A Pathway to Introductory Statistics

Jay Lehmann

College of San Mateo

PEARSON

Boston   Columbus   Indianapolis   New York   San Francisco
Amsterdam   Cape Town   Dubai   London   Madrid   Milan   Munich   Paris   Montréal   Toronto
Delhi   Mexico City   São Paulo   Sydney   Hong Kong   Seoul   Singapore   Taipei   Tokyo

Editorial Director: Chris Hoag
Editor in Chief: Michael Hirsch
Senior Acquisitions Editor: Dawn Giovanniello
Editorial Assistant: Megan Tripp
Program Team Lead: Karen Wernholm
Program Manager: Beth Kaufman
Project Team Lead: Peter Silvia
Project Manager: Tamela Ambush
Media Producer: Aimee Thorne
QA Manager, Assessment Content: Marty Wright
Senior Content Developer: John Flanagan
Executive Content Manager: Rebecca Williams
Associate Content Manager: Eric Gregg
Marketing Manager: Alicia Frankel
Marketing Assistant: Alexandra Habashi
Senior Technical Art Specialist: Joe Vetere
Manager Rights and Permissions: Gina M. Cheselka
Procurement Specialist: Carol Melville
Associate Director Art/Design: Andrea Nix
Program Design Lead/Cover Design: Beth Paquin
Composition: Integra
Cover Image: Mark Simms/Shutterstock

Library of Congress Cataloging-in-Publication Data
Lehmann, Jay.
   A pathway to introductory statistics / Jay Lehmann, College of San Mateo.
      pages cm
   ISBN 978-0-13-410717-2 — ISBN 0-13-410717-9
1. Mathematical statistics — Textbooks.   I. Title.
QA276.12.L45 2016
519.5 — dc23

                                                         2015005396

Screenshots from StatCrunch. Used by permission of StatCrunch.
Screenshots from Texas Instruments. Courtesy of Texas Instruments.

1 2 3 4 5 6 7 8 9 10—V003—19 18 17 16 15

www.pearsonhighered.com

ISBN-10:   0-13-410717-9
ISBN-13: 978-0-13-410717-2

**To Keri,
For carrying my aspirations
whenever I falter.**

Contents

B   USING STATCRUNCH   B-1

C   STANDARD NORMAL DISTRIBUTION TABLE   C-1

Preface

For a very long time, algebra has been viewed as an essential ingredient in a person's education. And for certain community college students such as STEM majors, this is definitely true. But recently, some instructors have begun to question whether the traditional algebra sequence best serves all students. Is it the ideal preparation for a career in political science? How about psychology? Social science? Probably not.

In addition to evaluating the long-range benefits of algebra, we should also assess the short-range ones. For some non-STEM majors, the only transferable math course they need to take is statistics. But is the traditional algebra sequence the best preparation for statistics? Without question, statistics students need to have a solid understanding of certain algebra concepts. But one would be hard-pressed to argue that factoring polynomials, completing the square, and solving complicated rational equations are the most important concepts to learn before embarking on a statistics course sequence.

It is not only the *content* of the traditional algebra sequence that is misaligned with statistics. It is also the *nature* of the activities. In algebra, much attention is devoted to manipulating symbols. Statistics focuses on analyzing situations, comparing measurements, and interpreting the meaning of concepts and results. Because the nature of the activities is so different, it is not surprising that many students enter introductory statistics unprepared.

A Pathway to Introductory Statistics is meant to serve non-STEM community college students better than a traditional algebra sequence. In particular, its main goals are to

- enhance students' ability to think statistically: analyze, compare, and interpret.
- address descriptive statistics, including the normal distribution and regression.
- empower students to discern good and bad practices of statistics.
- equip students with the algebra essential for success in introductory statistics.
- inspire students with exciting situations that are relevant to their careers.
- foster the use of technology to enhance, rather than replace, critical thinking.
- provide collaborative explorations in which students experience the joy of discovery.

Some or all of these goals might seem obvious. But the path that this text takes to achieve these goals may be a bit surprising. The following explanations will hopefully clarify this book's approach.

A Meaningful, Alternative Path This text contains the key concepts of descriptive statistics: experimental design, statistical diagrams, measures of center and spread, probability, the normal distribution, and regression. Teaching these topics along with the necessary algebra would certainly prepare students for an introductory statistics course better than the traditional algebra sequence. But to present the statistics concepts twice in the same manner—first in a *Pathway* course and again in an introductory statistics course—falls short of the sequence's highest potential. Teaching a concept from two perspectives rather than one provides students with a richer and broader learning experience.

But how can statistics be presented in a meaningful way that is different from its presentation in traditional statistics courses? There are actually many paths, but to discover the trailheads we must determine the foundational concepts with which statistics students struggle. Certainly one such concept is the normal curve, which lays the foundation for inferential statistics. How many introductory statistics students understand why the area of a region under the normal curve is equal to a probability? How many introductory statistics students understand how probability rules connect with finding such an area? And how many of them see that proportions, percentiles, and probabilities are closely related and understand why? Most instructors would agree, far too few.

What is compelling is that all three of these issues can be wonderfully addressed with one topic that is given short shrift in most, if not all, traditional statistics courses: density histograms. Because a normal curve can be viewed as a model that approximates a density histogram, students who have a firm grasp of density histograms can also gain a solid understanding of the three issues.

Many instructors' first reaction to this path is that density histograms are too difficult for students to comprehend. Actually, because density histograms are composed of rectangles, it is quite easy for students to compute areas and relate them to proportions, percentiles, and probabilities. *Pathway* takes full advantage of this by having students problem solve with rectangles in Chapter 1, construct and interpret density histograms in Chapter 3, reflect on how measures of center and spread are connected to density histograms in Chapter 4, and apply probability rules when working with density histograms in Chapter 5. After completing Chapters 1–5, the great majority of students will not only have a strong footing with the three issues mentioned earlier but also with related concepts such as probability and measures of center and spread.

Two Approaches: Acceleration versus Replacing Intermediate Algebra   In terms of sequencing courses, departments will use this text in one of two ways.

Some departments plan to accelerate their non-STEM students through their math programs by replacing elementary and intermediate algebra with *Pathway*. Some of these faculty feel that the traditional algebra sequence is an unnecessary obstacle for students whose careers will not depend on a significant portion of the sequence. Others feel that presenting algebra from a statistical perspective will engage students at a higher level and be more relevant to their careers.

And then some departments plan to use *Pathway* as an alternative to intermediate algebra. This means that their non-STEM students will first take elementary algebra and then enhance their knowledge of algebra by experiencing it from a statistical perspective. This will not only broaden students' understanding of algebra but may allow some departments to put greater emphasis on statistics because their students will have seen the necessary algebra once before.

The Big Picture   When the big picture is presented, students will have a map that tells them where the course's path is headed and how concepts connect. Once students have revisited many arithmetic concepts and a few simple algebra concepts in Chapter 1, they are ready for an overview of statistics in Chapter 2, which explores both good and bad experimental design. Unlike many statistics textbooks that then drop this crucial topic in subsequent chapters on descriptive statistics, *Pathway* encourages students to reflect on issues such as sampling error and sampling bias throughout the rest of the course.

At first glance, some reviewers wonder why the content of Chapter 6 was placed before the content of Chapters 7 and 8. After a closer look, they realize that Chapter 6's development of the four characteristics of an association (shape, strength, direction, and outliers) provides the big picture for the rest of this book. In fact, the four characteristics are further developed in a myriad of ways in subsequent chapters. A significant additional benefit to this organization is that Chapter 6 does not involve algebra. So, departments who want to heavily emphasize statistics can address all of Chapters 2–6 and, time permitting, pick and choose algebra topics from Chapters 7–10.

Compelling Modeling   Reviewers have praised the modeling in this text because the data sets are current, authentic, and compelling. And although a homework section's modeling exercises emphasize the concepts addressed in the section, investigations prompted by the "story" of an authentic situation are also embraced. It is in part due to these excursions "off the path" that make the modeling exercises come alive.

Judiciously Selected Algebra Topics   Some reviewers feel that this book contains too much algebra. Others think that the amount of algebra is just right. What is interesting is that almost all reviewers believe that for the most part, the only algebra that should be included are the concepts needed in an introductory statistics course. This suggests that instructors teach introductory statistics in different ways. For example, some instructors solve inequalities to derive confidence intervals. Others provide a more intuitive explanation. Some instructors solve equations to derive error formulas. Again, others get the idea across intuitively. And some do both.

With one possible exception, every algebra topic included in this text will be of service to *some* instructors who teach introductory statistics.

The one possible exception is the inclusion of functions. Although functions operate behind the scenes of introductory statistics courses, most textbooks do not make much, if any, use of function notation, language, and concepts. Nonetheless, keeping in mind that some departments will allow students to take *Pathway* instead of the traditional algebra sequence, functions have been included in this book in the hopes that any student who graduates from a community college will have an understanding and appreciation of a concept that is key for so much of mathematics.

Although exponential functions are definitely not included in most introductory statistics courses, they are arguably the second most important type of function (next to linear), and students can gain a significantly better understanding of linear modeling by comparing and contrasting the process with exponential modeling.

Arithmetic and Algebra Seen through a Statistics Lens   To better prepare students for Chapters 2–10, some of the arithmetic and simple algebra concepts in Chapter 1 are presented with statistics in mind. For example, Section 1.3 uses fractions as a springboard for proportions and the complement rule. In Section 1.7, students will evaluate statistics expressions and work with areas of rectangles that resemble density histograms.

Likewise, algebra concepts addressed in Chapters 7–10 have been developed from a statistics perspective. For example, rate of change is investigated in Section 7.2 before slope of a line is introduced. Evaluating linear functions in Section 7.4 is parlayed into using linear models to make predictions. And rather than have students work with geometry and science formulas, Section 8.4 requires students to solve probability and statistics formulas for a variable.

Group Explorations   Every section of *Pathway* contains at least one exploration that supports student investigation of a concept. Instructors can use explorations as **collaborative activities** during class time or as part of homework assignments. Section Opener Explorations are directed-discovery activities that are meant to be used at the start of class. Near the end of class, teams of students can work on additional explorations meant to deepen students' understanding of key concepts. Both types of explorations empower students to become active explorers of mathematics and can open the door to the wonder and beauty of the subject.

Balanced Raw Data and Visual Approach   Most statistics textbooks devote an entire chapter to constructing statistical diagrams but then make little use of such diagrams in homework exercises of subsequent chapters. This is unfortunate because students learn best when new concepts are integrated with previously learned ones. For example, to gain a solid understanding of the measures of center and spread, students should analyze some exercises that supply raw data and others that supply statistical diagrams. Throughout this book, homework sections contain a good balance of both types of exercises.

Technology   Back in the '80s, statistics students were expected to construct large numbers of statistical diagrams by hand and perform copious calculations with their calculators. Currently, most statistics instructors believe students should perform a limited number of such activities to get the idea and from then on use technology. The freed-up class time is devoted to enhancing students' ability to analyze authentic situations, compare measures of center and spread, and interpret concepts and results.

Pathway assumes students have access to technology. With so many packages to choose from, this text's technological support would be spread thin if it attempted to address all of them. This book focuses on the TI-84 graphing calculator and StatCrunch because the vast majority of community college instructors use one of these two technologies in their introductory statistics courses.

However, in the homework sections, the word *technology* is used rather than specifying the TI-84 or StatCrunch to accommodate classes using other technologies, unless an algebraic command specific to the TI-84 (such as *intersect*) is required.

Appendices A and B: TI-84 and StatCrunch Instructions   Appendices A and B contain instructions on how to use a TI-84 and StatCrunch, respectively. A subset of either appendix can serve as a tutorial early in the course. In addition, each time this text introduces a command from either technology, students are referred to a section of the appropriate appendix.

"Data" Icon   To support the appropriate use of technology, data sets in exercises and explorations that involve approximately 10 or more data values are available to download at MyMathLab and at the Pearson Downloadable Student Resources for Math and Statistics website: http://www.pearsonhighered.com/mathstatsresources. Such exercises are flagged in the text by the icon **DATA**.

Large Data Sets   It can make a significant, positive impression on students the first time they use technology to construct a histogram of about 100 observations when up to that point they have constructed histograms of only about 20 observations by hand. They are understandably struck by the ease, speed, and accuracy of using technology. But students can gain an even higher level of appreciation by using technology to describe a data set that consists of entries in thousands of rows and multiple columns.

Such an activity is especially relevant in today's age of big data. Although most *Pathway* students will not perform statistics in their career, some *will* work with large data sets. And as part of their general education, all students should have some sense of what statisticians do.

To meet this end, exercises that involve large data sets are sprinkled throughout this text. They directly follow the heading "Large Data Sets." Some of these data sets contain thousands of rows and tens of columns.

Hands-On Research   Even though every authentic data set in this book provides a source, some students still think that the data is fabricated. Having students find data sets themselves drives home the point that the concepts they are learning can truly be applied to real-life situations. Students begin to see that statistics can be used not only to inform but also to persuade.

To guide students in this process, this text contains exercises that direct students to analyze data found by online searches of blogs, newspapers, magazines, and scientific journals. These exercises are at the end of select homework sections, directly following the heading, "Hands-On Research."

Hands-On Projects   Compelling project assignments have been included near the end of most chapters. Some of the assignments are similar to the Hands-On Research exercises, but they are more extensive and challenging. These projects reinforce the idea that statistics is a powerful tool that can be used to analyze authentic situations. They are also an excellent opportunity for more in-depth writing assignments.

Some of the projects are about climate change and have been written at a higher reading level than the rest of this text to give students a sense of what it is like to perform research. Students will find that by carefully reading (and possibly rereading) the background information, they can comprehend the information and apply concepts they have learned in the course to make meaningful estimates about this compelling, current, and authentic situation.

Level of Difficulty   As was discussed earlier, some departments plan on using *Pathway* to accelerate non-STEM students through their math program. This is a worthy goal, provided it is done well. But some instructors have collapsed the notion of acceleration with making the course easier. The line of reasoning is that if certain students would not succeed in a traditional algebra course, then those students would not succeed in an alternative course that is just as challenging. This logic does not hold up because the nature of the two courses can differ greatly. We should not rob students of the knowledge and self-esteem that result from diligent study.

Furthermore, employers in search of college graduates certainly want a college degree to mean that students have succeeded at courses that are just as demanding as those in the past.

It is for these reasons that this text has been written to challenge students as much as they are challenged in traditional algebra courses. This is primarily achieved in two ways. First, exercises and projects require the interpretation of concepts and results, which causes significant growing pains in most students. Second, many exercises contain at least one part (often out of five parts) that challenges students to apply concepts in new ways.

Warnings   Throughout this text, the word WARNING in the margins flags paragraphs that describe common student misconceptions and the correct meanings or applications of concepts.

Tips for Success   Many sections close with practical study tips to help students succeed in the course. A complete list of these tips is included in the Index.

RESOURCES FOR INSTRUCTORS

Instructor's Resource Manual   This manual contains suggestions for pacing the course and creating homework assignments. It discusses how to incorporate technology and how to structure project assignments. The manual also contains section-by-section suggestions for presenting lectures and for undertaking the explorations in the text.

Instructor's Solutions Manual   This manual includes complete solutions to the even-numbered exercises in the homework sections of the text.

MyMathLab® Online Course (access code required)   MyMathLab from Pearson is the world's leading online resource in mathematics, integrating interactive homework, assessment, and media in a flexible, easy-to-use format.

MyMathLab helps individual **students succeed**.

- MyMathLab has a consistently positive impact on student retention, subsequent success, and overall achievement. MyMathLab can be successfully implemented in any environment—lab-based, hybrid, fully online, traditional.

- MyMathLab has a comprehensive online gradebook that automatically tracks your students' results on tests, quizzes, homework, and in the study plan. You can use the gradebook to quickly intervene if your students have trouble, or to provide positive feedback on a job well done.

MyMathLab provides **engaging experiences** that personalize, stimulate, and measure learning for each student.

- **Personalized Learning:** MyMathLab's personalized homework, and adaptive and companion study plan features allow your students to work more efficiently spending time where they really need to.

- **Exercises:** The homework and practice exercises in MyMathLab are correlated to the exercises in the textbook, and they regenerate algorithmically to give students unlimited opportunity for practice and mastery. Students receive immediate, helpful feedback when they work through each exercise.

- **Multimedia Learning Aids:** Exercises include guided solutions, sample problems, animations, videos, and eText access for extra help at point-of-use.

- **Learning Catalytics™:** MyMathLab now provides Learning Catalytics—an interactive student response tool that uses students' smartphones, tablets, or laptops to engage them in more sophisticated tasks and thinking.

- **StatCrunch™:** This MyMathLab course integrates the web-based statistical software, StatCrunch, within the online assessment platform so that students can easily analyze data sets from exercises and the text. In addition, this course includes access to www.statcrunch.com, a vibrant online community where users can access tens of thousands of shared data sets, create and conduct online surveys, perform complex analyses using the powerful statistical software, and generate compelling reports.

MyMathLab Accessibility: MyMathLab is compatible with the JAWS screen reader, and enables multiple-choice and free-response problem-types to be read and interacted with via keyboard controls and math notation input. MyMathLab also works with screen enlargers, including ZoomText, MAGic, and SuperNova. And all MyMathLab videos in this course are closed captioned. More information on this functionality is available at http://mymathlab.com/accessibility.

And, MyMathLab comes from an **experienced partner** with educational expertise and an eye on the future.

- Whether you are just getting started with MyMathLab, or have a question along the way, we're here to help.
- Contact your Pearson representative directly or at www.mymathlab.com.

TestGen   TestGen enables instructors to build, edit, print, and administer tests by using a computerized bank of questions developed to cover all the objectives of the text. TestGen is algorithmically based, allowing instructors to create multiple, but equivalent, versions of the same question or test with the click of a button. Instructors can also modify test-bank questions or add new questions. Tests can be printed or administered online. The software and test bank are available for download from Pearson Education's online catalogue.

PowerPoint Lecture Slides (download only)   Available through www.pearsonhighered.com or inside your MyMathLab course, these fully editable lecture slides include definitions, key concepts, and examples for use in a lecture setting and are available for each section of the text.

RESOURCES FOR STUDENTS

Interactive Video Lecture Series   This series provides students with extra help for each section of the textbook. The Lecture Series includes Interactive Lectures that highlight key examples and exercises for every section of the textbook.

Student's Solutions Manual   This manual contains the complete solutions to the odd-numbered exercises in the Homework sections of the text.

GETTING IN TOUCH

I would love to hear from you and would greatly appreciate receiving your comments regarding *Pathway*. If you have any questions, please ask them, and I will respond.

Jay Lehmann
MathNerdJay@aol.com

Acknowledgments

The commitment required to write a first textbook is substantial, but nothing compared to writing a second, or in this case, a fourth one because you know exactly what is in store. And so did Keri, my wife, who wholeheartedly supported me every step of the way, never flinching at the insane number of hours I spent in my author's cave.

Writing a textbook for a new course is both exhilarating and daunting, like driving along S-turns bordering cliffs with the lights dimmed. But Senior Acquisitions Editor Dawn Giovanniello turned on the high beams, connecting me with an amazing group of reviewers who were indispensable in helping me get the content and pedagogy right. Thanks, Dawn, for your idea to write *Pathway,* all your enthusiastic nudges to embark on the path, and your creative solutions to all the obstacles that popped up along the way.

I am greatly indebted to Editor in Chief Michael Hirsch, for sharing Dawn's vision and aligning Pearson to make the huge investment in this book.

Displaying a single name on the cover is deceptive. Tens of creative people helped build this text with great attention to detail. Project Manager Tamela Ambush did an incredible job coordinating many of them, always looking for ways to lighten my load. Thanks also to Integra-Chicago's Project Manager Valerie Iglar-Mobley, who expertly coordinated the rest of the team.

To the uber-supportive Senior Development Editor Elaine Page, for doing everything from helping me find the right word to inspiring me to make this text as cutting edge as possible. And thanks to Editorial Assistant Megan Tripp for responding to my e-mails within a nanosecond and thinking outside the box when assisting me in performing research.

If not for my colleague Ken Brown tirelessly responding to my countless inquiries, this textbook would have been out of step with current statistical practices. And to the equally generous Jon Freedman, who managed to crack me up even when replying to my most tedious e-mails, challenging me to present statistics in a way that would truly connect with students.

The quality of a textbook is only as good as its reviewer feedback. And *Pathway* received incredible reviews from a large number of passionate instructors, who often went beyond what was asked, ensuring that this book would not only meet the needs of students at their campuses, but at other colleges across the country. Deepest thanks goes to these fantastic reviewers:

Kate Acks, *Maui College*
Ken Anderson, *Chemeketa Community College*
Sasha Anderson, *Fresno City College*
Alvina Atkinson, *Georgia Gwinnett College*
Jannette Avery, *Monroe Community College*
Wayne Barber, *Chemeketa Community College*
Rosanne B. Benn, *Prince George's Community College*
Jack Bennett, *Ventura College*
Elena Bogardus, *Camden Community College*
Tony Bower, *St. Philip's College*
Joe Brenkert, *Front Range Community College*
Ronnie Brown, *University of Baltimore*
Jayalakshmi Casukhela, *Ohio State University at Lima*
Steven Cheng, *Quinsigamond CC*
Shawn Clift, *Eastern Kentucky University*
Michael Combs, *Bunker Hill Community College*
Eden Donahou, *Seminole State College of Florida*
Cynthia Ellis, *Indiana University–Purdue University Fort Wayne*
Mary Ann Esteban, *Kapiolani Community College*
Nancy Fees, *Northwest College*
Jon Freedman, *Skyline College*

David French, *Tidewater Community College*
Kim Ghiselin, *State College of Florida*
Dave Gilbert, *Santa Barbara City College*
Eric Gilbertsen, *Montana State University, Billings*
Ryan Girard, *Kauai Community College*
Lisa Green, *Middle Tennessee University*
Cheryl Gregory, *College of San Mateo*
Ryan Grossman, *Ivy Tech Community College of Indiana*
Edward Ham, *Bakersfield College*
Miriam Harris-Botzum, *Lehigh Carbon Community College*
Christy Hediger, *Lehigh Carbon Community College*
Bobbie Hill, *Coastal Bend College*
Steven Hal Huntsman, *City College of San Francisco*
Laura Iossi, *Broward College*
Sarah Isaksen, *University of Detroit Mercy*
Marilyn Jacobi, *Gateway Community College*
Yvette Janecek, *Blinn College, Brenham Campus*
Christopher Jett, *University of West Georgia*
Jonathan Kalk, *Kauai Community College*
Brian Karasek, *South Mountain Community College*
Cameron Kishel, *Columbus State Community College*
Alex Kolesnik, *Ventura College*
Lynne Kowski, *Raritan Valley Community College*
Kathryn Kozak, *Coconino Community College*
Julie Labbiento, *Lehigh Carbon Community College*
Marcia Lambert, *Pitt Community College*
Mary Margarita Legner, *Riverside City College*
Deb Lehman, *Columbia College*
Edith Lester, *Volunteer State Community College*
LaRonda Lowery, *Robeson College*
Christine Mac, *Front Range Community College*
Doug Mace, *Kirtland Community College*
Jason Malozzi, *Lehigh Carbon Community College*
Gayathri Manikandan, *El Camino College, Compton Community Educational Center*
Stacy Martig, *St. Cloud State University*
Nancy Matthews, *University of Oklahoma*
Sue McBride, *John Tyler Community College*
Judy McFarland, *Seminole State College of Florida*
Teresa McFarland, *Owensboro Community and Technical College*
Kim McHale, *Heartland Community College*
Andrea Nemeth, *California State University, Northridge*
Francis Kyei Nkansah, *Bunker Hill Community College*
Sue Norris, *Grinnell College*
Pat Rhodes, *Treasure Valley Community College*
Pat Riley, *Hopkinsville Community College*
Ruth Roberman, *South University*
Nicole Saporito, *Luzerne County Community College*
Ned Schillow, *Lehigh Carbon Community College*
Saliha Sha, *Ventura College*
Renee Shipp, *Jefferson Davis Community College*
Jenny Shotwell, *Central Texas College*
Joseph Spadaro, *Gateway Community College*
Brad Stetson, *Schoolcraft College*
Chairsty Stewart, *MSU Billings*
Marie St. James, *St. Clair County Community College*
Susan Twigg, *Wor-Wic Community College*
Mary Williams, *Roosevelt University*
Robin Williams, *Palomar College*
Cynthia Vanderlaan, *Indiana University–Purdue University Fort Wayne*
Cathleen Zucco-Teveloff, *Rider University*

Index of Applications

Performing Operations and Evaluating Expressions

Although the number of AIDS deaths in the United States greatly decreased from 37,787 deaths in 1996 to only 13,834 deaths in 2011, the number of AIDS deaths increased to 17,000 in 2012 (see Table 1). In Example 9 in Section 1.5, we will calculate how the number of deaths has changed in various years.

HIV infection can lead to contracting AIDS. In 2010, President Obama released the National HIV/AIDS Strategy to lower the number of new HIV infections. But how can we determine what will make a difference?

Table 1 Numbers of AIDS Deaths in the United States

Year	Number of AIDS Deaths
2006	15,564
2007	14,561
2008	16,084
2009	17,774
2010	15,529
2011	13,834
2012	17,000

Source: *Centers for Disease Control and Prevention*

In this course, we will address such questions by practicing statistics. We will learn to form *precise* questions such as "Does raising public awareness about the ways HIV can be contracted decrease the HIV infection rate?" Next, we will discuss how to develop a careful plan to answer a precise question, which will include taking a close look at how to collect the relevant information. Then, we will construct tables and diagrams and perform calculations to analyze the information. Throughout much of the course, we will determine what types of conclusions we can draw about questions raised.

In this chapter, we will discuss the arithmetic that forms the foundation of statistics.

▼1.1 Variables, Constants, Plotting Points, and Inequalities

Objectives

» Describe the meaning of *variable* and *constant.*

» Describe the meaning of *counting numbers, integers, rational numbers, irrational numbers, real numbers, positive numbers,* and *negative numbers.*

» Use a number line to describe numbers.

In this section, we will work with *variables* and *constants,* two extremely important building blocks of algebra and statistics. We will also discuss various types of numbers and how to describe numbers and pairs of numbers visually. Finally, we will compare the sizes of quantities.

Variables

In arithmetic, we work with numbers. In algebra and statistics, we work with *variables* as well as numbers.

▶ **Definition   Variable**

A **variable** is a symbol that represents a quantity that can vary.

1

» Graph data on a number line.

» Plot points on a coordinate system.

» Describe the meaning of *inequality symbols* and *inequality*.

» Graph an inequality and a compound inequality on a number line.

» Use inequality notation, interval notation, and graphs to describe possible values of a variable for an authentic situation.

» Describe a concept or procedure.

"The Ramsey account? Should be done any minute..."

For example, we can define h to be the height (in feet) of a specific child. Height is a quantity that varies: As time passes, the child's height will increase. So, h is a variable. When we say $h = 4$, we mean the child's height is 4 feet.

This definition of variable is typically used in algebra and will be extremely useful in this chapter. In Section 2.1, we will discuss a definition that is typically used in statistics. At the heart of both definitions is the fact that a variable describes something that can vary.

▶ **Example 1**   Using a Variable to Represent a Quantity

1. Let s be a car's speed (in miles per hour). What is the meaning of $s = 60$?
2. Let n be the number of people (in millions) who work from home at least once a week during normal business hours. For the year 2013, $n = 20$ (Source: *Telework Research Network*). What does that mean in this situation?
3. Let t be the number of years since 2010. What is the meaning of $t = 5$?

Solution

1. The speed of the car is 60 miles per hour.
2. In 2013, 20 million people worked from home at least once a week during normal business hours.
3. $2010 + 5 = 2015$; so, $t = 5$ represents the year 2015.

There are many benefits to using variables. For example, in Problem 2 of Example 1, we found that the simple equation "$n = 20$" means the same thing as the wordy sentence "20 million people worked from home at least once a week during normal business hours." Variables can help us describe some situations with a small amount of writing.

In Problem 3 of Example 1, we described the year 2015 by using $t = 5$. So, our definition of t allows us to use smaller numbers to describe various years—an approach that will be especially helpful in Chapters 6–10.

We will see other benefits of variables as we proceed through the course.

▶ **Example 2**   Using a Variable to Represent a Quantity

Choose a symbol to represent the given quantity. Explain why the symbol is a variable. Give two numbers that the variable can represent and two numbers that it cannot represent.

1. the weight (in pounds) of a baby at birth
2. the number of people who live in a two-bedroom house

Solution

1. Let w be the weight (in pounds) of a baby at birth. The weight of a baby at birth can vary, so w is a variable. For example, w can represent the numbers 6 and 8, because babies can weigh 6 or 8 pounds at birth. The variable w does not represent 0 or 300, because babies cannot weigh 0 or 300 pounds at birth!
2. Let n be the number of people who live in a two-bedroom house. The number of people who live in a two-bedroom house can vary, so n is a variable. For example, n can represent the numbers 2 and 3, because 2 or 3 people can live in a two-bedroom house. The variable n cannot represent the numbers 5000 or $\frac{1}{2}$, because 5000 people cannot live in a two-bedroom house and half of a person doesn't make sense.

In Problem 1 of Example 2, we stated that the units of w are pounds. Without stating the units of w, "$w = 10$" could mean the baby's weight was 10 ounces, 10 pounds, or 10 tons! In defining a variable, it is important to describe the variable's units.

Constants

A variable is a symbol—typically a letter—that represents a quantity that can vary. When we use a symbol to represent a quantity that does *not* vary, we call that symbol a *constant*. So, 2, 0, 4.8, and π are constants. The constant π is approximately equal to 3.14.

> ▶ **Definition** Constant
>
> A **constant** is a symbol that represents a specific number (a quantity that does *not* vary).

In the next example, we will compare the meanings of a variable and a constant while we consider the widths, lengths, and areas of some rectangles. The **area** (in square inches) of a flat surface is the number of square inches that it takes to cover the surface (see Fig. 1). **The area of a rectangle is equal to the rectangle's length times its width.**

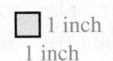
1 inch
1 inch

Figure 1 One square inch

▶ **Example 3** Comparing Constants and Variables

A rectangle has an area of 12 square inches. Let W be the width (in inches), L be the length (in inches), and A be the area (in square inches).

1. Sketch three possible rectangles of area 12 square inches.
2. Which of the symbols W, L, and A are variables? Explain.
3. Which of the symbols W, L, and A are constants? Explain.

Solution

1. We sketch three rectangles for which the width times the length is equal to 12 square inches (see Fig. 2).
2. The symbols W and L are variables since they represent quantities that vary.
3. The symbol A is a constant because in this problem the area does not vary—the area is always 12 square inches.

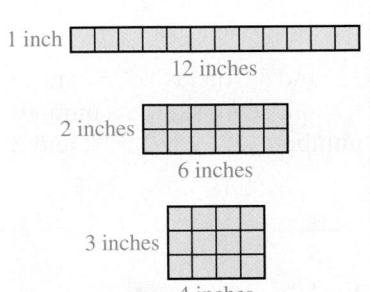
1 inch
12 inches

2 inches
6 inches

3 inches
4 inches

Figure 2 Three possible rectangles of area 12 square inches

Counting Numbers

When we describe people, it often helps to describe them in terms of certain categories, such as gender, ethnicity, and employment. In mathematics, it helps to describe numbers in terms of categories, too. We begin by describing the *counting numbers,* which are the numbers 1, 2, 3, 4, 5, and so on.

> ▶ **Definition** Counting numbers (natural numbers)
>
> The **counting numbers,** or **natural numbers,** are the numbers
>
> $$1, 2, 3, 4, 5, \ldots$$

The three dots mean that the pattern of the numbers shown continues without ending. In this case, the pattern continues with 6, 7, 8, and so on. When a list of numbers goes on forever, we say that there are an *infinite* number of numbers.

Integers

Next, we describe the *integers,* which include the counting numbers and other numbers.

> ▶ **Definition** Integers
>
> The **integers** are the numbers
>
> $$\ldots, -3, -2, -1, 0, 1, 2, 3, \ldots$$

The three dots on both sides mean that the pattern of the numbers shown continues without ending in both directions. In this case, the pattern continues with $-4, -5, -6$, and so on, and with 4, 5, 6, and so on.

If you write a check for more money than is in your checking account, you will have a negative balance. The balance -60 dollars is an integer.

The **positive integers** are the numbers 1, 2, 3,.... The **negative integers** are the numbers $-1, -2, -3, \ldots$. The integer 0 is neither positive nor negative. So, the integers consist of the counting numbers (which are positive integers), the negative integers, and 0.

The Number Line

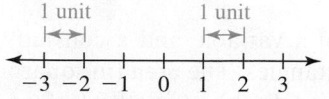

1 unit   1 unit

Figure 3 The number line

We can visualize numbers on a *number line* (see Fig. 3).

Each point (location) on the number line represents a number. The numbers increase from left to right. We refer to the distance between two consecutive integers on the number line as 1 *unit* (see Fig. 3).

▶ **Example 4**   Graphing Integers on a Number Line

Draw dots on a number line to represent the integers between -2 and 3, inclusive.

Solution

The integers between -2 and 3, inclusive, are $-2, -1, 0, 1, 2$, and 3. "Inclusive" means to include the first and last numbers, which in this case are -2 and 3. We sketch a number line and draw dots at the appropriate locations for the numbers $-2, -1, 0, 1, 2$, and 3 (see Fig. 4).

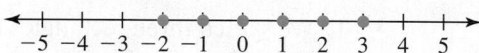

Figure 4 Graphing the numbers $-2, -1$, 0, 1, 2, and 3

When we draw dots on a number line, we say that we are "plotting points" or "graphing numbers."

WARNING   In Example 4, we worked with the integers between -2 and 3, inclusive: $-2, -1, 0, 1, 2$, and 3. Here are the integers between -2 and 3: $-1, 0, 1$, and 2. We did not include -2 or 3, because the word "inclusive" was not used. When working with such problems, it is important to check whether the word "inclusive" is used.

Rational Numbers

For a fraction $\frac{n}{d}$, we call n the **numerator** and d the **denominator.** The dash between the numerator and the denominator is the **fraction bar:**

$$\text{Numerator} \longrightarrow \quad \frac{n}{d} \quad \longleftarrow \text{Fraction bar}$$
$$\text{Denominator} \longrightarrow$$

A fraction can be used to describe a part of a whole. For example, consider the meaning of $\frac{5}{8}$ of a pizza. If we divide the pizza into 8 slices of equal area, 5 of the slices make up $\frac{5}{8}$ of the pizza (see Fig. 5).

The number $\frac{5}{8}$ is called a *rational number*.

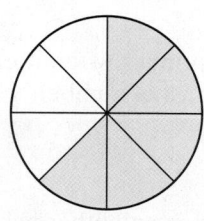

Figure 5 $\frac{5}{8}$ of a pizza

> ▶ **Definition** Rational numbers
>
> The **rational numbers** are the numbers that can be written in the form $\frac{n}{d}$, where n and d are integers and d is nonzero.

We specify that d is nonzero because, as we shall see later, division by zero does not make sense.

Here are some examples of rational numbers:

$$\frac{3}{7} \qquad \frac{-2}{5} \qquad 4 = \frac{4}{1}$$

Rational numbers include all the integers since any integer n can be written as $\frac{n}{1}$.

Irrational Numbers

There are numbers represented on the number line that are *not* rational. These numbers are called **irrational numbers.** An irrational number *cannot* be written in the form $\frac{n}{d}$, where n and d are integers and d is nonzero. The number $\sqrt{2}$ is the number greater than zero that we multiply by itself to get 2. The number $\sqrt{2}$ is an irrational number. Here are some more examples of irrational numbers:

$$\pi \qquad \sqrt{3} \qquad \sqrt{5}$$

We know that $\sqrt{9} = 3$, because $3 \times 3 = 9$. So $\sqrt{9} = 3 = \frac{3}{1}$. Therefore, $\sqrt{9}$ is rational (not irrational).

Decimals

The list price of an Xbox One console is $399.99, which is a decimal number.

Any rational number or irrational number can be written as a decimal number.

A rational number can be written as a decimal number that either terminates or repeats:

$$\frac{3}{4} = \underbrace{0.75}_{\text{terminates}} \qquad \frac{3}{11} = \underbrace{0.27272727\ldots}_{\text{repeats}}$$

We can use an overbar to write the repeating decimal $0.272727\ldots = 0.\overline{27}$.

An irrational number can be written as a decimal number that neither terminates nor repeats. It is impossible to write all the digits of an irrational number, but we can approximate the number by rounding. For example, earlier we *approximated* π by rounding to the second decimal place: $\pi \approx 3.14$.

Real Numbers

Recall that each point on the number line represents a number. We call all of the numbers represented by all of the points on the number line the *real numbers*.

> ▶ **Definition** Real numbers
>
> The **real numbers** are all of the numbers represented on the number line.

The real numbers are made up of the rational numbers and the irrational numbers. Here are some real numbers:

$$-1.8 \qquad -1 \qquad -\frac{7}{10} \qquad 0 \qquad 0.4 \qquad \frac{6}{5} \qquad \pi$$

We graph these real numbers in Fig. 6.

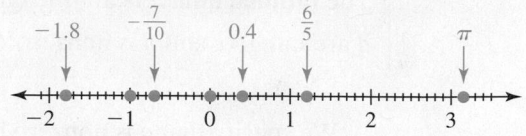

Figure 6 Graphing the real numbers -1.8, -1, $-\dfrac{7}{10}$, 0, 0.4, $\dfrac{6}{5}$, and π

We use an arrow to label points that do not fall on a labeled tick mark.

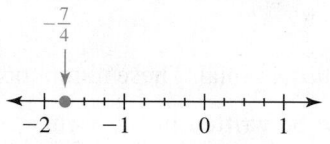

Figure 7 Graphing the number $-\dfrac{7}{4}$

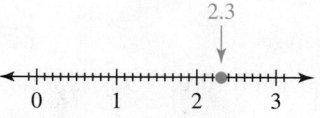

Figure 8 Graphing the number 2.3

▶ **Example 5**   Graphing Real Numbers on a Number Line

Graph the number on a number line.

1. $-\dfrac{7}{4}$ **2.** 2.3

Solution

1. We draw a number line so that the distance between tick marks is $\dfrac{1}{4}$ unit (see Fig. 7). To graph $-\dfrac{7}{4}$, we draw a dot at the seventh tick mark to the left of 0.

2. We draw a number line so that the distance between tick marks is $0.1 = \dfrac{1}{10}$ unit (see Fig. 8). To graph 2.3, we draw a dot at the third tick mark to the right of 2.

Figure 9 illustrates how the various types of numbers we have discussed so far are related. In particular, it shows that every counting number is an integer, every integer is a rational number, and every rational number is a real number. It also shows that **irrational numbers are the real numbers that are not rational.**

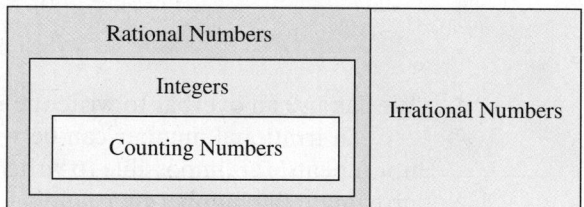

All numbers shown here are real numbers.

Figure 9 The real numbers

Graphing Data on a Number Line

Data are quantities or categories that describe people, animals, or things. For example, the following heights of six people, all in inches, are data: 64, 71, 75, 68, 71, and 69. The following genres of music of the top 5 grossing singles this week are data: pop, rock, pop, hip-hop, and country.

We often can get a better sense of data that are quantities by graphing them on a number line.

▶ **Example 6** Graphing Data

The total amount (in billions of dollars) of new federal student loans for the years 2008, 2009, 2010, 2011, and 2012 are 69, 87, 100, 107, and 108, respectively (Source: *College Board*). Let L be the total amount (in billions of dollars) of new federal student loans in a given year.

1. Graph the data.
2. Did the total amount of the loans increase, decrease, stay approximately constant, or none of these from 2008 to 2012, inclusive? Explain.
3. Did the *increases* in the total amounts of the loans increase, decrease, stay approximately constant, or none of these from 2008 to 2012, inclusive? Explain.

Solution

1. We sketch a number line and write "L" to the right of the number line and the units "Billions of dollars" underneath the number line (see Fig. 10). Because the data values are between 69 and 108, inclusive, we write the numbers 65, 70, 75, 80, 85, 90, 95, 100, 105, and 110 equally spaced on the number line. Then we graph the numbers 69, 87, 100, 107, and 108.

**Total Amounts of New Federal
Student Loans**

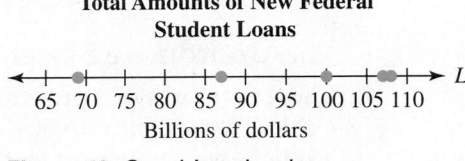

65 70 75 80 85 90 95 100 105 110
Billions of dollars

Figure 10 Graphing the data

2. From the opening paragraph, we know that the total amount of the loans is increasing. (From the graph alone, we cannot tell this, because the years are not included.)
3. As we look from left to right at the points plotted on the graph, we see that the distance between adjacent points decreases. This means that the increases in the total amounts of the loans decreased. That is, the jump from 69 to 87 is greater than the jump from 87 to 100, and so on.

WARNING   In Fig. 10, we wrote the numbers 65, 70, 75, 80, 85, 90, 95, 100, 105, and 110 on the number line. **When we write numbers on a number line, they should increase by a fixed amount and be equally spaced.**

Positive and Negative Numbers

The **negative numbers** are the real numbers less than 0, and the **positive numbers** are the real numbers greater than 0 (see Fig. 11).

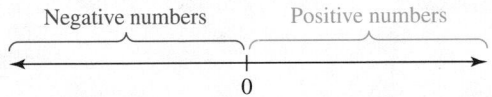

Negative numbers   Positive numbers

0

Figure 11 The location of the negative numbers and the positive numbers on the number line

Some examples of negative numbers are -13, -5.2, $-\frac{3}{4}$, and $-\sqrt{2}$. Some examples of positive numbers are 13, 5.2, $\frac{3}{4}$, and π. As we discussed earlier, the number 0 is neither positive nor negative.

We say that the *sign* of a negative number is negative and that the *sign* of a positive number is positive. To include zero, we define the **nonnegative numbers** as the positive numbers together with 0. Likewise, we define the **nonpositive numbers** as the negative numbers together with 0.

▶ **Example 7**   Graphing a Negative Quantity

A person bounces several checks and, as a result, is charged service fees. If b is the balance (in dollars) of the checking account, what value of b means the person owes $50? Graph the number on a number line.

Solution

Since the person *owes* money, the value of b is negative: $b = -50$. We graph -50 on a number line in Fig. 12.

Balance in a Checking Account

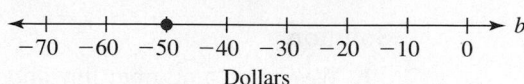

Figure 12 Graphing the number $b = -50$

So far, we have discussed how to describe the values of a *single* variable. Now we will discuss how to describe pairs of values of *two* variables.

The Coordinate System

The quiz scores of a student in one of the author's prestatistics classes are shown in Table 2. We define n to be the quiz number and s to be the quiz score (in points). From Table 2, we see the student's score on Quiz 1 was 7 points. So, when $n = 1$, $s = 7$. If we agree to write the quiz number first and the quiz score second, we can use the **ordered pair** $(1, 7)$ to mean that when $n = 1$, $s = 7$. We call each of the numbers in an ordered pair a **coordinate**. For $(1, 7)$ in this situation, we call 1 the *n-coordinate* and 7 the *s-coordinate*.

The ordered pair $(2, 6)$ means that when $n = 2$, $s = 6$. This indicates that the student's score on Quiz 2 was 6 points, which agrees with the second row of Table 2.

We graph the ordered pairs by using *two* number lines, which are called **axes** (singular: **axis**). To start, we draw a horizontal number line called the *n-axis* and a vertical number line called the *s-axis* (see Fig. 13). We refer to such a pair of axes as a **coordinate system.** The **origin** is the intersection point of the axes. The axes divide the coordinate system into four regions called **quadrants,** which we call Quadrants I, II, III, and IV. The quadrants do not include the axes.

Table 2 A Student's Quiz Scores

Quiz Number	Quiz Score (points)
n	*s*
1	7
2	6
3	9
4	8
5	9

Source: *J. Lehmann*

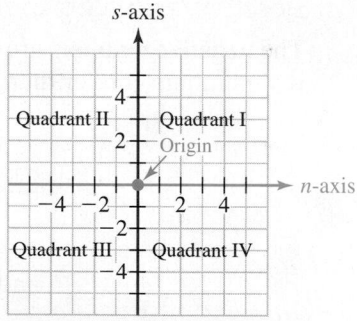

Figure 13 Coordinate system

Plotting Points on a Coordinate System

Next, we plot the ordered pair $(3, 9)$ shown in the third row of Table 2. To do so, we start at the origin, look 3 units to the right and 9 units up, and then draw a dot (see Fig. 14). In Fig. 15, we plot all the ordered pairs listed in Table 2.

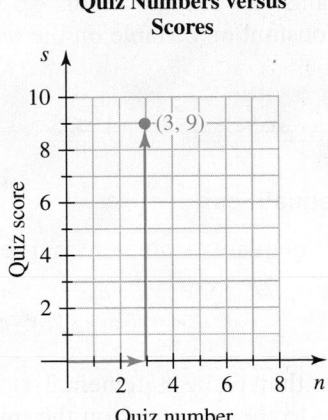

Figure 14 Plot (3, 9)

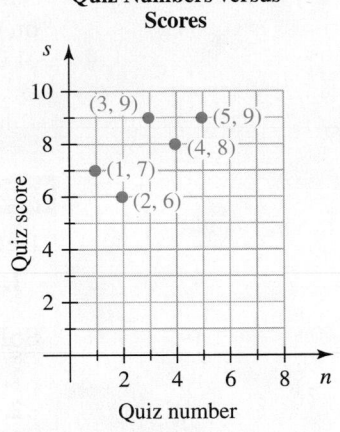

Figure 15 Plot the ordered pairs from Table 2

As we look at the plotted points in Fig. 15 from left to right, the points, in general, go upward. This means the quiz scores, in general, are increasing.

When we plot points that are not being used to describe authentic situations, we call the horizontal axis the *x-axis* and the vertical axis the *y-axis*. The ordered pair $(6, 3)$ means $x = 6$ and $y = 3$. So, the x-coordinate is 6 and the y-coordinate is 3.

▶ **Example 8** Plotting Points

Plot the points $(3, 4)$, $(-5, -3)$, $(-4, 2)$, and $(5, -4)$ on a coordinate system.

Solution

We plot the ordered pairs $(3, 4)$ and $(-5, -3)$ in Fig. 16, and we plot the ordered pairs $(-4, 2)$ and $(5, -4)$ in Fig. 17.

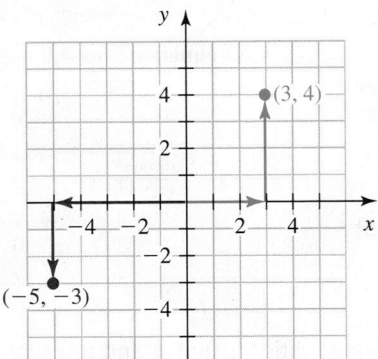

Figure 16 Plotting the ordered pairs (3, 4) and (−5, −3)

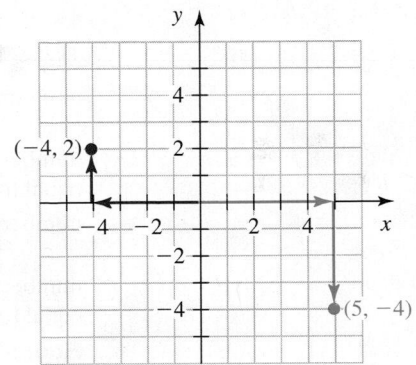

Figure 17 Plotting the ordered pairs (−4, 2) and (5, −4)

Inequality Symbols

In statistics, we often compare the sizes of two quantities. We can do this using the **inequality symbols** $<$, $\leq$, $>$, and $\geq$. Here are the meanings of these symbols and some examples of *inequalities*:

Symbol	Meaning	Examples of Inequalities
$<$	is less than	$2 < 5, 0 < 5, -6 < -1$
$\leq$	is less than or equal to	$4 \leq 7, 2 \leq 2, -3 \leq 0$
$>$	is greater than	$9 > 2, -4 > -6, 2 > 0$
$\geq$	is greater than or equal to	$8 \geq 3, 5 \geq 5, -2 \geq -8$

An **inequality** contains one of the symbols $<$, $\leq$, $>$, and $\geq$ with a constant or variable on one side and a constant or variable on the other side. Here are some more examples of inequalities:

$$x < -3 \qquad -4 \leq 5 \qquad 7 > 2 \qquad x \geq 6$$

▶ Example 9   Inequalities

Decide whether the inequality statement is true or false.

1. $3 \leq 6$ **2.** $-5 > -2$ **3.** $8 \geq 8$ **4.** $9 < 9$

Solution

1. Because 3 is less than 6, the statement $3 \leq 6$ is true.
2. Because -5 lies to the left of -2 on the number line, -5 is less than -2. So, -5 is *not* greater than -2, and the statement $-5 > -2$ is false.
3. Because 8 is equal to itself, the statement $8 \geq 8$ is true.
4. Because 9 is not less than itself, the statement $9 < 9$ is false.

Graphing an Inequality on a Number Line

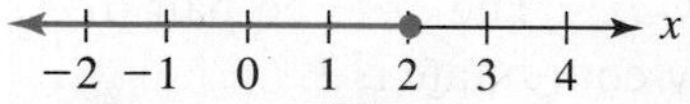

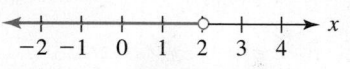

Figure 18 Graph of $x \leq 2$

Consider the inequality $x \leq 2$. This inequality says the values of x are less than or equal to 2. We can represent these values graphically on a number line by shading the part of the number line that lies to the left of 2 (see Fig. 18). We draw a *filled-in* circle at 2 to indicate that 2 is a value of x, too.

To graph the inequality $x < 2$, we shade the part of the number line that lies to the left of 2, but draw an *open* circle at 2 to indicate that 2 is *not* a value of x (see Fig. 19).

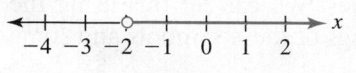

Figure 19 Graph of $x < 2$

We use **interval notation** to describe a set of numbers. For example, we describe the numbers greater than 3 by $(3, \infty)$. We describe the numbers greater than or equal to 3 by $[3, \infty)$. We describe the set of real numbers by $(-\infty, \infty)$. More examples of inequalities and interval notation are shown in Fig. 20.

In Words	Inequality	Graph	Interval Notation
numbers less than 3	$x < 3$	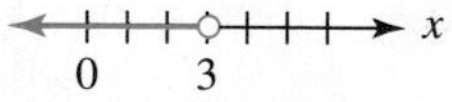	$(-\infty, 3)$
numbers less than or equal to 3	$x \leq 3$	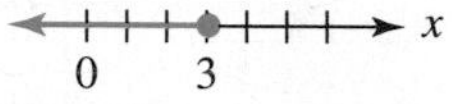	$(-\infty, 3]$
numbers greater than 3	$x > 3$	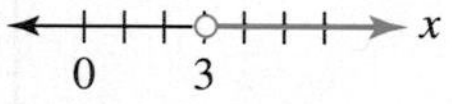	$(3, \infty)$
numbers greater than or equal to 3	$x \geq 3$	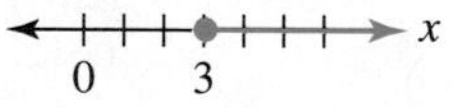	$[3, \infty)$

Figure 20 Words, inequalities, graphs, and interval notation

▶ Example 10   Graphing an Inequality

Write the inequality $x > -2$ in interval notation, and graph the values of x.

Solution

The inequality $x > -2$ means that the values of x are greater than -2. We describe these numbers in interval notation by $(-2, \infty)$. To graph the values of x, we shade the part of the number line that lies to the right of -2 and draw an open circle at -2 (see Fig. 21).

Figure 21 Graph of $x > -2$

Compound Inequalities

Now we will work with *compound inequalities in one variable,* such as $3 \leq x \leq 7$, which means the values of x are *both* greater than or equal to 3 *and* less than or equal to 7. In other words, all values of x are between 3 and 7, inclusive. To graph the solutions, we

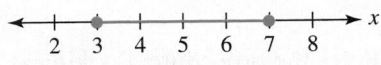

Figure 22 Graph of $3 \le x \le 7$

shade the part of the number line that lies between 3 and 7 (see Fig. 22). We draw filled-in circles at 3 and 7 to indicate that 3 and 7 are solutions, too.

We describe the numbers between 3 and 7, inclusive, in interval notation by [3, 7]. More examples of compound inequalities, with matching graphs and interval notation, are shown in Fig. 23.

In Words	Inequality	Graph	Interval Notation
Numbers between 1 and 3	$1 < x < 3$		$(1, 3)$
Numbers between 1 and 3, inclusive	$1 \le x \le 3$		$[1, 3]$
Numbers between 1 and 3, as well as 1	$1 \le x < 3$		$[1, 3)$
Numbers between 1 and 3, as well as 3	$1 < x \le 3$		$(1, 3]$

Figure 23 Words, inequalities, graphs, and interval notations

WARNING

We use notation such as $(3, 7)$ in two ways: When we work with one variable, the *interval* $(3, 7)$ is the set of numbers between 3 and 7; when we work with two variables, such as x and y, the *ordered pair* $(3, 7)$ means $x = 3$ and $y = 7$.

Describing Values of Variables for an Authentic Situation

When analyzing authentic situations, we will often use inequality notation, interval notation, and graphs to describe possible values of variables.

▶ **Example 11**   Describe Values of Variables and Graph a Compound Inequality

1. Let h be the height (in inches) of a woman, whose height is at least 64 inches. Describe her height using inequality notation, interval notation, and a graph.
2. A girl went for a run, traveling at most 4 miles. Let d be the distance (in miles) she ran. Describe the distance she ran using inequality notation, interval notation, and a graph.
3. Let w be the weight (in pounds) of a man. Interpret and graph the inequality $173 < w < 177$.

Solution

1. The phrase "at least 64" means greater than or equal to 64. So, we can describe the woman's height in inequality notation as $h \ge 64$ and in interval notation as $[64, \infty)$. To graph the values of h, we shade the part of the number line that lies to the right of 64 and draw a filled-in circle at 64 (see Fig. 24).
2. The phrase "at most 4" means less than or equal to 4. So, the distance traveled is less than or equal to 4 miles. Because the girl did in fact run, the distance traveled is also greater than 0 miles. So, we can describe the distance traveled in inequality notation as $0 < d \le 4$ and in interval notation as $(0, 4]$. To graph the values of d, we shade the part of the number line that lies between 0 and 4 (see Fig. 25). We draw an open circle at 0 and a filled-in circle at 4.
3. The inequality $173 < w < 177$ means the values of w are between 173 and 177. So, the man's weight is between 173 and 177 pounds. To graph the values of w, we shade the part of the number line that lies between 173 and 177 (see Fig. 26). We draw open circles at 173 and 177.

A Woman's Height

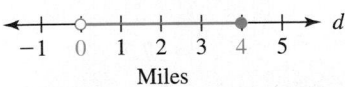

Inches

Figure 24 Graph of $h \ge 64$

A Girl's Running Distance

Miles

Figure 25 Graph of $0 < d \le 4$

A Man's Weight

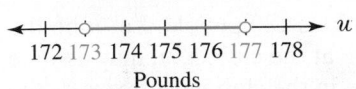

Pounds

Figure 26 Graph of $173 < w < 177$

Describing a Concept or Procedure

In some homework exercises, you will be asked to describe, in general, a concept or procedure.

Guidelines on Writing a Good Response

- Create an example that illustrates the concept or outlines the procedure. Looking at examples or exercises may jump-start you into creating your own example.
- Using complete sentences and correct terminology, describe the key ideas or steps of your example. You can review the text for ideas, but write your description in your own words.
- Describe also the concept or the procedure in general without referring to your example. It may help to reflect on several examples and what they all have in common.
- In some cases, it will be helpful to point out the similarities and the differences between the concept or the procedure and other concepts or procedures.
- Describe the benefits of knowing the concept or the procedure.
- If you have described the steps in a procedure, explain why it's permissible to follow these steps.
- Clarify any common misunderstandings about the concept, or discuss how to avoid making common mistakes when following the procedure.

▶ **Example 12** Responding to a General Question about a Concept

Describe the meaning of *variable*.

Solution

Let t be the number of hours that a person works at Starbucks. The symbol t is an example of a variable, because the value of t can vary. In general, a variable is a symbol that stands for an amount that can vary. A symbol that stands for an amount that does *not* vary is called a constant.

There are many benefits to using variables. We can use a variable to concisely describe a quantity; using the earlier definition of t, we see that the equation $t = 8$ means a person works at Starbucks for 8 hours. By using a variable, we can also use smaller numbers to describe various years.

In defining a variable, it is important to describe its units.

▶

Group Exploration

Reasonable values of a variable

1. Let u be the number of units (credits or hours) a student is currently taking at your college.

   a. Which of the following values of u are reasonable in this situation? Explain.

i. $u = 15$	**iv.** $u = 15.5$
ii. $u = -5$	**v.** $u = 15.1$
iii. $u = 200$	**vi.** $u = 0$

   b. Describe all of the real numbers that are reasonable values of u. Use a number line, a list of numbers, words, or some other way to describe these numbers.

2. A few months ago, a person bought a Porsche 911 Carrera Turbo for $160,700. It has a 17.7-gallon fuel tank. Let g be the amount of gasoline (in gallons) that is in the tank.

   a. Which of the following values of g are reasonable in this situation? Explain.

i. $g = 7$	**iv.** $g = 17.7$
ii. $g = 19$	**v.** $g = 0$
iii. $g = -4$	**vi.** $g = 10.392$

   b. Describe all of the real numbers that are reasonable values of g.

3. The legal capacity of a club is 180 people. Let n be the number of people who are at the club. You may assume that the number of people in the club never exceeds the legal limit. Describe all of the reasonable values of n.

▶ Tips for Success   Take Notes

It is always a good idea to take notes during classroom activities. Not only will you have something to refer to later when doing the homework, but also you will have something to help you prepare for tests. In addition, taking notes makes you become even more involved with the material, which will likely increase your understanding and retention of it.

Homework 1.1

For extra help ▶ MyMathLab®    Watch the videos in MyMathLab    Download the MyDashboard App

1. A(n) ____ is a symbol that represents a quantity that can vary.

2. A(n) ____ is a symbol that represents a specific number.

3. The ____ numbers are all of the numbers represented on the number line.

4. ____ are quantities or categories that describe people, animals, or things.

Respond to the questions in Exercises 5–12 by using complete sentences.

5. Let *n* be the number (in thousands) of fans who attend a Coldplay concert. What does $n = 25$ mean in this situation?

6. Let *p* be the percentage of Americans who own a gun, rifle, or pistol. For 2013, the value of *p* is 34 (Source: *Pew Research Center*). What does $p = 34$ mean in this situation?

7. Let *s* be the annual iPad® sales (in millions). The value of *s* is 70 for 2013 (Source: *Apple*). What does $s = 70$ mean in this situation?

8. Let *p* be the percentage of children ages 6–12 who participate in organized physical activity. The value of *p* is about 40 for 2013 (Source: *The Aspen Institute*). What does $p = 40$ mean in this situation?

9. Let *p* be a company's annual profit (in thousands of dollars). What does $p = -45$ mean in this situation?

10. Let *T* be the temperature (in degrees Fahrenheit). What does $T = -10$ mean in this situation?

11. Let *t* be the number of years since 2000. What does $t = 9$ mean in this situation?

12. Let *t* be the number of years since 2015. What does $t = -3$ mean in this situation?

For Exercises 13–20, choose a variable name for the given quantity. Give two numbers that the variable can represent and two numbers that it cannot represent.

13. The height (in inches) of a person

14. The amount of time (in hours) that a student prepares for an exam

15. The price (in dollars) of a video game

16. The number of students enrolled in a prestatistics class

17. The total time (in hours) a person works in a week

18. The temperature (in degrees Fahrenheit) in an oven

19. The annual salary (in thousands of dollars) of a person

20. The value (in thousands of dollars) of a new home

21. A rectangle has an area of 24 square inches. Let *W* be the width (in inches), *L* be the length (in inches), and *A* be the area (in square inches).

    a. Sketch three possible rectangles of area 24 square inches.
    b. Which of the symbols *W*, *L*, and *A* are variables? Explain.
    c. Which of the symbols *W*, *L*, and *A* are constants? Explain.

22. A rectangle has an area of 36 square feet. Let *W* be the width (in feet), *L* be the length (in feet), and *A* be the area (in square feet).

    a. Sketch three possible rectangles of area 36 square feet.
    b. Which of the symbols *W*, *L*, and *A* are variables? Explain.
    c. Which of the symbols *W*, *L*, and *A* are constants? Explain.

23. The length of a rectangle is 3 inches more than the width. Let *W* be the width (in inches), *L* be the length (in inches), and *A* be the area (in square inches).

    a. Sketch three possible rectangles of length 3 inches more than the width.
    b. Which of the symbols *W*, *L*, and *A* are variables? Explain.
    c. Which of the symbols *W*, *L*, and *A* are constants? Explain.

24. The length of a rectangle is twice the width. Let *W* be the width (in inches), *L* be the length (in inches), and *A* be the area (in square inches). [**Hint:** *Twice* means to multiply by 2.]

    a. Sketch three possible rectangles in which the length is twice the width.
    b. Which of the symbols *W*, *L*, and *A* are variables? Explain.
    c. Which of the symbols *W*, *L*, and *A* are constants? Explain.

Graph all of the given numbers on one number line.

25. $5, -2, 0, -3, 4, -1$

26. $-4, 1, -6, 2, 7, -3$

27. $-\dfrac{2}{3}, -1, \dfrac{7}{3}, 1, -\dfrac{5}{3}, 2$

28. $\dfrac{1}{4}, 0, -2, -\dfrac{5}{4}, \dfrac{9}{4}, 1$

29. $-2, 3.1, 1.2, -1.8, 0.5, 1$

30. $1, 0.2, -2.4, -0.7, 1.9, -1$

Graph the numbers on a number line.

31. Counting numbers between 3 and 8

32. Counting numbers between 1 and 5

33. Integers between -2 and 2, inclusive

34. Integers between -6 and 3, inclusive

35. Negative integers between -4 and 4

36. Positive integers between -4 and 4

Give three examples of the following types of numbers.

37. Negative integers less than -7

38. Integers that are not counting numbers

39. Rational numbers that are not integers

40. Rational numbers between 1 and 2

41. Irrational numbers between 1 and 10

42. Real numbers between -3 and -2

For Exercises 43–48, use points on a number line to describe the given values of a variable.

43. A student goes to a college for six semesters. Here are the numbers of units (credits or hours) taken per semester: $10, 12, 6, 9, 15, 14$. Let u be the number of units taken in one semester.

44. The percentages of airline flights that are on time for various years are $79\%, 82\%, 75\%, 77\%$, and 76%. Let p be the percentage of flights in a year that are on time.

45. The average annual lost time (in hours) due to traffic congestion on highways for various years is $22, 30, 24, 27$, and 32. Let L be the average annual lost time (in hours) due to traffic congestion.

46. The U.S. annual per person consumption of sports drinks (in gallons) for various years is $1.9, 2.5, 2.1, 2.3$, and 2.2. Let c be the per person consumption (in gallons per year) of sports drinks in a year.

47. The low temperatures (in degrees Fahrenheit) for three days in December in Chicago are $5°F$ above zero, $4°F$ below zero, and $6°F$ below zero. Let F be the low temperature (in degrees Fahrenheit) for one day.

48. Here are a company's annual profits and losses for various years: loss of $5 million, profit of $3 million, and loss of $8 million. Let p be the company's annual profit (in millions of dollars).

49. The revenue (in billions of dollars) of Apple in the years 2008, 2009, 2010, 2011, and 2012 is $38, 43, 65, 108$, and 157, respectively (Source: *Apple*). Let r be the annual revenue (in billions of dollars) of Apple.

  a. Use points on a number line to describe the given values of r.

  b. Did the annual revenue increase, decrease, stay approximately constant, or none of these between 2008 and 2012, inclusive? Explain.

  c. Did the annual *increases* in the annual revenue increase, decrease, stay approximately constant, or none of these between 2008 and 2012, inclusive? Explain.

50. The number of hours of video uploaded to YouTube per minute in the years 2009, 2010, 2011, 2012, and 2013 is $14, 25, 48, 73$, and 100, respectively (Source: *YouTube*). Let t be the number of hours of video uploaded to YouTube per minute.

  a. Use points on a number line to describe the given values of t.

  b. Did the number of hours of video uploaded to YouTube per minute increase, decrease, stay approximately constant, or none of these from 2009 to 2013? Explain.

  c. Did the *increases* in the number of hours of video uploaded to YouTube per minute increase, decrease, stay approximately constant, or none of these from 2009 to 2013? Explain.

51. The number (in millions) of smartphone users in the years 2010, 2011, 2012, 2013, and 2014 is $62, 93, 122, 145$, and 164, respectively (Source: *eMarketer*). Let n be the number (in millions) of smartphone users.

  a. Use points on a number line to describe the given values of n.

  b. Did the number of smartphone users increase, decrease, stay approximately constant, or none of these from 2010 to 2014? Explain.

  c. Did the *increases* in the number of smartphone users increase, decrease, stay approximately constant, or none of these from 2010 to 2014? Explain.

52. The number (in millions) of international tourists who travel to India in the years 2009, 2010, 2011, 2012, and 2013 is $5.1, 5.7, 6.2, 6.5$, and 6.6, respectively (Source: *Ministry of Tourism*). Let n be the number (in millions) of international tourists who travel to India in a year.

  a. Use points on a number line to describe the given values of n.

  b. Did the number of international tourists who traveled to India increase, decrease, stay approximately constant, or none of these from 2009 to 2013? Explain.

  c. Did the *increases* in the number of international tourists who traveled to India increase, decrease, stay approximately constant, or none of these from 2009 to 2013? Explain.

For Exercises 53–68, plot the given points in a coordinate system.

53. $(5, 1)$ **54.** $(2, 3)$

55. $(4, -2)$ **56.** $(3, -4)$

57. $(-5, 4)$ **58.** $(-1, 3)$

59. $(-3, -6)$ **60.** $(-5, -2)$

61. $(0, 2)$ **62.** $(0, -4)$

63. $(-3, 0)$ **64.** $(1, 0)$

65. $(2.5, -4.5)$ **66.** $(-3.5, 1.5)$

67. $(-1.3, -3.9)$ **68.** $(-2.4, -4.1)$

69. What is the x-coordinate of the ordered pair $(2, -4)$?

70. What is the y-coordinate of the ordered pair $(2, -4)$?

71. Find the coordinates of points A, B, C, D, E, and F shown in Fig. 27.

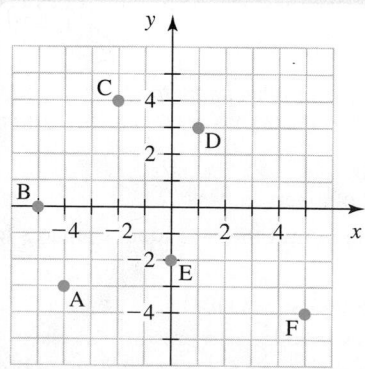

Figure 27 Exercise 71

72. Find the coordinates of points A, B, C, D, E, and F shown in Fig. 28.

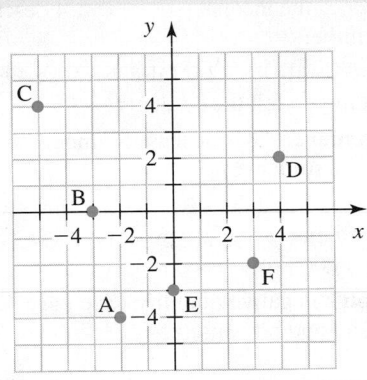

Figure 28 Exercise 72

Decide whether the given inequality is true or false.

73. $-3 > -5$

74. $-6 \leq -2$

75. $4 \geq 4$

76. $-5 < -5$

For Exercises 77–84, sketch the graph of the given inequality.

77. $x < 4$

78. $x < -5$

79. $x \geq -1$

80. $x \geq 1$

81. $x \leq -2$

82. $x \leq 4$

83. $x > 6$

84. $x > -3$

85. Use words, inequalities, graphs, and interval notation to complete Fig. 29.

In Words	Inequality	Graph	Interval Notation
		(graph: point at 4)	
numbers less than or equal to -2			
	$x > -5$		$(-\infty, 1)$

Figure 29 Exercise 85

86. Use words, inequalities, graphs, and interval notation to complete Fig. 30.

In Words	Inequality	Graph	Interval Notation
	$x \leq -6$		
numbers greater than 1			
		(graph: open circle at 5)	$[-4, \infty)$

Figure 30 Exercise 86

For Exercises 87–92, sketch the graph of the given inequality.

87. $2 \leq x \leq 4$

88. $1 \leq x \leq 6$

89. $-2 < x < 3$

90. $-5 < x < 1$

91. $-6 \leq x < -3$

92. $-5 < x \leq -2$

93. Use words, inequalities, graphs, and interval notation to complete Fig. 31.

In Words	Inequality	Graph	Interval Notation
numbers between 1 and 5			
			$[-5, -2)$
	$-2 < x \leq 4$	(graph: points at 0 and 4)	

Figure 31 Exercise 93

94. Use words, inequalities, graphs, and interval notation to complete Fig. 32.

In Words	Inequality	Graph	Interval Notation
			$(-3, 0)$
	$1 \leq x < 4$	(graph: open circle at -3, point at 1)	
numbers between -4 and -1, inclusive			

Figure 32 Exercise 94

95. Let w be the average daily coffee consumption (in ounces) of a person. Graph and interpret the inequality $w > 8$.

96. Let t be the time (in minutes) it takes for a student to complete a homework assignment. Graph and interpret the inequality $t \geq 30$.

97. Let h be the height (in inches) of a man who is at least 70 inches tall. Describe the man's height using inequality notation, interval notation, and a graph.

98. Let s be the running speed (in miles per hour) of a women who is running at a speed of at least 8 miles per hour. Describe the woman's running speed using inequality notation, interval notation, and a graph.

99. Let t be the length (in minutes) of a hip-hop concert that lasts at most 170 minutes. Describe the concert length using inequality notation, interval notation, and a graph.

100. Let G be the volume (in gallons) of gasoline in a car's gas tank in which there are at most 5 gallons. Describe the volume of gasoline in the tank using inequality notation, interval notation, and a graph.

101. Let w be the weight of a hamburger (in ounces) served at a fast-food restaurant. Graph and interpret the inequality $1 \leq w \leq 3$.

102. Let M be the average gas mileage (in miles per gallon) of a car on highways. Graph and interpret the inequality $35 < M < 40$.

103. Let d be the distance (in miles) of a person's work commute that is between 15 and 20 miles, inclusive. Describe the work commute distance using inequality notation, interval notation, and a graph.

104. Let w be the weight (in pounds) of a woman who weighs between 140 and 145 pounds. Describe the woman's weight using inequality notation, interval notation, and a graph.

Concepts

105. Let T be the temperature in degrees Fahrenheit.

    a. What value of T represents the temperature that is 5°F below zero?

    b. A student says that T represents only positive numbers and zero, because there is no negative sign. Is the student correct? Explain.

106. A student says the integers between 2 and 5 are the numbers 2, 3, 4, and 5. Is the student correct? Explain.

107. How is a variable different from a constant?

108. List five ordered pairs whose y-coordinate is 2. Then plot the ordered pairs in a coordinate system. What do you notice about the arrangement of the points? Explain why this makes sense.

109. How many numbers does the inequality $2 < x < 4$ describe? List three of those numbers.

110. A student says the inequality $4 \leq 4$ is false, because 4 is not less than 4. What would you tell the student?

111. A student says the sentence "x is at least 5", means $x < 5$. What would you tell the student?

112. List the various types of numbers discussed in this section and describe the meanings of each type. (See page 12 for guidelines on writing a good response.)

113. Describe how to graph a negative quantity. (See page 12 for guidelines on writing a good response.)

▼ 1.2 Expressions

Objectives

» Describe the meaning of *expression* and *evaluate an expression*.

» Use expressions to describe authentic quantities.

» Evaluate expressions.

» Translate English phrases to and from mathematical expressions.

» Evaluate expressions with more than one variable.

In this section, we will work with expressions—a very important concept in algebra and statistics.

Expressions

Addition, subtraction, multiplication, and division are examples of *operations*. In arithmetic, you performed operations with numbers. Since variables represent numbers, we can perform operations with variables, too.

▶ **Example 1** Using Operations with Variables and Numbers

Each employee at a small company receives a $500 bonus at the end of the year. For each employee's annual salary shown, find the employee's annual salary plus bonus.

    1. $28,000   **2.** $32,000   **3.** s dollars

Solution

    1. The employee's annual salary plus bonus is $28,000 + 500 = 28,500$ dollars.
    2. The employee's annual salary plus bonus is $32,000 + 500 = 32,500$ dollars.
    3. In Problems 1 and 2, we added the annual salary and $500, the bonus, to find the results. So, the employee's annual salary plus bonus (in dollars) is $s + 500$.

In Example 1, we took s to be an employee's annual salary and $s + 500$ to be the employee's annual salary plus bonus. We call s and $s + 500$ *expressions*.

▶ **Definition** Expression

An **expression** is a constant, a variable, or a combination of constants, variables, operation symbols, and grouping symbols, such as parentheses.

Here are some more examples of expressions:

$$t + 6 \qquad \pi \qquad L + W - 9 \qquad y \qquad 4 \qquad 5 \div (x + 2)$$

In Example 1, we used a variable to represent a quantity from an authentic situation. Sometimes we use variables to represent numbers in a math problem that is not being used to describe an authentic situation. In this case, we often use x for the variable. For example, we could let x represent a number. In this case, x could be *any* number.

To avoid confusing the multiplication symbol $\times$ and the variable name x, we use $\cdot$ or no operation symbol to indicate multiplication. For example, each of the following expressions describes multiplying 2 by 3:

$$2 \cdot 3 \qquad 2(3) \qquad (2)3 \qquad (2)(3)$$

And each of the following expressions describes multiplying 2 by k:

$$2 \cdot k \qquad 2k \qquad 2(k) \qquad (2)k \qquad (2)(k)$$

Using Expressions to Describe Authentic Quantities

We can use expressions to describe authentic quantities. In Example 2, we will find such an expression by noticing a pattern as we calculate values of a quantity.

▶ **Example 2** Describing a Quantity

A hot-dog stand sells hot dogs for $3 apiece. Find the total cost of buying the given number of hot dogs.

1. 2 hot dogs   **2.** 5 hot dogs   **3.** 8 hot dogs   **4.** n hot dogs

Solution

1. Two hot dogs cost $2(3) = 6$ dollars.
2. Five hot dogs cost $5(3) = 15$ dollars.
3. Eight hot dogs cost $8(3) = 24$ dollars.
4. In Problems 1–3, we found the total cost by multiplying the number of hot dogs by 3, the cost (in dollars) per hot dog. So, if there are n hot dogs, the total cost (in dollars) is $n(3)$. We can also write the expression as $3(n)$ or $3n$.

▶

In Example 3, we will use a table to help us find an expression that describes an authentic quantity.

▶ **Example 3** Using a Table to Find an Expression

An instructor adds 5 points to each student's test score. Find the new scores if the original scores are 60 points, 70 points, 80 points, and 90 points. Show the arithmetic to help you see the pattern. Organize the calculations in a table, and include an expression that stands for the new score if the original score is s points.

Solution

First, we construct Table 3. From the last row of the table, we see that the expression $s + 5$ represents the new score (in points) for a test with original score s points.

▶

Table 3 Original and New Test Scores

Original Score (points)	New Score (points)
60	60 + 5
70	70 + 5
80	80 + 5
90	90 + 5
s	$s + 5$

▶ **Example 4** Using a Table to Find an Expression

A person drives at a constant speed of 75 miles per hour. Find the distance traveled in 1, 2, 3, and 4 hours of driving at that speed. Show the arithmetic to help you see the pattern. Organize the calculations in a table, and include an expression that stands for the distance traveled in t hours.

Solution

First, we construct Table 4. From the last row of the table, we see that the expression $75t$ represents the distance traveled (in miles) in t hours.

▶

Table 4 Driving Times and Distances

Driving Time (hours)	Distance (miles)
1	$75 \cdot 1$
2	$75 \cdot 2$
3	$75 \cdot 3$
4	$75 \cdot 4$
t	$75 \cdot t$

Evaluating Expressions

In Example 4, we used $75t$ to describe the distance traveled (in miles) in t hours. This means if the driving time is 5 hours, the distance traveled is $75(5) = 375$ miles. To find the distance, we substituted 5 for t. We say we have *evaluated* the expression $75t$ for $t = 5$.

> ### Definition   Evaluate an expression
>
> We **evaluate an expression** by substituting a number for each variable in the expression and then calculating the result. If a variable appears more than once in the expression, the same number is substituted for that variable each time.

When we evaluate an expression, it is good practice to use parentheses each time a number is substituted for a variable. For example, here we evaluate $5x$ for $x = 3$:

$$5(3) = 15$$

This strategy will be especially helpful when we evaluate an expression for a negative number, which we will begin to do in Section 1.4.

> ### ▶ Example 5   Evaluating Expressions
>
> 1. In Example 1, we used s to represent an employee's annual salary (in dollars) and $s + 500$ to represent the employee's annual salary plus bonus (in dollars). Evaluate $s + 500$ for $s = 40,000$, and describe the meaning of the result.
> 2. In Example 2, we used n to represent the number of hot dogs bought and $n(3)$ to represent the total cost (in dollars) of n hot dogs. Evaluate $n(3)$ for $n = 4$, and describe the meaning of the result.

Solution

1. We substitute $40,000$ for s in $s + 500$:

$$(40,000) + 500 = 40,500$$

So, the annual salary plus bonus is $40,500.

2. We substitute 4 for n in $n(3)$:

$$(4)(3) = 12$$

So, the total cost of 4 hot dogs is $12.

Translating English Phrases to and from Expressions

In order to use mathematics to find results for authentic situations, we must translate from English to mathematics and vice versa. To do this, the following definitions are helpful:

> ### ▶ Definition   Product, factor, and quotient
>
> Let a and b be numbers. Then
>
> - The **product** of a and b is ab. We call a and b **factors** of ab.
> - The **quotient** of a and b is $a \div b$, where b is not zero.

For example, since $6 \cdot 3 = 18$, the number 18 is the product of 6 and 3 and the numbers 6 and 3 are factors of 18. The quotient of 6 and 3 is $6 \div 3 = 2$.

Here are some examples of English phrases or sentences and mathematical expressions that have the same meaning:

Operation	English Phrase or Sentence	Mathematical Expression
Addition	A number plus 3	$x + 3$
	The sum of a number and 3	$x + 3$
	The total of a number and 3	$x + 3$
	Add a number and 3.	$x + 3$
	3 more than a number	$x + 3$
	A number increased by 3	$x + 3$
Subtraction	A number minus 3	$x - 3$
	The difference of a number and 3	$x - 3$
	Subtract 3 from a number.	$x - 3$
	3 less than a number	$x - 3$
	A number decreased by 3	$x - 3$
Multiplication	Multiply 3 by a number.	$3x$
	3 times a number	$3x$
	The product of 3 and a number	$3x$
	Twice a number	$2x$
	One-third of a number	$\frac{1}{3}x$
Division	Divide a number by 3.	$x \div 3$
	The quotient of a number and 3	$x \div 3$
	The ratio of a number to 3	$x \div 3$

WARNING   To subtract 2 from 5, we write $5 - 2$, not $2 - 5$. Suppose you have \$5 and you take \$2 from the \$5. Then you have $5 - 2 = 3$ dollars left. So, subtracting 2 from 5 is $5 - 2$.

▶ **Example 6**   Translating from English to Mathematics

Let x be a number.

1. Translate the English phrase "The product of 2 and the number" into an expression.
2. Evaluate your result in Problem 1 for $x = 3$.
3. Evaluate your result in Problem 1 for $x = 7$.

Solution

1. The expression is $2x$.
2. $2(3) = 6$
3. $2(7) = 14$

▶

▶ **Example 7**   Translating from English to Mathematics

Let x be a number. Translate the English phrase or sentence into an expression. Then evaluate the expression for $x = 6$.

1. The quotient of the number and 3
2. Subtract the number from 8.

Solution

1. The expression is $x \div 3$. Next, we evaluate $x \div 3$ for $x = 6$:

$$(6) \div 3 = 2$$

2. The expression is $8 - x$. Next, we evaluate $8 - x$ for $x = 6$:

$$8 - (6) = 2$$

▶

▶ **Example 8**   Translating from Mathematics to English

Let x be a number. Translate the expression into an English phrase.

1. $6 - x$ **2.** $8x$

Solution

1. The difference of 6 and the number
2. The product of 8 and the number

▶

Expressions with More than One Variable

An expression may contain more than one variable. For example, let W be the width (in feet) and let L be the length (in feet) of a rectangle (see Fig. 33).

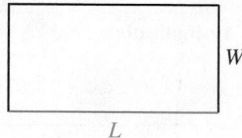

Figure 33 The length L and width W of a rectangle

Recall that the area of a rectangle is equal to the length times the width of the rectangle, so the area (in square feet) is equal to the expression LW. We can evaluate the expression LW for $L = 4$ and $W = 3$:

$$(4)(3) = 12$$

So, a 3-foot by 4-foot rectangle has an area of 12 square feet.

Note the power of algebra in that the expression LW *concisely* tells us how to find the area of *any* rectangle, no matter what its dimensions are.

▶ **Example 9**   Evaluating an Expression in Two Variables

If it takes a student T minutes to complete a test that has n questions, then $T \div n$ is the average time (in minutes) taken to respond to one question. Evaluate $T \div n$ for $T = 48$ and $n = 16$. What does the result mean in this situation?

Solution

We substitute 48 for T and 16 for n in the expression $T \div n$ and then calculate the result:

$$48 \div 16 = 3$$

If it takes a student 48 minutes to respond to 16 questions, the average response time is 3 minutes per question.

▶

▶ **Example 10**   Translating from English to Mathematics

Write the phrase as a mathematical expression, and then evaluate the result for $x = 8$ and $y = 4$.

1. The sum of x and y **2.** The quotient of x and y

Solution

1. The expression is $x + y$. Next, we evaluate $x + y$ for $x = 8$ and $y = 4$:

$$(8) + (4) = 12$$

2. The expression is $x \div y$. Next, we evaluate $x \div y$ for $x = 8$ and $y = 4$:

$$(8) \div (4) = 2$$

▶

Group Exploration

Expressions used to describe a quantity

Consider the expression $x + 2$. Suppose that a child has grown 2 inches within the last year. We could define x to be the child's height (in inches) last year, and then $x + 2$ would be the child's current height (in inches).

Describe a situation in which x represents a meaningful quantity and the expression given describes another meaningful quantity.

1. $x + 3$ **2.** $x - 4$

3. $3x$ **4.** $x \div 2$

For each of the four expressions, evaluate it for a reasonable value of x and describe the meaning of the result.

▶ **Tips for Success**   Study Time

For each hour of class time, study for at least two hours outside class. If your math background is weak, you may need to spend more time studying.

One way to study is to do what you are doing now: Read the text. Class time is a great opportunity to be introduced to new concepts and to see how they fit together with previously learned ones. However, there is usually not enough time to address details as well as a textbook can. In this way, a textbook can serve as a supplement to what you learn in class.

Homework 1.2

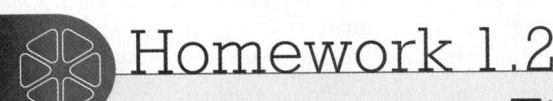

For extra help ▶ **MyMathLab®**   ▣ Watch the videos in MyMathLab   Download the MyDashboard App

1. A(n) _____ is a constant, a variable, or a combination of constants, variables, operation symbols, and grouping symbols, such as parentheses.

2. We _____ an expression by substituting a number for each variable in the expression and then calculating the result.

3. The product of a and b is _____.

4. For a rectangle with width W and length L, the area is equal to _____.

For Exercises 5–16, evaluate the expression for $x = 6$.

5. $x + 2$ **6.** $5 + x$ **7.** $9 - x$ **8.** $x - 4$

9. $7x$ **10.** $x(9)$ **11.** $x \div 3$ **12.** $30 \div x$

13. $x + x$ **14.** $x - x$ **15.** $x \cdot x$ **16.** $x \div x$

17. If a person buys n albums, the total cost is $9n$ dollars. Evaluate $9n$ for $n = 4$. What does your result mean in this situation?

18. If a student earns a total of T points on five tests, then $T \div 5$ is the student's average test score (in points). If a student earns a total of 440 points on five tests, what is the student's average test score?

19. For the period 2009–2013, if T is the number (in millions) of Comcast video subscribers, then $T - 10.3$ is approximately the number (in millions) of Time Warner Cable video subscribers (Source: *Comcast, Time Warner Cable*). There were about 21.7 million Comcast video subscribers in 2013. Estimate the number of Time Warner Cable video subscribers in 2013.

20. For the period 2005–2009, if M is the number (in millions) of male students in college, then $M + 2.68$ is approximately the number (in millions) of female students in college. There were about 8.77 million male students in college in 2009 (Source: *U.S. National Center for Education Statistics*). Estimate the number of female students in college in 2009.

21. Each student at a community college pays a student services fee of $20.

  a. Complete Table 5 to help find an expression that describes the total cost (in dollars) of tuition plus the services fee if a student pays t dollars for tuition. Show the arithmetic to help you see a pattern.

Table 5 Tuition and Total Cost

Tuition (dollars)	Total Cost (dollars)
400	
401	
402	
403	
t	

  b. Evaluate the expression you found in part (a) for $t = 417$. What does your result mean in this situation?

22. A person is driving 5 miles per hour over the speed limit.

  a. Complete Table 6 to help find an expression that describes the driving speed (in miles per hour) if the speed limit is

s miles per hour. Show the arithmetic to help you see a pattern.

Table 6 Speed Limit and Driving Speed

Speed Limit (miles per hour)	Driving Speed (miles per hour)
35	
40	
45	
50	
s	

b. Evaluate the expression you found in part (a) for $s = 65$. What does your result mean in this situation?

23. For the fall semester 2014, district residents at St. Louis Community College paid an enrollment fee of $101 per credit hour (unit). (Source: *St. Louis Community College*)

   a. Complete Table 7 to help find an expression that describes the total cost (in dollars) of enrolling in n credit hours of classes. Show the arithmetic to help you see a pattern.

Table 7 Credit Hours and Total Cost

Number of Credit Hours of Courses	Total Cost (dollars)
1	
2	
3	
4	
n	

   b. Evaluate the expression you found in part (a) for $n = 15$. What does your result mean in this situation?

24. Each share of Nike Inc stock was worth $93.64 on January 20, 2015 (Source: *Google Finance*).

   a. Complete Table 8 to help find an expression that describes the total value (in dollars) of n shares of the stock. Show the arithmetic to help you see a pattern.

Table 8 Number of Shares and Total Value

Number of Shares	Total Value (dollars)
1	
2	
3	
4	
n	

   b. Evaluate the expression you found in part (a) for $n = 7$. What does your result mean in this situation?

For Exercises 25–34, let x be a number. Translate the English phrase or sentence into a mathematical expression. Then evaluate the expression for x = 8.

25. The number plus 4   26. 8 minus the number

27. The quotient of the number and 2

28. Add 6 and the number.

29. Subtract 5 from the number.

30. 15 more than the number

31. The product of 7 and the number

32. The difference of the number and 7

33. 16 divided by the number

34. Multiply the number by 5.

Let x be a number. Translate the expression into an English phrase.

35. $x \div 2$ 36. $6 \div x$ 37. $7 - x$ 38. $x - 2$

39. $x + 5$ 40. $4 + x$ 41. $9x$ 42. $x(5)$

43. $x - 7$ 44. $x + 3$ 45. $x(2)$ 46. $x \div 5$

Evaluate the expression for x = 6 and y = 3.

47. $x + y$ 48. $y + x$ 49. $x - y$

50. xy 51. yx 52. $x \div y$

For Exercises 53–56, translate the phrase into a mathematical expression. Then evaluate the expression for x = 9 and y = 3.

53. The product of x and y 54. The sum of x and y

55. The difference of x and y 56. The quotient of x and y

57. If a car travels at a constant speed of r miles per hour for t hours, it will travel rt miles. Evaluate rt for $r = 62$ and $t = 3$. What does your result mean in this situation?

58. If a car can travel m miles on g gallons of gasoline, then the car's gas mileage is $m \div g$ miles per gallon. Evaluate $m \div g$ for $m = 240$ and $g = 12$. What does your result mean in this situation?

59. For the period 1992–2009, if E is the average verbal SAT score (in points) for a certain year, then the average math SAT score (in points) for that year is approximately $E + t$, where t is the number of years since 1992. The average verbal SAT score was 501 points in 2009. Estimate the average math SAT score in 2009.

60. Let b be the balance (in dollars) of a checking account. If a check is written for d dollars, then the new balance (in dollars) is $b - d$. Evaluate $b - d$ for $b = 3758$ and $d = 994$. What does your result mean in this situation?

Concepts

61. **a.** Evaluate $6 + x$ for $x = 1$, $x = 2$, and $x = 3$.
   b. Evaluate $6x$ for $x = 1$, $x = 2$, and $x = 3$.
   c. A student says that the expressions $6 + x$ and $6x$ are the same thing. What would you tell the student?

62. **a.** Evaluate $x + 2$ for $x = 4$, $x = 5$, and $x = 6$.
   b. Evaluate $2x$ for $x = 4$, $x = 5$, and $x = 6$.
   c. A student says that the expressions $x + 2$ and $2x$ are the same thing. What would you tell the student?

63. A person gets paid $10t$ dollars for t hours of work.
   a. Evaluate $10t$ for $t = 1$, $t = 2$, $t = 3$, and $t = 4$. Describe the meaning of these results.
   b. Refer to your results from part (a) to determine how much the person gets paid per hour. Explain.
   c. Compare your result from part (b) with the expression $10t$. What do you notice?

64. A person drives $50t$ miles in t hours.

    a. Evaluate $50t$ for $t = 1, t = 2, t = 3$, and $t = 4$. Describe the meaning of your results.

    b. Refer to your results from part (a) to determine at what speed the person is traveling. Explain.

    c. Compare your result from part (b) with the expression $50t$. What do you notice?

65. Compare the meaning of *variable* with the meaning of *expression*. (See page 12 for guidelines on writing a good response.)

66. Give an example of an expression containing a variable, and then evaluate it three times to get three different results.

67. Describe an authentic situation for the expression $8x$. Include a definition for the variable x in your description.

68. Describe an authentic situation for the expression $200 \div x$. Include a definition for the variable x in your description.

▼ 1.3 Operations with Fractions and Proportions; Converting Units

Objectives

» Describe the meaning of a fraction.

» Explain why division by zero is undefined.

» Describe the rules for $a \cdot 1$, $\dfrac{a}{1}$, and $\dfrac{a}{a}$.

» Perform operations with fractions.

» Find the prime factorization of a number.

» Simplify fractions.

» Find proportions.

» Convert units of quantities.

In this section, we will perform operations with fractions, which are used in numerous fields, including music, social science, business, engineering, and statistics.

Meaning of a Fraction

A fraction can be used to describe a part of a whole. For example, consider the meaning of $\dfrac{3}{4}$ of a pizza. If we divide the pizza into 4 slices of *equal* area, 3 of the slices make up $\dfrac{3}{4}$ of the pizza (see Fig. 34).

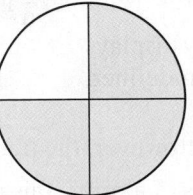

Figure 34 $\dfrac{3}{4}$ of a pizza

WARNING

Even though the orange region in Fig. 35 is 1 of 3 parts, it is *not* $\dfrac{1}{3}$ of the pizza, because the 3 parts do not have equal area. The orange region *is* equal to $\dfrac{1}{2}$ of the pizza, because it is 1 of 2 parts of equal area that make up the pizza (see Fig. 36).

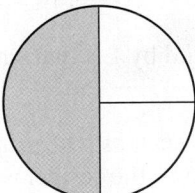

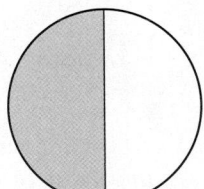

Figure 35 The 3 parts do not have equal area

Figure 36 The 2 parts have equal area, so the orange part is $\dfrac{1}{2}$ of the pizza

The fraction $\dfrac{a}{b}$ means $a \div b$. For example, $\dfrac{8}{4} = 8 \div 4 = 2$. So 8 quarters of pizza make 2 pizzas with 4 slices each (see Fig. 37).

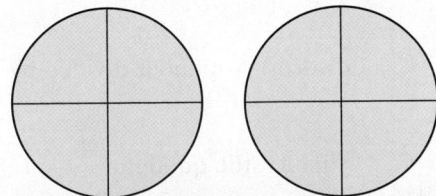

Figure 37 The 8 quarters of pizza make 2 pizzas

Division by Zero

We can think of division in terms of repeated subtraction. For example, $17 \div 5$ is equal to 3 with a remainder of 2 (try it). This means if we subtract 5 from 17 three times, the result is 2 (the remainder):

$$17 - 5 = 12, \qquad 12 - 5 = 7, \qquad 7 - 5 = 2$$

Note that the remainder, 2, is less than the divisor, 5.

As a matter of fact, the remainder must always be less than the divisor. This rule will help us see that division by 0 is undefined. For example, consider $8 \div 0$. No matter how many times we subtract 0 from 8, the result is always 8:

$$8 - 0 = 8, \qquad 8 - 0 = 8, \qquad 8 - 0 = 8, \text{ and so on}$$

If $8 \div 0$ is defined, the remainder would have to be the repeated result 8. Since the remainder must be less than the divisor, it is implied that 8 is less than 0, which is false. So, $8 \div 0$ is undefined. In fact, any number divided by 0 is undefined.

> **Division by Zero**
>
> The fraction $\dfrac{a}{b}$ is undefined if $b = 0$. Division by 0 is undefined.

For example, $\dfrac{6}{0}$ is undefined. If you use a calculator to divide by 0, the screen will likely display "Error," "ERR:," "E," or "ERR: Divide by 0" to indicate that division by 0 is undefined.

WARNING   However, the fraction $\dfrac{0}{6}$ *is* defined. In fact, $\dfrac{0}{6} = 0$. For example, if a person eats zero sixths of a pizza, this means that the person didn't eat any pizza.

Rules for $a \cdot 1$, $\dfrac{a}{1}$, and $\dfrac{a}{a}$

The products $4 \cdot 1 = 4$, $5 \cdot 1 = 5$, and $8 \cdot 1 = 8$ suggest the following property:

> **Multiplying a Number by 1**
>
> $$a \cdot 1 = a$$
>
> In words: A number multiplied by 1 is that same number.

When we write statements such as $a \cdot 1 = a$, we mean if we evaluate $a \cdot 1$ and a for *any* value of a in both expressions, the results will be equal. We say that the expressions $a \cdot 1$ and a are **equivalent expressions.**

The quotients $\dfrac{4}{1} = 4 \div 1 = 4$, $\dfrac{5}{1} = 5 \div 1 = 5$, and $\dfrac{8}{1} = 8 \div 1 = 8$ suggest the following property:

> **Dividing a Number by 1**
>
> $$\frac{a}{1} = a$$
>
> In words: A number divided by 1 is that same number.

Finally, the quotients $\dfrac{4}{4} = 4 \div 4 = 1$, $\dfrac{5}{5} = 5 \div 5 = 1$, and $\dfrac{8}{8} = 8 \div 8 = 1$ suggest the following property:

> **Dividing a Nonzero Number by Itself**
>
> If a is nonzero, then
>
> $$\frac{a}{a} = 1$$
>
> In words: A nonzero number divided by itself is 1.

The properties $a \cdot 1 = a$, $\frac{a}{1} = a$, and $\frac{a}{a} = 1$ (where a is nonzero) will help us when we work with fractions.

Multiplication of Fractions

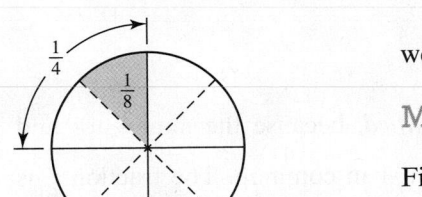

Figure 38 $\frac{1}{2}$ of $\frac{1}{4}$ of a pizza is $\frac{1}{8}$ of a pizza

Figure 38 illustrates that $\frac{1}{2}$ of $\frac{1}{4}$ of a pizza is $\frac{1}{8}$ of a pizza. We can calculate this result by finding the product $\frac{1}{2} \cdot \frac{1}{4}$:

$$\frac{1}{2} \cdot \frac{1}{4} = \frac{1 \cdot 1}{2 \cdot 4} = \frac{1}{8}$$

> **Multiplying Fractions**
>
> If b and d are nonzero, then
>
> $$\frac{a}{b} \cdot \frac{c}{d} = \frac{ac}{bd}$$
>
> In words: To multiply two fractions, write the numerators as a product and write the denominators as a product.

▶ **Example 1**   Finding the Product of Two Fractions

Find the product $\frac{2}{5} \cdot \frac{3}{7}$.

Solution

$$\frac{2}{5} \cdot \frac{3}{7} = \frac{2 \cdot 3}{5 \cdot 7} \qquad \text{Write numerators and denominators as products: } \frac{a}{b} \cdot \frac{c}{d} = \frac{ac}{bd}$$

$$= \frac{6}{35} \qquad \text{Find products.}$$

Prime Factorization

When we work with fractions, it can sometimes help to work with prime numbers.

> ▶ **Definition**   Prime number
>
> A **prime number,** or **prime,** is any counting number larger than 1 whose only positive factors are itself and 1.

Here are the first ten primes:

$$2, 3, 5, 7, 11, 13, 17, 19, 23, 29$$

Sometimes when we work with fractions, it is helpful to write a number as a product of primes. We call this product the **prime factorization** of the number.

▶ **Example 2**   Writing a Number as a Product of Primes

Write 54 as a product of primes.

Solution

$$54 = \underset{\downarrow\ \ \downarrow}{6} \cdot \underset{\downarrow\ \ \downarrow}{9} \qquad \textit{Write 54 as a product of two numbers.}$$
$$= 2 \cdot 3 \cdot 3 \cdot 3 \qquad \textit{Find prime factorizations of 6 and 9.}$$

The prime factorization of 54 is $2 \cdot 3 \cdot 3 \cdot 3$.

Simplifying Fractions

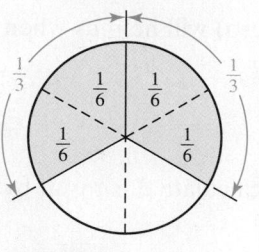

Figure 39 $\dfrac{4}{6} = \dfrac{2}{3}$

Figure 39 illustrates that $\dfrac{4}{6} = \dfrac{2}{3}$. We say $\dfrac{2}{3}$ is *simplified,* because the numerator and denominator do not have positive factors other than 1 in common. The fraction $\dfrac{4}{6}$ is not simplified, because the numerator and the denominator have a common factor of 2. To **simplify** a fraction, we write it as an equal fraction in which the numerator and the denominator do not have any common positive factors other than 1.

▶ **Example 3**   Simplifying Fractions

Simplify.

1. $\dfrac{4}{6}$ **2.** $\dfrac{30}{42}$

Solution

1. We begin to simplify $\dfrac{4}{6}$ by finding the prime factorizations of the numerator 4 and the denominator 6:

$$\frac{4}{6} = \frac{2 \cdot 2}{2 \cdot 3} = \frac{2}{2} \cdot \frac{2}{3} = 1 \cdot \frac{2}{3} = \frac{2}{3}$$

Our result matches with what we found in Fig. 39. By performing long division or using a calculator, we can check that both fractions are equal to the repeating decimal $0.\overline{6}$.

2. We begin to simplify $\dfrac{30}{42}$ by finding the prime factorizations of the numerator, 30, and the denominator, 42:

$$\frac{30}{42} = \frac{2 \cdot 3 \cdot 5}{2 \cdot 3 \cdot 7} = \frac{2 \cdot 3}{2 \cdot 3} \cdot \frac{5}{7} = 1 \cdot \frac{5}{7} = \frac{5}{7}$$

Simplifying fractions can make it easier to work out certain problems. Also, if two fractions are simplified, it is easy to tell whether they are equal. **If the result of an exercise is a fraction, simplify it.**

▶ **Simplifying a Fraction**

To simplify a fraction,
1. Find the prime factorizations of the numerator and denominator.
2. Find an equal fraction in which the numerator and the denominator do not have common positive factors other than 1 by using the property

$$\frac{ab}{ac} = \frac{a}{a} \cdot \frac{b}{c} = 1 \cdot \frac{b}{c} = \frac{b}{c}$$

where a and c are nonzero.

In Example 4, we will multiply two fractions and simplify the result.

▶ **Example 4**   Finding the Product of Two Fractions

Find the product $\dfrac{8}{9} \cdot \dfrac{15}{4}$.

Solution

$$\dfrac{8}{9} \cdot \dfrac{15}{4} = \dfrac{8 \cdot 15}{9 \cdot 4} \qquad \textit{Write numerators and denominators as products: } \dfrac{a}{b} \cdot \dfrac{c}{d} = \dfrac{ac}{bd}$$

$$= \dfrac{2 \cdot 2 \cdot 2 \cdot 3 \cdot 5}{3 \cdot 3 \cdot 2 \cdot 2} \qquad \textit{Find prime factorizations.}$$

$$= \dfrac{2 \cdot 5}{3} \qquad \textit{Simplify: } \dfrac{2 \cdot 2 \cdot 3}{2 \cdot 2 \cdot 3} = 1$$

$$= \dfrac{10}{3} \qquad \textit{Multiply.}$$

Division of Fractions

The **reciprocal** of $\dfrac{a}{b}$ is $\dfrac{b}{a}$. For example, the reciprocal of $\dfrac{3}{8}$ is $\dfrac{8}{3}$. We will need to find the reciprocal of a fraction when we divide two fractions.

Dividing Fractions

If b, c, and d are nonzero, then

$$\dfrac{a}{b} \div \dfrac{c}{d} = \dfrac{a}{b} \cdot \dfrac{d}{c}$$

In words: To divide by a fraction, multiply by its reciprocal.

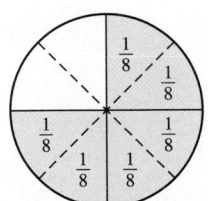

Figure 40 $\dfrac{3}{4}$ of a pizza divided into slices of size $\dfrac{1}{8}$ of the pizza gives 6 slices of pizza

Press $(\,3 \div 4\,) \div$
$(\,1 \div 8\,)$ ENTER.

Figure 41 Verify the work

▶ **Example 5**   Finding the Quotient of Two Fractions

Find the quotient $\dfrac{3}{4} \div \dfrac{1}{8}$.

Solution

$$\dfrac{3}{4} \div \dfrac{1}{8} = \dfrac{3}{4} \cdot \dfrac{8}{1} = \dfrac{3 \cdot 2 \cdot 2 \cdot 2}{2 \cdot 2 \cdot 1} = \dfrac{3 \cdot 2}{1} = 6$$

Our result makes sense, because $\dfrac{3}{4}$ of a pizza divided into slices, each of size $\dfrac{1}{8}$ of the pizza, gives 6 slices (see Fig. 40).

We use a TI-84 graphing calculator to check our work in Example 5 (see Fig. 41). From now on, we will refer to the calculator simply as "TI-84."

When you use a calculator to check work with fractions, it is good practice to enclose each fraction in parentheses. You will see the importance of using parentheses when we discuss the order of operations in Section 1.7.

To find the reciprocal of 6, we use the fact $6 = \dfrac{6}{1}$. So, the reciprocal of 6 is $\dfrac{1}{6}$.

▶ **Example 6**   Evaluating an Expression

Evaluate $\dfrac{a}{b} \div c$ for $a = 21$, $b = 2$, and $c = 3$.

Solution

We substitute $a = 21$, $b = 2$, and $c = 3$ into the expression $\dfrac{a}{b} \div c$:

$$\frac{(21)}{(2)} \div (3) = \frac{21}{2} \div \frac{3}{1} = \frac{21}{2} \cdot \frac{1}{3} = \frac{3 \cdot 7 \cdot 1}{2 \cdot 3} = \frac{7}{2}$$

▶

In Example 6, the result is $\dfrac{7}{2}$, which is an improper fraction (that is, the numerator is larger than the denominator). For *pure-math* exercises, if a fractional result is in improper form, we will leave it in that form. For exercises that involve an authentic situation, if a result is in improper form, we will write it as a mixed number. For example, we say that a car trip takes $3\dfrac{1}{2}$ hours rather than $\dfrac{7}{2}$ hours.

Addition of Fractions

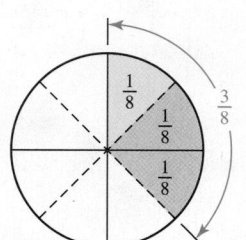

Figure 42 $\dfrac{1}{8}$ pizza plus $\dfrac{2}{8}$ pizza is $\dfrac{3}{8}$ pizza

Figure 42 illustrates that $\dfrac{1}{8}$ of a pizza plus $\dfrac{2}{8}$ of a pizza is equal to $\dfrac{3}{8}$ of a pizza. This illustration suggests that, to find the sum $\dfrac{1}{8} + \dfrac{2}{8}$, we add the numerators 1 and 2 and write the result, 3, over the common denominator, 8:

$$\frac{1}{8} + \frac{2}{8} = \frac{1 + 2}{8} = \frac{3}{8}$$

▶ **Adding Fractions with the Same Denominator**

If b is nonzero, then

$$\frac{a}{b} + \frac{c}{b} = \frac{a + c}{b}$$

In words: To add two fractions with the same denominator, add the numerators and write the result above the common denominator.

▶ **Example 7**   Adding Fractions with the Same Denominator

Find the sum $\dfrac{4}{15} + \dfrac{6}{15}$.

Solution

$$\frac{4}{15} + \frac{6}{15} = \frac{4 + 6}{15} = \frac{10}{15} = \frac{2 \cdot 5}{3 \cdot 5} = \frac{2}{3}$$

▶

Least Common Denominators

To find the sum $\dfrac{1}{4} + \dfrac{5}{6}$, in which the denominators of the fractions are different, we find an equal sum of fractions in which the denominators are equal. First, we list the multiples of 4 and the multiples of 6:

Multiples of 4:   4, 8, 12, 16, 20, 24, 28, 32, 36, . . .
Multiples of 6:   6, 12, 18, 24, 30, 36, 42, 48, 54, . . .

Common multiples of 4 and 6 are

$$12, 24, 36, \ldots$$

Note that 12 is the least (lowest) number in the list. We call it the least common multiple of 4 and 6. The **least common multiple (LCM)** of a group of numbers is the smallest number that is a multiple of *all* of the numbers in the group.

To find the sum $\dfrac{1}{4} + \dfrac{5}{6}$, we use the fact $\dfrac{a}{a} = 1$, where a is nonzero, to write an equal sum of fractions in which each denominator is equal to the LCM, 12:

$$\frac{1}{4} + \frac{5}{6} = \frac{1}{4} \cdot 1 + \frac{5}{6} \cdot 1 \qquad a = a \cdot 1$$

$$= \frac{1}{4} \cdot \frac{3}{3} + \frac{5}{6} \cdot \frac{2}{2} \qquad 1 = \frac{a}{a}$$

$$= \frac{3}{12} + \frac{10}{12} \qquad \begin{array}{l} \textit{Multiply numerators and multiply denominators:} \\ \dfrac{a}{b} \cdot \dfrac{c}{d} = \dfrac{ac}{bd} \end{array}$$

$$= \frac{13}{12} \qquad \begin{array}{l} \textit{Add numerators and keep common denominator:} \\ \dfrac{a}{b} + \dfrac{c}{b} = \dfrac{a+c}{b} \end{array}$$

We also call 12 the least common denominator of $\dfrac{1}{4}$ and $\dfrac{5}{6}$. The **least common denominator (LCD)** of a group of fractions is the LCM of the denominators of all of the fractions.

▶ **Example 8**   **Adding Fractions with Different Denominators**

Find the sum $\dfrac{5}{8} + \dfrac{5}{6}$.

Solution

We list multiples of 8 and multiples of 6:

> **Multiples of 8:**   8, 16, 24, 32, 40, 48, . . .
>
> **Multiples of 6:**   6, 12, 18, 24, 30, 36, . . .

The LCD is 24. We write an equal sum of fractions in which each denominator is 24:

$$\frac{5}{8} + \frac{5}{6} = \frac{5}{8} \cdot \frac{3}{3} + \frac{5}{6} \cdot \frac{4}{4} \qquad \text{LCD is 24.}$$

$$= \frac{15}{24} + \frac{20}{24} \qquad \begin{array}{l} \textit{Multiply numerators and multiply denominators:} \\ \dfrac{a}{b} \cdot \dfrac{c}{d} = \dfrac{ac}{bd} \end{array}$$

$$= \frac{35}{24} \qquad \begin{array}{l} \textit{Add numerators and keep common denominator:} \\ \dfrac{a}{b} + \dfrac{c}{b} = \dfrac{a+c}{b} \end{array}$$

We use a TI-84 to verify the work (see Fig. 43).

```
(5/8)+(5/6)
          1.458333333
35/24
          1.458333333
```

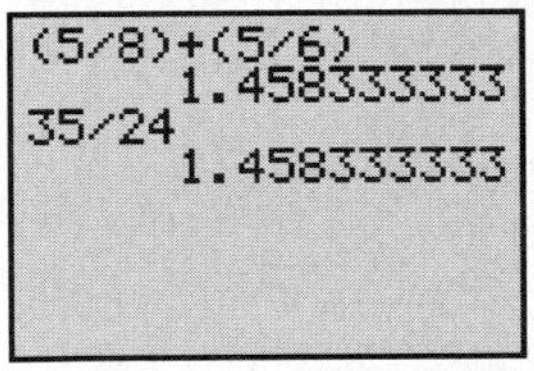

Figure 43 Verify the work

Subtraction of Fractions

The rule for subtracting two fractions with the same denominator is similar to the rule for adding such fractions, except we subtract the numerators.

> **Subtracting Fractions with the Same Denominator**

If b is nonzero, then

$$\frac{a}{b} - \frac{c}{b} = \frac{a - c}{b}$$

In words: To subtract two fractions with the same denominator, subtract the numerators and write the result above the common denominator.

> **Example 9**   Subtracting Fractions with the Same Denominator

Find the difference $\dfrac{5}{8} - \dfrac{3}{8}$.

Solution

$$\frac{5}{8} - \frac{3}{8} = \frac{5 - 3}{8} \qquad \text{\textit{Write numerators as a difference and keep common}}$$
$$\text{\textit{denominator:}} \ \frac{a}{b} - \frac{c}{b} = \frac{a - c}{b}$$

$$= \frac{2}{8} \qquad \text{\textit{Find difference.}}$$

$$= \frac{1}{4} \qquad \text{\textit{Simplify.}}$$

Subtracting fractions with different denominators is similar to adding them. The first step is to rewrite each fraction so that each denominator is the LCD.

> **Example 10**   Subtracting Fractions with Different Denominators

Find the difference $\dfrac{8}{9} - \dfrac{3}{5}$.

Solution

We list the multiples of 9 and the multiples of 5:

> **Multiples of 9:**   9, 18, 27, 36, 45, 54, 63, 72, 81, . . .
>
> **Multiples of 5:**   5, 10, 15, 20, 25, 30, 35, 40, 45, . . .

The LCD is 45. We now rewrite each fraction with the denominator 45:

$$\frac{8}{9} - \frac{3}{5} = \frac{8}{9} \cdot \frac{5}{5} - \frac{3}{5} \cdot \frac{9}{9} \qquad \text{\textit{LCD is 45.}}$$

$$= \frac{40}{45} - \frac{27}{45} \qquad \begin{array}{l}\text{\textit{Multiply numerators and multiply denominators:}}\\ \frac{a}{b} \cdot \frac{c}{d} = \frac{ac}{bd}\end{array}$$

$$= \frac{13}{45} \qquad \begin{array}{l}\text{\textit{Subtract numerators and keep common denominator:}}\\ \frac{a}{b} - \frac{c}{b} = \frac{a - c}{b}\end{array}$$

> **Adding (or Subtracting) Fractions with Different Denominators**

To add (or subtract) two fractions with different denominators, use the fact that $\dfrac{a}{a} = 1$, where a is nonzero, to write an equal sum (or difference) of fractions for which each denominator is the LCD.

Finding Proportions

If 3 of 4 bank tellers are women, then we say the *proportion* of tellers who are women is $\frac{3}{4} = 0.75$.

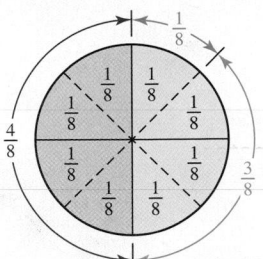

Figure 44 Sum of the proportions equals 1

> ▶ **Definition   Proportion**
>
> In statistics, a **proportion** is a fraction of the whole. A proportion can also be written as a decimal number.

Because a proportion is a fraction of the whole, **a proportion is always between 0 and 1, inclusive.** It is never negative or larger than 1. So, fractions such as $-\frac{3}{5}$ and $\frac{7}{2}$ are *not* proportions. Proportions are used a great deal in statistics.

In Fig. 44, a pizza is made up of 3 parts shaded green, blue, and orange. Here we add their proportions:

$$\frac{1}{8} + \frac{3}{8} + \frac{4}{8} = \frac{1 + 3 + 4}{8} = \frac{8}{8} = 1$$

So, the sum of the proportions equals 1, which stands for the *whole* pizza.

> ▶ **Sum of the Proportions Equals 1**
>
> If an object is made up of two or more parts, then the sum of their proportions equals 1.

The word "and" is used differently in mathematics than in English. To emphasize this difference, we will use capital letters for the mathematical version "AND." When "**AND**" is used with two categories, this means to consider the observations that the categories have in common. For example, the category of wealthy people AND Californians is the category of wealthy Californians.

When "**OR**" is used with two categories, this mean to consider the members in the categories all together. So, the category of wealthy people OR Californians is the category of all wealthy people together with all Californians.

Although the word "not" is used the same way in mathematics as in English, we will use the capital version "NOT" when performing mathematics so that we are consistent in using capital letters for the three key words "AND," "OR," and "NOT."

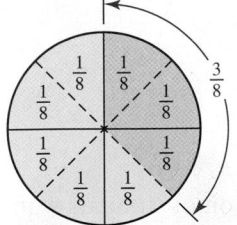

Figure 45 Find the proportion of the pizza NOT shaded orange

▶ **Example 11**   Find the Proportion of the Rest

Find the proportion of the pizza that is NOT shaded orange in Fig. 45.

Solution

We will use two methods to find the proportion of the pizza that is NOT orange.

Method 1   The part that is NOT orange consists of 5 blue slices that are each $\frac{1}{8}$ of the pizza. So, the proportion is $\frac{5}{8}$.

Method 2   The pizza is made up of the blue part and the orange part. So, the sum of their proportions equals 1. Therefore, we can find the proportion that is blue by subtracting the proportion that is orange, $\frac{3}{8}$, from 1:

$$1 - \frac{3}{8} = \frac{8}{8} - \frac{3}{8} \qquad \frac{a}{a} = 1$$

$$= \frac{5}{8} \qquad \frac{a}{b} - \frac{c}{b} = \frac{a-c}{b}$$

The proportion is $\frac{5}{8}$, which is the same result we found earlier.

Although Method 1 probably seems more straightforward than Method 2 in Example 11, we will use Method 2 to solve many problems in which Method 1 won't be an option.

▶ Proportion of the Rest

Let $\frac{a}{b}$ be the proportion of the whole that has a certain characteristic. Then the proportion of the whole that does *not* have that characteristic is

$$1 - \frac{a}{b}$$

▶ Example 12   Find the Proportion of the Rest

A student completes $\frac{7}{15}$ of a math assignment. What proportion of the assignment is not completed?

Solution

Because the proportion of the assignment completed is $\frac{7}{15}$, the proportion NOT completed is

$$1 - \frac{7}{15} = \frac{15}{15} - \frac{7}{15} \qquad \frac{a}{a} = 1$$

$$= \frac{8}{15} \qquad \frac{a}{b} - \frac{c}{b} = \frac{a-c}{b}$$

▶ Example 13   Find the Proportion of the Rest

At the City College of New York (CCNY) in fall 2013, the proportion of students who were African American was approximately $\frac{1}{5}$. The proportion of CCNY students who were Asian/Pacific Islander was approximately $\frac{1}{4}$ (Source: *CCNY*). Find the proportion of CCNY students who had other ethnicities.

Solution

We first find the proportion for African American and Asian/Pacific Islander ethnicities by adding the proportions for each group:

$$\frac{1}{5} + \frac{1}{4} = \frac{1}{5} \cdot \frac{4}{4} + \frac{1}{4} \cdot \frac{5}{5} \quad \text{LCD is 20.}$$

$$= \frac{4}{20} + \frac{5}{20} \quad \begin{array}{l}\text{Multiply numerators and multiply denominators:} \\ \frac{a}{b} \cdot \frac{c}{d} = \frac{ac}{bd}\end{array}$$

$$= \frac{9}{20} \quad \begin{array}{l}\text{Add numerators and keep common denominator:} \\ \frac{a}{b} + \frac{c}{b} = \frac{a+c}{b}\end{array}$$

Because the proportion for African American and Asian/Pacific Islander ethnicities was $\frac{9}{20}$, the proportion of students who were NOT those ethnicities is

$$1 - \frac{9}{20} = \frac{20}{20} - \frac{9}{20} \quad \frac{a}{a} = 1$$

$$= \frac{11}{20} \quad \frac{a}{b} - \frac{c}{b} = \frac{a-c}{b}$$

In Example 13, we found the proportion of CCNY students who had ethnicities other than African American and Asian/Pacific Islander is $\frac{11}{20}$. Because $11 \div 20 = 0.55$, we can also say the proportion is 0.55. When working with data values, we will usually write proportions as decimal numbers, because they are easier to perform calculations with and to compare to other proportions.

▶ **Example 14**   Using a Table to Find Proportions

On December 14, 2012, 20 children and 6 adults were shot at Sandy Hook Elementary School in Newtown, Connecticut. One week later, 1048 adults were surveyed about how closely they had been following the news about the shootings (see Table 9).

Table 9 Attention to News about School Shooting

	How Closely Following the News					
	Very Closely	Somewhat Closely	Not Too Closely	Not at All	No Opinion	Total
Number of People	519	384	104	21	20	1048

Source: *The Gallup Organization*

1. Find the proportion of those surveyed that
   a. followed the news very closely.
   b. did NOT follow the news very closely.
   c. followed the news very closely OR somewhat closely.

2. What does the response "no opinion" mean in this situation? Should those who responded that way be excluded from the study?

Solution

1. a. 519 out of 1048 adults followed the news very closely, so the proportion is $\frac{519}{1048} \approx 0.495$.

b. In Problem 1, we found that the proportion for following the news very closely is approximately 0.495. To find the proportion for NOT following the news very closely, we subtract 0.495 from 1: $1 - 0.495 = 0.505$. So, the proportion is approximately 0.505.

c. 519 of the adults followed the news very closely and 384 of the adults followed the news somewhat closely. We add the two values: $519 + 384 = 903$. Then we divide the result by the total number of people surveyed, 1048: $\frac{903}{1048} \approx 0.862$. So, the proportion is approximately 0.862.

2. The response "no opinion" does not describe how closely one follows the news. Perhaps "no opinion" means that a person did not want to respond to the survey question. Even though the meaning is unclear, the people who gave this response should *not* be excluded from the study. By including them, a researcher can be aware of the issue and might choose to look into the matter further.

In Examples 11–14, we described authentic situations using proportions, which are a special type of ratio. Now we will use ratios equal to 1 to convert units of quantities.

Converting Units of Quantities

Suppose we're ordering a rug measured in feet, but we measured the width and the length of the floor in inches. There is no need to remeasure: We can convert the units of a quantity to equivalent units by multiplying by ratios of units, where each ratio is equal to 1. For example, since there are 12 inches in 1 foot, the following ratio is equal to 1:

$$\frac{12 \text{ inches}}{1 \text{ foot}} = 1$$

The reciprocal is also equal to 1:

$$\frac{1 \text{ foot}}{12 \text{ inches}} = 1$$

Suppose the width of a floor is 110 inches. Here we convert from units of inches to feet:

$$\frac{110 \text{ inches}}{1} \cdot \frac{1 \text{ foot}}{12 \text{ inches}} \approx 9.2 \text{ feet}$$

When converting, we can eliminate "inches," because $\frac{\text{inches}}{\text{inches}} = 1$.

We can eliminate a pair of the same units even if one is in singular form and the other is in plural form. For example, $\frac{\text{feet}}{\text{foot}} = 1$. However, $\frac{\text{inch}^2}{\text{inch}} = \text{inch}$, not 1.

Some equivalent units are shown in Table 10.

Table 10 Equivalent Units

Length
1 inch = 2.54 centimeters
1 foot = 12 inches
1 yard = 3 feet
1 mile = 5280 feet
1 mile ≈ 1.61 kilometers

Volume
1 cup = 8 ounces
1 quart = 4 cups
1 quart ≈ 0.946 liter
1 gallon = 4 quarts

Weight
1 gram = 1000 milligrams
1 pound = 16 ounces

Time
1 year ≈ 365 days

▶ Example 15   Converting Units

Make the indicated unit conversions. Round the results to two decimal places for Problems 2 and 3.

1. The official height of a basketball hoop is 10 feet. What is its height in yards?
2. An electric 1974 Fender® Jazz Bass® is 46.25 inches long. What is the length in centimeters?
3. A person walks at a speed of 4 miles per hour. What is the person's speed in feet per second?

Solution

1. Since there are 3 feet in 1 yard, we can multiply 10 feet by $\dfrac{1 \text{ yard}}{3 \text{ feet}} = 1$. By doing so, we can eliminate "feet," because $\dfrac{\text{feet}}{\text{feet}} = 1$:

$$\frac{10 \text{ feet}}{1} \cdot \frac{1 \text{ yard}}{3 \text{ feet}} = \frac{10}{3} \text{ yards}$$

The official height of the hoop is $3\frac{1}{3}$ yards.

2. Since there are 2.54 centimeters in 1 inch, we multiply 46.25 inches by $\dfrac{2.54 \text{ centimeters}}{1 \text{ inch}} = 1$ so that the inches are eliminated:

$$\frac{46.25 \text{ inches}}{1} \cdot \frac{2.54 \text{ centimeters}}{1 \text{ inch}} \approx 117.48 \text{ centimeters}$$

The bass is approximately 117.48 centimeters long.

3. There are 5280 feet in 1 mile, 60 minutes in 1 hour, and 60 seconds in 1 minute. To convert, we multiply by ratios equal to 1. Again, we arrange the ratios so that the units we want to eliminate appear in one numerator and one denominator:

$$\frac{4 \text{ miles}}{1 \text{ hour}} \cdot \frac{5280 \text{ feet}}{1 \text{ mile}} \cdot \frac{1 \text{ hour}}{60 \text{ minutes}} \cdot \frac{1 \text{ minute}}{60 \text{ seconds}} \approx 5.87 \frac{\text{feet}}{\text{second}}$$

So, the person is walking at a speed of about 5.87 feet per second.

▶ **Converting Units**

To convert the units of a quantity,

1. Write the quantity in the original units.
2. Multiply by fractions equal to 1 so that the units you want to eliminate appear in one numerator and one denominator.

◤◢ Group Exploration

Illustrations of simplifying fractions and operations with fractions

Draw a picture of a pizza to show that the true statement makes sense. [**Hint:** See Figs. 38, 39, 40, and 42.]

1. $\dfrac{6}{8} = \dfrac{3}{4}$

2. $\dfrac{5}{8} + \dfrac{2}{8} = \dfrac{7}{8}$

3. $\dfrac{5}{6} - \dfrac{4}{6} = \dfrac{1}{6}$

4. $\dfrac{1}{2} \cdot \dfrac{1}{3} = \dfrac{1}{6}$

5. $\dfrac{2}{3} \div \dfrac{1}{6} = 4$

▶ **Tips for Success** **Get in Touch with Classmates**

It is wise to exchange phone numbers and e-mail addresses with some classmates. If any of you has to miss class, then you have someone to contact to find out what you missed and what homework was assigned.

Homework 1.3

For extra help ▶ MyMathLab®    Watch the videos in MyMathLab    Download the MyDashboard App

1. The fraction $\frac{a}{b}$ is undefined if _____ = 0.

2. The reciprocal of $\frac{a}{b}$ is ____.

3. In statistics, a proportion is a(n) ____ of the whole.

4. If an object is made up of two or more parts, then the sum of their proportions equals ____.

5. What is the denominator of $\frac{3}{7}$?

6. What is the numerator of $\frac{2}{5}$?

Write the number as a product of primes.

7. 20   8. 18   9. 36   10. 24

11. 45   12. 27   13. 78   14. 105

Simplify by hand.

15. $\frac{6}{8}$ 16. $\frac{10}{14}$ 17. $\frac{18}{30}$ 18. $\frac{27}{54}$

19. $\frac{5}{25}$ 20. $\frac{9}{81}$ 21. $\frac{20}{24}$ 22. $\frac{15}{18}$

Perform the indicated operation by hand.

23. $\frac{1}{3}\cdot\frac{2}{5}$ 24. $\frac{5}{7}\cdot\frac{4}{9}$ 25. $\frac{4}{5}\cdot\frac{3}{8}$ 26. $\frac{2}{3}\cdot\frac{5}{6}$

27. $\frac{5}{21}\cdot 7$ 28. $\frac{5}{12}\cdot 2$ 29. $\frac{5}{8}\div\frac{3}{4}$ 30. $\frac{7}{12}\div\frac{2}{3}$

31. $\frac{8}{9}\div\frac{4}{3}$ 32. $\frac{4}{7}\div\frac{8}{3}$ 33. $\frac{2}{3}\div 5$ 34. $\frac{4}{9}\div 2$

35. $\frac{5}{8}+\frac{1}{8}$ 36. $\frac{2}{15}+\frac{8}{15}$ 37. $\frac{11}{12}-\frac{7}{12}$ 38. $\frac{13}{18}-\frac{9}{18}$

39. $\frac{1}{4}+\frac{1}{2}$ 40. $\frac{1}{3}+\frac{5}{9}$ 41. $\frac{5}{6}+\frac{3}{4}$ 42. $\frac{3}{8}+\frac{1}{6}$

43. $4+\frac{2}{3}$ 44. $2+\frac{3}{7}$ 45. $\frac{7}{9}-\frac{2}{3}$ 46. $\frac{3}{4}-\frac{1}{2}$

47. $\frac{5}{9}-\frac{2}{7}$ 48. $\frac{5}{6}-\frac{4}{7}$ 49. $3-\frac{4}{5}$ 50. $1-\frac{9}{7}$

Perform the indicated operation by hand. If the fraction is undefined, say so.

51. $\frac{3172}{3172}$ 52. $\frac{62}{62}$ 53. $\frac{599}{1}$ 54. $\frac{215}{1}$

55. $\frac{842}{0}$ 56. $\frac{713}{0}$ 57. $\frac{0}{621}$ 58. $\frac{0}{798}$

59. $\frac{824}{631}\cdot\frac{631}{824}$ 60. $\frac{173}{190}\cdot\frac{190}{173}$ 61. $\frac{544}{293}-\frac{544}{293}$ 62. $\frac{345}{917}-\frac{345}{917}$

Evaluate the given expression for $w = 4$, $x = 3$, $y = 5$, and $z = 12$.

63. $\frac{w}{z}$ 64. $\frac{z}{x}$ 65. $\frac{x}{w}\div\frac{y}{z}$

66. $\frac{y}{z}\cdot\frac{w}{x}$ 67. $\frac{x}{w}-\frac{y}{z}$ 68. $\frac{y}{x}+\frac{y}{z}$

Use a calculator to compute. Round the result to two decimal places.

69. $\frac{19}{97}\cdot\frac{65}{74}$ 70. $\frac{67}{71}\cdot\frac{381}{399}$ 71. $\frac{684}{795}\div\frac{24}{37}$

72. $\frac{149}{215}\div\frac{31}{52}$ 73. $\frac{89}{102}-\frac{59}{133}$ 74. $\frac{614}{701}+\frac{391}{400}$

For Exercises 75 and 76, draw a picture of a pizza to show that the true statement makes sense.

75. $\frac{2}{8}=\frac{1}{4}$ 76. $\frac{1}{4}+\frac{2}{4}=\frac{3}{4}$

77. A rectangular plot of land has a length of $\frac{2}{5}$ mile and a width of $\frac{1}{4}$ mile. What is the area of this plot?

78. A rectangular picture has a width of $\frac{2}{3}$ foot and a length of $\frac{3}{4}$ foot. What is the area of this picture?

79. A person's playlist consists of hip-hop, rap, hard rock, and electronica songs. What is the sum of the playlist's proportions for the four types of songs?

80. A survey group consists of Republicans, Democrats, and Independents. What is the sum of the group's proportions for the three political parties?

81. The proportion of the disk in Fig. 46 that is yellow is $\frac{2}{9}$. Find the proportion of the disk that is green.

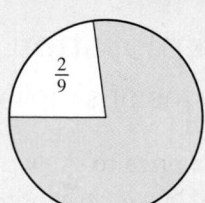

Figure 46 Exercise 81

82. The proportion of the disk in Fig. 47 that is blue is $\frac{2}{7}$. Find the proportion of the disk that is orange.

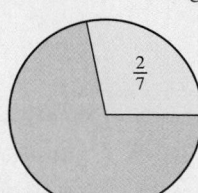

Figure 47 Exercise 82

83. The proportion of U.S. greenhouse-gas emissions that are from transportation is approximately $\frac{2}{7}$ (Source: *U.S. Environmental Protection Agency*). Find the proportion of U.S. greenhouse-gas emissions that are NOT from transportation. Write your result as a fraction.

84. The proportion of Twitter users who are Hispanic is approximately $\frac{3}{25}$ (Source: *Pew Research Center*). What proportion of Twitter users are NOT Hispanic? Write your result as a fraction.

85. The proportion of American adults who spend between 6 and 9.99 hours a day, inclusive, on digital devices is approximately $\frac{1}{3}$. The proportion of American adults who spend more than 9.99 hours on digital devices is approximately $\frac{2}{7}$ (Source: *Vision Council*). Find the proportion of American adults who spend at least 6 hours on digital devices. Write your result as a fraction.

86. The proportion of employees who spend between $101 and $200 per month, inclusive, on commuting to work is approximately $\frac{1}{5}$. The proportion of employees who spend more than $200 per month on commuting to work is approximately $\frac{1}{7}$ (Source: *Accounting Principals*). Find the proportion of employees who spend at least $101 per month on commuting to work. Write your result as a fraction.

87. The proportions of the disk in Fig. 48 that are orange and blue are $\frac{1}{4}$ and $\frac{1}{3}$, respectively. Find the proportion of the disk that is green.

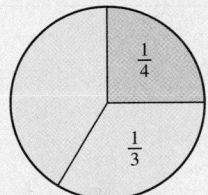

Figure 48 Exercise 87

88. The proportions of the disk in Fig. 49 that are orange and blue are $\frac{1}{2}$ and $\frac{1}{3}$, respectively. Find the proportion of the disk that is yellow.

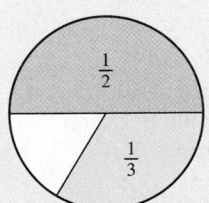

Figure 49 Exercise 88

89. At the Ann Arbor campus of the University of Michigan in fall 2013, the proportion of students who were Asian was approximately $\frac{2}{15}$. The proportion of students who were Caucasian was approximately $\frac{2}{3}$ (Source: *The University of Michigan*).

Find the proportion of students who had ethnicities other than Asian and Caucasian. Write your result as a fraction.

90. At Montclair State University in fall 2013, the proportion of undergraduates who were Hispanic was approximately $\frac{1}{5}$. The proportion of undergraduates who were Caucasian was approximately $\frac{2}{3}$ (Source: *Montclair State University*). Find the proportion of undergraduates who had ethnicities other than Caucasian and Hispanic. Write your result as a fraction.

91. For a prestatistics course, total course points are calculated by adding points earned on homework assignments, quizzes, tests, and the final exam. If the total of scores on tests is worth $\frac{1}{2}$ of the course points and the final exam score is worth $\frac{1}{4}$ of the course points, what proportion of the course points comes from homework assignments and quizzes?

92. A family spends $\frac{1}{3}$ of its income for the mortgage and $\frac{1}{6}$ of its income for food. What proportion of its income remains?

93. Undergraduate enrollments at Vanderbilt University in 2013 are shown in Table 11 for various schools within the university.

Table 11 Enrollments in Schools at Vanderbilt University

School	Enrollment
College of Arts and Science	4197
Blair School of Music	193
Divinity School	241
School of Engineering	1350
School of Nursing	881
Peabody College	1780
Division of Unclassified Studies	31
Total	8673

Source: *Vanderbilt University*

  a. Find the approximate proportion (rounded to the third decimal) of the Vanderbilt undergraduate students who were

  i. enrolled in the College of Arts and Science.

  ii. NOT enrolled in the College of Arts and Science.

  iii. enrolled in the Blair School of Music OR Peabody College.

  b. If the exact proportions for the seven schools in Table 11 were totaled, what would the result be? Explain.

94. The numbers of degrees awarded by Michigan University for the academic year 2012–2013 are shown in Table 12 for various types of degrees. Assume each student earned just one of the types of degrees shown in the table in the academic year 2012–2013.

Table 12 Degrees Awarded by Michigan University

Type of Degree	Number of Degrees Awarded
Bachelor's	3887
Master's	1402
Specialist	1
Doctoral	114
Total	5404

Source: *Michigan University*

a. Find the approximate proportion (rounded to the third decimal) of Michigan University students who were

   i. awarded a bachelor's degree in the academic year 2012–2013.

   ii. NOT awarded a bachelor's degree in the academic year 2012–2013.

   iii. awarded a master's degree OR a doctoral degree in the academic year 2012–2013.

b. If the exact proportions for the four types of degrees in Table 12 were totaled, what would the result be? Explain.

95. Some friends pay a total of $19 for a pizza. Each of the n friends pays an equal share of the cost. Complete Table 13 to help find an expression that describes the cost (in dollars) per person. Show the arithmetic to help you see a pattern.

Table 13 Cost per Person for the Pizza

Number of People	Cost per Person (dollars)
2	
3	
4	
5	
n	

96. A student drives her car 150 miles at a constant speed for t hours. Complete Table 14 to help find an expression that describes the car's speed (in miles per hour). Show the arithmetic to help you see a pattern.

Table 14 Speed of Car

Total Time (hours)	Speed (miles per hour)
2	
3	
4	
5	
t	

For Exercises 97–104, round approximate results to the second decimal place. Refer to the list of equivalent units in Table 10 on page 34 as needed.

97. The average height of a woman in the United States is 63.8 inches. What is that height in feet?

98. A German stein is 23 centimeters tall. What is the mug's height in inches?

99. A person buys 15 gallons of gasoline. How many liters of gasoline is that?

100. The speed limit on motorways in England is 113 kilometers per hour. What is the speed limit in miles per hour?

101. In Trader Giotto's Roasted Garlic Spaghetti Sauce, there are approximately 1.63 grams of salt in 1 pound of sauce. How many milligrams of salt are there in 1 ounce of sauce?

102. In Barbara's Puffins Cinnamon Cereal™, there are 42.5 milligrams of potassium in 1 ounce of cereal. How many grams of potassium are there in 1 pound of the cereal?

103. DairyCo, a UK dairy company, reports that the 2013 average yield of one of their cows is 7327 liters of milk per year. What is that yield in gallons per day?

104. A person drives at a speed of 25 meters per second. How fast is the person traveling in miles per hour?

Concepts

105. a. Perform the indicated operation.

   i. $\dfrac{5}{6} \cdot \dfrac{2}{3}$ **ii.** $\dfrac{5}{6} \div \dfrac{2}{3}$ **iii.** $\dfrac{5}{6} + \dfrac{2}{3}$ **iv.** $\dfrac{5}{6} - \dfrac{2}{3}$

b. Compare the methods you used to perform the operations in part (a). Describe how the methods are similar and how they are different.

106. a. Find each product.

   i. $\dfrac{2}{3} \cdot \dfrac{3}{2}$ **ii.** $\dfrac{4}{7} \cdot \dfrac{7}{4}$ **iii.** $\dfrac{1}{6} \cdot \dfrac{6}{1}$

b. On the basis of your results from part (a), use words to describe a property of a fraction and its reciprocal. Then describe the property in terms of variables.

107. A student tries to find the product $\dfrac{1}{2} \cdot \dfrac{1}{3}$:

$$\frac{1}{2} \cdot \frac{1}{3} = \left(\frac{1}{2} \cdot \frac{3}{3}\right) \cdot \left(\frac{1}{3} \cdot \frac{2}{2}\right)$$

$$= \frac{3}{6} \cdot \frac{2}{6}$$

$$= \frac{6}{36}$$

$$= \frac{1}{6}$$

What would you tell the student?

108. A student tries to find the product $2 \cdot \dfrac{3}{5}$:

$$2 \cdot \frac{3}{5} = \frac{6}{10} = \frac{3}{5}$$

Describe any errors. Then find the product correctly.

109. Why is division by 0 undefined? (See page 12 for guidelines on writing a good response.)

110. Explain why we do not add the denominators of two fractions when we add the fractions. (See page 12 for guidelines on writing a good response.)

▼1.4   Absolute Value and Adding Real Numbers

Objectives

» Find the opposite of a number.

» Find the absolute value of a number.

» Add two real numbers with the same sign.

» Add two real numbers with different signs.

» Add real numbers pertaining to authentic situations.

In this section, our main objective is to add real numbers.

The Opposite of a Number

Note that in Fig. 50 both the numbers -3 and 3 are 3 units from 0 on the number line, but they are on opposite sides of 0. We say that -3 is the *opposite* of 3, that 3 is the *opposite* of -3, and that -3 and 3 are *opposites*.

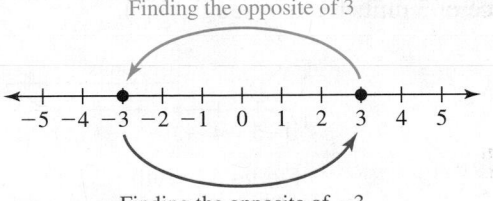

Finding the opposite of 3

Finding the opposite of -3

Figure 50 Finding the opposite of 3 and the opposite of -3

Two numbers are called **opposites** of each other if they are the same distance from 0 on the number line, but are on different sides of 0. We find the opposite of a number by writing a negative sign in front of the number. For example, the opposite of 3 is -3 (see Fig. 50).

Now consider this true statement:

The opposite of -3 is 3 (see Fig. 50).

In symbols, we write

The opposite of -3 is equal to   3.

$$-\quad(-3)\quad=\quad3$$

Here are some more examples of finding the opposite of a negative number:

$$-(-2) = 2$$
$$-(-7) = 7$$

We use a TI-84 to find $-(-7)$. See Fig. 51. We use the button $\boxed{-}$ for subtraction and the button $\boxed{(-)}$ for negative numbers and for taking opposites.

Press $\boxed{(-)}\boxed{(}\boxed{(-)}\boxed{7}\boxed{)}$ $\boxed{\text{ENTER}}$.

```
-(-7)
              7
```

Figure 51 Calculating $-(-7)$

WARNING

If we use the subtraction button $\boxed{-}$ to try to find $-(-7)$, a TI-84 will pull up the previous answer, and after we press ENTER, it will display an error message.

We can view $-(-7)$ as finding the opposite of -7 or as finding the opposite of the opposite of 7.

> ▶ **Finding the Opposite of the Opposite of a Number**
>
> $$-(-a) = a$$
>
> In words: The opposite of the opposite of a number is equal to that same number.

We use parentheses to separate two opposite symbols or an operation symbol and an opposite symbol.

▶ **Example 1**   Finding Opposites

Find the opposite.

1. $-(-5)$ **2.** $-(-(-8))$

Solution

1. $-(-5) = 5$ $-(-a) = a$

2. $-(-(-8)) = -(8)$ $-(-a) = a$

$\quad\quad\quad\quad = -8$ *Write without parentheses.*

Absolute Value

The *absolute value* of a number a, written $|a|$, is the distance the number a is from 0 on the number line.

> ▶ **Definition   Absolute value**
>
> The **absolute value** of a number is the distance the number is from 0 on the number line.

So $|-3| = 3$, because -3 is a distance of 3 units from 0, and $|3| = 3$, because 3 is a distance of 3 units from 0 (see Fig. 52).

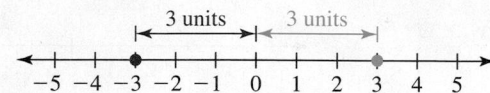

Figure 52 Both -3 and 3 are a distance of 3 units from 0

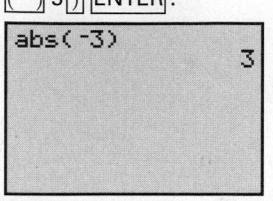

Press 2nd 0 ENTER
(−) 3) ENTER.

Figure 53 Calculating $|-3|$

On a TI-84, "abs" stands for absolute value. We find $|-3|$ in Fig. 53.

▶ **Example 2**   Finding Absolute Values of Numbers

Calculate.

1. $|2|$ **2.** $|-2|$ **3.** $-|2|$ **4.** $-|-2|$

Solution

1. $|2| = 2$, because 2 is a distance of 2 units from 0 (see Fig. 54).

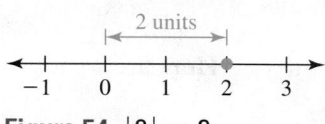

Figure 54 $|2| = 2$

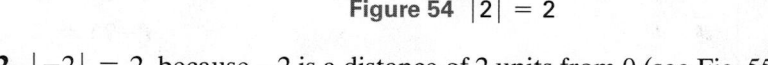

Figure 55 $|-2| = 2$

2. $|-2| = 2$, because -2 is a distance of 2 units from 0 (see Fig. 55).

3. $-|2| = -(2)$ $|2| = 2$

$\quad\quad\ = -2$ *Write without parentheses.*

4. $-|-2| = -(2)$ $|-2| = 2$

$\quad\quad\ \ = -2$ *Write without parentheses.*

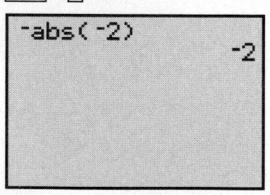

Press (−) 2nd 0 ENTER
(−) 2) ENTER.

Figure 56 Check that $-|-2| = -2$

We use a TI-84 to check that $-|-2| = -2$ (see Fig. 56).

Addition of Two Real Numbers with the Same Sign

Thinking about credit card balances or the number line can help us see how to add numbers with the same sign.

▶ **Example 3**   Finding the Sum of Two Numbers with the Same Sign

1. A person has a credit card balance of 0 dollars. If she uses her credit card to make two purchases, one for $2 and one for $5, what is the new balance?
2. Write a sum that is related to the computation in Problem 1.
3. Use a number line to illustrate the sum found in Problem 2.

Solution

1. By making purchases for $2 and $5, the person now owes $7. So, the new balance is -7 dollars.

2. Here is the sum:

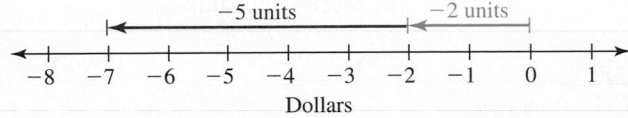

$$-2 + (-5) = -7$$

3. Using the number line, imagine moving 2 units to the left of 0 and then 5 more units to the left. Figure 57 illustrates that $-2 + (-5) = -7$.

Credit Card Balance

Figure 57 Illustration of $-2 + (-5) = -7$

In Example 3, we found that $-2 + (-5) = -7$. To get this result, we added the debts of 2 and 5 to get a total debt of 7. Note that 2 and 5 are the absolute values of -2 and -5. Note also that the result, -7, of the original sum has the same sign as both -2 and -5. These observations suggest the following procedure:

▶ Adding Two Numbers with the Same Sign

To add two numbers with the same sign,
1. Add the absolute values of the numbers.
2. The sum of the original numbers has the same sign as the sign of the original numbers.

▶ Example 4 Finding the Sum of Two Numbers with the Same Sign

Find the sum.

1. $-3 + (-6)$

2. $-\dfrac{1}{5} + \left(-\dfrac{3}{5}\right)$

Solution

1. First, we add the absolute values of the numbers -3 and -6: $3 + 6 = 9$. Since both -3 and -6 are negative, their sum is negative. So, $-3 + (-6) = -9$.

2. By adding the absolute values of the fractions, we have $\dfrac{1}{5} + \dfrac{3}{5} = \dfrac{4}{5}$. Since both original fractions are negative, their sum is negative. So,

$$-\frac{1}{5} + \left(-\frac{3}{5}\right) = -\frac{4}{5}$$

Press $(-)$ 3 $+$ $($ $(-)$ 6 $)$ ENTER.

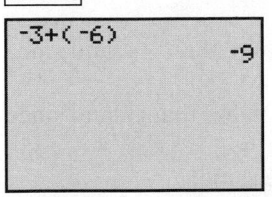

Figure 58 Calculating $-3 + (-6)$

After completing an exercise in this section's homework by hand, you can use a calculator to check your work. For example, we use a TI-84 to check our work for $-3 + (-6)$, Problem 1 in Example 4 (see Fig. 58).

Addition of Two Real Numbers with Different Signs

Thinking about the number line or exchanges of money can also help us see how to add numbers with different signs.

▶ **Example 5**   Finding the Sum of Two Numbers with Different Signs

1. A brother owes his sister $5. If he then pays her back $2, how much does he still owe her?
2. Write a sum that is related to your work in Problem 1.
3. Use a number line to illustrate the sum you found in Problem 2.

Solution

1. By owing his sister $5 and paying her back $2, the brother now owes his sister $3.
2. Here's the sum:

$$\underset{-5}{\underbrace{\text{Owe \$5}}} + \underset{2}{\underbrace{\text{Pay back \$2}}} = \underset{-3}{\underbrace{\text{Now owe \$3}}}$$

3. Using the number line, imagine moving 5 units to the left of 0 and then 2 units to the right of -5. Figure 59 illustrates that $-5 + 2 = -3$.

Balance in Sister's "Account"

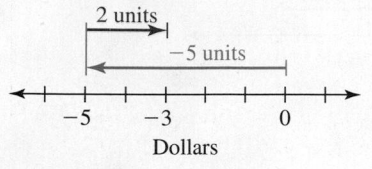

Figure 59 Illustration of $-5 + 2 = -3$

In Problem 2 in Example 5, we found that $-5 + 2 = -3$. We can get this result by first finding the difference of 5 and 2:

$$5 - 2 = 3$$

We can think of this operation as lowering a debt of 5 dollars by 2 dollars to get a debt of 3 dollars, so the result is -3. Note that the result, -3, has the same sign as -5, which has a larger absolute value than 2. These observations suggest the following procedure:

▶ **Adding Two Numbers with Different Signs**

To add two numbers with different signs,

1. Find the absolute values of the numbers. Then subtract the smaller absolute value from the larger absolute value.
2. The sum of the original numbers has the same sign as the original number with the larger absolute value.

▶ **Example 6**   Finding the Sum of Two Numbers with Different Signs

Find the sum.

1. $-4 + 7$ 2. $3 + (-9)$ 3. $-\dfrac{5}{6} + \dfrac{2}{3}$

Solution

1. First, we find that $7 - 4 = 3$. Since 7 has a larger absolute value than -4, and since 7 is positive, the sum is positive: $-4 + 7 = 3$.
2. First, we find that $9 - 3 = 6$. Since -9 has a larger absolute value than 3, and since -9 is negative, the sum is negative: $3 + (-9) = -6$.
3. First, we write the fractions so that the denominators are the same.

$$-\frac{5}{6} + \frac{2}{3} = -\frac{5}{6} + \frac{2}{3} \cdot \frac{2}{2} \quad \textit{LCD is 6.}$$

$$= -\frac{5}{6} + \frac{4}{6} \quad \textit{Multiply numerators and multiply}$$
$$\textit{denominators: } \frac{a}{b} \cdot \frac{c}{d} = \frac{ac}{bd}$$

Next, we subtract the smaller absolute value from the larger absolute value:

$$\frac{5}{6} - \frac{4}{6} = \frac{1}{6}$$

Since $-\dfrac{5}{6}$ has a larger absolute value than the fraction $\dfrac{4}{6}$, and since $-\dfrac{5}{6}$ is negative, the sum is negative:

$$-\frac{5}{6} + \frac{4}{6} = -\frac{1}{6}$$

We have discussed three ways to add real numbers: thinking in terms of money, the number line, and absolute value. **It is good practice to use one method to find a sum and then use another method (or a calculator) as a check.**

Adding Real Numbers in Authentic Situations

Knowing how to add real numbers is a useful skill when you work with quantities that can be negative, such as balances of checking accounts and temperature readings.

▶ **Example 7**   Applications of Adding Real Numbers

1. A person bounces several checks and is charged service fees such that the balance of the checking account is −90.75 dollars. If the person then deposits 300 dollars, what is the balance?
2. Three hours ago, the temperature was −11°F. If the temperature has increased by 5°F in the last three hours, what is the current temperature?

Solution

1. The balance is $-90.75 + 300$ dollars. To find this sum, we first find the difference $300 - 90.75 = 209.25$. Since 300 has a larger absolute value than −90.75 and since 300 is positive, the sum is positive: $-90.75 + 300 = 209.25$. So, the balance is $209.25.
2. The temperature is $-11 + 5$ degrees Fahrenheit. To find this sum, we first find the difference $11 - 5 = 6$. Since −11 has a larger absolute value than 5 and since −11 is negative, the sum is negative: $-11 + 5 = -6$. So, the current temperature is −6°F.

▶

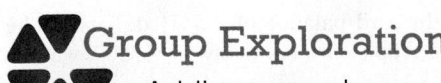 **Group Exploration**

Adding a number and its opposite

1. Evaluate $a + (-a)$ for the given values of a.
   a. $a = 2$ **b.** $a = 3$ **c.** $a = 5$

2. Evaluate $a + (-a)$ for the given values of a.
   [**Hint:** Use $-(-a) = a$.]
   a. $a = -2$ **b.** $a = -3$ **c.** $a = -5$

3. What do your results in Problems 1 and 2 suggest about $a + (-a)$?

▶ **Tips for Success   Make Good Use of This Text**

You can get more out of this course by making good use of the text. Before class, consider previewing the material for ten minutes. You can do this by reading the objectives and the boxed statements. Even if what you read doesn't make much sense to you, previewing will flag key concepts that you can focus on during class time.

After class, read the relevant section(s). When looking at each example, figure out how it goes from one step to the next.

Homework 1.4

For extra help ▶   Watch the videos in MyMathLab   Download the MyDashboard App

1. Two numbers are called _____ of each other if they are the same distance from 0 on the number line but are on different sides of 0.

2. The absolute value of a number is the distance the number is from _____ on the number line.

3. *True or False:* The sum of two negative numbers is always a negative number.

4. *True or False:* The sum of a negative number and a positive number is always a negative number.

Compute by hand.

5. $-(-4)$ 6. $-(-9)$ 7. $-(-(-7))$

8. $-(-(-2))$ 9. $|3|$ 10. $|6|$

11. $|-8|$ 12. $|-1|$ 13. $-|4|$

14. $-|5|$ 15. $-|-7|$ 16. $-|-9|$

Find the sum by hand.

17. $2 + (-7)$ 18. $5 + (-3)$ 19. $-1 + (-4)$

20. $-3 + (-2)$ 21. $7 + (-5)$ 22. $6 + (-9)$

23. $-8 + 5$ 24. $-3 + 4$ 25. $-7 + (-3)$

26. $-9 + (-5)$ 27. $4 + (-7)$ 28. $8 + (-2)$

29. $1 + (-1)$ 30. $8 + (-8)$ 31. $-4 + 4$

32. $-7 + 7$ 33. $12 + (-25)$ 34. $17 + (-14)$

35. $-39 + 17$ 36. $-89 + 57$

37. $-246 + (-899)$ 38. $-347 + (-594)$

39. $25,371 + (-25,371)$ 40. $127,512 + (-127,512)$

41. $-4.1 + (-2.6)$ 42. $-3.7 + (-9.9)$

43. $-5 + 0.2$ 44. $-0.3 + 7$

45. $2.6 + (-99.9)$ 46. $37.05 + (-19.26)$

47. $\dfrac{5}{7} + \left(-\dfrac{3}{7}\right)$ 48. $\dfrac{2}{5} + \left(-\dfrac{1}{5}\right)$

49. $-\dfrac{5}{8} + \dfrac{3}{8}$ 50. $-\dfrac{5}{6} + \dfrac{1}{6}$

51. $-\dfrac{1}{4} + \left(-\dfrac{1}{2}\right)$ 52. $-\dfrac{2}{3} + \left(-\dfrac{5}{6}\right)$

53. $\dfrac{5}{6} + \left(-\dfrac{1}{4}\right)$ 54. $\dfrac{2}{3} + \left(-\dfrac{3}{4}\right)$

For Exercises 55–60, use a calculator to find the sum. Round the result to two decimal places.

55. $-325.89 + 6547.29$ 56. $-7498.34 + 6435.28$

57. $-17,835.69 + (-79,735.45)$ 58. $-38,487.26 + (-83,205.87)$

59. $-\dfrac{34}{983} + \left(-\dfrac{19}{251}\right)$ 60. $-\dfrac{37}{642} + \left(-\dfrac{25}{983}\right)$

61. A person bounces several checks and is charged service fees such that the balance of the checking account is −75 dollars. If the person then deposits 250 dollars, what is the balance?

62. A person bounces several checks and is charged service fees such that the balance of the checking account is −112.50 dollars. If the person then deposits 170 dollars, what is the balance?

63. A check register is shown in Table 15. Find the final balance of the checking account.

Table 15   Check Register

Check No.	Date	Description of Transaction	Payment	Deposit	Balance
					−89.00
	7/18	Transfer		300.00	
3021	7/22	State Farm	91.22		
3022	7/22	MCI	44.26		
	7/31	Paycheck		870.00	

64. A check register is shown in Table 16. Find the final balance of the checking account.

Table 16   Check Register

Check No.	Date	Description of Transaction	Payment	Deposit	Balance
					−135.00
	2/31	Paycheck		549.00	
253	3/2	FedEx Kinko's	10.74		
	3/3	ATM	21.50		
254	3/7	Barnes and Noble	17.19		

65. A person has a credit card balance of −5471 dollars. If she sends a check to the credit card company for $2600, what is the new balance?

66. A person has a credit card balance of −2739 dollars. If he sends a check to the credit card company for $530, what is the new balance?

67. A student has a credit card balance of −3496 dollars. If he sends a check to the credit card company for $2500 and then uses his credit card to purchase a bicycle for $613 and a helmet for $24, what is the new balance?

68. A student has a credit card balance of −873 dollars. If she sends a check to the credit card company for $500 and then uses the card to buy a tennis racquet for $249 and a tennis outfit for $87, what is the new balance?

69. Three hours ago, it was −5°F. If the temperature has increased by 9°F in the last three hours, what is the current temperature?

70. Four hours ago, it was −12°F. If the temperature has increased by 8°F in the last four hours, what is the current temperature?

Concepts

71. If a is negative and b is negative, what can you say about the sign of $a + b$? Use a number line to show this property.

72. If a is positive, b is negative, and b is larger in absolute value than a, what can you say about the sign of $a + b$? Use a number line to show this property.

73. If $a + b = 0$, what can you say about a and b?

74. If $a + b$ is positive, what can you say about a and b?

75. a. Evaluate $-a$ for $a = -3$.
b. Evaluate $-a$ for $a = -4$.
c. Evaluate $-a$ for $a = -6$.
d. A student says that $-a$ represents only negative numbers because $-a$ has a negative sign. Is the student correct? Explain.

76. a. Evaluate $a + b$ for $a = -2$ and $b = 5$.
b. Evaluate $b + a$ for $a = -2$ and $b = 5$.
c. Compare your results from parts (a) and (b).
d. Evaluate $a + b$ and $b + a$ for $a = -4$ and $b = -9$, and then compare the results.
e. Evaluate $a + b$ and $b + a$ for values of your choosing for a and b, and then compare the results.
f. Is the statement $a + b = b + a$ true for all numbers a and b? Explain.

▼1.5   Change in a Quantity and Subtracting Real Numbers

Objectives

» Find the change in a quantity.

» Subtract real numbers.

» Find a change in elevation.

» Determine the sign of the change for an increasing or decreasing quantity.

In this section, we will discuss how to use subtraction to compute how much a quantity has changed. For example, we can compute the increase in a population of wolves or the decrease in the percentage of eligible voters who voted in an election.

Change in a Quantity

If the value of a stock increases from \$5 to \$8, we say the value *changed* by \$3. Finding the change in a quantity is a very important concept in mathematics and has many applications. A company is extremely focused on the change in its profits. During an operation, a surgeon keeps a close eye on the change in a patient's blood pressure. You probably care deeply about a change in your GPA.

▶ **Example 1**   Finding the Change in a Quantity

1. If a student's guitar collection increases from 2 guitars to 7 guitars, find the change in the number of guitars.
2. Write a difference that is related to the computation in Problem 1.

Solution

1. If the number of guitars increases from 2 guitars to 7 guitars, then the change in the number of guitars is 5 guitars (see Fig. 60).
2. Here is the difference:

Guitar Collection

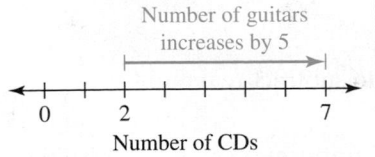

Figure 60 The change in the number of guitars is 5 guitars

Change in the number of guitars		Ending number of guitars		Beginning number of guitars
5	=	7	−	2

In Example 1, we found the change in the number of guitars by finding the difference of the ending number of guitars and the beginning number of guitars.

▶ **Change in a Quantity**

The change in a quantity is the ending amount minus the beginning amount:

Change in the quantity = Ending amount − Beginning amount

In Example 2, we will find the changes in a quantity from one year to the next.

Table 17 Average Movie Ticket Prices

Years	Average Price (dollars)
2008	7.18
2009	7.50
2010	7.89
2011	7.93
2012	7.96
2013	8.13

Source: *National Association of Theater Owners*

▶ **Example 2**   Finding Changes in a Quantity

Average movie ticket prices are shown in Table 17 for various years.

1. Find the change in the average movie ticket price from 2008 to 2009.
2. For the period 2008–2013, find each of the changes in the average movie ticket price from one year to the next.
3. From which year to the next did the average price increase the most?

Solution

1. We find the difference of the average price in 2009 (ending) and the average price in 2008 (beginning):

Ending average price (in dollars)		Beginning average price (in dollars)		Change in average price (in dollars)
7.50	−	7.18	=	0.32

So, the average ticket price from 2008 to 2009 changed by $0.32.

2. The changes in the average ticket price from one year to the next are listed in Table 18. The changes were found by computing the differences similar to the one found in Problem 1.

Table 18 Changes in Average Movie Ticket Prices from Year to Year

Years	Average Price (dollars)	Change in Average Price (dollars)
2008	7.18	
2009	7.50	7.50 − 7.18 = 0.32
2010	7.89	7.89 − 7.50 = 0.39
2011	7.93	7.93 − 7.89 = 0.04
2012	7.96	7.96 − 7.93 = 0.03
2013	8.13	8.13 − 7.96 = 0.17

Source: *National Association of Theater Owners*

3. The average ticket price changed by $0.39 from 2009 to 2010, the greatest change from any year to the next.

Subtraction of Real Numbers

Exploring the change in a quantity can help us see how to subtract real numbers.

▶ **Example 3**   Finding the Difference of Two Real Numbers

1. A college's enrollment decreases from 7 thousand students to 2 thousand students. What is the change in the enrollment?
2. Write a difference that is related to the computation in Problem 1.

Solution

1. Since the enrollment has decreased from 7 thousand students to 2 thousand students, the change is −5 thousand students. The change is negative because the enrollment is decreasing (see Fig. 61).
2. The change in the enrollment is the difference of the ending enrollment and the beginning enrollment:

Enrollment

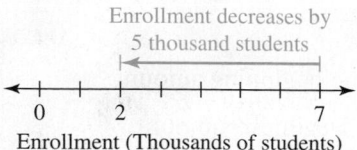

Enrollment decreases by 5 thousand students

Enrollment (Thousands of students)

Figure 61 Enrollment decreases from 7 thousand students to 2 thousand students

Ending enrollment (in thousands)		Beginning enrollment (in thousands)		Change in enrollment (in thousands)
2	−	7	=	−5

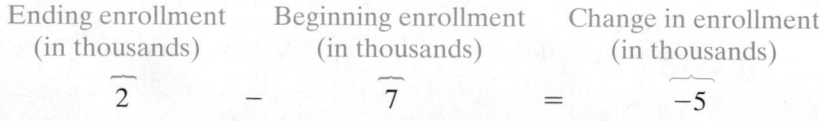

In Example 3, we found that

$$2 - 7 = -5$$

Note that $2 + (-7)$ gives the same result:

$$2 + (-7) = -5$$

This means

$$2 - 7 = 2 + (-7)$$

which suggests that subtracting a number is the same as adding the opposite of that number:

$$\overbrace{2 - 7}^{\text{Subtract 7.}} = \overbrace{2 + (-7)}^{\text{Add the opposite of 7.}}$$

▶ Subtracting a Real Number

$$a - b = a + (-b)$$

In words: To subtract a number, add its opposite.

To subtract real numbers, we first write the difference as a related sum and then find the sum.

▶ Example 4   Finding Differences of Real Numbers

Find the difference.

1. $4 - 6$

2. $\dfrac{2}{9} - \dfrac{5}{9}$

Solution

1. $\overbrace{4 - 6}^{\text{Subtract 6.}} = \overbrace{4 + (-6)}^{\text{Add the opposite of 6.}}$ $a - b = a + (-b)$

   $ = -2$ Add.

2. We begin by adding the opposite of $\dfrac{5}{9}$:

$$\frac{2}{9} - \frac{5}{9} = \frac{2}{9} + \left(-\frac{5}{9}\right) = -\frac{3}{9} = -\frac{1}{3}$$

Considering the change in a quantity can also help us see how to subtract a negative number.

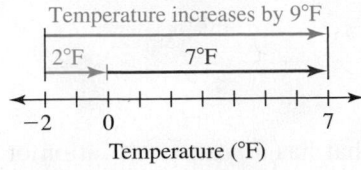

Temperature

Temperature increases by 9°F

2°F   7°F

−2   0   7

Temperature (°F)

Figure 62 Temperature increases by 9°F in going from −2°F to 7°F

▶ Example 5   Subtracting a Negative Number

1. The temperature increases from −2°F to 7°F. Find the change in temperature.
2. Write a difference that is related to the work in Problem 1.
3. Find the difference obtained in Problem 2 by using the rule $a - b = a + (-b)$.

Solution

1. Since the temperature increased from −2°F to 7°F, the change in temperature is 9°F (see Fig. 62).

2. The change in temperature is the ending temperature minus the beginning temperature:

Ending temperature   Beginning temperature   Change in temperature
(°F)   (°F)   (°F)
7   −   (−2)   =   9

3.

$$\overbrace{\text{Subtract}-2.}\quad\overbrace{\substack{\text{Add the opposite}\\\text{of }-2.\text{ (So, add 2.)}}}$$

$$7 - (-2) = 7 + 2 \qquad a - b = a + (-b)$$
$$= 9 \qquad \textit{Add.}$$

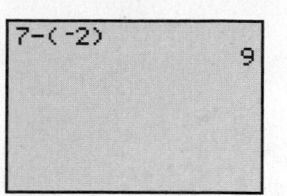

Press 7 $\boxed{-}$ $\boxed{(}$ $\boxed{(-)}$ 2 $\boxed{)}$ $\boxed{\text{ENTER}}$

```
7-(-2)
              9
```

Figure 63 Calculating
$7 - (-2)$

Note that to find the difference $7 - (-2)$, we add 2 and 7. This makes sense, because, in going from −2°F to 0°F, the temperature increases by 2°F, and in continuing from 0°F to 7°F, the temperature increases by another 7°F (see Fig. 62).

We use a TI-84 to check our work in Example 5 (see Fig. 63). Recall that we use the button $\boxed{-}$ for subtraction and the button $\boxed{(-)}$ for negative numbers and for taking opposites.

It is good practice to do homework exercises first by hand and then by using a calculator to check your hand results.

▶ **Example 6**   Subtracting a Negative Number

Find the difference.

1. $4 - (-6)$ **2.** $-9 - (-3)$

Solution

1.

$$\overbrace{\text{Subtract }-6.}\quad\overbrace{\substack{\text{Add the opposite}\\\text{of }-6.\text{ (So, add 6.)}}}$$

$$4 - (-6) = 4 + 6 = 10$$

2.

$$\overbrace{\text{Subtract }-3.}\quad\overbrace{\substack{\text{Add the opposite}\\\text{of }-3.\text{ (So, add 3.)}}}$$

$$-9 - (-3) = -9 + 3 = -6$$

▶ **Example 7**   Translating from English to Mathematics

Translate the phrase "the difference of a and b" into a mathematical expression. Then evaluate the expression for $a = 3$ and $b = -7$.

Solution

The expression is $a - b$. We substitute 3 for a and −7 for b in the expression $a - b$ and then find the difference:

$$(3) - (-7) = 3 + 7 \quad a - b = a + (-b)$$
$$= 10 \qquad \textit{Add.}$$

Change in Elevation

In Example 8, you will work with *elevation*. An object that has a *positive* elevation of 200 ft is 200 ft *above* sea level (see Fig. 64). An object that has a *negative* elevation of −200 ft is 200 feet *below* sea level (see Fig. 64).

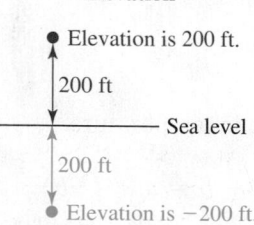

Elevation

• Elevation is 200 ft.

200 ft

——— Sea level

200 ft

• Elevation is −200 ft.

Figure 64 Elevations of 200 ft and −200 ft

▶ **Example 8** Finding a Change in Elevation

The Golden Gate Bridge has two towers that support the two main cables of the bridge (see Fig. 65). The top of each tower is at an elevation of 746 ft, and the foot of each tower is at an elevation of −136 ft (136 feet below sea level). Find the height of each tower.

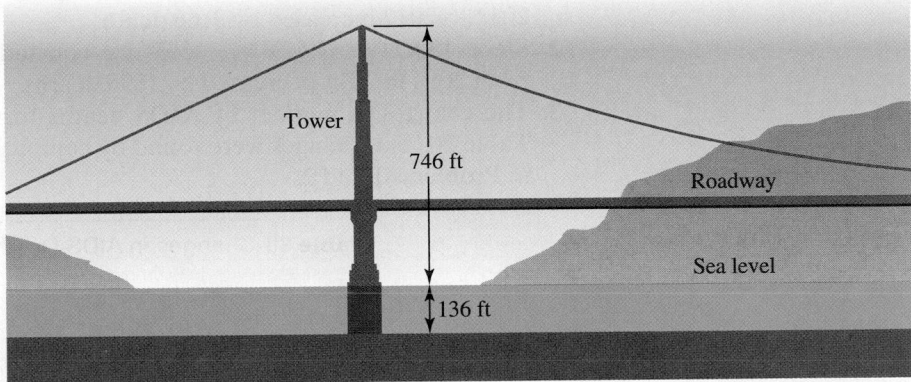

Figure 65 Golden Gate Bridge

Solution

We can find the height of each tower by computing the change in elevation from the bottom of each tower to the top:

$$\underset{\substack{\text{Top elevation}\\\text{(in feet)}}}{746} - \underset{\substack{\text{Bottom elevation}\\\text{(in feet)}}}{(-136)} = 746 + 136 \quad a - b = a + (-b)$$
$$= 882 \quad \text{Add.}$$

So, the height of each tower is 882 ft.

Changes of Increasing and Decreasing Quantities

An increasing quantity has a positive change. For instance, in Example 5, the temperature *increased* from −2°F to 7°F and the change in temperature was *positive* (9°F).

A decreasing quantity has a negative change. For instance, in Example 3, the college's enrollment *decreased* from 7 thousand students to 2 thousand students and the change in enrollment was *negative* (−5 thousand students).

> **Changes of Increasing and Decreasing Quantities**
>
> - An increasing quantity has a positive change.
> - A decreasing quantity has a negative change.

In Example 9, we will consider the meaning of a quantity with a positive or negative change.

Table 19 Numbers of AIDS Deaths in the United States

Year	Number of AIDS Deaths
2006	15,564
2007	14,561
2008	16,084
2009	17,774
2010	15,529
2011	13,834
2012	17,000

Source: *U.S. Centers for Disease Control and Prevention*

▶ **Example 9** Finding Changes in Quantities

The numbers of AIDS deaths in the United States are shown in Table 19 for various years.

1. Find the change in the number of AIDS deaths from 2011 to 2012. What does your result mean in terms of the number of AIDS deaths?
2. Find the change in the number of AIDS deaths from 2010 to 2011. What does your result mean in terms of the number of AIDS deaths?

3. Find the change in the number of AIDS deaths from one year to the next, beginning in 2006.

4. From what year to the next did the number of AIDS deaths decrease the most?

Solution

1. Since $17{,}000 - 13{,}834 = 3166$, we conclude that the number of AIDS deaths from 2011 to 2012 increased by 3166 deaths.

2. Since $13{,}834 - 15{,}529 = -1695$, we conclude that the number of AIDS deaths from 2010 to 2011 decreased by 1695 deaths.

3. The changes in number of AIDS deaths from one year to the next are listed in Table 20. The changes were found by computing differences similar to those found in Problems 1 and 2.

Table 20 Changes in AIDS Deaths from Year to Year

Year	Number of AIDS Deaths	Change in the Number of AIDS Deaths from Previous Year to Current Year
2006	15,564	
2007	14,561	$14{,}561 - 15{,}564 = -1003$
2008	16,084	$16{,}084 - 14{,}561 = 1523$
2009	17,774	$17{,}774 - 16{,}084 = 1690$
2010	15,529	$15{,}529 - 17{,}774 = -2245$
2011	13,834	$13{,}834 - 15{,}529 = -1695$
2012	17,000	$17{,}000 - 13{,}834 = 3166$

4. The smallest change is -2245, which is the change in the number of AIDS deaths from 2009 to 2010. So, the number of AIDS deaths decreased the most from 2009 to 2010.

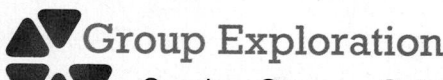

Group Exploration

Section Opener: Subtracting numbers

1. Find the difference $6 - 4$ and the sum $6 + (-4)$, and compare your results.

2. Find the difference $7 - 3$ and the sum $7 + (-3)$, and compare your results.

3. Find the difference $9 - 2$ and the sum $9 + (-2)$, and compare your results.

4. In Problems 1–3, for each difference, there is a related sum that gives the same result. Write $a - b$ as a sum.

5. In Problem 4, you wrote $a - b$ as a sum. Use this method to find the given difference.

   a. $8 - 3$ **b.** $2 - 5$ **c.** $-4 - 3$

▶ **Tips for Success**   Review Your Notes as Soon as Possible

How often do you get confused by class notes you wrote earlier the same day, even though the class activities made sense to you? If this happens a lot, review your notes as soon after class as possible. Even reviewing your notes for just a few minutes between classes will help. This will increase your likelihood of remembering what you learned in class and will give you the opportunity to add new comments to your notes while the class experience is still fresh in your mind.

Homework 1.5

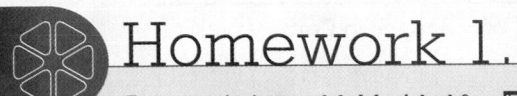

For extra help ▶ MyMathLab®   ▣ Watch the videos in MyMathLab   ⬇ Download the MyDashboard App

1. The _____ in a quantity is the ending amount minus the beginning amount.

2. To subtract a number, add its _____ .

3. *True or False:* A negative number minus a negative number equals a negative number.

4. *True or False:* A decreasing quantity has negative change.

Find the difference by hand.

5. $6 - 8$

6. $3 - 7$

7. $-1 - 5$

8. $-3 - 9$

9. $2 - (-7)$

10. $5 - (-1)$

11. $-3 - (-2)$

12. $-7 - (-3)$

13. $4 - 7$

14. $-4 - 7$

15. $4 - (-7)$

16. $-4 - (-7)$

17. $-3 - 3$

18. $-7 - 7$

19. $-54 - 25$

20. $-100 - 257$

21. $381 - (-39)$

22. $-1939 - (-352)$

23. $2.5 - 7.9$

24. $5.8 - 3.7$

25. $-6.5 - 4.8$

26. $-1.7 - 7.4$

27. $3.8 - (-1.9)$

28. $3.1 - (-3.1)$

29. $13.6 - (-2.38)$

30. $-159.24 - (-7.8)$

31. $-\dfrac{1}{3} - \dfrac{2}{3}$

32. $-\dfrac{1}{5} - \dfrac{4}{5}$

33. $-\dfrac{1}{8} - \left(-\dfrac{5}{8}\right)$

34. $-\dfrac{4}{9} - \left(-\dfrac{7}{9}\right)$

35. $\dfrac{1}{2} - \left(-\dfrac{1}{4}\right)$

36. $\dfrac{5}{12} - \left(-\dfrac{1}{6}\right)$

37. $-\dfrac{1}{6} - \dfrac{3}{8}$

38. $-\dfrac{2}{3} - \dfrac{2}{5}$

Perform the indicated operation by hand.

39. $-5 + 7$

40. $-3 + 9$

41. $-6 - (-4)$

42. $-4 - (-3)$

43. $\dfrac{3}{8} - \dfrac{5}{8}$

44. $-\dfrac{5}{6} + \dfrac{1}{6}$

45. $-4.9 - (-2.2)$

46. $-6.4 + 3.5$

47. $-2 + (-5)$

48. $-5 + (-8)$

49. $10 - 12$

50. $5 - 9$

For Exercises 51–56, use a calculator to perform the indicated operation. Round the result to two decimal places.

51. $-234.913 - 2893.26$

52. $-6178.39 - 52.387$

53. $29{,}643.52 - (-83{,}284.39)$

54. $83{,}451.6 - (-408.549)$

55. $-\dfrac{17}{89} - \dfrac{51}{67}$

56. $-\dfrac{49}{56} - \dfrac{85}{97}$

57. Three hours ago, the temperature was 7°F. If the temperature has decreased by 19°F in the last three hours, what is the current temperature?

58. Four hours ago, the temperature was −12°F. If the temperature has increased by 18°F in the last four hours, what is the current temperature?

59. Three hours ago, the temperature was −4°F. Now the temperature is 7°F. What is the change in temperature for the past three hours?

60. Four hours ago, the temperature was −2°F. Now the temperature is −13°F. What is the change in temperature for the past four hours?

61. Two hours ago, the temperature was 8°F. The temperature is now −4°F.

    a. What is the change in temperature for the past two hours?
    b. Estimate the change in temperature for the past hour.
    c. Explain why the estimate you found in part (b) may not be the actual change in temperature for the past hour.

62. Three hours ago, the temperature was −6°F. The temperature is now 9°F.

    a. What is the change in temperature for the past three hours?
    b. Estimate the change in temperature for the past hour.
    c. Explain why the estimate you found in part (b) may not be the actual change in temperature for the past hour.

63. The lowest elevation in the United States is at Death Valley, California (−282 ft), and the highest elevation is at the top of Mount McKinley, Alaska (20,320 ft). Find the change in elevation from Death Valley to Mount McKinley.

64. The lowest elevation on (dry) land in the world is at the edge of the Dead Sea, along the Israel-Jordan border (−1312 ft), and the highest elevation is at the top of Mount Everest, along the Nepal–Tibet border (29,035 ft). Find the change in the elevation from the Dead Sea to Mount Everest.

65. The U.S. presidential election in 2000 was the closest presidential race in the electoral vote since 1876. Yet, only a little over half of eligible voters chose to cast a vote (see Table 21).

Table 21 Presidential Election Voter Turnout

Year	Percent of Eligible Voters Who Voted
1980	59.2
1984	59.9
1988	57.4
1992	61.9
1996	54.2
2000	54.7
2004	63.8
2008	63.6
2012	57.5

Source: *U.S. Census Bureau, Current Population Study*

a. For the years listed in Table 21, find the changes in percent turnout from one presidential election to the next.
b. What was the greatest increase in percent turnout?

c. In 1993, in an attempt to increase the number of eligible voters, a "motor voter" law was passed that made voter registration a part of the process of applying for a driver's license. As a result, about 11 million new voters were registered. Compare the change in percent turnout between 1992 and 1996 with other changes you found in part (a). On the basis of the information in Table 21 alone, we cannot know for sure, but does it seem that many of these 11 million people voted? Explain.

66. In the 1930s, the gray wolf was hunted to near extinction across the western United States. In 1995, 14 wolves were reintroduced to Yellowstone National Park. In the following year, 17 more wolves were released into the park. By the end of 1996, there had been 20 births and 11 mortalities, leaving 40 wolves. The wolf population in Yellowstone National Park is shown in Table 22 for various years.

Table 22 Wolf Population

Year	Population
2005	118
2006	136
2007	171
2008	124
2009	96
2010	97
2011	98

Source: *National Park Service, Yellowstone National Park*

a. For the years listed in Table 22, find the changes in population from each year to the next.

b. From what year(s) to the next did the population increase the most? What is the change in population?

c. From what year(s) to the next did the population decrease the most? What is the change in population?

d. From 2005 to 2006, the change in the population was 18 wolves. Does that mean that there were 18 births? Explain.

67. The changes in Toyota Prius® U.S. sales (in thousands of cars) from one year to the next are shown in Table 23.

Table 23 Changes in Toyota Prius U.S. Sales

Years	Changes in Sales (thousands of cars)
2006–2007	74
2007–2008	−23
2008–2009	−19
2009–2010	1
2010–2011	−13
2011–2012	19
2012–2013	−2

Source: *National Renewable Energy Laboratory*

a. If 107 thousand cars were sold in 2006, what were the sales in 2013?

b. During which period(s) were sales increasing?

c. During which period(s) were sales decreasing?

68. The changes in corn harvests (in billions of bushels) in the United States from one year to the next are shown in Table 24.

Table 24 Changes in Corn Harvests

Years	Change in Corn Harvest (billions of bushels)
2008–2009	1.0
2009–2010	−0.6
2010–2011	−0.1
2011–2012	−1.6
2012–2013	3.1
2013–2014	0.3

Source: *Moebs Services*

a. If 12.1 billion bushels of corn were harvested in 2008, what was the amount of corn harvested in 2014?

b. During which period(s) were corn harvests increasing?

c. During which period(s) were corn harvests decreasing?

Evaluate the expression for $a = -5$, $b = 2$, and $c = -7$.

69. $a + b$ **70.** $a + c$ **71.** $a - b$

72. $c - a$ **73.** $b - c$ **74.** $b - a$

For Exercises 75–80, let x be a number. Translate the English phrase or sentence into a mathematical expression. Then evaluate the expression for $x = -5$.

75. −3 minus the number   **76.** The number decreased by 4

77. 8 less than the number   **78.** Subtract 5 from the number.

79. Subtract −2 from the number.

80. The difference of the number and −6

Concepts

81. A student tries to find the difference $7 - (-5)$:
$$7 - (-5) = 7 - 5 = 2$$
Describe any errors. Then find the difference correctly.

82. A student tries to find the difference $2 - 6$:
$$2 - 6 = 6 - 2 = 4$$
Describe any errors. Then find the sum correctly.

83. A quantity increases from amount a to amount b.

a. Find the change in the quantity.

   i. $a = 3, b = 5$ **ii.** $a = 1, b = 9$ **iii.** $a = 2, b = 7$

b. By referring to your work in part (a), explain why it makes sense that if a quantity increased, then the change will be positive.

84. A quantity decreases from amount a to amount b.

a. Find the change in the quantity.

   i. $a = 8, b = 2$ **ii.** $a = 9, b = 3$ **iii.** $a = 5, b = 1$

b. By referring to your work in part (a), explain why it makes sense that if a quantity decreased, then the change will be negative.

85. a. Evaluate $a - b$ for $a = 8$ and $b = 5$.

b. Evaluate $b - a$ for $a = 8$ and $b = 5$.

c. Compare your results from parts (a) and (b).

d. Evaluate $a - b$ and $b - a$ for $a = -2$ and $b = 4$, and compare your results.

e. Evaluate $a - b$ and $b - a$ for values of your choosing for a and b, and compare your results.

f. From your work on parts (a) through (e), what connection do you notice between $a - b$ and $b - a$?

86. If x and y are both negative, find the sign of $x - y$, if possible. If it is impossible to find the sign, explain why.

▼1.6 Ratios, Percents, and Multiplying and Dividing Real Numbers

Objectives

» Find the ratio of two quantities.

» Describe the meaning of *percent*.

» Convert percentages to and from decimal numbers.

» Use proportions and percentages to describe authentic situations.

» Find the percentage of a quantity.

» Multiply and divide real numbers.

» Describe which fractions with negative signs are equal to each other.

In this section, we will use ratios, percents, and proportions to describe various quantities. Ratios, percents, and proportions are important tools used in many fields, including business, political science, journalism, and statistics. We will also discuss how to multiply and divide real numbers.

The Ratio of Two Quantities

Recall that the ratio of a to b is the quotient $a \div b$. Usually, we write the ratio of a to b as the fraction $\dfrac{a}{b}$ or as $a:b$.

We can use a ratio to compare two quantities. For example, if a person has 6 cats and 2 dogs, then the ratio of cats to dogs is

$$\frac{6 \text{ cats}}{2 \text{ dogs}} = \frac{3 \text{ cats}}{1 \text{ dog}}$$

We say there are "3 cats to 1 dog." This means there are 3 cats per dog. Or we can say there are 3 times as many cats as dogs.

The ratio of 3 cats to 1 dog is an example of a unit ratio. A **unit ratio** is a ratio written as $\dfrac{a}{b}$ with $b = 1$ or as $a:b$ with $b = 1$.

▶ Example 1 Finding a Unit Ratio

In 2013, the average annual charge for tuition and fees was \$8070 at public four-year colleges and \$2792 at public two-year colleges (Source: *U.S. National Center for Education Statistics*). Find the unit ratio of the average annual charge at public four-year colleges to the average annual charge at public two-year colleges. What does the result mean?

Solution

We divide the average annual charge at public four-year colleges by the average annual charge at public two-year colleges:

$$\begin{array}{l} \text{public four-year colleges} \longrightarrow \\ \text{public two-year colleges} \longrightarrow \end{array} \frac{\$8070}{\$2792} \approx \frac{2.89}{1}$$

So, the average annual charge for tuition and fees at public four-year colleges is about 2.89 times the average annual charge for tuition and fees at public two-year colleges.

▶ Example 2 Comparing Ratios

The **median** of a group of numbers is the number in the middle (or the average of the two numbers in the middle) when the numbers are listed in order from smallest to largest. The median sales prices of existing homes and the median household incomes in 2011 are shown in Table 25 for four regions of the United States.

Table 25 Median Sales Prices of Existing Homes and Median Household Incomes

Region	Median Sales Price of Existing Homes (dollars)	Median Household Income (dollars)
Northeast	237,500	51,862
Midwest	135,400	49,549
South	135,200	42,590
West	201,300	53,367

Sources: *U.S. Census; Federal Reserve Bank of St. Louis*

Just think, our equity's growing by $100 per mile!

Welcome to California

OVER SIZE LOAD

1. Find the unit ratio of the median sales price of existing homes to the median household income in the Northeast. What does the result mean?
2. For each of the four regions, find the unit ratio of the median sales price of existing homes to the median household income. Taking into account the median household income of each region, list the regions in order of affordability of existing homes, from greatest to least.
3. A person believes that existing homes in the South are more affordable than in the Midwest, because the median price of existing homes is lower in the South than in the Midwest. What would you tell that person?

Solution

1. We divide the median sales price of existing homes in the Northeast by the median household income in the Northeast:

Median sales price of existing homes ⟶ $\dfrac{\$237{,}500}{\$51{,}862} \approx \dfrac{4.58}{1}$ ⟵ Median household income

So, the median sales price of an existing home in the Northeast is about 4.58 times the median household income in that region.

2. We find the unit ratios for each region by dividing the region's median sales price of existing homes by the region's median household income (see Table 26).

Table 26   Unit Ratios of Median Sales Prices of Existing Homes to Median Household Incomes

Region	Median Sales Price of Existing Homes (dollars)	Median Household Income (dollars)	Unit Ratio of Median Sales Price of Existing Homes to Median Household Income
Northeast	237,500	51,862	$\dfrac{237{,}500}{51{,}862} \approx \dfrac{4.58}{1}$
Midwest	135,400	49,549	$\dfrac{135{,}400}{49{,}549} \approx \dfrac{2.73}{1}$
South	135,200	42,590	$\dfrac{135{,}200}{42{,}590} \approx \dfrac{3.17}{1}$
West	201,300	53,367	$\dfrac{201{,}300}{53{,}367} \approx \dfrac{3.77}{1}$

The lower the unit ratio, the more affordable the existing homes are in the region. So, the regions, in order of affordability of existing homes, from greatest to least, are Midwest, South, West, Northeast.

3. Although the median sales price of homes in the South is a bit lower than that in the Midwest, the median income in the Midwest is so much higher than it is in the South that existing homes are actually more affordable in the Midwest. We can tell because the approximate unit ratio for the median sales price of existing homes to median income is smaller in the Midwest (2.73:1) than in the South (3.17:1). See Table 26.

In Example 2, to compare the affordability of existing homes, it was not enough to simply compare the median sales prices of existing homes. Instead, we compared unit ratios of median sales prices to median household incomes. In general, when comparing quantities, we will often need to compare the ratios of the quantities to yet other quantities in order to make a fair comparison. This is a key concept that we will discuss many times throughout the course.

Meaning of Percent

Suppose there are 53 women in a class of 100 students. Then the ratio of the number of women to the total number of students is $\dfrac{53}{100}$. We say that 53% of the students are women.

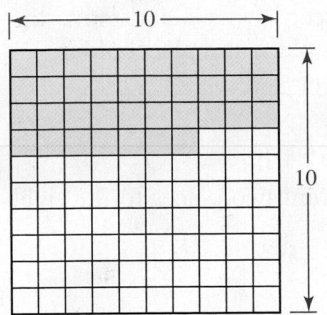

Figure 66 The area of the shaded region is 37% of the area of the large square

> **Definition   Percent**
>
> **Percent** means "for each hundred": $a\% = \dfrac{a}{100}$

For example, 37% means 37 for each 100 (the ratio $\dfrac{37}{100}$, or the unit ratio $\dfrac{0.37}{1}$). In Fig. 66, the area of the shaded region is 37% of the area of the large square, because 37 of 100 parts of equal area are shaded.

Converting Percentages to and from Decimal Numbers

Since 37% is the ratio $\dfrac{37}{100}$, 37% is 37 hundredths, or 0.37:

$$37\% = \frac{37}{100} = 0.\underbrace{\overset{\text{tenths place}}{3}\quad\overset{\text{hundredths place}}{7}}_{37\ \text{hundredths}}$$

So, to write 37% as a decimal number, first we remove the percent symbol. Then we divide 37 by 100, which is equivalent to moving the decimal point two places to the left:

$$37\% = 37.0\% = 0.37$$

two places to the left

To write 0.37 as a percentage, first we multiply 0.37 by 100, which is equivalent to moving the decimal point two places to the right. Then we insert a percent symbol:

$$0.37 = 37.0\% = 37\%$$

two places to the right

> **Converting Percentages to and from Decimal Numbers**
>
> - To write a percentage as a decimal number, remove the percent symbol and divide the number by 100 (move the decimal point two places to the left).
> - To write a decimal number as a percentage, multiply the number by 100 (move the decimal point two places to the right) and insert a percent symbol.

▶ **Example 3**   Converting Percentages and Decimal Numbers

Write each percentage as a decimal number, and write each decimal number as a percentage.

1. 86%   **2.** 7%   **3.** 0.125

Solution

1. To write 86% as a decimal number, we remove the percent symbol and move the decimal point two places to the left:

$$86\% = 86.0\% = 0.86$$

two places to the left

2. To write 7% as a decimal number, we remove the percent symbol and move the decimal point two places to the left, using 0 in the tenths place as a placeholder:

$$7\% = 7.0\% = 0.07$$

two places to the left

3. To write 0.125 as a percentage, we move the decimal point two places to the right and insert a percent symbol:

$$0.125 = 12.5\%$$

two places to the right

WARNING   From Problem 2 in Example 3, we see 7% is *not* equal to 0.7. Rather, 7% is equal to 0.07. Remember to move the decimal point *two* places to the left, using 0 in the tenths place as a placeholder.

Using Proportions and Percentages to Describe Authentic Situations

In this course, most of the fractions we work with are proportions. A key step of Example 4 will be to write a proportion as a percentage or to write a percentage as a proportion.

▶ **Example 4**   Using a Proportion or a Percentage to Describe an Authentic Situation

1. Approximately 25% of computers, electronics, and appliances were purchased online in 2013 (Source: *Kantar*). Use a proportion to describe this situation.
2. The proportion of employed mechanical engineers who are women is 0.05 (Source: *U.S. Labor Department*). Use a percentage to describe this situation.
3. Of 1821 adults surveyed, 382 adults said that it was extremely important to them that immigration reform is passed this year (Source: *Pew Research Center*). Use a percentage to describe this situation.

Solution

1. To write 25% as a proportion, we remove the percent symbol and move the decimal point two places to the left: 0.25. So, the proportion of computers, electronics, and appliances that were purchased online in 2013 is 0.25.
2. To write 0.05 as a percentage, we move the decimal point two places to the right and insert a percent symbol: 5%. So, the percentage of employed mechanical engineers who are women is 5%.
3. The proportion is $\dfrac{382}{1821} \approx 0.21$. To write 0.21 as a percentage, we move the decimal point two places to the right and insert a percent symbol: 21%. So, approximately 21% of the 1821 adults said it was extremely important to them that immigration reform is passed this year.

Percentage of a Quantity

How do we find the percentage of a quantity? For example, consider 75% of 4. That is the same as $\dfrac{75}{100}$ of 4. To find a fraction of a number, we *multiply* the fraction by that number:

$$\frac{75}{100} \text{ of } 4 = \frac{75}{100} \cdot 4 = \frac{3}{4} \cdot \frac{4}{1} = \frac{3}{1} = 3$$

So, using decimal notation, we find 75% of 4 by *multiplying* 0.75 by 4:

$$75\% \text{ of } 4 = 0.75(4) = 3$$

To see whether our result makes sense, we first form a large square made up of 4 medium-size squares of equal area (see Fig. 67). To find 75% of the four squares, we divide the large square into 100 small squares of equal area and shade 75 of them. The shaded region contains 3 of the 4 medium-size squares, which checks with our earlier computations.

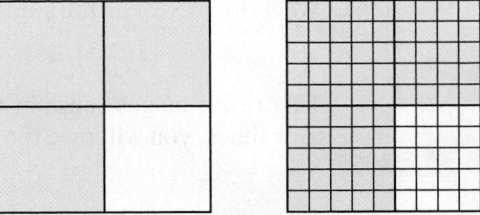

Figure 67 75% of 4 medium-size squares is made up of 75 small squares (in blue), or 3 medium-size squares (in blue)

▶ Finding the Percentage of a Quantity

To find the percentage of a quantity, multiply the decimal form of the percentage and the quantity.

▶ Example 5   Finding the Percentage of a Quantity

1. A person buys a Fender American Standard Jazz Bass for $1500 at Guitar Center in Indianapolis, Indiana, which has a sales tax of 7%. How much money is the sales tax?
2. At University of Oregon, 6.5% of 20,808 undergraduates were psychology majors in the fall semester of 2013 (Source: *University of Oregon*). How many psychology majors were there?

Solution

1. $0.07(1500) = 105$; so, the sales tax is $105.
2. $0.065(20{,}808) = 1352.52 \approx 1353$; so, there were 1353 psychology majors.

By definition, 100% means 100 for each 100. In other words, 100% of a quantity is *all* of the quantity. For example, 100% of 21 guitars is 21 guitars.

▶ One Hundred Percent of a Quantity

One hundred percent of a quantity is *all* of the quantity.

We will continue to work with ratios and percents as we discuss how to multiply and divide real numbers.

Multiplication of Two Real Numbers with Different Signs

We can think of multiplication as repeated addition. For example, 3(5) is equal to the sum of three 5s:

$$3(5) = 5 + 5 + 5 = 15$$

Also, 3(5) is equal to the sum of five 3s:

$$3(5) = 3 + 3 + 3 + 3 + 3 = 15$$

We can use the idea of repeated addition to help us find the product of two numbers with different signs.

▶ **Example 6**   Finding the Product of Two Numbers with Different Signs

Find the product.

1. $4(-2)$

Solution

1. We write $4(-2)$ as the sum of four -2s:
$$4(-2) = (-2) + (-2) + (-2) + (-2) = -8$$

This result makes sense in terms of money. If you borrow 2 dollars from a friend four times, you will owe the friend 8 dollars.

▶

In Example 6, we found the product of two numbers with different signs. Note that the result is negative:

$$\overbrace{4(-2)}^{\text{Different signs}} = \overbrace{-8}^{\text{Negative}}$$

▶ **Multiplying Two Numbers with Different Signs**

The product of two numbers that have different signs is negative.

▶ **Example 7**   Finding the Product of Two Numbers with Different Signs

Find the product.

1. $7(-4)$ **2.** $(-0.2)(0.3)$

Solution

1. Since the signs of 7 and -4 are different, their product is negative: $7(-4) = -28$.
2. Since the signs of -0.2 and 0.3 are different, their product is negative: $(-0.2)(0.3) = -0.06$.

▶

Multiplication of Two Real Numbers with the Same Sign

We have discussed how to multiply numbers with different signs. What if the signs are the same? To begin this investigation, consider the following pattern:

This factor decreases by 1.
$$\begin{array}{l}3(-5) = -15 \\ 2(-5) = -10 \\ 1(-5) = -5 \\ 0(-5) = 0\end{array}$$
The product increases by 5.

It turns out that this pattern continues. So, we have

This factor decreases by 1.
$$\begin{array}{l}-1(-5) = 5 \\ -2(-5) = 10 \\ -3(-5) = 15\end{array}$$
The product increases by 5.

Note that for each of the last three computations, the product of the two negative numbers is positive. This is, in fact, always true. Here we find another product of two negative numbers:

$$\overbrace{(-7)(-9)}^{\text{Same signs}} = \overbrace{63}^{\text{Positive}}$$

> **Multiplying Two Numbers with the Same Sign**
>
> The product of two numbers that have the same sign is positive.

▶ **Example 8**   Finding the Product of Two Numbers with the Same Sign

Find the product.

1. $-5(-6)$
2. $\left(-\dfrac{3}{2}\right)\left(-\dfrac{5}{7}\right)$

Solution

1. Since -5 and -6 have the same sign, their product is positive: $-5(-6) = 30$.

2. Since $-\dfrac{3}{2}$ and $-\dfrac{5}{7}$ have the same sign, their product is positive:

$$\left(-\frac{3}{2}\right)\left(-\frac{5}{7}\right) = \frac{15}{14}$$

In Fig. 68, we show a multiplication table for some specific numbers. In Fig. 69, we summarize the multiplication sign rules for all nonzero real numbers.

·	4	−4
2	8	−8
−2	−8	8

Figure 68 Multiplication table for 2, −2, 4, and −4

·	+	−
+	+	−
−	−	+

Figure 69 Multiplication table for all nonzero real numbers

▶ **Example 9**   Multiplying Real Numbers in an Authentic Situation

A person's credit card balance is -2340 dollars. If the person pays off 30% of the balance, what is the new balance?

Solution

If the person pays off 30% of the balance, then $100\% - 30\% = 70\%$ of the balance remains. We find 70% of -2340:

$$0.70(-2340) = -1638$$

The new balance is -1638 dollars.

Division of Real Numbers

We can get an idea of how to divide real numbers by writing multiplications as related divisions. For example, consider this statement:

$$2 \cdot 3 = 6 \text{ implies that } 6 \div 3 = 2.$$

We now write a similar statement for $(-2)(-3)$:

$$(-2)(-3) = 6 \text{ implies that } \overbrace{6 \div (-3)}^{\text{Different signs}} = \overbrace{-2}^{\text{Negative}}.$$

This statement suggests that the quotient of two numbers with different signs is negative.

Now consider the following statement:

$$2(-3) = -6 \text{ implies that } \overbrace{-6 \div (-3)}^{\text{Same signs}} = \overbrace{2}^{\text{Positive}}.$$

This statement suggests that the quotient of two numbers with the same sign is positive. Both statements suggest that the sign rules for dividing real numbers are similar to those for multiplying real numbers.

> ### ▶ Multiplying or Dividing Real Numbers
>
> The product or quotient of two numbers that have different signs is negative. The product or quotient of two numbers that have the same sign is positive.

▶ Example 10   Finding Quotients of Real Numbers

Find the quotient.

1. $-10 \div 2$

2. $-\dfrac{1}{6} \div \left(-\dfrac{3}{5}\right)$

Solution

Press $(-)$ $($ 1 $\div$ 6 $)$ $\div$ $(-)$ $($ 3 $\div$ 5 $)$ ENTER.

```
-(1/6)/-(3/5)
          .2777777778
5/18
          .2777777778
```

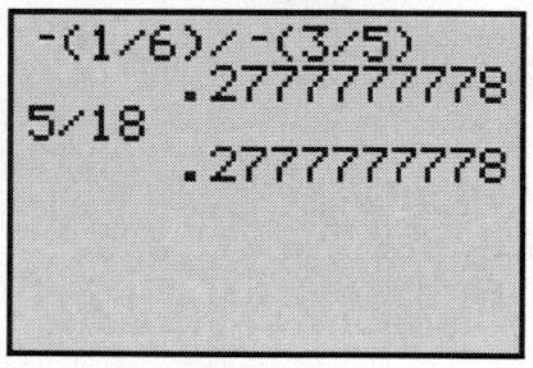

Figure 70 Verify the work

1. Since -10 and 2 have different signs, the quotient is negative: $-10 \div 2 = -5$. This makes sense in terms of money. If we divide a debt of $10 by 2, the result is a debt of $5.

2. The quotient of two negative numbers is positive. To find the result, we divide the absolute value of the fractions:

$$\frac{1}{6} \div \frac{3}{5} = \frac{1}{6} \cdot \frac{5}{3} = \frac{5}{18}$$

We use a TI-84 to check our work for Problem 2 in Example 10 (see Fig. 70).

▶ Example 11   Application of a Ratio of Two Real Numbers

A person has credit card balances of -3950 dollars on a Visa® account and -1225 dollars on a MasterCard® account.

1. Find the unit ratio of the Visa balance to the MasterCard balance.
2. If the person wishes to pay off both accounts gradually in the same amount of time, describe how the result in Problem 1 can help guide the person in making his next payment.

Solution

1. We divide the Visa balance by the MasterCard balance:

$$\frac{-3950}{-1225} \approx \frac{3.22}{1}$$

So, the Visa balance is about 3.22 times the MasterCard balance.

2. For each $1 the person pays to his MasterCard account, he should pay about $3.22 to his Visa account. (The ratio will need to be recalculated each month to take into account recent purchases, cash advances, and so on, as well as possible differences in interest rates on the cards.)

Equal Fractions with Negative Signs

Notice that $\dfrac{-6}{2} = -3$, $\dfrac{6}{-2} = -3$, and $-\dfrac{6}{2} = -3$. So we can write

$$\frac{-6}{2} = \frac{6}{-2} = -\frac{6}{2}$$

The positions of the negative signs in the three equal expressions suggest the following property.

> **Equal Fractions with Negative Signs**
>
> If $b \neq 0$, then
>
> $$\frac{-a}{b} = \frac{a}{-b} = -\frac{a}{b}$$

If the result of a computation is a negative fraction, we write the result in the form $-\dfrac{a}{b}$ rather than $\dfrac{-a}{b}$ or $\dfrac{a}{-b}$.

▲▼ Group Exploration

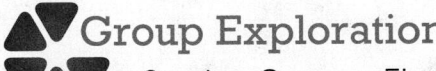

Section Opener: Finding the product of a positive number and a negative number

We can think of the multiplication of two counting numbers as a repeated addition. For example, we can think of $4(3)$ as adding four 3s:

$$4(3) = 3 + 3 + 3 + 3 = 12$$

We can use this idea to help us find the product of a positive number and a negative number.

1. Write each of the products that follow as a repeated sum. Then find the sum.
   - **a.** $3(-2)$ **b.** $5(-4)$ **c.** $7(-1)$

2. Are your results in Problem 1 positive or negative? What can you say about the product of a positive number and a negative number? If you are not sure, try some more multiplications.

3. Explain why the observation you made in Problem 2 makes sense. [**Hint:** What can you say about a negative number plus a negative number?]

> **▶ Tips for Success**   **Affirmations**
>
> Do you have difficulty with math? If so, do you ever tell yourself (or others) that you are not good at it? This is called *negative self-talk*. The more you say this, the more likely your subconscious will believe it—and you *will* do poorly in math.
>
> You can counteract years of negative self-talk by telling yourself with conviction that you are good at math.
>
> It might seem strange to state that something is true that hasn't happened yet, but it works! Such statements are called *affirmations*.

Homework 1.6

For extra help ▶   MyMathLab®   Watch the videos in MyMathLab   Download the MyDashboard App

1. To write a percentage as a decimal number, remove the percent symbol and divide the number by ____.

2. One hundred percent of a quantity is ____ of the quantity.

3. *True or False:* The product of two negative numbers is negative.

4. *True or False:* The quotient of two numbers that have different signs is negative.

For Exercises 5–12, write the percentage as a decimal number or write the decimal number as a percentage, as appropriate.

5. 63%   6. 91%   7. 0.08   8. 0.01

9. 9%   10. 4%   11. 0.052   12. 0.089

For Exercises 13–16, rewrite the sentence so that it contains a proportion.

13. In 2013, 12% of toys and sporting goods were purchased online (Source: *Kantar*).

14. In 2013, 89% of clothing was purchased in stores (Source: *Kantar*).

15. Approximately 7% of students' college costs are paid by their parents borrowing money (Source: *Sallie Mae*).

16. Approximately 3% of 18–49 year-old adults watch the network NBC (Source: *Nielsen Media Research*).

For Exercises 17–22, rewrite the sentence so that it contains a percentage. Round approximate percentages to the first decimal place.

17. In June 2014, the proportion of Americans who approved of the way President Obama was doing his job was 0.41 (Source: *WSJ/NBC News Polls*).

18. The proportion of 304 executives who said they would quit their job and be a stay-at-home parent if they could afford it was 0.37 (Source: *Accenture*).

19. The proportion of plastics that were recycled in 2012 was 0.088 (Source: *U.S. Environmental Protection Agency*).

20. Of the 41 million Americans who traveled at least 50 miles from home during Independence Day holiday weekend in 2014, the proportion who traveled by air was 0.076 (Source: *Newsroom.AAA.com*).

21. Of approximately 273 thousand people living with spinal-cord injuries, approximately 221 thousand of them are men (Source: *National Spinal Cord Injury Statistical Center*).

22. Out of 2048 surveyed adults who do not have a will, 287 adults do not have a will because they do not like thinking about death (Source: *Harris Poll*).

23. Find 35% of 7000 cars.

24. Find 67% of 4500 cars.

25. A person buys a refrigerator for $229.99 (not including sales tax) at Cummins Appliance in Baltimore, Maryland, which has a sales tax of 6%. How much money is the sales tax?

26. A person buys groceries for $125.35 (not including grocery tax) in Harmons Grocery in Salt Lake City, Utah, which has a groceries tax of 3%. How much money is the groceries tax?

27. At University of Arkansas, 83.2% of 26,237 students were undergraduates in fall semester, 2014 (Source: *University of Arkansas*). How many undergraduates were there?

28. At University of Iowa, 3.3% of 22,354 undergraduates were psychology majors in fall semester, 2014 (Source: *University of Iowa*). How many undergraduate psychology majors were there?

Perform the indicated operation by hand.

29. $-2(6)$ 30. $-5(4)$ 31. $-3(-6)$

32. $-8(-9)$ 33. $25 \div (-5)$ 34. $24 \div (-3)$

35. $-56 \div (-7)$ 36. $-1 \div (-1)$ 37. $-15(-37)$

38. $-124(-29)$ 39. $936 \div (-24)$ 40. $1008 \div (-21)$

41. $-0.2(-0.4)$ 42. $-0.3(-0.3)$ 43. $-0.06 \div 0.2$

44. $-0.12 \div 0.3$ 45. $\dfrac{36}{-4}$ 46. $\dfrac{9}{-3}$

47. $\dfrac{-32}{-8}$ 48. $\dfrac{-72}{-8}$ 49. $\dfrac{1}{2}\left(-\dfrac{1}{5}\right)$

50. $\dfrac{1}{3}\left(-\dfrac{7}{5}\right)$ 51. $\left(-\dfrac{4}{9}\right)\left(-\dfrac{3}{20}\right)$ 52. $\left(-\dfrac{7}{25}\right)\left(-\dfrac{5}{21}\right)$

53. $-\dfrac{3}{4} \div \dfrac{7}{6}$ 54. $-\dfrac{5}{7} \div \dfrac{15}{8}$

55. $-\dfrac{24}{35} \div \left(-\dfrac{16}{25}\right)$ 56. $-\dfrac{3}{8} \div \left(-\dfrac{9}{20}\right)$

Perform the indicated operation by hand.

57. $6 + (-9)$ 58. $-9 + (-4)$ 59. $-39 \div (-3)$

60. $-49 \div 7$ 61. $4 - (-2)$ 62. $-2 - 7$

63. $10(-10)$ 64. $-5(-9)$

65. $-\dfrac{8}{3} + \left(-\dfrac{5}{9}\right)$ 66. $-\dfrac{3}{8} - \left(-\dfrac{1}{10}\right)$

67. $\dfrac{9}{2}\left(-\dfrac{4}{21}\right)$ 68. $-\dfrac{22}{9} \div \left(-\dfrac{33}{18}\right)$

Simplify by hand.

69. $\dfrac{-16}{20}$ 70. $\dfrac{-15}{35}$ 71. $\dfrac{-18}{-24}$ 72. $\dfrac{-35}{-21}$

Use a calculator to perform the indicated operation. Round the result to two decimal places.

73. $-26.87(-381.572)$ 74. $-489.2(-8.39)$

75. $222.045 \div (-32.76)$ 76. $64.958 \div (-3.716)$

77. $-\dfrac{11}{18}\left(-\dfrac{15}{19}\right)$ 78. $-\dfrac{169}{175}\left(-\dfrac{64}{71}\right)$

79. $-\dfrac{59}{13} \div \dfrac{27}{48}$ **80.** $-\dfrac{75}{22} \div \dfrac{13}{48}$

For Exercises 81 and 82, write the ratio as a fraction.

81. the ratio of 6 to 8

82. the ratio of 9 to 15

83. There were 492 U.S. billionaires in 2014 and 261 U.S. billionaires in 2001 (Source: Forbes). Find the unit ratio of the number of U.S. billionaires in 2014 to the number of U.S. billionaires in 2001. Round to the second decimal place. What does your result mean in this situation?

84. During the week of January 12, 2012, the average number of viewers was 4.7 million viewers per day for the TV show *Good Morning America* and 2.2 million viewers for the competing *The Early Show* (Source: *Nielsen Media Research*). Find the unit ratio of the average number of viewers per day of *Good Morning America* to the average number of viewers per day of *The Early Show*. Round to the second decimal place. What does your result mean in this situation?

85. A recipe for roasted red-pepper pasta calls for 4 red bell peppers and 5 black olives. Calculate the given unit ratio. What does your result mean in this situation?

   a. The unit ratio of the number of red bell peppers to the number of black olives

   b. The unit ratio of the number of black olives to the number of red bell peppers

86. A recipe for beef stroganoff calls for 2 cups of sliced mushrooms and 4 cups of cooked noodles. Calculate the given unit ratio. What does your result mean in this situation?

   a. The unit ratio of the number of cups of sliced mushrooms to the number of cups of cooked noodles

   b. The unit ratio of the number of cups of cooked noodles to the number of cups of sliced mushrooms

87. The *full-time equivalent enrollment* (FTE enrollment) at a college is the number of full-time students it would take for their total credits (units or hours) to equal the total credits in which both part-time and full-time students combined are enrolled in one semester. The number of *full-time equivalent faculty* (FTE faculty) is the number of full-time faculty it would take to teach all the courses that are taught by both part-time and full-time faculty combined. The FTE enrollments and number of FTE faculty are shown in Table 27 for various colleges.

Table 27 FTE Enrollments and Numbers of FTE Faculty

College	FTE Enrollment	Number of FTE Faculty
Butler University	4168.0	335.1
St. Olaf College	2951.0	248.7
Stonehill College	2351.0	178.0
University of Massachusetts Amherst	17,016.2	982.4
Texas A&M University	37,682.49	1785.9

Sources: *Butler University, St. Olaf College, Stonehill College, University of Massachusetts, Texas A&M University*

Round all unit ratios to the second decimal place:

   a. Find the unit ratio of FTE enrollment at Texas A&M University to the FTE enrollment at St. Olaf College. What does your result mean in this situation?

   b. Find the unit ratio of the number of FTE faculty at the University of Massachusetts Amherst to the number of FTE faculty at Butler University. What does your result mean in this situation?

   c. Find the unit ratio of FTE enrollment to the number of FTE faculty at each of the colleges listed in Table 27.

   d. Which college listed in Table 27 has the largest ratio of FTE enrollment to the number of FTE faculty? Which has the smallest?

   e. A person believes that the ratio of FTE enrollment to the number of FTE faculty is lower at Stonehill College than at St. Olaf College, because Stonehill College has the lower FTE enrollment. Is that person correct? Explain.

88. The 2014 populations and land areas are shown in Table 28 for various states.

Table 28 Populations and Land Areas

State	Population	Land Area (square miles)
Alaska	736,732	571,951
California	38,802,500	155,959
Michigan	9,883,640	56,804
New Jersey	8,791,894	7417
New York	19,378,102	47,214

Sources: *U.S. Census Bureau; Infoplease*

Round all unit ratios to the second decimal place:

   a. Find the unit ratio of New York's population to New Jersey's population. What does your result mean in this situation?

   b. Find the unit ratio of Alaska's land area to California's land area. What does your result mean in this situation?

   c. The unit ratio of population to land area is called the *population density*. Find the population density of each state listed in Table 28.

   d. Which state listed in Table 28 has the greatest population density? Which has the least?

   e. A person believes that Michigan has a greater population density than New Jersey, because Michigan's population is more than New Jersey's population. Is that person correct? Explain.

89. A person has credit card balances of −4360 dollars on a Discover® account and −1825 dollars on a MasterCard® account.

   a. Find the unit ratio of the Discover balance to the MasterCard balance. Round the result to the second decimal place.

   b. If the person wishes to pay off both accounts gradually in the same amount of time, describe how the result you found in part (a) can help guide the person in making her next payment.

90. A person has credit card balances of −6810 dollars on a Visa® account and −2950 dollars on a Sears® account. Round the result to the second decimal place.

   a. Find the unit ratio of the Visa balance to the Sears balance.

   b. If the person wishes to pay off both accounts gradually in the same amount of time, describe how the result you found in part (a) can help guide the person in making his next payment.

91. A person's credit card balance is −3720 dollars. If the person pays off 15% of the balance, what is the new balance?

92. A person's credit card balance is -1590 dollars. If the person pays off 35% of the balance, what is the new balance?

93. A student has zero balance on a credit card. The student uses the credit card to buy 12.3 gallons of gasoline at a cost of $2.40 per gallon. What is the new balance?

94. A person has zero balance on a credit card. The person uses the credit card to buy three lamps at a cost of $89.50 per lamp. What is the new balance?

Concepts

95. a. Find the sum $-2 + (-4)$.

   b. Find the product $-2(-4)$.

   c. Consider the following statements:

   - Two negative numbers make a positive.

   - A negative number times a negative number is equal to a positive number.

   Which statement is clearer? Explain.

   d. Compare the sign rule for adding two negative numbers with the sign rule for multiplying two negative numbers.

96. a. Is $\dfrac{12}{-4}$ positive or negative? Explain.

   b. If a is positive and b is negative, is $\dfrac{a}{b}$ positive or negative? Explain.

   c. A student says that $\dfrac{a}{b}$ is positive because it has no negative signs. Is the student correct? Explain.

97. Which of the following fractions are equal? (There may be more than one pair of answers.)

$$\frac{a}{b} \quad \frac{-a}{b} \quad \frac{a}{-b} \quad -\frac{a}{b} \quad \frac{-a}{-b} \quad -\frac{-a}{-b} .$$

98. Discuss in terms of repeated addition why it makes sense that $3(-6)$ is negative.

99. If ab is negative, what can you say about a or b?

100. Some students believe $2x$ is greater than x for *all* real numbers.

   a. Find three values of x where $2x$ is greater than x.

   b. Find three values of x where $2x$ is less than x.

   c. Find one value of x where $2x$ is equal to x.

▼ 1.7 Exponents, Square Roots, Order of Operations, and Scientific Notation

Objectives

» Describe the meaning of *exponent*.

» Describe the meaning of the exponent zero.

» Describe the meaning of a negative-integer exponent.

» Describe the meaning of *square root* and *principal square root*.

» Approximate a principal square root.

» Use the rules for order of operations to perform computations and evaluate expressions.

» Use the rules for order of operations to make predictions.

» Find the area of part of an object.

» Use scientific notation.

In this section, we will discuss an operation called *exponentiation*. We will find the *principal square root* of a number and write numbers in *scientific notation*. We will also discuss the order in which we should perform various operations.

Exponents

The notation b^2 stands for $b \cdot b$. So, $7^2 = 7 \cdot 7 = 49$. The notation b^3 stands for $b \cdot b \cdot b$. So, $2^3 = 2 \cdot 2 \cdot 2 = 8$.

▶ **Definition Exponent**

For any counting number n,

$$b^n = \underbrace{b \cdot b \cdot b \cdot \ldots \cdot b}_{n \text{ factors of } b}$$

We refer to b^n as the **power**, the ***n*th power of *b***, or ***b* raised to the *n*th power**. We call b the **base** and n the **exponent**.

The expression 2^5 is a power. It is the 5th power of 2, or 2 raised to the 5th power. For 2^5, the base is 2 and the exponent is 5. Here, we label the base and the exponent of 2^5 and compute the power:

$$\overset{\text{Exponent}}{2^5} = \underbrace{2 \cdot 2 \cdot 2 \cdot 2 \cdot 2}_{5 \text{ factors of } 2} = 32$$

$\underset{\text{Base}}{}$

When we calculate a power, we say that we are performing **exponentiation.**

Notice that the notation b^1 stands for one factor of b, so $b^1 = b$.

Two powers of b have specific names. We refer to b^2 as the **square of b or b squared.** We refer to b^3 as the **cube of b or b cubed.**

For an expression of the form $-b^n$, we calculate b^n before taking the opposite. For example,

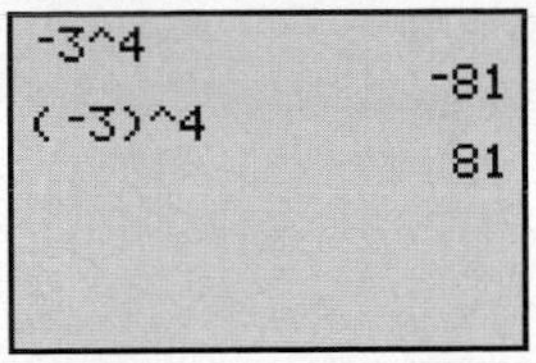

$$-3^4 = -(3^4) = -(3 \cdot 3 \cdot 3 \cdot 3) = -81$$

For -3^4, the base is 3. If we want the base to be -3, we enclose -3 in parentheses:

$$(-3)^4 = (-3)(-3)(-3)(-3) = 81$$

We use a TI-84 to check both computations (see Fig. 71).

Figure 71 Compute -3^4 and $(-3)^4$

▶ **Example 1** Calculating Expressions That Have Exponents

Perform the exponentiation.

1. 5^4 **2.** -3^2 **3.** $(-3)^2$ **4.** $\left(\dfrac{4}{5}\right)^3$

Solution

1. $5^4 = 5 \cdot 5 \cdot 5 \cdot 5 = 625$ The base is 5.

2. $-3^2 = -(3 \cdot 3) = -9$ The base is 3.

3. $(-3)^2 = (-3)(-3) = 9$ The base is -3.

4. $\left(\dfrac{4}{5}\right)^3 = \dfrac{4}{5} \cdot \dfrac{4}{5} \cdot \dfrac{4}{5} = \dfrac{64}{125}$ The base is $\dfrac{4}{5}$.

Zero as an Exponent

What is the meaning of 2^0? Computing powers of 2 can suggest the meaning:

$$
\begin{array}{l}
2^4 = 16 \\
2^3 = 8 \\
\text{The exponent decreases by 1.} \longrightarrow 2^2 = 4 \longleftarrow \text{The power is divided by 2.} \\
2^1 = 2 \\
2^0 = 1
\end{array}
$$

Each time we decrease the exponent by 1, the power is divided by 2. This pattern suggests that $2^0 = 1$.

Similar work with any other nonzero base b would suggest that $b^0 = 1$.

▶ **Definition** Zero exponent

For nonzero b,

$$b^0 = 1$$

So, $3^0 = 1$, $\left(\dfrac{2}{5}\right)^0 = 1$, and $(-4)^0 = 1$.

Negative Exponents

In statistics, we sometimes get results that a calculator will display using negative exponents. To learn the meaning of a negative exponent, we continue decreasing the exponent of 2 by 1 and dividing the power by 2, which is the same as multiplying by $\frac{1}{2}$:

$$2^2 = 4$$

$$2^1 = 2 \qquad 4 \cdot \frac{1}{2} = 2$$

$$2^0 = 1 \qquad 2 \cdot \frac{1}{2} = 1$$

$$2^{-1} = \frac{1}{2^1} \qquad 1 \cdot \frac{1}{2} = \frac{1}{2^1}$$

$$2^{-2} = \frac{1}{2^2} \qquad \frac{1}{2^1} \cdot \frac{1}{2} = \frac{1}{2^2}$$

$$2^{-3} = \frac{1}{2^3} \qquad \frac{1}{2^2} \cdot \frac{1}{2} = \frac{1}{2 \cdot 2} \cdot \frac{1}{2} = \frac{1}{2^3}$$

For each of the last three equations, the exponents (in red) are opposites. In general, $2^{-n} = \frac{1}{2^n}$, where n is a counting number. Next, we describe this pattern for any nonzero base.

▶ **Definition   Negative-integer exponent**

If n is a counting number and b is nonzero, then

$$b^{-n} = \frac{1}{b^n}$$

In words: To find b^{-n}, take its reciprocal and change the sign of the exponent.

▶ **Example 2**   Calculate Powers with Negative Exponents

Perform the exponentiation.

1. 5^{-2} **2.** 10^{-4}

Solution

1. $5^{-2} = \frac{1}{5^2}$ *Write power so that exponent is positive:* $b^{-n} = \frac{1}{b^n}$

$\qquad = \frac{1}{25}$ *Simplify.*

2. $10^{-4} = \frac{1}{10^4}$ *Write power so that exponent is positive:* $b^{-n} = \frac{1}{b^n}$

$\qquad = \frac{1}{10,000}$ *Simplify.*

▶

Principal Square Roots

What number squared is equal to 9? The numbers that "work" are -3 and 3:

$$(-3)^2 = 9 \qquad 3^2 = 9$$

We say that -3 and 3 are *square roots* of 9. A **square root** of a number a is a number we square to get a.

Of the square roots of 9, only the nonnegative square root, 3, is the *principal square root* of 9. Using symbols, we write $\sqrt{9} = 3$.

▶ **Definition   Principal square root**

If a is a nonnegative number, then $\sqrt{a}$ is the nonnegative number we square to get a. We call $\sqrt{a}$ the **principal square root** of a.

For example, $\sqrt{25} = 5$, since $5^2 = 25$. Also, $\sqrt{64} = 8$, because $8^2 = 64$.

The symbol "$\sqrt{}$" is called a **radical sign.** An expression under a radical sign is called a **radicand.** For $\sqrt{4x - 7}$, the radicand is $4x - 7$. A radical sign together with a radicand is called a **radical.** Here we label the radical sign and radicand of the radical $\sqrt{2x + 5}$:

$$\underbrace{\underset{\text{Radical}}{\underbrace{\sqrt{2x + 5}}}}\quad\overset{\text{Radical sign}}{\nearrow}$$
$$\longleftarrow \text{Radicand}$$

Here are some more radicals: $\sqrt{5}$, $\sqrt{x}$, $\sqrt{9x + 4}$.

An expression that contains a radical is called a **radical expression.** Here are some radical expressions that are used a lot in statistics:

$$\sqrt{5} \qquad \sqrt{n} \qquad \bar{x} + t\frac{s}{\sqrt{n}} \qquad \sqrt{\frac{s_1^2}{n_1} + \frac{s_2^2}{n_2}} \qquad \hat{p} - z\sqrt{\frac{\hat{p}(1 - \hat{p})}{n}}$$

WARNING   Is a square root of a negative number a real number? Consider $\sqrt{-9}$. Note that $\sqrt{-9} \neq -3$, because $(-3)^2$ is equal to 9, not -9. Since any number squared is nonnegative, we see that $\sqrt{-9}$ is not a real number. In general, **a square root of a negative number is not a real number.**

▶ **Example 3**   Finding Square Roots

Find the square root.

1. $\sqrt{49}$ **2.** $\sqrt{-49}$ **3.** $-\sqrt{49}$ **4.** $-\sqrt{-49}$

Solution

    1. $\sqrt{49} = 7$, because $7^2 = 49$.
    2. $\sqrt{-49}$ is not a real number, because the radicand -49 is negative.
    3. $-\sqrt{49} = -7$.
    4. $-\sqrt{-49}$ is not a real number, because the radicand -49 is negative.

Approximating Square Roots

Recall that a rational number is a number that can be written in the form $\frac{n}{d}$, where n and d are integers and d is nonzero. A **perfect square** is a number whose principal square root is rational. For example, 25 is a perfect square, because $\sqrt{25} = 5 = \frac{5}{1}$ is rational.

By squaring the integers from 0 to 15, we can find the integer perfect squares between 0 and 225, inclusive (see Table 29).

For a number that is not a perfect square, any square root of the number is not rational. Recall that we call such a number *irrational.* For example, $\sqrt{7}$ is irrational. We know that $\sqrt{7}$ is a number between 2 and 3, because $2^2 = 4$ and $3^2 = 9$. To use a TI-84 to get the estimate $\sqrt{7} \approx 2.645751311$, press [2nd] [$\sqrt{}$] 7 [)] [ENTER] (see Fig. 72).

Table 29 Perfect Squares

x	Perfect Square x^2
0	0
1	1
2	4
3	9
4	16
5	25
6	36
7	49
8	64
9	81
10	100
11	121
12	144
13	169
14	196
15	225

Figure 72 Estimating $\sqrt{7}$

▶ **Example 4**   Approximating Square Roots

State whether the square root is rational or irrational. If the square root is rational, find the (exact) value. If the square root is irrational, estimate its value by rounding to the second decimal place.

1. $\sqrt{19}$ **2.** $\sqrt{169}$

Solution

1. The number 19 is not a perfect square, so $\sqrt{19}$ is irrational. We use a TI-84 to compute $\sqrt{19} \approx 4.36$ (see Fig. 73).
2. The number 169 is a perfect square, so $\sqrt{169}$ is rational. In fact, $\sqrt{169} = 13$, because $13^2 = 169$.

Figure 73 Estimating $\sqrt{19}$

Order of Operations

We can establish the order of operations by using *grouping symbols,* such as parentheses ( ), brackets [ ], absolute-value symbols ‖, fraction bars, and square roots. We do operations that lie within grouping symbols before we perform other operations.

But does it matter in which order we perform operations? From the following calculations, it is clear that it *does* matter:

$$(3 + 2) \cdot 4 = 5 \cdot 4 = 20 \quad \textit{First add; then multiply.}$$
$$3 + (2 \cdot 4) = 3 + 8 = 11 \quad \textit{First multiply; then add.}$$

With a fraction such as $\dfrac{7 + 3}{3 - 1}$, the following use of parentheses is assumed:

$$\frac{7 + 3}{3 - 1} = \frac{(7 + 3)}{(3 - 1)} = \frac{10}{2} = 5$$

So, we compute both the numerator and the denominator before we divide.

Similarly, with a radical expression such as $\sqrt{9 + 16}$, the following use of parentheses is assumed:

$$\sqrt{9 + 16} = \sqrt{(9 + 16)} = \sqrt{25} = 5$$

WARNING   So, we compute the radicand before we take the square root.

It is a common error to write $\sqrt{9 + 16}$ as $\sqrt{9} + \sqrt{16}$, which is incorrect. In fact, $\sqrt{9} + \sqrt{16} = 3 + 4 = 7$, which is *not* equal to the correct value 5 we found earlier.

▶ **Example 5**   Performing Operations

Perform the indicated operations.

1. $(7 + 2)(3 - 8)$ **2.** $\dfrac{11 - 3}{1 - 5}$ **3.** $3\sqrt{2 + 14}$

Solution

1. $(7 + 2)(3 - 8) = (9)(-5) = -45$
2. $\dfrac{11 - 3}{1 - 5} = \dfrac{8}{-4} = -2$
3. $3\sqrt{2 + 14} = 3\sqrt{16} = 3(4) = 12$

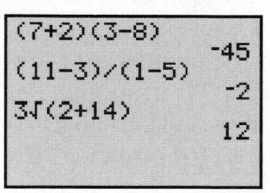

Figure 74 Verify the result

We use a TI-84 to check our results (see Fig. 74).

For an expression such as $3 + 2 \cdot 4$, where grouping symbols do not specify the order of operations for all operations in the expression, there is an understood order of operations.

> **Order of Operations**

We perform operations in the following order:

1. First, perform operations within parentheses or other grouping symbols, starting with the innermost group.
2. Then perform exponentiations.
3. Next, perform multiplications and divisions, going from left to right.
4. Last, perform additions and subtractions, going from left to right.

> **Example 6**   Performing Operations

Perform the indicated operations.

1. $9 - 8 \div 4$ **2.** $3 + 2^3 + 4(5)$ **3.** $2 - [7 + 4(3 - 5)]$

Solution

1.
$$9 - 8 \div 4 = 9 - 2 \quad \textit{Divide before subtracting.}$$
$$= 7 \quad \textit{Subtract.}$$

2.
$$3 + 2^3 + 4(5) = 3 + 8 + 4(5) \quad \textit{Perform exponentiation first: } 2^3 = 2 \cdot 2 \cdot 2 = 8$$
$$= 3 + 8 + 20 \quad \textit{Multiply before adding.}$$
$$= 31 \quad \textit{Add.}$$

3. The square brackets, "[" and "]," mean the same thing as parentheses.

$$2 - [7 + 4(3 - 5)] = 2 - [7 + 4(-2)] \quad \textit{Subtract within innermost parentheses.}$$
$$= 2 - [7 + (-8)] \quad \textit{Multiply.}$$
$$= 2 - [-1] \quad \textit{Add.}$$
$$= 2 + 1 \quad \textit{Simplify.}$$
$$= 3 \quad \textit{Add.}$$

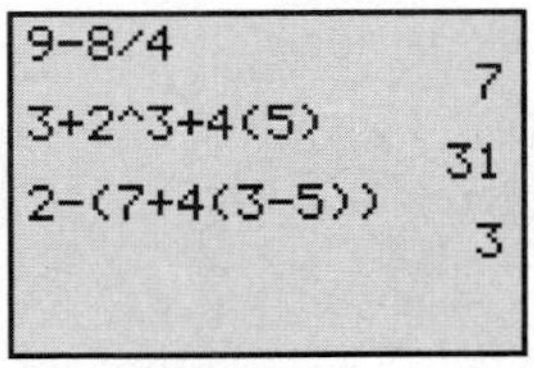

```
9-8/4
              7
3+2^3+4(5)
             31
2-(7+4(3-5))
              3
```

Figure 75 Verify the result

We use a TI-84 to verify our results (see Fig. 75).

There is a connection between the order of operations and the strengths of the operations. We explore the strengths of exponentiation, multiplication, and addition by performing these operations on a pair of 10s:

Operation	Computation with 10s
Exponentiation	$10^{10} = 10,000,000,000$
Multiplication	$10 \cdot 10 = 100$
Addition	$10 + 10 = 20$

Exponentiation is much more powerful than multiplication, which in turn is more powerful than addition. Since dividing by a number is the same as multiplying by the reciprocal of the number, division is as powerful as multiplication. Since subtracting a number is the same as adding the opposite of the number, subtraction is as powerful as addition. Here is a summary of the strengths of the operations:

Operation	Strength of Operation
Exponentiation	Most Powerful
Multiplication and Division	Next Most Powerful
Addition and Subtraction	Weakest

> **Order of Operations and the Strengths of Operations**

After we have performed operations in parentheses, the order of operations goes from the most powerful operation, exponentiation, to the next-most-powerful operations, multiplication and division, to the weakest operations, addition and subtraction.

Knowing the relationship between the order of operations and the strengths of the operations will likely help you remember the order of operations.

▶ Example 7   Performing Operations

Perform the indicated operations.

1. $\dfrac{(2 - 6)^2 + (4 - 6)^2 + (12 - 6)^2}{3 - 1}$ 2. $(47 - 41) - 2\sqrt{\dfrac{6^2}{9} + \dfrac{5^2}{5}}$

Solution

1.
$$\dfrac{(2 - 6)^2 + (4 - 6)^2 + (12 - 6)^2}{3 - 1} = \dfrac{(-4)^2 + (-2)^2 + 6^2}{2} \quad \text{Subtract within parentheses first.}$$

$$= \dfrac{16 + 4 + 36}{2} \quad \text{Perform exponentiations.}$$

$$= \dfrac{56}{2} \quad \text{Add.}$$

$$= 28 \quad \text{Divide.}$$

2.
$$(47 - 41) - 2\sqrt{\dfrac{6^2}{9} + \dfrac{5^2}{5}} = (47 - 41) - 2\sqrt{\dfrac{36}{9} + \dfrac{25}{5}} \quad \text{Perform exponentiations first.}$$

$$= (47 - 41) - 2\sqrt{4 + 5} \quad \text{Divide.}$$

$$= 6 - 2\sqrt{9} \quad \text{Work within parentheses and square root.}$$

$$= 6 - 2(3) \quad \text{Find the square root.}$$

$$= 6 - 6 \quad \text{Multiply.}$$

$$= 0 \quad \text{Subtract.}$$

▶ Example 8   Use a Calculator to Perform Operations

Use a calculator to perform the indicated operations for

$$\sqrt{\dfrac{(3.8 - 6.5)^2 + (8.6 - 6.5)^2 + (7.1 - 6.5)^2}{3 - 1}}.$$

Solution

We first use a TI-84 to compute the numerator of the radicand, which is 12.06 (see Fig. 76):

$$\sqrt{\dfrac{(3.8 - 6.5)^2 + (8.6 - 6.5)^2 + (7.1 - 6.5)^2}{3 - 1}} = \sqrt{\dfrac{12.06}{3 - 1}}$$

Next, we divide 12.06 by 3 − 1 = 2, which gives 6.03 (see Fig. 77):

$$\sqrt{\dfrac{12.06}{3 - 1}} = \sqrt{6.03}$$

Finally, we take the square root of 6.03, which gives approximately 2.5 (see Fig. 78).

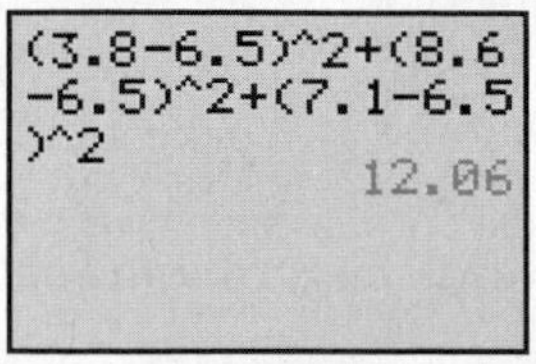

Figure 76 Compute numerator of radicand

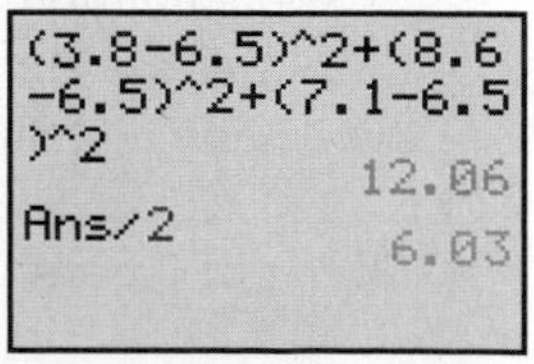

Figure 77 Divide by 2

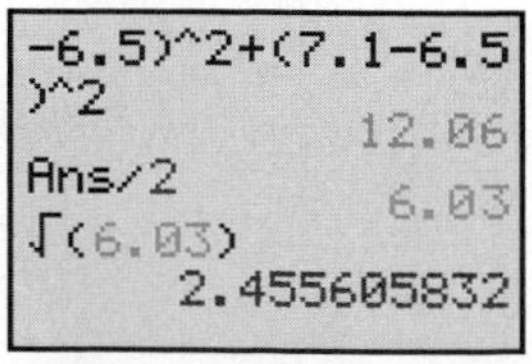

Figure 78 Compute square root

WARNING   In Example 8, we must make sure that we press ENTER before dividing by 2 (see Fig. 77).

The type of keystrokes we performed in Example 8 have to do with one of the most important concepts of statistics, which we will discuss in Section 4.2.

Expressions

Now that we know the order of operations, we can evaluate expressions that involve more than one operation.

We can use a combination of symbols and numbers to name a single variable. In Example 9, we will use the notation y_1 as the name of a single variable. We will do the same for the notations y_2, x_1, and x_2.

▶ **Example 9**   Evaluating an Expression

Evaluate $\dfrac{y_2 - y_1}{x_2 - x_1}$ for $y_2 = 4$, $y_1 = -2$, $x_2 = -5$, and $x_1 = 4$.

Solution

We begin by substituting 4 for y_2, -2 for y_1, -5 for x_2, and 4 for x_1:

$$\frac{(4) - (-2)}{(-5) - (4)} = \frac{6}{-9} \quad \text{Subtract.}$$

$$= -\frac{6}{9} \quad \frac{a}{-b} = -\frac{a}{b}$$

$$= -\frac{2}{3} \quad \text{Simplify.}$$

In Problem 1 of Example 10, we will work with the notation $\bar{x}$, which is the name of a key variable for statistics. We will discuss its meaning in Section 4.1.

▶ **Example 10**   Evaluate Expressions

1. Evaluate $\bar{x} - t\dfrac{s}{\sqrt{n}}$ for $\bar{x} = 30$, $t = 2$, $s = 12$, and $n = 16$.

2. Evaluate ab^x for $a = 5$, $b = 2$, and $x = -3$.

Solution

1. We substitute 30 for $\bar{x}$, 2 for t, 12 for s, and 16 for n in the expression $\bar{x} - t\dfrac{s}{\sqrt{n}}$:

$$30 - 2 \cdot \frac{12}{\sqrt{16}} = 30 - 2 \cdot \frac{12}{4} \quad \text{Find the square root.}$$
Divide, because the following
$$= 30 - 2 \cdot 3 \quad \text{parentheses are assumed: } 30 - 2 \cdot \left(\frac{12}{4}\right).$$
$$= 30 - 6 \quad \text{Multiply}$$
$$= 24 \quad \text{Subtract.}$$

2. We begin by substituting 5 for a, 2 for b, and -3 for x in the expression ab^x:

$$5(2)^{-3} = 5 \cdot \frac{1}{2^3} \quad b^{-n} = \frac{1}{b^n}$$

$$= 5 \cdot \frac{1}{8} \quad \text{Perform the exponentiation.}$$

$$= \frac{5}{1} \cdot \frac{1}{8} \quad a = \frac{a}{1}$$

$$= \frac{5}{8} \quad \frac{a}{b} \cdot \frac{c}{d} = \frac{ac}{bd}$$

▶ **Example 11**   Translating from English to Mathematics

Let x be a number. Translate the phrase "6 minus the product of -5 and the number" into a mathematical expression. Then evaluate the expression for $x = -3$.

Solution

First, we translate the given phrase into an expression:

$$\underbrace{6 -}_{\text{Six minus}} \quad \underbrace{(-5)x}_{\substack{\text{the product of } -5 \\ \text{and the number}}}$$

Then we substitute -3 for x in the expression $6 - (-5)x$ and perform the indicated operations:

$$6 - (-5)(-3) = 6 - 15 \quad \textit{Multiply before subtracting.}$$
$$= -9 \quad \textit{Subtract.}$$

Making Predictions

In Example 12, we will find an expression, which we will use to make a prediction.

▶ **Example 12**   Using a Table to Find an Expression

The number of cases worldwide of unruly passenger behavior on board aircraft was 2.7 thousand cases in 2009, and it increased by 1.4 thousand cases per year for the period 2009–2014 (Source: *International Air Transport Association*).

1. Let t be the number of years since 2009. For example, $t = 2$ stands for 2011, because 2011 is 2 years after 2009. Use a table to help find an expression that stands for the number (in thousands) of cases of unruly aircraft passenger behavior in the year that is t years since 2009.

2. Evaluate the expression you found in Problem 1 for $t = 5$. What does your result mean in this situation?

Solution

1. We construct Table 30. We show the arithmetic to help us see a pattern. From the last row of the table, we see that the number (in thousands) of cases of unruly aircraft passenger behavior can be represented by $1.4t + 2.7$.

Table 30 Numbers of Cases of Unruly Aircraft Passengers

Years since 2009	Number of Cases of Unruly Behavior (in thousands)
0	$1.4 \cdot 0 + 2.7$
1	$1.4 \cdot 1 + 2.7$
2	$1.4 \cdot 2 + 2.7$
3	$1.4 \cdot 3 + 2.7$
4	$1.4 \cdot 4 + 2.7$
t	$1.4 \cdot t + 2.7$

Source: *International Air Transport Association*

2. We substitute 5 for t in the expression $1.4t + 2.7$:

$$1.4(5) + 2.7 = 9.7$$

So, there were 9.7 thousand cases of unruly aircraft passenger behavior in 2014.

In Section 1.6, we found the percentage of a quantity. Now we will find the result of a percentage increase or percentage decrease in a quantity.

▶ Example 13   Solving a Percentage Problem

The attendance at Broadway shows was 12.3 million people in 2009, and it decreased by 2.9% in 2010 (Source: *National Arts Index*). What was the attendance in 2010?

Solution

The decrease in attendance from 2009 to 2010 was 0.029(12.3) million people. To find the attendance (in millions of people) in 2010, we subtract 0.029(12.3) from 12.3:

Attendance in 2009 (in millions)		Decrease in attendance (in millions)		Attendance in 2010 (in millions)
$\overbrace{12.3}$	$-$	$\overbrace{0.029(12.3)}$	$\approx$	$\overbrace{11.9}$

The attendance in 2010 was about 11.9 million people.

Finding the Area of Part of an Object

In statistics, we often find the area of part of an object in which the entire object has area 1.

▶ Example 14   Find the Area of Part of an Object

Use the fact that the total area of the object in Fig. 79 is 1 to find the area of the orange bar. The object has not been drawn to scale.

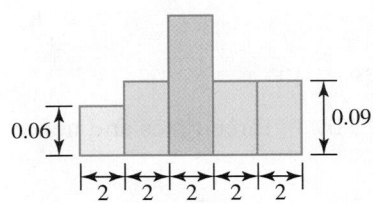

Figure 79 Find the area of the orange bar

Solution

We find the area of the orange bar by two methods.

Method 1   The blue bar has width 2 and height 0.06, so its area is $2(0.06) = 0.12$. Each of the green bars has width 2 and height 0.09, so the area of one green bar is $2(0.09) = 0.18$. We multiply this result by 3 to get the total area of all three green bars: $3(0.18) = 0.54$. Next, we find the total area of the blue and green bars: $0.12 + 0.54 = 0.66$. Because the area of the entire object is 1, we can find the area of the orange bar by subtracting 0.66 from 1: $1 - 0.66 = 0.34$.

Method 2   We use the logic in Method 1 to form a single expression that is equal to the area of the orange bar and then apply order of operations:

$$1 - [2(0.06) + 3(2(0.09))] = 1 - [2(0.06) + 3(0.18)] \quad \text{Work within innermost parentheses first.}$$

$$= 1 - [0.12 + 0.54] \quad \text{Multiply within parentheses.}$$

$$= 1 - 0.66 \quad \text{Add within parentheses.}$$

$$= 0.34 \quad \text{Subtract.}$$

So, the area is 0.34, which is the same result we found earlier.

Scientific Notation

We can use exponents to describe numbers in *scientific notation*. This will enable us to describe compactly a number whose absolute value is very large or very small. For example, the distance to Proxima Centauri, the nearest star other than the Sun, is 40,100,000,000,000 kilometers. Here we write the distance in scientific notation:

$$4.01 \times 10^{13}$$

The symbol "×" stands for multiplication.

As another example, 1 square inch is approximately 0.00000016 acre. Here we write 0.00000016 in scientific notation:

$$1.6 \times 10^{-7}$$

▶ Definition Scientific notation

A number is written in **scientific notation** if it has the form $N \times 10^k$, where k is an integer and the absolute value of N is between 1 and 10 or is equal to 1.

Here are more examples of numbers in scientific notation:

$$8.6 \times 10^{19} \quad 2.159 \times 10^8 \quad -4.23 \times 10^{-14} \quad 7.94 \times 10^{-97}$$

▶ Example 15 Writing Numbers in Standard Decimal Notation

Write the number in standard decimal form.

1. 3.52×10^4 **2.** 7×10^{-3} **3.** 9.48×10^{-4}

Solution

1. We simplify 3.52×10^4 by *multiplying* 3.52 by 10 four times and hence move the decimal point four places to the *right*:

$$3.52 \times 10^4 = 35200.0 = 35,200$$
$$\underbrace{\qquad\qquad}_{\text{four places to the right}}$$

2. Because

$$7 \times 10^{-3} = 7 \times \frac{1}{10^3} = \frac{7}{1} \times \frac{1}{10^3} = \frac{7}{10^3}$$

we see that we can simplify 7.0×10^{-3} by *dividing* 7.0 by 10 three times and hence move the decimal point three places to the *left*:

$$7.0 \times 10^{-3} = 0.007$$
$$\underbrace{\qquad\qquad}_{\text{three places to the left}}$$

3. We *divide* 9.48 by 10 four times and hence move the decimal point of 9.48 four places to the *left*:

$$9.48 \times 10^{-4} = 0.000948$$
$$\underbrace{\qquad\qquad}_{\text{four places to the left}}$$

Converting from Scientific Notation to Standard Decimal Notation

To write the scientific notation $N \times 10^k$ in standard decimal notation, we move the decimal point of the number N as follows:

- If k is *positive*, we multiply N by 10 k times and hence move the decimal point k places to the *right*.
- If k is *negative*, we divide N by 10 k times and hence move the decimal point k places to the *left*.

▶ Example 16 Writing Numbers in Scientific Notation

Write the number in scientific notation.

1. 8,459,000   **2.** 0.0000382

Solution

1. In scientific notation, we would have 8.459×10^k, but what is k? If we move the decimal point of 8.459 six places to the right, the result is 8,459,000. So, $k = 6$ and the scientific notation is 8.459×10^6.
2. In scientific notation, we would have 3.82×10^k, but what is k? If we move the decimal point of 3.82 five places to the left, the result is 0.0000382. So, $k = -5$ and the scientific notation is 3.82×10^{-5}.

Converting from Standard Decimal Notation to Scientific Notation

To write a number in scientific notation, count the number k of places that the decimal point must be moved so that the absolute value of the new number N is between 1 and 10 or is equal to 1:

- If the decimal point is moved to the left, then the scientific notation is written as $N \times 10^k$.

- If the decimal point is moved to the right, then the scientific notation is written as $N \times 10^{-k}$.

▶ Example 17   Writing Numbers in Scientific Notation

Write the number in scientific notation.

1. 778,000,000   *(Jupiter's average distance, in kilometers, from the Sun)*
2. 0.000012   *(the diameter, in meters, of a white blood cell)*

Solution

1. For 778,000,000, the decimal point needs to be moved eight places to the left so the new number N is between 1 and 10. Therefore, the scientific notation is 7.78×10^8.
2. For 0.000012, the decimal point needs to be moved five places to the right so the new number N is between 1 and 10. Therefore, the scientific notation is 1.2×10^{-5}.

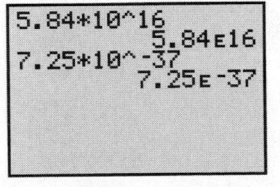

Figure 80 The numbers 5.84×10^{16} and 7.25×10^{-37}

Calculators express numbers in scientific notation so that the numbers "fit" on the screen. To represent 5.84×10^{16}, a TI-84 uses the notation 5.84 E 16, where E stands for exponent (of 10). It represents 7.25×10^{-37} as 7.25 E −37 (see Fig. 80).

▲▼ Group Exploration

Section Opener: Order of operations

If an expression has more than one operation, we can use parentheses to indicate which operation to do first.

1. Perform the indicated operations in $(2 + 3) \cdot 4$ by first doing the addition and then doing the multiplication.

2. Perform the indicated operations in $2 + (3 \cdot 4)$ by first doing the multiplication and then doing the addition.

3. Compare your results for Problems 1 and 2. Does it matter in which order we add and multiply?

For each object, determine whether the area is $(2 + 3) \cdot 4$ or $2 + (3 \cdot 4)$. Explain.

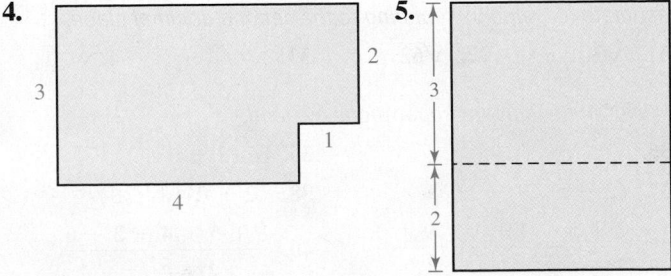

> ▶ Tips for Success   Practice Exams
>
> When studying for an exam (or quiz), try creating your own exam to take for practice. To create your exam, select several homework exercises from each section on which you will be tested. Choose a variety of exercises that address concepts your instructor has emphasized. Your test should include many exercises that are moderately difficult and some that are challenging. Completing such a practice test will help you reflect on important concepts and pin down what types of problems you need to study more.
>
>   It is a good idea to work on the practice exam for a predetermined period. Doing so will help you get used to a timed exam, build your confidence, and lower your anxiety about the real exam.
>
>   If you are studying with another student, you can each create a test and then take each other's test. Or you can create a test together and each take it separately.

 Homework 1.7

 For extra help ▶ MyMathLab®   Watch the videos in MyMathLab   Download the MyDashboard App

1. If n is a counting number and b is nonzero, then $b^{-n} = $ ____.

2. If a is a nonnegative number, then $\sqrt{a}$ is the nonnegative number we ____ to get a.

3. *True or False:* We perform all multiplications before we perform all divisions.

4. To write 3.56×10^{-4} in standard decimal notation, we move the decimal point k places to the ____ .

Perform the exponentiation by hand.

5. 4^3

6. 3^4

7. 2^5

8. 5^3

9. -8^2

10. -7^2

11. $(-8)^2$

12. $(-7)^2$

13. $\left(\dfrac{6}{7}\right)^2$

14. $\left(\dfrac{3}{5}\right)^3$

15. 8^0

16. $(-5)^0$

17. 2^{-3}

18. 4^{-2}

19. 3^{-4}

20. 6^{-1}

21. 10^{-5}

22. 10^{-3}

Find the square root by hand.

23. $\sqrt{4}$

24. $\sqrt{16}$

25. $-\sqrt{36}$

26. $-\sqrt{25}$

27. $\sqrt{-9}$

28. $\sqrt{-4}$

29. $-\sqrt{-25}$

30. $-\sqrt{-16}$

State whether the square root is rational or irrational. If the square root is rational, find the (exact) value. If the square root is irrational, estimate its value by rounding to the second decimal place.

31. $\sqrt{30}$

32. $\sqrt{62}$

33. $\sqrt{64}$

34. $\sqrt{81}$

Perform the indicated operations by hand.

35. $3 \cdot (5 - 1)$

36. $8 \cdot (2 - 6)$

37. $(2 - 5)(9 - 3)$

38. $(2 + 8)(3 - 8)$

39. $\dfrac{2 + 5 + 4 + 1 + 9 + 3}{6}$

40. $\dfrac{3 + 5 + 4 + 2 + 6}{5}$

41. $\dfrac{4 - (-6)}{-7 - 8}$

42. $\dfrac{1 - 9}{2 - (-4)}$

43. $4\sqrt{20 + 5}$

44. $2\sqrt{41 + 8}$

45. $-5 - 4 \cdot 3$

46. $1 + 9 \cdot (-4)$

47. $20 \div (-2) \cdot 5$

48. $-16 \div (-4) \cdot 2$

49. $-9 - 4 + 3$

50. $3 - 7 + 1$

51. $7 - 3(5 + 2)$

52. $2 - 4(9 - 6)$

53. $15 \div 3 - (2 - 7)(2)$

54. $6(2 + 3) - 5 \cdot 7$

55. $5 - [4 - 7(5 - 9)]$

56. $-3 - [6 + 2(4 - 8)]$

57. $\dfrac{7}{8} - \dfrac{3}{4} \cdot \dfrac{1}{2}$

58. $\dfrac{5}{6} + \dfrac{2}{3} \div \dfrac{2}{5}$

59. $2 + 5^2$

60. $8 - 3^2$

61. $-3(4)^2$

62. $8(-2)^3$

63. $(-1)^2 - (-1)^3$

64. $4^3 - (-4)^3$

65. $-4(-1)^2 - 2(-1) + 5$

66. $2(-4)^2 + 3(-4) - 7$

67. $8^2 + 2(4 - 8)^2 \div (-2)$

68. $(9 - 7)^2 \cdot (-3) - 2^4$

69. $\dfrac{(-3)^2 + (-2)^2 + 2^2 + 3^2}{4 - 1}$

70. $\dfrac{(-4)^2 + (-1)^2 + 1^2 + 4^2}{4 - 1}$

71. $\dfrac{(2 - 5)^2 + (4 - 5)^2 + (9 - 5)^2}{3 - 1}$

72. $\dfrac{(1 - 3)^2 + (2 - 3)^2 + (6 - 3)^2}{3 - 1}$

73. $(14 - 12) - 3\sqrt{\dfrac{4^2}{2} + \dfrac{2^2}{4}}$

74. $(30 - 40) + 2\sqrt{\dfrac{6^2}{9} + \dfrac{8^2}{2}}$

For Exercises 75–80, use a calculator to perform the indicated operations. Round your result to the second decimal place.

75. $\dfrac{15.9 + 21.8 + 19.2 + 20.4 + 27.9}{5}$

76. $\dfrac{84.7 + 82.9 + 89.3 + 80.1}{4}$

77. $\sqrt{\dfrac{0.24(1 - 0.24)}{30}}$

78. $\sqrt{\dfrac{0.35(1 - 0.35)}{50}}$

79. $\sqrt{\dfrac{(9.3 - 4.5)^2 + (1.8 - 4.5)^2 + (2.4 - 4.5)^2}{3 - 1}}$

80. $\sqrt{\dfrac{(5.8 - 6.2)^2 + (9.4 - 6.2)^2 + (3.4 - 6.2)^2}{3 - 1}}$

81. Evaluate $\bar{x} + zs$ for $\bar{x} = 20, z = -3, s = 4$.

82. Evaluate $\bar{x} + zs$ for $\bar{x} = 15, z = -2, s = 5$.

83. Evaluate $\dfrac{x - \bar{x}}{s}$ for $x = 2, \bar{x} = 8$, and $s = 3$.

84. Evaluate $\dfrac{x - \bar{x}}{s}$ for $x = 5, \bar{x} = 11$, and $s = 2$.

85. Evaluate $\dfrac{y_2 - y_1}{x_2 - x_1}$ for $x_1 = 1, x_2 = -6, y_1 = -10, y_2 = 3$.

86. Evaluate $\dfrac{y_2 - y_1}{x_2 - x_1}$ for $x_1 = -3, x_2 = -8, y_1 = -5, y_2 = -3$.

87. Evaluate $\bar{x} + t\dfrac{s}{\sqrt{n}}$ for $\bar{x} = 20, t = 3, s = 6$, and $n = 9$.

88. Evaluate $\bar{x} - t\dfrac{s}{\sqrt{n}}$ for $\bar{x} = 25, t = 2, s = 8$, and $n = 4$.

89. Evaluate ab^x for $a = 4, b = 3$, and $x = -2$.

90. Evaluate ab^x for $a = 8, b = 2$, and $x = -3$.

Use a calculator to help you evaluate

$\sqrt{\dfrac{(x_1 - \bar{x})^2 + (x_2 - \bar{x})^2 + (x_3 - \bar{x})^2}{n - 1}}$ *for the given values of* x_1,

$x_2, x_3, \bar{x},$ *and n. Round your result to the second decimal place.*

91. $x_1 = 2.1, x_2 = 5.8, x_3 = 10.4, \bar{x} = 6.1, n = 3$

92. $x_1 = 3.5, x_2 = 4.2, x_3 = 8.8, \bar{x} = 5.5, n = 3$

For Exercises 93–96, let x be a number. Translate the English phrase or sentence into a mathematical expression. Then evaluate the expression for x = −4.

93. 5 more than the product of −6 and the number

94. −3 minus the quotient of 8 and the number

95. Subtract 3 from the quotient of the number and −2.

96. The number plus the product of −5 and the number

97. Congressional pay in 2000 was $141.3 thousand, and it increased by approximately $3.6 thousand per year for the period 2000–2009 (Source: *Bureau of Labor Statistics*).

   a. Complete Table 31 to help find an expression that stands for the congressional pay (in thousands of dollars) at t years since 2000. For example, $t = 3$ stands for the year 2003, because 2003 is 3 years after 2000. Show the arithmetic to help you see a pattern.

Table 31 Congressional Pay

Years since 2000	Congressional Pay (thousands of dollars)
0	
1	
2	
3	
4	
t	

   b. Evaluate the expression you found in part (a) for $t = 7$. What does your result mean in this situation?

   c. Congress has had a pay freeze since 2009 due to a poor economy. The pay freeze might be lifted in 2015. If the congressional pay is increased by 1% in 2015, what will be the pay in that year?

98. The number of anti-government militia groups was 512 in 2009, and it increased by approximately 299 groups per year for the period 2009–2014. (Source: *The Southern Poverty Law Center*).

   a. Complete Table 32 to help find an expression that stands for the number of militia groups at t years since 2009. For example, $t = 2$ stands for the year 2011, because 2011 is 2 years after 2009. Show the arithmetic to help you see a pattern.

Table 32 Numbers of Militia Groups

Years since 2009	Number of Militia Groups
0	
1	
2	
3	
4	
t	

   b. Evaluate the expression you found in part (a) for $t = 5$. What does your result mean in this situation?

99. The population of Gary, Indiana, was about 102.7 thousand in 2000, and it decreased by about 2.0 thousand per year for the period 2009–2014 (Source: *U.S. Census Bureau*).

   a. Complete Table 33 to help find an expression that stands for the population of Gary (in thousands) at t years since 2000. Show the arithmetic to help you see a pattern.

Table 33 Populations of Gary

Years since 2000	Population (thousands)
0	
1	
2	
3	
4	
t	

   b. Evaluate the expression you found in part (a) for $t = 14$. What does your result mean in this situation?

100. The number of U.S. households with cable TV subscriptions was 61.8 million in 2009, and it decreased by approximately 1.5 million households for the period 2009–2014 (Source: *IHS*).

   a. Complete Table 34 to help find an expression that stands for the number (in millions) of U.S. households with cable TV subscriptions at *t* years since 2009. Show the arithmetic to help you see a pattern.

Table 34 Numbers of Households with Cable TV Subscriptions

Years since 2009	Number (in thousands)
0	
1	
2	
3	
4	
t	

   b. Evaluate the expression you found in part (a) for $t = 5$. What does your result mean in this situation?

101. U.S. wireless subscriber connections were 315.9 million in 2012, and they increased by 3.3% in 2013 (Source: *International Telecommunication Union*). What was the number of U.S. wireless subscriber connections in 2013?

102. Amazon.com's net sales were $74.5 billion in 2012, and they increased by 22% in 2013 (Source: Digital Book World). What were the net sales in 2013?

103. The worldwide revenue from recorded music was $15.6 billion in 2012, and it decreased by 3.9% in 2013 (Source: *International Federation of the Phonographic Industry*). What was the revenue in 2013?

104. U.S. revenue from newspapers was $38.6 billion in 2012, and it decreased by 2.6% in 2013 (Source: *Newspaper Association of America*). What was the revenue from newspapers in 2013?

For Exercises 105–108, use the fact that the total area of the object is 1 to find the area of the orange bar. The object has not been drawn to scale.

105. See Fig. 81.

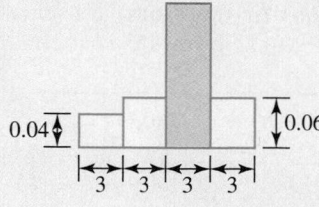

Figure 81 Exercise 105

106. See Fig. 82.

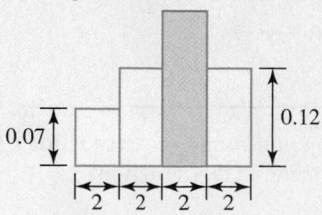

Figure 82 Exercise 106

107. See Fig. 83.

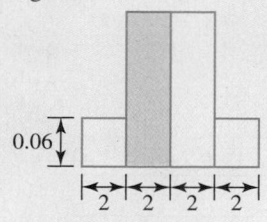

Figure 83 Exercise 107

108. See Fig. 84.

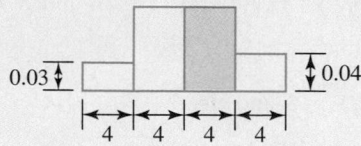

Figure 84 Exercise 108

Write the number in standard decimal form.

109. 4.9×10^4

110. 8.31×10^6

111. 8.59×10^{-3}

112. 6.488×10^{-5}

113. -2.95×10^{-4}

114. -8.7×10^{-2}

Write the number in scientific notation.

115. 45,700,000

116. 280,000

117. 0.0000659

118. 0.000023

119. −0.000001

120. −0.0004

For Exercises 121 and 122, numbers are displayed in TI-84's version of scientific notation. Write each number shown in the Y_1 column in standard decimal form.

121. See Fig. 85.

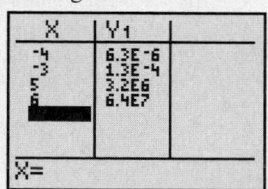

Figure 85 Exercise 121

122. See Fig. 86.

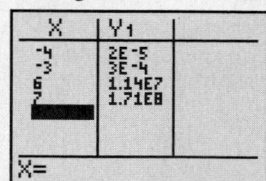

Figure 86 Exercise 122

For Exercises 123–126, the given sentence contains a number written in scientific notation. Write the number in standard decimal form.

123. The first evidence of life on Earth dates back to 3.6×10^9 years ago.

124. The Moon has an average distance from Earth of approximately 2.389×10^5 miles.

125. The hydrogen ion concentration in human blood is about 6.3×10^{-8} mole per liter.

126. The faintest sound humans can hear has an intensity of about 10^{-12} watt per square meter.

For Exercises 127–130, the given sentence contains a number (other than a date) written in standard decimal form. Write the number in scientific notation.

127. The tanker *Exxon Valdez* spilled about 10,080,000 gallons of oil in Prince William Sound, Alaska, in 1989.

128. The average distance from Earth to Alpha Centauri is about 25,000,000,000,000 miles.

129. The wavelength of violet light is about 0.00000047 meter.

130. One second is about 0.0000000317 year.

Concepts

131. A student tries to perform the indicated operations in $2(3)^2 + 2(3) + 1$:

$$2(3)^2 + 2(3) + 1 = 6^2 + 2(3) + 1$$
$$= 36 + 2(3) + 1$$
$$= 36 + 6 + 1$$
$$= 43$$

Describe any errors. Then perform the operations correctly.

132. A student tries to evaluate $x^2 + 4x + 5$ for $x = -3$:

$$-3^2 + 4(-3) + 5 = -9 - 12 + 5$$
$$= -21 + 5$$
$$= -16$$

Describe any errors. Then evaluate the expression correctly.

133. In Problem 2 of Example 5, we performed the indicated operations in $\dfrac{11 - 3}{1 - 5}$ by the following steps:

$$\frac{11 - 3}{1 - 5} = \frac{8}{-4} = -2$$

a. Perform the indicated operations in $(11 - 3) \div (1 - 5)$.

b. Perform the indicated operations in $11 - 3 \div 1 - 5$.

c. In using a calculator to simplify $\dfrac{11 - 3}{1 - 5}$, a student presses the following buttons:

$$11 \; \boxminus \; 3 \; \boxdiv \; 1 \; \boxminus \; 5$$

The result of this calculation is 3 rather than -2. Describe any errors. Then perform the operations correctly.

134. A student tries to perform the indicated operations in $16 \div 2 \cdot 4$:

$$16 \div 2 \cdot 4 = 16 \div 8 = 2$$

Describe any errors. Then perform the operations correctly.

135. a. Compute without using a calculator.

  i. $(-1)^2$ **ii.** $(-1)^3$
  iii. $(-1)^4$ **iv.** $(-1)^5$
  v. $(-1)^{87}$ **vi.** $(-1)^{596}$

  b. For what counting-number values of n is $(-1)^n$ equal to 1? Explain.

  c. For what counting-number values of n is $(-1)^n$ equal to -1? Explain.

136. Describe the order of operations in your own words.

Hands-On Projects

Stocks Project

Imagine that you have $5000 that you plan to invest in five stocks for one week. In this project, you will explore some possible outcomes of that investment.**

Collecting the Data

Use the Internet to select five stocks. For each stock, record the company name and the call letters of the stock. Here are some examples:

Gateway Computer has the call letters GTW.

Coca-Cola has the call letters KO.

EMC corporation has the call letters EMC.

Record the value of one share (the beginning share price) of each of the five stocks. Also, record how you distribute your $5000 investment. For example, you may invest all $5000 in one of the five stocks or $1000 in each of the five stocks, or you may opt for some other distribution of the money. The sum of your investments should equal or be close to $5000 by buying whole amounts of stock. Even if you do not invest money in some of the stocks, still record the information about all five stocks.

After one week, record the new value of one share (the ending share price) of each of the five stocks.

Analyzing the Data

1. Complete Table 35. The *profit* from a stock is the money you collect from the stock, minus the money you invested in the stock. What is the total profit from your $5000 investment?

Table 35 Five Stocks' Performances

Call Letters of Stock	Investment in Stock (dollars)	Beginning Share Price (dollars)	Number of Shares	Ending Share Price (dollars)	Money Collected from Stock (dollars)	Profit from Stock (dollars)

** Project suggested by Jim Ryan, State Center Community College District, Clovis Center, Clovis, CA.

2. Find the change in the share price of each of the five stocks. Which share price had the greatest change? The least?

3. The *percent change* of the value of a stock can be found by dividing the change in value of the stock by the original value of the stock and then converting the decimal result into percent form (by multiplying by 100). For example, suppose that a stock's value increases from $7 to $9. The change in value of this stock is its ending value minus its beginning value: $9 - 7 = 2$ dollars. Here we find the percent change in value:

$$\text{percent change} = \frac{\text{change in value}}{\text{beginning value}} \cdot 100$$

$$= \frac{9 - 7}{7} \cdot 100 = \frac{2}{7} \cdot 100 \approx 28.57$$

So, the percent change is about 28.57%.

Now find the percent change in value for each of your five stocks. Which stock had the greatest percent change? The least?

4. Among your five stocks, is there a pair of stocks for which one stock has the greater change but the other has the greater percent change? If yes, use these stocks to respond to the questions in parts (a)–(c). If no, then use the following values of fictional stocks A and B:

Stock A	Stock B
Increased from $4 to $5	Increased from $20 to $23

a. Find the profit earned from investing all of the $5000 in the stock with the larger change in value.

b. Find the profit earned from investing all of the $5000 in the stock with the larger percent change in value.

c. Which is the better measure of the growth of a stock, change in value or percent change in value? Explain.

5. a. Find the profit earned from each of the following scenarios:

  i. You invest the $5000 in the best-performing stock among the five stocks.

  ii. You invest the $5000 in the worst-performing stock among the five stocks.

  iii. You invest the $5000 by investing $1000 in each of the five stocks.

b. Describe the benefits and drawbacks to investing your money in a number of stocks rather than in just one stock.

Chapter Summary

Key Points of Chapter 1

Section 1.1   Variables, Constants, Plotting Points, and Inequalities

Variable	A **variable** is a symbol that represents a quantity that can vary.
Constant	A **constant** is a symbol that represents a specific number (a quantity that does *not* vary).
Counting numbers or natural numbers	The **counting numbers,** or **natural numbers,** are the numbers 1, 2, 3, 4, 5,....
Integers	The **integers** are the numbers ..., $-3, -2, -1, 0, 1, 2, 3, ...$
Rational numbers	The **rational numbers** are the numbers that can be written in the form $\frac{n}{d}$, where n and d are integers and d is nonzero.
Real numbers	The **real numbers** are all the numbers represented on the number line.
Irrational numbers	The **irrational numbers** are the real numbers that are not rational.
Data	**Data** are quantities or categories that describe people, animals, or things.
Negative numbers and positive numbers	The **negative numbers** are the real numbers less than 0, and the **positive numbers** are the real numbers greater than 0.
Coordinate	We call each of the numbers in an ordered pair a **coordinate**.
Inequality	An **inequality** contains one of the symbols $<, \leq, >$, and $\geq$ with a constant or variable on one side and a constant or variable on the other side.

Section 1.2   Expressions

Expression	An **expression** is a constant, a variable, or a combination of constants, variables, operation symbols, and grouping symbols, such as parentheses.
Evaluate an expression	We **evaluate an expression** by substituting a number for each variable in the expression and then calculating the result. If a variable appears more than once in the expression, the same number is substituted for that variable each time.

Section 1.3   Operations with Fractions and Proportions; Converting Units

Division by 0	The fraction $\frac{a}{b}$ is undefined if $b = 0$. Division by 0 is undefined.
Simplify a fraction	To simplify a fraction, 1. Find the prime factorizations of the numerator and denominator. 2. Find an equal fraction in which the numerator and the denominator do not have common positive factors other than 1 by using the property $$\frac{ab}{ac} = \frac{a}{a} \cdot \frac{b}{c} = 1 \cdot \frac{b}{c} = \frac{b}{c}$$ where a and c are nonzero.
Simplify results	If the result of an exercise is a fraction, simplify it.
Multiplying fractions	$\frac{a}{b} \cdot \frac{c}{d} = \frac{ac}{bd}$, where b and d are nonzero.
Dividing fractions	$\frac{a}{b} \div \frac{c}{d} = \frac{a}{b} \cdot \frac{d}{c}$, where b, c, and d are nonzero.
Adding fractions	$\frac{a}{b} + \frac{c}{b} = \frac{a + c}{b}$, where b is nonzero.
Subtracting fractions	$\frac{a}{b} - \frac{c}{b} = \frac{a - c}{b}$, where b is nonzero.
How to add or subtract two fractions with different denominators	To add (or subtract) two fractions with different denominators, use the fact that $\frac{a}{a} = 1$, where a is nonzero, to write an equal sum (or difference) of fractions for which each denominator is the LCD.
Proportion	In statistics, a **proportion** is a fraction of the whole. A proportion can also be written as a decimal number.
Value of a proportion	A proportion is always between 0 and 1, inclusive.
Sum of proportions	If an object is made up of two or more parts, the sum of their proportions equals 1.
Proportion of the rest	Let $\frac{a}{b}$ be the proportion of the whole that has a certain characteristic. Then the proportion of the whole that does NOT have that characteristic is $1 - \frac{a}{b}$.
Converting units	To convert the units of a quantity, 1. Write the quantity in the original units. 2. Multiply by fractions equal to 1 so the units you want to eliminate appear in one numerator and one denominator.

Section 1.4   Absolute Value and Adding Real Numbers

The opposite of the opposite of a number	$-(-a) = a$
Absolute value	The **absolute value** of a number is the distance that the number is from 0 on the number line.

Section 1.4   Absolute Value and Adding Real Numbers (*Continued*)

Adding two numbers with the same sign	To add two numbers with the same sign, **1.** Add the absolute values of the numbers. **2.** The sum of the original numbers has the same sign as the sign of the original numbers.
Adding two numbers with different signs	To add two numbers with different signs, **1.** Find the absolute values of the numbers. Then subtract the smaller absolute value from the larger absolute value. **2.** The sum of the original numbers has the same sign as the original number with the larger absolute value.

Section 1.5   Change in a Quantity and Subtracting Real Numbers

Change in a quantity	The change in a quantity is the ending amount minus the beginning amount: $$\text{Change in the quantity} = \text{Ending amount} - \text{Beginning amount}$$
Subtracting a number	To subtract a number, add its opposite: $$a - b = a + (-b)$$
Increasing quantity	An increasing quantity has a positive change.
Decreasing quantity	A decreasing quantity has a negative change.

Section 1.6   Ratios, Percents, and Multiplying and Dividing Real Numbers

Unit ratio	A **unit ratio** is a ratio written as $\frac{a}{b}$ with $b = 1$ or as $a:b$ with $b = 1$.
Percent	**Percent** means "for each hundred": $a\% = \frac{a}{100}$.
Writing a percentage as a decimal number	To write a percentage as a decimal number, remove the percent symbol and divide the number by 100 (move the decimal point two places to the left).
Writing a decimal number as a percentage	To write a decimal number as a percentage, multiply the number by 100 (move the decimal point two places to the right) and insert a percent symbol.
Finding the percentage of a quantity	To find the percentage of a quantity, multiply the decimal form of the percentage and the quantity.
One hundred percent	One hundred percent of a quantity is *all* of the quantity.
The product or quotient of two numbers	The product or quotient of two numbers that have different signs is negative. The product or quotient of two numbers that have the same sign is positive.
Equal fractions with negative signs	If $b \neq 0$, then $\frac{-a}{b} = \frac{a}{-b} = -\frac{a}{b}$.

Section 1.7 Exponents, Square Roots, Order of Operations, and Scientific Notation

Exponent	For any counting number n, $$b^n = \underbrace{b \cdot b \cdot b \cdot \ldots \cdot b}_{n \text{ factors of } b}$$ We refer to b^n as the **power,** the ***n*th power of *b*,** or ***b* raised to the *n*th power.** We call b the **base** and n the **exponent.**
Finding $-b^n$	For an expression of the form $-b^n$, we calculate b^n before taking the opposite.
Zero exponent	For nonzero b, $b^0 = 1$.

Section 1.7   Exponents, Square Roots, Order of Operations, and Scientific Notation (*Continued*)

Negative-integer exponent	If n is a counting number and b is nonzero, then $b^{-n} = \dfrac{1}{b^n}$.
Principal square root	If a is a nonnegative number, then $\sqrt{a}$ is the nonnegative number we square to get a. We call $\sqrt{a}$ the **principal square root** of a.
Order of operations	We perform operations in the following order: **1.** First, perform operations within parentheses or other grouping symbols, starting with the innermost group. **2.** Then perform exponentiations. **3.** Next, perform multiplications and divisions, going from left to right. **4.** Last, perform additions and subtractions, going from left to right.
Order of operations and the strengths of operations	After we have performed operations in parentheses, the order of operations goes from the most powerful operation, exponentiation, to the next-most-powerful operations, multiplication and division, to the weakest operations, addition and subtraction.
Scientific notation	A number is written in **scientific notation** if it has the form $N \times 10^k$, where k is an integer and the absolute value of N is between 1 and 10 or is equal to 1.
Converting from scientific notation to standard decimal notation	To write the scientific notation $N \times 10^k$ in standard decimal notation, we move the decimal point of the number N as follows: • If k is *positive*, we multiply N by 10 k times; hence, we move the decimal point k places to the *right*. • If k is *negative*, we divide N by 10 k times; hence, we move the decimal point k places to the *left*.
Converting from standard decimal notation to scientific notation	To write a number in scientific notation, count the number k of places that the decimal point must be moved so that the absolute value of the new number N is between 1 and 10 or is equal to 1: • If the decimal point is moved to the left, then the scientific notation is written as $N \times 10^k$. • If the decimal point is moved to the right, then the scientific notation is written as $N \times 10^{-k}$.

Chapter 1 Review Exercises

1. Let B be the total box office gross (in billions of dollars) from U.S. and Canada movie theaters. For 2014, the value of B is 10.02 (Source: *Rentrak Corporation*). What does that mean in this situation?

2. Choose a variable name for the percentage of students who are full-time students at a college. Give two numbers that the variable can represent and two numbers that it cannot represent.

3. Graph the numbers $-2, -\dfrac{3}{2}, 0, 1, \dfrac{5}{2}$, and 3 on a number line.

4. Graph the negative integers between -5 and 5 on a number line.

5. Here are a company's profits and losses for various years: profit of \$2 million, loss of \$4 million, loss of \$1 million, profit of \$3 million. Let p be the profit (in millions of dollars). Use points on a number line to describe the profits and losses of the company.

6. Plot the points $(2, 4), (-3, -1), (5, -2)$, and $(-4, 5)$ in a coordinate system.

7. Sketch the graph of the inequality $x < -3$. Also describe the inequality in words and in interval notation.

8. Let t be the time (in hours) a teenager spends playing a video game. Graph and interpret the inequality $2 \le t \le 5$.

9. The proportion of consumers who prefer to pay for purchases and other transactions with cash is $\dfrac{2}{5}$. The proportion of consumers who prefer to pay for purchases and other transactions with credit cards is $\dfrac{1}{6}$ (Source: *Federal Reserve Bank*). Find the proportion of consumers who prefer to pay for purchases and other transactions with cash OR credit cards.

10. The proportions of the disk in Fig. 87 that are orange and blue are $\frac{2}{3}$ and $\frac{1}{4}$, respectively. Find the proportion of the disk that is green.

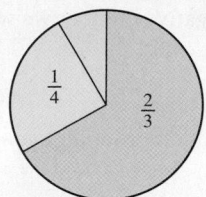

Figure 87 Exercise 10

11. Americans consume an average of 121 gallons of water per year (Source: *Wirthlin Worldwide*). What is this average in cups per day? Round your result to the second decimal place. Refer to the list of equivalent units in Table 10 on page 34 as needed.

Perform the indicated operations by hand.

12. $7 + (-10)$

13. $3 - (-5)$

14. $5(-9)$

15. $8 \div (-2)$

16. $-24 \div (10 - 2)$

17. $(2 - 6)(5 - 8)$

18. $\dfrac{5 + 2 + 7 + 6}{4}$

19. $\dfrac{2 - 8}{3 - (-1)}$

20. $-4 + 2(-6)$

21. $8 \div (-2) \cdot 5$

22. $2(4 - 7) - (8 - 2)$

23. $-14 \div (-7) - 3(1 - 5)$

24. $-5 - [3 + 2(1 - 7)]$

25. $4.2 - (-6.7)$

26. $\left(-\dfrac{8}{15}\right) \div \left(-\dfrac{16}{25}\right)$

27. $\dfrac{5}{9} - \left(-\dfrac{2}{9}\right)$

28. -8^2

29. $\left(\dfrac{3}{4}\right)^3$

30. 4^{-3}

31. $-\sqrt{49}$

32. $-6(3)^2$

33. $24 \div 2^3$

34. $(-2)^3 - 4(-2)$

35. $\dfrac{17 - (-3)^2}{5 - 4^2}$

36. $-3(2)^2 - 4(2) + 1$

37. $7^2 - 3(2 - 5)^2 \div (-3)$

38. $\dfrac{(2 - 4)^2 + (3 - 4)^2 + (7 - 4)^2}{3 - 1}$

For Exercises 39 and 40, use a calculator to compute. Round the result to two decimal places.

39. $-5.7 + 2.3^4 \div (-9.4)$

40. $\sqrt{\dfrac{0.15(1 - 0.15)}{200}}$

41. Simplify $\dfrac{-28}{-40}$ by hand.

42. A student has a credit card balance of −4789 dollars. If he sends a check to the credit card company for $800 and then uses his credit card to purchase a textbook for $102.99 and a notebook for $3.50, what is the new balance?

43. Three hours ago the temperature was 4°F. The temperature is now −8°F.
    a. What is the change in temperature for the past three hours?

b. Estimate the change in temperature for the past hour.
    c. Explain why your estimate in part (b) may not be the actual change in temperature for the past hour.

44. Contributions from individuals to major party nominees of presidential elections are shown in Table 36 for various years.

Table 36 Contributions from Individuals to Major Party Nominees of Presidential Elections

| Year | Contribution (millions of dollars) | |
	Democrat Nominee	Republican Nominee
1996	28.3	29.6
2000	33.9	91.3
2004	215	259
2008	454	220

Source: *Federal Election Commission*

a. Find the change in the annual individual contributions to the Democratic nominee from 1996 to 2000.
    b. Find the change in the annual individual contributions to the Republican nominee from 2004 to 2008.
    c. Over which four-year period was there the greatest change in annual individual contributions to the Democratic nominee? What was that change?
    d. Over which four-year period was there the greatest change in annual individual contributions to the Republican nominee? What was that change?

45. Text-messaging users sent or received an average of 41.5 messages per day in 2011 and sent or received an average of 29.7 messages per day in 2009 (Source: *The Pew Research Center's Internet & American Life Project*). Find the unit ratio of the number of messages sent or received per day in 2011 to the number of messages sent or received per day in 2009. What does your result mean?

46. Out of 1501 adults surveyed, 886 adults think Iran's nuclear program is the biggest threat to the United States (Source: *Pew Research Center*, USA Today). Describe this fact using a percentage, rounded to the first decimal place.

47. In a survey of 68,300 people aged at least 12 years, 7.3% of the people had used marijuana in the past month (Source: *National Institute on Drug Abuse*). How many of those surveyed used marijuana in the past month?

48. A person's credit card balance is −5493 dollars. If the person pays off 20% of the balance, what is the new balance?

Evaluate the expression for $a = 2$, $b = -5$, $c = -4$, and $d = 10$.

49. $-b - c^2$

50. $\dfrac{a - b}{c - d}$

51. Evaluate $\bar{x} - t\dfrac{s}{\sqrt{n}}$ for $\bar{x} = 30, t = 2, s = 12$, and $n = 9$.

For Exercises 52 and 53, let x be a number. Translate the English phrase into a mathematical expression. Then evaluate the expression for $x = -3$.

52. The number subtracted from −7

53. 1 plus the quotient of −24 and the number

54. If T is the total cost (in dollars) for a team to join a softball league and there are n players on the team, then $T \div n$ is the cost (in dollars) per player. Evaluate $T \div n$ for $T = 650$ and $n = 13$. What does your result mean in this situation?

55. A basement is flooded with 400 cubic feet of water. Each hour, 50 cubic feet of water is pumped out of the basement.

   a. Complete Table 37 to help find an expression that stands for the volume (in cubic feet) of water in the basement after water has been pumped out for t hours. Show the arithmetic to help you see a pattern.

Table 37 Volumes of Water

Time (hours)	Volume of Water (cubic feet)
0	
1	
2	
3	
4	
t	

   b. Evaluate the expression that you found in part (a) for $t = 7$. What does your result mean in this situation?

56. The number of Pfizer employees in 2012 was 90.0 thousand employees, and it had decreased 13.7% by 2013 (Source: *U.S. Securities and Exchange Commission*). What was the number of employees in 2013?

57. The object shown in Fig. 88 has not been drawn to scale. Use the fact that the total area is 1 to find the area of the orange bar.

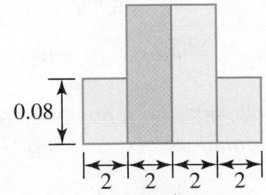

Figure 88 Exercise 57

58. Write 3.85×10^{-4} in standard decimal form.

59. Write 54,000,000 in scientific notation.

Chapter 1 Test

1. A rectangle has an area of 36 square feet. Let W be the width (in feet), L be the length (in feet), and A be the area (in square feet).

   a. Sketch three possible rectangles of area 36 square feet.

   b. Which of the symbols W, L, and A are variables? Explain.

   c. Which of the symbols W, L, and A are constants? Explain.

2. Graph the integers between -4 and 2, inclusive, on a number line.

3. The number of electric cars (in thousands) in use in the United States for various years is 4.5, 5.2, 7.0, 8.7, and 10.4. Let n be the number of electric cars (in thousands) in use. Use points on a number line to describe the given values of n.

4. Plot the points $(-5, 3)$ and $(-2, -4)$ in coordinate system.

5. Let c be the cost (in dollars) of a bicycle, which costs at least $1450. Describe the bicycle's cost using inequality notation, interval notation, and a graph.

6. In a 2014 survey, 1000 adults were asked, "Looking ahead, do you feel that torture of suspected terrorists can often be justified, sometimes justified, rarely justified, or never justified?" About $\frac{1}{6}$ of the adults said, "Justified," and $\frac{2}{5}$ of the adults said, "Sometimes justified" (Source: ABC News, Washington Post). What fraction of the adults gave other responses?

7. A TI-84 weighs 0.62 pound. What is the calculator's weight in ounces? Round your result to two decimal places. Refer to the list of equivalent units in Table 10 on page 34 as needed.

For Exercises 8–20, perform the indicated operations by hand.

8. $-3 + 9 \div (-3)$

9. $(4 - 2)(3 - 7)$

10. $\dfrac{4 - 7}{-1 - 5}$

11. $3\sqrt{32} + 4$

12. $5 - (2 - 10) \div (-4)$

13. $-20 \div 5 - (2 - 9)(-3)$

14. $0.4(-0.2)$

15. $-\dfrac{27}{10} \div \dfrac{18}{75}$

16. $-\dfrac{3}{10} + \dfrac{5}{8}$

17. 2^{-5}

18. $7 + 2^3 - 3^2$

19. $1 - (3 - 7)^2 + 10 \div (-5)$

20. $\dfrac{(-5)^2 + (-2)^2 + 2^2 + 5^2}{4 - 1}$

21. Simplify $\dfrac{84}{-16}$ by hand.

22. The chances of being audited by the Internal Revenue Service (IRS) are shown in Table 38 for various years.

Table 38 Tax Audit Rates

Year	Tax Audit Rate (number of audits per 1000 tax returns)
2001	5.8
2003	6.5
2005	9.7
2007	10.3
2009	10.3
2011	11.1
2013	9.6

Source: *Internal Revenue Service*

   a. Find the change in the tax audit rate from 2001 to 2003.

   b. Find the change in the tax audit rate from 2011 to 2013.

   c. For which two-year period was the change in the tax audit rate the most? What was that change?

23. Use a calculator to find

$$\sqrt{\dfrac{(6.2 - 4.1)^2 + (2.5 - 4.1)^2 + (3.6 - 4.1)^2}{3 - 1}}.$$

Round the result to two decimal places.

24. The average ticket price to major league baseball games was $9.14 in 1991 and $26.92 in 2012 (Source: *AP*). Find the unit ratio of the average ticket price in 2012 to the average ticket price in 1991. Round your result to the second decimal place. What does your result mean?

For Exercises 25 and 26, evaluate the expression for $a = -6$, $b = -2$, $c = 5$, and $d = -1$.

25. $ac - \dfrac{a}{b}$ **26.** $a + b^3 + c^2$

27. Evaluate $\dfrac{x - \bar{x}}{s}$ for $x = 4, \bar{x} = 10$, and $s = 2$.

For Exercises 28 and 29, let x be a number. Translate the English phrase into a mathematical expression. Then evaluate the expression for $x = -5$.

28. Twice the number minus the product of 3 and the number

29. 6 subtracted from the quotient of -10 and the number

30. U.S. Postal Service first-class mail volume was 91.7 billion pieces in 2008, and it decreased by about 4.9 billion pieces per year for the period 2008–2013 (Source: *U.S. Postal Service*).

    a. Complete Table 39 to help find an expression that stands for the U.S. Postal Service first-class mail volume (in billions of pieces) in the year that is *t* years since 2008. Show the arithmetic to help you see a pattern.

Table 39 U.S. Postal Service First-Class Mail Volume

Years since 2008	First-Class Mail Volume (billions of pieces)
0	
1	
2	
3	
4	
t	

    b. Evaluate the expression that you found in part (a) for $t = 5$. What does your result mean in this situation?

31. The number of married couples in which the mother is employed and the father is not in 2012 was 1.45 million, and it decreased by 3.3% by 2013 (Source: *Bureau of Labor Statistics*). What was the number of such couples in 2013?

32. Write 0.0000678 in scientific notation.

Designing Observational Studies and Experiments

Would monetary rewards motivate you to pass more classes? In a two-semester study, researchers divided 1019 low-income community college students who are parents into two groups. Each member of the first group received a $20 gift card. Each member of the second group received the gift card plus monetary rewards for enrolling in classes, staying enrolled in classes, and maintaining at least a C average. Students in the second group could be awarded as much as $2000 (Source: Paying for Performance: The Education Impacts of a Community College Scholarship Program for Low-Income Adults, *Lisa Barrow et al.*). In Exercise 14 of Homework 2.3, you will analyze key features of the study, including what the study's conclusion means.

In Chapter 1, we discussed many arithmetic and algebra concepts that will come in handy when performing statistics throughout the rest of the course. In this chapter, for example, we will use proportions to describe groups of things or people, such as the proportion of low-income students who maintain at least a C average.

Our main objective in this chapter will be to discuss two main types of statistical studies: observational studies and experiments. We will describe the benefits and drawbacks to both types and how to design them well.

▼ 2.1 Simple Random Sampling

Objectives

» Identify the *individuals*, the *variables*, and the *observations* of a study.

» Describe the five steps of *statistics*.

» Identify the *sample* and the *population* of a study.

» Identify *descriptive statistics* and *inferential statistics*.

» Select a *simple random sample*.

» Identify the *sampling bias*, the *nonresponse bias*, and the *response bias* of a study.

» Identify sampling and nonsampling error.

What is the GPA of a typical community college student? What is the lifetime of a typical LED television? Is an experimental drug an effective cure for AIDS? These are the sorts of questions that researchers attempt to answer by using statistics.

Individuals, Variables, and Observations

In order to investigate such questions, we must be clear on whom or what we are going to study. **Individuals** are the people or objects we want to learn about. For example, if we want to perform a study about community college students' IQs, then the individuals are the community college students. If we want to perform a study about how long LED televisions last, then the individuals are LED televisions.

▶ **Definition   Variable**

In statistics, a **variable** is a characteristic of the individuals to be measured or observed.

In a study of community college students, the variables might be GPA (in points), credits earned (in units), major, and age (in years). In a study of songs, the variables might be artist, genre, length (in seconds), lyrics, and price (in dollars).

You may have noticed that *variable* is defined differently in statistics than in algebra. Recall that in algebra, a variable is a *symbol* that represents a quantity that can vary (Section 1.1). In statistics, a variable is a *characteristic* of the individuals, not a symbol.

There is another difference. In algebra, a variable represents a *quantity* such as price (in dollars), but in statistics, a variable can also represent a characteristic such as music genre, which consists of *categories* such as hip-hop, rap, country, and alternative.

The data that we observe for a variable are called **observations**. For data the author has on the variable prices of statistics textbooks, some of the observations are $175.33, $225.76, and $71.25. For data the author has on the variable dog breed, some of the observations are French bulldog, beagle, and chihuahua.

▶ Example 1   Identifying Individuals, Variables, and Observations

The five movies with the largest worldwide gross receipts of all time are shown in Table 1.

Table 1 Movies with the Top 5 Worldwide Gross Receipts

Movie	Studio	Worldwide Gross Receipts (millions of dollars)	U.S. Gross Receipts (millions of dollars)
Avatar	Fox	2788	761
Titanic	Paramount Pictures	2187	659
Marvel's The Avengers	Buena Vista	1519	623
Harry Potter and the Deathly Hallows Part 2	Warner Brothers	1342	381
Frozen	Buena Vista	1215	409

Source: *Box Office Mojo*

1. Identify the individuals.
2. Identify the variables.
3. Identify the observations for each variable.

Solution

1. The individuals are the movies *Avatar, Titanic, Marvel's The Avengers, Harry Potter and the Deathly Hallows Part 2,* and *Frozen.*
2. The variables are studio, worldwide gross receipts (in millions of dollars), and U.S. gross receipts (in millions of dollars).
3. The observations for the variable studio are Fox, Paramount Pictures, Buena Vista, Warner Brothers, and Buena Vista. The observations for the variable worldwide gross receipts, all in millions of dollars, are 2788, 2187, 1519, 1342, and 1215. The observations for the variable U.S. gross receipts, all in millions of dollars, are 761, 659, 623, 381, and 409.

Five Steps of Statistics

Now that we have explored the meaning of a variable, we can discuss the meaning of *statistics,* which is the main point of this entire course.

▶ Definition   Statistics

Statistics is the practice of the following five steps:

1. **Raise a precise question about one or more variables.** The key word is precise. As Lewis Carroll once said, "If you don't know where you are going, any road will get you there."
2. **Create a plan to answer the question.** We must carefully design a plan so that our results are meaningful.

3. **Collect the data.** After observing individuals, measuring individuals, or asking people questions, we usually enter the data in software and fix any errors we can catch.
4. **Analyze the data.** Constructing tables and graphs can help us look for patterns. There are many procedures that involve calculations that can deepen our understanding of the variable(s).
5. **Draw a conclusion about the question.** Researchers often publish their findings. Learning something about one or more variables often raises new questions that can lead to more studies.

In Example 2, we will identify the five steps of performing statistics.

▶ **Example 2** Identifying the Five Steps of Statistics

Motivational enhancement therapy (*MET*) is a counseling process that aims to motivate drug abusers to stop using drugs. Some researchers wanted to test whether MET really works. Out of 70 drug abusers who participated in the study, 35 received the standard drug treatment provided at a hospital. The other 35 individuals received MET in addition to the standard drug treatment. After receiving treatment for 12 weeks, all 70 individuals took a questionnaire that measures a drug abuser's motivation to stop using drugs. The researchers concluded that MET effectively motivates drug abusers to stop using drugs. (Source: Motivational Enhancement Therapy for Substance Abusers: A Quasi Experimental Study, *Seema Rani et al.*). Describe the five steps of the study. If no details are given about a step, say so.

Solution

Here are the five steps:

1. **Raise a precise question about one or more variables.** The question the researchers addressed is whether MET effectively motivates drug abusers to stop using drugs.
2. **Create a plan to answer the question.** The researchers determined the number and identity of the people who would participate. They planned to divide those individuals into two groups of equal size. The researchers determined what treatments the groups would receive and for how long. Finally, they developed a questionnaire or selected an existing one.
3. **Collect the data.** After the two groups received their different treatments for 12 weeks, they completed the questionnaire, which measured their motivation to stop using drugs.
4. **Analyze the data.** No details are given about the analysis because it would not make sense at this point. In fact, you will spend the rest of this course as well as a statistics course learning about such an analysis.
5. **Draw a conclusion about the question.** The researchers concluded that MET effectively motivates drug abusers to stop using drugs. This conclusion raises questions. For example, would MET be effective if used only for 6 weeks? What proportion of drug abusers who participate in MET and the standard drug treatment actually stop abusing drugs for one year? For ten years?

Sample and Population

A key aspect of raising a precise question about one or more variables is to determine the group of individuals we want to study.

▶ **Definition** Population

A **population** is the *entire* group of individuals about which we want to learn.

In Example 2, the researchers raised and answered a question about all drug abusers, so the population is all drug abusers.

In some cases, a researcher will be able to collect data about *all* individuals in a population. A **census** is a study in which data are collected about *all* members of the population. If your instructor wants to know how well your class understands a concept and gives you a quiz, then your instructor is performing a census. In this case, the population is all the students in your class.

Once every ten years, the U.S. government performs the U.S. Census, which surveys all American adults about various characteristics including people's ages. In this study, the population is all American adults.

There is a reason that the U.S. Census is performed only once every ten years. Surveying over 200 million people requires a large number of people to perform the survey, takes a great amount of time, and costs an enormous amount of money.

If a pizza company would like to know what proportion of Americans prefer pepperoni pizza, then it would not be cost-effective to survey all Americans. Instead, it would survey a part of the population.

▶ **Definition　Sample**

A **sample** is the part of a population from which data are collected.

In the MET study, the sample is the 70 drug abusers who participated in the study.

Collecting data from a sample, rather than the entire population, saves time, money, and effort.

▶ **Example 3**　Identify the Variable, Sample, and Population of a Study

Researchers wanted to estimate the percentage of all American adults who think it should be legal for same-sex couples to marry. Of 1009 adults surveyed, 56% said yes. The researchers then made a conclusion about the proportion of all American adults who think it should be legal for same-sex couples to marry (Sources: ABC News, Washington Post).

1. Define a variable for the study.
2. Identify the population.
3. Identify the sample.

Solution

1. We define the variable same-sex marriage to represent all the "yes" responses and "no" responses to the question "Do you think it should be legal for same-sex couples to marry?"
2. The population is all American adults because that was the group the researchers wanted to learn about.
3. The sample is the 1009 adults who were surveyed because that is the group the researchers observed.

▶

Descriptive Statistics and Inferential Statistics

There are two main areas of statistics: descriptive statistics and inferential statistics. **Descriptive statistics** is the practice of using tables, graphs, and calculations about a sample to draw conclusions about only the sample. **Inferential statistics** is the practice of using information from a sample to draw conclusions about the *entire* population. When we perform inferential statistics, we call the conclusions **inferences**.

For example, if a researcher observes the eye colors of 1000 Americans and computes that 17% of the 1000 Americans have blue eyes, that conclusion is part of descriptive statistics. If the researcher uses the eye colors of the 1000 Americans to estimate the percentage of *all* Americans who have blue eyes, that conclusion is part of inferential statistics.

To be considered inferential statistics, a conclusion must go beyond the individuals from whom we collected data. So, if we collect data about *all* individuals in the population and draw a conclusion about the population, that is part of *descriptive* statistics.

In Chapters 3–10, we will focus on descriptive statistics. You will learn a lot about inferential statistics in a statistics course.

▶ **Example 4**   Identify the Variable, Sample, Population, and Conclusion of a Study

Researchers wanted to find out how a restricted diet would affect a Labrador Retriever's life-span. Of 48 Labrador Retrievers being observed, 24 were fed a normal diet and 24 were fed a restricted diet. The dogs' lifetimes were recorded. The study concluded that a typical Labrador Retriever on a restricted diet tends to live longer than a typical Labrador Retriever on a normal diet (Source: *Effects of Diet Restriction on Life Span and Age-Related Changes in Dogs*, *Richard D. Kealy, PhD, et al.*).

1. What question were the researchers trying to answer?
2. Identify the population.
3. Identify the sample.
4. Identify the conclusion. Is it part of descriptive or inferential statistics?

Solution

1. How does a restricted diet affect a Labrador Retriever's life-span?
2. The population is all Labrador Retrievers.
3. The sample is the 48 Labrador Retrievers.
4. The conclusion is that a typical Labrador Retriever on a restricted diet tends to live longer than a typical Labrador Retriever on a normal diet. The conclusion is part of inferential statistics because the study used information about a sample (48 Labrador Retrievers) to draw a conclusion about a population (all Labrador Retrievers).

Simple Random Sampling

When a study makes use of a sample, it is important that the sample represents the population well. Otherwise, any inferences that are made will be meaningless. But how can we select a sample that represents the population well? The key idea is to select a *simple random sample.*

▶ **Definition**   Simple random sample

A process of selecting a sample of size *n* is **simple random sampling** if every sample of size *n* has the same chance of being chosen. A sample selected by such a process is called a **simple random sample**.

There are many ways to select a simple random sample. For example, to randomly select 10 students from a class of 40 students, you could write each student's name on a slip of paper, mix the 40 slips in a hat, and select 10 slips.

If we allow an individual to be selected more than once, then we are **sampling with replacement.** If we do not allow an individual to be selected more than once, then we are **sampling without replacement.**

If the population size is large, it would not be practical to use slips of paper and a hat. Instead, technology can be used. The first step is to make a *frame*. A **frame** is a numbered list of all the individuals in the population. Then technology can generate random numbers, which would indicate the individuals to be included in the random sample.

▶ **Example 5**   Selecting a Simple Random Sample

1. Use a TI-84 to randomly select 5 of the following 10 students without replacement:

   Gabe, Tammy, Amanda, Nick, Victor, Mitch, Simone, Manak, Alex, Lisa

2. Use StatCrunch to randomly select 5 of the 10 students without replacement.
3. Compare the results of Problems 1 and 2.

Solution

1. First, we create a frame by numbering the names:

   1. Gabe, **2.** Tammy, **3.** Amanda, **4.** Nick, **5.** Victor, **6.** Mitch, **7.** Simone, **8.** Manak, **9.** Alex, **10.** Lisa

   Next, we use a TI-84 to generate random numbers between 1 and 10, inclusive. We will have the calculator generate 8 such numbers, rather than 5 numbers, in case there are repeats. The sample should not include a student more than once because we are sampling without replacement.

   We start by entering a number called a *seed*. A **seed** is a counting number that serves as a trigger for technology to generate random numbers. To select a seed, we can make up any counting number at all. Because this example was written on November 20 (11/20), we enter 1120 by storing it in the "rand" command (see Fig. 1). Then we use "randInt(" to generate the 8 random numbers 1, 7, 5, 6, 1, 2, 7, and 6 (see Fig. 2). For more detailed TI-84 instructions, see Appendix A.3.

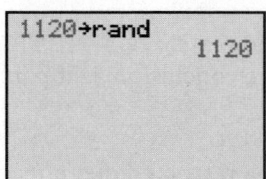

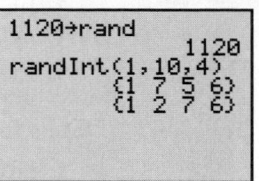

Figure 1 Store 1120 in the "rand" command

Figure 2 TI-84 generates 8 random numbers

   The first 5 different numbers are 1, 7, 5, 6, and 2, which correspond to the students Gabe, Simone, Victor, Mitch, and Tammy. So, our random sample consists of these five students.

2. We use the same frame and seed as in Problem 1 but now use StatCrunch to generate 8 random numbers between 1 and 10 (see Fig. 3). The 8 random numbers are 7, 9, 7, 2, 9, 1, 2, and 8 (see Fig. 4). For more detailed StatCrunch instructions, see Appendix B.1.

StatCrunch	Applets
Row	Discrete Uniform1
1	7
2	9
3	7
4	2
5	9
6	1
7	2
8	8

Figure 4 StatCrunch generates 8 random numbers

   The first 5 different numbers are 7, 9, 2, 1, and 8, which correspond to Simone, Alex, Tammy, Gabe, and Manak.

3. Different samples were selected by using a TI-84 and StatCrunch, even though the same seed was used. The two technologies must use different programs to generate random numbers.

Simulate Discrete Uniform

Number of rows and columns:

Rows: 8

Columns: 1

Discrete Uniform parameters:

Minimum: 1

Maximum: 10

Store samples:
- ⦿ Split across columns
- ○ Stacked with a sample id
- ○ Compute for each column (sample)
  --optional--
  e.g. mean("Discrete Uniform")

Column name(s):

Prefix: --optional--

Seeding:
- ○ Use dynamic seed
- ⦿ Use fixed seed
  Seed: 1120

Figure 3 StatCrunch settings

   In Example 5, we selected two different random samples from using a TI-84 and StatCrunch. We could select yet more different random samples by changing the seed. For example, using the seed 938 with a TI-84 would result in another different sample: Nick, Alex, Tammy, Simone, and Gabe (try it).

   In Example 6, we will explore how well a sample can represent a population if the sample size is large.

> **Example 6**   Estimating a Population Proportion with a Large Sample

In fall semester 2013, there were 19,773 students enrolled in LaGuardia Community College in Long Island City, New York. There were 11,310 female students and 8463 male students (Source: *LaGuardia Community College*).

1. What proportion of the students were female?
2. By using the numbers 1 through 11,310 to represent the female students and the numbers 11,311 through 19,773 to represent the male students, the author used technology to randomly select 1000 students without replacement. The random sample consisted of 585 female students and 415 male students. What proportion of the sample was female students?
3. If we did not know the proportion of the population who are female students and used the result we found in Problem 2 to estimate it, would that be part of descriptive statistics or inferential statistics? Explain.

Solution

1. Because there were 11,310 female students out of 19,773 students, the proportion of the population who are female is $\frac{11,310}{19,773} \approx 0.572$. This is called a *population proportion*.
2. Because there were 585 female students out of 1000 students, the proportion of the sample who are female is $\frac{585}{1000} \approx 0.585$. This is called a *sample proportion*.
3. Using the sample proportion 0.585 to estimate the population proportion 0.572 is part of inferential statistics.

In Example 6, the sample proportion 0.585 estimates the population proportion 0.572 quite well. The difference of the sample proportion and the population proportion is $0.585 - 0.572 = 0.013$, which we call the *sampling error*.

> **Definition   Sampling error**
>
> **Sampling error** is the error involved in using a sample to estimate information about a population due to randomness in the sample.

Recall that the benefit of using samples is that it can save us time, money, and effort. The somewhat bad news is that our inferences will have sampling error. The extremely good news is that **if random sampling is used to select a large enough sample, the sampling error will not be too large.**

Bias

For many studies, it can be quite challenging to select a sample that represents the population well.

> **Definition   Bias**
>
> A sampling method that consistently underemphasizes or overemphasizes some characteristic(s) of the population is said to be **biased.**

If a sampling method is biased, then any inferences made will be misleading. There are three types of bias in sampling:

1. *Sampling bias*
2. *Nonresponse bias*
3. *Response bias*

Sampling bias occurs if the sampling technique favors one group of individuals over another. To use an online survey to estimate the percentage of Americans who have a Facebook account is an example of sampling bias because people who go online are favored. In fact, people who never go online cannot participate in the poll. Because Facebook is an online service, the study will likely overestimate the percentage of Americans who use it. **The way to avoid sampling bias is to select a simple random sample.**

Nonresponse bias happens if individuals refuse to be part of the study or if the researcher cannot track down individuals identified to be in the sample. For example, there is almost always nonresponse bias with surveys because some people will refuse to take part in the survey. As another example, a person might be selected to be in a sample for a telephone survey, but the person may never answer the phone.

Response bias occurs if surveyed people's answers do not match with what they really think. For example, people might exaggerate how much money they earn, or a researcher might record the information incorrectly. A dishonest company might change customers' evaluations of their products.

Response bias can also result from the wording of questions. For example, compare the impact of the following two questions:

1. Do you brag about your past successes with others?
2. Do you inspire others by sharing your past successes?

Although both questions ask whether a subject shares about past successes, the first includes a negative judgment and the second includes a positive judgment. Each will likely generate response bias. It would be better to ask a neutral question, such as

> Do you share your past successes with others?

However, this is a yes/no question, which can create response bias. In general, people tend to say yes to yes/no questions to be polite (Source: *Leech, G. N., (1983). Principals of Pragmatics. London: Longman*). So, an even better question would be

> Do you share your past successes with others or keep them to yourself?

But even this question can be improved. The problem is that the order of the two choices might create response bias. We can fix that by asking half of the respondents the question as worded and asking the other half the question

> Do you keep your past successes to yourself or share them with others?

There is another problem with the question

> Do you inspire others by sharing your past successes?

If someone responds no, it will not be clear whether the person does not share past successes or the person does not think he or she inspires others when sharing past successes. Questions should address just one issue.

▶ Guidelines for Constructing Survey Questions

When constructing a survey question,

- Do not include judgmental words.
- Avoid asking a yes/no question.
- If the question includes two or more choices, switch the order of the choices for different respondents.
- Address just one issue.

▶ **Example 7** Identifying Bias

For each study, identify possible forms of bias. Also, discuss sampling error.

1. Throughout a week, a group of students surveys other students studying in the library and finds that 855 of 950 students study every day. The group concludes that 90% of all students at the college study every day.

2. SNAP is the U.S. food program for low-income individuals and families. A conservative television news station conducts a call-in survey, asking callers, "Should federal funding for SNAP be decreased so the national budget can be balanced?" Of the 1420 callers, 994 answered yes. The station concludes that 70% of all Americans think that the budget for SNAP should be decreased.

Solution

1. The sampling method favors students who study in the library over those that don't (sampling bias). Also, students busy studying for a test might refuse to take part in the survey (nonresponse bias). Finally, students who are surveyed might exaggerate how often they study (response bias). Even if the group of students had performed sampling without bias, there would still likely be sampling error, so the population percentage would probably not equal the sample percentage, although it would be close.

2. Because the station is conservative, its viewers will tend to be conservative, too. So, the survey favors Americans who are conservative (sampling bias). Also, the question is a yes/no question (response bias). Finally, the question includes two issues, reducing the SNAP budget and balancing the national budget (response bias). Even if the station had performed sampling without bias, there would still likely be sampling error, so the population percentage would probably not equal the sample percentage, although it would be close.

Sampling and Nonsampling Error

Recall that if random sampling is used correctly, then sampling error is the error that results from using a sample to estimate information about a population. But as we saw in Example 7, inferences can have error for other reasons.

▶ **Definition** Nonsampling Error

Nonsampling error is error from using biased sampling, recording data incorrectly, and analyzing data incorrectly.

▶ **Example 8** Identify Sampling and Nonsampling Errors

Identify whether sampling error, nonsampling error, or both have likely occurred.

1. An employee in human resources randomly selects some of the employees, and by referring to accurate records, he correctly computes that the proportion of sampled employees who have worked at the company at least 10 years is 0.37. But the population proportion is actually 0.36.

2. An employee in human resources randomly selects some of the employees and asks them whether they have ever called in sick when they were healthy. The proportion of the sample who say yes is 0.05, but the population proportion is actually 0.28.

Solution

1. There is no nonsampling error because the sampling is not biased, the data were recorded correctly, and the analysis was done correctly. The error $0.37 - 0.36 = 0.01$ is due to sampling error.

2. It is very likely that some employees did not admit they had called in sick when they were healthy. So, it is very likely that response bias occurred, which means non-sampling error occurred. It is also very likely that sampling error occurred because sampling error almost always occurs even when sampling is carried out correctly.

Through careful planning and execution, we can keep nonsampling error low or, in some cases, zero. There is no way to avoid sampling error. But by using large enough samples, we can keep the sampling error reasonably low.

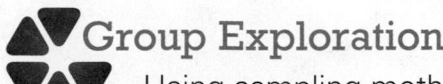 Group Exploration

Using sampling methods

1. Treating your classmates outside of your group as a population, decide on a characteristic about the population that interests your group and can be described by a proportion.

2. Form a question that you could ask of a sample to help you estimate the proportion you described in part (a). If you had any false starts in forming the question, describe how they would have resulted in response bias, if appropriate.

3. Draw a diagram that indicates the locations of the other groups of students in your classroom. Indicate which students are in which groups by referring to the students' names or using numbers to represent them.

4. Use simple random sampling on the students outside your group. Describe this process, including the seed that you chose.

Taking It One Step Further

5. Your professor may ask you to perform your survey. If so, describe the data you collected and any conclusions you can draw about the *sample*. Is this part of descriptive statistics or inferential statistics?

> ▶ **Tips for Success**   Math Journal
>
> Do you tend to make the same mistakes repeatedly throughout a math course? If so, it might help to keep a journal in which you list errors you have made on assignments, quizzes, and tests. For each error you list, include the correct solution as well as a description of the concept needed to solve the problem correctly. You can review this journal from time to time to help you avoid making these errors.

Homework 2.1

 For extra help ▶ MyMathLab®   ▣ Watch the videos in MyMathLab   Download the MyDashboard App

1. In statistics, a(n) ____ is a characteristic of the individuals to be measured or observed.

2. A(n) ____ is the part of a population from which data are collected.

3. ____ statistics is the practice of using information from a sample to draw conclusions about the entire population.

4. A sampling method that consistently underemphasizes or overemphasizes some characteristic(s) of the population is said to be ____ .

5. The volcanic eruptions with the greatest death tolls since 1900 are described in Table 2.

Table 2 Volcanic Eruptions

Volcano Name	Country	Year	Caused a Tsunami	Number of Deaths
Pelee	Martinique	1902	Yes	28,000
Kelut	Indonesia	1919	No	5110
Lamington	Papua New Guinea	1951	No	2942
El Chichon	Mexico	1982	No	1879
Ruiz	Colombia	1985	No	23,080

Source: *National Geophysical Data Center*

a. Identify the individuals.
b. Identify the variables.
c. Identify the observations for each variable.
d. Of the eruptions described in Table 2, which volcano caused the greatest number of deaths?
e. How many more people died in the 1902 Pelee eruption than in the 1982 El Chichon eruption?

6. Some Georgia inmates under the death sentence are described in Table 3.

Table 3 Some Georgia Death-Row Inmates

Name	County of Conviction	Race	Current Age (years)	Time Served (years)
Andrew Brannan	Laurens	Caucasian	66	14
Roger Collins	Houston	African American	55	37
Jerry Heidler	Toombs	Caucasian	37	15
Warren Hill	Lee	African American	54	23
Darryl Scott	Chatham	Caucasian	31	7

Source: *Georgia Department of Corrections*

a. Identify the individuals.
b. Identify the variables.
c. Identify the observations for each variable.
d. Calculate the inmates' ages when they were convicted.
e. Of the inmates included in Table 3, which one was youngest when convicted? What was the inmate's age?

7. Five statistics students' anonymous responses to a survey administered by the author are described in Table 4.

Table 4 Statistics Students' Anonymous Responses

Student	Major	Daily Exercise (minutes)	Statistics Study Time (hours)	Read Statistics Textbook
1	Psychology	0	4	Yes
2	Undecided	2	6.5	Yes
3	Film	30	8	Yes
4	Business	60	10	No
5	Journalism	60	5	No

Source: *J. Lehmann*

a. Identify the individuals.
b. Identify the variables.
c. Identify the observations for each variable.
d. For how long does Student 2 exercise daily, according to Table 4? Why does this seem odd? Describe two possible explanations for the observation.
e. Why is the variable Read Statistics Textbook vague?

8. Some missions to Mars are described in Table 5.

Table 5 Missions to Mars

Mission Name	Mission	Outcome	Cost (millions of dollars)	Launch Mass (pounds)
Mars Polar Lander	Lander	Failure	110	640
Opportunity	Rover	Success	400	408
Mars Reconnaissance Orbiter	Orbiter	Success	720	4810
Yinghuo-1	Orbiter	Failure	163	29,100
Mars Orbiter Mission	Orbiter	Success	74	2948

Source: *NASA*

a. Identify the individuals.
b. Identify the variables.
c. Identify the observations for each variable.
d. Compute the ratio of cost to launch mass (in millions of dollars per pound) for each mission. Round to the third decimal place.
e. Which mission has the least ratio of cost to launch mass?

9. In a poll of 1000 randomly selected American adults, 48% of respondents said that they strongly disapprove of the way Congress is doing its job. The study then made an inference about all American adults (Source: USA Today/*Bipartisan Policy Center*).
a. Define a variable for the study.
b. Identify the sample.
c. Identify the population.

10. In a poll of 1000 randomly selected likely voters, 48% of respondents said the Affordable Care Act goes too far. The study then made an inference about all likely voters (Source: *George Washington Battleground*).
a. Define a variable for the study.
b. Identify the sample.
c. Identify the population.

11. Do you think the drinking age should be lowered to age 18? In a telephone survey, 74% of 1013 randomly selected American adults said no. The study then made an inference about all American adults (Source: *The Gallup Organization*).
a. Define a variable for the study.
b. Identify the sample.
c. Identify the population.

12. In terms of solving the world's problems, do you think the United States does too much? In a survey of 1501 randomly selected American adults, 39% said yes. The study then made an inference about all American adults (Source: USA Today/*Pew Research Center*).
a. Define a variable for the study.
b. Identify the sample.
c. Identify the population.

13. In 2014, researchers surveyed 218 randomly selected non-Caucasians, asking whether they believed that police could protect them from violent crime. The percentage of those surveyed who said yes was 49%. The researchers then made an inference about all non-Caucasians (Source: *The Gallup Organization*).
a. Define a variable for the study.
b. Identify the sample.
c. Identify the population.

14. In 2014, researchers surveyed 776 randomly selected Caucasians, asking whether they believed that police could protect them from violent crime. The percentage of those surveyed who said yes was 60%. The researchers then made an inference about all Caucasians (Source: *The Gallup Organization*).
a. Define a variable for the study.
b. Identify the sample.
c. Identify the population.

15. After you graduate from college, do you plan to stay at your next full-time job for 3–5 years? In a survey of 13,127 randomly selected graduating college students, 43% said yes. Researchers then made an inference about all graduating college students (Source: *Achievers*).
a. Define a variable for the study.
b. Identify the sample.
c. Identify the population.

16. Researchers surveyed parents with college-bound teenagers ages 16 to 18 years, asking whether the parents will limit their children's choices of college based on cost. Of 1000 randomly selected individuals surveyed, 48% said yes. The researchers then made an inference about all parents with college-bound teenagers ages 16 to 18 years (Source: *Discover Student Loans*).
  a. Define a variable for the study.
  b. Identify the sample.
  c. Identify the population.

17. Researchers tested whether the flu shot helped hospital employees avoid getting the flu. A total of 179 hospital staff participated in the study. The researchers concluded that the flu shot does not help hospital staff avoid getting the flu (Source: Do Hospital Employees Benefit from the Influenza Vaccine? A Placebo-Controlled Clinical Trial, *S. Weingarten et al.*).
  a. What question were the researchers trying to answer?
  b. Identify the sample.
  c. Identify the population.
  d. Identify the conclusion. Is it part of descriptive or inferential statistics? Explain.

18. Simvastatin is a prescription drug used to lower blood cholesterol. However, researchers tested whether the drug also heals ulcers. A total of 66 ulcer patients were tested. The study concluded that the drug successfully heals ulcers (Source: Simvastatin as a Novel Therapeutic Agent for Venous Ulcers: A Randomized, Double-Blind, Placebo-Controlled Trial, *M. T. Evangelista et al.*).
  a. What question were the researchers trying to answer?
  b. Identify the sample.
  c. Identify the population.
  d. Identify the conclusion. Is it part of descriptive or inferential statistics? Explain.

19. Does music improve sleep quality? To find out, researchers studied 94 students ages 19 to 29 years. The students were divided into three groups. The first group listened to relaxing classical music for 45 minutes at bedtime for three weeks. The second group listened to an audiobook for the same periods. The researchers gave no instructions to the third group. To measure sleep quality, the individuals completed the Pittsburg Sleep Quality Index before, during, and after the three weeks. The researchers concluded that listening to relaxing classical music for 45 minutes at bedtime improves sleep quality in students ages 19 to 29 years (Source: Music Improves Sleep Quality in Students, *L. Harmat*).
  a. What question were the researchers trying to answer?
  b. Identify the sample.
  c. Identify the population.
  d. Identify the conclusion. Is it part of descriptive or inferential statistics? Explain.

20. People with autism have trouble relating to others. Some researchers wanted to test whether people with autism are less able to process social rewards than monetary rewards. There were 20 individuals in the study, 10 with autism and 10 without. All 20 individuals undertook a learning task that involved 200 attempts. For each success, they were shown either a photo of a person smiling or a picture of a dollar, which meant they would be paid a dollar later. Researchers concluded that autistic adults are less able to process social rewards than non-autistic

people (Source: Impaired Learning of Social Compared to Monetary Rewards in Autism, *Alice Lin et al.*).
  a. What question were the researchers trying to answer?
  b. Identify the sample.
  c. Identify the population.
  d. Identify the conclusion. Is it part of descriptive or inferential statistics? Explain.

21. Do low-skilled college algebra students learn more from completing online homework than textbook homework? To find out, researchers studied 75 low-skilled college algebra students at a large community college. Some of the students completed their homework using a textbook. The rest completed similar homework using an online homework system developed by the textbook publisher. The students' mathematical achievement was measured by a common final exam. The researchers concluded that low-skilled college algebra students learn more from completing online homework than textbook homework. (Source: The Effects of Online Homework on Achievement and Self-Efficacy of College Algebra Students, *David Shane Brewer*).
  a. What question were the researchers trying to answer?
  b. Identify the sample.
  c. Identify the population.
  d. Identify the conclusion. Is it part of descriptive or inferential statistics? Explain.

22. A researcher tried to determine whether women who are more sexually confident about themselves are more likely to achieve sexual satisfaction. She arranged for 45 women to take a 112-question anonymous online survey. The researcher concluded that the more sexually confident a woman is about herself, the more likely she will achieve sexual satisfaction (Source: Evaluating the Relationship between Women's Sexual Desire and Satisfaction from a Biopsychosocial Perspective, *Katherine J. Chartier*).
  a. What question were the researchers trying to answer?
  b. Identify the sample.
  c. Identify the population.
  d. Identify the conclusion. Is it part of descriptive or inferential statistics? Explain.

23. Use technology with seed 294 to randomly select 3 students without replacement from the frame:
  1. Becky, **2.** Nubia, **3.** Jabra, **4.** Alice, **5.** Paul, **6.** Jose, **7.** Dawn

24. Use technology with seed 836 to randomly select 4 students without replacement from the frame:
  1. Brenton, **2.** Kali, **3.** Mariah, **4.** May, **5.** Rani, **6.** Shea, **7.** Ken, **8.** Uriel

25. The favorite sports of students in one of the author's statistics classes are shown in Table 6.

Table 6 Students' Favorite Sports

1. Frank	Soccer	**7.** Harold	Football
2. Angelica	Football	**8.** Jason	None
3. Marissa	Football	**9.** Anas	Football
4. Gerardo	Baseball	**10.** Kat	None
5. Calvin	Basketball	**11.** Damon	Football
6. Keith	None	**12.** Amy	Soccer

Source: *J. Lehmann*

  a. Use technology with seed 49 to randomly select 6 students without replacement.

b. For the simple random sample you selected in part (a), find the proportion of students who prefer football. Write your result as a fraction. If your result is used to describe the sample, is this part of descriptive or inferential statistics? Explain.

c. Use technology with seed 87 to randomly select 6 students without replacement.

d. For the simple random sample you selected in part (c), find the proportion of students who prefer football. Write your result as a fraction.

e. Are your results in parts (b) and (d) equal? If yes, would all randomly selected samples of size 6 give the same results? If your results of parts (b) and (d) are not equal, explain why this happened.

26. The author surveyed some of his statistics students who drive, asking whether they run red lights. The students' responses are shown in the frame in Table 7. The survey was anonymous, so the names in the table are fictional.

Table 7 Students' Responses to "Do You Run Red Lights?"

Student	Response	Student	Response
1. Jeffrey	No	7. Taja	No
2. Phoebe	Yes	8. Samuel	Yes
3. Win	No	9. Joshua	Yes
4. Karen	Yes	10. Dulce	No
5. Paola	No	11. Monique	No
6. Arnold	Yes	12. Nathan	Yes

Source: *J. Lehmann*

a. Use technology with seed 25 to randomly select 5 students without replacement.

b. For the simple random sample you selected in part (a), find the proportion of students who run red lights. Write your result as a fraction. If your result is used to describe the sample, is this part of descriptive or inferential statistics? Explain.

c. Use technology with seed 76 to randomly select 5 students without replacement.

d. For the simple random sample you selected in part (c), find the proportion of students who run red lights. Write your result as a fraction.

e. Are your results in parts (b) and (d) equal? If yes, would all randomly selected samples of size 5 give the same results? If your results of parts (b) and (d) are not equal, explain why this happened.

27. The music preferences of students in one of the author's statistics classes are shown in Table 8.

Table 8 Students' Music Preferences

1. Ben	Hip-Hop	8. Andrew	Hip-Hop
2. Miles	Alternative Rock	9. Sean	Country
3. Jasmine	Rap	10. Charles	Hip-Hop
4. Jackson	Rap	11. Ryan	Country
5. Hallie	Metal	12. Hee-Sang	Electronic
6. Claudia	Electronic	13. Chris	Electronic
7. Pedram	Pop	14. Kellen	Metal

Source: *J. Lehmann*

a. Use technology with seed 95 to randomly select 6 students without replacement.

b. For the simple random sample you selected in part (a), find the proportion of students who prefer hip-hop. Write your result as a fraction.

c. For all 14 students shown in Table 8, find the proportion of students who prefer hip-hop. Write your result as a fraction.

d. Does your result from part (b) equal your result from part (c)? If no, is that due to sampling or nonsampling error?

e. If you were to select many simple random samples of size 6, would the proportion of students who prefer hip-hop for each sample equal the proportion of all 14 students who prefer hip-hop? If no, is that due to sampling or nonsampling error?

28. The author surveyed one of his statistics classes, asking students whether they thought it was more important to improve college students' success rates or to address climate change. The students' responses are shown in Table 9. The survey was anonymous, so the names in the table are fictional.

Table 9 Students' Priority of Success versus Climate Change

Student	Response	Student	Response
1. Aksana	Success	8. Luis	Climate
2. Patricia	Success	9. Jessica	Climate
3. Devin	Climate	10. Chris	Success
4. Julia	Success	11. Jose	Success
5. Gauri	Success	12. Fan	Success
6. Fiona	Success	13. Xiaowan	Success
7. Fadi	Climate	14. Dimitrios	Climate

Source: *J. Lehmann*

a. Use technology with seed 291 to randomly select 7 students without replacement.

b. For the simple random sample you selected in part (a), find the proportion of students who think it is more important to improve student success. Write your result as a fraction.

c. For all 14 student responses shown in Table 9, find the proportion of students who think it is more important to improve student success. Write your result as a fraction.

d. Does your result from part (b) equal your result from part (c)? If no, is that due to sampling or nonsampling error?

e. If you were to select many simple random samples of size 7, would the proportion of students who think it is more important to improve student success for each sample equal the proportion of all 14 students who think it is more important to improve student success? If no, is that due to sampling or nonsampling error?

29. In fall semester 2014, there were 27,972 students earning credit at a college in the Collin County Community College District in Texas. There were 9790 full-time students and 18,182 part-time students (Source: *Collin County Community College District*).

a. What proportion of the students were full-time? Round your result to the third decimal place.

b. By using the numbers 1 through 9790 to represent the full-time students and the numbers 9791 through 27,972 to represent the part-time students, the author used technology to randomly select 1000 students without replacement. The random sample consisted of 380 full-time students and 620 part-time students. What proportion of the sample were full-time students?

c. Does your result from part (b) equal your result from part (a)? If no, is that due to sampling or nonsampling error?

d. If we did not know the proportion of the population who are full-time students and used the result we found in part (b) to estimate it, would that be part of descriptive statistics or inferential statistics? Explain.

30. In fall semester 2013, there were 20,329 students at Diablo Valley College in Pleasant Hill, California. There were 10,571 female students and 9758 male students (Source: *Diablo Valley College*).

    a. What proportion of the students were female? Round your result to the third decimal place.

    b. By using the numbers 1 through 10,571 to represent the female students and the numbers 10,572 through 20,329 to represent the male students, the author used technology to randomly select 1000 students without replacement. The random sample consisted of 523 female students and 477 male students. What proportion of the sample were female students? Round your result to the third decimal place.

    c. Does your result from part (b) equal your result from part (a)? If no, is that due to sampling or nonsampling error?

    d. If we did not know the proportion of the population who are female students and used the result we found in part (b) to estimate it, would that be part of descriptive statistics or inferential statistics? Explain.

For Exercises 31–34, reword the question so that there is less chance of bias.

31. Do you drink smoothies?

32. Do you have a Facebook account?

33. Do you help society by paying taxes or do you not pay them?

34. Do you take care of yourself by exercising or do you not work out?

For Exercises 35–48, identify the type(s) of bias in the sampling method. Explain.

35. A student surveys the students in her calculus class and finds that 4 out of 35 students are engineering majors. She uses the data to predict the percentage of all students at her college who are engineering students.

36. A student surveys some students on campus in the evening and finds that 16 of 20 students are part-time students. He uses the data to predict the percentage of all students at his college who are part-time students.

37. A simple random sample of college students who are taking prestatistics at the same college are asked, "Do you read the textbook every day so you can get an A in your prestatistics course?" The pollster uses the data to predict the percentage of all college prestatistics students in the United States who read the textbook every day.

38. A simple random sample of American adults are asked, "Do you post messages on all your friends' Facebook pages every day because you like Facebook so much?" The pollster uses the data to predict the percentage of all American adults who post messages on their friends' Facebook pages every day.

39. A pollster surveys people attending a boxing match, asking them if they have ever struck their spouse. Of the 32 people surveyed, 5 say no and the others do not respond. The pollster uses the data to predict the percentage of people who strike their spouse.

40. A pollster randomly selects adults in a bar in the afternoon, asking them if they neglect their children. Of the 12 customers surveyed, 3 say no and the others do not respond. The pollster uses the data to predict the percentage of adults who neglect their children.

41. A simple random sample of customers who telephoned a company's customer service in the past 3 months were invited to take part in a survey. The 10% of the sample who agreed to take part were asked, "Did the representative you spoke with fully understand and resolve all your issues?" Of all customers who contacted the company's customer service in the past 3 months, the company predicts the percentage who felt the representative resolved all their issues.

42. A simple random sample of customers who purchased a company's product online in the past 4 months were invited by e-mail to respond to the question, "Did your purchase arrive on time and in good condition?" About 7% of those contacted responded. Of all customers who purchased the company's product online in the past 4 months, the company predicts the percentage who received their purchase on time and in good condition.

43. An automated telephone survey includes a question about a restaurant's food, directing the caller, "Press 1 for very satisfied and 5 for very dissatisfied. You may choose any number between 1 and 5, inclusive." For the next question about the restaurant's service, the caller is directed, "Press 1 for very dissatisfied and 5 for very satisfied. You may choose any number between 1 and 5, inclusive."

44. An automated telephone survey includes a question about a hotel's maid service, directing the caller, "Press 1 for very satisfied and 10 for very dissatisfied. You may choose any number between 1 and 10, inclusive." For the next question about the hotel's food service, the caller is directed, "Press 1 for very satisfied and 5 for very dissatisfied. You may choose any number between 1 and 5, inclusive."

45. At each building on campus, a pollster asks just the first student who enters the building, "Did you drive to campus?" The pollster uses the data to predict the percentage of all the college's students who drive to campus.

46. Once a minute, from 8 A.M. to 9 A.M. on a Monday, a used-car dealership owner records whether a Honda Civic is the first car to pass by on a highway. The owner uses the data to predict the percentage of all cars on the highway at any time on any day that are Honda Civics.

47. A liberal blog hosts an online poll, asking readers, "Should federal funding for education be increased so that crime is reduced?" The blogger uses the data to predict the percentage of all Americans who think that federal funding for education should be increased.

48. A health TV show runs a call-in survey, asking callers, "Should the government better support organic farmers so that the cancer rate of Americans can be lowered?" The show uses the data to predict the percentage of all Americans who think that the government should better support organic farmers.

49. In an often-used survey system, customers are asked to rate a company's product between 0 and 11 points, inclusive. Customers who chose scores between 0 and 6 points, inclusive, are interpreted to be unhappy with the product. People who selected either the score 7 or 8 points are interpreted to feel neutrally about the product. And customers who chose scores between 9 and 11 points, inclusive, are interpreted to be happy with the product.

    a. What real number is in the middle of the scores 0 through 11?

    b. On the basis of your result to part (a), does it make sense that people who selected the scores 7 or 8 are interpreted to feel neutrally about the product? Explain.

c. In your opinion, what would be a better way to interpret the scores?

50. A restaurant places cards on the dining tables, inviting customers to complete an online survey about their dining experience.

   a. Identify the type(s) of bias the survey has. Explain.
   b. Only 6% of customers complete the online survey. The restaurant decides to offer $3 off a customer's next meal if the customer completes the online survey. Which types of bias will likely increase? Which types will decrease? Explain.

Concepts

51. Describe the five steps of statistics. (See page 12 for guidelines on writing a good response.)

52. What are the benefits of using samples?

53. Give an example of how to use simple random sampling.

54. Describe sampling bias. Give an example of a study that might include sampling bias and explain how the study could be improved.

55. Describe nonresponse bias. Give an example of a study that might include nonresponse bias and how the study could be improved.

56. Describe response bias. Give an example of a study that might include response bias and how the study could be improved.

57. Compare the meanings of *descriptive statistics* and *inferential statistics*. Give an example of each.

58. Compare the meanings of *sampling error* and *nonsampling error*. Include in your comparison how careful planning and execution of a study will affect each type of error.

Hands-On Research

59. Google "Gallup" and click on the company's link. There you can find hundreds of surveys. Choose one that interests you. Briefly describe the survey question(s) and the results. Near the end of the article, there should be a section called "Survey Methods" or a similar title. The section should include a link that describes the sampling method in greater detail. Describe the sampling method and any possible bias in sampling.

▼ 2.2 Systematic, Stratified, and Cluster Sampling

Objectives

» Identify and explain *systematic sampling*.

» Identify and explain *stratified sampling*.

» Identify and explain *cluster sampling*.

» Compare sampling methods.

» Describe why *convenience sampling* and *voluntary response sampling* should never be used.

Recall that a simple random sample is beneficial because it usually represents a population well. However, for some studies, collecting a simple random sample is difficult or impossible. In this section, we will discuss other ways to randomly select individuals.

Systematic Sampling

On a certain day, a department store manager would like to get feedback from customers about its service. The manager has been disappointed with past e-mail surveys because the nonresponse rate was so high. He decides to try surveying customers as they leave the store, but there are too many customers to survey all of them. How should the manager select the sample?

It would be impossible to perform simple random sampling because the manager cannot predict who will shop in the store. So, he cannot create a frame. Instead, he decides to survey every 10th customer. To determine the first customer to be surveyed, the manager randomly selects a number between 1 and 10, inclusive. Say, 6. So, the pollster will approach the 6th customer, 16th customer, 26th customer, 36th customer, 46th customer, and so on. This is an example of *systematic sampling*.

▶ **Definition** Systematic sampling

To perform **systematic sampling**, we randomly select an individual out of the first k individuals and also select every kth individual after the first selected individual.

WARNING   Once k is chosen, the researcher must make sure that selecting every kth individual does not match with some pattern in the population. For example, if an appraiser records the market value of every 10th house along a street and there are 10 houses per block, the appraiser might record the value of only "corner houses," which might have different values than other houses.

▶ Example 1 Design a Plan to Perform Systematic Sampling

A grocery store owner wants to survey customers as they leave the store on a certain day. There are about 470 customers each day. The owner would like to get feedback from at least 40 customers. From past experience, she knows that if 60 customers are approached, about 40 of them will participate in the survey. Design a plan to select a sample.

Solution

First we divide 470 by 60: $470 \div 60 \approx 7.8$. If we round 7.8 up to 8 and approach every 8th person, the number of people approached would be only $470 \div 8 = 58.75$ people (58 people). But that would not meet the owner's goal to approach 60 people. If we round 7.8 *down* to 7 and approach every 7th customer, the number of people approached would be $470 \div 7 \approx 67$ people, which meets the owner's goal. So, every 7th customer should be approached.

To determine the first individual, we randomly select a number between 1 and 7, inclusive. Using the seed 58 (the current temperature outside the author's home when writing this paragraph) with a TI-84, we obtain the number 2 (see Fig. 5).

So, the pollster should approach the 2nd customer and every 7th customer after that:

$$2, 2 + 7 = 9, 9 + 7 = 16, 16 + 7 = 23, 23 + 7 = 30, \ldots$$

So, the pollster should approach the 2nd customer, the 9th customer, the 16th customer, the 23rd customer, the 30th customer, and so on.

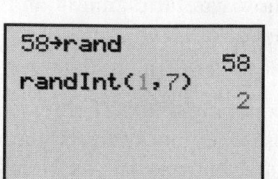

Figure 5 Randomly select a number between 1 and 7, inclusive

WARNING

In Example 1, we rounded 7.8 *down* to 7 so that we would approach at least 60 customers. **We should always round down when calculating k for systematic sampling.**

Stratified Sampling

Suppose that Dartmouth College wants to survey 100 students from its class of 2018, asking them whether they feel like they belong. Because some ethnic groups might feel differently than others, it is important that each ethnic group be well represented by the sample. An ethnic breakdown of the class of 2018 is shown in Table 10.

Because the percentages for Native American students, multi-racial students, and students who did not reveal their ethnicities are so small, a simple random sample of 100 students might not represent anyone in those groups. To make sure that all the groups in Table 10 are represented, we could poll 8 African Americans, 15 Asian Americans, 51 Caucasians, 9 Latinos, 4 Native Americans, 1 multi-racial student, 8 international students, and 4 students who did not reveal their ethnicities. We could perform simple random sampling for each group. This entire process is called *stratified sampling*.

Table 10 Ethnic Breakdown of Class of 2018 at Dartmouth College

Ethnicity	Percent
African American	8
Asian American	15
Caucasian	51
Latino	9
Native American	4
Multi-Racial	1
International	8
Non-Responder	4

Source: *Dartmouth College*

▶ **Definition   Stratified sampling**

To perform **stratified sampling**, a population is divided into subgroups called **strata** (singular: **stratum**) and simple random sampling is performed on each stratum. In each stratum, the individuals share some characteristic. The ratios of the number of individuals selected from each stratum are equal to the corresponding ratios of the stratum sizes.

Actually, if the corresponding ratios are not equal, there is a way to correct for this. You might learn about this in a statistics course.

One benefit of stratified sampling is that individuals in small strata are not excluded from the sample. Another benefit is that a smaller sample can be selected than by using simple random sampling because the sample will represent the population so well. This saves time, money, and effort.

▶ Example 2 Design a Plan to Perform Stratified Sampling

Researchers want to survey Stanford students about whether sexual harassment is prevalent in social settings on campus. Design a plan to use stratified sampling to survey 500 students.

Solution

Female, male, undergraduate, and graduate students might all have quite different impressions about sexual harassment on campus. So, we will choose the strata to be female undergraduates, male undergraduates, female graduate students, and male graduate students. Students within each strata might have similar impressions. The strata sizes at Stanford as of October 15, 2014, are shown in Table 11.

Table 11 Strata Sizes

Gender	Undergraduate	Graduate	Total
Female	3355	4693	8048
Male	3734	6687	10,421
Total	7089	11,380	18,469

Source: *Stanford University*

To calculate the proportion of students in each strata, we divide the number of students in each strata by the total number of students, 18,469 (see Table 12).

Table 12 Proportions of Students in the Four Strata

Gender	Undergraduate	Graduate	Total
Female	$3355 \div 18{,}469 \approx 0.182$	$4693 \div 18{,}469 \approx 0.254$	0.436
Male	$3734 \div 18{,}469 \approx 0.202$	$6687 \div 18{,}469 \approx 0.362$	0.564
Total	0.384	0.616	1.000

The four proportions add to 1, which checks.

Next, we compute the *number* of students to be sampled in each strata by multiplying the proportions found in Table 12 by the total desired sample size, 500 (see Table 13).

Table 13 Compute the Number of Students to Select from Each Strata

Gender	Undergraduate	Graduate	Total
Female	$0.182(500) = 91$	$0.254(500) = 127$	218
Male	$0.202(500) = 101$	$0.362(500) = 181$	282
Total	192	308	500

The numbers of students to be selected from the four strata add to 500, which checks. So, 91 female undergraduates, 101 male undergraduates, 127 female graduate students, and 181 male graduate students should be selected using simple random sampling on each strata.

WARNING The sum of the proportions for all the strata may not sum to exactly 1 due to rounding. Similarly, the sum of the numbers of students to be selected from all the strata may not sum exactly to the desired sample size due to rounding. We can prepare for this by increasing our total sample size goal a bit before calculating the sizes of the strata. If it turns out our sample size is a bit more than planned, this will not be a problem.

Cluster Sampling

Suppose that Mayor Kasim Reed of Atlanta wants to have pollsters survey 10,000 Atlanta residents in person about how well he is doing his job. Simple random sampling would require lots of driving, time, and money. So would systematic sampling and stratified sampling.

It would be easier to create a frame of all the blocks in Atlanta, perform simple random sampling to select some of the blocks, and survey all the adults who live on those blocks. This is an example of *cluster sampling*.

> ### ▶ Definition   Cluster sampling
>
> To perform **cluster sampling**, a population is divided into subgroups called **clusters**. Then simple random sampling is used to select some of the clusters. All of the individuals in those selected clusters form the sample.

▶ Example 3   Designing a Plan to Perform Cluster Sampling

The president of a college would like professors to survey 1000 of the college's 35,000 students about their awareness of various services available on campus. Although the president would like the survey to be administered during class time, she would prefer that it not disrupt all classes. Assuming that the smallest class size is 20 students, design a survey plan.

Solution

Simple random sampling, systematic sampling, and stratified sampling would disrupt a lot of classes. If each class is treated as a cluster, then cluster sampling would disrupt far fewer classes. The first step would be to create a frame of all classes. Then simple random sampling can be used to select the classes. To determine the number of classes that should be selected, we divide the desired sample size, 1000 students, by the smallest class size, 20 students: $1000 \div 20 = 50$ classes. All the students in those 50 classes should be surveyed.

The sample size will likely turn out to be more than the desired size of 1000 students because the minimum class size was used to calculate how many classes should be selected. To use a larger class size to perform the calculation runs the risk of ending up with fewer students than the desired sample size.

Comparing Sampling Methods

Now we will compare the four types of sampling methods we have discussed in Section 2.1 and this section. For simple random sampling, we randomly select individuals from a single frame. For example, in Fig. 6, we randomly select 5 students (shown in red) out of 22 students.

1. Noe	**6.** Janisa	**11.** Joe	**15.** Andrew	**19.** Alisa
2. Marcus	**7.** Steve	**12.** Edgar	**16.** Dean	**20.** Keith
3. Rose	**8.** Eddie	**13.** Mayra	**17.** Dustin	**21.** David
4. Reina	**9.** Trevor	**14.** Enzo	**18.** Seta	**22.** Lance
5. Erica	**10.** Owen			

Figure 6 Simple random sampling: Randomly select (shown in red) from just one frame

For systematic sampling, we randomly select an individual out of the first k individuals and then select every kth individual. For example, in Fig. 7, we select the 2nd individual and every 3rd individual after that (shown in red).

1. Mark	**2.** Ryan	**3.** Rob	**4.** Laura	**5.** Pete	**6.** Alice	**7.** James	**8.** Ruby	**9.** Joey	**10.** Jen	**11.** Angela

Figure 7 Systematic sampling: Select 2nd individual and every 3rd individual after that (shown in red)

WARNING   It is easy to confuse stratified sampling with cluster sampling. For stratified sampling, *all* strata are used but only *some* individuals in each strata are selected. For example, in Fig. 8, we perform simple random sampling (shown in red) on every strata. For cluster sampling, only *some* clusters are used but *all* individuals in those selected clusters are selected. For example, in Fig. 9, we randomly select clusters 2 and 5. All the individuals in those two clusters (shown in red) make up the sample.

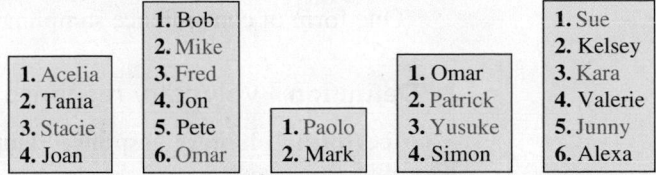

Figure 8 Stratified sampling: Perform simple random sampling (shown in red) on every strata

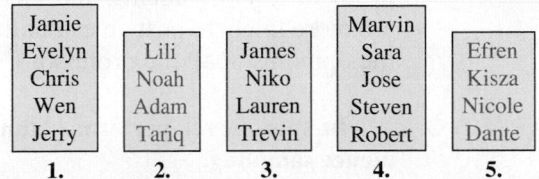

Figure 9 Cluster sampling: Randomly select Clusters 2 and 5. All the individuals (shown in red) in those two clusters make up the sample.

In Table 14, we summarize the requirements and the benefits of the sampling methods.

Table 14 Requirements and Benefits of Sampling Methods

Sampling Method	Requirement	Benefits
Simple Random	A frame of all individuals.	Works fine for telephone and e-mail surveys in which there's little risk of excluding anyone.
Systematic	Selecting every kth individual must not match with some pattern in the population.	No frame is required.
Stratified	In each strata, individuals are similar. A frame for individuals in each strata.	Individuals in small strata are not excluded from the sample. Can save time, money, and effort.
Cluster	A frame of clusters.	No frame of individuals is required. Can save time, money, and effort.

So far, we have discussed four useful sampling methods. There is another sampling method that is *not* helpful. We will identify it to make sure that we never use it.

Convenience Sampling and Voluntary Response Sampling

A student wants to estimate the proportion of all students at her college who work. To make things easy, she surveys students as they walk up to their cars in the campus lot one morning. This is an example of *convenience sampling*.

▶ Definition   Convenience sampling

To perform **convenience sampling**, we gather data that are easy to collect and do not bother with collecting them randomly.

WARNING   **Although convenience sampling is easy to perform, it should never be done because the sample will usually not represent the population well.** In other words, there will be sampling bias. Take the student who surveyed students in the parking lot. By polling only in the morning, she did not include any evening students, who might be more likely to work. And because the student polled only students who parked in the campus lot, she did not include students who walk or use public transportation. Those types of students might be more or less likely to work than students who drive to campus.

One form of convenience sampling is voluntary response sampling.

▶ Definition   Voluntary response sampling

To perform **voluntary response sampling**, we let individuals choose to be in the sample.

For example, the online survey Rate My Professors uses convenience sampling. The problem is that students who feel strongly (positively or negatively) about a professor are more likely to post an evaluation. So, the comments for a professor do not represent the opinions of all the professor's students well.

WARNING   **In sum, never perform voluntary response sampling or the more general convenience sampling.**

▶ Example 4   Identifying Sampling Methods

Identify whether the sampling method is simple random, systematic, stratified, cluster, or convenience. Explain.

1. A researcher randomly selects 20 Taco Bell locations and surveys all the employees at those locations.
2. A news station hosts a call-in survey about whether physician-assisted death should be legalized in all states.
3. A researcher randomly selects an LED TV out of the first 200 LED TVs on an assembly line and also selects every 200th LED TV after that.
4. In a study at a community college, 30 instructors are randomly selected from full-time instructors and 50 instructors are selected from part-time instructors.
5. The City Hall of Spring Hill, Kansas, creates a frame of its 5730 residents and randomly selects 60 residents.

Solution

1. Because all the employees at 20 clusters were surveyed, cluster sampling was performed.
2. A call-in survey would involve only those who choose to call in. So, the station performed voluntary response sampling, which is one type of convenience sampling. No type of convenience sample should ever be performed.
3. Because every 200th TV was selected, systematic sampling was performed.
4. Because instructors were randomly selected from (two) strata, stratified sampling was performed.
5. Because the residents were randomly selected from a single frame, simple random sampling was performed.

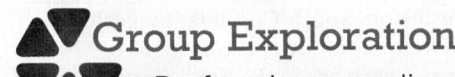

Group Exploration

Performing sampling methods

1. Draw a diagram that indicates the locations of the other groups of students in your classroom. Indicate which students are in which groups by referring to the students' names or using numbers to represent them.

2. Treating the other groups of students as clusters, perform cluster sampling. Describe this process, including the seed(s) that you used.

3. Treating the other groups of students as strata, perform stratified sampling. Describe this process, including the seed(s) that you used.

4. Perform systematic sampling on the students outside your group. Describe this process, including the seed(s) that you used.

▶ **Tips for Success**   **Use Your Instructor's Office Hours**

Helping students during office hours is part of an instructor's job. Keep in mind that your instructor wants you to succeed and hopes you take advantage of all opportunities to learn.

It is a good idea to come prepared to office visits. For example, if you are having trouble with a concept, attempt some related exercises and bring your work so your instructor can see where you are having difficulty. If you miss a class, it is helpful to read the material, borrow class notes, and try to complete assigned exercises before visiting your instructor so you get the most out of the visit.

Homework 2.2

For extra help ▶ MyMathLab®    Watch the videos in MyMathLab    Download the MyDashboard App

1. To perform systematic sampling, we randomly select an individual out of the first k individuals and also select every _____ individual after the first selected individual.

2. We should always round _____ when calculating k for systematic sampling.

3. *True or False:* For stratified sampling, all strata are used, but only some individuals in each strata are selected.

4. *True or False:* Convenience sampling is better than random sampling because it is easier to use.

For Exercises 5–18, identify whether the sampling method is simple random, systematic, stratified, cluster, or convenience. Explain.

5. Ten community colleges in the United States are randomly selected and all of the students at those colleges are surveyed.

6. A researcher creates a frame of all the blocks in Orlando, Florida. Then she randomly selects 40 of those blocks and surveys all the adult residents of those blocks.

7. A GameStop® manager randomly selects and surveys the 4th person who leaves the store and also surveys every 7th person who leaves the store after that.

8. A researcher randomly selects the 73rd car fuel tank on an assembly and every 100th tank after that.

9. To estimate the number of hours college students study per week, an instructor surveys his prestatistics class.

10. To estimate the proportion of Americans who use Instagram, a Costco employee surveys employees who work during her shift.

11. In a study at Harvard, 24 students are randomly selected from African American freshmen, 40 students are randomly selected from Asian American freshmen, 26 students are randomly selected from Latino freshmen, 106 students are randomly selected from Caucasian freshmen, 2 students are randomly selected from Native American freshmen, and 2 students are randomly selected from Pacific Islander freshmen.

12. In a nationwide study of registered voters conducted by *The New York Times*, 390 people are randomly selected out of those registered as Republicans, 430 people are randomly selected out of those registered as Democrats, and 180 people are randomly selected out of those registered as Independents.

13. Human Resources at Apple creates a frame of all its 80,300 employees and randomly selects some of the employees (Source: *Apple*).

14. A student writes the names of her classmates on slips of paper. Then she puts the slips in a hat, mixes them, and selects 5 of them without looking.

15. A wine grower divides her vineyard into 60 subsections, randomly selects 5 subsections, and measures the grape yields in those 5 subsections.

16. As hundreds of people wait in line for the new iPhone, a pollster surveys the 7th person and every 10th person after that.

17. In a study based on telephone interviews about Americans' confidence in the economy, 1518 telephone numbers were randomly selected from landline numbers and 1518 telephone numbers were randomly selected from cell phone numbers (Source: *The Gallup Organization*).

18. To assess its customers' satisfaction in the past month, the Hyatt Regency® Boston creates a frame of all its paying guests in the past month and randomly selects 250 guests to be telephoned for a survey.

19. A Wells Fargo bank manager wants to survey customers as they leave the store on a certain day. There are about 340 customers each day. The manager would like to get feedback from at least 50 customers by performing systematic sampling. From past experience, the manager knows that if 70 customers are approached, about 50 of them will participate in the survey.

   a. If every kth customer is approached, find k so that 70 customers are approached.

   b. To find which customer should be first approached, use 583 as a seed to find a random number between 1 and k, inclusive.

   c. List the first 5 customers who should be approached.

20. A Nordstrom manager wants to survey customers as they leave the store on a certain day. There are about 420 customers each day. The manager would like to get feedback from at least 30 customers by performing systematic sampling. From past experience, the manager knows that if 50 customers are approached, about 30 of them will participate in the survey.

   a. If every kth customer is approached, find k so that 50 customers are approached.

   b. To find which customer should be first approached, use 792 as a seed to find a random number between 1 and k, inclusive.

   c. List the first 5 customers who should be approached.

21. Human Resources at Genentech wants to perform a systematic sampling of its 12,300 employees (Source: *Genentech*). Using a frame of all its employees, it plans to survey 120 of them.

   a. If every kth employee is surveyed, find k so that 120 employees are surveyed.

   b. To find which employee should be first surveyed, use 264 as a seed to find a random number between 1 and k, inclusive.

   c. List the first 5 employees who should be surveyed.

22. Human Resources at Google wants to perform a systematic sampling of its 47,756 employees (Source: *Google*). Using a frame of all its employees, it plans to survey 150 of them.

   a. If every kth employee is surveyed, find k so that 150 employees are surveyed.

   b. To find which employee should be first surveyed, use 753 as a seed to find a random number between 1 and k, inclusive.

   c. List the first 5 employees who should be surveyed.

23. A researcher wants to select a sample of 70 full-time employees who work in either the police, fire, or judicial departments of Phoenix, Arizona, by performing stratified sampling. Of those three departments, 64% work in the police department, 33% work in the fire department, and 3% work in the judicial department (Source: *City-data.com*). How many full-time employees in each department should be surveyed?

24. A researcher wants to select a sample of 70 full-time employees who work in either the police, fire, or judicial departments of Charleston, South Carolina, by performing stratified sampling. Of those three departments, 62% work in the police department, 29% work in the fire department, and 9% work in the judicial department (Source: *City-data.com*). How many full-time employees in each department should be surveyed?

25. A researcher wants to select a sample of 40 students from the four private high schools in Green Bay, Wisconsin, by performing stratified sampling. The enrollments are shown in Table 15. How many students at each school should be included in the study?

Table 15 Enrollments at Private High Schools in Green Bay, Wisconsin

School	Enrollment
Notre Dame De La Baie Academy	730
Northeastern Wisconsin Lutheran High School	122
Bay City Baptist School	99
Beth Haven Academy	26

Source: *City-data.com*

26. A researcher wants to select a sample of 50 students from the four public high schools in Franklin, Tennessee, by performing stratified sampling. The enrollments are shown in Table 16. How many students at each school should be included in the study?

Table 16 Enrollments at Public High Schools in Franklin, Tennessee

School	Enrollment
Franklin High School	1936
Centennial High School	1466
Fred J. Page High School	899
Middle College High School	83

Source: *City-data.com*

27. The dean of the graduate fine arts school at University of Texas at Austin wants to select a sample of 120 students who applied to one of four music majors. The numbers of applicants for each major are shown in Table 17. If the dean wants to perform stratified sampling, how many students for each major should be included in the study?

Table 17 Numbers of Applicants to Graduate Music Majors

Major	Number of Applicants
Music and Human Learning	33
Music Performance	523
Music Theory	23
Musicology/Ethnomusicology	57

Source: *The University of Texas at Austin Graduate and International Admissions*

28. The dean of the graduate business administration school at University of Texas at Austin wants to select a sample of 100 students who applied to one of five majors. The numbers of applicants for each major are shown in Table 18. If the dean wants to perform stratified sampling, how many students for each major should be included in the study?

Table 18 Numbers of Applicants to Graduate Business Majors

Major	Number of Applicants
Accounting	85
Finance	368
Information Risk and Operations Management	109
Management	90
Marketing	83

Source: *The University of Texas at Austin Graduate and International Admissions*

29. Montana State University wants to survey its students about whether they think a 5% increase in next year's tuition is reasonable. Undergraduates who are given the Western Undergraduate Exchange (WUE) award pay lower tuition. Because WUE undergraduates, resident undergraduates, nonresident undergraduates, resident graduate students, and nonresident graduate students all pay different tuitions, the university plans to use these five types of students as strata. The strata sizes are shown in Table 19.

Table 19 Strata Sizes

Gender	Undergraduate	Graduate Student	Total
WUE Student	786	0	786
Resident	8653	1231	9884
Nonresident	3932	819	4751
Total	13,371	2050	15,421

Source: *Montana State University*

If the university wants to survey 900 students, how many students in each strata should be surveyed?

30. The University of North Carolina at Chapel Hill wants to survey its students about whether they think the dorms are in good enough condition. Because female undergraduates, male undergraduates, female graduate students, male graduate students, female professional students, and male professional students may have different opinions, the university plans to use these six types of students as strata. The strata sizes are shown in Table 20.

Table 20 Strata Sizes

Gender	Undergraduate	Graduate Student	Professional	Total
Female	10,588	4475	1421	16,484
Male	7762	3736	1153	12,651
Total	18,350	8211	2574	29,135

Source: *University of North Carolina*

If the university wants to survey 1200 students, how many students in each strata should be surveyed?

31. A 10-person environmental panel is to consist of Tennessee Senate members, who are shown in Table 21. Select a stratified sample of 8 Republicans and 2 Democrats, using 57 as the seed for the Republican strata and 82 as the seed for the Democrat strata.

Table 21 2015 Tennessee Senate Members

Republican		Democrat	
1. Beavers	14. Ketron	1. Burks	5. Kyle
2. Bell	15. Massey	2. Finney	6. Tate
3. Bowling	16. McNally	3. Ford	7. Yarbro
4. Campfield	17. Niceley	4. Harper	
5. Crowe	18. Norris		
6. Dickerson	19. Overbey		
7. Gardenhire	20. Ramsey		
8. Green	21. Southerland		
9. Gresham	22. Stevens		
10. Haile	23. Summerville		
11. Hensley	24. Tracy		
12. Johnson	25. Watson		
13. Kelsey	26. Yager		

Source: *Tennessee Senate*

32. A 9-person education panel is to be formed of Arizona Senate members, who are shown in Table 22. Select a stratified sample of 5 Republicans and 4 Democrats, using 649 as the seed for the Republican strata and 721 as the seed for the Democrat strata.

Table 22 2015 Arizona Senate Members

Republican		Democrat	
1. Pierce	10. Yarbrough	1. Dalessandro	8. Tovar
2. Ward	11. McCornish	2. Bedford	9. Hobbs
3. Crandell	12. Yee	3. Pancrazi	10. Ablesser
4. Melvin	13. Murphy	4. Begay	11. Taylor
5. Biggs	14. Burges	5. McGuire	12. Gallardo
6. Shooter	15. Reagan	6. Farley	13. Meza
7. Griffin	16. Worsley	7. Bradley	
8. Barto	17. Driggs		
9. Farnsworth			

Source: *Arizona Senate*

33. A researcher wants to test the conditioning of Major League Baseball players in the National League, which consists of the teams shown in Table 23. Each team has 25 active players.

Table 23 National League Baseball Teams

1. Braves	2. Marlins	3. Mets	4. Phillies
5. Nationals	6. Cubs	7. Reds	8. Brewers
9. Pirates	10. Cardinals	11. Diamondbacks	12. Rockies
13. Dodgers	14. Padres	15. Giants	

Source: *Major League Baseball*

The researcher wants to test 75 players using cluster sampling, where the clusters are the baseball teams. If 314 is used as a seed, which players will be included in the sample?

34. A researcher wants to test the conditioning of Major League Baseball players in the American League, which consists of the teams shown in Table 24. Each team has 25 active players.

Table 24 American League Baseball Teams

1. Orioles	2. Red Sox	3. Yankees	4. Rays
5. Blue Jays	6. White Sox	7. Indians	8. Tigers
9. Royals	10. Twins	11. Astros	12. Angels
13. Athletics	14. Mariners	15. Rangers	

Source: *Major League Baseball*

The researcher wants to test 75 players using cluster sampling, where the clusters are the baseball teams. If 829 is used as a seed, which players will be included in the sample?

35. If the samples you found in Exercises 33 and 34 are combined to represent all players in Major League Baseball, what type of sampling method is being used together with cluster sampling?

36. Due to a drought, the Lower Colorado River Authority (LCRA) wants to survey Colorado citizens about whether water deliveries should be cut off to rice farmers in the state. The LCRA divides the population into farmers and those who live in cities or suburbs. Cluster sampling is performed for farmers, and cluster sampling is performed for those who live in cities or suburbs. What type of sampling method is being used together with cluster sampling?

37. The Hamilton County Traffic Safety Partnership (HCTSP) is a collection of police departments in Hamilton County, Indiana. To combat drunk driving, the HCTSP set up a sobriety checkpoint in Hamilton County on August 22 and 23, 2014 (Source: *HCTSP*). Which of the four sampling methods was probably used? Explain. Describe how to perform the sampling.

38. Los Angeles City Hall wants to survey some of its residents in person, asking whether the tallest building in the world should be built in the city. Which of the four sampling methods discussed in this section would require the least money and effort? Explain. Describe how to perform the sampling.

39. The *Chicago Tribune* newspaper wants to survey Illinois residents about the job performance of Dick Durbin, Illinois Democratic Senator. The newspaper has access to phone numbers of most Republican, Democratic, and Independent residents in the state. Which of the four sampling methods should be used? Explain. Describe how to perform the sampling.

40. Barnes & Noble® wants to survey some of its customers about whether their online purchases arrived on time. The company has a frame of all of its online customers' e-mail addresses. Which of the four sampling methods should be used? Explain. Describe how to perform the sampling.

41. The president of Boston University wants some of the students to complete a written survey.

   a. If each class is treated as a cluster, describe how cluster sampling could be accomplished.

   b. If each ethnic group is treated as a strata, describe how stratified sampling could be accomplished.

   c. If the president wants the survey to be completed with a data collector present, which of the sampling methods described in parts (a) and (b) would be easier? Explain.

   d. Assume the president wants to find out if students feel like they belong at the college. If the sample size is small, which sampling method would probably give better results? Explain.

42. Kansas City Mayor Sly James wants data collectors to conduct in-person interviews with some Kansas City residents.

   a. If each city block is treated as a cluster, describe how cluster sampling could be accomplished.

   b. If Democrats, Republicans, Independents, and so on, are treated as strata, describe how stratified sampling could be accomplished.

   c. Which sampling method would be easier, cluster or stratified? Explain.

   d. Assume James wants to find out if residents are in favor of raising the sales tax to better fund schools. If the sample

size is small, which sampling method would probably give better results? Explain.

43. In 1990, the Supreme Court ruled that sobriety checkpoints are legal, but police departments must file a plan that describes how they will randomly select drivers for sobriety tests. Florida Highway Patrol troopers, Pasco Sheriff's deputies, and Tarpon Springs Police officers filed a plan for a sobriety checkpoint in New Port Richey, Florida, stating that they would pull over every third car (Source: *Tampa Bay's 10 News*).

   a. What type of sampling did the police department plan to use? Explain.

   b. A police officer's squad car video showed that 29 of 49 vehicles were pulled over at the New Port Richey checkpoint between 12:16 A.M. and 12:33 A.M. on December 17, 2011 (Source: *Sammis Law Firm*). What proportion of vehicles was pulled over? Round your result to the third decimal place.

   c. A driver arrested at the checkpoint for drunk driving contested the case in court. He argued that the police did not follow the law. Explain why it makes sense his case was dropped.

44. On July 2, 2006, the Forest Park Police Department set up a sobriety checkpoint in Forest Park, Illinois. For the first part of the evening—when traffic was heavier—the police stopped every fourth car. Later, when the traffic was lighter, the police stopped every third and fourth car (Source: *Forest Park Review*).

   a. What type of sampling method did the police use when the traffic was heavier? Explain.

   b. Later in the evening—when traffic was lighter—the police stopped every third and fourth car. Is this systematic sampling? Explain.

   c. Describe how the police could have performed systematic sampling when traffic was lighter and still stopped the same number of cars.

Concepts

45. Compare stratified sampling with cluster sampling.

46. If a researcher is unable to create a frame of the population, what type of sampling method is probably best? Explain.

47. Give an example not in the textbook in which cluster sampling is easier and less costly than simple random sampling.

48. Give an example not in the textbook in which stratified sampling would include at least one member of a strata but simple random sampling might not.

49. Give an example not in the textbook in which systematic sampling could be used but simple random sampling could not.

50. Describe the benefits/drawbacks of the four sampling methods discussed in this section.

Hands-On Research

51. Google "Gallup" and click on the company's link. There you can find hundreds of surveys. Choose one that interests you. Briefly describe the survey question(s) and the results. Near the end of the article, there should be a section called "Survey Methods" or a similar title. The section should include a link that describes the sampling method in greater detail. Did Gallup use simple, systematic, stratified, cluster, or some other type of sampling method? The company often works with a weighted sample. Research *weighted sample* and describe what this might mean for the study you selected.

▼2.3 Observational Studies and Experiments

Objectives

» Identify the following components of a good study: *treatment group, control group, placebo, single-blind and double-blind, random assignment,* and large-enough sample size.

» Identify *experiments* and *observational studies.*

» Identify *explanatory variables, response variables, association,* and *causation.*

» Identify *lurking variables* and *confounding variables.*

» Redesign an observational study into an experiment.

Components of a Good Study

In this section, we will determine key components of a good study.

Treatment and Control Groups

Suppose that a researcher randomly selects 100 adults who suffer from migraine headaches. Once a migraine headache begins, the subjects are instructed to take an experimental drug. Of the 100 adults, 60% report relief from migraines within one hour of taking the drug. What does this result tell us? Not much, because the 60 adults who found relief might have experienced that relief without taking the drug. Perhaps it was just the passage of time that gave them relief.

It would have been better if the 100-adult sample had been divided into two groups each of 50 adults. One group, called the *treatment group*, would take the drug, which is called the *treatment*. The other group, called the *control group*, would not take the drug. If 60% of the treatment group found relief but only 10% of the control group found relief, we would be one step closer to determining that the drug is effective, although we have more to discuss about performing a good study.

> ### ▶ Definition   Treatment group and control group
>
> Each **treatment group** in a study is a collection of individuals who receive a certain treatment (or have a certain characteristic of interest). The **control group** is a collection of individuals who do not receive any treatment (or do not have the characteristics of any treatment group).

When designing a study, it is important that each treatment group be as similar to the control group as possible, except for the characteristic of interest. For the migraine study, we want the treatment and control groups to be similar in all ways except that the treatment group takes the drug and the control group does not. Why is this so important? Suppose the groups were different in some other way. Say, most of the treatment group drinks small amounts of caffeinated beverages and most of the control group does not. For some people, small amounts of caffeine can relieve migraines (Source: *Mayo Clinic*). So, it would not be clear whether a greater proportion of the treatment group found relief than the control group because of the experimental drug or the caffeine.

▶ Example 1   Design a Study

To study the power of visualization, researchers plan to investigate whether college basketball players' ability to shoot free throws will improve if they imagine shooting free throws. Describe the design of a possible study, which includes a treatment group and a control group.

Solution

The researchers could have half of a college basketball team (the treatment group) visualize shooting free throws for 20 minutes every morning for two weeks. The other half of the team (the control group) would not perform visualizations in the mornings. All players would continue to participate in normal practices in the afternoons. After two weeks, both groups would shoot free throws, and the proportion of shots made by each group would be compared.

▶

Placebo Effect and Placebo

Even though a greater percentage of the treatment group had relief from migraines than the control group, the difference might not be due to the drug. It could be that many of the individuals in the treatment group found relief due to the psychological effect of believing they should feel better. This is an example of the *placebo effect*. In general, the **placebo effect** occurs when the characteristic of interest changes in individuals due to the individuals believing the characteristic should change.

In the migraine study, the researcher can deal with the placebo effect by having the control group take a harmless pill that has nothing to due with migraines, such as a sugar pill. The individuals in the treatment and control groups should be warned that some of them will be given fake pills, but they will not be told who will receive them. Such a pill is an example of a *placebo*. A placebo can also be a fake procedure. For example, in a recent study about a new knee surgical procedure, individuals in the control group underwent a fake surgery (Source: Arthroscopic Partial Meniscectomy versus Sham Surgery for a Degenerative Meniscal Tear, *Raine Sihvonen, MD*). In general, a **placebo** is a fake drug or procedure administered to the control group.

Single-Blind and Double-Blind Studies

If individuals in the migraine study knew which group they belonged to, they might behave differently. For example, individuals in the treatment group might not take as good care of themselves, thinking the drug would protect them. Or they might take better care of themselves because they would have more hope that they will be migraine-free. We can avoid this by *blinding* individuals in both groups. In a **single-blind study**, individuals do not know whether they are in the treatment group(s) or the control group.

Performing a single-blind migraine study with a sugar pill placebo would certainly improve the study. But there is another issue we must consider. If the researcher in touch with the individuals knows who is in the control group, she might communicate this to the individuals. For example, she might smile at those in the treatment group (because she might be helping them) and frown at the control group (out of guilt for giving them a fake pill). It would be better if the researcher did not know which individuals were in which group.

In a **double-blind study**, neither the individuals nor the researcher in touch with the individuals know who is in the treatment group(s) and who is in the control group. In the migraine study, this could be done by having one researcher randomly assign the individuals to the treatment and control groups (without telling the individuals). Then she could write the individuals' names on identical pill containers and fill each container with either the drug or sugar pills identical in appearance, according to the random assignment. Finally, another researcher who does not know the random assignment could give the containers to the individuals.

Random Assignment

Another issue about the migraine study is how the sample is divided into the treatment and control groups. The two groups should be formed so that they are as similar as possible. It would be bad practice to let the individuals themselves decide for two reasons. First, then the study would not be double-blind. Second, the individuals who choose to be in the treatment group might be more adventurous or have some other characteristics that would make it more or less likely that the drug would work. It is extremely important that we select the groups *randomly*.

> ▶ Definition   Random assignment
>
> **Random assignment** is the process of assigning individuals to the treatment group(s) and the control group randomly.

▶ **Example 2**   Perform Random Assignment

A researcher wants to test whether prestatistics students perform better on their homework if they do it while listening to hip-hop music. The researcher randomly selects the 16 students shown in Table 25. Use the seed 772 to randomly select 8 of the students for the treatment group. The other 8 students will be in the control group.

Table 25 Random Sample of 16 Students

1. Sofia	**2.** Trenton	**3.** Danielle	**4.** Gregory	**5.** Jacklyn	**6.** Jeriann	**7.** David	**8.** Anndee
9. Nicholas	**10.** Donovan	**11.** Amy	**12.** Cindy	**13.** Alejandra	**14.** Kevin	**15.** Jesse	**16.** Jaime

Solution

Recall that even with the same seed, different technologies may give different random numbers. Here, we enter the seed 772 into a TI-84 (see Fig. 10). Then we randomly select 15 numbers between 1 and 16, inclusive (see Fig. 11). We select 15 numbers rather than just 8 in case there are repeats.

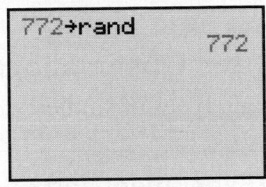

Figure 10 Store 772 in the "rand" command

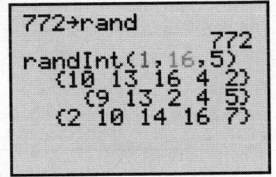

Figure 11 Generate 15 random numbers

The selected numbers are 10, 13, 16, 4, 2, 9, 13, 2, 4, 5, 2, 10, 14, 16, and 7. The first 8 numbers without repeats (in red) are 10, 13, 16, 4, 2, 9, 5, and 14. So, the following 8 students should be in the treatment group: Donavan, Alejandra, Jaime, Gregory, Trenton, Nicholas, Jacklyn, and Kevin. The other 8 students should be in the control group.

WARNING   It is important to keep the meaning of *random sampling* straight from *random assignment*. First, we perform random sampling by randomly selecting individuals *from the population* to be in a sample. Then we perform random assignment by randomly selecting individuals *from the sample* to be in the treatment and control groups (see Fig. 12).

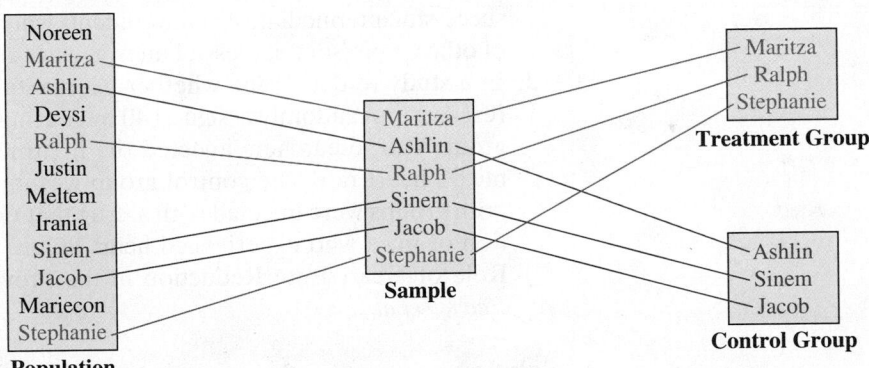

Figure 12 Random sampling followed by random assignment

Sample size

In Example 2, the sample size was only 16 students, so the example would be fairly easy to follow. In actual studies, it is often desirable to work with large sample sizes. The larger the sample size, the better the chance is that the sample will represent the population well. For example, when estimating the proportion of a population that has a certain characteristic, it is often desirable for the sample to be as large as 1000 individuals.

However, sometimes it would be too costly or unethical to use a large sample. For example, a small sample would have to be used for an expensive or risky new surgical procedure.

Also, in order to use certain statistical tests, the sample size must be large but not *too* large. You will learn more about good sample sizes in a statistics course.

Next, we summarize the components of a well-designed study.

> **Components of a Well-Designed Study**

In a well-designed study,

- There should be a control group and at least one treatment group.
- Individuals should be randomly assigned to the control and treatment group(s).
- The sample size should be large enough.
- A placebo should be used when appropriate.
- The study should be double-blind when possible. If this is impossible, then the study should be single-blind if possible.

Experiments and Observational Studies

There are two main types of studies: *experiments* and *observational studies*.

> **Definition   Experiment and observational study**

In an **experiment**, researchers determine which individuals are in the treatment group(s) and the control group, often by using random assignment. In an **observational study**, researchers do not determine which individuals are in the treatment group(s) and the control group.

> **Example 3**   Identifying an Experiment and an Observational Study

Identify whether the study is an experiment or an observational study. Discuss whether the components of a good study were used.

1. For five years, the author taught an innovative intermediate algebra course in which students learned by working in groups. Then the author compared the proportion of his successful intermediate algebra students who passed trigonometry with the proportion of other professors' successful intermediate algebra students who passed trigonometry.

2. In a study to determine whether heart rate reduction lowers heart failure in mice, researchers randomly assigned 40 mice to a treatment group and 8 mice to a control group. The researchers injected the treatment group with a drug that lowered the mice's heart rate. The control group was injected with a placebo. Then the mice in both groups were injected with a drug that typically causes heart failure. The proportion of mice who experienced heart failure was recorded for both groups (Source: *Role of Heart Rate Reduction in the Prevention of Experimental Heart Failure, Becher et al.*).

Solution

1. The author performed an observational study because he did not determine which students enrolled in his intermediate algebra class (the treatment group) and which students enrolled in other intermediate algebra classes (the control group).

   The study has several good components. It has treatment and control groups. A placebo is used in the sense that the control group still participated in an intermediate algebra course. The study is single-blind because the students did not know they were in a study. Although the sample size is not stated, it is likely pretty large because the study lasted for five years.

   There are three problems with the study. First, the students were not randomly assigned to the treatment and control groups. So, the two groups might not be similar. For example, if the author has a reputation for being challenging, weaker students may have avoided enrolling in his class. A second key problem is that the author may be better or worse at teaching than the professors teaching the control group. Finally, the study is not double-blind because the author knew that the students in his intermediate algebra courses were in the treatment group.

2. The researchers performed an experiment because the researchers determined which mice were in the treatment and control groups.

The study has many good components: There are treatment and control groups, a placebo was used, and there was random assignment. The study is single-blind because the mice were not aware of which group they were in.

Two components of the study are not clear: It is not stated whether the study is double-blind, and it is not clear whether the sizes of the treatment and control groups are large enough. You will learn to evaluate such group sizes in a statistics course.

Explanatory and Response Variables

For the visualization study in Example 1, we wanted to test whether visualization would improve basketball players' free-throw percentages. We call visualization the *explanatory variable* because we are investigating whether visualization might explain (improve) players' free-throw percentages. We say the free-throw percentage is the *response variable* because we are investigating whether players' free-throw percentages will respond to (be affected by) visualization.

> **Definition   Explanatory and response variables**
>
> In a study about whether a variable x explains (affects) a variable y,
> - We call x the **explanatory variable** (or **independent variable**).
> - We call y the **response variable** (or **dependent variable**).

WARNING   In the visualization study, calling visualization the explanatory variable does *not* mean for sure that visualization affects (explains) players' free-throw percentages. Although the researchers probably began the study because they were curious or even suspected visualization affects players' free-throw percentages, it is only by performing the study that they could find out. In general, **an explanatory variable may or may not turn out to affect (explain) the response variable.**

> **Example 4**   Identifying Components of a Study

Otis media with effusion (*OME*) is a collection of non-infected fluid in the middle ear space, mostly affecting children between 6 months and 3 years of age. It often causes hearing difficulties, loss of balance, and delayed speech development. In a double-blind study, researchers tested whether the drug mometasone in nasal spray form is effective for treatment of OME. Hundreds of children between the ages of 2 and 12 years were randomly assigned to receive mometasone nasal spray or a saline nasal spray for 6 months. The study concluded that mometasone nasal spray is an effective treatment of OME (Source: *A Double-Blind Randomized Placebo-Controlled Trial of Topical Intranasal Mometasone Furoate Nasal Spray in Children of Adenoidal Hypertrophy with Otitis Media with Effusion, R. Bhargava*).

1. Describe the treatment and control groups.
2. Is the study an experiment or an observational study? Explain.
3. Describe the sample and the population.
4. What are the explanatory and response variables?
5. What does it mean that the study is double-blind? How could that be accomplished?

Solution

1. The treatment group consists of the children in the study who were treated with mometasone. The control group consists of the children in the study who received saline nasal spray.
2. The study is an experiment because the children were randomly assigned to the treatment and control groups.

3. The sample consists of all the children in the treatment and control groups. The population consists of all children between the ages of 2 and 12 years who have OME.

4. The goal of the study was to determine whether mometasone is an effective treatment of OME. So, the variable of whether mometasone is used is the explanatory variable, and the variable of whether a patient is cured of OME is the response variable.

5. Double-blind means that neither the individuals nor the researcher in touch with the individuals know which individuals are in which group. To accomplish this, one researcher could randomly assign the individuals to the treatment and control groups (without telling the individuals). Then she could write the individuals' names on identical nasal spray containers and fill each container with either mometasone or saline solution, according to the random assignment. Finally, another researcher who does not know the random assignment could give the containers to the individuals.

Association versus Causation

There is an **association** between the explanatory and response variables if the response variable changes as the explanatory variable changes. For example, there is an association between the number of firefighters who fight a fire (explanatory variable) and the amount of fire damage that occurs (response variable). After all, the more firefighters that fight a fire, the larger the fire tends to be, so there tends to be more damage. However, that does *not* mean that an increase in the number of firefighters *causes* more damage. A larger fire causes more damage *and* the presence of more firefighters (see Fig. 13).

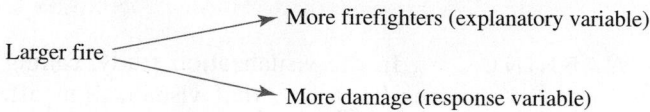

Figure 13 A larger fire causes more damage and the presence of more firefighters

If the change in the explanatory variable *causes* a change in the response variable, we say there is **causality**. When it comes to determining causality, experiments have a huge advantage over most observational studies.

▶ Determining Causality

- A well-designed experiment can determine whether there is causality between the explanatory and response variables.

- Most observational studies *cannot* determine whether there is causality between the explanatory and response variables. They can only determine whether there is an association between the two variables.

For example, because a much greater proportion of OME children were healed with mometasone nasal spray than with saline nasal spray and the study was a well-designed experiment, we can conclude that mometasone *causes* the healing of OME.

Why can we determine whether there is causality with an experiment but not with most observational studies? It is because we can perform random assignment with experiments but not with observational studies.

For example, consider the observational study about the innovative teaching method in Problem 1 of Example 3. It turns out that the proportion of students who passed trigonometry was larger in the treatment group than in the control group. However, in this observational study, we cannot conclude that the innovative teaching method *caused* the students in the treatment group to do better. It could be that the author is a demanding instructor and has a reputation for being challenging, so only stronger students enrolled in his course.

Lurking and Confounding Variables

If the author's characteristic of being demanding is the variable that really caused the treatment group to do better in trigonometry, we call the variable a *lurking variable*. This means that the author being demanding caused stronger students to enroll in the innovative course *and* for them to do better in trigonometry (see Fig. 14).

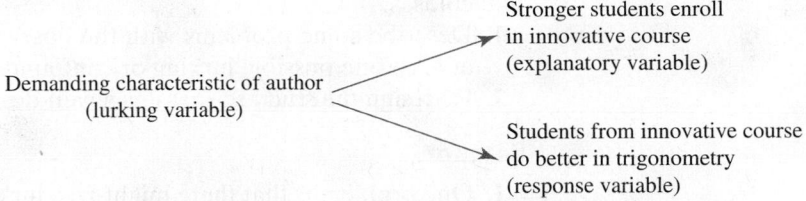

Demanding characteristic of author
(lurking variable)

Stronger students enroll
in innovative course
(explanatory variable)

Students from innovative course
do better in trigonometry
(response variable)

Figure 14 Demanding characteristic of author causes stronger students to enroll in the innovative course *and* for students from the innovative course to do better in trigonometry

> ▶ **Definition   Lurking variable**
>
> A **lurking variable** is a variable that causes both the explanatory and response variables to change during the study.

Lurking variables must be avoided. We can usually do so by using random assignment.

But even if we use random assignment, there is another type of variable that can cause confusion. Suppose for the innovative teaching study we randomly assign students to the innovative course and the control course. If a more effective professor teaches the innovative course, the students might do better in trigonometry because of the greater effectiveness of the professor, the innovative teaching method, or both. So, we would not be able to conclude that the innovative method caused the students to do better. The professor's greater effectiveness is called a *confounding variable* (see Fig. 15).

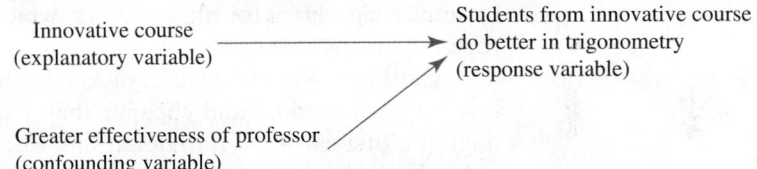

Innovative course
(explanatory variable)

Students from innovative course
do better in trigonometry
(response variable)

Greater effectiveness of professor
(confounding variable)

Figure 15 Students do better in trigonometry because of the innovative course, the greater effectiveness of the professor, or both

> ▶ **Definition   Confounding variable**
>
> A **confounding variable** is a variable other than the explanatory variable that causes or helps cause the response variable to change during the study.

We could address the issue of the professor's greater effectiveness by improving the experiment's design. For example, we could have the same professor teach both the innovative and control courses. But this means the two courses would have to be held at different times of day or on different days, which might be confounding variables. And the professor might teach the course more effectively for whichever course is taught second because she has had practice teaching the lesson. This illustrates a key concept: **A well-designed experiment requires careful planning so that there are as few confounding variables as possible.** Unfortunately, not all experiments can avoid all confounding variables. But there are ways to deal with that, which you can learn about in a higher-level statistics course.

Redesigning Observational Studies into Well-Designed Experiments

In Example 5, we will describe the problems with a study and how to improve it.

> ▶ **Example 5**   Redesign an Observational Study into a Well-Designed Experiment

A researcher wants to determine whether taking vitamin C helps people avoid getting the flu and the common cold. She randomly selects 100 people and asks them whether they take vitamin C and how often they had the flu or a cold in the past year. The researcher analyzes the responses and concludes that vitamin C helps people avoid the flu and colds.

1. Describe some problems with the observational study. Include in your description at least one possible lurking or confounding variable and identify which type it is.
2. Redesign the study so that it is a well-designed experiment.

Solution

1. One problem is that there might be a lurking variable. Individuals who take vitamin C might take good care of themselves in other ways, such as eating a healthy diet, getting enough sleep, getting a flu shot, and exercising. So, taking good care of oneself might both cause a person to take vitamin C and cause a person to avoid getting sick. Another problem is that there might be a confounding variable. The placebo effect of taking vitamin C might help individuals avoid getting sick. Yet another problem is that there might be response bias. Individuals might give false information because they have forgotten times when they were sick.

2. We can address the lurking variable (taking good care of oneself) by randomly assigning individuals to the treatment group and the control group. We can address the confounding variable (placebo effect) by giving the control group a placebo (sugar pill) and making the study single-blind. We could improve the experiment further by making it double-blind. We could address the response bias (forgetting) by having all individuals keep a daily journal of their health.

Even though we usually cannot claim causation when using observational studies, they are still useful. One reason is that for some studies, an experiment is not an option because it would be unethical. For example, consider studies about the harmful effects of smoking. To run a 15-year experiment, individuals in the treatment group would have to smoke cigarettes for all that time, which might give them lung cancer. That would certainly be unethical.

Another reason that observational studies are useful is that they are often easier, less time-consuming, and cheaper than experimental studies. An observational study might be first carried out to determine whether there is an association. If that turns out to be the case, researchers might then perform the experiment.

◢◣ Group Exploration ────────────────────

Designing an experiment

1. Treating your classmates outside of your group as a population, design an experiment that includes a treatment and control group. Your design should also include the following components or an explanation why a certain component would not be possible or appropriate: random sample, random assignment, placebo, and double-blind.

2. Draw a diagram that indicates the locations of the other groups of students in your classroom. Indicate which students are in which groups by referring to their names or using numbers to represent them.

3. Treating the other groups of students as clusters, perform cluster sampling. Describe this process, including the seed that you used.

4. Randomly assign students in the sample to the treatment and control groups. Describe this process, including the seed that you used.

Taking It One Step Further

5. Your professor may ask you to carry out the experiment. If so, describe the data you collected and any conclusions you can draw about the *sample*.

> **Tips for Success**   **Study in a Test Environment**
>
> Do you ever feel that you understand your homework assignments yet perform poorly on quizzes and tests? If so, you may not be studying enough to be ready to solve problems *in a test environment*. For example, although it is a good idea to refer to your lecture notes when you are stumped on a homework exercise, you must continue to solve similar exercises until you can solve them *without* referring to your lecture notes (unless your instructor uses open-notebook tests). The same idea applies to using the help buttons on MyMathLab, looking up answers in the back of the textbook, referring to examples in the textbook, getting help from someone, or any other form of support.
>
> Getting support to help you learn math is a great idea. Just make sure you spend the last part of your study time completing exercises without such support. One way to complete your study time would be to make up a practice quiz or test for you to do in a given amount of time.

Homework 2.3

For extra help ▶ MyMathLab®    Watch the videos in MyMathLab    Download the MyDashboard App

1. The _____ group is a collection of individuals who do not receive any treatment (or do not have the characteristics of any treatment group).

2. In a _____ study, neither the individuals nor the researcher in touch with the individuals know who is in the treatment group(s) and who is in the control group.

3. *True or False:* Random assignment means that we randomly select individuals from the population to be in the sample.

4. A(n) _____ variable is a variable that causes both the explanatory and response variables to change during the study.

5. In a double-blind study, researchers tested whether small doses of caffeine would improve highly trained male cyclists' times in a hot environment (95°F, 25% humidity). The individuals were 9 highly trained male cyclists who rode a stationary bicycle in a hot climate chamber on three separate occasions. The first session was a familiarization session. On the second session, some athletes took a caffeine pill and the rest took a placebo pill of Metamucil. On the third session, each athlete took the other pill. For each of the second and third sessions, the cyclists were randomly assigned to the treatment and control groups. The study concluded that a small dose of caffeine improves a highly trained male cyclist's time in a hot environment (Source: *Effect of Caffeine on Cycling Time-Trial Performance in the Heat, Nathan W. Pitchford et al.*).

   a. Describe the treatment and control groups.
   b. Is the study an experiment or an observational study? Explain.
   c. What does random assignment mean in this study? How could this be accomplished?
   d. Describe the sample and the population.

6. In a double-blind study, 50 older adults who had difficulty keeping their balance and/or were slow walkers were randomly assigned to one of three groups. The first group received 4 weeks of training designed to improve balance and walking speed. The second group performed the same physical training but also practiced mental tasks at the same time. They were instructed to give equal attention to the physical and mental tasks. The third group participated in the same activities as the second group but spent half the session focused on the physical tasks and half the session focused on the mental tasks. The researchers concluded that the training methods used for the second and third groups are more effective than the method used for the first group at improving walking speed when an older person is performing a mental task at the same time (Source: *Effects of Single-Task versus Dual-Task Training on Balance Performance in Older Adults: A Double-Blind, Randomized Controlled Trial, Patima Silsupadol et al.*).

   a. Describe the treatment and control groups.
   b. Is the study an experiment or an observational study? Explain.
   c. What does random assignment mean in this study? How could this be accomplished?
   d. Describe the sample and the population.

7. Refer to Exercise 5 for a description of a study about caffeine.

   a. Why was the placebo a Metamucil pill rather than a sugar pill?
   b. What does it mean that the study is double-blind? How could that be accomplished?
   c. What are the explanatory and response variables?
   d. What is the conclusion of the study? Does this mean there is causality or only an association between the two variables? Explain.
   e. The cyclists were instructed to avoid caffeine for at least 12 hours prior to each session. The *half-life* of caffeine in the bloodstream is 6 hours, which means that caffeine will be cut in half every 6 hours. Drinking 4 cups of coffee 12 hours before the trial would be equivalent to drinking how much coffee at the start of the session? Would that be a problem? Explain.

8. Refer to Exercise 6 on page 119 for a description of a study about walking speed.

   a. What are the explanatory and response variables?

   b. What is the conclusion of the study? Does this mean there is causality or only an association between the two variables? Explain.

   c. The researchers also concluded that only the first group increased their self-reported confidence when performing daily activities. On the basis that the researchers felt the activities (mental and physical) the second and third groups performed were much more difficult than the activities (only physical) the first group performed, what might be the reason only the first group's confidence increased?

9. In 2009, a law was passed in New Zealand banning the use of handheld cell phones while driving. In 2012, researchers wanted to determine if a smaller proportion of drivers use handheld cell phones now that the law was in effect. To find out, data collectors stood at the side of three streets in Wellington, New Zealand, and recorded the proportion of drivers who drove by while talking on their handheld cell phones. By referring to a 2006 study, the researchers obtained the proportion of drivers in Auckland, New Zealand, who talked on handheld cell phones while driving before the law. They concluded that the proportion of drivers who use handheld cell phones while driving has decreased since the law was passed (Source: Mobile Phone Use while Driving after a New Law: Observational Study, *Drury Christopher et al.*).

   a. Is such a study an experiment or an observational study? Explain.

   b. What are the explanatory and response variables?

   c. What is the conclusion of the study? Does this mean there is causality or only an association between the two variables? Explain.

   d. Were the 2006 and 2012 studies performed in the same city? Why might this be a problem?

   e. How many years apart were the two studies? Why might this be a problem?

10. In a 1999 study, researchers attempted to determine whether the death rate of older adults who have had a major bone fracture is higher than the death rate for older adults who have never had a major fracture. Researchers reviewed records of all people at least 60 years old who lived in Dubbo, Australia, between 1989 and 1994. The study concluded that the death rate of older adults who have had a major fracture is higher than the death rate for older adults who have never had a major fracture (Source: Mortality after All Major Types of Osteoporotic Fracture in Men and Women: An Observational Study, *J. R. Center*).

   a. The title of the journal article says that the study is observational. Why does that make sense?

   b. Explain why it would be unethical to perform an experiment about this topic.

   c. Describe the sample and the population.

   d. What are the explanatory and response variables?

   e. What is the conclusion of the study? Does this mean there is causality or only an association between the two variables? Explain.

11. In a 2011 study, researchers explored whether there is an association between presidential TV advertising and voters' opinions of the candidates. The researchers analyzed data about the Bush-Gore 2000 presidential election, including local ratings of TV shows in which presidential advertisements were run and interviews conducted in 2000 by National Annenberg Election Survey. The study concluded that a voter who had been exposed to more advertisements for a certain candidate tended to have a more positive opinion about that candidate (Source: Identifying the Persuasive Effects of Presidential Advertising, *Gregory A. Huber et al.*).

   a. Is the study an experiment or an observational study? Explain.

   b. What are the explanatory and response variables?

   c. In states where the presidential race was close, the candidates had a lot of "on-the-ground" interaction with voters, hoping to improve voters' opinions of them. Even if the race was not close in neighboring states, the states often still received the local TV presidential advertising from the state with the close race. Why did the researchers analyze only voters in such neighboring states where the race was *not* close?

   d. What is the conclusion of the study? Does this mean there is causality or only an association between the two variables? Explain.

12. Researchers wanted to see whether children who take the drug levodroprophizine recover better from cough than children who take other cough syrups. The individuals were 161 children who visited one of four doctors from February 1, 2010, to April 30, 2010, for an acute cough. The doctors prescribed levodroprophizine to 101 children and other cough syrups to the other 60 children. Parents completed a questionnaire about the frequency and intensity of their children's cough before taking one of the drugs. Six days later, the parents completed the questionnaire again. The researchers concluded that the children who took levodroprophizine recovered better from cough than the children who took other cough syrups (Source: An Observational Study on Cough in Children: Epidemiology, Impact on Quality of Sleep and Treatment Outcome, *Francesco De Blasio et al.*).

   a. The title of the journal article says the study is observational. Why does that make sense?

   b. Explain why it would be unethical to use a placebo in this study.

   c. Describe the sample and the population.

   d. What is the conclusion of the study? Does this mean there is causality or only an association between the two variables? Explain.

   e. In the journal article, two of the researchers report that they are employees of the company that manufactures levodroprophizine. Why is it important that this is stated in the article?

13. Approximately 85% of patients diagnosed with multiple sclerosis have attacks—often called relapses—of neurologic problems. The relapses are followed by periods during which symptoms improve partially or fully. This form of the disease is called *relapsing-remitting multiple sclerosis (RRMS)*. In a double-blind study, researchers tested whether the drug natalizumab reduces the annual relapse rate for RRMS patients.

A total of 942 RRMS patients ages 18 to 50 years, inclusive, were randomly assigned to receive the drug or a placebo by intravenous infusion for 120 weeks. The study concluded that the drug reduces the relapse rate (Source: A Randomized, Placebo-Controlled Trial of Natalizumab for Relapsing Multiple Sclerosis, *Chris H. Polman, MD et al.*).

a. Describe the treatment and control groups.

b. Is the study an experiment or an observational study? Explain.

c. What does random assignment mean in this study? How could this be accomplished?

d. Describe the sample and the population.

14. Would monetary rewards motivate you to pass more classes? In a two-semester study, researchers randomly assigned 1019 low-income community college students who were parents into two groups. Each member of the first group received a $20 gift card. Each member of the second group received the gift card plus monetary rewards for enrolling in classes, staying enrolled in classes, and maintaining at least a C average. Students in the second group could be awarded as much as $2000. The researchers concluded that monetary rewards increase the number of credits earned by low-income community college students who are parents (Source: Paying for Performance: The Education Impacts of a Community College Scholarship Program for Low-Income Adults, *Lisa Barrow et al.*).

a. Describe the treatment and control groups.

b. Is the study an experiment or an observational study? Explain.

c. What does random assignment mean in this study? How could this be accomplished?

d. Describe the sample and the population.

15. Refer to Exercise 13 on page 120 for a description of a study about the drug natalizumab.

a. What does it mean that the study is double-blind? How could it be accomplished?

b. What are the explanatory and response variables?

c. What is the conclusion of the study? Does this mean there is causality or only an association between the two variables? Explain.

d. In the journal article, many of the researchers reported that they have received consulting fees, lecture fees, and grant support from the company that manufactures natalizumab. Why is it important that this is stated in the article?

16. Refer to Exercise 14 for a description of a study about offering college students monetary rewards.

a. Explain why it would be impossible to use a placebo in this study.

b. Explain why it would be impossible to redesign this study to be double-blind.

c. What are the explanatory and response variables?

d. What is the conclusion of the study? Does this mean there is causality or only an association between the two variables? Explain.

e. Three students in the control group were mistakenly given monetary rewards. Why is it important this was stated in the article?

17. A 2001 study reviewed data from a previous study of 7316 Danish adults who completed a questionnaire about their diet. They determined that there were two dietary patterns: a high-fiber diet (frequent consumption of wholemeal bread, fruits, and vegetables) and a Western diet (frequent consumption of meat products, potatoes, white bread, butter, and lard). The 2001 study found that those on a high-fiber diet have lower death rates than those on the Western diet (Source: Dietary Patterns and Mortality in Danish Men and Women: A Prospective Observational Study, *M. Osler*).

a. The title of the journal article says the study is observational. Why does that make sense?

b. Explain why it would be unethical to perform an experiment about this topic.

c. Describe the sample and the population.

d. What are the explanatory and response variables?

e. What is the conclusion of the study? Does this mean there is causality or only an association between the two variables? Explain.

18. According to Illinois law, a crosswalk is the part of the roadway at an intersection that is in line with at least one sidewalk. A crosswalk need not have markings on the road. In 2010, a law was passed in Illinois that requires motorists to come to a complete stop for pedestrians in crosswalks. Researchers wanted to determine whether motorists are more likely to follow the "Must Stop" law for marked crosswalks than for unmarked ones. A trial consisted of a researcher standing with one foot in the street, which the study claims is a clear indication that the researcher wanted to cross the crosswalk. Data was collected at 52 sites—some marked crosswalks and some unmarked—throughout Chicago. At each site, 4 trials were completed. The study found that motorists are more likely to follow the Must Stop law at marked crosswalks than at unmarked ones. (Source: *Active Transportation Alliance*).

a. Is the study an experiment or an observational study? Explain.

b. Describe the sample and the population.

c. What are the explanatory and response variables?

d. What is the conclusion of the study? Does it have to do with causality or only association? Explain.

e. Can we assume the conclusion is also true for motorists in Prairie City, Illinois, which has population 379 and is surrounded by farmland? Explain.

19. The 12 smartphones in Table 26 have been randomly selected for the sample of an experiment. Use the seed 994 to randomly select 6 of the phones for the treatment group. The other 6 phones will be in the control group. Is this process an example of random assignment? Explain.

Table 26 Smartphones Randomly Selected for a Sample

1. ZTE Grand X	7. Kyocera DuraForce
2. Microsoft Lumia 535	8. HTC Desire 510
3. CAT B15Q	9. Motorola Droid Turbo
4. Nokia Lumia 735	10. Google Nexus 6
5. Huawei Ascend Y550	11. LG G3 Vigor
6. Posh Memo S580	12. Meizu MX4

20. The 8 ultrabooks in Table 27 have been randomly selected for the sample of an experiment. Use the seed 274 to randomly select 4 of the ultrabooks for the treatment group. The other 4 ultrabooks will be in the control group. Is this process an example of random assignment? Explain.

Table 27 Ultrabooks Randomly Selected for a Sample

1. Toshiba Portégé Z10t	**5.** Dell Latitude 7440
2. Acer Aspire S7	**6.** Dell XPS 13
3. Lenovo ThinkPad X240	**7.** HP Spectre 13 Ultrabook
4. Samsung ATIV Book 9	**8.** Asus VivoBook

21. The 10 cars in Table 28 have been randomly selected for the sample of an experiment. Use the seed 469 to randomly select 5 of the cars for the treatment group. The other 5 cars will be in the control group. Is this process an example of random assignment? Explain.

Table 28 Cars Randomly Selected for a Sample

1. Hyundai Elantra	**6.** Audi 55
2. Chrysler 300	**7.** Mercedes-Benz ML350
3. KIA Sportage	**8.** Volkswagen Jetta
4. Mini Cooper S Countryman	**9.** Lexus LS 460 L
5. Chevrolet Camaro	**10.** Chevrolet Silverado 15

22. The 12 public, 4-year colleges in Table 29 have been randomly selected for the sample of an experiment. Use the seed 528 to randomly select 6 of the colleges for the treatment group. The other 6 colleges will be in the control group. Is this process an example of random assignment? Explain.

Table 29 Colleges Randomly Selected for a Sample

1. Temple University	**7.** Angelo State University
2. Oakland University	**8.** Palm Beach State College
3. Lander University	**9.** Fitchburg State University
4. Langston University	**10.** SUNY College at Oneonta
5. Boise State College	**11.** Virginia Military Institute
6. Eastern Illinois University	**12.** Kean University

23. A weight-loss company wants to test whether its weight-loss program works. They identify 30 new members, who agree to attend weekly meetings and weigh in at the meetings. They also pay 30 nonmembers to weigh in once a week with the understanding that they will avoid the company's weight-loss meetings. After 3 months, a greater proportion of the 30 members lose weight than the nonmembers. The company advertises that its program works.

a. Describe the problems with the weight-loss study. Include in your description at least one possible lurking or confounding variable and identify which type it is.

b. Redesign the study so that it is a well-designed experiment.

24. A student wants to test whether exercise helps people quit smoking. She surveys adults, asking whether they ever smoked, whether they ever quit smoking, and whether they exercise. The student concludes that exercise helps people quit smoking.

a. Describe the problems with the student's study. Include in your description at least one possible lurking or confounding variable and identify which type it is.

b. Redesign the study so that it is a well-designed experiment.

25. A student wants to compare the effectiveness of the laundry detergents Tide® and Wisk® Deep Clean. He washes some stained towels with Tide in a Kenmore® washing machine and washes other stained towels with Wisk Deep Clean in a Samsung® washing machine. After inspecting the washed towels, the student concludes that Tide is more effective than Wisk Deep Clean.

a. Describe the problems with the student's study. Include in your description at least one possible lurking or confounding variable and identify which type it is.

b. Redesign the study so that it is a well-designed experiment.

26. A high school humanities teacher wants to determine whether monetary awards will motivate students to create better projects. The teacher assigns a project about community spirit. He says that the student with the best project will win a $25 prize awarded by the mayor and be featured in a front-page article in the local newspaper. After grading the completed projects, the teacher concludes that monetary awards motivate students to create better projects.

a. Describe the problems with the teacher's study. Include in your description at least one possible lurking or confounding variable and identify which type it is.

b. Redesign the study so that it is a well-designed experiment.

27. A weight-lifting trainer wants to determine if a new protein shake helps weight lifters bench-press more weight. To find out, she interviews weight lifters at a local fitness center, asking them how much they can bench-press and whether they drink the shake. She finds out that the lifters who drink the shake can bench-press more, so she concludes that the shake works.

a. Describe the problems with the trainer's study. Include in your description at least one possible lurking or confounding variable and identify which type it is.

b. Redesign the study so that it is a well-designed experiment.

28. *Popular Mechanic*™ wants to determine if a new fuel additive improves gas mileage. The magazine hosts an online survey, asking car owners whether they use the fuel additive and what their car's gas mileage is. Because car owners who use the fuel additive reported better gas mileages, the magazine concludes that the fuel additive works.

a. Describe the problems with the magazine's study. Include in your description at least one possible lurking or confounding variable and identify which type it is.

b. Redesign the study so that it is a well-designed experiment.

29. A pharmaceutical company has designed an experimental drug in pill form that is supposed to treat severe acne within 6 months. Design an experiment to test whether the drug works.

30. A pharmaceutical company has designed an experimental drug in pill form that is supposed to help adults who suffer from insomnia sleep better. Design an experiment to test whether the drug works.

31. A track coach wants to know whether having his long-distance runners train with a 10-pound weight belt for one month of daily workouts will improve their 2-mile running times. Design an experiment to help the coach.

32. If students wear earphones and play recordings of their professor's lectures while they sleep, will they perform better on tests? Design an experiment to test this.

Concepts

33. Describe how a study's sample, population, treatment group, and control group are related. Also draw a figure to illustrate the relationships.

34. A student says that stratified sampling and random assignment are the same thing because both involve simple random sampling for parts of the sample. What would you tell the student?

35. Compare the meaning of *association* and *causality*.

36. What is the key difference in the designs of an experiment and an observational study? What does that difference have to do with the fact that we can determine causality between explanatory and response variables with experiments but not with observational studies?

37. Compare the meaning of *placebo* and *placebo effect*. Give an example of each that is not in the textbook.

38. Describe a lurking variable. Give a possible example that is not in the textbook.

39. What is a double-blind study? What is the benefit of such a study?

40. A student says any study in which the individuals are objects is either a single-blind study or a double-blind study. Is the student correct?

41. A student says that if a study concludes that there is an association between the explanatory and response variables, then there must be causation because *explanatory variable* means the variable *explains* things. What would you tell the student?

Hands-On Projects

Survey about Proportions Project

In this project, you will perform a survey about characteristics of people that can be described using proportions. Recall that there are five steps of performing statistics. What follows are those five steps and suggestions on how to take those steps.

Raise a precise question about one or more variables

1. What would you like to know about others that can be described using proportions? To gain a thorough understanding, you may need to find out about several related characteristics of others. If so, what are those characteristics?

2. What is the population?

3. Construct the exact wording of your question(s). Explain how you have tried to avoid response bias.

4. Do you think individuals will be comfortable responding honestly to your survey? If no, how could you address this?

Create a plan to answer the question

5. How many individuals do you plan on surveying?

6. Which of the four sampling methods will you use? Why is it the best choice, given your limitations of time and money?

7. Do you think that sampling bias will be an issue for your study? If so, how can you improve your design? How would you further address sampling bias if you had more time and money?

8. Do you think that nonresponse bias will be an issue for your study? If so, how can you improve your design?

How would you further address nonresponse bias if you had more time and money?

9. How will the individuals' responses be recorded?

10. Describe where and when you will collect the data.

11. If your sampling method requires any calculations to determine whom to poll, perform them. Describe which technology and which seeds you will use to generate random numbers, if appropriate.

Collect the data

12. Collect the data and include them in your report.

Analyze the data

13. Summarize the data. Use a table, if appropriate.

14. Calculate proportions of the individuals' characteristics you are interested in.

Draw a conclusion about the question

15. Using proportions and complete sentences, describe what you have learned about the *sample*.

16. Will your results in Problem 15 also be true for the population? Explain.

Online Report Project

In this project, you will find an experiment's online report and analyze it. Recall that an experiment must have a treatment group, control group, and random assignment.

Searching for a Report

To find an experiment's report, perform an online search using keywords such as "control group," "randomly assigned," "double-blind," and "placebo." Use several of these words or even all of them. You can further steer the search by using yet one more keyword about some general topic. Experiments with placebos are often used to test pharmaceutical drugs, so you could pick some health-related topic, such as cancer or diabetes. But you can try other topics, such as music or sports, and see what comes up. Your project is not required to be double-blind or have a placebo.

Experiments' reports usually begin with an abstract, which is a concise description of the experiment. Because abstracts are packed with so much information, they can be difficult to understand, so reading the full report can clarify the components of the experiment. If the full report is not shown, look for a button that might allow you to view or download it. If you cannot access the full report, you might choose to search for another experiment's report, but there is often enough information in the abstract to get a good sense of the experiment after reading and *rereading* it many times.

Analyzing the Experiment

Respond to the following questions. **If the report does not give details to fully answer a question about a component of the experiment, then describe what the component might be and how the researchers might have accomplished it.**

1. State the title and the first-listed author of the report.
2. Describe the treatment and control groups.
3. Explain why the study is an experiment (and not an observational study).
4. What does random assignment mean in this study? How was it accomplished?
5. Describe the sample and the population.
6. Was a placebo used? If yes, describe as much as the article reveals about it.

7. Is the study double-blind? If yes, how was that accomplished?
8. What are the explanatory and response variables?
9. What is the conclusion of the study? Does this mean there is causality or only an association between the two variables? Explain.
10. How could a person or a company benefit from the conclusion of the experiment? (Even if the conclusion is that the explanatory variable and the response variable have no association, that can often still be useful.)

Taking It One Step Further

11. The workplaces of the researchers are usually stated near the beginning of the report or in a footnote. To find them, you might have to click on a button called "author information" or "author affiliations." List the researchers' workplaces.
12. Does it seem likely that any of the researchers might have something to gain by the report's conclusion? For example, if the experiment tested the effectiveness of a drug, check whether any of the researchers work at the pharmaceutical company that manufactures the drug. You can probably find this out by searching for the drug and the company in the same search.
13. Near the end of most reports is a section titled "Discussion." If you have access to the full report, read this section. What did you find interesting or unusual? What did you learn? If the researchers described how the experiment is related to other experiments, describe this. If the researchers described what could be studied next, describe this.

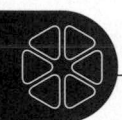 # Chapter Summary

Key Points of Chapter 2

Section 2.1　Simple Random Sampling

Individual	**Individuals** are the people or objects we want to learn about.
Variable	In statistics, a **variable** is a characteristic of the individuals to be measured or observed.
Observation	The data that we observe for a variable are called **observations**.
Statistics	**Statistics** is the practice of the following five steps: 1. **Raise a precise question about one or more variables.** 2. **Create a plan to answer the question.** 3. **Collect the data.** 4. **Analyze the data.** 5. **Draw a conclusion about the question.**

Section 2.1 Simple Random Sampling (*Continued*)

Population	A **population** is an *entire* group of individuals that we want to learn about.
Sample	A **sample** is the part of a population from which data are collected.
Descriptive statistics	**Descriptive statistics** is the practice of using tables, graphs, and calculations about a sample to draw conclusions about only the sample.
Inferential statistics	**Inferential statistics** is the practice of using information from a sample to draw conclusions about the *entire* population.
Inference	When we perform inferential statistics, we call the conclusions **inferences**.
Simple random sample	A process of selecting a sample of size *n* is **simple random sampling** if every sample of size *n* has the same chance of being chosen. A sample selected by such a process is called a **simple random sample**.
Frame	A **frame** is a numbered list of all the individuals in the population.
Sampling error	**Sampling error** is the error involved in using a sample to estimate information about a population due to randomness in the sample.
Bias	A sampling method that consistently underemphasizes or overemphasizes some characteristic(s) of the population is said to be **biased**.
Sampling bias	**Sampling bias** occurs if the sampling technique favors one group of individuals over another.
Nonresponse bias	**Nonresponse bias** happens if individuals refuse to be part of the study or if the researcher cannot track down individuals identified to be in the sample.
Response bias	**Response bias** occurs if surveyed people's answers do not match with what they really think.
Guidelines for constructing survey questions	When constructing a survey question, • Do not include judgmental words. • Avoid asking a yes/no question. • If the question includes two or more choices, switch the order of the choices for different respondents. • Address just one issue.
Nonsampling error	**Nonsampling error** is error from using biased sampling, recording data incorrectly, and analyzing data incorrectly.

Section 2.2 Systematic, Stratified, and Cluster Sampling

Systematic sampling	To perform **systematic sampling**, we randomly select an individual out of the first *k* individuals and also select every *k*th individual after the first selected individual.
Rounding when calculating *k* for systematic sampling	We should always round down when calculating *k* for systematic sampling.
Stratified sampling	To perform **stratified sampling**, a population is divided into subgroups called **strata** (singular: **stratum**) and simple random sampling is performed on each stratum. In each stratum, the individuals share some characteristic. The ratios of the number of individuals selected from each stratum are equal to the corresponding ratios of the stratum sizes.
Cluster sampling	To perform **cluster sampling**, a population is divided into subgroups called **clusters**. Then simple random sampling is used to select some of the clusters. All of the individuals in those selected clusters form the sample.

Section 2.2   Systematic, Stratified, and Cluster Sampling (*Continued*)

Requirements and benefits of sampling methods	Sampling Method	Requirement	Benefits
	Simple Random	A frame of all individuals.	Works fine for telephone and e-mail surveys in which there's little risk of excluding anyone.
	Systematic	Selecting every *k*th individual must not match with some pattern in the population.	No frame is required.
	Stratified	In each strata, individuals are similar. A frame for individuals in each strata.	Individuals in small strata are not excluded from the sample. Can save time, money, and effort.
	Cluster	A frame of clusters.	No frame of individuals is required. Can save time, money, and effort.

Convenience sampling	To perform **convenience sampling**, we gather data that are easy to collect and do not bother with collecting them randomly. Convenience sampling should never be performed.
Voluntary response sampling	To perform **voluntary response sampling**, we let individuals choose to be in the sample. Voluntary response sampling should never be performed.

Section 2.3   Observational Studies and Experiments

Treatment group and control group	Each **treatment group** in a study is a collection of individuals who receive a certain treatment (or have a certain characteristic of interest). The **control group** is a collection of individuals who do not receive any treatment (or do not have the characteristics of any treatment group).
Similar treatment and control group	When designing a study, it is important that each treatment group be as similar to the control group as possible, except for the characteristic of interest.
Placebo effect	The **placebo effect** occurs when the characteristic of interest changes in individuals due to the individuals believing the characteristic should change.
Placebo	A **placebo** is a fake drug or procedure administered to the control group.
Single-blind study	In a **single-blind study**, individuals do not know whether they are in the treatment group(s) or the control group.
Double-blind study	In a **double-blind study**, neither the individuals nor the researcher in touch with the individuals know who is in the treatment group(s) and who is in the control group.
Random assignment	**Random assignment** is the process of assigning individuals to the treatment group(s) and the control group randomly.
Components of a well-designed study	In a well-designed study, • There should be a control group and at least one treatment group. • Individuals should be randomly assigned to the control and treatment group(s). • The sample size should be large enough. • A placebo should be used when appropriate. • The study should be double-blind when possible. If this is impossible, then the study should be single-blind, if possible.
Experiment and observational study	In an **experiment**, researchers determine which individuals are in the treatment group(s) and the control group, often by using random assignment. In an **observational study**, researchers do not determine which individuals are in the treatment group(s) and the control group.
Explanatory and response variables	In a study about whether a variable *x* explains (affects) a variable *y*, • We call *x* the **explanatory variable** (or **independent variable**). • We call *y* the **response variable** (or **dependent variable**).

Section 2.3 Observational Studies and Experiments (*Continued*)

Association	There is an **association** between the explanatory and response variables if the response variable changes as the explanatory variable changes.
Causality	If the change in the explanatory variable *causes* a change in the response variable, we say there is **causality**.
Determining causality	• A well-designed experiment can determine whether there is causality between the explanatory and response variables. • Most observational studies *cannot* determine whether there is causality between the explanatory and response variables. They can only determine whether there is an association between the two variables.
Lurking variable	A **lurking variable** is a variable that causes both the explanatory and response variables to change during the study.
Confounding variable	A **confounding variable** is a variable other than the explanatory variable that causes or helps cause the response variable to change during the study.

Chapter 2 Review Exercises

1. Some Middle East countries are described in Table 30.

Table 30 Some Middle East Countries

Country	Government	Population (millions)	2012 Military Expenditure (billion dollars)	Oil Production (billion barrels per day)
Bahrain	Monarchy	1.3	0.92	0.05
Iraq	Republic	31.9	5.69	2.99
Israel	Republic	7.7	15.54	0.20
Kuwait	Monarchy	2.7	5.95	2.69
Saudi Arabia	Monarchy	26.9	54.22	9.90

Source: *Stockholm International Peace Research Institute*

   a. Identify the individuals.
   b. Identify the variables.
   c. Identify the data for each variable.
   d. Compute the ratio of 2012 military expenditure to oil production (in dollars per barrel) for each country. Round your results to the third decimal place.
   e. Which country has the greatest ratio of 2012 military expenditure to oil production?

2. In a telephone survey, 54% of 500 American adults said they experience a lot of happiness and enjoyment. The study then made an inference about all American adults (Source: *The Gallup Organization*).

   a. Define a variable for the study.
   b. Identify the sample.
   c. Identify the population.

3. The favorite movie genres of students in one of the author's statistics classes are shown in Table 31.

Table 31 Students' Favorite Movie Genres

Student	Genre	Student	Genre
1. Antoine	Comedy	**7.** Dante	Drama
2. Nicholas	Horror	**8.** Jose	Comedy
3. Sandra	Comedy	**9.** Ruben	Drama
4. Mario	Comedy	**10.** Jacob	Thriller
5. John	Drama	**11.** Alyssa	Drama
6. Chelsea	Thriller	**12.** Sabrina	Comedy

Source: *J. Lehmann*

   a. Use technology with seed 43 to randomly select 6 students without replacement.
   b. For the simple random sample you found in part (a), find the proportion of students who prefer comedies. If your result is used to describe the sample, is this part of descriptive or inferential statistics? Explain.
   c. For all 12 students shown in Table 31, find the proportion of students who prefer comedies.
   d. Does your result from part (b) equal your result from part (c)?
   e. If you were to select many simple random samples of size 6, would the proportion of students who prefer comedies for each sample equal the proportion of all 12 students who prefer comedies? Explain.
   f. In general, if two researchers perform the same study but with different simple random samples of the same size, will their inferences necessarily be the same? Explain.

4. Reword the following question so that there is less chance of bias: "Do you help your professor out by asking lots of questions or do you keep silent?"

For Exercises 5 and 6, identify the type(s) of bias in the sampling method. Explain.

5. A militia group hosts an online poll at their website, asking readers whether balancing the national budget is a top priority. The militia group uses the data to predict the percentage of all Americans who think balancing the budget is a top priority.

6. A pollster surveys adults in the financial district of Chicago, asking them their annual salary. Of 100 adults approached, 55 refuse to take part in the survey. The pollster uses the data to predict the proportion of Chicagoans who earn over $100 thousand per year.

7. Recall that there are three types of bias: sampling, response, and nonresponse. For each type, give an example of a study that might include that type of bias and how the study could be improved.

For Exercises 8–12, identify whether the sampling method is simple random, systematic, stratified, cluster, or convenience.

8. A researcher creates a frame of all the blocks in Chicago. Then he randomly selects 50 of the blocks and surveys all the adult residents of those blocks.

9. Human Resources at General Motors creates a frame of all its U.S. employees and randomly selects some of the employees.

10. A pollster surveys people who walk by a busy intersection in New York City, asking whether state taxes should be raised to fund improvements in schools.

11. In a study based on telephone interviews about whether racism is the most important issue facing the United States today, 402 telephone numbers were randomly selected from landline numbers and 402 telephone numbers were randomly selected from cell phone numbers (Source: *The Gallup Organization*).

12. A Macy's® manager randomly selects and surveys the 5th person who leaves the store and also surveys every 8th person who leaves the store after that.

13. Human Resources at Sony® wants to perform a systematic sampling of its 105,000 employees (Source: *Sony*). Using a frame of all its employees, they plan to survey 800 of them.

   a. If every *k*th employee is surveyed, find *k* so that 800 employees are surveyed.

   b. To find which employee should be first surveyed, use 297 as a seed to find a random number between 1 and *k*.

   c. List the first five employees that should be surveyed.

14. A researcher wants to select a sample of 80 employees at Boeing by performing stratified sampling. The employments are shown in Table 32. How many employees from each group should be included in the study?

Table 32 Enrollments by Group at Boeing

Group	Employment (thousands)
Commercial Airplanes	82
Defense, Space, and Security	57
Corporate	30
Engineering, Operations, and Technology	19
Shared Services Group	8
Other	3

Source: *Boeing*

15. A 5-person energy panel is to consist of Connecticut Senate members, who are shown in Table 33. Select a stratified sample of 3 Democrats and 2 Republicans, using 337 as the seed for the Democrat strata and 629 as the seed for the Republican strata.

Table 33 Connecticut Senate Members

Democrat		Republican	
1. Fonfara	12. Meyer	1. Kissel	8. Witkos
2. Coleman	13. Bartolomeo	2. Markley	9. Kelly
3. LeBeau	14. Slossberg	3. McLachian	10. Boucher
4. Cassano	15. Hartley	4. McKinney	11. Chapin
5. Bye	16. Crisco	5. Welch	12. Kane
6. Gerratana	17. Maynard	6. Linares	13. Fasano
7. Doyle	18. Osten	7. Guglielmo	14. Frantz
8. Holder-Winfield	19. Stillman		
9. Looney	20. Musto		
10. Ayala	21. Duff		
11. Leone	22. Williams		

Source: *Connecticut Senate*

16. Atlanta City Hall wants to survey some of its residents in person, asking whether the city's public transportation is satisfactory. Which of the four sampling methods discussed in this chapter would require the least money and effort? Explain. Describe how to perform the sampling.

17. In a double-blind study, researchers tested whether the antidepressant drug Lu AA21004 lowered depression in adults with major depressive disorder (MDD). A total of 560 MDD adults were randomly assigned to 3 treatment groups of different Lu AA21004 dosages or a placebo group. Those in the treatment groups were given the drug in pill form for 8 weeks. The individuals' levels of depression were measured using a questionnaire called the Hamilton Rating Scale for Depression (HRSD). The study found that the drug successfully lowers MDD adults' HRSD scores (Source: A Randomized, Double-Blind, Placebo-Controlled 8-Week Trial of the Efficacy and Tolerability of Multiple Doses of Lu AA21004 in Adults with Major Depressive Disorder, *Neven Henigsber, MD, et al.*).

   a. Describe the treatment and control groups.

   b. Is the study an experiment or an observational study? Explain.

   c. What does random assignment mean in this study? How could this be accomplished?

   d. Describe the sample and the population.

18. Refer to Exercise 17 for a description of a study about the drug Lu AA21004.

   a. The placebo is not described in the journal article, but what might it be?

   b. What does it mean that the study is double-blind? How could it be accomplished?

   c. What are the explanatory and response variables?

   d. What is the conclusion of the study? Does this mean there is causality or only an association between the two variables? Explain.

   e. Some experts criticize the HRSD because it places more emphasis on insomnia than on suicidal thoughts and gestures (Source: *Bagby, R. M. et al. (2004). "The Hamilton Depression Rating Scale: Has the Gold Standard Become a Lead Weight?," American Journal of Psychiatry 161(12): 2163–77*). Why is this important to consider?

19. Researchers wanted to determine whether mothers with eating disorders expressed more negative emotions toward their first-born infants during mealtimes than mothers without eating disorders. Researchers observed how both types of mothers interacted with their first-born infants during mealtimes. The researchers found that mothers with eating disorders express more negative emotions toward their first-born infants during mealtimes than mothers without eating disorders (Source: *An Observational Study of Mothers with Eating Disorders and Their Infants, A. Stein, et al.*).

   a. The title of the journal article says that the study is observational. Why does that make sense?

   b. Explain why it would be impossible to use random assignment in this study.

   c. Describe the sample and the population.

   d. What are the explanatory and response variables?

   e. What is the conclusion of the study? Does it have to do with causality or only association? Explain.

20. The 10 private, 4-year colleges in Table 34 have been randomly selected for the sample of an experiment. Use the seed 929 to randomly select 5 of the colleges for the treatment group. The other 5 colleges will be in the control group. Is this process an example of random assignment? Explain.

Table 34 4-Year, Private Colleges Randomly Selected for a Sample

1. Mills College	6. Nichols College
2. Campbell University	7. Elon University
3. Rider University	8. Villanova University
4. Wellesley College	9. University of Mount Union
5. Urbana University	10. Columbia College

21. The math center coordinator at a college wants to show that the center helps students get better grades in math classes. Two weeks after the semester is over, she researches students' grades and finds that a greater proportion of students who attended the math center passed their math classes that semester than those that did not attend the center. The person concludes that the center helps students get better grades in math classes.

   a. Describe the problems with the coordinator's study. Include in your description at least one possible lurking or confounding variable.

   b. Redesign the study so that it is a well-designed experiment.

22. A pharmaceutical company has designed an experimental drug in pill form that is supposed to treat baldness within 8 months. Design an experiment to test whether the drug works.

Chapter 2 Test

1. The numbers of workers and the numbers of workers in unions are shown in Table 35 for various states.

Table 35 Numbers of Workers and Numbers of Workers in Unions

State	Region	Number of Workers (thousands)	Number of Workers in Unions (thousands)
Delaware	East	370	38
Hawaii	West	549	121
Mississippi	South	1040	38
Texas	South	10,877	518
Wisconsin	Midwest	2569	317

Source: *Bureau of Labor Statistics*

   a. Identify the individuals.

   b. Identify the variables.

   c. Identify the data for each variable.

   d. Compute the percentage of workers in unions for each state shown in Table 35. Round your results to the first decimal place.

   e. Which state in Table 35 has the largest percentage of workers in unions?

2. In a sample consisting of 2011 American adults, 18% of the adults intend to buy wearable technology in the next 12 months. The researchers then made an inference about all American adults (Source: *Ipsos*).

   a. Define a variable for the study.

   b. Identify the sample.

   c. Identify the population.

3. A simple random sample of customers who purchased products at Amazon.com® in the past 3 months were invited by e-mail to respond to the question, "Was our website easy to use, and did you find everything you needed?" About 8% of those contacted responded. Of all customers who purchased products at Amazon.com in the past 3 months, the company predicts the percentage who found the website easy to use and found everything they needed. Identify the type(s) of bias the study has. Explain.

4. Use technology with seed 564 to randomly select 4 students without replacement from the frame:

     1. Isabel, **2.** Liliana, **3.** Michael, **4.** Brianna,
     5. Jamie, **6.** Lisa, **7.** Jared, **8.** Lance, **9.** Dan

5. An apple farmer divides his orchard into 50 subsections, randomly selects 8 subsections, and measures the apple yield in those 8 subsections. Is the sampling method simple random, systematic, stratified, cluster, or convenience?

6. A Gap manager wants to survey customers as they leave the store on a certain day. There are about 500 customers each day. The manager would like to get feedback from at least 60 customers by performing systematic sampling. From past experience, the manager knows that if 80 customers are approached, about 60 of them will participate in the survey.

   a. If every kth customer is approached, find k so that 80 customers are approached.

   b. To find which customer should be first approached, use 212 as a seed to find a random number between 1 and k.

   c. List the first five customers that should be approached.

7. Georgia Tech wants to survey its students about whether they think Bobby Dodd Stadium should be renovated. Because female undergraduates, male undergraduates, female graduate students, and male graduate students may have different opinions, the university plans to use these four types of students as strata. The strata sizes are shown in Table 36.

Table 36 Strata Sizes

Gender	Undergraduate	Graduate Student	Total
Female	4833	1792	6625
Male	9725	5121	14,846
Total	14,558	6913	21,471

Source: *Georgia Tech*

If the university wants to survey 500 students, how many students in each strata should be surveyed?

8. By measuring the glycated hemoglobin level in the bloodstream, a doctor can determine what a patient's average blood sugar level has been in past months. The glycated hemoglobin level of untreated type 2 diabetes patients is too high. Between March 2008 and March 2009, researchers tested whether the drug ipragliflozin decreased glycated hemoglobin levels in Japanese patients with type 2 diabetes at 39 sites in Japan. The drug is in pill form. In the double-blind study, the 361 patients were randomly assigned to 4 treatment groups of different ipragliflozin dosages or a placebo group. The study concluded that the drug successfully lowers glycated hemoglobin levels in Japanese patients with type 2 diabetes (Source: Randomized, Placebo-Controlled, Double-Blind, Glycemic Control Trial of Novel Sodium-Dependent Glucose Cotransporter 2 Inhibitor Ipragliflozin in Japanese Patients with Type 2 Diabetes Mellitus, *Atsunori Kashiwagi et al.*).

a. Describe the treatment and control groups.

b. Is the study an experiment or an observational study? Explain.

c. What does random assignment mean in this study? How could this be accomplished?

d. Describe the sample and the population.

9. Refer to Exercise 8 for a description of a study about the drug ipragliflozin.

a. The placebo is not described in the journal article, but what might it be?

b. What does it mean that the study is double-blind? How could it be accomplished?

c. What are the explanatory and response variables?

d. What is the conclusion of the study? Does this mean there is causality or only an association between the two variables? Explain.

e. In the journal article, three of the researchers report that they are employees of the company that manufactures ipragliflozin. Why is it important that this is stated in the article?

10. A researcher wants to determine if college basketball players score more points if they run every day. To find out, she surveys players on one college team and finds out that a greater proportion of those who run every day score over 10 points than those that don't run every day. The researcher advises coaches across the country to have their players run every day.

a. Describe the problems with the researcher's study. Include in your description at least one possible lurking or confounding variable.

b. Redesign the study so that it is a well-designed experiment.

11. The owner of car dealership wants to know whether a workshop about emotions would increase the number of cars a salesperson sells per month. Design an experiment to help him.

Graphical and Tabular Displays of Data

Do you have body piercings and tattoos? Table 1 compares class standings with types of body art for 490 undergraduate students surveyed at a state university in the Southwest. In Exercise 15 of Homework 3.2, you will determine whether the juniors were more likely to have both body piercings and tattoos than the sophomores.

Table 1 Class Standing versus Types of Body Art

Class Standing	Body Piercing Only	Tattoos Only	Both Body Piercing and Tattoos	No Body Art	Total
Freshman	61	7	14	86	168
Sophomore	43	11	10	64	128
Junior	20	9	7	43	79
Senior	21	17	23	54	115
Total	145	44	54	247	490

Source: Contemporary College Students and Body Piercing, *Myrna L. Armstrong, Ed. D., et al.*

In Chapter 2, we discussed how to use random sampling to collect data. In this chapter, we will use tables and diagrams to give us a much better sense of the data. We will use proportions to further analyze the data. For certain types of data, we will also identify unusual observations and investigate the *shape*, *center*, and *spread* of diagrams. Finally, we will carefully analyze misleading diagrams so that we understand what they really mean.

▼ 3.1 Frequency Tables, Relative Frequency Tables, and Bar Graphs

Objectives

» Identify *categorical variables* and *numerical variables*.

» Construct and interpret *frequency tables* and *relative frequency tables*.

» Construct and interpret *frequency bar graphs* and *relative frequency bar graphs*.

» Describe the meanings of *AND* and *OR*.

» Use a relative frequency bar graph to find proportions.

» Interpret *multiple bar graphs*.

Categorical Variables and Numerical Variables

What is your favorite movie genre? Do you love thrillers? Or do you get more excited about comedies? Favorite movie genre is an example of a *categorical variable* because it consists of names of groups (genres) of movies, such as action, comedy, and drama.

▶ **Definition** Categorical Variable

A **categorical variable** (or **qualitative variable**) consists of names or labels of groups of individuals.

How much does the movie theater closest to you charge for an adult ticket? Adult movie ticket price is an example of a *numerical variable* because it consists of measurable quantities such as $10.50, $11.01, and $9.50 (Source: *Fandango*).

▶ **Definition** Numerical Variable

A **numerical variable** (or **quantitative variable**) consists of measurable quantities that describe individuals.

▶ **Example 1**   Identifying Categorical and Numerical Variables

Identify whether the variable is categorical or numerical.

1. Age of a person at a hip-hop concert
2. Academic major of a community college student
3. Maximum speed of a car (in miles per hour)
4. ZIP code of a home in Arkansas
5. A variable consisting of zeroes and ones, where 0 stands for a man and 1 stands for a woman

Solution

1. The ages of people who attend a hip-hop concert are numerical measures such as 19 years and 25 years. So, the variable is numerical.
2. Majors are categories of community college students such as business majors and psychology majors. So, the variable is categorical.
3. Maximum speeds of cars are numerical measures such as 108 miles per hour and 231 miles per hour. So, the variable is numerical.
4. ZIP codes identify groups of homes. So, the variable is categorical.
5. The numbers 0 and 1 represent the categories men and women, respectively. So, the variable is categorical.

▶

WARNING   In Problems 4 and 5 of Example 1, it is a common error to think that the variables are numerical because the variables involve numbers. But we are *not* measuring a home to obtain its ZIP code, so ZIP code is *not* numerical. Similarly, we are *not* measuring a characteristic of a man or woman to get the result 0 or 1, respectively. So, the variable is *not* numerical.

All variables are either categorical or numerical. If a variable is not categorical, then it is numerical. If a variable is not numerical, then it is categorical.

We will use very different methods to analyze categorical and numerical variables. In this section and Section 3.2, we will focus on categorical variables. In Sections 3.3 and 3.4, we will focus on numerical variables.

Frequency and Relative Frequency Tables

The author surveyed one of his statistics classes about favorite movie genres. The observations are shown in Table 2.

Table 2   Favorite Movie Genres

Drama	Comedy	Action	Drama	Horror	Comedy	Comedy
Thriller	Comedy	Drama	Other	Comedy	Horror	Comedy
Action	Other	Thriller	Other	Comedy	Thriller	Other
Horror	Other	Comedy	Action	Comedy	Thriller	Comedy
Drama	Comedy	Other	Drama	Comedy	Action	Comedy
Comedy	Action	Comedy	Horror	Comedy	Action	Horror

Source: *J. Lehmann*

Because there are so many observations, it is hard to take in all the information. It will help to summarize the data. First, we determine that the variable favorite movie genre is categorical, where the categories are drama, comedy, action, and so on. Second, we can find each category's *frequency*, which is the number of observations in that category. For example, the Drama category has 5 observations (see Table 2), so its frequency is 5.

▶ **Definition   Frequency of a category**

The **frequency of a category** is the number of observations in that category.

A **frequency table** for a categorical variable is a table that lists all the categories and their frequencies. We will construct such a table for the movie data in Example 2.

▶ **Example 2**   Constructing a Frequency Table

Construct a frequency table of favorite movie genres selected by the 42 students (see Table 2).

Solution

We record the different categories in Table 3 and draw a mark across from a category and under the second column "Tally" each time an observation falls in that category. Then for each tally, we count the marks and record the number under the third column "Frequency."

Table 3 Frequency Table

Category	Tally	Frequency
Drama	‖‖‖	5
Action	‖‖‖ ‖	6
Thriller	‖‖‖‖	4
Comedy	‖‖‖ ‖‖‖ ‖‖‖ ‖	16
Horror	‖‖‖	5
Other	‖‖‖ ‖	6
Total		42

To check our work, we add all the frequencies, which gives 42 observations, which is correct (see the bottom row, far right, of Table 3).

The categories of movies together with their frequencies are an example of a *frequency distribution*.

▶ **Definition**   Frequency distribution of a categorical variable

The **frequency distribution of a categorical variable** is the categories of the variable together with their frequencies.

Was the drama category a popular choice for the class? We can determine this by finding the fraction of the observations that were in that category. Because there were 5 such observations out of 42, the fraction is $\frac{5}{42}$. Recall from Section 1.3 that because $\frac{5}{42}$ describes a part of a whole, we also call it a proportion. We also say the *relative frequency* of the drama category is $\frac{5}{42}$.

▶ **Definition**   Relative frequency

The **relative frequency of a category** is given by

$$\frac{\text{frequency of the category}}{\text{total number of observations}}$$

In words, the relative frequency of a category is the proportion of all the observations that fall in that category.

In short, **a relative frequency is a proportion**.

A **relative frequency table** of a categorical variable is a list of the categories and their relative frequencies. We will construct such a table for the movie data in Example 3.

▶ Example 3   Constructing a Frequency and Relative Frequency Table

Table 4 displays the frequencies of the movie categories.

Table 4 Frequency Table

Category	Tally	Frequency
Drama	卌	5
Action	卌	6
Thriller	||||	4
Comedy	卌 卌 卌		16
Horror	卌	5
Other	卌		6
Total		42

1. What is the relative frequency of the action category? What does it mean in this situation?
2. Insert a column in Table 4 for the relative frequencies of the categories.
3. Which category has the largest relative frequency? What is that relative frequency? What does it mean in this situation?

Solution

1. The relative frequency of the action category is the frequency of this category (6) divided by the total number of observations (42): $\frac{6}{42} = \frac{3}{21}$. Because $\frac{3}{21} \approx 0.143$, we conclude that 14.3% of the class prefers action movies.
2. We compute the relative frequencies of each category and list them in the fourth column of Table 5.

Table 5 Frequency and Relative Frequency Table

Category	Tally	Frequency	Relative Frequency
Drama	卌	5	$\frac{5}{42} \approx 0.119$
Action	卌		6	$\frac{6}{42} \approx 0.143$
Thriller	||||	4	$\frac{4}{42} \approx 0.095$
Comedy	卌 卌 卌		16	$\frac{16}{42} \approx 0.381$
Horror	卌	5	$\frac{5}{42} \approx 0.119$
Other	卌		6	$\frac{6}{42} \approx 0.143$
Total		42	$\frac{42}{42} = 1$

To check our work, we add all the relative frequencies, which gives 1 (see the bottom row, far right, of Table 5). This makes sense because the sum of all the parts gives the whole (1).

3. From Table 5, we see that the comedy category has the largest relative frequency: $\frac{16}{42} = \frac{8}{21}$. Because $\frac{8}{21} \approx 0.381$, we conclude that 38.1% of the class prefers comedies.

In Problem 2 of Example 3, we calculated some relative frequencies. **In general, we agree to round relative frequencies and proportions to the third decimal place.**

The categories of movies together with their relative frequencies are an example of a *relative frequency distribution*.

> **Definition** Relative frequency distribution of a categorical variable
>
> The **relative frequency distribution of a categorical variable** is the categories of the variable together with their relative frequencies.

Although it is good to practice constructing a few frequency and relative frequency tables by hand to get the idea, we usually use technology to create them. Figure 1 displays such a table for the movie distribution constructed by StatCrunch. For StatCrunch instructions, see Appendix B.3.

Frequency and relative frequency table results for Movie:

Movie	Frequency	Relative Frequency
Drama	5	0.11904762
Action	6	0.14285714
Thriller	4	0.095238095
Comedy	16	0.38095238
Horror	5	0.11904762
Other	6	0.14285714

Figure 1 Frequency and relative frequency table of movie distribution

In Problem 2 of Example 3, we found that the sum of the relative frequencies of all the movie categories is equal to 1. This makes sense because the sum of all the parts is equal to the whole (Section 1.3).

> ▶ **Sum of Relative Frequencies**
>
> For a categorical variable, the sum of the relative frequencies of all the categories is equal to 1.

WARNING   In Problem 3 of Example 3, we found that 38.1% of the 42 statistics students surveyed preferred comedies. Recall from Chapter 2 that this does *not* mean that 38.1% of *all* statistics students in the United States prefer comedies for two reasons. First, the sample was not selected randomly. Second, even if the sample had been selected randomly, there would likely be sampling error.

Frequency and Relative Frequency Bar Graphs

Although frequency tables contain very useful information, it is often easier to mentally process the information if it is presented in graphical form. A **frequency bar graph** is a graph that uses heights of bars to describe frequencies of categories.

▶ **Example 4**   Constructing a Frequency Bar Graph

Construct a frequency bar graph of the movie distribution.

Solution

To begin, we list the categories on the horizontal axis and write the variable "Favorite movie genre" below the categories (see Fig. 2). Because the frequencies range from 4 to 16, we write 0, 5, 10, 15, and 20 equally spaced on the vertical axis and "Frequency" above the numbers. Then we draw bars above the categories so that the tops of the bars line up with the appropriate frequencies on the vertical axis. For example, because the drama category has frequency 5, the top of the red bar lines up with 5 on the vertical axis.

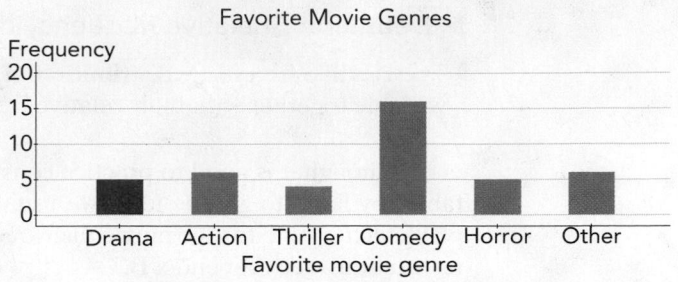

Figure 2 Frequency bar graph of movie distribution

WARNING

In Example 4, we wrote the numbers 0, 5, 10, 15, and 20 on the vertical axis of the bar graph (see Fig. 2). Recall from Section 1.1 that when we write numbers on a number line, they should increase by an equal amount and be equally spaced.

Just as with frequency tables, it is useful to practice constructing a few frequency bar graphs by hand, but we usually use technology to create them. In fact, the frequency bar graph in Fig. 2 was constructed using StatCrunch. For StatCrunch instructions, see Appendix B.4.

There is another useful way to describe a distribution. A **relative frequency bar graph** is a graph that uses heights of bars to describe relative frequencies of categories.

▶ Example 5   Constructing a Relative Frequency Bar Graph

Construct a relative frequency bar graph of the movie distribution.

Solution

We construct a relative frequency bar graph much like we construct a frequency bar graph, but we write relative frequencies on the vertical axis (see Fig. 3). Because the approximate relative frequencies in Table 5 range from 0.095 to 0.381, we write 0.10, 0.20, 0.30, and 0.40 equally spaced on the vertical axis. We draw bars above the categories so that the tops of the bars line up with the appropriate relative frequencies on the vertical axis. We also write the relative frequencies above the bars, although this is not done in statistical articles. The relative frequencies will often be shown in this book so that you can find precise answers.

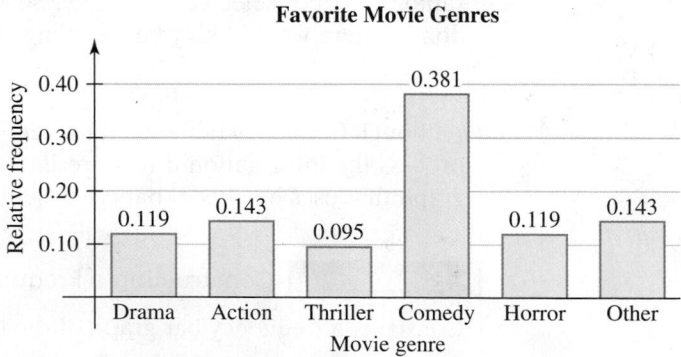

Figure 3 Relative frequency bar graph of movie distribution

Although the relative frequencies will often be shown in bar graphs in this book so that you can find accurate answers, you should be able to estimate them if they are not shown. Figure 4 displays a relative frequency bar graph of the movie distribution constructed with StatCrunch. You should be able to estimate the relative frequencies to the second decimal place. Try it, and then compare your estimates with the relative frequencies shown in Fig. 3.

Figure 4 StatCrunch relative frequency bar graph of movie distribution

For StatCrunch instructions on constructing relative frequency bar graphs, see Appendix B.4.

Meanings of AND and OR

If we know the relative frequencies for the categories full-time students and business majors, we might want to find the relative frequency for the category of all the students in the two categories together. You might think we could use the word "and" to describe the new category, but the word "and" is used differently in mathematics than in English. To emphasize this difference, we will use capital letters for the mathematical version: "AND." When "**AND**" is used with two categories, this means to consider the observations that the categories have in common. For example, the category of full-time students AND business majors is the category of full-time business majors.

When "**OR**" is used with two categories, this means to consider the observations in the categories all together. So, the category of full-time students OR business majors is the category of all full-time students together with all business majors.

Figure 5 contains five students' names. The left circle contains female names. The right circle contains names that start with "M." To find any names that are female names AND start with "M," we identify any names that lie where the circles overlap: Mary. To find any names that are female names OR start with "M," we identify the names in either circle: Keri, Jenny, Mary, Matt, and Mark.

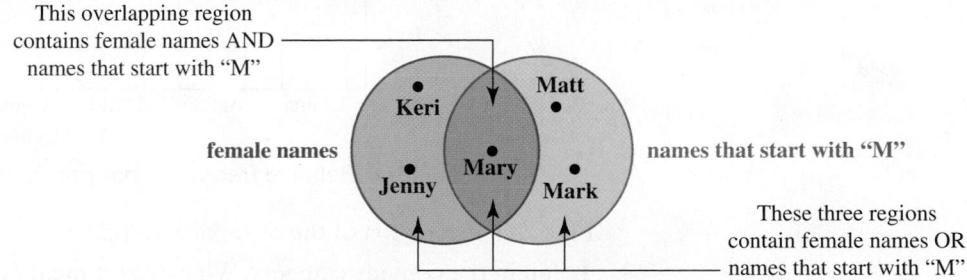

Figure 5 Meanings of "AND" and "OR"

> **Example 6** Identifying Categories Involving AND or OR

The dates of February 2015 are shown in Table 6.

Table 6 Dates of February 2015

Sun	Mon	Tues	Wed	Thurs	Fri	Sat
1	2	3	4	5	6	7
8	9	10	11	12	13	14
15	16	17	18	19	20	21
22	23	24	25	26	27	28

1. Find the dates that are Thursdays.
2. Find the dates in the third week.
3. Find the dates that are in the third week AND are Thursdays.
4. Find the dates that are in the third week OR are Thursdays.

Solution

1. From the yellow column, we see the dates are 5, 12, 19, and 26.
2. From the green row, we see the dates are 15, 16, 17, 18, 19, 20, and 21.
3. The date that the green row and the yellow column have in common is 19.
4. The dates in the green row together with the dates in the yellow column are 5, 12, 15, 16, 17, 18, 19, 20, 21, and 26.

Using a Relative Frequency Bar Graph to Find Proportions

In Example 7, we will use the concept that a relative frequency is a proportion along with the meanings of OR and AND to find some proportions. We will also find the proportion of observations that do NOT fall in a category, using capital letters as we do for AND and OR.

> **Example 7** Using a Relative Frequency Bar Graph to Find Proportions

A relative frequency bar graph of the movie distribution is shown in Fig. 6.

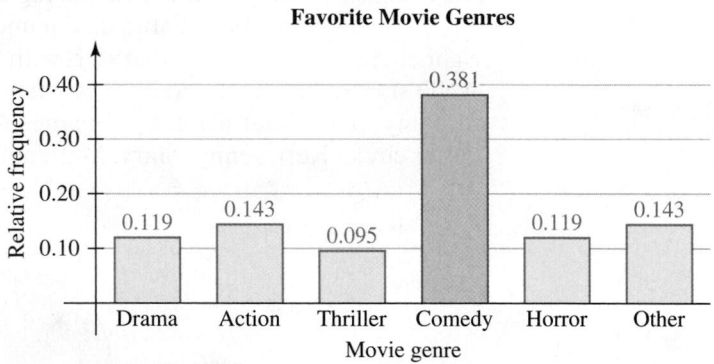

Figure 6 Relative frequency bar graph of movie distribution

Find the proportion of the observations that

1. fall in the comedy category. What does it mean in this situation?
2. do NOT fall in the comedy category.
3. fall in the thriller category OR fall in the horror category.
4. fall in the thriller category AND fall in the horror category.

Solution

1. The relative frequency of the comedy category is 0.381 (see the orange bar). Because a relative frequency is a proportion, we conclude the proportion is 0.381. This means 38.1% of the 42 students prefer comedy movies.

2. We discuss two methods of finding the proportion.

   Method 1 We can find the proportion of observations that do NOT fall in the comedy category by adding up the relative frequencies of all the *other* categories:

   $$0.119 + 0.143 + 0.095 + 0.119 + 0.143 = 0.619$$

   Method 2 Because the relative frequencies of all the categories of any categorical variable always add to 1, we can find the proportion of observations that do NOT fall in the comedy category by subtracting 0.381 from 1:

   $$1 - 0.381 = 0.619$$

   So, the proportion is 0.619, which is the same result we found by method 1.

3. To find the proportion of observations that fall in the thriller category OR fall in the horror category, we add the relative frequencies of the categories thriller and horror:

   $$0.095 + 0.119 = 0.214$$

   So, the proportion is 0.214.

4. The distribution describes each student's *favorite* movie genre, so no student chose both the categories thriller AND horror. So, the proportion is 0.

By Method 2 in Problem 2 of Example 7, we used the fact that the relative frequency of the comedy category is 0.381 to find the proportion of observations that do NOT fall in the comedy category by subtracting 0.381 from 1:

$$1 - 0.381 = 0.619$$

Note that we have used the proportion-of-the-rest property from Section 1.3.

Proportion of the Rest

Let $\dfrac{a}{b}$ be the proportion of the whole that has a certain characteristic. Then the proportion of the whole that does NOT have that characteristic is

$$1 - \dfrac{a}{b}$$

Multiple Bar Graphs

We can compare two or more categorical variables visually. A **multiple bar graph** is a graph that has two or more bars for each category of the variable described on the horizontal axis (see Fig. 7).

▶ Example 8 Interpreting a Multiple Bar Graph

In a survey in 2012, 1960 adults were asked the following question: "Generally speaking, do you usually think of yourself as a Republican, Democrat, Independent, or what?" The results of the survey are described by the multiple bar graph in Fig. 7.

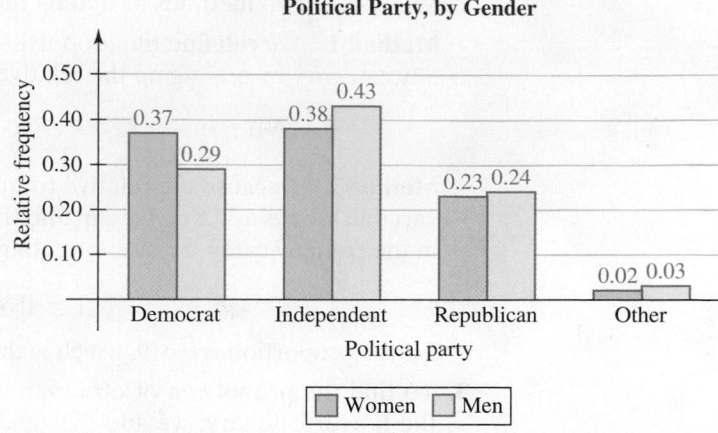

Figure 7 Multiple bar graph of political party distribution
(**Source:** *General Social Survey*)

1. What proportion of women thought of themselves as Democrats?
2. Which political party did the greatest proportion of men choose?
3. Compare the proportion of women who thought of themselves as Independents to the proportion of men who thought of themselves as Independents.
4. A total of 1081 women and 879 men responded to the survey. Were there more women or men who thought of themselves as Independents? How is this possible, given there was a smaller proportion of women who thought of themselves as Independents than men?
5. On the basis of the multiple bar graph, a student concludes that 24% of *all* American men are Republicans. What would you tell the student?

Solution

1. The orange bar above "Democrat" has height 0.37, so the proportion of women who thought of themselves as Democrats is 0.37.
2. The tallest blue bar is above "Independent." The height of the bar is 0.43. So, the proportion of men who thought of themselves as Independents is 0.43—the largest proportion of men for any political party.
3. The orange bar above "Independent" has height 0.38, so the proportion of women who thought of themselves as Independents is 0.38. In Problem 2, we found that the proportion of men who thought of themselves as Independents is 0.43. Therefore, a smaller proportion of women thought of themselves as Independents than men.
4. The number of women who thought of themselves as Independents was $0.38(1081) \approx 411$ women. The number of men who thought of themselves as Independents was $0.43(879) \approx 378$ men. So, more women thought of themselves as Independents than men. This is possible because more women responded to the survey than men.
5. The blue bar above "Republican" has height 0.24, so we can conclude that 24% of the men *surveyed* are Republicans. But we *cannot* conclude 24% of *all* American men are Republicans. Even though the General Social Survey is well designed, there would still likely be sampling error. The percentage for all men might actually be a bit lower, say 23%, or a bit higher, say 26%.

See Appendix B.4 for instructions on how to use StatCrunch to construct a multiple bar graph.

In the survey described in Example 8, more women thought of themselves as Independents than men, but that's just because more women responded to the study. In fact, we found that a smaller proportion of women thought of themselves as Independents than men. In general, **for a multiple bar graph, it is more meaningful for the vertical axis to describe relative frequencies rather than frequencies**.

The multiple bar graph in Fig. 7 is the combination of two relative frequency bar graphs: one for the political parties of women and one for the political parties of men. So, the relative frequencies for the political parties for women should add to 1, which is true:

$$0.37 + 0.38 + 0.23 + 0.02 = 1$$

The relative frequencies for the political parties of men should also add to 1:

$$0.29 + 0.43 + 0.24 + 0.03 = 0.99 \approx 1$$

The relative frequencies add to 0.99, rather than 1, due to rounding.

Group Exploration
Relative frequencies

A total of 2508 adults were surveyed on how the income gap between the rich and the poor may have changed in the past 10 years. Their responses are described by the relative bar graph in Fig. 8.

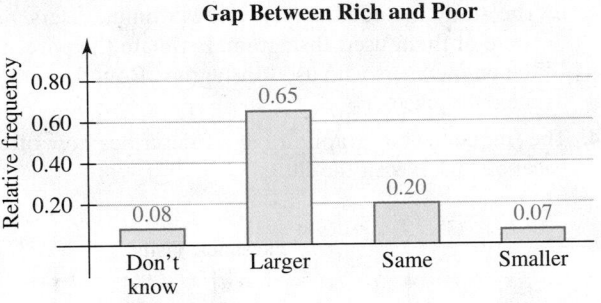

Figure 8 Gap between rich and poor
(**Source:** *Pew Research Center*)

1. Add the relative frequencies of the four categories. What does your result mean in this situation?

2. Explain why the sum of the relative frequencies of all the categories of a categorical variable equals 1 for *any* categorical variable.

3. Most states have adopted Common Core standards, which are academic standards for mathematics and English language arts/literacy for grades up through high school. The relative frequency bar graph in Fig. 9 summarizes the responses of 1000 adults who were surveyed about whether they support the standards. What proportion of those surveyed are strongly opposed?

4. The relative frequency above the bar for those that strongly support is missing on purpose. Although you could estimate the relative frequency by inspecting the vertical axis, explain how you can find the exact relative frequency by using the concept you described in Problem 2. Then find the relative frequency.

5. What proportion of those surveyed strongly oppose the standards OR somewhat oppose the standards? What proportion of those surveyed strongly support the standards OR somewhat support the standards? Are most of the 1000 people surveyed opposed or in support of the standards? Explain.

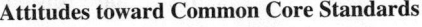

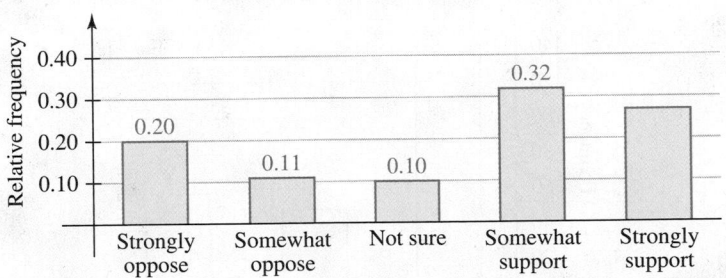

Figure 9 Common Core standards
(**Source:** *WSJ/NBC*)

> ▶ **Tips for Success**   **Study with a Classmate**
>
> It can be helpful to meet with a classmate and discuss what happened in class that day. Not only can you ask questions of each other, but you will learn just as much by explaining concepts to each other. Explaining a concept to someone else forces you to clarify your own understanding of the concept.

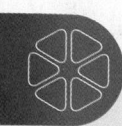

Homework 3.1

For extra help ▶ **MyMathLab®**  Watch the videos in MyMathLab    Download the MyDashboard App

1. A(n) ____ variable consists of names or labels of groups of individuals.

2. A(n) ____ variable consists of measurable quantities that describe individuals.

3. The ____ of a category is the number of observations in that category.

4. The ____ frequency of a category is the proportion of all the observations that fall in that category.

For Exercises 5–12, identify whether the variable is categorical or numerical.

5. The price (in dollars) of a song on iTunes®

6. The height (in inches) of a person

7. A person's ski level (beginner, intermediate, or advanced)

8. A person's ethnicity

9. The possible observations of a variable are 0 and 1, where 0 stands for a new car and 1 stands for a used car.

10. The area code of a phone number

11. The maximum sound level (in decibels) at a rap show

12. The weight (in pounds) of a cat

13. Do you have an Instagram account? If so, how often do you visit the site? Data for 247 Instagram users are described by the frequency bar graph in Fig. 10.

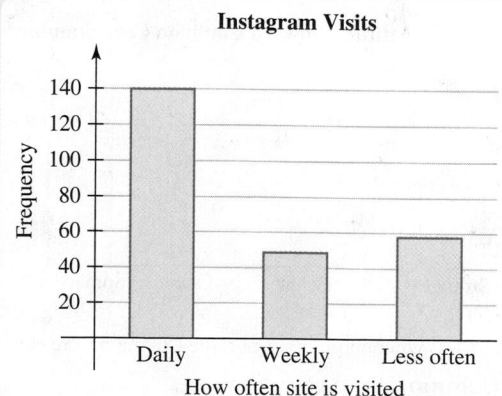

Figure 10 Exercise 13
(**Source:** *Pew Research Center*)

a. What variable is described by the bar graph? Is it categorical or numerical?

b. Estimate the number of Instagram users in the study who visited the site daily.

c. Estimate the number of Instagram users in the study who visited the site *at least* weekly.

d. Estimate the number of Instagram users in the study who did NOT visit Instagram daily.

e. The study included a total of 1445 online users, but only some of them used Instagram. Estimate the percentage of the online users who used Instagram. Round your result to the ones place.

14. The frequency bar graph in Fig. 11 describes how often 1026 Facebook users visit the site.

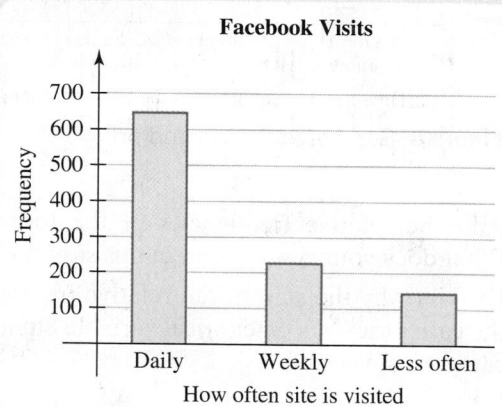

Figure 11 Exercise 14
(**Source:** *Pew Research Center*)

a. What variable is described by the bar graph? Is it categorical or numerical? Explain.

b. Estimate the number of Facebook users in the study who visited the site less than once per week.

c. Estimate the number of Facebook users in the study who visited the site *at least* weekly.

d. Estimate the number of Facebook users in the study who did NOT visit the site daily.

e. The study included a total of 1445 online users, but only some of them used Facebook. Estimate the percentage of the online users who used Facebook. Round your result to the ones place.

15. Of 2050 American tech employees who took part in an online survey, 389 said they had stolen at least one object from their work. The reasons why they said they stole are described by the bar graph in Fig. 12.

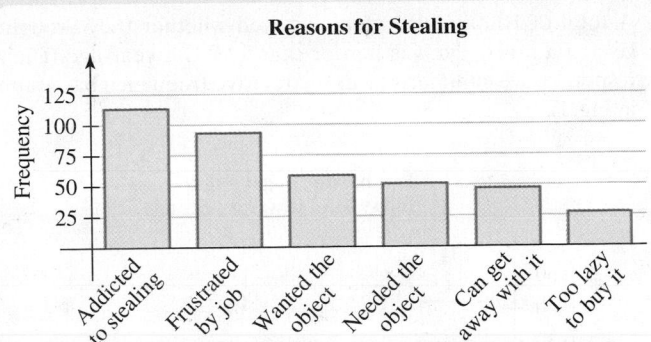

Figure 12 Exercise 15
(**Source:** *GETVOIP*)

a. Estimate the number of the tech employees who said they stole because they needed the object.
b. Estimate the number of the tech employees who said they stole because they wanted the object OR needed the object.
c. Estimate the number of the tech employees who did NOT say they stole because they are addicted to stealing.
d. Estimate the *relative frequency* of the tech employees who said they stole because they were too lazy to buy it. Although we will usually round relative frequencies to the third decimal place, round your result to the second decimal place.
e. The individuals who participated in the online survey were selected from a group of tech employees who volunteered to take part in such online surveys. If the population is all tech employees, what type(s) of bias in sampling likely occurred?

16. A total of 1014 American adults were asked, "How worried are you that you or someone in your family will become a victim of terrorism?" The adults' responses are described by the bar graph in Fig. 13.

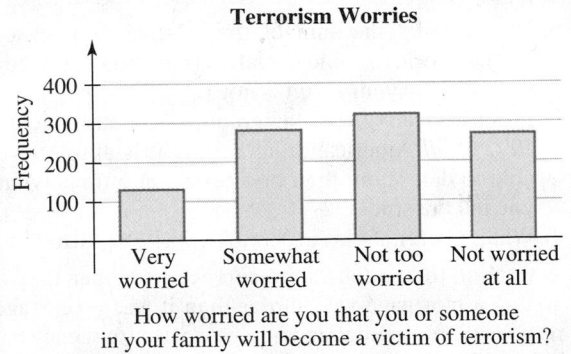

Figure 13 Exercise 16
(**Source:** *CNN/ORC Poll*)

a. Estimate the number of those surveyed who are not worried at all about them or someone in their family becoming a victim of terrorism.

b. Estimate the number of those surveyed who are somewhat worried OR not too worried about them or someone in their family becoming a victim of terrorism.
c. Estimate the number of those surveyed who are NOT very worried about them or someone in their family becoming a victim of terrorism.
d. Estimate the *relative frequency* of those surveyed who are very worried about them or someone in their family becoming a victim of terrorism. Although we will usually round relative frequencies to the third decimal place, round your result to the second decimal place.

For Exercises 17–24, refer to the dates of February 2015 shown in Table 7. Find all dates that fall in the given category.

17. Tuesdays
18. Fridays
19. The second week
20. The fourth week
21. Tuesdays OR the second week
22. Fridays OR the fourth week.
23. Tuesdays AND the second week
24. Fridays AND the fourth week

Table 7 Dates of February 2015

Sun	Mon	Tues	Wed	Thurs	Fri	Sat
1	2	3	4	5	6	7
8	9	10	11	12	13	14
15	16	17	18	19	20	21
22	23	24	25	26	27	28

25. At the website Lending Club®, customers can either borrow money or fund money for others' loans. When someone applies for a loan, Lending Club grades the potential loan from A to G, where A means the potential loan has very low risk and G means it has very high risk. A loan with grade A will have a much lower interest rate than a loan with grade G. The grades of loans funded in September 2014 are described by the bar graph in Fig. 14.

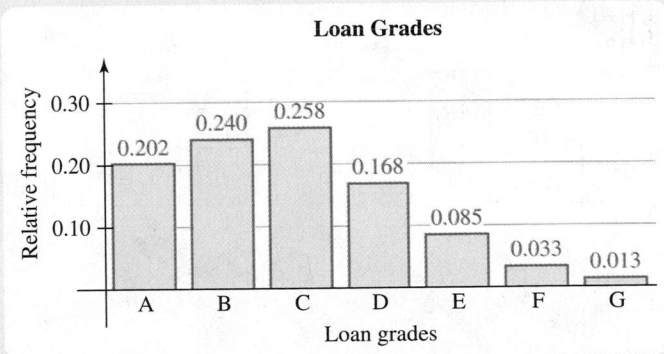

Figure 14 Exercise 25
(**Source:** *Lending Club*)

a. Loans of which grade were most frequently funded?
b. What proportion of the loans were B loans OR C loans?
c. What proportion of the loans were NOT G loans?
d. What proportion of the loans were more risky than D loans?

e. A total of 10,256 loans were funded at Lending Club in September 2014. How many of those were A loans?

26. The relative frequency bar graph in Fig. 15 describes 2234 Internet users' reading habits of print books and e-books.

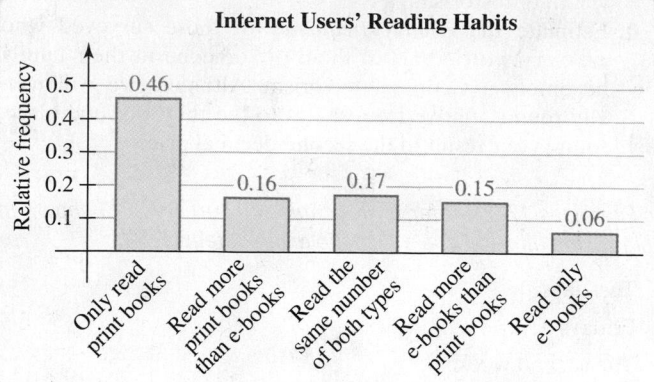

Figure 15 Exercise 26
(**Source:** *The Harris Poll*)

a. Find the proportion of the surveyed Internet users who read only print books.

b. Find the proportion of the surveyed Internet users who read more e-books than print books OR read only print books.

c. Find the proportion of the surveyed Internet users who do NOT only read print books.

d. Find the *number* of those surveyed who read more e-books than print books.

e. The individuals who participated in the survey were selected from a group of Internet users who volunteered to take part in such online surveys. If the population is all Internet users, what type(s) of bias in sampling likely occurred?

27. A total of 2647 single adults, ages 18–59, took part in an online survey by responding to the following question: "Do you believe it is acceptable to date more than one person at a time?" Their responses are summarized in the relative frequency bar graph in Fig. 16.

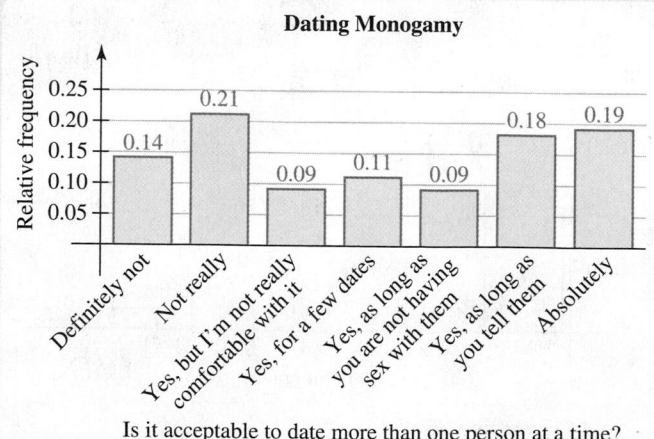

Figure 16 Exercise 27
(**Source:** *Quonundrums*)

Find the proportion of those surveyed who

a. thought it was absolutely acceptable to date more than one person at a time.

b. did NOT think it was absolutely acceptable to date more than one person at a time.

c. thought it was definitely not acceptable to date more than one person at a time OR not really acceptable to date more than one person at a time.

28. A total of 1000 adults were surveyed whether they thought getting a mortgage was harder than it was a year ago. Their responses are summarized in the relative frequency bar graph in Fig. 17.

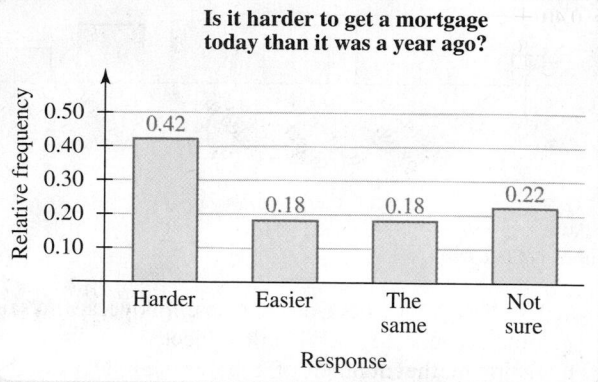

Figure 17 Exercise 28
(**Source:** *loanDepot*)

Find the proportion of those surveyed who

a. thought it was harder to get a mortgage at the time of the survey than it was one year before.

b. did NOT think it was harder to get a mortgage at the time of the survey than it was one year before.

c. thought it was easier to get a mortgage at the time of the survey than one year before OR thought it was the same difficulty.

29. A total of 2647 single adults, ages 18–59, took part in an online survey by responding to the following question: "Do you believe it is acceptable to date more than one person at a time?" Their responses are summarized in the relative frequency bar graph in Fig. 16.

a. What is the *number* of adults in the study who thought it was acceptable to date more than one person for a few dates?

b. Explain why the sum of the relative frequencies of all the categories should equal 1. Then find the actual total. Explain why your result is not 1.

c. From inspecting the bar graph, a student concludes that 19% of *all* American adults think it is absolutely acceptable to date more than one person at a time. What would you tell the student?

d. What type(s) of bias in sampling likely occurred? Explain.

30. A total of 1000 adults were surveyed whether they thought getting a mortgage was harder than it was a year ago. Their responses are summarized in the relative frequency bar graph in Fig. 17.

a. Find the sum of the relative frequencies of all the categories. Why does your result make sense?

b. Construct a *frequency* table of the data.

c. From inspecting the bar graph, a student concludes that 18% of *all* American adults think that it's easier to get a mortgage than it was a year ago. What would you tell the student?

d. Individuals were contacted by landlines and cell phones. What type(s) of bias in sampling likely occurred? Explain.

31. In 2012, 709 women and 586 men were asked how much confidence they had in Congress. The results of the survey are described by the multiple bar graph in Fig. 18.

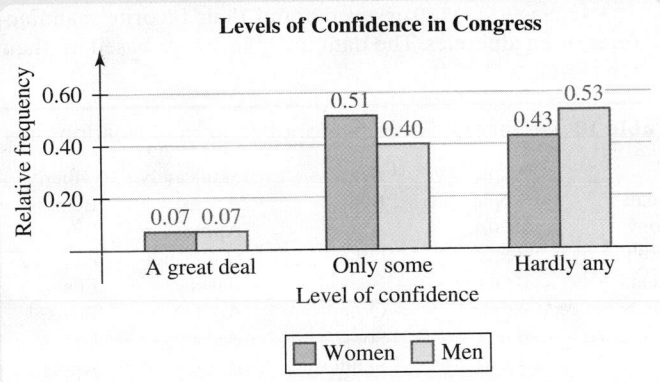

Figure 18 Exercise 31
(**Source:** *General Social Survey*)

a. Which gender of those surveyed tended to have more confidence in Congress? Explain.

b. What proportion of men surveyed did NOT have a great deal of confidence in Congress?

c. How many women had only some confidence in Congress?

d. From inspecting the multiple bar graph, a student concludes that 53% of *all* American men have hardly any confidence in Congress. What would you tell the student?

e. A student says that the same *number* of women as men had a great deal of confidence in Congress because the relative frequency for each gender is 0.07. What would you tell the student?

32. In 1989, fathers who stayed at home were surveyed about why they stayed at home. A similar survey was carried out in 2012. The fathers' responses and the relative frequencies are described by the multiple bar graph in Fig. 19 for both years.

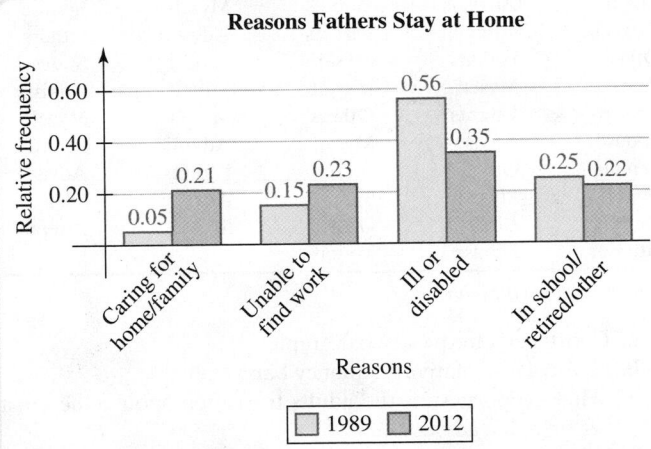

Figure 19 Exercise 32
(**Source:** *Pew Research Center*)

a. What percentage of fathers in the 2012 study stayed at home because they were unable to find work?

b. From inspecting the multiple bar graph, a student concludes that 35% of *all* American fathers who stayed at

home in 2012 stayed at home because they were ill or disabled. What would you tell the student?

c. Why did most fathers in the 1989 study stay at home? What percentage of fathers in the 1989 study stayed home for that reason?

d. The number of fathers who stayed at home increased from 1.1 million in 1989 to 2.0 million in 2012. Although the number of *all* fathers increased from 1989 to 2012, that increase alone cannot account for the large percentage growth in fathers who stayed at home. Assuming the live-at-home fathers surveyed represent all live-at-home fathers really well, what are the other reasons for the increase in stay-at-home fathers?

e. Why is it important that the vertical axis of the multiple bar graph describes relative frequencies and not frequencies?

33. Table 8 shows the top 5 cosmetic surgeries in 2013, their frequencies (in number of thousands of times performed in 2013), and their relative frequencies.

Table 8 Top 5 Cosmetic Surgeries in 2013

Procedure	Frequency (in thousands)	Relative Frequency
Liposuction	364	0.318
Breast Augmentation	313	0.273
Eyelid Surgery	161	0.140
Tummy Tuck	160	0.140
Nose Surgery	148	0.129
Total	1146	1.000

Source: *American Society for Aesthetic Plastic Surgery*

a. Construct a frequency bar graph.

b. Construct a relative frequency bar graph.

c. Are the bar graphs you constructed examples of descriptive statistics or inferential statistics? Explain.

d. Find the proportion of cosmetic surgeries performed that involved liposuction OR a tummy tuck.

e. Find the proportion of the top 5 cosmetic surgeries performed that did NOT involve eyelid surgery.

34. Table 9 shows the top 6 bands played on alternative rock station Live 105 on April 18, 2014. The table includes the bands' frequencies (numbers of times played that day) and their relative frequencies.

Table 9 Top 6 Bands Played on Live 105

Band	Frequency	Relative Frequency
Arctic Monkeys	16	0.205
The Black Keys	15	0.192
Fitz and the Tantrums	14	0.179
Bastille	11	0.141
Cage the Elephant	11	0.141
The Neighborhood	11	0.141
Total	78	0.999

Source: *Live 105*

a. Construct a frequency bar graph.

b. Construct a relative frequency bar graph.

c. Are the bar graphs you constructed examples of descriptive statistics or inferential statistics? Explain.

d. Of the songs played on Live 105 on April 18, 2014, that were recorded by the bands listed in Table 9, what proportion were recorded by Arctic Monkeys OR Cage the Elephant?

e. On the basis of the relative frequencies totaling 0.999, rather than 1, a student concludes that nonsampling error happened. What would you tell the student?

35. DATA Here are the countries of the top 15 men in the 2013 ING® New York City Marathon (Source: *New York City Marathon*):

Kenya	Ethiopia	South Africa	Kenya
Kenya	Japan	Uganda	Kenya
Kenya	Italy	Japan	Uganda
U.S.	U.S.	France	

a. Construct a frequency and relative frequency table of the data. Use the category "Other" for Ethiopia, South Africa, Italy, and France.

b. Construct a frequency bar graph.

c. Construct a relative frequency bar graph of the data.

d. What percentage of the top 15 men are NOT from the United States?

e. What percentage of the top 15 men are from Kenya OR Uganda?

36. DATA Here are the countries of the top 22 women in the 2013 Chicago Marathon (Source: *Chicago Marathon*):

Kenya	Ethiopia	Kenya	Ethiopia	U.S.
U.S.	U.S.	Ethiopia	U.S.	U.S.
U.S.	U.S.	U.S.	U.S.	U.S.
U.S.	U.S.	U.S.	U.S.	U.S.
U.S.	U.S.			

a. Construct a frequency and relative frequency table.

b. Construct a frequency bar graph.

c. Construct a relative frequency bar graph.

d. What percentage of the top 22 women are NOT from the United States?

e. What percentage of the top 22 women are from Kenya OR the United States?

37. DATA The author surveyed one of his prestatistics classes, asking, "If you could have one superpower, what would it be?" Here are the students' responses:

Mind read	Fly	Fly	Other
Telekinesis	Fly	Other	Teleport
Other	Other	Telekinesis	Fly
Teleport	Teleport	Invisible	Other
Other	Invisible	Fly	Other
Mind read	Other	Fly	Other

a. Construct a frequency bar graph.

b. Construct a relative frequency bar graph.

c. What proportion of the students chose flying?

d. What proportion of the students did NOT choose invisibility?

e. What proportion of the students chose telekinesis OR mind reading?

38. DATA The author surveyed one of his statistics classes about favorite music genres. Here are students' responses:

Other	R & B	Hip-hop	Hip-hop
Rap	Hip-hop	Country	Hip-hop
Other	Hip-hop	Hip-hop	Other
Hip-hop	Hip-hop	Other	Electronic
R & B	Electronic	Rap	Country

a. Construct a frequency bar graph.

b. Construct a relative frequency bar graph.

c. What proportion of the students prefer electronic music?

d. What proportion of the students do NOT prefer electronic music?

e. What proportion of the students prefer hip-hop OR rap?

39. DATA Teenagers were surveyed about their favorite manufacturer of headphones. The data in Table 10 are based on their responses.

Table 10 Teenagers' Favorite Manufacturers of Headphones

Beats	Apple	Beats	Skullcandy	Other
Beats	Apple	Beats	Apple	Beats
Sony	Beats	Apple	Apple	Bose
Skullcandy	Beats	Other	Skullcandy	Apple
Beats	Beats	Beats	Beats	Beats
Beats	Apple	Other	Apple	Beats
Apple	Beats	Sony	Beats	Beats
Other	Apple	Skullcandy	Beats	Apple
Beats	Beats	Beats	Apple	Beats
Skullcandy	Beats	Apple	Bose	Other

Source: *Piper Jaffray & Co.*

a. Construct a frequency bar graph.

b. Construct a relative frequency bar graph.

c. What proportion of the teenagers prefer Skullcandy headphones?

d. What proportion of the teenagers do NOT prefer Skullcandy headphones?

e. For the surveyed teenagers, which is the most popular manufacturer of headphones? Is it easier to tell this by inspecting your frequency bar graph or Table 10? Explain.

40. DATA American adults were surveyed about whom they trust most for financial advice. The data in Table 11 are based on the survey's results.

Table 11 Whom American Adults Trust the Most for Financial Advice

Myself	Advisor	Parents	Myself	Spouse
Advisor	Others	Parents	Advisor	Others
Others	Parents	Myself	Spouse	Spouse
Parents	Myself	Myself	Myself	Advisor
Others	Parents	Others	Myself	Myself
Spouse	Advisor	Myself	Advisor	Parents
Parents	Others	Myself	Parents	Advisor
Myself	Myself	Spouse	Advisor	Myself
Parents	Advisor	Others	Myself	Parents
Advisor	Spouse	Myself		

Source: *Springleaf Financial*

a. Construct a frequency bar graph.

b. Construct a relative frequency bar graph.

c. What proportion of the adults trust their spouse the most for financial advice?

d. What proportion of the adults do NOT trust their spouse the most for financial advice?

e. For the surveyed adults, whom is an adult more likely to trust most for financial advice, parents or a financial advisor? Is it easier to tell this by inspecting your frequency bar graph or Table 11? Explain.

Concepts

41. Compare the meanings of a categorical variable and a numerical variable. Give an example that is not in the textbook of each type of variable. (See page 12 for guidelines on writing a good response.)

42. The possible observations of a variable are 0 and 1, where 0 stands for a full-time student and 1 stands for a part-time student. A student says that the variable is numerical because 0 and 1 are numbers. What would you tell the student? (See page 12 for guidelines on writing a good response.)

43. Explain why the vertical axis of a multiple bar graph should describe relative frequency rather than frequency.

44. Is it possible to calculate the relative frequency of a category by working with a frequency bar graph? Explain. Is it possible to calculate the frequency of a category by working with a relative frequency bar graph?

45. Explain why a relative frequency cannot be larger than 1.

46. Explain why the sum of the relative frequencies of all the categories of a categorical variable is equal to 1.

47. What is a frequency distribution? Give an example that is not in the textbook.

48. What is the meaning of relative frequency? Give an example that is not in the textbook.

49. Describe the meaning of AND in mathematics. Give an example not included in the textbook.

50. Describe the meaning of OR in mathematics. Give an example not included in the textbook.

Hands-On Research

51. Find a bar graph in a blog, newspaper, magazine, or journal.

  a. For each variable the bar graph describes, state its definition.

  b. What is the point of the article? Does the bar graph support the point? Explain.

52. Collect data that involves a categorical variable. Your study should include 20 individuals.

  a. Define the variable.

  b. If you performed a survey, give the exact wording of your question(s). If you did not perform a survey, describe how you made your observations.

  c. Provide the data you collected.

  d. Use a bar graph to describe the data.

  e. What conclusion(s) can you draw about the 20 individuals?

  f. If the 20 individuals were a sample, what group of individuals could be the population? In this case, would your sampling have bias? If so, which type(s)? Explain. Would it be OK to draw the same conclusion(s) about the population that you drew about the sample?

▼ 3.2 Pie Charts and Two-Way Tables

Objectives

» Construct and interpret *pie charts*.

» Construct and interpret *two-way tables*.

» Use a two-way table to compute proportions.

Pie Charts

The distribution of a categorical variable can be described by a **pie chart**, which is a disk where slices represent the categories. The proportion of the total area for one slice is equal to the relative frequency for the category represented by the slice. The relative frequencies are usually written as percentages.

Recall from Section 3.1 that the author surveyed one of his statistics classes about favorite movie genres. A pie chart of the movie distribution constructed with StatCrunch is shown in Fig. 20.

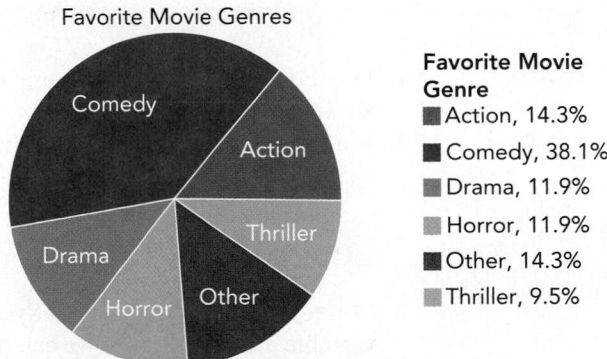

Figure 20 Pie chart of movie distribution

Pie charts are difficult to draw by hand, so we will use technology to construct them. For StatCrunch instructions, see Appendix B.5.

▶ **Example 1**   Construct and Interpret a Pie Chart

A total of 273 children were surveyed about what job they would want to do. The jobs and the percentages of the children who voted for them are shown in Table 12.

Table 12 Top Jobs Children Want to Do When They Grow Up

Job	Percent
Spy/Agent	16
Veterinarian	13
Professional Athlete	12
Movie Star	10
Video Game Designer	8
Doctor	6
Other	35

Source: *National Geographic Kids survey*

1. Use technology to construct a pie chart of the distribution.
2. Find the proportion of the observations that fall in the spy category.
3. Find the proportion of the observations that do NOT fall in the spy category.
4. Find the proportion of the observations that fall in the athlete category OR fall in the movie-star category.
5. On the basis of the pie chart, a student concludes that 6% of all American children want to be a doctor when they grow up. What would you tell the student?

Solution

1. We use StatCrunch to construct the pie chart shown in Fig. 21.

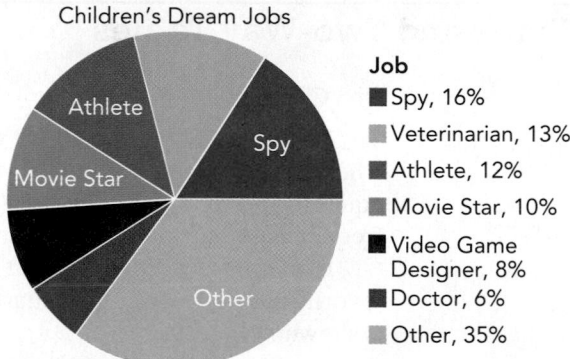

Figure 21 Pie chart of children's dream jobs

2. From the purple slice in the pie chart, we see 16% of the children want to be a spy. So, the proportion is 0.16.
3. We discuss two methods of finding the proportion.

   Method 1   We can find the proportion of observations that do NOT fall in the spy category by adding up the relative frequencies of all the *other* categories:

   $$0.13 + 0.12 + 0.10 + \mathbf{0.08} + 0.06 + 0.35 = 0.84$$

   Method 2   Because the relative frequencies of all the categories of any categorical variable always add to 1, we can find the proportion of observations that do NOT fall in the spy category by subtracting 0.16 from 1:

   $$1 - 0.16 = 0.84$$

   So, the proportion is 0.84, which is the same result we found by Method 1.

4. To find the proportion of observations that fall in the athlete category OR fall in the movie-star category, we add the relative frequencies of the categories athlete and movie star:

$$0.12 + 0.10 = 0.22$$

5. Although it is true that 6% of the 273 children surveyed want to be a doctor, we *cannot* conclude that 6% of *all* American children want to be a doctor. If the survey was not carried out well, there would be nonsampling error. And even if the study was performed well, there would be sampling error.

Two-Way Tables

In Section 3.1, we used a multiple bar graph to compare two categorical variables. We can also compare two categorical variables using a *two-way table*. A **two-way table** is a table in which frequencies correspond to two categorical variables. The categories of one variable are listed vertically on the left side of the table, and the categories of the other variable are listed along the top.

For example, consider the following situation. In a survey of one of the author's statistics classes, students anonymously recorded their gender and whether they had read at least one novel in the past year. A few of the results are shown in Table 13.

Table 13 Genders and Novel Readers

Gender	Read Novel in Past Year?
Male	Yes
Male	No
Female	Yes
Female	No
Male	No
Female	Yes
Female	Yes

Source: *J. Lehmann*

There are four possible pairs: "Male"–"Yes" (in gray), "Male"–"No" (in blue), "Female"–"Yes" (in red), and "Female"–"No" (in green). We summarize the paired data using tally marks in Table 14. For example, for the first data pair, "Male"–"Yes" (in gray), we make a gray tally mark so it lies in the bottom row and far-right column of Table 14.

Table 14 Genders and Novel Readers

Gender	Did Not Read Novel	Read Novel
Female	|	|||
Male	||	|

Source: *J. Lehmann*

Once we have completed the tally, we record the frequencies for each of the four types of paired data and total all rows and columns (see Table 15).

Table 15 Genders and Novel Readers

Gender	Did Not Read Novel	Read Novel	Total
Female	1	3	4
Male	2	1	3
Total	3	4	7

Source: *J. Lehmann*

Figure 22 displays a StatCrunch two-way table of the novel data. For StatCrunch instructions, see Appendix B.6.

Contingency table results:
Rows: Gender
Columns: Novel

	Did Not Read Novel	Read Novel	Total
Female	1	3	4
Male	2	1	3
Total	3	4	7

Figure 22 StatCrunch two-way table of novel data

Using Two-Way Tables to Compute Proportions

In Example 2, we will use a two-way table to find some proportions.

▶ **Example 2** Using a Two-Way Table to Find Proportions

Table 16 summarizes the responses from all 42 students who participated in the survey about whether they had read a novel in the past year.

Table 16 Genders and Novel Readers

Gender	Did Not Read Novel	Read Novel	Total
Female	6	19	25
Male	6	11	17
Total	12	30	42

Source: *J. Lehmann*

1. How many of the students read a novel in the past year?
2. What proportion of the students did not read a novel in the past year?
3. What proportion of the women read a novel in the past year?
4. What proportion of the men read a novel in the past year?

Solution

1. At the bottom of the green column in Table 16, we see that 30 students read a novel in the past year.
2. At the bottom of the yellow column, we see that 12 students did not read a novel. There are a total of 42 students, so the proportion of students who did not read a novel in the past year is $\frac{12}{42} \approx 0.286$. Therefore, approximately 28.6% of the students did not read a novel in the past year.
3. To find *any* proportion of women, we use data only in the green row of Table 17. We see 19 of 25 women read a novel in the past year. So, the proportion is $\frac{19}{25} = 0.76$. Therefore, 76% of the women read a novel in the past year.

Table 17 Data about Women

Gender	Did Not Read Novel	Read Novel	Total
Female	6	19	25
Male	6	11	17
Total	12	30	42

Source: *J. Lehmann*

4. To find *any* proportion of men, we use data only in the yellow row of Table 18. We see 11 of 17 men read a novel in the past year. So, the proportion is $\frac{11}{17} \approx 0.647$. Therefore, approximately 64.7% of the men read a novel in the past year.

Table 18 Data about Men

Gender	Did Not Read Novel	Read Novel	Total
Female	6	· 19	25
Male	6	11	17
Total	12	30	42

Source: *J. Lehmann*

▶ **Example 3**   Using a Two-Way Table to Find Proportions

In 2012, a total of 1824 adults were asked the following question: "Do you favor or oppose the death penalty for persons convicted of murder?" Table 19 compares the adults' responses with their ethnicities.

Table 19 Opinion on Death Penalty versus Ethnicity

Favor or Oppose the Death Penalty	African American	Caucasian	Other	Total
Favor	128	953	108	1189
Oppose	140	414	81	635
Total	268	1367	189	1824

Source: *General Social Survey*

1. Find the proportion of adults in the survey who oppose the death penalty OR are Caucasian.
2. Find the proportion of adults in the survey who oppose the death penalty AND are Caucasian.
3. Find the proportion of Caucasians who oppose the death penalty.
4. Find the proportion of African Americans who oppose the death penalty.
5. A student says that Caucasians in the study are more likely to oppose the death penalty than African Americans in the study because more of the Caucasians (414) oppose the death penalty than the African Americans (140). What would you tell the student?

Solution

1. The surveyed adults who oppose the death penalty are described by the green row, and the surveyed Caucasians are described by the yellow column. So, the number of adults who oppose the death penalty OR are Caucasian is the sum of all the numbers in either the green row or the yellow column, but we do not count 414 twice: $953 + 140 + 414 + 81 = 1588$. There are a total of 1824 adults, so the proportion is $\frac{1588}{1824} \approx 0.871$.

2. The surveyed adults who oppose the death penalty are described by the green row, and the surveyed Caucasians are described by the yellow column. So, the number of adults who oppose the death penalty AND are Caucasian is in *both* the green row and the yellow column. That number is 414 adults (out of 1824 adults). So, the proportion is $\frac{414}{1824} \approx 0.227$.

3. Because we are to consider only Caucasians, we use only the numbers in the yellow column. Because 414 out of 1367 Caucasians oppose the death penalty, the proportion is $\frac{414}{1367} \approx 0.303$.

4. Because we are to consider only African Americans, we use only the numbers in the yellow column of Table 20. Because 140 out of the 268 African Americans oppose the death penalty, the proportion is $\frac{140}{268} \approx 0.522$.

Table 20 Opinion on Death Penalty versus Ethnicity

Favor or Oppose the Death Penalty	African American	Caucasian	Other	Total
Favor	128	953	108	1189
Oppose	140	414	81	635
Total	268	1367	189	1824

5. Although it is true that more of the surveyed Caucasians opposed the death penalty than the surveyed African Americans, a lot more Caucasians were surveyed than African Americans. In fact, from our results in Problems 3 and 4, we see the proportion of Caucasians who oppose the death penalty (0.303) is less than the proportion of African Americans who oppose the death penalty (0.522). So, the Caucasians were less likely to oppose the death penalty than the African Americans.

WARNING   In Problem 1 of Example 3, we added all the numbers in either the green row or the yellow column of Table 19, but we did not count 414 twice. In general, when finding a proportion that involves an OR statement, avoid double-counting.

WARNING   It might seem like the results of Problems 2 and 3 in Example 3 should be equal, but the groups of people we are to consider are different. Study these two problems carefully until you are confident that you understand what is going on.

Group Exploration

Association versus causation

In a survey of 56,629 American adults, the religious preferences for four age groups are shown in Table 21.

1. Find the proportion of the adults ages 19 to 30 years who are Protestants. Do the same for the adults ages 31 to 40 years. Compare your results.

Table 21 Religious Preference versus Age

| Religious Preference | Age Group (years) | | | | |
	19–30	31–40	41–55	56–89	Total
Protestant	6650	6473	8636	11,137	32,896
Catholic	3733	3248	3666	3924	14,571
Other	803	857	898	701	3259
None	2181	1497	1290	935	5903
Total	13,367	12,075	14,490	16,697	56,629

Source: *General Social Survey*

2. A student says that the adults ages 19 to 30 years are more likely to be Protestants than the adults ages 31 to 40 years because there are more Protestants ages 19 to 30 years (6650) than there are Protestants ages 31 to 40 years (6473). What would you tell the student?

3. Find the proportion of the adults ages 41 to 55 years who are Protestants. Do the same for adults ages 56 to 89 years.

4. What pattern do you notice in the four proportions you found in Problems 1 and 3?

5. On the basis of the pattern you described in Problem 4, a student concludes that as Catholics grow older, they tend to become Protestants. What would you tell the student? Include in your response a discussion about sample, population, inference, association, and causality.

6. Find four other proportions that show there is another pattern. What does the pattern mean in this situation?

Homework 3.2

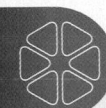

For extra help ▶ **MyMathLab®**  Watch the videos in MyMathLab  Download the MyDashboard App

1. In a pie chart, the proportion of the total area for one slice is equal to the ____ frequency for the category represented by the slice.

2. The sum of all the percentages indicated by a pie chart is equal to ____.

3. In a two-way table, we compare two ____ variables.

4. *True or False:* When using a two-way table to find a proportion that involves an OR statement, avoid double-counting.

5. From 2008 to 2012, college students were surveyed about how they spend their time on weekdays. Their responses and the percentages are described by the pie chart in Fig. 23.

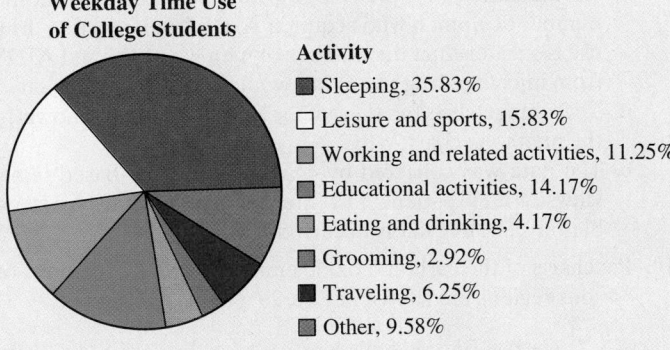

Weekday Time Use of College Students

Activity
- Sleeping, 35.83%
- Leisure and sports, 15.83%
- Working and related activities, 11.25%
- Educational activities, 14.17%
- Eating and drinking, 4.17%
- Grooming, 2.92%
- Traveling, 6.25%
- Other, 9.58%

Figure 23 Exercise 5
(**Source:** *Bureau of Labor Statistics*)

a. What variable is described by the pie chart? Is it categorical or numerical? Explain.
b. What proportion of a weekday does a typical college student spend NOT doing educational activities?
c. What proportion of a weekday does a typical college student spend grooming OR traveling?
d. Estimate the number of hours a typical college student spends doing educational activities Monday through Friday. Round to the ones place.

e. On the basis of the pie chart, a student says that a typical college student spends 14.17% of the *weekend* doing educational activities. What would you tell the student?

6. In Exercise 32 of Homework 3.1, you analyzed reasons why fathers stayed at home in 2012. Their responses and the percentages are described by the pie chart in Fig. 24.

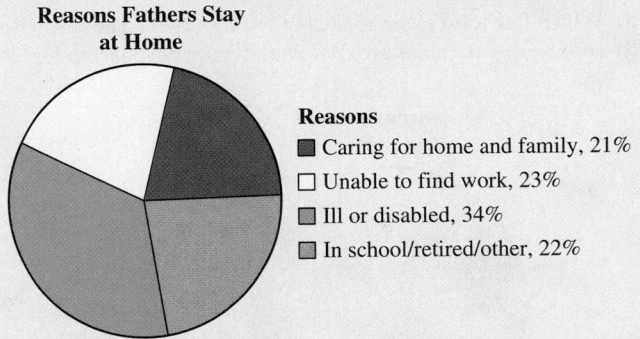

Reasons Fathers Stay at Home

Reasons
- Caring for home and family, 21%
- Unable to find work, 23%
- Ill or disabled, 34%
- In school/retired/other, 22%

Figure 24 Exercise 6
(**Source:** *Pew Research Center*)

a. What variable is described by the pie chart? Is it categorical or numerical? Explain.
b. A father might stay at home for multiple reasons. For example, a father might stay at home because he is ill and because he can't find work. How can we tell from the pie chart that for each father, only one reason has been described?
c. Find the proportion of fathers surveyed whose main reason was caring for his home and family.
d. Find the proportion of fathers surveyed whose main reason was NOT caring for his home and family.
e. Find the proportion of fathers surveyed whose main reason was they were unable to find work OR were caring for their home and family.

7. The percentages of murders in 2010 carried out by using various types of weapons are described by the pie chart in Fig. 25.

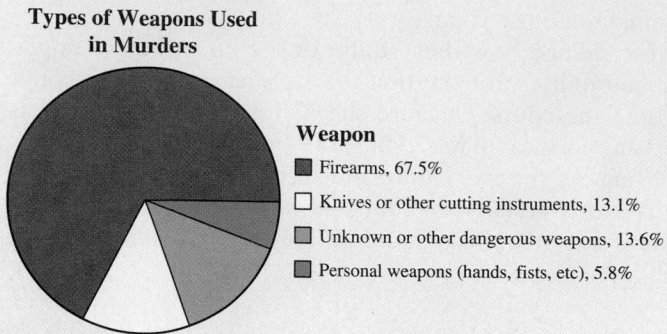

Types of Weapons Used in Murders

Weapon
- ■ Firearms, 67.5%
- □ Knives or other cutting instruments, 13.1%
- ■ Unknown or other dangerous weapons, 13.6%
- ■ Personal weapons (hands, fists, etc), 5.8%

Figure 25 Exercise 7
(**Source:** *FBI*)

a. Find the percentage of murders that were carried out using firearms OR knives or other cutting instruments.
b. If a murder victim was both stabbed with a knife and shot with a handgun, how do we know that only one of the two weapons was counted as the murder weapon in the pie chart?
c. Construct a relative frequency bar graph that describes the data.
d. Does a pie chart contain more, less, or the same information as a relative frequency bar graph? Explain.

8. Which fast food chain makes the best burger? The responses of surveyed Americans are described by the pie chart in Fig. 26.

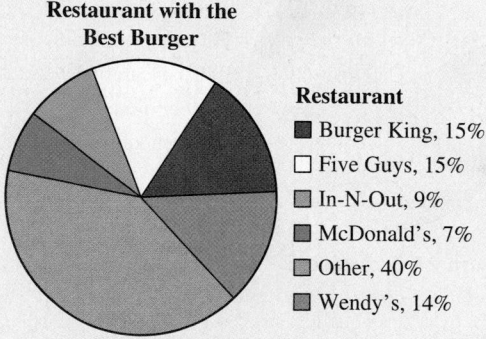

Restaurant with the Best Burger

Restaurant
- ■ Burger King, 15%
- □ Five Guys, 15%
- ■ In-N-Out, 9%
- ■ McDonald's, 7%
- ■ Other, 40%
- ■ Wendy's, 14%

Figure 26 Exercise 8
(**Source:** *YouGov*)

a. Find the percentage of respondents who said Five Guys® OR In-N-Out® have the best burgers.
b. All the restaurants included in the "other" category have percentages less than McDonald's®. What is the minimum number of restaurants in the "other" category?
c. Construct a relative frequency bar graph that describes the data.
d. Does a pie chart contain more, less, or the same information as a bar graph? Explain.

9. The most likely ways (transmission routes) women and men acquired HIV and the percentages are described by the pie charts in Figs. 27 and 28 for women and men, respectively.

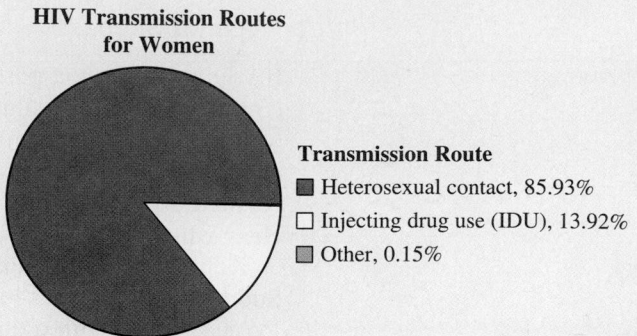

HIV Transmission Routes for Women

Transmission Route
- ■ Heterosexual contact, 85.93%
- □ Injecting drug use (IDU), 13.92%
- ■ Other, 0.15%

Figure 27 Transmission routes of HIV for women
(**Source:** *Centers for Disease Control and Prevention*)

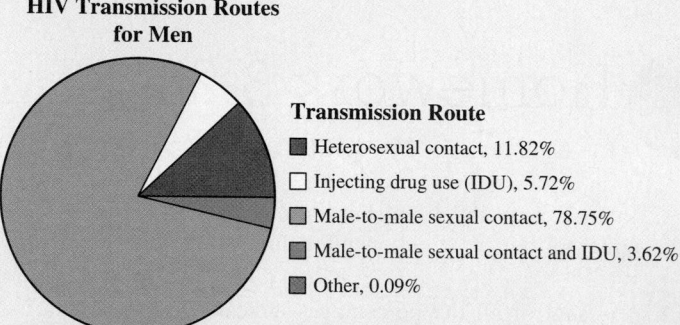

HIV Transmission Routes for Men

Transmission Route
- ■ Heterosexual contact, 11.82%
- □ Injecting drug use (IDU), 5.72%
- ■ Male-to-male sexual contact, 78.75%
- ■ Male-to-male sexual contact and IDU, 3.62%
- ■ Other, 0.09%

Figure 28 Transmission routes of HIV for men
(**Source:** *Centers for Disease Control and Prevention*)

a. What percentage of the women who acquired HIV were injecting drug users?
b. What percentage of men who acquired HIV were injecting drug users? [**Hint:** Read the descriptions of all the slices carefully.]
c. On the basis of the pie charts, a student concludes that the number of women who acquired AIDS from injecting drug use is greater than the number of men who acquired AIDS from injecting drug use. What would you tell the student?
d. The orange slice for the "other" category is not shown in the men's pie chart. Why?
e. The data was collected by confidential name-based interviews of HIV patients just after diagnosis. What type(s) of bias in sampling likely occurred?

10. Pie charts of the carbon dioxide emissions and populations of various regions are shown in Figs. 29 and 30, respectively.

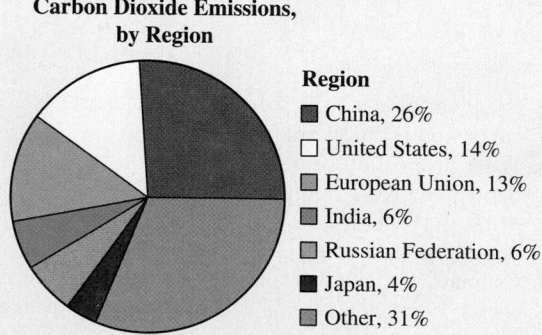

Carbon Dioxide Emissions, by Region

Region
- ■ China, 26%
- □ United States, 14%
- ■ European Union, 13%
- ■ India, 6%
- ■ Russian Federation, 6%
- ■ Japan, 4%
- ■ Other, 31%

Figure 29 Carbon dioxide emissions of regions
(**Source:** *Carbon Dioxide Information Analysis Center*)

Populations, by Region

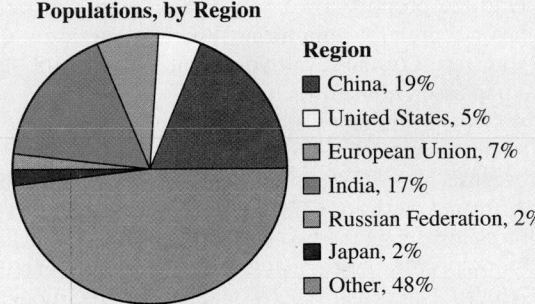

Region
- ■ China, 19%
- □ United States, 5%
- ■ European Union, 7%
- ■ India, 17%
- ■ Russian Federation, 2%
- ■ Japan, 2%
- ■ Other, 48%

Figure 30 Populations of regions
(**Source:** *Organization for Economic Cooperation and Development*)

a. Of the six regions specified in the pie charts, which region emits the most carbon dioxide?

b. Of the six regions specified in the pie charts, which region has the largest population?

c. What percentage of worldwide carbon dioxide emissions does the United States emit? What percentage of world population lives in the United States? Why could one argue that the United States emits more than its fair share of carbon dioxide?

d. On the basis of the argument described in part (c), what other regions can you say for sure emit more than their fair share of carbon dioxide?

e. Explain why your list of regions in part (d) may not be complete.

11. A total of 2022 American adults were surveyed about their opinion on the degree to which political leaders talk about their faith and prayer. Their responses are summarized in Table 22.

Table 22 Opinions on the Degree to Which Political Leaders Talk about Their Faith and Prayer

Responses	Frequency
Too Much	601
Too Little	821
Right Amount	460
Not Sure	140
Total	2022

Source: *Pew Research Center*

a. Construct a pie chart. Round the percentages to the first decimal place.

b. What proportion of those surveyed think that political leaders talk too much about their faith and prayer?

c. What proportion of those surveyed think that political leaders talk the right amount OR too much about faith and prayer?

d. What proportion of those surveyed do NOT think political leaders talk the right amount about faith and prayer?

e. On the basis of the information in Table 22, a student concludes that if politicians want to gain more popularity with Americans, they should talk more about their faith and prayer because the response "too little" has the largest frequency. Identify two problems with the student's logic.

12. A total of 8719 American adults were asked, "How often do you unplug from technology?" Their responses are summarized in Table 23.

Table 23 Responses to "How often do you unplug from technology?"

Responses	Frequency
Never	3749
A few times a year	1482
Once a month	349
Once a week	523
A few times a week	872
Daily	1744
Total	8719

Source: The Wall Street Journal

a. Construct a pie chart. Round the percentages to the first decimal place.

b. What proportion of those surveyed never unplug from technology?

c. What proportion of those surveyed do NOT unplug from technology daily?

d. What proportion of those surveyed unplug from technology a few times a year OR once a month.

e. On the basis of the frequencies in Table 23, a student concludes that 17% of *all* Internet users unplug from technology a few times per year. What would you tell the student?

13. A total of 2452 American adults responded to a survey about their income class and their general health level. Table 24 summarizes their responses.

Table 24 Income Class versus General Health Level

	Health in General			Total
	Excellent	Good	Poor	
Upper Income	223	228	51	502
Middle Income	412	592	283	1287
Lower Income	127	268	268	663
Total	762	1088	602	2452

Source: *Pew Research Center*

a. What proportion of the lower-income adults have excellent health?

b. What proportion of the middle-income adults have excellent health?

c. What proportion of the upper-income adults have excellent health?

d. On the basis of finding the results for parts (a), (b), and (c), a student concludes that an increase in income *causes* an increase in health. Describe two problems with the student's logic.

e. To form the sample, the researchers randomly selected some landline-telephone users and randomly selected some cell-phone users. What type of sampling method is this?

14. A total of 2452 American adults responded to a survey about their income level and their general level of health. Table 24 summarizes their responses.

a. What proportion of the adults who have poor health are lower-income?

b. What proportion of the lower-income adults have poor health?

c. What proportion of the middle-income adults have poor health?

d. A student says that the middle-income adults are more likely to have poor health than the lower-income adults because more of the middle-income adults (283) have poor health than the lower-income adults (268). What would you tell the student?

e. The survey included questions about topics other than income and health. Of the people who participated in the study, 56 refused to reveal their income class and/or level of health. What type(s) of bias in sampling likely occurred? Explain.

15. Table 25 compares class standings with types of body art for 490 undergraduate students surveyed at a state university in the Southwest.

Table 25 Class Standing versus Types of Body Art

Class Standing	Body Piercing Only	Tattoos Only	Both Body Piercing and Tattoos	No Body Art	Total
Freshman	61	7	14	86	168
Sophomore	43	11	10	64	128
Junior	20	9	7	43	79
Senior	21	17	23	54	115
Total	145	44	54	247	490

Source: *Contemporary College Students and Body Piercing, Myrna L. Armstrong, Ed. D., et al.*

a. Find the proportion of those surveyed who have tattoos only.

b. Find the proportion of sophomores surveyed who have both body piercings and tattoos.

c. Find the proportion of juniors surveyed who have both body piercings and tattoos. Compare the result to your result in part (b).

d. A student says that the surveyed sophomores are more likely than the surveyed juniors to have both body piercing and tattoos because more sophomores (10) have both body piercing and tattoos than juniors (7). What would you tell the student?

e. On the basis of the two-way table, a student concludes that 11% of *all* college students in the United States have both body piercings and tattoos. What would you tell the student?

16. A total of 1220 adults were asked the following question: "Some people say that because of past discrimination, African Americans should be given preference in hiring and promotion. Others say that such preference in hiring and promotion of African Americans is wrong because it discriminates against Caucasians. What about your opinion—are you for or against preferential hiring and promotion of African Americans?" Table 26 compares the adults' responses with their ethnicities.

Table 26 Race versus Preference in Hiring African Americans

	Favor Preference	Against Preference	Total
Caucasian	134	795	929
African American	77	103	180
Other	21	90	111
Total	232	988	1220

Source: *General Social Survey*

a. Find the proportion of surveyed Caucasians who favored preference in hiring African Americans.

b. Find the proportion of surveyed African Americans who favored preference in hiring African Americans. Compare your result to the result you found in part (a). Why is this comparison not surprising?

c. A student says that Caucasians in the study are more likely to favor preference than African Americans in the study because more Caucasians (134) favored preference than African Americans (77). What would you tell the student?

d. From inspecting the two-way table, a student concludes that 81.0% of *all* American adults are against preference in hiring African Americans. What would you tell the student?

e. The survey question is quite long (see the opening paragraph of this exercise). Why do you think the first two sentences were included? Discuss whether it would have been better if the phrase "because it discriminates against Caucasians" had been changed to "because it discriminates against those who are not African American."

17. Table 27 compares the genders of patients who died from oral cancer and the types of oral cancer.

Table 27 Types of Oral Cancer versus Gender

	Tongue	Mouth	Other	Total
Women	690	720	380	1790
Men	1380	1080	1260	3720
Total	2070	1800	1640	5510

Source: *National Center for Health Statistics*

a. What proportion of those who died from tongue cancer were men?

b. What proportion of oral cancer deaths were from tongue cancer?

c. What proportion of women who died of oral cancer did NOT die from mouth cancer?

d. What proportion of those who died from oral cancer were women OR patients who died from mouth cancer?

e. What proportion of those who died from oral cancer were women AND patients who died from mouth cancer?

18. A total of 2397 American adults responded to a survey about their income class and whether their financial situation is better, worse, or no different than before the recession started in December 2007. Table 28 summarizes their responses.

Table 28 Income Class versus whether Financial Situation Is Better, Worse, or No Different than Before Recession

	Financial Situation			
	Better	Worse	No Different	Total
Upper Income	213	172	106	491
Middle Income	412	541	296	1249
Lower Income	161	389	107	657
Total	786	1102	509	2397

Source: *Pew Research Center*

a. What proportion of those surveyed have financial situations that are no different than before the recession?

b. What proportion of the upper-income adults' financial situations are better than before the recession started?

c. What proportion of the adults are middle-income OR have financial situations that are worse than before the recession started?

d. What proportion of the adults are middle-income AND have financial situations that are worse than before the recession started?

e. The survey included questions about topics other than income and financial situation. Of the people who participated in the study, 111 refused to reveal their income class and/or whether their financial situation had changed since the start of the recession. What type(s) of bias in sampling likely occurred? Explain.

19. **DATA** The author surveyed students in one of his statistics classes about whether they smoke and whether they run red traffic lights (see Table 29). The survey was anonymous, so the students' names are fictional.

Table 29 Smoking versus Running Red Lights

Student	Smoke?	Run Lights?	Student	Smoke?	Run Lights?
1. Chris	No	No	6. Victoria	No	No
2. Ada	No	Yes	7. Joseph	No	No
3. Spencer	Yes	No	8. Kristen	Yes	Yes
4. Stephen	No	Yes	9. Rachel	No	No
5. Alan	No	No	10. Sue	No	No

Source: *J. Lehmann*

Summarize the data by completing Table 30.

Table 30 Smoking versus Running Red Lights

	Run Lights	Don't Run Lights	Total
Smoke			
Don't Smoke			
Total			

20. **DATA** In a survey of one of the author's statistics classes, 42 students recorded whether they were working. A few of the results are shown in Table 31.

Table 31 Gender versus Working

Student	Gender	Working?
1. Charysse	Female	Yes
2. Peggy	Female	Yes
3. Arthur	Male	Yes
4. Jorge	Male	No
5. Marco	Male	Yes
6. Jenna	Female	No
7. Karin	Female	Yes
8. Brandon	Male	Yes
9. Marisha	Female	No
10. Holly	Female	Yes

Source: *J. Lehmann*

Summarize the data by completing Table 32.

Table 32 Gender versus Working

	Working	Not Working	Total
Female			
Male			
Total			

21. In Exercise 19, you analyzed 10 students' responses to a survey about whether they smoked and whether they ran red lights. Table 33 shows the responses of all 23 students who anonymously responded to both questions.

Table 33 Smoking versus Running Red Lights

	Run Lights	Don't Run Lights	Total
Smoke	1	3	4
Don't Smoke	3	16	19
Total	4	19	23

Source: *J. Lehmann*

a. What proportion of the students run red lights OR smoke?

b. What proportion of the students run red lights AND smoke?

c. What proportion of the students run red lights, given that they smoke?

d. What proportion of the students run red lights, given that they do not smoke? Compare your result to the result you found in part (c).

e. A total of 5 students did not say whether they run red lights and 1 other student did not say whether he or she smokes. What type(s) of bias in sampling likely occurred? Assuming each of the 6 students did at least one of those things, how would combining the data about those 6 students with the data about the other 23 students affect your result to part (a)? Explain why there is not enough information to know how your results to parts (b), (c), and (d) would be affected.

22. In Exercise 20, you analyzed 10 students' responses to a survey about their gender and whether they were working. Table 34 shows the responses of all 42 students.

Table 34 Gender versus Working

	Working	Not Working	Total
Female	18	7	25
Male	11	6	17
Total	29	13	42

Source: *J. Lehmann*

a. What proportion of the male students were working?

b. What proportion of the female students were working? Compare your result to the result you found in part (a).

c. What proportion of the students were female AND were working?

d. A student says that the results for parts (b) and (c) should be equal because in both parts we are interested in female students who were working. What would you tell the student?

e. What proportion of the students were female OR were working?

Concepts

For Exercises 23–28, choose all the following tables and diagrams that would be appropriate ways to describe the given data: frequency and relative frequency table, frequency bar graph, relative frequency bar graph, multiple bar graph, pie chart, two-way table.

23. A statistician compares the political party affiliation and type of work of 96 adults.

24. The grades of 27 prestatistics students are recorded.

25. Each of 50 teenagers state their favorite music genre.

26. A researcher compares the gender and eye color of 79 children.

27. A researcher compares the ethnicity of individuals in an experiment and whether they are in the treatment group or control group.

28. The breed of each dog in an animal shelter is recorded.

29. When analyzing data described by a two-way table, is it more meaningful to compare frequencies or relative frequencies? Explain.

30. The percentages of world population who have unfavorable views on various religious groups are shown in Table 35. Why don't the percentages add up to 100%? Would it be appropriate to use a pie chart to describe the data? Explain.

Table 35 Percentages of World Population Who Have Unfavorable Views on Various Religious Groups

Religious Group	Percent
Muslims	24
Jews	21
Hindus	18
Christians	15
Buddhists	14

Source: *First International Resources*

31. Compare the ease of constructing and analyzing a pie chart and a relative frequency bar graph. Which diagram is easier to construct by hand? by using technology? Which diagram makes it easier to compare the relative frequencies of two categories? Which diagram better illustrates how the frequency of a category compares to the whole? Explain.

32. The number of car thefts in 2013 for the top 5 most-stolen cars are shown in Table 36. What other piece of information do you need in order to construct a pie chart for *all* car thefts?

Table 36 Numbers of Stolen Cars

Car	Number of Stolen Cars
Honda Accord®	53,995
Honda Civic®	45,001
Chevrolet Silverado®	27,809
Ford F-150®	26,494
Toyota Camry®	14,420

Source: *National Insurance Crime Bureau*

Hands-On Research

33. Find a bar graph, pie chart, or two-way table in a blog, newspaper, magazine, or journal.

   a. Define each variable.
   b. What is the point of the article? Does the diagram or table support that point? Explain.
   c. Could one or both of the other diagrams or table have been used instead? If so, identify which one(s) and determine which of all the diagrams and table would best support the point of the article. Explain.

34. Collect data that involves one or two categorical variables. Your study should include 20 individuals.

   a. For each variable, state the definition.
   b. If you performed a survey, give the exact wording of your question(s). If you did not perform a survey, describe how you made your observations.
   c. Provide the data you collected.
   d. Use a statistical diagram or two-way table to describe the data.
   e. What conclusion(s) can you draw about the 20 individuals?
   f. If the 20 individuals were a sample, what group of individuals could be the population? In this case, would your sampling have bias? If so, which type(s)? Explain. Would it be OK to draw the same conclusion(s) about the population that you drew about the sample?

▼ 3.3 Dotplots, Stemplots, and Time-Series Plots

Objectives

» Identify *discrete variables* and *continuous variables*.

» Construct and interpret *dotplots.*

» Identify *outliers* of a distribution.

» Find *percentiles* of a distribution.

» Measure the center of a distribution.

» Construct and interpret *stemplots.*

» Construct and interpret *time-series plots.*

Have you ever taken a test and wondered how well you did compared to others in the class? In this section, we will construct various graphs that can help us answer such questions.

Discrete and Continuous Variables

In Sections 3.1 and 3.2, we used tables and diagrams to describe categorical variables. In this section, we will use tables and diagrams to describe numerical variables. There are two types of numerical variables: *discrete* and *continuous*.

A discrete variable has gaps between successive, possible values. For example, the number of friends a person has is a discrete variable. Possible values of the variable are 0, 1, 2, 3, ... Because a person cannot have a fraction of a friend, the variable cannot take on values between 0 and 1, between 1 and 2, and so on.

▶ **Definition   Discrete variable**

A **discrete variable** is a variable that has gaps between successive, possible values.

A continuous variable can take on any value between two possible values. For example, the height (in inches) of a person is a continuous variable. Not only are the heights 66 inches and 67 inches possible values, but every height between them, such as 66.71285 inches, is also possible.

> **Definition**   Continuous variable
>
> A **continuous variable** is a variable that can take on any value between two possible values.

▶ **Example 1**   Identifying Discrete and Continuous Variables

Identify whether the variable is discrete or continuous.

1. the amount of time (in seconds) it takes a person to run 100 meters
2. the number of times a person has traveled to the Grand Canyon

Solution

1. The time (in seconds) it takes a person to run 100 meters is a continuous variable because the variable can take on any value between two possible values. For example, not only are 10 seconds and 11 seconds possible times, but the time 10.59329 seconds is also possible.
2. The number of times a person has traveled to the Grand Canyon is a discrete variable because there are gaps between successive values. Possible values are 0, 1, 2, 3, ... The variable cannot take on values between 0 and 1, between 1 and 2, and so on. For example, a person cannot travel to the Grand Canyon 3.2 times.

WARNING   When identifying a variable as discrete or continuous, we consider the possible values of the variable *before* rounding. So, even though 100-meter-run times might be rounded to the second decimal place, the variable running time is continuous.

Identifying whether a variable is discrete or continuous and to what decimal place the data may have been rounded can help us select a type of diagram to describe the data.

Dotplots

The students' scores on the first test in one of the author's applied calculus classes are shown in Table 37.

Table 37 Scores on Test 1 (points)

68	80	38	83	73	98	75	95	93	88
85	60	80	85	78	85	90	100	98	75
70	62	93	65	100	88	93	95	65	95
85	88	68	90	100					

Source: *J. Lehmann*

Because there are so many scores, it's hard to take in all the information. We can learn about the data by constructing a special graph called a *dotplot*. To construct a **dotplot**, for each observation, we plot a dot above the number line, stacking dots as necessary.

▶ **Example 2**   Constructing and Interpreting a Dotplot

1. Construct a dotplot of the test scores shown in Table 37.
2. What observation occurred the most?
3. Each student who scored at most 69 points did not pass the test. How many students did not pass the test?

4. Each student who scored at least 90 points received a grade of A. What proportion of students earned an A on the test?

5. Identify any observations that are quite a bit smaller or larger than the other observations. What are some possible reasons that this happened?

Solution

1. Plotting points on a dotplot is a lot like plotting points on a number line, but we draw the dots above the number line (see Fig. 31). The first step is to determine that the lowest score is 38 points and the three highest scores are all 100 points (see Table 37). So, we write the numbers 35, 40, 45, …, 100 equally spaced on the number line and write the units "Points" below the numbers. Then for each data value, we draw a dot above the value of the observation, stacking dots if the same value occurs more than once. All the dots should be the same size.

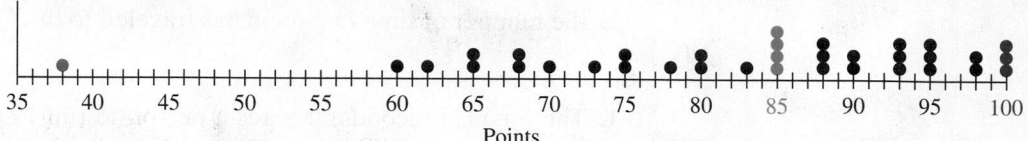

Figure 31 Dotplot of test scores

2. The most number of dots above a number is 4 dots, which are above the test score 85 (see the four green dots in Fig. 31). So, the observation 85 points occurred the most.

3. The phrase "at most 69 points" means less than or equal to 69 points. There are 7 dots to the left of 69 points (see the blue dots in Fig. 32). So, 7 students did not pass the test.

4. The phrase "at least 90 points" means 90 or more points. There are 13 dots at 90 points or to the right of 90 points (see the red dots in Fig. 32). So, 13 out of 35 students earned As on the test. The proportion is $\dfrac{13}{35} \approx 0.371$.

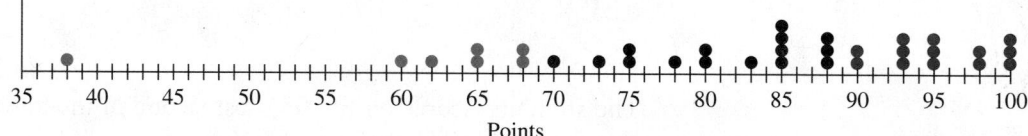

Figure 32 Dotplot of test scores

5. The score 38 points (see the blue dot in Fig. 33) is quite a bit lower than the other scores. The student who earned that score might have suffered from math anxiety, have a weak math background, or have skipped studying for the test.

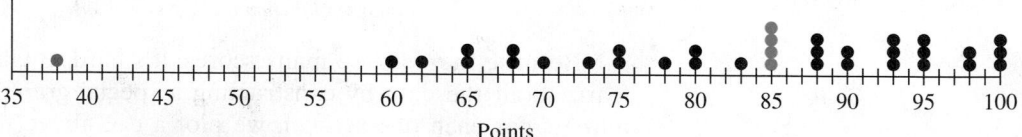

Figure 33 Dotplot of test scores

WARNING   In Example 2 we wrote the numbers 35, 40, 45, …, 100 on the number line of the dotplot (see Fig 33). Recall from Section 1.1 that when we write numbers on a number line, they should increase by an equal amount and be equally spaced.

In Example 2 we used a dotplot to describe test scores, which is a discrete variable. **We tend to use dotplots to describe data values of discrete variables, but they can be used to describe data values of continuous variables, too.** If a data set is large, each dot can represent a certain equal number of observations. This should be clearly stated in the diagram.

In Fig. 33, we see that there are 4 scores equal to 85 points (see the green dots). We say that the score 85 points has *frequency* 4.

> **Definition** Frequency of an observation
>
> The **frequency of an observation** of a numerical variable is the number of times the observation occurs in the group of data.

The *frequency distribution* of the test scores is all the test scores together with their frequencies.

> **Definition** Frequency distribution of a numerical variable
>
> The **frequency distribution of a numerical variable** is the observations together with their frequencies.

Although it can be helpful to construct one or two dotplots by hand to fully understand them, we will usually construct dotplots using technology. A dotplot of the test scores constructed by StatCrunch is shown in Fig. 34. For StatCrunch instructions, see Appendix B.7.

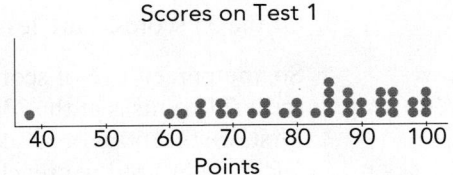

Figure 34 Dotplot of test scores

Outliers

In Example 2, we determined that the test score 38 points is quite different (smaller) than the other scores. We say that 38 points is an *outlier*. In general, an **outlier** is an observation that is quite a bit smaller or larger than the other observations. We will define an outlier more precisely in Section 4.3.

Percentiles

Suppose you take an aptitude test and your score on managing others is 27 points. The score alone tells us nothing. Even if we were told that the test is out of 40 points, we would not know how to interpret your score because we would not know how points were assigned. But if we were told your score is greater than or equal to approximately 98% of others' scores (and less than approximately 2% of others' scores), we would know that you have strong management skills. We say 27 points is at the *98th percentile* (see Fig. 35).

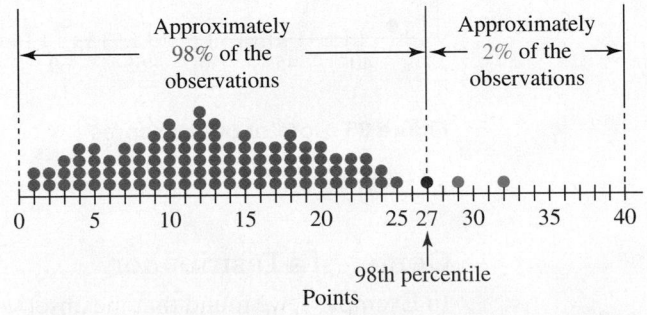

Figure 35 27 points is at the 98th percentile

> ▶ Definition   Percentile
>
> The **kth percentile** of some data is a value (not necessarily a data value) that is greater than or equal to approximately $k\%$ of the observations and is less than approximately $(100 - k)\%$ of the observations.

▶ **Example 3**   Percentiles

A dotplot of the Test 1 scores is shown in Fig. 36.

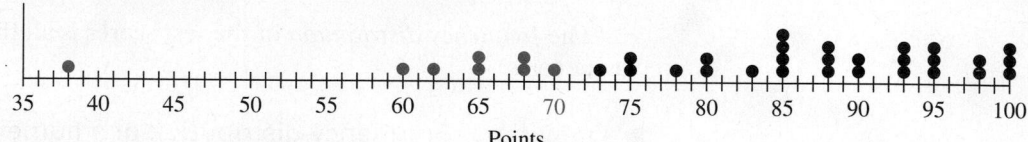

Figure 36 Dotplot of test scores

1. Find the percentile of the test score 70 points.
2. Find the 50th percentile.
3. Suppose the author decides that the test he gave was too difficult. In each past applied calculus class, his cutoff for a C on the first test was approximately at the 15th percentile. He decides to set the cutoff at the 15th percentile. Find its value.

Solution

1. Of the 35 scores, 8 are less than or equal to 70 points (see the blue dots in Fig. 36).

   So, the percentage of scores less than or equal to 70 is $\dfrac{8}{35} \approx 0.23 = 23\%$. So, the score 70 points is at the 23rd percentile.

2. First, we find 50% of 35 scores: $0.50(35) \approx 18$ scores. Then we count the dots from left to right until we reach the 18th dot, which is at 85 points (see the red dots in Fig. 37). So, 85 points is at the 50th percentile.

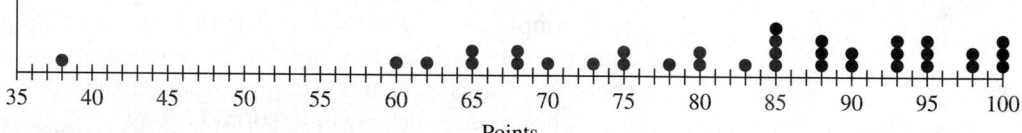

Figure 37 Dotplot of test scores

3. First, we find 15% of the 35 scores: $0.15(35) \approx 5$ scores. Then we count the dots from left to right until we reach the 5th dot, which is at 65 points (see the 5 blue dots in Fig. 38). So, the cutoff should be 65 points.

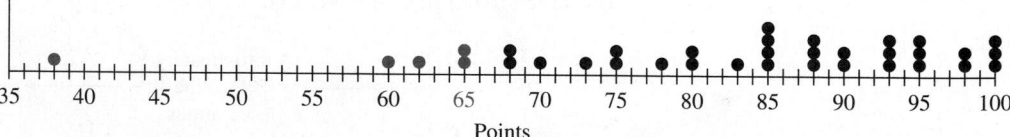

Figure 38 Dotplot of test scores

Center of a Distribution

In Example 3, we found that the observation 85 points is at the 50th percentile of the Test 1 distribution. So, about 50% (half) of the observations are less than or equal to 85 points and about 50% (half) of the observations are greater than 85. This suggests that the 50th percentile (85 points) is a reasonable measure of the center. In Section 4.1 we will discuss various methods to measure the center that often give different results, but **for this chapter we will always use the 50th percentile to measure the center of a distribution.**

Stemplots

A **stemplot** (or **stem-and-leaf plot**) breaks up each data value into two parts: the **leaf**, which is the rightmost digit, and the **stem**, which is the other digits. For the data value 375, the leaf is 5 and the stem is 37. A stemplot of 375 is shown in Fig. 39.

To create a stemplot of the data values 375 and 379, we list both of the leaves, 5 and 9, but only write the stem, 37, once (see Fig. 40).

Stem (tens)	Leaf (ones)
37	5

Figure 39 Stemplot of 375

Stem (tens)	Leaf (ones)
37	5 9

Figure 40 Stemplot of 375 and 379

A stemplot of the data values 375, 379, 384, 391, 398, and 416 is shown in Fig. 41.

Stem (ones)	Leaf (tenths)
37	5 9
38	4
39	1 8
40	
41	6

Figure 41 Stemplot of 375, 379, 384, 391, 398, and 416

▶ **Example 4**   Constructing a Stemplot

Construct a stemplot for the test scores we analyzed in Example 2 (see Table 38).

Table 38 Scores on Test 1 (points)

68	80	38	83	73	98	75	95	93	88
85	60	80	85	78	85	90	100	98	75
70	62	93	65	100	88	93	95	65	95
85	88	68	90	100					

Solution

The first data value in Table 38 is 68. The leaf is the rightmost digit, 8, and the stem is the other number, 6. Because the numbers range from 38 to 100, the stems range from 3 to 10, inclusive. We construct a stemplot for 68 and the other data values in Fig. 42.

Then we construct a final stemplot, where in each row we list the leaves from smallest to largest (see Fig. 43).

Stem (tens)	Leaf (ones)
3	8
4	
5	
6	8 0 2 5 5 8
7	3 5 8 5 0
8	0 3 8 5 0 5 5 8 5 8
9	8 5 3 0 8 3 3 5 5 0
10	0 0 0

Figure 42 Stemplot of scores on Test 1

Stem (tens)	Leaf (ones)
3	8
4	
5	
6	0 2 5 5 8 8
7	0 3 5 5 8
8	0 0 3 5 5 5 5 8 8 8
9	0 0 3 3 3 5 5 5 8 8
10	0 0 0

Figure 43 Stemplot of scores on Test 1 with leaves arranged from smallest to largest

The numbers are fairly bunched up in the stemplot in Fig. 43. We can address this by constructing a *split stem*. A **split stem** is a stemplot which lists the leaves from 0 to 4 in one row and lists the leaves from 5 to 9 in the next. A StatCrunch split stem of the Test 1 scores is shown in Fig. 44. The split stem offers a more detailed view of the spread. See Appendix B.8 for StatCrunch instructions.

Variable: Test 1

Decimal point is 1 digit(s) to the right of the colon.
Leaf unit = 1

```
 3 :  8
 4 :
 4 :
 5 :
 5 :
 6 :  02
 6 :  5588
 7 :  03
 7 :  558
 8 :  003
 8 :  5555888
 9 :  00333
 9 :  55588
10 :  000
```

Figure 44 StatCrunch split stem of scores on Test 1

Remember that we rarely construct diagrams such as dotplots and stemplots by hand. Usually we use technology to do this, especially when there are a lot of observations.

In Example 4, we used a stemplot to describe test scores, which we've determined is a discrete variable. Stemplots work well with discrete variables, but they also work fine with continuous variables, provided the data values have all been rounded to the same decimal place. In Example 5, we will work with female college students' shot-put distances, which is a continuous variable. The observations have been rounded to the first decimal place.

▶ **Example 5**   Constructing a Stemplot

In a 2013 study, researchers compared the strength of 28 female college shot-putters with their personal best shot-put distances. The distances are shown in Table 39.

Table 39 Personal Best Shot-Put Distances (meters)

18.6	15.8	17.0	10.2	12.7	13	13.9	12.4	17.9	15.5
15.8	16.8	13.0	17.8	14.8	15.5	16.1	16.2	17.1	15.9
15.8	16.7	17.6	16.8	6.4	18.2	19.2	13.2		

Source: A Pilot Study Exploring the Quadratic Nature of the Relationship of Strength to Performance among Shot Putters, *L. W. Judge et al.*

1. Construct a stemplot.
2. If there were exactly one outlier, which observation would it be?
3. At what percentile is the observation 17.1 meters?
4. Measure the center of the distribution by finding the 50th percentile.

Solution

1. For the first observation, 18.6, the leaf is 6 and the stem is 18. We construct a stemplot for 18.6 and the other observations in Fig. 45. Then we write the leaves from smallest to largest in Fig. 46.
2. If there were exactly one outlier, then it would be 6.4 meters because it is quite a bit smaller than the other observations.

3. Of the 28 observations, 22 are less than or equal to 17.1 meters (see the blue leaves in Fig. 46). So, the percentage of observations less than or equal to 17.1 meters is $\frac{22}{28} \approx 0.79 = 79\%$. So, 17.1 meters is at the 79th percentile.

4. First, we find 50% of 28 observations: $0.50(28) = 14$. Then we count the (blue) leaves in Fig. 47 from left to right and top to bottom (like we read a book) until we reach the 14th leaf, which is 8. The stem for the leaf 8 is 15. So, the 50th percentile is 15.8. The observation 15.8 appears to be in the center of the distribution, which checks.

Stem (ones)	Leaf (tenths)	Stem (ones)	Leaf (tenths)	Stem (ones)	Leaf (tenths)
6	4	6	4	6	4
7		7		7	
8		8		8	
9		9		9	
10	2	10	2	10	2
11		11		11	
12	7 4	12	4 7	12	4 7
13	0 9 0 2	13	0 0 2 9	13	0 0 2 9
14	8	14	8	14	8
15	8 5 8 5 9 8	15	5 5 8 8 8 9	15	5 5 8 8 8 9
16	8 1 2 7 8	16	1 2 7 8 8	16	1 2 7 8 8
17	0 9 8 1 6	17	0 1 6 8 9	17	0 1 6 8 9
18	6 2	18	2 6	18	2 6
19	2	19	2	19	2

Figure 45 Stemplot of shot puts

Figure 46 Stemplot of shot puts

Figure 47 Stemplot of shot puts

WARNING When constructing a stemplot, it is important to remember to indicate the units of the stems and leaves. After all, there's a big difference between earning 98 points and 9.8 points on a test!

Stemplots are awkward to use if there is a large number of observations because the stemplot will be really tall, really wide, or both. In sum, **stemplots work best with a small number of observations whose variable is discrete or continuous with rounded values.**

Time-Series Plots

In Section 1.1, we described the relationship between two variables by plotting points in a coordinate system. To construct a **time-series plot**, we plot points in a coordinate system where the horizontal axis represents time and the vertical axis represents some other quantity, and we draw line segments to connect each pair of successive dots (see Fig. 48).

▶ **Example 6** Constructing a Time-Series Plot

Total annual U.S. revenues of products with gluten-free labels are shown in Table 40 for various years.

Table 40 Total Annual U.S. Revenues of Products with Gluten-Free Labels

Year	Total Annual U.S. Revenue (billions of dollars)
2010	12
2011	13
2012	17
2013	20
2014	23

Source: *Nielsen*

1. Construct a time-series plot of the data.
2. Describe how the total annual U.S. revenue of products with gluten-free labels has changed over the years.

Solution

1. For a time-series plot, we always describe the time (in this situation calendar years) on the horizontal axis (see Fig. 48). Because the earliest year is 2010 and the latest year is 2014, we write 2010, 2011, 2012, 2013, and 2014 equally spaced on the horizontal axis and write the units "Year" below them. Because the smallest total annual revenue is $12 billion and the largest total annual revenue is $23 billion, we write the numbers 10, 12, 14, 16, 18, 20, 22, and 24 equally spaced on the vertical axis and write the units "Billions of dollars" along the vertical axis. Then for the data in the first row of Table 40, we plot a dot above 2010 and across from 12 (see the red dot). In a similar way, we plot a dot for each of the remaining rows in the table. Finally, we use line segments to connect each pair of successive dots.

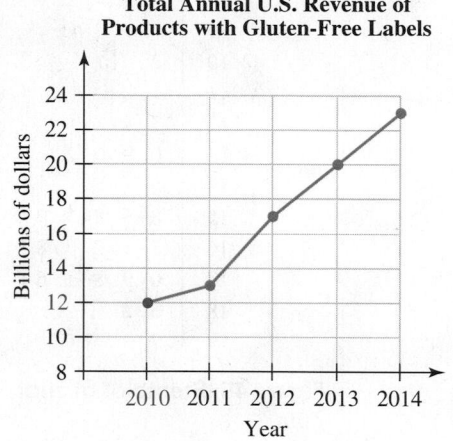

Figure 48 A time-series plot of the gluten-free data

2. The total annual U.S. revenue of products with gluten-free labels has increased greatly over time.

Even though the total annual U.S. revenue of products with gluten-free labels greatly increased for the years shown in Table 40, we *cannot* assume the it will continue to increase. A new factor could come into play at any moment that would cause the total annual revenue to decrease.

WARNING   If a variable has increased (or decreased) throughout a period, we *cannot* assume it will continue to increase (or decrease) after that period. We also *cannot* assume the variable increased (or decreased) before the period.

StatCrunch and TI-84 time-series plots of the gluten data are shown in Figs. 49 and 50, respectively. See Appendixes B.9 and A.10 for StatCrunch instructions and TI-84 instructions, respectively.

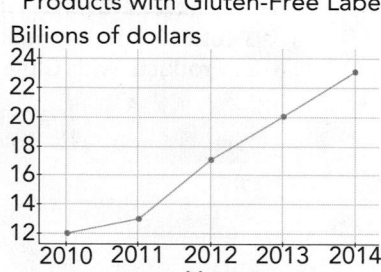

Figure 49 StatCrunch time-series plot of gluten-free data

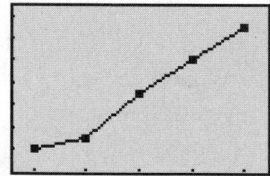

Figure 50 TI-84 time-series plot of gluten-free data

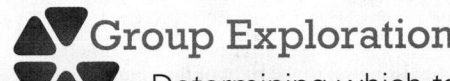

Group Exploration

Determining which table or diagram to use

For the following problems, refer to the following types of diagrams and table: relative frequency bar graph, multiple bar graph, pie chart, two-way table, dotplot, stemplot, and time-series plot. Determine which of these can be used to describe

1. a single categorical variable. Explain.
2. two categorical variables. Explain.
3. a single numerical variable. Explain.
4. two numerical variables. Explain.
5. a single numerical variable that has thousands of observations. Explain.

▶ **Tips for Success**   Visualize

To help prepare themselves mentally and physically for competition, many exceptional athletes visualize themselves performing well at their event many times throughout their training period. For example, a runner training for the 100-meter dash might imagine getting set in the starting blocks, taking off right after the gun goes off, being in front of the other runners, and so on, right up until the moment of breaking the tape at the finish line.

   You can do visualizations, too. Visualize doing all the things you feel you need to do to succeed in the course. If you do this regularly, you will have better follow-through with what you intend to do. You will also feel more confident about succeeding.

Homework 3.3

For extra help ▶ **MyMathLab®**   Watch the videos in MyMathLab   Download the MyDashboard App

1. A(n) _____ variable is a variable that has gaps between successive, possible values.

2. A(n) _____ variable is a variable that can take on any value between two possible values.

3. The _____ percentile of some data is a value that is greater than or equal to approximately 45% of the observations and is less than approximately 55% of the observations.

4. A reasonable measure of the center of a distribution is the _____ percentile.

For Exercises 5–12, identify whether the variable is discrete or continuous.

5. the number of times last year a person went out to eat

6. the number of dogs a person has

7. the sound level (in decibels) of a hip-hop concert

8. the speed (in miles per hour) of an airplane

9. the amount of time (in seconds) it takes a person to construct a histogram

10. the weight (in ounces) of a 14-inch cheese pizza

11. the price (in dollars) of a skateboard

12. the number of strings on a musical instrument

13. The asking prices (in thousands of dollars) of some four-bedroom homes in Akron, Ohio, are described by the dotplot in Fig. 51.

Asking Prices of Four-Bedroom Homes

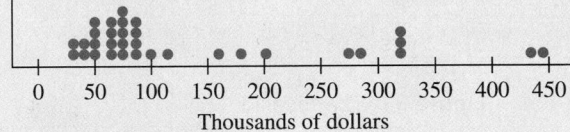

Figure 51 Exercise 13
(**Source:** *Zillow*)

a. A buyer can afford to pay up to $150 thousand. Assuming sellers will not lower their asking prices, what proportion of the homes can the buyer afford?

b. What proportion of the homes have asking prices between $250 thousand and $350 thousand?

c. If there were two outliers, estimate their values.

d. It turns out that the two homes with asking prices you identified in part (c) also have square footages that are outliers. Why does this make sense?

e. The data set does not include the asking prices of foreclosed homes. If the population were all types of four-bedroom homes, what type of bias in sampling occurred? If the asking prices of foreclosed homes had been included, how would the dotplot most likely be different?

14. The square footages of some three-bedroom homes for sale in Akron, Ohio, are described by the dotplot in Fig. 52.

Square Footage of Three-Bedroom Homes

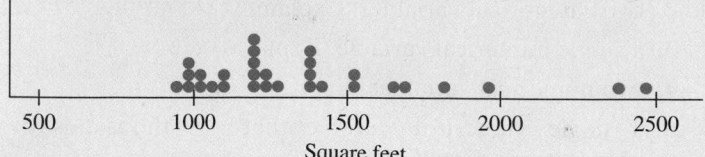

Figure 52 Exercise 14
(**Source:** *Zillow*)

   a. A family wants a home with at least 1500 square feet. What proportion of the homes meet this condition?
   b. What proportion of the homes have square footages between 1500 and 2000 square feet?
   c. If there were two outliers, estimate their values.
   d. It turns out that the two homes with square footages you identified in part (c) also have *asking prices* that are outliers. Why does this make sense?
   e. For the 5 dots stacked above 1200 square feet, four of the observations are each 1200 square feet but the other observation is actually 1224 square feet. Why didn't StatCrunch plot one of the five dots to the right of the other four?

15. The percentages of residents who are Protestant for each of the 50 states are described by the stemplot in Fig. 53.

Variable: Protestant

Decimal point is 1 digit(s) to the right of the colon.
Leaf unit = 1

```
1 : 1
1 :
2 : 0
2 : 99
3 : 14
3 : 78899
4 : 224
4 : 5668999
5 : 012223344
5 : 5668889
6 : 023
6 : 8
7 : 0113
7 : 56677
```

Figure 53 Exercise 15
(**Source**: *The Gallup Organization*)

   a. What is the frequency of the observation 58%? What does it mean in this situation?
   b. Alabama and Mississippi have the largest Protestant percentages of all 50 states. What is that percentage?
   c. Use the 50th percentile to measure the center of the distribution.
   d. If there were one outlier, what would it be? Guess which state the observation describes. [**Hint:** A lot of Mormons live there.]

16. Students in one of the author's prestatistics classes anonymously estimated the author's age (in years). Their responses are described by the stemplot in Fig. 54.

Variable: Age

Decimal point is 1 digit(s) to the right of the colon.
Leaf unit = 1

```
3 : 0
3 : 8
4 : 333
4 : 55556789
5 : 00223
5 : 5677
6 :
6 : 5
```

Figure 54 Exercise 16
(**Source:** *J. Lehmann*)

   a. Use the 50th percentile to measure the center.
   b. The author's actual age is 52 years. Find the error in the estimate you found in part (a) by finding the difference of your estimate and the actual age.
   c. Give two possible reasons why the 50th percentile does not equal the actual age.
   d. Construct a dotplot of the data.
   e. Assume there is exactly one outlier. On the basis of the stemplot, why does the observation 65 years *appear* to be the outlier? On the basis of the dotplot, why does the observation 30 years *appear* to be the outlier? If there is just one outlier, what is it? Explain.
   f. If the author performed the survey to measure the youthfulness of his physical appearance, give at least one reason why it was good practice to perform the survey on the first day of class rather than later on.

17. A time-series plot of the numbers of Radio Shack® stores is shown in Fig. 55.

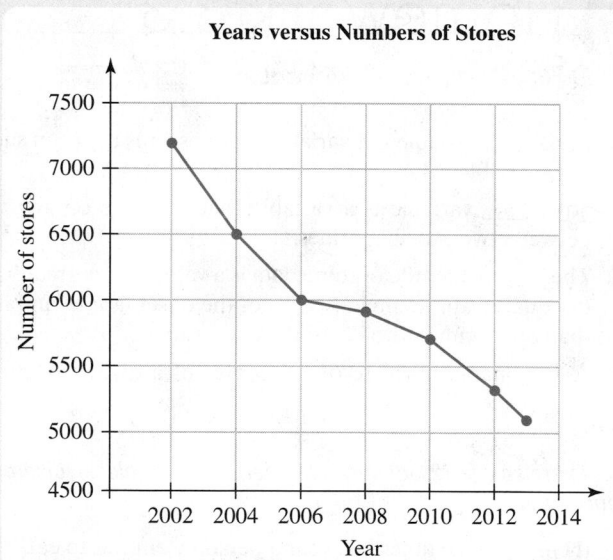

Figure 55 Exercise 17
(**Source:** *Radio Shack*)

   a. Estimate the number of stores in 2004.
   b. Estimate when there were 6000 stores.
   c. Has the number of stores increased, stayed approximately constant, decreased, or none of these from 2002 to 2013?
   d. A student estimates that there were 5100 stores in 2013 and that the number of stores had been decreasing by about 200 stores per year. He concludes that there were 5100 − 200 = 4900 stores in 2014. What would you tell the student?

e. In March, 2014, Radio Shack announced its plans to close as many as 1100 stores. Does this support your response to part (d)? Explain.

18. A time-series plot of college students' average spending (in dollars) on course materials is shown in Fig. 56.

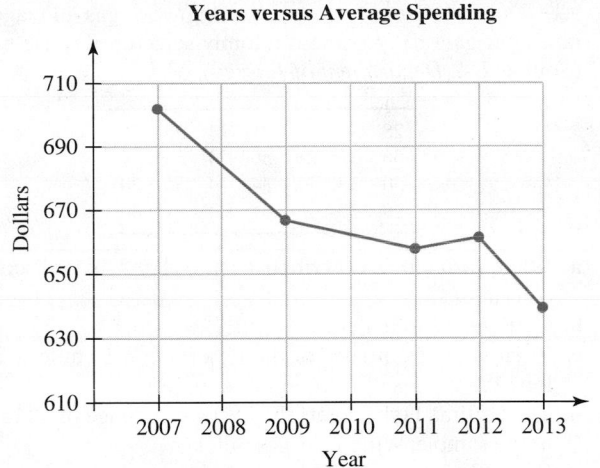

Figure 56 Exercise 18

(**Source:** *National Association of College Stores*)

a. Estimate the average spending on course materials in 2009.

b. Estimate when the average annual spending on course materials was $640.

c. Estimate the change in average annual spending on course materials from 2012 to 2013.

d. Has the average annual spending on course materials generally increased, stayed approximately constant, generally decreased, or none of these from 2007 to 2013?

e. Textbook prices have increased an average of 6% per year for the past decade (Source: *The Wall Street Journal*). State at least two possible reasons how college students' average spending on course materials could generally decrease while textbook prices increase.

19. **DATA** The following data are the percentages of adults who exercise often for the Midwestern states (Source: *The Gallup Organization*).

51	49	53	52	51	56
52	56	56	49	52	54

a. Construct a dotplot.

b. What is the frequency of the observation 52%? What does it mean in this situation?

c. Indiana and Ohio have the lowest exercise rates of the Midwestern states. What are their exercise rates?

d. For how many Midwestern states is the exercise rate at least 54%?

e. Find the proportion of states that have an exercise rate of at most 52%.

20. **DATA** The following data are the percentages of Americans who exercise often for the Western states (Source: *The Gallup Organization*).

60	53	55	60	62	58	60
55	57	58	54	56	54	

a. Construct a dotplot.

b. What is the frequency of the observation 60%? What does it mean in this situation?

c. Hawaii has the largest exercise rate of the Western states. What is the exercise rate?

d. For how many states is the exercise rate at most 54%?

e. Find the proportion of states that have an exercise rate of at least 60%.

21. **DATA** The author surveyed the students in one of his algebra classes, asking how many hours they spend watching TV shows, movies, and videos on a weekday. Here are their responses (Source: *J. Lehmann*):

10	4	3	2	0	30	2	1
2	4	2	4	0	1	1	2
6	4	2	5	6	4	4	3

a. Is the observation 30 hours an outlier? What might the student have thought the question was? What type of bias is this? Because we cannot be sure what the observation means, remove it from the sample for parts (c) through (d).

b. One student wrote "4–10." What value would represent this response well? Include the value in the sample for parts (c) through (d).

c. Construct a dotplot.

d. What proportion of the observations are between 2 and 4 hours, inclusive?

e. What proportion of the observations are less than 3 hours?

22. **DATA** The author surveyed the students in one of his algebra classes, asking them their ages. Here are their responses (Source: *J. Lehmann*):

18	18	24	20	19	19	17	22	18
18	20	28	20	20	18	19	19	21
25	20	24	21	21	20	22	19	

a. Construct a dotplot.

b. What proportion of the observations are between 20 and 22 years, inclusive?

c. What proportion of the observations are at least 24 years?

d. What proportion of the observations are greater than 24 years?

e. If there were exactly one outlier, what would it be? What does it mean in this situation?

23. **DATA** If a restaurant in Alameda County, California, receives a food safety grade of yellow, that means two or more major food safety violations were identified and corrected during an inspection. The following data are the numbers of restaurants in Alameda County, California, that received yellow grades per day between October 1, 2014, and January 13, 2015 (Source: *Alameda County Data Sharing Initiative*).

1	2	1	0	5	2	2	1	1	2
0	5	0	1	4	1	4	5	1	8
5	2	3	1	6	6	5	6	3	0
3	4	3	3	4	5	4	4	1	3
2	6	2	1	4	2	3	1	4	4

a. What variable is described by the data? Is it discrete or continuous?

b. Construct a dotplot.

c. What is the frequency of the observation 6 restaurants receiving yellow grades per day? What does it mean in this situation?

d. What observation has the greatest frequency? What is that frequency? What does the frequency mean in this situation?

e. Use the 50th percentile to measure the center of the distribution.

24. **DATA** Muhammad Ali is considered by many to be one of the greatest heavyweight boxers in history. The following data are the numbers of rounds in his wins from November 1966 to September 1978 (Source: *Ali Enterprises LLC*).

3	15	7	3	15	12	12
7	15	12	7	11	7	7
12	12	12	12	8	15	11
15	14	5	15	5	15	15
15	15					

a. What variable is described by the data? Is it discrete or continuous?

b. Construct a dotplot.

c. Use the 50th percentile to measure the center of the distribution.

d. Each boxing match is limited to a certain number of rounds, although it can end earlier from a knockout or a technical knockout. For the rounds shown in blue, the matches could last a maximum of 12 rounds. For the rounds shown in red, the matches could last a maximum of 15 rounds. Explain how the round limits tie into the shape of the dotplot.

e. In 1976, Ali won all 4 of his matches. On the basis of only the provided data, what is the least possible total number of rounds Ali boxed in those 4 wins? What is the greatest possible total number?

25. **DATA** The following data are the numbers of miles of bike lanes per square mile for the 10 cities with the most miles of bike lanes (Source: *Alliance of Biking and Walking*).

2.6	3.4	1.3	2.9	4.3
4.1	3.7	1.5	4.2	2.7

a. Construct a stemplot.

b. How many of the 10 cities have at most 3.0 miles of bike lanes per square mile?

c. Find the proportion of the 10 cities that have at least 4.1 miles of bike lanes per square mile.

d. Find the proportion of the 10 cities that have less than 3.7 miles of bike lanes per square mile.

e. Philadelphia has more miles of bike lanes per square mile than any other city. How many miles of bike lanes per square mile is that?

26. **DATA** The following data are the U.S. sales (in millions of albums) of the top 10 best-selling Christmas albums (Source: *Billboard*).

3.4	5.3	3.5	7.3	3.7
5.4	3.4	3.7	5.7	3.5

a. Construct a stemplot.

b. *Miracle: The Holiday Album* by Kenny G is the best-selling Christmas album. What are its sales?

c. *Miracle: The Holiday Album* is approximately 36 minutes long. If all the albums sold were played one at a time, one right after the other, for how many years would the music play? There are 365 days in one year.

d. Find the number of the 10 albums that have sales of at least 5.4 million albums.

e. Find the proportion of the 10 albums that have sales of at most 3.5 million albums.

27. **DATA** The following data are the highway gas mileages (in miles per gallon) of some randomly selected Kia car models (Source: *U.S. Department of Energy*).

28	39	29	37	24	35	26
36	31	24	36	24	34	34
30	30	31	34	24	30	29
29	37	28	31	34	26	40

a. What variable is described by the data? Is it discrete or continuous?

b. Construct a split stem.

c. What is the frequency of the observation 30 miles per gallon? What does it mean in this situation?

d. The Optima Hybrid gets the best gas mileage of all the cars in the sample. What is its gas mileage?

e. Use the 50th percentile to measure the center of the distribution.

28. **DATA** The following data are the highway gas mileages (in miles per gallon) of some randomly selected Dodge car models (Source: *U.S. Department of Energy*).

40	25	23	36	27	23	33
24	25	30	31	29	27	25
26	27	27	27	34	31	25
25	31	41	24	35	31	23

a. What variable is described by the data? Is it discrete or continuous?

b. Construct a split stem.

c. What is the frequency of the observation 27 miles per gallon? What does it mean in this situation?

d. The Dart Aero gets the best gas mileage of all the cars in the sample. What is its gas mileage?

e. Use the 50th percentile to measure the center of the distribution.

29. **DATA** The following data are the top 50 men's times (in minutes) in the 2013 ING New York City Marathon (Source: *ING New York City Marathon*).

148	128	133	129	130	130	150
131	155	132	152	155	142	152
153	132	137	143	145	131	155
152	133	150	144	144	138	143
147	152	153	146	143	154	152
148	131	150	150	132	146	144
155	149	131	145	151	154	143
142						

a. Construct a split stem.

b. What is the frequency of the observation 131 minutes? What does it mean in this situation?

c. Which observation has the greatest frequency? What is the frequency?

d. How many of the observations are between 140 and 149 minutes, inclusive?

e. How many of the observations are greater than 149 minutes?

30. DATA The following data are the top 50 women's times (in minutes) in the 2013 ING New York City Marathon (Source: *ING New York City Marathon*).

161	163	179	179	180	146	149
176	149	150	177	177	179	148
177	177	160	158	177	172	160
166	167	170	171	150	172	173
173	174	175	175	161	164	176
152	155	155	176	177	161	176
178	180	145	178	148	148	149
181						

a. Construct a split stem.
b. What is the frequency of the observation 148 minutes? What does it mean in this situation?
c. Which observation has the greatest frequency? What is the frequency?
d. How many of the observations are between 160 and 169 minutes, inclusive?
e. How many of the observations are less than 160 minutes?

31. DATA Open Streets Initiatives temporarily close streets to motorized traffic and open them for jogging, bicycling, and other physical activities. The numbers of large cities with Open Streets Initiatives are shown in Table 41 for various years.

Table 41 Numbers of Large Cities with Open Streets Initiatives

Year	Number
2006	1
2007	2
2008	8
2009	7
2010	12
2011	22
2012	30

Source: *Alliance for Biking and Walking*

a. Construct a time-series plot.
b. Have the number of large cities with Open Streets Initiatives generally increased, decreased, stayed approximately constant, or none of these?
c. Find the greatest change in the number of large cities with Open Streets Initiatives from one year to the next.
d. Find the change in the number of large cities with Open Streets Initiatives from 2006 to 2012. If the number of large cities with Open Streets Initiatives changes by that same amount from 2012 to 2018, what will it be in 2018? Do you have much faith that this will turn out to be true? Explain.

32. DATA The percentages of U.S. households with only a cell phone (no landline phone) are shown in Table 42 for various years.

Table 42 Percentages of U.S. Households with Only a Cell Phone

Year	Percent
2003	2
2005	8
2007	13
2009	22
2011	31
2013	39

Source: *Centers for Disease Control and Prevention*

a. Construct a time-series plot.
b. Has the percentage of U.S. households with only a cell phone increased, decreased, stayed approximately constant, or none of these?
c. Find the greatest change for a 2-year period in the percentage of U.S. households with only a cell phone.
d. Find the change in the cell phone–only percent from 2003 to 2013. If the cell phone–only percent changes by the same amount from 2013 to 2023, what will it be in 2023? Do you have much faith that this will turn out to be true? Explain.

33. DATA The number of deaths from collisions at highway-railroad crossings are shown in Table 43 for various years.

Table 43 Numbers of Deaths from Collisions at Highway-Railroad Crossings

Year	Number of Deaths
1990	698
1995	579
2000	425
2005	359
2010	260
2013	251

Source: *Federal Railroad Administration*

a. Construct a time-series plot.
b. Has the number of deaths from collisions per year increased, decreased, stayed approximately constant, or none of these?
c. Find the change in the number of deaths from collisions from 2010 to 2013. What does it mean in this situation?
d. Find the change in the number of deaths from collisions from 2005 to 2013. If the number of deaths changes by the same amount from 2013 to 2021, what will it be in 2021? Do you have much faith that this will turn out to be true? Explain.

34. DATA The violent crime rates (in number of violent crimes per 100,000 people) in the United States are shown in Table 44 for various years.

Table 44 U.S. Violent Crime Rates

Year	Violent Crime Rate (crimes per 100,000 people)
1990	730
1995	685
2000	507
2005	469
2010	405
2012	387

Source: *FBI*

a. Construct a time-series plot.
b. Has the violent crime rate increased, decreased, stayed approximately constant, or none of these?
c. Find the change in the violent crime rate from 2010 to 2012. What does it mean in this situation?
d. Find the change in the violent crime rate from 2000 to 2012. If the violent crime rate changes by the same amount from 2012 to 2024, what will it be in 2024? Do you have much faith that this will turn out to be true? Explain.

35. The ages (in years) of 1098 people stopped (on foot or in a vehicle) by police in New York City on December 1, 2012, are described by the dotplot in Fig. 57 and the stemplot in Fig. 58, although the right part of the stemplot has been cut off (truncated).

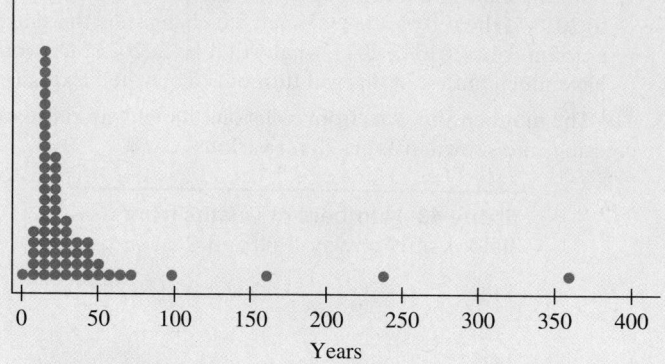

Stopped by Police Up to 21 values per dot

Years

Figure 57 Dotplot of the ages of people stopped
(**Source:** *New York Police Department*)

Decimal point is 1 digit(s) to the right of the colon.

```
0 : 1113
0 : 9
1 : 333333444444444444444444444444
1 : 5555555555555555555555555555555555555555555
2 : 00000000000000000000000000000000000000000
2 : 555555555555555555555555555555555555556666
3 : 00000000000000000000001111111111111111111
3 : 555555555555555556666666666666677777777777
4 : 000000000000011122222222233333333344444444
4 : 5555555555555566666677777777777888888888
5 : 001111111222222222344444
5 : 5555555666666677789999
6 : 0000011222223
6 : 59
7 : 1
```

High: 76, 99, 99, 160, 236, 360

Figure 58 Truncated stemplot of the ages of people stopped
(**Source:** *New York Police Department*)

a. In order to save vertical space, each dot in the dotplot may represent up to 21 ages. On the basis of the dotplot alone, estimate the outliers.

b. To the right of "High," the truncated stemplot identifies the outliers. Compare the outliers and their frequencies with your result in part (a).

c. Which data are ridiculous or questionable in this situation? Explain. What type of bias in sampling occurred? [**Hint:** Consider non-outliers as well as outliers.]

d. Here are some actions that might be taken on each of the outliers, questionable data, and ridiculous data:

- Remove the data value.
- Review the original report completed by the police officer and check the age recorded.
- Interview the police officer who completed the report about the age of the suspect.

Which actions would you recommend be taken, and in which order should they be done so that the researcher's and police officers' time is not wasted?

36. Estimates of the numbers of militants who died from 74 drone strikes in Pakistan in 2012 and 2013 are described by the dotplots in Fig. 59. The numbers of militants who died for certain and the most possible numbers of militants who died are both described.

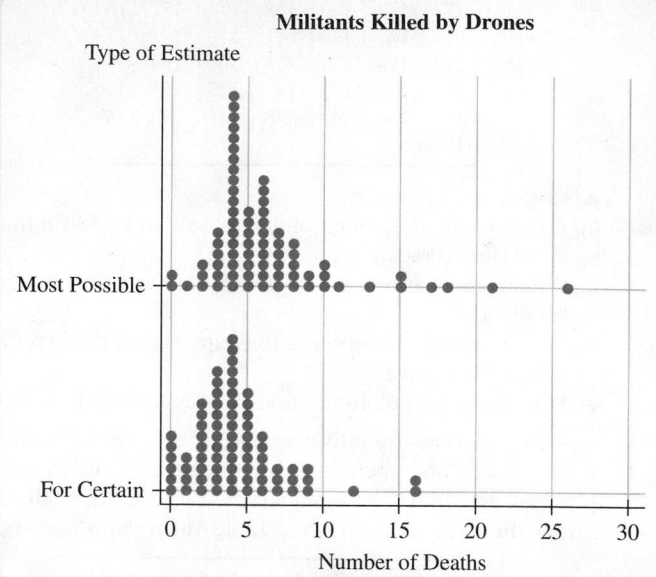

Figure 59 Exercise 36
(**Source:** *New America Foundation*)

a. For each distribution, use the 50th percentile to measure the center.

b. Find the difference of the center for the most-possible distribution and the center for the for-certain distribution. What does it mean in this situation? Why does it make sense that your result is positive?

c. What was the greatest possible number of militants killed in one drone strike? What was the greatest number of militants killed in one drone strike for certain? Explain why the difference of your two results may *not* be the number of militant deaths we are unsure about in a particular strike.

Large Data Sets

37. ▼**DATA** Access the data about the United Airlines departure delays, which are available at MyMathLab and at the Pearson Downloadable Student Resources for Math & Stats website.

a. Construct a dotplot of the departure delays.

b. Estimate the smallest observation. What does it mean in this situation?

c. Estimate the largest observation. What does it mean in this situation?

d. Estimate the observation that has the greatest frequency. What does that mean in this situation?

e. The 150 observations that are at least 39 minutes can be considered outliers. What does that mean in this situation?

38. ▼**DATA** Access the data about tuitions and fees at 4-year colleges and universities, which are available at MyMathLab and at the Pearson Downloadable Student Resources for Math & Stats website.

a. Construct a dotplot of the tuitions and fees at *public*, 4-year colleges and universities and a dotplot of the tuition and fees at *private, not-for-profit*, 4-year colleges and universities. The two dotplots should share one horizontal axis.

b. Which type of 4-year institution tends to have larger tuition and fees, public colleges and universities or private, not-for-profit colleges and universities?

c. Estimate the smallest and largest tuition and fees for 4-year, public colleges and universities. Find the difference of the largest value and the smallest value.

d. Estimate the smallest and largest tuition and fees for 4-year, private, not-for-profit colleges and universities. Find the difference of the largest value and the smallest value.

e. Compare the results you found in parts (c) and (d). What does the comparison mean in this situation?

Concepts

39. A student surveys 16 students who work and constructs a dotplot of the numbers of hours they work (see Fig. 60). On the basis of the dotplot, the student concludes that there are no outliers.

Hours of Work

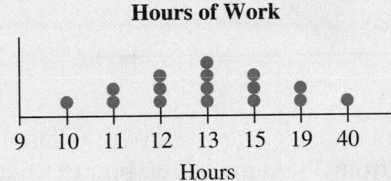

Hours

Figure 60 Exercise 39

a. Explain why the dotplot is incorrect and construct a correct one.

b. Is the student correct that there are no outliers? Explain.

c. In general, would making the student's type of error really matter much? Explain.

40. A student tries to construct a time-series plot to describe U.S. imports (in millions of pounds) of quinoa (see Fig. 61). The student concludes that the data points lie on a straight line.

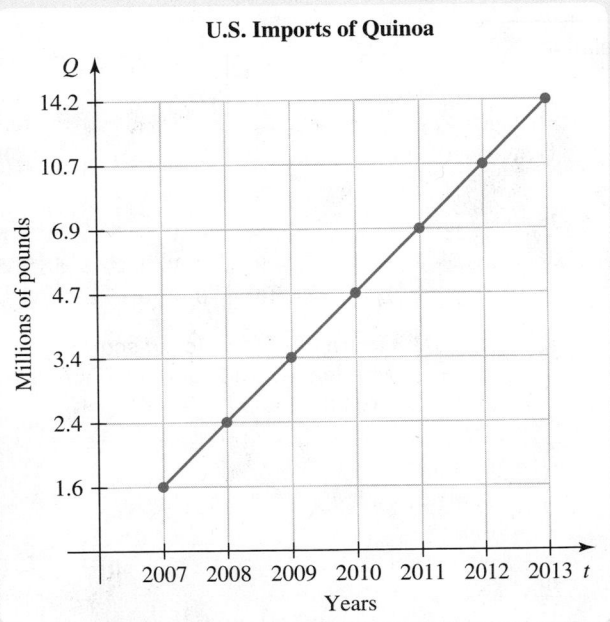

U.S. Imports of Quinoa

Figure 61 Exercise 40
(**Source:** *Datamyne*)

a. Explain why the time-series plot is incorrect and construct a correct one.

b. Is the student correct that the data points lie on a straight line? Explain.

c. In general, would making the student's type of error really matter much? Explain.

41. From which type of diagram is it easier to tell the values of a variable, a dotplot or a stemplot? Explain.

42. At the Rockville campus of Montgomery College in Maryland, fall 2013 enrollment was 16,441 students (Source: *Montgomery College*). Explain why a stemplot would not be the best type of diagram to describe the ages of the students.

43. **DATA** The numbers of countries requiring picture warnings on cigarette packages are shown in Table 45 for various years.

Table 45 Numbers of Countries Requiring Picture Warnings on Cigarette Packages

Year	Number
2010	34
2011	40
2012	55
2013	63
2014	70
2015	77

Source: *Canadian Cancer Society*

a. Construct a time-series plot.

b. We can find the *average rate of change* of the number of countries requiring picture warnings on cigarette packages by dividing the change in the number of such countries for the period 2010–2015 by the change in time for the same period. Find the average rate of change.

c. For which 1-year periods does your result in part (b) underestimate the change in the number of countries requiring picture warnings on cigarette packages?

d. For which 1-year periods does your result in part (b) overestimate the change in the number of countries requiring picture warnings on cigarette packages?

44. **DATA** The following data are students' scores (in points) on a test in one of the author's intermediate algebra classes (Source: *J. Lehmann*).

73	85	63	97	71	64	95
83	34	84	38	53	88	53
85	80	95	84	70	35	77
65	91	58	72	61	46	86
81	92	83	73			

Assume the author used the following cutoffs for grades:

- A: 85th percentile
- B: 50th percentile
- C: 30th percentile
- D: 10th percentile

Find the frequencies of the grades A, B, C, D, and F.

▼ 3.4 Histograms

Objectives

» Construct and interpret *frequency and relative frequency tables*.

» Construct and interpret *frequency histograms* and *relative frequency histograms*.

» Interpret *density histograms*.

» Describe the *shape* of a distribution.

» Describe the *spread* of a distribution.

» Describe the meaning of *model*.

Frequency and Relative Frequency Tables

We can construct frequency and relative frequency tables for numerical variables much like we do for categorical variables.

▶ **Example 1**   Constructing a Frequency and Relative Frequency Table

Construct a frequency and relative frequency table for the Test 1 distribution that we worked with in Section 3.3 (see Table 46).

Table 46 Scores on Test 1 (points)

68	80	38	83	73	98	75	95	93	88
85	60	80	85	78	85	90	100	98	75
70	62	93	65	100	88	93	95	65	95
85	88	68	90	100					

Solution

The first step is to form categories of numbers called *classes*. We often use between 5 and 10 classes. Because the test scores range from 38 points to 100 points, we use the following classes: 30–39 points, 40–49 points, 50–59 points, …, 100–109 points (see Table 47). If we think of the classes as categories, then we can find the tallies, frequencies, and relative frequencies of the classes in much the same way as we construct frequency and relative frequency tables for categorical variables (Section 3.1). For example, because 6 of the 35 observations fall in the class 60–69 (see the blue numbers in Table 46), the relative frequency of the class is $\frac{6}{35}$.

In Section 3.3, we found a stemplot for the Test 1 distribution (see Fig. 62). Note that each stem acts as a class in Table 47 and the stem's number of leaves equals the frequency of the class.

Table 47 Frequency and Relative Frequency Table

Class	Tally	Frequency	Relative Frequency
30–39	|	1	$\frac{1}{35} \approx 0.029$
40–49		0	$\frac{0}{35} = 0$
50–59		0	$\frac{0}{35} = 0$
60–69	ⅢⅢ |	6	$\frac{6}{35} \approx 0.171$
70–79	ⅢⅢ	5	$\frac{5}{35} \approx 0.143$
80–89	ⅢⅢ ⅢⅢ	10	$\frac{10}{35} \approx 0.286$
90–99	ⅢⅢ ⅢⅢ	10	$\frac{10}{35} \approx 0.286$
100–109	|||	3	$\frac{3}{35} \approx 0.086$
Total		35	$\frac{35}{35} \approx 1.001$

Stem (tens)	Leaf (ones)
3	8
4	
5	
6	0 2 5 5 8 8
7	0 3 5 5 8
8	0 0 3 5 5 5 5 8 8 8
9	0 0 3 3 3 5 5 5 8 8
10	0 0 0

Figure 62 Stemplot of scores on Test 1 with leaves arranged from smallest to largest

To check our work, we add all the frequencies, which gives 35 observations, which is correct. We also add all the relative frequencies, which gives 1.001, which is approximately 1. We did not get the correct value, 1, due to round-off error. Recall that these are the same types of checks we did for frequency and relative frequency tables for distributions of categorical variables in Section 3.1.

In Example 1, the first class is 30–39. We say that 30 is the *lower class limit* and 39 is the *upper class limit*. In general, for a class $[a, b]$, the **lower class limit** is a and the **upper class limit** is b.

To find the **class width** of a class, we subtract the lower class limit of a class from the lower class limit of the next class. In Example 1, the first two classes are 30–39 and 40–49. So, the class width is $40 - 30 = 10$ points.

In Example 1, we found the frequencies and the relative frequencies of classes.

> ▶ **Definition**   Frequency of a class and relative frequency of a class
>
> The **frequency of a class** is the number of observations in the class. The **relative frequency of a class** is the proportion of the observations in the class.

When using classes, the **frequency distribution of a numerical variable** is the classes together with their frequencies and the **relative frequency distribution of a numerical variable** is the classes together with their relative frequencies.

In Section 3.1, we found that the sum of the relative frequencies of all the categories of a categorical variable is equal to 1. A similar property is true for numerical variables.

> ▶ **Sum of Relative Frequencies**
>
> For a numerical variable, the sum of the relative frequencies of all the classes is equal to 1.

In Table 47 on page 174, recall that the class widths are equal to 10 points. Because of this choice of classes, the frequencies are equal to the number of leaves in the stemplot in Fig. 62. However, we don't always use classes of width 10 units. We are free to use any class width we want, provided we use that same width for all the classes.

Frequency Histograms and Relative Frequency Histograms

Although dotplots and stemplots have their uses, the most popular way to view a distribution is by constructing a graph called a *histogram*. We will construct two types of histograms in Example 2 (see Figs. 63 and 64). Part of the reason histograms are so popular is that **histograms can be used with any numerical variable (discrete or continuous) and any number of observations.**

▶ **Example 2**   Constructing Histograms

1. Use Table 47 on page 174 to help construct a frequency histogram for the test score distribution.
2. Use Table 47 to help construct a relative frequency histogram for the test score distribution.

Solution

1. Because the numbers $30, 40, 50, \ldots$, and 100 are the lower class limits, we write these numbers as well as 110 equally spaced on the horizontal axis (see Fig. 63). Because the largest frequency is 10, we write the numbers 0, 2, 4, 6, 8, and 10 equally spaced

on the vertical axis. Next, we write "Points" and "Frequency" on the appropriate axes. Then we draw rectangles whose widths are 10 (the class widths) and whose heights are the class frequencies.

2. The process of drawing a relative histogram is similar to drawing a frequency histogram, but we handle the vertical axis differently (see Fig. 64). Because the largest relative frequency is approximately 0.286, we write the numbers 0.05, 0.10, 0.15, ..., 0.30 equally spaced on the vertical axis. Then we write "Relative frequency" on that axis.

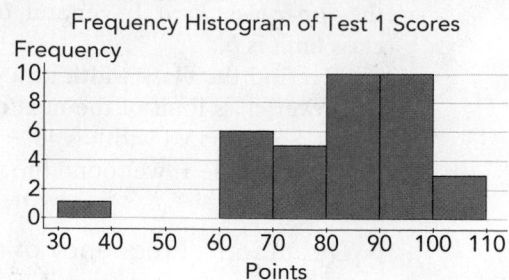

Figure 63 Frequency histogram of Test 1 Scores

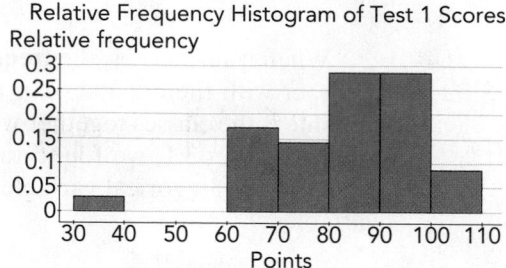

Figure 64 Relative frequency histogram of Test 1 scores

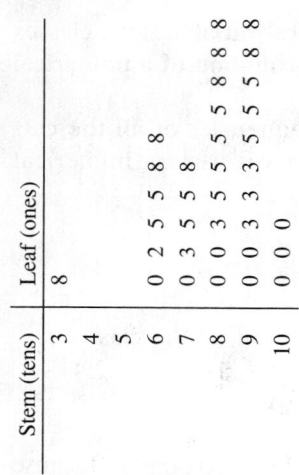

Figure 65 Stemplot of Test 1 distribution

If we rotate the stemplot of the Test 1 distribution clockwise by 90 degrees (see Fig. 65), the shape of the result looks similar to the frequency histogram in Fig. 63.

The histograms in Figs. 63 and 64 were drawn using StatCrunch. The frequency histogram in Fig. 66 was drawn using a TI-84. See Appendixes B.10 and A.5 for StatCrunch and TI-84 instructions, respectively.

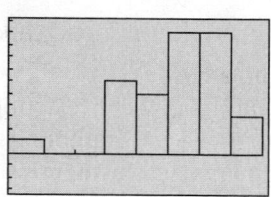

Figure 66 TI-84 frequency histogram

WARNING   Even though histograms and bar graphs look similar, histogram bars can touch, but bar-graph bars never touch.

Nonetheless, we can use relative frequency histograms to estimate proportions in much the same way as we used relative frequency bar graphs to estimate proportions in Section 3.1.

▶ **Example 3**   Use a Relative Frequency Histogram to Find Proportions

Figure 67 displays a relative frequency histogram describing the numbers of wildfires per year in California for the period 1933–2010.

Numbers of Wildfires per Year in California

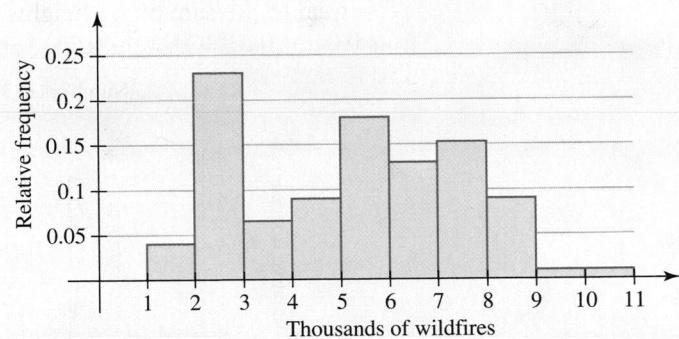

Figure 67 Wildfires in California

(**Source:** *California Department of Forestry and Fire Protection*)

1. For each class of data shown in the histogram, estimate the relative frequency.
2. Estimate the proportion of years in the period 1933–2010 when the number of wildfires per year were

   a. between 6000 and 6999, inclusive.
   b. less than 4 thousand.
   c. at least 4 thousand.

Solution

1. We estimate the relative frequencies and write them above the bars of the histogram in Fig. 68.

Numbers of Wildfires per Year in California

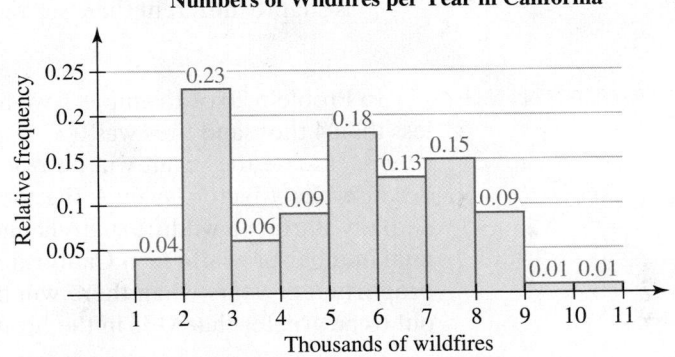

Figure 68 Estimate the relative frequencies

2. **a.** The proportion of years in the period 1933–2010 when there were between 6000 and 6999 wildfires, inclusive, per year is equal to the height of the orange rectangle in Fig. 69. So, the proportion is 0.13.

Numbers of Wildfires per Year in California

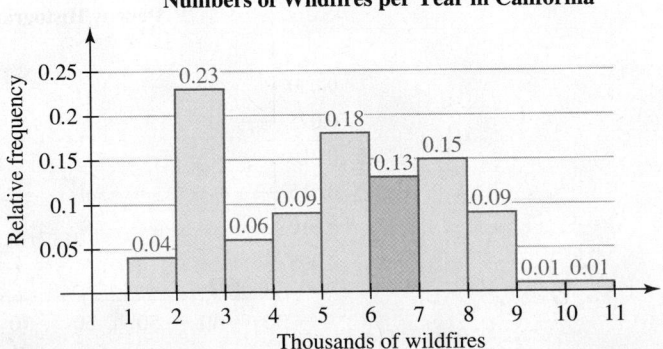

Figure 69 Years when there were between 6 thousand and 7 thousand wildfires

b. The proportion of years when there were less than 4 thousand wildfires per year is equal to the sum of the heights of the orange bars in Fig. 70. So, the proportion is $0.04 + 0.23 + 0.06 = 0.33$.

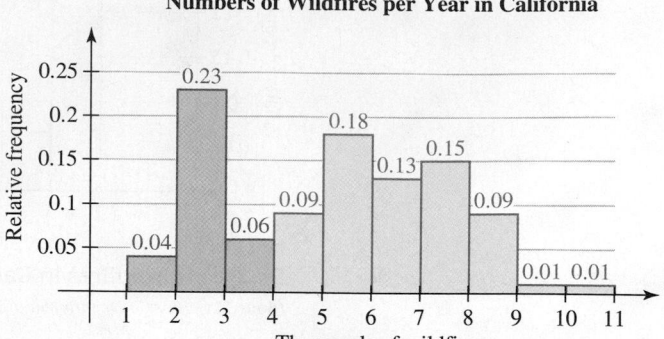

Figure 70 Years when there were less than 4 thousand wildfires

c. We discuss two ways of finding the proportion.

Method 1 The proportion of years when there were at least 4 thousand wildfires is equal to the sum of the heights of the blue rectangles in Fig. 70. So, the proportion is $0.09 + 0.18 + 0.13 + 0.15 + 0.09 + 0.01 + 0.01 = 0.66$.

Method 2 The sum of all the estimated relative frequencies in Fig. 70 should be approximately 1. So, the proportion of years when there were at least 4 thousand fires per year (blue bars) is approximately equal to 1 minus the proportion of years when there was less than 4 thousand wildfires (orange bars), which we found to be 0.33 in Problem 3b. So the proportion is $1 - 0.33 = 0.67$. Our result is slightly different than our result by Method 1 due to round-off error.

WARNING   In Problem 2b of Example 3, we found that the proportion of years when there were less than 4 thousand fires was 0.33. This result is true only for the period 1933–2010. We *cannot* assume the result will be true for the next 100 years or even the next 10 years because many factors such as the weather and people's awareness of the risks of accidentally starting a wildfire may change. In fact, a little research would show that the annual number of wildfires in California has been generally decreasing since 1980. So, the proportion of years when there will be less than 4 thousand fires next year might turn out to be greater than 0.33 in the future.

Interpreting Density Histograms

For a relative frequency histogram, the height of each bar is the relative frequency of the bar's class. For a **density histogram**, the vertical axis has units called **density** so that the *area* of each bar is the relative frequency of the bar's class. A density histogram of the Test 1 distribution is shown in Fig. 71. Each number inside a bar is the area of the bar, which is also the relative frequency of the bar's class (see Fig. 72).

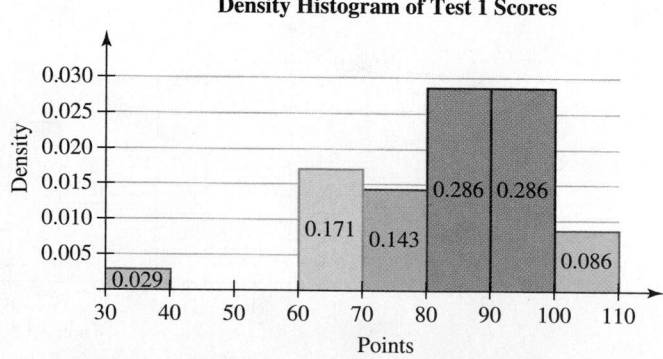

Figure 71 Density histogram of Test 1 scores

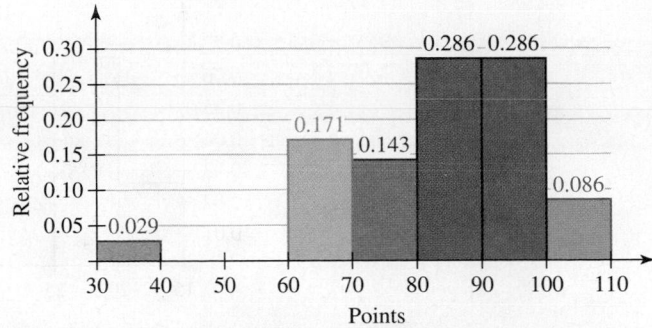

Relative Frequency Histogram of Test 1 Scores

Figure 72 Relative frequency histogram of Test 1 scores

To check this, recall from Section 1.1 that the area of a rectangle is equal to the rectangle's length times its width (or its width times its height). In Fig. 73, we calculate the area of the (green) bar for the class 60–69 points. The result 0.171 equals the relative frequency of the class 60–69 points (see the number 0.171 written inside the green bar in Fig. 71 and just above the green bar in Fig. 72).

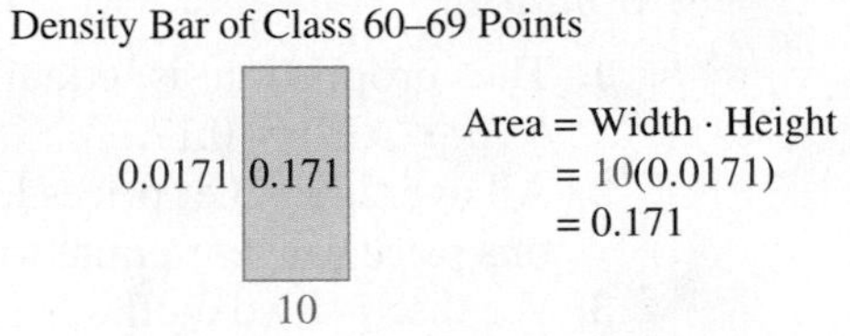

Density Bar of Class 60–69 Points

$$0.0171 \quad 0.171$$

Area = Width · Height
= 10(0.0171)
= 0.171

10

Figure 73 Area of density bar is 0.171

By comparing the density histogram with the relative frequency histogram, we see that the histograms look the same, except for the scaling on the vertical axis. If the horizontal axis remains the same, a density histogram will always look the same as the corresponding relative frequency histogram, except for the scaling on the vertical axis.

When we wish to add relative frequencies, it is easier to work with density histograms because it is easier to picture adding areas of bars rather than adding heights of bars with relative frequency histograms.

Earlier in this section, we found that the sum of the relative frequencies of all the classes of a numerical variable is equal to 1. Because the areas of the bars of a density histogram are equal to the relative frequencies of the bars' classes, this means the total area of the bars is equal to 1.

To check this, we add the areas of all the bars in the density histogram in Fig. 71:

$$0.029 + 0.171 + 0.143 + 0.286 + 0.286 + 0.086 = 1.001 \approx 1$$

We did not get the correct result of 1 due to rounding error, just like what happened with our check of the relative frequencies in Table 47.

> **Area of Bars of a Density Histogram**

The following statements are true for a density histogram.

- The area of each bar is equal to the relative frequency of the bar's class.
- The total area of the bars is equal to 1.

> **Example 4**   Using a Density Histogram to Find Proportions

The average prices of 2014 Major League Baseball® (MLB) tickets at the stadiums are described by the density histogram in Fig. 74.

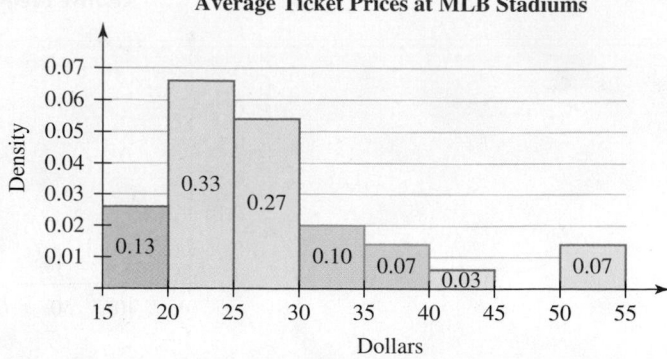

Figure 74 Average ticket prices at MLB stadiums
(**Source:** *Team Marketing Report*)

Find the proportion of stadiums whose average price of 2014 MLB tickets is

1. between $30 and $39.99, inclusive.
2. less than $20.
3. at least $20.

Solution

1. The proportion is equal to the total area of the green rectangles, which is 0.10 + 0.07 = 0.17.
2. All average ticket prices less than $20 are between $15 and $20. The proportion for this price range is equal to the area of the orange bar, which is 0.13.
3. We discuss two methods of finding the proportion.

   Method 1 The proportion of stadiums whose average price is at least $20 is equal to the sum of the areas of the blue bars and green bars. So, the proportion is 0.33 + 0.27 + 0.10 + 0.07 + 0.03 + 0.07 = 0.87.

   Method 2 The sum of the areas of all the bars is equal to 1. So, the proportion of stadiums whose average price is at least $20 is equal to 1 minus the proportion of stadiums whose average price is less than $20, which we found to be 0.13 in Problem 2. So, the proportion is 1 − 0.13 = 0.87.

WARNING   For the density histogram in Fig. 74, the relative frequency of the class 30–34.99 is the *area* of the bar, 0.10, not the *height* of the bar, 0.02, which is the density. In general, when using a density histogram to determine a relative frequency (proportion), make sure you refer to the area of a bar, not the height.

See Appendix B.10 for instructions on how to use StatCrunch to construct density histograms.

▶ Example 5   Working with Percentiles

In Example 4, we worked with the average ticket prices at MLB stadiums.

1. Estimate the percentile of a $20 average ticket price. Round to the nearest dollar.
2. The average ticket price at Nationals Park, home of the Washington Nationals, is at the 83rd percentile. Estimate the average ticket price. Round to the nearest dollar.
3. Identify the class that contains the 50th percentile, which is a reasonable measure of the center.

Solution

1. From the orange bar of the density histogram in Fig. 75, we see that 13% of observations are less than $20. So, $19.99 is at the 13th percentile. Rounding to the nearest dollar, we conclude the approximate average ticket price at Nationals Park is $20.

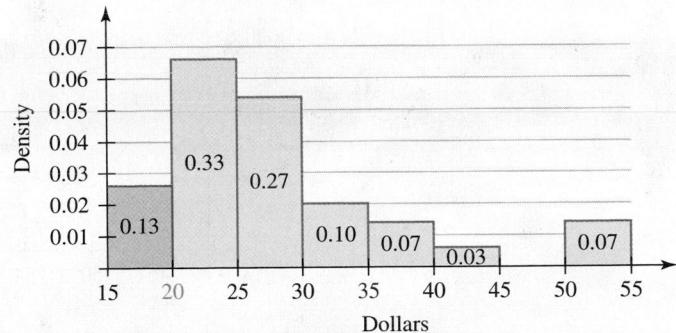

Figure 75 Average ticket prices at MLB stadiums

2. We add the areas of the bars from left to right until we reach 0.83:

$$0.13 + 0.33 = 0.46$$
$$0.13 + 0.33 + 0.27 = 0.73$$
$$0.13 + 0.33 + 0.27 + 0.10 = 0.83$$

So, 0.83 is equal to the sum of the areas of the first four rectangles. From Fig. 76, we see that the percentage of stadiums that charge less than $35 is 83%. So, Nationals Park must charge an average price of approximately $35 for 2014 MLB tickets.

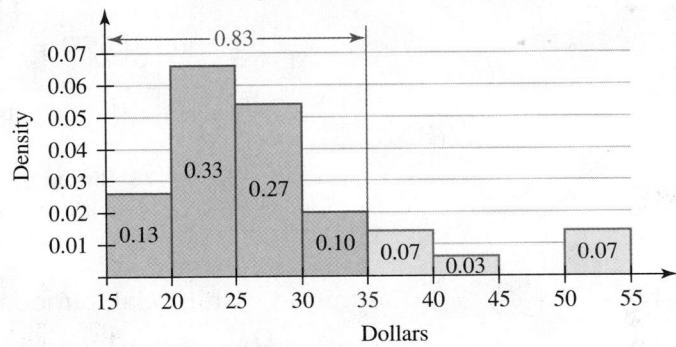

Figure 76 Average ticket prices at MLB stadiums
(**Source:** *Team Marketing Report*)

3. We add the areas of the bars from the left until we reach or go beyond 0.50:

$$0.13 + 0.33 = 0.46$$
$$0.13 + 0.33 + 0.27 = 0.73$$

Because 0.50 is between 0.46 and 0.73, we conclude that the 50th percentile is in the class 25–29.99 dollars. So, a typical ticket price is between $25 and $29.99.

Shape of a Distribution

Do you think that the area where you live is getting better as a place to live? The percentages of state residents who think so are described by the histogram in Fig. 77 for the 50 states. Because the histogram is in the shape of a single mound, we say the distribution is *unimodal*.

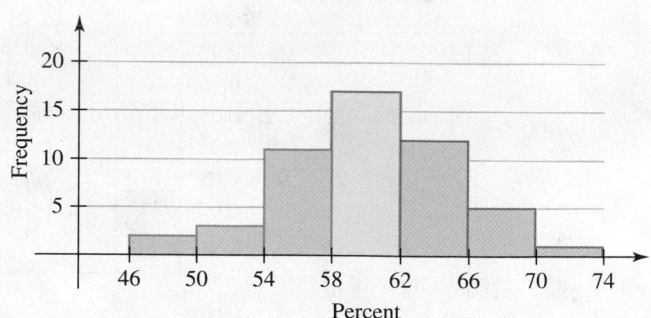

Figure 77 Unimodal distribution
(**Source:** *The Gallup Organization*)

Figure 78 displays a histogram of the top 100 male finishers and top 100 female finishers in the 2013 ING New York City Marathon. We say the distribution is *bimodal* because there are two mounds. It makes sense that there are two mounds because men tend to run faster than women.

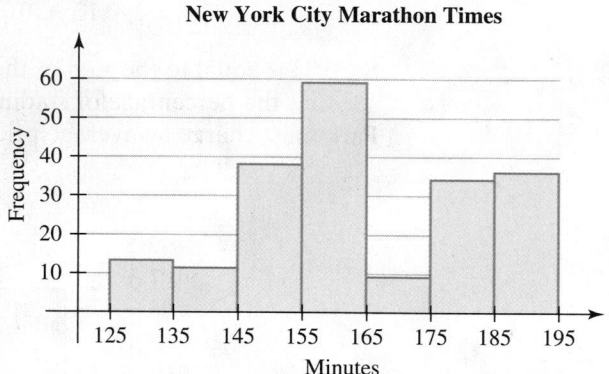

Figure 78 Bimodal distribution
(**Source:** *New York City Marathon*)

> **Definition**   Unimodal, bimodal, and multimodal distributions
>
> A distribution is **unimodal** if it has one mound, **bimodal** if it has two mounds, and **multimodal** if it has more than two mounds.

The students' scores on the second test of one of the author's statistics classes are described by the histogram in Fig. 79. Because there is one mound, the distribution is unimodal.

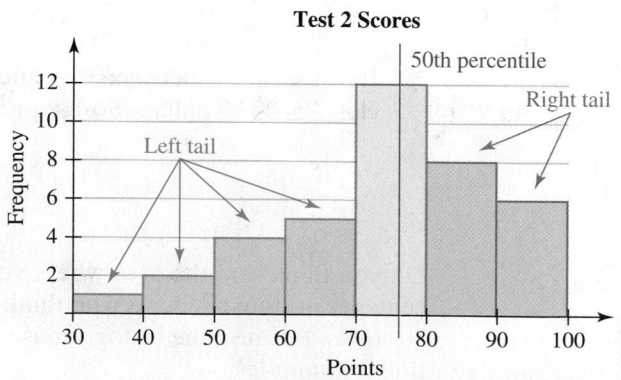

Figure 79 Skewed-left distribution
(**Source:** *J. Lehmann*)

In general, for a unimodal distribution, the **left tail** is the part of the histogram to the left of the 50th percentile and the **right tail** is the part of the histogram to the right of the 50th percentile (see Fig. 79).

Because the left tail (in green) of the Test 2 distribution is longer than the right tail (in orange), we say the distribution is *skewed left*. The right tail is short because the maximum possible test score is 100 points.

Figure 80 displays a histogram of all National Football League® players' ages in 2014. Because the right tail (in orange) is longer than the left tail (in green), we say the distribution is *skewed right*. The left tail is short because players must be three years removed from high school.

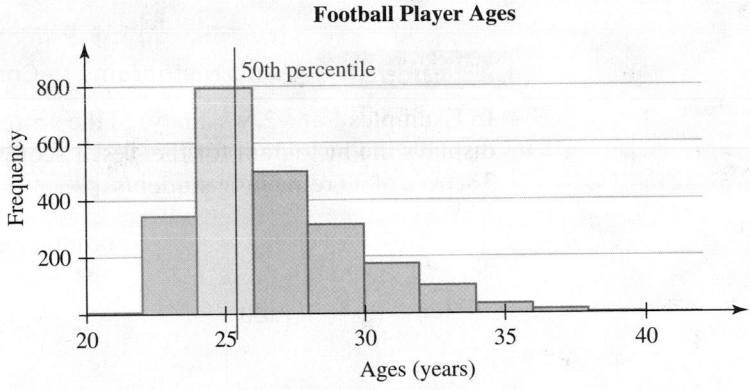

Figure 80 Skewed-right distribution
(**Source:** *National Football League Players Association*)

Finally, we return to the distribution about a living area getting better (see Fig. 81). Because the left tail (in green) and the right tail (in orange) are roughly mirror images, we say that the distribution is *symmetric*.

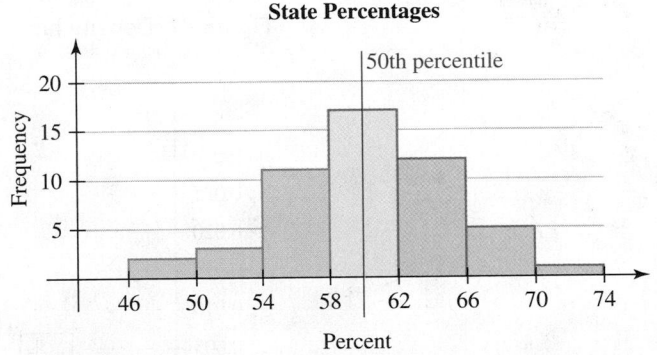

Figure 81 Symmetric distribution
(**Source:** *The Gallup Organization*)

▶ **Definition   Skewed-left, skewed-right, and symmetric distributions**

- If the left tail of a unimodal distribution is longer than the right tail, then the distribution is **skewed left** (see Fig. 79).
- If the right tail of a unimodal distribution is longer than the left tail, then the distribution is **skewed right** (see Fig. 80).
- If the left tail of a distribution is roughly the mirror image of the right tail, the distribution is **symmetric** (see Fig. 81).

Notice that for the three unimodal distributions in Figs. 79, 80, and 81, the 50th percentile lies in the class with the greatest frequency. In general, **an observation at the 50th percentile (the approximate center) of a unimodal distribution tends to be a typical observation.**

Spread of a Distribution

Is Sioux Falls, South Dakota, very spread out? That depends on which city we compare it to. On the one hand, the land area of Sioux Falls is over 8 times greater than the land area of Cambridge, Massachusetts. On the other hand, it is less than 4% of the area of Anchorage, Alaska.

In Example 6, we will compare the spreads of two distributions.

▶ **Example 6**   Using Histograms to Compare Two Groups of Data

In Examples 1 and 2, we analyzed the scores on Test 1 for 35 calculus students. Figure 82 displays the histogram for the Test 1 scores, and Fig. 83 displays the histogram for Test 3 scores of 30 remaining students.

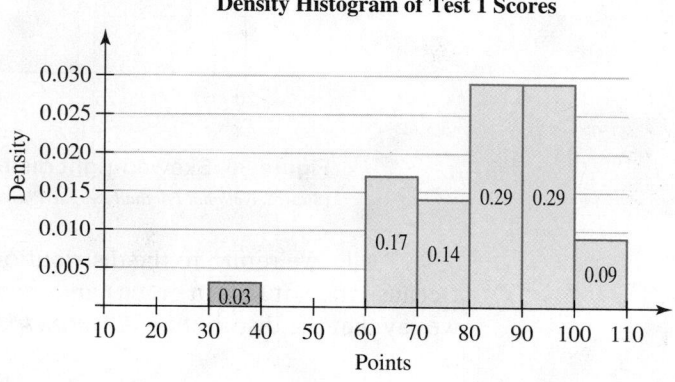

Figure 82 Density histogram of Test 1 scores

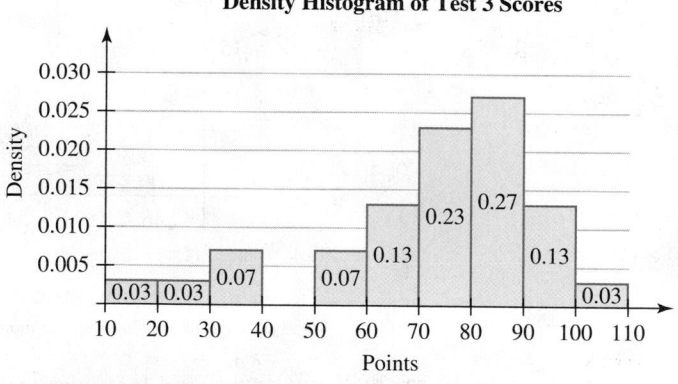

Figure 83 Density histogram of Test 3 scores
(**Source:** *J. Lehmann*)

1. For the two distributions identify any outliers.
2. Compare the shapes of the two distributions.
3. Compare the centers of the two distributions. What does that mean in this situation?
4. Compare the spreads of the two distributions. What does that mean in this situation?
5. Give three possible reasons why the distributions are different.

Solution

1. For Test 1, there is at least one outlier in the class 30–39 points (see the orange bar in Fig. 82). Because $0.03(35) \approx 1$, we conclude there is exactly one outlier. So, one student did quite a bit worse than the other students. Test 3 does not have any outliers.

2. One could argue that the Test 1 distribution is bimodal, but the mound for the class 60–69 points is not very significant. So, we conclude that the distribution is unimodal and skewed left. The Test 3 distribution is unimodal and even more skewed left than the Test 1 distribution.

3. For Test 1, the 50th percentile is in the class 80–89 points (try it). For Test 3, the 50th percentile is in the class 70–79 points. So, the center is lower for the Test 3 distribution than for the Test 1 distribution. This suggests that a typical score on Test 3 is lower than a typical score on Test 1.

4. The Test 3 distribution is more spread out than the Test 1 distribution. This occurred because some students continued to earn high scores but many students did worse.

5. The students might have done worse on Test 3 because the course material had become more challenging, the author wrote harder questions for Test 3, or the students didn't study as much for Test 3. In fact, it could be that two or all three of these events occurred.

In Example 6, we compared four characteristics of the test distributions. When working with distributions that involve a numerical variable, we will often determine the characteristics in the same order as we did in Example 6.

> **Order of Determining the Four Characteristics of a Distribution with a Numerical Variable**
>
> We often determine the four characteristics of a distribution with a numerical variable in the following order:
>
> **1.** Identify all outliers.
>   **a.** For outliers that stem from errors in measurement or recording, correct the errors, if possible. If the errors cannot be corrected, remove the outliers.
>   **b.** For other outliers, determine whether they should be analyzed in a separate study.
> **2.** Determine the shape. If the distribution is bimodal or multimodal, determine whether subgroups of the data should be analyzed separately.
> **3.** Measure and interpret the center.
> **4.** Describe the spread.

For now, the best way to describe the spread of a distribution is to compare it to the spreads of other distributions. In Chapter 4, we will discuss how to *measure* spread. We will also discuss new computational methods to measure the center and identify outliers.

Model

Even though the histograms of the Test 1 and Test 2 distributions do not precisely describe each individual test score, they are still useful in determining characteristics of the distributions such as shape. The histograms are examples of *models*.

> **Definition   Model**
>
> A **model** is a mathematical description of an authentic situation. We say the description *models* the situation.

For an authentic situation that involves a numerical variable, we can model the situation using a stemplot, a dotplot, or a histogram. But which one is the best? Although each diagram has its advantages, histograms are used most often. However, because it is so easy to use technology to construct diagrams, we will sometimes view more than one type of diagram to gain a more complete understanding of the data.

In Table 48, we summarize the diagrams that can be used with various types of variables and the benefits of the diagrams.

Table 48 Types of Variables Needed for Diagrams and Benefits of Diagrams

Diagram	Types of Variables	Benefits
Frequency Bar Graph	One categorical variable	Compare frequencies of categories.
Relative Frequency Bar Graph	One categorical variable	Compare a part to the whole.
Multiple Bar Graph	Two categorical variables	Compare a part to the whole.
Pie Chart	One categorical variable	Compare a part to the whole.
Two-Way Table	Two categorical variables	Compare a part to the whole.
Dotplot	One numerical variable	Describe individual values for a small or medium number of observations.
Stemplot	One numerical variable	Describe individual values for a small number of observations.
Frequency Histogram	One numerical variable	Compare the frequencies of classes.
Relative Frequency Histogram	One numerical variable	Compare a part to the whole.
Density Histogram	One numerical variable	Compare a part to the whole.
Time-Series Plot	Two numerical variables	Find the association between two variables.

 Group Exploration

Comparing histograms with different class widths

Figure 84 contains three histograms of the Test 1 scores with class widths of 1, 10, and 20.

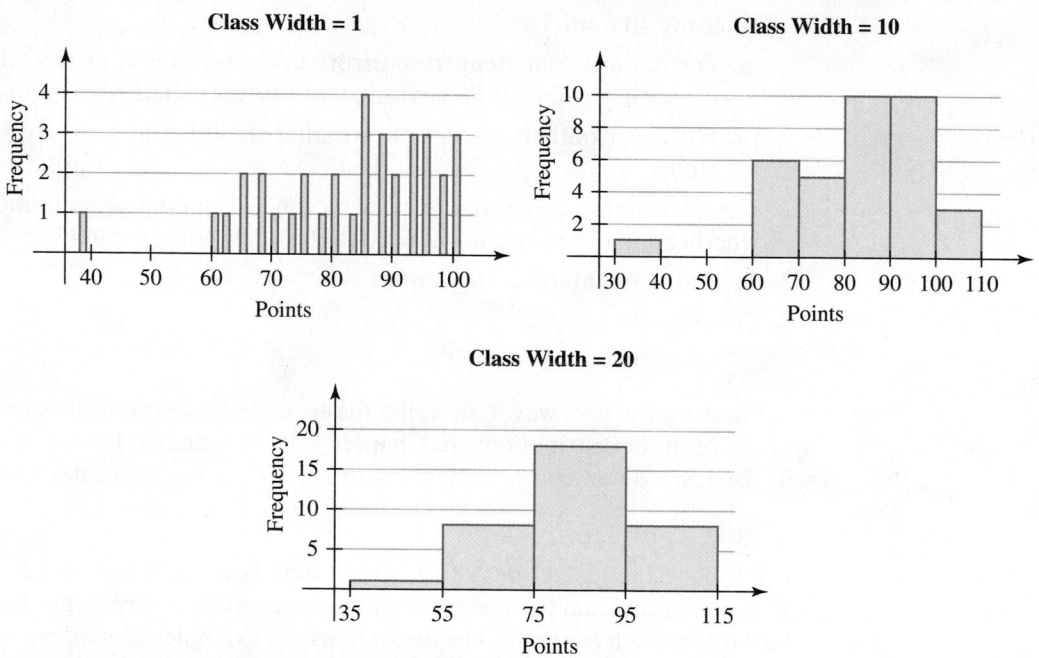

Figure 84 Frequency histograms of Test 1 scores using class widths 1, 10, and 20

For which histogram(s) is the following true?

1. It can be determined that there is an outlier.

2. The value of the outlier can be estimated quite well.

3. From quick inspection, you can determine the shape of the distribution.

4. From quick inspection, you can identify the class that contains the 50th percentile.

5. From careful work, the 50th percentile can be estimated quite well.

▶ **Tips for Success**   **Get the Most Out of Working Exercises**

If you work an exercise by referring to a similar example in your notebook or in the text, it is a good idea to try the exercise again without referring to your source of help. If you need to refer to your source of help to solve the exercise a second time, consider trying the exercise a third time without help. When you finally complete the exercise without help, reflect on which concepts you used to work the exercise, where you had difficulty, and what the key idea was that opened the door of understanding for you.

A similar strategy can be applied after using help features on MyMathLab or getting help from a student, an instructor, or a tutor.

If this sounds like a lot of work, it is! But it is well worth it. Although it is important to complete each assignment, it is also important to learn as much as possible while progressing through it.

Homework 3.4

For extra help ▶   **MyMathLab®**    Watch the videos in MyMathLab    Download the MyDashboard App

1. For a numerical variable, the sum of the relative frequencies of all the classes is equal to _____ .

2. For a(n) _____ histogram, the area of a bar is equal to the relative frequency of the bar's class.

3. A(n) _____ distribution has two mounds.

4. If the right tail of a unimodal distribution is longer than the left tail, the distribution is _____ _____ .

5. The author surveyed some of the students in one of his statistics classes about the distances (in miles) they commute to college. Their responses are described by the histogram shown in Fig. 85.

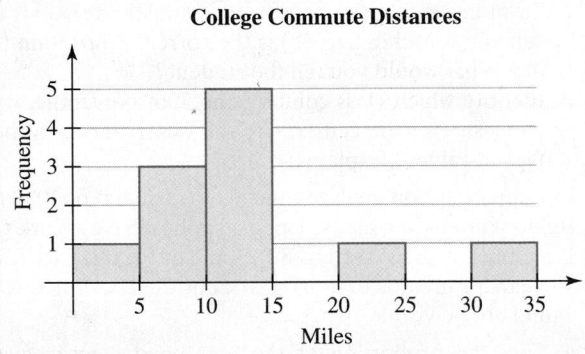

Figure 85 Exercise 5
(**Source:** *J. Lehmann*)

a. What variable is described by the histogram? Is it discrete or continuous?

b. How many of the commute distances are between 17 and 38 miles?

c. How many of the commute distances are at least 10 miles?

d. What proportion of the commute distances are less than 15 miles?

e. Create some data that would have the histogram shown in Fig. 85.

6. The author surveyed some of the students in one of his statistics classes about the numbers of hours they work per week. Their responses are described by the histogram shown in Fig. 86.

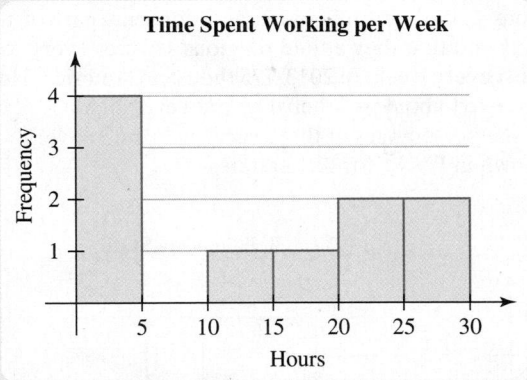

Figure 86 Exercise 6
(**Source:** *J. Lehmann*)

a. What variable is described by the histogram? Is it discrete or continuous?

b. How many of the students work between 9 and 25 hours per week?

c. How many of the students work at least 15 hours per week?

d. What proportion of the students work less than 5 hours per week?

e. Create some data that would have the histogram shown in Fig. 86.

7. A histogram of the numbers of murders per state in 2012 is shown in Fig. 87 for the 50 states.

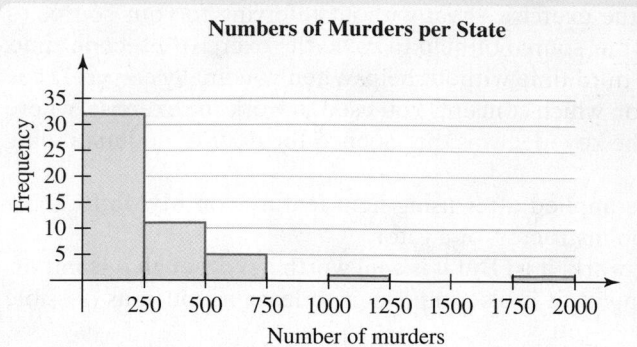

Figure 87 Exercise 7
(**Source:** *FBI*)

a. If there were two outliers, then estimate what they would be. For each of your estimates, find the largest possible error.
b. Describe the shape of the distribution.
c. Estimate the number of states that had between 250 and 749 murders, inclusive.
d. Identify which class contains the 50th percentile, which is a measure of the center. Are the observations in the class typical values? Explain.
e. California and Texas had the largest and the next-to-largest numbers of murders in 2012, respectively. California and Texas also had the largest and the next-to-largest populations in that year, respectively. Explain why we *cannot* conclude that Californians and Texans had the greatest chance of being murdered in 2012. What would be a better measure of comparing the risk of being murdered?

8. The Gallup Organization® classifies Americans as very religious if they say religion is an important part of their daily lives and that they attend religious services every week or almost every week. In 2013, 175 thousand American adults were surveyed about whether they are very religious. A histogram of the percentages of those surveyed who are very religious is shown in Fig. 88 for the 50 states.

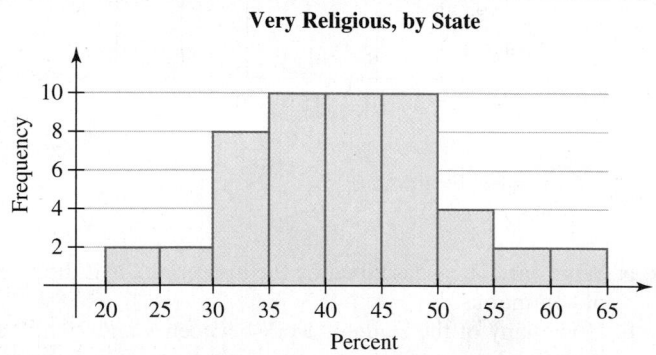

Figure 88 Exercise 8
(**Source:** *The Gallup Organization*)

a. Describe the shape of the distribution.
b. Estimate the number of states where at least 50% of those surveyed are very religious.

c. Mississippi and Utah have the highest percentage of those surveyed who say they are very religious. Estimate the percentage for each of those two states.
d. Identify the class that contains the 50th percentile, which is a measure of the center. Are the observations in the class typical values? Explain.
e. Why might your result from part (d) be a poor measurement of the center of the distribution for *all* residents in the United States. What type of bias in sampling is that?

9. In Exercise 35 of Homework 3.3, you analyzed the ages of 1098 people (on foot or in a vehicle) stopped by police in New York City on December 1, 2012. Without being able to view the police reports or interview the police officers, the author did his best to clean up the data. The ages (in years) of the remaining 1089 people are described by the relative frequency histogram in Fig. 89. Some of the ages are less than 10 years or greater than 65 years, but the bars for those ages would not be visible.

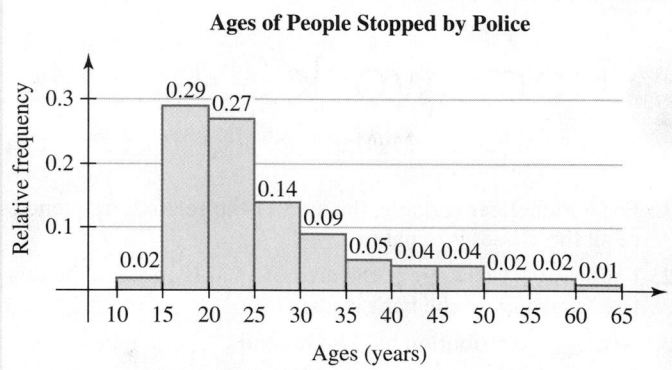

Figure 89 Exercise 9
(**Source:** *New York Police Department*)

a. Describe the shape of the distribution. What does it mean in this situation?
b. Estimate the proportion of people stopped by police who were at most 19 years of age.
c. From inspecting the histogram, a student says that the result you found in part (b) is the correct proportion for *any* day. What would you tell the student?
d. Identify which class contains the 50th percentile, which is a measure of the center. Are the observations in the class typical values? Explain.

10. In Exercise 9, you analyzed the ages (in years) of 1089 people (on foot or in a vehicle) stopped by police in New York City on December 1, 2012 (see Fig. 89). Some of the ages are less than 10 years or greater than 65 years, but the bars for those ages would not be visible.

a. Find the proportion of those stopped whose ages were between 15 and 19 years, inclusive.
b. There was just one child under the age of 10 years who was stopped. Find the proportion of those stopped whose ages were at most 29 years.
c. Find the proportion of those stopped whose ages were at least 20 years.
d. On the basis of the histogram, a student concludes that on *any* day, 27% of those who are stopped are between the ages of 20 and 24 years, inclusive. What would you tell the student?

11. The *approval rating* of the President is the percentage of people surveyed who believe the President is doing a good job. The histogram in Fig. 90 displays President Obama's daily approval ratings in 2009.

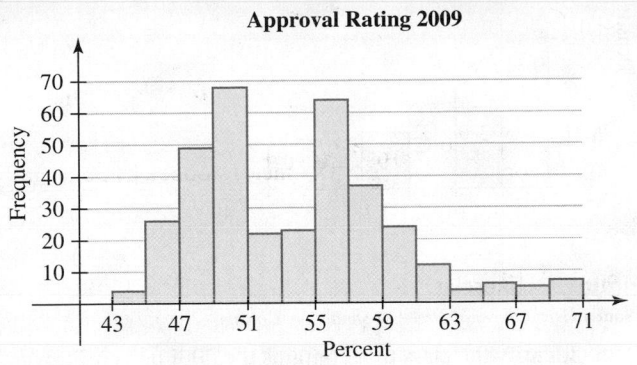

Figure 90 Exercise 11
(**Source:** *Rasmussen Reports*)

a. Describe the shape of the distribution. What does this mean in this situation?

b. The approval rating 54% is at the 50th percentile, which is a measure of the center. The center of a unimodal distribution tends to be a typical value. Is that true for this situation? Explain.

c. Figures 91 and 92 display histograms of President Obama's approval ratings for the first half of 2009 and the second half of 2009, respectively. Is the 50th percentile for the first half of 2009 less than, greater than, or approximately equal to the 50th percentile for the second half of 2009? What does that mean in this situation?

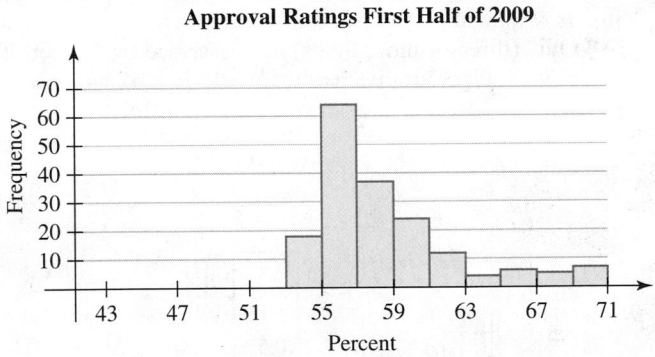

Figure 91 President Obama's daily approval ratings for the first half of 2009

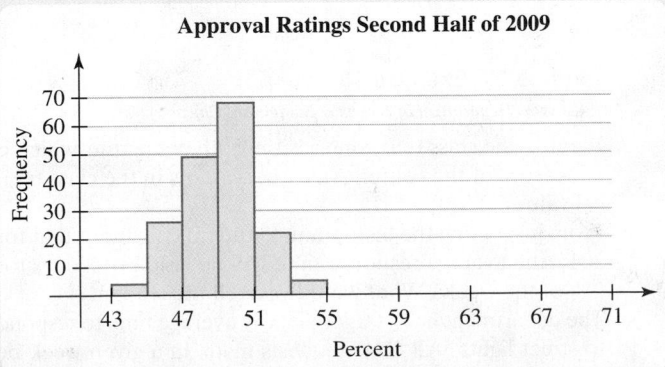

Figure 92 President Obama's daily approval ratings for the second half of 2009

d. Compare the spreads of the distributions for the two halves of 2009.

e. How do the distributions for the two halves of 2009 explain why the distribution for all of 2009 is bimodal?

12. The salaries (in millions of dollars) of football players on the Eagles and Steelers teams are described by the relative frequency histograms in Fig. 93.

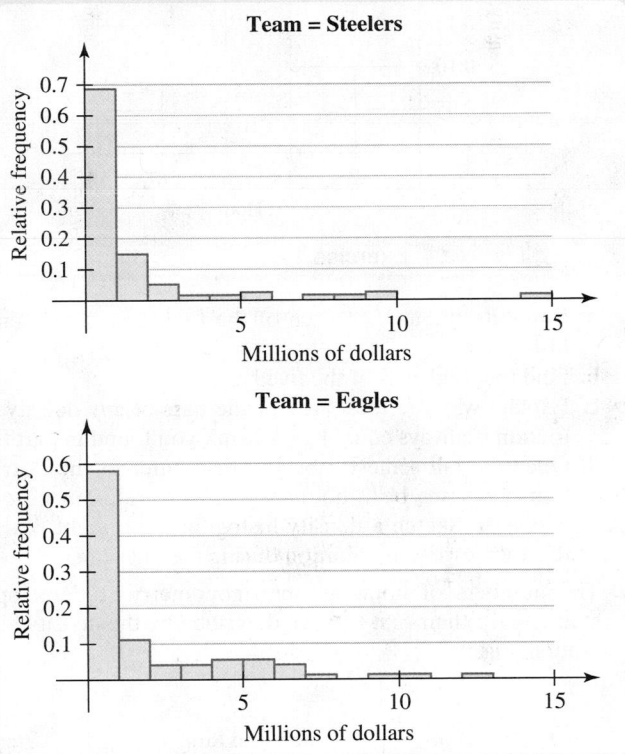

Figure 93 Exercise 12
(**Source:** *NFL*)

a. Estimate the proportion of the Steelers who earn less than $1 million. Estimate the proportion of the Eagles who earn less than $1 million.

b. Estimate the proportion of the Steelers who earn at least $1million. Estimate the proportion of the Eagles who earn at least $1million.

c. A student says the Steelers' distribution is more heavily skewed to the right than the Eagles' distribution because the top-paid player of the Steelers is paid more than the top-paid player of the Eagles. What would you tell the student?

d. Make an educated guess which team tends to pay its players more. Explain why it's not possible to know for sure on the basis of just the histograms.

e. The Steelers have 10 more players on their payroll than the Eagles. Explain why it is better to compare relative frequency histograms rather than frequency histograms in order to respond to part (d).

13. The numbers of hours some college students spend watching television during the summer are described by the density histogram in Fig. 94.

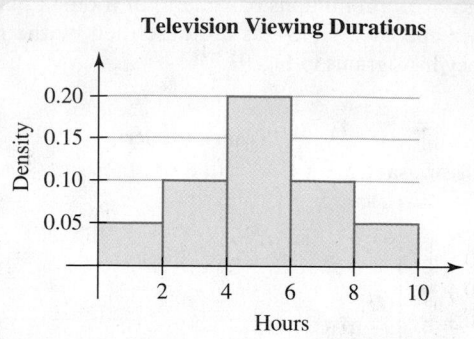

Figure 94 Exercise 13

a. Compute the area of each of the five bars in the density histogram.

b. Find the total area of the five bars.

c. Explain why the total area of the bars of *any* density histogram is always equal to the result you found in part (b).

d. Once the fall semester begins, the center of the distribution decreases by 2 hours and the distribution becomes narrower. Sketch a density histogram that might describe the TV viewing distribution during the fall semester.

14. The numbers of hours some trigonometry students spent studying for their first test are described by the density histogram in Fig. 95.

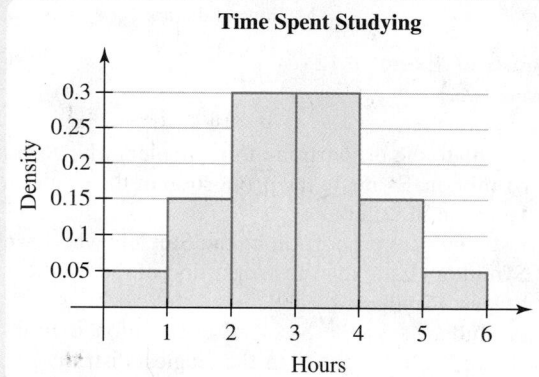

Figure 95 Exercise 14

a. Compute the area of each of the six bars in the density histogram.

b. Find the total area of the six bars.

c. Explain why the total area of the bars of *any* density histogram is always equal to the result you found in part (b).

d. For the second test, the center of the distribution increases by 2 hours and the distribution becomes more spread out. Sketch a density histogram that might describe the study-time distribution for the second test.

15. The average numbers of days it took Chicago's Department of Streets & Sanitation to respond to requests to fix potholes are described by the density histogram in Fig. 96 for each of the weeks in 2012 and 2013.

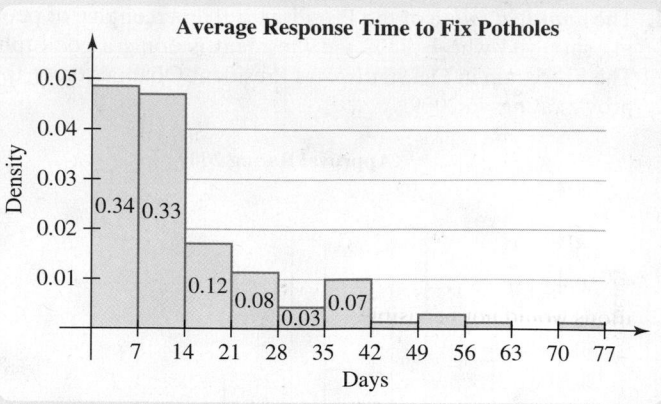

Figure 96 Exercise 15

(**Source:** *Department of Streets & Sanitation, Chicago*)

a. Identify the class that contains the 50th percentile, which is a measure of the center. Are the observations in the class typical values? Explain.

b. For what proportion of the weeks was the average response time at least 42 days? Some of the bars in the density histogram were left blank on purpose.

c. From inspecting the histogram, a student concludes that in 2015, the average response time will be less than 21 days for 79% of the weeks. What would you tell the student?

d. The department has a target that the average time to respond to pothole requests made in a given week be less than 7 days. Find the proportion of observations that are less than 7 days.

e. Suppose the department decides to change the target to be less than 14 days. Find the proportion of observations that are less than 14 days.

16. The average numbers of days it took Chicago's Department of Streets & Sanitation to respond to requests to fix street lights "All Out" (three or more lights) are described by the density histogram in Fig. 97 for each of the weeks in 2012 and 2013.

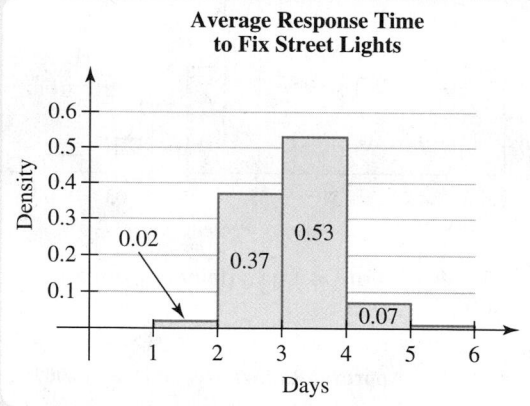

Figure 97 Exercise 16

(**Source:** *Department of Streets & Sanitation, Chicago*)

a. Identify the class that contains the 50th percentile, which is a measure of the center. Are observations in the class typical values? Explain.

b. From inspecting the histogram, a student concludes that for 2015, the average response time will be less than 2 days for 2% of the weeks. What would you tell the student?

c. The department has a target that the average time to respond to street lights "All Out" requests made in a given week be less than 4 days. Find the proportion of observations that are less than 4 days.

d. For what proportion of the weeks was the average response time at least 4 days? One of the bars in the density histogram was left blank on purpose.

e. Suppose the department changes the target to be less than 3 days. Find the proportion of observations that are less than 3 days.

17. The 2011–2012 tuitions (in thousands of dollars) of 4-year, private, not-for-profit universities and colleges are described by the density histogram in Fig. 98. Some of the tuitions are greater than 45 thousand dollars, but the bars for those tuitions would not be visible.

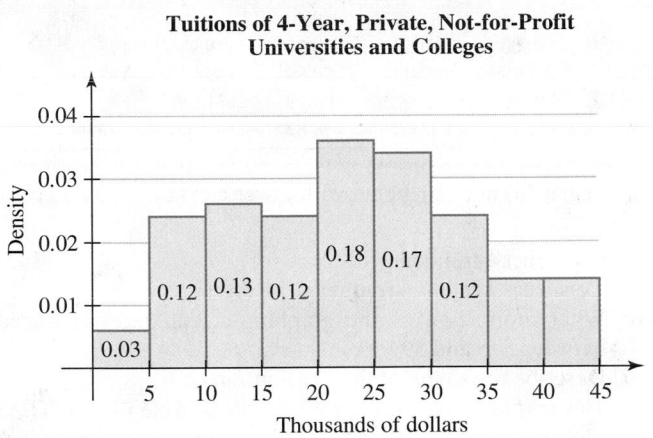

Figure 98 Exercise 17
(**Source:** *U.S. Department of Education*)

a. What variable is described by the histogram? Is it discrete or continuous?

b. Identify the class that contains the 50th percentile, which is a measure of the center. Are observations in the class typical values? Explain.

c. What proportion of the institutions had tuitions less than $30 thousand?

d. What proportion of the institutions had tuitions at least $35 thousand? Some of the bars in the density histogram were left blank on purpose.

e. Find the proportion of the institutions that had tuitions between $35,000.00 and $39,999.99, inclusive. [**Hint:** Use the result you found in part (d) to help.]

18. The 2011–2012 tuitions (in thousands of dollars) of 2-year, public colleges are described by the density histogram in Fig. 99. Some of the tuitions are greater than 8 thousand dollars, but the bars for those tuitions would not be visible.

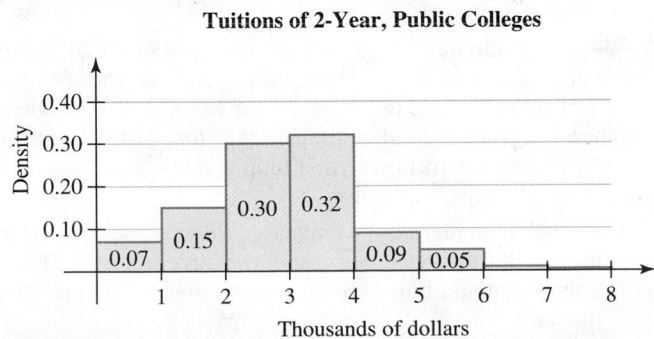

Figure 99 Exercise 18
(**Source:** *U.S. Department of Education*)

a. What variable is described by the histogram? Is it discrete or continuous?

b. Estimate the proportion of the colleges that had tuitions between $3000.00 and $3999.99, inclusive.

c. What proportion of the colleges had tuitions less than $6 thousand?

d. Identify the class that contains the 50th percentile, which is a measure of the center. Are observations in the class typical values? Explain.

e. What proportion of the colleges had tuitions at least $6 thousand? Some of the bars in the density histogram were left blank on purpose.

19. The 2011–2012 tuitions (in thousands of dollars) of 4-year, private, not-for-profit universities and colleges are described by the density histogram in Fig. 98. Some of the tuitions are greater than 45 thousand dollars, but the bars for those tuitions would not be visible.

a. Estimate the percentile of Florida Memorial University's tuition, which was $15 thousand.

b. Estimate the percentile of Morningside College's tuition, which was $25 thousand.

c. The tuition at Rabbinical College of Ohr Shimon Yisroel is at the 15th percentile. Estimate the tuition. Round to the nearest dollar.

d. The tuition at Worcester Polytechnic Institute is at the 40th percentile. Estimate the tuition. Round to the nearest dollar.

20. The 2011–2012 tuitions (in thousands of dollars) of 2-year, public colleges are described by the density histogram in Fig. 99. Some of the tuitions are greater than 8 thousand dollars, but the bars for those tuitions would not be visible.

a. At Cossatot Community College of the University of Arkansas, the tuition was about $2 thousand. Estimate the percentile for the tuition.

b. Estimate the percentile of Highline Community College's tuition, which was $4 thousand.

c. The tuition at New Mexico State University–Carlsbad is at the 7th percentile. Estimate the tuition. Round to the nearest dollar.

d. The tuition at Hennepin Technical College is at the 93rd percentile. Estimate the tuition. Round to the nearest dollar.

21. ⏚ Here are the percentages of Americans who are obese for the Northeastern and Midwestern states (Source: *Centers for Disease Control and Prevention*):

25.6	28.4	22.9	27.3	24.6	23.6
29.1	25.7	23.7	28.1	31.4	30.4
29.9	31.1	25.7	29.6	28.6	29.7
30.1	28.1	29.7			

For parts (a), (b) and (c), begin with lower class limit 22 and class width 2.

a. Construct a frequency and relative frequency table.

b. Construct a frequency histogram.

c. Construct a relative frequency histogram.

d. Describe the shape of the distribution.

e. Most of the obesity rates (in blue) of the Northeastern states are less than the obesity rates (in red) of the Midwestern states. How does that tie in with the shape of the distribution of the obesity rates for all 21 Northeastern and Midwestern states?

22. **DATA** Here are the percentages of Americans who are obese for 11 randomly selected Southern states and 9 randomly selected Western states (Source: *Centers for Disease Control and Prevention*):

29.1	31.3	34.7	27.6	29.6	31.1
29.2	33.8	33.0	34.5	26.9	25.7
20.5	23.6	24.3	27.1	27.3	24.3
26.8	24.6				

For parts (a), (b), and (c), begin with lower class limit 20 and class width 2.

a. Construct a frequency and relative frequency table.
b. Construct a frequency histogram.
c. Construct a relative frequency histogram.
d. Describe the shape of the distribution.
e. All but one of the obesity rates (in red) of the Southern states are greater than the obesity rates (in blue) of the Western states. A student says the distribution of all 20 obesity rates must be bimodal because the obesity rate for the Southern and the Western states are so different. What would you tell the student?

23. **DATA** The author surveyed students in one of his prestatistics classes about the number of texts they send per weekday. Here are their responses (Source: *J. Lehmann*):

50	4	100	30	100	50	20
25	50	15	300	35	5	100
40	30	6	10	100	50	25
5	200	50	75	2	90	250
4	50					

For parts (a) and (b), begin with lower class limit 0 and class width 50.

a. Construct a frequency histogram.
b. Construct a relative frequency histogram.
c. What proportion of the students send at least 100 texts per day?
d. Describe the shape of the distribution.
e. One student sends 1000 texts per weekday. If a frequency histogram with the same classes you used to construct the frequency histogram in part (a) were constructed that included this outlier, how would it compare with the frequency histogram you constructed in part (a)?

24. **DATA** The author surveyed Facebook users in one of his statistics classes about the number of friends they have on Facebook. Here are their responses (Source: *J. Lehmann*):

30	1000	300	500	150	300
500	600	1100	200	200	1523
500	7	0	200	700	1200
250	600	200	500	450	200
600	40	1300	1300	756	

For parts (b) and (c), begin with lower class limit 0 and class width 200.

a. One student wrote "100–200." What value would represent this response well? Include the value in the sample for parts (b) through (e).

b. Construct a frequency histogram.
c. Construct a relative frequency histogram.
d. Estimate the proportion of the surveyed Facebook users who have between 400 and 800 Facebook friends.
e. Estimate the proportion of the surveyed Facebook users who have at most 800 Facebook friends.

25. **DATA** Here are the prices (in dollars) of some randomly selected new graphing calculators (Source: *Amazon.com*):

90.00	91.50	95.80	125.44	136.26
132.99	74.95	37.99	136.99	59.99
96.50	49.25	101.99	137.81	60.00
129.00	81.95	69.50	94.99	93.99
159.00	69.50	89.37	179.99	115.00
199.99	169.99	59.99	159.95	199.95
110.63	95.97	44.49	131.98	33.99
132.21	132.00	138.82	139.99	49.99

For parts (a) and (b), begin with lower class limit 25 and class width 25.

a. Construct a frequency histogram.
b. Construct a relative frequency histogram.
c. What proportion of the graphing calculators are priced between $75 and $99.99, inclusive?
d. Describe the shape of the distribution.
e. The graphing calculator with the price $115.00 is a TI-83 Plus. Explain why its price can be considered somewhat unusual.

26. **DATA** Here are the prices (in dollars) of some randomly selected smartphones (Source: *Amazon.com*):

95.99	79.99	119.99	79.99	139.99
84.95	139.99	164.00	79.45	249.00
89.99	95.99	175.99	64.98	147.00
70.99	48.43	49.99	175.99	82.95
299.99	68.74	98.89	134.12	137.95
179.72	176.00	39.99	136.00	299.99

For parts (a) and (b), begin with lower class limit 25 and class width 25.

a. Construct a frequency histogram.
b. Construct a relative frequency histogram.
c. Describe the shape of the distribution.
d. What proportion of the smartphones have prices between $100 and $149.99, inclusive?
e. Which class has the largest relative frequency? What is that relative frequency? What does that mean in this situation?

Large Data Sets

27. **DATA** Access the data about the gas mileages of all 2014 cars, which are available at MyMathLab and at the Pearson Downloadable Student Resources for Math & Stats website. When constructing relative frequency histograms for parts (a) and (b), begin with lower class limit 5 miles per gallon and class width 5 miles per gallon.

a. Describe the highway gas mileages (in miles per gallon) by using a dotplot, a stemplot, and a relative frequency histogram. Compare how helpful the diagrams are in analyzing the data.

b. Construct a relative frequency histogram that describes the city gas mileages (in miles per gallon) of the cars. Compare the spreads of the city-gas-mileage distribution and the highway-gas-mileage distribution.

c. Find the class that contains the 50th percentile of the city gas mileages. Find the class that contains the 50th percentile of the highway city gas mileages. Compare your two results. What does this mean in this situation?

d. Construct a dotplot of the highway gas mileages where the mileages are grouped by the number of cylinders. As the number of cylinders increases, what tends to happen to the highway gas mileage? Can you assume there is causality? Explain.

e. A student says front-wheel drive is more fuel efficient than rear-wheel drive, which is more fuel efficient than four-wheel drive. To investigate, construct a dotplot of the highway gas mileages where the mileages are grouped by the drive description. Is the student correct? Explain.

28. **DATA** Access the data about cruise ships, which are available at MyMathLab and at the Pearson Downloadable Student Resources for Math & Stats website.

a. Construct a pie chart that describes the distribution of the companies. What proportion of the cruise ships are owned by Royal Caribbean, Carnival, and Princess? The 3 companies are what proportion of all the companies? Compare the two proportions you found and what this means in this situation.

b. Construct a relative frequency histogram that describes the lengths (in hundreds of feet) of the cruise ships. Begin with lower class limit 2 and class width 1. Describe the shape of the length distribution.

c. Construct a relative frequency histogram that describes the crew sizes (in hundreds). Begin with lower class limit 0 and class width 2. There are two outliers. Describe them.

d. Ignoring the two outliers, determine the shape of the crew-size distribution. Is the shape the same as the length distribution's? Why does this make sense?

e. For the crew-size distribution, identify the class that contains the 50th percentile. Do the same for the length distribution. Use your two results to help you estimate the number of crew members needed per foot.

Concepts

29. **DATA** A student measures the temperatures on 9 days in a row, all in degrees Fahrenheit, and lists them from smallest to largest:

$$40, 60, 60, 67, 67, 67, 68, 68, 69$$

Then she constructs a frequency histogram (see Fig. 100) and concludes that the distribution is symmetric.

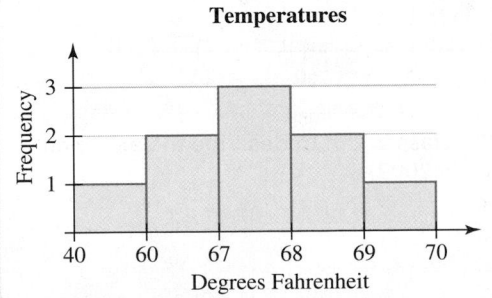

Figure 100 Exercise 29

a. Explain why the student's histogram is incorrect and construct a correct one.

b. Is the student correct that the distribution is symmetric? Explain.

c. For the horizontal axis of a histogram, is it important that the numbers increase by an equal amount and be equally spaced? Explain.

30. Explain why the sum of the heights of the bars of a relative frequency histogram equals 1.

31. Describe at least one advantage of using a stemplot over a histogram with equal class widths of 5.

32. Explain why if we describe a frequency distribution using a histogram with equal class widths of 1, the histogram is about the same thing as a dotplot.

33. Why do we often search for two meaningful parts of a bimodal distribution?

34. Explain how the frequency histogram of a bimodal distribution might appear to be describing a unimodal distribution if the equal class widths are too large.

35. Sketch a frequency histogram that describes a skewed-right distribution, which has an approximate center 35 and an outlier of 95.

36. Sketch a frequency histogram that describes a skewed-left distribution, which has an approximate center 70 and an outlier of 20.

37. For a density histogram, the area of a bar is equal to the relative frequency of the bar's class. Why can't there be a density bar graph?

38. Thirty-five students take a prestatistics test worth 100 points. Draw a density histogram of the test scores (in points) with 0–100 points as its one and only class.

For Exercises 39–42, identify whether the distribution of the given variable is skewed left, skewed right, or symmetric. Explain.

39. The year on a penny in circulation

40. The age (in years) of a car in use

41. The height (in inches) of a women

42. The price (in dollars) of gasoline at a gas station in Atlanta, Georgia, on May 1, 2015

For Exercises 43–48, choose all the following diagrams and tables that would be an appropriate way to describe the given data: frequency and relative frequency table, relative frequency bar graph, multiple bar graph, pie chart, two-way table, dotplot, stemplot, time-series plot, and relative frequency histogram. Do not include diagrams or tables that would be possible but awkward.

43. The maximum speeds (in miles per hour) of 895 Porsche Boxsters are recorded. The speeds are rounded to the second decimal place.

44. The colors of cars are recorded.

45. For each of 70 adults, a researcher compares whether the adult has a tattoo and whether the adult smokes.

46. A researcher compares the annual Home Depot revenues (in billions of dollars) to the years from 2000 to 2014.

47. For each of 30 intermediate algebra courses offered in a semester at a community college, the head of the department records the number of students who passed the course.

48. A statistician compares the genders of 1000 young adults and their favorite types of music.

Hands-On Research

49. Find a dotplot, stemplot, time-series plot, or histogram in a blog, newspaper, magazine, or journal.

   a. For each variable the diagram illustrates, state its definition including units and whether the variable is discrete or continuous.

   b. What is the point of the article? Does the diagram support that point? Explain.

   c. Could one or more of the other diagrams have been used instead? If so, identify which ones and determine which of all the diagrams would best support the point of the article. Explain.

   d. If the diagram describes one numerical variable, describe the shape, center, spread, and any outliers of the distribution. If the diagram is a time-series plot, what happens to one variable as the other variable increases?

50. Collect data that involve one or two numerical variables. Your study should include 20 individuals.

   a. For each variable, state its definition including units.

   b. Provide the data you collected.

   c. Use a statistical diagram to illustrate the data.

   d. What conclusion(s) can you draw about the 20 individuals?

   e. If the 20 individuals were a sample, what group of individuals could be the population? In this case, would your sampling have bias? If so, which type(s)? Explain. Would it be OK to draw the same conclusion(s) about the population that you drew about the sample?

▼ 3.5 Misleading Graphical Displays of Data

Objectives

» Explain why a certain class width of a histogram can emphasize or de-emphasize an aspect of a distribution.

» Explain why the starting value of the vertical axis of a bar graph or a time-series plot can emphasize or de-emphasize an aspect of a distribution.

» Explain why nonuniform scaling can be misleading.

» Explain why *three-dimensional graphs* can be misleading.

The way that a statistical diagram is designed can greatly affect people's impression of the information, especially if the diagram is not studied carefully.

The Impact of the Class Width of a Histogram

In Example 1, we will determine how a certain class width of a histogram can be misleading.

▶ **Example 1**   Determine How a Class Width Can Be Misleading

In Exercise 11 of Homework 3.4, you worked with the daily approval ratings of President Obama. Figure 101 displays a histogram of the daily approval ratings in 2009 using a smaller class width (2 percentage points), and Fig. 102 displays a histogram using a larger class width (5 percentage points).

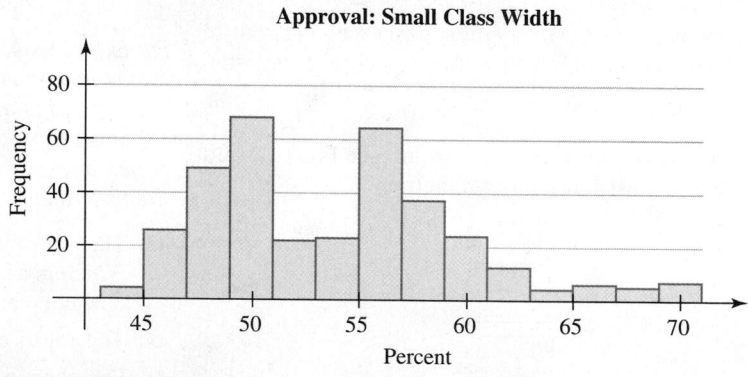

Figure 101 Using a smaller class width to describe the approval rating of President Obama in 2009

(**Source:** *Rasmussen Reports*)

Approval: Large Class Width

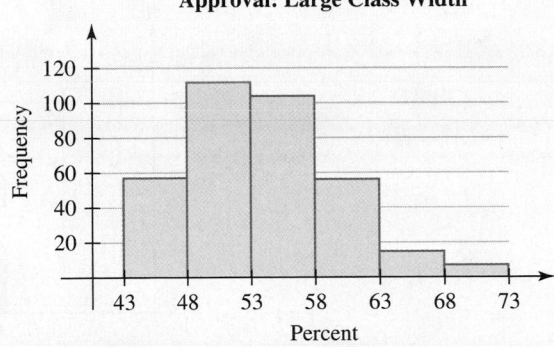

Figure 102 Using a larger class width to describe the approval rating of President Obama in 2009

Which of these histograms is misleading? Explain.

Solution

The histogram with the smaller class width reveals that the distribution is bimodal. This would point us in the correct direction of discovering that President Obama's approval rating was high in the first half of 2009 and low in the second half (see Exercise 11 of Homework 3.4). The histogram with the larger class width is misleading because it suggests that the distribution is unimodal, which is false. If someone wanted to hide that the approval rating dropped in 2009, he or she might use a larger class width, but that would be dishonest.

WARNING   **When viewing a histogram, keep in mind that a certain choice of class width can emphasize or de-emphasize certain aspects of the distribution.**

The Impact of the Starting Value of the Vertical Axis of a Bar Graph or Time-Series Plot

The differences in values described by the vertical axis of a bar graph can be de-emphasized by starting the vertical axis at a nonzero value.

▶ **Example 2** Determine the Impact of Starting a Vertical Axis at a Nonzero Value

The percentages of veterans who used GI Bill benefits from 2002 through 2013 to complete educations ranging from vocational training to postgraduate are described by the bar graphs in Figs. 103 and 104 for various service branches.

Education Completion Rates

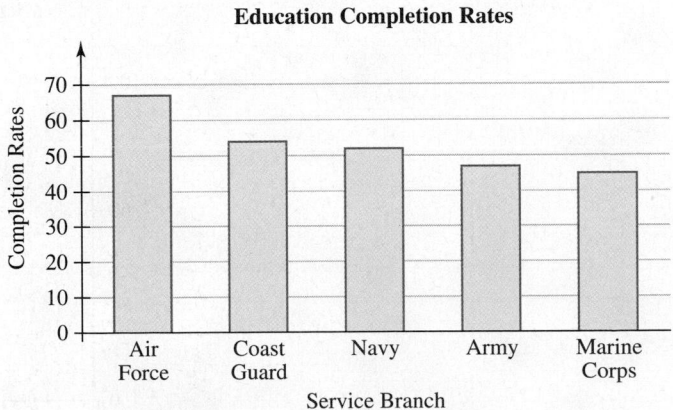

Figure 103 Vertical axis starting at 0%

(**Source:** *Student Veterans of America*)

Education Completion Rates

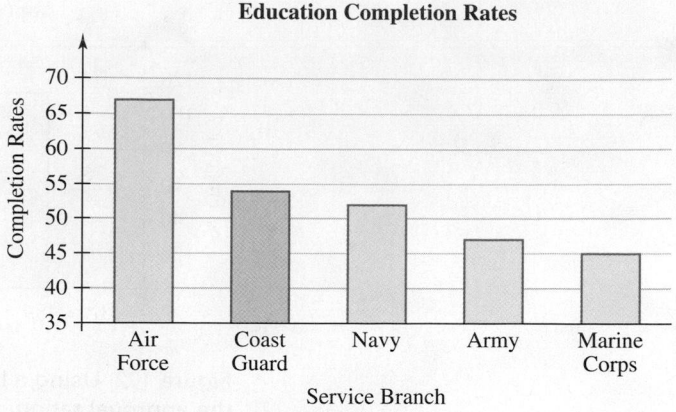

Figure 104 Vertical axis starting at 35%

1. If the Air Force wants to convince potential recruits how much better the education completion rate is for Air Force veterans than for veterans of the other four service branches, which bar graph would it display? Explain.
2. If the Navy wants to convince potential recruits that the education completion rate for Navy veterans is very close to the rate for Coast Guard veterans, which bar graph would it display? Explain.
3. From which bar graph can you better estimate the education completion rate for Coast Guard veterans? Explain. Estimate the rate.

Solution

1. The bar graph in Fig. 103 emphasizes the *sizes* of the education completion rates, whereas the bar graph in Fig. 104 emphasizes the *differences* in the rates. So, the Air Force would display the bar graph in Fig. 104.
2. The Navy would display Fig. 103 because the bars for the Coast Guard and the Navy appear to have almost the same height.
3. We can better estimate the education completion rate for Coast Guard veterans using the bar graph in Fig. 104 because the numbers on the vertical axis go up by smaller amounts than in the bar graph in Fig. 103. The orange bar above "Coast Guard" has height approximately 54%. So, the education completion rate for Coast Guard veterans is approximately 54%.

▶ Example 3   Determine the Impact of Starting a Vertical Axis at a Nonzero Value

The annual revenues of Microsoft® are described by the time-series plots in Figs. 105 and 106 for various years. If Microsoft wants to emphasize the growth of its revenue, which time-series plot would it want to display?

Microsoft Annual Revenues

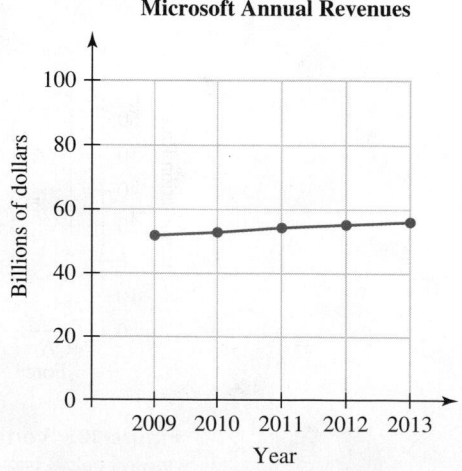

Figure 105 Vertical axis starting at $0 billion
(**Source:** *Microsoft Corporation*)

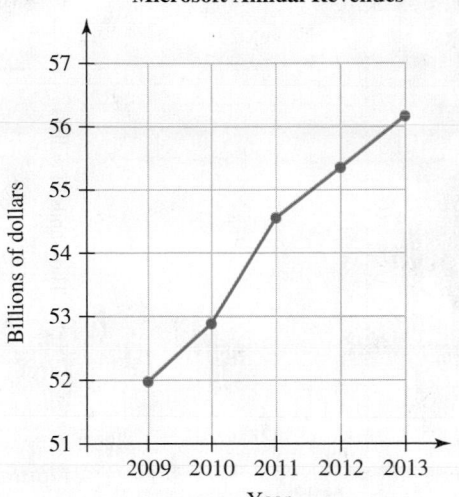

Figure 106 Vertical axis starting at $51 billion

Solution

The time-series plot in Fig. 105 emphasizes the *sizes* of the annual revenues, whereas the time-series plot in Fig. 106 emphasizes *increases* in the annual revenue. So, if Microsoft wanted to emphasize the growth of its revenue, the company would choose to display the time-series plot in Fig. 106 (the one whose vertical axis starts with a nonzero value).

WARNING   There is nothing mathematically wrong with having the vertical axis of a time-series plot start at a nonzero value. But be aware that **if the vertical axis of a time-series plot does not start at 0, the changes in the variable described by that axis are being emphasized.**

In some cases, the starting value of the vertical axis of a time-series plot might be 0 so that the changes in the variable described by that axis are de-emphasized.

▶ Example 4   Determine the Impact of Starting a Vertical Axis at 0

Northwestern University tuitions are described by the time-series plots in Figs. 107 and 108 for various years, where the starting year of the academic year is shown on the horizontal axis. If Northwestern wants to de-emphasize how much its tuition has increased, which time-series plot would it want to display?

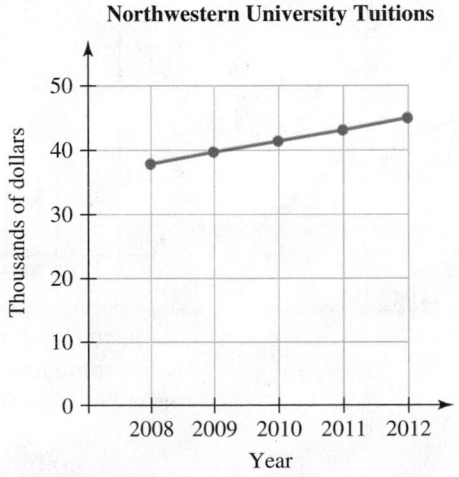

Figure 107 Vertical axis starting at $0 thousand
(**Source:** *Northwestern University*)

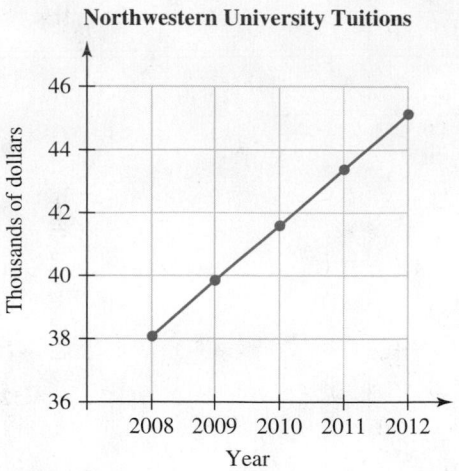

Northwestern University Tuitions

Figure 108 Vertical axis starting at $36 thousand

Solution

The times series plot in Fig. 107 emphasizes the *sizes* of the tuitions, whereas the time-series plot in Fig. 108 emphasizes *increases* in tuition. So, if Northwestern wants to de-emphasize its increases in tuition, it would want to display the time-series plot in Fig. 107 (the one where the vertical axis starts at 0).

WARNING   **If the vertical axis of a time-series plot starts at 0, be aware that the changes in the variable described by that axis are being de-emphasized.**

Nonuniform Scaling Can Be Misleading

When using categories of years for a bar graph, there is nothing mathematically wrong with having the years increase by different amounts, but it can be misleading.

▶ **Example 5**   Determining the Impact of Using Categories of Years That Increase by Different Amounts

The revenues (in billions of dollars) of the Walt Disney Company® are described by the bar graphs in Figs. 109 and Fig. 110 for various years. If Walt Disney wanted to emphasize its growth in revenue, which bar graph would be more convincing? Why is that bar graph misleading? What type of graph would contain the same information but not be misleading?

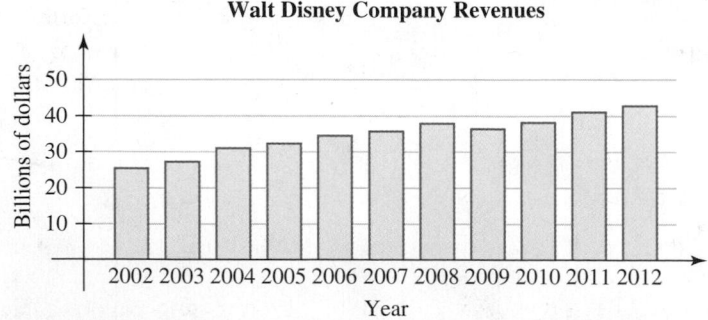

Walt Disney Company Revenues

Figure 109 Walt Disney revenues of years that increase by the same amount

(**Source:** *Walt Disney Company*)

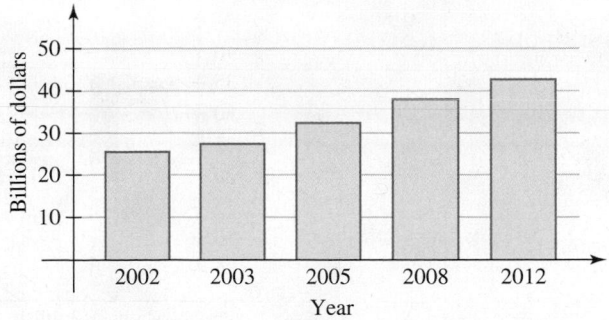

Figure 110 Walt Disney revenues of years that increase by different amounts

In Fig. 110, the revenues seem to increase a lot more than in Fig. 109, because the years used for the categories increase by larger and larger amounts. So, the Walt Disney Company could make its financial progress seem much better by using Fig. 110, but this would be misleading. A clear way to describe such data is with a time-series plot. A time-series plot for the Walt Disney revenues, where the years increase by equal amounts, is shown in Fig. 111.

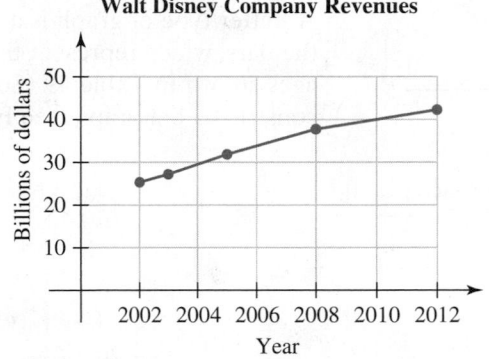

Figure 111 A time-series plot of the Walt Disney Company revenues

The time-series plot has the same overall steepness as the bar graph in Fig. 109, as it should.

WARNING **If the categories for a bar graph are various years and those years do not increase by the same amount, the bar graph can be misleading. A time-series plot would be a better graph to use.**

Three-Dimensional Graphs Can Be Misleading.

Although three-dimensional graphs can look nice, they can be misleading.

▶ **Example 6** Determine How a Three-Dimensional Graph Is Confusing

A total of 2059 randomly selected adults were asked which school subject has been the most valuable to them in their lives. Their responses are described by the three-dimensional graph in Fig. 112.

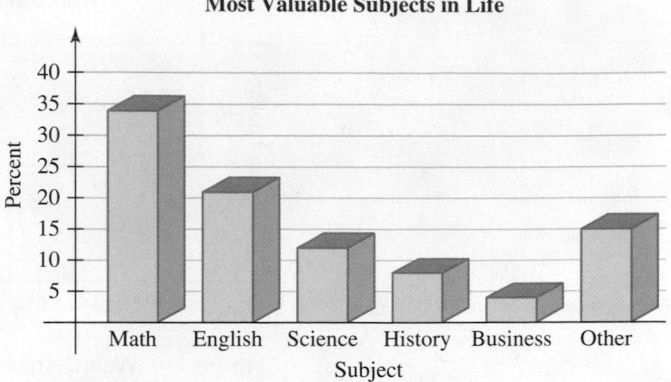

Figure 112 Three-dimensional graph of most valuable subjects
(**Source:** *The Gallup Organization*)

1. What is confusing about this graph?
2. Determine a better type of graph and construct it.
3. Which subject received the most votes? What percentage of adults voted for it?

Solution

1. Because of the three-dimensional perspective, it is difficult to tell how the tops of the boxes line up with the scaling on the vertical axis.
2. A better type of graph is a bar graph because it is easier to estimate the heights of the bars, which represent the percentages. A little research would give the percentages shown in Table 49 (Source: *The Gallup Organization*). We use these data to construct a bar graph (see Fig. 113).

Table 49 Most Valuable Subjects in Life

Subject	Percent
Math	34
English	21
Science	12
History	8
Business	4
Other	15

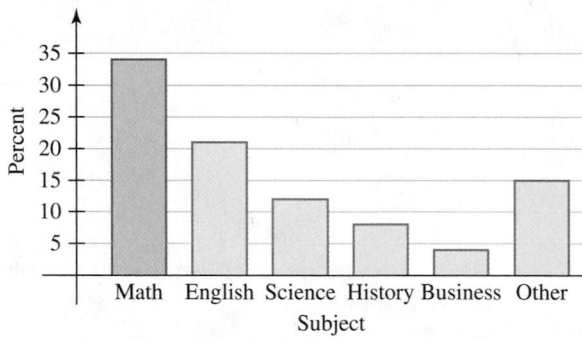

Figure 113 Bar graph of most valuable subjects

3. The highest bar in the bar graph is the orange one, which represents the percentage of adults who voted for math. The height of the bar is approximately 34, which checks with the data in Table 49. So, about 34% of the adults thought math was the most valuable subject to them in their lives.

Now that we've discussed various types of misleading diagrams, take your time sizing up the axes and other aspects of a diagram to make sure you understand what it really means.

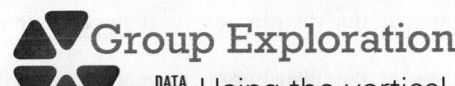 Group Exploration

DATA Using the vertical axis to persuade

The amounts of farmland in the United States are shown in Table 50 for various years.

Table 50 Amounts of Farmland in the United States

Year	Amount of Farmland (millions of acres)
1993	968
1995	961
2000	945
2005	928
2010	920
2012	915

Source: *Agricultural Statistics Board*

1. Suppose an agricultural interest group wants to emphasize how much the acreage of farmland has decreased in hopes of increasing subsidies for farmers. Construct a time-series plot that will support the group's argument.

2. Suppose a city planner wants to de-emphasize how much the acreage of farmland has decreased in hopes of replacing some farmland with a subdivision. Construct a time-series plot that will support the planner's argument.

3. For the time-series plot that you created in Problem 1, extend the line segment that connects the data points for 2010 and 2012 to the right so you can predict the amount of farmland in 2020. Then repeat this process, but use the line segment that connects the data points for 2000 and 2005.

4. Which of the two predictions you made in Problem 3 would better serve the agricultural interest group? the city planner? Explain.

5. Do you have much faith that either of the predictions you made in Problem 3 will turn out to be true? Explain.

 Homework 3.5

For extra help ▶ MyMathLab®   ▤ Watch the videos in MyMathLab   ◑ Download the MyDashboard App

1. *True or False:* If the vertical axis of a time-series plot starts at 0, the changes in the variable described by that axis are being emphasized.

2. *True or False:* If the vertical axis of a time-series plot does not start at 0, the changes in the variable described by that axis are being emphasized.

3. If the categories for a bar graph are various years and those years do not increase by the same amount, a(n) ____ plot would be a better graph to use.

4. Graphs in ____ dimensions can be misleading.

5. The histograms in Fig. 114 and Fig. 115 describe the high temperatures (in degrees Fahrenheit) in San Mateo, California, on each day of May 2014. Fig. 114 uses class widths of 3°F, and Fig. 115 uses class widths of 7°F.

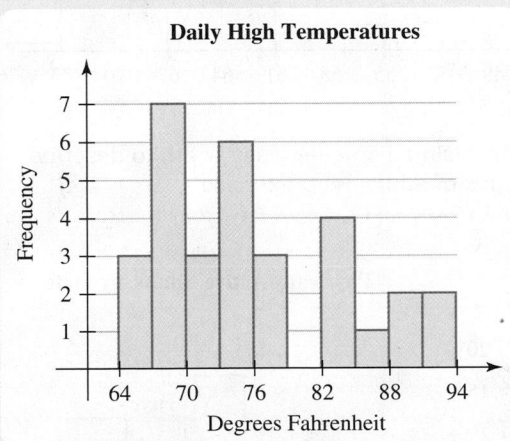

Figure 114 Using a smaller class width
(**Source:** *The Weather Channel*)

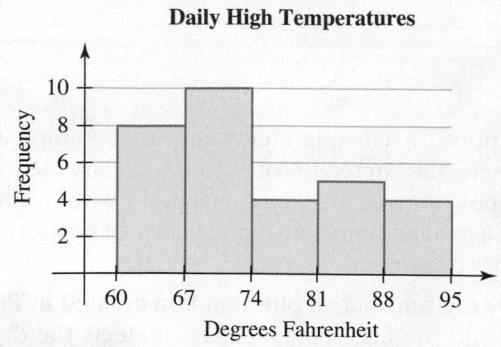

Figure 115 Using a larger class width

a. In San Mateo in May, there tend to be two types of weather: cool, foggy days and hot, sunny days. Which of the two histograms better reflects this?

b. With which histogram can you determine the number of days when the high temperature was between 76°F and 79°F, inclusive? How many days is that?

c. With which histogram can you better estimate the largest high temperature in May? State the class that contains that temperature.

d. If a home buyer tells her realtor that she prefers sunny days with a high temperature of 80°F, which histogram would make San Mateo's weather seem more attractive to the home buyer? Explain. Would it be misleading for the realtor to show the client only that histogram?

6. The histograms in Fig. 116 and Fig. 117 describe the percentages of adults who exercise for each of the 50 states. Figure 116 uses class widths of 3 percentage points, and Fig. 117 uses class widths of 4 percentage points.

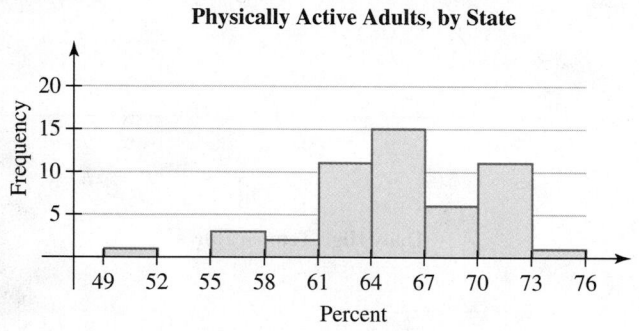

Figure 116 Using a smaller class width to describe percentages of adults who exercise
(**Source:** *Centers for Disease Control and Prevention*)

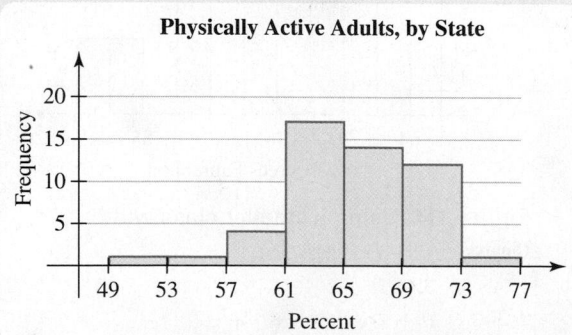

Figure 117 Using a larger class width to describe percentages of adults who exercise

a. Describe the shape of the distribution as it appears in Fig. 116.

b. Describe the shape of the distribution as it appears in Fig. 117.

c. It turns out that the center of the exercise distribution for the West is about 6 percentage points greater than the center of the exercise distribution for the rest of the country (Source: *Centers for Disease Control and Prevention*). Which of the two histograms suggests this? Explain.

d. Tennessee has the lowest percentage of adults who exercise. Estimate that percentage.

e. With which histogram can you determine the number of states where the percentage of adults who exercise is at least 70% AND less than 73%? Find that number of states.

7. For this exercise, we define the *resale rate* of a car to be the percentage of the sticker price the car is worth 5 years later. The resale rates of the cars with the 5 best resale rates are described by the bar graphs in Figs. 118 and 119.

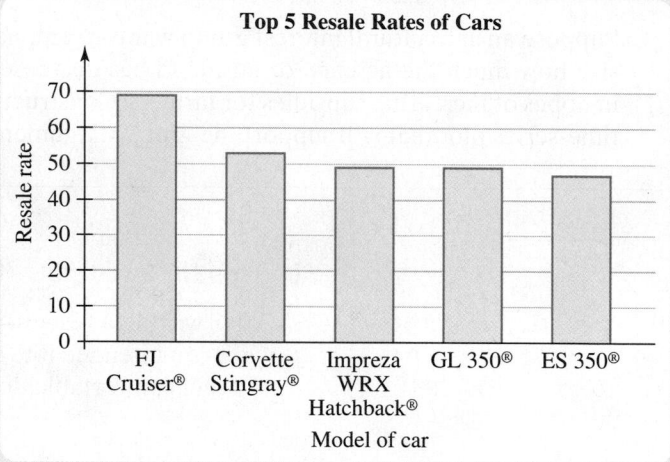

Figure 118 Vertical axis starting at 0%
(**Source:** *Kiplinger's Personal Finance/Kelly Blue Book*)

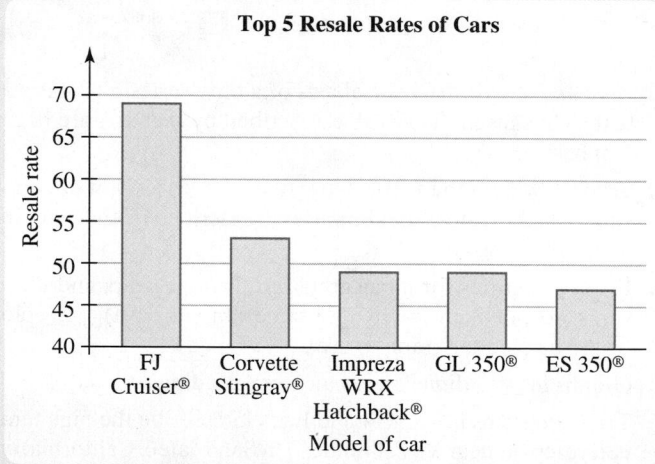

Figure 119 Vertical axis starting at 40%

a. If Toyota wants to convince potential customers how much better the resale rate of its FJ Cruiser is than the other four cars described in the bar graphs, which bar graph would it display? Explain.

b. If Lexus wants to convince potential customers that the resale rate of its ES 350 is very close to the resale rate of the GL 350, which bar graph would it display? Explain.

c. From which bar graph can you better estimate the re-sale rate for a Corvette Stingray? Explain. Estimate the Stingray's resale rate.

d. In Boston, Massachusetts, the base price of a 2014 Impreza WRX Hatchback is $25,549. Estimate the resale *value* (the dollar value in 5 years).

e. In Glen Ellyn, Illinois, the 2014 base price of an ES 350 is $71,005 and the 2014 base price of an FJ Cruiser is $27,230. Find the difference in base prices of the ES 350 and the FJ Cruiser. Then estimate the difference in the resale values of the two cars. Compare the two differences.

8. The percentages of commuters who walk to work are de-scribed by the bar graphs in Figs. 120 and 121 for the five cities with the highest percentages.

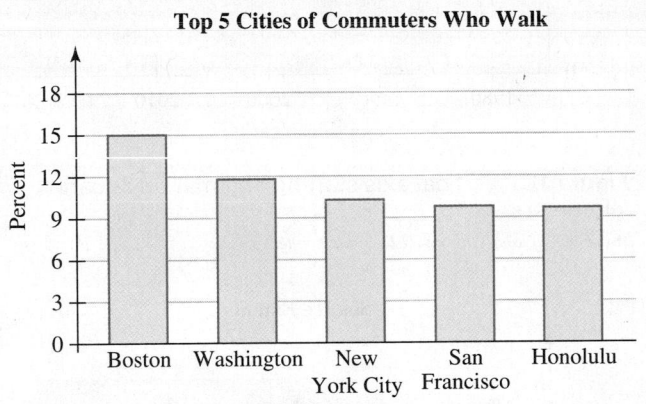

Figure 120 Vertical axis starting at 0%
(**Source:** *Alliance for Biking and Walking 2014 Benchmarking Report*)

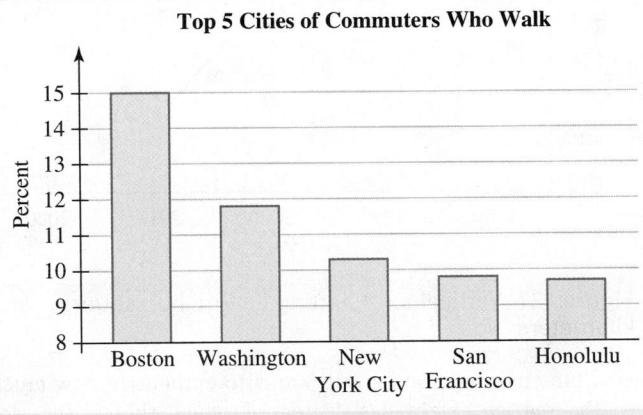

Figure 121 Vertical axis starting at 8%

a. If Boston's mayor's office wants to convince people that the percentage of commuters who walk is much higher in its city than in any other city, which bar graph would it display? Explain.

b. If San Francisco's mayor's office wants to promote its city as being almost as much of a walking-commuter city as Washington, which bar graph would it display? Explain.

c. From which bar graph are you more likely to better estimate the percentage for San Francisco? Explain. Estimate the percent.

d. There are approximately 467.9 thousand commuters in Honolulu (Source: *U.S. Census Bureau*). Estimate the *number* of commuters who walk to work in Honolulu.

e. There are approximately 3.64 million commuters in New York City and 0.36 million commuters in Boston (Source: *U.S. Census Bureau*). For each city, estimate the percentage of commuters who walk to work and estimate the number of commuters who walk to work. Which pair of results would New York City's mayor's office highlight if it wants to promote its city as being more of a walking-commuter city than Boston?

9. University of Massachusetts Boston tuitions (in thousands of dollars) are described by the time-series plots in Figs. 122 and 123 for various years, where the starting year of each aca-demic year is shown on the horizontal axis.

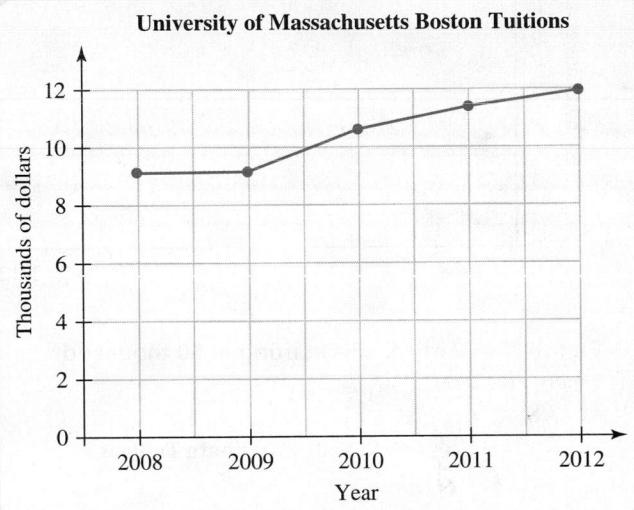

Figure 122 Vertical axis starting at $0 thousand
(**Source:** *Massachusetts Department of Education*)

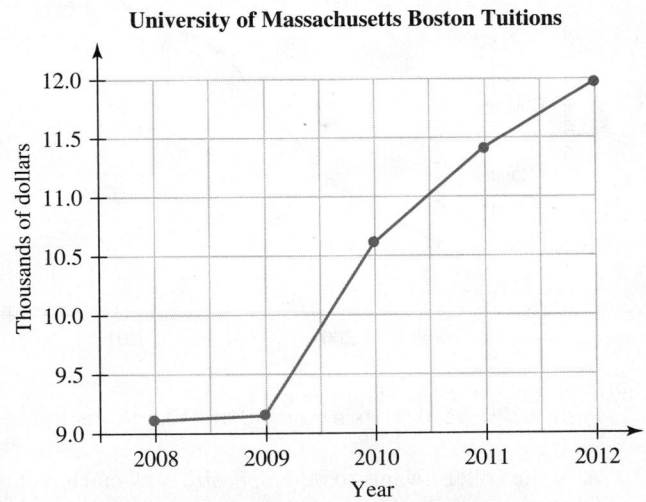

Figure 123 Vertical axis starting at $9 thousand

a. If the university wants to de-emphasize how much its tuition has increased, which time-series plot would it want to display? Explain.

b. From which time-series plot can you better estimate the tuition in 2011? Explain. What is that tuition?

c. From which year to the next did the tuition change the most? Estimate that change.

d. Estimate the change in tuition from 2008 to 2012. If the tuition were to change that same amount from 2012 to 2016, what would the tuition be in 2016? Do you have much faith that your result will turn out to be true? Explain.

10. Freshman tuitions (in thousands of dollars) at Penn State Harrisburg are described by the time-series plots in Figs. 124 and 125 for various years, where the starting year of each academic year is shown on the horizontal axis.

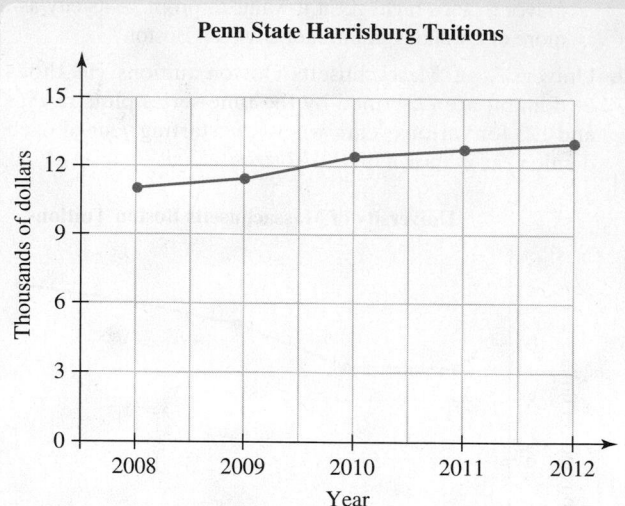

Figure 124 Vertical axis starting at $0 thousand
(**Source:** *Penn State*)

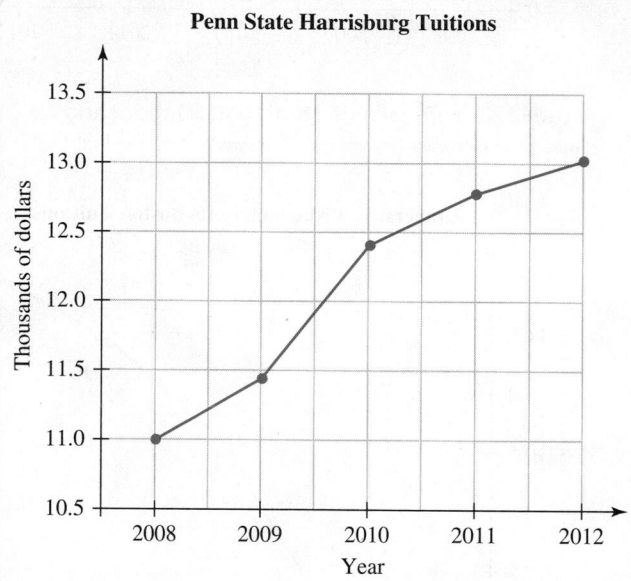

Figure 125 Vertical axis starting at $10.5 thousand

a. If the college wants to de-emphasize how much its tuition has increased, which time-series plot would it want to display? Explain.
b. From which time-series plot can you better estimate the tuition in 2011? What is that tuition?
c. From which year to the next did the tuition change the most? Estimate that change.
d. Estimate the change in tuition from 2008 to 2012. If the tuition were to change that same amount from 2012 to 2016, what would the tuition be in 2016? Do you have much faith that your result will turn out to be true? Explain.

11. The *sea ice extent* is the area of the ocean where at least 15% of the surface is frozen. The average sea ice extents (in million square kilometers) of the Northern Hemisphere are described

by the time-series plots in Figs. 126 and 127 for the months of September in various years.

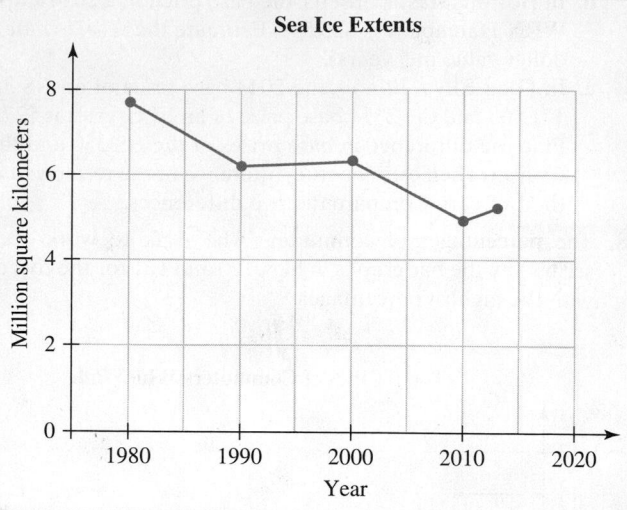

Figure 126 Vertical axis starting at 0 million square kilometers
(**Source:** *National Snow and Ice Data Center*)

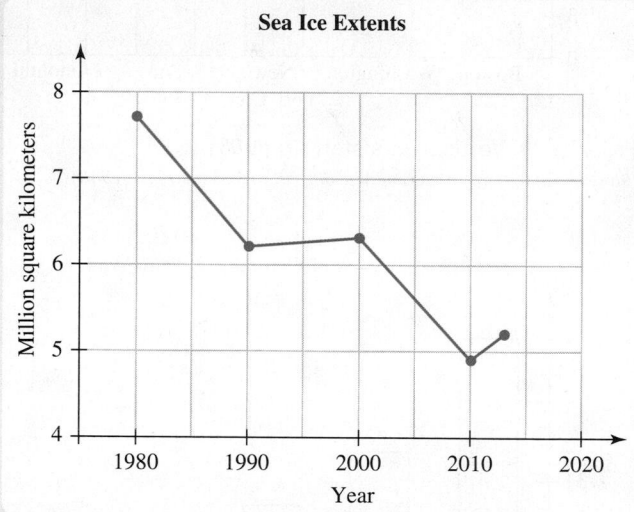

Figure 127 Vertical axis starting at 4 million square kilometers

a. If an environmental group wants to emphasize how much the sea ice extent has decreased, which time-series plot would it display? Explain.
b. If an oil company wants to de-emphasize how much the sea ice extent has decreased, which time-series plot would it display? Explain.
c. From which time-series plot are you more likely to better estimate the sea ice extent in 2010? Explain. Estimate the sea ice extent in that year.
d. Estimate the change in the sea ice extent from 2000 to 2010. If the sea ice extent were to change that same amount from 2010 to 2020, what would be the sea ice extent in 2020? Do you have much faith that your result will turn out to be true? Explain.
e. The time-series plots describe the average sea ice extents in September, when the sea ice extents are the smallest. For the years 1980, 1990, 2000, 2010, and 2013, the average

sea ice extents in March are larger than in September by approximately 9.4 million square kilometers. Construct a time-series plot that describes the average sea ice extents in March for the period 1980–2013. Use scaling on the vertical axis that emphasizes the decrease in sea ice extent.

12. The United States' market shares (percentages) of world manufacturing are described by the time-series plots in Figs. 128 and 129 for various years.

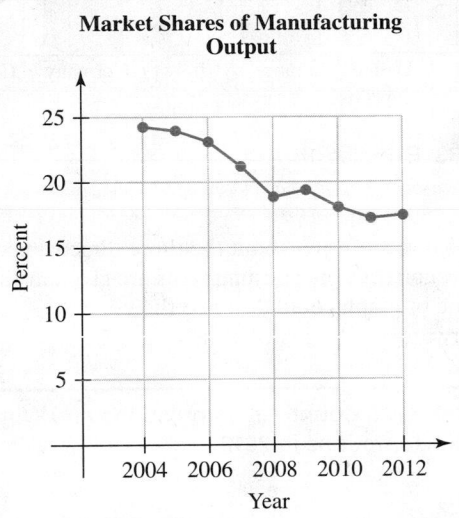

Figure 128 Vertical axis starting at 0%
(**Source:** *United Nations*)

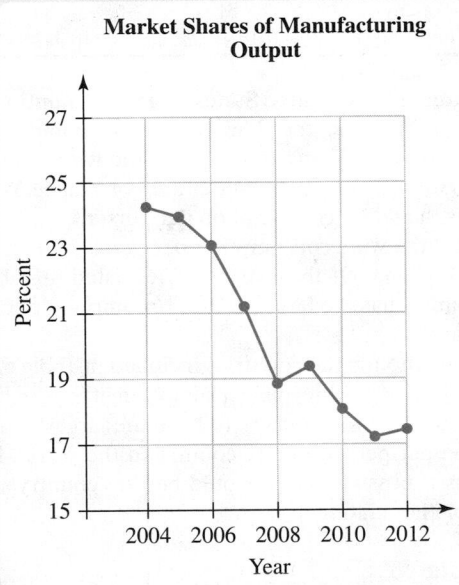

Figure 129 Vertical axis starting at 15%

a. If the Obama administration wants to de-emphasize how much the U.S. market share of world manufacturing has decreased, which time-series plot would it display? Explain.

b. If the Republican Party wants to emphasize how much the market share has decreased, which time-series plot would it display? Explain.

c. From which time-series plot are you more likely to better estimate the market share in 2010? Explain. Estimate the market share in that year.

d. Estimate the change in market share from 2004 to 2012. If the market share were to change that same amount from 2012 to 2020, what would be the market share in 2020? Do you have much faith that your result will turn out to be true? Explain.

e. World manufacturing output was $11.4 trillion in 2012 (Source: *United Nations*). Estimate the U.S. manufacturing output in that year.

13. The annual revenues (in billions of dollars) of Nike® are described by the bar graph in Fig. 130 for various years.

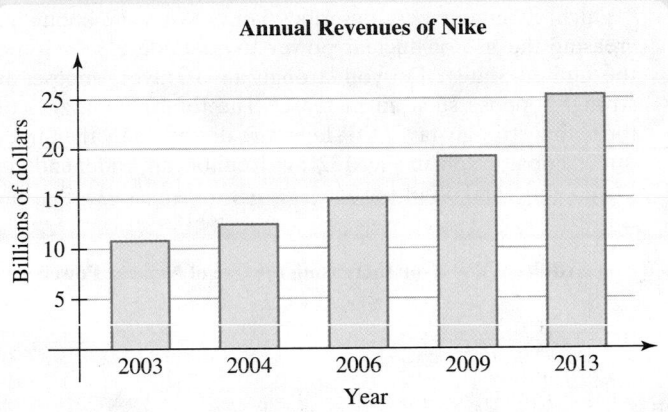

Figure 130 Exercise 13
(**Source:** *Wikinvest*)

a. Explain why the bar graph is misleading.

b. Construct a time-series plot of the data. Explain why this graph is not misleading.

c. If someone does not look carefully, which graph makes it seem like the annual revenue is increasing by greater and greater amounts, the bar graph or the time-series plot? Explain.

d. Estimate the revenue in 2013.

e. Estimate how much the annual revenue increased by from 2003 to 2013. If it increased by the same amount from 2013 to 2023, what would the annual revenue be in 2023? Do you have much faith that this prediction will turn out to be true? Explain.

14. The annual revenues (in billions of dollars) of eBay® are described by the bar graph in Fig. 131 for various years.

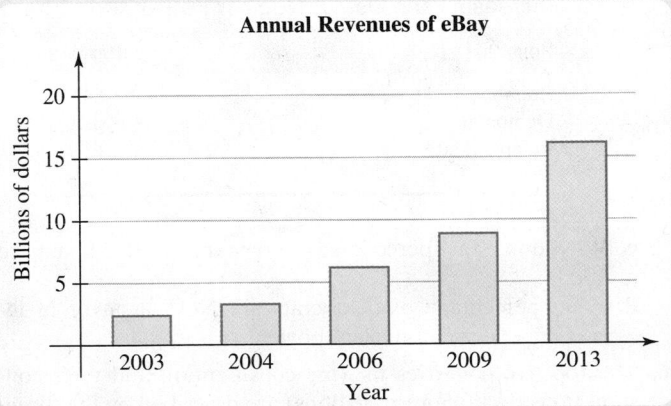

Figure 131 Exercise 14
(**Source:** *Wikinvest*)

a. Explain why the bar graph is misleading.

b. Construct a time-series plot of the data. Explain why this graph is not misleading.

c. If someone does not look carefully, which graph makes it seem like the annual revenue is increasing by greater and greater amounts, the bar graph or the time-series plot? Explain.

d. Estimate the revenue in 2013.

e. Estimate how much the annual revenue increased from 2003 to 2013. If it increased by the same amount from 2013 to 2023, what would the annual revenue be in 2023? Do you have much faith that this prediction will turn out to be true? Explain.

15. A total of 470 adults were asked the following question: "Which statement best describes your own view about increasing the use of nuclear power to generate electricity in the United States? Do you strongly favor, favor, oppose, or strongly oppose such an increase? The total percentages of those that strongly favor OR favor are described by the three-dimensional graph in Fig. 132 for Republican, Independent, Democratic, and other parties.

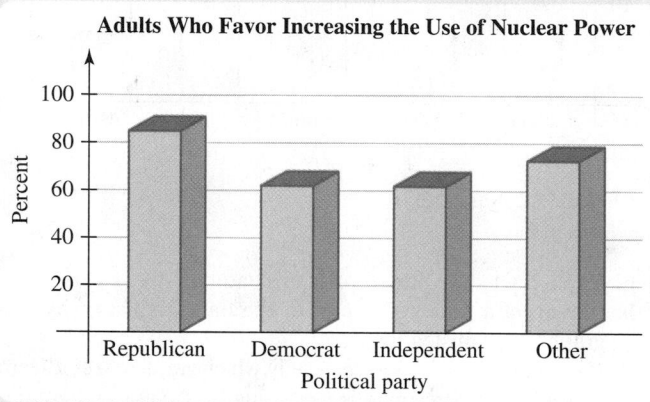

Figure 132 Exercise 15
(**Source:** *General Social Survey*)

a. What is confusing about the three-dimensional graph?

b. The correct percentages are given in Table 51. What type of graph could describe the percentages in a straightforward way? Construct such a graph.

Table 51 Percentages of Adults Who Favor Increasing the Use of Nuclear Power

Political Party	Percent
Republican	84.9
Democrat	62.2
Independent	61.9
Other	72.8

c. Why don't the percentages shown in Table 51 sum to 100%?

d. What percentage of democrats are NOT in favor of increasing the use of nuclear power?

16. The top five countries in wine consumption and their consumptions (in millions of gallons) are described by the three-dimensional graph in Fig. 133.

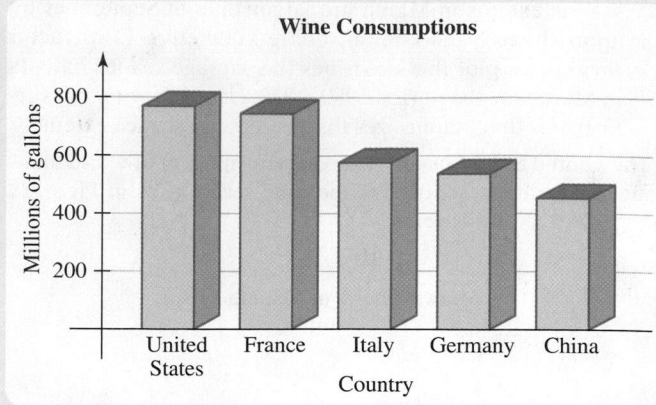

Figure 133 Exercise 16
(**Source:** *International Organization of Vine and Wine*)

a. What is confusing about the three-dimensional graph?

b. The correct wine consumptions are given in Table 52. What type of graph could describe the percentages in a straightforward way? Construct such a graph.

Table 52 Countries' Populations and Wine Consumptions in 2013

Country	Population (millions)	Wine Consumption (millions of gallons)
United States	316	770
France	66	745
Italy	61	576
Germany	81	536
China	1350	444

c. Although the United States' wine consumption was greater than France's, the United States' population is much greater than France's. Calculate the wine consumption *per person* for the United States and for France. Which country has the larger consumption per person?

d. Find the wine consumption per person for Italy, Germany, and China. Of the five countries listed in Table 52, which country has the largest wine consumption per person? The least?

e. Even though the countries included in Table 52 are the top five countries in wine consumption, it is possible, in theory, for some other country to have the largest wine consumption per person of any country in the world. Explain why this is possible. What would be that country's largest possible population?

Concepts

17. For a time-series plot, what is emphasized if the vertical axis starts at a nonzero value?

18. For a bar graph or a time-series plot, why is it important that the numbers written on the vertical axis increase by the same amount and are equally spaced?

19. If a distribution is bimodal, explain why a frequency histogram would make the distribution appear unimodal if a large enough class width is used. Should such a class width be used? Explain.

20. If the categories for a bar graph are various years and those years do not increase by the same amount, explain why the bar graph can be misleading. What would be a better type of graph to use?

21. Explain why three-dimensional graphs can be misleading. What would be a better type of graph to use?

Hands-On Research

22. Find a statistical diagram in a blog, newspaper, or magazine that emphasizes or de-emphasizes an aspect of the distribution or is misleading. If you cannot find one, then select a straightforward statistical diagram and change it so that it emphasizes or de-emphasizes an aspect of the distribution or is misleading. Describe how the diagram might persuade people to think or act a certain way.

Hands-On Projects

DATA Student Loan Default Project

Table 53 describes the tuitions of 4-year, public colleges and students who have defaulted on the Federal Perkins Loans, which means they have not made payment for at least 240 days.

Analyzing the Data

1. The 20 colleges described in Table 53 were selected by creating a frame of all 4-year, public colleges and performing simple random sampling. Describe what this means as if you were explaining it to a student who has not studied statistics.

2. At each college, for borrowers who entered repayment status in 2010–2011, find the proportion who defaulted by 2012. We will refer to this variable as the *2012 one-year default rate*.

3. Construct a frequency histogram for the 2012 one-year default rate. Describe the shape of the distribution. Identify the class that contains the 50th percentile, which is a measure of the center. What does it mean in this situation?

4. A student thinks that the 2012 one-year default rate is higher for colleges with higher tuitions. To get started in helping the student investigate this, complete Table 54.

Table 53 4-year Public College Tuitions and Student Loan Defaults

College	Tuition	Number of Borrowers Who Entered Repayment Status in 2010–2011	Number of Borrowers from Previous Column in Default in 2012	Total Number of Borrowers in Default	Total Outstanding Principal on Loans in Default
University of Arkansas at Pine Bluff	5724	16	7	217	166,389
University of Minnesota–Morris	12,549	67	2	36	103,186
University of Virginia Main Campus	12,216	590	38	429	1,573,645
Weber State University	4768	325	6	438	1,110,692
College of Charleston	9918	111	17	297	437,961
University of Hawaii Maui College	2550	49	15	71	126,740
Massachusetts College of Art and Design	10,400	101	26	172	297,554
Illinois State University	12,726	363	18	772	1,545,783
The College of New Jersey	14,378	141	21	470	1,091,519
George Mason University	9620	312	28	191	383,295
University of Louisiana at Lafayette	5374	123	17	672	2,879,239
Georgia College & State University	8618	160	14	47	146,007
University of Rhode Island	12,450	540	117	1074	2,241,281
University of Oregon	9310	827	81	831	2,119,849
SUNY College at Cortland	6942	213	7	161	204,449
Chadron State College	5600	53	3	33	81,982
Indiana University–South Bend	6728	189	27	237	312,585
California State University–Monterey Bay	5963	3	3	3	7581
University of the District of Columbia	7244	33	15	477	747,211
Georgia Southwestern State University	4914	58	11	78	156,888

Source: *U.S. Department of Education*

Table 54 Tuitions versus 2012 One-Year Default Rates

	2012 One-Year Default Rate Is Less than 15%	2012 One-Year Default Rate Is at Least 15%	Total
Tuition is less than $8000			
Tuition is at least $8000			
Total			

5. Of colleges with tuitions less than $8000, find the proportion that have a 2012 one-year default rate at least 15%.

6. Of colleges with tuitions at least $8000, find the proportion that have a 2012 one-year default rate at least 15%.

7. Is the student correct? That is, for the 20 colleges in the sample, is the 2012 one-year default rate higher for colleges with higher tuitions? Can we make the same conclusion for *all* 4-year, public colleges? Explain.

8. The *outstanding principal* is the part of the original amount borrowed that has not been paid back. For each college, estimate the outstanding principal per defaulted loan.

9. Construct a histogram of the outstanding principal per defaulted loan. Identify the class that contains the 50th percentile, which is a measure of the center. What does it mean in this situation?

Chapter Summary

Key Points of Chapter 3

Section 3.1   Frequency Tables, Relative Frequency Tables, and Bar Graphs

Categorical variable	A **categorical variable** (or **qualitative variable**) consists of names or labels of groups of individuals.
Numerical variable	A **numerical variable** (or **quantitative variable**) consists of measurable quantities that describe individuals.
Frequency of a category	The **frequency of a category** is the number of observations in that category.
Frequency distribution of a categorical variable	The **frequency distribution of a categorical variable** is the categories of the variable together with their frequencies.
Relative frequency of a category	The **relative frequency of a category** is given by $$\frac{\text{frequency of the category}}{\text{total number of observations}}$$
Relative frequency distribution of a categorical variable	The **relative frequency distribution of a categorical variable** is the categories of the variable together with their relative frequencies.
Sum of relative frequencies	For a categorical variable, the sum of the relative frequencies of all the categories is equal to 1.
Frequency bar graph	A **frequency bar graph** is a graph that uses heights of bars to describe frequencies of categories.
Relative frequency bar graph	A **relative frequency bar graph** is a graph that uses heights of bars to describe relative frequencies of categories.
AND	When **AND** is used with two categories, this means to consider the observations that the categories have in common.
OR	When **OR** is used with two categories, this means to consider the observations in the categories all together.
Multiple bar graph	A **multiple bar graph** is a graph that has two or more bars for each category of the variable described on the horizontal axis.

Section 3.2   Pie Charts and Two-Way Tables

Pie chart	The distribution of a categorical variable can be described by a **pie chart**, which is a disk where slices represent the categories. The proportion of the total area for one slice is equal to the relative frequency for the category represented by the slice.
Two-way table	A **two-way table** is a table in which frequencies correspond to two categorical variables. The categories of one variable are listed vertically on the left side of the table, and the categories of the other variable are listed along the top.

Section 3.3   Dotplots, Stemplots, and Time-Series Plots

Discrete variable	A **discrete variable** is a variable that has gaps between successive, possible values.
Continuous variable	A **continuous variable** is a variable that can take on any value between two possible values.
Dotplot	To construct a **dotplot**, for each observation, we plot a dot above the number line, stacking dots as necessary.
Frequency of an observation	The **frequency of an observation** of a numerical variable is the number of times the observation occurs in the group of data.
Frequency distribution of a numerical variable	The **frequency distribution of a numerical variable** is the observations together with their frequencies.
Outlier	An **outlier** is an observation that is quite a bit smaller or larger than the other observations.
kth percentile	The **kth percentile** of some data is a value (not necessarily a data value) that is greater than or equal to approximately $k\%$ of the observations and is less than approximately $(100 - k)\%$ of the observations.
Measuring the center	The 50th percentile can be used to measure the center of a distribution.
Stemplot	A **stemplot** (or **stem-and-leaf plot**) breaks up each data value into two parts: the **leaf**, which is the rightmost digit, and the **stem**, which is the other digits.
Time-series plot	To construct a **time-series plot**, we plot points in a coordinate system where the horizontal axis represents time and the vertical axis represents some other quantity, and we draw line segments to connect each pair of successive dots.

Section 3.4   Histograms

Frequency of a class	The **frequency of a class** is the number of observations in the class.
Relative frequency of a class	The **relative frequency of a class** is the proportion of the observations in the class.
Frequency distribution of a numerical variable	When using classes, the **frequency distribution of a numerical variable** is the classes together with their frequencies.
Relative frequency distribution of a numerical variable	When using classes, the **relative frequency distribution of a numerical variable** is the classes together with their relative frequencies.
Sum of relative frequencies	For a numerical variable, the sum of the relative frequencies of all the classes is equal to 1.
Density histogram	For a **density histogram**, the vertical axis has units called **density** so that the area of each bar is the relative frequency of the bar's class.
Area of bars of density histogram	The following statements are true for a density histogram. • The area of each bar is equal to the relative frequency of the bar's class. • The total area of the bars is equal to 1.
Unimodal, bimodal, and multi-modal distributions	A distribution is **unimodal** if it has one mound, **bimodal** if it has two mounds, and **multimodal** if it has more than two mounds.

Section 3.4   Histograms (*Continued*)

Left and right tails	For a unimodal distribution, the **left tail** is the part of the histogram to the left of the 50th percentile and the **right tail** is the part of the histogram to the right of the 50th percentile.
Skewed-left, skewed-right, and symmetric distributions	• If the left tail of a unimodal distribution is longer than the right tail, then the distribution is **skewed left**. • If the right tail of a unimodal distribution is longer than the left tail, then the distribution is **skewed right**. • If the left half of a distribution is roughly the mirror image of the right half, the distribution is **symmetric**.
Typical observation	An observation at the 50th percentile (approximate center) of a unimodal distribution tends to be a typical observation.

Order of determining the four characteristics of a distribution with a numerical variable	We often determine the four characteristics of a distribution with a numerical variable in the following order:
	1. Identify all outliers.
	    a. For outliers that stem from errors in measurement or recording, correct the errors if possible. If the errors cannot be corrected, remove the outliers.
	    b. For other outliers, determine whether they should be analyzed in a separate study.
	2. Determine the shape. If the distribution is bimodal or multimodal, determine whether subgroups of the data should be analyzed separately.
	3. Measure and interpret the center.
	4. Describe the spread.

Model	A **model** is a mathematical description of an authentic situation. We say the description *models* the situation.

Types of variables needed for diagrams and benefits of diagrams	Diagram	Types of Variables	Benefits
	Frequency Bar Graph	One categorical variable	Compare frequencies of categories.
	Relative Frequency Bar Graph	One categorical variable	Compare a part to the whole.
	Multiple Bar Graph	Two categorical variables	Compare a part to the whole.
	Pie Chart	One categorical variable	Compare a part to the whole.
	Two-Way Table	Two categorical variables	Compare a part to the whole.
	Dotplot	One numerical variable	Describe individual values for a small or medium number of observations.
	Stemplot	One numerical variable	Describe individual values for a small number of observations.
	Frequency Histogram	One numerical variable	Compare the frequencies of classes.
	Relative Frequency Histogram	One numerical variable	Compare a part to the whole.
	Density Histogram	One numerical variable	Compare a part to the whole.
	Time-Series Plot	Two numerical variables	Find the association between two variables.

Section 3.5   Misleading Graphical Displays of Data

Impact of class width on a histogram	When viewing a histogram, keep in mind that a certain choice of class width can emphasize or de-emphasize certain aspects of the distribution.
Impact of a histogram's vertical axis not starting at 0	If the vertical axis of a time-series plot does not start at 0, the changes in the variable described by that axis are being emphasized.
Impact of a histogram's vertical axis starting at 0	If the vertical axis of a time-series plot starts at 0, be aware that the changes in the variable described by that axis are being de-emphasized.
Impact of a bar graph's categories not increasing by the same amount	If the categories for a bar graph are various years and those years do not increase by the same amount, the bar graph can be misleading. A time-series plot would be a better graph to use.

Chapter 3 Review Exercises

For Exercises 1 and 2, identify whether the variable is categorical or numerical. Explain.

1. the distance (in miles) a student drives to college

2. a person's favorite flavor of ice cream

3. Adults were surveyed about how often they are late to work. Their responses are summarized in the relative frequency bar graph in Fig. 134.

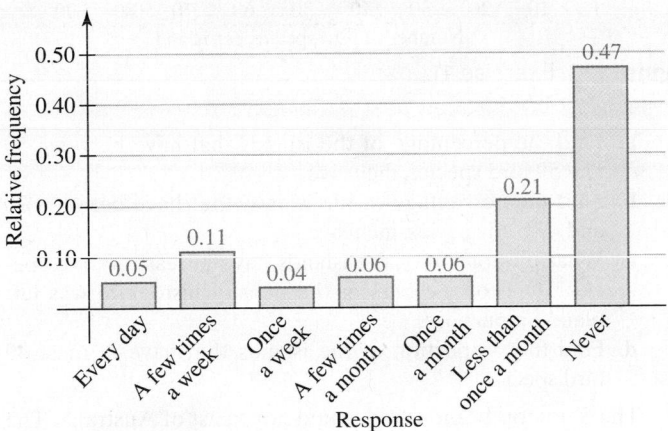

How Often a Worker Is Late to Work

Figure 134 Exercise 3
(**Source:** *YouGov*)

 a. Find the proportion of those surveyed who say they are late to work every day.

 b. Find the proportion of those surveyed who say they are NOT late to work every day.

 c. Find the proportion of those surveyed who say they are late to work less than once a month OR never.

 d. Find the proportion of those surveyed who say they are late to work *at least* once a week.

 e. Explain why there is probably response bias. If there were no response bias, would the relative frequency for the "Never" category likely be larger or smaller?

4. **DATA** The following data are the genres of the top 20 U.S. music singles during the week of December 14, 2014 (Source: *Top40-Charts.com*).

Country	Alternative	Pop	Pop	Pop
Pop	Pop	Country	Rap	Pop
Hip-hop	Pop	Hip-hop	Soul	Pop
Hip-hop	Pop	Country	Pop	Pop

 a. Construct a frequency and relative frequency table.

 b. Construct a frequency bar graph.

 c. Construct a relative frequency bar graph.

 d. What proportion of the observations are NOT country?

 e. What proportion of the observations are hip-hop OR rap?

5. In a 2014 study, some Americans were asked which fast-food restaurant has the best fries. Their responses and the relative frequencies are described by the pie chart in Fig. 135.

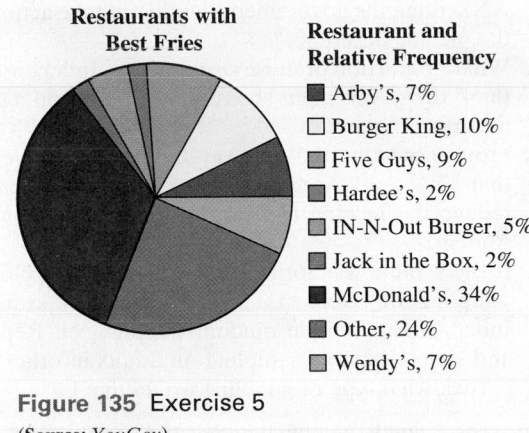

Restaurants with Best Fries

Restaurant and Relative Frequency

■ Arby's, 7%
☐ Burger King, 10%
■ Five Guys, 9%
■ Hardee's, 2%
■ IN-N-Out Burger, 5%
■ Jack in the Box, 2%
■ McDonald's, 34%
■ Other, 24%
■ Wendy's, 7%

Figure 135 Exercise 5
(**Source:** *YouGov*)

 a. Find the proportion of those surveyed who said the restaurant with the best fries was Wendy's.

 b. Find the proportion of those surveyed who said the restaurant with the best fries was NOT Wendy's.

 c. Find the proportion of those surveyed who said the restaurant with the best fries was Arby's OR McDonald's.

 d. If each restaurant in the "Other" category received at most 2% of the votes, what is the *least* possible number of restaurants that received votes in the survey?

6. A total of 28,468 adults were asked whether they thought the U.S. government should reduce the income differences between the rich and the poor. Table 55 compares the adults' responses with their political party affiliations.

Table 55 Opinion about Income Difference versus Political Party Affiliation

	Reduce Difference	No Opinion	Take No Action	Total
Democrat	5895	2085	2333	10,313
Independent	4849	2179	3212	10,240
Republican	2344	1390	3779	7513
Other	169	68	165	402
Total	13,257	5722	9489	28,468

Source: *General Social Survey*

 a. What proportion of those surveyed think the government should take no action about the income difference?

 b. What proportion of those surveyed are Independents?

 c. What proportion of the Democrats surveyed think the government should reduce the income difference?

 d. What proportion of the Republicans surveyed think the government should reduce the income difference? Compare your result with the result you found in part (c).

 e. What proportion of the surveyed adults who think the government should reduce the income difference are Republicans?

 f. A student says the results for parts (d) and (e) should be equal because both parts involve Republicans who think the government should reduce the income difference. What would you tell the student?

7. A total of 28,468 adults were asked whether they thought the U.S. government should reduce the income differences between the rich and the poor. Table 55 compares the adults' responses with their political party affiliation.

a. What proportion of those surveyed are NOT Republicans?
b. What proportion of those surveyed are Independents AND think the government should not take action against the income difference?
c. What proportion of those surveyed are Independents OR think the government should not take action against the income difference?
d. From inspecting the two-way table, a student concludes that 47% of *all* Americans think the government should reduce the income difference. What would you tell the student?
e. If the sample was formed by performing simple random sampling on Democrats, simple random sampling on Independents, simple random sampling on Republicans, and simple random sampling on adults in other political parties, what type of sampling would that be?

For Exercises 8 and 9, identify whether the variable is discrete or continuous.

8. the length (in centimeters) of an alligator
9. the number of musicians in a band
10. The surface-wave magnitude scale, M_s, measures the size of an earthquake. The M_s readings of earthquakes in the United States in 1985 are described by the histogram in Fig. 136.

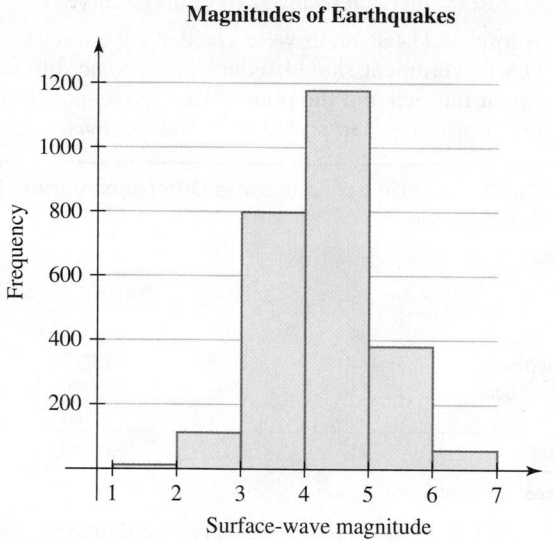

Magnitudes of Earthquakes

Figure 136 Exercise 10
(**Source:** *NOAA*)

a. Describe the shape of the distribution.
b. Identify the class that contains the 50th percentile, which is a measure of the center. What does it mean in this situation?
c. Estimate the number of U.S. earthquakes with M_s readings that were at least 3 AND less than 5. Round to the hundreds place.
d. Estimate the number of U.S. earthquakes that had M_s readings greater than or equal to 4. Round to the hundreds place.
e. California is known for having lots of earthquakes. In fact, there were 862 earthquakes in California in 1985. Estimate the percentage of U.S. earthquakes that happened in California in 1985.

11. The Solomon Islands are located northeast of Australia. The numbers of bird species per island are described by the density histogram in Fig. 137.

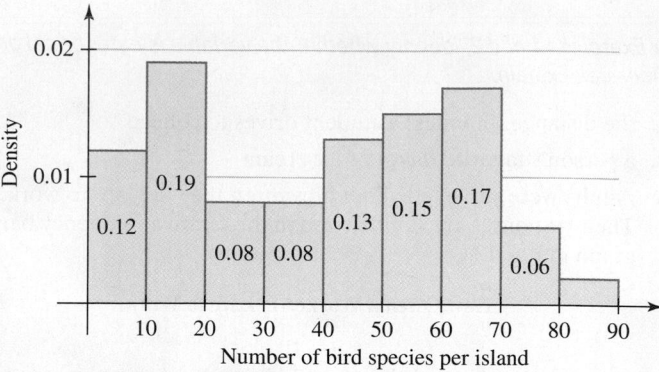

Birds on Solomon Islands

Figure 137 Exercise 11
(**Source:** *solomon.dat*)

a. Find the percentage of the islands that have between 50 and 59 bird species, inclusive.
b. Find the percentage of the islands that have between 30 and 49 bird species, inclusive.
c. What proportion of the islands have at least 60 bird species? One of the bars in the density histogram was left blank on purpose.
d. Find the proportion of the islands that have at most 49 bird species.

12. The Solomon Islands are located northeast of Australia. The numbers of bird species per island are described by the density histogram in Fig. 137.

a. Describe the shape of the distribution.
b. Identify the class that contains the 50th percentile, which is a measure of the center.
c. Find the percentile of Malaita Island's 69 bird species.
d. The number of bird species on Samarai Island is at the 31st percentile. Estimate the number of bird species on the island.

13. **DATA** The following data are the percentages of adult residents in the Western and Midwestern states who think life is getting better (Source: *The Gallup Organization*).

66	62	60	65	62	63	63	56	56
59	68	60	61	55	59	65	65	56
67	58	68	57	53	71	60		

a. Construct a dotplot of the distribution.
b. Find the proportion of Western and Midwestern states in which between 54% and 64% of adult residents think life is getting better.
c. Find the proportion of Western and Midwestern states in which at most 56% of adult residents think life is getting better.
d. Find the proportion of Western and Midwestern states in which at least 65% of adult residents think life is getting better.

14. **DATA** The following data are annual earnings (in millions of dollars) by the top 10 best paid DJs in 2014 (Source: *Forbes*).

21	16	17	30	28
66	28	23	17	22

a. Construct a stemplot of the distribution.
b. If there were one outlier, what would it be? What does it mean in this situation?
c. Find the 30th percentile. What does it mean in this situation?

d. Use the 50th percentile to measure the center.

e. Find the percentile of the observation $30 million. What does it mean in this situation?

15. **DATA.** Lego's revenues (in billions of dollars) are shown in Table 56 for various years.

Table 56 Lego's Revenues

Year	Revenue (billions of dollars)
2010	1.0
2011	1.3
2012	1.6
2013	1.8
2014	2.0

Source: *The Lego Group*

a. Construct a time-series plot.

b. Has the revenue increased, stayed approximately constant, decreased, or none of these? Explain.

c. Find the change in revenue from 2013 to 2014. What does it mean in this situation?

d. How much did the revenue change from 2010 to 2014? If the revenue changes by the same amount from 2014 to 2018, what will the revenue be in 2018? Do you have much faith that this will turn out to be true? Explain.

16. **DATA.** The following data are the numbers of threatened species of mammals for some randomly selected African countries (Source: *International Union for Conservation of Nature*).

11	7	9	11	38	4	8	13	5
11	31	23	7	19	10	33	15	10
16	21	12	29	2	18	114	8	12

For parts (a), (b), and (c), begin with lower class limit 0 and class width 10.

a. Construct a frequency and relative frequency table. Find relative frequencies to the third decimal place.

b. Construct a frequency histogram.

c. Construct a relative frequency histogram.

d. Describe the shape of the distribution.

e. Identify the class that contains the 50th percentile, which is a measure of the center.

f. If there were one outlier, what would it be?

17. The numbers (in thousands) of cars stolen in 2013 for the four most-stolen car models are described by the bar graphs in Figs. 138 and 139.

Number of Cars Stolen

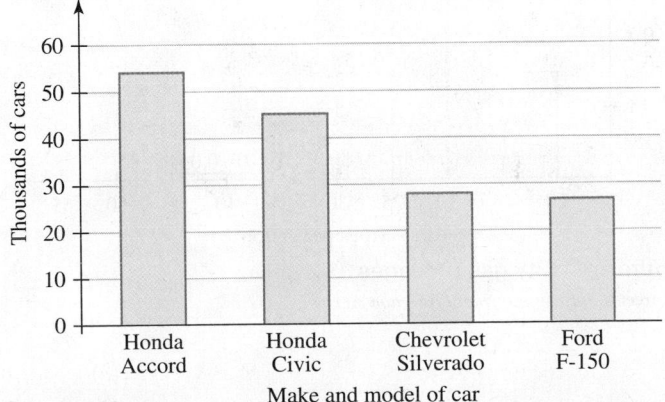

Figure 138 Vertical axis starting at 0 thousand cars

(**Source:** *NICB*)

Number of Cars Stolen

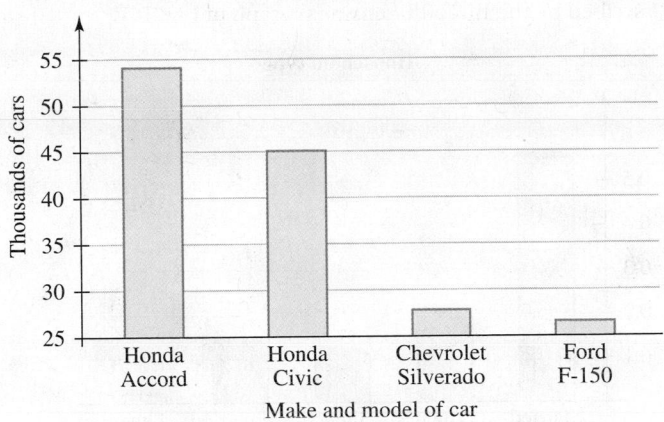

Figure 139 Vertical axis starting at 25 thousand cars

a. If a Chevrolet salesperson wants to convince a potential customer that Silverados are stolen much less often than Honda Civics and Accords, which bar graph would she refer to? Explain.

b. If a Honda salesperson wants to convince a potential customer that the Accord and Civic are not stolen much more often than the Chevrolet Silverado and the Ford F-150, which bar graph would he refer to? Explain.

c. From which bar graph can you better estimate the number of Honda Accords stolen? Explain.

d. Estimate how many more Honda Civics were stolen than Chevrolet Silverados.

18. The numbers (in thousands) of charter schools are described by the bar graph in Fig. 140 for various years.

Charter Schools

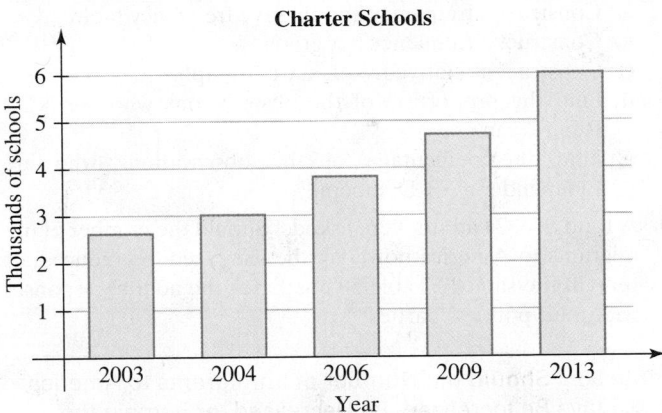

Figure 140 Exercise 18

(**Source:** *National Alliance for Public Charter Schools*)

a. Why is the graph in Fig. 140 misleading?

b. Construct a time-series plot of the data. Explain why this graph is not misleading.

c. If someone does not look carefully, which graph makes it seem like the number of charter schools is increasing by greater and greater amounts, the bar graph or the time-series plot? Explain.

d. Estimate the number of charter schools in 2013.

e. Estimate how much the number of charter schools increased from 2003 to 2013. If it increases by the same amount from 2013 to 2023, what would the number of charter schools be in 2023? Do you have much faith that this prediction will turn out to be true? Explain.

19. The household types of first-time buyers of homes are described by the three-dimensional graph in Fig. 141.

Household type

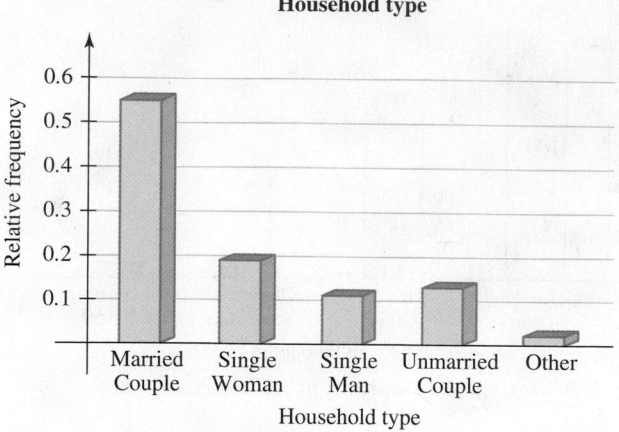

Figure 141 Exercise 19
(**Source:** *National Association of Realtors®*)

a. Why is the graph in Fig. 141 confusing?

b. The correct relative frequencies are given in Table 57. What type of graph could describe the relative frequencies in a straightforward way? Construct such a graph.

Table 57 Relative Frequencies of Household Types of First-Time Home Buyers

Household Type	Relative Frequency
Married Couple	0.55
Single Woman	0.19
Single Man	0.11
Unmarried Couple	0.13
Other	0.02

c. What proportion of first-time buyers were *not* married couples?

d. What percentage of first-time buyers were single women OR single men?

Chapter 3 Test

1. Is the cost (in dollars) of a car a numerical variable or a categorical variable? Explain.

2. ^{DATA} Here are the political parties of 15 students in one of the author's statistics classes (Source: *J. Lehmann*):

Republican	Independent	None	Libertarian	Democrat
Republican	Democrat	Democrat	Independent	Republican
None	Republican	Democrat	Democrat	Democrat

a. Construct a frequency and relative frequency table.

b. Construct a frequency bar graph.

c. Construct a relative frequency bar graph.

d. Find the proportion of the observations who are NOT Republican.

e. Find the percentage of the observations who are Independent OR Democrat.

3. A total of 8824 adults were asked, "Should the number of immigrants to America nowadays be increased, be reduced, or remain the same?" Table 58 compares the adults' responses with their political parties.

Table 58 "Should the Number of Immigrants to America Nowadays Be Increased, Be Decreased, or Remain the Same?"

Response	Democrats	Independents	Republicans	Total
Increased	420	425	178	1023
Remain the Same	1167	1234	702	3103
Reduced	1528	1679	1491	4698
Total	3115	3338	2371	8824

Source: *General Social Survey*

a. What proportion of those surveyed are Republican OR think that the number of immigrants should be reduced?

b. What proportion of those surveyed are Republican AND think that the number of immigrants should be reduced?

c. What proportion of the surveyed Democrats think that the number of immigrants should be increased?

d. What proportion of the surveyed Independents think that the number of immigrants should be increased?

e. A student says that one of the surveyed Independents is more likely to think that the number of immigrants should be increased than one of the surveyed Democrats because more Independents (425) said that the number of immigrants should be increased than Democrats (420). What would you tell the student?

4. Is the high temperature (in degrees Fahrenheit) of Orlando, Florida, a discrete or continuous variable? Explain.

5. The city and highway gas mileages (in miles per gallon) of all cars are described by the histograms in Figs 142 and Fig 143, respectively.

City Gas Mileages

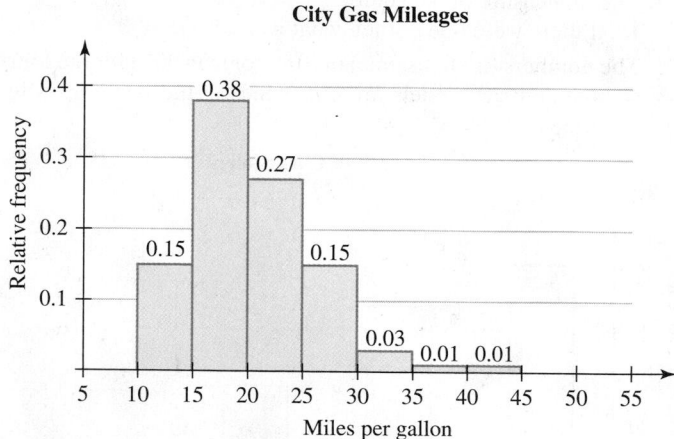

Figure 142 City gas mileages
(**Source:** *U.S. Environmental Protection Agency*)

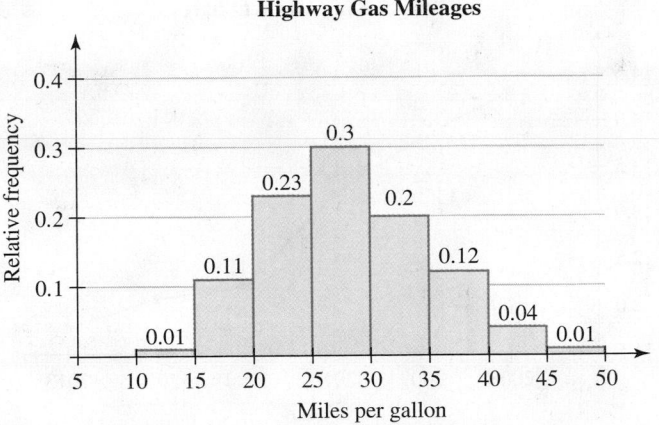

Figure 143 Highway gas mileages

(**Source:** *U.S. Environmental Protection Agency*)

a. Describe the shapes of the distributions.
b. Compare the spreads of the distributions. What does your comparison mean in this situation?
c. For city gas mileages, which class contains the 50th percentile? For highway gas mileages, which class contains the 50th percentile?
d. Use the results you found in part (c) to help you estimate the difference of the 50th percentile for highway gas mileages and the 50th percentile for city gas mileages. What does it mean in this situation?
e. The city and highway gas mileages of all cars are described by just one histogram in Fig. 144. Explain why the distribution is unimodal even though the 50th percentiles for the city distribution and the highway distribution are quite different.

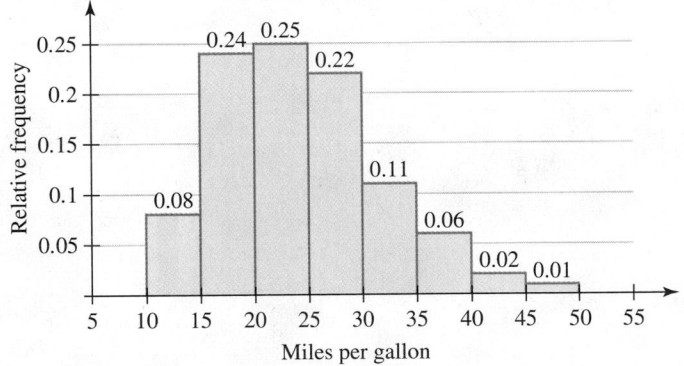

Figure 144 City and highway gas mileages

f. Explain why your response to part (e) suggests that a population with a unimodal distribution may consist of two or more subgroups that are quite different.

6. New York City taxi fares (in dollars) for the month of January 2013 are described by the density histogram in Fig. 145. There were fares over $60, but the bars in the density histogram would not be visible.

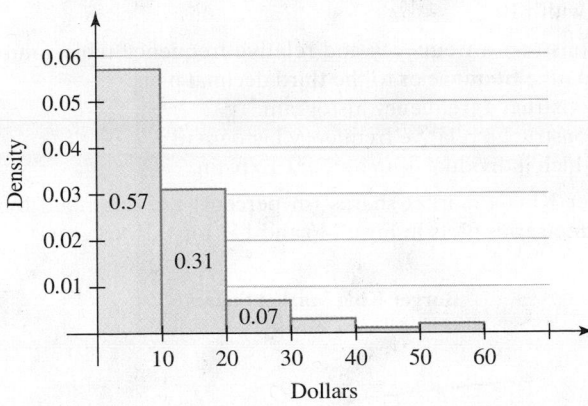

Figure 145 Exercise 6

(**Source:** *Taxi and Limousine Commission*)

a. Describe the shape of the distribution.
b. The largest fare was $460. Is that an outlier? Explain.
c. Find the proportion of the fares that were between $10 and $29.99, inclusive.
d. What proportion of the fares were at most $29.99?
e. There were 1,048,576 fares in New York City during January 2013. How many of the fares were at least $30?

7. New York City taxi fares (in dollars) for the month of January 2013 are described by the density histogram in Fig. 145. There were fares over $60, but the bars in the density histogram would not be visible.

a. Estimate the percentile of a $20 fare.
b. Estimate the fare at the 57th percentile. Round your result to the ones place.
c. There were 1,048,576 fares in New York City during January 2013. Estimate the total money collected in January from all fares that were less than $30.

8. **DATA** The following data are the ages (in years) of the U.S. Supreme Court Justices (Source: *Supreme Court of the United States*).

64	76	81	54	78
59	78	60	66	

a. Construct a stemplot of the ages.
b. Describe the shape of the distribution.
c. Use the 50th percentile to measure the center.
d. What proportion of the justices are at least 78 years in age?
e. Find the percentile of the observation 60 years, which is Justice Sotomayor's age.

9. **DATA** Here are the damages (in billions of dollars) from the top 32 most destructive hurricanes (Source: *WeatherUnderground*):

6	12	4	7	11	46	6	4
4	9	16	10	10	8	3	8
6	28	2	16	3	20	106	11
12	6	6	21	3	9	7	5

For parts (a), (b), and (c), begin with lower class limit 0 and class width 10.

   a. Construct a frequency and relative frequency table. Find relative frequencies to the third decimal place.
   b. Construct a frequency histogram.
   c. Construct a relative frequency histogram.
   d. Which individual is an outlier? Explain.

10. Burger King's market shares (in percent) are described by the time-series plots in Figs. 146 and 147 for various years.

Burger King Market Shares

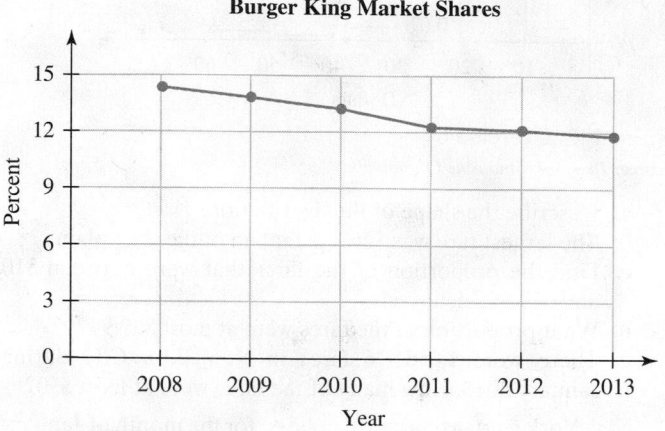

Figure 146 Vertical axis starting at 0%
(**Source:** *Technomic Top 500 Reports*)

Burger King Market Shares

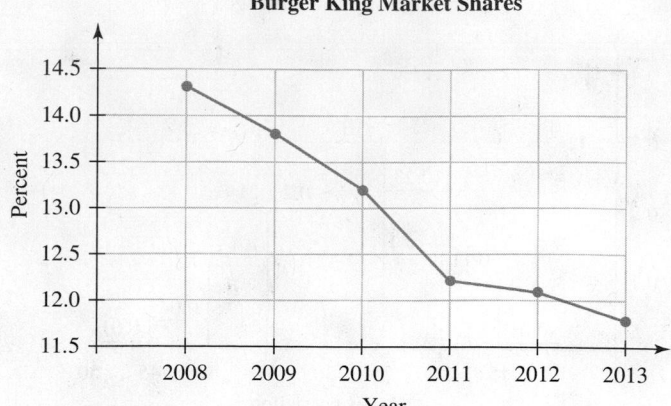

Figure 147 Vertical axis starting at 11.5%

   a. If Burger King wants to de-emphasize how much its market share has decreased, which time-series plot would it want to refer to? Explain.
   b. From which time-series plot can you better estimate the market share in 2010? What is that market share?
   c. From which year to the next did the market share decrease the most? Estimate that decrease.
   d. Estimate the change in market share from 2008 to 2013. If the market share were to change that same amount from 2013 to 2018, what would the market share be in 2018? Do you have much faith that your result will turn out to be true? Explain.

Summarizing Data Numerically

A household suffers from *food insecurity* if at some point in the year the household eats less, goes hungry, or eats less nutritional meals because there is not enough money for food. The percentages of households suffering from food insecurity are shown in Table 1 for the West North Central and Pacific states. In Example 3 of Section 4.2, we will measure the centers and spreads of the two distributions and use the results to make meaningful comparisons.

Table 1 Food Insecurities for West North Central and Pacific States

West North Central State	Food Insecurity (percent of households)	Pacific State	Food Insecurity (percent of households)
Iowa	12.7	Alaska	14.0
Kansas	14.8	California	16.2
Minnesota	10.7	Hawaii	14.2
Missouri	17.1	Oregon	16.7
Nebraska	13.4	Washington	15.0
North Dakota	7.7		
South Dakota	12.3		

Source: *Feeding America; U.S. Department of Agriculture*

In Chapter 3, we used the 50th percentile to measure the center of a distribution. In this chapter, we will use two more methods to measure the center. We will also discuss three ways to measure the spread of a distribution. All of these methods will empower us to make meaningful comparisons of two or more groups of data such as the changes in weight by people who took part in one of the following four diets: Atkins®, Zone®, Weight Watchers®, and Ornish® (see Exercise 44 of Homework 4.2). We will also construct a diagram called a *boxplot* that illustrates the center and spread of a distribution.

▼4.1 Measures of Center

Objectives

» Compute the *median* of some data.

» Describe the meaning of *sigma notation*.

» Compute the *arithmetic mean* of some data.

» Compare the means of two groups of data.

Recall that when investigating the distribution of a numerical variable, we take note of its shape, center, spread, and unusual characteristics such as outliers. In this section, we will focus on the center, which up until now we have measured using the 50th percentile. In this section, we will take a closer look at this technique. We will also discuss two other measures of the center and determine which measure should be used for various types of distributions.

Median

Because the 50th percentile is such an important measure of the center, we give it a special name: the *median*. To remember the name, it might help to think of the phrase *median strip*, which is the strip of land between the lanes of opposing traffic.

» Compare the mean and the median of some data.

» Find the means and medians of bimodal and multimodal distributions.

» Find the *mode* of some data.

▶ **Definition   Median**

The **median** of some data values is the 50th percentile.

About half (50%) of the data values are less than the median and about half (50%) of the data values are greater than it. We use the symbol M to stand for the median. In Chapter 3, we discussed one way to find the 50th percentile (the median). We will now discuss another way. For most large data sets, the two methods will give the same result or results that are very close to each other. For small data sets, finding the 50th percentile can be a bit tricky, but the new method will make it clear what to do.

▶ **Finding the Median of a Distribution**

To find the median of some data values, first list the observations from smallest to largest.

- If the number of observations is odd, then the median is the middle observation.
- If the number of observations is even, then the median is the average of the two middle observations.

▶ **Example 1**   Comparing the Medians of Two Distributions

The minimum wages (in dollars per hour) of the Pacific states and the Mountain states as of April 3, 2014, are shown in Table 2.

Table 2 Minimum Wages for Pacific and Mountain States

Pacific State	Minimum Wage (dollars)	Mountain State	Minimum Wage (dollars)
Alaska	7.75	Arizona	7.90
California	8.00	Colorado	8.00
Hawaii	7.25	Idaho	7.25
Oregon	9.10	Montana	7.90
Washington	9.32	Nevada	8.25
		New Mexico	7.50
		Utah	7.25
		Wyoming	5.15

Source: *National Conference of State Legislatures*

1. Find the median minimum wage of the Pacific states. What does it mean in this situation?
2. Find the median minimum wage of the Mountain states. What does it mean in this situation?
3. Compare the results in Problems 1 and 2. What does this mean in this situation?

Solution

1. First we list the minimum wages of the Pacific states from smallest to largest:

$$7.25, 7.75, 8.00, 9.10, 9.32$$

Because there are five observations, which is odd, the median is the middle observation, 8.00. So, $M = 8.00$. This means that about half of the minimum wages are less than $8.00 and about half of the minimum wages are greater than $8.00.

To find the median using a TI-84, we enter the five observations in the stat list editor (see Fig. 1). Then we use "1-Var Stats" and press the downward arrow three times to find the median 8 (see Fig. 2). For TI-84 instructions, see Appendix A.6.

We can also find the median 8 by using StatCrunch (see Fig. 3). For StatCrunch instructions, see Appendix B.11.

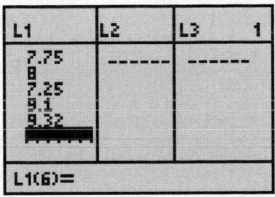

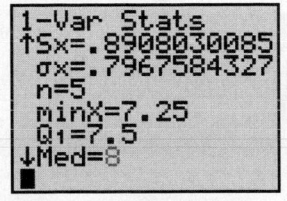

Summary statistics:

Column	n	Median
Wage	5	8

Figure 1 Enter data into stat list editor of TI-84

Figure 2 Use TI-84 to compute the median

Figure 3 Use StatCrunch to compute the median

2. First we list the minimum wages of the Mountain states from smallest to largest:

$$5.15, 7.25, 7.25, 7.50, 7.90, 7.90, 8.00, 8.25$$

Because there are eight observations, which is even, we average the two middle observations, 7.50 and 7.90:

$$\frac{7.50 + 7.90}{2} = 7.70$$

So, the median is $7.70. We can write $M = 7.70$. This means that half of the minimum wages are less than $7.70 and half of the minimum wages are greater than $7.70.

3. The median minimum wage of the Pacific states, $8.00, is more than the median wage of the Mountain states, $7.70. This means that the minimum wage in Pacific states tends to be more than the minimum wage in Mountain states.

In Example 1, we found that the median of the minimum wage of the Pacific states ($8.00) is larger than the median of the minimum wage of the Mountain states ($7.70), even though there are fewer Pacific states than Mountain states. This makes sense, because the median is determined by only the middle observation(s); it is unaffected by the other observations or the number of observations in the data set.

Sigma Notation

Before discussing a second way to measure the center, it will be helpful to discuss a notation that stands for the total of two or more numbers. To begin, consider the following situation. A student in one of the author's statistics classes earned the following points on the first six quizzes: 8, 10, 8, 9, 7, and 9. In order to measure the center of the scores, a first step would be to find their sum:

$$8 + 10 + 8 + 9 + 7 + 9 = 51$$

So, the student earned a total of 51 points on the quizzes.

In order to come up with notation we can use to describe the total points for any student in the class, we start by letting x_i stand for the ith quiz score. So x_1 is the first score, x_2 is the second score, and so on. The letter i is called the *subscript* of x_i. We can write the total score as

$$x_1 + x_2 + x_3 + \cdots + x_6$$

Instead of writing $x_1 + x_2 + x_3 + \cdots + x_6$, we can use *sigma notation* Σx_i:

$$\Sigma x_i = x_1 + x_2 + x_3 + \cdots + x_6$$

The symbol Σ is the Greek letter capital sigma.

▶ Definition   Summation notation

Let $x_1, x_2, x_3, \ldots, x_n$ be some data values. The **summation notation** Σx_i stands for the sum of the data values:

$$\Sigma x_i = x_1 + x_2 + x_3 + \cdots + x_n$$

Table 3 Top Five Foreign Aid Donors

Country	Foreign Aid (billions of dollars)
United States	31.5
Britain	17.9
Germany	14.1
Japan	11.8
France	11.4

Source: *Organization for Economic Cooperation and Development*

▶ **Example 2**   Sigma Notation

The top five foreign aid donors to poor countries and the amounts (in billions of dollars) donated are shown in Table 3.

Let x_i be the amount of foreign aid (in billions of dollars) listed in the ith row of Table 3.

1. Find the values of $x_1, x_2, x_3, x_4,$ and x_5.
2. Find Σx_i. What does it mean in this situation?

Solution

1. From Table 3, we see that $x_1 = 31.5, x_2 = 17.9, x_3 = 14.1, x_4 = 11.8,$ and $x_5 = 11.4$.
2. To find Σx_i, we add the values of each x_i:

$$\Sigma x_i = 31.5 + 17.9 + 14.1 + 11.8 + 11.4 = 86.7$$

So, the total aid donated by the top five donors is $86.7 billion.

To find the sum using a TI-84, we first enter the five observations in the stat list editor (see Fig. 4). Then we use "1-Var Stats" to compute the sum 86.7 (see Fig. 5). For TI-84 instructions, see Appendix A.6.

We can also find the sum 86.7 by using StatCrunch (see Fig. 6). For StatCrunch instructions, see Appendix B.11.

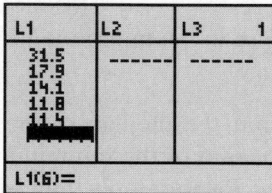

Figure 4 Enter data into stat list editor of TI-84

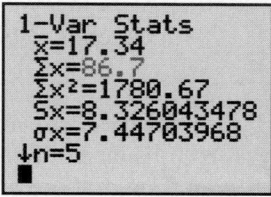

Figure 5 Use TI-84 to compute the sum

Summary statistics:

Column	n	Sum
Aid	5	86.7

Figure 6 Use StatCrunch to compute the sum

Arithmetic Mean

Earlier in this section, we found the sum of six quiz scores 8, 10, 8, 9, 7, and 9:

$$\Sigma x_i = 8 + 10 + 8 + 9 + 7 + 9 = 51$$

One measure of the center is the *arithmetic mean*. There are different types of means, but because the only type we will work with is the arithmetic mean, we will refer to it simply as the *mean*.

To find the mean, we divide the total quiz score by the number of scores, which is 6:

$$\text{mean} = \frac{8 + 10 + 8 + 9 + 7 + 9}{6} = \frac{51}{6} = 8.5$$

We say the mean score is 8.5 points.

In Fig. 7, we see that the mean score 8.5 points is a reasonable measure of the center.

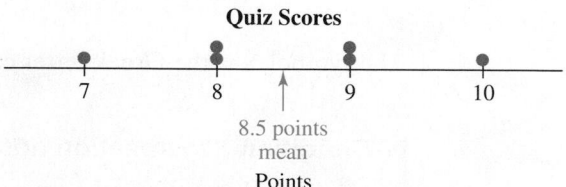

Figure 7 Mean quiz score is 8.5 points

Although the mean is not equal to any of the quiz scores, it is fairly close to all the values, so we say it is a typical value.

Imagine that the horizontal axis in Fig. 7 is a seesaw and the dots are bowling balls that all weigh the same. The mean would be the balance point.

We use the symbol $\bar{x}$ (pronounced "x bar") to stand for the mean.

▶ **Definition** Mean

The **arithmetic mean** (or **mean**) of n data values $x_1, x_2, x_3, \ldots, x_n$ is given by

$$\bar{x} = \frac{\Sigma x_i}{n}$$

In words, the mean is the sum of the observations divided by the number of observations.

Notice that the mean score, 8.5 points, includes the units points. **Remember to always include the units of the mean.**

A **formula** is an equation that contains two or more variables. So, the equation $\bar{x} = \frac{\Sigma x_i}{n}$ is an example of a formula. Many formulas are used in statistics.

Table 4 Minimum Wages for New England States

State	Minimum Wage (in dollars)
Connecticut	8.70
Maine	7.50
Massachusetts	8.00
New Hampshire	7.25
Rhode Island	8.00
Vermont	8.73

Source: *National Conference of State Legislatures*

▶ **Example 3** Calculating the Mean of Some Data

The minimum wages (in dollars per hour) for the New England states as of April 3, 2014, are shown in Table 4.

Find the mean of the minimum wages of the New England states.

Solution

$\bar{x} = \dfrac{\Sigma x_i}{n}$ *Mean formula*

$= \dfrac{8.70 + 7.50 + 8.00 + 7.25 + 8.00 + 8.73}{6}$ *Add the 6 observations; divide by 6, the number of observations.*

$= \dfrac{48.18}{6}$

$= 8.03$

The mean is $8.03.

We will see later in this section that the mean is not always a good measure of the center. The following property describes when we can count on the mean to measure the center well.

▶ **Mean Is a Measure of the Center**

If a distribution is unimodal and approximately symmetric, the mean is a reasonable measure of the center. In this case, we say the mean is a typical value.

Because the number of observations in Example 3 is so small, it is difficult to determine whether the distribution is unimodal and approximately symmetric. In Example 4, the number of observations will be large enough to easily determine whether the distribution is unimodal and approximately symmetric.

Table 5 Statistics Students' Course Loads

5	3	3	4	4
6	6	3	4	5
4	6	4	5	4
5	4	5	5	5
4	4	3	5	4
3	5	3	5	3
4	3	4	5	4
5	4	1	4	6
4	2			

Source: *J. Lehmann*

▶ **Example 4**   Finding and Interpreting the Mean of Some Data

The author surveyed one of his statistics classes. The students' course loads (number of classes per semester) are shown in Table 5.

1. Determine whether the mean should be a reasonable measure of the center of the distribution.
2. Compute the mean. What does it tell us in this situation?
3. A student calculates the mean to be 4.2 classes and concludes that a typical course load at the college where the survey was performed is 4 classes. What would you tell the student?

Solution

1. A histogram of the course-load distribution is shown in Fig. 8. The mean should be a reasonable measure of the center, because the distribution is unimodal and approximately symmetric.

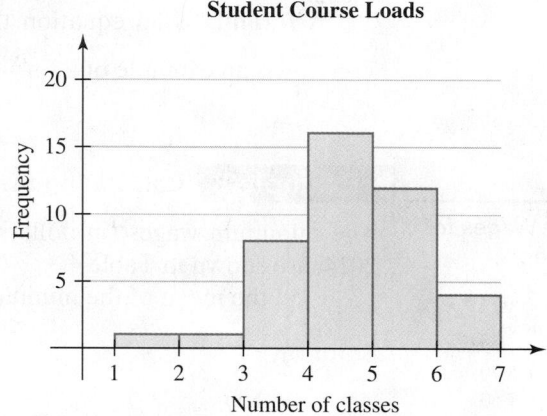

Student Course Loads

Figure 8 Course loads of 42 statistics students

2. $\bar{x} = \dfrac{\Sigma x_i}{n}$ *Mean formula*

$= \dfrac{5 + 3 + 3 + 4 + 4 + \cdots + 2}{42}$ *Add the 42 observations; divide by 42, the number of observations.*

$= \dfrac{175}{42}$

≈ 4.2

The mean is approximately 4.2 classes, which is a measure of the center of the course-load distribution. This suggests that a typical course load for a student in the survey is 4 classes.

To find the mean using a TI-84, we perform the same steps as we would to find the sum. First, we enter the 42 observations in the stat list editor (see Fig. 9 to view the first seven observations). Then we use "1-Var Stats" to compute the approximate mean 4.166666667 (see Fig. 10). For TI-84 instructions, see Appendix A.6.

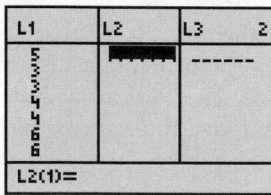

Figure 9 Enter data into stat list editor of TI-84

Figure 10 Use TI-84 to compute the mean

We can also find the approximate mean 4.1666667 by using StatCrunch (see Fig. 11). For StatCrunch instructions, see Appendix B.11.

Summary statistics:

Column	n	Mean
Classes	42	4.1666667

Figure 11 Use StatCrunch to compute the mean

3. Although 4 classes is a typical course load for the surveyed students, it might not be a typical load for *all* students at the college. For example, it turns out that the statistics class in which the surveyed students attended was offered during the daytime, which tends to attract more full-time students than in the evening. There may be many other reasons why the typical course load for the surveyed students is not a typical course load for all students at the college.

In Problem 2 of Example 4, we rounded the mean to the first decimal place: 4.2 classes. The number 4.2 has one more decimal place than the data, which were given to the ones place. **When computing the mean or other measures of data, we will round to one more decimal place than the data.**

Comparing the Means of Two Groups of Data

It can be difficult to compare two groups of data, because there is so much information to process. Using the means to compare the distributions' centers can help bring things into focus.

▶ **Example 5**   Comparing the Means of Two Groups of Data

The prices (in dollars) of hot dogs at each of the stadiums for the baseball teams in the National League Central (NLC) and the National League West (NLW) are shown in Table 6.

Table 6 Prices of Hot Dogs at Baseball Stadiums for the NLC and NLW

NLC Team	Hot Dog Price (dollars)	NLW Team	Hot Dog Price (dollars)
Chicago Cubs	5.50	Arizona Diamondbacks	2.75
Cincinnati Reds	1.00	Colorado Rockies	4.75
Milwaukee Brewers	3.50	Los Angeles Dodgers	5.50
Pittsburgh Pirates	3.25	San Diego Padres	4.00
St. Louis Cardinals	4.25	San Francisco Giants	5.25

Source: *Team Marketing Report*

1. Determine whether the mean hot dog price of the NLC distribution should be a reasonable measure of the center of the distribution. Do the same for the NLW distribution.
2. Find the mean hot dog price for the NLC.
3. Find the mean hot dog price for the NLW.
4. What do your results in Problems 2 and 3 mean in this situation?

Solution

1. Histograms of the hot-dog-price distributions of the NLC and the NLW are shown in Fig. 12. Because both distributions are unimodal and approximately symmetric, the mean prices should be reasonable measures of the centers.

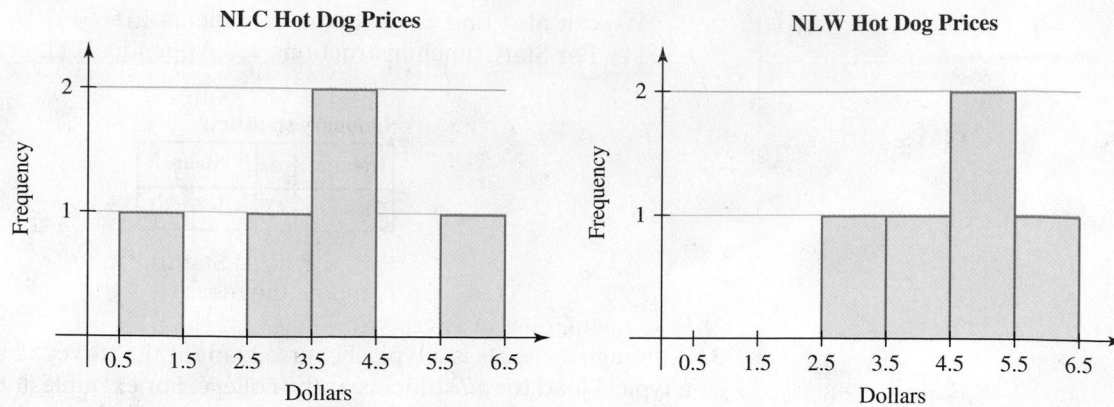

Figure 12 Prices of hot dogs in the NLC and NLW

2. We use the symbol $\bar{x}_{\text{NLC}}$ to stand for the mean of the NLC distribution:

$$\bar{x}_{\text{NLC}} = \frac{\Sigma x_i}{n} \qquad \textit{Mean formula}$$

$$= \frac{5.50 + 1.00 + 3.50 + 3.25 + 4.25}{5} \qquad \textit{Add the 5 observations; divide by 5, the number of observations.}$$

$$= \frac{17.50}{5}$$

$$= 3.50$$

The mean price of a hot dog in the NLC is $3.50.

3. We use the symbol $\bar{x}_{\text{NLW}}$ to stand for the mean of the NLW distribution:

$$\bar{x}_{\text{NLW}} = \frac{\Sigma x_i}{n} \qquad \textit{Mean formula}$$

$$= \frac{2.75 + 4.75 + 5.50 + 4.00 + 5.25}{5} \qquad \textit{Add the 5 observations; divide by 5; the number of observations.}$$

$$= \frac{22.25}{5}$$

$$= 4.45$$

The mean price of a hot dog in the NLW is $4.45.

   We can use StatCrunch to find the mean price $3.50 for the NLC and the mean price $4.45 for the NLW (see Fig. 13).

Summary statistics:

Column	n	Mean
NLC	5	3.5
NLW	5	4.45

Figure 13 Using StatCrunch to find the means

4. The means $3.50 and $4.45 are reasonable measures of the centers of the NLC and NLW distributions, respectively (see Fig. 12). The typical price of a hot dog for the NLC is almost one dollar ($0.95) less than the typical price of a hot dog for the NLW.

Comparing the Mean and the Median

Although the mean and the median both measure the center of a distribution, they often give different results. In Example 6, we will explore how much the mean and the median are affected by outliers.

▶ **Example 6**   Comparing the Mean and the Median

The following data are the savings (in thousands of dollars) of five adults: 2, 10, 8, 7, and 14.

1. Find the mean savings.
2. Find the median savings.
3. Suppose that the adult who had $14 thousand in savings wins $900 thousand (after taxes) from the lottery. So, the adult's new savings is $914 thousand. Find the mean savings of the five adults after the lottery win.
4. Find the median savings of the five adults after the lottery win.
5. Describe how much the outlier $914 thousand affected the mean and the median.

Solution

1.
$$\bar{x} = \frac{2 + 10 + 8 + 7 + 14}{5} = \frac{41}{5} = 8.2$$

So, the mean is $8.2 thousand dollars.

2. To find the median, we list the savings from smallest to largest:

$$2, 7, 8, 10, \text{ and } 14.$$

Because the number of observations is 5, an odd number, the median is the middle observation, $M = 8$. So, the median is $8 thousand.

3.
$$\bar{x} = \frac{2 + 10 + 8 + 7 + 914}{5} = \frac{941}{5} = 188.2$$

So, the mean is $188.2 thousand.

4. To find the median, we list the savings from smallest to largest:

$$2, 7, 8, 10, \text{ and } 914.$$

Because the middle number is 8, the median is $M = 8$. So, the median is $8 thousand.

5. The outlier $914 thousand caused the mean to increase from $8.2 thousand to $188.2 thousand, which is a large increase. The outlier did not affect the median at all.

▶

In Example 6, we found that the mean was greatly affected by the outlier $914 thousand. This makes sense because increasing the savings of $14 thousand to the outlier $914 thousand adds $900 thousand to the numerator of the mean. In general, we say the mean is **sensitive to outliers**, because it is strongly affected by outliers (either small or large).

We also found that the median was unaffected by the outlier $914 thousand. This makes sense because only the middle number $8 thousand determined the median. In general, we say the median is **resistant to outliers**, because it is not strongly affected by outliers (either small or large).

▶ **The Effect of Outliers on the Mean and the Median**

- The mean is sensitive to outliers.
- The median is resistant to outliers.

▶ **Example 7**   Comparing the Mean and the Median

The durations (in seconds) of wood roller coaster rides in Ohio are shown in Table 7.

Table 7 Duration of Wood Roller Coaster Rides in Ohio

Name of Roller Coaster	Park	Duration (seconds)
Teddy Bear	Stricker's Grove	63
Fairly Odd Coaster	Kings Island	90
Sea Dragon	Jungle Jack's Landing	90
Tornado	Stricker's Grove	93
Blue Streak	Cedar Point	105
Racer	Kings Island	120
Son of Beast	Kings Island	140
Mean Streak	Cedar Point	193
Beast	Kings Island	250

Source: *Roller Coaster DataBase*

1. Find the mean duration.
2. Find the median duration.
3. Construct a frequency histogram and indicate the mean and the median on it. Describe the shape.
4. Which measures the center better, the mean or the median? Explain.
5. Explain why it makes sense that the mean is larger than the median.

Solution

1. $\bar{x} = \dfrac{63 + 90 + 90 + 93 + 105 + 120 + 140 + 193 + 250}{9} \approx 127.1$

   So, the mean duration is approximately 127.1 seconds.

2. Because the number of roller coasters is 9, an odd number, the median is the middle observation, $M = 105$ (see Table 7). So, the median is 105 seconds, which is the duration of Blue Streak.

3. We construct a frequency histogram (see Fig. 14) and indicate the mean 127.1 and the median 105 on it. The distribution is skewed right.

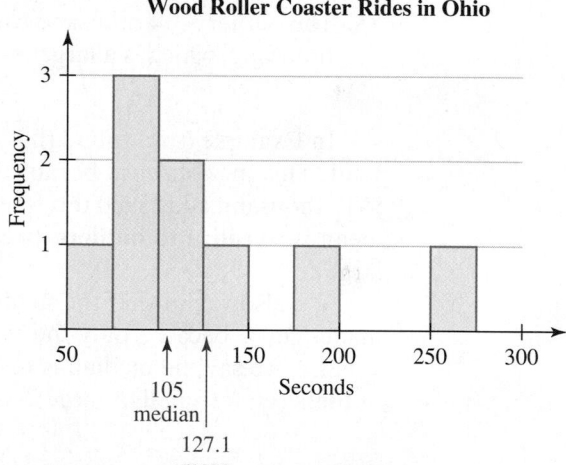

Figure 14 Durations of wood roller coaster rides in Ohio

4. The median 105 seconds is closer to the classes with larger frequencies than the mean 127.1 seconds (see Fig. 14). So, the median measures the center better than the mean.

5. In general, the mean is sensitive to outliers and the median is resistant to outliers. On the basis of the large outlier 250 seconds, it makes sense that the mean is larger than the median. The other observations in the longer right tail also contribute to making the mean larger than the median.

WARNING   Some students think that for a unimodal distribution, the mean and the median are always in the class with the largest frequency. However, this is *not* always true. For example, consider the roller coaster distribution, where both the mean and the median are *not* in the class with largest frequency 3 (see Fig. 14 on page 226).

In Example 7, we concluded that for the skewed-right duration distribution, the mean is greater than the median and the median is a better measure of the center. This is true for most skewed-right distributions.

By considering the mean's sensitivity to outliers and the median's resistance to outliers, we can also sort out how the mean and median are related for skewed-left and symmetric distributions.

> ### How the Shape of a Distribution Affects the Mean and the Median
>
> - If a distribution is skewed left, the mean is usually less than the median and the median is usually a better measure of the center (see Fig. 15).
> - If a distribution is symmetric, the mean is approximately equal to the median and both are reasonable measures of the center (see Fig. 16).
> - If a distribution is skewed right, the mean is usually greater than the median and the median is usually a better measure of the center (see Fig. 17).

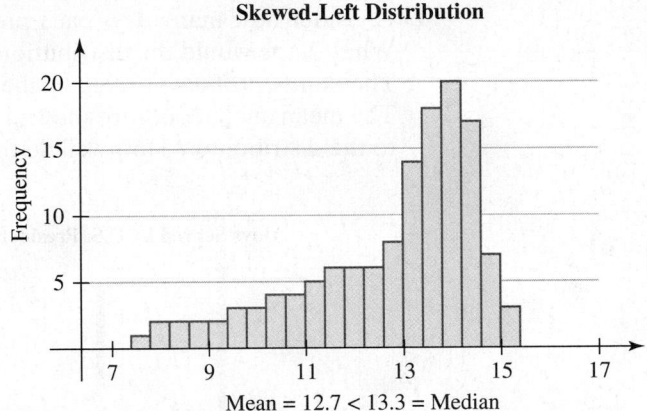

Figure 15 Skewed-left distribution

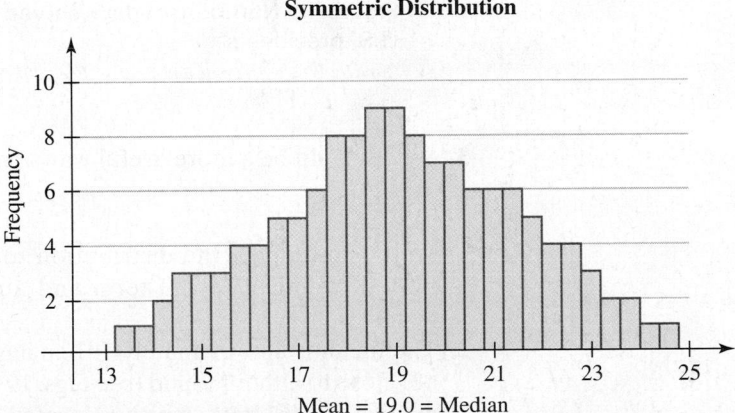

Figure 16 Symmetric distribution

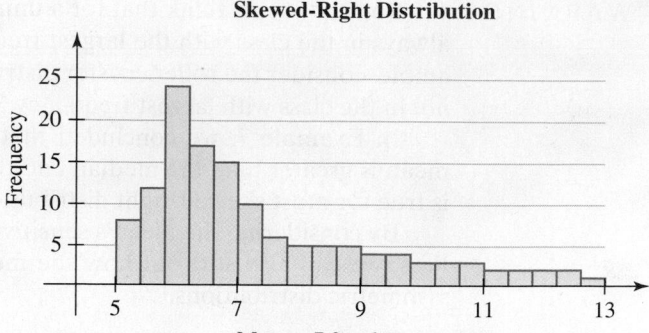

Mean = 7.3 > 6.7 = Median

Figure 17 Skewed-right distribution

Finding Means and Medians of Bimodal and Multimodal Distributions

Recall that a bimodal distribution has two mounds and a multimodal distribution has three or more mounds. Also recall that for such distributions, it is often useful to separate the data into two or more groups.

▶ **Example 8**   Finding the Mean and Median of a Bimodal Distribution

1. Consider how many days each president of the United States has served in office. What shape would the distribution have?
2. The numbers of days served by the presidents are described by the dotplot in Fig. 18. The mean and median are shown in Fig. 19. How do the mean and the median relate to the distribution? How well do they describe a typical observation?

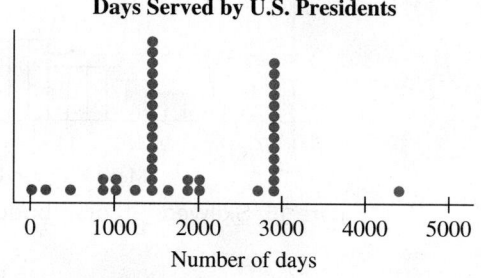

Figure 18 Numbers of days served by U.S. presidents

(**Source:** *World Almanac and Book of Facts 2014*)

Summary statistics:

Column	Mean	Median
Length (days)	1910.5581	1461

Figure 19 Numbers of days served by U.S. presidents

3. What would be a more useful way to analyze this situation?

Solution

1. We would expect the distribution to be bimodal, because many presidents have served exactly one full term and almost as many have served exactly two full terms.
2. The mean is approximately 1910.6 days, which lies between the two mounds and is not close to either mound (see Figs. 19 and 20). So, it is not a typical observation. The median is 1461 days, which is located at one mound but not the other (see Figs. 19 and 20). So, it represents one type of typical value but not the other.

Days Served by U.S. Presidents

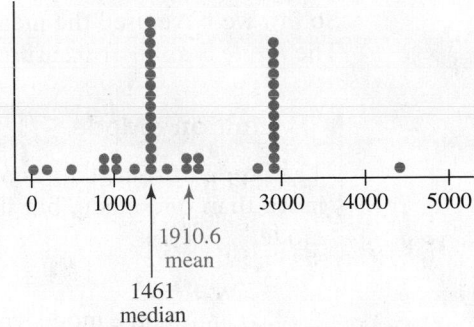

Figure 20 President dotplot with median and mean

3. It would be more useful to separate the data into three groups: presidents who served at most one term, presidents who served two terms (but not more), and presidents who served four terms. (Franklin D. Roosevelt is the only president in the last group; he served three full terms and died during his fourth term.) The dotplots for these three groups are shown in Fig. 21.

Days Served by U.S. Presidents
Up to 2 values per dot

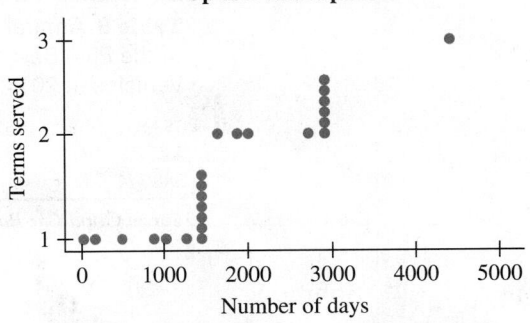

Figure 21 Numbers of days served by U.S. presidents, separated into three groups

Because the first and second groups are both skewed left, it makes sense to use the medians (and not the means) to measure the centers. Using StatCrunch, we can find that the median for the one-term presidents is 1460.5 days and the median for the two-term presidents is 2922 days (see Fig. 22).

Summary statistics for Length (days):
Group by: Terms Served

Terms Served	Mean	Median
1	1188.4091	1460.5
2	2579.35	2922
4	4422	4422

Figure 22 Numbers of days served by U.S. presidents, separated into three groups

In Example 8, we separated a bimodal distribution into groups and described the centers of the groups using medians. In general, it is sometimes useful to separate bimodal and multimodal distributions into two or more groups and use means or medians to describe the centers of the groups.

Mode

So far, we have used the mean and the median to measure the center of a distribution. The *mode* is another measure of the center.

> ▶ **Definition   Mode**
>
> The **mode** of some data is an observation with the greatest frequency. There can be more than one mode, but if all the observations have frequency 1, then there is no mode.

We can find the mode when working with categorical variables as well as numerical variables.

▶ **Example 9** Finding the Mode

Table 8 Numbers of Tweets Sent Daily

0	1	0	0	1
0	0	0	4	0

Source: *J. Lehmann*

For each data set, find the mode.

1. The author surveyed his statistics students about the number of tweets they send daily on average. The responses of some randomly selected students are shown in Table 8.
2. The annual sales of the five best-selling vehicles in 2014 are shown in Table 9.
3. The countries of the winners of the women's Olympic 100-meter run from 1984 to 2012 are shown in Table 10.

Table 9 Annual Sales of the Five Best-Selling Vehicles in 2014

753,851	529,755
439,789	428,606
388,374	

Source: *Good Car Bad Car*

Table 10 Countries of Winners of Women's Olympic 100-Meter Run

United States	United States
United States	France
Bahamas	Jamaica
Jamaica	United States

Source: *International Olympic Committee results database*

Summary statistics:

Column	n	Mode
Tweets	10	0

Figure 23 Using StatCrunch to find the mode

Solution

1. The observation 0 tweets has the greatest frequency (7). So, the mode is 0 tweets. We can also use StatCrunch to find that 0 is the mode (see Fig. 23).
2. Because each observation has a frequency of 1, there is no mode.
3. The observation United States has the greatest frequency (4). So, the mode is the observation United States.

The mean and/or the median often measure the center better than the mode. For example, consider the following situation. The author surveyed students in one of his statistics classes about how many times per week they eat at fast-food restaurants. A dotplot and the mean, median, and mode are shown in Fig. 24.

Fast-Food Restaurants

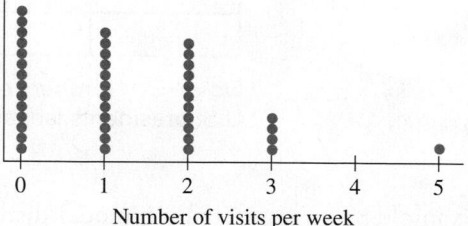

Summary statistics:

Column	n	Mean	Median	Mode
Visits	42	1.2142857	1	0

Figure 24 Dotplot and mean, median, and mode of the fast-food distribution
(**Source:** *J. Lehmann*)

The mode (0 visits) is not a reasonable measure of the center, because 0 is the smallest observation. The median (1 visit) is a much better measure of the center. The mean 1.2142857 rounded to the ones place gives the same (reasonable) result.

When working with numerical variables, the mean and the median are used much more often than the mode. However, the mode has one clear advantage over the other two measures: It can be used when working with categorical variables.

Group Exploration

Comparing the mean and median

1. Construct a frequency histogram for the data in Table 11.

Table 11 Some Data

10	11	12	12	13	13	14	14
14	15	15	15	15	15	15	16
16	16	17	17	18	18	19	20

2. Describe the shape of the distribution.

3. Compute the mean and median. Compare them.

4. Insert three numbers that are larger than 20 into the group of data. Then construct a frequency histogram and describe the shape of the distribution. Next, compute the mean and median and compare them.

5. Replace the three numbers you inserted with three numbers that are less than 10. Then construct a frequency histogram and describe the shape of the distribution. Next, compute the mean and median and compare them.

6. Summarize your findings. Your summary should describe how the mean and median are related for skewed-left, symmetric, and skewed-right distributions.

▶ **Tips for Success**   Complete the Rest of the Assignment

If you have spent a good amount of time trying to solve an exercise but cannot, consider going on to the next exercise in the assignment. You may find that the next exercise involves a different concept or involves a more familiar situation. You may even find that, after completing the rest of the assignment, you are able to complete the exercise(s) you skipped. One explanation of this phenomenon is that you may have learned or remembered some concept in a later exercise that relates to the exercise with which you were struggling.

Homework 4.1

For extra help ▶ MyMathLab®    Watch the videos in MyMathLab    Download the MyDashboard App

1. The median is resistant to _____ .

2. The mean is a measure of the _____ .

3. If a distribution is skewed left, then the mean is usually _____ than the median.

4. *True or False:* If a distribution is skewed, then the mean measures the center better than the median.

For Exercises 5 and 6, the given sentence describes the value of the mean or median number of days it took Chicago's Department of Streets & Sanitation to respond to requests to remove graffiti during the week of May 21–27, 2012 (Source: Department of Streets & Sanitation, Chicago). Use the appropriate symbol to describe the value.

5. The mean number of days is 7.41 days.

6. The median number of days is 3 days.

7. The six most popular cosmetic surgeries and the numbers (in thousands) of those surgeries performed in 2013 are shown in Table 12.

Table 12 Numbers of Cosmetic Surgeries Performed in 2013

Procedure	Frequency (thousands)
Breast Augmentation	290
Nose Reshaping	221
Eyelid Surgery	216
Liposuction	200
Facelift	133
Tummy Tuck	112

Source: *American Society of Plastic Surgeons*

Let n be the number of types of procedures listed in Table 12, and let x_i be the number of surgeries (in thousands) in 2013 for the procedure listed in the ith row of the table.

a. Find the values of $x_1, x_2, x_3, x_4, x_5,$ and x_6.

b. Find Σx_i. What does it mean in this situation?

c. Find $\dfrac{\Sigma x_i}{n}$. What does it mean in this situation?

8. The names and the costs (in billions of dollars) of the six costliest U.S. hurricanes are shown in Table 13.

Table 13 Costs of Six Costliest U.S. Hurricanes

Name	Cost (billions of dollars)
Katrina	105.8
Andrew	45.6
Ike	27.8
Wilma	20.6
Ivan	19.8
Charley	15.8

Source: *Weather Underground*

Let n be the number of hurricanes listed in Table 13, and let x_i be the cost (in billions of dollars) of the hurricane listed in the ith row of the table.

a. Find the values of $x_1, x_2, x_3, x_4, x_5,$ and x_6.

b. Find Σx_i. What does it mean in this situation?

c. Find $\dfrac{\Sigma x_i}{n}$. What does it mean in this situation?

9. The top four grossing superhero movies and their grosses (in millions of dollars adjusted to 2014 ticket prices) are shown in Table 14.

Table 14 Grosses of Superhero Movies

Movie	Gross (millions of dollars)
Marvel's The Avengers	641
The Dark Knight	622
Spider-Man	580
Batman	526

Source: *Box Office Mojo*

Let n be the number of movie titles listed in Table 14, and let x_i be the gross (in millions of 2014 dollars) of the movie listed in the ith row of the table.

a. Find the values of $x_1, x_2, x_3,$ and x_4.

b. Find Σx_i. What does it mean in this situation?

c. Find $\dfrac{\Sigma x_i}{n}$. What does it mean in this situation?

10. Five of the most highly paid CEOs in 2013 and their compensations (in millions of dollars) are shown in Table 15. Steve Ells and Monty Moran are co-CEOs of Chipotle.

Table 15 CEO Compensations

CEO	Company	Compensation (millions of dollars)
Philippe Dauman	Viacom	37.2
Robert Iger	Walt Disney	34.3
David Cote	Honeywell International	25.4
Steve Ells	Chipotle	25.1
Monty Moran	Chipotle	24.4

Source: USA Today *research*

Let n be the number of CEOs listed in Table 15, and let x_i be the compensation (in millions of dollars) of the CEO listed in the ith row of the table.

a. Find the values of $x_1, x_2, x_3, x_4,$ and x_5.

b. Find Σx_i. What does it mean in this situation?

c. Find $\dfrac{\Sigma x_i}{n}$. What does it mean in this situation?

Find the mean and median of the data. Use the correct symbols.

11. 12, 7, 9, 2, 10

12. 9, 15, 4, 14, 13, 10, 12

13. 31, 22, 40, 33, 30, 16, 35, 25

14. 21, 10, 31, 17, 23, 24, 26, 22, 27, 19

For Exercises 15 and 16, find the mean, median, and mode of the data.

15. 8, 2, 6, 6, 8, 3, 9, 6, 2, 3

16. 5, 1, 0, 1, 4, 3, 5, 1, 2, 2, 0

17. The following data are the numbers (in millions) of downloads of the top five most downloaded songs in the United States in 2013 (Source: *Nielsen SoundScan*). Find the mean and the median of the numbers of downloads.

4.7	6.1	5.5	6.5	4.4

18. The following data are the numbers (in millions) of times a song has been streamed for the top five most streamed songs in the United States in 2013 (Source: *Nielsen SoundScan*). Find the mean and the median of the numbers of times a song has been streamed.

490	280	257	189	171

For Exercises 19–22, refer to the given histogram to determine whether the mean is less than, greater than, or approximately equal to the median. Explain.

19. See Fig. 25
20. See Fig. 26
21. See Fig. 27
22. See Fig. 28

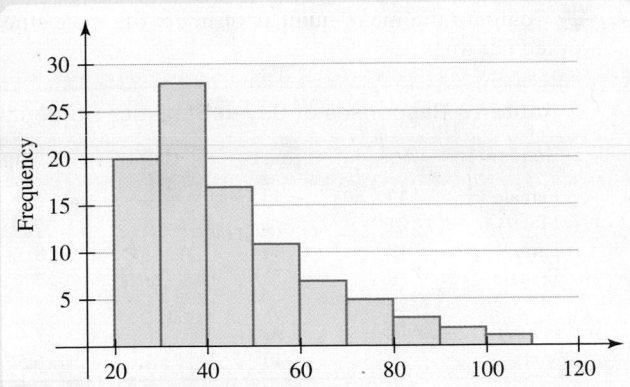

Figure 25 Exercise 19

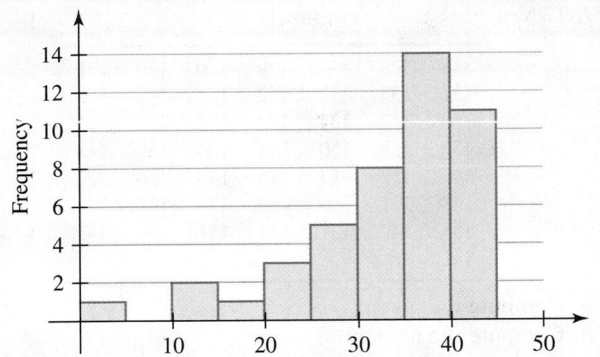

Figure 26 Exercise 20

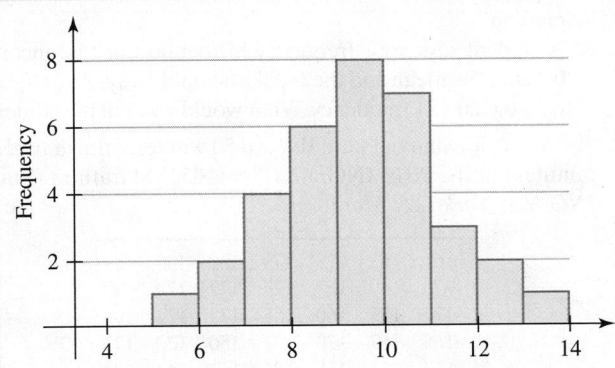

Figure 27 Exercise 21

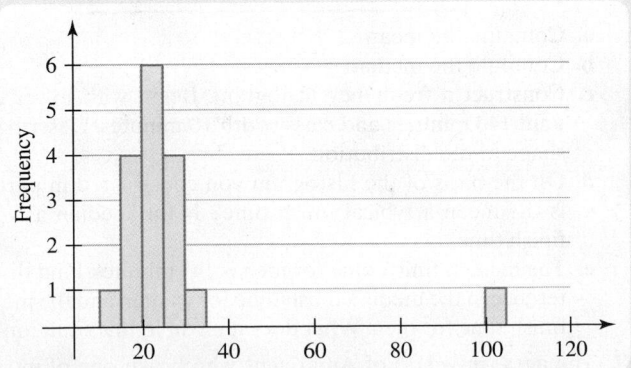

Figure 28 Exercise 22

For Exercises 23–26, match the given information to the appropriate histogram in Fig. 29.

23. mean: 68, median: 75   **24.** mean: 81, median: 75

25. mean: 45, median: 45   **26.** mean: 75, median: 75

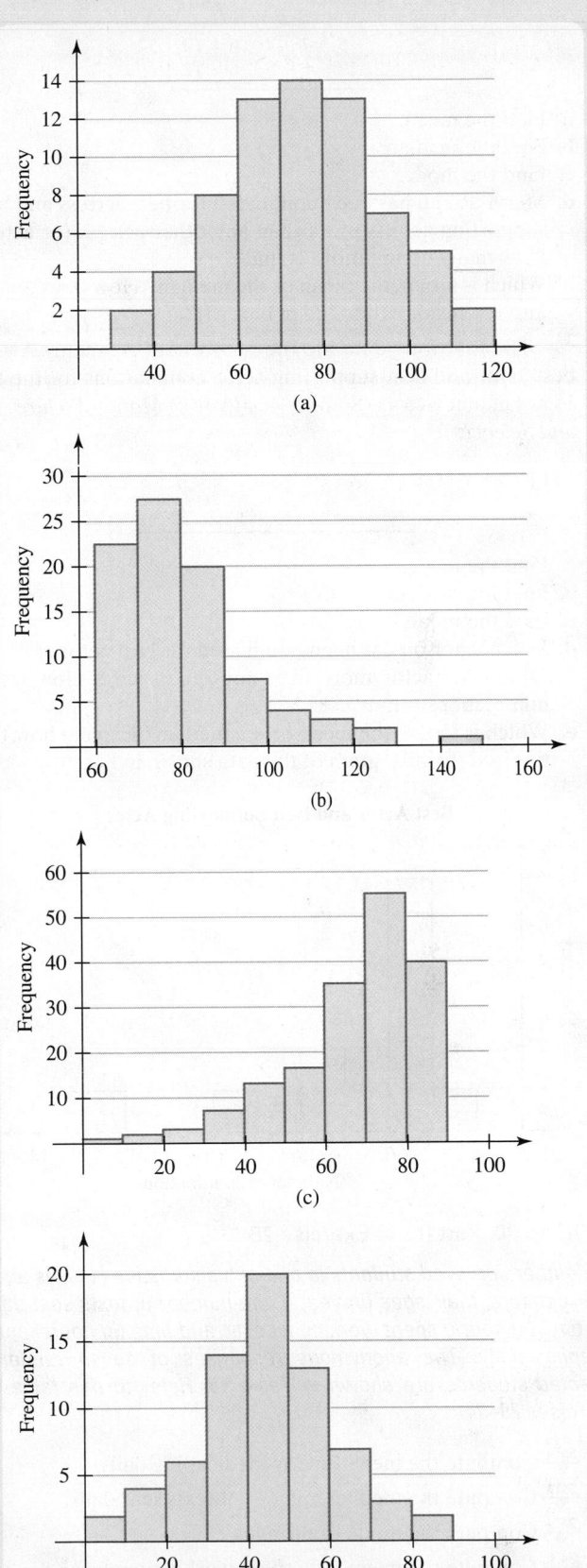

Figure 29 Exercises 23–26

27. DATA The following data are the numbers of Academy Award best-actress and best-supporting-actress nominations for the top 10 nominated actresses (Source: *Academy of Motion Picture Arts and Sciences*).

5	7	7	6	10
19	6	5	6	12

a. Find the mean.
b. Find the median.
c. Find the mode.
d. Meryl Streep has been nominated for best actress and best supporting actress more than any other actress (or actor). How many nominations is that?
e. Which is larger, the mean or the median? How does this tie into the result you found in part (d)?

28. DATA The following data are the numbers of Academy Award best-actor and best-supporting-actor nominations for the top 15 nominated actors (Source: *Academy of Motion Picture Arts and Sciences*).

5	8	5	4	6	5	7	8
9	12	5	5	7	5	6	

a. Find the mean.
b. Find the median.
c. Find the mode.
d. Jack Nicholson has been nominated for best actor and best supporting actor more than any other actor. How many nominations is that?
e. Which is larger, the mean or the median? Explain how this ties into the histogram of the data shown in Fig. 30.

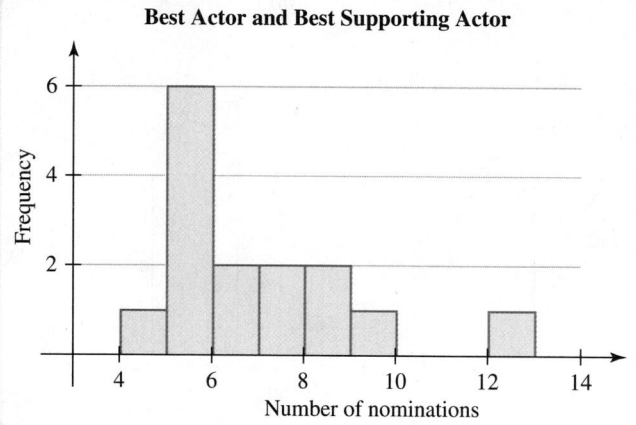

Figure 30 Part (e) of Exercise 28

The author surveyed students in one of his statistics classes about their genders, their ages (in years), the number of texts sent daily, the time (in hours) spent working weekly, and time (in hours) spent sleeping daily. The anonymous responses of seven randomly selected students are shown in Table 16. Refer to this table for Exercises 29–34.

29. DATA Compute the mean time spent sleeping daily.
30. DATA Compute the median number of texts sent daily.
31. DATA Compute the mode of the ages.
32. DATA Compute the mean time spent working per week.
33. DATA Compute the median number of texts the *female* students sent daily.

34. DATA Compute the mean number of hours the *male* students worked per week.

Table 16 Responses of Seven Statistics Students

Gender	Age	Texts	Work	Sleep
Female	19	50	10	8.5
Male	21	1	16	7.5
Female	19	300	12	8
Female	19	5	15	7
Male	20	0	11	7
Female	19	90	20	7
Male	20	40	24	6.5

Source: *J. Lehmann*

35. DATA The following data are the top 50 men's finish times (in minutes) in the 2013 ING New York City Marathon (Source: *ING New York City Marathon*).

148	128	133	129	130	130	150
131	155	132	152	155	142	152
153	132	137	143	145	131	155
152	133	150	144	144	138	143
147	152	153	146	143	154	152
148	131	150	150	132	146	144
155	149	131	145	151	154	143
142						

a. Compute the mean.
b. Compute the median.
c. Construct a frequency histogram. Begin with lower class limit 120 minutes and class width 10 minutes.
d. On the basis of the histogram you constructed in part (c), explain why it makes sense that the mean is less than the median.
e. A student says your frequency histogram must be incorrect, because the mean and the median should always be in the class with the largest frequency. What would you tell the student?

36. DATA The following data are the top 50 women's finish times (in minutes) in the 2013 ING New York City Marathon (Source: *ING New York City Marathon*).

161	163	179	179	180	146	149
176	149	150	177	177	179	148
177	177	160	158	177	172	160
166	167	170	171	150	172	173
173	174	175	175	161	164	176
152	155	155	176	177	161	176
178	180	145	178	148	148	149
181						

a. Compute the mean.
b. Compute the median.
c. Construct a frequency histogram. Begin with lower class limit 140 minutes and class width 10 minutes. Describe the shape of the distribution.
d. On the basis of the histogram you constructed in part (c), is the mean a typical finish time? Is the median a typical finish time?
e. The median finish time for men is 145 minutes. Find the difference in the median finish time for women and the median finish time for men. What does it mean in this situation?

37. The ages (in years) of Americans who broke one of more of their fingers from 2006 to 2007 are described by the density histogram in Fig. 31. The ages are rounded to the ones place.

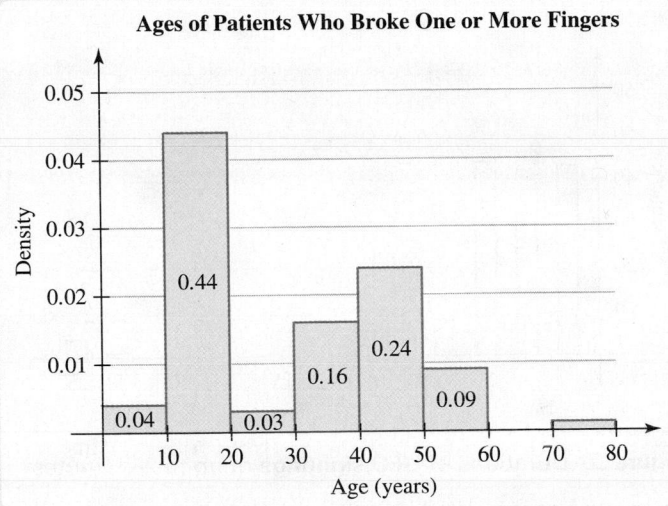

Figure 31 Ages of patients who broke one or more fingers
(Source: *NAMCS, NHAMCS*)

a. Identify which class contains the median.
b. Estimate the percentile for the age 19 years.
c. Estimate the 91st percentile.
d. Describe the shape of the distribution. Why might the distribution have that shape?
e. Discuss how the distribution could be divided into two groups and identify the class that contains the median for each group.

38. The weights (in pounds) of type 2 diabetes patients from 2006 to 2007 are described by the density histogram in Fig. 32. The weights are rounded to the ones place.

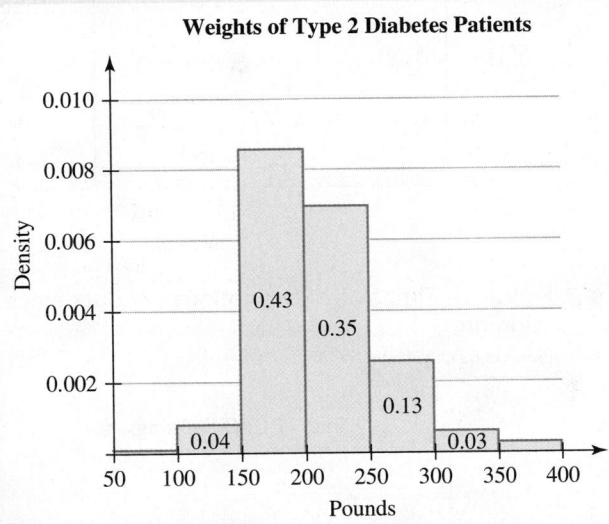

Figure 32 Weights of type 2 diabetes patients
(Source: *NAMCS, NHAMCS*)

a. Describe the shape of the distribution. Which is larger, the mean or the median? Which is a better measure of the center? Explain.
b. Identify which class contains the median.
c. A student says the density histogram must not have been constructed correctly, because the median should aways be in the class with largest density. What would you tell the student?
d. Estimate the percentile for the weight 249 pounds.
e. Estimate the 95th percentile.

39. The temperatures (in degrees Fahrenheit) in Indianapolis, Indiana, at various times of day and night on June 1, 2014, are described by the histogram in Fig. 33. The temperatures are rounded to the first decimal place.

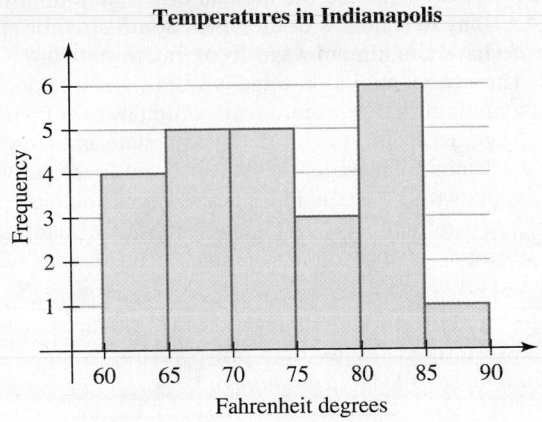

Figure 33 Temperatures in Indianapolis on June 1, 2014
(Source: *Weatherbase*)

a. Identify which class contains the median.
b. A student says the result you found in part (a) must also contain the median temperature for June 2, 2014, because that was only one day after the temperatures were recorded. What would you tell the student?

40. The temperatures (in degrees Fahrenheit) in Phoenix, Arizona, at various times of day and night on June 1, 2014, are described by the histogram in Fig. 34. The temperatures are rounded to the first decimal place.

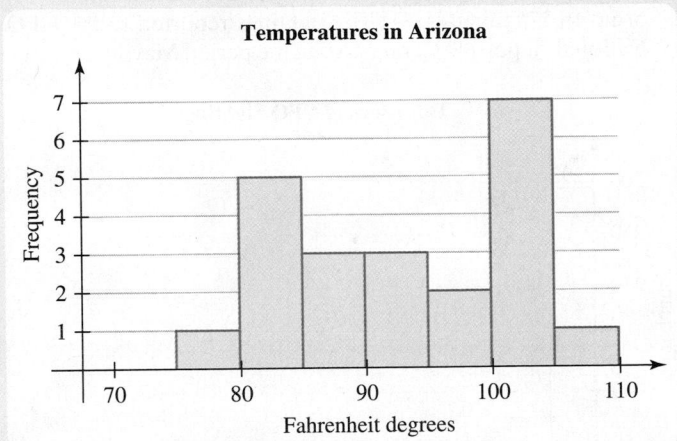

Figure 34 Temperatures in Phoenix on June 1, 2014
(Source: *Weatherbase*)

a. Identify which class contains the median.
b. A student says your result in part (a) must also contain the median temperature for June 2, 2014, because that was only one day after the temperatures were recorded. What would you tell the student?

41. **DATA** The following data are the minimum wages (in dollars per hour) according to *state* law for the South Atlantic states as of April 3, 2014 (Source: *National Conference of State Legislatures*). The word *none* has been entered for South Carolina, which does not have a minimum wage according to state law.

7.25	7.25	7.93	none
5.15	7.25	7.25	7.25

a. A student would like to measure the center of the minimum wages according to state law of the South Atlantic states. He plans on using a minimum wage of 0 dollars for South Carolina. Discuss whether this is appropriate.

b. Find the mean and the median of the minimum wages according to state law of the seven South Atlantic states that *do* have a minimum wage according to state law.

c. The *federal* minimum wage is $7.25. If a state doesn't have a minimum wage according to state law or if the minimum wage according to state law for a state is below the federal minimum wage, then the federal level automatically applies to that state. Find the mean and median of the *effective* minimum wages of all eight South Atlantic states.

d. Compare your two results in part (b) with your two results in part (c).

42. **DATA.** Here are the minimum wages (in dollars per hour) for Midwestern states as of April 3, 2014 (Source: *National Conference of State Legislatures*):

8.25	7.50	7.25	7.25	7.25	7.25
7.25	7.95	7.40	7.25	7.25	7.25

a. Find the mean and median of the minimum wages for the Midwestern states.

b. If 12 people earn minimum wages in the Midwest, one person per state, would, $12\bar{x}$, $12M$, or neither be the total income earned by the 12 people if each person works one hour? Explain. Find that total income.

c. If P people earn minimum wages in the Midwest, would $P\bar{x}$, PM, or neither be the total income earned by the P people if each person works one hour? Explain.

43. The histogram in Fig. 35 describes the durations (in minutes) of up to 270 minutes of UFO sightings reported to the UFO National Reporting Center during the period May 14–31, 2014.

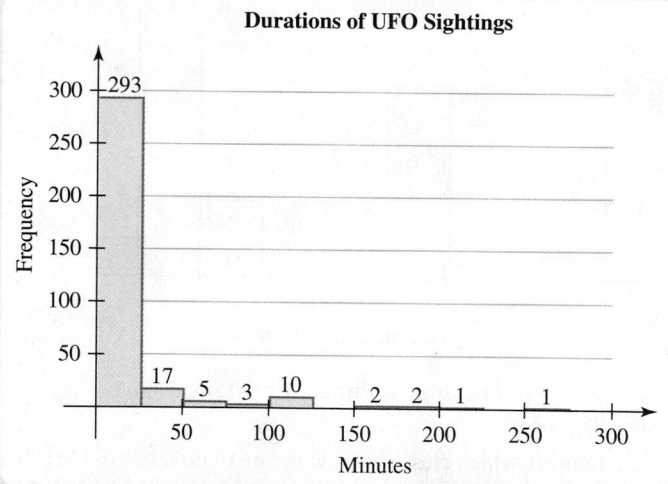

Figure 35 Durations of UFO sightings of up to 270 minutes
(**Source:** *UFO National Reporting Center*)

a. Describe the shape of the distribution.

b. Which is larger, the mean or the median? Which is a better measure of the center? Explain.

c. The durations of up to 25 minutes of UFO sightings are described by the histogram in Fig. 36. Describe the shape of the distribution.

d. Guess why there are spikes at the following times (in minutes): 5, 10, 15, 20, and 25. What type(s) of bias occurred?

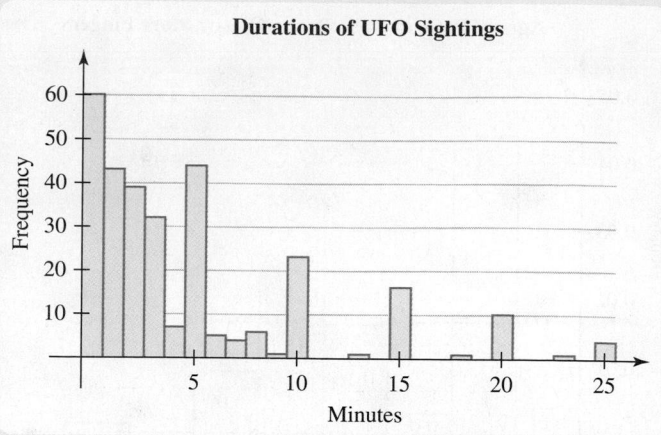

Figure 36 Durations of UFO sightings of up to 25 minutes

e. By inspecting both Figs. 35 and 36, is it reasonable to assume that the mean is even greater than the median than if one were to inspect only Fig. 35? Explain.

44. The times of UFO sightings in the United States reported to the UFO National Reporting Center during the period May 14–31, 2014, are described by the histograms in Fig. 37 (using number of hours since midnight) and Fig. 38 (using number of hours since noon).

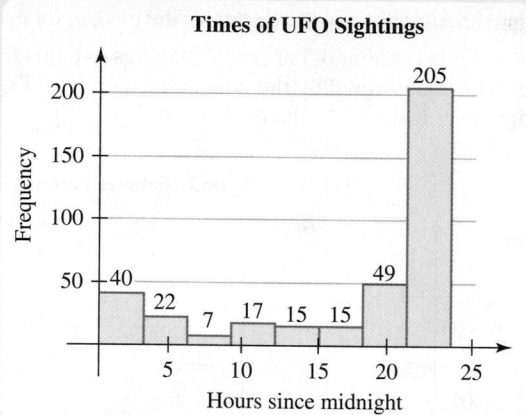

Figure 37 Times of UFO sightings in hours since midnight
(**Source:** *UFO National Reporting Center*)

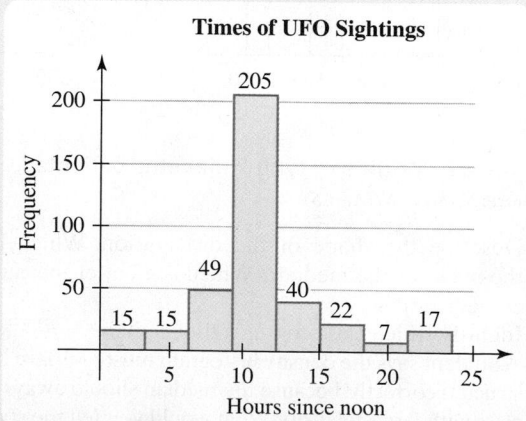

Figure 38 Times of UFO sightings in hours since noon

a. Explain why one distribution is unimodal and the other is bimodal. Which of the two distributions is somewhat misleading?

b. The means and medians for both distributions are shown in Fig. 39. A student thinks that the values for the medians must be wrong, because he thinks the values should be 12 hours apart, not 11 hours. What would you tell the student?

Summary statistics:

Column	n	Mean	Median
Hours since midnight	370	16.910811	21
Hours since noon	370	10.489189	10

Figure 39 Means and medians for the two distributions

c. How is it possible that the means of the two distributions differ by only approximately 6.4 hours when the medians differ by so much more (11 hours)?

d. During what three-hour time slot were there the most sightings? Do you get the same result by inspecting the two distributions?

e. Why were there so many sightings during the time slot you found in part (d)? A UFO believer explains that alien aircraft tend to approach the United States then. What is another explanation? What type(s) of bias likely occurred?

45. Histograms of the ages (in years) of the winners of best actress and best actor at the Academy Awards upon winning the awards are shown in Fig. 40.

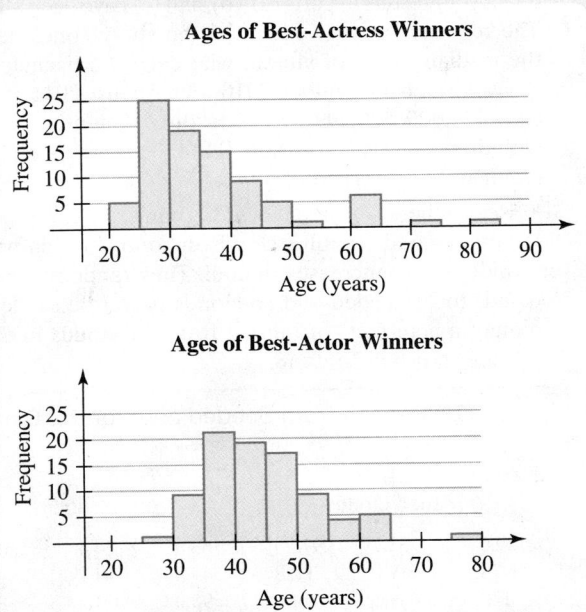

Ages of Best-Actress Winners

Ages of Best-Actor Winners

Figure 40 Ages of the winners of best actress and best actor

(**Source:** *Academy of Motion Picture Arts and Sciences*)

a. Is the distribution of ages of winners of best actress skewed left, skewed right, or symmetric? Which is larger, the mean or the median? Which is a better measure of the center? Explain.

b. Which class contains the median age of the winners of best actress?

c. Which class contains the median age of the winners of best actor?

d. It is a common belief that the typical age of an actor who is highly successful is greater than the typical age of an actress who is highly successful. Is this true? Explain. The phrase *highly successful* can have many meanings. In your explanation, clarify how you are using the phrase.

e. From inspecting the histograms, a student concludes that for next year's Academy Awards, the best-actress winner will be younger than the best-actor winner. What would you tell the student?

46. The amounts of time (in minutes) between eruptions of Old Faithful geyser during the first week of August 2011 are described by the histogram in Fig. 41.

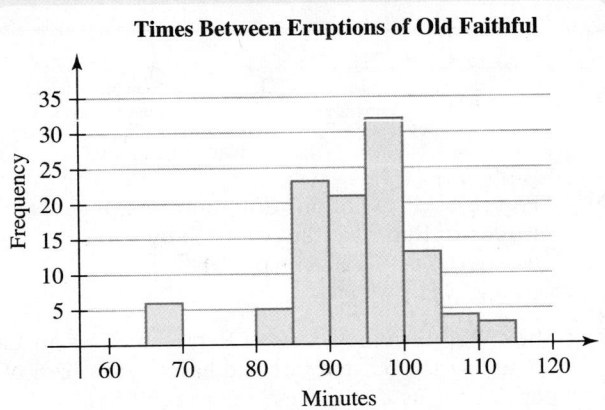

Times Between Eruptions of Old Faithful

Figure 41 Old Faithful

(**Source:** *The Geyser Observation and Study Association*)

a. It is a common belief that Old Faithful erupts on the hour every hour (Source: *The Geyser Observation and Study Association*). Is this true for the first week of August 2011?

b. Old Faithful has two types of eruptions: short and long. The long eruptions are more common and last over 4 minutes. The short eruptions last about 2.5 minutes. After a short eruption, it takes a relatively short time until the next eruption. After a long eruption, it takes a relatively long time until the next eruption (Source: *The Geyser Observation and Study Association*). The distribution of times between eruptions could be viewed as unimodal with some outliers between 65 minutes and 70 minutes, or the distribution could be viewed as bimodal. Which viewpoint is probably better? Explain.

c. Assuming the distribution is bimodal, discuss how the distribution could be divided into two groups and identify the class that contains the median for each group.

d. Estimate the number of eruptions that occurred during the week.

47. Histograms of the 2010 populations (in millions) and the numbers of seats in the House of Representatives, both by state, are shown in Fig. 42.

Population by State

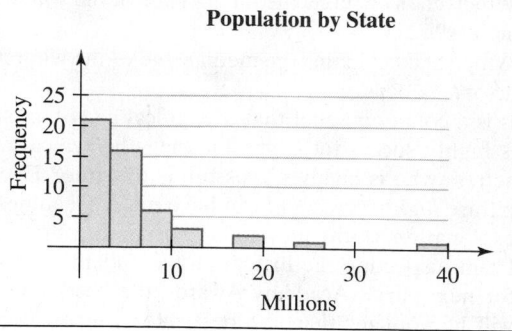

House Seats by State

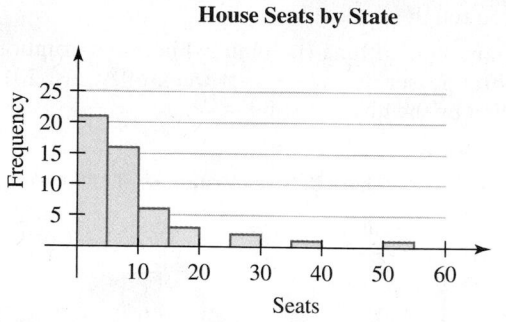

Figure 42 Populations and numbers of seats in the House of Representatives, both by state
(**Source:** *U.S. Census Bureau*)

a. The 435 seats in the House of Representatives are divided up among the 50 states on the basis of the sizes of their populations as of the most recent census (in 2010). Why does it follow that the two histograms in Fig. 42 should have the same shape?

b. For each distribution, which is larger, the mean or the median? Explain.

c. The U.S. population was 308 million in 2010. Find the ratio of the 435 seats to the U.S. population (in millions). Round to the second decimal place. Use your result to find the number of representatives in Pennsylvania, whose 2010 population was 12.7 million.

d. The means of the population distribution and seat distribution are 6.16 million and 8.7 seats, respectively. Show that the mean state population (in millions) times the ratio you found in part (c) is approximately equal to the mean number of House seats in a state.

e. For the population distribution, the mean is 6.16 million and the median is 4.44 million. For the seat distribution, the mean is 8.7 seats and the median is 6 seats. Compute the ratio of the mean to the median for both the population distribution and the seat distribution. Round your results to the second decimal place. Explain why it makes sense that they are approximately equal.

48. **DATA** Gallaudet University offers a liberal education and career development for deaf and hard-of-hearing students. A researcher sent surveys to 334 alumni who had graduated from Gallaudet between December 2010 and August 2011. Of 176 alumni who completed at least part of the survey, 134 were employed, and of those 134 alumni, 111 revealed their salary (in dollars). See Table 17.

Table 17 Salaries of Recent Graduates of Gallaudet University

Salary Range (dollars)	Highest Degree Earned		
	Bachelor's	Master's	Beyond Master's
0–9,999	3	0	1
10,000–19,999	4	1	0
20,000–29,999	8	1	0
30,000–39,999	11	4	0
40,000–49,999	10	18	1
50,000–59,999	2	8	7
60,000–69,999	0	5	8
70,000–79,999	1	9	1
80,000–89,999	1	1	1
90,000–99,999	0	0	2
100,000 or more	0	0	3

Source: *Gallaudet University*

a. Which class contains the median salary of respondents to the survey who earned a bachelor's degree?

b. Which class contains the median salary of respondents to the survey who earned a master's degree?

c. Which class contains the median salary of respondents to the survey who earned a degree beyond a master's?

d. Explain why it makes sense that the median described in part (a) is smaller than the median described in part (b), which is smaller than the median described in part (c).

e. Describe at least two possible reasons why 23 respondents to the survey did not reveal their salaries. What type of bias occurred? If those 23 people had revealed their salaries, how might your results to parts (a), (b), and (c) have been affected?

f. The researcher who conducted the study concluded that the median salary of alumni who earned a bachelor's degree between December 2010 and August 2011 is in the class $20,000–$29,999. Describe three potential problems with drawing this conclusion. [**Hint:** See the opening paragraph and parts (a) and (e).]

49. **DATA** Researchers wanted to determine whether seeding (treating) isolated cumulus clouds in south Florida with silver iodide smoke increases rainfall. They randomly assigned 26 clouds to be seeded and 26 clouds to not be seeded. The amounts (in acre-feet) of rainfall from the clouds in the two groups are shown in Table 18.

Table 18 Rainfall from Seeded and Control Clouds

Rainfall (acre-feet)					
Seeded Clouds			Control Clouds		
129.6	32.7	978.0	163.0	24.4	1202.6
1697.8	119.0	302.8	345.5	41.1	4.9
17.5	489.1	4.1	95.0	321.2	81.2
703.4	2745.6	200.7	87.0	4.9	28.6
40.6	31.4	92.4	147.8	21.7	17.3
334.1	274.7	115.3	244.3	36.6	29.0
7.7	1656.0	430.0	11.5	830.1	68.5
274.7	198.6	255.0	1.0	26.3	372.4
242.5	118.3		47.3	26.1	

Source: A Bayesian Analysis of a Multiplicative Treatment Effect in Weather Modification, *Simpson, Alsen, and Eden.*

a. Construct two dot plots that share the same horizontal axis to describe the rainfall from the two groups of clouds.

b. On the basis of the shapes of the two distributions, would it be better to use the means or the medians to measure the centers? Explain.

c. On the basis of your response to part (b), measure the centers of the two distributions.

d. Did one of the seeded clouds tend to have greater, equal, or less rainfall than one of the control clouds?

e. In general, we cannot be certain that seeding clouds in south Florida increases rainfall. Is that because of possible sampling bias, sampling error, both, or neither? Explain.

50. **DATA** Researchers wanted to determine whether calcium can lower the systolic blood pressure of African American men. They randomly assigned 21 men to two groups. Each member of one group took a calcium supplement for 12 weeks, and each member of the other group took a placebo during the same period. The systolic blood pressures (in mmHg) of the men before and after the 12-week period are shown in Table 19.

Table 19 Levels of Systolic Blood Pressure in 21 African American Men

Systolic Blood Pressure (mmHg)					
Calcium Group			Placebo Group		
Before	After	Difference	Before	After	Difference
107	100	7	123	124	−1
110	114		109	97	
123	105		112	113	
129	112		102	105	
112	115		98	95	
111	116		114	119	
107	106		119	114	
112	102		112	114	
136	125		110	121	
102	104		117	118	
			130	133	

Source: Blood Pressure and Metabolic Effects of Calcium Supplementation in Normotensive White and Black Men, *Lyle, Roseann M., et al.*

a. Complete the third column of Table 19 by finding the difference of the first and second columns. For example, the entry 7 was found by computing $107 - 100 = 7$.

b. Complete the sixth column of Table 19 by finding the difference of the fourth and fifth columns. For example, the entry −1 was found by computing $123 - 124 = -1$.

c. Compute the mean of the differences in the third column. What does it mean in this situation?

d. Compute the mean of the differences in the sixth column. What does it mean in this situation?

e. Did calcium tend to lower the systolic blood pressure of individuals in the treatment group?

51. Refer to Exercise 49 on page 238 for a description of a study about seeding clouds.

a. Is the study an experiment or an observational study? Explain.

b. What are the explanatory and response variables?

c. What does random assignment mean in this study? How could this be accomplished?

d. Describe the sample and the population.

e. After performing a statistical test, the study concluded that seeding isolated cumulus clouds increases rainfall. Does this mean there is causality or only an association between the two variables? Explain.

52. Refer to Exercise 50 for a description of a study about using calcium to lower systolic blood pressure.

a. Is the study an experiment or an observational study? Explain.

b. What are the explanatory and response variables?

c. What does random assignment mean in this study? How could this be accomplished?

d. Describe the sample and the population.

e. After performing a statistical test, the study concluded that calcium lowers systolic blood pressure in African American men. Does this mean there is causality or only an association between the two variables? Explain.

Large Data Sets

53. **DATA** At the website Lending Club®, customers can either borrow money or fund money for others' loans. Access the data about the loans funded at Lending Club in September, 2014, which are available at MyMathLab and at the Pearson Downloadable Student Resources for Math & Stats website.

a. Construct a relative frequency histogram of the loan amounts for those who rent their home, beginning with lower class limit 0 dollars and class width 5000 dollars. Do the same for those who own their home. Also, do the same for those who have a mortgage on a home.

b. For each of the three distributions described in part (a), determine the shape.

c. For each of the three distributions, determine whether the mean or the median is the better measure of the center. Explain.

d. For each of the three distributions, find the mean and median.

e. For each of the three distributions, which is larger, the mean or the median? How does this relate to your response to part (b)?

f. Compare the medians for the three distributions. What does your comparison mean in this situation? In your opinion, why does this make sense? [**Hint:** It may help to consider the typical ages, employment, and needs of the three types of borrowers as well as whether they can put something up for collateral (something of value they promise to give up if they fail to pay the loan).]

54. **DATA** Access the data about worldwide volcanic eruptions, which are available at MyMathLab and at the Pearson Downloadable Student Resources for Math & Stats website.

a. Construct a pie chart of the countries where volcanic eruptions occurred. In which three countries did almost half of the volcanic eruptions occur?

b. Construct a histogram of the numbers of deaths, beginning with lower class limit 0 deaths and class width 1000 deaths.

c. Describe the shape of the death distribution.

d. Which is a better measure of the center for the death distribution, the mean or the median? Explain.

e. Find and compare the mean and the median of the death distribution. Why is one value so much larger than the other?

f. What does the median number of deaths tell you about this situation?

Concepts

55. Find two groups of five numbers that both have a mean of 5.

56. Find two groups of seven numbers that both have a median of 10.

57. The median salary of employees at a company is $50 thousand. If all the employees receive a $2 thousand raise, what is the new median salary?

58. The mean weight of some people in an exercise class is 157 pounds. If each person loses 10 pounds, what is their new mean weight?

59. The mean height of a group of 9 men is 6 feet. If a 6-foot tall man joins the group, will the mean height of the group decrease, increase, or stay the same? Explain.

60. The median salary of employees at a company is $60 thousand. If a person is hired and paid a salary of $60 thousand, will the median salary decrease, increase, or stay the same? Explain.

61. The following data are the savings (in thousands of dollars) of five adults: 6, 9, 5, 1, and 13.

    a. Find the mean savings.

    b. Find the median savings.

    c. Suppose that the adult who had $13 thousand in savings wins $800 thousand (after taxes) from the lottery. So, the adult's new savings is $813 thousand. Find the mean savings of the five adults after the lottery win.

    d. Find the median savings of the five adults after the lottery win.

    e. Describe how much the outlier $813 thousand affected the mean and the median.

62. **DATA** The author surveyed the students in one of his statistics classes, asking how many Facebook friends they have. Here are some of their randomly selected responses (Source: *J. Lehmann*):

1000	756	900	150	200	700	250
600	200	2400	450	600	200	40
1300	300	500	150	300	500	

    a. Compute the mean and the median. Compare the two results.

    b. Construct a frequency histogram. Begin with lower class limit 0 friends and class width 500 friends. Explain how its shape ties in with the results you found in part (a).

    c. To find the *10% trimmed mean*, we first list the observations from smallest to largest. Then we remove the bottom 10% of the observations and the top 10% of the observations. Finally, we compute the mean of the remaining observations. Find the 10% trimmed mean for the Facebook friend data.

    d. Compare the 10% trimmed mean and the mean that you found in part (a).

    e. Which is more resistant to outliers, the mean or the trimmed mean? Explain.

63. Which would tend to be larger, the mean weight of 20,000 randomly selected cats or the mean weight of 5 randomly selected human adults? Explain.

64. Which would tend to be larger, the mean height of 10 randomly selected professional NBA basketball players or the mean height of 10,000 randomly selected high-school basketball players? Explain.

65. Compare how the mean and the median are computed. Compare their meanings. (See page 12 for guidelines on writing a good response.)

66. Describe distributions in which the mean measures the center better than the median, the mean and the median give approximately equal results, and neither the mean nor the median measures the center well. (See page 12 for guidelines on writing a good response.)

Hands-On Research

67. Collect data that involve a numerical variable. Your study should include 20 individuals.

    a. Define the variable, including units.

    b. Provide the data you collected.

    c. Use a statistical diagram to describe the distribution.

    d. Describe the shape of the distribution.

    e. On the basis of the shape of the distribution, which is a better measure of the center, the mean or the median? Explain. Find that measure of the center. Is it a typical value for your sample?

▼ 4.2 Measures of Spread

Objectives

» Compute the *range* of some data.

» Compute the *standard deviation* of some data.

» Compare the means and standard deviations of two groups of data.

» Apply the *Empirical Rule* to some data.

» Determine whether an observation is unusual.

» Compare the range and the standard deviation.

» Compute the *variance* of some data.

Recall that key elements of the distribution of a numerical variable are its shape, center, spread, and unusual characteristics such as outliers. In this section, we focus on spread. So far, we have compared the spreads of two distributions by inspecting diagrams such as histograms. Now we will discuss three methods that involve making *computations* rather than inspecting a diagram.

Range

The students in one of the author's statistics classes were asked how many people live in their households, including themselves. The sizes of the households for five of those students are 2, 3, 4, 6, and 10. One way to describe the spread of the five numbers is the *range*, which is equal to the largest observation, 10, minus the smallest observation, 2:

$$\text{range} = 10 - 2 = 8 \text{ people}$$

Note that we included the units "people" for the range. Just like for the mean, median, and mode, **it is important that we state the units for the range.**

The range is the distance between the smallest and largest observations (see Fig. 43).

Household Sizes

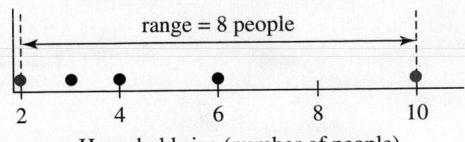

Household size (number of people)

Figure 43 The range is 8 people

We use the symbol **R** to stand for the range.

> **Definition** Range
>
> The **range** of some data values is given by
>
> $$R = \text{largest observation} - \text{smallest observation}$$

▶ **Example 1** Calculating the Range

Vehicle sales in April 2014 are shown in Table 20 for each automaker.

Table 20 Vehicle Sales in April 2014

Company	Sales	Company	Sales
General Motors	254,076	KIA	53,676
Chrysler	178,652	Nissan	103,934
Ford Motor	210,355	SUBARU	40,083
Honda	132,456	Toyota	199,660
Volkswagen	46,715	Hyundai	66,107

Source: *MotorIntelligence.com*

Find the range in sales and explain what it means in this situation.

Solution

From Table 20, we see that the smallest sales were 40,083 vehicles (by SUBARU) and the largest sales were 254,076 vehicles (by General Motors). Next, we calculate the range:

$$R = \text{largest sales} - \text{smallest sales} = 254,076 - 40,083 = 213,993$$

So, the range is 213,993 vehicles. This suggests that the spread of the vehicle sales of the automakers is quite large.

Summary statistics:

Column	n	Range
Sales	10	213993

Figure 44 The range is 213,993 vehicles

We can use StatCrunch to find that the range of the vehicle data is 213,993 vehicles (see Fig. 44). For StatCrunch instructions, see Appendix B.11.

Standard Deviation

One problem with the range is that no matter how many observations there are, the range takes into account only two observations, the smallest and largest ones. For example, suppose a group of five students from a business class have household sizes of 2, 6, 6, 6, and 10. The range of this data is $10 - 2 = 8$ people, which is the same as the range we found for the statistics students, even though the business student data have less spread than the statistics student data (see Fig. 45).

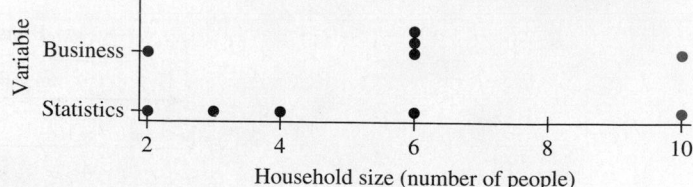

Figure 45 Household sizes for statistics students and business students

In order to find a better way to measure the spread of the statistics student data 2, 3, 4, 6, and 10, it would be helpful to know the mean, which is $\bar{x} = 5$ (try it). One way to take into account *all* of the statistics student data would be to find the distances between each of the observations and the mean (5). From our work with the range, we know that if we subtract two numbers in the correct order, the result is the distance between the numbers, but using the expression $\Sigma(x_i - \bar{x})$ doesn't work, because some of the differences give the *opposite* of distances:

$$\Sigma(x_i - \bar{x}) = (2 - 5) + (3 - 5) + (4 - 5) + (6 - 5) + (10 - 5)$$
$$= -3 + (-2) + (-1) + 1 + 5$$
$$= 0$$

The result is 0, because the total distance for observations to the right of the mean minus the total distance for observations to the left of the mean equals 0. This makes sense, because the mean is a measure of the center.

To avoid getting negative differences, we square each difference:

$$\Sigma(x_i - \bar{x})^2 = (2 - 5)^2 + (3 - 5)^2 + (4 - 5)^2 + (6 - 5)^2 + (10 - 5)^2$$
$$= (-3)^2 + (-2)^2 + (-1)^2 + 1^2 + 5^2$$
$$= 40$$

One problem with the expression $\Sigma(x_i - \bar{x})^2$ is that it can be large even for observations that are close together if there are a lot of observations. It would be better to compute the *mean* of the squared differences, so we should divide by the number of observations, $n = 5$. However, for the ways in which we will be measuring spread, it turns out to be more useful to divide by $n - 1 = 4$:

$$\frac{\Sigma(x_i - \bar{x})^2}{n - 1} = \frac{40}{4} = 10$$

We say the *variance* is 10 people squared. The units are people squared, because we squared the differences, which each have units of people. In this course, it is preferable to work with units of just people (not squared) so the units of spread and center are the same. To make this happen, we take the square root of the expression $\dfrac{\Sigma(x_i - \bar{x})^2}{n - 1}$ and round to one more decimal place than the data:

$$\sqrt{\frac{\Sigma(x_i - \bar{x})^2}{n - 1}} = \sqrt{10} \approx 3.2$$

We say the *standard deviation* is 3.2 people.

We use the letter s to stand for the standard deviation.

▶ Definition   Standard deviation

The **standard deviation** of n data values $x_1, x_2, x_3, \ldots, x_n$ is given by

$$s = \sqrt{\frac{\Sigma(x_i - \overline{x})^2}{n - 1}}$$

In words, the standard deviation is the square root of all the following: the sum of the squared distances between the data values and the mean, all divided by 1 less than the number of data values.

Yes, the formula is complicated, but keep in mind the main point: **The standard deviation measures the spread. The greater the spread, the greater the standard deviation will be.**

▶ **Example 2**   Finding the Standard Deviation

The levels of cholesterol (in milligrams) in the equivalent of $\frac{1}{8}$ of a 14" cheese pizza are shown in Table 21 for four pizza companies. Find the standard deviation of the cholesterol levels. Even though California Pizza Kitchen's "Five Cheese + Fresh Tomato" pizza has a topping other than cheese, we include it in the group of data, because tomato does not have cholesterol. Find the standard deviation for the data.

Table 21 Levels of Cholesterol in the Equivalent of $\frac{1}{8}$ of a 14" Pizza

Company	Name of Pizza	Cholesterol (milligrams)
Pizza Hut	Thin 'N Crispy Cheese®	30
Little Caesar's	Cheese	20
California Pizza Kitchen	Five Cheese + Fresh Tomato	41
Papa Murphy's	Cheese	21

Source: *The companies*

Solution

First, we find the mean cholesterol level:

$$\overline{x} = \frac{\Sigma x_i}{n} = \frac{30 + 20 + 41 + 21}{4} = \frac{112}{4} = 28$$

So, the mean cholesterol level is 28 milligrams.

Next, we list the values of x_i and $\overline{x} = 28$ in the first and second columns, respectively, of Table 22. In the third and fourth columns, we find the values of $x_i - \overline{x}$ and $(x_i - \overline{x})^2$, respectively.

Table 22 Finding Values of $(x_i - \overline{x})^2$

x_i	$\overline{x}$	$x_i - \overline{x}$	$(x_i - \overline{x})^2$
30	28	$30 - 28 = 2$	$2^2 = 4$
20	28	$20 - 28 = -8$	$(-8)^2 = 64$
41	28	$41 - 28 = 13$	$13^2 = 169$
21	28	$21 - 28 = -7$	$(-7)^2 = 49$
		$\Sigma(x_i - \overline{x}) = 0$	$\Sigma(x_i - \overline{x})^2 = 286$

Then we check that the sum of the values in the third column is 0, which is the case. So, we move forward and find the sum of the values in the fourth column, which gives $\Sigma(x_i - \overline{x})^2 = 286$.

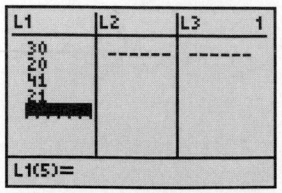

Figure 46 Enter the data

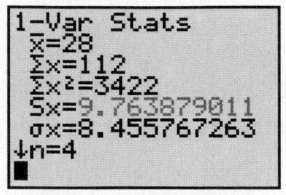

Figure 47 Compute the standard deviation

Summary statistics:

Column	n	Mean	Std. dev.
Cholesterol	4	28	9.763879

Figure 48 Using StatCrunch to compute the standard deviation

Finally, we substitute 286 for $\Sigma(x_i - \bar{x})^2$ in the standard deviation formula, compute, and round to one more decimal place than the data:

$$s = \sqrt{\frac{\Sigma(x_i - \bar{x})^2}{n-1}} = \sqrt{\frac{286}{4-1}} = \sqrt{\frac{286}{3}} \approx 9.7639$$

So, the standard deviation is approximately 9.8 milligrams.

In Example 2, we found the standard deviation for some pizza data. We can also find the standard deviation by using a TI-84. We perform the same steps as we would to find the mean. First, we enter the four observations in the stat list editor (see Fig. 46). Then we use "1-Var Stats" to compute the standard deviation, which is given as "$S_x = 9.763879011$" (see Fig. 47). If we round this value to the first decimal place, we get 9.8 milligrams, which is equal to the value we found in Example 2. For TI-84 instructions, see Appendix A.6.

We can also use StatCrunch to find the standard deviation, which is listed under "Std. dev." (see Fig. 48). If we round the displayed value 9.763879 to the first decimal place, we get 9.8 milligrams, which is equal to the value we found in Example 2. For StatCrunch instructions, see Appendix B.11.

Comparing the Means and Standard Deviations of Two Groups of Data

In Section 4.1, we compared the centers of two distributions by comparing their means. In addition to comparing the centers, we can compare the spreads by comparing the standard deviations.

▶ **Example 3**   Comparing Means and Standard Deviations of Two Distributions

A household suffers from *food insecurity* if at some point in the year the household eats less, goes hungry, or eats less nutritional meals because there is not enough money for food. The percentages of households suffering from food insecurity are shown in Table 23 for the West North Central and Pacific states.

Table 23 Food Insecurities for West North Central and Pacific States

West North Central State	Food Insecurity (percent of households)	Pacific State	Food Insecurity (percent of households)
Iowa	12.7	Alaska	14.0
Kansas	14.8	California	16.2
Minnesota	10.7	Hawaii	14.2
Missouri	17.1	Oregon	16.7
Nebraska	13.4	Washington	15.0
North Dakota	7.7		
South Dakota	12.3		

Source: *Feeding America; U.S. Department of Agriculture*

The means and standard deviations of the two distributions are shown in Fig. 49. What do they mean in this situation?

Summary statistics:

Column	n	Mean	Std. dev.
West North Central	7	12.671429	2.9836858
Pacific	5	15.22	1.196662

Figure 49 Food insecurity

Solution

The mean for the West North Central states (12.67%) is less than the mean for the Pacific states (15.22%), so a West North Central state tends to have a lower percentage of households with food insecurity than a Pacific state. However, the standard deviation for the West North Central states (2.98%) is more than twice the standard deviation for the Pacific states (1.20%), so the spread of the West North Central distribution is much larger than the spread of the Pacific distribution. In fact, the spread is so much larger for the West North Central distribution, one of the West North Central states (Missouri) has a larger percentage of households with food insecurity (17.1%) than every one of the Pacific states.

We verify these conclusions by inspecting the dotplot shown in Fig. 50.

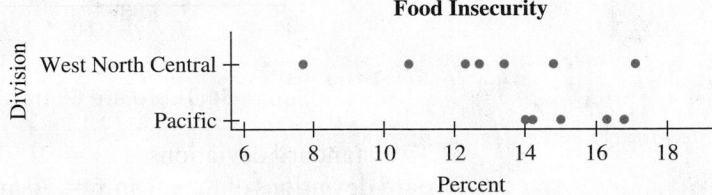

Figure 50 Percentages of households with food insecurity

Example 3 is a good reminder that even though it is quite useful to compare the means (centers) of two distributions, it is also important to compare the standard deviations (spreads).

Empirical Rule

Suppose your car runs out of gasoline at the 7 mile mark on a highway and there is a gas station within 3 miles of you. Where might the gas station be? From Fig. 51, we see the gas station is between the $7 - 3 = 4$ mile mark and the $7 + 3 = 10$ mile mark. We will use this type of logic in Example 4.

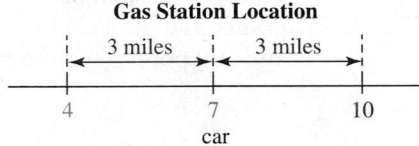

Figure 51 The 4 mile mark and the
10 mile mark are both 3 miles from the car

▶ **Example 4**　Discovering the Empirical Rule

Suppose that 100 statistics students' scores (in points) on a test are described by the dotplot shown in Fig. 52 and the information in Fig. 53.

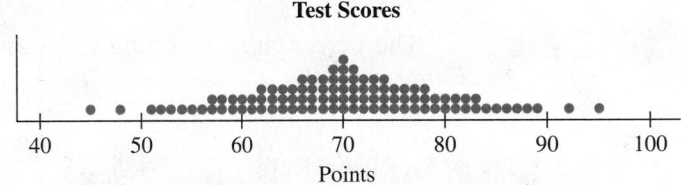

Figure 52 Dotplot of test scores

1. Describe the distribution.
2. Find the percentage of the scores that are within one standard deviation of the mean.
3. Find the percentage of the scores that are within two standard deviations of the mean.
4. Find the percentage of the scores that are within three standard deviations of the mean.

Solution

1. The distribution is unimodal and symmetric (see Fig. 52). The mean is $\bar{x} = 70$ points, and the approximate standard deviation is $s = 10$ points (see Fig. 53).

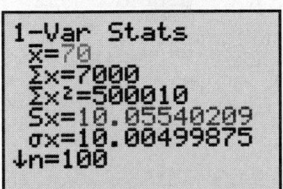

Figure 53 Test scores

2. The scores within $s = 10$ points of $\bar{x} = 70$ points are the scores between $70 - 10 = 60$ points and $70 + 10 = 80$ points. There are 68 such scores (see the red dots in Fig. 54). Since there are a total of 100 scores, the percentage is 68%.

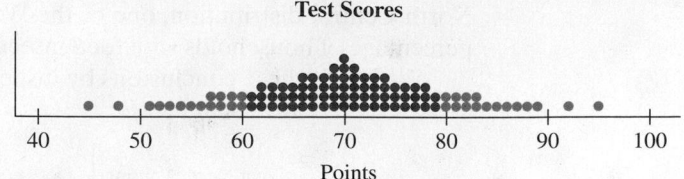

Figure 54 There are 68 (red) dots between 60 and 80

3. Two standard deviations is $2s = 2(10) = 20$ points. So, the scores within two standard deviations of the mean $\bar{x} = 70$ are the scores between $70 - 2(10) = 50$ points and $70 + 2(10) = 90$ points. There are 96 such scores (see the red dots in Fig. 55). The percentage is 96%.

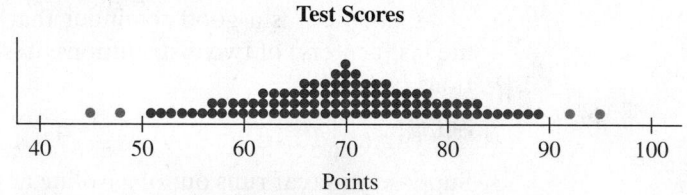

Figure 55 There are 96 (red) dots between 50 and 90

4. Three standard deviations is $3s = 3(10) = 30$ points. So, the scores within three standard deviations of the mean $\bar{x} = 70$ are the scores between $70 - 3(10) = 40$ points and $70 + 3(10) = 100$ points. All of the 100 scores are between 40 points and 100 points (see the red dots in Fig. 56). The percentage is 100%.

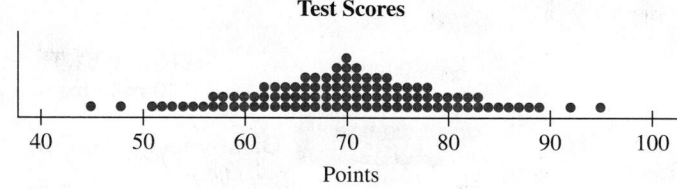

Figure 56 There are 100 (red) dots between 40 and 100

The percentages we found in Example 4 are quite close to the ones stated in the *Empirical Rule.*

▶ **Empirical Rule**

If a distribution is unimodal and symmetric, then the following statements are true.

- Approximately 68% of the observations lie within one standard deviation of the mean. So, approximately 68% of the observations lie between $\bar{x} - s$ and $\bar{x} + s$.
- Approximately 95% of the observations lie within two standard deviations of the mean. So, approximately 95% of the observations lie between $\bar{x} - 2s$ and $\bar{x} + 2s$.
- Approximately 99.7% of the observations lie within three standard deviations of the mean. So, approximately 99.7% of the observations lie between $\bar{x} - 3s$ and $\bar{x} + 3s$.

Recall that for a density histogram, the areas of the bars are equal to proportions. For the test-score distribution, this means that the total area of the bars within one standard deviation of the mean is approximately 0.68 (see the orange bars in Fig. 57). It also means that the total area of the bars within two standard deviations of the mean is approximately 0.95 (see the orange bars in Fig. 58). And it means that the total area of the bars within three standard deviations of the mean is approximately 0.997 (see the orange bars in Fig. 59).

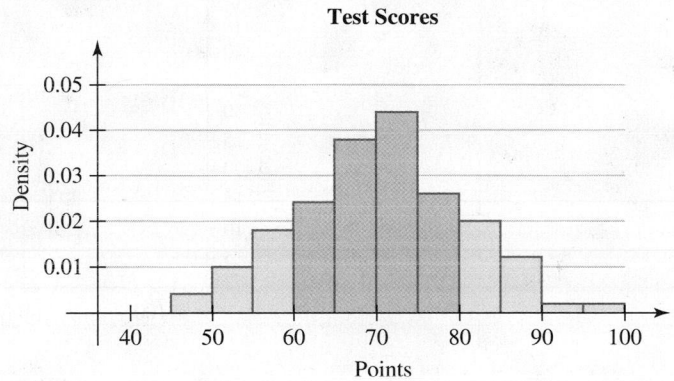

Figure 57 The total area of the (orange) bars within *one* standard deviation of the mean is approximately 0.68

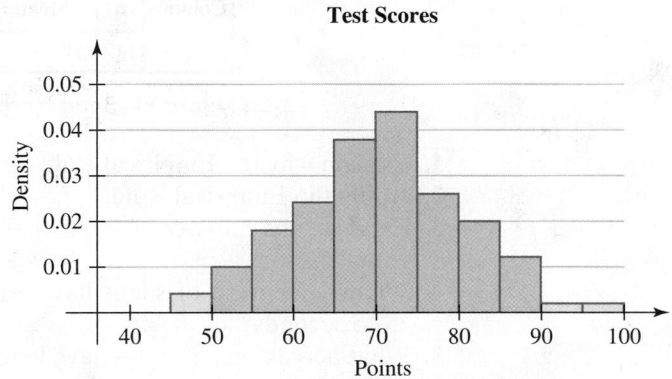

Figure 58 The total area of the (orange) bars within *two* standard deviations of the mean is approximately 0.95

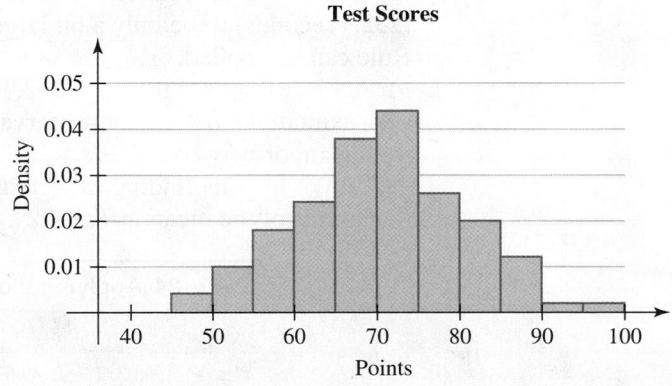

Figure 59 The total area of the (orange) bars within *three* standard deviations of the mean is approximately 0.997

WARNING   Before applying the Empirical Rule to a particular distribution, we must make sure the distribution is unimodal and symmetric. Other distributions can have quite different percentages.

▶ Example 5   Using the Empirical Rule

Live 105® is a San Francisco Bay Area alternative rock radio station. The lengths (in seconds) of 149 songs played on Friday, April 18, 2014, are described by the histogram in Fig. 60 and the information in Fig. 61. If a song was played multiple times, its length was included in the data set just once.

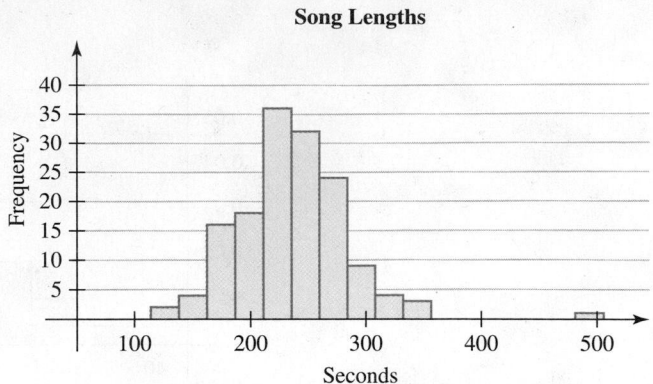

Song Lengths

Figure 60 Histogram of song lengths
(**Source:** *Live 105; iTunes Store*)

Summary statistics:

Column	n	Mean	Std. dev.	Median	Min	Max
Length	149	232.92617	49.634079	231	110	499

Figure 61 Song lengths

1. Explain why the Empirical Rule can be applied.
2. Apply the Empirical Rule.
3. Estimate the *number* of songs that have lengths between 133.7 seconds and 332.1 seconds.
4. What percentage of songs have lengths less than 183.3 seconds OR greater than 282.5 seconds?
5. What percentage of songs have lengths greater than 282.5 seconds?

Solution

1. For the data other than the outlier song length 499 seconds (see "Max" in Fig. 61), the distribution is symmetric and unimodal. Because the outlier has caused the mean (232.9 seconds) to be only a bit larger than the median (231 seconds), the Empirical Rule can be applied.
2. From Fig. 61, we see that $\bar{x} = 232.9$ and $s = 49.6$. According to the Empirical Rule, approximately 68% of the observations will lie within one standard deviation of the mean, or between $\bar{x} - s = 232.9 - 49.6 = 183.3$ and $\bar{x} + s = 232.9 + 49.6 = 282.5$. We list this finding as well as the results for within two and three standard deviations of the mean in Table 24.

Table 24 Applying the Empirical Rule

Interval	Interval	Interval	Percent
$\bar{x} \pm s$	232.9 ± 49.6	$(183.3, 282.5)$	68%
$\bar{x} \pm 2s$	$232.9 \pm 2(49.6)$	$(133.7, 332.1)$	95%
$\bar{x} \pm 3s$	$232.9 \pm 3(49.6)$	$(84.1, 381.7)$	99.7%

The proportions of all the songs that have lengths within one, two, and three standard deviations of the mean are equal to the total areas of the orange bars in the density histograms in Figs. 62, 63, and 64, respectively.

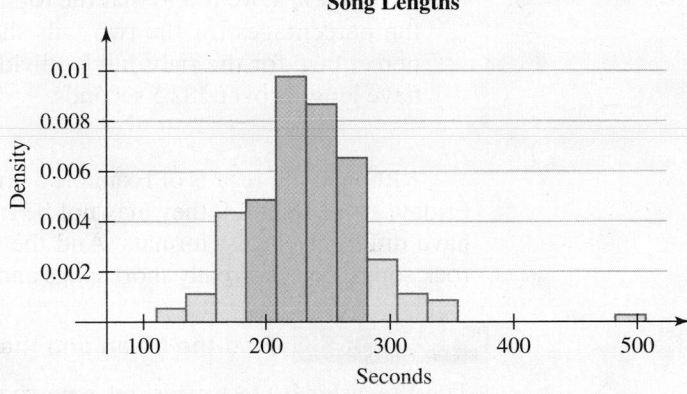

Figure 62 The proportion (approximately 0.68) of all the songs that have lengths within *one* standard deviation of the mean is equal to the total area of the orange bars

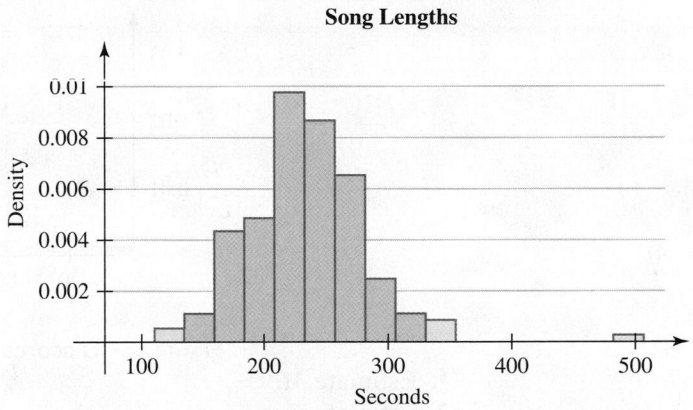

Figure 63 The proportion (approximately 0.95) of all the songs that have lengths within *two* standard deviations of the mean is equal to the total area of the orange bars

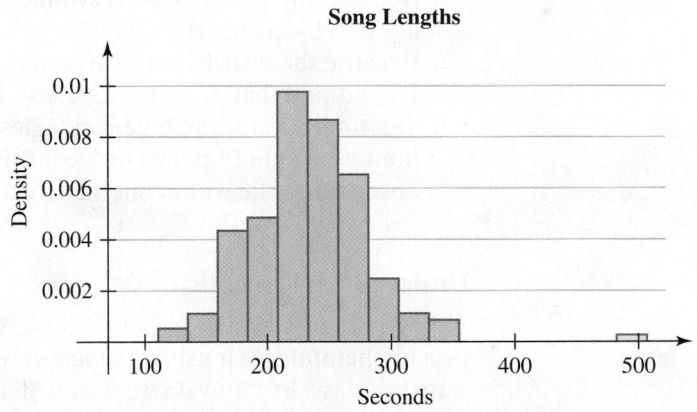

Figure 64 The proportion (approximately 0.997) of all the songs that have lengths within *three* standard deviations of the mean is equal to the total area of the orange bars

3. Approximately 95% of the 149 song lengths lie in the interval (133.7, 332.1). See Table 24 on page 248. So, approximately $0.95(149) \approx 142$ songs have lengths between 133.7 seconds and 332.1 seconds.

4. Approximately 68% of songs have lengths between 183.3 seconds and 282.5 seconds (see Table 24 on page 248). That leaves $100 - 68 = 32\%$ for the total area of the blue tails shown in Fig. 62. So, approximately 32% of the songs have lengths less than 183.3 seconds OR greater than 282.5 seconds

5. In Problem 4, we found that the total area of the blue tails in Fig. 62 is 32%. Because the percentages for the two tails should be approximately equal, we can find the percentage for the right tail by dividing by 2: $32 \div 2 = 16\%$. So, 16% of the songs have lengths over 282.5 seconds.

Although the results of Example 5 are likely true for the songs played on Live 105 on Friday, April 18, 2014, they may not be true for songs played on other days, which might have different playlist formats. And the results are certainly not true for all alternative rock songs, because really short songs and really long songs don't tend to get airplay.

▶ **Example 6**   Find the Mean and the Standard Deviation of a Distribution

The Wechsler IQ test measures a person's intelligence. A density histogram of the IQs (in points) of some people is displayed in Fig. 65. Some of the bars have been left blank on purpose.

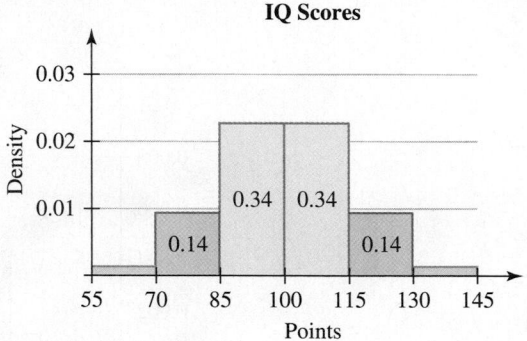

Figure 65 IQ scores

1. Estimate M.
2. Estimate $\bar{x}$.
3. Estimate s.

Solution

1. Because the distribution is symmetric, the median is at the vertical line where the left half meets the right half. So, $M = 100$.
2. Because the distribution is symmetric, the mean is equal to the median. In Problem 1, we found that $M = 100$. So, $\bar{x} = 100$.
3. The total area of the blue rectangles is 0.68, which means that 68% of the observations lie within 15 points of the mean 100 points. By the Empirical Rule, 68% of the observations lie within one standard deviation of the mean. So, $s = 15$.

Unusual Observations

A key part of statistics is determining whether an observation is unusual. This is especially helpful when using *hypothesis testing*, which you will learn about in your next statistics class. But how do we decide if an observation is unusual?

If a distribution is unimodal and symmetric, the answers lies with the Empirical Rule, which states that 95% of the observations are within 2 standard deviations of the mean. This means that only 5% of the observations are more than 2 standard deviations away from the mean. So, **if an observation is more than two standard deviations away from the mean, we refer to the observation as unusual.**

▶ **Example 7**   Determining whether an Observation Is Unusual

In Example 5, we found that some songs played on Live 105 on April 18, 2014, have song lengths with a mean of 232.9 seconds and a standard deviation of 49.6 seconds. One of the songs is *Lazy Eye* (by Silversun Pickups), which lasts for 354 seconds. Is the song's length unusual?

Solution

In Example 5, we found that song lengths between 133.7 seconds and 332.1 seconds are within 2 standard deviations of the mean. Since *Lazy Eye* lasts for 354 seconds, which is outside the interval (133.7, 332.1), we conclude that the song length is unusual for songs played on Live 105 on April 18, 2014.

Range versus Standard Deviation

If a distribution is skewed, the median is usually a better measure of the center than the mean (Section 4.1). And if we are not using the mean to measure the center, then it doesn't make sense to use standard deviation to measure the spread, because the mean is part of the formula for standard deviation. Instead, we use the range to measure the spread.

> **Summary of Measuring Center and Spread**
>
> - If a distribution is unimodal and symmetric, then we usually use the mean to measure the center and the standard deviation to measure the spread.
> - If a distribution is skewed, then we usually use the median to measure the center and the range to measure the spread.

Variance

In Example 4, the students' test scores have approximate standard deviation 10 points. If we square the standard deviation, the result is 100 points^2, which we call the *variance*. We use the notation s^2 to stand for the variance.

> ### Definition   Variance
>
> The **variance** is the square of the standard deviation.

▶ Example 8   Variance

In Example 2, we analyzed the cholesterol levels in the equivalent of $\frac{1}{8}$ of a 14" cheese pizza for some pizza companies. We calculated the standard deviation of the cholesterol levels to be 9.7639 milligrams. Find the variance.

Solution

To find the variance we square both sides of $s = 9.7639$:

$$s^2 = 9.7639^2 \approx 95.3$$

So, the variance is 95.3 milligrams^2.

WARNING   Note that in Example 8, the units of the variance are milligrams^2, not just milligrams.

In Section 4.1 and this section, we have worked with many measures of distributions: mean, median, mode, range, standard deviation, and variance. It is important to keep their meanings and symbols straight. The meanings and symbols are summarized in Table 25.

Table 25 Meanings and Symbols for the Mean, Median, Range, Standard Deviation, and Variance

Measure	What It Measures	Symbol
mean	center	$\bar{x}$
median	center	M
mode	center	(no standard symbol)
range	spread	R
standard deviation	spread	s
variance	spread	s^2

Group Exploration

ᴅᴀᴛᴀ. Empirical Rule: To Be Used Just After Standard Deviation Is Defined

1. Construct a frequency histogram for the data in Table 26. Begin with lower class limit 9 and class width 2.

Table 26 Some Data

9	11	12	12	13	13	14	14
14	15	15	15	15	15	15	16
16	16	17	17	18	18	19	21

2. Describe the shape of the distribution. Should the mean or the median be used to measure the center?

3. Compute the mean and the standard deviation.

4. Compute $\bar{x} - s$ and $\bar{x} + s$. Then count how many of the observations are between these two values. We say we are finding the number of observations within one standard deviation of the mean.

5. Find the *percentage* of observations that lie within one standard deviation of the mean. [**Hint:** Divide your result in Problem 4 by the total number of observations. Then write your result in percentage form.]

6. Find the percentage of observations that lie within *two* standard deviations of the mean. [**Hint:** Repeat Problems 4 and 5 but work with $\bar{x} - 2s$ and $\bar{x} + 2s$.]

7. Find the percentage of observations that lie within *three* standard deviations of the mean. [**Hint:** Repeat Problems 4 and 5 but work with $\bar{x} - 3s$ and $\bar{x} + 3s$.]

8. Because the distribution is unimodal and symmetric, a rule called the Empirical Rule states that your results to Problems 5, 6, and 7 should be approximately 68%, 95%, and 99.7%, respectively. Did that happen?

▶ **Tips for Success**   Review Material

At various times throughout this course, you can improve your understanding of statistics by reviewing material that you have learned so far. Your review should include solving problems, redoing explorations, and reexamining concepts and techniques from previous sections.

Homework 4.2

For extra help ▶ **MyMathLab®**   Watch the videos in MyMathLab   Download the MyDashboard App

1. The standard deviation measures the ____ of a distribution.

2. If a distribution is unimodal and symmetric, then 68% of the observations lie within ____ standard deviation(s) of the mean.

3. If an observation is more than ____ standard deviation(s) away from the mean, we refer to the observation as unusual.

4. The ____ is the square of the standard deviation.

For Exercises 5–8, the given sentence describes the value of the mean, median, standard deviation, or variance of the numbers of drone strikes per month in Pakistan in 2013 (Source: New America Foundation). Use the appropriate symbol to describe the value.

5. The mean number of strikes per month is 2.2 strikes.

6. The variance of the number of strikes per month is 2.3 strikes².

7. The standard deviation of the number of strikes per month is 1.5 strikes.

8. The median number of strikes per month is 2 strikes.

For Exercises 9–12, find the mean, median, range, standard deviation, and variance for the given data. Use the correct symbols.

9. 10, 4, 7, 6, 7, 8

10. 7, 1, 5, 12, 5

11. 18, 3, 12, 15, 24, 5, 28, 15

12. 24, 13, 25, 39, 20, 23, 4, 23, 31, 18

13. The following data are the numbers (in millions) of downloads of the top five most downloaded songs in the United States in 2013 (Source: *Nielsen SoundScan*). Find the range and the standard deviation of the numbers of downloads.

4.7	6.1	5.5	6.5	4.4

14. The following data are the numbers (in millions) of times a song has been streamed for the top five most streamed songs in the United States in 2013 (Source: *Nielsen SoundScan*).

Find the range and the standard deviation of the numbers of times a song has been streamed.

490	280	257	189	171

The author surveyed students in one of his statistics classes about their genders, their ages (in years), the number of texts sent daily, the number of hours spent working weekly, and the time (in hours) spent sleeping daily. The anonymous responses of seven randomly selected students are shown in Table 27. Refer to this table for Exercises 15–20.

15. **DATA** Compute the standard deviation of the number of hours worked weekly.

16. **DATA** Compute the variance of the amount of time spent sleeping daily.

17. **DATA** Compute the range of the number of texts sent daily.

18. **DATA** Compute the range of the ages.

19. **DATA** Compute the standard deviation of the number of hours the *male* students worked weekly.

20. **DATA** Compute the variance of the ages of the *female* students.

Table 27 Responses of Seven Statistics Students

Gender	Age	Texts	Work	Sleep
Female	19	50	10	8.5
Male	21	1	16	7.5
Female	19	300	12	8
Female	19	5	15	7
Male	20	0	11	7
Female	19	90	20	7
Male	20	40	24	6.5

Source: *J. Lehmann*

21. **DATA** Here are the 2013 tuitions (in dollars) of 10 randomly selected 2-year, public colleges (Source: *U.S. Department of Education*):

1772	1752	3485	1138	2328
1854	3598	3722	4680	3030

a. Find the mean tuition.

b. Find the standard deviation of the tuitions.

c. The mean tuition of 10 randomly selected 2-year, *private* colleges is $13,475. The standard deviation is $4863. Which sample of colleges has the lower typical tuition, the 10 public colleges or the 10 private colleges? Explain.

d. Which sample of colleges has tuitions with less spread, the 10 public colleges or the 10 private colleges? Explain. See part (c).

e. On the basis of comparing only the standard deviations of the two samples, a student concludes that the tuitions of all 2-year, public colleges have less spread than the tuitions of all 2-year, private colleges. What would you tell the student?

22. **DATA** Here are the 2013 tuitions (in dollars) of 10 randomly selected 4-year, public colleges (Source: *U.S. Department of Education*):

7238	8570	2550	4255	6180
6238	9864	7327	8706	3875

a. Find the mean tuition.

b. Find the standard deviation of the tuitions.

c. The mean tuition of 10 randomly selected 4-year, *private* colleges is $20,750. The standard deviation is $10,065. Which sample of colleges has the higher typical tuition, the 10 public colleges or the 10 private colleges? Explain.

d. Which sample of colleges has tuitions with more spread, the 10 public colleges or the 10 private colleges? Explain. See part (c).

e. On the basis of comparing only the standard deviations of the two samples, a student concludes that the tuitions of all 4-year, private colleges have more spread than the tuitions of all 4-year, public colleges. What would you tell the student?

For Exercises 23–26, match the given information to the appropriate histogram in Fig. 66.

23. mean: 40, standard deviation: 2.5

24. mean: 40, standard deviation 12.4

25. median: 70, range: 7.6

26. median: 70, range: 38

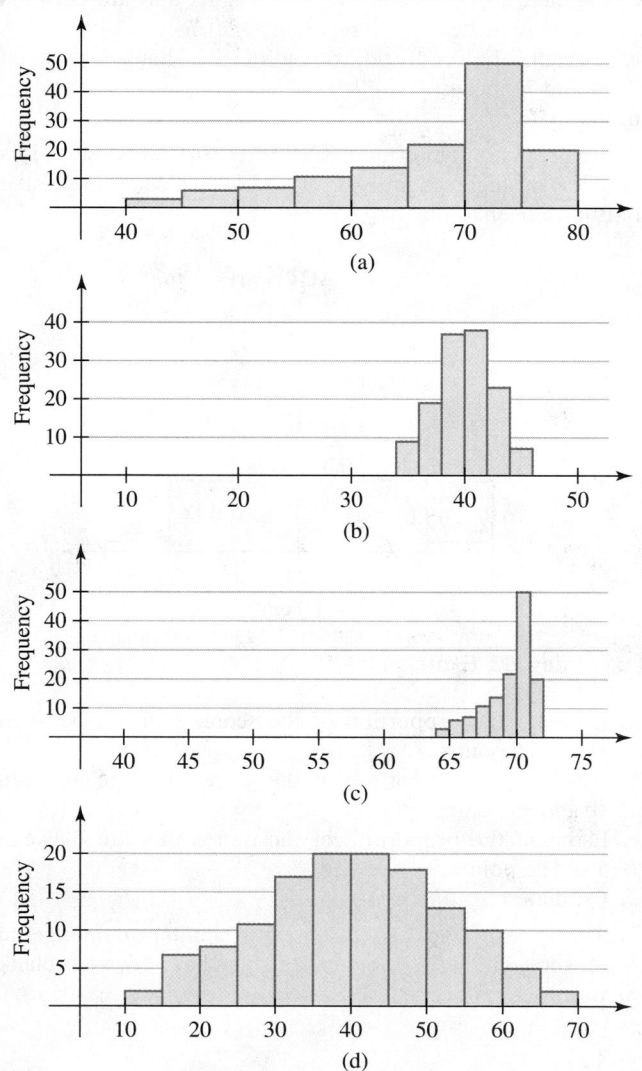

Figure 66 Exercises 23–26

27. A density histogram of the heights (in inches) of women at a large company is displayed in Fig. 67. Some of the bars have been left blank on purpose.

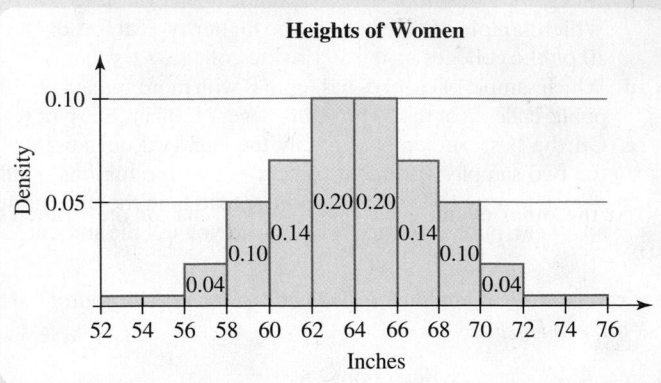

Figure 67 Exercise 27

a. Estimate the proportion of the heights that are between 60 and 68 inches.
b. Estimate the proportion of the heights that are between 56 and 72 inches.
c. Estimate the proportion of the heights that are between 52 and 76 inches.
d. Estimate s.

28. A density histogram of the ACT scores (in points) of seniors at a certain high school is displayed in Fig. 68. Some of the bars have been left blank on purpose.

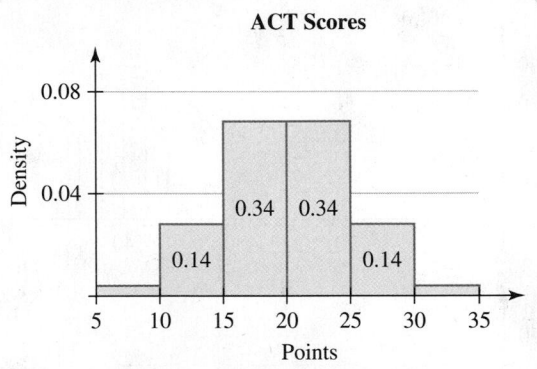

Figure 68 Exercise 28

a. Estimate the proportion of the scores that are between 15 and 25 points.
b. Estimate the proportion of the scores that are between 10 and 30 points.
c. Estimate the proportion of the scores that are between 5 and 35 points.
d. Estimate s.

29. A density histogram of the scores (in points) on an algebra test is shown in Fig. 69. Some of the bars have been left blank on purpose.

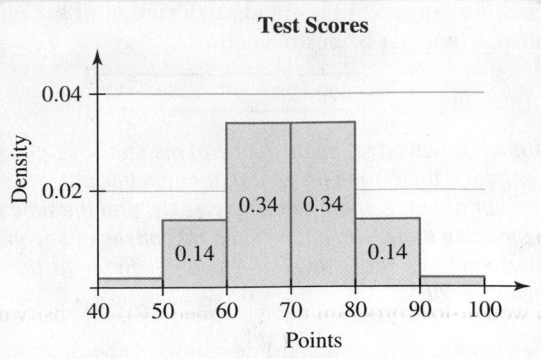

Figure 69 Exercise 29

a. Estimate M.
b. Estimate $\bar{x}$.
c. Estimate s.
d. Estimate s^2.

30. A density histogram of the ages (in years) of people who attend a movie is displayed in Fig. 70. Some of the bars have been left blank on purpose.

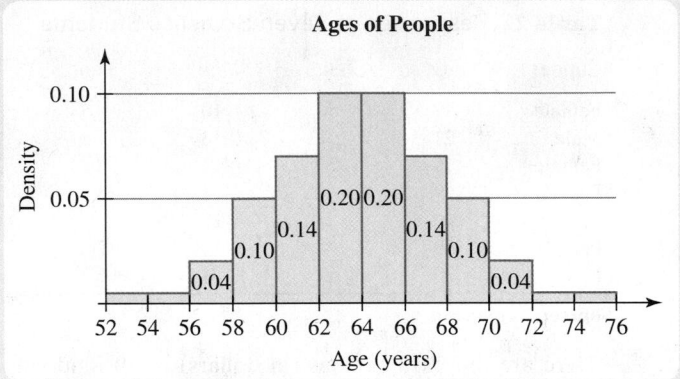

Figure 70 Exercise 30

a. Estimate M.
b. Estimate $\bar{x}$.
c. Estimate s.
d. Estimate s^2.

31. The scores (in points) on a first prestatistics test are described by the density histogram in Fig. 71.

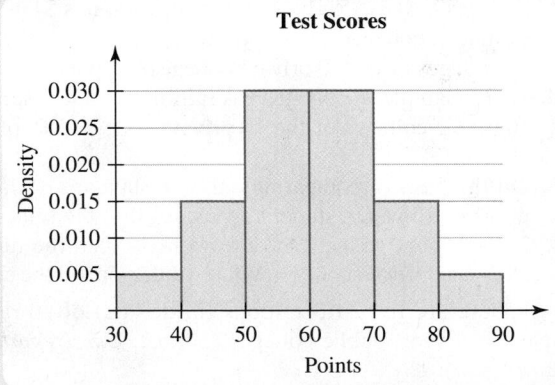

Figure 71 Exercise 31

a. Compute the area of each of the six bars in the density histogram.

b. Find the total area of the six bars.

c. Explain why the total area of the bars of *any* density histogram is always equal to your result in part (b).

d. On the second test, the mean is 10 points higher and the standard deviation is smaller than on the first test. Construct a density histogram that might describe the scores on the second test.

32. The weights (in pounds) of some people who have just started a weight-loss program are described by the density histogram in Fig. 72.

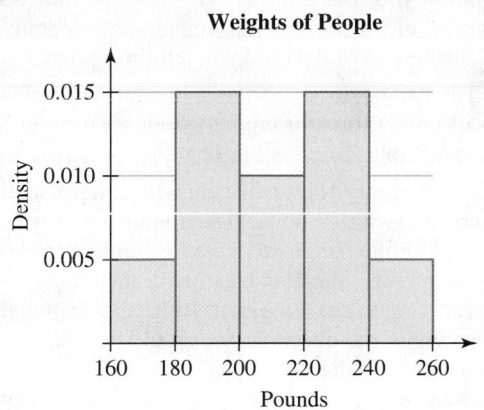

Figure 72 Exercise 32

a. Compute the area of each of the five bars in the density histogram.

b. Find the total area of the five bars.

c. Explain why the total area of the bars of *any* density histogram is always equal to your result in part (b).

d. Three months later, the mean weight is 20 pounds less, but the standard deviation is greater. Construct a density histogram that might describe the people's new weights.

33. A student who usually does very well in math is trying to decide whether to take Professor A or Professor B for statistics. The two professors have similar teaching styles and the same grading policies. Last semester, the distribution of Professor A's students' scores on all tests had mean 70% and standard deviation 10%, and the distribution of Professor B's students' scores on all tests had mean 70% and standard deviation 5%.

a. With which professor would the student probably get higher test scores? Explain.

b. Is it possible the student would actually get higher tests scores with the other professor? Give at least two reasons why or why not.

34. A student who usually does very poorly in math is trying to decide whether to take Professor A or Professor B for statistics. The two professors have similar teaching styles and the same grading policies. Last semester, the distribution of Professor A's students' scores on all tests had mean 80% and standard deviation 10%, and the distribution of Professor B's students' scores on all tests had mean 80% and standard deviation 3%.

a. With which professor would the student probably get higher test scores? Explain.

b. Is it possible the student would actually get higher tests scores with the other professor? Give at least two reasons why or why not.

35. The finish times (in minutes) for runners in three-mile courses at Events A and B are shown in Fig. 73. At each event, the women's mean time was 19 minutes and the men's mean time was 17 minutes. At one event, the standard deviations for the women's times and the men's times were both 0.5 minute. At the other event, the standard deviations for the women's times and the men's times were both 2.5 minutes.

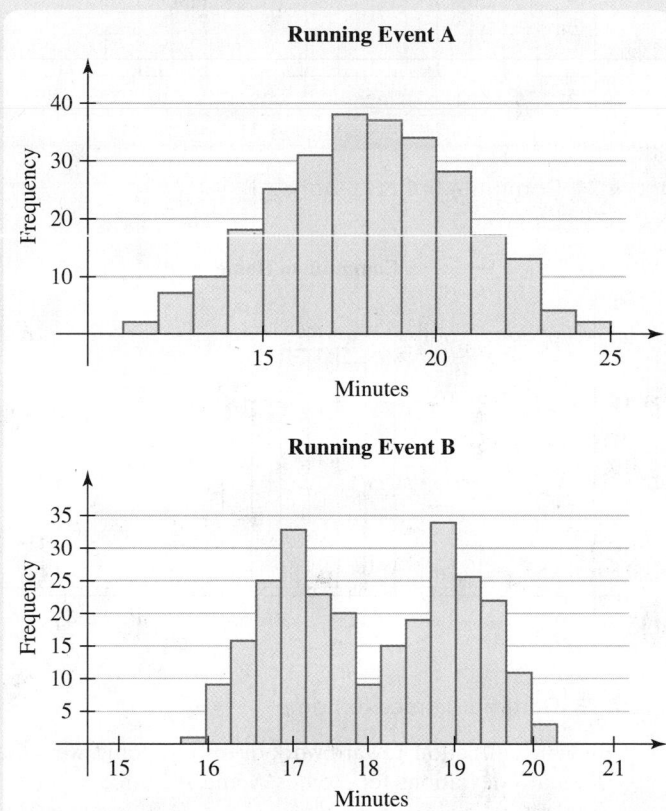

Figure 73 Three-mile times for Events A and B

a. At which Event, A or B, were the standard deviations for women and men both 0.5 minutes?

b. At which Event, A or B, were the standard deviations for women and men both 2.5 minutes?

c. Refer to the given means and standard deviations to explain why the distribution for Event A is unimodal but the distribution for Event B is bimodal.

d. When determining how different women's times are from men's times, explain why it is not enough to simply compare the mean times.

36. A person wants to know whether Route A or Route B is the better commute to work. He also wants to know which route is the better commute home. He times the commutes (in minutes) to work by both routes many times (see Fig. 74) and also times the return trip home by both routes many times (see Fig. 75). In each direction (toward work and toward home), the mean commute time by Route A is 25 minutes and the mean commute time by

Route B is 29 minutes. In one direction, the standard deviations for Routes A and B are both 5 minutes. In the other direction, the standard deviations for Routes A and B are both 1 minute.

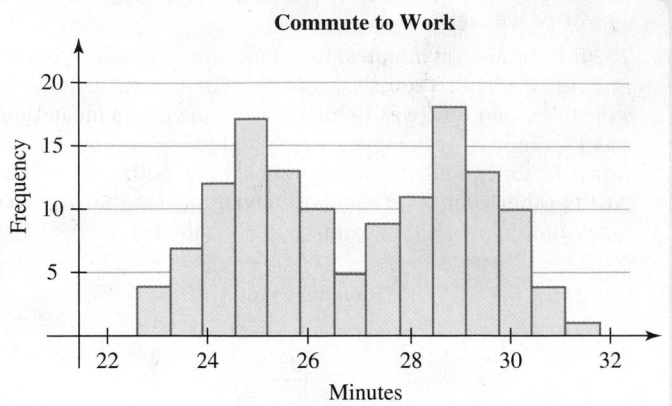

Figure 74 Commute times to work

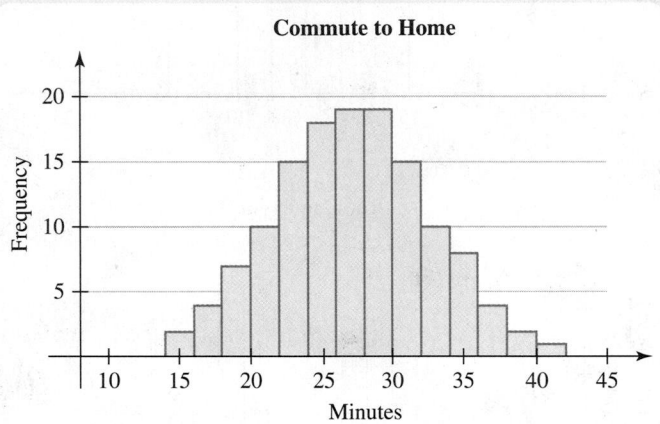

Figure 75 Commute times to home

a. In which direction, toward work or toward home, were the standard deviations for Routes A and B both 5 minutes? Explain.

b. In which direction, toward work or toward home, were the standard deviations for Routes A and B both 1 minute? Explain.

c. Refer to the given means and standard deviations to explain why the distribution for commuting to work is bimodal but the distribution for commuting to home is unimodal.

d. When determining how different the commute times for Route A are from the commute times for Route B, explain why it is not enough to simply compare the mean times.

37. **DATA** The author surveyed one of his statistics classes. The numbers of units (hours or credits) students were enrolled in are shown in Table 28.

Table 28 Numbers of Units

16	10	11	13	13	15	15
13	15	17	13	17	13	14
11	14	12	14	14	16	12
15	11	13	10	13	10	15
15	16	8	13	16	14	12.5
14	18	4	13	11	13	5

Source: *J. Lehmann*

a. Construct a histogram. Begin with lower class limit 4 units and class width 2 units. Describe the shape of the distribution. Can the Empirical Rule be applied?

b. Compute the mean and standard deviation.

c. According to the Empirical Rule, approximately what percentage of the observations should lie within one standard deviation of the mean?

d. Count the actual number of observations shown in Table 28 that lie within one standard deviation of the mean. What percentage is that of the 42 students? Compare the percentage with the result you found in part (c). [**Hint:** You could use a stemplot to help perform the count.]

e. A student concludes that the mean and the standard deviation you found in part (b) describe the center and the spread of the numbers of units for *all* students enrolled at the college. What would you tell the student?

38. **DATA** The author surveyed one of his statistics classes. The numbers of units (hours or credits) students were enrolled in are shown in Table 28.

a. Construct a histogram. Begin with lower class limit 4 units and class width 2 units. Describe the shape of the distribution. Can the Empirical Rule be applied?

b. Compute the mean and standard deviation.

c. According to the Empirical Rule, approximately what percentage of the observations should lie within two standard deviations of the mean?

d. Count the actual number of observations shown in Table 28 that lie within two standard deviations of the mean. What percentage is that of the 42 students? Compare the percentage with the result you found in part (c). [**Hint:** You could use a stemplot to help perform the count.]

39. **DATA** Here are the sales (in billions of dollars) of Pfizer®'s and AstraZeneca®'s top-ten-selling drugs (Source: *Pfizer and AstraZeneca*):

Pfizer			AstraZeneca		
2.9	1.2	4.0	0.6	3.5	1.0
1.9	2.3	1.1	1.3	5.6	0.8
3.8	1.4	1.2	1.1	3.9	0.7
4.6			0.9		

a. Construct a histogram for the Pfizer data and a histogram for the AstraZeneca data. For each histogram, begin with lower class limit 0.5 billion dollars and class width 0.5 billion dollars. Describe the distributions.

b. What should be used to measure the center of each data set, the mean or the median? Explain.

c. Find the median sales of Pfizer's top-ten-selling drugs. Also find the median sales of AstraZeneca's top-ten-selling drugs.

d. Which company has the larger median sales? What does that mean in this situation?

e. In January 2014, Pfizer tried to buy AstraZeneca. Explain why it makes sense that Pfizer might be in a better position than AstraZeneca to buy the other company.

40. **DATA** The sales (in billions of dollars) of Pfizer's and AstraZeneca's top-ten-selling drugs are shown in Exercise 39.

a. Construct a histogram for the Pfizer data and a histogram for the AstraZeneca data. For each histogram, begin with lower class limit 0.5 billion dollars and class width 0.5 billion dollars. Describe the distributions.

b. What should be used to measure the spread of each data set, the range or the standard deviation? Explain.

c. Find the range of the sales of Pfizer's top-ten-selling drugs. Also find the range of the sales of AstraZeneca's top-ten-selling drugs.

d. Which company has the larger range of sales? What does that mean in this situation?

e. By referring to your result in part (d), explain why it makes sense that even though Pfizer has the larger median sales, AstraZeneca has the best-selling drug of the two companies. What are the sales of that drug?

41. In one of the author's statistics classes, 27 students responded anonymously to the following two questions: "How many friends did you have during your senior year in high school? How many of those people are you still in touch with and consider to be friends?" For each student, the author computed the *friend-retention rate* (*FRR*), which is the percentage of the student's high-school friends that the student was still friends with. The distribution of FRRs is unimodal and symmetric with mean 42.7% and standard deviation 20.5%.

a. Two students had no friends in high school. Explain why it makes sense that the author did not compute FRRs for those two students.

b. Out of the 25 FRRs, approximate how many were between 22.2% and 63.2%.

c. Out of the 25 FRRs, approximate how many were between 1.7% and 83.7%.

d. One student was still friends with all of his or her high school friends. What is that student's FRR? Is that an unusual FRR? Explain.

e. From reading the opening paragraph of this exercise, a student concludes that the mean FRR for all college students is 42.7%. What would you tell the student?

42. In Example 5, we analyzed the lengths of songs played on the alternative rock radio station Live 105. The lengths (in seconds) of all songs on albums by bands typically featured on Live 105 are described by the histogram in Fig. 76 and the information in Fig. 77. Some of the songs on the albums have been played on Live 105, but many have not.

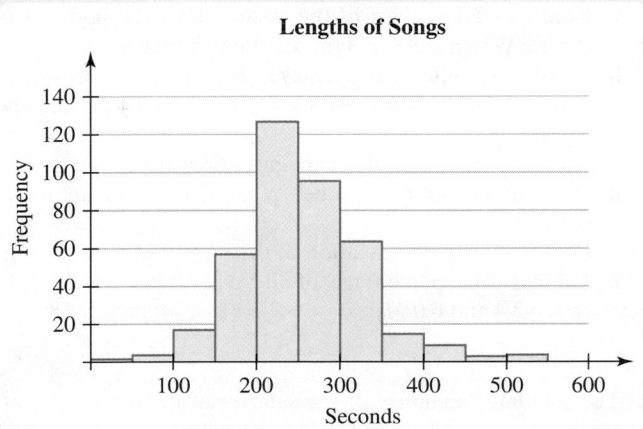

Figure 76 Histogram of song lengths
(**Source:** *Live 105; iTunes Store*)

Summary statistics:

Column	n	Mean	Std. dev.
Time	398	253.09045	74.579911

Figure 77 Song lengths

a. Explain why the Empirical Rule can be applied.

b. What interval contains approximately 95% of the song lengths?

c. Would the interval you found in part (b) contain 95% of the song lengths of *all* types of music? Explain.

d. In Example 5, we found that songs played on Live 105 have a mean length of 232.9 seconds and standard deviation 49.6 seconds. By comparing these values to the mean and standard deviation of song lengths described in Fig. 77, explain why this suggests that the radio station does not tend to play songs that are too short or too long.

e. In Example 5, we found that 95% of the songs played on Live 105 have lengths that lie in the interval (133.7, 332.1). Is the interval wider or narrower than the interval you found in part (b)? What does this mean? How does this tie in with your explanation in part (d)?

43. The numbers of calories in slices of stuffed-crust pizza and Thin 'N Crispy pizza are shown in Table 29 for various types of pizza at Pizza Hut. For each type of pizza, the difference in calories for stuffed-crust and Thin 'N Crispy slices is also shown.

Table 29 Numbers of Calories in One Slice ($\frac{1}{8}$ of a 14" Pizza) at Pizza Hut

Type of Pizza	Stuffed-Crust Pizza Calories	Thin 'N Crispy Pizza Calories	Paired Difference of Stuffed-Crust Pizza Calories and Thin 'N Crispy Pizza Calories
Cheese Only	310	260	310 − 260 = 50
Pepperoni	310	280	310 − 280 = 30
Supreme	380	320	380 − 320 = 60
Super Supreme	400	320	400 − 320 = 80
Chicken Supreme	310	250	310 − 250 = 60
Meat Lover's®	440	390	440 − 390 = 50
Pepperoni Lover's®	400	350	400 − 350 = 50
Veggie Lovers'®	300	240	300 − 240 = 60
Ultimate Cheese Lover's®	340	280	340 − 280 = 60

Source: *Pizza Hut*

a. StatCrunch can be used to find the information shown in Fig. 78. Compare the mean of the stuffed-crust data and the mean of the Thin 'N Crispy data. Referring to your knowledge of pizza, explain why your comparison makes sense.

Summary statistics:

Column	n	Mean	Std. dev.
Stuffed	9	354.44444	51.505124
Thin	9	298.88889	49.860918
Difference	9	55.555556	13.333333

Figure 78 Caloric content in one slice of Pizza Hut pizza

b. Subtract the mean of the Thin 'N Crispy data from the mean of the stuffed-crust data. Round your result to the first decimal place. Compare the result and the mean (rounded to the first decimal place) of the paired differences. By referring to how the means are calculated, explain why this makes sense.

c. Subtract the standard deviation of the Thin 'N Crispy data from the standard deviation of the stuffed-crust data. Round your result to the first decimal place. Compare the result and the standard deviation (rounded to the first decimal place) of the paired differences.

d. Compare the standard deviation of the paired differences and the standard deviation of the stuffed-crust data. Also compare the standard deviation of the paired differences and the standard deviation of the Thin 'N Crispy data. On the basis of both comparisons, if someone is trying to lose weight, which is more important: deciding between stuffed-crust pizza and Thin 'N Crispy pizza or deciding which toppings to order?

44. Table 30 shows the mean changes in weight (in pounds) by people who took part in one of four popular diets. Forty people were randomly assigned to each of the diets. The mean changes were found after 2 months, 6 months, and 12 months.

Table 30 Participants' Mean Changes in Weight by Four Popular Diets

| Duration | Mean Change in Weight (in pounds) | | | |
	Atkins®	Zone®	Weight Watchers®	Ornish®
2 months	−3.6	−3.8	−3.5	−3.6
6 months	−3.2	−3.4	−3.5	−3.6
12 months	−2.1	−3.2	−3.0	−3.3

Source: Comparison of the Atkins, Ornish, Weight Watchers, and Zone Diets for Weight Loss and Heart Disease Risk Reduction: A Randomized Trial, *by Dansinger, Gleason, Griffith, Selker, and Schaefer*

a. Which of the four groups of people had the largest mean decrease in weight after 2 months?

b. Which of the four groups of people had the largest mean decrease in weight after 12 months?

c. For each diet, compare the mean change in weight after 2 months with the mean change in weight after 12 months. What do these comparisons mean in this situation?

d. For the first 2 months, the researchers asked the participants to follow their diet plan to the best of their ability. But after that, the researchers encouraged the participants to follow their diet as much or little as they wanted. Explain how this *might* explain what you observed in part (c). What else might be the reason?

e. The standard deviations of the changes in weight (in pounds) of the participants are shown in Table 31.

Table 31 Standard Deviations of Changes in Weight by Four Popular Diets

| Duration | Standard Deviation of the Change in Weight (in pounds) | | | |
	Atkins	Zone	Weight Watchers	Ornish
2 months	3.3	3.6	3.8	3.4
6 months	4.9	5.7	5.6	6.7
12 months	4.8	6.0	4.9	7.3

A student concludes that almost all the Ornish participants lost more weight than the Zone participants after 12 months, because the mean decrease in weight was larger for the Ornish participants. What would you tell the student, considering that the standard deviations shown in Table 31 are relatively large?

Large Data Sets

45. **DATA.** At the website Lending Club®, customers can either borrow money or fund money for others' loans. Access the data about the loans funded at Lending Club in September, 2014, which are available at MyMathLab and at the Pearson Downloadable Student Resources for Math & Stats website.

a. Construct a relative frequency histogram of the loan amounts for those who rent their home, beginning with lower class limit 0 dollars and class width 5000 dollars. Do the same for those who own their home. Finally, do the same for those who have a mortgage on a home.

b. On the basis of shape, which is the better measure of the spread for each of the three distributions described in part (a), the standard deviation or the range? Explain.

c. Find and compare the ranges of the three distributions. What does your comparison mean in this situation?

d. For each histogram you constructed in part (a), mark the location of the mean and the median. For each distribution, do the mean and median appear to be close together or far apart?

e. On the basis of your response to part (d), explain why it might actually be reasonable to compare the standard deviations of the three distributions even though they are skewed.

f. Find and compare the standard deviations of the three distributions.

g. Compare the spreads of the three distributions. Your comparison should refer to your responses in parts (c) and (f) as well as the histograms you constructed in part (a). What does your comparison mean in this situation?

46. **DATA.** Access the data about earthquakes, which are available at MyMathLab and at the Pearson Downloadable Student Resources for Math & Stats website.

a. Construct a pie chart of the states where earthquakes occurred. Which state had the most earthquakes?

b. Construct a relative frequency histogram of the earthquake magnitudes, beginning with lower class limit 1 M_s and class width 1 M_s.

c. Describe the shape of the magnitude distribution.

d. Find the mean magnitude. What does it mean in this situation?

e. Find the standard deviation of the magnitudes.

f. Estimate the percentage of the magnitudes that are between 2.4 and 6.0 M_s.

Concepts

47. The standard deviation of the salaries of the employees at a company is $15 thousand. If all the employees receive a $2 thousand raise, will the standard deviation decrease, increase, or stay the same?

48. The standard deviation of the weights of some people in an exercise class is 12 pounds. If each person loses 8 pounds, will the standard deviation decrease, increase, or stay the same?

49. The standard deviation of a group of numbers is 5. If the smallest number is decreased by 2 and the largest number is increased by 2, will the standard deviation decrease, increase, or stay the same? Explain.

50. The variance of a group of numbers is 80. If the smallest number is decreased by 20 and the largest number is increased by 20, will the variance decrease, increase, or stay the same? Explain.

51. The distribution of the heights of a group of women has mean 66 inches and standard deviation 2 inches. If a 66-inch-tall woman joins the group, will the standard deviation of the heights decrease, increase, or stay the same? Explain.

52. The distribution of salaries of employees at a company has mean $60 thousand and standard deviation $20 thousand. If a person is hired and paid a salary of $60 thousand, will the standard deviation of the salaries decrease, increase, or stay the same? Explain.

53. The numbers 4, 8, 10, 10, 12, and 16 have standard deviation 4. Use this information to find another group of six numbers that have standard deviation 4.

54. The numbers 6, 12, 15, 15, 18, and 24 have mean 15 and standard deviation 6. Use this information to find another group of six numbers that have mean 15 and standard deviation less than 6.

55. Which would tend to be larger, the standard deviation of the weights of 30,000 cats or the standard deviation of the weights of 5 randomly selected human adults? Explain.

56. Which would tend to be larger, the standard deviation of the prices of 7 randomly selected new homes or the standard deviation of the prices of 30,000 randomly selected new cars?

57. A distribution of tests scores is unimodal and symmetric with mean $\bar{x} = 60$ points and standard deviation $s = 10$ points.

  a. Estimate the percentage of scores that are between 30 and 90 points.

  b. How many standard deviations apart are the scores 30 and 90 points?

  c. Use parts (a) and (b) to explain why if there are no outliers, it makes sense that the range of the scores is approximately $6s$.

58. In general, if a distribution is unimodal and symmetric with no outliers, the range is approximately $6s$ (see Exercise 57). Explain why this suggests that the standard deviation of the distribution is approximately $\dfrac{R}{6}$, where R is the range.

59. In 2013, 4579 runners completed the Great Cow Harbor 10K Run in Northport, New York. The youngest finisher was 7 years old and the oldest finisher was 82 years old. The age distribution is unimodal and symmetric (Source: *The Great Cow Harbor 10-Kilometer Run, Inc.*).

  a. What is the range of the ages?

  b. In Exercise 58, you found that if a distribution is unimodal and symmetric with no outliers, then the standard deviation is approximately $\dfrac{R}{6}$, where R is the range. Estimate the standard deviation of the ages.

  c. The actual standard deviation of the ages is 12.4 years. Find the error of the result you found in part (b). (The error is the estimated standard deviation minus the actual standard deviation.)

60. Recall from Example 3 that a household suffers from *food insecurity* if at some point in the year the household eats less, goes hungry, or eats less nutritional meals because there is not enough money for food. The distribution of percentages of households suffering from food insecurity in a U.S. state is unimodal and symmetric with no outliers. North Dakota has the smallest percentage (7.7%), and Mississippi has the largest percentage (22.3%).

  a. What is the range of the food insecurity percentages of the 50 states?

  b. In Exercise 58, you found that if a distribution is unimodal and symmetric with no outliers, then the standard deviation is approximately $\dfrac{R}{6}$, where R is the range. Estimate the standard deviation of the food insecurity percentages of the 50 states.

  c. The actual standard deviation is 2.7 percentage points. Find the error of the result you found in part (b). (The error is the estimated standard deviation minus the actual standard deviation.)

61. An investor plans to invest in a stock in such a way that if the stock value increases the investor will earn money and if the stock value decreases the investor will lose money. The more the stock value changes, the more money the investor will earn or lose. The investor is trying to decide between investing in a certain small-cap stock and a certain blue-chip dividend stock. Both stocks currently have the same value. Typically, the daily values of a small-cap stock have greater standard deviation than the daily values of a blue-chip dividend stock.

  a. Which stock has higher risk? Explain.

  b. Even though the stock you found in part (a) has higher risk, some investors still choose to invest in that type of stock. Why?

Hands-On Research

62. Visit the website www.amazon.com.

  a. Select a certain product and describe it. Then find and record 20 prices of the product.

  b. Compute the mean and standard deviation of the prices.

  c. What percentage of the prices lie within one standard deviation of the mean?

  d. What percentage of the prices lie within two standard deviations of the mean?

  e. What percentage of the prices lie within three standard deviations of the mean?

  f. Compare the results you found in parts (c), (d), and (e) with the Empirical Rule.

  g. Construct a histogram that describes the price distribution. How does its shape tie in with the comparisons you made in part (f)?

▼4.3 Boxplots

Objectives

» Find the *quartiles* of some data.

» Find the *interquartile range* of some data.

» Construct a *boxplot* to describe a distribution without outliers.

» Identify outliers.

» Construct a *boxplot* to describe a distribution with one or more outliers.

» Compare the boxplots of two groups of data.

In Section 4.2, we discussed three measures of the spread of a distribution. In this section, we will discuss another such measure. Then we will incorporate this measure into a diagram that will enable us to both visualize and measure the spread. Plus, the diagram will indicate the median as well as any outliers that the distribution might have.

Quartiles

In Sections 4.1 and 4.2, we discussed that for skewed data, we usually measure the center by the median and measure the variation by the range. However, one problem with the range is that it is based on only two observations, the smallest and largest ones. We can fine-tune our measure of spread by finding the *quartiles*.

▶ **Definition**   **First quartile, second quartile, and third quartile**

The *first quartile*, *second quartile*, and *third quartile* are the 25th, 50th, and 75th percentiles, respectively.

The quartiles break up observations into quarters (see Fig. 79).

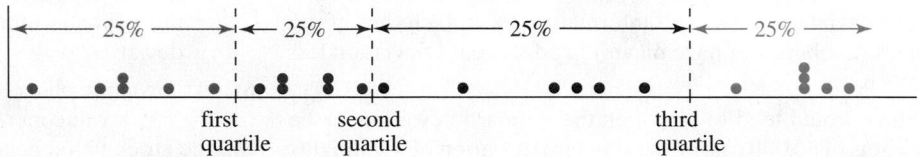

Figure 79 Quartiles

Because the second quartile is the 50th percentile, it is also the median.

We use the symbols Q_1, Q_2, and Q_3 to stand for the first, second, and third quartiles, respectively.

▶ **Example 1**   Finding Quartiles

Students in one of the author's statistics classes were surveyed about the number of hours they study per week for the course. Here are the anonymous responses (in hours) of ten of the students: 5, 7, 2, 3, 4, 10, 17, 4, 28, and 6.

1. Which is a better measure of the center, the mean or the median?
2. Find Q_2 (the median).
3. Find Q_1.
4. Find Q_3.
5. Use technology to verify the results in Problems 2–4.
6. Describe the meaning of the results of Problems 2–4.

Solution

1. Because the distribution is skewed (see Fig. 80), the median is a better measure of the center than the mean.
2. To find Q_2 (the median), we list the observations from smallest to largest:

$$2, 3, 4, 4, 5, 6, 7, 10, 17, 28$$

Because there are ten observations, which is even, we average the two middle observations, 5 and 6:

$$\frac{5 + 6}{2} = 5.5$$

So, Q_2 is 5.5 hours.

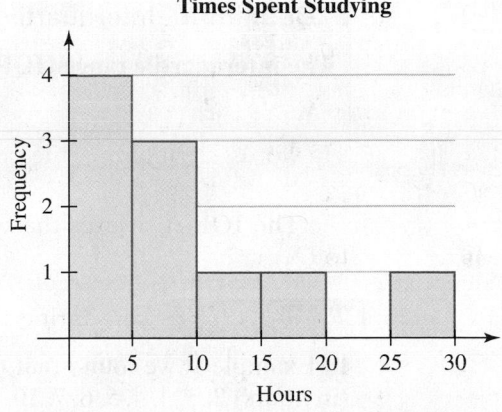

Figure 80 Study times of statistics students

3. To find Q_1, we list the observations that are less than the median:

$$2, 3, 4, 4, 5$$

Q_1 is the 25th percentile, which is the middle of the lower 50% of the observations. So, Q_1 is 4 hours.

4. To find Q_3, we list the observations that are greater than the median:

$$6, 7, 10, 17, 28$$

Q_3 is the 75th percentile, which is the middle of the upper 50% of the observations. So, Q_3 is 10 hours.

5. To use a TI-84 or StatCrunch to find the quartiles, we take the usual steps to find the median (see Fig. 81). We see that $Q_1 = 4$, median $= 5.5$, and $Q_3 = 10$, which checks with our results in Problems 2–4.

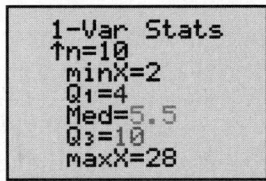

Summary statistics:

Column	n	Median	Q1	Q3
Study Time	10	5.5	4	10

Figure 81 Compute the quartiles

6. Approximately 25% of the observations lie in each of the following groups of study times: less than 4 hours, between 4 and 5.5 hours, between 5.5 hours and 10 hours, and greater than 10 hours.

In Example 1, we found the quartiles for some data with an even number of observations. **If some data have an odd number of data values, do not include the median in the lower half of the data when finding Q_1. Likewise, do not include the median in the upper half of the data when finding Q_3.**

WARNING   Although a TI-84 follows this practice when the number of observations is any odd number, StatCrunch follows this practice only when the number of observations is one of certain odd numbers (3, 7, 11, 15, 19, . . .).

Interquartile Range

In Section 4.2, we discussed three measures of spread: range, standard deviation, and variance. One problem with all three measures is that they are sensitive to outliers. A measure of spread that is resistant to outliers is the *interquartile range*.

> ▶ Definition   Interquartile Range
>
> The **interquartile range** (**IQR**) is given by
>
> $$IQR = Q_3 - Q_1$$
>
> In words, the interquartile range is equal to the difference of Q_3 and Q_1.

The IQR measures the spread of the middle 50% of the observations (from Q_1 to Q_3).

▶ **Example 2**   Comparing the Range and Interquartile Range

In Example 1, we found that $Q_1 = 4$, $Q_2 = 5.5$, and $Q_3 = 10$ for the study distribution (in hours) 2, 3, 4, 4, 5, 6, 7, 10, 17, and 28.

  1. Calculate the IQR.
  2. Calculate the range.
  3. What do the IQR and the range mean in this situation?
  4. Discuss whether the IQR and the range are sensitive to the outlier 28 hours.

Solution

  1. We substitute 4 for Q_1 and 10 for Q_3 in the formula $IQR = Q_3 - Q_1$:

$$IQR = Q_3 - Q_1 = 10 - 4 = 6 \text{ hours}$$

  We can use StatCrunch to find the IQR 6 hours (see Fig. 82).

Summary statistics:

Column	n	Q1	Q3	IQR
Study	10	4	10	6

Figure 82 Compute the IQR

  2. The smallest observation is 2, and the largest observation is 28. We find the difference of 28 and 2:

$$\text{range} = \text{largest observation} - \text{smallest observation} = 28 - 2 = 26 \text{ hours}$$

  3. The relatively small value of the IQR (6 hours) suggests that the middle 50% (from Q_1 to Q_3) has small spread, which is true (see Fig. 83). The relatively large range (26 hours) suggests that there might be at least one outlier, which is true. The study time 28 hours is an outlier (see Fig. 83).

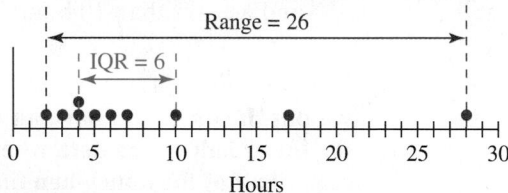

Figure 83 IQR and Range

  4. The range (26 hours) is relatively large due to the outlier 28 hours. The IQR (6 hours) is relatively small, partly because it is unaffected by the outlier 28 hours. So, the range is sensitive to the outlier and the IQR is resistant to it.

In general, **the IQR is resistant to outliers and the range is sensitive to outliers.**

Boxplots That Describe Distributions with No Outliers

A **boxplot** is a diagram that allows us to visualize the IQR and the range. We begin by considering boxplots that describe distributions with no outliers. The author surveyed the students in one of his statistics classes about the number of minutes they exercise per day. A boxplot of the exercise distribution is shown in Fig. 84. It consists of a line segment (called a **whisker**) from the smallest observation (0) to $Q_1 = 15$, a box from $Q_1 = 15$ to $Q_3 = 90$, and another whisker from $Q_3 = 90$ to the largest observation (150). There is also a vertical line segment in the box at $Q_2 = 48$ (the median).

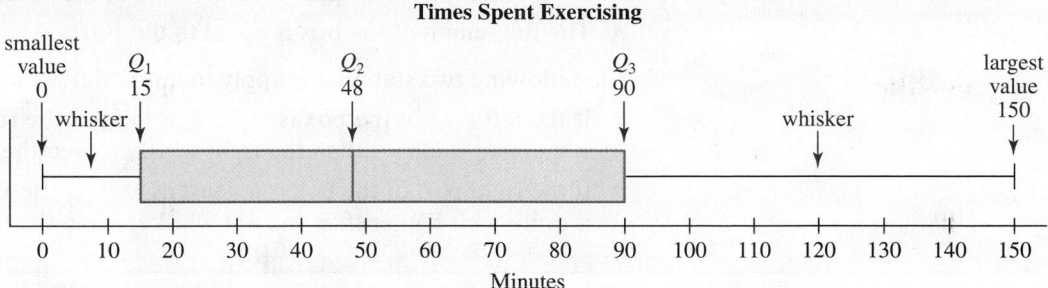

Figure 84 Boxplot of the exercise distribution

The boxplot consists of four parts: the left whisker, the left part of the box, the right part of the box, and the right whisker. Each part represents 25% of the observations (see Fig. 85). The full length of the box is equal to the IQR (75).

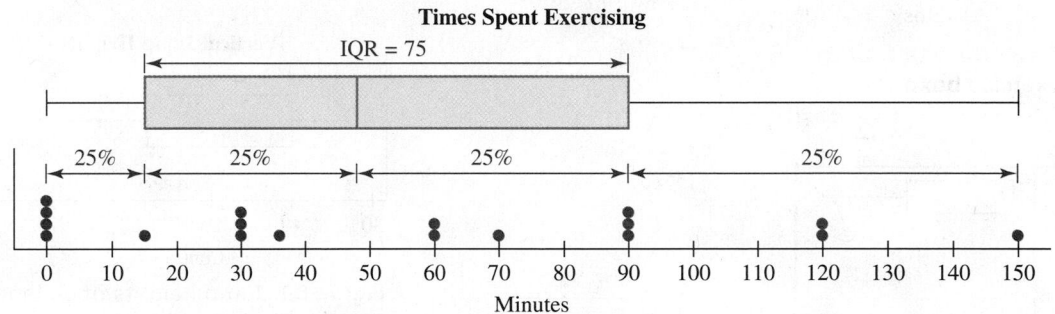

Figure 85 The IQR and percentages for the boxplot of the exercise distribution

WARNING   If one whisker is longer than the other, that does *not* mean more observations are represented by the longer whisker. It means that the observations represented by the longer whisker are more spread out than the approximately equal number of observations represented by the shorter whisker. The blue dots in Fig. 85 illustrate this.

From the histogram in Fig. 86, we see the exercise distribution is skewed right. The boxplot in Fig. 85 confirms this, because the right part of the box is a bit longer than the left part and the right whisker is much longer than the left whisker.

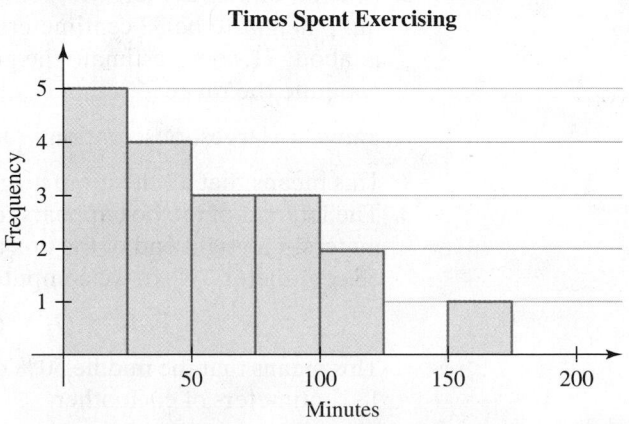

Figure 86 Histogram for exercise durations

> **Meaning of Boxplots that Describe Distributions without Outliers**

- A boxplot that describes a distribution without outliers consists of four parts: the left whisker, the left part of the box, the right part of the box, and the right whisker. Each part represents approximately 25% of the observations.
- The full length of the box is equal to the IQR.

The following two statements apply to unimodal distributions:

- If the left part of the box is at least as long as the right part and the left whisker is quite a bit longer than the right whisker, then the distribution is skewed left.
- If the right part of the box is at least as long as the left part and the right whisker is quite a bit longer than the left whisker, then the distribution is skewed right.

We can use StatCrunch to construct a boxplot of the exercise distribution (see Fig. 87). A boxplot constructed with a TI-84 is shown in Fig. 88. For StatCrunch and TI-84 instructions, see Appendices B.12 and A.7, respectively.

Times Spent Exercising

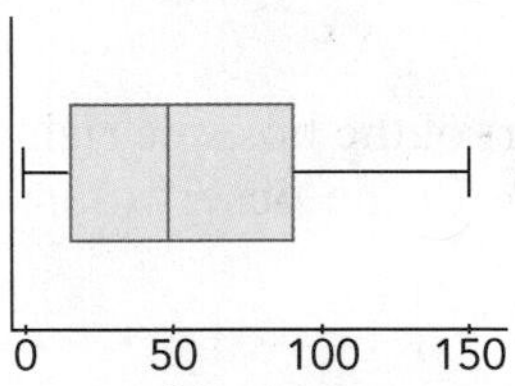

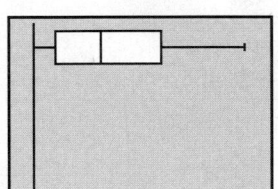

Figure 87 StatCrunch exercise boxplot

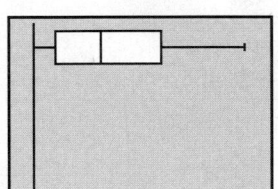

Figure 88 TI-84 exercise boxplot

▷ **Example 3**   Analyzing a Boxplot

The jump heights (in centimeters) of 40 college soccer players are described by the boxplot in Fig. 89.

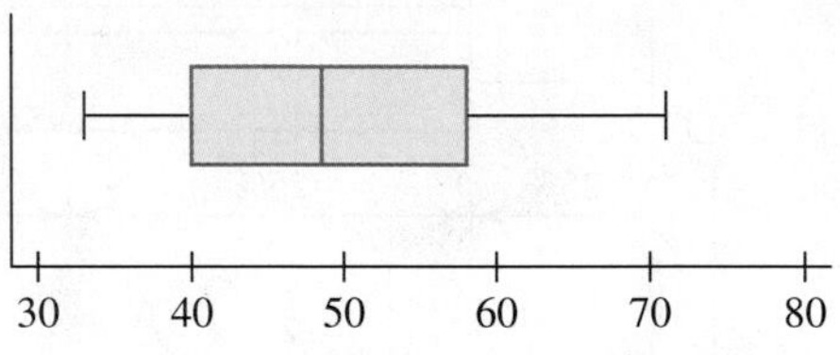

Vertical Jump Heights

Figure 89 Jump heights of college soccer players
(**Source:** Triple-Hop Distance as a Valid Predictor of Lower Limb Strength and Power, *R. T. Hamilton et al.*

1. Is the distribution of jump heights symmetric, skewed left, or skewed right?
2. Estimate the range of jump heights. What does it mean in this situation?
3. Estimate the IQR. What does it mean in this situation?

Solution

1. Because the right side of the box is a bit longer than the left side and the right whisker is longer than the left whisker, we conclude the distribution is skewed right.
2. The left end of the left whisker appears to be at about 33, so we estimate the least jump height to be 33 centimeters. The right end of the right whisker appears to be at about 71, so we estimate the greatest jump height to be 71 centimeters. Next, we compute the range:

   range = largest observation − smallest observation = 71 − 33 = 38 centimeters

   This means that all the jump heights are within 38 centimeters of each other.
3. The left end of the box appears to be at about 40, so we estimate that Q_1 is 40 centimeters. The right end of the box appears to be at about 58, so we estimate that Q_3 is 58 centimeters. Next, we compute the IQR:

$$IQR = Q_3 - Q_1 = 58 - 40 = 18 \text{ centimeters}$$

   This means that the middle 50% of the jump heights (between Q_1 and Q_3) are within 18 centimeters of each other.

Outliers

So far, we have been vague about the definition of an outlier. We will soon give a precise definition. First, we need to define the fences of a distribution.

▶ **Definition   Fences**

The **left fence** and the **right fence** of some data values are given by

- left fence $= Q_1 - 1.5 \, \text{IQR}$
- right fence $= Q_3 + 1.5 \, \text{IQR}$

For the study distribution, we found that $Q_1 = 4$ hours, $Q_3 = 10$ hours, and IQR $= 6$. Here we find the fences:

- left fence $= Q_1 - 1.5 \, \text{IQR} = 4 - 1.5(6) = 4 - 9 = -5$
- right fence $= Q_3 + 1.5 \, \text{IQR} = 10 + 1.5(6) = 10 + 9 = 19$

We indicate the fences with red dashed lines in Fig. 90.

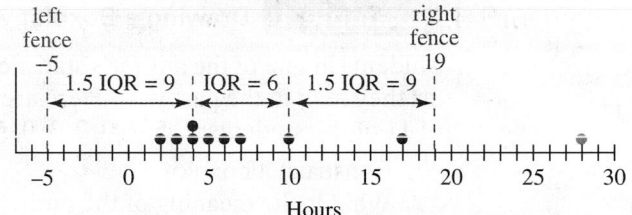

Figure 90 The fences are at -5 and 19.

Now that we have defined the fences, we can define an outlier more precisely.

▶ **Definition   Outlier**

An **outlier** is a data value that is less than the left fence or greater than the right fence. We say that the outlier lies *outside* the fences.

It makes sense that an outlier would lie outside (rather than inside) the fences because an outlier should be far from the center.

For the study distribution, we identify 28 hours as an outlier because it is greater than the right fence 19 (see the green dot in Fig. 90). There are no other outliers, because all the other study times are between the left fence, -5 hours, and the right fence, 19 hours.

Boxplots That Describe Distributions with Outliers

Now we will discuss how to construct a boxplot for a distribution with one or more outliers. A boxplot of the study distribution is shown in Fig. 91. The box is constructed in the usual way. But the right whisker runs from Q_3 to the largest observation that is *not* an outlier. Similarly, the left whisker runs from Q_1 to the smallest observation that is *not* an outlier, but because there are no small outliers, the left whisker extends to the smallest observation 2. We plot dots to represent outliers, so a (red) dot is plotted at 28.

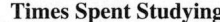

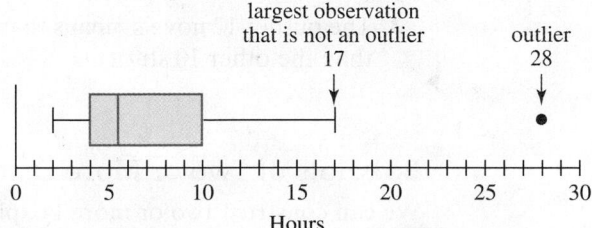

Figure 91 A boxplot with one outlier at 28 hours

We can use StatCrunch to construct a boxplot that displays the outlier of the study distribution (see Fig. 92). A similar boxplot drawn with a TI-84 is shown in Fig. 93. For StatCrunch and TI-84 instructions, see Appendices B.12 and A.7, respectively.

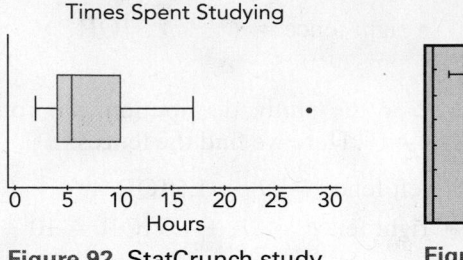

Times Spent Studying

Figure 92 StatCrunch study boxplot with outlier

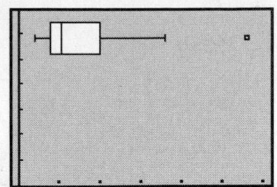

Figure 93 TI-84 study boxplot with outlier

▶ **Example 4**   Drawing a Boxplot with an Outlier

Students in one of the author's statistics classes were surveyed about the number of novels they read in the past year. Here are the anonymous responses (in numbers of novels) of 11 of the students: 2, 5, 2, 0, 2, 3, 0, 5, 6, 4, and 12.

1. Construct a boxplot.
2. What is the meaning of the outlier?

Solution

1. First, we list the observations from smallest to largest:

$$0, 0, 2, 2, 2, 3, 4, 5, 5, 6, 12$$

Next, we can find that $Q_1 = 2$, $Q_2 = 3$, and $Q_3 = 5$ (try it). So, IQR $= Q_3 - Q_1 = 5 - 2 = 3$. Then we find the fences:

- left fence $= Q_1 - 1.5\,\text{IQR} = 2 - 1.5(3) = -2.5$
- right fence $= Q_3 + 1.5\,\text{IQR} = 5 + 1.5(3) = 9.5$

All the observations are between -2.5 and 9.5, except the observation 12 novels, which is an outlier.

   We take the usual steps to construct the boxplot, except that the right end of the right whisker stops at the second-largest observation, 6 (see Fig. 94). We draw a dot at 12 to indicate the outlier.

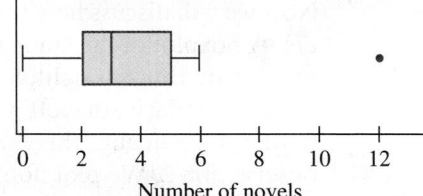

Number of Novels Read in Past Year

Figure 94 Boxplot of novels distribution

2. The outlier 12 novels means that the student read many more novels in the past year than the other 10 students.

Boxplots of Two or More Distributions

We can construct two or more boxplots that share a horizontal axis to help us compare two or more distributions.

▶ Example 5   Comparing Boxplots

The distributions of ages (in years) of all 835 Major League Baseball players and all 2282 National Football League players in 2014 are described by the boxplots in Fig. 95.

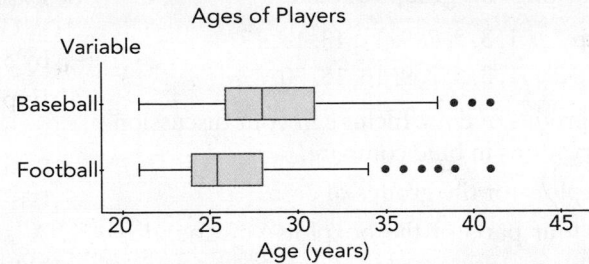

Figure 95 Boxplots of ages of professional baseball and football players

(**Source:** *Major League Baseball Players Association; National Football League Players Association*)

1. Compare the centers of the two distributions. What does this mean in this situation?
2. Estimate Q_1 and Q_3 for both of the distributions. What do they mean in this situation?
3. A student estimates that the ages of the youngest baseball player(s) and the youngest football player(s) are both 21 years. Also, he estimates that the ages of the oldest baseball player(s) and the oldest football player(s) are both 41 years. He concludes that baseball players tend to be the same age as football players. What would you tell the student?
4. The student discussed in Problem 3 computes that the ranges are $41 - 21 = 20$ years for both baseball and football players. He concludes that the spread of ages is the same for baseball and football players. What would you tell the student?

Solution

1. Because both distributions have outliers, it is better to measure the centers with the medians, which are resistant to outliers. The median is larger for baseball (28 years) than for football (26 years). This suggests that baseball players tend to be older than football players. This makes sense, because football is more physically demanding than baseball.
2. The value of Q_1 is larger for baseball (26 years) than for football (24 years). The value of Q_3 is larger for baseball (31 years) than for football (28 years). These values suggest that baseball players tend to be older than football players, which matches our conclusion in Problem 1.
3. Although the student's comparisons appear to be correct, there is more to the story than just comparing the ages of the youngest players and comparing the ages of the oldest players. Our comparisons of the quartiles in Problems 1 and 2 give a more complete picture, suggesting that baseball players tend to be older than football players.
4. Although the student's calculations of the ranges are correct, the ranges take into account only the ages of the youngest and oldest players. Each of the four main parts of the baseball boxplot (two whiskers and two parts of the box) is longer than each of the respective parts of the football boxplot. Not counting the outliers, this suggests that the remaining baseball players' ages are more spread out than the remaining football players' ages.

We can use StatCrunch to construct two or more boxplots that share a horizontal axis. In fact, the boxplots shown in Fig. 95 were constructed with StatCrunch. See Appendix B.12 for StatCrunch instructions.

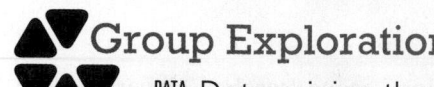

Group Exploration

DATA Determining the meaning of a boxplot

Consider the following two groups of data:

Group 1   1, 3, 5, 7, 9, 11, 13, 15, 17
Group 2   1, 3, 5, 7, 9, 16, 18, 20, 22

1. Compare the groups of data. Include in your discussion how the observations in blue compare.

2. Construct boxplots for the groups of data.

3. Compare the four parts of the boxplots you found in Problem 2.

4. Explain why adding 5 to each of the blue observations in Group 1 did not affect the values of the minimum observation, Q_1, and the median.

5. Explain why adding 5 to each of the blue observations in Group 1 increased both Q_3 and the maximum value by 5.

6. Explain why adding 5 to each of the blue observations in Group 1 increased the length of the right part of the box but did not affect the length of the right whisker.

7. A student says that there are fewer observations between the median and Q_3 for Group 1 than for Group 2, because the left part of the box is shorter for Group 1 than for Group 2. What would you tell the student?

▶ **Tips for Success**   **Relax during a Test**

If you get flustered during a test, close your eyes, take a couple of deep breaths, and think about something pleasant or nothing at all for a moment. This short break from the test might give you some perspective and help you relax.

Homework 4.3

For extra help ▶ **MyMathLab®**   Watch the videos in MyMathLab   Download the MyDashboard App

1. The third quartile is the ____ percentile.

2. The ____ ____ measures the spread of the middle 50% of the observations.

3. The left whisker represents the observations up to the ____ percentile.

4. *True or False:* The IQR is sensitive to outliers.

For Exercises 5–8, match the given histogram to the appropriate boxplot in Fig. 100.

5. See Fig. 96

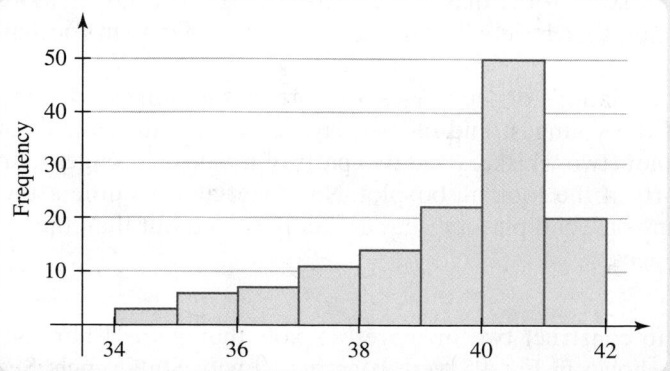

Figure 96 Exercise 5

6. See Fig. 97

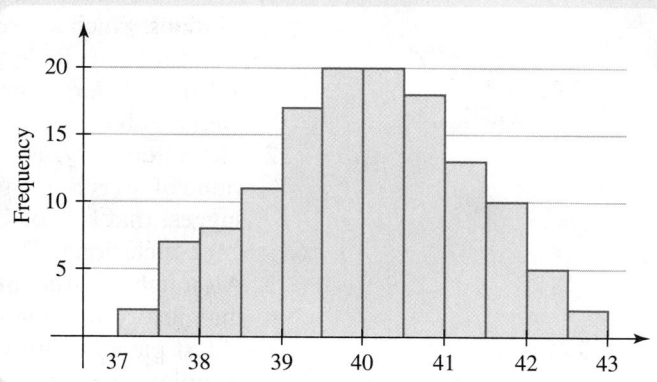

Figure 97 Exercise 6

7. See Fig. 98

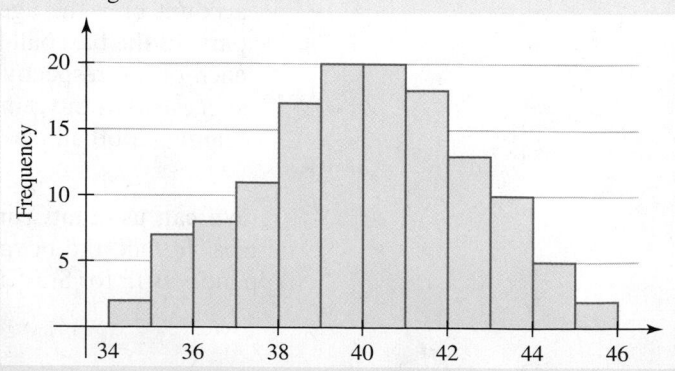

Figure 98 Exercise 7

8. See Fig. 99

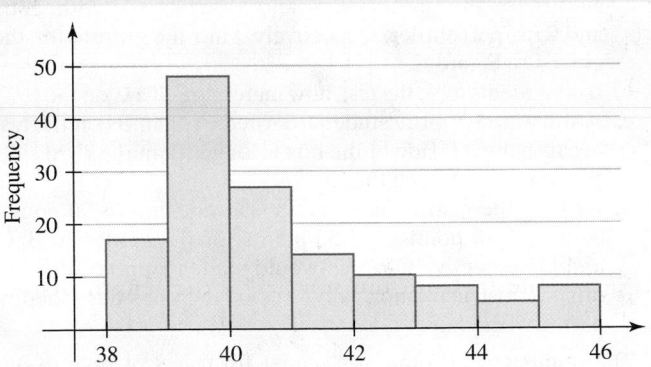

Figure 99 Exercise 8

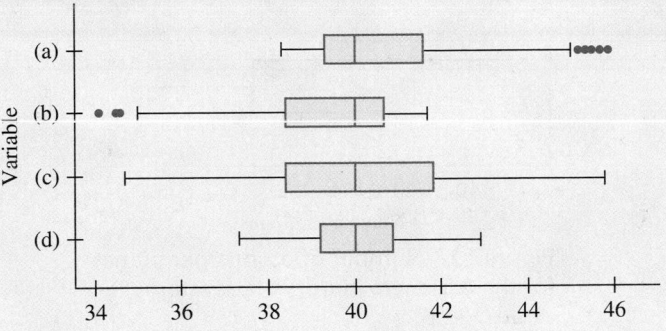

Figure 100 Exercises 5–8

9. The 2013 mean personal incomes (in thousands of dollars) of the 50 states are described by the boxplot in Fig. 101.

Mean Personal Incomes

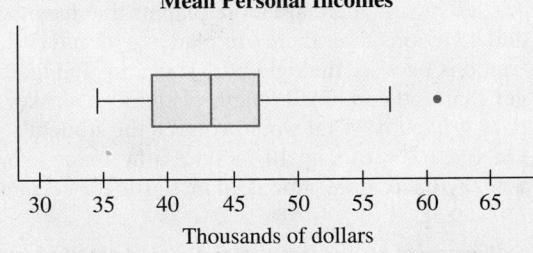

Thousands of dollars

Figure 101 Boxplot of mean incomes of the 50 states
(**Source:** *Bureau of Economic Analysis*)

a. Estimate the mean personal income of Connecticut, which has the largest mean personal income of all 50 states.
b. Estimate the percentage of states that have mean personal incomes less than $39 thousand.
c. Estimate the percentage of states that have mean personal incomes greater than or equal to $39 thousand.
d. Add the results you found in parts (b) and (c). Explain why the result makes sense.
e. Estimate the *number* of states that have mean personal incomes between $39 thousand and $47 thousand.

10. The mean prices (in dollars) of regular gasoline in the 50 states on April 27, 2014, are described by the boxplot in Fig. 102.

Mean Prices of Regular Gasoline

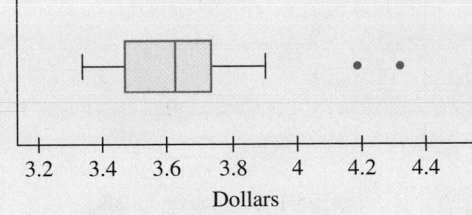

Dollars

Figure 102 Mean prices of regular gasoline in the 50 states
(**Source:** *GasBuddy.com*)

a. Estimate the mean price of regular gasoline in Hawaii, which has the highest mean price of the 50 states.
b. What percentage of states have a mean regular gasoline price that is at least $3.76?
c. What percentage of states have a mean regular gasoline price that is less than $3.76?
d. Add the results you found in parts (b) and (c). Explain why the sum should be 100%.
e. Estimate the *number* of states that have mean regular gasoline prices between $3.48 and $3.76.

11. At the website Lending Club®, customers can either borrow money or fund money for others' loans. When someone applies for a loan, Lending Club grades the potential loan from A to G, where A means the potential loan has very low risk and G means it has very high risk. A loan with grade A will have a much lower interest rate than a loan with grade G. The interest rates of loans funded at Lending Club in September 2014 are described by the boxplot in Fig. 103.

Interest Rates for Loans

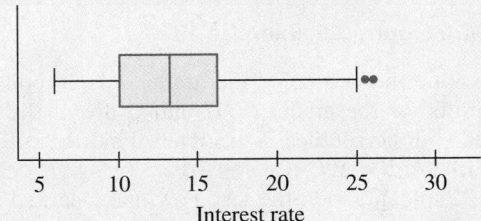

Interest rate

Figure 103 Exercise 11
(**Source:** *Lending Club*)

a. Given that the distribution is unimodal, determine whether it is skewed left, symmetric, or skewed right. What does this mean in this situation?
b. Estimate the 75th percentile. What does it mean in this situation?
c. Estimate the percentile of a 10.2% interest rate. What does it mean in this situation?
d. There are 186 outliers. Estimate the largest one.
e. C loans and D loans have interest rates between 12.29% and 17.86%. Is the percentage of C loans together with D loans probably less than, equal to, or greater than 25%?

12. The amounts (in thousands of dollars) of loans funded at Lending Club in September 2014 are described by the boxplot in Fig. 104.

Lending Club Loan Amounts

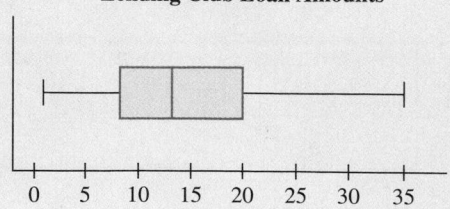

Figure 104 Exercise 12
(**Source:** *Lending Club*)

a. Given that the distribution is unimodal, determine whether it is skewed left, symmetric, or skewed right. What does this mean in this situation?
b. Estimate the 75th percentile of the loan amounts. What does it mean in this situation?
c. Estimate the percentile of a $8.3 thousand loan. What does it mean in this situation?
d. Estimate the largest loan.
e. Estimate the IQR. What does it mean in this situation?

13. The scores (in points) on a test are described by the boxplot in Fig. 105.

Test Scores

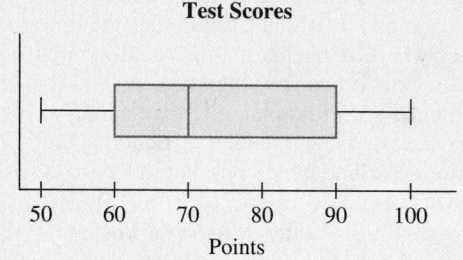

Figure 105 Exercise 13

a. Assume the cutoff for the grade D is 50 points and the cutoffs for the grades A, B, and C are at the 75th, 50th, and 25th percentiles, respectively. Find the cutoffs for the grades A, B, and C.
b. If 32 students took the test, how many earned a B on the test?
c. A student says more students earned Bs than Cs on the test because the right side of the box is longer than the left side. What would you tell the student?
d. Assume the instructor normally uses the cutoffs 90 points, 80 points, 70 points, and 50 points for the grades A, B, C, and D, respectively. Which would students prefer, the cutoffs the instructor normally uses or the ones described in part (a)? Explain.

14. The scores (in points) on a test are described by the boxplot in Fig. 106.

Test Scores

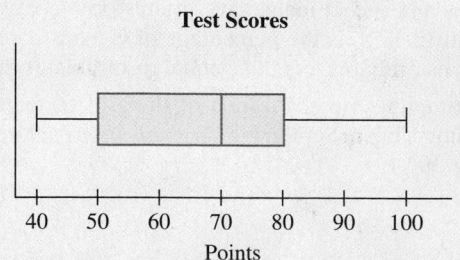

Figure 106 Exercise 14

a. Assume the cutoff for the grade D is 40 points and the cutoffs for the grades A, B, and C are at the 75th, 50th, and 25th percentiles, respectively. Find the cutoffs for the grades A, B, and C.
b. If 40 students took the test, how many earned a C on the test?
c. A student says more students earned Cs than Bs on the test because the left side of the box is longer than the right side. What would you tell the student?
d. Assume the instructor normally uses the cutoffs 90 points, 80 points, 70 points, and 50 points for the grades A, B, C, and D, respectively. Which would students prefer, the cutoffs the instructor normally uses or the ones described in part (a)? Explain.

15. The numbers of moons per planet for the 8 planets in our solar system are described by the boxplot in Fig. 107.

Number of Moons per Planet

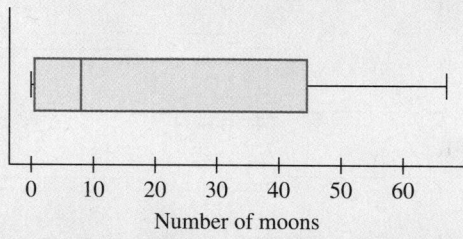

Figure 107 Number of moons per planet for the 8 planets in our solar system
(**Source:** *NASA*)

a. Given that the distribution is unimodal, is it skewed left, skewed right, or symmetric? Which is a better measure of the center, the mean or the median? Explain.
b. Estimate the median. What does it mean in this situation?
c. Estimate Q_3. What does it mean in this situation?
d. A student says there are more planets that have between 8 and 45 moons than there are planets that have fewer than 8 moons, because the right part of the box in Fig. 107 is longer than the combined length of the left whisker and left part of the box. What would you tell the student?
e. On the basis of Fig. 107 alone, how many planets can you say have more moons than Earth for certain? [**Hint:** $Q_1 = 0.5$]

16. The *diameter* of a planet is the distance through the center of the planet from one point on the equator to the opposite side. The diameters (in thousands of kilometers) of the 8 planets in our solar system are described by the boxplot in Fig. 108.

Diameters

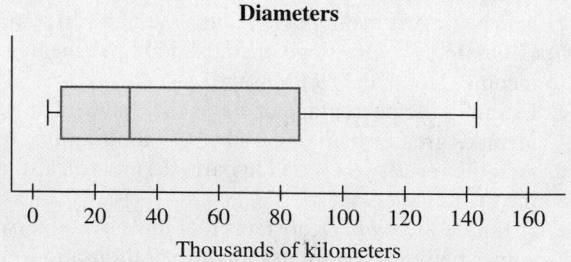

Figure 108 Diameters of the 8 planets in our solar system
(**Source:** *NASA*)

a. Given that the distribution is unimodal, is it skewed left, skewed right, or symmetric? Which is a better measure of the center, the mean or the median? Explain.

b. Estimate the median diameter. What does it mean in this situation?

c. Estimate Q_1. What does it mean in this situation?

d. A student says there are more planets that have diameters between 31 thousand kilometers and 86 thousand kilometers than there are planets that have diameters less than 31 thousand kilometers, because the right part of the box in Fig. 108 is longer than the combined length of the left whisker and left part of the box. What would you tell the student?

e. The diameter of Earth is approximately 13 thousand kilometers. On the basis of Fig. 108 alone, how many planets can you say have diameters larger than Earth's for certain?

17. Find some ages (in years) of people that can be described by the boxplot shown in Fig. 109.

Ages

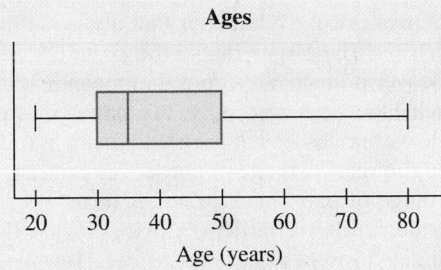

Figure 109 Exercise 17

18. Find some incomes (in thousands of dollars) that can be described by the boxplot shown in Fig. 110.

Incomes

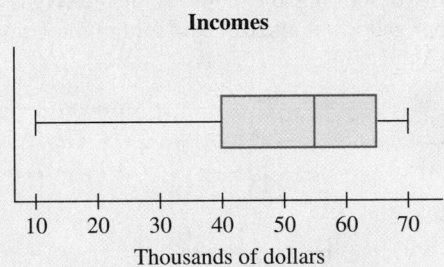

Figure 110 Exercise 18

19. **DATA** The author surveyed 11 of the students in one of his statistics classes. Here are the one-way distances (in miles) they drove to college (Source: *J. Lehmann*):

10	6	7	6	5	13
20	3	3	7	30	

a. Find the smallest observation, the quartiles, and the largest observation.

b. Find any outliers.

c. Construct a boxplot.

d. Given that the distribution is unimodal, is it skewed left, skewed right, or symmetric? Explain.

e. Which is a better measure of the center, the mean or the median? Explain.

20. **DATA** In one of the author's statistics classes, students were asked how many people live in their households, including themselves. Here are the household sizes (in number of people) for 12 of those students (Source: *J. Lehmann*):

6	3	4	3	7	4
3	4	3	2	10	3

a. Find the smallest observation, the quartiles, and the largest observation.

b. Find any outliers.

c. Construct a boxplot.

d. Given that the distribution is unimodal, is it skewed left, skewed right, or symmetric? Explain.

e. Which is a better measure of the center, the mean or the median? Explain.

21. **DATA** Here are the durations (in seconds) of wood roller coaster rides in Ohio (Source: *Roller Coaster DataBase*):

63	90	90	93	105
120	140	193	250	

a. Construct a boxplot.

b. What is Q_3? What does it mean in this situation?

c. What is the IQR? What does it mean in this situation?

d. Are there any outliers? If yes, find them.

e. If the result you found in part (b) were used to estimate the third quartile of both wood and steel roller coasters in Ohio, what type of bias would that be?

22. **DATA** Here are the 2013 tuitions (dollars) of 10 randomly selected 2-year, public colleges (Source: *U.S. Department of Education*):

2960	2904	1111	2760	1704
2640	3000	3780	3342	4511

a. Construct a boxplot.

b. Given that the distribution is unimodal, is it skewed left, skewed right, or symmetric? Explain.

c. Which would be a better measure of the center, the mean or the median? Explain.

d. What is the median? What does it mean in this situation?

e. What is the IQR? What does it mean in this situation?

23. **DATA** The following data are the top 50 women's times (in minutes) in the 2013 ING New York City Marathon (Source: *ING New York City Marathon*).

161	163	179	179	180	146	149
176	149	150	177	177	179	148
177	177	160	158	177	172	160
166	167	170	171	150	172	173
173	174	175	175	161	164	176
152	155	155	176	177	161	176
178	180	145	178	148	148	149
181						

a. Construct a boxplot.

b. What is the median? What does it mean in this situation?

c. What is the IQR? What does it mean in this situation?

d. Which has greater range, the running times in the top 25% of the sample or the running times in the bottom 25% of the sample? Explain.

e. A student says there are the same number of observations in the class 160–169 minutes as there are in the class 170–179 minutes, because each class has width 10 minutes. What would you tell the student? Refer to the boxplot in your response.

24. ⬇️**DATA** The following data are the top 50 men's times (in minutes) in the 2013 ING New York City Marathon (Source: *ING New York City Marathon*).

148	128	133	129	130	130	150
131	155	132	152	155	142	152
153	132	137	143	145	131	155
152	133	150	144	144	138	143
147	152	153	146	143	154	152
148	131	150	150	132	146	144
155	149	131	145	151	154	143
142						

a. Construct a boxplot.
b. What is the median? What does it mean in this situation?
c. What is the IQR? What does it mean in this situation?
d. Which has greater range, the times of the runners in the top 25% of the sample or the times of the runners in the bottom 25% of the sample? Explain.
e. A student says there are the same number of observations in the class 135–145 minutes as there are in the class 145–155 minutes, because each class has width 10 minutes. What would you tell the student? Refer to the boxplot in your response.

25. Some data with smallest observation 17 and largest observation 78 are described by the density histogram in Fig. 111. Construct a boxplot that describes the data.

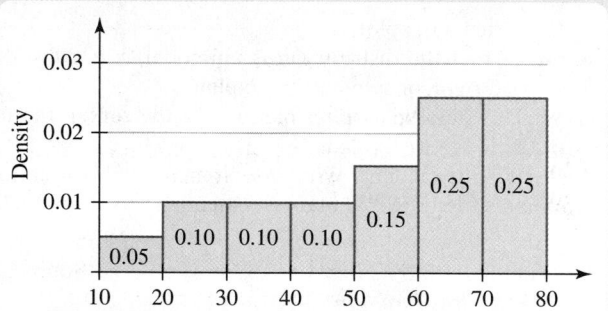

Figure 111 Exercise 25

26. Some data with smallest observation 7 and largest observation 34 are described by the density histogram in Fig. 112. Construct a boxplot that describes the data.

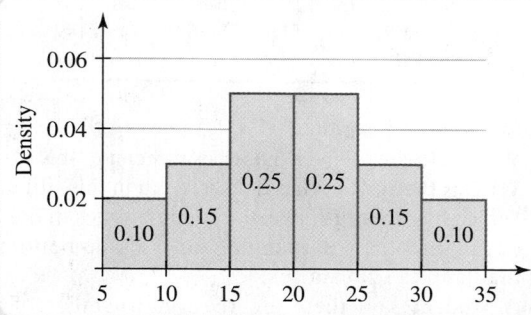

Figure 112 Exercise 26

27. Amazon.com's prices (in dollars) of new hardcover and paperback elementary statistics textbooks on May 18, 2014, are described by the boxplots in Fig. 113.

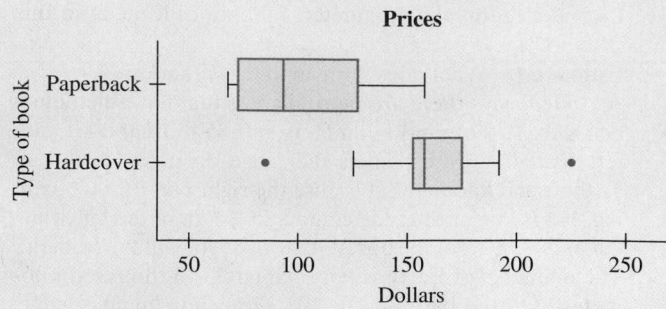

Figure 113 Prices of new hardcover and paperback elementary statistics textbooks
(**Source:** *Amazon.com*)

a. Which has a larger median, the paperback prices or the hardcover prices? What does that mean in this situation?
b. Which has the larger range, the paperback prices or the hardcover prices? What does that mean in this situation?
c. Which has the larger IQR, the paperback prices or the hardcover prices? What does that mean in this situation?
d. Even though the range of hardcover prices is greater than the range of paperback prices, it turns out that the standard deviation of hardcover prices is less than the standard deviation of paperback prices. How is this possible?
e. From inspecting the boxplots, a student concludes that the range of hardcover prices is greater than the range of paperback prices for *all* elementary statistics textbooks. What would you tell the student?

28. Boxplots describing the highway and city gas mileages (in miles per gallon) of all 2014 cars sold in the United States are shown in Fig. 114.

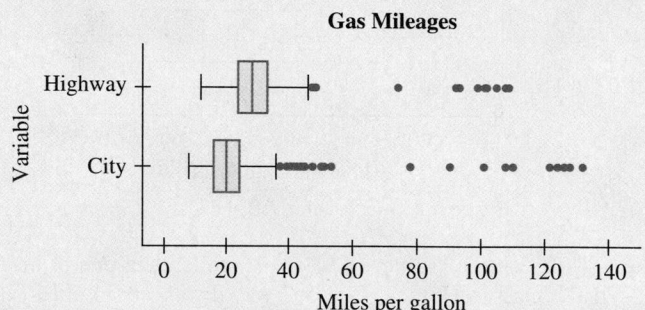

Figure 114 Gas mileages of all 2014 cars
(**Source:** *U.S. Department of Energy*)

a. Which has the larger median, the highway distribution or the city distribution? What does that mean in this situation?
b. Which has the larger Q_3, the highway distribution or the city distribution? What does that mean in this situation?
c. The Honda Accord Plug-in Hybrid has the best city gas mileage. Estimate its gas mileage. On the basis of your responses to parts (a) and (b), explain why it is surprising that the car's city gas mileage is better than every car's highway gas mileage including its own (105 miles per gallon).
d. Even though the IQRs are approximately equal for the two distributions, the standard deviation of the city distribution (11.7 miles per gallon) is larger than the standard deviation of the highway distribution (9.9 miles per gallon). How is this possible?

e. Most or all of the outliers are gas mileages of green cars such as hybrid cars. Explain why it might be better to analyze green cars separately from other cars.

29. In 2011, Forbes surveyed 16,000 alumni (class of 2006) at more than 100 2-year business schools. Of the 30% who responded, Forbes computed the alumni's returns on investment from attending the schools. Fig. 115 displays boxplots of the annual tuitions (in thousands of dollars) of the 74 schools with the best returns on investment, the median pre-MBA annual salaries for those schools' alumni, and the median post-MBA annual salaries.

Tuitions

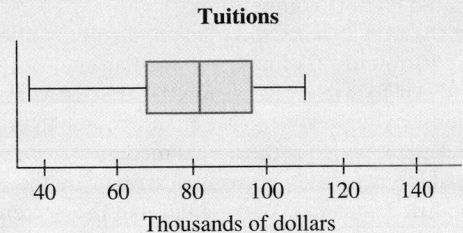

Pre-MBA Salaries

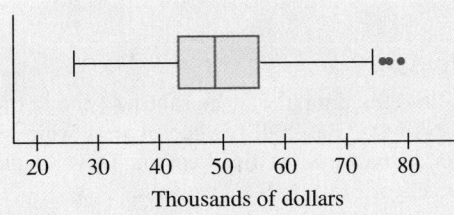

Post-MBA Salaries

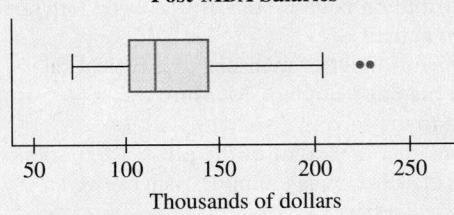

Figure 115 Class of 2006 business-school tuitions, pre-MBA salaries, and post-MBA salaries
(**Source:** *Forbes*)

a. Given that the distribution of post-MBA salaries is unimodal, is it skewed left, skewed right, or symmetric? Which is a better measure of the center, the mean or the median? Explain.

b. Estimate the median pre-MBA salary and the median post-MBA salary. What do they mean in this situation?

c. Find the difference of the median post-MBA salary and the median pre-MBA salary. What does it mean in this situation?

d. Estimate the median tuition.

e. Estimate a student's *break-even point*. That is, counting from the first day a person attends business school, estimate how long it will take the person to make up for paying two years of tuition plus not working for two years. Assume the person pays the full tuition and that salaries do not increase.

30. In 2013, Forbes surveyed 17,000 alumni (class of 2008) at more than 100 2-year business schools. Of the 27% who responded, Forbes computed the alumni's returns on investment from

attending the schools. Fig. 116 displays boxplots of the annual tuitions (in thousands of dollars) of the 70 schools with the best returns on investment, the median pre-MBA annual salaries for those schools' alumni, and the median post-MBA annual salaries.

Tuitions

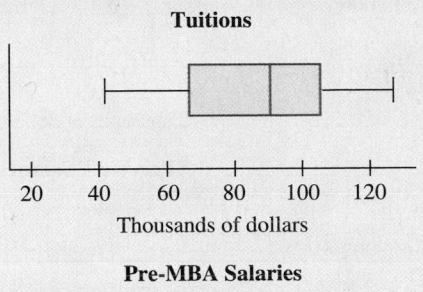

Pre-MBA Salaries

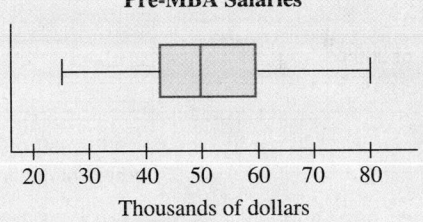

Post-MBA Salaries

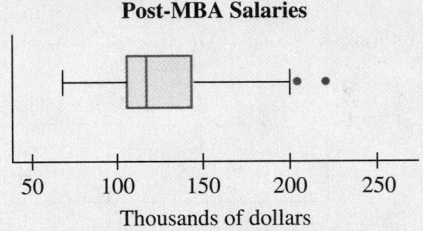

Figure 116 Class of 2008 business-school tuitions, pre-MBA salaries, and post-MBA salaries
(**Source:** *Forbes*)

a. Given that the distribution of post-MBA salaries is unimodal, is it skewed left, skewed right, or symmetric? Which is a better measure of the center, the mean or the median? Explain.

b. Estimate the median pre-MBA salary and the median post-MBA salary. What do they mean in this situation?

c. Find the difference of the median post-MBA salary and the median pre-MBA salary. What does it mean in this situation?

d. Estimate the median tuition.

e. Estimate a student's *break-even point*. That is, counting from the first day a person attends business school, estimate how long it will take the person to make up for paying two years of tuition plus not working for two years. Assume the person pays the full tuition and that salaries do not increase.

31. In Exercises 29 and 30, you analyzed the tuitions, the pre-MBA salaries, and the post-MBA salaries of the classes of 2006 and 2008, respectively, of the business schools with the best returns on investment. Boxplots comparing these observations (in thousands of dollars) for the two classes are shown in Fig. 117.

Tuitions

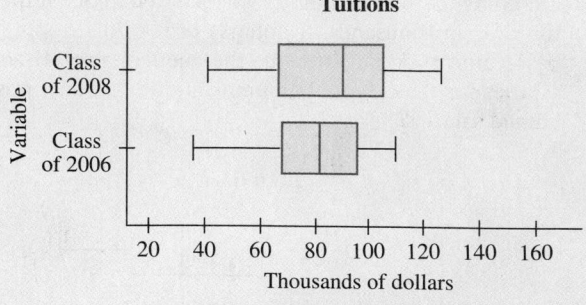

Pre-MBA Salaries

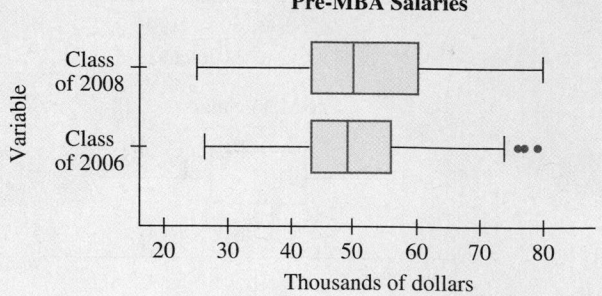

Post-MBA Salaries

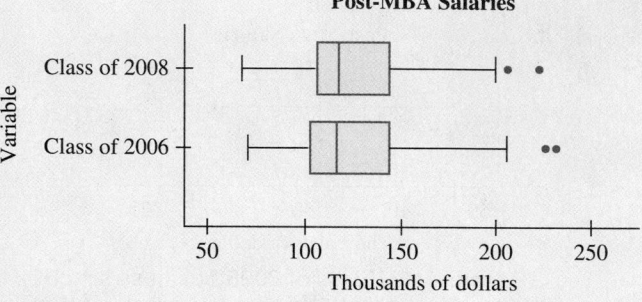

Figure 117 Classes of 2006 and 2008 business-school tuitions, pre-MBA salaries, and post-MBA salaries
(**Source:** *Forbes*)

a. In comparing the class of 2006 to the class of 2008, describe any changes in the median tuition, median pre-MBA salary, and median post-MBA salary.

b. In part (e) of both Exercises 29 and 30, you estimated the students' break-even points for the classes of 2006 and 2008, respectively. Refer to part (a) to help explain why the break-even point was larger for the class of 2008 than for the class of 2006.

32. Muhammad Ali is considered by many to be one of the greatest heavyweight boxers in history. Boxplots of the percentages of maximum allowed rounds it took Ali to win his first 28 wins and his last 28 wins are shown in Fig. 118.

Percentages of Maximum Allowed Rounds

Figure 118 Percentages of maximum allowed rounds it took Ali to win boxing matches
(**Source:** *Muhammad Ali Enterprises LLC*)

a. The boxplot for the last 28 wins does not have a right whisker. How is this possible, assuming the boxplot has been drawn correctly?

b. The boxplot for the last 28 wins does not include a vertical line (to indicate the median) inside the box. How is this possible, assuming the boxplot has been drawn correctly?

c. The median of the last-28-wins distribution turns out to be 100%. Is this median less than, greater than, or equal to the median for the first-28-wins distribution? Discuss how some or all of the following might explain why this occurred: Ali's age, Ali's physical condition, Ali's experience, and the level of competition.

d. For the first 28 wins, the median maximum allowed rounds was 10 rounds. Estimate the median number of rounds it took Ali to win a match for the first 28 wins. Determine whether your result is less than, greater than, or equal to 5.5 rounds, which is the actual median.

e. In addition to Ali's 56 wins described by the boxplots, Ali also lost 5 matches. When analyzing the percentages of the maximum allowed rounds it took for a match to be over, why should Ali's 5 losses be analyzed separately from his 56 wins?

Large Data Sets

33. **DATA** Access the data about the salaries (in millions of dollars) of the National Baseball League players, which are available at MyMathLab and at the Pearson Downloadable Student Resources for Math & Stats website.

a. Construct a boxplot of the average salaries. Given that the distribution is unimodal, is it skewed left, skewed right, or symmetric?

b. Which is a better measure of a typical salary, the mean or the median? Explain. Measure the center. Round your result to the second decimal place.

c. Construct a boxplot of the players' experience. Given that the distribution is unimodal, is it skewed left, skewed right, or symmetric?

d. Which is a better measure of a typical player's experience, the mean or the median? Explain. Measure the typical experience.

e. Use the results you found in parts (b) and (d) to estimate the total earnings so far of a player with a typical salary and a typical amount of experience. If such a player's performance has been improving over his career, is your estimate likely an underestimate or an overestimate? Explain.

34. **DATA** Access the data about the salaries (in millions of dollars) of the National Baseball League players, which are available at MyMathLab and at the Pearson Downloadable Student Resources for Math & Stats website.

a. Construct a pie chart of the positions of the players.

b. What proportion of the players are pitchers? The abbreviations SP, RP, and CL stand for starting pitchers, relief pitchers, and closers (a pitcher who specializes in finishing a game), respectively.

c. For each position, compute the median salary. Which position has the largest median salary?

d. Compute the median salaries and salary ranges for third basemen (3B) and the median salaries and salary ranges for catchers (C). If a player has typical skills for both positions, at which of the two positions would he tend to earn a larger salary?

e. If a player is a better third baseman and catcher than most players in those positions, at which of the two positions would he tend to earn a larger salary?

Concepts

35. If the median of some data is 50 and the IQR is 0, what are the values of Q_1 and Q_3?

36. If the IQR of some data is equal to the range and $Q_1 = 7$, what can you conclude about the smallest observation?

37. Forty students in a prestatistics class take a test. The IQR of the scores is 20 points and $Q_1 = 65$. If the instructor adds 5 points to everyone's score, what are the new values of IQR, Q_1, and Q_3?

38. On October 1, the share values (in dollars) of 20 different green stocks have IQR $= 43$ and $Q_3 = 60$. If each of the share values falls by $5, what is the new IQR, Q_1, and Q_3?

39. On March 1, a group of 10 people signed up for a weight-loss program. During the next two months, the lightest person lost 10 pounds, the heaviest person gained 5 pounds, and everyone else kept their same weight.

  a. Did the range of the weights decrease, increase, or stay the same over the two-month period? Explain.

  b. Did the IQR of the weights decrease, increase, or stay the same over the two-month period? Explain.

40. At 1 P.M., a meteorologist records the temperatures of seven locations. One hour later, the temperature of the coldest location is 1°F less, the temperature of the hottest location is 2°F more, and the temperatures of the other locations stayed the same.

  a. Did the range of the temperatures decrease, increase, or stay the same over the one-hour period? Explain.

  b. Did the IQR of the temperatures decrease, increase, or stay the same over the one-hour period? Explain.

41. A student says that if one whisker of a boxplot is twice as long as the other whisker, then the longer whisker represents twice as many observations as the shorter whisker. What would you tell the student?

42. The IQR of the salaries of employees at company A is larger than the IQR of the salaries of employees at company B. A student concludes that the standard deviation of the salaries at company A is larger than the standard deviation of the salaries at company B. What would you tell the student?

43. Some data have $Q_1 = 30$ and $Q_3 = 50$.

  a. Describe all possible values of the median.

  b. Describe all possible values of the range.

44. Some data have $Q_1 = 6$ and $Q_3 = 10$.

  a. A student says the mean must be between 6 and 10. Is the student correct? If yes, explain. If no, give an example of some data that show the student is incorrect.

b. A student says the range must be at least 4. Is the student correct? If yes, explain. If no, give an example of some data that show the student is incorrect.

45. The author surveyed the Facebook users in one of his statistics classes, asking how many Facebook friends they had. The observations ranged from 0 to 2400 friends, with $Q_1 = 200$, $Q_2 = 500$, and $Q_3 = 900$. A student constructed the boxplot shown in Fig. 119. Why is the boxplot incorrect? Construct a correct one.

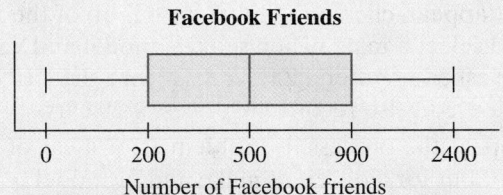

Figure 119 Exercise 45

(**Source:** *J. Lehmann*)

46. The following data are the savings (in thousands of dollars) of six adults: 3, 7, 8, 10, 13, and 15.

  a. Find the standard deviation of the savings.

  b. Find the IQR of the savings.

  c. Suppose that the adult who had $15 thousand in savings wins $700 thousand (after taxes) from the lottery. So, the adult's new savings is $715 thousand. Find the standard deviation of the six adults after the lottery win.

  d. Find the IQR of the savings of the six adults after the lottery win.

  e. Describe how much the outlier $715 affected the standard deviation and the IQR.

For Exercises 47–50, choose all the following diagrams and tables that would be an appropriate way to describe the given data: frequency and relative frequency table, relative frequency bar graph, multiple bar graph, pie chart, two-way table, dotplot, stemplot, time-series plot, relative frequency histogram, and boxplot. Do not include diagrams or tables that would be possible but awkward.

47. The majors of college students are recorded.

48. For a sample of 40 Olympic swimmers, a researcher measures how long (in seconds) each swimmer can hold his or her breath.

49. The lifespans (in years) of 750 Golden Retrievers are recorded.

50. A statistician compares the political parties of 1200 American adults and their ethnicities.

Hands-On Projects

Comparison Shopping of Cars Project

To begin, visit the website www.edmunds.com.

1. If a pop-up window appears, select a certain make, model, and year of a *used* car. If a pop-up window does not appear, click on "Make" at the top of the web page and select a make. Then select a model and year. If you are asked to enter a ZIP code, record the ZIP code that you used. Also record the prices of 20 cars.

2. Repeat the process in Problem 1 for a car of the same year but for a different make and/or model.

3. For each price distribution, construct a frequency histogram and describe the shape.

4. For each distribution, use the shape to help you determine which is the better measure of the center, the mean or the median. Explain.

5. Measure the centers of the two distributions and compare the results. What does your comparison tell you about the situation?

6. For each distribution, use the shape to help you determine which is the better measure of the spread, the standard deviation or the range. Explain.

7. Measure the spreads of the two distributions and compare the results. What does your comparison tell you about the situation?

8. When most people compare prices of two types of cars that are different makes and/or models, they probably compare typical prices but do not think about the spreads of the two price distributions. If two price distributions have typical prices that are approximately equal, explain why the spreads of the distributions could be an important consideration.

9. Are the spreads of the two price distributions you researched an important consideration? Explain.

10. If you were to shop for only one of the two types of cars you researched, which type would that be? Explain.

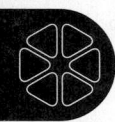

Chapter Summary

Key Points of Chapter 4

Section 4.1   Measures of Center

Median	The **median** of some data values is the 50th percentile.
Finding the median of a distribution	To find the median of some data values, first list the observations from smallest to largest. • If the number of observations is odd, then the median is the middle observation. • If the number of observations is even, then the median is the average of the two middle observations.
Summation notation	Let $x_1, x_2, x_3, \ldots, x_n$ be some data values. The **summation notation** Σx_i stands for the sum of the data values: $$\Sigma x_i = x_1 + x_2 + x_3 + \cdots + x_n$$
Mean	The **arithmetic mean** (or **mean**) of n data values $x_1, x_2, x_3, \ldots, x_n$ is given by $$\bar{x} = \frac{\Sigma x_i}{n}$$
Mean is a measure of the center	If a distribution is unimodal and approximately symmetric, the mean is a reasonable measure of the center. In this case, we say the mean is a typical value.
The effect of outliers on the mean and the median	• The mean is sensitive to outliers. • The median is resistant to outliers.

Section 4.1   Measures of Center (*Continued*)

How the shape of a distribution affects the mean and the median	• If a distribution is skewed left, the mean is usually less than the median and the median is usually the better measure of the center. • If a distribution is symmetric, the mean is approximately equal to the median and both are reasonable measures of the center. • If a distribution is skewed right, the mean is usually greater than the median and the median is usually the better measure of the center.
Mode	The **mode** of some data is an observation with the greatest frequency. There can be more than one mode, but if all the observations have frequency 1, then there is no mode.

Section 4.2   Measures of Spread

Range	The **range** of some data values is given by $$R = \text{largest observation} - \text{smallest observation}$$
Standard deviation	The **standard deviation** of n data values $x_1, x_2, x_3, \ldots, x_n$ is given by $$s = \sqrt{\dfrac{\sum (x_i - \bar{x})^2}{n-1}}$$
Empirical Rule	If a distribution is unimodal and symmetric, then the following statements are true. • Approximately 68% of the observations lie within one standard deviation of the mean. So, approximately 68% of the observations lie between $\bar{x} - s$ and $\bar{x} + s$. • Approximately 95% of the observations lie within two standard deviations of the mean. So, approximately 95% of the observations lie between $\bar{x} - 2s$ and $\bar{x} + 2s$. • Approximately 99.7% of the observations lie within three standard deviations of the mean. So, approximately 99.7% of the observations lie between $\bar{x} - 3s$ and $\bar{x} + 3s$.
Unusual observation	If an observation is more than two standard deviations away from the mean, we refer to the observation as unusual.
Summary of measuring center and spread	• If a distribution is unimodal and symmetric, then we usually use the mean to measure the center and the standard deviation to measure the spread. • If a distribution is skewed, then we usually use the median to measure the center and the range to measure the spread.
Variance	The **variance** is the square of the standard deviation.

Section 4.3   Boxplots

First quartile, second quartile, and third quartile	The *first quartile*, *second quartile*, and *third quartile* are the 25th, 50th, and 75th percentiles, respectively.
Quartile notation	We use the symbols Q_1, Q_2, and Q_3 to stand for the first, second, and third quartiles, respectively.
Finding Q_1 and Q_3 for an odd number of observations	If some data have an odd number of data values, do not include the median in the lower half of the data when finding Q_1. Likewise, do not include the median in the upper half of the data when finding Q_3.
Interquartile range	The **interquartile range (IQR)** is given by $$\text{IQR} = Q_3 - Q_1$$
Outliers' impact on the IQR and the range	The IQR is resistant to outliers, and the range is sensitive to outliers.

Section 4.3   Boxplots (*Continued*)

Meaning of boxplots that describe distributions without outliers	• A boxplot that describes a distribution without outliers consists of four parts: the left whisker, the left part of the box, the right part of the box, and the right whisker. Each part represents approximately 25% of the observations. • The full length of the box is equal to the IQR. The following two statements apply to unimodal distributions: • If the left part of the box is at least as long as the right part and the left whisker is quite a bit longer than the right whisker, then the distribution is skewed left. • If the right part of the box is at least as long as the left part and the right whisker is quite a bit longer than the left whisker, then the distribution is skewed right.
Fences	The **left fence** and the **right fence** of some data values are given by • left fence $= Q_1 - 1.5$ IQR • right fence $= Q_3 + 1.5$ IQR
Outlier	An **outlier** is a data value that is less than the left fence or greater than the right fence. We say that the outlier lies *outside* the fences.

Chapter 4 Review Exercises

1. An apple farmer randomly selects some apples from his orchard. The mean weight is 8 ounces. Use the appropriate symbol to represent the result.

2. Air pollution can be measured using the units *average particulate pollution* (*PM2.5*). Air pollution levels over 10 PM2.5 are considered hazardous to health by the World Health Organization (WHO). The names and the levels of pollution (in PM2.5) for the five worst U.S. cities are shown in Table 32.

Table 32 Air Pollution for the Five Worst U.S. Cities

City	Air Pollution (PM2.5)
Bakersfield, CA	18.2
Merced, CA	18.2
Fresno, CA	17.0
Hanford, CA	16.2
Los Angeles, CA	15.3

Source: *American Lung Association; WHO*

Let n be the number of cities listed in Table 32, and let x_i be the air pollution (in PM2.5) of the city listed in the ith row of the table.

a. Find the values of $x_1, x_2, x_3, x_4,$ and x_5.
b. Find Σx_i. What does it mean in this situation?
c. Find $\dfrac{\Sigma x_i}{n}$. What does it mean in this situation?

3. Find the mean, median, and mode of the data 57, 25, 80, 56, 25, 62, 40, and 93.

4. **DATA** The following data are the 2013 monthly fuel consumptions (in millions of gallons) by United Airlines (Source: *U.S. Department of Transportation*).

117	107	126	122	128	135
138	138	121	127	119	125

a. Find the mean.
b. Find the median.
c. Find the mode.
d. Is the mode a reasonable measure of the center? Explain. [**Hint:** If you are unsure, construct a dotplot.]
e. United Airlines' fuel consumption in December 2013 was 125 million gallons. Air travel is very intense near Christmas. Does that cause United Airlines' fuel consumption in December to be larger than a typical monthly fuel consumption? Explain.

5. The values (in billions of dollars) of the 73 worldwide startup companies worth at least $1 billion are described by the relative frequency histogram in Fig. 120. The values are rounded to the first decimal place.

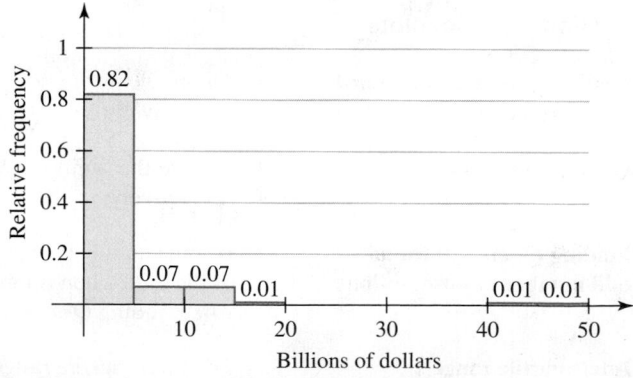

Figure 120 Exercise 5
(**Source:** *Dow Jones VentureSource and the Wall Street Journal*)

a. Describe the shape of the distribution.
b. Which is larger, the mean or the median? Which is the better measure of the center? Explain.
c. Which class contains the median?
d. Which is the better measure of the spread, the standard deviation or the range? Explain.

e. On the basis of the relative frequency histogram and the fact that the companies are worth at least $1 billion, what is the largest possible range? Round your result to the ones place.

6. The values (in billions of dollars) of the worldwide startup companies worth at least $1 billion are described by the relative frequency histogram in Fig. 120 on page 278. The values are rounded to the first decimal place.

   a. Estimate the percentile of the value of Dropbox, which is worth $10 billion.

   b. Palantir is at the 96th percentile. Estimate its value.

   c. On the basis of the relative frequency histogram, what is the largest possible value of Q_3?

   d. On the basis of the relative frequency histogram and the fact that the companies are worth at least $1 billion, what is the largest possible value of the IQR?

   e. The companies Xiaomi and Uber are worth $46 billion and $41.2 billion, respectively. Use the results you found in parts (c) and (d) to determine whether the values of the two companies are outliers.

7. The mean of a group of eight numbers is 13. If the smallest number is decreased by 2 and the largest number is increased by 2, will the mean decrease, increase, or stay the same? Explain.

8. The median of a group of six numbers is 85. If the smallest number is decreased by 10 and the largest number is increased by 20, will the median decrease, increase, or stay the same? Explain.

9. Find the mean, median, range, standard deviation, and variance of the data 48, 45, 34, 42, 55, 43, and 66.

The author surveyed students in one of his statistics classes about their genders, the number of units they were taking that semester, the number of hours they spent online per day, and the number of novels they read in the past year. The anonymous responses of six randomly selected students are shown in Table 33. Refer to this table for Exercises 10–12.

10. ᴰᴬᵀᴬ Compute the standard deviation of the number of units the students were taking that semester.

11. ᴰᴬᵀᴬ Compute the variance of the number of novels the *male* students read in the past year.

12. ᴰᴬᵀᴬ Compute the range of the number of hours spent online per day.

Table 33 Responses of Six Statistics Students

Gender	Units	Online	Novels
Male	17	4	1
Male	14	5	0
Female	11	2	3
Male	15	5	6
Male	15	4	1
Female	13	3	0

Source: *J. Lehmann*

For Exercises 13–16, match the given information to the appropriate histogram in Fig. 121.

13. mean: 70, standard deviation: 12.4

14. mean: 70, standard deviation: 2.5

15. median: 40, range: 7.6

16. median: 40, range: 38

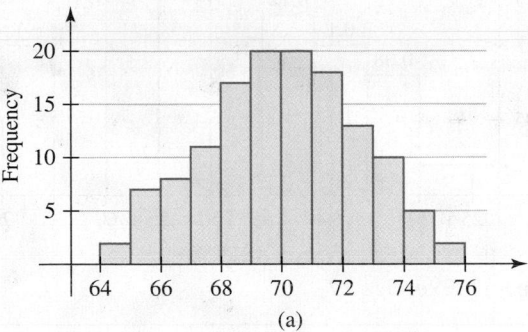

(a)

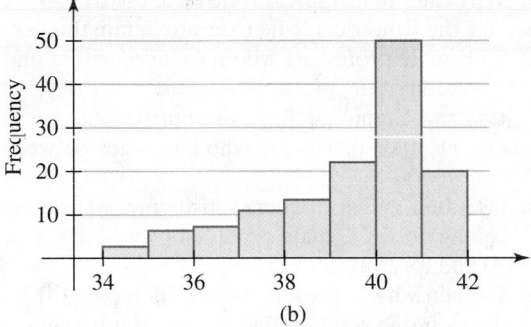

(b)

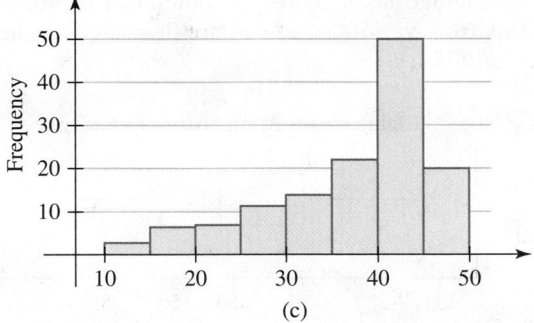

(c)

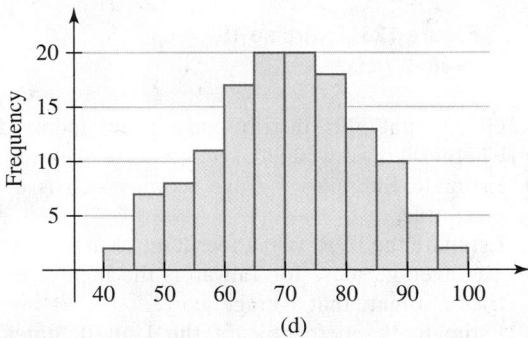

(d)

Figure 121 Exercises 13–16

17. The ages (in years) of tenured and tenure-eligible (TTE) male professors at 4-year colleges and universities are described by the relative frequency histogram in Fig. 122. The mean age is 51.3 years, and the standard deviation is 12.0 years.

Ages of Male Professors

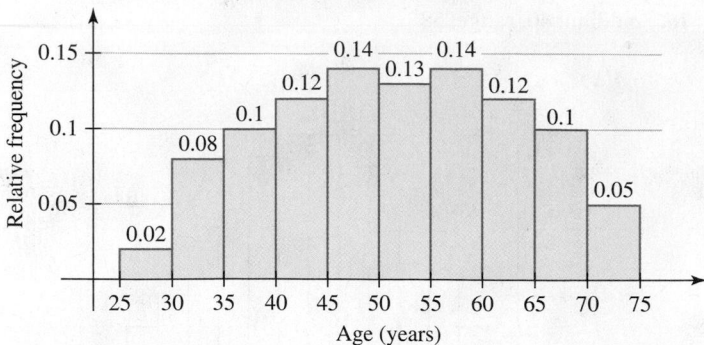

Figure 122 Exercise 17
(**Source:** *CBMS*)

a. Why can the Empirical Rule be applied?
b. Use the Empirical Rule to approximate the percentage of TTE male professors who have ages within one standard deviation of the mean.
c. Use the Empirical Rule to approximate the percentage of TTE male professors who have ages between 27.3 and 75.3 years.
d. Use the relative frequency histogram to estimate the percentage of TTE male professors who have ages between 30 and 75 years.
e. Explain why the result you found in part (d) implies that the estimate you found in part (c) is too low.

18. The average math scores (in points) of eighth-grade students from various countries are described by the boxplot in Fig. 123.

Eighth Grade Math Scores, by Country

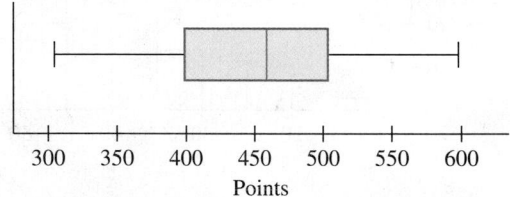

Figure 123 Exercise 18
(**Source:** *TIMSS*)

a. Given that the distribution is unimodal, determine whether it is skewed left, symmetric, or skewed right.
b. Estimate Bahrain's average score, which is at the 25th percentile.
c. Estimate the IQR. What does it mean in this situation?
d. The average score for Taiwan is the highest of all countries. Estimate that average score.
e. Estimate the percentile of the United States' average score, which is 508 points.

19. **DATA** Here are the number (in millions) of viewers of the 10 most-watched TV series finales (Source: USA Today, *Reuters*, TV Guide, *ABC*):

44.4	52.5	40.2	33.7	80.4
50.7	105.9	35.5	76.3	36.3

a. Construct a boxplot.
b. Given that the distribution is unimodal, determine whether it is skewed left, symmetric, or skewed right. What does this mean in this situation?
c. Which is a better measure of the center, the mean or the median? Explain.
d. Estimate the median by inspecting your boxplot. What does it mean in this situation?
e. A student says there are fewer observations between 34 million and 36 million viewers than there are observations between 76 million and 106 million viewers because the left whisker is shorter than the right whisker. What would you tell the student?

20. **DATA** Here are the numbers of times the alternative songs and hip-hop songs in the top 100 were listened to at Last.fm during the week ending on February 15, 2015 (Source: *Last.fm*):

Alternative			Hip-Hop		
1526	3206	5360	1534	5817	2365
3820	1436	1886	5056	4147	1825
5136	2856	5215	2591	1466	3591
1995	1883	3359	1581	1691	2501
1985	1410	1417	2045	1706	1830
4925	3395	1419	6117	1779	4792
4865	3161	1544	1565	1399	1617
1467	1384	1527	3491	5407	1945
5495	3887	3067	5407	5934	1400
1446	2038	4398	4941	4981	1804
1867	1460	1494	1453	3900	2566
3732	3891	3591			
1617	4047	4845			
3078	1408	3843			
2680	3607	3976			

a. Construct boxplots of the alternative distribution and the hip-hop distribution that share the same horizontal axis.
b. Was the genre of the song that had the most listens alternative or hip-hop?
c. For each distribution, compute the median. Compare the two results. What does your comparison mean in this situation?
d. For each distribution, compute the IQR. Compare the two results. What does your comparison mean in this situation?
e. For each distribution, compute Q_3. Compare the two results. What does your comparison mean in this situation?

Chapter 4 Test

1. Refer to the histogram shown in Fig. 124 to determine whether the mean is less than, greater than, or approximately equal to the median. Explain.

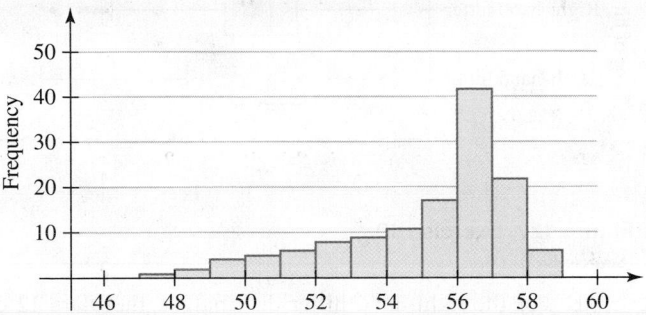

Figure 124 Exercise 1

2. The ages (in years) of tenured and tenure-eligible (TTE) female professors at 4-year colleges and universities are described by the relative frequency histogram in Fig. 125. The ages are rounded to the ones place.

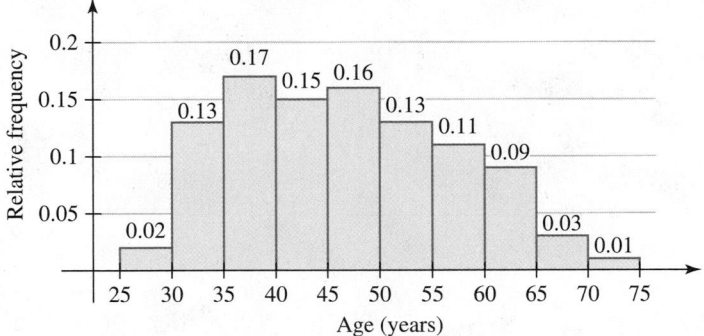

Figure 125 Exercise 2
(**Source:** *CBMS*)

    a. Describe the shape of the distribution.

    b. Which is larger, the mean or the median? Which is the better measure of the center? Explain.

    c. Which class contains the median?

    d. The median age of TTE male professors lies in the class 50–55 years. Compare this with the result you found in part (c). What does your comparison mean in this situation?

    e. A female professor who is 64 years in age is at what percentile?

3. The mean income of a group of nine people is $40,000. If a person is added to the group but the mean income of the ten people stays the same, what is the new person's income?

4. Find the mean, median, mode, range, standard deviation, and variance of the data 92, 27, 55, 48, 80, 41, 62, 80, and 16.

5. ᴰᴬᵀᴬ Here are the sales (in thousands) of the best-selling vinyl albums in 2014 (Source: *Nielsen Soundscan*):

31.8	33.6	42.1	58.7	27.8
31.7	38.2	86.7	37.8	34.2

    a. Find the mean sales.

    b. Find the standard deviation of the sales.

    c. Find the variance of the sales.

6. The scores on a first test are unimodal and symmetric with mean 75 points and standard deviation 8 points.

    a. Find the approximate percentage of scores within three standard deviations of the mean.

    b. Find the approximate percentage of scores between 67 points and 83 points.

    c. If 40 students took the test, then find the *number* of students who scored between 59 points and 91 points.

    d. On the second test, the mean is 70 points and the standard deviation is 10 points. Compare the center of the first-test scores and the center of the second-test scores. Also, compare the spread of the first-test scores and the spread of the second-test scores.

7. The standard deviation of the prices of some skateboards is $14. Use the appropriate symbol to represent the result.

8. The 2012 infant mortality rates (deaths per 1000 births) of 195 countries are described by the boxplot in Fig. 126.

Infant Mortality Rate, by Country

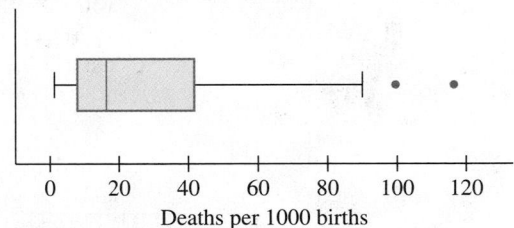

Deaths per 1000 births

Figure 126 Exercise 8
(**Source:** *UNICEF; the Human Mortality Database*)

    a. Given that the distribution is unimodal, determine whether it is skewed left, symmetric, or skewed right. What does this mean in this situation?

    b. Jordan's infant mortality rate is at the 50th percentile. Estimate its infant mortality rate.

    c. Sierra Leone has the largest infant mortality rate. Estimate its infant mortality rate. Is it an outlier?

    d. Estimate the percentile of the Saudi Arabia infant mortality rate, which is 7.4 deaths per 1000 births.

    e. There were 3,932,181 births in the United States in 2012. Estimate the number of U. S. infant deaths in 2012.

9. ᴰᴬᵀᴬ Here are the average numbers of tornados per month in July from 1989 to 2013 for states in which the average is at least 2 tornados per month (Source: *NOAA*):

3	15	4	20	2	2	3	10
2	3	17	6	24	2	5	4
20	2	2	14	3	15	7	2
12	4	10	8	2			

    a. Is the variable numerical or categorical?

    b. Construct a boxplot.

c. Given that the distribution is unimodal, determine whether it is skewed left, symmetric, or skewed right. Explain.

d. The average number of tornados in July in Michigan is at the 50th percentile. What is that average number?

e. What does the fact that there is no left whisker mean in this situation?

10. On September 20, 2014, the Chicago Cubs baseball team played the Los Angeles Dodgers. The heights of pitches in which the batters did not make contact with the ball were determined by a system of cameras.

a. Six of the observations were negative. Discuss various options of what could be done with those observations and then determine the best course of action.

b. The author was unable to find out why six of the observations were negative. So, he made a big assumption that the six pitches hit the ground before reaching the batter. On the basis of this assumption, he replaced each of the negative observations with 0 feet and then constructed boxplots for the pitch heights (in feet) for right-hand hitters and for left-hand hitters (see Fig. 127). Estimate the 75th

percentile for the left-hand-hitter distribution. What does it mean in this situation?

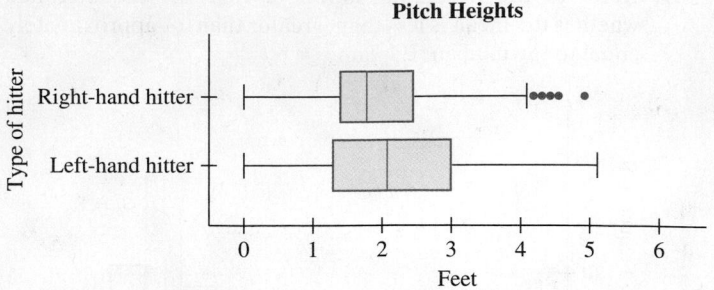

Figure 127 Exercise 10
(**Source:** *Brooks Baseball*)

c. For the right-hand-hitter distribution, there are 12 outliers. Explain why there are only 5 dots in the boxplot.

d. Which distribution has the smaller median? What does that mean in this situation?

e. Which distribution has the smaller IQR? What does that mean in this situation?

Computing Probabilities

Table 1 compares the numbers (in thousands) of students who started college at a 4-year private, a 4-year public, or a 2-year public institution in fall 2007 with their educational status as of spring 2013 (six years later). The category "Completed a degree or certificate" includes students who eventually attended different colleges or earned a

Table 1 College Students' Educational Status after 6 Years

Educational Status as of Spring 2013	Numbers of Students (in thousands)			
	4-Year Private	4-Year Public	2-Year Public	Total
Completed a degree or certificate	325	637	342	1304
Still enrolled at any college	43	150	162	355
Not enrolled in any college	78	217	353	648
Total	446	1004	857	2307

Source: *National Student Clearinghouse Research Center*

different degree than first intended. In Exercise 33 of Homework 5.3, you will investigate how likely it was for a student in a certain type of institution to complete a degree or certificate by spring 2013.

In Chapters 3 and 4, we used relative frequencies to describe data. In this chapter, we will use relative frequencies to describe the chance that something will occur. We will in turn use this concept to explore whether a claim made by a company is true. We will also explore whether there might be a connection between *events*, such as a person smiling and the person's age (see Exercise 35 of Homework 5.3). Finally, we will describe some distributions using a special type of curve called a *normal curve*, which is often called a "bell-shaped curve."

▼ 5.1 Meaning of Probability

Objectives

» Compute and interpret *probabilities*.

» Use *simulation* to estimate a probability.

WARNING

Probability

Recall from Section 2.1 that **if we randomly select one item from a group of items, each item has the same chance of being selected.** For example, we can randomly select heads or tails by flipping a fair coin because each outcome (heads or tails) has an equal chance of occurring.

Even though flipping a coin is a random selection of heads or tails, this does *not* mean that if we flip a coin over and over, the outcomes will go back and forth between heads and tails with no streaks. For example, here are the outcomes of the author flipping a coin 20 times, listed in order from left to right:

$$T\ T\ T\ T\ H\ T\ T\ H\ T\ H\ H\ T\ T\ T\ H\ T\ H\ T\ T$$

Does it surprise you that there is a streak of 5 tails at the start? Streaks can be quite long if a coin is flipped a large number of times. For example, if a coin were flipped 1 million times, there would likely be a streak of about 20 heads or 20 tails.

We call a process that selects items randomly a **random experiment**. Flipping a coin and rolling a die are both random experiments. Randomly selecting 1 student from a class of 35 students is also a random experiment.

If we were to flip a coin an infinite number of times, how often would we expect to get heads? Because there are two possible outcomes (heads and tails) for each toss and those two outcomes are equally likely to occur, we would expect to get heads about $\frac{1}{2}$ the time. We say the *probability* of getting heads is $\frac{1}{2}$.

▶ **Definition   Probability**

The **probability** of an outcome from a random experiment is the relative frequency of the outcome if the experiment were run an infinite number of times.

We can describe probabilities in fraction or decimal form. Both forms are common.

▶ **Example 1**   Finding Probabilities

An experiment consists of rolling a six-sided die (numbered 1 through 6) once. Find and interpret the probability of rolling a 4.

Solution

Because 4 is 1 of 6 equally likely outcomes, we conclude that the probability is $\frac{1}{6}$. If we were to roll the die an infinite number of times, we would roll a 4 exactly $\frac{1}{6}$ of the time. Keeping to what is physically possible, if we were to roll the die a *large number* of times, we would roll a 4 *approximately* $\frac{1}{6}$ of the time.

◀

WARNING

A probability does not tell us what to expect if we run an experiment a *small* number of times. For example, the probability $\frac{1}{2}$ for getting heads does not mean that if we toss a coin only 20 times we would expect to get heads 10 of the 20 tosses. In fact, in the author's experiment, there were only 6 heads:

$$T\ T\ T\ T\ H\ T\ T\ H\ T\ H\ H\ T\ T\ T\ H\ T\ H\ T\ T$$

Using Simulation to Estimate a Probability

The relative frequency of heads for the author's 20 flips of the coin is $\frac{6}{20} = 0.3$, which is not a good estimate of the probability of getting heads: $\frac{1}{2} = 0.5$. But what if we flipped the coin a large number of times?

Instead of flipping a coin by hand, which can be time consuming, we could investigate by using a TI-84 to randomly select from the numbers 0 and 1. The number 0 could stand for tails, and the number 1 could stand for heads. Using technology to imitate a random experiment is called a **simulation**.

We can also perform a coin-flipping simulation using StatCrunch, which we will do in Example 2.

> ▶ **Example 2** Using Relative Frequencies to Estimate a Probability

Use StatCrunch to simulate flipping a coin the given number of times. Discuss how well the relative frequency of heads estimates the probability of getting heads.

   1. 5 times   **2.** 100 times   **3.** 1100 times   **4.** 10,100 times

Solution

   1. The outcomes are H T T H H (see Fig. 1).

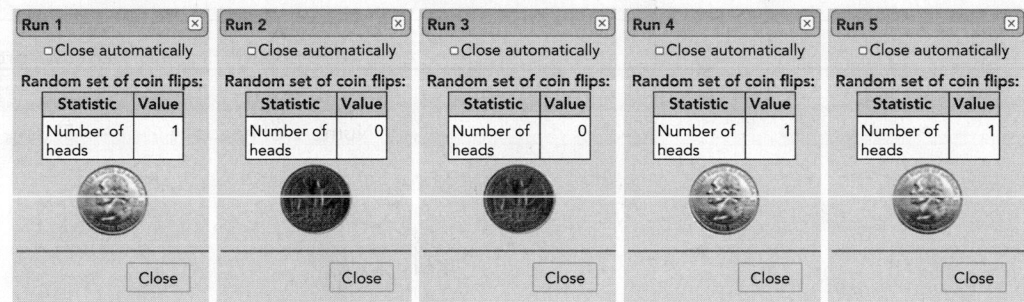

Figure 1 The outcomes of five tosses

Because there are 3 heads out of 5 tosses, the relative frequency of heads is $\frac{3}{5} = 0.6$.

In Fig. 2, "Count," "Total," and "Proportion" give the frequency of heads (3), the total number of flips (5), and the relative frequency (0.6), respectively. The relative frequency 0.6 is also described by the height of the right end of the curve (the zigzag line). The heights of the points on the curve above 1, 2, 3, and 4 on the Run # axis describe the relative frequency of heads after 1, 2, 3, and 4 flips, respectively.

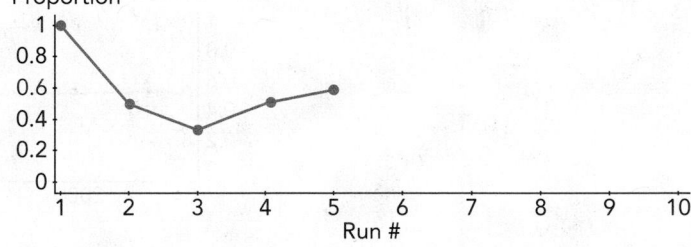

# of heads for flipping 1 coin w/ prob. of heads= 0.5

Event	Count	Total	Proportion
Number of heads = 1	3	5	0.6

Figure 2 Relative frequency of heads from 5 tosses

The relative frequency 0.6 is not equal to the probability 0.5, but it is not a terrible estimate, considering the coin was flipped so few times.

   2. We simulate flipping a coin an additional 95 times, which makes a total of 100 times (see Fig. 3). The relative frequency 0.58 is a bit better estimate of the probability 0.5 than we found with just 5 flips.

   3. We simulate flipping a coin an additional 1000 times, which makes a total of 1100 times (see Fig. 4). The relative frequency 0.5209 is a much better estimate of the probability 0.5 than we found with just 100 flips, but we can do better.

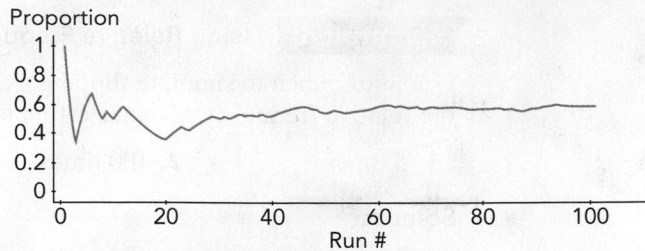

Figure 3 Relative frequency of heads from 100 tosses

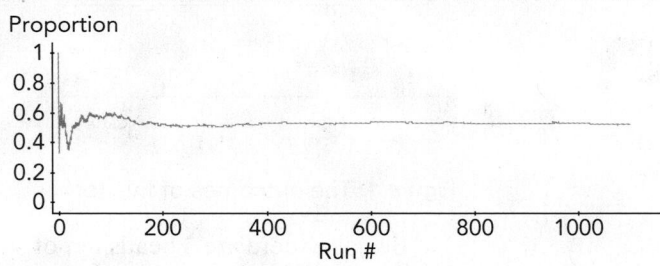

Figure 4 Relative frequency of heads from 1100 tosses

4. We simulate flipping a coin an additional 9000 times, which makes a total of 10,100 times (see Fig. 5). The relative frequency 0.5013 is a very good estimate of the probability 0.5.

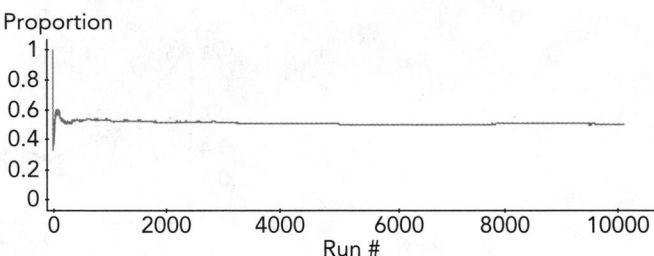

Figure 5 Relative frequency of heads from 10,100 tosses

Even after simulating 1100 runs of the coin experiment, the relative frequency of heads (0.5209) was not that close to the probability 0.5. But with 10,100 runs, the relative frequency 0.5013 was very close to the probability 0.5. This suggests that **by running a random experiment a large number of times, we can use relative frequency to estimate probability**, which is true.

But why bother estimating a probability when we can find it exactly by using logic? The reason is that some probabilities involve so many factors we cannot find them using logic.

For example, it is too challenging to determine the probability that a house will catch fire in the next year using logic alone. But we can estimate it by computing the relative frequency of homes in the United States that caught fire in the past year.

Or even better, we could take into account the climate and whether someone in the household smokes. For example, we could estimate the probability that a smoker household in Boston will experience fire damage in the next year by finding the relative frequency of smoker households in Boston that caught fire last year.

Equally Likely Probability Formula

In Example 3, we will find another probability for the die experiment. This will help us find a formula that we can use to compute probabilities.

▶ **Example 3**   Finding Probabilities

An experiment consists of rolling a six-sided die once. Find and interpret the probability of rolling an even number.

Solution

The even numbers on a die are the numbers 2, 4, and 6. Because those numbers are 3 of 6 equally likely outcomes, we conclude that the probability is $\frac{3}{6} = \frac{1}{2}$. This means that if we roll the die a large number of times, we would roll an even number about $\frac{1}{2}$ of the time.

◀

For the random experiment of rolling a die once, the possible outcomes are the numbers 1, 2, 3, 4, 5, and 6. We say the *sample space* is the group of outcomes 1, 2, 3, 4, 5, and 6.

▶ **Definition   Sample space**

The **sample space** of a random experiment is the group of all possible outcomes.

For the experiment of flipping a coin once, the sample space is the group of outcomes heads and tails.

▶ **Definition   Event**

An **event** is some of the outcomes in the sample space, all of them, or none of them.

For the die experiment, each of the following is an event:

- the evens 2, 4, and 6
- the number 3
- the numbers 1, 2, 3, 4, 5, and 6

We usually use a capital letter to stand for an event. For example, we could let E stand for the even numbers on a die. We write $P(E)$ to stand for the probability of the event E. In other words, $P(E)$ stands for the probability of getting an even number. On the basis of our work in Example 3, we can write $P(E) = \frac{1}{2}$. Another option is to replace E with the even numbers themselves: $P(2, 4, 6) = \frac{1}{2}$.

In order to find this probability, we first wrote it as $\frac{3}{6}$ by dividing the number of even numbers (3) by the total number of possible outcomes (6). This strategy works in general, provided that all the possible outcomes are equally likely.

> **Equally Likely Probability Formula**
>
> If the sample space of a random experiment consists of n equally likely outcomes and an event E consists of m of those outcomes, then
>
> $$P(E) = \frac{m}{n}$$
>
> In other words, if the sample space of a random experiment consists of a finite number of equally likely outcomes, then the probability of an event is equal to the number of outcomes in the event divided by the number of outcomes in the sample space.

WARNING   If the outcomes of a random experiment are not equally likely, then we *cannot* use the equally likely probability formula. For example, suppose someone spins the spinner shown in Fig. 6 once. Even though the random experiment consists of 3 outcomes, the probability of getting a 1 is *not* $\frac{1}{3}$. The correct probability is $\frac{1}{2}$, because the area of the region for the outcome 1 is half the total area.

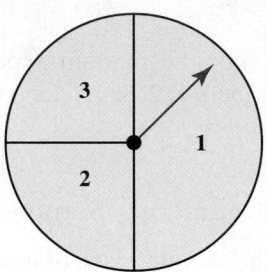

Figure 6 The probability of the spinner landing on 1 is $\frac{1}{2}$

▶ **Example 4**   Calculating Probabilities

For a group of 10 students in one of the author's algebra classes, there are 2 nursing majors, 4 business majors, 1 architect major, and 3 undecided majors. None of the students are double-majors. Assume one student is randomly selected from the group. Use N for nursing, B for business, A for architect, and U for undecided.

   1. Find $P(N)$.
   2. Construct a frequency and relative frequency table of the students' majors.
   3. Find $P(B)$, $P(A)$, and $P(U)$.
   4. Find $P(\text{philosophy})$.
   5. Find the probability of randomly selecting a nursing, business, architect, OR undecided major.
   6. Find $P(N) + P(B) + P(A) + P(U)$. Why does the result make sense?

Solution

   1. There are 2 nursing majors out of the 10 students, so the probability is $\frac{2}{10} = 0.2$.

   2. The frequency and relative frequency table is shown in Table 2. The row for nursing contains the relative frequency (probability) 0.2 that we found in Problem 1.

   3. From the far right column of Table 2, we see that $P(B) = 0.4$, $P(A) = 0.1$, and $P(U) = 0.3$.

   4. There are no philosophy majors, so the probability is $\frac{0}{10} = 0$.

Table 2 Students' Majors

Major	Frequency	Relative Frequency
Nursing	2	$\frac{2}{10} = 0.2$
Business	4	$\frac{4}{10} = 0.4$
Architect	1	$\frac{1}{10} = 0.1$
Undecided	3	$\frac{3}{10} = 0.3$
Total	10	1

5. The event consists of $2 + 4 + 1 + 3 = 10$ of the 10 students, so the probability is $\frac{10}{10} = 1$.

6. We add all four probabilities:

$$P(N) + P(B) + P(A) + P(U) = \frac{2}{10} + \frac{4}{10} + \frac{1}{10} + \frac{3}{10} = \frac{10}{10} = 1$$

The results makes sense, because we are adding the relative frequencies of all the majors, which must equal 1.

Properties of Probability

In Example 4, we found that $P(\text{philosophy}) = 0$ because none of the 10 students were philosophy majors. In general, an **impossible event** does not contain any outcomes of the sample space. **The probability of an impossible event is 0.**

In Example 4, we found that the probability of randomly selecting a nursing, business, architect, OR undecided major is 1. In general, a **sure event** (or **certain event**) contains all the outcomes in the sample space. **The probability of a sure event is 1.**

All the probabilities that we found in Example 4 are between 0 and 1, inclusive. In general, **the probability of an event is between 0 and 1, inclusive.**

In Example 4, we say N, B, A, and U are *single-outcome* events. We found that

$$P(N) + P(B) + P(A) + P(U) = 1$$

In general, **the sum of the probabilities of all the single-outcome events in the sample space is equal to 1.**

> ### Probability Properties
>
> - The probability of an impossible event is equal to 0.
> - The probability of a sure event is equal to 1.
> - The probability of an event is between 0 and 1, inclusive.
> - The sum of the probabilities of all the single-outcome events in the sample space is equal to 1.

WARNING   If you ever calculate a probability to be negative or greater than 1, you should realize that you've made an error.

▶ Example 5 Probability Properties

1. A student determines that the probability of randomly selecting a classmate who owns a cat is -0.4. What would you tell the student?

2. For each of the ethnicities shown in Table 3, a counselor at a college calculates the probability of randomly selecting a student at the college of that ethnicity. What would you tell the counselor?

Table 3 Probabilities of Ethnicities

Ethnicity	Probability
Hispanic	0.35
African-American	0.27
Asian	0.49
Caucasian	0.26
Other	0.12

Solution

1. A probability cannot be negative, so the student must have made a data-entry or calculation error.
2. We add the probabilities: $0.35 + 0.27 + 0.49 + 0.26 + 0.12 = 1.49$. The sum should be equal to 1. The error is too large to be due to rounding. A data-entry or calculation error must have happened.

In Example 4, 2 out of 10 students are nursing majors. We can interpret this in two ways:

1. The proportion of the 10 students who are nursing majors is $\frac{2}{10} = 0.2$.

2. The probability of randomly selecting a student (from the 10 students) who is a nursing major is $\frac{2}{10} = 0.2$.

In general, if a situation involves a proportion or a probability, it can be interpreted to involve the other as well.

Using a Pie Chart to Find Probabilities

In Section 3.2, we used a pie chart to find proportions, which we now know can be interpreted to be probabilities.

> **Example 6**   Using a Pie Chart to Find Probabilities

The pie chart in Fig. 7 describes the percentages of 1445 online adults who use various numbers of social media websites.

Number of Social Media Sites Used

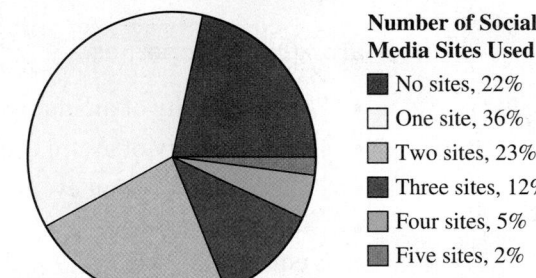

Number of Social Media Sites Used

- ■ No sites, 22%
- □ One site, 36%
- ■ Two sites, 23%
- ■ Three sites, 12%
- ■ Four sites, 5%
- ■ Five sites, 2%

Figure 7 Number of Social Media Sites Used
(**Source:** *Pew Research Center*)

1. Find the probability that a person randomly selected from the survey uses
   a. exactly three sites.
   b. at least three sites.
   c. at most three sites.

2. From inspecting the pie chart, a student concludes that 36% of *all* Americans use just one site. What would you tell the student?

Solution

1. **a.** From the pie chart, we see that 12% of the online adults use three sites. So, the relative frequency is 0.12, and the probability is also 0.12.
    b. At least three sites means three, four, or five sites, so we add the percentages for those categories: 12% + 5% + 2% = 19%. The probability is 0.19.
    c. At most three sites means zero, one, two, or three sites, so we add the percentages for those categories: 22% + 36% + 23% + 12% = 93%. The probability is 0.93.
2. Although it is true that 36% of the online adults surveyed use just one site, we *cannot* conclude that 36% of *all* Americans use just one site. The online adults surveyed might not represent all online adults well and would probably represent all adults (online and non-online) even worse.

▶

In Example 6, we could have let X be the number of social media sites used by a person randomly selected from the survey. So, $X = 2$ would represent randomly selecting someone who uses 2 social media sites. We say X is a *random variable*.

▶ **Definition   Random variable**

A **random variable** is a numerical measure of an outcome from a random experiment. We often use a capital letter such as X to stand for a random variable.

▶ **Example 7**   Finding Probabilities

Let X be the outcome of rolling a six-sided die once. Find the given probability.

1. $P(X = 4)$.   2. $P(X \leq 4)$ 3. $P(X > 4)$
4. $P(3 \leq X \leq 6)$ 5. $P(3 < X < 6)$

Solution

1. The equation $X = 4$ means that the outcome is 4, which is 1 out of 6 equally likely outcomes, so $P(X = 4) = \dfrac{1}{6}$.

2. $P(X \leq 4) = P(X \text{ is less than or equal to } 4)$
    $= P(X = 1, 2, 3, 4)$

    The outcomes 1, 2, 3, and 4 are 4 out of 6 equally likely outcomes, so $P(X \leq 4) = \dfrac{4}{6} = \dfrac{2}{3}$.

3. $P(X > 4) = P(X \text{ is greater than } 4)$
    $= P(X = 5, 6)$

    The outcomes 5 and 6 are 2 out of 6 equally likely outcomes, so $P(X > 4) = \dfrac{2}{6} = \dfrac{1}{3}$.

4. $P(3 \leq X \leq 6) = P(X \text{ is between 3 and 6, inclusive})$
    $= P(X = 3, 4, 5, 6)$

    The outcomes 3, 4, 5, and 6 are 4 out of 6 equally likely outcomes, so

    $$P(3 \leq X \leq 6) = \frac{4}{6} = \frac{2}{3}$$

5. $P(3 < X < 6) = P(X \text{ is between 3 and 6})$
    $= P(X = 4, 5)$

    The outcomes 4 and 5 are 2 out of 6 equally likely outcomes, so

    $$P(3 < X < 6) = \frac{2}{6} = \frac{1}{3}$$

▶

Using a Density Histogram to Find Probabilities

Recall that the areas of the bars of a density histogram are proportions, which we now know can be interpreted as probabilities.

▶ **Example 8**   Using a Density Histogram to Find Probabilities

The density histogram in Fig. 8 describes the number of mammal species threatened in each country in the world. Some countries have at least 80 threatened mammal species, but the bars would not be visible.

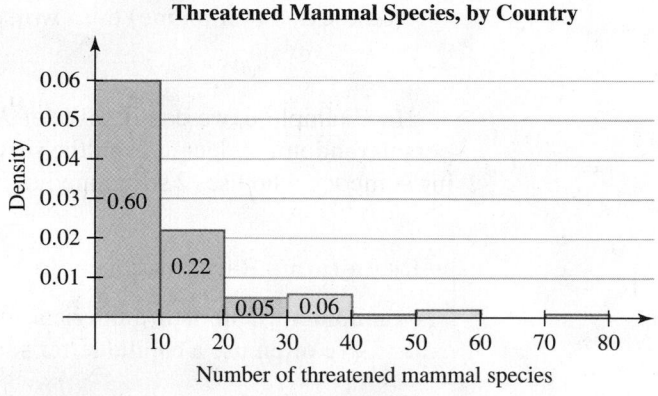

Figure 8 Threatened mammal species, by country
(**Source:** *The IUCN Red List of Threatened Species^{TM}*)

Let *X* be the number of threatened mammal species in a randomly selected country. Find the given probability.

1. $P(20 \leq X \leq 29)$
2. $P(X < 20)$
3. $P(X \geq 20)$

Solution

1. The compound inequality $20 \leq X \leq 29$ means that a randomly selected country has between 20 and 29 threatened mammal species, inclusive. The proportion of countries that have that many threatened mammal species is equal to the area of the green bar, which is 0.05 (see Fig. 8). So, $P(20 \leq X \leq 29) = 0.05$.

2. The inequality $X < 20$ means that a randomly selected country has less than 20 threatened mammal species. The proportion of countries that have that many threatened mammal species is equal to the total area of the two orange bars:

$$0.60 + 0.22 = 0.82$$

So, $P(X < 20) = 0.82$.

3. The inequality $X \geq 20$ means that a randomly selected country has 20 or more threatened mammal species. The sum of the areas of all the bars (including ones that would not be visible) is equal to 1. So, the proportion of countries that have 20 or more threatened mammal species is equal to 1 minus the proportion of countries that have less than 20 threatened mammal species, which we found to be 0.82 in Problem 2. So, the proportion is $1 - 0.82 = 0.18$. Thus, $P(X \geq 20) = 0.18$.

In Example 8, we found that the area of the green bar of the density histogram in Fig. 8 is equal to 0.05. We can interpret the area 0.05 in two ways:

1. The proportion of countries that have between 20 and 29 threatened mammal species, inclusive, is 0.05.

2. The probability of randomly selecting a country that has between 20 and 29 threatened mammal species, inclusive, is equal to 0.05.

> **Interpreting the Area of a Bar of a Density Histogram**

We can interpret the area of a bar of a density histogram as

- the proportion of observations that lie in the bar's class.
- the probability of randomly selecting an observation that lies in the bar's class.

Using the Empirical Rule to Find Probabilities

Recall from Section 4.2 that if a distribution is unimodal and symmetric, then we can apply the Empirical Rule.

> **Example 9**   Using the Empirical Rule to Find Probabilities

The Wechsler IQ test measures a person's intelligence. The distribution of IQ scores is unimodal and symmetric. The test is designed so that the mean score is 100 points and the standard deviation is 15 points (Source: *Essentials of WAIS-IV Assessment, Elizabeth Lichtenberger et al.*). Find the probability of randomly selecting a person who has an IQ between 70 and 130 points.

Solution

Because the distribution is unimodal and symmetric, we can apply the Empirical Rule (see Table 4). From the table, we see that 95% of the IQ scores are between 70 and 130 points. So, the probability is 0.95.

Table 4 Applying the Empirical Rule

Interval	Interval	Interval	Percent
$\bar{x} \pm s$	100 ± 15	$(85, 115)$	68%
$\bar{x} \pm 2s$	$100 \pm 2(15)$	$(70, 130)$	95%
$\bar{x} \pm 3s$	$100 \pm 3(15)$	$(55, 145)$	99.7%

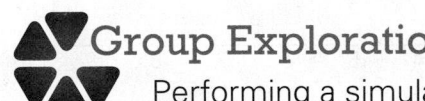

Group Exploration

Performing a simulation

In this exploration, you will use technology to simulate rolling a six-sided die.

1. What is the probability of getting a 4?

2. Simulate rolling a die for each of the numbers of rolls displayed in Table 5. Record the relative frequencies of the outcome 4 in the second column of the table.

Table 5 Relative Frequencies of the Outcome 4

Number of Rolls	Relative Frequency of the Outcome 4	Absolute Value of Difference of Relative Frequency and Probability
10		
100		
1000		
2000		
4000		
6000		
8000		
10,000		

Also, for each relative frequency, compute the absolute value of the difference of the relative frequency and the probability you found in Problem 1 and record it in the third column of the table.

3. After how many rolls is the relative frequency very close to the probability?

4. Does the absolute value of the difference of the relative frequency and the probability decrease each time the number of rolls increases?

5. In any six-sided-die random experiment, after how many rolls can you be certain that the relative frequency of the outcome 4 will be within 0.001 of the probability of getting a 4?

> **Tips for Success   Choose a Good Time and Place to Study**
>
> To improve your effectiveness at studying, consider taking stock of when and where you are best able to study. Tracy, a student who lives in a sometimes-distracting household, completes her assignments at the campus library just after she attends her classes. Gerome, a morning person, gets up early so that he can study before classes. Being consistent in the time and location for studying can help form a good habit.

Homework 5.1

For extra help ▶ MyMathLab®    Watch the videos in MyMathLab    Download the MyDashboard App

1. The sum of the probabilities of all the single-outcome events in the sample space is equal to ____ .

2. The probability of an impossible event is equal to ____ .

3. A(n) ____ variable is a numerical measure of an outcome from a random experiment.

4. We can interpret the area of a bar of a(n) ____ histogram as the probability of randomly selecting an observation that lies in the bar's class.

5. The simulations of flipping a coin 5 times and an additional 10,000 times are shown in Figs. 9 and 10, respectively.

# of heads for flipping 1 coin w/ prob. of heads= 0.5

Event	Count	Total	Proportion
Number of heads = 1	2	5	0.4

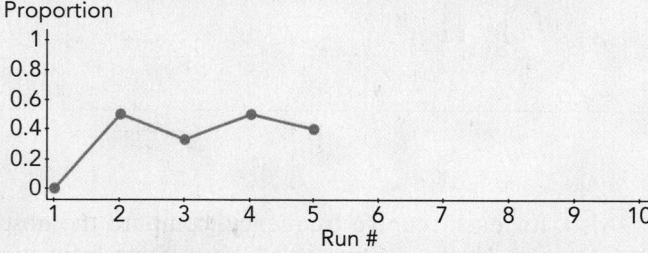

Figure 9 Relative frequency of heads from 5 tosses

# of heads for flipping 1 coin w/ prob. of heads= 0.5

Event	Count	Total	Proportion
Number of heads = 1	4967	10005	0.4965

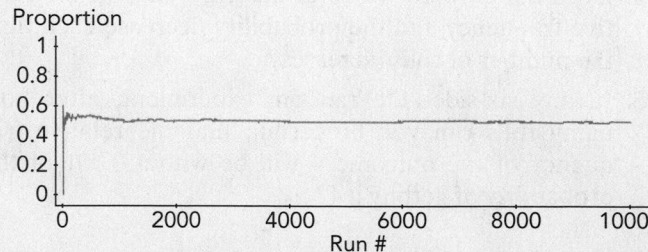

Figure 10 Relative frequency of heads from 10,005 tosses

a. Determine the first five outcomes of the simulated experiment.

b. What was the relative frequency of heads after 5 flips of the coin? Is this a good estimate of the probability of getting heads from flipping a coin once? Explain.

c. What was the relative frequency of heads after 10,005 flips of the coin? Is this a good estimate of the probability of getting heads from flipping a coin once? Explain.

6. The simulations of rolling a six-sided die 5 times and an additional 10,000 times are shown in Figs. 11 and 12, respectively.

Sum of die with 1 die that has 6 sides

Event	Count	Total	Proportion
Sum of 1 rolls = 4	1	5	0.2

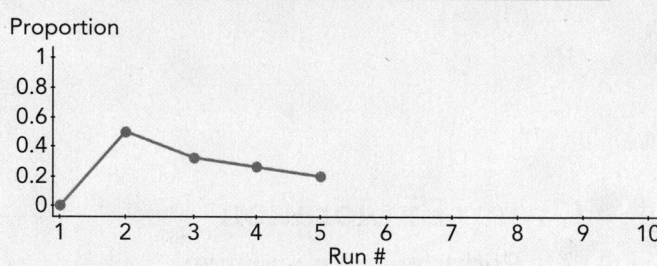

Figure 11 Relative frequency of the outcome 4 from 5 rolls

Sum of die with 1 die that has 6 sides

Event	Count	Total	Proportion
Sum of 1 rolls = 4	1677	10005	0.1676

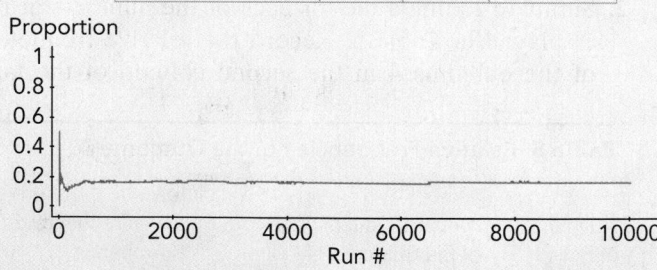

Figure 12 Relative frequency of the outcome 4 from 10,005 rolls

a. In which of the first five rolls did a 4 occur?

b. What was the relative frequency of the outcome 4 after 5 rolls of the die? Is this a good estimate of the probability of rolling a 4 from rolling a die once? Explain.

c. What was the relative frequency of the outcome 4 after 10,005 rolls of the die? Is this a good estimate of the probability of rolling a 4 from rolling a die once? Explain.

For Exercises 7–10, suppose there are 12 employees working at a Mexican restaurant on a Saturday night: 3 cooks, 2 bus people, 5 waiters, 1 hostess, and 1 bartender. An employee of the restaurant is randomly selected. Find the given probability in fraction form.

7. $P(\text{a waiter})$

8. $P(\text{a cook})$

9. $P(\text{an employee at the restaurant})$

10. $P(\text{President Abraham Lincoln})$

11. A student guesses the answer for a true-or-false test question. What is the probability (in fraction form) that the student guesses the correct answer?

12. A student guesses the answer for a multiple-choice test question with 5 choices. What is the probability (in fraction form) that the student guesses the correct answer?

In Exercises 13–18, you will analyze the gambling game roulette in which a ball is dropped onto a revolving wheel that has 38 numbered slots: the numbers 1–36, a zero, and a double zero (see Fig. 13). There are 18 red slots, 18 black slots, and 2 green slots:

Red: 1, 3, 5, 7, 9, 12, 14, 16, 18, 19, 21, 23, 25, 27, 30, 32, 34, 36

Black: 2, 4, 6, 8, 10, 11, 13, 15, 17, 20, 22, 24, 26, 28, 29, 31, 33, 35

Green: 0, 00

Consider the random experiment of dropping the ball onto the wheel once. Find the given probability in fraction form.

13. $P(5)$ **14.** $P(37)$ **15.** $P(\text{black})$

16. $P(\text{green})$ **17.** $P(\text{yellow})$ **18.** $P(999)$

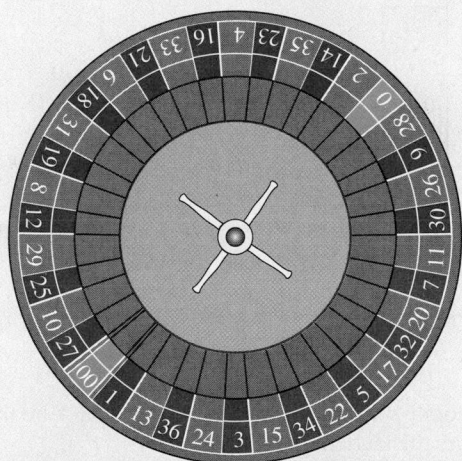

Figure 13 A roulette wheel (Exercises 13–18)

19. Suppose a person bets \$10 on red in the gambling game roulette described for Exercises 13–18. If the ball lands on a red number, the person will receive the \$10 she bet plus another \$10. If the ball lands on a black or a green number, she will lose the \$10 she bet. If she were to play the game 10,000 times over the course of her life, would she most likely win or lose money overall? Refer to probabilities to help explain.

20. A person plays a gambling game in which a six-sided die is rolled once. The person bets \$5 on the numbers 1 and 2. If

the number 1 or 2 is rolled, the person will get the \$5 he bet plus another \$5. If the number 3, 4, or 5 is rolled, the person will lose the \$5 he bet. If the number 6 is rolled, the person will get back the \$5 he bet but won't get additional money. If he were to play the game 10,000 times over the course of his life, would he most likely win or lose money overall? Refer to probabilities to help explain.

For Exercises 21–30, suppose someone spins the spinner shown in Fig. 14 once. Find the given probability in fraction form.

21. $P(1)$ **22.** $P(2)$ **23.** $P(3)$ **24.** $P(4)$

25. $P(\text{an odd number})$ **26.** $P(\text{an even number})$

27. $P(\text{at least 2})$ **28.** $P(\text{at least 3})$

29. $P(\text{at most 2})$ **30.** $P(\text{at most 3})$

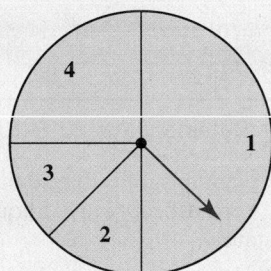

Figure 14 A spinner (Exercises 21–30)

31. Table 6 shows all the genres of music on a student's playlist. For each of the genres, the student calculates the probability of randomly selecting a song from her playlist that is in that genre. What would you tell the student?

Table 6 Probabilities of Music Genres

Genre	Probability
Hip-Hop	0.2
Rap	0.2
Electronica	0.1
Other	0.1

32. A professor grades a prestatistics test and writes the numbers of students who earned the grades A, B, C, D, and F on the board. For each grade, a student computes the probability of randomly selecting a student's test that has that grade (see Table 7). What would you tell the student?

Table 7 Probabilities of Grades on Test

Grade	Probability
A	0.3
B	0.6
C	0.4
D	0.1
F	0.1

33. Identify any of the following numbers that cannot be probabilities. Explain.

$$\frac{9}{2} \qquad 0.999 \qquad \frac{3}{5} \qquad -0.65 \qquad 0.0001$$

34. Identify any of the following numbers that cannot be probabilities. Explain.

$$0.75 \quad 0 \quad -0.37 \quad 1.29 \quad 0.897$$

For Exercises 35–40, one of the dates of June 2016 is randomly selected (see Fig. 15). Find the given probability in fraction form.

35. $P(\text{Monday})$

36. $P(\text{Thursday})$

37. $P(\text{weekend})$

38. $P(\text{weekday})$

39. $P(\text{in the third week})$

40. $P(\text{in the fifth week})$

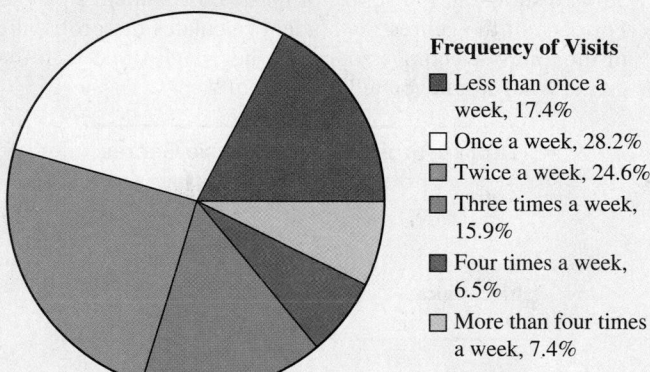

Sun	Mon	Tues	Wed	Thurs	Fri	Sat
			1	2	3	4
5	6	7	8	9	10	11
12	13	14	15	16	17	18
19	20	21	22	23	24	25
26	27	28	29	30		

Figure 15 Dates of June 2016 (Exercises 35–40)

41. A total of 1487 randomly selected adults who are at least 22 years in age were asked, "How frequently do you visit quick service restaurants?" Their responses are described by the pie chart in Fig. 16.

Frequency of Visits to Quick Service Restaurants

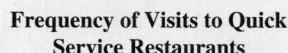

Frequency of Visits

- ■ Less than once a week, 17.4%
- □ Once a week, 28.2%
- ▨ Twice a week, 24.6%
- ▨ Three times a week, 15.9%
- ▨ Four times a week, 6.5%
- ▨ More than four times a week, 7.4%

Figure 16 Exercises 41 and 42
(**Source:** *Deloitte*)

a. Find the probability that a person randomly selected from the survey visits quick service restaurants 4 times a week.

b. Find the probability that a person randomly selected from the survey visits quick service restaurants at least 3 times a week.

c. Find the probability that a person randomly selected from the survey visits quick service restaurants less than 3 times a week.

d. To be included in the study, a person had to meet two criteria. Within 30 days prior to the study,
- the person must have visited a quick service restaurant at least 4 times or a casual restaurant at least twice.
- the person must have visited quick service restaurants at least as many times as casual restaurants.

On the basis of these criteria, what can you say about the 17.4% of respondents who said they visit quick service restaurants less than once a week?

42. Some randomly selected adults who are at least 22 years in age were asked, "How frequently do you visit quick service restaurants?" Their responses are described by the pie chart in Fig. 16.

a. Find the probability that a person randomly selected from the survey visits quick service restaurants 3 times a week.

b. Find the probability that a person randomly selected from the survey visits quick service restaurants at most twice a week.

c. Find the probability that a person randomly selected from the survey visits quick service restaurants more than twice a week.

d. On the basis of the pie chart, a student concludes that 6.5% of all teenagers eat at quick service restaurants 4 times a week. What would you tell the student?

For Exercises 43–52, let X be the outcome of rolling a six-sided die once. Find the given probability in fraction form.

43. $P(X = 2)$

44. $P(X = 5)$

45. $P(X > 2)$

46. $P(X \geq 4)$

47. $P(X \leq 3)$

48. $P(X < 5)$

49. $P(3 \leq X \leq 6)$

50. $P(2 \leq X \leq 4)$

51. $P(1 < X < 6)$

52. $P(2 < X < 5)$

53. The author surveyed students in one of his statistics classes about the number of hours they worked per week. The density histogram in Fig. 17 describes their responses.

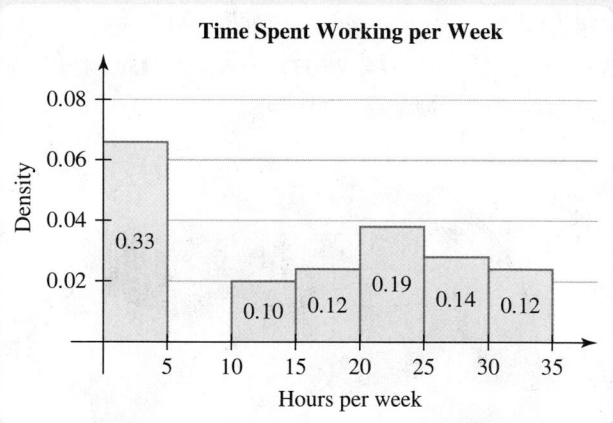

Figure 17 Exercises 53 and 54
(**Source:** *J. Lehmann*)

For a randomly selected student in the class, let X be the number of hours the student works per week.

a. Find $P(X \geq 25)$.

b. Find $P(X < 20)$.

c. Find $P(15 \leq X < 30)$.

d. Describe the shape of the distribution.

e. Rather than use the mean or the median to estimate the center of the entire distribution, explain why it would be more meaningful to separate the data into two groups and estimate the center of each group.

54. The author surveyed students in one of his statistics classes about the number of hours they worked per week. The density histogram in Fig. 17 describes their responses. For a randomly selected student in the class, let X be the number of hours the student works per week.

a. Find $P(X \geq 20)$.
b. Find $P(X < 15)$.
c. Find $P(10 \leq X < 25)$.
d. From inspecting the density histogram, a student concludes that 33% of all statistics students in the United States work less than 5 hours per week. What would you tell the student?
e. A counselor advises that part-time students work less than 25 hours per week and that full-time students work less than 15 hours per week. What percentage of the class did not follow this guideline for certain?

55. The mean numbers of days it took Chicago's Department of Streets & Sanitation to respond to requests to fix potholes are described by the density histogram in Fig. 18 for each of the weeks in 2012 and 2013.

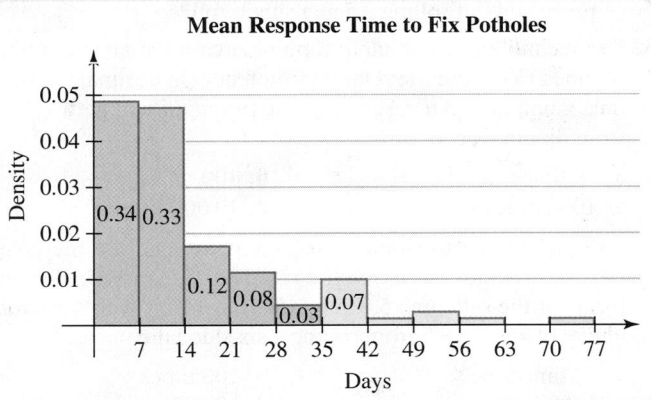

Figure 18 Exercise 55
(**Source:** *Department of Streets & Sanitation, Chicago*)

a. Interpret the area of the orange region in the density histogram as a proportion.
b. Interpret the area of the orange region in the density histogram as a probability.
c. Find the probability that the mean response time for a randomly selected pothole is less than 14 days.
d. Find the probability that the mean response time for a randomly selected pothole is between 7 and 27 days, inclusive.
e. Find the probability that the mean response time for a randomly selected pothole is at least 21 days.

56. The average prices (in dollars) of 2014 Major League Baseball (MLB) tickets at the stadiums are described by the density histogram in Fig. 19.

a. Interpret the area of the orange region in the density histogram as a proportion.
b. Interpret the area of the orange region in the density histogram as a probability.
c. Find the probability that the average ticket price at a randomly selected MLB stadium is at most $24.99.
d. Find the probability that the average ticket price at a randomly selected MLB stadium is at least $35.
e. Find the probability that the average ticket price at a randomly selected MLB stadium is between $20 and $39.99, inclusive.

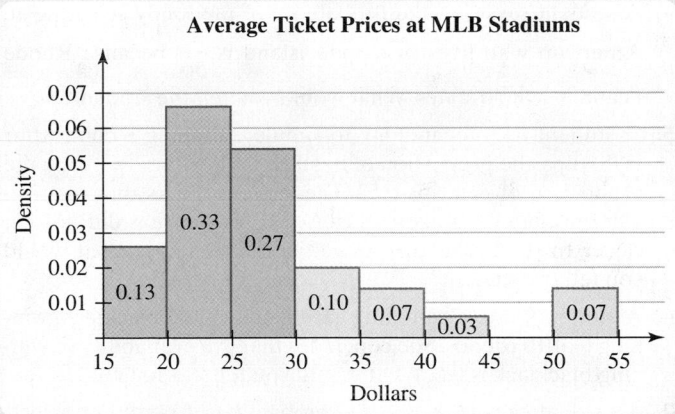

Figure 19 Exercise 56
(**Source:** *Team Marketing Report*)

57. A total of 940 healthy, pregnant women between 20 and 34 years of age, inclusive, participated in a study. The distribution of their lengths of pregnancy is unimodal and symmetric with mean 281 days and standard deviation 8 days (Source: The Length of Human Pregnancy as Calculated by Ultrasonographic Measurement of the Fetal Biparietal Diameter, *H. Kieler et al.*). Estimate the probability that a woman randomly selected from the study had a pregnancy that lasted between

a. 265 and 297 days.
b. 257 and 305 days.
c. 273 and 289 days.

58. In a study of U.S. veteran patients undergoing a first liver transplant, the distribution of their ages is unimodal and symmetric with mean 54 years and standard deviation 6 years (Source: Impact of the MELD Score on Waiting Time and Disease Severity in Liver Transplantation in United States Veterans, *Jawad Ahmad et al.*). Estimate the probability that the age of a veteran randomly selected from the study is between

a. 36 and 72 years.
b. 48 and 60 years.
c. 42 and 66 years.

59. The proportion of Americans who live in Ohio is approximately 0.04 (Source: *U.S. Census Bureau*). Interpret this situation using a probability.

60. For the class of 2015 at Harvard Business School, the proportion of MBA students who are minorities is 0.25 (Source: *Harvard Business School*). Interpret this situation using a probability.

61. The probability of randomly selecting a Hispanic person among Instagram users is 0.17 (Source: Wall Street Journal). Interpret this situation using a proportion.

62. The probability that a randomly selected American adult spends 10 or more hours daily on digital devices is 0.28 (Source: *Vision Council*). Interpret this situation using a proportion.

Concepts

63. A student says that the probability of a traffic light being green when you reach an intersection is $\frac{1}{3}$, because the light can be one of three colors: green, yellow, or red. What would you tell the student?

64. A student says that the probability of randomly selecting an American who lives in Rhode Island is $\frac{1}{50}$, because Rhode Island is 1 of 50 states. What would you tell the student?

65. A student uses technology to simulate flipping a coin. After 1000 flips, the proportion of heads is 0.51. After 2000 flips, the proportion of heads is 0.52. The student thinks that the simulation must be broken, because the proportion did not get closer to the probability of getting heads (0.5). What would you tell the student?

66. A student plays a gambling card game called blackjack, winning 6 of 10 rounds. She concludes that the probability of winning blackjack is 0.6. What would you tell the student?

67. A student calculates that the probability of randomly selecting a thriller novel from his bookcase is $\frac{125}{17}$. What would you tell the student?

68. A student applies to Princeton University. It is so hard to get in, she estimates that the probability she will be accepted is -0.2. What would you tell the student?

69. A car insurance company wants to estimate the probability that a 20-year-old American driver will get in a car accident in the next year. Describe how relative frequency could be used to make such an estimate.

70. A life insurance company wants to estimate the probability that a 60-year-old American who smokes will die in the next year. Describe the research and the calculation(s) that would need to be performed to use relative frequency to make such an estimate.

71. Give an example of a probability. Explain how it can be interpreted as a proportion. (See page 12 for guidelines on writing a good response.)

72. Give an example of a proportion. Explain how it can be interpreted as a probability. (See page 12 for guidelines on writing a good response.)

Hands-On Research

73. a. Flip a coin 20 times and record the outcomes.
   b. Compute the relative frequency (in decimal form) of heads. Compare your result with the *probability* of getting heads from flipping a coin once.
   c. If you were to flip the coin a million times, how would the relative frequency of heads compare with the probability of getting heads from flipping a coin once?

74. a. Roll a six-sided die 10 times and record the outcomes.
   b. Compute the relative frequency (in decimal form) of the outcome 5. Compare your result with the *probability* of rolling a 5 in a single roll.
   c. If you were to roll the die a million times, how would the relative frequency of the outcome 5 compare with the probability of rolling a 5 in a single roll?

75. Use technology to simulate flipping a coin the given number of times. Compute the relative frequency (in decimal form) of tails. Compare your result with the *probability* of getting tails from flipping a coin once.

   a. 10 times   **b.** 100 times
   c. 1000 times   **d.** 10,000 times

76. Use technology to simulate rolling a six-sided die the given number of times. Compute the relative frequency (in decimal form) of the outcome 5. Compare your result with the *probability* of getting a 5 from rolling a six-sided die once.

   a. 10 times   **b.** 100 times
   c. 1000 times   **d.** 10,000 times

▼ 5.2 Complement and Addition Rules

Objectives

» Use the *complement rule* to find probabilities.

» Use the addition rule for disjoint events to find probabilities.

» Use the general addition rule to find probabilities.

Complement Rule

In Chapters 3 and 4, we found proportions of people who were NOT in certain categories. In Example 1, we will find the *probability* of randomly selecting a person who is NOT in a certain category.

> ▶ **Example 1** Find the Probability of Randomly Selecting a Person Who Is NOT in a Certain Category

In a group of 10 students, there are 4 accounting majors, 2 biology majors, 1 kinesiology major, and 3 undecided majors. Use *A* for accounting, *B* for biology, *K* for kinesiology, and *U* for undecided. Find the probability of randomly selecting

  1. an accounting major.
  2. NOT an accounting major.

Solution

  1. $P(accounting) = P(A) = \dfrac{4}{10} = 0.4$

  2. We discuss two methods of finding the probability.

Method 1   The majors that are non-accounting are 2 biology, 1 kinesiology, and 3 undecided. So, there are $2 + 1 + 3 = 6$ non-accounting majors out of 10 students. Therefore, the probability is $\dfrac{6}{10} = 0.6$.

Method 2 $P(A) = 0.4$, and the probabilities of $A, B, K,$ and U add to 1. So, we can find the probability of NOT selecting an accounting major by subtracting 0.4 from 1:

$$1 - 0.4 = 0.6$$

So, the probability is 0.6, which is the same result we found earlier.

In Example 1, we found that the probability of selecting a non-accounting major. We say selecting a non-accounting major is the *complement* of selecting an accounting major.

> **Definition   Complement**
>
> The **complement** of an event E is the event that consists of all the outcomes not in E. We write "NOT E" to stand for the complement of E.

For the experiment of randomly selecting an adult, the complement of selecting a woman is selecting a man.

We can represent an event and its complement with a figure called a *Venn diagram* (see Fig. 20). A **Venn diagram** consists of one or more shaded circles that are all bordered by a rectangle. The region inside the rectangle represents the sample space, and the region inside each circle represents an event.

Consider the experiment of rolling a six-sided die once, and let the event E stand for the outcomes 1 and 2. In Fig. 20, E is represented by the blue-shaded circle with "1" and "2" inside it. The complement of E is represented by the orange region with "3," "4," "5," and "6" in it.

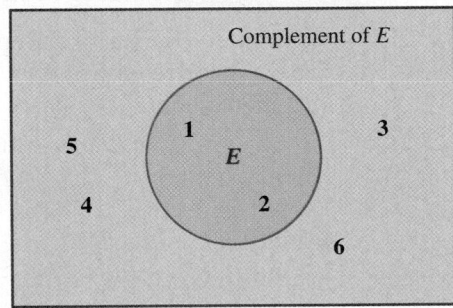

Figure 20 An Event E (in blue) and the complement of E (in orange)

In Example 1, we found that

$$P(\text{NOT } A) = 1 - 0.4$$

In Example 1, we also found that $P(A) = 0.4$. Substituting $P(A)$ for 0.4 gives

$$P(\text{NOT } A) = 1 - P(A)$$

This equation suggests the following rule.

> **Complement Rule**
>
> For any event E,
>
> $$P(\text{NOT } E) = 1 - P(E)$$
>
> In words, the probability of the complement of E is equal to 1 minus the probability of E.

▶ **Example 2**   Using the Complement Rule to Find Probabilities

Consider the experiment of randomly selecting a student in a prestatistics class. The probability of selecting a student with blue eyes is 0.3. Find the probability of selecting a student who does not have blue eyes.

Solution

We let B stand for the outcome of a person with blue eyes. So, $P(B) = 0.3$. We apply the complement rule:

$$P(\text{NOT } B) = 1 - P(B) = 1 - 0.3 = 0.7$$

So, the probability of selecting a student who does not have blue eyes is 0.7.

Addition Rule for Disjoint Events

Before we discuss the next probability rule, we need to define *disjoint events*.

▶ **Definition   Disjoint events**

Two events are **disjoint** (or **mutually exclusive**) if they have no outcomes in common.

For example, consider randomly selecting a person in a town. The event of selecting a teenager and the event of selecting a senior citizen are disjoint.

In Example 3, we will find a probability having to do with disjoint events.

▶ **Example 3**   Finding Probabilities Involving Disjoint Events

A committee consists of 5 Republicans, 3 Democrats, and 2 Independents. One person is randomly selected from the committee. Use R for Republican, D for Democrat, and I for Independent.

  1. Are the events R and D disjoint? Explain.
  2. Find $P(R)$.
  3. Find $P(D)$.
  4. Find $P(R \text{ OR } D)$.
  5. Find $P(R) + P(D)$.
  6. Compare the results of Problems 4 and 5.

Solution

  1. The events are disjoint because Republicans and Democrats have no members in common.

  2. There are 5 Republicans out of 10 members, so $P(R) = \dfrac{5}{10} = 0.5$.

  3. There are 3 Democrats out of 10 members, so $P(D) = \dfrac{3}{10} = 0.3$.

  4. The total number of Republicans and Democrats is $5 + 3 = 8$. So, $P(R \text{ OR } D) = \dfrac{8}{10} = 0.8$.

  5. In Problems 2 and 3, we found that $P(R) = 0.5$ and $P(D) = 0.3$, respectively. So, $P(R) + P(D) = 0.5 + 0.3 = 0.8$.
  6. The results of Problems 4 and 5 are equal.

In Example 3, we found that $P(R \text{ OR } D)$ and $P(R) + P(D)$ are equal:

$$P(R \text{ OR } D) = P(R) + P(D)$$

This is true for any two *disjoint* events, R and D.

▶ **Addition Rule for Disjoint Events**

If E and F are disjoint events, then

$$P(E \text{ OR } F) = P(E) + P(F)$$

In words, if two events have no outcomes in common, then the probability of one event OR the other event occurring is equal to the sum of their probabilities.

A Venn diagram can help us visualize this rule. We return to the disjoint events R and D, where R stands for Republican and D stands for Democrat (see Fig. 21). The 5 dots in the circle for R stand for the 5 Republicans, the 3 dots in D stand for the 3 Democrats, and the 2 dots outside both circles stand for the 2 Independents.

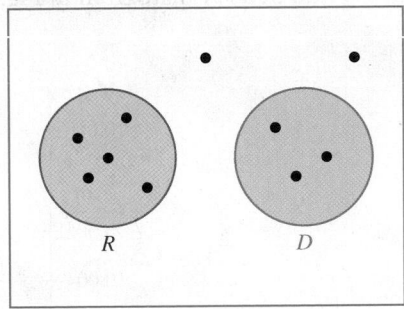

Figure 21 Venn diagram of two disjoint events

Because there are 5 outcomes in R out of 10 outcomes, $P(R) = \dfrac{5}{10} = 0.5$. Since there are 3 outcomes in D out of 10 outcomes, $P(D) = \dfrac{3}{10} = 0.3$. Because there are 8 outcomes in R OR D out of 10 outcomes, $P(R \text{ OR } D) = \dfrac{8}{10} = 0.8$. Now we verify the addition rule for disjoint events:

$$P(R \text{ OR } D) \stackrel{?}{=} P(R) + P(D) \quad \text{Addition rule for disjoint events}$$
$$0.8 \stackrel{?}{=} 0.5 + 0.3 \quad \text{Substitute the probabilities.}$$
$$0.8 \stackrel{?}{=} 0.8$$
$$\text{true}$$

▶ **Example 4** Using the Addition Rule for Disjoint Events

A six-sided die is rolled once. Use E to stand for an even-number outcome and F to stand for the outcome 5. Find the given probabilities.

1. $P(E)$ **2.** $P(F)$ **3.** $P(E \text{ OR } F)$

Solution

1. The even-number outcomes are 2, 4, and 6. Because 3 of the 6 outcomes are even,
$$P(E) = \frac{3}{6} = \frac{1}{2}.$$

2. The outcome 5 is 1 of 6 outcomes. So, $P(F) = \dfrac{1}{6}$.

3. Because an outcome cannot be even and the number 5, the events E and F are disjoint. So, we can use the addition rule for disjoint events to find $P(E \text{ OR } F)$:

$$P(E \text{ OR } F) = P(E) + P(F) \quad \text{Addition rule for disjoint events}$$

$$= \frac{3}{6} + \frac{1}{6} \quad \text{Substitute the probabilities found in Problems 1 and 2.}$$

$$= \frac{4}{6} \quad \frac{a}{b} + \frac{c}{b} = \frac{a+c}{b}$$

$$= \frac{2}{3} \quad \text{Simplify.}$$

So, $P(E \text{ OR } F) = \frac{2}{3}$.

▶ Example 5   Using the Addition Rule for Disjoint Events

The weights (in pounds) of type 2 diabetes patients from 2006 to 2007 are described by the density histogram in Fig. 22. The weights are rounded to the ones place.

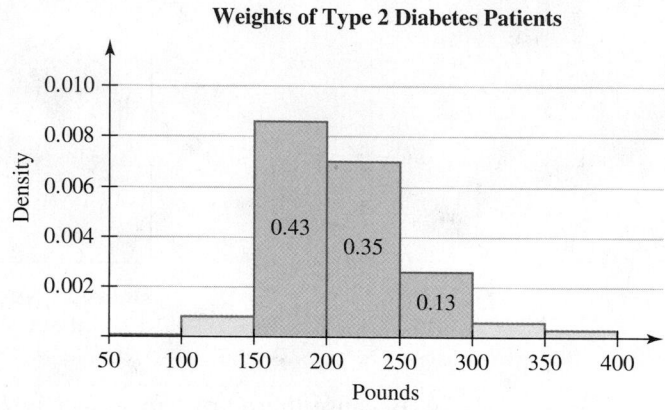

Weights of Type 2 Diabetes Patients

Figure 22 Weights of type 2 diabetes patients
(**Source:** *NAMCS, NHAMCS*)

1. Find the probability that a patient randomly selected from the study would weigh between 150 and 299 pounds, inclusive.
2. Find the probability that a patient randomly selected from the study would weigh less than 150 pounds OR at least 300 pounds.
3. Find the probability that a patient randomly selected from the study would weigh between 150 and 199 pounds, inclusive, OR between 250 and 299 pounds, inclusive.

Solution

1. The proportion of the observations between 150 and 299 pounds, inclusive, is equal to the sum of the areas of the orange bars. So, the proportion is $0.43 + 0.35 + 0.13 = 0.91$. Therefore, the probability that a patient randomly selected from the study would weigh between 150 and 299 pounds, inclusive, is 0.91.
2. The probability of the given event is equal to the total area of the blue bars, but the areas of those bars are not given. So, we apply the complement rule and conclude that the total area of the blue bars is equal to 1 minus the total area of the orange bars, which we found in Problem 1. Using symbols, we write

$P(\text{less than 150 pounds OR at least 300 pounds})$

$= 1 - P(\text{between 150 and 299 pounds, inclusive}) \quad \text{Complement rule}$

$= 1 - 0.91 \quad \text{Substitute the result 0.91 from Problem 1.}$

$= 0.09$

So, the probability is 0.09.

3. The event between 150 and 199 pounds, inclusive, and the event between 250 and 299 pounds, inclusive, are disjoint because a patient cannot have a weight in both classes. From the far-left orange rectangle, we see that $P(\text{between 150 and 199 pounds, inclusive}) = 0.43$. From the far-right orange rectangle, we see that $P(\text{between 250 and 299 pounds, inclusive}) = 0.13$. We use these results after applying the addition rule for disjoint events:

$P(\text{between 150 and 199 pounds, inclusive, OR between 250 and 299 pounds, inclusive})$

$= P(\text{between 150 and 199 pounds, inclusive})$

$\quad + P(\text{between 250 and 299 pounds, inclusive})$ *Addition rule for disjoint events*

$= 0.43 + 0.13$

$= 0.56$

So, the probability is 0.56.

General Addition Rule

In Example 6, we will explore finding probabilities that involve non-disjoint events. This will help us discover another probability rule.

▶ **Example 6**   Finding Probabilities That Involve Non-Disjoint Events

A total of 1287 randomly selected adults were asked the following question: "Do you agree or disagree? Gay couples should have the right to marry one another." Their responses and their religious preferences are summarized in Table 8.

Table 8 Gay Couples' Right to Marry versus Religious Preference

	Protestant	Catholic	Jewish	Other	None	Total
Agree	230	171	12	46	167	626
Neither Agree nor Disagree	71	36	1	22	24	154
Disagree	308	94	2	60	43	507
Total	609	301	15	128	234	1287

Source: *General Social Survey*

An adult is randomly selected from those surveyed. Use D to stand for an adult who disagrees and C to stand for a Catholic. Find the given probability.

  1. $P(D)$
  2. $P(C)$
  3. $P(D \text{ AND } C)$
  4. $P(D \text{ OR } C)$

Solution

  1. From the far-right column, we see 507 of the 1287 surveyed adults disagree with the statement. So, $P(D) = \dfrac{507}{1287} \approx 0.394$.

  2. From the bottom row, we see 301 of the 1287 surveyed adults are Catholic. So, $P(C) = \dfrac{301}{1287} \approx 0.234$.

  3. The adults who disagreed with the statement are described by the yellow row in Table 9, and the adults who are Catholics are described by the green column. So, the number of adults who disagreed AND are Catholic is in *both* the yellow row and the green column. The number is 94 out of 1287 adults, so $P(D \text{ AND } C) = \dfrac{94}{1287} \approx 0.073$.

Table 9 Gay Couples' Right to Marry versus Religious Preference

	Protestant	Catholic	Jewish	Other	None	Total
Agree	230	171	12	46	167	626
Neither Agree nor Disagree	71	36	1	22	24	154
Disagree	308	94	2	60	43	507
Total	609	301	15	128	234	1287

4. The 507 adults who disagreed with the statement are described by the yellow row in Table 9, and the 301 Catholics are described by the green column in the same table. So, the number of adults who disagreed OR are Catholic is the sum of all the numbers in either the yellow row or the green column, but we do not count 94 twice: $507 + 301 - 94$. There are a total of 1287 adults, so

$$P(D \text{ OR } C) = \frac{507 + 301 - 94}{1287} = \frac{714}{1287} \approx 0.555$$

In Problem 1 of Example 6, we found that $P(D) \approx 0.394$. Similar to finding approximate proportions, **we will round approximate probabilities to the third decimal place.**

In Problem 4 of Example 6, we found that

$$P(D \text{ OR } C) = \frac{507 + 301 - 94}{1287}$$

We can write the fraction as the sum and difference of three fractions:

$$P(D \text{ OR } C) = \frac{507}{1287} + \frac{301}{1287} - \frac{94}{1287}$$

In Problems 1–3 of Example 6, we found that the fractions $\frac{507}{1287}, \frac{301}{1287},$ and $\frac{94}{1287}$ are $P(D)$, $P(C)$, and $P(D \text{ AND } C)$, respectively. Substituting the probabilities for the fractions in the equation gives

$$P(D \text{ OR } C) = P(D) + P(C) - P(D \text{ AND } C)$$

This equation suggests the following rule.

▶ **General Addition Rule**

For any events E and F,

$$P(E \text{ OR } F) = P(E) + P(F) - P(E \text{ AND } F)$$

In words, the probability of one event OR the other event occurring is equal to the sum of their probabilities minus the probability of one event AND the other event occurring.

A Venn diagram can help us visualize this rule. Suppose we roll a six-side die once, where event E consists of the outcomes 1, 2, and 3, and event F consists of the outcomes 3, 4, and 5 (see Fig. 23). The region formed by combining the two circles' regions represents $E \text{ OR } F$, and the football-shaped overlapping region represents $E \text{ AND } F$.

To find $P(E \text{ OR } F)$, we see from the Venn diagram that simply finding the sum $P(E) + P(F)$ would double-count $P(E \text{ AND } F)$, so we subtract $P(E \text{ AND } F)$ from the sum:

$$P(E \text{ OR } F) = P(E) + P(F) - P(E \text{ AND } F)$$

This is the general addition rule.

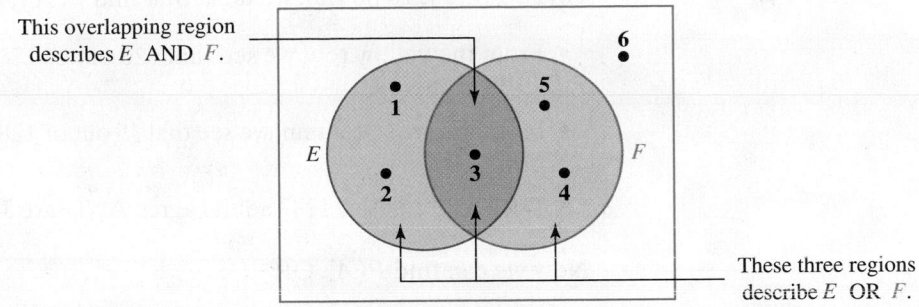

Figure 23 Venn diagram of two non-disjoint events

We can use the events described by the Venn diagram to verify the general addition rule:

$$P(E \text{ OR } F) \stackrel{?}{=} P(E) + P(F) - P(E \text{ AND } F) \quad \textit{General addition rule}$$

$$\frac{5}{6} \stackrel{?}{=} \frac{3}{6} + \frac{3}{6} - \frac{1}{6} \quad \textit{Compute the probabilities.}$$

$$\frac{5}{6} \stackrel{?}{=} \frac{5}{6} \qquad \frac{a}{b} + \frac{c}{b} = \frac{a+c}{b}; \frac{a}{b} - \frac{c}{b} = \frac{a-c}{b}$$

true

▶ **Example 7** Using the General Addition Rule

A six-sided die is rolled once. Find the probability of rolling a number that is at least 3 OR is even.

Solution

Let T be an outcome that is at least 3 $(3, 4, 5, 6)$. Let E be an even outcome $(2, 4, 6)$. Because the outcomes 4 and 6 are in both T and E, the events are *not* disjoint. So, we must use the general addition rule to find $P(T \text{ OR } E)$:

$$P(T \text{ OR } E) = P(T) + P(E) - P(T \text{ AND } E) \quad \textit{General addition rule}$$

$$= \frac{4}{6} + \frac{3}{6} - \frac{2}{6} \quad \textit{Substitute } P(T) = \frac{4}{6}, P(E) = \frac{3}{6}, \textit{ and } P(T \text{ AND } E) = \frac{2}{6}.$$

$$= \frac{5}{6} \qquad \frac{a}{b} + \frac{c}{b} = \frac{a+c}{b}; \frac{a}{b} - \frac{c}{b} = \frac{a-c}{b}$$

▶ **Example 8** Using the General Addition Rule

We return to the survey about gay couples' right to marry versus religious preference (see Table 10). Find the probability of randomly selecting a surveyed adult who agrees OR is Jewish. Round the result to the third decimal place.

Table 10 Gay Couples' Right to Marry versus Religious Preference

	Protestant	Catholic	Jewish	Other	None	Total
Agree	230	171	12	46	167	626
Neither Agree nor Disagree	71	36	1	22	24	154
Disagree	308	94	2	60	43	507
Total	609	301	15	128	234	1287

Source: *General Social Survey*

Solution

We use A to stand for agree and J to stand for Jewish. The number of adults who agree AND are Jewish is 12, which is in both the yellow row and the green column in Table 10.

Because A and J are non-disjoint, we must use the general addition rule to find $P(A$ OR $J)$. To do this, we must first find $P(A), P(J),$ and $P(A$ AND $J)$:

- From the yellow row, we see that 626 out of 1287 adults agree. So, $P(A) = \dfrac{626}{1287}$.

- From the green column, we see that 15 out of 1287 adults are Jewish. So, $P(J) = \dfrac{15}{1287}$.

- Because 12 out of 1287 adults agree AND are Jewish, $P(A$ AND $J) = \dfrac{12}{1287}$.

Now we can find $P(A$ OR $J)$:

$$
\begin{aligned}
P(A \text{ OR } J) &= P(A) + P(J) - P(A \text{ AND } J) &&\text{\textit{General addition rule}}\\
&= \frac{626}{1287} + \frac{15}{1287} - \frac{12}{1287} &&\text{\textit{Substitute the probabilities found earlier.}}\\
&= \frac{629}{1287} &&\frac{a}{b} + \frac{c}{b} = \frac{a+c}{b}; \frac{a}{b} - \frac{c}{b} = \frac{a-c}{b}\\
&\approx 0.489
\end{aligned}
$$

So, the probability of randomly selecting a surveyed adult who agrees OR is Jewish is approximately 0.489.

▶

Suppose events E and F are disjoint. What would happen if we were to use the general addition rule to find $P(E$ OR $F)$? Here we investigate:

$$
\begin{aligned}
P(E \text{ OR } F) &= P(E) + P(F) - P(E \text{ AND } F) &&\text{\textit{General addition rule}}\\
&= P(E) + P(F) - 0 &&\text{\textit{P(E AND F) = 0}}\\
& &&\text{\textit{because E and F are disjoint.}}\\
&= P(E) + P(F)
\end{aligned}
$$

Note that the result is the addition rule for disjoint events:

$$P(E \text{ OR } F) = P(E) + P(F)$$

So, if the general addition rule is used with two disjoint events, the result will be the same as using the addition rule for disjoint events.

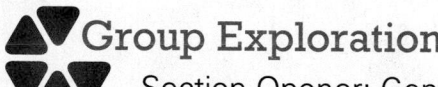

Group Exploration

Section Opener: General addition rule

A random selection of American adults who were planning on watching the Super Bowl were asked whether they were more interested in watching the game or the commercials. Table 11 compares their responses and their genders.

Table 11 Super Bowl Game versus Commercials

	Women	Men	Total
Game	116	156	272
Commercials	82	42	124
Total	198	198	396

Source: *The Gallup Organization*

Assume one respondent is randomly selected. Let G stand for the event that the person's main interest is the game and W stand for the event that the person is a woman.

1. Find $P(G)$.
2. Find $P(W)$.
3. Find $P(G$ AND $W)$.
4. Find $P(G$ OR $W)$.
5. Why is $P(G) + P(W)$ larger than $P(G$ OR $W)$? Refer to the data in Table 11 in your response.
6. How can you use $P(G), P(W),$ and $P(G$ AND $W)$ to find $P(G$ OR $W)$?
7. Use your response to Problem 6 to help you write an equation in which the expression on the left side is $P(G$ OR $W)$ and the expression on the right side involves $P(G), P(W),$ and $P(G$ AND $W)$.

Homework 5.2

1. The complement of an event E is the event that contains all the outcomes NOT in ____.

2. For any event E, $P(\text{NOT } E) =$ _____.

3. If E and F are disjoint events, then $P(E \text{ OR } F) =$ _____.

4. If E and F are not disjoint events, then $P(E \text{ OR } F) =$ _____.

5. If $P(D) = 0.7$, find $P(\text{NOT } D)$.

6. If $P(R) = 0.2$, find $P(\text{NOT } R)$.

7. If A and B are disjoint events in which $P(A) = 0.3$ and $P(B) = 0.4$, find $P(A \text{ OR } B)$.

8. If M and R are disjoint events in which $P(M) = 0.7$ and $P(R) = 0.2$, find $P(M \text{ OR } R)$.

9. If K and W are events in which $P(K) = 0.4$, $P(W) = 0.8$, and $P(K \text{ AND } W) = 0.3$, find $P(K \text{ OR } W)$.

10. If C and D are events in which $P(C) = 0.5$, $P(D) = 0.4$, and $P(C \text{ AND } D) = 0.2$, find $P(C \text{ OR } D)$.

For Exercises 11–16, suppose there are 8 members of a rock band: 2 guitarists, 1 lead singer, 3 backup singers, 1 bass player, and 1 drummer. No one both sings and plays an instrument. A person is randomly selected from the band. Find the given probability in fraction form.

11. $P(\text{a guitarist})$

12. $P(\text{a backup singer})$

13. $P(\text{NOT a guitarist})$

14. $P(\text{NOT a backup singer})$

15. $P(\text{a guitarist OR a singer})$

16. $P(\text{a stringed instrument player OR a backup singer})$

In Exercises 17–24, you will analyze the gambling game roulette in which a ball is dropped onto a revolving wheel that has 38 numbered slots: the numbers 1–36, a zero, and a double zero. There are 18 red slots, 18 black slots, and 2 green slots:

Red: 1, 3, 5, 7, 9, 12, 14, 16, 18, 19, 21, 23, 25, 27, 30, 32, 34, 36

Black: 2, 4, 6, 8, 10, 11, 13, 15, 17, 20, 22, 24, 26, 28, 29, 31, 33, 35

Green: 0, 00

Consider the random experiment of dropping the ball onto the wheel once. Find the given probability in fraction form.

17. $P(\text{NOT green})$

18. $P(\text{NOT black})$

19. $P(\text{black OR green})$

20. $P(\text{red OR black})$

21. $P(\text{red AND at most 10})$

22. $P(\text{black AND at most 7})$

23. $P(\text{black OR at least 30})$

24. $P(\text{red OR at least 28})$

For Exercises 25–32, a person rolls a six-sided die once. Find the given probability in fraction form.

25. $P(\text{NOT even})$

26. $P(\text{NOT odd})$

27. $P(\text{even AND at least 3})$

28. $P(\text{odd AND at most 3})$

29. $P(\text{even OR at least 3})$

30. $P(\text{odd OR at most 3})$

31. $P(\text{even AND an odd number})$

32. $P(\text{less than 2 AND greater than 4})$

For Exercises 33–40, suppose someone spins the spinner shown in Fig. 24 once. Find the given probability in fraction form.

33. $P(2)$

34. $P(4)$

35. $P(\text{NOT 2})$

36. $P(\text{NOT 4})$

37. $P(\text{NOT even})$

38. $P(\text{NOT odd})$

39. $P(\text{odd OR at most 2})$

40. $P(\text{even OR at least 3})$

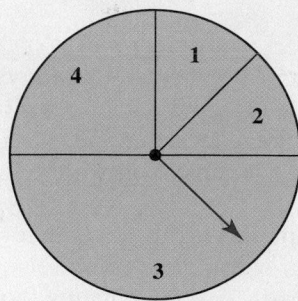

Figure 24 A spinner (Exercises 33–40)

41. The percentages of weapons used in murders in 2010 are described by the pie chart in Fig. 25.

Types of Weapons Used in Murders

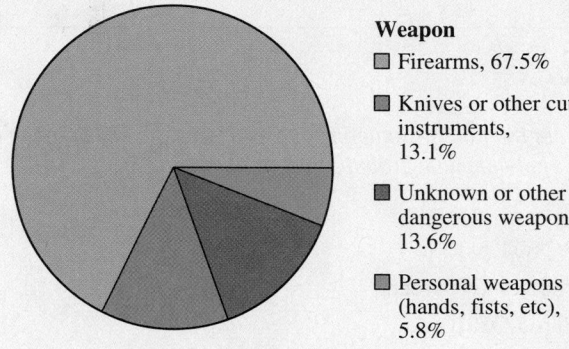

Weapon

☐ Firearms, 67.5%

◼ Knives or other cutting instruments, 13.1%

◼ Unknown or other dangerous weapons, 13.6%

☐ Personal weapons (hands, fists, etc), 5.8%

Figure 25 Exercises 41 and 42
(**Source:** *FBI*)

a. Find the probability that a weapon randomly selected from those used in murders in 2010 was a firearm.

b. Find the probability that a weapon randomly selected from those used in murders in 2010 was NOT a firearm.

c. Find the probability that a weapon randomly selected from those used in murders in 2010 was a firearm OR a personal weapon.

d. From inspecting the pie chart, a student concludes that in any year, 5.8% of weapons used in murders are personal weapons. What would you tell the student?

42. The percentages of weapons used in murders in 2010 are described by the pie chart in Fig. 25.

a. Find the probability that a weapon randomly selected from those used in murders in 2010 was a personal weapon.

b. Find the probability that a weapon randomly selected from those used in murders in 2010 was NOT a personal weapon.

c. Find the probability that a weapon randomly selected from those used in murders in 2010 was a personal weapon OR a knife or other cutting instrument.

d. Explain why the following statements do *not* have the same meaning. Which one is correct?

- In 2010, 13.1% of murders involved knives or other cutting instruments.

- In 2010, 13.1% of weapons used in murders were knives or other cutting instruments.

43. A total of 18,525 randomly selected adults were contacted by phone and asked to what degree social media influenced their purchasing decisions. Their responses are summarized by the pie chart in Fig. 26.

Social Media Influence on Purchasing Decisions

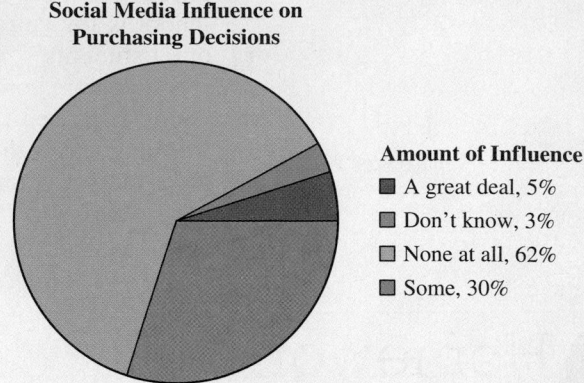

Amount of Influence

◼ A great deal, 5%

◼ Don't know, 3%

◼ None at all, 62%

◼ Some, 30%

Figure 26 Exercises 43 and 44
(**Source:** *The Gallup Organization*)

Suppose an adult from the study is randomly selected. Consider the following events, which describe to what degree social media influences the adult's purchasing decisions.

- *G*: a great deal
- *D*: don't know

a. Find $P(G)$.

b. Find $P(\text{NOT } G)$.

c. Find $P(G \text{ OR } D)$.

d. The percentages in the pie chart are based on the adults' perceptions. Suppose that 10% of those who said they were not influenced at all were actually influenced to some extent and that 10% of those who said they were influenced to some extent were actually influenced a great deal. What would be the actual percentages for the four categories of amount of influence? Round your results to the ones place.

e. What type(s) of sampling bias likely occurred? Explain.

44. A total of 18,525 adults were asked to what degree social media influenced their purchasing decisions. Their responses are summarized by the pie chart in Fig. 26. Suppose an adult from the study is randomly selected. Consider the following events, which describe to what degree social media influences the adult's purchasing decisions.

- *N*: none at all
- *S*: some

a. Find $P(S)$.

b. Find $P(\text{NOT } S)$.

c. Find $P(S \text{ OR } N)$.

d. From inspecting the pie chart, a student concludes that 62% of all American adults think that social media has no influence at all on their purchasing demands. What would you tell the student?

45. In 2013, 4579 runners completed the Great Cow Harbor 10K Run in Northport, New York. Their ages (in years) are described by the density histogram in Fig. 27.

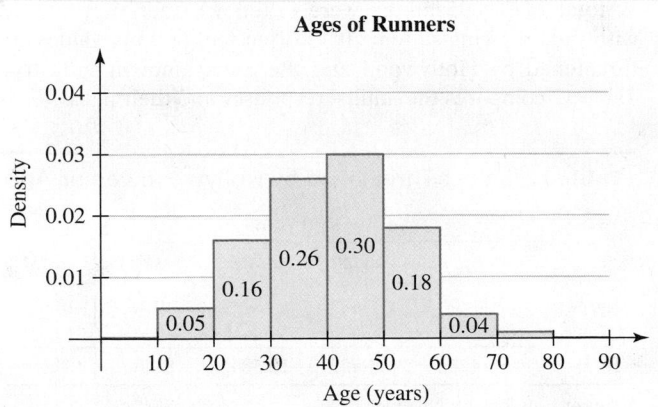

Figure 27 Exercise 45

(**Source:** *The Great Cow Harbor 10-Kilometer Run, Inc.*)

a. Find the probability that a randomly selected runner's age was between 10 and 69 years, inclusive.

b. Find the probability that a randomly selected runner's age was at most 9 years OR at least 70 years. Some of the bars have been left blank on purpose.

c. Find the probability that a randomly selected runner's age was between 20 and 29 years, inclusive, OR between 50 and 59 years, inclusive.

d. From inspecting the density histogram, a student concludes that 5% of all Americans who run are between 10 and 19 years in age, inclusive. What would you tell the student?

e. The ages of the runners were self-reported. What type of bias likely occurred? Explain.

46. In 1994, California passed a three-strikes sentencing law mandating that defendants convicted of a third felony should serve a state prison term of 25 years to life. The density histogram in Fig. 28 describes the ages (in years) of third-strike felons in California's state prison population as of September 30, 2012.

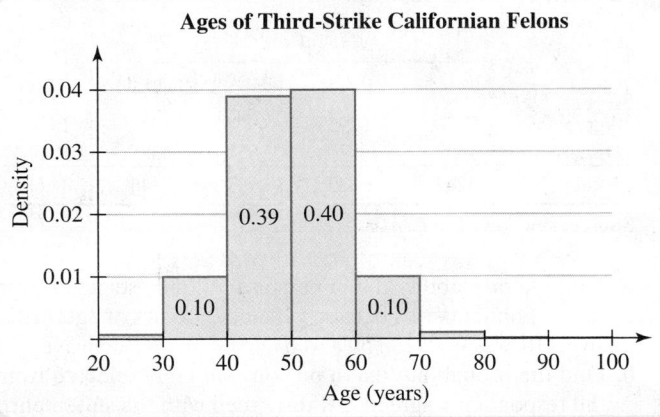

Figure 28 Exercise 46

(**Source:** *California Department of Corrections and Rehabilitation*)

a. Find the probability that a third-strike Californian felon randomly selected on September 30, 2012, was between 30 and 69 years of age, inclusive.

b. Find the probability that a third-strike Californian felon randomly selected on September 30, 2012, was younger than 30 years OR at least 70 years of age. Some of the bars have been left blank on purpose.

c. Find the probability that a third-strike Californian felon randomly selected on September 30, 2012, was between 30 and 39 years of age, inclusive, OR between 60 and 69 years of age, inclusive.

d. Nearly 120,000 inmates are jailed in California prisons meant to hold 84,000 inmates (Source: Wall Street Journal). On November 6, 2012, California passed a proposition allowing third-strike felons to apply for reduced sentences, provided that their third offense was not violent or serious. More than 2800 prisoners may be able to reduce their sentences (Source: *California Department of Corrections and Rehabilitations*). If all 2800 prisoners were released, would that do much to reduce the overcrowding in California prisons?

47. The numbers of minutes that United Airlines flights were delayed are described by the density histogram in Fig. 29 for the 1219 United flights that departed from O'Hare Airport in Chicago during the period October 2–7, 2013. Some flights were delayed at least 150 minutes, but the bars would not be visible. The data are rounded to the ones place.

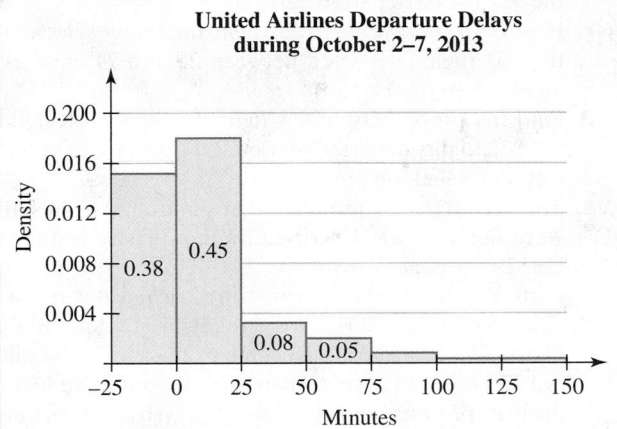

Figure 29 Exercise 47

(**Source:** *Bureau of Transportation Statistics*)

a. The minimum observation is −15 minutes. What does it mean in this situation?

b. Find the probability that a flight randomly selected from the 1219 flights departed early.

c. Find the probability that a flight randomly selected from the 1219 flights departed between 25 and 74 minutes late, inclusive.

d. Find the probability that a flight randomly selected from the 1219 flights departed on time OR late. Some of the bars were left blank on purpose.

e. From inspecting the density histogram, a student concludes that 38% of United flights departing from O'Hare Airport tomorrow will depart early. What would you tell the student?

48. The numbers of minutes that United Airlines flights were delayed are described by the density histogram in Fig. 30 for the 922 United flights that departed from O'Hare Airport in Chicago during the period December 22–28, 2013. Some flights were delayed at least 175 minutes, but the bars would not be visible. The data are rounded to the ones place.

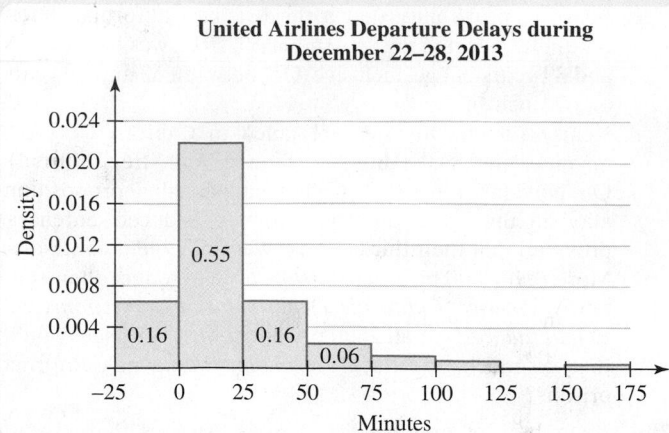

Figure 30 Exercise 48
(**Source:** *Bureau of Transportation Statistics*)

a. The minimum observation is −8 minutes. What does it mean in this situation?

b. Find the probability that a flight randomly selected from the 922 flights departed early.

c. Find the probability that a flight randomly selected from the 922 flights departed between 25 and 74 minutes late, inclusive.

d. Find the probability that a flight randomly selected from the 922 flights departed on time OR late. Some of the bars were left blank on purpose.

e. The numbers of minutes that United Airlines flights were delayed are described by the density histogram in Fig. 29 on page 309 for the 1219 United flights that departed from O'Hare Airport in Chicago during the period October 2–7, 2013. Were flights more delayed during October 2–7 or during December 22–28? Explain. If you did Exercise 47, your explanation should include a comparison of your results in parts (a), (b), and (c) of that exercise and your results of parts (a), (b), and (c) of this exercise, respectively.

For Exercises 49–56, one of the dates of June 2016 is randomly selected (see Fig. 31). Find the given probability in fraction form.

49. $P(\text{NOT a Wednesday})$

50. $P(\text{NOT in the third week})$

51. $P(\text{in the second week OR in the fifth week})$

52. $P(\text{a Tuesday OR a Thursday})$

53. $P(\text{in the second week AND a Tuesday})$

54. $P(\text{in the third week AND a Saturday})$

55. $P(\text{in the first week OR a Thursday})$

56. $P(\text{in the second week OR a Sunday})$

Sun	Mon	Tues	Wed	Thurs	Fri	Sat
			1	2	3	4
5	6	7	8	9	10	11
12	13	14	15	16	17	18
19	20	21	22	23	24	25
26	27	28	29	30		

Figure 31 Dates of June 2016 (Exercises 49–56)

57. A total of 34,695 adults were asked whether they agreed with the following statement: "I often feel that my values are threatened by Hollywood and the entertainment industry." Table 12 compares the adults' responses and their ages.

Table 12 Values Threatened by Hollywood versus Age

	Ages (in years)				
	18–29	30–49	50–64	over 64	Total
Agree	1400	4847	4823	3910	14,980
Disagree	2842	6976	5661	4236	19,715
Total	4242	11,823	10,484	8146	34,695

Source: *Pew Research Center*

a. Find the probability that a person randomly selected from the respondents was between 30 and 49 years of age, inclusive, OR was over 64 years of age.

b. Find the probability that a person randomly selected from the respondents agreed OR disagreed with the statement.

c. Find the probability that a person randomly selected from the respondents agreed AND disagreed with the statement.

d. Find the probability that a person randomly selected from the respondents agreed with the statement OR was between 18 and 29 years of age, inclusive.

e. Find the probability that a person randomly selected from the respondents agreed with the statement AND was between 18 and 29 years of age, inclusive.

58. A total of 34,695 adults were asked whether they agreed with the following statement: "Evolution is the best explanation for the origins of human life on earth." Table 13 compares the adults' responses and their ages.

Table 13 Evolution Is the Best Explanation of Human Life on Earth versus Age

	Ages (in years)				
	18–29	30–49	50–64	over 64	Total
Agree	2333	5793	4927	3258	16,311
Disagree	1909	6030	5557	4888	18,384
Total	4242	11,823	10,484	8146	34,695

Source: *Pew Research Center*

a. Find the probability that a person randomly selected from the respondents was between 18 and 29 years of age, inclusive, OR was between 50 and 64 years of age, inclusive.

b. Find the probability that a person randomly selected from the respondents agreed OR disagreed with the statement.

c. Find the probability that a person randomly selected from the respondents agreed AND disagreed with the statement.

d. Find the probability that a person randomly selected from the respondents disagreed with the statement OR was between 50 and 64 years of age, inclusive.

e. Find the probability that a person randomly selected from the respondents disagreed with the statement AND was between 50 and 64 years of age, inclusive.

59. A total of 803 adults were asked whether they prefer to drink coffee, tea, or neither. Table 14 compares the adults' responses and the regions of the country where they live.

Table 14 Drink Preference versus Region of the Country

	Northeast	Midwest	South	West	Total
Coffee	101	102	169	89	461
Tea	40	52	106	42	240
Neither	11	39	28	24	102
Total	152	193	303	155	803

Source: *YouGov*

Let S stand for a person who lives in the South, C stand for a person who prefers coffee, and T stand for a person who prefers tea. Assuming a person is randomly selected from the 803 adults, find the given probability.

a. $P(\text{NOT } T)$
b. $P(C \text{ AND } T)$
c. $P(C \text{ OR } T)$
d. $P(S \text{ AND } T)$
e. $P(S \text{ OR } T)$

60. A total of 803 adults were asked whether they prefer to drink coffee, tea, or neither. Table 14 compares the adults' responses and the regions of the country where they live. Let N stand for a person who lives in the Northeast, W stand for a person who lives in the West, and C stand for a person who prefers coffee. Assuming a person is randomly selected from the 803 adults, find the given probability.

a. $P(\text{NOT } C)$
b. $P(N \text{ AND } W)$
c. $P(N \text{ OR } W)$
d. $P(N \text{ AND } C)$
e. $P(N \text{ OR } C)$

Concepts

61. Let E be an event in a sample space S. What is the complement of the complement of E? Explain.

62. If A and B are disjoint events and B and C are disjoint events, does that guarantee that A and C are disjoint? Explain.

63. Give an example of two disjoint events E and F in which $P(E \text{ OR } F) = 1$.

64. Give an example of two disjoint events E and F in which $P(E \text{ OR } F) < 1$.

65. Let E and F be events for some random experiment.

a. Can $P(E \text{ AND } F)$ ever be larger than $P(E)$? If yes, give an example. If no, explain.
b. Can $P(E \text{ AND } F)$ ever be equal to $P(E)$? If yes, give an example. If no, explain.

66. Let E and F be events for some random experiment.

a. Can $P(E \text{ OR } F)$ ever be less than $P(E)$? If yes, give an example. If no, explain.
b. Can $P(E \text{ OR } F)$ ever be equal to $P(E)$? If yes, give an example. If no, explain.

67. If E and F are disjoint events, then $P(E \text{ OR } F) = P(E) + F(F)$. But if E and F are non-disjoint events, then we must use the more complicated rule $P(E \text{ OR } F) = P(E) + F(F) - P(E \text{ AND } F)$. Use a Venn diagram to help explain why this makes sense.

68. Find a general addition rule for three events. [**Hint:** Draw a Venn diagram.]

5.3 Conditional Probability and the Multiplication Rule for Independent Events

Objectives

» Find and interpret *conditional probabilities.*

» Determine whether two events are *independent* or *dependent.*

» Use the multiplication rule for independent events to find probabilities.

Conditional Probability

Let's define a man to be tall if his height is at least 76 inches. The probability that we would randomly select a tall man from all American men is small: approximately 0.02. Here, the sample space is all American men. But what if we narrow the sample space to just NBA basketball players? The probability that we would randomly select a tall man from NBA basketball players is much larger: approximately 0.73 (Source: *NBA*). We say 0.73 is the probability that we would select a tall man, *given that the man is an NBA player*. Using symbols, we write $P(\text{tall} \mid \text{NBA player}) = 0.73$. This is an example of a *conditional probability*.

▶ **Definition Conditional probability**

If E and F are events, then the **conditional probability** $P(E \mid F)$ is the probability that E occurs, given that F occurs.

For the notation $P(E \mid F)$, the sample space has been narrowed to event F.

WARNING   It is important to keep straight the difference in meanings of $P(E \text{ AND } F)$ and $P(E \mid F)$. For example, $P(\text{tall AND NBA player})$ is the probability of randomly selecting a tall, NBA player from *all Americans*, which is very close to 0. But $P(\text{tall} \mid \text{NBA player})$ is the probability of randomly selecting a tall, NBA player from *just NBA players*, which is fairy large (0.73). In other words, the sample spaces for the two probabilities are very different. Figures 32 and 33 show that difference.

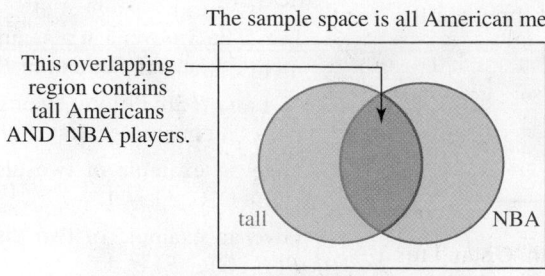

Figure 32 For $P(\text{tall AND NBA player})$, the sample space is all Americans

Figure 33 For $P(\text{tall} \mid \text{NBA player})$, the sample space is narrowed to just NBA players

WARNING

Some students think that $P(\text{tall} \mid \text{NBA player})$ must be smaller than $P(\text{tall})$ because the sample space of $P(\text{tall} \mid \text{NBA player})$ is just NBA players whereas the sample space of $P(\text{tall})$ is so much larger (all American men). But actually, $P(\text{tall} \mid \text{NBA player}) = 0.73$ is much larger than $P(\text{tall}) = 0.02$.

In general, if E and F are events for a random experiment, $P(E \mid F)$ might be larger, smaller, or equal to $P(E)$. We must consider the situation to determine what is true.

▶ Example 1   Finding Conditional Probabilities

A total of 989 randomly selected adults were asked whether they are interested in international issues. Table 15 compares their responses and their ages.

Table 15 Interest in International Issues versus Age

	Age Group (years)				
	18–30	31–40	41–55	56–89	Total
Very interested	17	36	67	106	226
Moderately interested	81	69	128	184	462
Not at all interested	77	65	79	80	301
Total	175	170	274	370	989

Source: *General Social Survey*

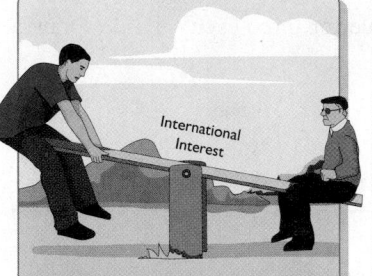

Suppose a person is randomly selected from the survey.

1. Find $P(\text{very interested})$. What does it mean in this situation?
2. Find $P(\text{very interested} \mid \text{ages } 56-89 \text{ years})$. What does it mean in this situation?
3. Compare the results you found in Problems 1 and 2. What does your comparison mean in this situation?

Solution

1. From the yellow column in Table 15, we see that 226 out of 989 adults are very interested in international issues. So, $P(\text{very interested}) = \dfrac{226}{989} \approx 0.229$. This means that the probability of randomly selecting a respondent who is very interested in international issues is 0.229.
2. We are to focus on the data for adults ages 56–89 years. Those adults are described by the green column in Table 15. Because 106 of 370 of those adults are very interested in international issues, $P(\text{very interested} \mid \text{ages } 56-89 \text{ years}) = \dfrac{106}{370} \approx 0.286$. This means that the probability of randomly selecting a respondent ages 56–89 years who is very interested in international issues is 0.286.
3. The result we found in Problem 2 is larger than the one we found in Problem 1. This means that an adult randomly selected from the respondents who are between 56 and 89 years of age, inclusive, is more likely to be very interested in international issues than an adult randomly selected from all those surveyed.

We must think carefully about whether comparing two or more specific conditional probabilities is meaningful. We will explore this issue in Example 2.

▶ **Example 2**   Thinking Critically about Conditional Probabilities

In Example 1, we worked with the data shown in Table 16.

Table 16 Interest in International Issues versus Age

	Age Group (years)				
	18–30	31–40	41–55	56–89	Total
Very interested	17	36	67	106	226
Moderately interested	81	69	128	184	462
Not at all interested	77	65	79	80	301
Total	175	170	274	370	989

Source: *General Social Survey*

1. Find the probability that an adult randomly selected from the survey is between 56 and 89 years of age, inclusive, given the person is not at all interested in international issues.
2. Find the probability that an adult randomly selected from the survey is between 31 and 40 years of age, inclusive, given the person is not at all interested in international issues.
3. Compare the results you found in Problems 1 and 2. What does your comparison mean in this situation?

Solution

1. From the yellow row in Table 16, we see that 80 adults ages 56–89 years are not at all interested in international issues out of 301 adults who are not at all interested in international issues. So, the probability is $\dfrac{80}{301} \approx 0.266$.

2. From the yellow row, we see that 65 adults ages 31–40 years are not at all interested in international issues out of 301 adults who are not at all interested in international issues. So, the probability is $\dfrac{65}{301} \approx 0.216$.

3. The result we found in Problem 1 is larger than the result we found in Problem 2. That is because more adults ages 56–89 years were in the study (370) than adults ages 31–40 years (170). In fact, the proportion of adults ages 56–89 years who are not at all interested in international issues is $\dfrac{80}{370} \approx 0.216$, which is actually *less* than the proportion of adults ages 31–40 years who are not at all interested in international issues $\left(\dfrac{65}{170} \approx 0.382 \right)$.

WARNING   In Example 1, we compared two conditional probabilities, which was quite instructive. In Example 2, we compared two other conditional probabilities about the same situation, but the comparison was not instructive. The main point is that care should be taken to understand the meaning of probabilities before jumping to conclusions by comparing them.

▶ **Example 3**   Finding a Conditional Probability

A person rolls a six-sided die once. Find the probability that the outcome is

1. a 6, given the number is even.
2. an even number, given the number is at least 4.

Solution

1. The sample space is narrowed to just the even numbers 2, 4, and 6. Because the outcome 6 is 1 of those 3 equally likely outcomes, the probability is $\dfrac{1}{3}$.

2. The sample space is narrowed to just the numbers that are at least 4: 4, 5, and 6. Of those numbers, 4 and 6 are the only even numbers. Because the desired event is 2 out of 3 equally likely outcomes, the probability is $\frac{2}{3}$.

Independent Events and Dependent Events

In Example 1, we found that a person randomly selected from the study who is between 56 and 89 years of age, inclusive, is more likely to be very interested in international issues than an adult randomly selected from all those surveyed:

$$P(\text{very interested} \mid \text{ages } 56{-}89) = 0.286$$
$$P(\text{very interested}) = 0.229$$

Because the probabilities are not equal we say the event very interested in international issues and the event ages 56–89 are *dependent events*.

Not all events are dependent. For the experiment of randomly selecting an adult, the probability that we select a very intelligent person (IQ over 130) is 0.023. The probability that we select a very intelligent person, given the person is a man, is also 0.023. This makes sense because research has shown that men are just as likely to be very intelligent as all people. Using symbols, we can write

$$P(\text{very intelligent} \mid \text{man}) = P(\text{very intelligent})$$

Because there is equality, we say the event very intelligent and the event man are *independent events*.

> **Definition**   Independent events and dependent events
>
> Two events E and F are **independent** if $P(E \mid F) = P(E)$. The events are **dependent** if $P(E \mid F)$ and $P(E)$ are *not* equal.

WARNING   Students often think that independent events are disjoint events, but this is not true. Note that the independent events men and very intelligent people are *not* disjoint events because very intelligent men are in both events.

> ▶ Example 4   Determining whether Two Events Are Independent

Determine whether the given events E and F are independent.

1. Experiment: An American adult is randomly chosen.
Event E: The person loses weight.
Event F: The person lowers his or her daily caloric intake.
2. Experiment: A person flips a coin and rolls a six-sided die.
Event E: The coin lands on tails.
Event F: The person rolls a 3.

Solution

1. A person randomly selected from those who lower their daily caloric intake is more likely to lose weight than a person randomly selected from all adults. So, $P(\text{lose weight} \mid \text{lower daily caloric intake})$ and $P(\text{lose weight})$ are not equal. Therefore, the events lose weight and lower daily caloric intake are dependent.
2. The outcome of the diet does not affect the outcome of the coin. The probability of getting tails is 0.5 no matter what number is rolled. So, $P(\text{tails} \mid \text{roll a 3}) = 0.5 = P(\text{tails})$. Therefore, getting tails is independent of rolling a 3.

WARNING   If two events are dependent, this does not necessarily mean that one event *causes* the other. For example, even though we found in Example 1 that adults ages 56–89 years are more likely to be very interested in international issues than all adults in the study, we cannot say that aging is the cause. It could be that the adults ages 56–89 years felt the same way about international issues as they did when they were between 18 and 55 years of age. Perhaps they shared a nationwide experience when they were teenagers that created a strong interest in international issues. A study could be designed to try to sort this out.

▶ **Example 5**   Determine whether Two Events Are Independent

A total of 1310 adults were surveyed about whether they had a great deal of confidence, only some confidence, or hardly any confidence in the press (newspapers and magazines). Table 17 compares their responses and their ages.

Table 17 Confidence in Press

Confidence	Caucasian	African American	Hispanic	Other	Total
A great deal	56	27	17	16	116
Only some	370	90	83	33	576
Hardly any	439	83	75	21	618
Total	865	200	175	70	1310

Source: *General Social Survey*

1. Find the probability that a person randomly selected from the survey
   a. had a great deal of confidence in the press.
   b. had a great deal of confidence in the press, given the person is Hispanic.
2. Determine whether having a great deal of confidence in the press is independent of being Hispanic. What does that mean in this situation?

Solution

1. a. From the yellow column in Table 17, we see that 116 of 1310 adults had a great deal of confidence in the press. So, the probability is $\frac{116}{1310} \approx 0.089$.

   b. We are to focus on the data for Hispanics, which are described by the green column in Table 17. Because 17 of 175 Hispanics had a great deal of confidence in the press, the probability is $\frac{17}{175} \approx 0.097$.

2. Although the probabilities we found in Problem 1 are not equal, they are fairly close. Loosely speaking, we could say having a great deal of confidence in the press is "almost independent" of being Hispanic. More importantly, whatever dependency might exist, it is so small it is probably not worth investigating further.

▶

Multiplication Rule for Independent Events

If we flip a coin twice, what is the probability of getting two tails? First, we take note that getting tails on the first flip and getting tails on the second flip are independent events because the flips do not affect each other. Then we list the possible outcomes:

$$\text{TT} \quad \text{TH} \quad \text{HT} \quad \text{HH}$$

Each of the outcomes is equally likely, and getting two tails (TT) is 1 of those 4 outcomes. So,

$$P(\text{tails on the first flip AND tails on the second flip}) = \frac{1}{4}$$

We can find the same result by considering the experiment in stages. If we were to flip an infinite number of coins each once, we would get tails in $P(\text{tails on the first flip}) = \dfrac{1}{2}$ of the tosses. And if we were to flip the coins that landed tails a second time, we would get tails in $P(\text{tails on the second flip}) = \dfrac{1}{2}$ of those tosses. So, for two tosses of all the coins, we would expect to get two tails $\dfrac{1}{2}$ of $\dfrac{1}{2}$ the pairs of tosses. So,

$$
\begin{aligned}
P(\text{tails on the first flip AND} \\
\text{tails on the second flip}) &= \frac{1}{2} \cdot \frac{1}{2} \\
&= P(\text{tails on the second flip}) \cdot P(\text{tails on the first flip}) \\
&= P(\text{tails on the first flip}) \cdot P(\text{tails on the second flip})
\end{aligned}
$$

This result suggests the following rule.

▶ Multiplication Rule for Independent Events

If E and F are independent events, then

$$P(E \text{ AND } F) = P(E) \cdot P(F)$$

In words, if two events are independent, then the probability of one event AND the other event occurring is equal to the product of their probabilities.

▶ Example 6   Using the Multiplication Rule for Independent Events

1. A person flips a coin and rolls a six-sided die. What is the probability that the person rolls a 4 AND the coin lands on heads?
2. Three coins are flipped. What is the probability that all the outcomes are tails?

Solution

1. We let F stand for rolling a 4 and let H stand for the coin landing on heads. The events F and H do not affect each other, so the events are independent. Thus, we can use the multiplication rule for independent events:

$$
\begin{aligned}
P(F \text{ AND } H) &= P(F) \cdot P(H) \quad \text{\small Multiplication rule for independent events} \\
&= \frac{1}{6} \cdot \frac{1}{2} \quad\quad \text{\small } P(F) = \frac{1}{6}; P(H) = \frac{1}{2} \\
&= \frac{1}{12}
\end{aligned}
$$

2. The three outcomes of tails are independent events, so we can use the multiplication rule for independent events:

$$
\begin{aligned}
P(\text{first is tails AND second is} \\
\text{tails AND third is tails}) &= P(\text{first is tails}) \cdot P(\text{second is tails}) \cdot P(\text{third is tails}) \\
&= \frac{1}{2} \cdot \frac{1}{2} \cdot \frac{1}{2} \\
&= \frac{1}{8}
\end{aligned}
$$

In Example 7, we will use the multiple rule for independent events to draw a comparison that might surprise you.

▶ Example 7 Using the Multiplication Rule for Independent Events

A person flips a coin 8 times.

1. Find the probability of alternating between getting heads and tails, beginning with heads: HTHTHTHT.
2. Find the probability of getting all heads.
3. Compare the results you found in Problems 1 and 2.

Solution

1. The events of alternating heads and tails, beginning with heads, are independent. So, we can use the multiplication rule for independent events:

$$P(\text{HTHTHTHT}) = P(\text{H}) \cdot P(\text{T}) \cdot P(\text{H}) \cdot P(\text{T}) \cdot P(\text{H}) \cdot P(\text{T}) \cdot P(\text{H}) \cdot P(\text{T})$$

$$= \frac{1}{2} \cdot \frac{1}{2} \cdot \frac{1}{2} \cdot \frac{1}{2} \cdot \frac{1}{2} \cdot \frac{1}{2} \cdot \frac{1}{2} \cdot \frac{1}{2} \quad P(H) = \frac{1}{2}; P(T) = \frac{1}{2}$$

$$= \frac{1}{256}$$

2. The events of getting heads every time are independent. Therefore, we can use the multiplication rule for independent events:

$$P(\text{HHHHHHHH}) = P(\text{H}) \cdot P(\text{H}) \cdot P(\text{H}) \cdot P(\text{H}) \cdot P(\text{H}) \cdot P(\text{H}) \cdot P(\text{H}) \cdot P(\text{H})$$

$$= \frac{1}{2} \cdot \frac{1}{2} \cdot \frac{1}{2} \cdot \frac{1}{2} \cdot \frac{1}{2} \cdot \frac{1}{2} \cdot \frac{1}{2} \cdot \frac{1}{2} \quad P(H) = \frac{1}{2}$$

$$= \frac{1}{256}$$

3. The results are equal. So, it is just as likely to flip 8 heads in a row as it is to alternate between flipping heads and tails, beginning with heads.

Suppose a person flips a coin 7 times and gets all heads. What is the probability of getting another head? There are two ways to interpret the question:

1. What is the probability of getting 8 heads in a row?
2. What is the probability of getting a head, given that the first 7 outcomes were all heads?

For interpretation (1), we found in Problem 1 of Example 7 that the probability is quite low: $\frac{1}{256} \approx 0.00391$.

For interpretation (2), the probability is conditional. We assume the first 7 heads have already occurred. We are to find the probability of getting heads from *one* toss, which is $\frac{1}{2} = 0.5$. After all, the coin does not "remember" that it came up heads the previous 7 times.

WARNING   Many students confuse interpretations (1) and (2). Make sure you understand the meaning of each.

WARNING   Before using the multiplication rule for independent events, we must check that the events are truly independent. For example, consider the random experiment of rolling a six-sided die *once*. The event rolling a 2 and the event rolling an even number are *not* independent. After all, rolling a 2 definitely affects whether an even number has been rolled because a 2 *is* even. Here we check that the multiplication rule for independent events does *not* work for the two dependent events:

$$P(2 \text{ AND even}) \overset{?}{=} P(2) \cdot P(\text{even})$$ *Multiplication rule for independent events*

$$P(2) \overset{?}{=} \frac{1}{6} \cdot \frac{3}{6}$$ $P(2 \text{ AND even}) = P(2)$ *because* 2 *is even.*

$$\frac{1}{6} \overset{?}{=} \frac{1}{12}$$ $\frac{1}{6} \cdot \frac{3}{6} = \frac{1}{6} \cdot \frac{1}{2} = \frac{1}{12}$

false

In a statistics course, you will learn about the *general* multiplication rule, which works for dependent events.

In Example 8, we will see that sometimes it is useful to use a combination of the multiplication and complement rules.

▶ **Example 8**   Using the Multiplication and Complement Rules

For the spring semester 2014, 57% of students at Bellevue College were female (Source: *Bellevue College*). If 4 students were randomly selected with replacement from the college during that semester, what is the probability that at least 1 of the students would have been female?

Solution

To find $P(\text{at least 1 female})$ directly, we would have to find

$$P(1 \text{ female}) + P(2 \text{ females}) + P(3 \text{ females}) + P(4 \text{ females})$$

But the first three probabilities would be complicated to find. Fortunately, there is an easier way. We start by using the complement rule:

$$P(\text{at least 1 female}) = 1 - P(\text{no females})$$ *Complement rule*
$$= 1 - P(\text{all males})$$
$$= 1 - P(\text{first male AND second male AND third male AND fourth male})$$

The events of selecting the male students are independent because selecting a male student with replacement does not affect the probability of selecting another. So, we can use the multiplication rule for independent events:

$$= 1 - P(\text{first male}) \cdot P(\text{second male}) \cdot P(\text{third male}) \cdot P(\text{fourth male})$$
$$= 1 - (0.43) \cdot (0.43) \cdot (0.43) \cdot (0.43)$$ $P(male) = 1 - P(NOT \ male) =$
$$= 0.966$$ $1 - P(female) = 1 - 0.57 = 0.43$

So, the probability that at least 1 out of the 4 students would have been female is approximately 0.966.

▶

Next, we will summarize all the probability rules that we have discussed in this chapter so far.

> **Summary of Probability Rules**
>
> Let E and F be events.
> - $P(\text{NOT } E) = 1 - P(E)$ Complement rule
> - If E and F are disjoint, then $P(E \text{ OR } F) = P(E) + P(F)$.   Addition rule for disjoint events
> - $P(E \text{ OR } F) = P(E) + P(F) - P(E \text{ AND } F)$ General addition rule
> - If E and F are independent, then $P(E \text{ AND } F) = P(E) \cdot P(F)$.   Multiplication rule for independent events

Group Exploration
Conditional probability

For this exploration, you will need to work with only one other student.

1. Have your partner pick a counting number between 1 and 4, inclusive, but your partner should *not* tell you the number. What is the probability that you can guess the number?

2. Try to guess the number. Record whether your guess was correct.

3. Repeat Problems 1 and 2 twenty times. Each time, record whether your guess was correct.

4. Compute the relative frequency of correct guesses. Compare your result with the probability you found in Problem 1.

5. This time *you* should pick a counting number between 1 and 4, inclusive. Do *not* tell your partner the number, but let your partner know whether the number is odd or even. What is the probability that your partner can guess the number? Explain why your result is a conditional probability.

6. Have your partner try to guess the number. Record whether your partner's guess was correct.

7. Repeat Problems 5 and 6 twenty times. Each time, record whether your partner's guess was correct.

8. Compute the relative frequency of correct guesses. Compare your result with the probability you found in Problem 5.

9. Explain why it makes sense that the probability you found in Problem 1 is half of the probability that you found in Problem 5.

10. If a number is randomly selected from the counting numbers between 1 and 4, inclusive, is the number being 3 independent of the number being odd?. Explain.

▶ **Tips for Success**   **Do Five New Study Activities**

Have you ever had trouble with mathematics? Did you try out a new study activity such as getting a tutor, but it did not seem to help? Sometimes employing one new behavior is not enough to cross the threshold to success. Incorporating five such strategies can greatly enhance your chances.

Imagine taking part in five new activities such as asking more questions in class, doing extra problems, forming a study group, visiting your professor's office hours, and taking practice exams. When you get involved with so many activities, the benefits of one activity tend to increase the benefits of all of the others. It also sends a message to your psyche that you mean business about passing your prestatistics course, which can do wonders.

Homework 5.3

For extra help ▶     Watch the videos in MyMathLab   Download the MyDashboard App

1. If E and F are events, then the ____ probability $P(E|F)$ is the probability that E occurs, given that F occurs.

2. Two events E and F are independent if $P(E|F) =$ _____.

3. If E and F are independent events, then $P(E \text{ AND } F) =$ _____.

4. If E and F are disjoint events, then $P(E \text{ AND } F) =$ _____.

For Exercises 5 and 6, suppose there are 8 members of a rock band: 2 guitarists, 1 lead singer, 3 backup singers, 1 bass player, and 1 drummer. No one both sings and plays an instrument. Find the probability (in fraction form) that a person randomly selected from the 8 band members:

5. is a backup singer, given the person sings.

6. plays guitar, given the person plays an instrument.

For Exercises 7–12, a person rolls a six-sided die once. When finding a probability, write it in fraction form.

7. Find $P(3|\text{no more than }5)$

8. Find $P(4|\text{no less than }3)$

9. Find $P(\text{even number}|\text{at least }3)$

10. Find $P(\text{odd number}|\text{at most }5)$

11. Is a number being even independent of a number being at least 3? Explain.

12. Is a number being odd independent of a number being at most 5? Explain.

In Exercises 13–20, you will analyze the gambling game roulette in which a ball is dropped onto a revolving wheel that has 38 numbered slots: the numbers 1–36, a zero, and a double zero. There are 18 red numbers, 18 black numbers, and 2 green numbers:

Red: 1, 3, 5, 7, 9, 12, 14, 16, 18, 19, 21, 23, 25, 27, 30, 32, 34, 36

Black: 2, 4, 6, 8, 10, 11, 13, 15, 17, 20, 22, 24, 26, 28, 29, 31, 33, 35

Green: 0, 00

Consider the random experiment of dropping the ball onto the wheel once. When finding a probability, write it in fraction form.

13. Find $P(0 \mid \text{green})$.

14. Find $P(5 \mid \text{red})$.

15. Find $P(\text{odd} \mid \text{red})$.

16. Find $P(\text{even} \mid \text{black})$.

17. Find $P(\text{greater than } 29 \mid \text{black})$.

18. Find $P(\text{less than } 5 \mid \text{red})$.

19. Are the events red and black independent?

20. Are the events red and black disjoint?

For Exercises 21–24, suppose someone spins the spinner shown in Fig. 34 once. Find the probability (in fraction form) that the outcome is

21. a 3, given the number is at least 3

22. a 1, given the number is at most 2

23. a 1, given the number is odd

24. a 4, given the number is even

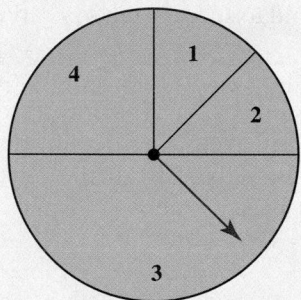

Figure 34 A spinner
(Exercises 21–24)

For Exercises 25–28, refer to the dates of June 2016 shown in Fig. 35.

25. Find the probability (in fraction form) that a randomly selected date from the month is a Tuesday, given the date is in the fifth week.

26. Find the probability (in fraction form) that a randomly selected date from the month is in the second week, given it is a Thursday.

27. Are the events Monday and in the second week independent? Explain.

28. Are the events Wednesday and in the fifth week independent? Explain.

Sun	Mon	Tues	Wed	Thurs	Fri	Sat
			1	2	3	4
5	6	7	8	9	10	11
12	13	14	15	16	17	18
19	20	21	22	23	24	25
26	27	28	29	30		

Figure 35 Dates of June 2016 (Exercises 25–28)

29. Using W to stand for the event workers ages 18–34 years and L to stand for the event bring a packed lunch to work, use probability notation to describe that 45% of workers ages 18–34 years bring a packed lunch to work (Source: *Accounting Principals*). Your result should include W, L, the equal sign "=," and other symbols.

30. Using C to stand for the event current cigarette smoker and D to stand for the event smoke every day, use probability notation to describe that 78.4% of current cigarette smokers smoke every day (Source: *Centers for Disease Control and Prevention*). Your result should include C, D, the equal sign "=," and other symbols.

31. A total of 34,695 adults were asked whether they agreed with the following statement: "I often feel that my values are threatened by Hollywood and the entertainment industry." Table 18 compares the adults' responses and their ages.

Table 18 Values Threatened by Hollywood versus Age

	Ages (in years)				
	18–29	30–49	50–64	over 64	Total
Agree	1400	4847	4823	3910	14,980
Disagree	2842	6976	5661	4236	19,715
Total	4242	11,823	10,484	8146	34,695

Source: *Pew Research Center*

a. Find the probability that a person randomly selected from the 34,695 adults was between 30 and 49 years of age, inclusive, given the person agreed with the statement.

b. Find the probability that a person randomly selected from the 34,695 adults was over 64 years of age, given the person agreed with the statement.

c. By inspecting the results you found in parts (a) and (b), a student concludes that a surveyed adult between 30 and 49 years of age, inclusive, was more likely to agree with the statement than a surveyed adult over 64 years of age. Is the student correct? If yes, explain. If no, find two appropriate conditional probabilities and compare them to explain.

32. A total of 34,695 adults were asked whether they agreed with the following statement: "Evolution is the best explanation for the origins of human life on earth." Table 19 compares the adults' responses and their ages.

Table 19 Evolution Is the Best Explanation of Human Life on Earth versus Age

	Ages (in years)				Total
	18–29	30–49	50–64	over 64	
Agree	2333	5793	4927	3258	16,311
Disagree	1909	6030	5557	4888	18,384
Total	4242	11,823	10,484	8146	34,695

Source: *Pew Research Center*

a. Find the probability that a person randomly selected from the 34,695 adults was between 18 and 29 years of age, inclusive, given the person agreed with the statement.

b. Find the probability that a person randomly selected from the 34,695 adults was between 30 and 49 years of age, inclusive, given the person agreed with the statement.

c. By inspecting the results of parts (a) and (b), a student concludes that a surveyed adult between 30 and 49 years of age, inclusive, was more likely to agree with the statement than a surveyed adult between 18 and 29 years of age, inclusive. Is the student correct? If yes, explain. If no, find two appropriate conditional probabilities and compare them to explain.

33. Table 20 compares the numbers (in thousands) of students who started college at a 4-year private, a 4-year public, or a 2-year public institution in fall 2007 and their educational status as of spring 2013 (six years later). The category "Completed a degree or certificate" includes students who eventually attended different colleges or earned a different degree than first intended.

Table 20 College Students' Educational Status after 6 Years

Educational Status as of Spring 2013	Numbers of Students (in thousands)			
	4-Year Private	4-Year Public	2-Year Public	Total
Completed a degree or certificate	325	637	342	1304
Still enrolled at any college	43	150	162	355
Not enrolled in any college	78	217	353	648
Total	446	1004	857	2307

Source: *National Student Clearinghouse Research Center*

a. Find the probability that a student randomly selected from the study began studying at a 2-year public institution in fall 2007, given the student was still enrolled at any college during spring 2013.

b. Find the probability that a student randomly selected from the study completed a degree or certificate by spring 2013, given the student began at a 4-year private institution in fall 2007.

c. Find the probability that a student randomly selected from the study completed a degree or certificate by spring 2013, given the student began at a 2-year public institution in fall 2007.

d. Compare the results you found in parts (b) and (c). What does your comparison mean in this situation?

e. A student says that for students who began college in fall 2007, 2-year public institutions were more effective than 4-year private institutions, because more 2-year public students (342 thousand) completed a degree or certificate by spring 2013 than 4-year private students (325 thousand). What would you tell the student?

34. A random selection of 969 married adults were asked about their level of happiness. Table 21 compares their responses and their genders.

Table 21 Level of Happiness versus Gender

	Very Happy	Fairly Happy	Not too Happy	Total
Female	215	247	38	500
Male	183	243	43	469
Total	398	490	81	969

Source: *General Social Survey*

a. Find the probability that a person randomly selected from the 969 married adults is fairly happy, given the person is a woman.

b. Find the probability that a person randomly selected from the 969 married adults is fairly happy, given the person is a man.

c. Compare the results you found in parts (a) and (b). What does your comparison mean in this situation?

d. A student says married women in the study are more likely to be fairly happy than married men in the study because more married women (247) were fairly happy than married men (243). What would you tell the student?

35. Researchers observed 15,824 individuals in malls, schools, stadiums, and other public places, recording whether each person smiled within 10 seconds. Table 22 compares their observations and the researchers' estimates of the individuals' ages.

Table 22 Smiling versus Age Group

	Age Group (years)					Total
	0–10	11–20	21–40	41–60	over 60	
Smiling	1131	1748	1691	937	522	6029
Not smiling	1187	2020	2955	2124	1509	9795
Total	2318	3768	4646	3061	2031	15,824

Source: *Frequency of Public Smiling across the Life Span,* Mark S. Chapell

Suppose one participant from the study is randomly selected.

a. Find $P(\text{smile} \mid \text{ages } 0{-}10 \text{ years})$.

b. Find $P(\text{smile} \mid \text{ages } 11{-}20 \text{ years})$.

c. Find $P(\text{smile} \mid \text{ages } 21{-}40 \text{ years})$.

d. Find $P(\text{smile} \mid \text{ages } 41{-}60 \text{ years})$.

e. Find $P(\text{smile} \mid \text{over age } 60 \text{ years})$.

f. Describe a key pattern in the results you found in parts (a) through (e). What does the pattern mean in this situation? Can causation be assumed? Explain.

g. If the population is all Americans, what type(s) of bias likely occurred? Explain.

36. Between August 1973 and September 1974, 957 randomly selected adults in Washington County, Maryland, were interviewed about whether they wore seat belts and whether they had undergone a dental checkup in the past two years. Their responses are summarized in Table 23. During that period, no state required car occupants to wear seat belts.

Table 23 Wore a Seat Belt versus Underwent a Dental Checkup

Dental Checkup	Wear Seat Belt		
	Yes	No	Total
Within past two years	428	332	760
More than two years ago	90	107	197
Total	518	439	957

Source: *What Kinds of People Do Not Use Seat Belts?, Knud J. Helsing et al.*

Suppose one individual from the study is randomly selected.

a. Find $P(\text{wear seat belt} \mid \text{dental checkup within two years})$.

b. Find $P(\text{wear seat belt} \mid \text{dental checkup more than two years ago})$.

c. Compare the results you found in parts (a) and (b). What does your comparison mean in this situation?

d. On the basis of the results you found in parts (a) and (b), a student concludes that undergoing frequent dental checkups causes a person to wear a seat belt. What would you tell the student? Include a possible lurking variable in your response.

e. The sampling was done by randomly selecting a household in a frame of all the households in the county and then selecting every kth household. An individual to be interviewed was randomly selected from each household. Identify the sampling method used to select the *households*.

f. Some individuals were not included in the study because they did not own cars. Other individuals refused to participate in the study. Individuals who participated in the study self-reported whether they wore seat belts and when they last had a dental checkup. Identify likely type(s) of bias.

For Exercises 37–40, determine whether the given events E and F are independent or dependent.

37. Experiment: A student is randomly selected from a prestatistics class.
Event E: The student earns an A on the first test.
Event F: The student studies at least ten hours for the first test.

38. Experiment: An American adult is randomly selected.
Event E: The person earns more than $100,000 per year.
Event F: The person has a bachelor's degree.

39. Experiment: A high school senior is randomly selected.
Event E: The student gets accepted to Harvard on March 1.
Event F: The student drinks a cup of coffee on March 1.

40. Experiment: An American adult is randomly selected.
Event E: The person wears contact lenses.
Event F: The person's favorite ice cream flavor is chocolate.

41. If D and W are independent events in which $P(D) = 0.2$ and $P(W) = 0.3$, find $P(D \text{ AND } W)$.

42. If H and Y are independent events in which $P(H) = 0.9$ and $P(Y) = 0.5$, find $P(H \text{ AND } Y)$.

43. If five coins are flipped, what is the probability (in fraction form) that all the outcomes are heads?

44. If a six-sided die is rolled three times, what is the probability (in fraction form) that all the outcomes are fives?

45. In a survey of 2000 American adults, 18% of men have experienced street verbal harassment (Source: *Stop Street Harassment*). If three men are randomly selected with replacement from the sample, what is the probability that they all have experienced street harassment? Round your result to five decimal places.

46. In a survey of 2000 American adults, 57% of women have experienced street verbal harassment (Source: *Stop Street Harassment*). If four women are randomly selected with replacement from the sample, what is the probability that they all have experienced street harassment? Round your result to three decimal places.

47. A student guesses the answers to four multiple-choice test questions that each have five choices. What is the probability (in fraction form) that the student guesses the correct answers for all four questions?

48. A student guesses the answers to six true-or-false test questions. What is the probability (in fraction form) that the student guesses the correct answers for all six questions?

49. In a study of 1004 randomly selected adults, 37% of the sample described themselves as middle class (Source: *Robert Morris University Polling Institute*).

a. If four participants of the study are randomly selected with replacement, what is the probability that all four describe themselves as middle class? Round your result to five decimal places.

b. If four participants of the study are randomly selected with replacement, what is the probability that at least one participant describes himself or herself as middle class? Round your result to five decimal places.

c. The randomly selected participants were asked to complete an online survey. What type(s) of bias likely occurred?

50. In an online survey of 1.2 million Americans, 34% of the sample say McDonald's makes the best French fries of all fast-food chains (Source: *YouGov*).

a. If three participants of the study are randomly selected with replacement, what is the probability that all three participants say McDonald's makes the best French fries of all fast-food chains? Round your result to five decimal places.

b. If three participants of the study are randomly selected with replacement, what is the probability that at least one participant says McDonald's makes the best French fries of all fast-food chains? Round your result to five decimal places.

51. In a survey of 34,525 randomly selected adults, 18.1% of the sample smoke cigarettes (Source: *Centers for Disease Control and Prevention*).

a. If five participants of the study are randomly selected with replacement, find the probability that all five participants smoke cigarettes. Round your result to five decimal places.

b. If five participants of the study are randomly selected with replacement, find the probability that at least one of the participants smokes cigarettes. Round your result to five decimal places.

c. Of those contacted, 38.8% refused to participate in the survey. The participants' smoking behaviors were self-reported. What type(s) of bias probably occurred?

52. On the basis of physical examinations of approximately 40,000 Americans at least 12 years of age, researchers determined that 20.3% of the sample had hearing loss (Source: *Johns Hopkins Medicine*).

a. If four individuals of the study are randomly selected with replacement, find the probability that all four individuals have hearing loss. Round your result to five decimal places.

b. If four individuals of the study are randomly selected with replacement, find the probability that at least one of the individuals has hearing loss. Round your result to five decimal places.

53. A total of 34,695 adults were asked whether they agreed with the following statement: "I often feel that my values are threatened by Hollywood and the entertainment industry." Table 24 compares the adults' responses and their ages.

Table 24 Values Threatened by Hollywood versus Age

	Ages (in years)				
	18–29	30–49	50–64	over 64	Total
Agree	1400	4847	4823	3910	14,980
Disagree	2842	6976	5661	4236	19,715
Total	4242	11,823	10,484	8146	34,695

Source: *Pew Research Center*

a. Find the probability that a person randomly selected from the 34,695 adults agreed with the statement.

b. Find the probability that a person randomly selected from the 34,695 adults agreed with the statement, given the person was between 18 and 29 years of age, inclusive.

c. For those surveyed, is agreeing with the statement independent of being between 18 and 29 years of age, inclusive? Explain.

54. A total of 34,695 adults were asked whether they agreed with the following statement: "Evolution is the best explanation for the origins of human life on earth." Table 25 compares the adults' responses and their ages.

Table 25 Evolution Is the Best Explanation of Human Life on Earth versus Age

	Ages (in years)				
	18–29	30–49	50–64	over 64	Total
Agree	2333	5793	4927	3258	16,311
Disagree	1909	6030	5557	4888	18,384
Total	4242	11,823	10,484	8146	34,695

Source: *Pew Research Center*

a. Find the probability that a person randomly selected from the 34,695 adults agreed with the statement.

b. Find the probability that a person randomly selected from the 34,695 adults agreed with the statement, given the person was between 18 and 29 years of age, inclusive.

c. For those surveyed, is agreeing with the statement independent of being between 18 and 29 years of age, inclusive? Explain.

55. A random selection of 1481 adults were asked whether upper-income people pay too much, too little, or their fair share of taxes. Table 26 compares their responses and their political affiliations.

Table 26 Opinion on Taxing Upper-Income People versus Political Party

Response	Republican	Democrat	Total
Fair share	235	123	358
Too much	75	38	113
Too little	177	738	915
Don't know	48	47	95
Total	535	946	1481

Source: *Pew Research Center*

One person is randomly selected from the sample. Let F be the event that the person thinks upper-income people pay their fair share of taxes. Let R be the event that the person is a Republican.

a. Find $P(F)$.
b. Find $P(F|R)$.
c. Are the events F and R independent? Explain.
d. Some people refused to respond to the survey. What type of bias is this?

56. A random selection of 1481 adults were asked whether upper-income people pay too much, too little, or their fair share of taxes. Table 26 compares their responses and their political affiliations. One person is randomly selected from the sample. Let L be the event that the person thinks upper-income people pay too little taxes. Let D be the event that the person is a Democrat.

a. Find $P(L)$.
b. Find $P(L|D)$.
c. Are the events F and D independent? Explain.
d. On the basis of your response to part (c), a student concludes that registering as a Democrat causes a person to think that upper-income people pay too little taxes. What would you tell the student?

Concepts

57. A six-sided die is rolled four times.

a. Find the probability (in fraction form) of rolling a 2, then a 6, then a 4, and then a 1.
b. Find the probability (in fraction form) of rolling all 6s.
c. Compare the results you found in parts (a) and (b).
d. A student says the result you found in part (b) should be less than the result you found in part (a) because it is less likely to get all 6s than four different numbers. What would you tell the student?

58. A coin is flipped six times.

a. Find the probability (in fraction form) of alternating between getting tails and heads, beginning with tails: THTHTH.
b. Find the probability (in fraction form) of getting all tails.
c. Compare the results you found in parts (a) and (b).
d. A student says the result you found in part (b) should be less than the result you found in part (a) because it is less likely to get all tails than three tails. What would you tell the student?

59. A coin is flipped seven times.

a. What is the probability (in fraction form) of getting all tails?
b. What is the probability (in fraction form) of getting a tail on the seventh toss, given the first six tosses were all tails?
c. Compare the results you found in parts (a) and (b).
d. A student says the results you found in parts (a) and (b) should be equal because both are the probability of getting all tails. What would you tell the student?

60. A six-sided die is rolled five times.

a. What is the probability (in fraction form) of getting all 3s?
b. What is the probability (in fraction form) of getting a 3 on the fifth roll, given the first four rolls were all 3s?
c. Compare the results you found in parts (a) and (b).

d. A student says the results you found in parts (a) and (b) should be equal because both are the probability of getting all 3s. What would you tell the student?

61. Let E be an event in a sample space S. Find $P(E|E)$.

62. Let E and F be disjoint events. Find $P(E|F)$.

63. Let E and F be events. A students says $P(E|F)$ and $P(F|E)$ are always equal, because in both cases we are to find the probability that an observation in both E and F occurs. What would you tell the student?

64. Let E and F be events. A student says that $P(E|F)$ and $P(E \text{ AND } F)$ are always equal, because in both cases we are to find the probability that an observation in both E and F occurs. What would you tell the student?

65. Let E and F be events for some random experiment.

    a. Can $P(E|F)$ ever be larger than $P(E)$? If yes, give an example. If no, explain.

    b. Can $P(E|F)$ ever be less than $P(E)$? If yes, give an example. If no, explain.

66. A gambler plays a game in which a six-sided die is rolled. If the person guesses correctly which number will come up, he wins \$4. If he guesses incorrectly, he loses \$1. For the previous five rolls of the die, the outcome was the number 2 each time. For the next round, the gambler decides to guess that the number 3 will be rolled, because it is so unlikely that a 2 will come up yet again. What would you tell the gambler?

67. A total of 41 students in one of the author's prestatistics classes completed an anonymous survey, responding to the questions "Do you read the course textbook?" and "Which do you think is more important, improving the success rates of students or stopping climate change?" The author found that for the 41 students, reading the course textbook and identifying success rates of students as more important than climate change are dependent (Source: *J. Lehmann*). For the 41 students, does that mean that believing that the success rates of students are more important than climate change *causes* students to read the course textbook? Explain.

68. Compare the meanings of *disjoint events* and *independent events*.

69. Give an example of two independent events. Explain why they are independent.

70. Give an example of two dependent events. Explain why they are dependent.

▼ 5.4 Finding Probabilities for a Normal Distribution

Objectives

» Describe the meaning of a *normal curve* and a *normal distribution*.

» Compute and interpret z-scores.

» Use z-scores and tables to find probabilities for a normal distribution.

» Use technology to find probabilities for a normal distribution.

Normal Curve

In Example 5 of Section 4.2, we analyzed the lengths (in seconds) of songs played on Friday, April 18, 2014, on Live 105, which is a San Francisco Bay Area alternative rock radio station (see Fig. 36). The song lengths are rounded to the ones place.

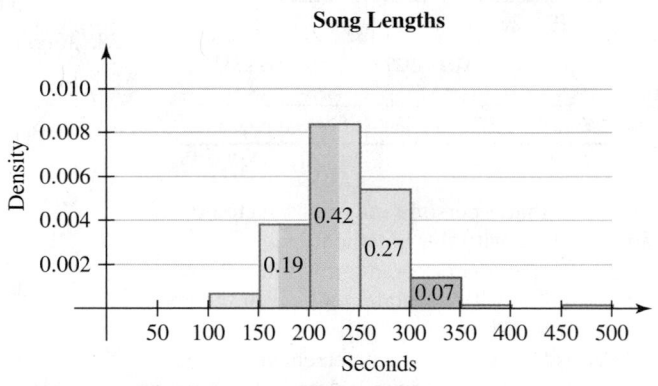

Figure 36 Density histogram of song lengths
(**Source:** *Live 105; iTunes Store*)

Recall that we can find the probability for any of the classes in a density histogram. For example, the probability of randomly selecting a song length between 300 and 349 seconds, inclusive, is 0.07 (see the orange bar in the density histogram). But how can we find the probability for another span of numbers, such as song lengths between 170 and 230 minutes? One option is to add the areas of portions of rectangles (see the green region). Another option is to use a graph called a normal curve, which is often called a "bell-shaped curve" (see the blue curve in Fig. 37).

The shape of the normal curve is similar to the shape of the density histogram in Fig. 37. The shape of the normal curve is also similar to the shape of a density histogram with smaller class size, such as the density histogram with class size 10 in Fig. 38.

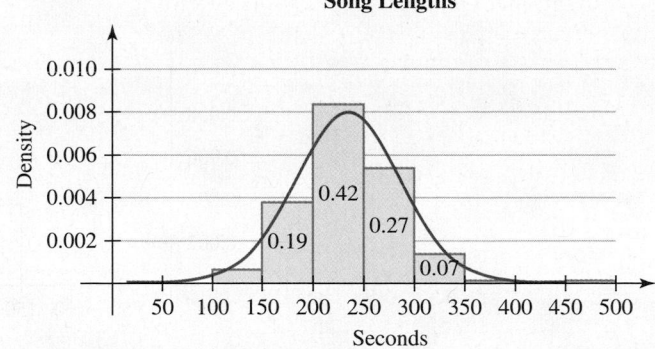

Figure 37 Normal curve and density histogram with class size 50

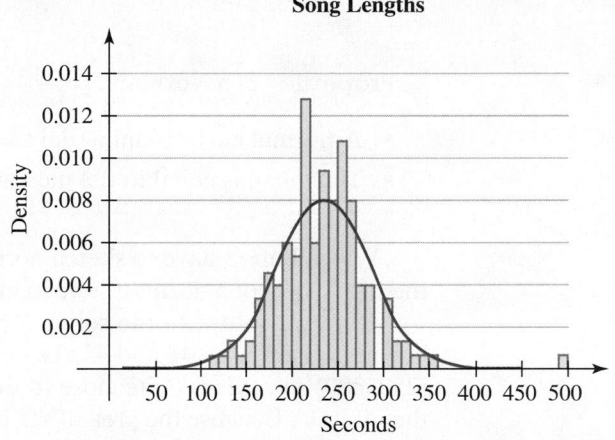

Figure 38 Normal curve and density histogram with class size 10

Recall that the Empirical Rule works better for some unimodal, symmetric distributions than others. The better that the Empirical Rule works, the better the normal curve will describe the distribution.

When a normal curve describes a distribution perfectly, we say the distribution is **normal** or **normally distributed.** When a normal curve describes a distribution fairly well (but not exactly), we say the distribution is **approximately normal** or **approximately normally distributed.**

Many authentic distributions are approximately normally distributed. Some examples are SAT scores, heights of women, heights of men, lifetimes of batteries, and ages of death-row prisoners. Recall that a model is a mathematical description of an authentic situation. So, a normal curve is a model.

In Figs. 39 and 40, we see that the mean $\bar{x} = 232.93$ is the center of both the normal curve and the density histogram. Similarly to histograms, because the left half of the curve (in green) and the right half (in blue) are mirror images, we say the curve is *symmetric*. And because the curve is symmetric, the mean is equal to the median.

Because the curve has one mound, we say it is *unimodal.*

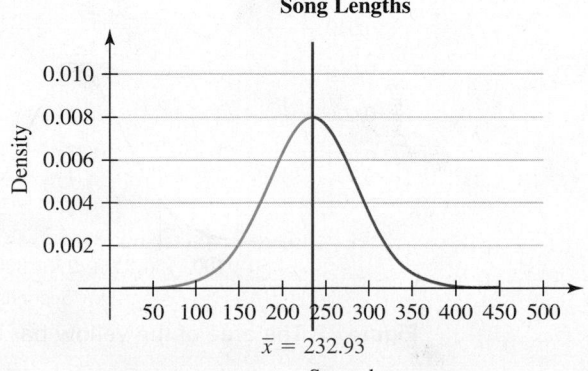

Figure 39 The center of the normal curve is $\bar{x} = 232.93$

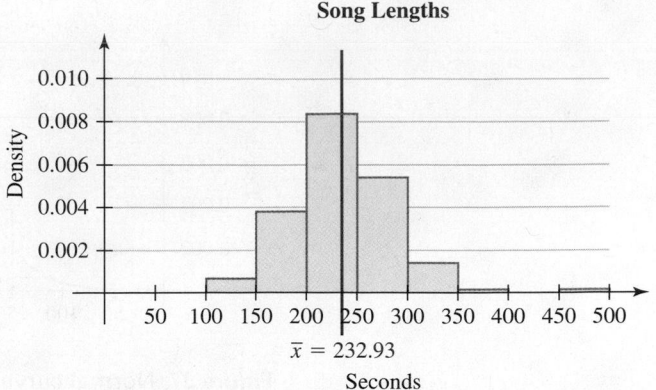

Figure 40 The center of the density histogram is $\bar{x} = 232.93$

▶ **Properties of a Normal Curve**

- A normal curve is unimodal and symmetric (see Fig. 39).
- The mean is equal to the median. Both are the center of the curve (see Fig. 39).

We will often have to sketch normal curves in this section and Section 5.5. Keeping the properties of a normal curve in mind can help us draw the curves well.

Here we return to the song-length distribution and compare the areas of the yellow regions in Figs. 41 and 42. The area (0.28) under the normal curve for the interval [250, 300) seconds is quite close to the area 0.27 of the bar in the density histogram for the interval. Because the area of the bar is equal to the probability of randomly selecting a song length (in seconds) in the interval [250, 300), we conclude that the area under the normal curve estimates the probability well. In fact, if the song-length distribution were *exactly* normal, then the area under the curve would give the *exact* probability.

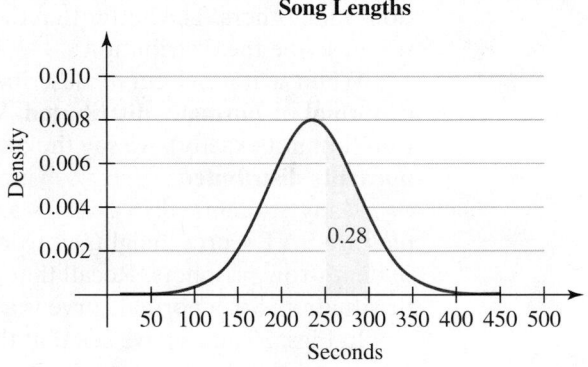

Figure 41 The area under the curve for the interval [250, 300] is 0.28

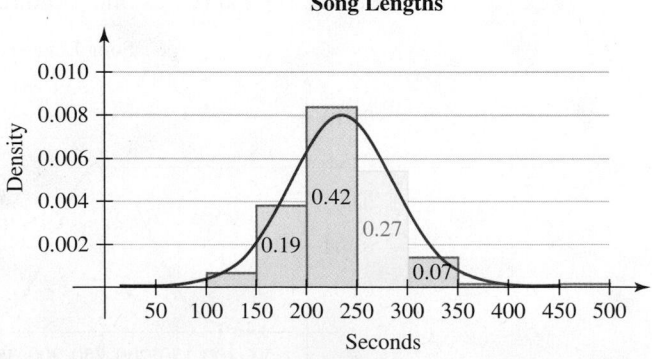

Figure 42 The area of the yellow bar is 0.27

> **Area-Probability Equality Property**
>
> The area under a normal curve for an interval is equal to the probability of randomly selecting an observation that lies in the interval (see Fig. 43).

In short, an area under the normal curve is a probability.

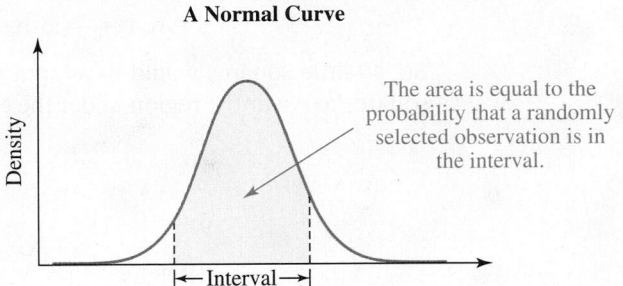

Figure 43 Area under the normal curve is a probability

We can see why the areas of the yellow regions in Figs. 41 and 42 are so close in value by inspecting Fig. 44. Although the normal curve overcounted the area by including the orange region, it also undercounted the area by missing the yellow region, and these two errors approximately cancel each other.

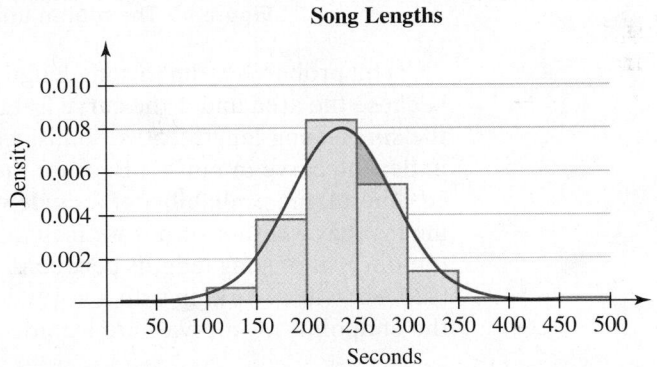

Figure 44 The overcounted area (in orange) is approximately equal to the undercounted area (in yellow)

In Fig. 45, the total overcounted area (see the orange regions) is equal to the total undercounted area (see the yellow regions). So, the area under the curve is equal to the total area of the bars, which is 1. In general, the area under *any* normal curve is equal to 1.

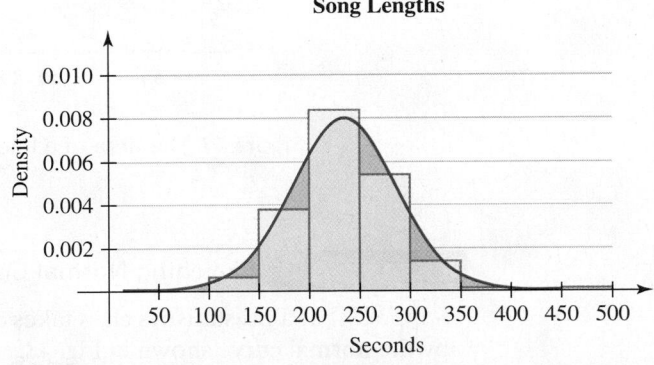

Figure 45 The total overcounted area (in orange) is equal to the total undercounted area (in yellow)

> **Total Area under the Normal Curve**
>
> The total area under a normal curve is equal to 1.

To visualize this property, we start by computing the area of one of the little squares in the grid in Fig. 46:

$$\text{Area} = \text{Width} \cdot \text{Height} = 25(0.001) = 0.025$$

So, 40 little squares would have area $40(0.025) = 1$. And it *does* appear to take 40 little squares to cover the region under the curve (try estimating it).

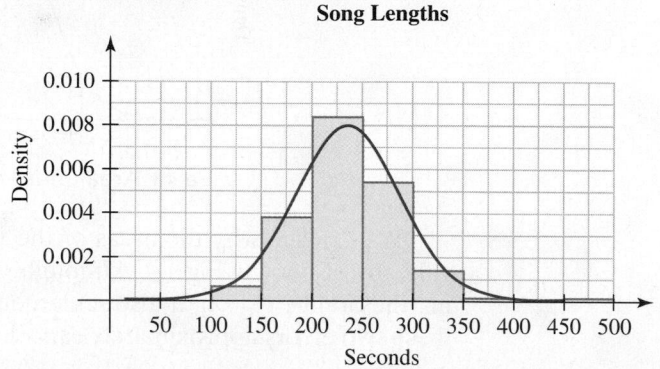

Figure 46 The region under the curve consists of 40 little squares

The probability that a song length (in seconds) is in the interval [250, 300] is 0.28 because the area under the curve is 0.28. To find the probability of randomly selecting the *single* song length 250 seconds, we need to find the area of the red line segment under the curve in Fig. 47. Because the area of a line segment is 0, the probability is 0. In general, **the probability of a single value of a normally distributed variable is 0.** This means that whether or not we include the endpoints 250 seconds and 300 seconds for the interval of song lengths between 250 and 300 seconds, we will get the same result, 0.28. So, when working with a normal curve, we don't have to be picky about whether the endpoints of intervals are included.

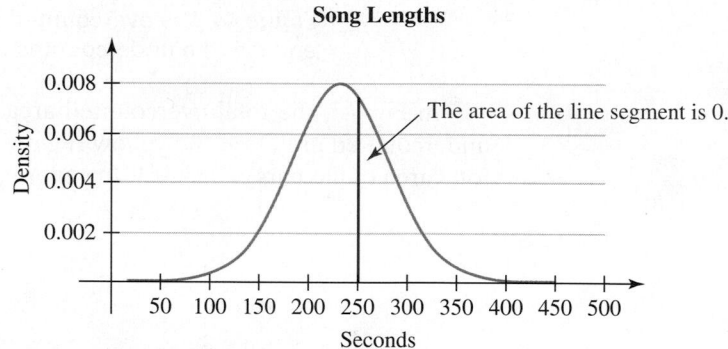

Figure 47 The area of a line segment is 0

▶ Example 1   Sketching Normal Curves

Suppose that a prestatistics class takes a first test and students' scores are described well by the normal curve shown in Fig. 48.

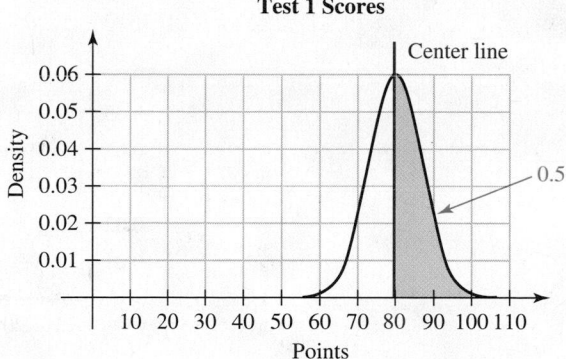

Figure 48 Scores on Test 1

1. What is the probability that a randomly selected student scored more than 80 points?
2. On Test 2, the mean score is 20 points less than the mean score on Test 1, but the standard deviation is the same. Assuming the scores are still normally distributed, sketch the curve that describes the scores.
3. On Test 3, the mean score is the same as on Test 2, but the standard deviation is larger. Assuming the scores are still normally distributed, sketch a curve that describes the scores.

Solution

1. The area to the right of 80, the median, is half of the total area, 1 (see the green region in Fig. 48). So, the probability is 0.5.
2. Because the standard deviation hasn't changed, the shape and spread of the Test 1 and Test 2 distributions are exactly the same. But because the mean is 20 points lower, we move the curve for Test 1 20 points to the left to get the curve for Test 2 (see Fig. 49).
3. Since the means for Test 2 and Test 3 are equal, we do not move the curve for Test 2 left or right (see Fig. 50). But because the standard deviation is larger for Test 3, we sketch the curve wider. By drawing the curve wider, we must also make it flatter, because the area under the curve must still equal 1. As a check, we could calculate that each little square in the grids of both Figs. 49 and 50 is 0.1 (try it). So, 10 little squares should cover the region below each curve, which is true.

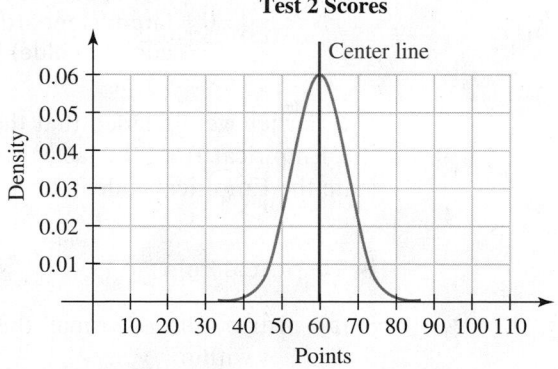

Figure 49 Scores on Test 2

Test 3 Scores

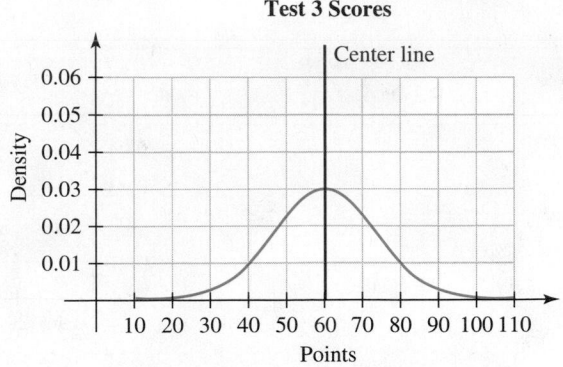

Figure 50 Scores on Test 3

Imagine pressing down on a mound of salt on a table. As it becomes flatter, it will become wider, because the volume of the salt does not change (see Fig. 51). Similarly, in Problem 3 of Example 1, the normal curve for the test distribution with larger standard deviation was both wider and flatter than the normal curve for the test distribution with smaller standard deviation. This makes sense, because the area under both curves is 1.

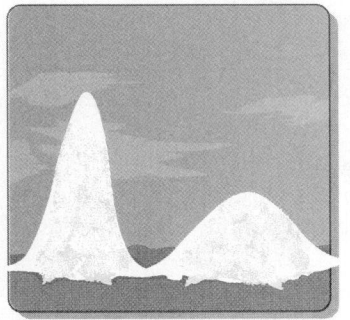

Figure 51 Pressing down on a mound of salt results in a flatter and wider mound

> **Comparing Normal Curves with Different Standard Deviations**
>
> A normal curve for a distribution with larger standard deviation will be wider and flatter than a normal curve for a distribution with smaller standard deviation (see Fig. 52).

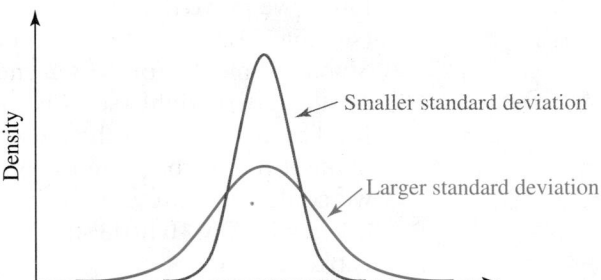

Figure 52 The normal curve (in green) for a distribution with larger standard deviation is wider and flatter than the normal curve (in blue) for a distribution with smaller standard deviation

Earlier we discussed that there is a connection between a normal distribution and the Empirical Rule. In fact, the Empirical Rule is based on normal curves. Next, we state the Empirical Rule in terms of probabilities as well as percentages.

> **Empirical Rule**
>
> If a distribution is normal, then the probability that a randomly selected observation lies within
> - one standard deviation of the mean is approximately 0.68. So, about 68% of the data lie between $\bar{x} - s$ and $\bar{x} + s$ (see Fig. 53).
> - two standard deviations of the mean is approximately 0.95. So, about 95% of the data lie between $\bar{x} - 2s$ and $\bar{x} + 2s$ (see Fig. 54).
> - three standard deviations of the mean is approximately 0.997. So, about 99.7% of the data lie between $\bar{x} - 3s$ and $\bar{x} + 3s$ (see Fig. 55).

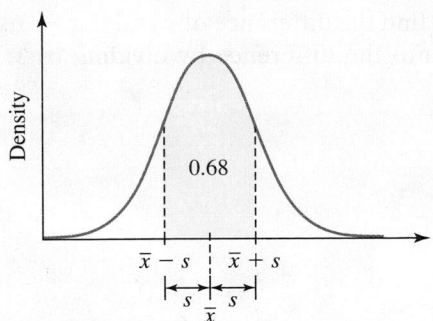

Figure 53 The probability that a randomly selected observation lies within one standard deviation of the mean is approximately 0.68

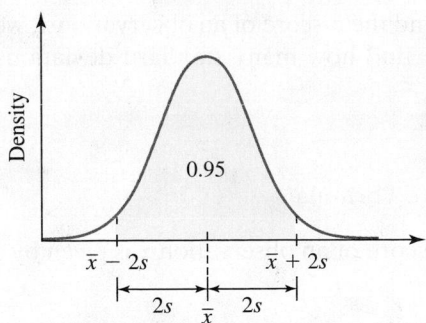

Figure 54 The probability that a randomly selected observation lies within two standard deviations of the mean is approximately 0.95

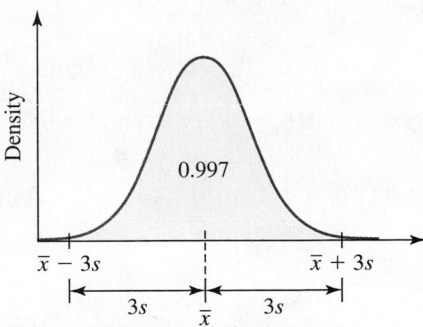

Figure 55 The probability that a randomly selected observation lies within three standard deviations of the mean is approximately 0.997

Computing and Interpreting z-Scores

The Empirical Rule tells us the proportion of observations that lie within one, two, or three standard deviations of the mean. But how can we compute the number of standard deviations that an observation is from the mean?

Consider the Wechsler IQ test, which measures a person's intelligence. People's IQ scores are normally distributed with mean 100 points and standard deviation 15 points (Source: *Essentials of WAIS-IV Assessment, Elizabeth Lichtenberger et al.*).

A student has an IQ of 130 points. Here we find the difference of the student's IQ, 130, and the mean, 100:

$$130 - 100 = 30$$

So, the student scored 30 points above the mean (see Fig. 56).

Next, we find the number of standard deviations, 15 points, that divide into 30 points:

$$\frac{130 - 100}{15} = \frac{30}{15} = 2$$

So, the student scored 2 standard deviations above the mean (see Fig. 57). We say the *z-score* is 2 and write $z = 2$.

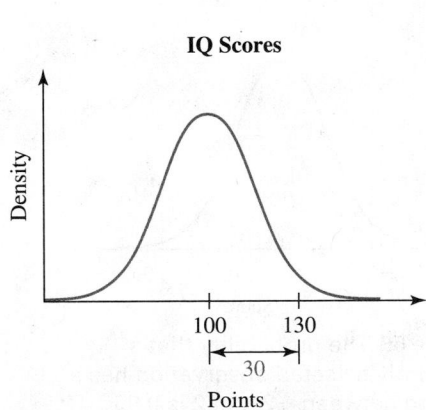

Figure 56 The student scored 30 points above the mean

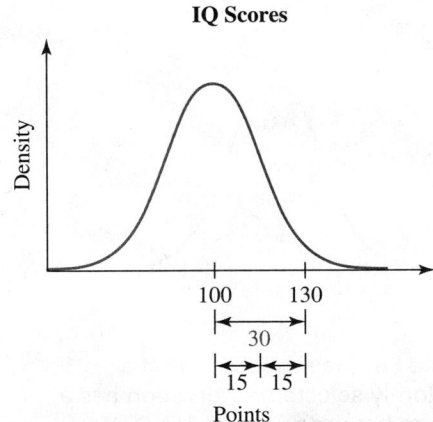

Figure 57 The student scored 2 standard deviations above the mean

▶ **Definition** z-Score

The **z-score** of an observation is the number of standard deviations that the observation is from the mean. If the observation lies to the left of the mean, then its z-score is negative. If the observation lies to the right of the mean, then its z-score is positive.

To find the z-score of an observation x, we first find the difference of x and $\bar{x}$: $x - \bar{x}$. Then we find how many standard deviations go into the difference by dividing by s:

$$z = \frac{x - \bar{x}}{s}.$$

---

z-Score Formula

The z-score of an observation x is given by

$$z = \frac{x - \bar{x}}{s}$$

▶ **Example 2   Compute and Interpret a Z-Score**

The 2014 draft picks for NBA basketball teams have heights that are approximately normally distributed with mean 79.1 inches and standard deviation 3.0 inches (Source: *nbadraft.net*). Shabazz Napier was the shortest 2014 draft pick with a height of 72 inches. Find the z-score for 72 inches. What does it mean?

Solution

We substitute 72 for x, 79.1 for $\bar{x}$, and 3.0 for s in the formula $z = \dfrac{x - \bar{x}}{s}$:

$$z = \frac{72 - 79.1}{3.0} \approx -2.37$$

The z-score is -2.37, which means that Napier's height is 2.37 standard deviations less than the mean.

Because the z-score of an observation is the number of standard deviations that the observation is from the mean, we can restate the Empirical Rule in terms of z-scores. So, if a distribution is normally distributed, then

- 68% of its z-scores lie between -1 and 1 (see Fig. 58),
- 95% of its z-scores lie between -2 and 2 (see Fig. 59),
- and 99.7% of its z-scores lie between -3 and 3 (see Fig. 60).

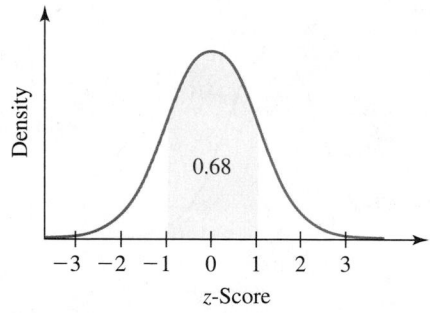

Figure 58 The probability that a randomly selected observation has a z-score between -1 and 1 is 0.68

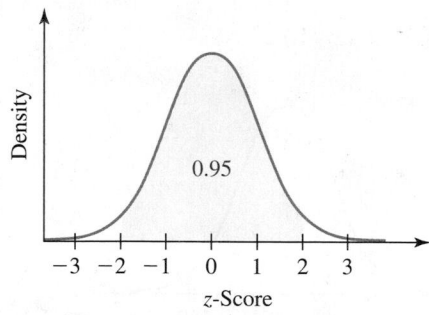

Figure 59 The probability that a randomly selected observation has a z-score between -2 and 2 is 0.95

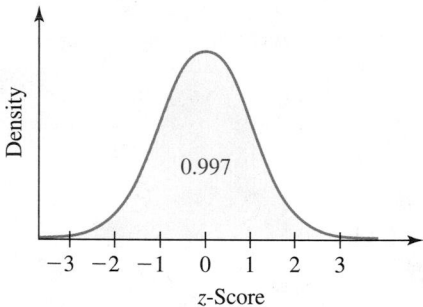

Figure 60 The probability that a randomly selected observation has a z-score between -3 and 3 is 0.997

Standard Normal Distribution

If a distribution is normally distributed, then the distribution of the observations' z-scores is also normally distributed with mean 0 and standard deviation 1. We call the distribution of z-scores the **standard normal distribution**.

Using z-Scores and Tables to Find Probabilities for a Normal Distribution

To find a probability for a normal distribution, we can compute an observation's z-score, which we then use to look up the area under the standard normal curve in Table 1 in Appendix C.

▶ **Example 3**   Find a Probability for the Standard Normal Distribution

Find the probability that a z-score randomly selected from the standard normal distribution is less than -1.35.

Solution

First, we sketch the standard normal curve, which has mean 0 and standard deviation 1. Next, we mark the z-score -1.35 on the horizontal axis and use orange to shade the region to the left of -1.35 (see Fig. 61).

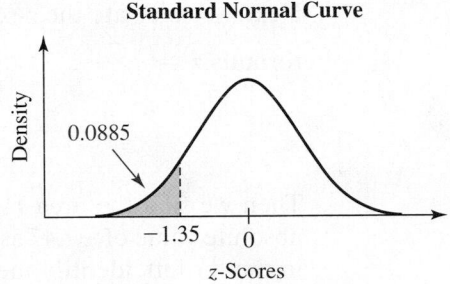

Figure 61 z-Scores less than -1.35

Then we find the area by using Table 1 in Appendix C. To do this, we write the absolute value of -1.35 as $1.35 = 1.3 + 0.05$ and identify the yellow row with **−1.3** in the far left, identify the green column with **.05** at the top, and find the number that's in both that row and column, which is 0.0885 (see Table 27, which is just two portions of Table 1 in Appendix C). The number 0.0885 describes the area to the *left* of the z-score -1.35, which is what we are interested in.

Table 27 Looking Up the z-Score -1.35

z	.00	.01	.02	.03	.04	.05	.06	.07	.08	.09
−3.4	.0003	.0003	.0003	.0003	.0003	.0003	.0003	.0003	.0003	.0002
−3.3	.0005	.0005	.0005	.0004	.0004	.0004	.0004	.0004	.0004	.0003
−3.2	.0007	.0007	.0006	.0006	.0006	.0006	.0006	.0005	.0005	.0005
−1.6	.0548	.0537	.0526	.0516	.0505	.0495	.0485	.0475	.0465	.0455
−1.5	.0668	.0655	.0643	.0630	.0618	.0606	.0594	.0582	.0571	.0559
−1.4	.0808	.0793	.0778	.0764	.0749	.0735	.0721	.0708	.0694	.0681
−1.3	.0968	.0951	.0934	.0918	.0901	**.0885**	.0869	.0853	.0838	.0823
−1.2	.1151	.1131	.1112	.1093	.1075	.1056	.1038	.1020	.1003	.0985

So, the probability is 0.0885.

▶ **Example 4**   Using z-Scores and Tables to Find Probabilities

Earlier in this section, we worked with the Wechsler IQ test, which measures a person's intelligence. Recall that people's IQ scores are normally distributed with mean 100 points and standard deviation 15 points. Let X be the IQ (in points) of a randomly selected person. Find the probability that a randomly selected person has an IQ

1. less than 78 points.
2. greater than 117 points.

Solution

1. We first sketch a normal curve with mean 100, mark the number 78 on the horizontal axis, and use orange to shade the region to the left of 78 (see Fig. 62).

IQ Scores

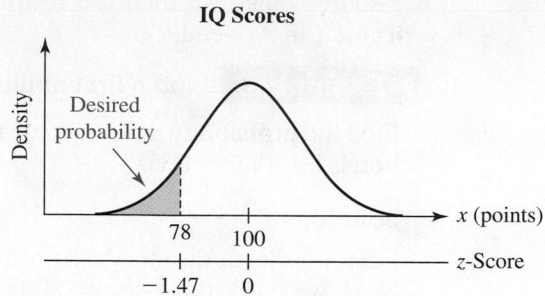

Figure 62 IQ scores less than 78 points

Next, we calculate the z-score by substituting 78 for x, 100 for $\bar{x}$, and 15 for s in the formula $z = \dfrac{x - \bar{x}}{s}$:

$$z = \frac{78 - 100}{15} = \frac{-22}{15} \approx -1.47$$

Then we find the area by using Table 1 in Appendix C. To do this, we write the absolute value of -1.47 as $1.47 = 1.4 + 0.07$ and identify the yellow row with $-\mathbf{1.4}$ in the far left, identify the green column with $\mathbf{.07}$ at the top, and find the number that's in both that row and column, which is 0.0708 (see Table 28, which is just two portions of Table 1 in Appendix C). The number 0.0708 describes the area of the orange region to the *left* of 78 points, which is what we are interested in.

Table 28 Looking Up the z-Score -1.47

z	.00	.01	.02	.03	.04	.05	.06	.07	.08	.09
−3.4	.0003	.0003	.0003	.0003	.0003	.0003	.0003	.0003	.0003	.0002
−3.3	.0005	.0005	.0005	.0004	.0004	.0004	.0004	.0004	.0004	.0003
−3.2	.0007	.0007	.0006	.0006	.0006	.0006	.0006	.0005	.0005	.0005
−1.6	.0548	.0537	.0526	.0516	.0505	.0495	.0485	.0475	.0465	.0455
−1.5	.0668	.0655	.0643	.0630	.0618	.0606	.0594	.0582	.0571	.0559
−1.4	.0808	.0793	.0778	.0764	.0749	.0735	.0721	**.0708**	.0694	.0681
−1.3	.0968	.0951	.0934	.0918	.0901	.0885	.0869	.0853	.0838	.0823

So, the probability is 0.0708. The probability is less than 0.5, which checks with the orange region having area less than half of the area under the entire curve.

2. We first sketch a normal curve with mean 100, mark the number 117 on the horizontal axis, and use blue to shade the region to the right of 117 (see Fig. 63).

IQ Scores

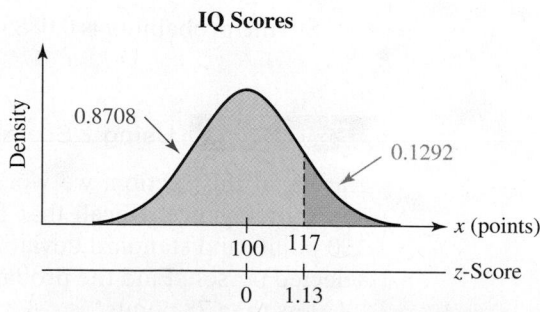

Figure 63 IQ scores greater than 117 points

Next, we calculate the z-score by substituting 117 for x, 100 for $\bar{x}$, and 15 for s in the formula $z = \dfrac{x - \bar{x}}{s}$:

$$z = \frac{117 - 100}{15} = \frac{17}{15} \approx 1.13$$

Then we use Table 1 in Appendix C to find an area. To do this, we write $1.13 = 1.1 + 0.03$ and identify the green row with **1.1** in the far left, identify the yellow column with **.03** at the top, and find the number that's in both that row and column, which is 0.8708 (see Table 29, which is just two portions of Table 1 in Appendix C).

Table 29 Looking Up the z-Score 1.13

z	.00	.01	.02	.03	.04	.05	.06	.07	.08	.09
0.0	.5000	.5040	.5080	.5120	.5160	.5199	.5239	.5279	.5319	.5359
0.1	.5398	.5438	.5478	.5517	.5557	.5596	.5636	.5675	.5714	.5753
0.2	.5793	.5832	.5871	.5910	.5948	.5987	.6026	.6064	.6103	.6141
0.9	.8159	.8186	.8212	.8238	.8264	.8289	.8315	.8340	.8365	.8389
1.0	.8413	.8438	.8461	.8485	.8508	.8531	.8554	.8577	.8599	.8621
1.1	.8643	.8665	.8686	**.8708**	.8729	.8749	.8770	.8790	.8810	.8830
1.2	.8849	.8869	.8888	.8907	.8925	.8944	.8962	.8980	.8997	.9015

The number 0.8708 describes the area (in orange) to the *left* of 117 points, which is *not* what we are interested in (see Fig. 63). Because the area under the entire curve is 1, we can find the area (in blue) to the *right* of 117 by finding 1 minus 0.8708:

$$1 - 0.8708 = 0.1292$$

So, the probability is 0.1292, which we indicate in Fig. 63. The probability is less than 0.5, which checks with the blue region having area less than half of the area under the entire curve.

WARNING   In Problem 2 of Example 4, we used a table of areas to find the area 0.8708, which is the area to the *left* of 117 points on the IQ test. In order to find the area to the *right* of 117 points, we subtracted 0.8708 from 1. Remember that **the entries in the area table are areas to the *left* of a value. We have to subtract an entry from 1 to find the area to the right.**

▶ **Example 5**   Using z-Scores and Tables to Find Probabilities

Recall from Example 2 that the 2014 draft picks for NBA basketball teams have heights that are approximately normally distributed with mean 79.1 inches and standard deviation 3.0 inches. Find the probability that a randomly selected draft pick has height between 75 inches and 83 inches.

Solution

We first sketch a normal curve with mean 79.1, mark the numbers 75 and 83 on the horizontal axis, and use orange to shade the region between 75 and 83 (see Fig. 64).

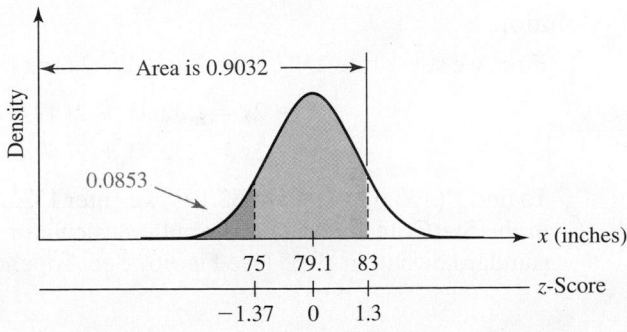

Figure 64 Heights between 75 and 83 inches

Next, we calculate the z-score of 75 by substituting 75 for x, 79.1 for $\bar{x}$, and 3.0 for s in the formula $z = \dfrac{x - \bar{x}}{s}$:

$$z = \frac{75 - 79.1}{3.0} \approx -1.37$$

And we calculate the z-score of 83 by substituting 83 for x, 79.1 for $\bar{x}$, and 3.0 for s in the formula $z = \dfrac{x - \bar{x}}{s}$:

$$z = \frac{83 - 79.1}{3.0} = 1.3$$

By looking up $z = -1.37$ in Table 1 in Appendix C, we find that the area (in blue) to the left of 75 is 0.0853, which we indicate in Fig. 64. By looking up $z = 1.3$, we find that the area (in blue OR orange) to the left of 83 is 0.9032, which we also indicate in Fig. 64. To find the area of the orange region, we find the difference of the area to the left of 83 and the area to the left of 75:

$$0.9032 - 0.0853 = 0.8179$$

So, the approximate probability is 0.8179.

Using Technology to Find Probabilities for a Normal Distribution

We can find probabilities for a normal distribution using technology rather than using a table. We will continue to round probabilities to the fourth decimal place.

In Example 6, we will use StatCrunch to find probabilities. In Example 7, we will use a TI-84 to find probabilities.

▶ **Example 6**   Using StatCrunch to Find Probabilities for a Normal Distribution

We can use StatCrunch to find the mean and the standard deviation of the song-length distribution (see Fig. 65).

Summary statistics:

Column	Mean	Std. dev.
Song Length	232.92617	49.634079

Figure 65 Mean and standard deviation of the song-length distribution

Let X be a song length (in seconds) randomly selected from the song-length data set.

1. Use StatCrunch to find the probability that a song length randomly selected from the data lies within 2 standard deviations of the mean. Compare your result with the Empirical Rule.
2. Use StatCrunch to find the probability that a randomly selected song length is at most 195 seconds.
3. Use StatCrunch to find the probability that a randomly selected song length is at least 260 seconds.

Solution

1. First, we substitute 232.93 for $\bar{x}$ and 49.63 for s in the expressions $\bar{x} - 2s$ and $\bar{x} + 2s$:

$$x - 2s = 232.93 - 2(49.63) = 133.67$$
$$x + 2s = 232.93 + 2(49.63) = 332.19$$

To find $P(133.67 \le X \le 332.19)$, we enter 133.67 and 332.19 in the appropriate boxes in the StatCrunch normal distribution calculator, along with the mean 232.93 and the standard deviation 49.63 (see Fig. 66). See Appendix B.13 for StatCrunch instructions.

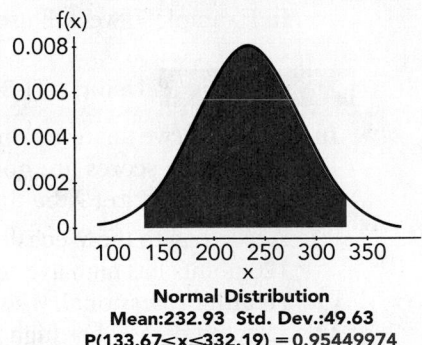

Figure 66 Calculate the probability

From the StatCrunch output, we see that $P(133.67 \leq x \leq 332.19) \approx 0.95$, which is what the Empirical Rule predicts.

2. Recall that the phrase *at most* means *less than or equal to*:

$$P(\text{at most 195 seconds}) = P(\text{less than or equal to 195 seconds})$$
$$= P(X \leq 195)$$

So, we enter 195 in the appropriate box in the StatCrunch normal distribution calculator, along with the mean 232.93 and the standard deviation 49.63 (see Fig. 67). See Appendix B.13 for StatCrunch instructions.

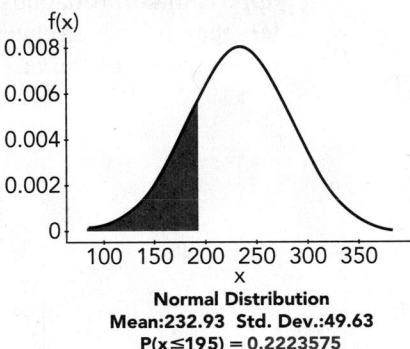

Figure 67 Calculate the probability

From the StatCrunch output, we see that $P(X \leq 195) \approx 0.2224$.

3. Recall that the phrase *at least* means *greater than or equal to*:

$$P(\text{at least 260 seconds}) = P(\text{greater than or equal to 260 seconds})$$
$$= P(X \geq 260)$$

So, we enter 260 in the appropriate box in the StatCrunch normal distribution calculator, along with the mean 232.93 and the standard deviation 49.63 (see Fig. 68). See Appendix B.13 for StatCrunch instructions.

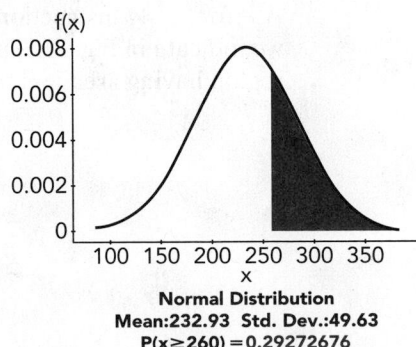

Figure 68 Calculate the probability

From the StatCrunch output, we see that $P(X \geq 260) \approx 0.2927$.

In Example 7, we will use a TI-84 to find probabilities for a normal distribution.

▶ Example 7   Using a TI-84 to Find Probabilities for a Normal Distribution

In Example 4, we analyzed the Wechsler IQ test, which measures a person's intelligence. Recall that the scores are normally distributed with mean 100 points and standard deviation 15 points. Let X be the IQ (in points) of a randomly selected person.

1. According to the Wechsler classification, a person with an IQ score between 110 and 120 points has high average intelligence (Source: *Practical Resources for the Mental Health Professional, Weiss et al.*). Use a TI-84 to find the probability that a randomly selected person has high average intelligence.
2. Use a TI-84 to find the probability that a randomly selected person has an IQ less than 80 points.
3. Use a TI-84 to find the probability that a randomly selected person has an IQ greater than 80 points.

Solution

1. We first sketch a normal curve with mean 100 points, mark the numbers 110 and 120 on the horizontal axis, and use blue to shade the region between 110 and 120 (see Fig. 69). To use "normalcdf" on a TI-84, we enter 110 for "lower," 120 for "upper," 100 for the (population) mean μ, and 15 for the (population) standard deviation σ (see Fig. 70). See Appendix A.8 for TI-84 instructions. From Fig. 70, we see that the approximate probability is 0.1613, which we indicate in Fig. 69. The probability is less than 0.5, which checks with the blue region having area less than half of the area under the entire curve.

Figure 70 Calculate the probability

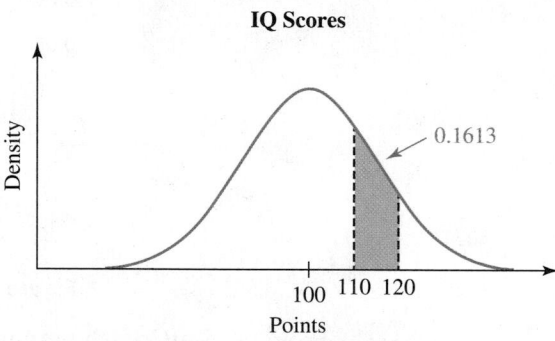

Figure 69 IQ scores between 110 and 120 points

2. We first sketch a normal curve with mean 100 points and mark the number 80 on the horizontal axis (see Fig. 71). Because we want to find the probability that a randomly selected person has an IQ *less* than 80 points, shade the region (in orange) to the *left* of 80. To use "normalcdf" on a TI-84, we enter −999999 for "lower," 80 for "upper," 100 for the (population) mean μ, and 15 for the (population) standard deviation σ (see Fig. 72). We use −999999 because there is no left value. See Appendix A.8 for TI-84 instructions. From Fig. 72, we see that $P(X < 80) \approx 0.0912$, which we indicate in Fig. 71. The probability is less than 0.5, which checks with the orange region having area less than half of the area under the entire curve.

Figure 72 Calculate the probability

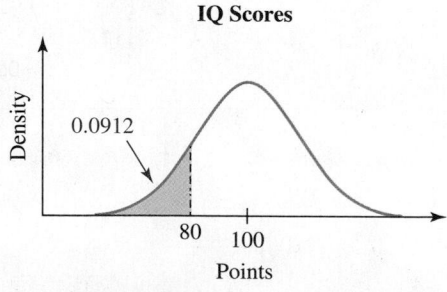

Figure 71 IQ scores at most 80 points

3. We will discuss two methods to find the probability.

Method 1 We first sketch a normal curve with mean 100 points and mark the number 80 on the horizontal axis (see Fig. 73). Because we want to find the probability that a randomly selected person has an IQ *greater* than 80 points, we shade the region (in green) to the *right* of 80. To use "normalcdf" on a TI-84, we enter 80 for "lower," 999999 for "upper," 100 for the (population) mean μ, and 15 for the (population) standard deviation σ (see Fig. 74). We use 999999 because there is no right value. See Appendix A.8 for TI-84 instructions. From Fig. 74, we see that $P(X > 80) \approx 0.9088$, which we indicate in Fig. 73. The probability is greater than 0.5, which checks with the green region having area greater than half of the area under the entire curve.

Figure 74 Calculate the probability

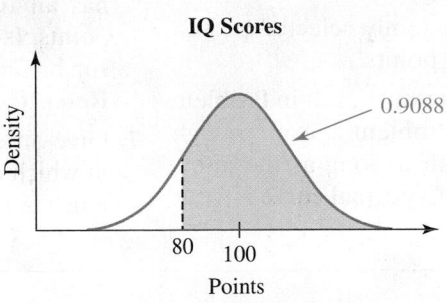

Figure 73 IQ scores at most 80 points

Method 2 In Problem 2 we found that the probability for IQ scores less than 80 points is approximately 0.0912. Because the area under the entire curve is 1, we can find the area for IQ scores greater than 80 points by finding 1 minus 0.0912: $1 - 0.0912 = 0.9088$. This is the same result we found by Method 1.

In Problem 2 of Example 7, we found that for the normal curve shown in Fig. 75, the area of the orange region is approximately 0.0912. We can interpret the area 0.0912 in two ways:

1. The probability of randomly selecting a person with an IQ less than 80 points is 0.0912.

2. The proportion of people with an IQ less than 80 points is 0.0912.

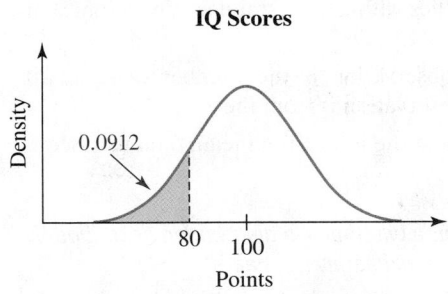

Figure 75 IQ scores at most 80 points

> **Interpreting the Area under a Normal Curve for an Interval**

We can interpret the area A under the normal curve for an interval as
- the probability that a randomly selected observation will lie in the interval.
- the proportion of observations that lie in the interval.

Group Exploration

Symmetry of a normal curve

Recall from Example 4 that scores on the Wechsler IQ test are normally distributed with mean 100 points and standard deviation 15 points.

1. Sketch the IQ normal curve.

2. Find the probability that a randomly selected person has an IQ between 100 and 120 points.

3. Find the probability that a randomly selected person has an IQ between 100 and 110 points.

4. Explain why it makes sense that your result in Problem 3 is *not* half of your result in Problem 2, even though the interval in Problem 3 is half as long as the interval in Problem 2. Refer to the IQ normal curve in your explanation.

5. Find the probability that a randomly selected person has an IQ between 90 and 110 points. Is the probability that a randomly selected person has an IQ between 100 and 110 half as much? Explain. Refer to the IQ normal curve in your explanation.

6. Find the probability that a randomly selected person has an IQ less than 80 points OR greater than 120 points. Is the probability that a randomly selected person has an IQ less than 80 points half as much? Explain. Refer to the IQ normal curve in your explanation.

7. Give an example other than those in this exploration in which one probability involving IQ scores is half as much as another.

▶ **Tips for Success**   **Stick with It**

If you are having difficulty doing an exercise, do not panic! Reread the exercise and reflect on what you have already sorted out about the problem—what you know and where you need to go. Your solution to the problem may be just around the corner.

Homework 5.4

For extra help ▶ **MyMathLab®**   ▦ Watch the videos in MyMathLab   ◆ Download the MyDashboard App

1. The total area under the normal curve is equal to ____ .

2. If a distribution is normal, then the probability that a randomly selected observation lies within ____ standard deviation(s) of the mean is approximately 0.997.

3. The *z*-score of an observation is the number of standard deviations that the observation is from the ____ .

4. If an observation lies to the left of the mean, then its *z*-score is ____ .

For Exercises 5–8, determine whether the distribution described by the density histogram is approximately normal.

5. See Fig. 76.

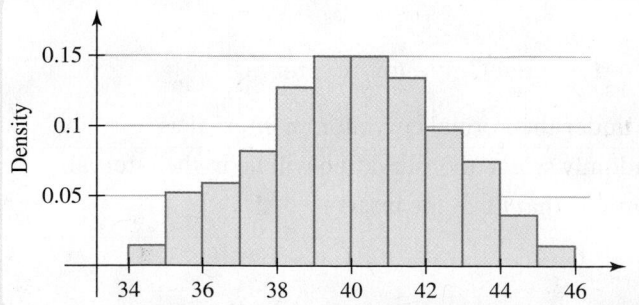

Figure 76 Exercise 5

6. See Fig. 77.

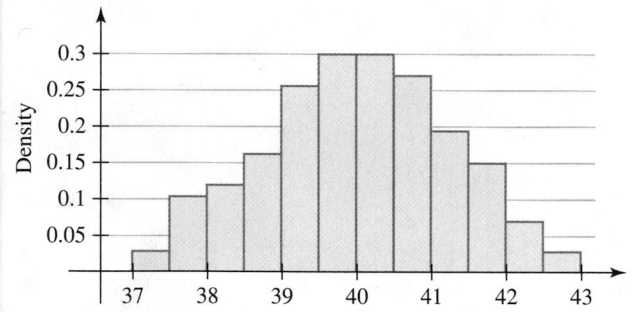

Figure 77 Exercise 6

7. See Fig. 78.

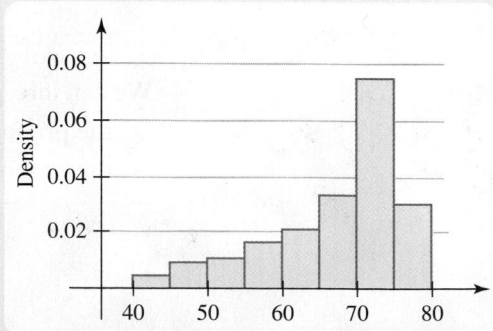

Figure 78 Exercise 7

8. See Fig. 79.

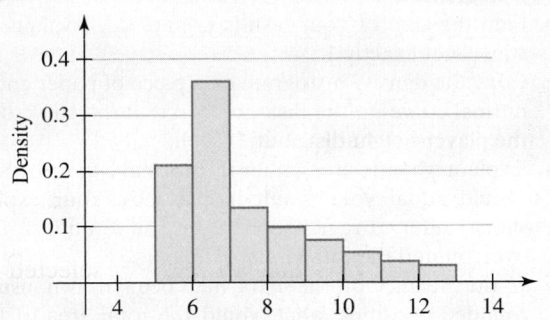

Figure 79 Exercise 8

For Exercises 9–12, match the given normal curve to the appropriate histogram in Fig. 84.

9. See Fig. 80.

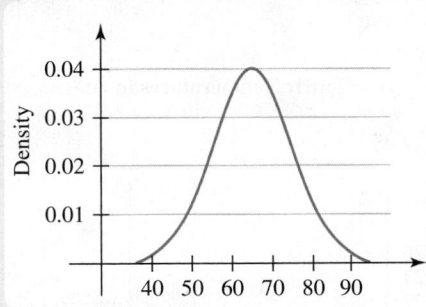

Figure 80 Exercise 9

10. See Fig. 81.

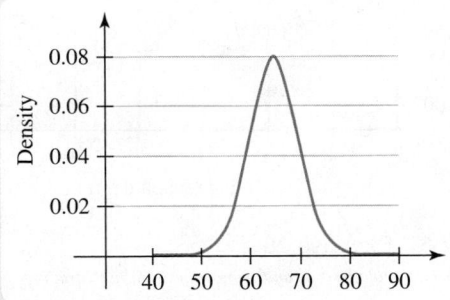

Figure 81 Exercise 10

11. See Fig. 82.

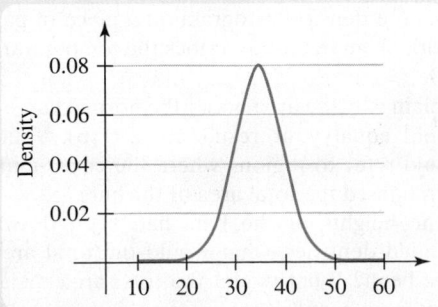

Figure 82 Exercise 11

12. See Fig. 83.

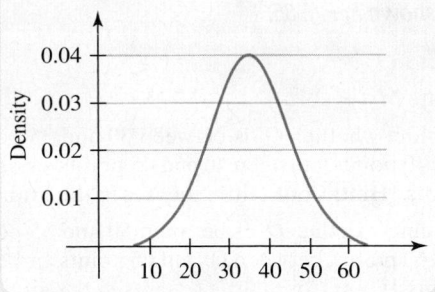

Figure 83 Exercise 12

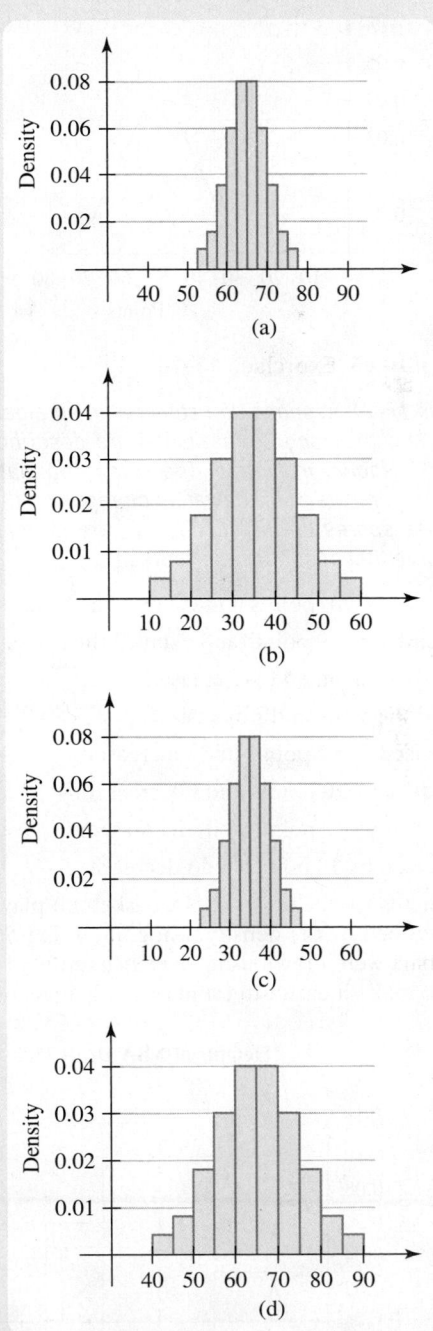

(a)

(b)

(c)

(d)

Figure 84 Exercises 9–12

For Exercises 13–16, suppose that students in an algebra class take a first test and their scores (in points) are described well by the normal curve shown in Fig. 85.

13. Estimate $\bar{x}$.

14. Estimate M.

15. Determine whether Q_1 is between 30 and 35 points, between 35 and 40 points, between 40 and 45 points, or between 45 and 50 points. [**Hint:** Count little squares in the grid.]

16. Determine whether Q_3 is between 50 and 55 points, between 55 and 60 points, between 60 and 65 points, or between 65 and 70 points. [**Hint:** Count little squares in the grid.]

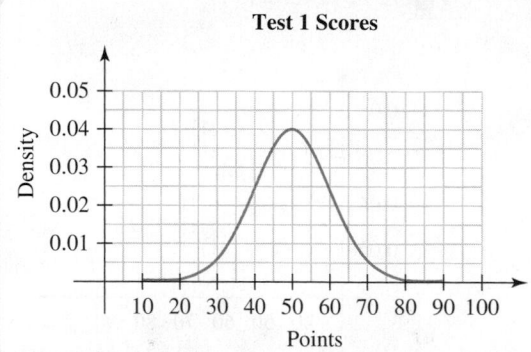

Figure 85 Exercises 13–16

In Exercises 17–24, suppose that students in an algebra class take a first test and their scores (in points) are described well by the normal curve shown in Fig. 85. The given information describes how the mean and standard deviation changed on the second test. Assuming the scores on the second test are normally distributed, sketch a curve that describes the scores.

17. $\bar{x}$ increased by 10 points and s stayed the same.

18. $\bar{x}$ decreased by 10 points and s stayed the same.

19. $\bar{x}$ stayed the same and s decreased.

20. $\bar{x}$ stayed the same and s increased.

21. $\bar{x}$ decreased by 10 points and s increased.

22. $\bar{x}$ increased by 10 points and s decreased.

23. $\bar{x}$ increased by 10 points and s increased.

24. $\bar{x}$ decreased by 10 points and s decreased.

25. The heights (in inches) of NBA basketball player draft picks are described by the density histogram in Fig. 86. The heights of the bars were drawn using densities rounded to the second place to make it easier to estimate the heights of the bars.

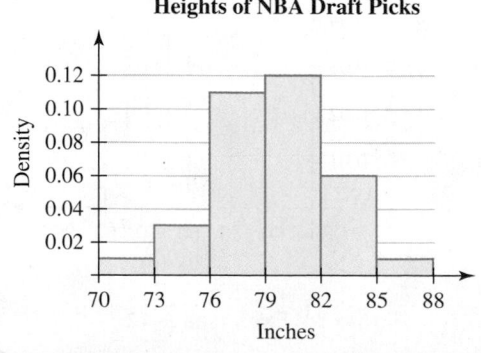

Figure 86 Exercise 25

(**Source:** *nbadraft.net*)

a. Estimate the area of each of the six bars of the density histogram.

b. Find the sum of your results in part (a). Explain why the sum is not exactly 1.

c. Copy the density histogram on a piece of paper and draw a normal curve on the histogram so that the curve describes the player-height distribution well.

d. Explain why the area under the normal curve you sketched should equal your result in part (b). Your explanation should refer to regions where the curve undercounted or overcounted the total area of the bars.

e. If the heights of the bars had been drawn using non-rounded densities, what would the total area of the bars have been? What would the total area under the normal curve have been?

26. The hourly temperatures (in degrees Fahrenheit) at Eppley Airfield Airport in Omaha, Nebraska, in May 2014 are described by the density histogram in Fig. 87. The heights of the bars were drawn using densities rounded to one of the values shown on the vertical axis to make it easier to estimate the heights of the bars.

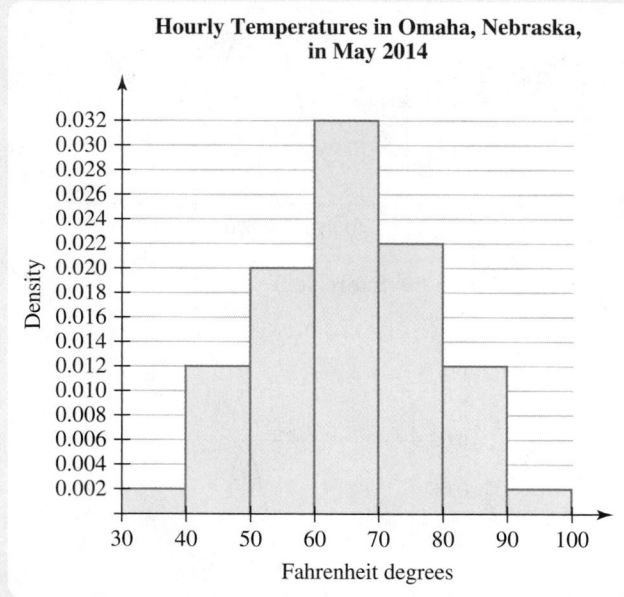

Figure 87 Exercise 26

(**Source:** *National Oceanic and Atmospheric Administration*)

a. Find the area of each of the seven bars of the density histogram.

b. Find the sum of your results in part (a). Explain why the sum is not exactly 1.

c. Copy the density histogram on a piece of paper and draw a normal curve that describes the temperature distribution well.

d. Explain why the area under the normal curve you sketched should equal your result in part (b). Your explanation should refer to regions where the curve undercounted or overcounted the total area of the bars.

e. If the heights of the bars had been drawn using non-rounded densities, what would the total area of the bars have been? What would the total area under the normal curve have been?

27. At the start of this section, we analyzed the lengths of songs played on Friday, April 18, 2014, on Live 105. The lengths of the songs are approximately normally distributed with mean 232.93 seconds and standard deviation 49.63 seconds

(Source: *Live 105*). Find the *z*-score of the length of Coldplay's song "Magic," which is 285 seconds long. Round your result to two decimal places. What does it mean in this situation?

28. Recall that scores on the Wechsler IQ test are normally distributed with mean 100 points and standard deviation 15 points. Kim Ung-yong, a Korean child prodigy, has an IQ of 210. By age 4 years, he was solving integral calculus problems (Source: *"What Ever Became of Geniuses?,"* Time). Find the *z*-score of his IQ. Round your result to two decimal places. What does it mean in this situation?

29. The hourly temperatures at Eppley Airfield Airport in Omaha, Nebraska, in May 2014 are approximately normally distributed with mean 64.4°F and standard deviation 12.2°F (Source: *National Oceanic and Atmospheric Administration*). Find the *z*-score for the lowest temperature, which was 34°F. Round your result to two decimal places. What does it mean in this situation?

30. The age distribution of inmates on death row in Alabama is approximately normal with mean 43.0 years and standard deviation 9.7 years (Source: *Alabama Department of Corrections*). The age of one of the inmates is 23 years. Find the *z*-score of his age. Round your result to two decimal places. What does it mean in this situation?

For Exercises 31–42, find the probability that a z-score randomly selected from the standard normal distribution meets the given condition.

31. The *z*-score is less than −1.93.

32. The *z*-score is less than −2.38.

33. The *z*-score is less than 3.30.

34. The *z*-score is less than 0.40.

35. The *z*-score is greater than −0.69.

36. The *z*-score is greater than −2.62.

37. The *z*-score is greater than 2.15.

38. The *z*-score is greater than 1.96.

39. The *z*-score is between −2.50 and 1.40.

40. The *z*-score is between −3.10 and −1.80.

41. The *z*-score is between −1.73 and 0.

42. The *z*-score is between 0.25 and 2.74.

43. Smoke alarms with a UL label have been certified to last 10 years (Source: *Underwriters Laboratories*). Assume the lifetimes of smoke alarms with a UL label are normally distributed with mean 10.8 years and standard deviation 0.3 year. Find the probability that a randomly selected smoke alarm with a UL label will last

   a. between 10 and 11 years.
   b. greater than 11 years.
   c. less than 10.6 years.
   d. less than the certified 10 years. Why is it important that your result is so close to 0?

44. A researcher measured the lifetimes of some zinc-air batteries when used for hearing aids. For the Activair® 13 HPX batteries, the mean lifetime was 252.4 hours and the standard deviation was 18.4 hours. For the Rayovac® 13 AE battery, the mean lifetime was 272.1 hours and the standard deviation

was 5.4 hours (Source: Measurement with Various Loads on Three Sizes of Zinc-Air Batteries for Hearing Aids, *Björn Hagerman*). Assume the lifetimes of the tested Activair batteries are normally distributed. Also assume the lifetimes of the tested Rayovac batteries are normally distributed.

   a. Find the probability that a Activair battery randomly selected from those tested lasted more than 285 hours.
   b. Find the probability that a Rayovac battery randomly selected from those tested lasted more than 285 hours.
   c. Explain how it is possible for your result in part (a) to be greater than your result in part (b), even though the mean lifetime of the tested Activair batteries was less than the mean lifetime of the tested Rayovac batteries. Sketch two normal curves to help explain.

45. Recall that scores on the Wechsler IQ test are normally distributed with mean 100 points and standard deviation 15 points.

   a. According to the Wechsler classification, a person with an IQ score between 120 and 130 points has superior intelligence (Source: Practical Resources for the Mental Health Professional, *Weiss et al.*). Find the probability that a randomly selected person has superior intelligence.
   b. According to the Wechsler classification, a person with an IQ score less than 70 points has extremely low intelligence. Find the probability that a randomly selected person has extremely low intelligence.
   c. Find the probability that a randomly selected person does NOT have extremely low intelligence.

46. Recall that scores on the Wechsler IQ test are normally distributed with mean 100 points and standard deviation 15 points. If a prisoner has a low enough IQ, it is against the law to sentence the prisoner to death. Florida used to have an inflexible cutoff of 70 points to decide whether prisoners were too intellectually disabled to be executed. But on May 27, 2014, the Supreme Court ruled that Florida and nine other states need to have a flexible cutoff, like other states do. A cutoff as high as 75 points is used for some prisoners, depending on their behavior and history.

   a. Find the probability that a randomly selected person (not necessarily a prisoner) has an IQ less than 70 points.
   b. Find the probability that a randomly selected person (not necessarily a prisoner) has an IQ less than 75 points.
   c. As of the ruling, there were 392 Florida prisoners on death row. Assume there would have been 9 additional prisoners on death row, but their IQs were lower than 70. Estimate the greatest number of prisoners that could be removed from death row due to the ruling.
   d. Describe any assumptions that you made in calculating your estimate in part (c).

47. For students who graduated from high school in 2013, their SAT math scores were approximately normally distributed with a mean of 514 points and a standard deviation of 118 points (Source: *College Board*).

   a. If a student is randomly selected from those who graduated from high school in 2013 and took the SAT, what is the probability that the student scored between 600 and 700 points on the math sections?
   b. The mean SAT math score for psychology majors was 490 points. Find the percentile for 490 points. A student says

that all psychology majors's SAT math scores are at the percentile you found. What would you tell the student?

c. The mean SAT math scores for social-science majors and engineering majors were 553 points and 580 points, respectively. Find the percentiles for 553 points and 580 points. What is surprising about your results?

48. For students who graduated from high school in 2013, the ACT math scores are approximately normally distributed with mean 20.9 points and standard deviation 5.3 points (Source: *ACT, Inc.*).

a. If a freshman at Thomas More College (TMC) scored 17 points on the math sections of the ACT, the student was required to take a one-semester beginning algebra course. What is the z-score of 17 points? Why does it make sense that your result is negative?

b. If a freshman at TMC scored at most 16 points on the math sections of the ACT, the student was required to take a *two*-semester beginning algebra course. What proportion of students who took the ACT scored at most 16 points?

c. A student says your result in part (c) is the proportion of all TMC students who had to take the two-semester beginning algebra course. What would you tell the student?

49. For students who graduated from high school in 2013, their SAT math scores had a mean of 514 points and a standard deviation of 118 points (Source: *College Board*).

a. Assuming the SAT math scores are normally distributed, find the percentiles for the following scores (in points): 400, 600, and 800.

b. Table 30 displays some of the percentiles by assuming normality and all of the *actual* percentiles for various SAT math scores. Use the results you found in part (a) to fill in the blanks.

Table 30 Percentiles by Assuming Normality and Actual Percentiles for Various SAT Math Scores

Math Score	Percentile (Assuming Normality)	Actual Percentile
300	3	3
400		16
500	45	45
600		75
700	94	93
800		99

Source: *College Board*

c. Compare the percentiles assuming normality and the actual percentiles in your completed Table 30. Does your comparison suggest that the SAT math scores are normally distributed, approximately normally distributed, or not at all normally distributed? Explain.

50. For students who graduated from high school in 2013, their ACT English scores had a mean of 20.2 points and a standard deviation of 6.5 points (Source: *ACT, Inc.*).

a. Assuming the ACT English scores are normally distributed, find the percentiles for the following scores (in points): 15, 25, and 35.

b. Table 31 displays some of the percentiles by assuming normality and all of the *actual* percentiles for various ACT English scores. Use your results of part (a) to fill in the blanks.

Table 31 Percentiles by Assuming Normality and Actual Percentiles for Various ACT English Scores

English Score	Percentile (Assuming Normality)	Actual Percentile
10	6	7
15		26
20	49	52
25		79
30	93	92
35		100

Source: *College Board*

c. Compare the percentiles assuming normality and the actual percentiles in your completed Table 31. Does your comparison suggest that the ACT English scores are normally distributed, approximately normally distributed, or not at all normally distributed? Explain.

51. HDL is the "good" cholesterol. The higher a person's HDL level, the lower the person's risk of heart attack and stroke will be. In a study of 1751 women in their twenties, the women's HDL cholesterol levels are approximately normally distributed with mean 55 mg/dl and standard deviation 14.8 mg/dl (Source: *NHANES*).

a. A woman has low HDL (higher risk) if her HDL reading is less than 50 mg/dl (Source: *American Heart Association*). Find the probability that a woman randomly selected from the 1751 women has low HDL.

b. A woman has high HDL (lower risk) if her HDL reading is greater than 60 mg/dl (Source: *American Heart Association*). Find the probability that a woman randomly selected from the 1751 women has high HDL.

c. Find the probability that a woman randomly selected from the 1751 women has an HDL reading between 40 and 60 mg/dl.

d. Find the *number* of women in the study who have HDL readings between 40 and 60 mg/dl.

52. HDL is the "good" cholesterol. The higher a person's HDL level, the lower the person's risk of heart attack and stroke will be. In a study of 1541 men in their twenties, the men's HDL cholesterol levels are approximately normally distributed with mean 47 mg/dl and standard deviation 12.5 mg/dl (Source: *NHANES*).

a. A man has low HDL (higher risk) if his HDL reading is less than 40 mg/dl (Source: *American Heart Association*). Find the probability that a man randomly selected from the 1541 men has low HDL.

b. A man has high HDL (lower risk) if his HDL reading is greater than 60 mg/dl (Source: *American Heart Association*). Find the probability that a man randomly selected from the 1541 men has high HDL.

c. Find the probability that a man randomly selected from the 1541 men has an HDL reading between 40 and 60 mg/dl.

d. Find the *number* of men in the study who have HDL readings between 40 and 60 mg/dl.

53. The risk of coronary disease for diabetes patients is related not only to the mean glucose level in the bloodstream but also to the standard deviation. (Source: *"Understanding the Ups and Downs of Blood Glucose," Irl B. Hirsch, MD.*).

a. A diabetes patient we will call Frank posted online that his blood glucose has mean 93.1 mg/dL and standard deviation 41.5 mg/dL. Dr. Irl B. Hirsch recommends that diabetes patients regulate their glucose levels so that the standard deviation is at most $\frac{1}{2}$ the mean. Has Frank met this recommendation?

b. Dr. Hirsch says that, ideally, the standard deviation of blood glucose levels should be at most $\frac{1}{3}$ the mean. Has Frank met this stricter goal?

c. Frank posted that 33% of his glucose blood readings were between 79 and 121 mg/dL. What would the percentage have been if his readings were normally distributed? Recall that the mean is 93.1 mg/dL and the standard deviation is 41.5 mg/dL.

d. Glucose levels tend to be low just before meals and high just after meals. If Frank measured his blood glucose levels only just before meals and just after meals, would the data he collected be unimodal, bimodal, or multimodal? Would the data be normally distributed? Explain why that might account for the error in the result you found in part (c).

54. The lower a person's bone mineral density (BMD), the greater the person's risk of a bone fracture will be. BMD levels of 75-year-old Caucasian women's hips are approximately normally distributed with mean 740 mg/cm² and standard deviation 129 mg/cm² (Source: *NHANES*).

a. Find the probability of randomly selecting a 75-year-old Caucasian woman whose hip has a BMD at least 908 mg/cm².

b. Find the probability of randomly selecting a 75-year-old Caucasian woman whose hip has a BMD between 572 and 636 mg/cm².

c. If the z-score of an adult's hip BMD reading is less than −1.5, then this suggests that bone loss has happened for reasons other than aging, such as thyroid abnormalities, malabsorption, alcoholism, smoking, and the use of certain medications. Find the probability of randomly selecting a Caucasian woman who has experienced bone loss for reasons other than aging.

55. A total of 940 healthy, pregnant women between 20 and 34 years of age participated in a study. Their lengths (in days) of pregnancy are approximately normally distributed (see the normal curve in Fig. 88).

Lengths of Pregnancy

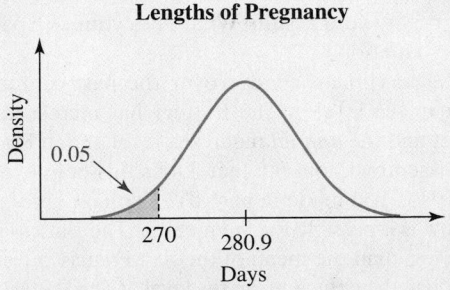

Figure 88 Exercise 55

(**Source:** The Length of Human Pregnancy as Calculated by Ultrasonographic Measurement of the Fetal Biparietal Diameter, *H. Kieler et al.*)

a. Interpret the area of the orange region in Fig. 88 as a probability.

b. Interpret the area of the orange region in Fig. 88 as a proportion.

c. Interpret the area of the orange region in Fig. 88 as a percentile.

d. The researchers defined a *preterm delivery* as a pregnancy lasting at most 259 days. Women with preterm deliveries were excluded from the study. If these women had been included in the study, how would the mean pregnancy length have been affected? How would the standard deviation have been affected?

56. In a study of U.S. veteran patients undergoing a first liver transplant, the ages (in years) of patients are approximately normally distributed (see the normal curve in Fig. 89).

Ages of Veteran Patients Undergoing a First Liver Transplant

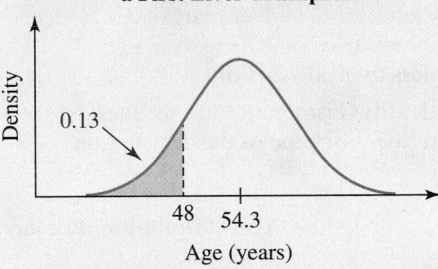

Figure 89 Exercise 56

(**Source:** Impact of the MELD Score on Waiting Time and Disease Severity in Liver Transplantation in United States Veterans, *Jawad Ahmad et al.*)

Interpret the area of the orange region in Fig. 89 as a

a. probability.

b. proportion.

c. percentile.

57. In a study of African American and Caucasian girls of ages 9 and 10 years, the girls recorded their diets in a diary for three days. The researcher then computed the daily salt intake. The distribution of daily salt intakes for the African American girls has a mean of 3073 mg and a standard deviation of 1072 mg. The distribution of daily salt intakes for the Caucasian girls has a mean of 2803 mg and a standard deviation of 833 mg (Source: *World Health Organization*). Assume the distributions of daily salt intakes for the African American girls and for the Caucasian girls are approximately normal.

a. Find the probability that a girl randomly selected from the African American girls had a daily salt intake of at least 3500 mg.

b. Find the probability that a girl randomly selected from the Caucasian girls had a daily salt intake of at least 3500 mg.

c. Refer to the given means and/or standard deviations to explain why it makes sense that the result you found in part (a) is greater than the result you found in part (b).

58. In a study of African American and Caucasian girls of ages 9 and 10 years, the girls recorded their diets in a diary for three days. The researcher then computed the daily salt intake. The distribution of daily salt intakes for the African American girls has a mean of 3073 mg and a standard deviation of 1072 mg. The distribution of daily salt intakes for the Caucasian girls has a mean of 2803 mg and a standard deviation of 833 mg (Source: *World Health Organization*). Assume the distributions of daily salt intakes for the African American girls and for the Caucasian girls are approximately normal.

a. Find the probability that a girl randomly selected from the African American girls had a daily salt intake of at most 1500 mg.

b. Find the probability that a girl randomly selected from the Caucasian girls had a daily salt intake of at most 1500 mg.

c. Explain how it is possible for the result you found in part (a) to be greater than the result you found in part (b), even though the mean intake of salt per day was larger for the African American girls than the Caucasian girls.

59. The age distribution of prisoners on death row in Florida has mean 48.4 years and standard deviation 11.0 years (Source: *Florida Department of Corrections*).

a. Assuming the distribution is normally distributed, find the probability that a randomly selected death-row prisoner in Florida is

   i. between 40 and 60 years of age.
   ii. between 30 and 70 years of age.
   iii. less than 30 years of age.

b. A density histogram of the ages is shown in Fig. 90. Describe the shape of the distribution.

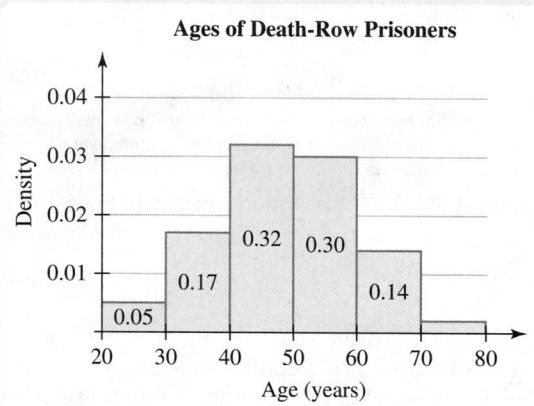

Ages of Death-Row Prisoners

Figure 90 Exercise 59
(**Source:** *Florida Department of Corrections*)

c. Use the density histogram to estimate the probability that a randomly selected death-row prisoner in Florida is

   i. between 40 and 60 years of age.
   ii. between 30 and 70 years of age.
   iii. less than 30 years of age.

d. Compare the results you found in part (c) with the results you found in part (a). On the basis of your comparison and your response to part (b), determine whether the distribution is approximately normal.

60. In 2013, 4579 runners completed the Great Cow Harbor 10K Run in Northport, New York. Their age distribution has mean 40.2 years and standard deviation 12.4 years (Source: *The Great Cow Harbor 10-Kilometer Run, Inc.*).

a. Assuming the distribution is normally distributed, find the probability that a randomly selected runner's age was

   i. between 40 and 50 years.
   ii. between 30 and 60 years.
   iii. between 20 and 70 years.

b. A density histogram of the ages is shown in Fig. 91. Describe the shape of the distribution.

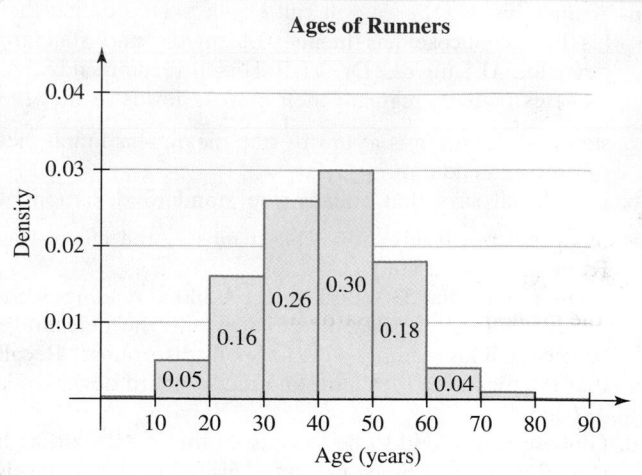

Ages of Runners

Figure 91 Exercise 60
(**Source:** *The Great Cow Harbor 10-Kilometer Run, Inc.*)

c. Use the density histogram to estimate the probability that a randomly selected runner's age was

   i. between 40 and 50 years.
   ii. between 30 and 60 years.
   iii. between 20 and 70 years.

d. Compare the results you found in part (c) with the results you found in part (a). On the basis of your comparison and your response to part (b), determine whether the distribution is approximately normal.

Large Data Sets

61. **DATA** Access the data about sea levels, which are available at MyMathLab and at the Pearson Downloadable Student Resources for Math & Stats website. Did you know that the sea level varies around the Earth and is increasing at different rates? For example, the sea level along the West Coast of the United States is greater than the sea level along the East Coast.

a. Construct relative frequency histograms of the sea levels since 1900 at The Battery (in New York) and at San Francisco.

b. Describe the shapes of The Battery and San Francisco distributions.

c. Compute the means of the two distributions. Compare your results. What does your comparison mean in this situation?

d. Compute the standard deviations of the two distributions. Compare your results. What does your comparison mean in this situation?

e. Due to climate change over the past century, the *annual* mean sea level at The Battery has increased by about 2.1 feet and the *annual* mean sea level at San Francisco has increased by about 0.6 feet. Does this help explain the comparison you made in part(d)? Explain.

f. How is it possible for the mean of The Battery distribution to be less than the mean of the San Francisco distribution even though the annual mean sea level at The Battery has increased in the past century by more than 3 times as much as at San Francisco?

62. **DATA** Access the data about lengths of alternative rock songs, which are available at MyMathLab and at the Pearson Downloadable Student Resources for Math & Stats website. At the start of this section, we analyzed the lengths of songs played on the alternative rock radio station Live 105. You will now analyze the lengths (in seconds) of all songs on albums by bands typically featured on Live 105. Some of the songs on the albums have been played on Live 105, but many have not.

a. Construct a relative frequency histogram of the song lengths, beginning with lower class limit 0 seconds and class width 50 seconds. Display the relative frequencies above the bars if the technology you are using has this feature.

b. Describe the shape of the distribution.

c. Compute the mean and the standard deviation.

d. Assuming the distribution is approximately normal, use the mean and the standard deviation you found in part (c) to estimate the proportion of song lengths that are between 200 seconds and 300 seconds.

e. Find the *actual* proportion of song lengths that are between 200 seconds and 300 seconds. Compare the result to the one you found in part (d). Does your comparison suggest that there are too few or too many observations near the mean for the distribution to be exactly normal? Explain.

f. Assuming the distribution is approximately normal, use the mean and the standard deviation you found in part (c) to estimate the proportion of song lengths that are between 100 seconds and 400 seconds. Then find the *actual* proportion of song lengths that are between 100 seconds and 400 seconds. Compare your two results.

Concepts

63. Explain why the area under a normal curve is equal to 1. Include a comparison to a density histogram in your explanation.

64. For a normal distribution, explain why the probability of randomly selecting an observation that is greater than the mean is equal to 0.5.

65. Explain why a normal curve for a distribution with larger standard deviation is wider and flatter than a normal curve for a distribution with smaller standard deviation.

66. Suppose some data are approximately normally distributed. Sketch a standard normal curve to explain why so few

observations (approximately 4.55%) of the data values lie outside 2 standard deviations of the mean.

67. If a father's height is 1 standard deviation greater than his son's, how much larger is his height's z-score than his son's?

68. Describe how to compute a z-score. What does it mean?

69. The wait times of 2444 patients at 18 health care clinics in the Wake Forest University Baptist Medical Center located in Winston-Salem, North Carolina, were studied. The mean wait time was 21.0 minutes with standard deviation 14.7 minutes (Source: The Relationship Between Patient's Perceived Wait Time and Office-Based Practice Satisfaction, *Fabian Camacho, MS, et al.*).

a. Explain why the wait-time distribution cannot be normal or even approximately normal. [**Hint:** Find what the proportion of *negative* wait times would be if the distribution were normal.]

b. On the basis of the mean, standard deviation, and your explanation in part (a), determine whether the wait-time distribution is skewed left, skewed right, or symmetric.

c. For the purposes of the study, it was reasonable to exclude patients who had wait times greater than 75 minutes. A total of 91 such patients were excluded. How would the mean wait time and standard deviation have changed if the 91 patients had been included in the study? Explain.

70. The wait times of 4534 patients at an emergency department in a hospital in Akron, Ohio, were studied. The mean wait time was 63.7 minutes with standard deviation 62.4 minutes (Source: Posted Emergency Department Wait Times Are Not Always Accurate, *Nicholas Jouriles, MD, et al.*).

a. Explain why the wait-time distribution cannot be normal or even approximately normal. [**Hint:** Find what the proportion of *negative* wait times would be if the distribution were normal.]

b. On the basis of the mean, standard deviation, and your explanation in part (a), determine whether the wait-time distribution is skewed left, skewed right, or symmetric. Explain.

5.5 Finding Values of Variables for Normal Distributions

Objectives

» Find an observation from its z-score.

» Use z-scores to find the value of a normally distributed variable from a probability.

» Use technology to find the value of a normally distributed variable from a probability.

» Use z-scores to compare the relative size of two observations from different groups of data.

» Identify unusual observations.

» Investigate whether a claim is true.

In Section 5.4, we used the normal curve to find a probability for a given interval. In this section, we will go backward. That is, we will find an interval from a probability.

Finding an Observation from Its z-Score

Suppose the scores on a test are approximately normally distributed with mean 70 points and standard deviation 10 points. We can find the z-score for a given test score, but can we find a test score from its z-score?

For example, we will find the test score that has $z = 2$, which means the test score is 2 standard deviations (10) greater than the mean, 70. So, the test score is $2(10) = 20$ points greater than the mean, 70. Therefore, the test score is $70 + 2(10) = 90$ points (see Fig. 92).

We can perform similar logic to find a formula. If z is the z-score of an observation, then the observation is zs away from the mean $\bar{x}$ (to the left if z is negative and to the right if z is positive). So, the observation is $\bar{x} + zs$.

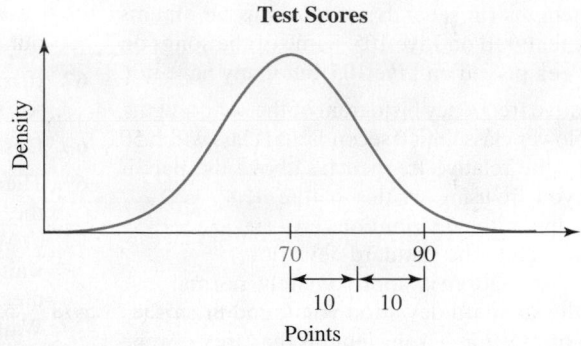

Figure 92 The test score is $70 + 2(10) = 90$

> **Finding the Value of an Observation from Its z-Score**
>
> An observation x with z-score z is given by
>
> $$x = \bar{x} + zs$$
>
> In words, the observation is equal to the mean plus the product of the z-score and the standard deviation.

▶ **Example 1** Find an Observation from Its z-Score

Recall that the scores on the Wechsler IQ test are normally distributed with mean 100 points and standard deviation 15 points. Film director Quentin Tarantino is reported to have a z-score of $z = 4$ (Source: Chicago Tribune). What is his IQ score?

Solution

We substitute 100 for $\bar{x}$, 4 for z, and 15 for s in the formula $x = \bar{x} + zs$:

$$x = 100 + 4(15) = 160$$

Quentin Tarantino's IQ is 160 points.

Using a z-Score to Find a Value of a Variable from a Percentage

Now that we know how to find an observation from its z-score, we are ready to find a value of a variable that is normally distributed from a percentage.

▶ **Example 2** Using a z-Score to Find a Value of a Variable from a Percentage

A professor gives a test to her calculus students. The scores are approximately normally distributed with mean 75 points and standard deviation 9 points. The professor decides to give As to approximately 10% of the students but not less than 10%. Find the cutoff score for an A.

Solution

We sketch a normal curve with mean 75 points. Because As are given to high scores, we shade (using blue) the far right 10% of the total area (see Fig. 93). We let c be the cutoff for an A and write it below the left edge of the shaded region.

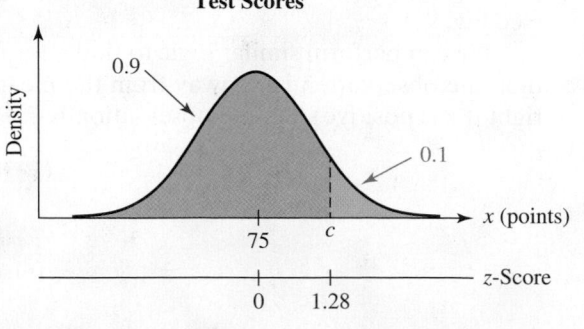

Figure 93 Top 10% of students earn As

Because the area (in blue) to the right of c has area 0.1, the area (in orange) to the left of c has area $1 - 0.1 = 0.9$. Next, we find the value in the middle of Table 1 in Appendix C that is closest to 0.9, which is 0.8997 (see Table 32). Then we find the number in the far left of the yellow row (1.2) and the number at the top of the blue column (.08) to form the z-score 1.28.

Table 32 Looking Up area 0.9

z	.00	.01	.02	.03	.04	.05	.06	.07	.08	.09
0.0	.5000	.5040	.5080	.5120	.5160	.5199	.5239	.5279	.5319	.5359
0.1	.5398	.5438	.5478	.5517	.5557	.5596	.5636	.5675	.5714	.5753
0.2	.5793	.5832	.5871	.5910	.5948	.5987	.6026	.6064	.6103	.6141
0.9	.8159	.8186	.8212	.8238	.8264	.8289	.8315	.8340	.8365	.8389
1.0	.8413	.8438	.8461	.8485	.8508	.8531	.8554	.8577	.8599	.8621
1.1	.8643	.8665	.8686	.8708	.8729	.8749	.8770	.8790	.8810	.8830
1.2	.8849	.8869	.8888	.8907	.8925	.8944	.8962	.8980	.8997	.9015
1.3	.9032	.9049	.9066	.9082	.9099	.9115	.9131	.9147	.9162	.9177

Finally, we substitute 75 for $\bar{x}$, 1.28 for z, and 9 for s in the formula $c = \bar{x} + zs$:

$$c = 75 + 1.28(9) = 86.52$$

If the professor were to round 86.52 points up to 87 points, less than 10% of the students would get As (about 9%). By rounding down to 86 points, at least 10% of the students would get As (about 11%). So, the cutoff should be 86 points.

WARNING In order to use the table of areas in Example 2, we looked for an area approximately equal to 0.9, which is the area to the *left* of c, not the area to the right. **Remember that whenever you use the table of areas, you must look for the area to the *left* of a value.**

Using Technology to Find a Value of a Variable from a Percentage

Next, we will redo Example 2 using a TI-84.

▶ **Example 3** Use a TI-84 to Find a Value of a Variable from a Percentage

In Example 2, we worked with test scores that are approximately normally distributed with mean 75 points and standard deviation 9 points. We found a cutoff for an A so that approximately 10% of the students would get As but not less than 10%. Now find the result using a TI-84.

Solution

We sketch a normal curve with mean 75 points. Because As are given to high scores, we shade (using blue) the far right 10% of the total area (see Fig. 94). We let c be the cutoff for an A and write it below the left edge of the blue region. To use "invNorm" on a TI-84, we must enter the area to the *left* of c, which is $1 - 0.1 = 0.9$ (see the orange region in Fig. 94). So, we enter 0.9 for "area," 75 for the (population) mean μ, and 9 for the (population) standard deviation σ (see Fig. 95). For TI-84 instructions, see Appendix A.9. From Fig. 95, we see the cutoff score is 86.5339641 points.

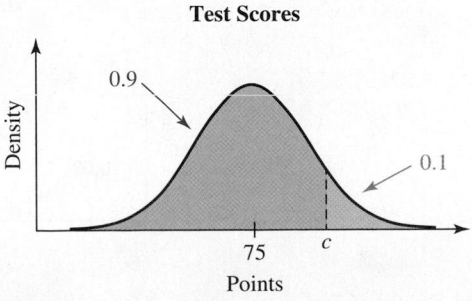

Test Scores

Figure 94 Shade the far right 10% of the total area

```
       invNorm
area:0.9
µ:75
σ:9
Paste
```

```
invNorm(0.9,75,▸
          86.5339641
```

Figure 95 Find the cutoff score

If the professor were to round the result up to 87 points, less than 10% of the students would get As (about 9%). By rounding down to 86 points, at least 10% of the students would get As (about 11%). So, the cutoff should be 86 points. The cutoff 86 points is larger than the mean 75 points, which checks with our sketch in Fig. 94, where we wrote c to the right of the mean. The result is the same as the result we found in Example 2.

WARNING

In order to use invNorm in Example 3, we entered 0.9, which is the area to the *left* of c, not the area to the right. **Remember that whenever you use invNorm, you must enter the area to the *left* of a value.**

In Example 3, we used a TI-84 to find the value of a normally distributed variable from a percentage. In Example 4, we will use StatCrunch to perform a similar task.

▶ **Example 4**   Using StatCrunch to Find a Value of a Variable from a Percentage

A researcher measured the lifetimes of some Activair® 13 HPX batteries when used for hearing aids. The mean lifetime was 252.4 hours, and the standard deviation was 18.4 hours (Source: Measurement with Various Loads on Three Sizes of Zinc-Air Batteries for Hearing Aids, *Björn Hagerman*). Assume all Activair 13 HPX batteries are normally distributed with the same mean and same standard deviation as those tested.

1. Explain why it would be unwise for the company to advertise that its batteries last 252.4 hours.
2. What should the company advertise the lifetime of its batteries to be so that 98% of the batteries would last at least that long?

Solution

1. Because the median (and mean) is equal to 252.4 hours, 50% of the batteries would last less than 252.4 hours (see Fig. 96). So, half of the batteries would last less than advertised. This could be bad for business, because the company might receive lots of complaints from customers and might run into trouble for false advertising.

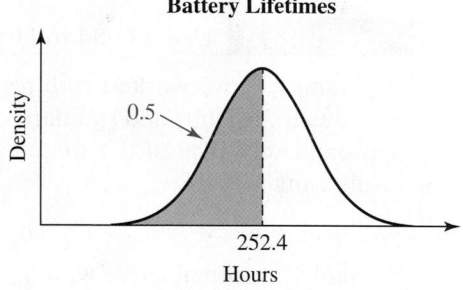

Figure 96 Shade (in orange) to the left of 252.4 hours

2. We sketch the same normal curve, but we use orange to shade only the far left 2% of the total area (see Fig. 97). We let c be the lifetime that should be advertised and write it on the horizontal axis. To use StatCrunch, we enter the mean 252.4 hours,

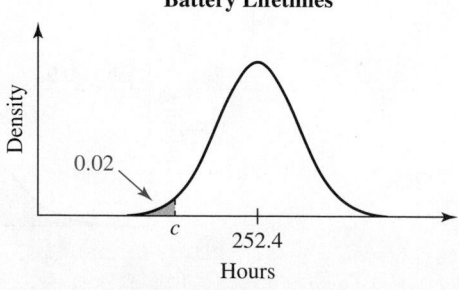

Figure 97 Shade (in orange) the far left 2% of the total area

the standard deviation 18.4 hours, and the probability 0.02 (see Fig. 98). See Appendix B.14 for StatCrunch instructions. From the StatCrunch output, we see that the lifetime that should be advertised is 214.6 hours. The result is less than the mean 252.4 hours, which checks with our sketch in Fig. 97, where we wrote c to the left of the mean.

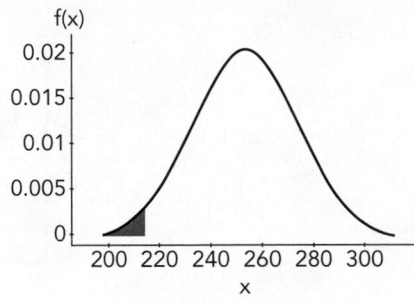

Normal Distribution
Mean:252.4   Std. Dev.:18.4
P(x≤214.61102) = 0.02

Figure 98 Find the lifetime that should be advertised

If the company were to round to the ones place, it would be better to round *down* to 214 hours, even though 6 is in the first decimal place of 214.6. If the company rounded up, the percentage of batteries that would last the advertised amount would be a bit less than 98%.

Use z-Scores to Compare the Relative Sizes of Two Observations

Suppose two students take tests on the same material, but one test is harder than the other. How can we fairly determine who did relatively better? We can factor in the difficulty of the tests by comparing the z-scores of the student's test scores.

▶ **Example 5**   Comparing *z*-Scores of Two Observations from Different Distributions

Maria and Roberto took tests on probability in two different sections of prestatistics. Maria scored 91 points on a test with mean 77 points and standard deviation 6 points. Roberto scored 80 points on a test with mean 68 points and standard deviation 4 points. The test scores on each test are approximately normally distributed.

1. Find the z-score for Maria's test score. What does it mean in this situation?
2. Find the z-score for Roberto's test score. What does it mean in this situation?
3. Assuming a typical student in one class knows the material as well as a typical student in the other class, determine whether Maria did relatively better than Roberto.

Solution

1. We substitute 91 for x, 77 for $\bar{x}$, and 6 for s in the z-score formula:

$$z = \frac{x - \bar{x}}{s} = \frac{91 - 77}{6} = \frac{14}{6} \approx 2.33$$

The z-score is 2.33, which means that Maria's test score is 2.33 standard deviations greater than the mean of her section's scores.

2. We substitute 80 for x, 68 for $\bar{x}$, and 4 for s in the z-score formula:

$$z = \frac{x - \bar{x}}{s} = \frac{80 - 68}{4} = \frac{12}{4} = 3$$

The *z*-score is 3, which means that Roberto's test score is 3 standard deviations greater than the mean of his section's scores.

3. Although Maria's score is larger than Roberto's, the smaller mean and smaller standard deviation on Roberto's test suggest that his test was harder than Maria's (see Fig. 99). So, comparing the actual scores is not a fair comparison.

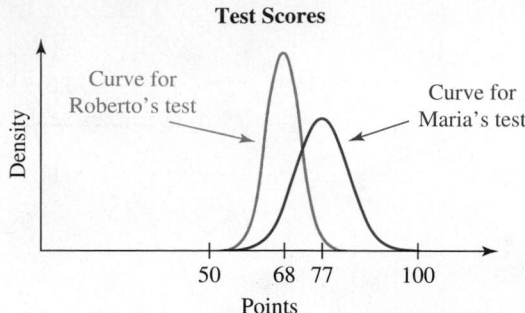

Figure 99 Roberto's test was harder than Maria's

Because Roberto's *z*-score is larger than Maria's, we conclude that Roberto outperformed a typical student in his class by more than Maria outperformed a typical student in her class. Since it's given that a typical student in one class knows the material as well a typical student in the other class, we conclude that Roberto did relatively better than Maria.

Identifying Unusual Observations

Identifying observations that are unlikely or unusual is a key tool in statistics. In section 4.2, we decided that an observation more than 2 standard deviations away from the mean is unusual. For data that are normally or approximately normally distributed, we will use the more precise cutoff of 1.96 standard deviations.

> **Definition   Unusual observation**
>
> Assume that some data are normally or approximately normally distributed. An observation is **unusual** if it is more than 1.96 standard deviations away from the mean (see Fig. 100).

In other words, an observation is unusual if its *z*-score is less than -1.96 OR greater than 1.96. So, the unusual observations lie under the blue regions in Fig. 100. By using Table 1 in Appendix C, we can show that the area of each of the blue regions in Fig. 100 is equal to 0.025. So, the total area is $2(0.025) = 0.05$, which means 5% of the observations are unusual. The number 1.96 might seem like an strange cutoff, but it leads to the simple percentage 5%.

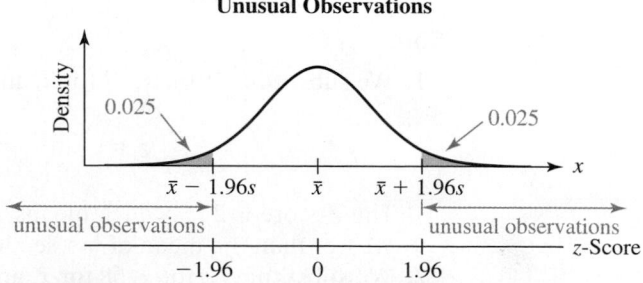

Figure 100 Unusual observations are more than 1.96 standard deviations away from the mean

Depending on the situation, statisticians might use a cutoff other than 1.96 standard deviations away from the mean. But to be consistent, we will always use 1.96 in this course.

▶ **Example 6** Determine Whether an Event Is Unusual

In a study of 1541 randomly selected young men (ages 20–29 years), the young men's HDL cholesterol levels are approximately normally distributed with mean 47 mg/dl and standard deviation 12.5 mg/dl (Source: *NHANES*).

1. Suppose that a young man in the study has an HDL cholesterol level of 20 mg/dl. Is that an unusual observation?
2. Find the probability of randomly selecting a young man from the study with an HDL cholesterol less than or equal to 20 mg/dl.
3. Explain why it makes sense that your result in Problem 2 is less than 0.025.

Solution

1. We substitute 20 for x, 47 for $\bar{x}$, and 12.5 for s in the formula $z = \dfrac{x - \bar{x}}{s}$:

$$z = \frac{20 - 47}{12.5} = -2.16$$

Because the z-score is less than -1.96, we conclude that the young man's 20 mg/dl level is unusual.

2. We draw a normal curve with mean 47 mg/dl, write 20 on the horizontal axis, and use orange to shade the region to the left of 20 (see Fig. 101). We use the z-score -2.16 and Table 1 in Appendix C to find that the area of the orange region is 0.0154. So, the probability is 0.0154, which we indicate in Fig. 101.

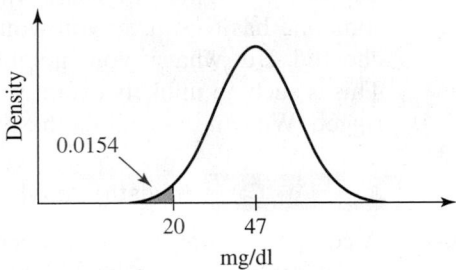

Figure 101 The area of the orange region is 0.0154

3. Because the observation's z-score -2.16 is less than the cutoff -1.96, the area of the orange region 0.0154 is less than the area to the left of $z = -1.96$, which is 0.025 (see Fig. 102).

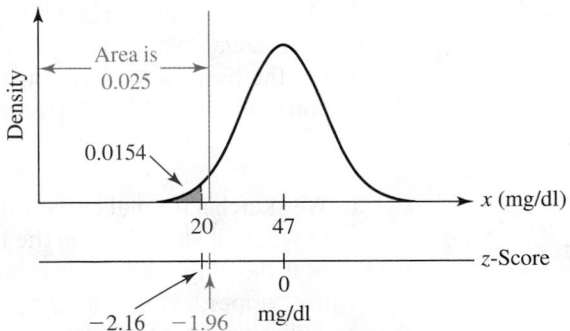

Figure 102 The area of the orange region, 0.0154, is less than 0.025

Our work in Example 6 suggests that there are two ways to determine if an observation is unusual, which is true in general.

> **Determining Whether an Observation Is Unusual**

An observation of a normal or an approximately normal distribution is unusual if either of the following equivalent statements is true:

- The observation's z-score is less than -1.96 OR greater than 1.96 (see Fig. 103).
- The probability of randomly selecting the observation OR one on the same side of the mean but even more extreme is less than 0.025.

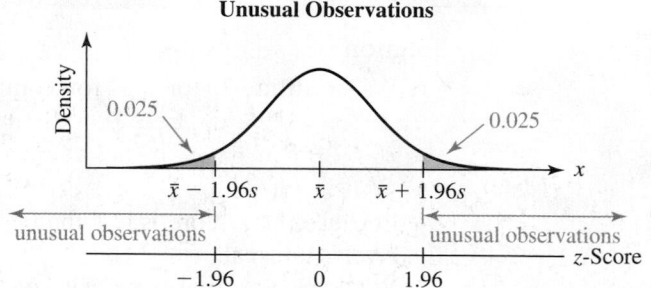

Figure 103 Unusual observations are more than 1.96 standard deviations away from the mean

Investigating whether a Claim Is True

Suppose a neighbor of yours wins the lottery. Although this is certainly a rare event, someone has to win, so you would probably not suspect that your neighbor somehow cheated. But what if your neighbor continues to win the next ten lotteries in a row? This is such an unlikely event, you would probably suspect that the lottery is somehow rigged. We will think along these lines in the next example.

▶ **Example 7** Investigate whether a Company's Claim Is True

A company claims that its low-salt potato chips contain only 85 mg of sodium per serving on average. An independent company randomly selects and tests a bag of the low-salt potato chips, which turns out to contain 104 mg of sodium per serving.

1. Assuming the sodium levels per serving of the low-salt chips are approximately normally distributed with mean 85 mg and standard deviation 5 mg, find the probability that a randomly selected bag would have at least 104 mg of sodium per serving.
2. If the assumptions you made in Problem 1 are true, is the bag containing 104 mg of sodium per serving an unusual event?
3. Describe a scenario in which the bag containing 104 mg of sodium per serving is not an unusual event.
4. On the basis of the bag containing 104 mg of sodium per serving, do the assumptions you made in Problem 1 seem reasonable? Should a more thorough investigation occur?

Solution

1. We sketch a normal curve with mean 85 milligrams (mg). We write 104 a bit to the left of where it should be on the horizontal axis so we can see the blue region to the right of it (see Fig. 104). To use "normalcdf" on a TI-84, we enter 104 for "lower," 999999 for "upper," 85 for the (population) mean μ, and 5 for the (population) standard deviation σ (see Fig. 105). From Fig. 105, we see that the probability is approximately 7×10^{-5}, or 0.00007, which we indicate in Fig. 104. The probability is very close to 0, which checks with the blue region being so small. In fact, if we were to write 104 in its correct position (a bit more to the right), the blue region would be so small it would not be visible.
2. The probability 0.00007 that we found in Problem 1 is much less than 0.025, so the bag that contains 104 mg of sodium per serving is certainly an unusual event, provided the assumptions we made in Problem 1 are correct.

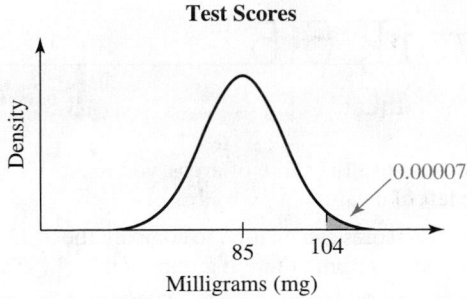

Figure 104 Shade to the right of 104

Figure 105 Find the probability

3. The bag that contains 104 mg of sodium per serving might not be an unusual event if the assumptions we made in Problem 1 are false. For example, if the mean sodium per serving was actually 104 mg, then a bag that contains 104 mg of sodium per serving would be a typical event, not an unusual one.

4. If the assumptions are true, then the event of randomly selecting a bag of chips that has 104 mg of sodium is extremely unlikely. In fact, because $0.00007 \approx 0.0001$, we would expect to select such a bag only about 1 time in 10,000 random selections. Yet this event occurred in just 1 try. So, we suspect that one of the assumptions is false. For example, the mean sodium per serving might actually be more than the advertised 85 mg. A more thorough investigation should occur.

 Group Exploration

Grading on a curve

A professor gives a 100-point test to his prestatistics students. The scores are approximately normally distributed with mean 76 points and standard deviation 8 points.

1. The professor typically uses the cutoffs (in points) 90, 80, 70, and 60 for the grades A, B, C, and D, respectively. By this grading method, find the percentages of students who would get the grades A, B, C, D, and F.

2. The professor is considering **grading on a curve**, which means to use the normal curve to compute the grade cutoffs. The professor decides to give As to the top 10%, Bs to the next 20%, Cs to the next 40%, Ds, to the next 20%, and Fs to the lowest 10%. Find the grade cutoffs for each of the grades A, B, C, and D.

3. Compare the grade cutoffs for each of the grades A, B, C, and D of the two grading techniques.

4. For what test scores would students get a higher grade by grading on a curve? For what test scores would students get a lower grade?

5. State a mean and standard deviation of a test-score distribution in which no student's grade would lower by grading on the curve.

6. State a mean and standard deviation of a test-score distribution in which no student's grade would improve by grading on the curve.

7. Students typically want their professors to grade on a curve. What would you tell such students?

> **Tips for Success**   Scan Test Problems
>
> When you take a test, scan the test problems quickly, pick the ones with which you feel most comfortable, and complete those problems first. By doing so, you will warm up and gain confidence, and you may do better on the rest of the test. Also, you will probably have a better idea of how to allot your time on the remaining problems.

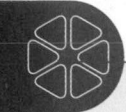

Homework 5.5

For extra help ▶  MyMathLab®    Watch the videos in MyMathLab   Download the MyDashboard App

1. *True or False:* Whenever you use the table of areas, you must look for the area to the left of a value.

2. *True or False:* The values of z-scores can be used to compare the relative sizes of two observations from different groups of data.

3. Assume some data are approximately normally distributed. An observation is unusual if it is more than ____ standard deviations away from the mean.

4. If an observation's z-score is greater than the number ____, then it is unusual.

5. The 2014 draft picks for NBA basketball teams have heights that are approximately normally distributed with mean 79.1 inches and standard deviation 3.0 inches (Source: *nbadraft.net*). The z-score of the tallest 2014 draft pick, Walter Tavares, is 2.63. What is Tavares's height? Round to the ones place.

6. The hourly temperatures at Eppley Airfield Airport in Omaha, Nebraska, in May 2014 are approximately normally distributed with mean 64.4°F and standard deviation 12.2°F (Source: *National Oceanic and Atmospheric Administration*). The z-score of the highest temperature is 2.59. What is the highest temperature? Round to the ones place.

7. Near the start of Section 5.4, we analyzed the lengths of songs played on Friday, April 18, 2014, on Live 105. The lengths of the songs are approximately normally distributed with mean 232.93 seconds and standard deviation 49.63 seconds (Source: *Live 105*). The shortest song played was "Fell in Love with a Girl" by the White Stripes. The song's length has a z-score of −2.48. What is the length of the song? Round to the ones place.

8. The age distribution of inmates on death row in Alabama is approximately normal with mean 43.0 years and standard deviation 9.7 years (Source: *Alabama Department of Corrections*). The z-score of the youngest inmate is −2.16. What is his age? Round to the ones place.

9. A professor gives a test to a trigonometry class. The scores are approximately normally distributed with mean 78 points and standard deviation 7 points.

    a. The professor usually uses 90 points as the cutoff for an A. If the professor uses this cutoff, what percentage of the students would get As?

    b. If the professor decides to give As to approximately 8% of the students but not less than 8%, what should the cutoff score for an A be?

10. A professor gives a test to an algebra class. The scores are approximately normally distributed with mean 67 points and standard deviation 11 points.

    a. The professor usually uses 87 points as the cutoff for an A. If the professor uses this cutoff, what percentage of the students would get As?

    b. If the professor decides to give As to approximately 7% of the students but not less than 7%, what should the cutoff score for an A be?

11. A prestatistics class takes a test in which the scores are approximately normally distributed with mean 80 points and standard deviation 7 points. The professor is debating whether to use

90 points or the 90th percentile as the cutoff for As. Which cutoff would allow more students to get As? Explain.

12. A statistics class takes a test in which the scores are approximately normally distributed with mean 82 points and standard deviation 6 points. The professor is debating whether to use 88 points or the 88th percentile as the cutoff for As. Which cutoff would allow more students to get As? Explain.

13. American women's heights are approximately normally distributed with mean 63.8 inches and standard deviation 4.2 inches. American men's heights are approximately normally distributed with mean 69.4 inches and standard deviation 4.7 inches (Source: *National Health Statistics Report*).

    a. At what doorway height would 95% of men be able to pass through without having to duck? Round to the ones place.

    b. What percentage of women would have to duck to pass through a doorway at the height you found in part (a)? Round to the second decimal place.

    c. The height of a doorway is typically 80 inches. What percentage of men can pass through such a doorway without ducking their head? Round to the second decimal place.

    d. For each gender, the mean height increased over time. For each of parts (a), (b), and (c), determine whether the correct answer to the question in the future will be larger, smaller, or impossible to know from the given information. Assume that for each gender, the standard deviation of heights will be constant.

14. American women's heights are approximately normally distributed with mean 63.8 inches and standard deviation 4.2 inches. American men's heights are approximately normally distributed with mean 69.4 inches and standard deviation 4.7 inches (Source: *National Health Statistics Report*).

    a. What height is at the 30th percentile for women's heights?

    b. What height is at the 30th percentile for men's heights?

    c. If there were the same number of women as men in the United States, what would be the mean height of all American adults?

    d. In reality, there are more women than men in the United States. Is your result in part (c) an underestimate or an overestimate of the mean height of all American adults?

15. The age distribution of inmates on death row in Alabama is approximately normal with mean 43.0 years and standard deviation 9.7 years (Source: *Alabama Department of Corrections*).

    a. What age is at the 10th percentile?

    b. What age is at the 90th percentile?

    c. According to the normal-curve model, what percentage of death-row inmates are under 18 years of age?

    d. The death penalty is not allowed for those under the age of 18 at the time of their crime. If the death penalty were allowed for those people, how many would be on death row in Alabama, according to the normal curve model? There are 196 inmates on death row in Alabama.

16. The age distribution of prisoners on death row in Florida has mean 48.4 years and standard deviation 11.0 years (Source: *Florida Department of Corrections*).

a. What age is at the 10th percentile?

b. What age is at the 90th percentile?

c. According to the normal-curve model, what percentage of death-row inmates are under 18 years of age?

d. The death penalty is not allowed for those under the age of 18 at the time of their crime. If the death penalty were allowed for those people, how many would be on death row in Florida, according to the normal-curve model? There are 394 inmates on death row in Florida.

17. The distribution of the weights of cereal in boxes of a specific brand of cereal is approximately normally distributed with mean 12 ounces and standard deviation 0.15 ounce.

a. Explain why it would be unwise for the cereal company to advertise that each box contains 12 ounces of cereal.

b. What should the company advertise the weight of the cereal in each box to be so that 99% of the boxes contain at least that much cereal? Round to the first decimal place.

18. For a specific brand of bread, the distribution of levels of fiber per loaf is approximately normally distributed with mean 6 g and standard deviation 0.2 g.

a. Explain why it would be unwise for the bread company to advertise that each loaf contains 6 g of fiber.

b. What should the company advertise the level of fiber per loaf to be so that 97% of the loaves contain at least that much fiber? Round to the first decimal place.

19. One way to measure a person's reading fluency is to have him or her read an unpracticed passage for one minute. The researcher then counts the number of words read correctly, which is called *words correct per minute* (*WCPM*). In a study at a private college in the Northeast, the WCPMs of participants were approximately normally distributed with mean 189.16 words and standard deviation 22.50 words (Source: *Assessment of Reading Rate in Postsecondary Students, Lewandowski, Lawrence J., et al.*). Round your results to the first decimal place.

a. Find Q_1.

b. Find Q_3.

c. Find the IQR.

d. Of the original group of students recruited for the study, 15% were excluded due to learning disabilities, attention disorders, or speaking English as a second language. For each of parts (a), (b), and (c), determine whether your result would most likely have been larger, smaller, or the same if the students who were excluded had been allowed to participate. [**Hint:** For part (c), think about the shape of the normal curve.]

20. One way to measure a person's reading fluency is to have them read an unpracticed passage for one minute. The researcher then counts the number of words read correctly, which is called *words correct per minute* (*WCPM*). The WCPMs for eighth-graders are approximately normally distributed with mean 151 words and standard deviation 45 words (Source: *Oral Reading Fluency Norms: A Valuable Assessment Tool for Reading Teachers, Hasbrouck, Jan, and Tindal, Gerald A.*).

a. Find the 50th percentile for the eighth-grader WCPM distribution.

b. The authors of the article cited in the opening paragraph "recommend that a [WCPM] score falling within 10 words above or below the 50 percentile should be interpreted as within the normal, expected, and appropriate range." Find the percentage of eighth-graders' WCPMs that fall in the normal, expected, and appropriate range.

c. Assume that readers who do not score in the range described in part (b) are either slow readers or fast readers. On the basis of your result in part (b), what percentage of eighth-graders are slow readers? Fast readers? Do these percentages seem reasonable? Explain.

d. Describe an interval of WCPM scores that contains the middle 80% of the scores.

21. Students A and B took tests on probability in two different sections of prestatistics. Student A scored 81 points on a test with mean 73 points and standard deviation 3 points. Student B scored 92 points on a test with mean 74 points and standard deviation 8 points.

a. Find the z-score for Student A's test score. What does it mean in this situation?

b. Find the z-score for Student B's test score. What does it mean in this situation?

c. Which student did relatively better? Explain.

22. Students A and B took tests on probability in two different sections of prestatistics. Student A scored 89 points on a test with mean 74 points and standard deviation 7 points. Student B scored 78 points on a test with mean 66 points and standard deviation 4 points.

a. Find the z-score for Student A's test score. What does it mean in this situation?

b. Find the z-score for Student B's test score. What does it mean in this situation?

c. Which student did relatively better? Explain.

23. For students who graduated from high school in 2013, the means and standard deviations of SAT and ACT math test scores are shown in Table 33. Assume the distributions are normally distributed. Many students who apply to top colleges take both tests and submit their better performance in their applications. If a student scored 675 points on the math sections of the SAT and 29 points on the math sections of the ACT, which is the relatively better score? Explain.

Table 33 Means and Standard Deviations for 2013 SAT and ACT Math Scores

Test	Mean	Standard Deviation
SAT	514	118
ACT	20.9	5.3

Source: College Board and ACT, Inc.

24. For students who graduated from high school in 2013, the means and standard deviations of SAT and ACT math test scores are shown in Table 33. Assume the distributions are normally distributed. Many students who apply to top colleges take both tests and submit their better performance in their applications. If a student scored 410 points on the math sections of the SAT and 15 points on the math sections of the ACT, which is the relatively better score? Explain.

25. In a study of 1541 men in their twenties, the men's HDL cholesterol levels are approximately normally distributed with mean 47 mg/dl and standard deviation 12.5 mg/dl. In a study of 823 men in their fifties, the men's HDL cholesterol levels are approximately normally distributed with mean 44 mg/dl and standard deviation 13.9 mg/dl (Source: *NHANES*).

a. Suppose one of the men in their twenties has an HDL reading of 42 mg/dl. Find the z-score of the reading relative to the men in their twenties. What does it mean in this situation?

b. Suppose one of the men in their fifties has an HDL reading of 40 mg/dl. Find the z-score of the reading relative to the men in their fifties. What does it mean in this situation?

c. HDL is the "good" cholesterol. The higher a person's HDL level, the lower the person's risk of heart attack and stroke will be. Doctors sometimes judge a characteristic of a person's health by comparing it to others of the same age. Limiting such a comparison to just those in the study, determine which of the two men described in parts (a) and (b) has the relatively better HDL reading.

d. How is it possible that of the two men described in parts (a) and (b), the man who has the higher HDL reading has the lower z-score?

26. In a study of 1744 women in their thirties, the women's HDL cholesterol levels are approximately normally distributed with mean 54 mg/dl and standard deviation 15.0 mg/dl. In a study of 955 women in their fifties, the women's HDL cholesterol levels are approximately normally distributed with mean 57 mg/dl and standard deviation 17.2 mg/dl (Source: *NHANES*).

a. Suppose one of the women in their thirties has an HDL reading of 60 mg/dl. Find the z-score of the reading relative to the women in their thirties. What does it mean in this situation?

b. Suppose one of the women in their fifties has an HDL reading of 62 mg/dl. Find the z-score of the reading relative to the women in their fifties. What does it mean in this situation?

c. HDL is the "good" cholesterol. The higher a person's HDL level, the lower the person's risk of heart attack and stroke will be. Doctors sometimes judge a characteristic of a person's health by comparing it to others of the same age. Limiting such a comparison to just those in the study, determine which of the two women described in parts (a) and (b) has the relatively better HDL reading.

d. How is it possible that of the two women described in parts (a) and (b), the woman who has the higher HDL reading has the lower z-score?

27. The lower a person's bone mineral density (BMD), the greater the person's risk of a bone fracture will be. BMD levels of 45-year-old African American women's hips are approximately normally distributed with mean 1034 mg/cm^2 and standard deviation 160 mg/cm^2 (Source: *NHANES*).

a. A 45-year-old African American woman's hip has BMD 986 mg/cm^2. What is the z-score of the BMD level? What does it mean in this situation?

b. Another 45-year-old African American woman's hip has BMD 666 mg/cm^2. What is the z-score of the BMD level? What does it mean in this situation?

c. For negative z-scores, the risk of a hip fracture doubles for each decrease of 1 in the z-score (Source: *American Academy of Orthopedic Surgeons*). Compare the risks of bone fracture for the women described in parts (a) and (b).

28. The lower a person's bone mineral density (BMD), the greater the person's risk of a bone fracture will be. BMD levels of

45-year-old Caucasian women's hips are approximately normally distributed with mean 920 mg/cm^2 and standard deviation 136 mg/cm^2 (Source: *NHANES*).

a. A 45-year-old Caucasian woman's hip has a BMD of 893 mg/cm^2. What is the z-score of the BMD level? What does it mean in this situation?

b. Another 45-year-old Caucasian woman's hip has a BMD of 485 mg/cm^2. What is the z-score of the BMD level? What does it mean in this situation?

c. For negative z-scores, the risk of a hip fracture doubles for each decrease of 1 in the z-score (Source: *American Academy of Orthopedic Surgeons*). Compare the risks of bone fracture for the women described in parts (a) and (b).

29. Recall that scores on the Wechsler IQ test are normally distributed with mean 100 points and standard deviation 15 points.

a. A student's IQ is at the 95th percentile. Find the student's IQ. Is it an unusual IQ?

b. Mensa is a high-IQ society with members in more than 100 countries. Membership is open to people who have attained a score within the upper 2 percent of the general population on an approved intelligence test, such as the Wechsler IQ test (Source: *Mensa International Limited*). What is the cutoff score on the Wechsler IQ test in order to qualify for Mensa? Do all Mensa members have unusual IQs?

c. A person scores 130 points on the Wechsler IQ test. Is that an unusual score? It is high enough to qualify for Mensa?

30. Recall that scores on the Wechsler IQ test are normally distributed with mean 100 points and standard deviation 15 points.

a. A student's IQ is at the 87th percentile. Find the student's IQ. Is it an unusual IQ? Explain.

b. Alexis Martin, who just turned 3 years old, has an IQ of 160 points. She learned to read when she was just 2 years old (Source: New York Daily News). What is her z-score? Does she have an unusual IQ? Explain.

c. Actress Geena Davis reportedly has an IQ of 140 points (Source: *CBS News*). Sir Isaac Newton, who shares credit with Gottfried Leibniz for inventing calculus, has been estimated to have had an IQ of 190 points (Source: *livescience. com*). Find the percentiles for Davis's and Newton's IQs. Do your equal results mean that Davis is as intelligent as Newton was? Explain.

31. American women's heights are approximately normally distributed with mean 63.8 inches and standard deviation 4.2 inches. American men's heights are approximately normally distributed with mean 69.4 inches and standard deviation 4.7 inches (Source: *National Health Statistics Report*).

a. Tulsa Cambage and Phoenix Griner are the two tallest women currently playing in the Women's National Basketball Association. They are 80 inches tall (Source: *WNBA.com*). Find the z-score of 80 inches for the women's heights distribution.

b. Hasheem Thabeet is the tallest man currently playing in the men's National Basketball Association. He is 87 inches tall (Source: *NBA*). Find the z-score of 87 inches for the men's heights distribution.

c. Who has the relatively larger height, Thabeet or Cambage and Griner?

d. Do Cambage and Griner have an unusual height for women? How about Thabeet for men? Explain.

32. American women's heights are approximately normally distributed with mean 63.8 inches and standard deviation 4.2 inches. American men's heights are approximately normally distributed with mean 69.4 inches and standard deviation 4.7 inches (Source: *National Health Statistics Report*).

a. Tamara De Treaux played the role of the alien in the movie 'E.T., the Extra-Terrestrial.' She was 31 inches tall (Source: *The New York Times Company*). Find the z-score of 31 inches for the women's height distribution. What does it mean in this situation?

b. Verne Troyer played the role of Mini-Me in the 'Austin Powers' movie series. He is 32 inches tall (Source: *IMDb*). Find the z-score of 32 inches for the men's height distribution. What does it mean in this situation?

c. Who has the relative smaller height, De Treaux or Troyer?

d. Does De Treaux have an unusual height for women? How about Troyer for men? Explain.

33. For students who graduated from high school in 2013, their SAT math scores are approximately normally distributed with a mean of 514 points and a standard deviation of 118 points (Source: *College Board*).

a. A student who graduated from high school in 2013 scored 704 points on the math sections of the SAT. Is that an unusual score? Explain.

b. A student who graduated from high school in 2013 scored 779 points on the math sections of the SAT. Is that an unusual score? Explain.

c. For students who graduated from high school in 2013, find the cutoff for unusually high scores on the math sections of the SAT. Round to the ones place.

34. For students who graduated from high school in 2013, their ACT math scores are approximately normally distributed with mean 20.9 points and standard deviation 5.3 points (Source: *ACT, Inc.*).

a. Is 25 points an unusual score on the math sections of the ACT? Explain.

b. Is 35 points an unusual score on the math sections of the ACT? Explain.

c. Find the cutoff for unusual scores greater than the mean on the math sections of the ACT. Round to the ones place.

35. A company claims that its low-salt tortillas contain only 130 mg of sodium per serving on average. An independent company randomly selects a package of the tortillas, which turns out to contain 141 mg of sodium per serving.

a. Assuming the sodium levels per serving of the low-salt tortillas are approximately normally distributed with mean 130 mg and standard deviation 4 mg, find the probability that a randomly selected package of the tortillas would have at least 141 mg of sodium per serving.

b. If the assumptions you made in part (a) are true, is the package containing 141 mg of sodium per serving an unusual event?

c. Describe a scenario in which the package containing 141 mg of sodium per serving is not an unusual event.

d. On the basis of the package containing 141 mg of sodium per serving, do the assumptions you made in part (a) seem reasonable? Should a more thorough investigation occur?

36. A company claims that its low-fat yogurt contains only 3 g of fat per cup on average. An independent company randomly selects a cup of the yogurt, which turns out to contain 3.5 g of fat.

a. Assuming the fat levels per cup of the yogurt are approximately normally distributed with mean 3 g and standard deviation 0.2 g, find the probability that a randomly selected cup of the yogurt would have at least 3.5 g of fat.

b. If the assumptions you made in part (a) are true, is the cup containing 3.5 g of fat an unusual event?

c. Describe a scenario in which the cup containing 3.5 g of fat is not an unusual event.

d. On the basis of the cup containing 3.5 g of fat, do the assumptions you made in part (a) seem reasonable? Should a more thorough investigation occur?

37. A pizza restaurant claims that its mean delivery time is 30 minutes. An independent company times a randomly selected delivery, which takes 40 minutes.

a. Assuming the delivery times are approximately normally distributed with mean 30 minutes and standard deviation 4 minutes, find the probability that a randomly selected delivery would take at least 40 minutes.

b. If the assumptions you made in part (a) are true, is the 40-minute delivery an unusual event?

c. Describe a scenario in which the 40-minute delivery is not an unusual event.

d. On the basis of the 40-minute delivery, do the assumptions you made in part (a) seem reasonable? Should a more thorough investigation occur?

38. A car company claims that its only hybrid car has a highway gas mileage of 40 mpg on average. An independent company randomly selects one of the cars, which turns out to have a gas mileage of 38 mpg.

a. Assuming the gas mileages of the cars are approximately normally distributed with mean 40 mpg and standard deviation 0.6 mpg, find the probability that a randomly selected hybrid car manufactured by the company has a gas mileage of at most 38 mpg.

b. If the assumptions you made in part (a) are true, is the randomly selected car's gas mileage of 38 mpg an unusual event?

c. Describe a scenario in which the randomly selected car's gas mileage of 38 mpg is not an unusual event.

d. On the basis of the randomly selected car having a gas mileage of 38 mpg, do the assumptions you made in part (a) seem reasonable? Should a more thorough investigation occur?

39. A company claims that its high-fiber bread contains 5 g of fiber per serving on average. An independent company randomly selects a loaf of the bread, which turns out to contain 4.6 g of fiber per serving.

a. Assuming the fiber levels per serving of the bread are approximately normally distributed with mean 5 g and standard deviation 0.15 g, find the probability that a randomly selected loaf of the bread would have at most 4.6 g of fiber per serving.

b. If the assumptions you made in part (a) are true, is the loaf containing 4.6 g of fiber per serving an unusual event?

c. Describe a scenario in which the loaf containing 4.6 g of fiber per serving is not an unusual event.

d. On the basis of the package containing 4.6 g of fiber per serving, do the assumptions you made in part (a) seem reasonable? Should a more thorough investigation occur?

40. A company claims that its high-protein shake contains 30 g of protein per serving on average. An independent company randomly selects a package of the shake, which turns out to contain 28.6 g of protein per serving.

a. Assuming the protein levels per serving of the shake are approximately normally distributed with mean 30 g and standard deviation 0.5 g, find the probability that a randomly selected package of the shake would have at most 28.6 g of protein per serving.

b. If the assumptions you made in part (a) are true, is the package containing 28.6 g of protein per serving an unusual event?

c. Describe a scenario in which the package containing 28.6 g of protein per serving is not an unusual event.

d. On the basis of the package containing 28.6 g of protein per serving, do the assumptions you made in part (a) seem reasonable? Should a more thorough investigation occur?

Concepts

41. A husband is taller than his wife. A student says the husband's relative height (compared with other men) is therefore greater than the women's relative height (compared with other women). What would you tell the student?

42. Suppose you scored fewer points on a prestatistics test than a friend of yours, who is in another section. Your friend concludes that your z-score (based on your section's scores) must be less than hers (based on her section's scores). What would you tell your friend?

43. Recall that an observation from a normal distribution is unusual if it is more than 1.96 standard deviations away from the mean. Use this fact to show that 5% of all observations are unusual.

44. Recall that a boxplot consists of four parts: the left whisker, the left part of the box, the right part of the box, and the right whisker. Assume a boxplot is constructed for a normal distribution.

a. Is the left whisker longer, shorter, or the same length as the right whisker? Explain.

b. Is the left part of the box longer, shorter, or the same length as the right part of the box? Explain.

c. Is the right part of the box longer, shorter, or the same length as the right whisker? Explain.

45. For a normal distribution with known mean and standard deviation, given the percentile of an observation, we can find the observation. Given an observation, we can find its percentile. Which steps of the two methods are the same? Similar but different? Completely different?

46. Describe two methods to determine whether an observation is unusual. In your opinion, which method is easier? Explain.

47. If observation x_1 is larger than observation x_2 from the same distribution, is the z-score of x_1 necessarily larger than the z-score of x_2? Explain.

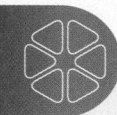

Hands-On Projects

Heights of Adults Project

In this project you will analyze adults' heights. Ask your instructor whether you should find all the data yourself or whether you should share data with others in the class.

1. When performing a study on adults' heights, explain why it is good practice to analyze women's heights separately from men's heights. Ask your instructor about how many individuals should be included in your study.

2. Here are some possible methods to estimate the height of a person:

- Ask the person.
- Refer to the person's driver's license.
- Use a tape measure to measure his or her height.

Describe the advantages and disadvantages of each method. Your description should include a discussion about any type(s) of likely bias. Determine and state which method you will use.

3. Determine what type of random sampling method or combination of methods you will use. Explain why you made this choice.

4. Collect the data. If you are working with others, share the data with each other. Include the data in your report.

5. For each gender, construct a relative frequency histogram.

6. For each distribution, describe the shape. What does it mean in this situation?

7. For each distribution, compute the mean and the standard deviation.

8. Compare the means of the two distributions. What does your comparison mean in this situation?

9. Compare the standard deviations of the two distributions. Are the two values approximately equal? Does this make sense? Explain.

10. Use the Empirical Rule to find the proportion of the women's heights that should be within one standard deviation of the mean if the distribution is normal. Compare your result with the actual proportion for your sample.

11. Use the Empirical Rule to find the proportion of the women's heights that should be within two standard deviations of the mean if the distribution is normal. Compare your result with the actual proportion for your sample.

12. Respond to Problems 10 and 11 again, but this time for men's heights.

13. American women's heights are approximately normally distributed with mean 63.8 inches and standard deviation 4.2 inches. American men's heights are approximately normally distributed with mean 69.4 inches and standard deviation 4.7 inches (Source: *National Health Statistics Report*). Compare these values and the ones you found. If any of the values are different, is this due to sampling bias, response bias, nonresponse bias, and/or sampling error?

14. How would each distribution be different if children were included in the study? Draw a rough sketch.

Chapter Summary

Key Points of Chapter 5

Section 5.1   Meaning of Probability

Random experiment	We call a process that selects items randomly a **random experiment.**
Probability	The **probability** of an outcome from a random experiment is the relative frequency of the outcome if the experiment were run an infinite number of times.
Simulation	Using technology to imitate a random experiment is called a **simulation.**
Using relative frequency to estimate probability	By running a random experiment a large number of times, we can use relative frequency to estimate probability.
Sample space	The **sample space** of a random experiment is the group of all possible outcomes.
Event	An **event** is some of the outcomes in the sample space, all of them, or none of them.
Equally likely probability formula	If the sample space of a random experiment consists of n equally likely outcomes and an event E consists of m of those outcomes, then $P(E) = \dfrac{m}{n}$.
Impossible event	An **impossible event** does not contain any outcomes of the sample space.
Sure event	A **sure event** (or **certain event**) contains all the outcomes in the sample space.
Probability properties	• The probability of an impossible event is equal to 0. • The probability of a sure event is equal to 1. • The probability of an event is between 0 and 1, inclusive. • The sum of the probabilities of all the single-outcome events in the sample space is equal to 1.
Random variable	A **random variable** is a numerical measure of an outcome from a random experiment. We often use a capital letter such as X to stand for a random variable.
Interpreting the area of a bar of a density histogram	We can interpret the area of a bar of a density histogram as • the proportion of observations that lie in the bar's class. • the probability of randomly selecting an observation that lies in the bar's class.

Section 5.2   Complement and Addition Rules

Complement	The **complement** of an event E is the event that consists of all the outcomes not in E. We write "NOT E" to stand for the complement of E.
Venn diagram	A **Venn diagram** consists of one or more shaded circles that are all bordered by a rectangle. The region inside the rectangle represents the sample space, and the region inside each circle represents an event.
Complement rule	For any event E, $P(\text{NOT } E) = 1 - P(E)$.
Disjoint events	Two events are **disjoint** (or **mutually exclusive**) if they have no outcomes in common.
Addition rule for disjoint events	If E and F are disjoint events, then $P(E \text{ OR } F) = P(E) + P(F)$.
General addition rule	For any events E and F, $P(E \text{ OR } F) = P(E) + P(F) - P(E \text{ AND } F)$.

Section 5.3   Conditional Probability and the Multiplication Rule for Independent Events

Conditional probability	If E and F are events, then the **conditional probability** $P(E|F)$ is the probability that E occurs, given that F occurs.
Independent events and dependent events	Two events E and F are **independent** if $P(E|F) = P(E)$. The events are **dependent** if $P(E|F)$ and $P(E)$ are *not* equal.
Multiplication rule for independent events	If E and F are independent events, then $P(E \ \text{AND} \ F) = P(E) \cdot P(F)$.

Section 5.4   Finding Probabilities for a Normal Distribution

Normal	When a normal curve describes a distribution perfectly, we say the distribution is **normal** or **normally distributed**. When a normal curve describes a distribution fairly well (but not exactly), we say the distribution is **approximately normal** or **approximately normally distributed**.
Properties of a normal curve	• A normal curve is unimodal and symmetric. • The mean is equal to the median. Both are the center of the curve.
Area-probability equality property	The area under a normal curve for an interval is equal to the probability of randomly selecting an observation that lies in the interval.
Total area under the normal curve	The total area under a normal curve is equal to 1.
Probability of a single value	The probability of a single value of a normally distributed variable is 0.
Comparing normal curves with different standard deviations	A normal curve for a distribution with larger standard deviation will be wider and flatter than a normal curve for a distribution with smaller standard deviation.
Empirical Rule	If a distribution is normal, then the probability that a randomly selected observation lies within • one standard deviation of the mean is approximately 0.68. So, about 68% of the data lie between $\bar{x} - s$ and $\bar{x} + s$. • two standard deviations of the mean is approximately 0.95. So, about 95% of the data lie between $\bar{x} - 2s$ and $\bar{x} + 2s$. • three standard deviations of the mean is approximately 0.997. So, about 99.7% of the data lie between $\bar{x} - 3s$ and $\bar{x} + 3s$.
z-Score formula	The z-score of an observation x is given by $z = \dfrac{x - \bar{x}}{s}$.
Standard normal distribution	If a distribution is normally distributed, then the distribution of the observations' z-scores is also normally distributed with mean 0 and standard deviation 1. We call the distribution of z-scores the **standard normal distribution.**
Interpreting the area under a normal curve for an interval	We can interpret the area A under the normal curve for an interval as • the probability that a randomly selected observation will lie in the interval. • the proportion of observations that lie in the interval.

Section 5.5   Finding Values of Variables for Normal Distributions

Finding the value of an observation from its z-score	An observation x with z-score z is given by $x = \bar{x} + zs$.
Unusual observation	Assume that some data are approximately normally distributed. An observation is **unusual** if it is more than 1.96 standard deviations away from the mean.
Determining whether an observation is unusual	An observation of a normal or an approximately normal distribution is unusual if either of the following equivalent statements is true: • The observation's z-score is less than -1.96 OR greater than 1.96. • The probability of randomly selecting the observation OR one on the same side of the mean but even more extreme is less than 0.025.

Chapter 5 Review Exercises

1. A student plans to buy one lottery ticket. The student says that the probability he will win is $\frac{1}{2}$ because one of two things will happen: He will either win or lose. What would you tell the student?

In Exercises 2–10, you will analyze the gambling game roulette in which a ball is dropped onto a revolving wheel that has 38 numbered slots: the numbers 1–36, a zero, and a double zero. There are 18 red slots, 18 black slots, and 2 green slots:

Red: 1, 3, 5, 7, 9, 12, 14, 16, 18, 19, 21, 23, 25, 27, 30, 32, 34, 36

Black: 2, 4, 6, 8, 10, 11, 13, 15, 17, 20, 22, 24, 26, 28, 29, 31, 33, 35

Green: 0, 00

Consider the random experiment of dropping the ball onto the wheel once. When finding a probability, write it in fraction form.

2. $P(9)$

3. $P(\text{odd})$

4. $P(\text{NOT red})$

5. $P(\text{blue})$

6. $P(\text{red OR green})$

7. $P(\text{black AND at most } 9)$

8. $P(\text{red OR at least } 26)$

9. Are the events green and black disjoint?

10. Are the events odd and black independent?

11. A total of 801 adults were asked whether they prefer to drink coffee, tea, or neither. Table 34 compares the adults' responses with their ages.

Table 34 Drink Preference versus Age

	Age (years)				
	18–29	30–44	45–64	over 64	Total
Coffee	67	121	174	97	459
Tea	67	66	85	24	242
Neither	29	33	30	8	100
Total	163	220	289	129	801

Source: *YouGov*

Suppose that one person in the sample is randomly selected.

a. Find $P(\text{coffee} \mid 18–29 \text{ years})$.

b. Find $P(\text{coffee} \mid 30–44 \text{ years})$.

c. Find $P(\text{coffee} \mid 45–64 \text{ years})$.

d. Find $P(\text{coffee} \mid \text{over 64 years})$.

e. What pattern do you notice in parts (a) through (d)? What does it mean in this situation? Can causation be assumed? Explain.

12. A total of 801 adults were asked whether they prefer to drink coffee, tea, or neither. Table 34 compares the adults' responses with their ages.

a. Find the probability that a person randomly selected from the study prefers tea.

b. Find the probability that a person randomly selected from the study prefers tea, given the person is between 18 and 29 years of age, inclusive.

c. For individuals in the study, is preferring tea independent of being between 18 and 29 years of age, inclusive? Explain.

d. Find the probability that a person randomly selected from the study prefers coffee AND is between 45 and 64 years of age, inclusive.

e. Find the probability that a person randomly selected from the study prefers tea OR is over 64 years of age.

13. In a survey of 1000 randomly selected adults, 68% of the respondents think that the government should require parents to have their children vaccinated (Source: *YouGov*).

a. If five individuals of the study are randomly selected, find the probability that all five individuals think that the government should require parents to have their children vaccinated.

b. If five individuals of the study are randomly selected, find the probability that at least one of the individuals think the government should require parents to have their children vaccinated.

14. The age distribution of inmates on death row in Alabama is approximately normal with mean 43.0 years and standard deviation 9.7 years (Source: *Alabama Department of Corrections*). The age of one of the inmates is 35 years. Find the z-score of his age. What does it mean in this situation?

15. Find the probability that a z-score randomly selected from the standard normal distribution is between -1.7 and 2.5.

16. The ages of Americans who died from alcohol poisoning are normally distributed with mean 46.6 years and standard deviation 12.5 years (Source: *Centers for Disease Control and Prevention*).

a. Find the proportion of alcohol-poisoning fatalities in which the victims were under the minimum drinking age (21 years).

b. Find the proportion of alcohol-poisoning fatalities in which the victims were between 50 and 60 years of age.

c. Find the proportion of alcohol-poisoning fatalities in which the victims were senior citizens (at least 65 years of age).

17. The hourly temperatures at Eppley Airfield Airport in Omaha, Nebraska, in May 2014 are approximately normally distributed with mean 64.4°F and standard deviation 12.2°F (Source: *National Oceanic and Atmospheric Administration*). The z-score of the lowest temperature is -2.49. What is the lowest temperature? Round your result to the ones place.

18. A prestatistics class takes a test in which the scores are approximately normally distributed with mean 82 points and standard deviation 5 points. The professor is trying to decide whether to use 90 points as the cutoff for As or to give As to students who scored above the 90th percentile. By which grading method would more students get As?

19. For students who graduated from high school in 2013, their SAT math scores are approximately normally distributed with a mean of 514 points and a standard deviation of 118 points. Their ACT math scores are approximately normally distributed with mean 20.9 points and standard deviation 5.3 points (Source: *College Board; ACT, Inc.*).

a. A student who graduated from a high school in Texas in 2013 scored 765 points on the math sections of the SAT. Is that an unusual score? Explain.

b. A student who graduated from high school in Florida in 2013 scored 33 points on the math sections of the ACT. Is that an unusual score? Explain.

c. Which student performed better on the math sections of their test, the student from Texas or Florida? Explain.

20. A company claims that its cheddar cheese contains 3.5 g of fat per serving on average. An independent company randomly selects a package of the cheese, which turns out to contain 3.8 g of fat per serving.

a. Assuming the fat levels per serving of the cheese are approximately normally distributed with mean 3.5 g and standard deviation 0.12 g, find the probability that a randomly selected package of cheese would have at least 3.8 g of fat per serving.

b. If the assumptions you made in part (a) are true, is the package containing 3.8 g of fat per serving an unusual event?

c. Describe a scenario in which the package containing 3.8 g of fiber per serving is not an unusual event.

d. On the basis of the package containing 3.8 g of fiber per serving, do the assumptions you made in part (a) seem reasonable? Should a more thorough investigation occur?

Chapter 5 Test

For Exercises 1–6, a six-sided die is rolled once. When finding a probability, write it in fraction form.

1. $P(3)$

2. $P(\text{greater than } 4)$

3. $P(\text{even AND at most } 3)$

4. $P(\text{odd OR at least } 4)$

5. $P(\text{odd} \mid \text{less than } 4)$

6. Is rolling a number less than 5 independent of rolling an even number?

7. In a study of patients with coronary heart disease, 605 patients were randomly assigned to follow a Mediterranean-type diet or the step 1 American Heart Association (AHA) prudent diet for four years. Table 35 describes whether the individuals were healthy (other than having coronary heart disease), had a nonfatal heart-related episode, had cancer, or had died during the four years.

Table 35 Diet versus Health

	Healthy	Nonfatal Heart-Related Episode	Cancer	Died	Total
Mediterranean	273	8	7	14	302
AHA	237	25	17	24	303
Total	510	33	24	38	605

Source: *Mediterranean Dietary Pattern in a Randomized Trial: Prolonged Survival and Possible Reduced Cancer Rate,* de Lorgeril M. et al.

a. Find the probability that a patient randomly selected from the study had a nonfatal heart-related episode.

b. Find the probability that a patient randomly selected from the study had a nonfatal heart-related episode OR died.

c. Find the probability that a patient randomly selected from the study had cancer OR was on the Mediterranean diet.

d. Find the probability that a patient randomly selected from the study had cancer, given he or she was on the step 1 AHA prudent diet.

e. For individuals in the study, is being healthy during the four-year study independent of being on the Mediterranean diet?

8. For students who graduated from high school in 2013, their SAT math scores were approximately normally distributed with a mean of 514 points and a standard deviation of 118 points (Source: *College Board*).

a. If a student is randomly selected from those who graduated from high school in 2013 and took the SAT, what is the probability that the student scored between 500 and 600 points on the math sections?

b. If a student is randomly selected from those who graduated from high school in 2013 and took the SAT, what is the probability that the student scored less than 350 points on the math sections?

c. If a student is randomly selected from those who graduated from high school in 2013 and took the SAT, what is the probability that the student scored more than 450 points on the math sections?

9. The ages of male professors are approximately normally distributed with mean 51.3 years and standard deviation 12.0 years (Source: *Statistical Abstract of Undergraduate Programs in the Mathematical Programs in the Mathematical Sciences in the United States*). Estimate the age of a professor whose age is at the 90th percentile. Round your result to the ones place.

10. The Trends in International Mathematics and Science Study (TIMSS) measures the mathematical achievement of eighth-graders of various countries (Source: *TIMSS*).

a. The scores of countries in 2007 are approximately normally distributed with mean 454.1 points and standard deviation 74.2 points. Is the United States' score of 508 points unusual?

b. The scores of countries in *2011* are approximately normally distributed with mean 490.5 points and standard deviation 72.0 points. Is the United States' score of 541 points unusual?

c. Relative to other countries, did the United States perform better in 2007 or 2011? Explain.

11. A company claims that its low-salt potato chips contain only 85 mg of sodium per serving on average. An independent company randomly selects a package of the potato chips, which turns out to contain 93 mg of sodium per serving.

a. Assuming the sodium levels per serving of the low-salt potato chips are approximately normally distributed with mean 85 mg and standard deviation 3 mg, find the probability that a randomly selected package of the tortillas would have at least 93 mg of sodium per serving.

b. If the assumptions you made in part (a) are true, is the package containing 93 mg of sodium per serving an unusual event?

c. Describe a scenario in which the package containing 93 mg of sodium per serving is not an unusual event.

d. On the basis of the package containing 93 mg of sodium per serving, do the assumptions you made in part (a) seem reasonable? Should a more thorough investigation occur?

6

Describing Associations of Two Variables Graphically

Think about the last concert you attended. What was the ticket price? Was it worth it? The mean ticket price for the top-50-grossing concert tours has increased greatly (see Table 1). In Example 2 of Section 6.3, we will estimate the mean ticket price in 2010.

In this chapter, we will discuss how to describe the relationship between two quantities that occur in an authentic situation. For example, we will compare the ages of Americans with the percentages of Americans who support same-sex marriage.

Table 1 Mean Ticket Prices for Top-50-Grossing Concert Tours

Year	Mean Ticket Price (dollars)
1998	33
2001	47
2004	59
2008	67
2011	85

Source: *Pollstar*

▼ 6.1 Scatterplots

Objectives

» Identify explanatory and response variables.

» Construct a scatterplot.

» Determine the direction of an association.

Recall from Section 3.3 that a time-series plot describes the relationship between a numerical variable and time. In this section, we will construct and analyze a graph called a *scatterplot*, which describes the relationship between two numerical variables in which the variables do not necessarily represent time.

Explanatory and Response Variables

Suppose we want to investigate whether the number of times a teacher encourages her students affects her students' test performances. Let x be the number of times a teacher encourages a student per class period, and let y be a student's test score (in points). Recall that we call x the *explanatory* variable, because we are investigating whether the number of encouragements might explain (affect) a student's score (Section 2.3). Also recall that we call y the *response variable*, because we are investigating whether students will respond to (be affected by) encouragement (Section 2.3).

▶ **Definition**   **Explanatory and response variables**

In a study about whether a variable x explains (affects) a variable y,

- We call x the **explanatory variable** (or **independent variable**).
- We call y the **response variable** (or **dependent variable**).

Although StatCrunch uses the terminologies *independent variable* and *dependent variable*, we will avoid using them to avoid confusion with independent events, which we discussed in Section 5.3.

Explanatory and response variables can be categorical, but in this chapter we will assume they are numerical.

WARNING

In the encouragement study, calling n the explanatory variable does not mean for sure that encouragement affects (explains) a student's test performance. After all, we'd have to carry out the study to find out. In general, an explanatory variable may or may not turn out to affect (explain) the response variable.

▶ **Example 1**   Identifying Explanatory and Response Variables

For each situation, identify the explanatory variable and the response variable.

1. Let w be the number of minutes some shirts are washed, and let p be the percentage of stains that are removed.
2. Let p be the percentage of people who think that a man of height h inches is attractive.
3. A car is traveling at speed s (in mph) on a dry asphalt road, and the brakes are suddenly applied. Let d be the stopping distance (in feet).

Solution

1. A study would explore to what extent the washing time explains the percentage of stains that are removed. So, the study would test to what extent w explains p. Therefore, w is the explanatory variable and p is the response variable. (It would *not* make sense to explore to what extent the percentage of stains removed affects the washing time.)
2. A researcher would investigate to what extent the height of a man explains (affects) the percentage of people who find the man attractive. So, the researcher would test to what extent h explains p. Therefore, h is the explanatory variable and p is the response variable. (It would *not* make sense to investigate to what extent the percentage of people who find a man attractive affects the man's height.)
3. A researcher would test to what extent the traveling speed explains (affects) the stopping distance. So, the researcher would test to what extent s explains d. Therefore, s is the explanatory variable and d is the response variable. (It would *not* make sense to test to what extent the stopping distance affects the traveling speed.)

▶

In Section 1.1, we plotted ordered pairs such as (2, 5). Recall that if we are working with the variables x and y, then (2, 5) means that when $x = 2$, $y = 5$. But what if we are working with variables other than x and y? Which variable's value do we list first? In general, **for an ordered pair (a, b), we write the value of the explanatory variable in the first (left) position and the value of the response variable in the second (right) position.**

▶ **Example 2**   Describing Authentic Situations with Ordered Pairs

1. Let m be the amount of money (in dollars) a pollster offers each individual to complete a survey, and let c be the completion rate (the percentage of individuals who complete the survey). For a survey with a monetary incentive of $5, 14% of individuals who are contacted complete the survey. Express this as an ordered pair.
2. Let b be a person's systolic blood pressure reading (in mm Hg), and let s be the number of times the person smiles daily. An individual who smiles 25 times daily has a systolic blood pressure reading of 117 mm Hg. Express this as an ordered pair.

Solution

1. A monetary incentive of $5 and a completion rate of 14% are described by $m = 5$ and $c = 14$. A researcher would explore to what extent a monetary incentive

explains the completion rate. So, m is the explanatory variable and c is the response variable. Because we always list the value of the explanatory variable first and the response variable second, the ordered pair is $(5, 14)$.

2. Smiling 25 times daily and having a systolic blood pressure of 117 mm Hg are described by $s = 25$ and $b = 117$. A statistician would investigate to what extent the number of times a person smiles daily explains a person's systolic blood pressure. So, s is the explanatory variable and b is the response variable. Because we always list the value of the explanatory variable first and the response variable second, the ordered pair is $(25, 117)$.

In Example 3, we will interpret ordered pairs.

> **Example 3**   Interpreting Ordered Pairs

1. Let w be the weight (in pounds) of a person, and let h be the person's height (in inches). What does the ordered pair $(66, 142)$ mean in this situation?
2. Let c be the amount of caffeine (in mg) a person ingests daily, and let s be the mean number of hours the person sleeps per night. What does the ordered pair $(300, 7)$ mean in this situation?

Solution

1. A researcher would investigate to what extent a person's height explains a person's weight. So, h is the explanatory variable and w is the response variable. The ordered pair $(66, 142)$ means that $h = 66$ and $w = 142$. A person's height is 66 inches, and the person's weight is 142 pounds.
2. A statistician would test to what extent caffeine explains the mean number of hours a person sleeps. So, c is the explanatory variable and s is the response variable. The ordered pair $(300, 7)$ means that $c = 300$ and $s = 7$. An individual ingests 300 mg of caffeine daily and sleeps an average of 7 hours per night.

Recall that when working with the variables x and y and the ordered pair $(2, 5)$, we can describe the ordered pair in a table (Section 1.1). We do so by writing the x-coordinate 2 in the first column and the y-coordinate 5 in the second column. Recall that for a coordinate system, the x-coordinate 2 is described by the horizontal axis and the y-coordinate 5 is described by the vertical axis. We follow a similar practice when working with other variables.

> ### Columns of Tables and Axes of Coordinate Systems

Assume that an authentic situation can be described by using two variables. Then
- For tables, the values of the explanatory variable are listed in the first column and the values of the response variable are listed in the second column (see Table 2).
- For coordinate systems, the values of the explanatory variable are described by the horizontal axis and the values of the response variable are described by the vertical axis (see Fig. 1).

Table 2 Position of the Variables

Explanatory Variable	Response Variable
*	*
*	*
*	*

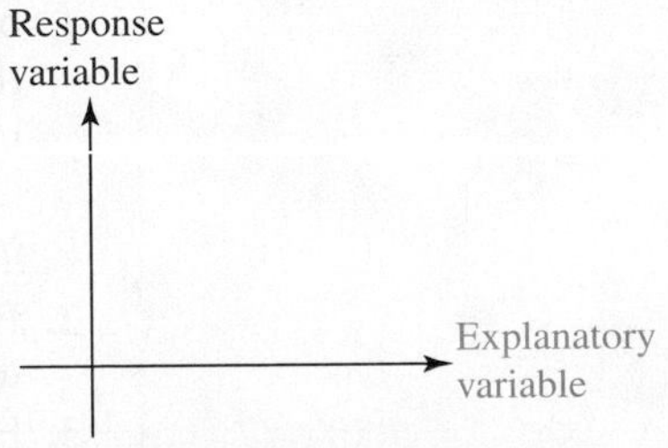

Figure 1 Positions of the variables

Constructing Scatterplots

A coordinate system with plotted ordered pairs is called a **scatterplot**. We will construct one in Example 4.

▶ **Example 4**   Construct a Scatterplot

The mean ticket prices for the Super Bowl are shown in Table 3 for various years. Let p be the mean ticket price (in dollars) and n be the Super Bowl number.

1. Construct a scatterplot.
2. For the Super Bowls described in Table 3, which Super Bowl had the highest mean ticket price? What was that price?
3. Describe any patterns you see in the prices.

Table 3 Mean Ticket Prices for the Super Bowl

Super Bowl Number	Mean Ticket Price (dollars)
1 (I)	12
5 (V)	15
10 (X)	20
15 (XV)	40
20 (XX)	75
25 (XXV)	150
30 (XXX)	250
35 (XXXV)	325
40 (XL)	650
45 (XLV)	900
48 (XLVIII)	1500

Source: *NFL.com*

Solution

1. A scatterplot of the data is shown in Fig. 2. It makes sense to think of n as the explanatory variable, because a researcher would investigate whether the Super Bowl number explains the mean ticket price (and not the other way around). So, we let the horizontal axis be the n-axis and let the vertical axis be the p-axis. Note that we write the variable names "n" and "p" and the units "Super Bowl number" and "Dollars" on the appropriate axes.

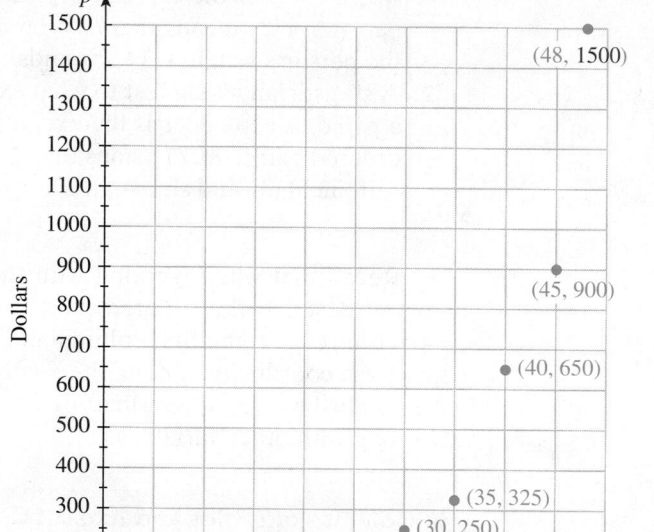

Figure 2 Scatterplot of mean Super Bowl ticket prices

Recall from Section 1.1 that when we write numbers on an axis, they should increase by a fixed amount and be equally spaced. Because the Super Bowl numbers are between 1 and 48, inclusive, we write the numbers 5, 10, 15,…, 50 equally spaced on the n-axis. Because the mean prices shown in Table 3 are between \$12 and \$1500, inclusive, we write the numbers 100, 200, 300,…, 1500 on the p-axis.

Then we plot the data points. For example, the ordered pair (1, 12) indicates that in Super Bowl I, the mean ticket price was \$12.

2. From Table 3 and the scatterplot in Fig. 2, we see that the highest mean ticket price was \$1500 in Super Bowl XLVIII.

3. From Table 3 and the scatterplot in Fig. 2, we see that mean ticket prices have increased.

Just as with other statistical diagrams, it is useful to practice constructing some scatterplots by hand, but we usually use technology to construct them. A scatterplot of the Super Bowl data constructed by StatCrunch is shown in Fig. 3. For StatCrunch instructions, see Appendix B.15.

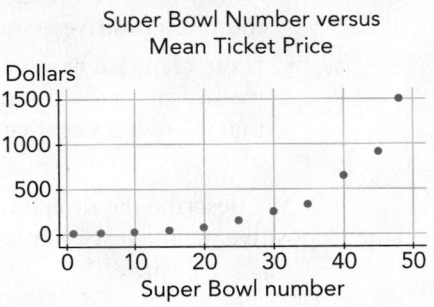

Figure 3 StatCrunch scatterplot of Super Bowl data

Recall that there is an **association** between the explanatory and response variables if the response variable changes as the explanatory variable changes (Section 2.3). For instance, in Example 4, we determined that as the Super Bowl number increases, the mean ticket price increases.

WARNING   However, an increase in the Super Bowl number does not *cause* an increase in ticket price. The true causes are probably increasing popularity, increasing cost of the event, and inflation.

An association of *numerical* explanatory and response variables is often called a **correlation**. However, we will continue to use the word *association* as a reminder that the concepts about associations we discussed in Section 2.3 still apply.

Direction of an Association

Because the mean ticket price *increases* as the Super Bowl number increases, we say there is a *positive association*.

Figure 4 displays a scatterplot, which compares the percentage of adults who exercise with the percentage of adults who are obese, for each of the 50 states, Puerto Rico, and District of Columbia.

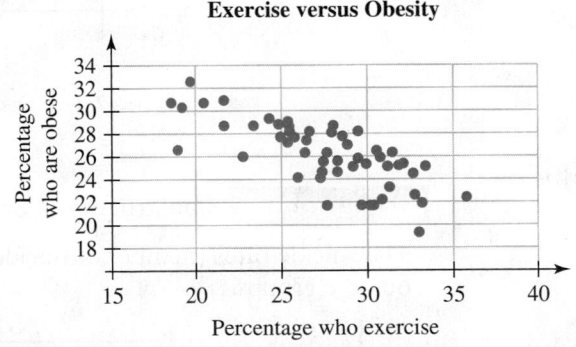

Figure 4 Exercise versus obesity
(**Source:** *Centers for Disease Control and Prevention*)

The scatterplot has a less clear-cut pattern than the Super Bowl scatterplot. We say it has more *scatter*. Even though there is quite a bit of scatter, we can still observe that the percentage of adults who are obese tends to *decrease* as the percentage of adults who exercise increases. We say there is a *negative association*.

▶ Definition   Positive and negative association

Assume two numerical variables are the explanatory and response variables of a study.

- If the response variable tends to increase as the explanatory variable increases, we say the variables are **positively associated** (or **positively correlated**) and that there is a **positive association** (or **positive correlation**). See Fig. 5.
- If the response variable tends to decrease as the explanatory variable increases, we say the variables are **negatively associated** (or **negatively correlated**) and that there is a **negative association** (or **negative correlation**). See Fig. 6.

We describe the **direction** of an association by determining whether the association is positive, negative, or neither (see Figs. 5, 6, and 7, respectively).

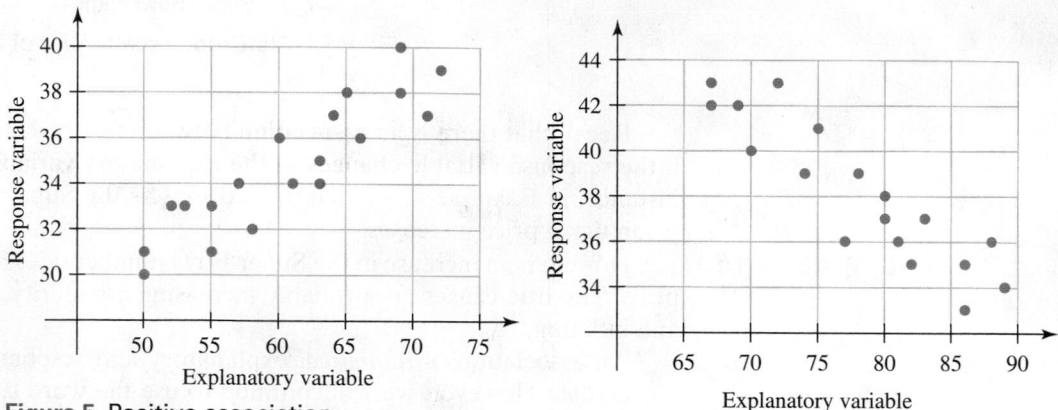

Figure 5 Positive association

Figure 6 Negative association

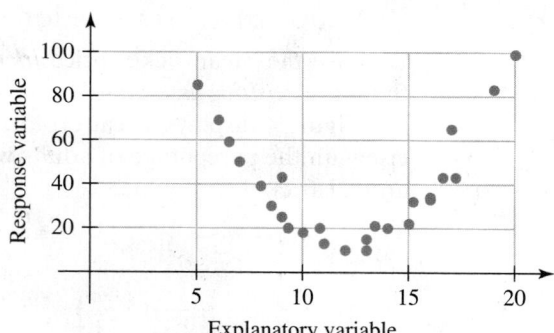

Figure 7 The association is neither positive nor negative

▶ Example 5   Constructing a Scatterplot and Analyzing the Association

The suicide rates (number of suicides per 100,000 people) are shown in Table 4 for various age groups.

Table 4 Ages versus Suicide Rates

Age Group (years)	Age Used to Represent Age Group (years)	Number of Suicides per 100,000 People
35–39	37	15
40–44	42	17
45–49	47	19
50–54	52	20
55–59	57	19
60–64	62	16

Source: *Centers for Disease Control and Prevention*

Let *s* be the suicide rate (number of suicides per 100,000 people) for adults who are *a* years of age.

1. Construct a scatterplot by hand.
2. Use technology to construct a scatterplot.
3. Determine whether the association is positive, negative, or neither.

Solution

1. A look at the first row of Table 4 suggests that we use $a = 37$ to stand for the age group from 35 to 39 years. The age 37 is the mean of the ages 35 and 39. (Try it.) Likewise, we will use 42 to stand for the age group from 40 to 44 years, and so on.

    A scatterplot of the data is shown in Fig. 8. It makes sense to think of *a* as the explanatory variable because a researcher would explore whether age explains the suicide rate (and not the other way around). So, we let the horizontal axis be the *a*-axis and let the vertical axis be the *s*-axis. Note that we write the variable names *a* and *s* and the units *Age in years* and *Number of suicides per 100,000 people* on the appropriate axes.

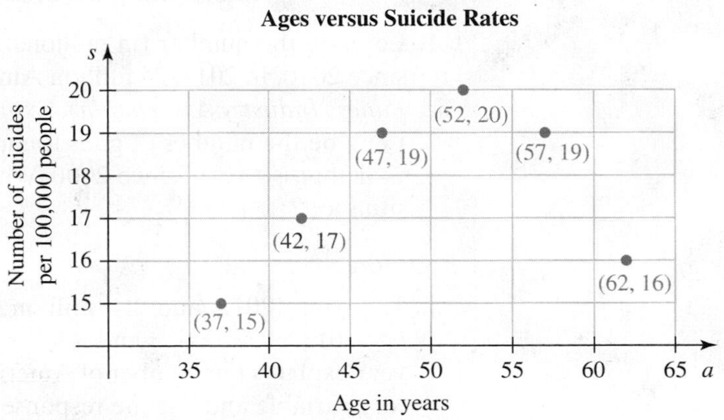

Ages versus Suicide Rates

Figure 8 Ages versus suicide rates

Because the ages used to stand for the age groups are between 37 and 62 years, inclusive, we write the numbers 35, 40, 45, 50, 55, 60, and 65 equally spaced on the *a*-axis. Because the suicide rates shown in Table 4 are between 15 and 20 suicides per 100,000 people, inclusive, we write the numbers 15, 16, 17, 18, 19, and 20 equally spaced on the *s*-axis.

Then we plot the data points. The ordered pair (37, 15) means that, for the age group from 35 to 39 years, the suicide rate is 15 suicides per 100,000 people.

2. We use a TI-84 to construct a scatterplot in Fig. 9. For TI-84 instructions, see Appendix A.10.
3. From Table 4 and Fig. 8, we see the association is neither positive nor negative.

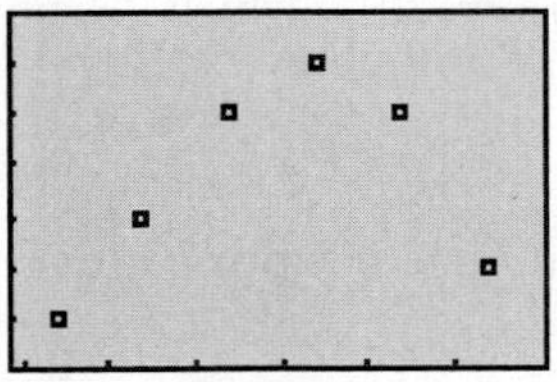

Figure 9 A TI-84 scatterplot

In Example 6, we will define a variable to represent time.

▶ **Example 6** Defining a Variable for Time

Let *t* be the number of years since 1990. Find the values of *t* that represent the years 1990, 1996, 1999, 2000, and 2003.

Solution

We can represent 1990 by $t = 0$, because 1990 is 0 years after 1990. We can represent 1996 by $t = 6$, because 1996 is 6 years after 1990. We list the value of *t* for each of the years 1990, 1996, 1999, 2000, and 2003 in Table 5.

Table 5 Values of *t*

Year	Years since 1990 *t*		
1990	0	because	1990 − 1990 = 0
1996	6		1996 − 1990 = 6
1999	9		1999 − 1990 = 9
2000	10		2000 − 1990 = 10
2003	13		2003 − 1990 = 13

The values of *t* in Table 5 are much smaller numbers than the years they represent. When working with authentic situations, we will often perform calculations that involve years. Using definitions similar to the one in Example 6 will enable us to perform those calculations with smaller numbers. It is also easier to label the axes of a coordinate system with smaller numbers.

▶ **Example 7** Interpreting an Ordered Pair

1. Let *n* be the number (in millions) of American adults who practice yoga at *t* years since 2010. In 2013, 24 million American adults practiced yoga (Source: *Sports and Fitness Industry Association*). Express this as an ordered pair.
2. Let *g* be the number of guns found among airline passengers at U.S. airports in the year that is *t* years since 2000. What does the ordered pair (14, 2122) mean in this situation?

Solution

1. The year 2013 and 24 million Americans practicing yoga are described by $t = 2013 - 2010 = 3$ and $n = 24$. A statistician would explore to what extent the year explains the number of American adults who practice yoga. So, *t* is the explanatory variable and *n* is the response variable. Because we always list the explanatory variable first and the response variable second, the ordered pair is $(3, 24)$.
2. A researcher would investigate to what extent the year explains the number of guns found among airline passengers. So, *t* is the explanatory variable and *g* is the response variable. The ordered pair $(14, 2122)$ means that $t = 14$ and $g = 2122$. There were 2122 guns found among airline passengers in $2000 + 14 = 2014$, which is true (Source: *Transportation Security Administration*).

▶ **Example 8** Constructing a Scatterplot and Analyzing the Association

Atlantic City gambling revenues are shown in Table 6 for various years.

Table 6 Years versus Atlantic City Gambling Revenues

Year	Revenue (billions of dollars)
2008	4.7
2009	4.0
2010	3.6
2011	3.3
2012	3.0
2013	2.9

Source: *New Jersey Division of Gaming Entertainment*

Let *r* be the annual revenue (in billions of dollars) at *t* years since 2005.

1. Construct a scatterplot.
2. Determine whether the association is negative, positive, or neither.

Table 7 Values of t and r

t (years since 2005)	r (billions of dollars)
3	4.7
4	4.0
5	3.6
6	3.3
7	3.0
8	2.9

Solution

1. First, we list the values of t and r in Table 7. For example, $t = 3$ stands for 2008 because 2008 is 3 years after 2005.

   A scatterplot of the data is shown in Fig. 10. Because t is the explanatory variable and r is the response variable, we let the horizontal axis be the t-axis and let the vertical axis be the r-axis.

Figure 10 Gambling revenue scatterplot

2. From Table 6 and Fig. 10, we see that the revenue has been decreasing since 2008. So, there is a negative association.

Group Exploration

Analyzing points below, above, and on a line containing points with equal coordinates

The scatterplot in Fig. 11 compares the scores of Test 2 and Test 3 for 35 calculus students taught by the author. Each dot represents a pair of scores for exactly one student.

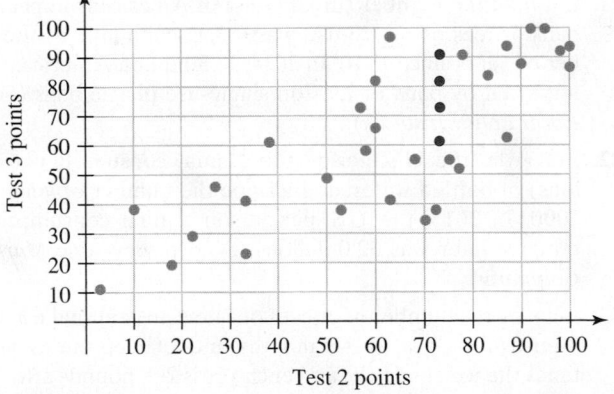

Figure 11 Scores of Test 2 and Test 3

1. Is the association positive, negative, or neither? What does that mean in this situation?

2. On the scatterplot, imagine a line that goes through the points (20, 20), (40, 40), (60, 60), (80, 80), and (100, 100). Or use a strand of thread, a stretched rubber band, or the edge of a clear ruler to form the line. Which data points lie on the line? What do they mean in this situation?

3. Give the coordinates of three points that lie above the line. For your first point, determine whether the corresponding student scored higher on Test 2 or Test 3. Is the same true for your other two points? the other points above the line?

4. Give the coordinates of three points that lie below the line. For your first point, determine whether the corresponding student scored higher on Test 2 or Test 3. Is the same true for your other two points? the other points below the line?

5. Two students have equal total scores for Test 2 and Test 3. The data point for one student is above the line, and the data point for the other student is below the line. Which student shows more promise? Explain.

6. A student who scored 73 points on Test 2 was absent for Test 3. Predict what the student's Test 3 score would have been by computing the mean Test 3 score for the students (represented by the red dots) who scored 73 points on Test 2.

> ▶ Tips for Success   Take a Break
>
> Have you ever had trouble solving a problem but returned to the problem hours later and found it easy to solve? By taking a break, you can return to the exercise with a different perspective and renewed energy. You've also given your unconscious mind a chance to reflect on the problem while you take your break. You can strategically take advantage of this phenomenon by allocating time to complete your homework assignment at two different points in your day.

Homework 6.1

For extra help ▶ **MyMathLab®**   Watch the videos in MyMathLab   Download the MyDashboard App

1. For an ordered pair, we write the value of the _____ variable in the first position.

2. For a coordinate system, the values of the _____ variable are described by the vertical axis.

3. There is a(n) _____ between the explanatory and response variables if the response variable changes as the explanatory variable changes.

4. If the response variable tends to decrease as the explanatory variable increases, we say there is a(n) _____ association.

For Exercises 5–16, identify the explanatory and the response variables. Which variable should be described by the horizontal axis of a coordinate system? How about the vertical axis?

5. Let n be the number of hours a student studies for a quiz, and let s be the student's score (in points) on the quiz.

6. Let n be the number of office parties an employer hosts per year, and let p be the percentage of employees who say they enjoy working at the company.

7. Let h be the height (in inches) of a girl, and let a be the age (in years) of the girl.

8. Let p be the percentage of colleges that would accept a student whose grade point average (GPA) is G points.

9. Let I be a student's IQ (in points), and let G be the student's GPA (in points) in college.

10. Let h be the height (in inches) of a professional basketball player, and let P be the player's mean points scored per game.

11. Let c be a person's cholesterol level (in mg/dl) after ingesting d milligrams of an experimental cholesterol drug each day for one month.

12. Let p be the percentage of men at age a years who have gray hair.

13. Let T be the increase in height (in feet) of a tree one year after n experimental fertilizer stakes were driven into the ground 10 feet from the tree's trunk.

14. A person cooks a potato in an oven for an hour and then removes the potato and allows it to cool. Let t be the number of minutes since the potato was removed from the oven, and let F be the temperature (in degrees Fahrenheit) of the potato.

15. Let t be the number of seconds after a baseball is hit upward, and let h be the baseball's height (in feet).

16. Let p be the percentage of people at age a years who own a computer.

For Exercises 17–22, describe the situation with an ordered pair. Which variable should be described by the horizontal axis of a coordinate system? How about the vertical axis?

17. Let T be the total number of T-shirts (in thousands) a hip-hop group sells at C concerts per year. A group sells a total of 29.4 thousand T-shirts at 98 concerts in a year.

18. Let p be the asking price (in thousands of dollars) of a home with b bedrooms. The asking price of a 3-bedroom home is $152 thousand.

19. Let B be the bonus (in thousands of dollars) a car dealership pays a salesperson if the salesperson makes a quota of selling at least 15 cars in a month, and let n be the number of salespeople who make the quota. A dealership offers a bonus of $3 thousand for making the quota, and 4 salespeople make the quota.

20. A person goes for a run. Let s be the runner's speed (in miles per hour), and let p be the runner's pulse rate (in beats per minute). When the runner's speed is 10 miles per hour, his pulse rate is 160 beats per minute.

21. Let n be the number (in millions) of Americans impacted by bans or fees on single-use plastic bags, and let t be the number of years since 2010. In 2014, 18.3 million Americans were impacted by bans or fees on single-use plastic bags (Source: *Earth Policy Institute*).

22. Let w be the U.S. per-person annual consumption (in gallons) of bottled water, and let t be the number of years since 2000. In 2013, the U.S. per-person annual consumption of bottled water was 32.0 gallons (Source: *Beverage Marketing Company*).

23. Let p be the number of ounces of a new protein shake a weight lifter drinks daily for six months, and let n be the number of times the weight lifter can bench-press 200 pounds after those six months. What does the ordered pair (16, 8) mean in this situation?

24. Let s be the mean number of times a television salesperson compliments customers per day, and let c be the mean number of televisions the salesperson sells per month. What does the ordered pair (72, 39) mean in this situation?

25. Let p be the percentage of Americans at age A years who say they volunteer. What does the ordered pair (21, 38) mean in this situation?

26. Let *n* be the number of pimples a teenager has on his face after ingesting *d* milligrams of an experimental acne medication daily for three months. What does the ordered pair (30, 2) mean in this situation?

27. Let *p* be the percentage of Americans who stream television shows at least once a month at *t* years since 2005. What does the ordered pair (9, 45) mean in this situation?

28. Let *r* be Google's annual revenue (in billions of dollars) at *t* years since 2010. What does the ordered pair (4, 66) mean in this situation?

29. Let *p* be the percentage of Americans who believe travel websites do a good job of presenting travel choices at *t* years since 2010. What does the ordered pair (−1, 33) mean in this situation?

30. Let *p* be the percentage of Americans who are satisfied with the size and power of major corporations at *t* years since 2010. What does the ordered pair (−2, 35) mean in this situation?

For Exercises 31–34 refer to the given scatterplot to determine whether the association is positive, negative, or neither.

31. See Fig. 12.

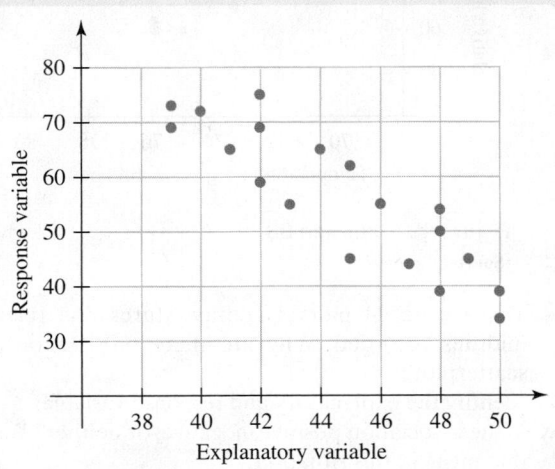

Figure 12 Exercise 31

32. See Fig. 13.

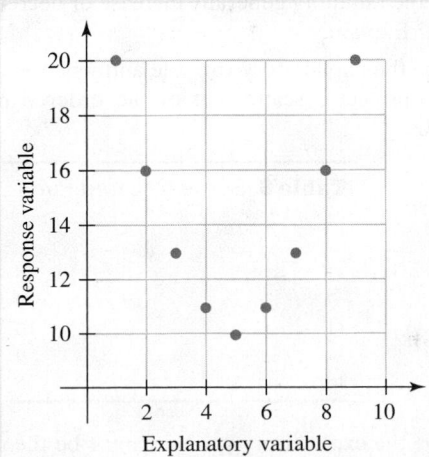

Figure 13 Exercise 32

33. See Fig. 14.

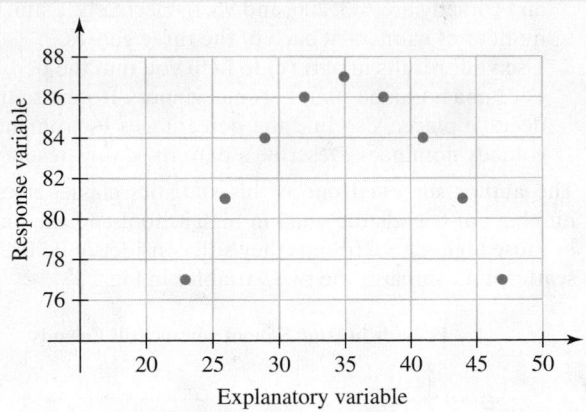

Figure 14 Exercise 33

34. See Fig. 15.

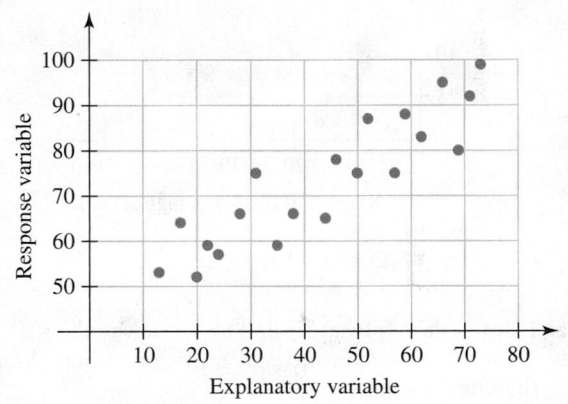

Figure 15 Exercise 34

35. The numbers of nominees for best picture at the Academy Awards and the numbers of winners are compared in the scatterplot in Fig. 16 for 17 genres. The genres were determined by the description of the movies at IMDb. If a nominated or winning movie is described as being a combination of genres, then the movie was counted for each of those genres.

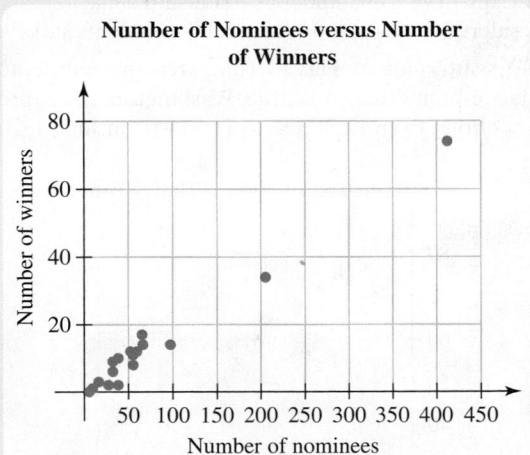

Figure 16 Exercise 35
(**Source:** *Collider*)

a. Identify the explanatory and response variables.

b. Is the association positive, negative, or neither? What does that mean in this situation?

c. The number of nominations for the genres drama, romance, and comedy are 413, 206, and 98, respectively. Estimate the number of winners of each of the three genres.

d. Use your results in part (c) to help you find the percentage of drama nominees that became winners. Round to the first decimal place. Also find the percentages for romance and comedy nominees. Describe a pattern in your results.

36. The author surveyed one of his statistics classes about the numbers of friends they had in high school and the numbers of those high school friends they still consider to be friends. A scatterplot compares the two variables in Fig. 17.

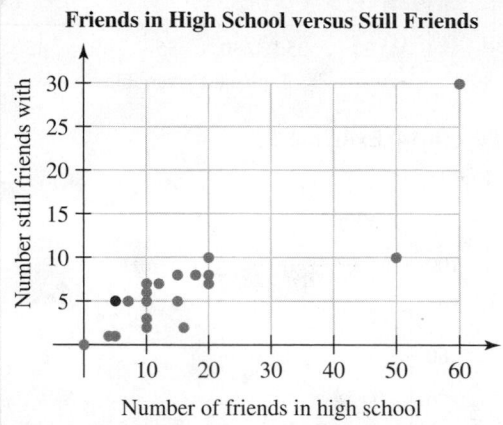

Figure 17 Exercise 36
(**Source:** *J. Lehmann*)

a. Identify the explanatory and response variables.

b. Is the association positive, negative, or neither? What does that mean in this situation?

c. For the student who had 60 friends in high school, estimate how many of those people the student still considers to be friends. Do the same for the student who had 18 friends in high school. Finally, do the same for the student who had 7 friends in high school.

d. For each of the students discussed in part (c), find the percentage of the friends the student still considers to be friends. Also, do the same for the student represented by the red dot. Explain why this student's friendship behavior could be considered the most unusual of all the students surveyed.

37. The scatterplot in Fig. 18 compares the temperatures and relative humidities in Seattle, Washington, for approximately every hour, from 12:53 A.M. to 11:53 P.M. on June 1, 2014.

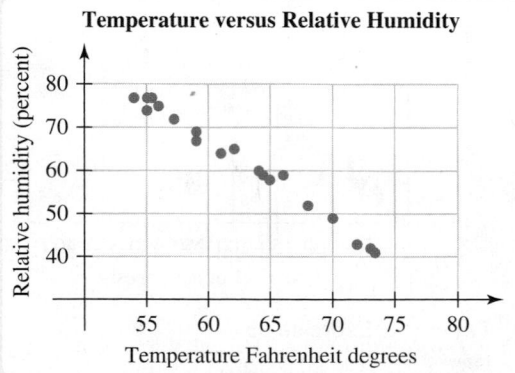

Figure 18 Exercise 37
(**Source:** *Weatherbase*)

a. There were 24 pairs of temperatures and relative humidities recorded. Why are there only 19 dots in the scatterplot?

b. Identify the explanatory and response variables.

c. Is the association positive, negative, or neither? What does that mean in this situation?

d. On the basis of the scatterplot, a student concludes that the relative humidity generally decreased throughout the day. What would you tell the student?

e. The temperature decreased from 12:53 A.M. to 4:53 A.M. Did the humidity generally increase or decrease in that period? Explain.

38. The scatterplot in Fig. 19 compares the temperatures and relative humidities in Atlanta, Georgia, for every hour, from 12:52 A.M. to 11:52 P.M. on June 1, 2014.

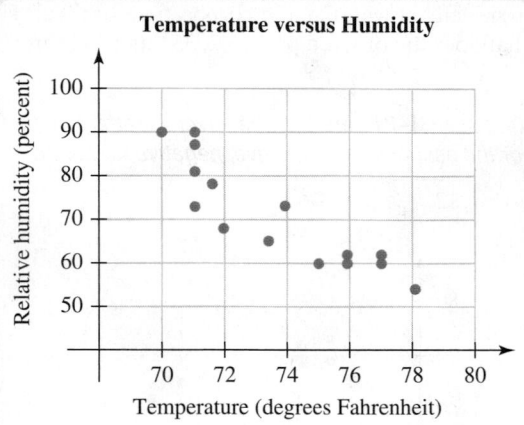

Figure 19 Exercise 38
(**Source:** *Weatherbase*)

a. There were 24 pairs of temperatures and relative humidities recorded. Why are there only 15 dots in the scatterplot?

b. Identify the explanatory and response variables.

c. Is the association positive, negative, or neither? What does that mean in this situation?

d. On the basis of the scatterplot, a student concludes that the relative humidity generally decreased throughout the day. What would you tell the student?

e. The temperature d ecreased from 12:52 A.M. to 4:53 A.M. Did the humidity generally increase or decrease in that period? Explain.

39. Let *x* be the explanatory variable and *y* be the response variable. Construct a scatterplot of the ordered pairs listed in Table 8.

Table 8 Some Ordered Pairs

x	y
2	5
7	9
11	10
14	9
16	5

40. Let *x* be the explanatory variable and *y* be the response variable. Construct a scatterplot of the ordered pairs listed in Table 9.

Table 9 Some Ordered Pairs

x	y
−5	2
−4	3
−1	5
4	9
11	15

41. <u>DATA</u> The number of pages in each of the books in the Harry Potter series is listed in Table 10.

Table 10 Numbers of Pages in the Books of the Harry Potter Series

Book Number	Number of Pages
1	309
2	341
3	435
4	734
5	870
6	652
7	784

Let p be the number of pages and b be the corresponding book number.

a. Identify the explanatory and response variables.
b. Which variable should be described by the horizontal axis of a coordinate system, p or b? How about the vertical axis?
c. Construct a scatterplot.
d. Which book has the greatest number of pages?
e. From which book to the next did the number of pages increase the most? Explain how you can tell this by inspecting your scatterplot.

42. <u>DATA</u> The mean life-spans of various denominations of bills are shown in Table 11. For example, the mean life-span of a $1 bill is 1.5 years before it is taken out of circulation due to wear and tear.

Table 11 Mean Life-spans of Denominations of Bills

Value of Bill (dollars)	Life-span (years)
1	1.5
5	2
10	3
20	4
50	9

Source: *Federal Reserve System*

Let L be the mean life-span (in years) of a bill that is worth d dollars.

a. Identify the explanatory and response variables.
b. Which variable should be described by the horizontal axis of a coordinate system, L or d? How about the vertical axis?
c. Construct a scatterplot.
d. Is the association positive, negative, or neither? What does that mean in this situation? Explain why this makes sense.
e. Each year, many more $1 bills are printed than $50 bills. Give at least two reasons why this makes sense.

43. <u>DATA</u> Tiger Woods's golf tournament earnings and numbers of wins are shown in Table 12 for various years.

Table 12 Tiger Woods's Tournament Earnings and Numbers of Wins

Year	Tournament Earnings (millions of dollars)	Wins
2006	9.9	8
2007	10.9	7
2008	5.8	4
2009	10.5	6
2010	1.3	0
2011	0.7	0
2012	6.1	3
2013	8.6	5
2014	0.1	0

Source: *PGA Tour*

Let E be Tiger Woods's tournament earnings (in millions of dollars) in the year that is t years since 2000. For example, $t = 0$ represents 2000, and $t = 6$ represents 2006.

a. Identify the explanatory and response variables.
b. Construct a scatterplot.
c. For which of the years shown in Table 12 were Woods's tournament earnings the least? What were those earnings?
d. For which of the years shown in Table 12 were Woods's tournament earnings the greatest? What were those earnings?
e. For the years shown in Table 12, did Woods have his greatest tournament earnings in the same year as his greatest number of wins? Give two reasons why this could be possible.

44. <u>DATA</u> The mean price per barrel of crude oil and the total fuel cost for the global airline industry are shown in Table 13 for various years.

Table 13 Mean Price per Barrel of Crude Oil and Total Fuel Cost for the Global Airline Industry

Year	Mean Price per Barrel of Crude Oil (dollars)	Total Airline Fuel Cost (billions of dollars)
2007	73	133
2008	99	187
2009	62	123
2010	79	138
2011	111	174

Source: *IATA*

Let C be the total airline fuel cost (in billions of dollars) in the year that is t years since 2005. For example, $t = 0$ represents 2005, and $t = 2$ represents 2007.

a. Identify the explanatory and response variables.
b. Construct a scatterplot that compares the years and the total airline fuel costs.
c. For which of the years shown in Table 13 was the total airline fuel cost the least? What was that cost?

d. For which of the years shown in Table 13 was the total airline fuel cost the greatest? What was that cost?

e. For the years shown in Table 13, was the total fuel cost the greatest in the same year that the mean price per barrel was the greatest? Explain why this is possible.

45. DATA The total costs to individuals and companies to prepare their taxes are shown in Table 14 for various years.

Table 14 Total Costs to Individuals and Companies to Prepare Taxes

Year	Total Cost (billions of dollars)
1990	80
1995	108
2000	172
2005	260
2010	368
2015	420

Source: *Tax Foundation*

Let c be the total annual cost (in billions of dollars) to individuals and companies to prepare taxes at t years since 1990.

a. Construct a scatterplot.

b. Is the association positive, negative, or neither? What does that mean in this situation?

c. The mean price of all goods and services increased by a factor of 1.80 from 1990 to 2015 (Source: *Bureau of Labor Statistics*). Did the total cost to individuals and companies to prepare taxes grow by a smaller, equal, or greater factor for that period?

46. DATA The mean hourly manufacturing pay is shown in Table 15 for various years.

Table 15 Mean Hourly Manufacturing Pay

Year	Mean Hourly Pay (dollars)
1970	3.24
1980	7.15
1990	10.78
2000	14.32
2010	18.61
2015	19.71

Source: *Bureau of Labor Statistics*

Let p be the mean hourly manufacturing pay (in dollars) at t years since 1970.

a. Construct a scatterplot.

b. Is the association positive, negative, or neither? What does that mean in this situation?

c. Estimate the *annual* pay for a specific employee who worked 40 hours per week for 50 weeks in 2010 at a car plant. Give at least two reasons why your estimate may be in error.

d. The mean price of all goods and services increased by a factor of 6.05 from 1970 to 2010 (Source: *Bureau of Labor Statistics*). Did the mean hourly manufacturing pay fall behind, approximately keep up with, or grow faster than the increase in the mean price of goods and services for that period? Explain.

47. DATA The numbers of automobile accidents per 1000 licensed drivers per year are shown in Table 16 for various age groups.

Table 16 Automobile Accidents

Age Group (years)	Age Used to Represent Age Group (years)	Accident Rate (number of accidents per 1000 licensed drivers per year)
16	16	190.3
17	17	163.2
18	18	142.9
19	19	127.8
20–29	24.5	91.4
30–39	34.5	54.7
40–49	44.5	43.9
50–59	54.5	36.4
60–69	64.5	31.3
over 69	75	32.1

Source: *National Highway Traffic Safety Administration*

Let r be the automobile accident rate (number of accidents per 1000 licensed drivers per year) for licensed drivers at age a years.

a. Construct a scatterplot.

b. Which age group shown in Table 16 has the lowest accident rate?

c. Which age group shown in Table 16 has the highest accident rate?

d. Between what two consecutive drivers' ages does there seem to be the greatest change in the accident rate? Explain why we can't be sure this is true, because of the way the data are described in Table 16.

e. Many states put limits on teenage driving. For example, some states do not allow 16-year-old drivers to drive at night. Some states require parental supervision at all times. Why do you think these regulations were adopted?

48. DATA The percentages of Americans of various age groups who are ordering more takeout food than they did two years ago are shown in Table 17.

Table 17 Percentages of Americans Who Are Ordering More Takeout Food than They Did Two Years Ago

Age Group (years)	Age Used to Represent Age Group (years)	Percent
18–24	21.0	34
25–34	29.5	31
35–44	39.5	27
45–54	49.5	17
55–64	59.5	15
over 64	70.0	7

Source: *National Restaurant Association Survey*

Let p be the percentage of Americans at age a years who are ordering more takeout food than they did two years ago.

a. Construct a scatterplot.

b. Is the association positive, negative, or neither? What does that mean in this situation?

c. Which of the points in your scatterplot is highest? What does that mean in this situation?

d. Which of the points in your scatterplot is lowest? What does that mean in this situation?

49. DATA Several inventions are listed in Table 18, along with the years they were invented and how long it took for one-quarter of the U.S. population to use them ("mass use").

Table 18 Number of Years until Inventions Reached Mass Use

Invention	Year Invented	Years until Mass Use
Electricity	1873	46
Telephone	1876	35
Gasoline-Powered Automobile	1886	55
Radio	1897	31
Television	1923	29
Microwave Oven	1953	36
VCR	1965	13
Personal Computer	1975	16
Mobile Phone	1985	11
CD Player	1985	8
World Wide Web	1991	7
DVD Player	1997	5

Source: *Newsweek*

Let M be the number of years elapsed until an invention reached mass use if it was invented at t years since 1870.

a. Construct a scatterplot.
b. Is the association positive, negative, or neither? What does that mean in this situation? In your opinion, why does that make sense?
c. Does the point for the microwave oven fit the overall pattern you described in part (b)? Explain.
d. For a while after the microwave oven was invented, many people feared it would cause radiation poisoning, blindness, or impotence. Discuss the impact of these fears in terms of your response to part (c).
e. Explain why the point for the gasoline-powered automobile does not fit the overall pattern you described in part (b). Why do you think this happened? We say that the point is an *outlier*.

50. DATA The percentages of adults of various age groups who approve of single men raising children on their own are shown in Table 19.

Table 19 Percentages of Adults Who Approve of Single Men Raising Children on Their Own

Age Group (years)	Age Used to Represent Age Group (years)	Percentage
18–34	26.0	81
35–44	39.5	73
45–54	49.5	73
55–64	59.5	66
over 64	70	47

Source: *Taylor Nelson Sofres*

Let p be the percentage of adults at age a years who approve of single men raising children on their own.

a. Construct a scatterplot.
b. Is the association positive, negative, or neither? What does that mean in this situation?
c. Which age group shown in Table 19 has the most faith in single men raising children on their own?

d. Which age group shown in Table 19 has the least faith in single men raising children on their own?

51. DATA Data about existing home sales in area 232 of Palo Alto, California, in 2007 are shown in Table 20.

Table 20 Existing Home Sales in Area 232 of Palo Alto, California

Street	List Price (thousands of dollars)	Sales Price (thousands of dollars)	Beds	Baths	Square Footage (square Feet)
Barron Av	2150	2400	4	3	2331
La Jennifer Wy	1195	1652	3	2	1351
Kendall Av	1295	1600	3	2	1745
Josina Av	1395	1715	3	2	2204
Magnolia Av	898	1000	2	1	996
Timlott Ln	1289	1500	3	2	1802
Irven Ct	999	1120	3	1	1024
Wisteria Ln	1199	1150	3	2.5	1750
Kendall	2845	2800	5	4.5	2958
Matadero Av	1595	1755	5	3	2123
Driscoll Pl	900	1035	3	3.5	1836
Laguna Wy	2495	2450	5	3	3105
Los Robles Av	959	1070	2	1	1114

Source: *Peninsula Realtors*

Let F be the square footage of a home, and let P be the home's sales price (in thousands of dollars).

a. Construct a scatterplot that describes the association between square footage and sales price.
b. Is the association positive, negative, or neither? What does that mean in this situation?
c. State the street of the home that sold for the least price per square foot. Explain how sizing up the scatterplot can help narrow your search so you do not have to perform so many calculations.
d. State the street of the home that sold for the greatest price per square foot. Explain how sizing up the scatterplot can help narrow your search so you do not have to perform so many calculations.

52. DATA Data about existing home sales in area 232 of Palo Alto, California, in 2007 are shown in Table 20. A *list price* is the price an owner originally requested before negotiations began. Let L be the list price (in thousands of dollars) of a home, and let S be the home's sales price (in thousands of dollars).

a. Construct a scatterplot that describes the association between list prices and sales prices.
b. On the scatterplot, imagine a line that goes through the points (1000, 1000), (1500, 1500), (2000, 2000), and (2500, 2500). Or use a strand of thread, a stretched rubber band, or the edge of a clear ruler to form the line. If a home's data point lies on the line, what can we say about the list price and sales price?
c. Does the line from part (b) come close to few, most, or all of the data points in the scatterplot? For each home whose data point is close to the line, what can we say about the list price and the sales price?

d. Determine the street of the home that sold for the most over the list price. Explain how sizing up the scatterplot can help narrow your search so you do not have to perform so many calculations.

e. Determine the street of the home that sold for the most under the list price. Explain how sizing up the scatterplot can help narrow your search so you do not have to perform so many calculations.

53. _{DATA} Data about existing home sales in area 232 of Palo Alto, California, in 2007 are shown in Table 20 on page 379. Let x be the number of bedrooms in a home, and let y be the number of baths in the home.

a. Construct a scatterplot that describes the association between number of bedrooms and number of baths.

b. Is the association positive, negative, or neither? What does that mean in this situation?

c. The house on Wisteria Ln has 2.5 bathrooms. What does it mean to have half a bathroom?

d. Find the mean number of baths for homes that have 2 bedrooms. Do the same for 3 bedrooms, 4 bedrooms, and 5 bedrooms.

e. Use your results in part (d) to help you construct a scatterplot that describes the association between the number of bedrooms and the *mean* number of baths. Compare the scatterplot to the one you constructed in part (a).

54. _{DATA} Data about existing home sales in area 232 of Palo Alto, California, in 2007 are shown in Table 20 on page 379. Let B be the number of bedrooms in a home, and let S be the home's sales price (in thousands of dollars).

a. Construct a scatterplot that describes the association between number of bedrooms and sales prices.

b. Is the association positive, negative, or neither? What does that mean in this situation?

c. Find the mean sales price for homes that have 2 bedrooms. Do the same for 3 bedrooms, 4 bedrooms, and 5 bedrooms.

d. Use your results in part (c) to help you construct a scatterplot that describes the association between the number of bedrooms and the *mean* sales prices. Compare the scatterplot to the one you constructed in part (a).

Concepts

55. On the basis of 18 LCD TVs, the mean annual costs to operate LCD TVs are shown in Table 21 for various sizes.

Table 21 Sizes of LCD TVs and Mean Annual Operating Costs

Size (inches)	Mean Annual Cost (dollars)
32	21
42	30
46	39
52	58

Source: *Columbia ISA*

A student lets C be the mean annual cost (in dollars) to run a LCD TV of size S inches. The student tries to construct a scatterplot (see Fig. 20). Describe all the errors made by the student, and draw a correct scatterplot.

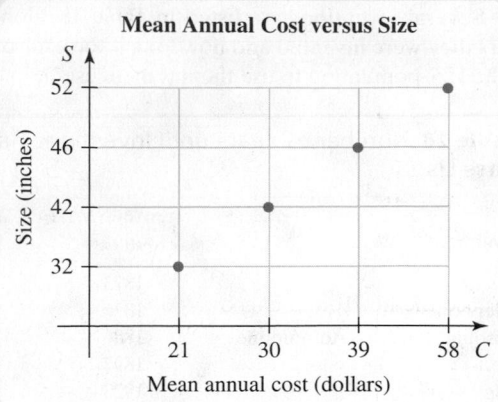

Figure 20 Exercise 55

56. In a study, the explanatory variable is the number of milligrams of an experimental drug that a person ingests per day, and the response variable is the person's blood pressure. Because the number of milligrams of the drug is the *explanatory variable*, does that mean that the drug definitely lowers blood pressure? Explain.

57. The author surveyed his algebra students about their favorite music genres. The students' responses are summarized in the bar graph in Fig. 21. A student says there is a negative association between favorite music genre and frequency. What would you tell the student?

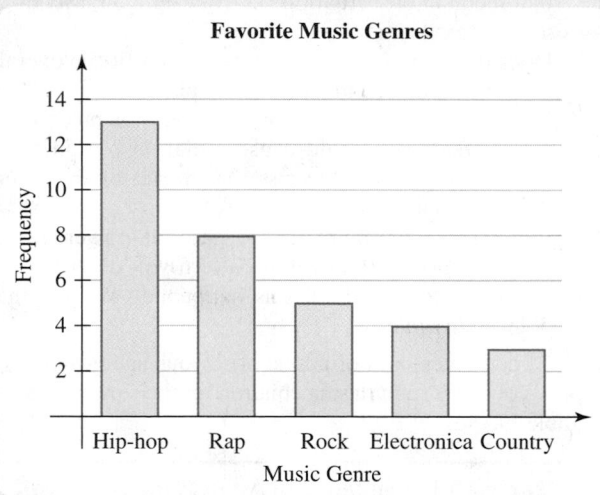

Figure 21 Exercise 57
(**Source:** *J. Lehmann*)

58. Determine whether a scatterplot can describe the given types of variables. If it is not possible, explain why by describing how we construct axes of a scatterplot.

a. two categorical variables

b. a categorical variable and a numerical variable

c. two numerical variables

For Exercises 59–62, choose all the following diagrams and tables that would be an appropriate way to describe the given data: frequency and relative frequency table, relative frequency bar graph, multiple bar graph, pie chart, two-way table, dotplot, stemplot, scatterplot, relative frequency histogram, and boxplot. Do not include diagrams or tables that would be possible but awkward.

59. A researcher compares the breeds of dogs and their coat colors.

60. For a sample of 1250 executives, a researcher records the number of texts they send in one day.

61. The list prices (in dollars) of toasters are compared with their lifetimes (in years).

62. Students' favorite music genres are recorded.

63. Assume the numerical variables x and y are negatively associated and the numerical variables y and z are also negatively associated. Determine whether the variables x and z are positively associated, negatively associated, or neither.

64. Suppose the horizontal axis of a coordinate system describes a variable x, the vertical axis describes a variable y, and the variables are negatively associated. Determine whether the variables would still be negatively associated if the variables are swapped—that is, the horizontal axis describes y and the vertical axis describes x.

65. Give an example of two variables not in the textbook that are positively associated.

66. Give an example of two variables not in the textbook that are negatively associated.

67. A students says an association must be one of two things: positive or negative. What would you tell the student?

68. Give an example of two variables not in the textbook that are neither positively associated nor negatively associated.

69. Compare the meaning of *explanatory variable* with the meaning of *response variable*. (See page 12 for guidelines on writing a good response.)

70. Compare the meaning of *positively associated* with the meaning of *negatively associated*. (See page 12 for guidelines on writing a good response.)

6.2   Determining the Four Characteristics of an Association

Objectives

» Determine the shape of an association.

» Determine the strength of an association.

» Compute and interpret the correlation coefficient.

» Determine whether there are any outliers and what to do with them.

» Determine the four characteristics of an association in the correct order.

» Explain why strong (or weak) association does not guarantee causation.

In Section 6.1, we analyzed the direction of an association. That is one of four characteristics of an association. The other three are *shape*, *strength*, and *outliers*. We will take a close look at each of these three new characteristics one at a time and then determine all four characteristics for an association described in Example 2. We will begin this section by discussing the shape of an association.

Shape of an Association

Figure 22 displays a scatterplot that describes the association between lengths of cruise ships and the crew sizes. The points in the scatterplot lie near a line.

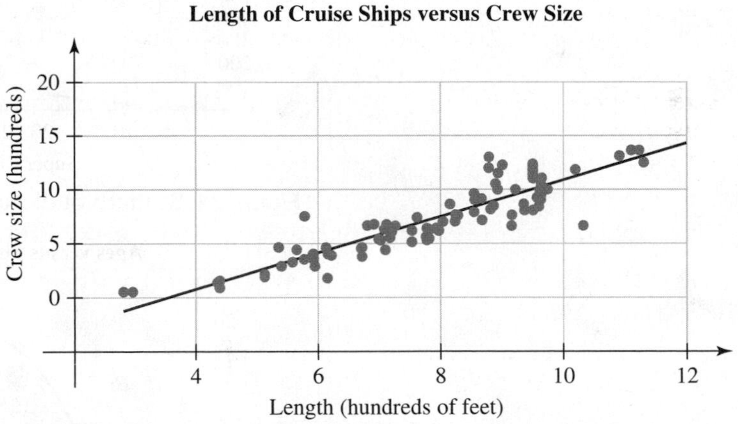

Figure 22 Scatterplot for cruise data
(**Source:** *True Cruise*)

In general, if the points of a scatterplot lie close to (or on) a line, we say the variables are **linearly associated** and that there is a **linear association**. Another example of linear association is shown in Fig. 23. The scatterplot describes the linear association between temperature and relative humidity in Phoenix, Arizona, for every hour on June 1, 2014.

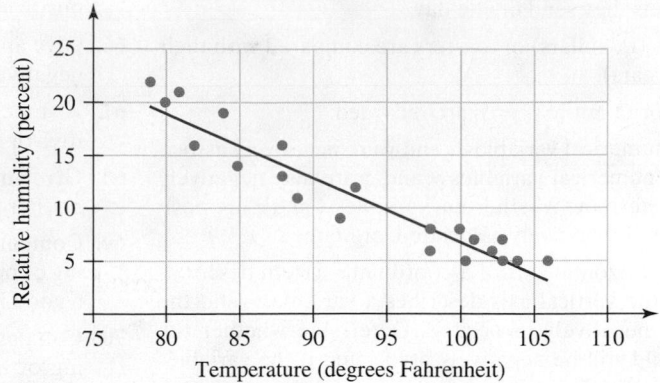

Figure 23 Linear association
(**Source:** *Weatherbase*)

In Example 4 in Section 6.1, we analyzed the Super Bowl scatterplot (see Fig. 24). The points lie close to (or on) a curve, but the curve is not a line. We will devote Chapter 10 to investigating this type of curve. In general, if the points of a scatterplot lie close to (or on) a curve that is not a line, we say there is a **nonlinear association**. Another example of a nonlinear association is the association between ages and suicide rates that we investigated in Example 5 in Section 6.1 (see Fig. 25).

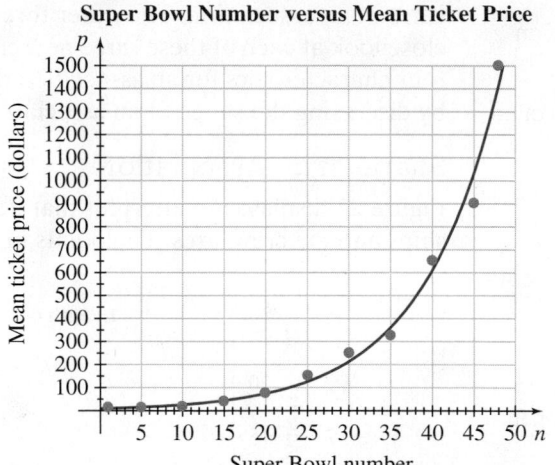

Figure 24 Scatterplot of mean Super Bowl ticket prices

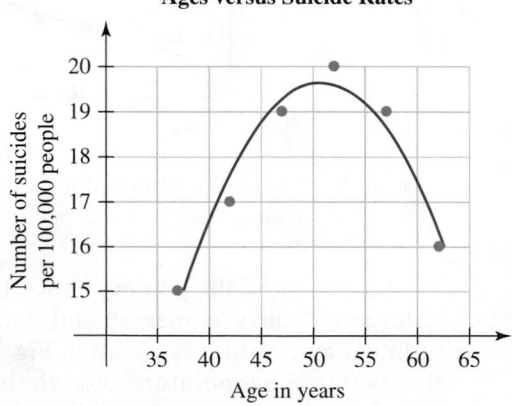

Figure 25 Scatterplot suicide data

Figure 26 displays a scatterplot comparing the ages and the finish times of 4578 participants of the Great Cow Harbor 10-Kilometer Run in Northport, New York. There is so much scatter, no curve comes close to all the points. In general, if no curve comes close to all the points of a scatterplot, we say there is **no association**.

Ages versus Finish Times

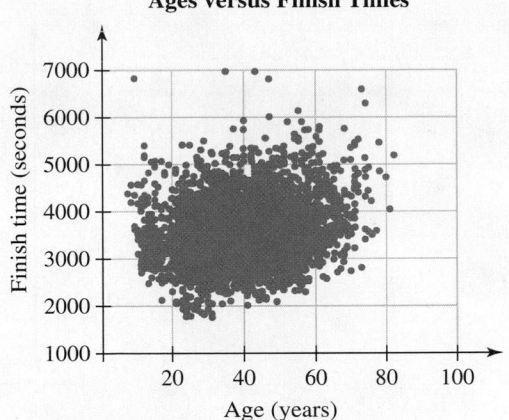

Figure 26 No association
(**Source:** *The Great Cow Harbor 10-Kilometer Run, Inc.*)

Most people would probably expect that for adults, the older the runner is, the larger the finish time tends to be. But the scatterplot does not support this. So, even though there is no association, we *have* learned something.

For any pair of numerical variables, there is either a linear association, a nonlinear association, or no association. We will refer to the **shape** of an association as being one of these three types.

Strength of an Association

If a curve passes through all the points of a scatterplot, we say there is an **exact association** with respect to the curve. If a curve comes quite close to all the points, we say there is a **strong association** with respect to the curve. If a curve comes somewhat close to all the points, we say there is a **weak association** with respect to the curve. The definitions for strong and weak associations are vague and require us to make a judgment call.

Determining the strength of an association between two numerical variables is similar to determining the spread of a distribution involving one numerical variable.

▶ **Example 1**   Interpreting a Scatterplot

Figure 27 shows a scatterplot comparing carbohydrates and calories for 29 pizzas made by six of the leading pizza companies. A scatterplot comparing carbohydrates and fat for the same pizzas is shown in Fig. 28.

Carbohydrates versus Calories

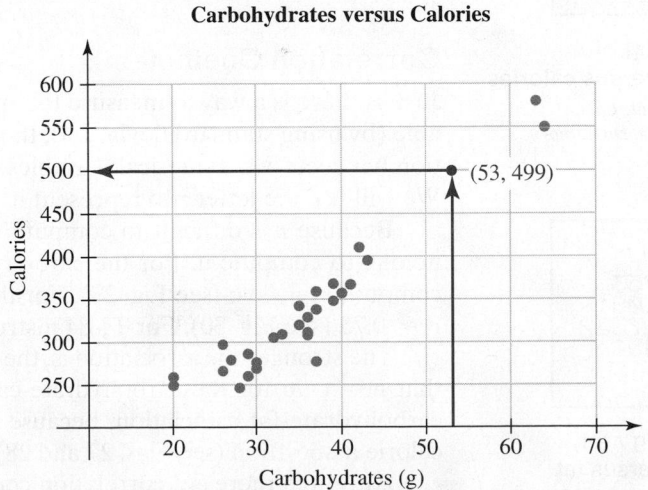

Figure 27 Carbohydrates versus Calories
(**Source:** *Domino's, Little Caesar's, Papa John's, Pizza Hut, DiGiorno Frozen, Kashi Frozen*)

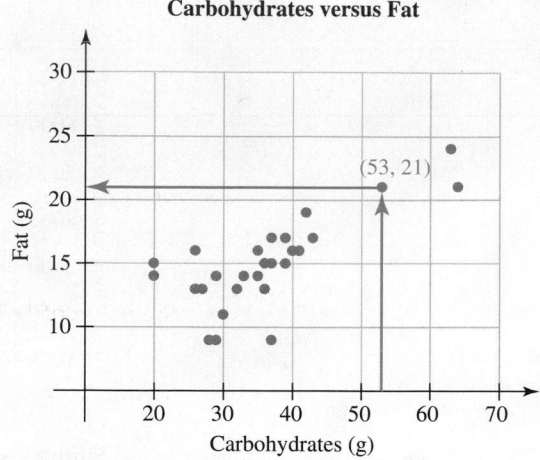

Figure 28 Carbohydrates versus fat
(**Source:** *Domino's, Little Caesar's, Papa John's, Pizza Hut, DiGiorno Frozen, Kashi Frozen*)

1. For each of the scatterplots, describe the shape of the association.
2. Compare the strengths of the two associations.
3. For each of the scatterplots, identify whether the association is positive, negative, or neither.
4. Estimate the caloric content of the pizza with 53 g of carbohydrates.
5. Estimate the fat content of the pizza with 53 g of carbohydrates.

Solution

1. For both scatterplots, the points lie near a line. So, the associations are both linear.
2. The points in the scatterplot in Fig. 27 lie closer to a line than the points in the scatterplot in Fig. 28. So, the association between carbohydrates and calories is stronger than the association between carbohydrates and fat.
3. For both scatterplots, the response variable increases as the number of grams of carbohydrates increases. So, the associations are both positive.
4. We start at 53 on the horizontal axis (see Fig. 27) and look up to the (red) dot. Then we look to the left of the dot and determine that the number of calories is approximately 500 g.
5. We start at 53 on the horizontal axis (see Fig. 28) and look up to the (green) dot. Then we look to the left of the dot and determine that the fat content is approximately 21 g.

Figure 29 Finding *r* for carbohydrates versus calories
(**Source:** *Domino's, Little Caesar's, Papa John's, Pizza Hut, DiGiorno Frozen, Kashi Frozen*)

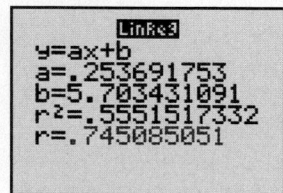

Figure 30 Finding *r* for carbohydrates versus fat
(**Source:** *Domino's, Little Caesar's, Papa John's, Pizza Hut, DiGiorno Frozen, Kashi Frozen*)

Correlation Coefficient

Just as there is a way to measure the spread of a distribution for a single numerical variable (by using standard deviation), there is a way to measure the strength of the association between two numerical variables. The measure is called the *correlation coefficient*. We will use the letter r to represent it.

Because r is difficult to compute with a nonstatistical calculator, we will use technology to compute it. For the carbohydrate-calorie association, we can use a TI-84 to compute $r \approx 0.96$ (see Fig. 29). For the carbohydrate-fat association, we compute that $r \approx 0.75$ (see Fig. 30). For TI-84 instructions, See Appendix A.12.

The stronger the association is, the larger the absolute value of r is. So, it makes sense that $r \approx 0.96$ for the carbohydrate-calorie association is larger than $r \approx 0.75$ for the carbohydrate-fat association, because the association is stronger for the carbohydrate-calorie association (see Figs. 27 and 28).

Although there are correlation coefficients for nonlinear associations, we will work only with the correlation coefficient that measures the strength of *linear* associations in Chapters 6–9.

> **Properties of the Linear Correlation Coefficient**
>
> Assume r is the linear correlation coefficient for the association between two numerical variables. Then
> - The values of r are between -1 and 1, inclusive.
> - If r is positive, then the variables are positively associated.
> - If r is negative, then the variables are negatively associated.
> - If $r = 0$, there is no *linear* association.
> - The larger the value of $|r|$, the stronger the linear association will be.
> - If $r = 1$, then the points lie exactly on a line and the association is positive.
> - If $r = -1$, then the points lie exactly on a line and the association is negative.

Figures 31–38 display the values of r for various associations.

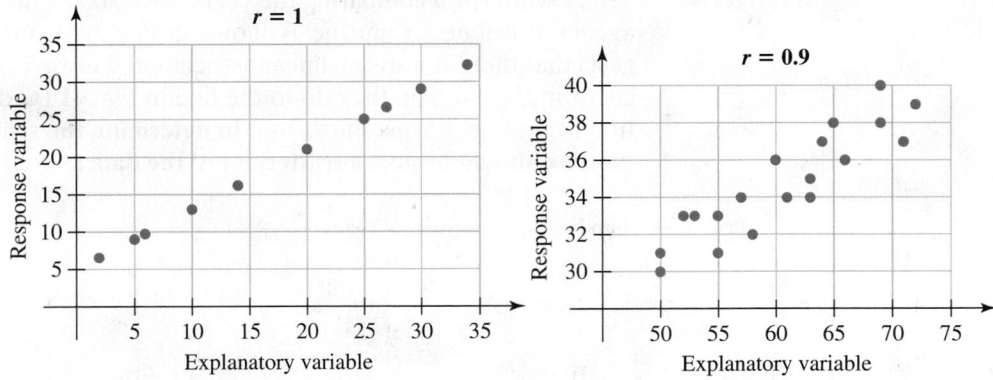

Figure 31 Exact linear association   **Figure 32** Strong linear association

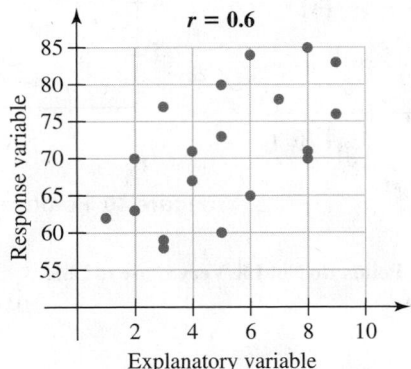

Figure 33 Weak linear association

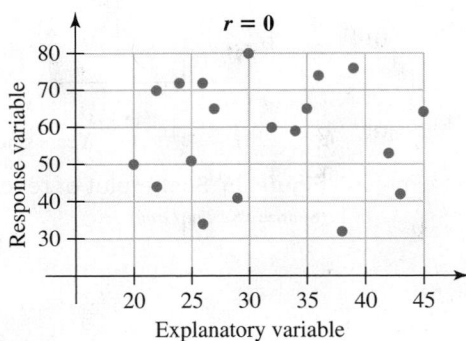

Figure 34 No linear association

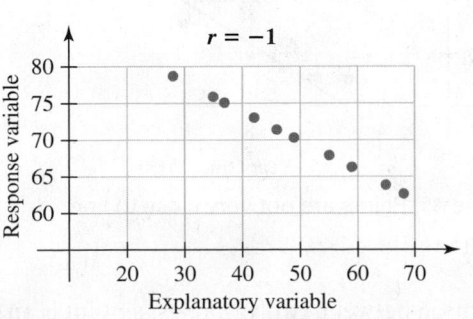

Figure 35 Exact linear association

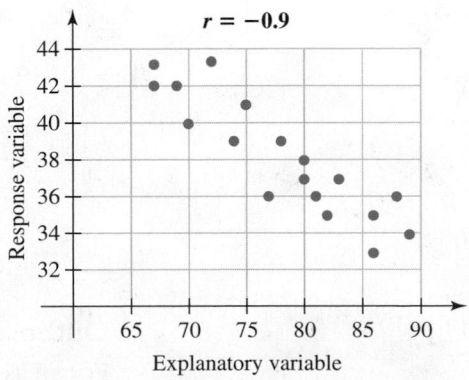

Figure 36 Strong linear association

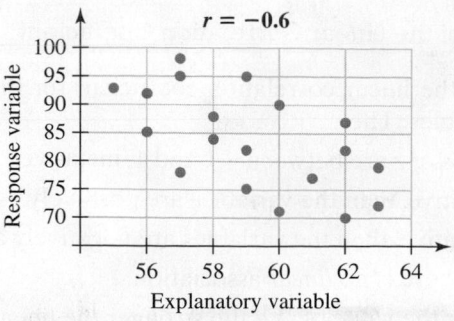

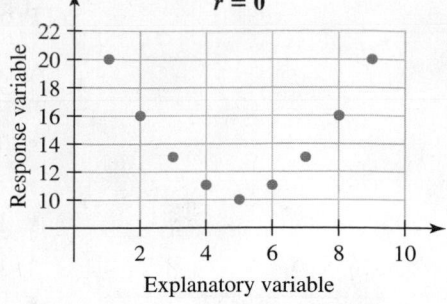

Figure 37 Weak linear association   **Figure 38** No linear association

WARNING

In Fig. 38, $r = 0$ tells us that there is no *linear* association. However, there is a strong *nonlinear* association. In general, if $r = 0$, then there is no linear association, but there may be a nonlinear association.

WARNING

A scatterplot comparing the years since 2000 with the revenues of the online travel agency Priceline Group Inc. is shown in Fig. 39. It turns out that $r \approx 0.96$, which suggests that there is a strong linear association. However, the points lie much closer to the curve in Fig. 40 than they do to the line in Fig. 41 (and any other line). So, the association is nonlinear. This shows that **to determine the strength and the type of an association, we should inspect a scatterplot of the data as well as compute r.**

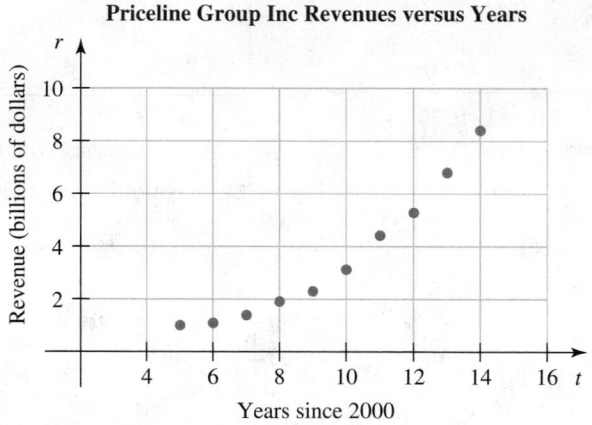

Figure 39 Scatterplot of revenues
(**Source:** *Priceline Group Inc.*)

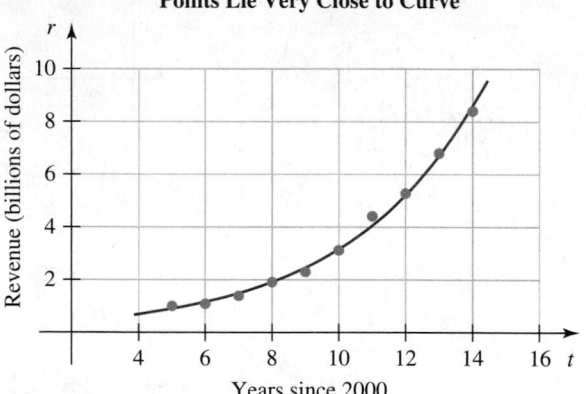

Figure 40 Points are very close to curve

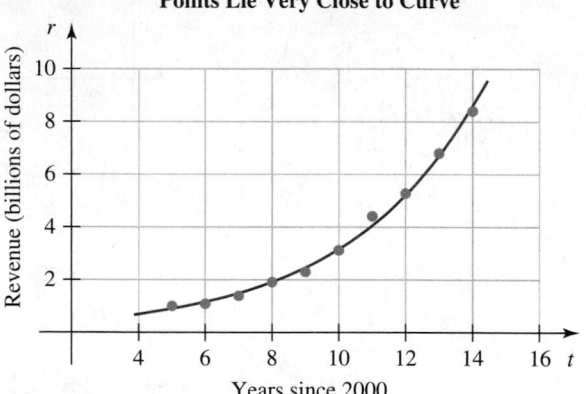

Figure 41 Points are not very close to line

Outliers

For an association between two variables, a point is an outlier if it does not fit the overall pattern of the points in the scatterplot. For example, the red dot in Fig. 42 is an outlier.

A Scatterplot with an Outlier

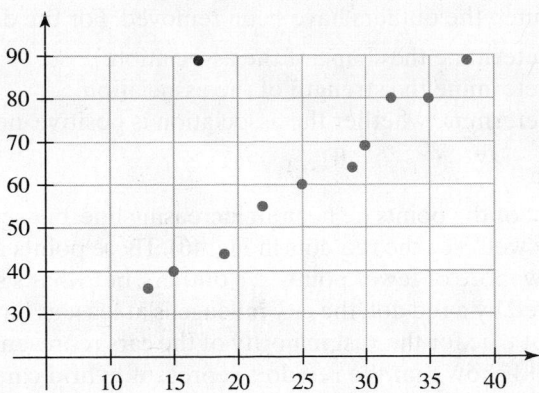

Figure 42 The red dot is an outlier

Similar to our work with one-variable distributions, if an outlier is due to error in measurement or recording, the coordinate(s) should be corrected, if possible, or the point should be excluded from the study. Sometimes a group of outliers will indicate a need for a separate study.

WARNING

The correlation coefficient is very sensitive to outliers. For example, the scatterplot in Fig. 43 shows no association. The value of r is 0, which makes sense. But by adding the outlier (45, 40), r is increased all the way up to 0.95 (see Fig. 44). Because there is still no association, the value of r is very misleading. Once again, we see that to determine the strength of an association, we should inspect a scatterplot of the data as well as compute r.

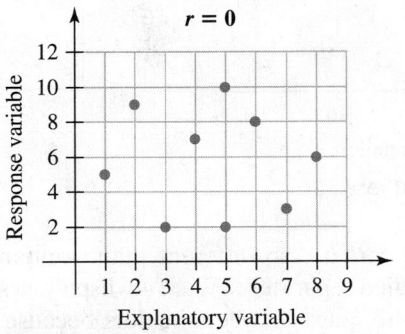

Figure 43 No association and $r = 0$

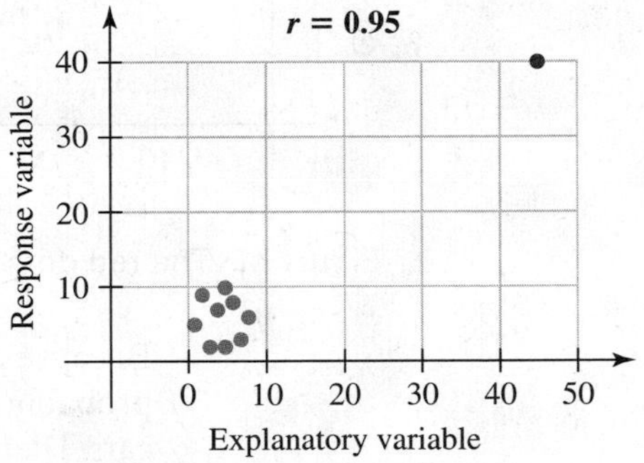

Figure 44 By adding an outlier, there is still no association but r is now 0.95.

▶ Example 2   Determine the Four Characteristics of an Association

A scatterplot comparing the highway mileages and city mileages of 1195 types of cars is displayed in Fig. 45.

Highway Mileage versus City Mileage

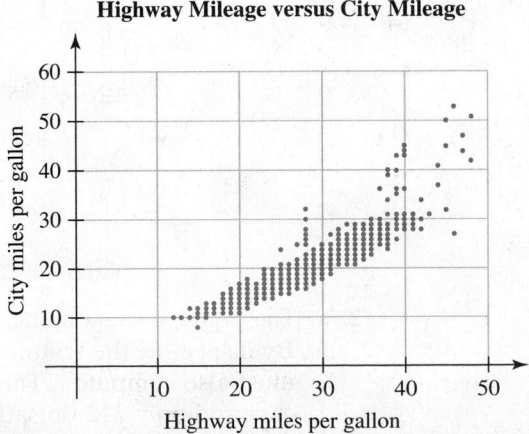

Figure 45 Highway mileage versus city mileage
(**Source:** *U.S. Environmental Protection Agency*)

1. Identify the outliers. Should they be removed? If yes, for what purpose?
2. Assume the outliers have been removed. For the data that remain,
    a. determine the shape of the association.
    b. determine the strength of the association.
    c. determine whether the association is positive, negative, or neither.

Solution

1. Most of the points lie near an increasing line, but several points do not fit the pattern very well (see the red dots in Fig. 46). These points are outliers. One could argue that a few more or fewer points are outliers, but what's striking is that for each car represented by a red dot, the city mileage is at least as large as the highway mileage, which is not true for the vast majority of the cars represented by blue dots. A little research would show that the red dots represent hybrid cars. Yet more research would show that all the red dots in Fig. 47 represent hybrid cars.

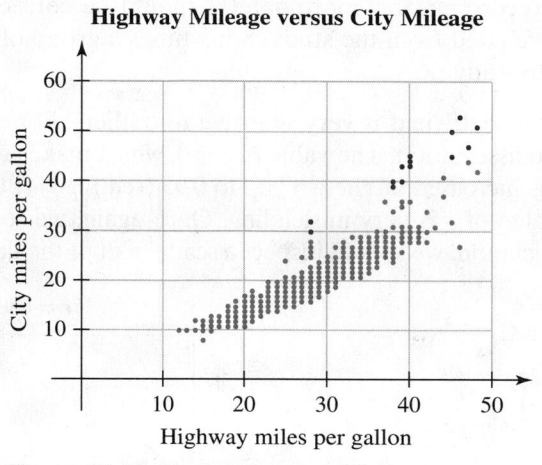

Figure 46 The red dots are outliers

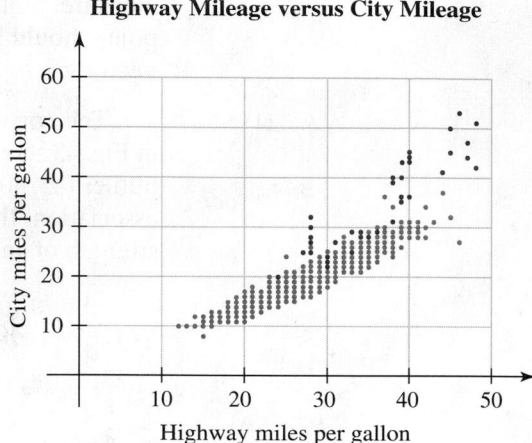

Figure 47 The red dots represent hybrid cars

Because hybrid cars are so different than standard cars, their gas mileages should probably be studied separately. Figure 48 displays a scatterplot for only the non-hybrid cars. There is more space between the dots, because the axes have been rescaled.

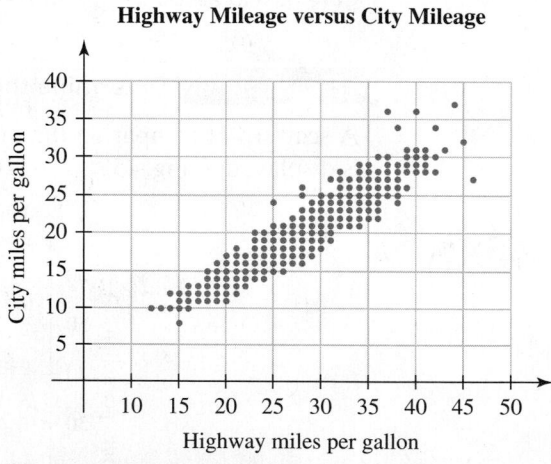

Figure 48 Scatterplot with hybrid-car points removed

2. a. The points lie near a line, so the association is linear.
    b. By inspecting the scatterplot, the association appears to be quite strong. But we must also compute r. This is especially important, because with 1159 remaining cars and only 212 dots, this means each dot represents approximately 5.5 cars on average. Some dots may represent many more cars than others, so we can't completely trust our eyes. StatCrunch, which uses a capitol R to stand for the

correlation coefficient, computes r to be approximately 0.95 (see Fig. 49). This checks with our visual determination that the linear association is strong. For StatCrunch instructions, see Appendix B.16.

Simple linear regression results:
Dependent Variable: City
Independent Variable: Highway
City = -0.88894966 + 0.74600579 Highway
Sample size: 1157
R (correlation coefficient) = 0.94829949
R-sq = 0.89927193
Estimate of error standard deviation: 1.5558934

Figure 49 $r \approx 0.95$

c. As highway mileage increases, city mileage tends to increase, so the association is positive. This checks with $r = 0.95$ being positive.

Order of Determining the Four Characteristics of an Association

In Example 2, we described four key characteristics of an association. The order in which we determine the characteristics matters.

> ### ▶ Order of Determining the Four Characteristics of an Association
>
> We determine the four characteristics of an association in the following order:
>
> 1. Identify all outliers.
>   **a.** For outliers that stem from errors in measurement or recording, correct the errors if possible. If the errors cannot be corrected, remove the outliers.
>   **b.** For other outliers, determine whether they should be analyzed in a separate study.
> 2. Determine the shape of the association.
> 3. If the shape is linear, then on the basis of r and the scatterplot, determine the strength. If the shape is nonlinear, then on the basis of the scatterplot, determine the strength.
> 4. Determine the direction. In other words, determine whether the association is positive, negative, or neither.

Strong (or Weak) Association Does Not Guarantee Causation

A scatterplot comparing the U.S. per-person consumption of margarine (in pounds) and the divorce rate (number of divorces per 1000 people) in Maine is shown in Fig. 50.

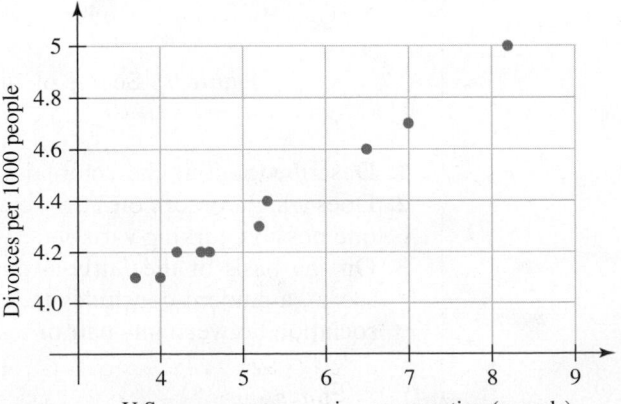

Margarine Consumption versus Maine Divorce Rate

Figure 50 Margarine consumption versus divorce rate in Maine
(**Source:** *USDA, U.S. Census Bureau*)

The linear association appears to be very strong, and *r* turns out to be 0.99, which checks. However, the very strong association does not mean that a change in the U.S per-person consumption of margarine will *cause* a change in the divorce rate in Maine! Recall that we say there is *no causation* between U.S. per-person margarine consumption and the divorce rate in Maine (Section 2.3).

For another example, it turns out that as ice cream consumption increases, drowning deaths increase. In other words, there is a positive association between ice cream consumption and drowning deaths. However, ice cream consumption does not *cause* drowning deaths.

So why is there a positive association? The answer lies with temperature. During the hot summer months, ice cream consumption increases, and drowning deaths increase, too, because people are more likely to go swimming when it's hot out. Recall from Section 2.3 that this means temperature is a lurking variable.

> ▶ **Definition** Lurking variable
>
> A **lurking variable** is a variable that causes both the explanatory and response variables to change during the study.

WARNING

In general, **a strong association between two variables does not guarantee that a change in the explanatory variable will cause a change in the response variable.**

So how do we know when strong association implies causation? Recall from Section 2.3 that we can determine this only by performing an experiment. Recall that a key feature of an experiment is that researchers *randomly assign* individuals to the treatment and control groups.

▶ **Example 3** Determining Possible Lurking Variables

The scatterplot in Figure 51 compares the scores of Test 4 and Test 5 for a calculus course taught by the author.

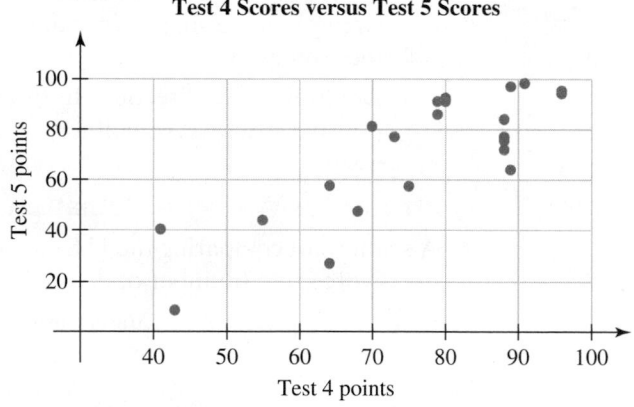

Test 4 Scores versus Test 5 Scores

Figure 51 Scores of Test 4 and Test 5
(**Source:** *J. Lehmann*)

1. Describe the four characteristics of the association.
2. Does a higher score on Test 4 cause a higher score on Test 5? If no, describe at least one possible lurking variable.
3. On the basis of the fairly strong, positive association between Test 4 and Test 5 scores, a student concludes that there must have been a fairly strong, positive association between any pair of tests for the course. What would you tell the student?

Solution

1. • **Outliers:** There are no outliers.
   • **Shape:** The points appear to come close to a line, so the association is linear.

- **Strength:** We compute that r is 0.81 (see Fig. 52). Because r is fairly close to 1 and the points in the scatterplot appear to lie fairly close to a line, we conclude that the association is fairly strong.

> **Simple linear regression results:**
> Dependent Variable: Test 5 points
> Independent Variable: Test 4 points
> Test 5 points = -25.428804 + 1.2469308 Test 4 points
> Sample size: 24
> R (correlation coefficient) = 0.81134401
> R-sq = 0.6582791
> Estimate of error standard deviation: 14.044018

Figure 52 Computing r

- **Direction:** As Test 4 points increase, Test 5 points tend to increase as well, so the association is positive. This checks with $r = 0.81$ being positive.

2. A student's score on Test 4 does not directly affect the student's score on Test 5. There are many possible lurking variables, including a student's study habits, the consistency in the level of difficulty of the tests, Test 4's impact on a student's motivation or confidence, the extent to which Test 5's concepts build on Test 4's, and a student's strength in algebra, which is a useful tool in calculus.

3. The fairly strong, positive association between Test 4 and Test 5 scores tells us nothing about the other test scores. In fact, it turns out that although there is a positive association between Test 1 and Test 5 scores, the association is much weaker ($r = 0.58$) than the association for Test 4 and Test 5 scores ($r = 0.81$). The impacts of the various lurking variables described in our solution to Problem 2 might explain why.

Group Exploration

 Impact of transforming data on the correlation coefficient

1. Construct a scatterplot and compute r for the ordered pairs described in Table 22.

Table 22 Values of x and y

Explanatory Variable x	Response Variable y
3	12
4	17
5	18
6	23
7	24
8	29

2. Transform the data in Table 22 as instructed. Then construct a scatterplot and compute r. How do the scatterplot and r compare to your results in Problem 1? Why does this make sense?

    a. Add 4 to all the x values.
    b. Subtract 10 from all the y values.
    c. Multiply all the x values by 5.
    d. Treat y as the explanatory variable and x as the response variable.
    e. Change the signs of all the y values.

3. Find a group of six ordered pairs that has the same value of r as the ordered pairs in Table 22, where one of the ordered pairs is (9, 24).

4. Find another group of six ordered pairs that meets the conditions stated in Problem 3.

> ▶ Tips for Success   Write a Summary
>
> Consider writing, after each class meeting, a summary of what you have learned. Your summaries will increase your understanding, as well as your memory, of concepts and procedures and will also serve as a good reference for quizzes and exams.

Homework 6.2

 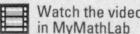

1. If the points of a scatterplot lie close to (or on) a line, we say there is a(n) _____ association.

2. If *r* is positive, then there is a(n) _____ association.

3. If *r* = _____, the points lie exactly on a line and the association is negative.

4. *True or False:* A strong association between two variables means that a change in the explanatory variable will cause a change in the response variable.

For Exercises 5–8, refer to the given scatterplot to determine whether there is a linear association, a nonlinear association, or no association.

5. See Fig. 53.

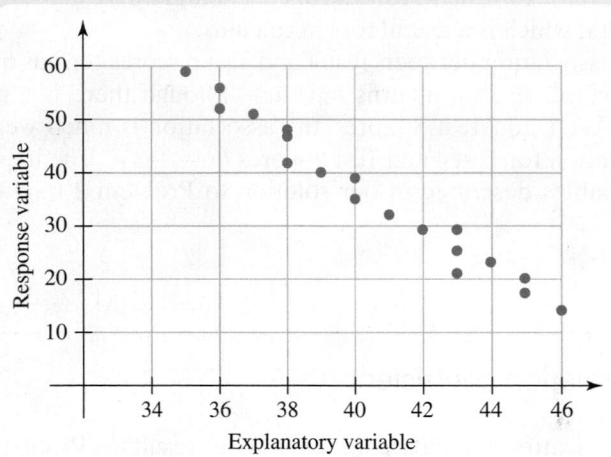

Figure 53 Exercise 5

6. See Fig. 54.

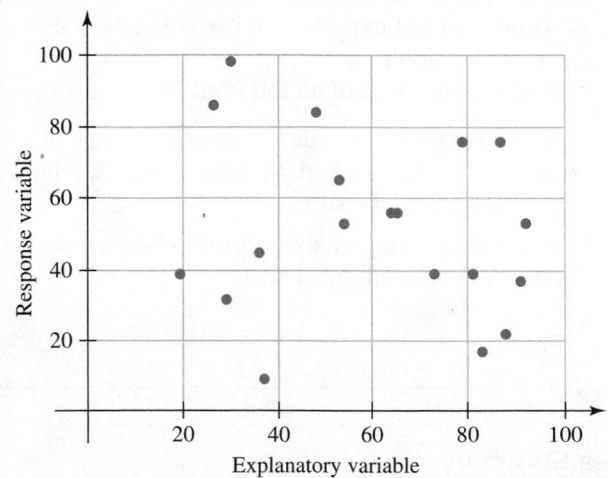

Figure 54 Exercise 6

7. See Fig. 55.

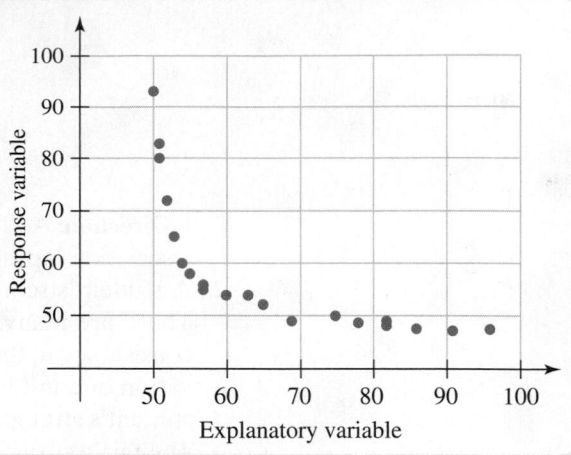

Figure 55 Exercise 7

8. See Fig. 56.

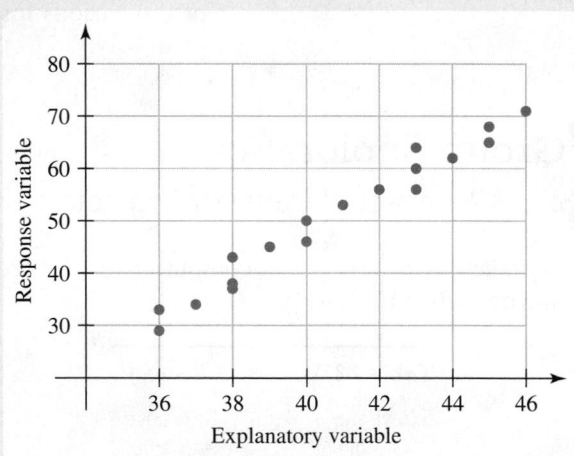

Figure 56 Exercise 8

For Exercises 9–14, match the given information to the appropriate scatterplot in Fig. 57.

9. *r* = 0.6

10. *r* = 0

11. *r* = −1

12. *r* = −0.9

13. *r* = 0.9

14. *r* = 1

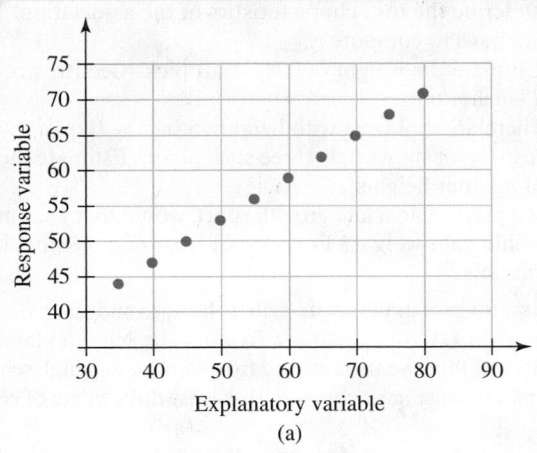

(a)

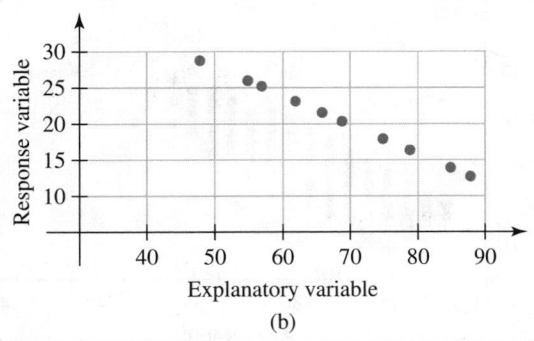

(b)

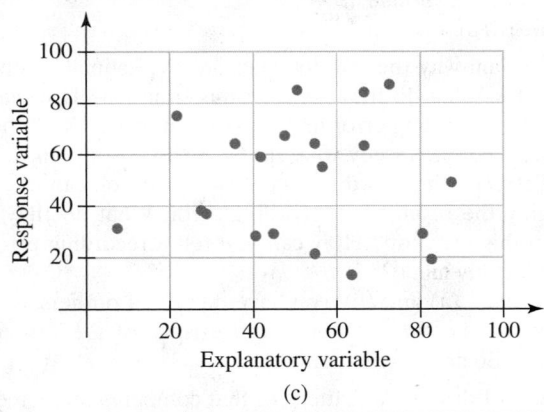

(c)

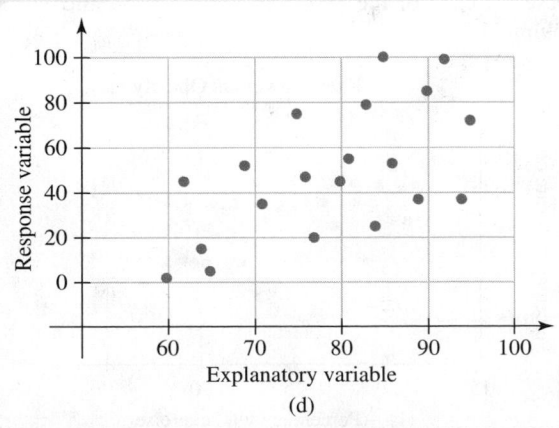

(d)

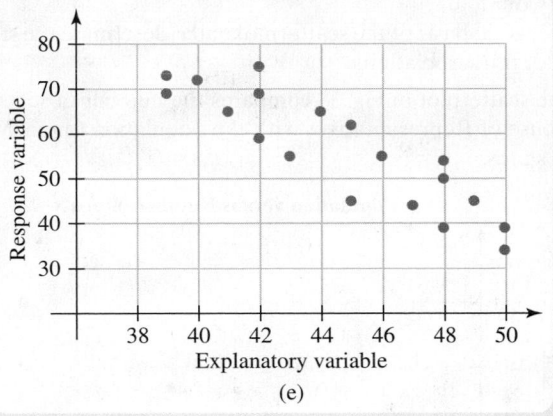

(e)

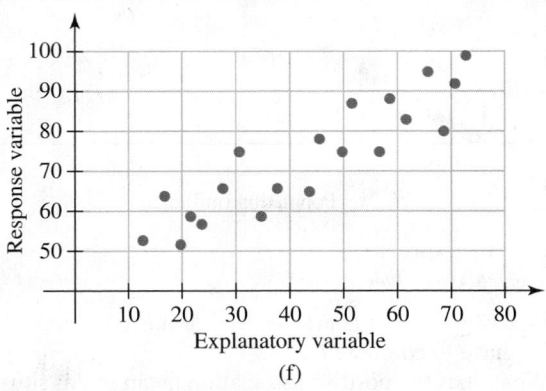

(f)

Figure 57 Exercises 9–14

15. Some ordered pairs are shown in Table 23. Let x be the explanatory variable and y be the response variable.

Table 23 Values of x and y

x	y
1	2
3	6
4	4
6	8
7	10

a. Construct a scatterplot.
b. Is there a linear association, nonlinear association, or no association?
c. Compute r.
d. On the basis of the scatterplot and r, determine the strength of the association.

16. Some ordered pairs are shown in Table 24. Let x be the explanatory variable and y be the response variable.

Table 24 Values of x and y

x	y
0	18
2	16
2	13
4	11
6	8

a. Construct a scatterplot.
b. Is there a linear association, nonlinear association, or no association?

c. Compute *r*.

d. On the basis of the scatterplot and *r*, determine the strength of the association.

17. The scatterplot in Fig. 58 compares the number of seats in the House of Representatives with the population for each of the 50 states.

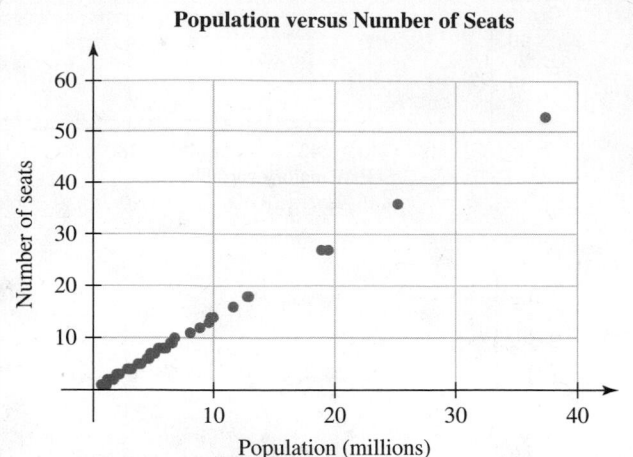

Figure 58 Exercise 17

(**Source:** *U.S. Census Bureau*)

a. Describe the four characteristics of the association. You do not have to compute *r*.

b. What does the positive association mean in this situation?

c. Estimate the number of seats in California, New York, and Michigan, which have populations of approximately 37 million, 19 million, and 10 million, respectively.

d. Find the ratio of number of seats to population (in millions) for each of California, New York, and Michigan. Round to the first decimal place. Are the ratios approximately equal?

e. In the *Great Compromise of 1787*, large and small states agreed that in the House of Representatives, the number of seats for each state would be based on the state's population. They also agreed that in the Senate, there would be exactly two Senators from each state. Which type of states were in favor of the system used for the House of Representatives, large or small? How about the Senate?

18. The heights and weights of the 60 players picked in the 2014 draft for NBA basketball teams are described by the scatterplot in Fig. 59.

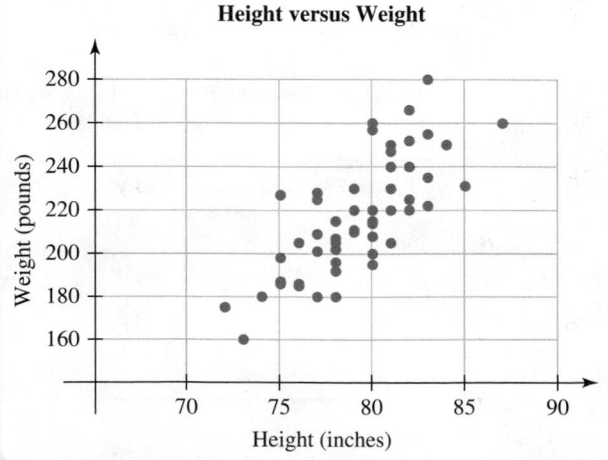

Figure 59 Exercise 18

(**Source:** *nbadraft.net*)

a. Describe the four characteristics of the association. You do not have to compute *r*.

b. Estimate the weight of 2014 draft pick Alec Brown, who is 85 inches tall.

c. There are 4 players with height 75 inches. Explain why the scatterplot shows only three such points. Estimate the range of the four heights.

d. If a player had a late growth spurt, would that guarantee he would gain weight? Describe at least one possible lurking variable.

19. The scatterplot in Fig. 60 describes the ages and years of experience of NFL football players. To enter the NFL, a player must be at least three years removed from completing high school, or the player must have completed at least three years of college.

Figure 60 Exercise 19

(**Source:** *NFL*)

a. Explain why the red dot is an outlier. Estimate its coordinates. What do they mean in this situation? How can you tell a recording error must have been made? If all the data were part of a study, what should be done about the outlier?

b. Estimate the coordinates of one other point that is probably the result of a recording error. What do they mean in this situation? How can you tell a recording error was probably made?

c. In parts (a) and (b), you have described outliers. Now describe the three other characteristics of the association. You do not have to compute *r*.

20. Figure 61 displays a scatterplot that compares the percentage of adults who exercise with the percentage of adults who are obese for each of the 50 states, Puerto Rico, and District of Columbia.

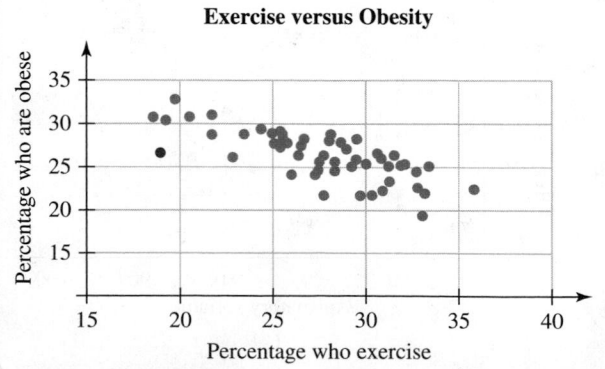

Figure 61 Exercise 20

(**Source:** *Centers for Disease Control and Prevention*)

a. Explain why the red dot in the scatterplot might be considered an outlier. What does this mean in this situation? The dot represents Puerto Rico. Why might this make sense?

b. In part (a), you analyzed a possible outlier. Describe the other three characteristics of an association. You do not have to compute *r*.

c. There are 5 states in which approximately 33% of the residents exercise. For each of those 5 states, estimate the percentage of residents who are obese. Then find the range of those values. Does the range check with your analysis of the strength of the association in part (b)? Explain.

d. On the basis of the scatterplot, a student concludes that exercise and obesity are negatively associated in Niger, which has had large-scale famine in recent years. What would you tell the student?

21. **DATA** A planet's *orbital velocity* is the speed that a planet travels around the Sun. The orbital velocities and the mean distances from the Sun of the 8 planets in our solar system are shown in Table 25.

Table 25 Planets' Orbital Velocities and Distances from the Sun

Planet	Mean Distance from Sun (million km)	Orbital Velocity (km/s)
Mercury	57.9	47.4
Venus	108.2	35.0
Earth	149.6	29.8
Mars	227.9	24.1
Jupiter	778.6	13.1
Saturn	1433.5	9.7
Uranus	2872.5	6.8
Neptune	4495.0	5.4

Source: *NASA*

Let *v* be the orbital velocity (in km/s) of a planet, and let *d* be the planet's mean distance (in million km) from the Sun.

a. Construct a scatterplot. Treat a planet's mean distance from the Sun as the explanatory variable and the planet's orbital velocity as the response variable.

b. Describe the four characteristics of the association.

c. The value of *r* is −0.77, which confuses a student, who thinks *r* should be −1, because the association is so strong. What would you tell the student?

d. Find Earth's orbital velocity in miles per hour. Round to the ones place. [**Hint:** There is 0.621 mile in 1 km.]

e. The association is actually true for all bodies that travel around the Sun. Pluto is 5870 million km from the Sun. Is Pluto's orbital velocity greater than or less than Neptune's orbital velocity of 5.4 km/s? Explain.

22. **DATA** *Escape velocity* is the initial velocity that an object needs in order to break free of a planet's gravitational pull. The mass of a planet can be used to help find the weight of an object on that planet. The escape velocities and the masses of 7 of the 8 planets in our solar system are shown in Table 26.

Table 26 Planets' Escape Velocities and Masses

Planet	Mass (10^{24} kg)	Escape Velocity (km/s)
Mercury	0.3	4.3
Mars	0.6	5.0
Earth	6.0	11.2
Uranus	86.8	21.3
Neptune	102	23.5
Saturn	568	35.5
Jupiter	1898	59.5

Source: *NASA*

Let *V* be the escape velocity (in km/s) of a planet, and let *M* be the planet's mass (in 10^{24} kg).

a. Assume you want to investigate whether a planet's mass explains its escape velocity. Construct a scatterplot.

b. Describe the four characteristics of the association.

c. Find Earth's escape velocity in miles per hour. [**Hint:** There is 0.621 mile in 1 km.]

d. Venus's mass is 4.9×10^{24} kg. Use two other planets' escape velocities to form an interval to estimate Venus's escape velocity.

e. Find the ratio of Jupiter's mass to Neptune's mass. Find the ratio of Jupiter's escape velocity to Neptune's escape velocity. Is the ratio of masses less than, equal to, or greater than the ratio of escape velocities? Explain how this matches with the shape of the association by referring to your scatterplot.

23. The scatterplot in Fig. 62 compares the prices of hot dogs and soft drinks at all the Major League Baseball (MLB) stadiums.

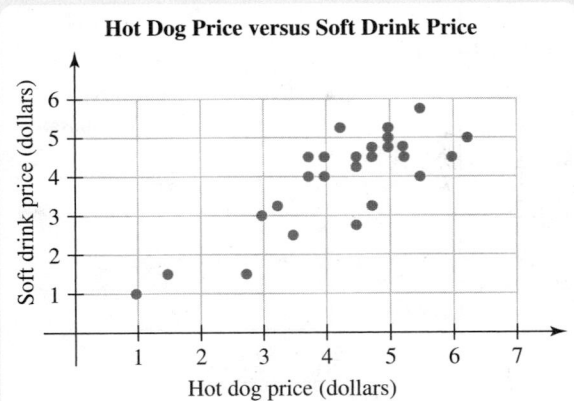

Figure 62 Exercise 23
(**Source:** *Team Marketing Report*)

a. Describe the four characteristics of the association. You do not have to compute *r*.

b. The correlation coefficient is 0.83. Does this support your analysis of the strength of the association in part (a)? Explain.

c. Because the association is positive, a student concludes that if a stadium raises the price of hot dogs, then it will raise the price of soft drinks. What would you tell the student?

d. A researcher suspects that there is a lurking variable, which is the costliness of a city. She reasons that if she is right, then there should be a positive, linear association between prices of any pair of concessions available at the stadiums. Figure 63 shows a scatterplot that compares the prices of hot dogs and beers at all the MLB stadiums. The value of *r* is 0.27. Do the scatterplot and *r* support the researcher's suspicion? Explain.

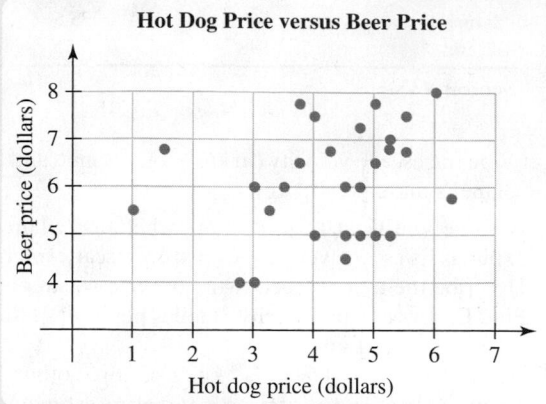

Figure 63 Exercise 23(d)
(**Source:** *Team Marketing Report*)

24. Figure 64 displays a scatterplot that compares the mean lengths with the mean heights of 51 types of dinosaurs. There are two main categories (called *orders*) of dinosaurs: Saurischians and Ornithischians. Saurischians had a pelvis also found in modern lizards, whereas Ornithischians had a pelvis that was more bird-like.

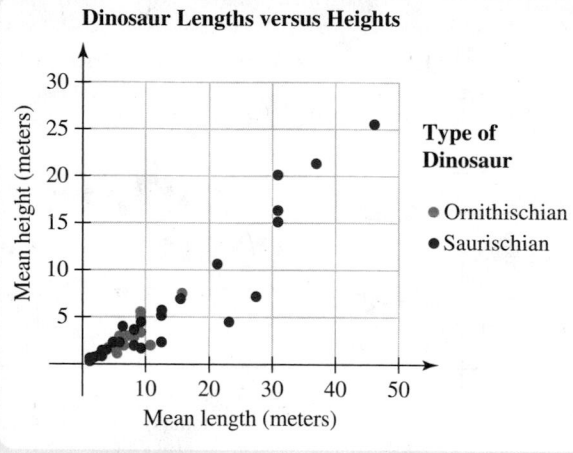

Figure 64 Exercise 24
(**Source:** *DinoDatabase.com*)

Let *L* be the mean length and *H* be the mean height, both in meters, of a dinosaur.

a. Describe the four characteristics of the Saurischian association. You do not have to compute *r*.

b. Describe the four characteristics of the Ornithischian association. You do not have to compute *r*.

c. For the Saurischian association, *r* = 0.94. For the Ornithischian association, *r* = 0.75. Compare these results with your responses to parts (a) and (b).

d. The mean length (45.7 meters) of a Seismosaurus is an outlier for the distribution of mean lengths of the 51 types of dinosaurs. The mean height (25.6 meters) of a Seismosaurus is an outlier for the distribution of mean heights of the 51 types of dinosaurs. Is the data point (45.7, 25.6) an outlier for the association of lengths and heights of dinosaurs? Explain.

e. If the association of lengths and heights of *individual* dinosaurs were analyzed rather than the association of mean lengths and mean heights of types of dinosaurs, how would the values of the Saurischian and Ornithischian correlation coefficients change? Explain.

25. A football fan says Super Bowl games should be hosted only at indoor stadiums and at outdoor stadiums with warm January climates so the games will all be high-scoring. The scatterplot in Fig. 65 describes the temperature on the field and the total score from both teams in each of the 48 Super Bowls. The StatCrunch computation of *r* is shown in Fig. 66. Do the scatterplot and *r* support the fan's assumption that warmer temperatures on the field lead to higher-scoring games? Explain.

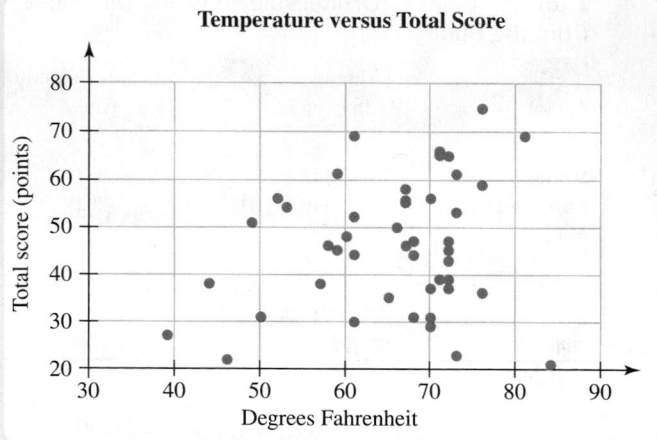

Figure 65 Exercise 25
(**Source:** *Pro Football Hall of Fame*)

Simple linear regression results:
Dependent Variable: Total points
Independent Variable: Fahrenheit Degrees
Total points = 27.958316 + 0.27682566 Fahrenheit Degrees
Sample size: 48
R (correlation coefficient) = 0.19835823
R-sq = 0.039345988
Estimate of error standard deviation: 13.229974

Figure 66 StatCrunch output

26. The governor of a certain state says parents should exercise more to set a good example for their teenagers. The percentages of parents who exercise and the percentages of teenagers who exercise are compared by the scatterplot in Fig. 67 for the 40 states in which the data were available. The StatCrunch computation of *r* is shown in Fig. 68. Do the scatterplot and *r* support the governor's assumption that a change in parents' exercise habits will lead to a change in their teenagers' exercise habits? Explain.

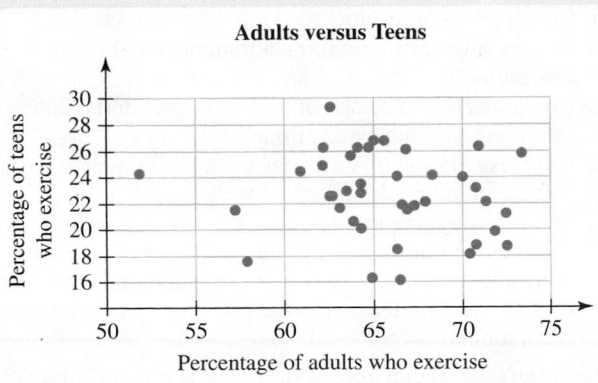

Adults versus Teens

Figure 67 Exercise 26
(**Source:** *Centers for Disease Control and Prevention*)

Simple linear regression results:
Dependent Variable: Teens
Independent Variable: Adults
Teens = 29.307489 − 0.10048686 Adults
Sample size: 40
R (correlation coefficient) = − 0.14359581
R-sq = 0.020619757
Estimate of error standard deviation: 3.1514132

Figure 68 StatCrunch output

27. **DATA** A study was based on the records of 6789 patients in 369 hospitals in which the patients' hearts stopped because of conditions that could have been reversed by an electrical shock from a defibrillator. The percentages of patients who survived at least until being discharged from the hospital are shown in Table 27 for various delays in the hospital staff giving the shock.

Table 27 Delays in Using Defibrillator versus Percentages of Patients Who Survived

Delay (minutes)	Percent
1	39
2	38
3	34
4	23
5	25
6	18
7	14

Source: *New England Journal of Medicine*

Let p be the percentage of patients who survived a delay of d minutes before receiving treatment from a defibrillator.

a. Construct a scatterplot.
b. Is there a linear association, a nonlinear association, or no association?
c. Compute r. On the basis of the scatterplot and r, describe the strength of the association.
d. Expert guidelines say the shock should be given within 2 minutes after the heart stops. Explain why this cutoff makes sense.
e. Hospital staffs took longer than 2 minutes for 30% of their patients. Estimate the number of patients in the study who died in the hospital due to the delay being more than 2 minutes. To simplify the calculations, assume that all delays more than 2 minutes took 4 minutes.

28. **DATA** The probability that a randomly selected newborn is a boy is actually more than $\frac{1}{2}$. The number of newborn boys for every 100 newborn girls is called the *sex ratio at birth*. The *birth order* of a first child is 1, the birth order of a second child is 2, and so on. The sex ratios at birth are shown in Table 28 for various birth orders.

Table 28 Birth Orders versus Sex Ratios at Birth

Birth Order	Sex Ratio at Birth
1	105.8
2	105.1
3	104.8
4	104.4
5	104.0
6	103.4

Source: *U.S. Centers for Disease Control and Prevention*

Let r be the sex ratio at birth for newborns whose birth order is n.

a. Construct a scatterplot.
b. Is there a linear association, a nonlinear association, or no association?
c. Compute r. On the basis of the scatterplot and r, determine the strength of the association.
d. Find the *probability* that a randomly selected firstborn is a boy.
e. China limits the number of babies a couple can have to one or two babies. Chinese parents tend to want to have boys rather than girls (Source: *NBC News*). The sex ratio at birth in China in 2010 was 120 (Source: *United Nations*). Give a possible reason why the sex ratio at birth might have been so high.

29. **DATA** The tank and aircraft productions in Germany and the United States during World War II are shown in Table 29.

Table 29 Tank and Aircraft Productions in Germany and the United States

Year	Country	Production Tanks	Production Aircraft
1939	Germany	247	8295
1940	Germany	1643	10,862
1941	Germany	3790	12,401
1942	Germany	6180	15,409
1943	Germany	12,063	24,807
1944	Germany	19,002	40,593
1945	Germany	3932	7540
1939	United States		2141
1940	United States	331	6068
1941	United States	4052	19,433
1942	United States	24,997	47,836
1943	United States	29,497	85,898
1944	United States	17,565	96,318
1945	United States	11,968	46,001

Source: *The National WWII Museum;* World War Two Tanks, *George Forty, 1995;* No Simple Victory: World War II in Europe, 1939–1945, *Norman Davies, 2008;* Atlas of Tank Warfare: From 1916 to the Present Day, *Stephen Hart, 2012*

Let T be a country's annual tank production, and let A be the country's annual aircraft production. Treat T as the explanatory variable and A as the response variable.

a. Construct a scatterplot that describes the association between tank and aircraft productions in Germany. Do the same for the United States. Show both scatterplots in the same coordinate system if you know how to use colors or shapes to identify which points are for which country.

b. Is there a linear association, a nonlinear association, or no association between tank and aircraft production in Germany? How about the United States?

c. For the Germany association, compute r. On the basis of the scatterplot and the value of r, determine the strength of the Germany association. Do the same for the United States.

d. For a given tank production, which country tended to produce more aircraft, Germany or the United States?

30. **DATA** The numbers of math degrees granted by Northwestern University are shown in Table 30 for various years.

Table 30 Years versus Numbers of Math Degrees Granted

Year	Number of Math Degrees	Year	Number of Math Degrees
2003	42	2009	46
2004	32	2010	81
2005	44	2011	81
2006	51	2012	72
2007	53	2013	78
2008	73		

Source: *Northwestern University*

Let n be the number of math degrees granted by Northwestern University in the year that is t years since 2000.

a. Construct a scatterplot.

b. Is there a linear association, a nonlinear association, or no association?

c. Compute r. On the basis of the scatterplot and r, determine the strength of the association.

d. Is the association positive, negative, or neither? What does that mean in this situation?

e. What is the drop in math majors from 2008 to 2009? Give at least one reason why this result stands out.

31. **DATA** Women's winning times in Olympic 500-meter speed skating are shown in Table 31 for various years.

Table 31 Years versus Women's 500-Meter Speed Skating Times

Year	Winning Time (seconds)	Year	Winning Time (seconds)
1972	43.33	1994	39.25
1976	42.76	1998	38.21
1980	41.78	2002	37.375
1984	41.02	2006	38.285
1988	39.10	2010	38.050
1992	40.33	2014	37.35

Source: *Universal Almanac*

Let w be the women's winning time (in seconds) at t years since 1970.

a. Construct a scatterplot.

b. Is there a linear association, a nonlinear association, or no association?

c. Compute r. On the basis of the scatterplot and r, determine the strength of the association.

d. Is the association positive, negative, or neither? What does that mean in this situation?

e. The faster Olympic competitors skate, the more innovative their training programs and nutrition plans must be to skate yet faster. Refer to some of the data in Table 31 that suggest that the linear association might not continue for much longer.

32. **DATA** Men's 400-meter record times are shown in Table 32 for various years.

Table 32 Years versus Men's 400-Meter Run Record Times

Year	Record Time (seconds)	Year	Record Time (seconds)
1900	47.8	1950	45.8
1916	47.4	1960	44.9
1928	47.0	1968	43.86
1932	46.2	1988	43.29
1941	46.0	1999	43.18

Source: *International Association of Athletics Federations*

Let r be the men's 400-meter record time (in seconds) at t years since 1900.

a. Construct a scatterplot.

b. Is there a linear association, a nonlinear association, or no association?

c. Compute r. On the basis of the scatterplot and r, determine the strength of the association.

d. Is the association positive, negative, or neither? What does that mean in this situation?

e. The lower the record is, the more innovative runners' training programs and nutrition plans must be to shave yet more tenths of seconds off the record. Explain what is surprising about the change in the record time from 1950 to 1999 as compared with the change in the record time from 1900 to 1950.

33. **DATA** The percentages of Americans who believe marriages between same-sex couples should be recognized by the law as valid are shown in Table 33 for various age groups.

Table 33 Ages versus Percentages of Americans Who Support Same-Sex Marriage

Age Group (years)	Age Used to Represent Age Group (years)	Percent
18–29	23.5	71
30–39	34.5	56
40–49	44.5	49
50–59	54.5	46
60–69	64.5	42
70–79	74.5	32
over 79	85.0	21

Source: *The Gallup Organization*

Let p be the percentage of Americans at age a years who believe marriages between same-sex couples should be recognized by the law as valid.

a. Construct a scatterplot.

b. Is there a linear association, a nonlinear association, or no association?

c. Compute *r*. On the basis of your scatterplot and *r*, determine the strength of the association.

d. Is the association positive, negative, or neither? What does that mean in this situation?

e. On the basis of your response to part (d), a student concludes that as adults grow older, they tend to switch from being in support of same-sex marriage to being against it. What would you tell the student?

34. DATA The percentages of adult women who are overweight (including obese) are shown in Table 34 for various age groups.

Table 34 Ages versus Percentages of Adult Women Who Are Obese

Age Group (years)	Age Used to Represent Age Group (years)	Percent
20–34	27.0	55.2
35–44	39.5	62.4
45–54	49.5	70.5
55–64	59.5	75.1
65–74	69.5	73.8
75 and over	80.0	62.4

Source: *U.S. Centers of Disease and Protection*

Let *p* be the percentage of women at age *a* years who are obese.

a. Construct a scatterplot.

b. Is there a linear association, a nonlinear association, or no association?

c. Is the association positive, negative, or neither?

d. On the basis of your response to part (c), a student concludes that once women reach 59.5 years in age, they lose weight as they grow older. What would you tell the student?

e. Describe at least one possible lurking variable. Explain.

35. The *sea ice extent* is the area of the ocean with at least 15% sea ice. The association between years since 1970 and the mean sea ice extent in the northern hemisphere in *March* is described by the scatterplot in Fig. 69. The association between years since 1970 and the mean sea ice extent in the northern hemisphere in *September* is described by the scatterplot in Fig. 70.

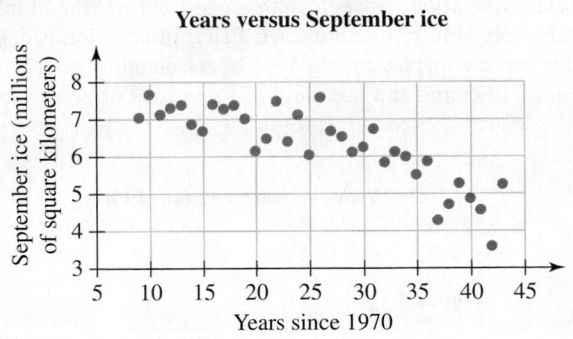

Figure 70 Years since 1970 versus September ice extent

(**Source:** *National Snow and Ice Data Center*)

a. Is there a positive association, a negative association, or no association between the years since 1970 and March sea ice extent? What does that mean in this situation?

b. Is there a positive association, a negative association, or no association between the years since 1970 and September sea ice extent? What does that mean in this situation?

c. Estimate the mean loss of sea ice extent from March to September for the period 1979–2013.

d. The association between years since 1970 and the *difference* in sea ice extent in March and September is described by the scatterplot in Fig. 71. Does the scatterplot check with your response to part (c)? Explain.

e. The association between the sea ice extent in March and the sea ice extent in September is described by the scatterplot in Fig. 72. Is there a positive association, a negative association, or no association? What does that mean in this situation? Explain how this connects to your responses to parts (c) and (d).

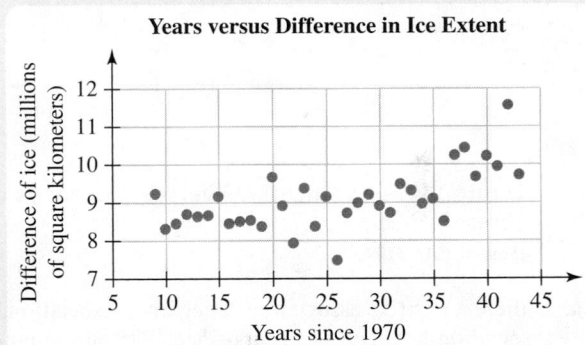

Figure 71 Difference in sea ice extent in March and September

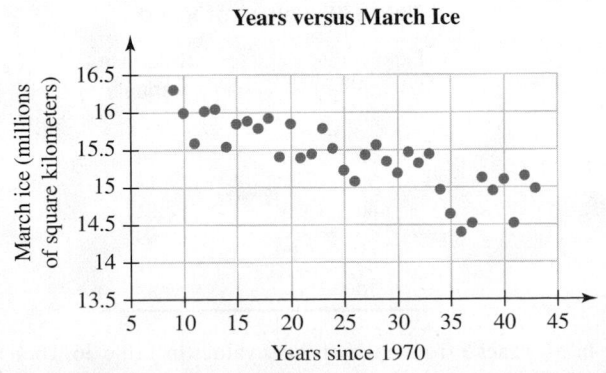

Figure 69 Years since 1970 versus March ice extent

(**Source:** *National Snow and Ice Data Center*)

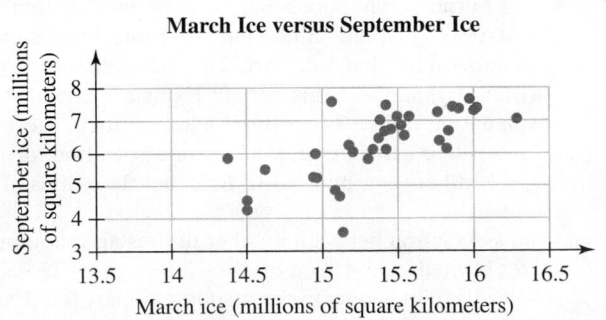

Figure 72 March ice extent versus September ice extent

36. The association between years since 1995 and the numbers of fires in California within CAL FIRE jurisdiction is described by the scatterplot in Fig. 73. The association between years since 1995 and the numbers of thousands of acres burned is described by the scatterplot in Fig. 74.

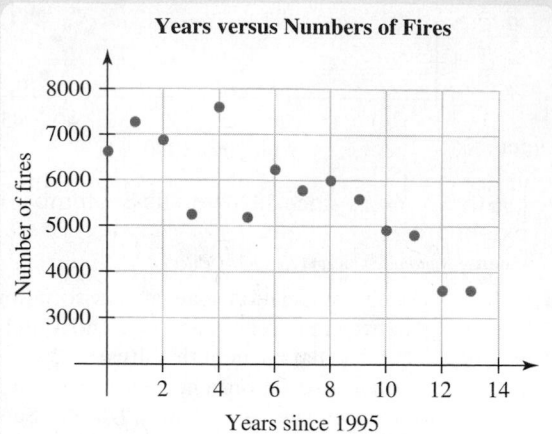

Figure 73 Years since 1995 versus numbers of fires
(**Source:** *CAL FIRE*)

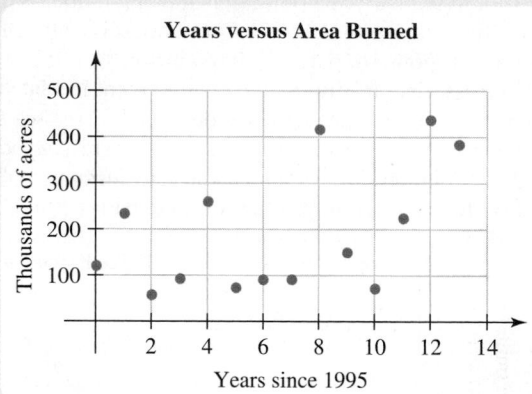

Figure 74 Years since 1995 versus thousands of acres burned
(**Source:** *CAL FIRE*)

a. Is there a positive association, a negative association, or no association between the years since 1995 and numbers of fires? What does that mean in this situation?

b. Is there a positive association, a negative association, or no association between the years since 1995 and thousands of acres burned? What does that mean in this situation?

c. Most people would think that the more fires there are, the more acres that will burn. Do your responses to parts (a) and (b) support this belief? Explain. Include in your explanation a discussion about whether there most likely is a positive association, a negative association, or no association between number of fires and thousands of acres burned.

d. The association between number of fires and thousands of acres burned is described by the scatterplot in Fig. 75. Does the scatterplot support your response to part (c)? Explain.

e. On the basis of the scatterplot in Fig. 75, a student concludes that there is a weak negative association between number of fires and thousands of acres burned. The student

says this means more fires should be set so that the area burned will be less. What would you tell the student?

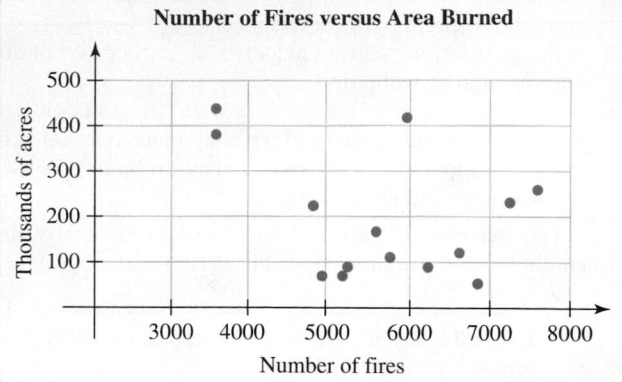

Figure 75 Number of fires versus thousands of acres burned

Concepts

37. a. **DATA** Construct a scatterplot and compute *r* for the association of the variables *x* and *y* described in Table 35.

Table 35 Values of *x* and *y*	
Explanatory Variable *x*	Response Variable *y*
2	17
3	19
4	30
5	31
6	39
7	44

b. Add 10 to each of the *x* values in Table 35. Then construct a scatterplot and compute *r*. How do the scatterplot and *r* compare to your results in part (a)? Why does this make sense?

c. Add 3 to each of the *y* values in Table 35. Then construct a scatterplot and compute *r*. How do the scatterplot and *r* compare to your results in part (a)? Why does this make sense?

38. a. **DATA** Find *r* for the association of the variables *x* and *y* described in Table 36.

Table 36 Values of *x* and *y*	
Explanatory Variable *x*	Response Variable *y*
20	96
30	94
40	91
50	89
60	91
70	88

b. Subtract 5 from each of the *x* values in Table 36. Then construct a scatterplot and compute *r*. How do the scatterplot and *r* compare to your results in part (a)? Why does this make sense?

c. Subtract 2 from each of the y values in Table 36. Then construct a scatterplot and compute r. How do the scatterplot and r compare to your results in parts (a) and (b)? Why does this make sense?

39. The six ordered pairs in Table 37 describe an association that has $r \approx -0.9489971718$. Find six other ordered pairs that describe an association that has the same value of r, where one of the ordered pairs is $(10, 60)$. [**Hint:** See Exercise 37.]

Table 37 Values of x and y

Explanatory Variable	Response Variable
x	y
3	54
4	41
5	43
6	25
7	28
8	15

40. The six ordered pairs in Table 38 describe an association that has $r \approx 0.9257954372$. Find six other ordered pairs that describe an association that has the same value of r, where one of the ordered pairs is $(20, 30)$. [**Hint:** See Exercise 38.]

Table 38 Values of x and y

Explanatory Variable	Response Variable
x	y
23	31
24	46
25	42
26	63
27	57
28	75

41. A student says that the larger the absolute value of r is, the greater the probability that a change in the explanatory variable will cause a change in the response variable. What would you tell the student?

42. A student says that if there is a weak association between two variables, then a change in the explanatory variable might not cause a change in the response variable, but if the absolute value of r is equal to 1, then a change in the explanatory variable will cause a change in the response variable. What would you tell the student?

43. A student says that if $r = 0$ for two variables, then there is no association between the variables. What would you tell the student?

44. State the four characteristics of an association and describe each one.

45. In what order should the four characteristics of an association be determined? Why should they be found in that order?

46. What is a lurking variable?

Hands-On Research

47. In this exercise, you will explore the association of adults' heights and their arm spans.

a. The association for women is different than for men. Which gender will you study? Develop a plan to select 20 adults of that gender. Describe any type(s) of bias that might occur.

b. To measure an individual's arm span, the person should stand flat against a wall with arms stretched out as far as possible. Measure the heights and arm spans of 20 individuals.

c. Construct a scatterplot.

d. Describe the four characteristics of the association. Compute and interpret r as part of your analysis.

e. There are medical conditions such as spondyloepiphyseal dysplasia that hold the spine back from growing. There are other medical conditions such as Marfan's syndrome that cause arms to grow excessively (Source: *Katrina Parker, MD, previously with Dept. of Pediatrics, University of Texas Medical Branch. Now at Morehouse School of Medicine, Atlanta, Georgia. Core Concepts © 2009*). How could children with such conditions be identified?

▼ 6.3 Modeling Linear Associations

Objectives

» Use a line to model an association between two numerical variables.

» Use a linear model to make estimates and predictions.

» Describe the meaning of *input* and *output*.

» Find errors in estimations.

In Section 6.2, we determined whether an association between two numerical variables was linear. In this section, we will use a line to describe a linear association and use the line to predict values of the response variable.

Modeling a Linear Association

▶ **Example 1** Using a Line to Model an Association

The mean ticket prices for the top-50-grossing concert tours are shown in Table 39 for various years. Let p be the mean ticket price (in dollars) at t years since 1995.

1. Construct a scatterplot of the data.

2. Is there a linear association, a nonlinear association, or no association?

» Describe the meaning of *interpolate, extrapolate,* and *model breakdown.*

» Find intercepts of a line.

» Find intercepts of a linear model.

» Modify a model.

Table 39 Mean Ticket Prices for Top-50-Grossing Concert Tours

Year	Mean Ticket Price (dollars)
1998	33
2001	47
2004	59
2008	67
2011	85

Source: *Pollstar*

Table 40 Using Values of *t* to Stand for the Years

Number of Years since 1995 *t*	Mean Ticket Price (dollars) *p*
3	33
6	47
9	59
13	67
16	85

Figure 77 Compute $r \approx 0.99$

3. Compute r. On the basis of r and the scatterplot, determine the strength of the association.

4. Draw a line that comes close to the points of the scatterplot.

Solution

1. First, we list values of t and p in Table 40. For example, $t = 3$ represents 1998, because 1998 is 3 years after 1995; and $t = 6$ represents 2001, because 2001 is 6 years after 1995.

Next, we sketch a scatterplot in Fig. 76. Because t is the explanatory variable and p is the response variable, we let the horizontal axis be the t-axis and let the vertical axis be the p-axis.

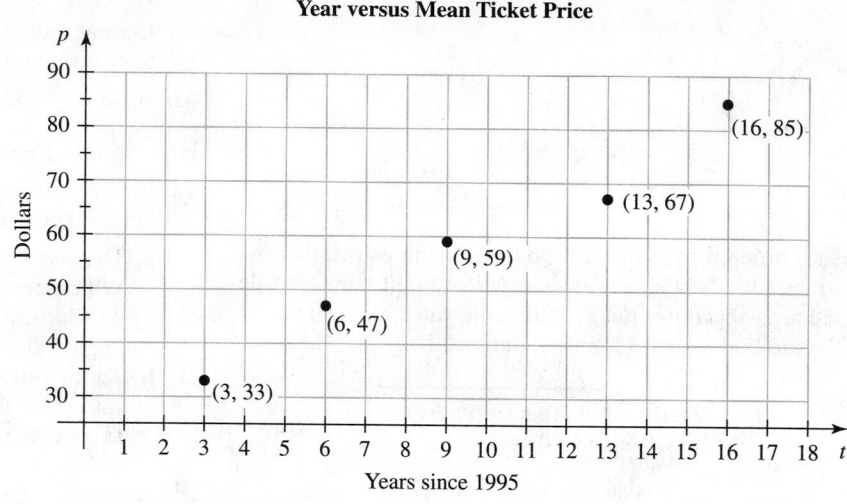

Figure 76 Mean ticket price scatterplot and model

2. The points appear to lie very close to a line, so the association is linear.

3. We use a TI–84 to compute $r \approx 0.99$ (see Fig. 77). Because r is so close to 1 and because the points in the scatterplot appear to lie so close to a line, we conclude that the association is very strong.

4. In Fig. 78, we sketch a line that comes close to the points of the scatterplot.

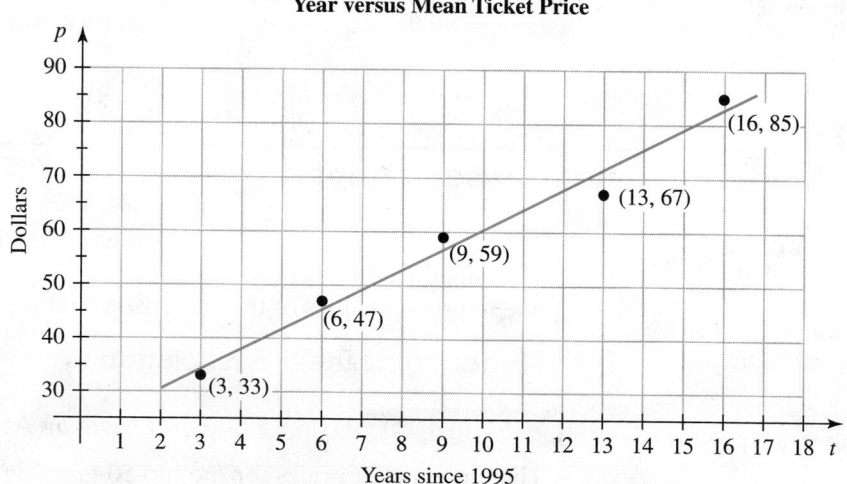

Figure 78 Mean ticket price scatterplot and model

The line needn't contain any of the points, but it should come close to all of them. Figure 79 shows that many such lines are possible.

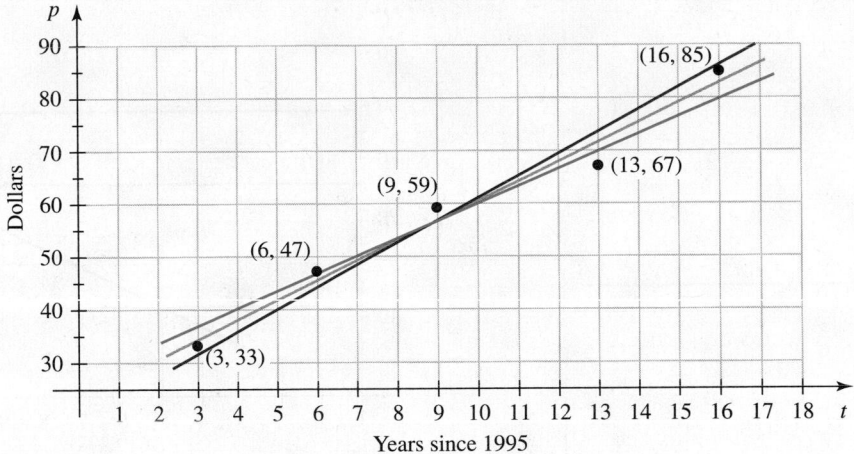

Figure 79 A few of the many reasonable linear models

In Fig. 78, we sketched a line that describes the mean-ticket-price situation. However, this description is not exact. For example, the line does not describe exactly what happened in the years 1998, 2001, 2004, 2008, or 2011, because the line does not contain any of the data points. However, the line does come pretty close to these data points, so it suggests pretty good approximations for those years.

The process of choosing a line to represent the association between time and mean ticket prices is an example of *modeling.*

▶ **Definition**   Model

A **model** is a mathematical description of an authentic situation. We say the description *models* the situation.

We call the line in Fig. 78 a *linear model.* In Chapter 10, we will model certain nonlinear associations with a *nonlinear* model. The term "model" is being used in much the same way as it is used in "airplane model." Just as an airplane designer can use the behavior of an airplane model in a wind tunnel to predict the behavior of an actual airplane, a linear model can be used to predict what might happen in a situation in which two variables are linearly associated.

▶ **Definition**   Linear model

A **linear model** is a nonvertical line that describes the association between two quantities in an authentic situation.

Using a Linear Model to Make Estimates and Predictions

Because all of the ticket-price data points lie close to our linear model in Fig. 78, it seems reasonable that data points for the years between 1998 and 2011 that are not shown in Table 39 might also lie close to the line.

▶ **Example 2**   Using a Linear Model to Make Estimates

1. Use the linear model shown in Fig. 78 to estimate the mean ticket price in 2000.
2. Use the linear model to estimate the mean ticket price in 2010.
3. Use the model to estimate in which year the mean ticket price was $65.

Solution

1. The year 2000 corresponds to $t = 5$, because $2000 - 1995 = 5$. To find the mean ticket price in 2000, we locate the point on the linear model where the t-coordinate is 5 (see the blue arrows in Fig. 80). The p-coordinate of that point is 42. So, according to the model, the mean ticket price in 2000 was approximately $42.

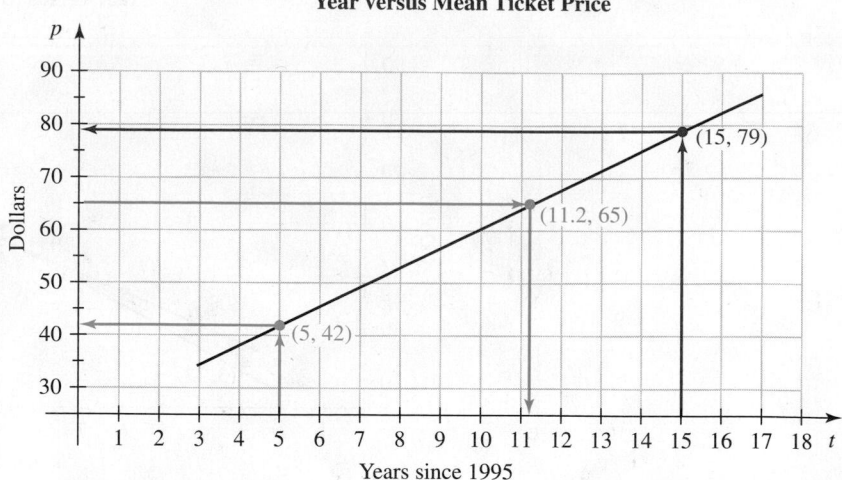

Figure 80 Mean-ticket-price model

We verify our work by checking that our result is consistent with the values shown in Table 41. Since the mean ticket price was $33 in 1998 and $47 in 2001, it follows that the mean ticket price in 2000 probably would be between $33 and $47, which checks with our result of $42.

2. The year 2010 corresponds to $t = 15$, because $2010 - 1995 = 15$. To find the mean ticket price in 2010, we locate the point on the linear model where the t-coordinate is 15 (see the red arrows in Fig. 80). The p-coordinate of that point is 79. So, according to the model, the approximate mean ticket price in 2010 was $79. This result is consistent with the values in Table 41.

3. To find the year when the mean ticket price was $65, we locate the point on the linear model where the p-coordinate is 65 (see the green arrows in Fig. 80). The approximate t-coordinate of that point is 11.2. So, according to the linear model, the mean ticket price in about $1995 + 11 = 2006$ was $65. This result is consistent with the values in Table 41.

Table 41 Mean Ticket Prices

Year	Mean Ticket Price (dollars)
1998	33
2001	47
2004	59
2008	67
2011	85

We construct a scatterplot of data to determine whether there is a linear association. If so, we draw a line that comes close to the data points and use the line to make estimates and predictions.

In Problem 3 of Example 2, we estimated a value of the explanatory variable ($t = 11.2$) that corresponds to a value of the response variable ($p = 65$). In statistics, we don't usually find a value of the explanatory variable for a given value of the response variable, but we certainly do this in algebra. So, to practice algebra, we will continue to find values of the explanatory variable in the rest of the course.

WARNING   It is a common error to try to find a line that contains the greatest number of points. However, our goal is to find a line that comes close to *all* of the data points. For example, even though model 1 in Fig. 81 does not contain any of the data points shown, it fits the complete group of data points much better than does model 2, which contains three data points.

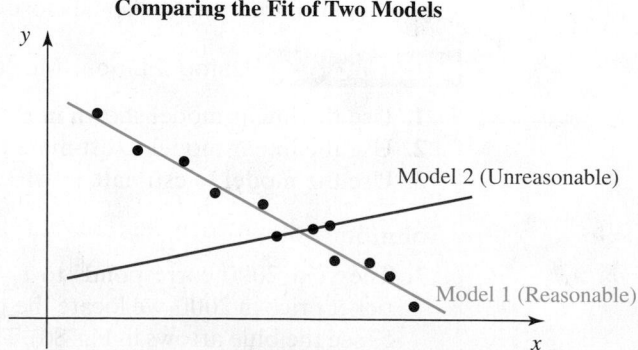

Figure 81 Comparing the fit of two models

Input and Output

In Problem 1 of Example 2, we found that when the value of the explanatory variable t is 5, the corresponding value of the response variable p is 42. We say the *input* 5 leads to the *output* 42. The blue arrows in Fig. 82 show the action of the input $t = 5$ leading to the output $p = 42$.

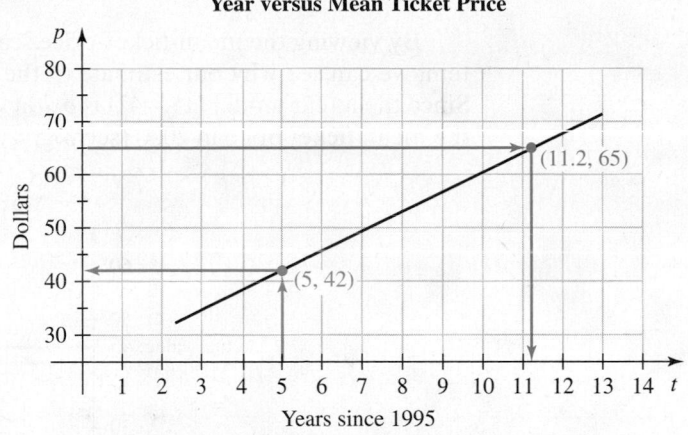

Year versus Mean Ticket Price

Figure 82 Mean-ticket-price model

> ## Definition   Input, output
>
> An **input** is a permitted value of the *explanatory* variable that leads to at least one **output**, which is a permitted value of the *response* variable.

For a value to be permitted, it must make physical sense and be defined. For instance, in Example 2, the value -50 is not a permitted value of the variable p, because it does not make sense for the mean ticket price to be -50 dollars. Later in the course we will discuss values that are not permitted for mathematical reasons.

Sometimes we will go "backward," from an output back to an input. For instance, in Problem 3 of Example 2, we found that the output $p = 65$ originates from the input $t = 11.2$. The green arrows in Fig. 82 show the action of going backward from the output $p = 65$ to the input $t = 11.2$.

Errors in Estimations

WARNING   It is a common error to confuse the meaning of data points and points that lie on a linear model. Data points are accurate descriptions of an authentic situation. Points on a model may or may not be accurate descriptions.

For example, the data in Table 41 are accurate values for the mean ticket prices for the given years. For the linear model in Fig. 78 (page 402), some points on the line describe the situation well, but some points on the line do not. The advantage of using the linear model is that we can estimate the mean ticket price for years other than those in Table 41.

The **error** in an estimate is the amount by which the estimate differs from the actual value. For an overestimate, the error is positive. For an underestimate, the error is negative. If the estimate is equal to the actual value, then the error is 0.

> ▶ **Example 3**   Calculating Errors
>
> 1. In Problem 1 of Example 2, we estimated that the mean ticket price was $42 in 2000. The actual mean ticket price was $45. Calculate the error in the estimate.
> 2. In Problem 2 of Example 2, we estimated that the mean ticket price was $79 in 2010. The actual mean ticket price was $75. Calculate the error in the estimate.

Solution

1. Since $45 - 42 = 3$ and we underestimated the actual mean price, the error is -3 dollars.
2. Since $79 - 75 = 4$ and we overestimated the actual mean price, the error is 4 dollars.

By viewing the mean-ticket-price scatterplot and model in the same coordinate system, we can see why our estimate of the mean ticket price in 2000 is an underestimate. Since the linear model at $(5, 42)$ is *below* the data point $(5, 45)$, the model *underestimates* the mean ticket price in 2000 (see Fig. 83).

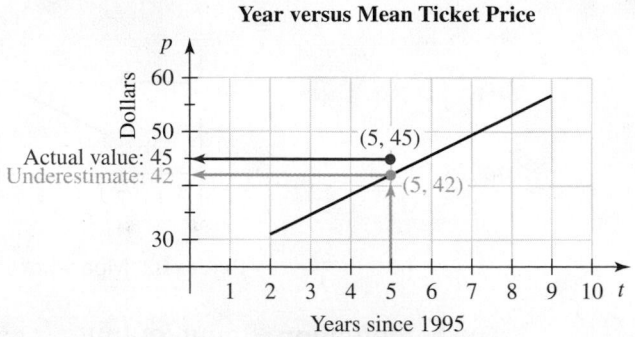

Figure 83 Comparing the data point and the model for 2000

We can also see why our estimate of the mean ticket price in 2010 is an overestimate. Since the linear model at $(15, 79)$ is *above* the data point $(15, 75)$, the model *overestimates* the mean ticket price in 2010 (see Fig. 84).

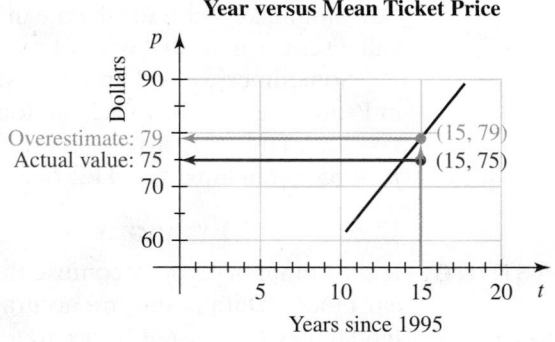

Figure 84 Comparing the data point and the model for 2010

Interpolation, Extrapolation, and Model Breakdown

To draw the ticket-price model in Fig. 78 (page 402), we used a scatterplot consisting of data points representing various years from 1998 to 2011. In Fig. 85, we draw that portion of the model in blue, and we draw the rest of the model in red. When we use the blue portion of the model to make estimates, we are performing *interpolation*. When we use the red portions of the model, we are performing *extrapolation*.

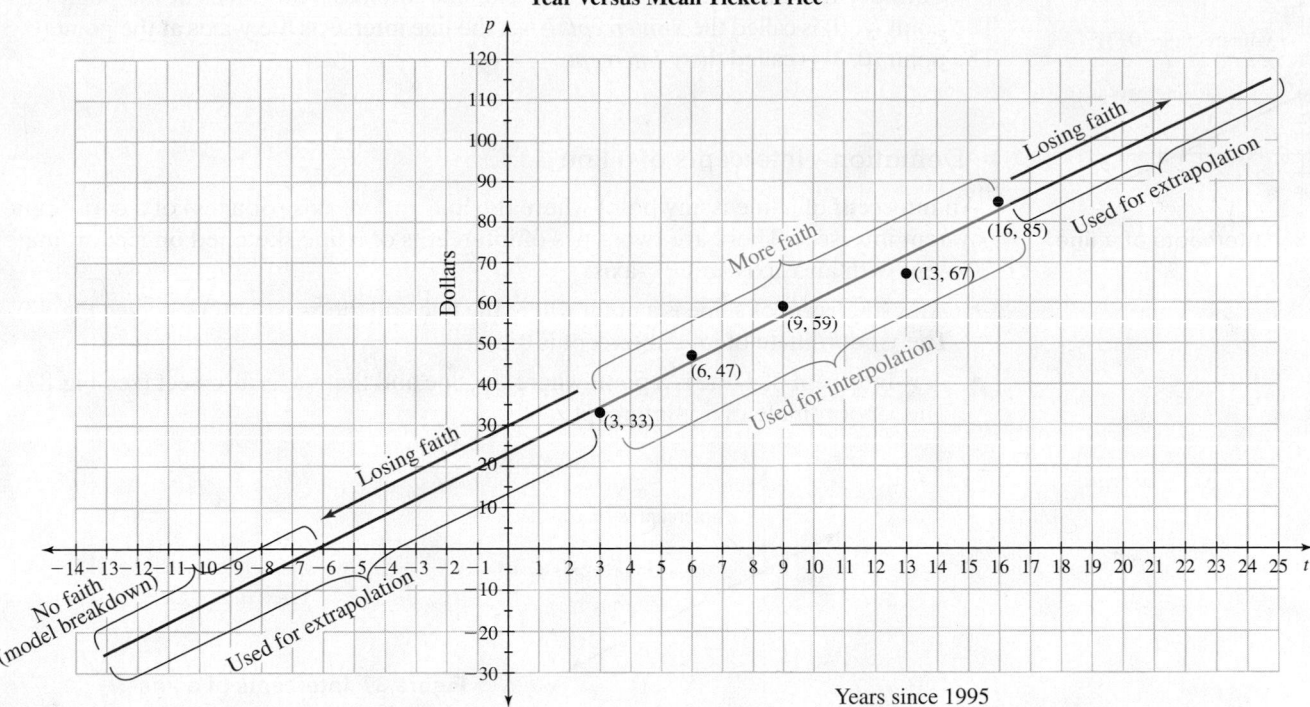

Figure 85 Interpolation versus Extrapolation

> ▶ **Definition**   Interpolation, extrapolation
>
> For a situation that can be described by a model whose explanatory variable is x,
>
> - We perform **interpolation** when we use a part of the model whose x-coordinates are between the x-coordinates of two data points.
> - We perform **extrapolation** when we use a part of the model whose x-coordinates are not between the x-coordinates of any two data points.

Although we could get large errors from interpolating with the ticket-price model, we have more faith in our results from interpolating than from extrapolating. That's because the blue portion of the model comes close to several known data points, whereas we have no idea whether the red portion of the model comes close to *any* data points.

When we extrapolate, our faith declines more and more as we stray farther and farther from the blue portion of the ticket-price model. In fact, we have no faith in the portion of the model in 1989 and before then because the model estimates nonpositive ticket prices for these years. We say *model breakdown* has occurred in 1989 and before then.

> ▶ **Definition**   Model breakdown
>
> When a model gives a prediction that does not make sense or an estimate that is not a good approximation, we say **model breakdown** has occurred.

WARNING   In sum, when we perform extrapolation, we have little or no faith in our result. In other words, we run the risk of model breakdown occurring.

Intercepts of a Line

Recall from Section 1.1 that when we plot points that are not being used to describe authentic situations, we let the horizontal axis be the x-axis and the vertical axis be the y-axis. Therefore, x is the explanatory variable and y is the response variable.

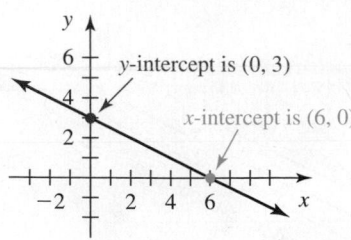

Figure 86 Intercepts of a line

Consider the line sketched in Fig. 86. The line intersects the x-axis at the point $(6, 0)$. The point $(6, 0)$ is called the *x-intercept*. Also, the line intersects the y-axis at the point $(0, 3)$. The point $(0, 3)$ is called the *y-intercept*.

▶ **Definition   Intercepts of a line**

An **intercept** of a line is any point where the line and an axis (or axes) of a coordinate system intersect. There are two types of intercepts of a line sketched on a coordinate system with an x-axis and a y-axis:

- An **x-intercept** of a line is a point where the line and the x-axis intersect (see Fig. 87). The y-coordinate of an x-intercept is 0.
- A **y-intercept** of a line is a point where the line and the y-axis intersect (see Fig. 87). The x-coordinate of a y-intercept is 0.

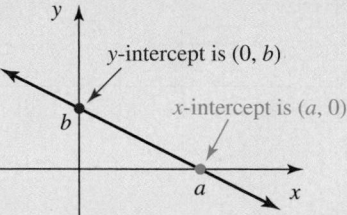

Figure 87 Intercepts of a line

▶ **Example 4**   Finding Intercepts and Coordinates

Refer to Fig. 88 for the following problems.

1. Find the x-intercept of the line.
2. Find the y-intercept of the line.
3. Find y when $x = 4$.
4. Find x when $y = -2$.

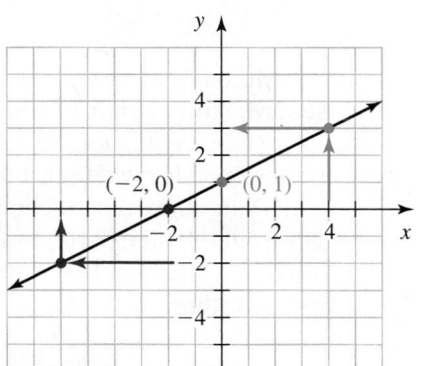

Figure 88 Problems 1–4 of Example 4

Solution

1. The line and the x-axis intersect at $(-2, 0)$. So, the x-intercept is $(-2, 0)$.
2. The line and the y-axis intersect at $(0, 1)$. So, the y-intercept is $(0, 1)$.
3. The blue arrows in Fig. 88 show that the input $x = 4$ leads to the output $y = 3$. So, $y = 3$ when $x = 4$.
4. The red arrows in Fig. 88 show that the output $y = -2$ originates from the input $x = -6$. So, $x = -6$ when $y = -2$.

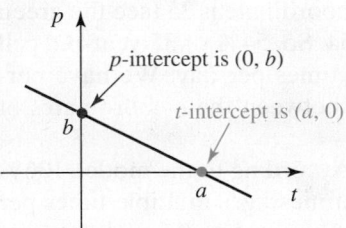

Figure 89 Intercepts of a linear model

Intercepts of a Linear Model

Suppose a linear model describes the association between two variables t and p, where t is the explanatory variable. Then the t-intercept is a point where the line and the t-axis intersect, and the p-intercept is a point where the line and the p-axis intersect (see Fig. 89).

▶ **Example 5**   Intercepts of a Model

The percentages of cell phone users who send or receive text messages multiple times per day are shown in Table 42 for various age groups.

Table 42 Percentages of Cell Phone Users Who Send or Receive Text Messages Multiple Times per Day

Age Group (years)	Age Used to Represent Age Group (years)	Percent
18–24	21.0	76
25–34	29.5	63
35–44	39.5	42
45–54	49.5	37
55–64	59.5	17

Source: *Edison Research and Arbitron*

Let p be the percentage of cell phone users at age a years who send or receive text messages multiple times per day.

1. Draw a model that describes the association between a and p.
2. Predict the percentage of 35-year-old cell phone users who send or receive text messages multiple times per day. Did you perform interpolation or extrapolation?
3. Find the p-intercept. What does it mean in this situation? Did you perform interpolation or extrapolation?
4. Find the a-intercept. What does it mean in this situation? Did you perform interpolation or extrapolation?

Solution

1. We begin by viewing the positions of the data points in the scatterplot (see Fig. 90). It appears a and p are linearly associated, so we sketch a line that comes close to the data points.

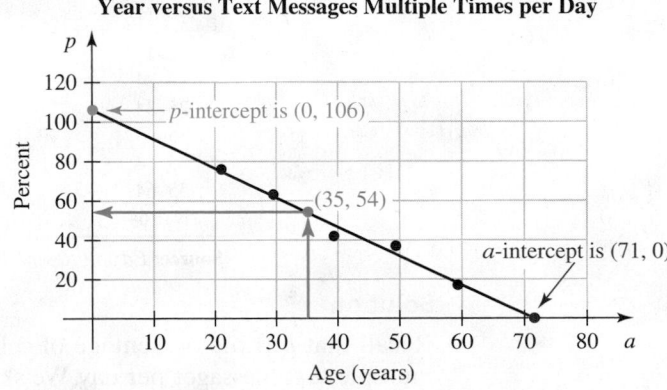

Year versus Text Messages Multiple Times per Day

Figure 90 Intercepts of the text message model

2. We locate the point on the linear model where the a-coordinate is 35 (see the green arrows in Fig. 90). The p-coordinate of that point is 54. So, 54% of 35-year-old cell phone users send or receive text messages multiple times per day. We have performed interpolation because the a-coordinate 35 is between the a-coordinates of the data points (29.5, 63) and (39.5, 42).

3. The p-intercept is (0, 106), or $p = 106$, when $a = 0$. According to the model, 106% of newborns who use cell phones send or receive text messages multiple times per day. We have performed extrapolation because the a-coordinate 0 is not between the a-coordinates of two data points. Model breakdown has occurred for two reasons: Percentages cannot be larger than 100% in this situation, and newborns cannot send or receive text messages.

4. The a-intercept is (71, 0), or $p = 0$, when $a = 71$. According to the model, no 71-year-old cell phone users send or receive text messages multiple times per day. We have performed extrapolation because the a-coordinate 71 is not between the a-coordinates of two data points. Model breakdown has occurred because some 71-year-old cell phone users send or receive text messages multiple times per day.

Modifying a Model

When model breakdown occurs, it is time to modify our model or possibly rethink our modeling process. A different model might give more reasonable predictions. It could be helpful to gather more data to check our choice of model.

▶ **Example 6**   Modifying a Model

Additional research yields the data shown in the first and last rows of Table 43. Use this data and the following assumptions to modify the model we found in Example 5:

- Children 3 years old and younger do not send or receive text messages multiple times per day.

- The percentage of cell phone users who send or receive text messages levels off at 5% for users over 80 years in age.

- The age of the oldest cell phone users is 116 years.

Table 43 Percentages of Cell Phone Users Who Send or Receive Text Messages Multiple Times per Day

Age Group (years)	Age Used to Represent Age Group (years)	Percent
12–17	14.5	75
18–24	21.0	76
25–34	29.5	63
35–44	39.5	42
45–54	49.5	37
55–64	59.5	17
over 64	70.0	7

Source: *Edison Research and Arbitron*

Solution

Recall that p is the percentage of cell phone users at age a years who send or receive multiple text messages per day. We sketch a scatterplot of the data in Table 43, and taking into account the three assumptions, we draw a model that comes close to the data points (see Fig. 91).

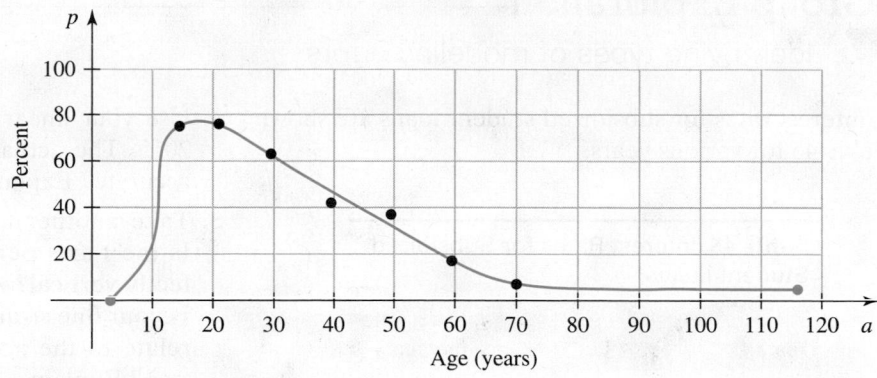

Year versus Text Messages Multiple Times per Day

Figure 91 Modified text message model

Your results for homework exercises will likely be different from the answers provided near the end of this textbook for two possible reasons. First, there will be many reasonable linear models to choose from, unless the association is exactly linear. Second, it is impossible to do a perfect job of sketching models and estimating coordinates of points. However, if you do a careful job, your results should be close to those in the textbook.

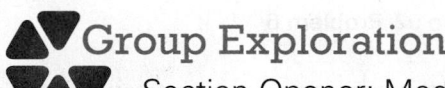

Group Exploration

Section Opener: Modeling a linear association

Total amounts of overdraft fees collected by banks are shown in Table 44 for various years.

Table 44 Total Amount of Overdraft Fees Collected by Banks

Year	Amount of Overdraft Fees (billions of dollars)
2001	22.2
2003	27.1
2005	29.7
2007	34.1
2009	37.1

Source: *Moebs Services*

Let A be the total amount (in billions of dollars) of overdraft fees collected in the year that is t years since 2000.

1. Construct a scatterplot by hand.
2. Draw a (straight) line that comes close to all of the data points in your scatterplot.
3. Use your line to estimate the total amount of fees collected in 2008. [**Hint:** On the line, locate the point whose t-coordinate is 8.]
4. Use your line to estimate when the total amount of fees collected was $32 billion.
5. Find the point where the line intersects the A-axis. What does it mean in this situation?
6. On July 1, 2010, new overdraft rules went into effect for debit and ATM card users. Now a bank must ask permission to apply its standard overdraft practices, including charging overdraft fees. For customers who do not give permission, their transactions will be declined if there is not enough money in their accounts. Use your line to estimate the total amount of fees collected in 2011. Compare your result with $31.6 billion, which is the actual amount. Explain why it is not surprising that your result is an overestimate.

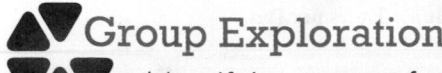

Group Exploration

Identifying types of modeling errors

The interest rates for subsidized student loans are shown in Table 45 for various years.

Table 45 Interest Rates for Subsidized Student Loans

Years	Interest Rate (percent)
2008	6.8
2009	6.0
2010	5.7
2011	4.5
2012	3.4

Source: *New America Foundation*

Here you will explore possible causes of error for estimates and predictions based on a linear model for the interest rate data.

1. Let r be the interest rate (percent) for subsidized student loans at t years since 2005. Construct a scatterplot by hand.

2. Draw a linear model on your scatterplot.

3. Use your linear model to estimate the interest rate in 2012. What is the actual interest rate? Calculate the error in your estimate for 2012.

4. Use your linear model to predict the interest rate in 2015. The actual interest rate is 4.7%. Is your result accurate? Explain.

5. Take another look at your sketch from Problem 2. Is the t-axis perfectly horizontal and the r-axis perfectly vertical? Are the scalings of both axes precise? Is your line straight? How might these considerations relate to the accuracy of an estimation or a prediction? Explain.

6. What are the coordinates of point S plotted in Fig. 92? Do you think you have found the correct first decimal place (tenths place) for these coordinates? How about the second decimal place?

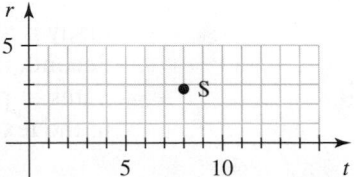

Figure 92 Problem 6

7. Problems 3–6 of this exploration suggest several possible causes of error for estimates and predictions based on a linear model. Describe the possible causes of error.

▶ **Tips for Success**   **The Value of Learning Mathematics**

Imagine someone who is a math expert. Are you imagining a person wearing broken glasses that are taped together, a pocket protector, and wrinkled, unfashionable clothing—in other words, a math nerd? Of course, this stereotype does not accurately describe most mathematicians. But perhaps it is because of the stereotype of a math nerd that many students don't want to become "too good" at mathematics.

Learning mathematics will not transform you into a nerd. Rather, it will transform you into a more educated, well-rounded person. Learning mathematics will equip you to be more effective in many lines of work, as well as when you do math-intensive activities such as investing or filling out a tax return. Most people have a high respect for mathematicians due to their commitment to a useful and challenging subject.

Homework 6.3

1. A(n) _____ model is a nonvertical line that describes the association between two quantities in an authentic situation.

2. An input is a permitted value of the _____ variable that leads to at least one output.

3. For a situation that can be described by a model whose explanatory variable is x, we perform _____ when we use a part of the model whose x-coordinates are not between the x-coordinates of any two data points.

4. When a model gives a prediction that does not make sense or an estimate that is not a good approximation, we say model _____ has occurred.

For Exercises 5–10, refer to Fig. 93.

5. Find y when $x = 4$.

6. Find y when $x = 2$.

7. Find x when $y = 1$.

8. Find x when $y = 5$.

9. What is the x-intercept of the line?

10. What is the y-intercept of the line?

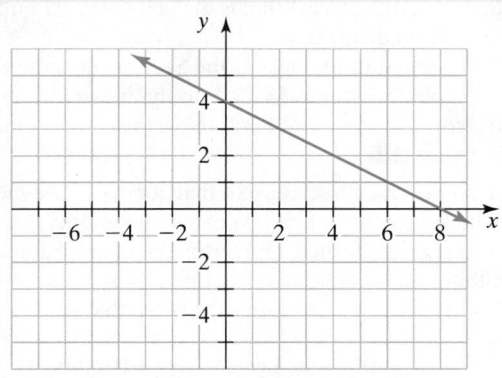

Figure 93 Exercises 5–10

For Exercises 11–16, refer to Fig. 94.

11. Find y when $x = -3$.

12. Find y when $x = -6$.

13. Find x when $y = 1$.

14. Find x when $y = 0$.

15. What is the y-intercept of the line?

16. What is the x-intercept of the line?

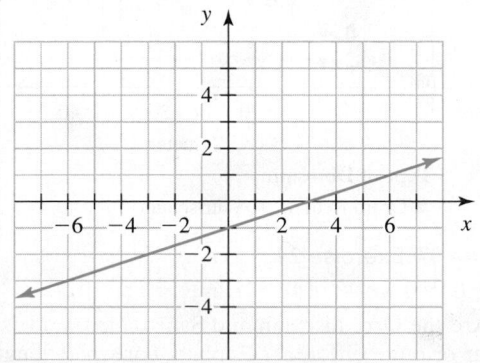

Figure 94 Exercises 11–16

17. Some ordered pairs are listed in Table 46.

Table 46 Some Ordered Pairs

x	y
1	13
3	9
5	8
7	4
9	2

a. Construct a scatterplot by hand.

b. Is there a linear association, a nonlinear association, or no association?

c. Draw a line that comes close to the points in your scatterplot.

d. Which point on your line has x-coordinate 8?

e. Which point on your line has y-coordinate 6?

18. Some ordered pairs are listed in Table 47.

Table 47 Some Ordered Pairs

x	y
2	4
3	6
5	6
7	11
10	13

a. Construct a scatterplot by hand.

b. Is there a linear association, a nonlinear association, or no association?

c. Draw a line that comes close to the points in your scatterplot.

d. Which point on your line has x-coordinate 6?

e. Which point on your line has y-coordinate 12?

19. Some ordered pairs are listed in Table 48.

Table 48 Some Ordered Pairs

x	y
−8	−5
−5	−3
−2	4
1	5
3	9

a. Construct a scatterplot by hand.

b. Is there a linear association, a nonlinear association, or no association?

c. Draw a line that comes close to the points in your scatterplot.

d. What is the x-intercept of your line?

e. What is the y-intercept of your line?

20. Some ordered pairs are listed in Table 49.

Table 49 Some Ordered Pairs

x	y
−10	4
−7	1
−3	−3
−1	−2
2	−5

a. Construct a scatterplot by hand.
b. Is there a linear association, a nonlinear association, or no association?
c. Draw a line that comes close to the points in your scatterplot.
d. What is the x-intercept of your line?
e. What is the y-intercept of your line?

21. Let *n* be the number (in thousands) of ride-related injuries at fixed-site amusement parks in the year that is *t* years since 2000. A scatterplot of some data and a linear model are sketched in Fig. 95.

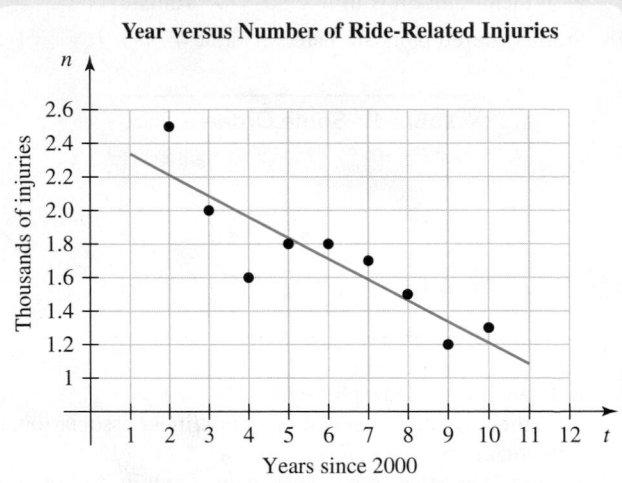

Figure 95 Exercises 21 and 22
(**Source:** *National Safety Council*)

a. Use the linear model to estimate the number of ride-related injuries in 2004.
b. What was the actual number of ride-related injuries in 2004?
c. Is your result in part (a) an underestimate or an overestimate? Explain how you can tell this from the graph of the scatterplot and the sketch of the model. Calculate the error in the estimate.

22. Refer to Exercise 21, including Fig. 95.

a. Use the linear model to estimate the number of ride-related injuries in 2002.
b. What was the actual number of ride-related injuries in 2002?
c. Is your result in part (a) an underestimate or an overestimate? Explain how you can tell this from the graph of the scatterplot and the sketch of the model. Calculate the error in the estimate.

23. The scatterplot and linear model in Fig. 96 describe the prices of hot dogs and soft drinks at all Major League Baseball (MLB) stadiums.

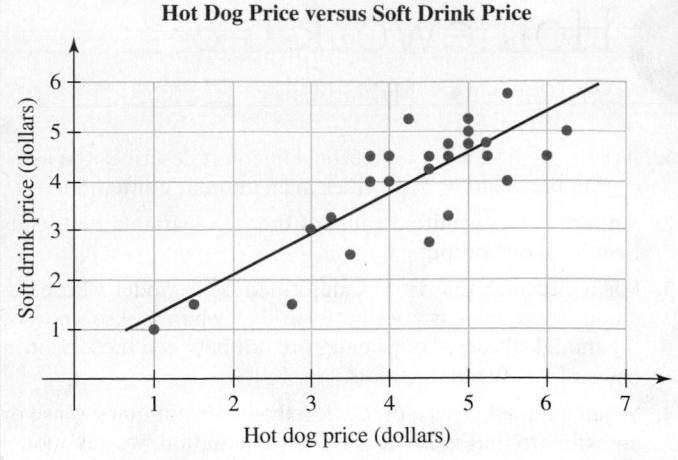

Figure 96 Exercise 23
(**Source:** *Team Marketing Report*)

a. Use the linear model to predict the soft drink price at an MLB stadium that charges $4 for a hot dog.
b. Use the linear model to predict the hot dog price at an MLB stadium that charges $5 for a soft drink.
c. At Miller Park, where the Milwaukee Brewers play, the price of a hot dog is $3.50. Use the linear model to predict the soft drink price. What is the actual price? Find the error in your prediction.
d. At Busch Stadium, where the St. Louis Cardinals play, the price of a hot dog is $4.25. Use the linear model to predict the soft drink price. What is the actual price? Find the error in your prediction.

24. Figure 97 displays a scatterplot that compares the mean lengths and mean heights of 51 types of dinosaurs. There are two main categories (called *orders*) of dinosaurs: Saurischians and Ornithischians. Saurischians had a pelvis also found in modern lizards, whereas Ornithischians had a pelvis that was more bird-like.

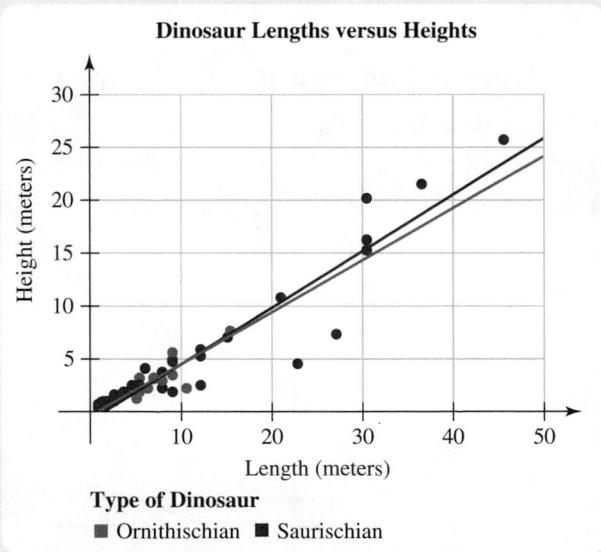

Figure 97 Exercise 24
(**Source:** *DinoDatabase.com*)

a. Are the Ornithischian and Saurischian models quite similar or quite different? Explain. For which lengths of dinosaurs is it reasonable to compare the models?
b. Suppose a paleontologist discovers the fossils of a new type of Saurischian, which she estimates has length 30 meters.

Use the Saurischian model to estimate the height. Do you have much faith in your estimate? Explain.

c. Suppose a paleontologist discovers the fossils of a new type of Ornithischian, which he estimates has length 30 meters. Use the Ornithischian model to estimate the height. Do you have much faith in your estimate? Explain.

d. Use the Saurischian model to estimate the heights of Saurischians that are 10 meters, 20 meters, and 30 meters in length.

e. Use the results you found in part (d) to find three ratios of height to length for Saurischians. Round each ratio to the first decimal place and compare them. How could a paleontologist use your result to quickly estimate the height of a Saurischian for a given length?

25. A scatterplot and linear model comparing the highway mileages and city mileages of 1159 conventional cars (neither hybrid nor electric) are displayed in Fig. 98.

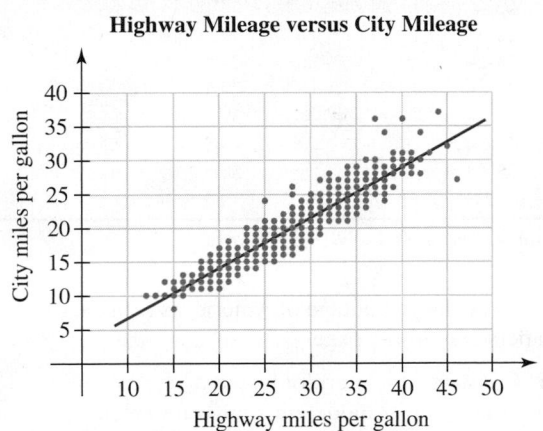

Figure 98 Exercise 25
(**Source:** *U.S. Environmental Protection Agency*)

a. Use the linear model to predict the city mileage for a car with highway mileage 20 miles per gallon.

b. Use the linear model to predict the highway mileage for a car with city mileage 20 miles per gallon.

c. Use the linear model to predict the city mileage for a car with highway mileage 25 miles per gallon. Find the error in this prediction for the car(s) with highway mileage 25 miles per gallon and smallest city mileage.

d. Use the linear model to predict the city mileage for a car with highway mileage 40 miles per gallon. Find the error in this prediction for the car(s) with highway mileage 40 miles per gallon and largest city mileage.

26. The scatterplot and linear model in Fig. 99 describe the ages and years of experience of NFL football players.

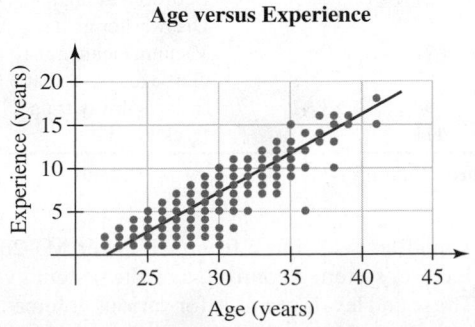

Figure 99 Exercise 26
(**Source:** *NFL*)

a. Use the linear model to predict the years of experience of a 30-year-old player.

b. Use the linear model to predict the age of a player with 5 years of experience.

c. Use the linear model to predict the years of experience for a 35-year-old player. Find the error in this prediction for 35-year-old players with the least experience.

d. Use the linear model to predict the years of experience for a 25-year-old player. Find the error in this prediction for 25-year-old players with the most experience.

27. The total numbers of animal and plant species in the United States that are listed as endangered or threatened are shown in Table 50 for various years.

Table 50 Numbers of Endangered or Threatened Species

Year	Number of Species Listed
1985	384
1990	596
1995	962
2000	1244
2005	1264
2011	1967
2013	1519

Source: *U.S. Fish and Wildlife Service*

Let n be the number of species that are listed as endangered or threatened at t years since 1980.

a. Construct a scatterplot by hand.

b. Is there a linear association, a nonlinear association, or no association?

c. Draw a linear model on your scatterplot.

d. Estimate when 800 species were listed.

e. Estimate the number of species that were listed in 2008.

28. The percentages of Americans who say there should be a ban on the possession of handguns are shown in Table 51 for various years.

Table 51 Percentages of Americans Who Say There Should Be a Ban on the Possession of Handguns

Year	Percent
1991	43
1993	39
2000	36
2005	35
2010	29
2014	26

Source: *The Gallup Organization*

Let p be the percentage of Americans who say there should be a ban on the possession of handguns at t years since 1990.

a. Construct a scatterplot by hand.

b. Is there a linear association, a nonlinear association, or no association?

c. Draw a linear model on your scatterplot.

d. Use your model to estimate the percentage of Americans in 2012 who said there should be a ban on the possession of handguns.

e. Use your model to estimate when 31% of Americans said there should be a ban on the possession of handguns.

29. Due to improved technology and public-service campaigns, the number of collisions at highway–railroad crossings per year has declined since 1992 (see Table 52).

Table 52 Numbers of Collisions at Highway–Railroad Crossings

Year	Number of Collisions (thousands)
1992	4.9
1995	4.6
2000	3.5
2005	3.1
2010	2.1
2014	2.3

Source: *Federal Railroad Administration*

Let n be the number of collisions (in thousands) for the year that is t years since 1990.

a. Construct a scatterplot by hand.
b. Draw a linear model on your scatterplot.
c. Use your linear model to estimate the number of collisions in 2012. Did you perform interpolation or extrapolation?
d. Use your linear model to predict in which year there will be 1.0 thousand collisions. Did you perform interpolation or extrapolation? Do you have much faith in your result? Explain.
e. Find the t-intercept. What does it mean in this situation? Do you have much faith in your result? Explain.

30. Repeat Exercise 29, but let n be the number of collisions (in thousands) for the year that is t years *since 1985*. Which of your responses for this exercise are the same as those for Exercise 29? Explain why it makes sense that these responses are the same. Explain why it makes sense that the other responses are different.

31. The percentages of Americans who believe marriages between same-sex couples should be recognized by the law as valid are shown in Table 53 for various age groups.

Table 53 Percentages of Americans Who Support Same-Sex Marriage

Age Group (years)	Age Used to Represent Age Group (years)	Percent
18–29	23.5	71
30–39	34.5	56
40–49	44.5	49
50–59	54.5	46
60–69	64.5	42
70–79	74.5	32
over 79	85.0	21

Source: *The Gallup Organization*

Let p be the percentage of Americans at age a years who believe marriages between same-sex couples should be recognized by the law as valid.

a. Create a scatterplot by hand.
b. Draw a linear model on your scatterplot.

c. Use your linear model to predict at what age 65% of Americans believe same-sex couples should be recognized by the law as valid.
d. Find the p-intercept of your linear model. What does it mean in this situation?
e. Find the a-intercept of your linear model. What does it mean in this situation?

32. The percentages of Americans who currently have a personal profile page on a social networking website such as Facebook are shown in Table 54 for various age groups.

Table 54 Percentages of Americans Who Currently Have a Personal Profile Page on a Social Networking Website

Age Group (years)	Age Used to Represent Age Group (years)	Percent
12–17	14.5	78
18–24	21.0	77
25–34	29.5	65
35–44	39.5	51
45–54	49.5	35
55–64	59.5	31
over 64	70.0	13

Source: *Edison Research and Arbitron*

Let p be the percentage of Americans at age a years who have a personal profile page.

a. Construct a scatterplot by hand.
b. Draw a linear model on your scatterplot.
c. Use your linear model to predict at what age 70% of Americans have a profile page.
d. Find the p-intercept of your linear model. What does it mean in this situation? Do you have much faith in your result? Explain.
e. Find the a-intercept of your linear model. What does it mean in this situation? Do you have much faith in your result? Explain.

33. The loudness of sound can be measured by using a *decibel scale*. Some examples of sounds at various sound levels are listed in Table 55.

Table 55 Examples of Sound Levels

Sound Level (decibels)	Example
0	Faintest sound heard by humans
27	Bedroom at night
50	Dishwasher next room
71	Vacuum cleaner at 10 ft
82	Garbage disposal at 3 ft
98	Inside subway train
111	Rock band

Source: Caltrans Noise Manual, M. M. Hatano

The sound level of music from a Pioneer MT-2000® stereo-CD-receiver system is controlled by the system's volume number. The sound levels of music for various volume numbers are shown in Table 56.

Table 56 Sound Levels of Music Played by a Stereo

Volume Number	Sound Level (decibels)
6	60
8	66
10	69
12	74
14	78
16	82
18	86
20	90

Source: *J. Lehmann*

Let *S* be the sound level (in decibels) for a volume number *n*.

a. Construct a scatterplot by hand.
b. Is there a linear association, a nonlinear association, or neither?
c. Draw a linear model on your scatterplot.
d. Use your model to predict the sound level when the volume number is 19.
e. Use your model to predict for what volume number the sound level is comparable to that of a vacuum cleaner at 10 ft (see Table 55).

34. The temperature at which water boils (the *boiling point*) depends on elevation: The higher the elevation, the lower the boiling point will be. At sea level, water boils at 212°F; at an elevation of 10,000 meters, water boils at about 151°F. Boiling points are listed in Table 57 for various elevations.

Table 57 Boiling Points of Water

Elevation (in thousands of meters)	Boiling Point (°F)
0	212
1	205
2	200
5	181
10	151
15	123

a. Let *B* be the boiling point (in degrees Fahrenheit) at an elevation of *E* thousand meters. Construct a scatterplot by hand.
b. Draw a linear model on your scatterplot.
c. Mount Everest, the highest mountain in the world, reaches 8850 meters at its peak. Predict the boiling point of water at the peak. Did you perform interpolation or extrapolation? Explain.
d. The *Armstrong limit* is the altitude at which water boils at body temperature (about 98.6°F). So, at that altitude, the saliva in a person's mouth would boil. Predict the Armstrong limit. Did you perform interpolation or extrapolation? Explain.
e. Extrapolation is bad practice, unless there is some underlying principle that tells us a model describes the situation well for values outside the scope of the data. The Armstrong limit is reportedly between 18.9 and 19.4 thousand meters (Source: *NASA*). Is the result you found in part (d) between these values?

35. The discounts State Farm® offers to customers who have insured a vehicle with the company are shown in Table 58 for various years of loyalty.

Table 58 Years of Loyalty to State Farm versus Loyalty Discounts

Years	Discount (percent)
0	0
1	3
3	9
4	12
6	18

Source: *State Farm Insurance*

Let *p* be the discount (in percent) for customers who have insured a vehicle with the company for *t* years.

a. Construct a scatterplot by hand and draw a model on the scatter plot.
b. Use your model to predict the discount for 10 years of loyalty. Did you perform interpolation or extrapolation?
c. It turns out that for years of loyalty beyond 6 years, the discount is still 18%. Construct a scatterplot of the data for years of loyalty from 0 to 10 years.
d. Compute the error in the prediction you made in part (b). Explain why the error is so large.
e. Use the scatterplot that you constructed in part (c) to modify the model.

36. The discounts State Farm offers to customers who have insured a vehicle with the company are shown in Table 58 for various years of loyalty.

Let *p* be the discount (in percent) for customers who have insured a vehicle with the company for *t* years.

a. Construct a scatterplot by hand, and draw a linear model on the scatterplot.
b. Use the model to predict the number of years of loyalty it takes to reach a 6% discount.
c. A customer's car has been insured by State Farm for 5 years. Without the loyalty discount, the customer would have to pay $120 per month. Use the model to predict how much money the customer actually has to pay per month.
d. It turns out that a linear model does in fact predict the correct discount for every year of loyalty up to 6 years. Up until now, State Farm used to offer no discount for up to 2 years of loyalty, a 10% discount for 3 to 5 years of loyalty, and an 18% discount for at least 6 years of loyalty. For which years of loyalty is the new policy the better deal?
e. A customer's car has been insured by State Farm for 6 years. Without any loyalty discounts, the customer would have had to pay $100 per month for the past 6 years. Find the total amount of money the customer would have saved if the new policy had been in effect rather than the old policy.

37. DATA The percentages of Americans who are satisfied with the way things are in the United States are shown in Table 59 for various years.

Table 59 Percentages of Americans Who Are Satisfied

Year	Percent
1992	21
1993	28
1994	33
1995	32
1996	39
1997	49
1998	60
1999	59

Source: *The Gallup Organization*

Let p be the percentage of Americans at t years since 1990 who are satisfied with the way things are.

a. Construct a scatterplot by hand and draw a linear model on the scatterplot.
b. Use your model to estimate the percentage of Americans who were satisfied in 2006. Did you perform interpolation or extrapolation?
c. Data for the years 2000–2011 are shown in Table 60. Construct a scatterplot of the data for the years 1992–2011.

Table 60 Percentages of Americans Who Are Satisfied

Year	Percent
2000	60
2002	52
2004	43
2006	31
2008	20
2011	11

Source: *The Gallup Organization*

d. Compute the error in the estimation for 2006 that you made in part (b). Explain why the error in your estimate is so large.
e. Use the scatterplot that you constructed in part (c) to modify the model.

38. The percentages of Americans living below the poverty level are shown in Table 61 for various years.

Table 61 Percentages of Americans Living Below the Poverty Level

Year	Percent
1993	15.1
1994	14.5
1995	13.8
1996	13.7
1997	13.3
1998	12.7
1999	11.9
2000	11.3

Source: *U.S. Census Bureau*

Let p be the percentage of Americans living below the poverty level at t years since 1990.

a. Construct a scatterplot by hand.
b. Draw a linear model on your scatterplot.
c. Use your model to estimate the percentage in 2012. Did you perform interpolation or extrapolation?
d. Data for the years 2002–2012 are shown in Table 62. Construct a scatterplot of the data by hand for the years 1993–2012.

Table 62 Percentages of Americans Living Below the Poverty Level

Year	Percent
2002	12.1
2004	12.7
2006	13.3
2008	13.2
2010	15.3
2012	15.0

Source: *U.S. Census Bureau*

e. Compute the error in the estimation for 2012 that you made in part (c). Explain why the estimate is so inaccurate.
f. Use the scatterplot that you constructed in part (d) to modify the model.

39. The annual profits of Alaska Air Group are shown in Table 63 for various years.

Table 63 Annual Profits of Alaska Air Group

Year	Annual Profit (millions of dollars)
2002	−68
2003	−31
2005	55
2007	92
2009	89
2011	245

Source: *Alaska Air Group*

Let p be the annual profit (in millions of dollars) at t years since 2002.

a. Without graphing, estimate the coordinates of the t-intercept for a line that comes close to the data points. What does that point mean in this situation? If you don't see how to estimate the coordinates, construct a scatterplot of the data first.
b. Without graphing, estimate the coordinates of the p-intercept for a line that comes close to the data points. What does that point mean in this situation? If you don't see how to estimate the coordinates, construct a scatterplot of the data first.

40. The *windchill* (or *windchill factor*) is a measure of how cold you feel as a result of being exposed to wind. Table 64 provides some data on windchills for various temperatures when the wind speed is 10 mph.

Table 64 Windchills for a 10-mph Wind

Temperature (°F)	Windchill (°F)
−15	−35
−10	−28
−5	−22
5	−10
10	−4
15	3
20	9
25	15

Source: *National Weather Service Forecast Office*

Let w be the windchill (in degrees Fahrenheit) corresponding to a temperature of t degrees Fahrenheit when the wind speed is 10 mph.

a. Without graphing, estimate the coordinates of the t-intercept for a line that comes close to the data points. What does that point mean in this situation? If you don't see how to estimate the coordinates, construct a scatterplot of the data first.

b. Without graphing, estimate the coordinates of the w-intercept for a line that comes close to the data points. What does that point mean in this situation? If you don't see how to estimate the coordinates, construct a scatterplot of the data first.

41. **DATA** President Obama's statewide approval ratings and Democratic candidates' shares of votes in Senate races in 2010 and 2012 are compared in Table 65 for 12 randomly selected Senate races in recent years.

Table 65 President Obama's Statewide Approval Ratings versus Democratic Candidates' Shares of Votes in Senate Races

President Obama's Statewide Approval Rating (percent)	Democratic Candidate's Share of Votes in a Senate Race (percent)
42	22
51	60
56	60
40	34
46	48
55	53
48	47
48	53
37	48
44	41
66	75
59	71

Source: *The Gallup Organization, Clerk of the House of Representatives*

Let P be President Obama's statewide approval rating (in percent) and D be a Democratic candidate's share of votes (in percent) in a Senate race, both for the same state.

a. Construct a scatterplot by hand. Assume P is the explanatory variable.

b. Describe the four characteristics of the association. Compute and interpret r as part of your analysis.

c. Draw a linear model on your scatterplot.

d. President Obama's approval rating in March 2014 was 33% in South Dakota, which had one Democratic senator: Tim Johnson. Assuming Senator Johnson ran for reelection in 2014 and President Obama's approval rating didn't change, use your model to predict Senator Johnson's share of votes. On the basis of your result, explain why his seat was viewed to be at risk.

e. Did you perform interpolation or extrapolation in part (d)? Do you have much faith in your result? Explain.

42. **DATA** The mean annual U.S. per-person consumptions of lower fat and skim milk and the mean annual U.S. per-person consumptions of whole milk are shown in Table 66 for various years.

Table 66 Mean Annual U.S. Per-Person Consumptions of Lower Fat and Skim Milk, Whole Milk

Year	Mean Annual Consumption (gallons per person)	
	Lower Fat and Skim Milk	Whole Milk
1955	2.9	33.5
1965	3.7	28.8
1975	8.1	21.7
1985	12.3	14.3
1995	15.3	8.6
2005	14.1	7.0
2010	14.8	5.6

Source: *USDA/Economic Research Service*

Let L be the mean annual consumption of lower fat and skim milk, and let W be the mean annual consumption of whole milk, both in gallons per person.

a. Compare the values of L with the values of W by constructing a scatterplot by hand. Assume L is the explanatory variable.

b. Describe the four characteristics of the association. Compute and interpret r as part of your analysis.

c. Draw a linear model on your scatterplot.

d. Assuming time has no effect on the data, predict the mean annual per-person consumption of whole milk if the mean annual consumption of lower fat and skim milk is 10 gallons per person.

e. Describe at least one possible lurking variable.

43. **DATA** The heights, lengths, and maximum speeds of 11 randomly selected U.S. wooden roller coasters are shown in Table 67.

Table 67 Heights, Lengths, and Maximum Speeds of Wooden Roller Coasters

Name	Height (feet)	Length (feet)	Maximum Speed (mph)
Ghost Rider	36	1382	56
Grizzly	28	991	55
Terminator Salvation	29	877	50
Ride of Prey	23	792	50
Cornball Express	17	640	45
Voyage	50	1964	67
Judge Roy Scream	22	814	53
Texas Giant	44	1500	62
American Eagle	39	1417	66
Viper	30	1054	50
Son of Beast	66	2143	79

Source: *Roller Coaster DataBase*

Let H be a roller coaster's height (in feet), and let S be the roller coaster's maximum speed (in mph).

a. Compare the heights and maximum speeds of the roller coasters by constructing a scatterplot by hand.
b. Describe the four characteristics of the association. Compute and interpret r as part of your analysis.
c. Draw a linear model on your scatterplot.
d. Use the model to predict the maximum speed of a 40-foot-tall wooden roller coaster.

44. **DATA** The heights, lengths, and maximum speeds of 11 randomly selected U.S. wooden roller coasters are shown in Table 67 on page 419. Let H be a roller coaster's height (in feet), and let L be the roller coaster's length (in feet).

a. Compare the values of H with the values of L by constructing a scatterplot by hand. Assume H is the explanatory variable.
b. Describe the four characteristics of the association. Compute and interpret r as part of your analysis.
c. Draw a linear model on your scatterplot.
d. Use the model to predict the length of a 40-foot-tall wooden roller coaster.

45. **DATA** The heights, lengths, and maximum speeds of 11 randomly selected U.S. wooden roller coasters are shown in Table 67 on page 419. Let S be a roller coaster's maximum speed (in mph), and let L be the roller coaster's length (in ft).

a. Compare the values of S with the values of L by constructing a scatterplot by hand. Assume S is the explanatory variable.
b. Describe the four characteristics of the association. Compute and interpret r as part of your analysis.
c. Draw a linear model on your scatterplot.
d. Use the model to predict the length of a wooden roller coaster with maximum speed 60 mph.

46. **DATA** The pedestrian and bicyclist fatality rates in 10 randomly selected large cities are shown in Table 68.

Table 68 Pedestrian and Bicyclist Fatality Rates

City	Pedestrian Fatality Rate (number of deaths per 10,000 walking commuters)	Bicyclist Fatality Rate (number of deaths per 10,000 bicycling commuters)
Baltimore	7	5
Charlotte	22	18
Dallas	26	18
El Paso	25	8
Jacksonville	42	33
Las Vegas	17	11
New Orleans	10	6
Portland	6	1
Raleigh	17	3
Wichita	17	0

Source: *Fatality Analysis Reporting System*

Let p be a city's pedestrian fatality rate (in number of deaths per 10,000 walking commuters), and let b be the city's bicyclist fatality rate (in number of deaths per 10,000 bicycling commuters).

a. Construct a scatterplot by hand. Assume the pedestrian fatality rate is the explanatory variable.
b. Describe the four characteristics of the association. Compute and interpret r as part of your analysis.
c. Draw a linear model on your scatterplot.
d. Use the model to predict the bicyclist fatality rate of Oklahoma City, which has a pedestrian fatality rate of 20 deaths per 10,000 walking commuters.
e. Describe at least one lurking variable.

Large Data Sets

47. **DATA** Access the data about head sizes versus brain weights, which are available at MyMathLab and at the Pearson Downloadable Student Resources for Math & Stats website. Treat head size as the explanatory variable and brain weight as the response variable.

a. Construct a scatterplot that describes the association between the head sizes and the brain weights for the women. Do the same for the men. Show both scatterplots in the same coordinate system if you know how to use colors or shapes to identify which points are for which gender.
b. Describe the four characteristics of the women's association. Compute and interpret r as part of your analysis. Do the same for the men's association.
c. How is the women's association similar to the men's? How is the women's association different?
d. Print the scatterplots you constructed in part (a) and draw linear models for the women's and men's associations on the scatterplots. Compare the two models.
e. The author plunged his head into a kitchen sink filled with water. The excess water drained into an adjacent sink. It took 18.4 cups of water to replace the displaced water. Predict the weight of the author's brain in pounds. Round your result to the first decimal place. There are 236.6 cm^3 in 1 cup, and there is 0.002205 pound in 1 gram.
f. The data were collected for a study published in 1905. Do you have much faith in the result you found in part (e)? Explain.

48. **DATA** Access the data about the San Mateo Real Estate, which are available at MyMathLab and at the Pearson Downloadable Student Resources for Math & Stats website.

a. Construct a scatterplot that describes the association between the square footages and the sales prices of the homes.
b. Describe the four characteristics of the association. Compute and interpret r as part of your analysis.
c. Print the scatterplot you constructed in part (a) and draw a linear model on the scatterplot.
d. Predict the sales price of a home with 2000 square feet.
e. Predict the square footage of a home with sales price $2.9 million.

Concepts

49. When modeling a situation in which the variables are linearly associated, different students may all do good work yet not get the same results. Draw a scatterplot and at least two reasonable linear models to show how this is possible.

50. Which is more desirable, finding a linear model that contains several, but not all, data points or finding a linear model that does not contain any data points but comes close to all data points? Include in your discussion some sketches of scatterplots and linear models.

51. When using a line to model a situation, do we usually have more faith in a result obtained by interpolation or extrapolation? Explain.

52. A student comes up with a shortcut for modeling a situation. Instead of plotting all of the given data points, the student plots only two of the data points and draws a line that contains the two chosen points. Give an example to illustrate what can go wrong with this shortcut.

53. A person collects data by doing research. If the data points lie exactly on a line, will all points on the line describe the situation exactly? Explain.

54. Let t be the number of years since 2010. A student believes $t = -2$ is an example of model breakdown because time cannot be negative. What would you tell the student?

55. **a.** Sketch a nonvertical line in a coordinate system. Find any outputs for the given input. State how many outputs there are for that single input.

    i. the input 2   **ii.** the input 4
    iii. the input -3

    b. For your line, a single input leads to how many outputs? Explain.

    c. For *any* nonvertical line, a single input leads to how many outputs? Explain.

56. A student says the x-intercept of the ordered pair $(-3, 4)$ is -3. Is the student correct? Explain.

57. A student says the y-intercept of a line is $(5, 0)$. Is the student correct? Explain.

58. Explain why the x-coordinate of a y-intercept of a line is 0.

59. Are there any lines for which the x-intercept is the same point as the y-intercept? If yes, sketch such a line, and what is that point? If no, explain why not.

60. Sketch three distinct lines that all have the same x-intercept.

61. In your own words, describe the meaning of *linear model*.

62. Describe how to find a linear model for a situation and how to use the model to make estimates and predictions.

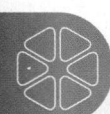

Hands-On Projects

DATA Climate Change Project

Many scientists are greatly concerned that the mean temperature of the surface of Earth has increased since 1900 (see Table 69).

Table 69 Mean Surface Temperatures of Earth

Year	Mean Temperature (degrees Fahrenheit)	Year	Mean Temperature (degrees Fahrenheit)
1900	57.1	1960	57.2
1905	56.8	1965	57.0
1910	56.6	1970	57.3
1915	57.0	1975	57.1
1920	56.9	1980	57.6
1925	56.9	1985	57.3
1930	57.1	1990	57.8
1935	57.0	1995	57.9
1940	57.3	2000	57.8
1945	57.3	2005	58.3
1950	56.9	2010	58.3
1955	57.0	2014	58.4

Source: *NASA–GISS*

Although it may not seem that the mean temperatures shown in Table 69 have increased much, many scientists believe an increase as small as 3.6°F could be a dangerous climate change.* Global warming would cause the extinction of plants and animals, lead to severe water shortages, create more extreme weather events, increase the number of heat-related illnesses and deaths, and melt glaciers, which would raise ocean levels and, thus, submerge coastlands.

Despite these alarming predictions, not all experts are concerned. Robert Mendelsohn, an environmental economist at Yale University, argues that "global warming will increase agricultural production in the northern half of the United States" and that "the southern half will be able to maintain its current level of production." Even from a global perspective, he believes,[†] the benefits of global warming will offset the damages.[†]

To test such theories, Peter S. Curtis, an ecologist at Ohio State University, and colleagues ran experiments that showed that increased carbon dioxide levels do in fact increase plant growth. However, the nutritional value of the produce was lower—so much lower that the increase in growth did not make up for the decrease in nutrition.[‡]

Some other theories suggest benefits to global warming. The scientific report *Impacts of a Warming Arctic* points out that it will be easier to extract oil from the Arctic due to less extensive and thinner sea ice. However, the study also says that oil spills are more difficult to clean up in icy seas than in open waters and that many species

---

* "Meeting the Climate Challenge: Recommendations of the International Climate Change Taskforce," The Institute for Public Policy Research/The Center for American Progress/The Australian Institute, January 2005.

[†] From *The Impact of Climate Change on the United States Economy,* R. Mendelsohn and J. Neumann (eds.), Cambridge University Press.
[‡] "Plant Reproduction under Elevated CO_2 Conditions," L. M. Jablonski, X. Wang, and P. S. Curtis, *New Phytologist* (156):9–26, 2002.

would suffer from such spills. In addition, potential structural problems such as broken pipelines could mean that the costs outweigh the benefits.[§]

A study done by economist Thomas Gale Moore at Stanford University suggests that a 4.5°F increase in temperature could reduce deaths in the United States by 40,000 per year and that medical costs might be reduced by at least $20 billion annually. Moore also points out that most people prefer warmer climates.[¶]

Most scientists do not share Moore's perspective. On a global scale, they believe warmer climates could increase the spread of diseases such as malaria and, thus, increase the global death rate and medical costs.

Although there may be some relatively small and short-lived benefits to global warming, the vast majority of scientists agree that global warming is already taking its toll and that further warming would bring catastrophic results.

Glaciologists report that, over the past century, glaciers around the globe have been melting.[‖] Biologists note that many species throughout the world have changed their habitats in search of cooler climates.[**] One species, the golden toad, was not able to migrate and as a result has become extinct due to heat stress.[††]

Looking to the future, an international study, the most comprehensive analysis of its kind, predicts that 15 to 37 percent of all species of plants and animals—well over a million species—will become extinct by 2050.[*] Klaus Toepfer, head of the United Nations Environment Programme (UNEP), said, "If one million species become extinct… it is not just the plant and animal kingdoms and the beauty of the planet that will suffer. Billions of people, especially in the developing world, will suffer too as they rely on nature for such essential goods and services as food, shelter and medicines."[†]

Analyzing the Situation

1. Discuss those theories that describe the benefits of global warming and whether they are likely correct.

2. What are some possible costs of global warming? In your opinion, do the possible costs outweigh the possible benefits? Explain.

3. Let F be the mean surface temperature of Earth (in degrees Fahrenheit) at t years since 1900. Construct a scatterplot of the data in Table 69.

[§]Report of the Arctic Climate Impact Assessment, Cambridge University Press, 2004.
[¶]"In sickness or in health: The Kyoto Protocol versus global warming," Hoover Institution, Stanford University, August 2000.
[‖]National Snow and Ice Data Center, 2003.
[**]T. L. Root et al., "Fingerprints of global warming on wild animals and plants," January 2, 2003, *Nature,* 421:57–60.
[††]J. A. Pounds et al., "Biological response to climate change on a tropical mountain," 1999, *Nature (London),* 398(6728):611–615.
[*]C. D. Thomas et al., "Extinction risk from climate change," January 8, 2004, *Nature,* 427:145–148.
[†]Reported by UNEP (United Nations Environment Programme), January 8, 2004.

4. On the basis of your scatterplot, in what year is it first clear that global warming is occurring? Explain. Also, explain why it makes sense that the first World Climate Conference convened in 1979.

5. Are the variables t and F linearly associated for the years 1900–2014? Are the variables linearly associated for the years 1965–2014? Explain.

6. In Problem 5, you determined whether there is a linear association for the years 1965–2014. Now describe the three other characteristics of the association for the years 1965–2014. Compute and interpret r as part of your analysis.

7. Print the scatterplot and draw a linear model on the scatterplot for the years 1965–2014.

8. Use the model to estimate the mean global temperature in 2008. Have you performed interpolation or extrapolation? How much faith do you have in your result? Explain.

9. Use the model to predict the mean global temperature in 2020. Have you performed interpolation or extrapolation? How much faith do you have in your result? Explain.

Volume Project

In this lab, you will explore the relationship between the volume of some water in a cylinder and the height of the water. Check with your instructor whether you should collect your own data or use the data listed in Table 70.

Table 70 Heights of Water in a Cylinder with Radius 4.45 Centimeters

Height (centimeters)	Volume (ounces)
0	0
0.9	2
1.9	4
2.9	6
3.8	8
4.8	10
5.7	12

Source: *J. Lehmann*

Materials

You will need the following items:

- A "perfect" cylinder (the diameter of the top should equal the diameter of the base) that can hold at least 8 ounces of water
- At least 8 ounces of water
- A $\frac{1}{4}$-cup measuring cup
- A ruler

Recording of Data

Pour $\frac{1}{4}$ cup (2 ounces) of water into the cylinder, and measure the height of the water, using units of centimeters. Then continue adding $\frac{1}{4}$ cup of water and measuring the height after you have added each $\frac{1}{4}$ cup until there is at least 8 ounces of water in the cylinder. Also, measure the height of the cylinder in units of centimeters.

Analyzing the Data

1. Display your data in a table similar to Table 70. If you are using the data in Table 70, the height of the cylinder is 12 centimeters.

2. Let V be the volume of water (in ounces) in the cylinder when the height is h centimeters. Assume that h is the explanatory variable. Construct a scatterplot of the data.

3. Draw a linear model on your scatterplot.

4. What is the V-intercept of your model? What does it mean in this situation?

5. Use the model to estimate the volume of water when the height of the water is 3 centimeters.

6. Use the model to estimate the height of 7 ounces of water in the cylinder.

7. What is the height of the cylinder? Use this height and the model to estimate the maximum amount of water that the cylinder can hold.

8. Indicate on your graph of the model where model breakdown occurs. Also, describe in words when model breakdown occurs.

Linear Graphing Project: Topic of Your Choice

Your objective in this lab is to use a linear model to describe some authentic situation. Choose a situation that has not been discussed in this text. Your first task will be to find some data, which you could find by online searches of blogs, newspapers, magazines, and scientific journals. Or you can conduct a survey or a physical experiment. Choose something that interests you!

Analyzing the Situation

1. What two variables did you explore?

2. Which variable is the explanatory variable? Which variable is the response variable? Explain.

3. Describe how you found your data. If you conducted a survey or an experiment, provide a careful description with specific details of how you conducted your survey or experiment. If you didn't conduct a survey or an experiment, state the source of your data.

4. Include a table of your data.

5. Construct a scatterplot.

6. Describe the four characteristics of the association. Compute and interpret r as part of your analysis. (If the variables are not linearly associated, find some variables that are.)

7. Draw a linear model on your scatterplot.

8. Choose a value for your explanatory variable. On the basis of your chosen value, use your model to predict a value for your response variable. Describe what your result means in the situation.

9. Choose a value for your response variable. On the basis of your chosen value, use your model to predict a value for your explanatory variable. Describe what your result means in the situation.

10. Comment on your project experience.

    a. For example, you might address whether this project was enjoyable, insightful, and so on.

    b. Were you surprised by any of your findings? If so, which ones?

    c. How would you improve your process for this project if you were to do it again?

    d. How would you improve your process if you had more time and money?

Chapter Summary

Key Points of Chapter 6

Section 6.1   Scatterplots

Explanatory and response variables	In a study about whether a variable x explains (affects) a variable y, • We call x the **explanatory variable** (or **independent variable**). • We call y the **response variable** (or **dependent variable**).
Values of an ordered pair	For an ordered pair (a, b), we write the value of the explanatory variable in the first (left) position and the value of the response variable in the second (right) position.
Columns of tables and axes of coordinate systems	Assume that an authentic situation can be described by using two variables. Then • For tables, the values of the explanatory variable are listed in the first column and the values of the response variable are listed in the second column. • For coordinate systems, the values of the explanatory variable are described by the horizontal axis and the values of the response variable are described by the vertical axis.

Section 6.1   Scatterplots (*Continued*)

Scatterplot	A coordinate system with plotted ordered pairs is called a **scatterplot**.
Association	There is an association between the explanatory and response variables if the response variable changes as the explanatory variable changes.
Positive and negative association	Assume two numerical variables are the explanatory and response variables of a study. • If the response variable tends to increase as the explanatory variable increases, we say the variables are **positively associated** (or **positively correlated**) and that there is a **positive association** (or **positive correlation**). • If the response variable tends to decrease as the explanatory variable increases, we say the variables are **negatively associated** (or **negatively correlated**) and that there is a **negative association** (or **negative correlation**).
Direction	We describe the **direction** of an association by determining whether the association is positive, negative, or neither.

Section 6.2   Determining the Four Characteristics of an Association

Linear association	If the points of a scatterplot lie close to (or on) a line, we say the variables are **linearly associated** and that there is a **linear association**.
Nonlinear association	If the points of a scatterplot lie close to (or on) a curve that is not a line, we say there is a **nonlinear association**.
Exact association	If a curve passes through all the points of a scatterplot, we say there is an **exact association** with respect to the curve.
Strong association	If a curve comes quite close to all the points of a scatterplot, we say there is a **strong association** with respect to the curve.
Weak association	If a curve comes somewhat close to all the points of a scatterplot, we say there is a **weak association** with respect to the curve.
Properties of the linear correlation coefficient	Assume r is the linear correlation coefficient for the association between two numerical variables. Then • The values of r are between -1 and 1, inclusive. • If r is positive, then the variables are positively associated. • If r is negative, then the variables are negatively associated. • If $r = 0$, there is no *linear* association. • The larger the value of $|r|$, the stronger the linear association will be. • If $r = 1$, then the points lie exactly on a line and the association is positive. • If $r = -1$, then the points lie exactly on a line and the association is negative.
Determining the strength of an association	To determine the strength and the type of an association, we should inspect a scatterplot of the data as well as compute r.
Order of determining the four characteristics of an association	We determine the four characteristics of an association in the following order: 1. Identify all outliers. 　a. For outliers that stem from errors in measurement or recording, correct the errors if possible. If the errors cannot be corrected, remove the outliers. 　b. For other outliers, determine whether they should be analyzed in a separate study. 2. Determine the shape of the association. 3. If the shape is linear, then on the basis of r and the scatterplot, determine the strength. If the shape is nonlinear, then on the basis of the scatterplot, determine the strength. 4. Determine the direction. In other words, determine whether the association is positive, negative, or neither.
Strong association does not guarantee causation	A strong association between two variables does not guarantee that a change in the explanatory variable will cause a change in the response variable.

Section 6.3   Modeling Linear Associations

Model	A **model** is a mathematical description of an authentic situation. We say the description *models* the situation.
Linear model	A **linear model** is a nonvertical line that describes the association between two quantities in an authentic situation.
Constructing scatterplots and making estimates and predictions	We construct a scatterplot of data to determine whether there is a linear association. If so, we draw a line that comes close to the data points and use the line to make estimates and predictions.
Input and output	An **input** is a permitted value of the *explanatory* variable that leads to at least one **output**, which is a permitted value of the *response* variable.
Interpolation and extrapolation	For a situation that can be described by a model whose explanatory variable is x, <ul><li>We perform **interpolation** when we use a part of the model whose x-coordinates are between the x-coordinates of two data points.</li><li>We perform **extrapolation** when we use a part of the model whose x-coordinates are not between the x-coordinates of any two data points.</li></ul>
Model Breakdown	When a model gives a prediction that does not make sense or an estimate that is not a good approximation, we say **model breakdown** has occurred.
Intercept	An **intercept** of a line is any point where the line and an axis (or axes) of a coordinate system intersect. There are two types of intercepts of a line sketched on a coordinate system with an x-axis and a y-axis: <ul><li>An **x-intercept** of a line is a point where the line and the x-axis intersect. The y-coordinate of an x-intercept is 0.</li><li>A **y-intercept** of a line is a point where the line and the y-axis intersect. The x-coordinate of a y-intercept is 0.</li></ul>

Chapter 6 Review Exercises

For Exercises 1 and 2, identify the explanatory and the response variables. Which variable should be described by the horizontal axis of a coordinate system? How about the vertical axis?

1. Let s be the mean salary (in dollars) for people with t years of education.

2. Let c be the mean number of calories that a person consumes daily, and let w be the person's mean weight (in pounds).

For Exercises 3 and 4, describe the situation with an ordered pair. Which variable should be described by the horizontal axis of a coordinate system? How about the vertical axis?

3. Let M be the number of miles a person runs weekly, and let T be the person's best marathon time (in minutes). A marathoner runs 110 miles weekly, and her best marathon time is 139 minutes.

4. Let s be the number (in thousands) of Starbucks stores, and let t be the number of years since 2010. There were 21.4 thousand Starbucks stores in 2014 (Source: *FactSet*).

5. Let G be the gas mileage (in miles per gallon) of a car while it has been driven at M miles per hour. What does the ordered pair (64, 42) mean in this situation?

6. Let n be the number of U.S. billionaires at t years since 2000. What does the ordered pair (14, 1645) mean in this situation?

7. Let x be the explanatory variable and y be the response variable. Construct a scatterplot of the ordered pairs listed in Table 71.

Table 71 Some Ordered Pairs

x	y
2	20
4	15
7	13
11	16
13	21

8. **DATA** *Sepsis* is the presence of harmful bacteria in a patient's tissue. The numbers of Medicare sepsis patients discharged between October 1, 2012, and September 30, 2013, are shown in Table 72 for various lengths of stays.

Table 72 Numbers of Medicare Sepsis
Patients Discharged

Length of Stay (days)	Number of Sepsis Patients Discharged
1	37
5	43
10	67
15	83
19	90
20	931
25	279
30	113
35	73
40	27

Source: *Centers for Medicare and Medicaid Services; Medicare claims data*

Let *d* be the number of Medicare sepsis patients discharged after staying for *n* days.

a. Construct a scatterplot.
b. Is the association positive, negative, or neither?
c. Which of the points in your scatterplot is lowest? What does that mean in this situation?
d. Which of the points in your scatterplot is highest? What does that mean in this situation?
e. Under Medicare rules for most sepsis patients, hospitals get paid small amounts for patients who have short stays and a larger amount for patients who stay 20 days. But hospitals do not get additional monies if patients stay more than 20 days. Explain why the data suggest hospitals are keeping some patients longer than necessary.

9. **DATA** The *gauge* of an extension cord measures the diameter (thickness) of the wires inside. The maximum rates (in amperes) at which electric current can safely flow through 16-gauge extension cords are shown in Table 73 for various lengths of cords. Not following these guidelines can cause a fire from overheating.

Table 73 16-Gauge Extension Cord Lengths versus Safe Maximum Rates of Electric Current

Length (feet)	Safe Maximum Rate of Electric Current (amperes)
25	9.55
50	6.05
75	4.25
100	2.75
150	1

Source: *Milwaukee Tool*

Let *A* be the maximum number of amperes that can be safely delivered by a 16-gauge extension cord that is *L* feet long.

a. Identify the explanatory and response variables.
b. Which variable should be described by the horizontal axis of a coordinate system? How about the vertical axis?
c. Construct a scatterplot.
d. Is the association positive, negative, or neither? What does that mean in this situation?

e. A Milwaukee Screwdriver Power Unit® requires 6.5 amperes. Select all the lengths of 16-gauge extension cords shown in Table 73 that can be safely used with the screwdriver.

For Exercises 10–12, match the given information to the appropriate scatterplot in Fig. 100.

10. *r* = −0.6
11. *r* = −0.9
12. *r* = 1

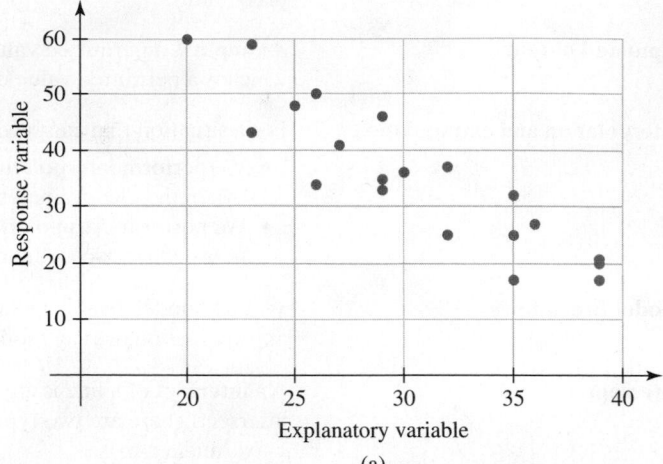
(a)

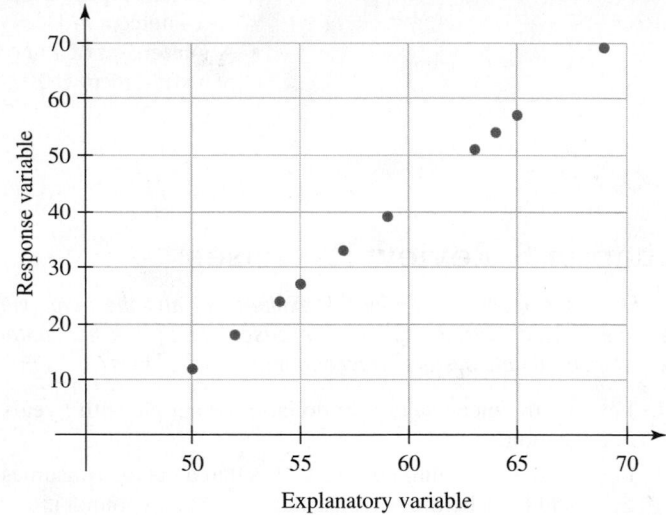
(b)

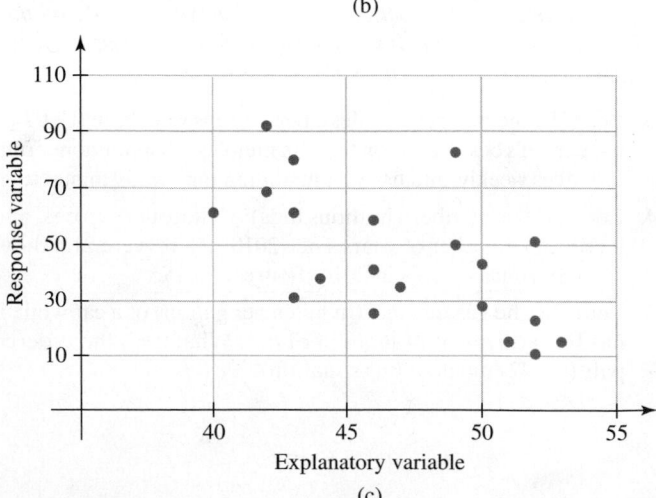
(c)

Figure 100 Exercises 10–12

13. DATA The median sales prices of vacation and investment homes are shown in Table 74 for various years.

Table 74 Median Sales Prices of Vacation and Investment Homes

| Year | Median Sales Price (thousands of dollars) | |
	Vacation Home	Investment Home
2007	195	150
2008	150	108
2009	169	105
2010	150	94
2011	121	100
2012	150	115
2013	169	130
2014	150	125

Source: *National Association of Realtors*®

Let *V* be the median sales price (in thousands of dollars) of vacation homes in a certain year, and let *I* be the median sales price (in thousands of dollars) of investment homes in the same year.

a. Construct a scatterplot that describes the association between *V* and *I*. Treat *V* as the explanatory variable.

b. Describe the four characteristics of the association. Compute and interpret *r* as part of your analysis.

c. A student concludes that an increase in the median sales price of vacation homes causes an increase in the median sales price of investment homes. What would you tell the student?

d. Describe a possible lurking variable. Explain.

For Exercises 14–17, refer to Fig. 101.

14. Find *y* when *x* = −2.

15. Find *x* when *y* = 1.

16. What is the *x*-intercept of the line?

17. What is the *y*-intercept of the line?

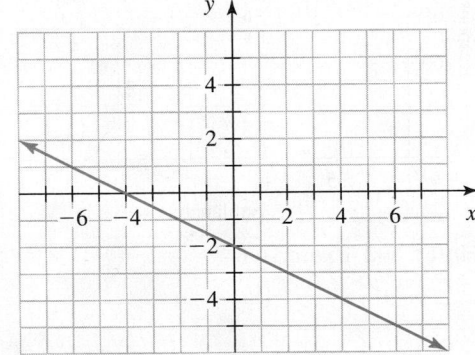

Figure 101 Exercises 14–17

18. Some ordered pairs are listed in Table 75.

Table 75 Some Ordered Pairs

x	y
1	12
2	10
5	8
8	6
9	3

a. Construct a scatterplot by hand.

b. Is there a linear association, a nonlinear association, or neither?

c. Draw a line that comes close to the points in your scatterplot.

d. Find *y* when *x* = 4.

e. Find *x* when *y* = 7.

19. DATA After gaining enough experience to advance through a first salary schedule (A), police officers at the Los Angeles Police Department are paid annual salaries as described in Table 76 (Schedule 2). As police officers gain more experience, they receive step increases.

Table 76 Steps versus Annual Salaries (Schedule 2)

Step	Annual Salary (dollars)
2	67,442
3	71,243
4	75,168
6	83,791
7	88,427

Source: *City of Los Angeles Personnel Department*

Let *s* be the annual salary (in dollars) for step *n*.

a. Construct a scatterplot by hand.

b. Describe the four characteristics of the association. Compute and interpret *r* as part of your analysis.

c. Draw a linear model on your scatterplot.

d. Predict the salary at step 5. Have you performed interpolation or extrapolation? Do you have much faith in your result?

e. Suppose police officers negotiated adding Step 8 to Schedule 2. Predict the annual salary at Step 8. Have you performed interpolation or extrapolation? Do you have much faith in your result?

20. Willie Mays, with all-around talent, was one of the greatest baseball players of all time. The numbers of stolen bases by Mays are shown in Table 77 for various years.

Table 77 Numbers of Stolen Bases by Willie Mays

Year	Number of Stolen Bases
1956	40
1957	38
1958	31
1960	25
1961	18
1962	18
1963	8

Source: *The Sports Encyclopedia: Baseball 2004, D. S. Neft et al., 2004, St. Martin's Press, NY.*

Let *n* be the number of stolen bases by Mays in the year that is *t* years since 1955.

a. Construct a scatterplot by hand.
b. Draw a linear model on your scatterplot.
c. Estimate the number of bases Mays stole in 1959. Have you performed interpolation or extrapolation? The actual number is 27 bases. Find the error in your estimate.

d. What is the *n*-intercept? What does it mean in this situation? In 1955, Mays stole 24 bases. Has model breakdown occurred?
e. What is the *t*-intercept? What does it mean in this situation? In 1965, Mays stole 9 bases. Has model breakdown occurred?
21. A student says the correlation coefficient *r* is the probability that some data points lie near a line. What would you tell the student?

Chapter 6 Test

1. Let *p* be the percentage of Americans at age *a* years who own a home. Identify the explanatory and the response variables. Which variable should be described by the horizontal axis of a coordinate system? How about the vertical axis?

2. Let *c* be the total cost (in dollars) of *n* tickets to an alt-J concert. The total cost of 6 tickets was $255 at the Greek Theater at University of California, Berkeley, on April 16, 2015 (Source: *Ticketmaster*). Describe the situation with an ordered pair. Which variable should be described by the horizontal axis of a coordinate system? How about the vertical axis?

3. Let *s* be the annual salary (in millions of dollars) of basketball player LeBron James at *t* years since 2010. What does the ordered pair (4, 72.3) mean in this situation?

4. DATA The mean amounts of savings various age groups believe are enough to have at retirement are shown in Table 78.

Table 78 Mean Savings Believed Necessary to Have at Retirement

Age Group (years)	Age Used to Represent Age Group (years)	Mean Savings Believed Necessary for Retirement (millions of dollars)
20–29.99	25	4.3
30–39.99	35	2.4
40–49.99	45	2.5
50–59.99	55	3.4
60–69.99	65	4.0

Source: *Transamerica Center for Retirement Studies*

Let *s* be the mean amounts of savings (in millions of dollars) a person at age *a* years believes are enough to have at retirement.

a. Identify the explanatory and response variables.
b. Construct a scatterplot.
c. Which point is highest in your scatterplot? What does it mean in this situation?
d. Which point is lowest in your scatterplot? What does it mean in this situation?
e. In terms of people's opinions about having enough savings to retire, which age group in Table 78 does not fit the pattern of the other age groups? Explain.

For Exercises 5–7, refer to the given scatterplot to determine whether there is a linear association, a nonlinear association, or no association.

5. See Fig. 102.

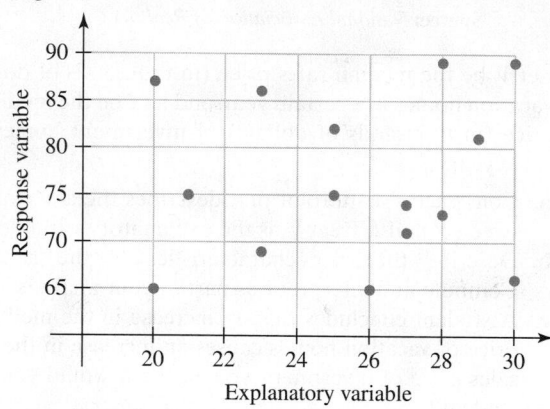

Figure 102 Exercise 5

6. See Fig. 103.

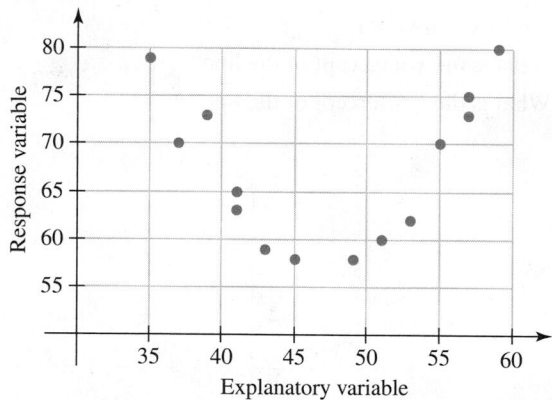

Figure 103 Exercise 6

7. See Fig. 104

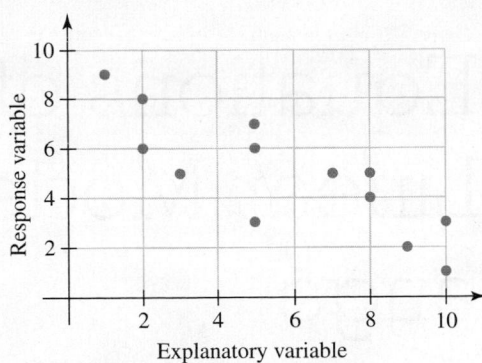

Figure 104 Exercise 7

8. **DATA** The percentages of women and men who have high blood pressure are shown in Table 79 for various age groups.

Table 79 Percentages of Women and Men Who Have High Blood Pressure

| Age Group (years) | Age Used to Represent Age Group (years) | Percent | |
		Women	Men
20–34	27	6.7	9.1
35–44	39.5	17.6	24.4
45–54	49.5	34.0	37.7
55–64	59.5	52.0	52.0
65–74	69.5	70.8	63.9
over 74	80	80.1	72.1

Source: *National Center for Health Statistics; National Heart, Lung, and Blood Institute*

Let W and M be the percentages of women and men, respectively, at age a years who have high blood pressure.

a. Construct a scatterplot that describes the association between ages and blood pressures for the women. Do the same for the men. Show both scatterplots in the same coordinate system if you know how to use colors or shapes to identify which points are for which gender.

b. Describe the four characteristics of the women's association. Compute and interpret r as part of your analysis.

c. Describe the four characteristics of the men's association. Compute and interpret r as part of your analysis.

d. On the basis of your responses to parts (b) and (c), a student concludes that aging causes high blood pressure. What would you tell the student?

e. On the basis of the ages you used to represent the age groups, at what age are women as likely as men to have high blood pressure? For people younger than that age, which gender is more likely to have high blood pressure? How about people older than that age?

9. Some ordered pairs are listed in Table 80.

Table 80 Some Ordered Pairs

x	y
1	23
3	17
6	10
8	9
9	4

a. Construct a scatterplot by hand.
b. Is there a linear association, a nonlinear association, or neither?
c. Draw a line that comes close to the points in your scatterplot.
d. What is the x-intercept of your line?
e. What is the y-intercept of your line?

10. The number of spacecraft, rocket bodies, mission-related debris, and fragmentation debris orbiting the Earth has increased greatly since the beginning of the Space Age in 1957. The numbers of such debris larger than 10 cm are shown in Table 81 for various years.

Table 81 Numbers of Space Debris

Year	Number of Space Debris (thousands)
1960	0.02
1970	2.6
1980	5.5
1990	7.5
2000	9.8
2005	10.2

Source: *NASA Orbital Debris Program Office*

Let n be the number (in thousands) of space debris at t years since 1960.

a. Construct a scatterplot by hand.
b. Draw a linear model on your scatterplot.
c. Use your model to estimate the number of space debris in 1995. Have you performed interpolation or extrapolation?
d. Use your model to estimate the number of space debris in 2010. Have you performed interpolation or extrapolation? Do you have much faith in your estimate? Explain.
e. The actual number of space debris in 2010 was 16.0 thousand. A huge number of space debris was caused by collisions of spacecraft in 2007 and 2009. Use the result you found in part (d) to help you estimate the number of space debris caused by the two collisions. Describe any assumptions you have made.

7 Graphing Equations of Lines and Linear Models; Rate of Change

Worldwide iPhone sales have greatly increased since 2009 (see Table 1). In fact, the sales were so large in 2014 that you will compute quite an impressive number of iPhones that were sold *per hour* in the last three months of that year (see Exercise 47 of Homework 7.1).

In Chapter 6, we used a line to describe the association between two quantities that are linearly associated. In this chapter, we will discuss how to describe such an association by a symbolic statement called an *equation*.

Table 1 Worldwide Sales of iPhones in the Last Three Months of Various Years

Year	Worldwide Sales (millions)
2009	8.7
2010	16.2
2011	37.0
2012	47.8
2013	51.0

Source: *Pew Research Center's Internet & American Life Project surveys*

We will discuss how to use an equation to sketch a line. We will also describe the steepness of a line and how that steepness is related to how quickly one quantity changes in relation to another, such as how quickly jewelry sales have increased in the United States over time. Finally, we will describe some associations by using an extremely important concept called a *function*.

▼ 7.1 Graphing Equations of Lines and Linear Models

Objectives

» For an equation in two variables, determine the meaning of *solution, satisfy,* and *solution set.*

» Describe the meaning of the *graph* of an equation.

» Graph equations of the form $y = mx + b$.

» Describe the meaning of b in equations of the form $y = mx + b$.

» Graph equations of the form $y = b$ and $x = a$.

» Describe the Rule of Four for equations.

» Graph an equation of a linear model.

In Chapter 6, we sketched linear models on scatterplots of data so we could make predictions. Unfortunately, errors usually result from not drawing lines perfectly straight and not estimating coordinates of points well. To avoid such errors, it turns out we can describe a linear model with an *equation* that we can use to make predictions. Here are some examples of equations that describe linear models:

$$T = 575C + 15 \qquad s = 1.86t - 8.26 \qquad p = -0.15t + 61.8$$

In this section, we will determine the line that is described by such an equation.

Solutions, Satisfying Equations, and Solution Sets

Consider the equation $y = x + 4$. Let's find y when $x = 3$:

$$y = x + 4 \quad \text{Original equation}$$
$$y = 3 + 4 \quad \text{Substitute 3 for x.}$$
$$= 7 \quad \text{Add.}$$

So, $y = 7$ when $x = 3$. Recall from Section 1.1 that the ordered-pair notation $(3, 7)$ is shorthand for saying that when $x = 3$, $y = 7$.

For the equation $y = x + 4$, we found that $y = 7$ when $x = 3$. This means that the equation $y = x + 4$ becomes a true statement when we substitute 3 for x and 7 for y:

$$y = x + 4 \quad \text{\textit{Original equation}}$$
$$7 \stackrel{?}{=} 3 + 4 \quad \text{\textit{Substitute 3 for x and 7 for y.}}$$
$$7 \stackrel{?}{=} 7 \quad \text{\textit{Add.}}$$
$$\text{true}$$

We say that $(3, 7)$ is a *solution* of the equation $y = x + 4$ and that $(3, 7)$ *satisfies* the equation $y = x + 4$.

A *set* is a container. Much as an egg carton contains eggs, a *solution set* contains solutions.

> **Definition** *Solution, satisfy,* and *solution set* of an equation in two variables
>
> An ordered pair (a, b) is a **solution** of an equation in terms of x and y if the equation becomes a true statement when a is substituted for x and b is substituted for y. We say (a, b) **satisfies** the equation. The **solution set** of an equation is the set of all solutions of the equation.

▶ **Example 1** Identifying Solutions of an Equation

1. Is $(2, 1)$ a solution of $y = 3x - 5$?
2. Is $(4, 9)$ a solution of $y = 3x - 5$?

Solution

1. We substitute 2 for x and 1 for y in the equation $y = 3x - 5$:

$$y = 3x - 5 \quad \text{\textit{Original equation}}$$
$$1 \stackrel{?}{=} 3(2) - 5 \quad \text{\textit{Substitute 2 for x and 1 for y.}}$$
$$1 \stackrel{?}{=} 6 - 5 \quad \text{\textit{Multiply before subtracting.}}$$
$$1 \stackrel{?}{=} 1 \quad \text{\textit{Subtract.}}$$
$$\text{true}$$

So, $(2, 1)$ is a solution of $y = 3x - 5$.

2. We substitute 4 for x and 9 for y in the equation $y = 3x - 5$:

$$y = 3x - 5 \quad \text{\textit{Original equation}}$$
$$9 \stackrel{?}{=} 3(4) - 5 \quad \text{\textit{Substitute 4 for x and 9 for y.}}$$
$$9 \stackrel{?}{=} 12 - 5 \quad \text{\textit{Multiply before subtracting.}}$$
$$9 \stackrel{?}{=} 7 \quad \text{\textit{Subtract.}}$$
$$\text{false}$$

So, $(4, 9)$ is *not* a solution of $y = 3x - 5$.

▶

Definition of Graph

Next we will learn how to *graph* an equation. As a first step, we plot some solutions of an equation in the next example.

▶ **Example 2** Plotting Some Solutions of an Equation

Find five solutions of $y = 2x - 1$, and plot them in the same coordinate system.

Solution

To find solutions, we are free to choose *any* values we'd like to substitute for x, but it's a good idea to pick integers close to or equal to 0 so the solutions are easy to plot. For example, here we substitute 0, 1, and 2 for x:

$$y = 2(0) - 1 \qquad\qquad y = 2(1) - 1 \qquad\qquad y = 2(2) - 1$$
$$= 0 - 1 \qquad\qquad\qquad = 2 - 1 \qquad\qquad\qquad = 4 - 1$$
$$= -1 \qquad\qquad\qquad\quad = 1 \qquad\qquad\qquad\quad = 3$$

Solution: $(0, -1)$ Solution: $(1, 1)$ Solution: $(2, 3)$

Next we substitute -2 and -1 for x:

$$y = 2(-2) - 1 \qquad\qquad\qquad y = 2(-1) - 1$$
$$y = -4 - 1 \qquad\qquad\qquad\quad y = -2 - 1$$
$$y = -5 \qquad\qquad\qquad\qquad\quad y = -3$$

Solution: $(-2, -5)$ Solution: $(-1, -3)$

We organize our findings in Table 2. In Fig. 1, we plot the five solutions.

Table 2 Solutions of $y = 2x - 1$

x	y
-2	$2(-2) - 1 = -5$
-1	$2(-1) - 1 = -3$
0	$2(0) - 1 = -1$
1	$2(1) - 1 = 1$
2	$2(2) - 1 = 3$

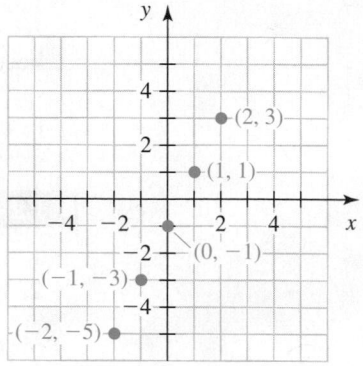

Figure 1 Five solutions of $y = 2x - 1$

In Example 2, we plotted five solutions of $y = 2x - 1$. Note that a line contains these five points (see Fig. 2). It turns out *every* point on the line represents a solution of the equation $y = 2x - 1$. For example, in Fig. 3, we see that the point $(3, 5)$ lies on the line, and we can show that the ordered pair $(3, 5)$ does satisfy the equation $y = 2x - 1$:

$$y = 2x - 1 \qquad \text{\textit{Original equation}}$$
$$5 \stackrel{?}{=} 2(3) - 1 \qquad \text{\textit{Substitute 3 for x and 5 for y.}}$$
$$5 \stackrel{?}{=} 6 - 1 \qquad \text{\textit{Multiply before subtracting.}}$$
$$5 \stackrel{?}{=} 5 \qquad \text{\textit{Add.}}$$
$$\text{true}$$

It also turns out points that do not lie on the line represent ordered pairs that do *not* satisfy the equation. For example, by Fig. 3, we see that the point $(4, 2)$ does not lie on the line, and we can show that the ordered pair $(4, 2)$ does not satisfy the equation $y = 2x - 1$:

$$y = 2x - 1 \qquad \text{\textit{Original equation}}$$
$$2 \stackrel{?}{=} 2(4) - 1 \qquad \text{\textit{Substitute 4 for x and 2 for y.}}$$
$$2 \stackrel{?}{=} 8 - 1 \qquad \text{\textit{Multiply before subtracting.}}$$
$$2 \stackrel{?}{=} 7 \qquad \text{\textit{Subtract.}}$$
$$\text{false}$$

We call the line in Fig. 3 the *graph* of the equation $y = 2x - 1$.

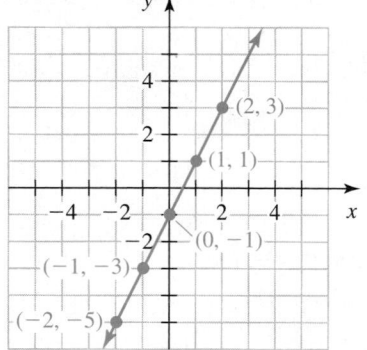

Figure 2 The line contains solutions of $y = 2x - 1$ found in Example 2

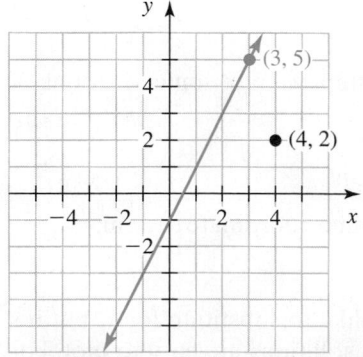

Figure 3 $(3, 5)$ lies on the line, but $(4, 2)$ does not lie on the line

▶ Definition   Graph

The **graph** of an equation in two variables is the set of points that correspond to all solutions of the equation.

The graph of an equation in two variables is a visual description of the solutions of the equation. Every point on the graph represents a solution of the equation. Every point *not* on the graph represents an ordered pair that is *not* a solution.

Graphs of Equations of the Form $y = mx + b$

Directly after Example 2, we found that the graph of the equation $y = 2x - 1$ is a line. The equation $y = 2x - 1$, or $y = 2x + (-1)$, is of the form $y = mx + b$, where $m = 2$ and $b = -1$. It turns out that for any equation of the form $y = mx + b$, where m and b are constants, the graph is a line.

> **Graph of $y = mx + b$**
>
> The graph of an equation of the form $y = mx + b$, where m and b are constants, is a line.

Here are some equations whose graphs are lines:

$$y = 3x + 7, \quad y = -4x - 5, \quad y = -4x, \quad y = x + 3, \quad y = 2$$

The equation $y = -4x$ is of the form $y = mx + b$ because we can write it as $y = -4x + 0$. The equation $y = 2$ is of the form $y = mx + b$ because we can write it as $y = 0x + 2$.

▶ Example 3   Graphing an Equation

Sketch the graph of $y = -2x + 3$. Also, find the y-intercept.

Solution

Since $y = -2x + 3$ is of the form $y = mx + b$, the graph is a line. Although we can sketch a line from as few as two points, we plot a third point as a check. If the third point is not in line with the other two, then we know that we have computed or plotted at least one of the solutions incorrectly.

To begin, we calculate three solutions of $y = -2x + 3$ in Table 3. We use 0, 1, and 2 as values of x because they correspond to points that are easy to plot. Then we plot the three points and sketch the line through them (see Fig. 4).

We use the TI-84 command ZStandard followed by ZSquare to verify our graph (see Fig. 5). See Appendices A.14, A.15, and A.17 for TI-84 instructions.

Table 3 Solutions of $y = -2x + 3$

x	y
0	$-2(0) + 3 = 3$
1	$-2(1) + 3 = 1$
2	$-2(2) + 3 = -1$

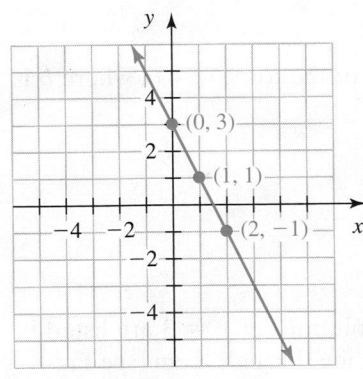

Figure 4 Graph of $y = -2x + 3$

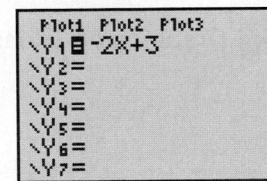 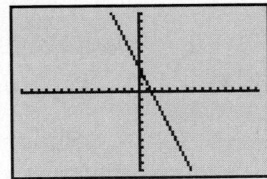

Figure 5 Graph of $y = -2x + 3$

From Table 3 and Fig. 4, we see that the y-intercept is $(0, 3)$.

▶ Example 4   Graphing an Equation

Sketch the graph of $y = \frac{3}{2}x - 2$. Also, find the y-intercept.

Solution

In Table 4, we use 0 and multiples of 2 as values of x to avoid fractional values of y. Since $y = \frac{3}{2}x - 2$ is of the form $y = mx + b$, the graph is a line. We plot the points that correspond to the solutions we found and sketch the line that contains them in Fig. 6.

Table 4 Solutions of $y = \frac{3}{2}x - 2$

x	y
0	$\frac{3}{2}(0) - 2 = -2$
2	$\frac{3}{2}(2) - 2 = 1$
4	$\frac{3}{2}(4) - 2 = 4$

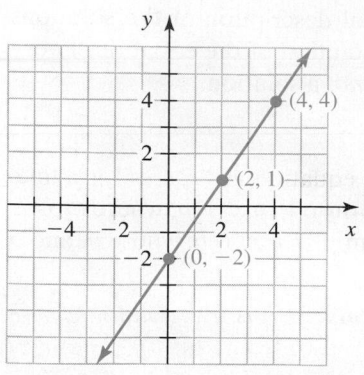

Figure 6 Graph of $y = \dfrac{3}{2}x - 2$

We use ZDecimal to verify our graph (see Fig. 7). See Appendices A.14, A.15, and A.17 for TI-84 instructions.

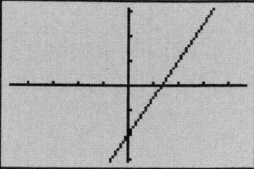

Figure 7 Graph of $y = \dfrac{3}{2}x - 2$

From Table 4 and Fig. 6, we see that the y-intercept is $(0, -2)$.

The Meaning of b for an Equation of the Form $y = mx + b$

Recall from Section 6.3 that the x-coordinate of a y-intercept is 0. For a line with an equation of the form $y = mx + b$, the y-intercept is $(0, b)$. For instance, in Example 3, we found that the line $y = -2x + 3$ has y-intercept $(0, 3)$. In Example 4, we found that the line $y = \dfrac{3}{2}x - 2$ has y-intercept $(0, -2)$.

Now consider any equation of the form $y = mx + b$. Substituting 0 for x gives

$$y = m(0) + b = 0 + b = b$$

which shows that the y-intercept is $(0, b)$.

> **y-Intercept of the Graph of $y = mx + b$**
>
> The graph of an equation of the form $y = mx + b$ has y-intercept $(0, b)$.

For the line $y = -5x + 9$, the y-intercept is $(0, 9)$, and for the line $y = 8x - 4$, the y-intercept is $(0, -4)$.

Equations of the Form $y = b$ and $x = a$

In Example 5, we will explore how to graph an equation of the form $y = b$, where b is a constant.

Table 5 Solutions of $y = 3$

x	y
-2	3
-1	3
0	3
1	3
2	3

▶ **Example 5**   Graphing an Equation of the Form $y = b$

Sketch the graph of $y = 3$.

Solution

Note that y must be 3, but x can have any value. Some solutions of $y = 3$ are listed in Table 5. We plot the corresponding points and sketch the line through them (see Fig. 8). The graph of $y = 3$ is a horizontal line.

We can use a TI-84 to verify our graph (see Fig. 9).

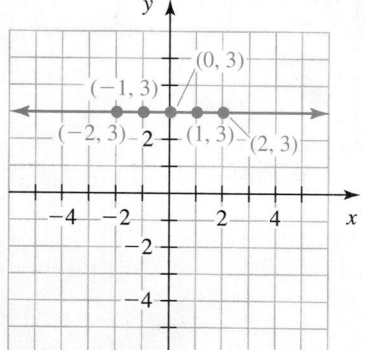

Figure 8 Graph of $y = 3$

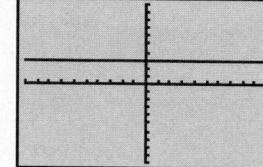

Figure 9 The graph of $y = 3$

In Example 5, we saw that the graph of the equation $y = 3$ is a horizontal line. Any equation that can be written in the form $y = b$, where b is a constant, has a horizontal line as its graph.

Table 6 Solutions of $x = 2$

x	y
2	-2
2	-1
2	0
2	1
2	2

▶ **Example 6** Graphing an Equation of the Form $x = a$

Sketch the graph of $x = 2$.

Solution

Note that x must be 2, but y can have any value. Some solutions of $x = 2$ are listed in Table 6. We plot the corresponding points and sketch the line through them (see Fig. 10). The graph of $x = 2$ is a vertical line.

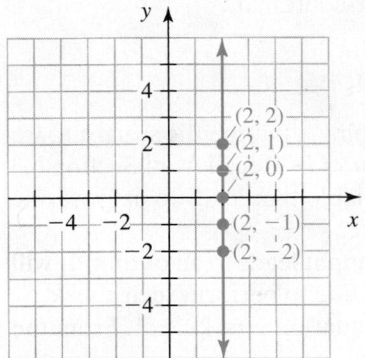

Figure 10 Graph of $x = 2$

In Example 6, we saw that the graph of the equation $x = 2$ is a vertical line. Any equation that can be written in the form $x = a$, where a is a constant, has a vertical line as its graph.

▶ **Equations for Horizontal and Vertical Lines**

If a and b are constants, then

- The graph of $y = b$ is a horizontal line (see Fig. 11).
- The graph of $x = a$ is a vertical line (see Fig. 12).

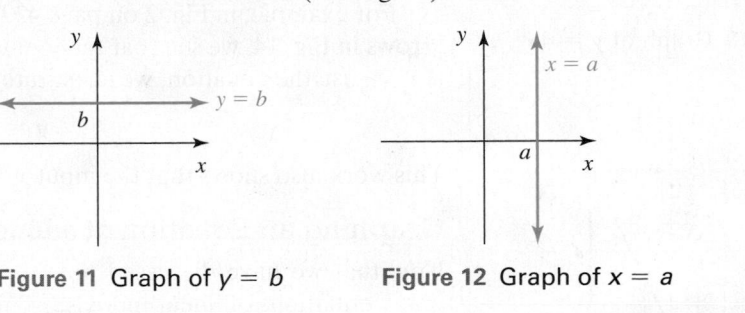

Figure 11 Graph of $y = b$ **Figure 12** Graph of $x = a$

For example, the graphs of the equations $y = 5$ and $y = -2$ are horizontal lines. The graphs of the equations $x = 6$ and $x = -4$ are vertical lines.

▶ **Equations Whose Graphs Are Lines**

If an equation can be put into the form

$$y = mx + b \quad \text{or} \quad x = a$$

where m, a, and b are constants, then the graph of the equation is a line. We call such an equation a **linear equation in two variables**.

Any equation that can be put into the form $x = a$ has a vertical line as its graph. Any equation that can be put into the form $y = mx + b$ has a nonvertical line as its graph.

Rule of Four for Equations

We can describe the solutions of an equation in two variables in four ways. For instance, in Example 4, we described the solutions of the equation $y = \frac{3}{2}x - 2$ by using the equation and a graph (see Fig. 6). We also described some of the solutions by using a table (see Table 4). Finally, we can describe the solutions verbally: For each solution, the y-coordinate is three-halves of the x-coordinate minus 2.

▶ **Rule of Four for Solutions of an Equation**

We can describe some or all of the solutions of an equation in two variables with:

1. an equation,   **2.** a table,   **3.** a graph, or   **4.** words.

These four ways to describe solutions are known as the **Rule of Four**.

Table 7 Solutions of $y = 4x$

x	y
-1	$4(-1) = -4$
0	$4(0) = 0$
1	$4(1) = 4$

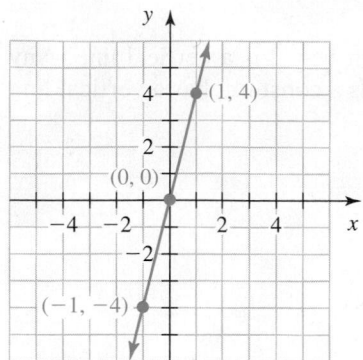

Figure 13 Graph of $y = 4x$

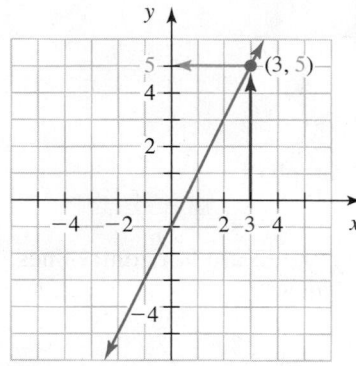

Figure 14 The input $x = 3$ leads to the output $y = 5$

▶ **Example 7**   Describing Solutions by Using the Rule of Four

1. List some solutions of $y = 4x$ by using a table.
2. Describe the solutions of $y = 4x$ by using a graph.
3. Describe the solutions of $y = 4x$ by using words.

Solution

1. We list three solutions in Table 7.
2. We plot the solutions listed in Table 7 and sketch the line through them (see Fig. 13).
3. For each solution, the y-coordinate is four times the x-coordinate.

Recall from Section 6.3 that when we are not describing an authentic situation, we use x as the *explanatory variable* and y as the *response variable*. Recall from Section 6.3 that an *input* is a permitted value of the explanatory variable that leads to at least one *output,* which is a permitted value of the response variable.

In Section 6.3, we used a sketched line to see how an input leads to an output. It will often be easier and more efficient to use an *equation* of a line to perform such a task.

For example, in Fig. 2 on page 432, we graphed the equation $y = 2x - 1$. From the arrows in Fig. 14, we see that the input $x = 3$ leads to the output $y = 5$.

To use the equation, we substitute 3 for x in $y = 2x - 1$:

$$y = 2(3) - 1 = 5$$

This work also shows that the input $x = 3$ leads to the output $y = 5$.

Graphing an Equation of a Linear Model

Now that we have discussed how to graph linear equations in two variables, we can graph equations of linear models.

▶ **Example 8**   Graphing an Equation of a Linear Model

In the United Kingdom, nervous dental patients are sometimes given a sedative to calm them. To administer the sedative, the dentist performs a *cannulation*, which involves inserting a thin tube into a vein in a patient's hand. In an experiment involving 20 individuals, researchers tested whether the amount of pain the individuals experienced during cannulation explains the level of anxiety that the individuals predicted they would experience by undergoing cannulation again. After undergoing cannulations, the individuals rated their experiences of pain from 0 to 100 and their predicted anxieties about undergoing cannulations again from 0 to 100. The patients' ratings are shown in Table 8.

Table 8 Pain Ratings versus Predicted-Anxiety Ratings

Pain Rating	Anxiety Rating	Pain Rating	Anxiety Rating	Pain Rating	Anxiety Rating	Pain Rating	Anxiety Rating
37	50	23	13	29	35	69	68
63	70	38	35	34	30	14	3
8	2	60	20	41	48	12	10
10	10	38	30	75	70	49	43
70	81	42	42	8	32	48	43

Source: Anaesthetics: A Randomised, Double-Blind, Placebo-Controlled, Comparative Study of Topical Skin Analgesics and the Anxiety and Discomfort Associated with Venous Cannulation, *A. F. Speirs et al.*

Let a be the anxiety rating a patient predicts for undergoing another cannulation after having undergone one with a pain rating of p.

1. Identify the explanatory and response variables.
2. Construct a scatterplot.
3. Describe and interpret the direction of the association.
4. Graph the model $a = 0.91p + 1.71$ on the scatterplot.
5. Does the model come close to the data points?
6. Use the model to predict the anxiety rating a patient would predict about undergoing another cannulation after having undergone one with a pain rating of 70.

Solution

1. Researchers wanted to test whether the pain rating explains the predicted-anxiety rating, so p is the explanatory variable and a is the response variable.
2. We construct a scatterplot in Fig. 15.

Pain versus Predicted Anxiety

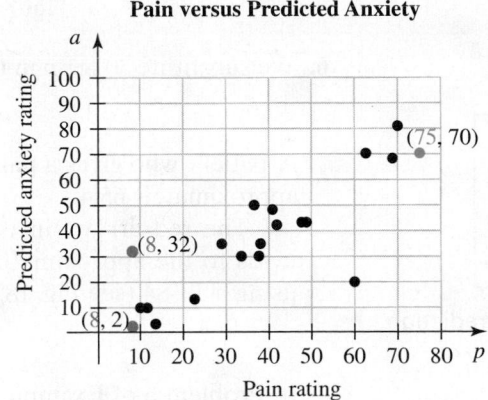

Figure 15 Pain-anxiety scatterplot

3. The association is positive, which means the larger the pain rating, the larger the predicted-anxiety rating tends to be.
4. The p-coordinates of the data points are between 8 and 75, inclusive (see Table 8 and Fig. 15). The line should extend a bit beyond the data on both the left-hand and right-hand sides. So, for values to substitute for p, we choose 5 and 80. We also choose 60 as a check. We list the solutions in Table 9. Then we plot the (red) points and draw the linear model from (5, 6) to (80, 75) in Fig. 16.

Table 9 Approximate Solutions of $a = 0.91p + 1.71$

p	a
5	6
60	56
80	75

Pain versus Predicted Anxiety

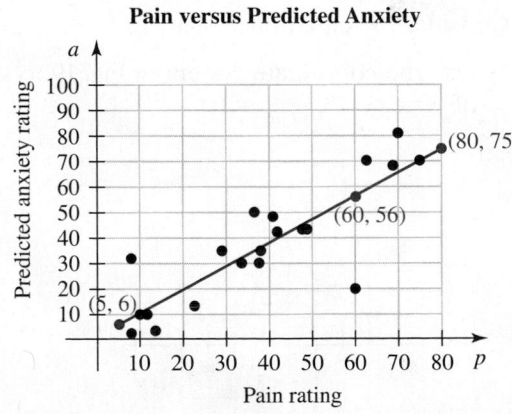

Figure 16 Pain-anxiety scatterplot and model

5. The model comes fairly close to the data points except for (60, 20), which is an outlier (see Fig. 17). The outlier represents a patient who gave a relative high pain rating (60) but predicted a relatively low anxiety rating (20).

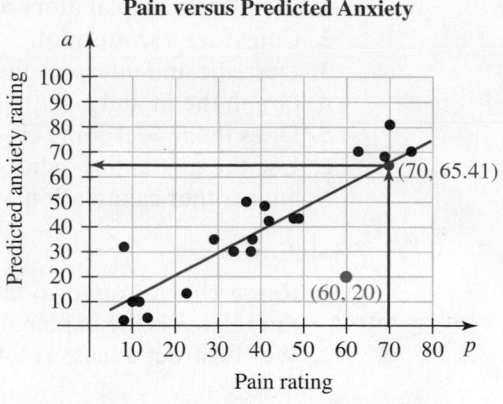

Figure 17 Pain-anxiety scatterplot and model

6. We substitute 70 for p in the equation $a = 0.91p + 1.71$:

$$a = 0.91(70) + 1.71 \approx 65.4$$

A patient who gives a pain rating of 70 tends to predict their anxiety rating will be approximately 65.4.

   The red arrows in the scatterplot in Fig. 17 confirm that the input $p = 70$ leads to the approximate output $a = 65.4$. We can also verify our prediction by using a TI-84 (see Fig. 18). See Appendix A.20 for TI-84 instructions.

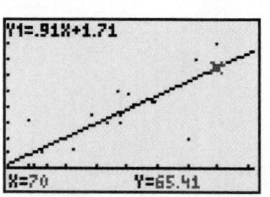

Figure 18 Verify the prediction ▶

In Problem 6 of Example 8, our predicted value 65.4 for an anxiety rating has one more decimal place than the data values for anxiety ratings. In general, we will round predicted values for a response variable to one more decimal place than the data values for the response variable.

In Example 8, we used an equation of a linear model to make a prediction. But how do we find such an equation? We will discuss one way in Section 7.3.

▲▼ Group Exploration
Solutions of an equation

Consider the equation $y = x - 3$.

1. Use the coordinate system in Fig. 19 to sketch a graph of $y = x - 3$.

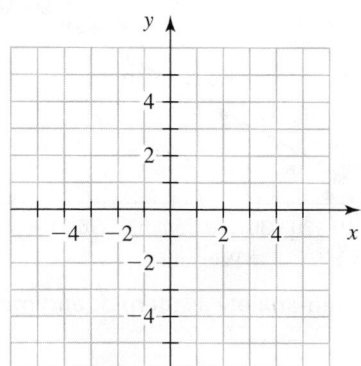

Figure 19 Sketch a graph of $y = x - 3$

2. Pick three points that lie on the graph of $y = x - 3$. Do the coordinates of these points satisfy the equation $y = x - 3$?

3. Pick three points that do not lie on the graph of $y = x - 3$. Do the coordinates of these points satisfy the equation $y = x - 3$?

4. Which ordered pairs satisfy the equation $y = x - 3$? There are too many to list, but describe them in words. [**Hint:** You should say something about the points that do or do not lie on the line.]

5. The graph of an equation is sketched in Fig. 20. Which of the points A, B, C, D, E, and F represent ordered pairs that satisfy the equation?

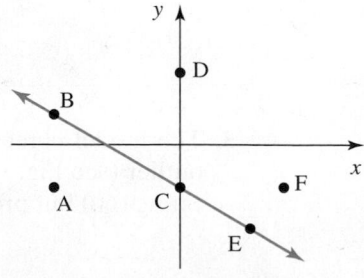

Figure 20 Problem 5

Homework 7.1

1. When an equation in terms of x and y becomes a true statement when a is substituted for x and b is substituted for y, we say (a, b) ____ the equation.

2. The ____ of an equation in two variables is the set of points that correspond to all solutions of the equation.

3. The graph of an equation of the form $y = mx + b$, where m and b are constants, is a(n) ____.

4. The graph of $x = a$, where a is a constant, is a(n) ____ line.

Which of the given ordered pairs satisfy the given equation?

5. $y = 2x - 4$ $(-3, -10), (1, -3), (2, 0)$

6. $y = 4x - 12$ $(-2, -20), (1, -8), (3, 0)$

7. $y = -3x + 7$ $(-1, 4), (0, 7), (4, -5)$

8. $y = -5x + 8$ $(-2, 3), (0, 8), (3, -7)$

Find the y-intercept. Also, graph the equation by hand.

9. $y = x + 2$ 10. $y = x + 4$ 11. $y = x - 4$

12. $y = x - 6$ 13. $y = 2x$ 14. $y = 5x$

15. $y = -3x$ 16. $y = -4x$ 17. $y = x$

18. $y = -x$ 19. $y = \frac{1}{3}x$ 20. $y = \frac{1}{2}x$

21. $y = -\frac{5}{3}x$ 22. $y = -\frac{3}{2}x$ 23. $y = 2x + 1$

24. $y = 3x + 2$ 25. $y = 5x - 3$ 26. $y = 4x - 1$

27. $y = -3x + 5$ 28. $y = -2x + 4$ 29. $y = -2x - 3$

30. $y = -4x - 2$ 31. $y = \frac{1}{2}x - 3$ 32. $y = \frac{1}{3}x - 2$

33. $y = -\frac{2}{3}x + 1$ 34. $y = -\frac{4}{3}x + 5$

Graph the equation by hand.

35. $x = 3$ 36. $x = 6$ 37. $y = 1$ 38. $y = 5$

39. $y = -2$ 40. $y = -4$ 41. $x = -1$ 42. $x = -3$

43. $x = 0$ 44. $y = 0$

45. The numbers of Winter Olympic medal events are shown in Table 10 for various years.

Table 10 Numbers of Winter Olympic Medal Events

Year	Number of Medal Events
1988	46
1992	57
1994	61
1998	68
2002	78
2006	84
2010	86
2014	98

Source: *International Olympic Committee*

Let n be the number of Winter Olympic medal events at t years since 1980.

a. Identify the explanatory and response variables.

b. Construct a scatterplot by hand.

c. Is the association positive, negative, or neither? What does that mean in this situation?

d. Graph the model $n = 1.88t + 33.75$ by hand on the scatterplot. Does the line come close to the data points?

e. Use the model to estimate the number of medal events in 2010. Did you perform interpolation or extrapolation? Calculate the error in the estimate.

46. The number of priests in the world are shown in Table 11 for various years.

Table 11 Numbers of Priests

Year	Number of Priests (thousands)
1985	57
1990	52
1995	49
2000	46
2005	41
2010	40
2014	38

Source: *Official Catholic Directory*

Let n be the number of priests (in thousands) at t years since 1980.

a. Identify the explanatory and response variables.

b. Construct a scatterplot by hand.

c. Is the association positive, negative, or neither? What does that mean in this situation?

d. Graph the model $n = -0.65t + 59.07$ by hand on the scatterplot. Does the line come close to the data points?

e. Use the model to estimate the number of priests in 2011. Did you perform interpolation or extrapolation?

47. Worldwide sales of iPhones® are shown in Table 12 for the last three months of various years.

Table 12 Worldwide Sales of iPhones in the Last Three Months of Various Years

Year	Worldwide Sales (millions)
2009	8.7
2010	16.2
2011	37.0
2012	47.8
2013	51.0

Source: *Apple*

Let s be worldwide iPhone sales (in millions) for the last three months of the year that is t years since 2005.

a. Identify the explanatory and response variables.

b. Construct a scatterplot by hand.

c. Graph the model $s = 11.62t - 37.58$ by hand on the scatterplot. Does the line come close to the data points?

d. Use the model to estimate worldwide sales of iPhones for the last three months of 2014. The actual sales were 74.5 million iPhones. Give a possible explanation of why analysts' predictions of the sales turned out to be less than 74.5 million iPhones (Source: The Wall Street Journal).

e. Compute the number of iPhones that were sold per hour throughout the day and night in the last three months of 2014 on average.

48. A researcher recorded the rate at which a cricket chirps for various temperatures. A random selection of the data is shown in Table 13.

Table 13 Temperatures versus Chirp Rates

Temperature (°F)	Chirp Rate (number of chirps per 13 seconds)
57	18
73	36
66	33
81	44
51	13
59	24
69	29

Source: The Sound of Crickets: Using Evidence-Based Reasoning to Measure Temperature Using Cricket Chirps, *Peggy LeMone,* The Science Teacher, *Vol. 76, No. 8, November 2009*

Let R be the chirp rate (number of chirps per 13 seconds) when the temperature is F degrees Fahrenheit.

a. Identify the explanatory and response variables.

b. Construct a scatterplot by hand.

c. Is the association positive, negative, or neither? What does that mean in this situation?

d. Graph the model $R = 1.02F - 38.03$ by hand on the scatterplot. Does the line come close to the data points?

e. Use the model to predict the chirp rate when the temperature is 70°F.

49. The operating costs per trip of bus services and their mean fares, both in dollars, are shown in Table 14 for some randomly selected fixed-route bus services.

Table 14 Operating Costs per Trip and Mean Fares

Bus Service	Operating Cost per Trip (dollars)	Mean Fare (dollars)
CATS (Charlotte)	3.75	0.79
Port Authority (Pittsburgh)	5.00	1.23
LA Metro (Los Angeles)	2.60	0.68
Sound Transit (Puget Sound)	7.30	1.59
SMART (Southeast Michigan)	6.90	1.14
ITA (Jacksonville)	5.20	0.80
VIA (San Antonio)	2.90	0.49

Source: *Federal Transit Administration*

Let C be the operating cost per trip and F be the mean fare, both in dollars, for a fixed-route bus service.

a. For the association between C and F, identify the explanatory and response variables.

b. Construct a scatterplot by hand that describes the association between C and F.

c. Is the association positive, negative, or neither? What does that mean in this situation? Why does this make sense?

d. Graph the model $F = 0.18C + 0.11$ by hand on the scatterplot. Does the model come close to the data points?

e. Use the model to predict the mean fare of a fixed-route bus service with operating cost $4.25.

50. The numbers (in thousands) of vehicle-train collisions and injuries are shown in Table 15 for various years.

Table 15 Numbers of Vehicle-Train Collisions and Injuries

Year	Number of Collisions (thousands)	Number of Injuries (thousands)
1981	9.5	3.3
1985	7.1	2.7
1990	5.7	2.4
1995	4.6	1.9
2000	3.5	1.2
2005	3.1	1.1
2010	2.1	0.9
2014	2.3	0.8

Source: *Federal Railroad Administration*

Let I be the number (in thousands) of injuries from vehicle-train collisions in a year when there were C collisions (in thousands).

a. For the association between C and I, identify the explanatory and response variables.

b. Construct a scatterplot by hand that describes the association between C and I.

c. Is the association positive, negative, or neither? What does that mean in this situation? Why does this make sense?

d. Graph the model $I = 0.36C + 0.10$ by hand on the scatterplot. Does the model come close to the data points?

e. Use the model to predict the number of injuries in a year when there were 5 thousand vehicle-train collisions.

Concepts

51. Recall that we can describe some or all of the solutions of an equation in two variables with an equation, a table, a graph, or words.

a. Describe three solutions of $y = 2x - 3$ by using a table.

b. Describe the solutions of $y = 2x - 3$ by using a graph.

c. Describe the solutions of $y = 2x - 3$ by using words.

52. Recall that we can describe some or all of the solutions of an equation in two variables with an equation, a table, a graph, or words.

a. Describe three solutions of $y = -4x + 5$ by using a table.

b. Describe the solutions of $y = -4x + 5$ by using a graph.

c. Describe the solutions of $y = -4x + 5$ by using words.

53. a. For the equation $y = 3x + 1$, find all outputs for the input $x = 2$. State how many outputs there are for that single input.

b. For $y = 3x + 1$, how many outputs originate from any single input? Explain.

c. Give an example of an equation of the form $y = mx + b$. Using your equation, find all outputs for the input $x = 3$. State how many outputs there are for that single input.

d. For your equation, how many outputs originate from any single input? Explain.

e. For *any* equation of the form $y = mx + b$, how many outputs originate from a single input? Explain.

54. a. Graph $y = 2x - 4$ by hand.

b. For the equation $y = 2x - 4$, find all outputs for the input $x = 3$. Explain by using arrows on your graph in part (a). State how many outputs there are for that single input.

c. For $y = 2x - 4$, how many outputs originate from any single input? Explain in terms of drawing arrows.

d. Give an example of an equation of the form $y = mx + b$. Graph your equation by hand.

e. Using your equation, find all outputs for the input $x = 2$. Explain by using arrows on your graph in part (d). State how many outputs there are for that single input.

f. For your equation, how many outputs originate from any single input? Explain in terms of drawing arrows.

g. For *any* equation of the form $y = mx + b$, how many outputs originate from a single input? Explain in terms of drawing arrows.

55. a. Graph the equation by hand. Find all x-intercepts and y-intercepts.

  i. $y = 3x$ **ii.** $y = -2x$ **iii.** $y = \dfrac{2}{5}x$

b. What are the intercepts of the graph of an equation of the form $y = mx$, where $m \neq 0$?

56. Find the intersection point of the lines $y = 4x$ and $y = -5x$. Try to do this without graphing.

57. The graph of an equation is sketched in Fig. 21. Create a table of ordered-pair solutions of this equation. Include at least five ordered pairs.

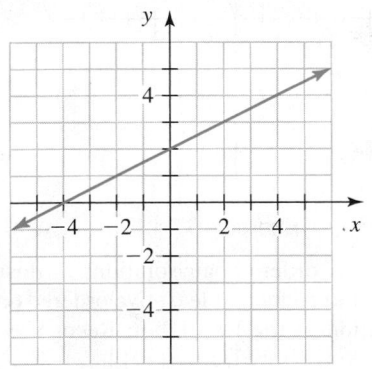

Figure 21
Exercise 57

58. The graph of an equation is sketched in Fig. 22. Create a table of ordered-pair solutions of this equation. Your table should contain at least five ordered pairs.

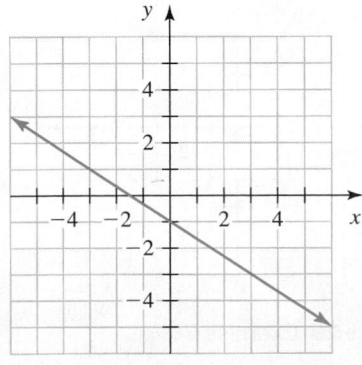

Figure 22
Exercise 58

For Exercises 59–66, refer to the graph sketched in Fig. 23.

59. Find y when $x = -4$.   **60.** Find y when $x = 0$.

61. Find y when $x = 2$.   **62.** Find y when $x = -2$.

63. Find x when $y = -1$.   **64.** Find x when $y = 0$.

65. Find x when $y = 2$.   **66.** Find x when $y = 3$.

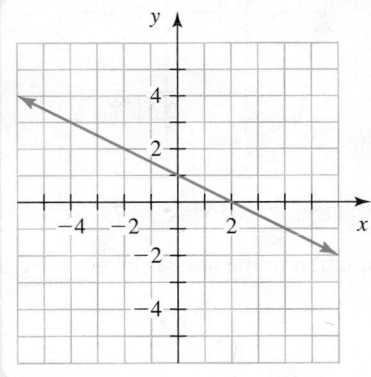

Figure 23 Exercises 59–66

67. The graph of an equation is sketched in Fig. 24. Which of the points A, B, C, D, E, and F represent ordered pairs that satisfy the equation?

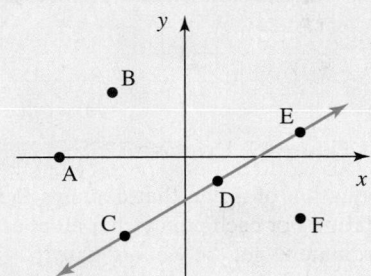

Figure 24 Exercise 67

68. The graphs of equations 1 and 2 are sketched in Fig. 25.

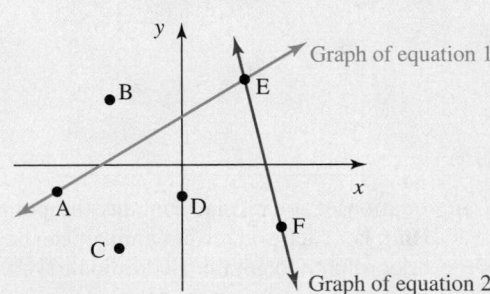

Figure 25 Exercise 68

For each part, decide which one or more of the points A, B, C, D, E, and F represent ordered pairs that

a. satisfy equation 1.

b. satisfy equation 2.

c. satisfy both equations.

d. do not satisfy either equation.

69. Find a solution of $y = x + 2$ that lies in Quadrant II. How many solutions of this equation are in Quadrant II?

70. Find a solution of $y = x - 4$ that lies in Quadrant III. How many solutions of this equation are in Quadrant III?

71. Find an equation of the line sketched in Fig. 26.

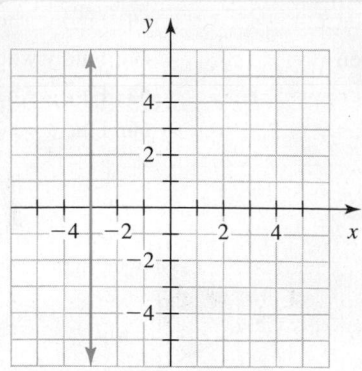

Figure 26 Exercise 71

72. Find an equation of the line sketched in Fig. 27.

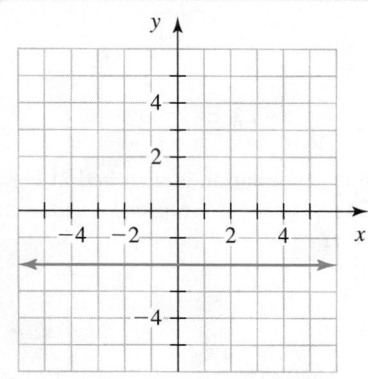

Figure 27 Exercise 72

73. Find an equation of a line that contains the points listed in Table 16. [**Hint:** For each point, what number can be added to the x-coordinate to get the y-coordinate?]

Table 16 Points on a Line
(Exercise 73)

x	y
0	3
1	4
2	5
3	6
4	7

74. Find an equation of a line that contains the points listed in Table 17. [**Hint:** For each point, what number can be subtracted from the x-coordinate to get the y-coordinate?]

Table 17 Points on a Line
(Exercise 74)

x	y
0	−1
1	0
2	1
3	2
4	3

75. Find an equation of a line that contains the points listed in Table 18.

Table 18 Points on a Line
(Exercise 75)

x	y
0	0
1	1
2	2
3	3
4	4

76. Find an equation of a line that contains the points listed in Table 19.

Table 19 Points on a Line
(Exercise 76)

x	y
0	0
1	−1
2	−2
3	−3
4	−4

77. The graph of an equation is sketched in Fig. 28.

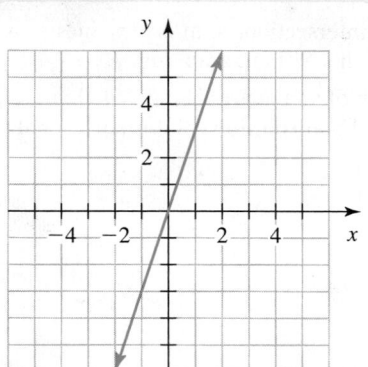

Figure 28 Exercise 77

a. Create a table of ordered-pair solutions of this equation. Your table should contain at least five ordered pairs.
b. Find an equation of the line. [**Hint:** Recognize a pattern from the table you created in part (a).]

78. The graph of an equation is sketched in Fig. 29.

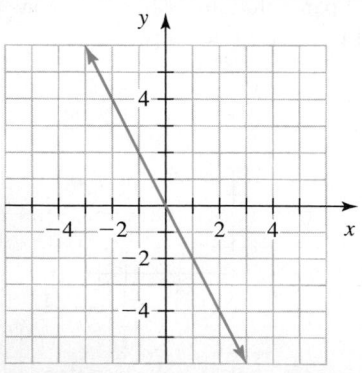

Figure 29 Exercise 78

a. Create a table of ordered-pair solutions of this equation. Your table should contain at least five ordered pairs.

b. Find an equation of the line. [**Hint:** Recognize a pattern from the table you created in part (a).]

79. Graph $x + y = 5$ by hand. [**Hint:** Assume that the graph is a line. Think of pairs of numbers whose sum is 5.]

80. Graph $x - y = 5$ by hand. [**Hint:** Assume that the graph is a line. Think of pairs of numbers whose difference is 5.]

81. Why does the graph of the equation $y = mx + b$ have y-intercept $(0, b)$?

82. Find an ordered pair that satisfies both of the equations $y = x + 1$ and $y = -2x + 4$. [**Hint:** Graph the equations.]

83. Assume a is a constant. Is the graph of $x = a$ a horizontal line, a vertical line, or neither? Explain.

84. Assume b is a constant. Is the graph of $y = b$ a horizontal line, a vertical line, or neither? Explain.

85. Does every line have an x-intercept? If yes, explain why. If no, give an equation of a line that doesn't have one.

86. Give an example of an equation of the form $y = mx + b$, where m and b are constants. Graph the equation by hand.

87. Describe how to graph an equation of the form $y = mx + b$. (See page 12 for guidelines on writing a good response.)

88. In your own words, describe the Rule of Four for equations. (See page 12 for guidelines on writing a good response.)

▼ 7.2 Rate of Change and Slope of a Line

Objectives

» Calculate the rate of change of a quantity with respect to another quantity.

» Explain the connection between constant rate of change and an exact linear association.

» Use a run and the corresponding rise of a linear model to find a rate of change.

» Describe the connection between the direction of an association and rate of change.

» Calculate the slope of a line.

» Explain why the slope of an increasing line is positive and the slope of a decreasing line is negative.

» Explain why the absolute value of the slope of a line measures the steepness of the line.

» Explain why the slope of a horizontal line is zero and the slope of a vertical line is undefined.

In Chapter 6 and this chapter, we have discussed the direction of an association. For instance, in Example 8 of Section 6.1, we found that there is a negative association between years and gambling revenues in Atlantic City. That is, we determined that the annual revenue is declining, but we did not estimate by how much per year.

In this section, we will determine *how quickly* the response variable changes as the explanatory variable increases. For instance, in Example 5, we will estimate how quickly the number of large ships lost at sea per year declined for the period 2006–2014.

Calculating Rate of Change

Suppose the temperature increased *steadily* by 6°F in the past 3 hours. We can compute how much the temperature changed *per hour* by finding the unit ratio of the change in temperature (6°F) to the change in time (3 hours):

$$\frac{6°F}{3 \text{ hours}} = \frac{2°F}{1 \text{ hour}}$$

So, the temperature increased by 2°F per hour. This is an example of a *rate of change*. We say the rate of change of temperature with respect to time is 2°F per hour. The rate of change is a *constant* because the temperature increased *steadily*.

Here are some other examples of rates of change:

• The number of friends on a student's Facebook account increases by 10 friends per month.

• The revenue of a company decreases by $3 million per year.

• A college charges $70 per unit (hour or credit).

▶ **Example 1**   Finding Rates of Change

1. An airplane's altitude increases steadily by 10,000 feet over a 5-minute period. Find the rate of change of altitude with respect to time.

2. The temperature increased steadily from 60°F at 6 A.M. to 72°F at 10 A.M. Find the rate of change of temperature with respect to time.

Solution

1. We find the unit ratio of the change in altitude (10,000 feet) to the change in time (5 minutes):

$$\frac{\text{change in altitude}}{\text{change in time}} = \frac{10{,}000 \text{ feet}}{5 \text{ minutes}} \qquad \textit{Find ratio.}$$

$$= \frac{2000 \text{ feet}}{1 \text{ minute}} \qquad \textit{Find unit ratio.}$$

So, the airplane climbed at a rate of 2000 feet per minute.

2. We find the unit ratio of the change in temperature to the change in time:

$$\frac{\text{change in temperature}}{\text{change in time}} = \frac{72°F - 60°F}{10{:}00 - 6{:}00} \qquad \begin{array}{l}\textit{Change in a quantity is ending}\\ \textit{amount minus beginning amount.}\end{array}$$

$$= \frac{12°F}{4 \text{ hours}} \qquad \textit{Subtract.}$$

$$= \frac{3°F}{1 \text{ hour}} \qquad \textit{Find unit ratio.}$$

So, the temperature increased 3°F per hour.

▶ **Formula for Rate of Change**

Suppose that a quantity y changes steadily from y_1 to y_2 as a quantity x changes steadily from x_1 to x_2. Then the **rate of change** of y with respect to x is the ratio of the change in y to the change in x:

$$\frac{\text{change in } y}{\text{change in } x} = \frac{y_2 - y_1}{x_2 - x_1}$$

We often refer to rate of change *with respect to time* simply as "rate of change."

▶ **Example 2**   Finding Rates of Change

1. The number of postal employees declined approximately steadily from 583,908 employees in 2010 to 491,017 employees in 2013 (Source: *U.S. Postal Service*). Find the approximate rate of change of the number of postal employees.
2. The total cost of 12 karate classes and an enrollment fee is $158. The total cost of 20 karate classes and the same enrollment fee is $230. The charge per class is the same, regardless of the number of classes for which you pay. Find the rate of change of the total cost with respect to the number of classes.

Solution

1. $$\frac{\text{change in number of employees}}{\text{change in time}}$$

$$= \frac{491{,}017 \text{ employees} - 583{,}908 \text{ employees}}{\text{year } 2013 - \text{year } 2010} \qquad \begin{array}{l}\textit{Change in a quantity is ending}\\ \textit{amount minus beginning amount.}\end{array}$$

$$= \frac{-92{,}891 \text{ employees}}{3 \text{ years}} \qquad \textit{Subtract.}$$

$$\approx \frac{-30{,}963.7 \text{ employees}}{1 \text{ year}} \qquad \textit{Find unit ratio.}$$

The rate of change of the number of postal employees was about −30,963.7 employees per year. The number of employees had a yearly decline of about 30,963.7 employees. Our result is an approximation because we rounded and because the number of postal employees declined *approximately* steadily.

2. To be consistent in finding the signs of the changes, we assume that the number of classes increases from 12 to 20 and that the total cost increases from \$158 to \$230:

$$\frac{\text{change in total cost}}{\text{change in number of classes}} = \frac{230 \text{ dollars} - 158 \text{ dollars}}{20 \text{ classes} - 12 \text{ classes}}$$

Change in a quantity is ending amount minus beginning amount.

$$= \frac{72 \text{ dollars}}{8 \text{ classes}}$$

Subtract.

$$= \frac{9 \text{ dollars}}{1 \text{ class}}$$

Find unit ratio.

The rate of change of the total cost with respect to the number of classes is \$9 per class. So, the cost of each class is \$9. Our result is exact because it is given that the charge per class is the same.

Connection between Constant Rate of Change and an Exact Linear Association

▶ **Example 3** Analyzing an Association That Involves a Constant Rate of Change

Suppose that a student travels at 50 miles per hour on a road trip. Let d be the distance (in miles) that the student can drive in t hours.

1. Identify the explanatory and response variables.
2. Construct a table to describe the association between t and d for driving times of 0, 1, 2, 3, 4, and 5 hours.
3. Construct a scatterplot for driving times of 0, 1, 2, 3, 4, and 5 hours.
4. Describe the four characteristics of the association. Compute and interpret r as part of your analysis. If there is an association, draw an appropriate model in the scatterplot.

Solution

1. By letting d be the distance that the student can drive in t hours, we are viewing time to be what explains the distance traveled. So, the explanatory variable is t and the response variable is d.
2. At time 0 hours, the distance traveled is 0 miles. After driving 50 miles per hour for 1 hour, the distance traveled is 50 miles. After 2 hours, the distance traveled is $2 \cdot 50 = 100$ miles. After 3 hours, the distance traveled is $3 \cdot 50 = 150$ miles. We summarize these findings as well as the ordered pairs for 4 and 5 miles in Table 20.
3. We construct a scatterplot in Fig. 30.
4. There are no outliers. The association is exactly linear and positive (see Fig. 31). The value of r is 1, which checks with the association being exactly linear and positive.

Table 20 Time versus Distance

Time (hours)	Distance (miles)
0	0
1	50
2	100
3	150
4	200
5	250

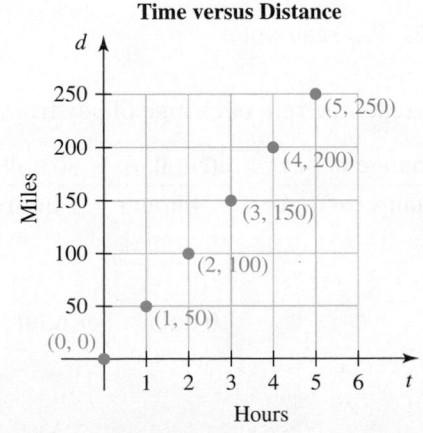

Figure 30 Car scatterplot

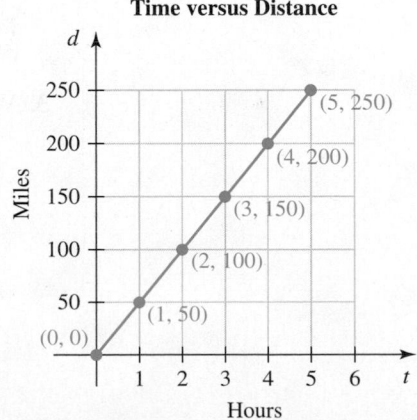

Figure 31 Car model and scatterplot

In Example 3, we found that the association between time and distance for a car traveling at a *constant* rate (50 miles per hour) is exactly linear.

> **Constant Rate of Change Implies an Exact Linear Association**
>
> If the rate of change of one variable with respect to another variable is constant, then there is an exact linear association between the variables.

▶ **Example 4**   Finding Rates of Change of a Linear Association

The amounts of money a student is paid are shown in Table 21 for various numbers of hours worked.

Let p be the student's pay (in dollars) for working t hours.

1. Identify the explanatory and response variables.
2. Describe the four characteristics of the association. Compute and interpret r as part of your analysis.
3. Find the rate of change of pay from 3 to 4 hours.
4. Find the rate of change of pay from 0 to 5 hours. Compare the result with the result you found in Problem 3.

Table 21 Time versus Pay

Time (hours)	Pay (dollars)
0	0
1	10
2	20
3	30
4	40
5	50

Solution

1. The amount of time the student works explains the student's pay. So, the explanatory variable is t and the response variable is p.
2. We construct a scatterplot in Fig. 32. There are no outliers. The association is exactly linear and positive (see Fig. 33). The value of r is 1, which checks with the association being exactly linear and positive.

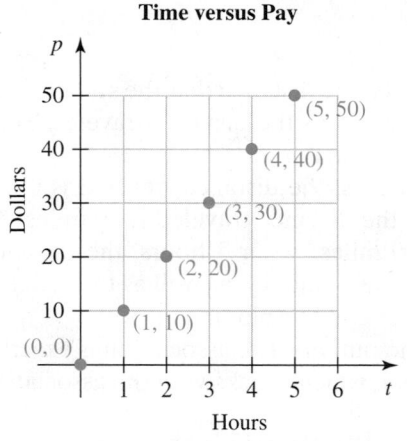

Figure 32 Pay scatterplot

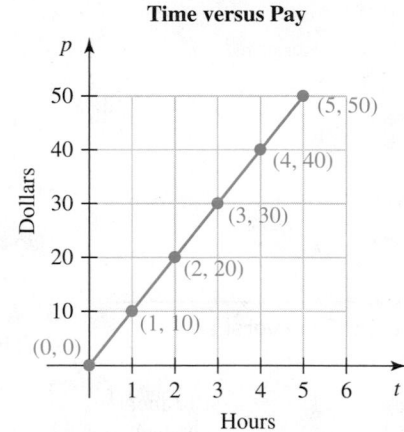

Figure 33 Pay model and scatterplot

3. We calculate the rate of change of pay from $t = 3$ to $t = 4$:

$$\frac{\text{change in pay}}{\text{change in time}} = \frac{40 \text{ dollars} - 30 \text{ dollars}}{4 \text{ hours} - 3 \text{ hours}} \quad \textit{Change in a quality is ending amount minus beginning amount.}$$

$$= \frac{10 \text{ dollars}}{1 \text{ hour}} \quad \textit{Subtract.}$$

$$= 10 \text{ dollars per hour} \quad \textit{Divide.}$$

4. We calculate the rate of change of pay from $t = 0$ to $t = 5$:

$$\frac{\text{change in pay}}{\text{change in time}} = \frac{50 \text{ dollars} - 0 \text{ dollars}}{5 \text{ hours} - 0 \text{ hours}} \qquad \textit{Change in a quantity is ending amount minus beginning amount.}$$

$$= \frac{50 \text{ dollars}}{5 \text{ hours}} \qquad \textit{Subtract.}$$

$$= \frac{10 \text{ dollars}}{1 \text{ hours}} \qquad \textit{Find unit ratio.}$$

$$= 10 \text{ dollars per hour} \qquad \textit{Divide.}$$

The result is equal to the result we found in Problem 3.

In Example 4, we found that for the exact linear association between time and pay, the rate of change of pay is equal (to $10 per hour) for two different periods. In fact, we could show that the rate of change of pay is $10 per hour for *any* period when the student works. In other words, the rate of change of pay is constant.

> **An Exact Linear Association Implies Constant Rate of Change**
>
> If there is an exact linear association between two variables, then the rate of change of one variable with respect to the other is constant.

Use Rise and Run to Compute Rate of Change

In Problem 3 of Example 4, we calculated the rate of change of pay from $t = 3$ to $t = 4$:

$$\frac{\text{change in pay}}{\text{change in time}} = \frac{10 \text{ dollars}}{1 \text{ hour}}$$

We interpret this graphically in Fig. 34. To go from point $(3, 30)$ to point $(4, 40)$, we look 1 hour to the right and then look 10 dollars up. So, the horizontal change, called the *run*, is 1 hour and the vertical change, called the *rise*, is 10 dollars. The rate of change of pay is equal to the ratio of the rise to the run:

$$\text{rate of change} = \frac{\text{rise}}{\text{run}} = \frac{10 \text{ dollars}}{1 \text{ hour}}$$

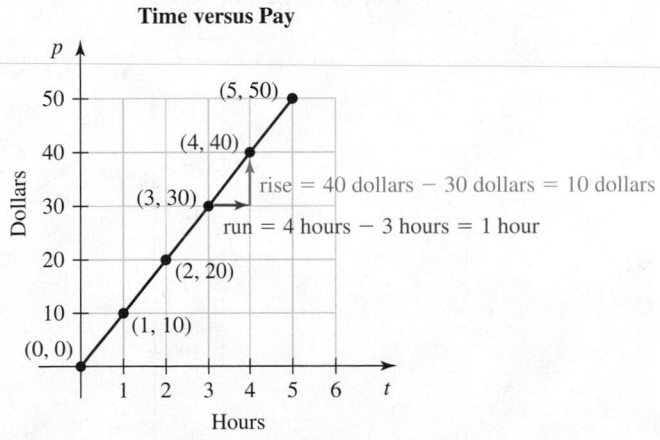

Time versus Pay

Figure 34   rate of change $= \dfrac{\text{rise}}{\text{run}} = \dfrac{10 \text{ dollars}}{1 \text{ hour}}$

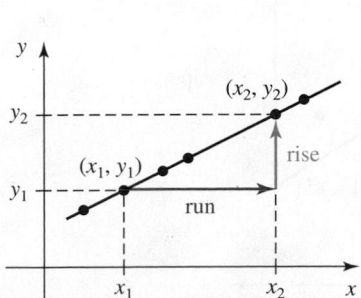

Figure 35 The *rise* and *run* between any two points

For any linear model, the **run** is the horizontal change and the **rise** is the vertical change in going from one point on the line to another point on the line (see Fig. 35).

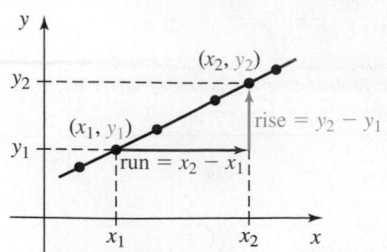

Figure 36 The rate of change of y with respect to x is
$$\frac{y_2 - y_1}{x_2 - x_1} = \frac{rise}{run}$$

> ### Using Run and Rise to Find Rate of Change
>
> Assume there is an exact linear association between an explanatory variable x and a response variable y and (x_1, y_1) and (x_2, y_2) are two distinct points of the linear model. Then the rate of change of y with respect to x is the ratio of the rise to the run in going from point (x_1, y_1) to point (x_2, y_2):
>
> $$\text{rate of change of } y \atop \text{with respect to } x = \frac{y_2 - y_1}{x_2 - x_1} = \frac{rise}{run}$$
>
> See Fig. 36.

▶ **Example 5**   Estimating Rates of Change

1. A *carat* is a unit of weight for precious stones and pearls. The scatterplot and the model in Fig. 37 describe the association between the weights (in carats) of some diamonds and their value (in Singapore dollars). Is the association positive, negative, or neither? Estimate the rate of change of value with respect to number of carats.

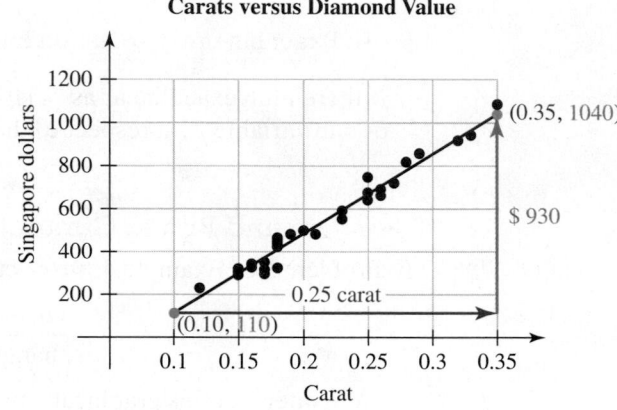

Figure 37 Diamond model
(**Source:** *Straits Times*)

2. The scatterplot and the model in Fig. 38 describe the association between years and the number of ships lost at sea due to fire, collision, storm, and machine breakdown. Is the association positive, negative, or neither? Estimate the rate of change of number of large ship losses.

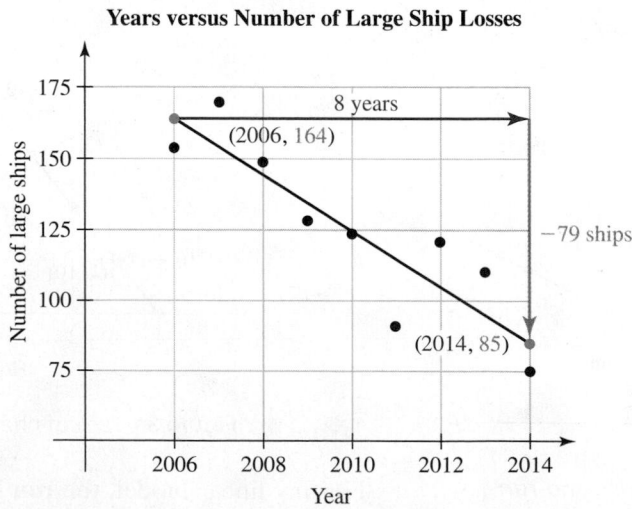

Figure 38 Ship model
(**Source:** *Lloyd's List Intelligence Casualty Statistics*)

Solution

1. The association is positive. To estimate the rate of change, we begin by estimating that the ends of the line segment are the ordered pairs $(0.10, 110)$ and $(0.35, 1040)$. See the green dots in the scatterplot in Fig. 37. Then using the formula for rate of change with $(x_1, y_1) = (0.10, 110)$ and $(x_2, y_2) = (0.35, 1040)$, we have

$$\text{rate of change} = \frac{y_2 - y_1}{x_2 - x_1} = \frac{1040 \text{ dollars} - 110 \text{ dollars}}{0.35 \text{ carats} - 0.10 \text{ carats}} = \frac{930 \text{ dollars}}{0.25 \text{ carats}}$$
$$= 3720 \text{ dollars per carat}$$

Because the association is not exactly linear, our result is approximate. So, if one diamond weighs 1 carat more than another diamond, it will cost approximately \$3720 more than the other diamond.

2. The association is negative. To estimate the rate of change, we begin by estimating that the ends of the line segment are the ordered pairs $(2006, 164)$ and $(2014, 85)$. See the green dots in the scatterplot in Fig. 38. Then using the formula for rate of change with $(x_1, y_1) = (2006, 164)$ and $(x_2, y_2) = (2014, 85)$, we have

$$\text{rate of change} = \frac{y_2 - y_1}{x_2 - x_1} = \frac{85 \text{ ships} - 164 \text{ ships}}{\text{year } 2014 - \text{year } 2006} = \frac{-79 \text{ ships}}{8 \text{ years}}$$
$$\approx -9.9 \text{ ships per year}$$

Because we rounded and more significantly because the association is not exactly linear, our result is approximate. We conclude that the number of large ship losses had a yearly decline of approximately 9.9 ships.

WARNING   In Problem 1 of Example 5, we found that if one diamond weighs 1 carat more than another diamond, it will cost approximately \$3720 more than the other diamond. This does *not* mean that the diamonds cost approximately \$3720 per carat. In fact, our estimate that a 0.1-carat diamond costs \$110 means that it costs $110 \div 0.1 = 1100$ dollars per carat. And our estimate that a 0.35-carat diamond costs \$1040 means that it costs $1040 \div 0.35 \approx 2971$ dollars per carat, which is a lot more than 1100 dollars per carat.

Connection between the Direction of an Association and Rate of Change

In Problem 1 of Example 5, we found that the diamond association is *positive* and the rate of change of the value of the diamonds with respect to their weight is *positive*. In Problem 2 of Example 5, we found that the ship-loss association is *negative* and the rate of change of ship losses is *negative*.

▶ **Connection between the Direction of an Association and Rate of Change**

- If an association between two variables is positive, then the approximate rate of change of one variable with respect to the other is positive (see Fig. 39).
- If an association between two variables is negative, then the approximate rate of change of one variable with respect to the other is negative (see Fig. 40).

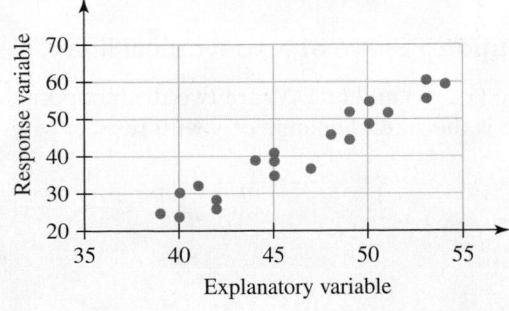

Positive Association and Positive Approximate Rate of Change

Figure 39 Positive association and positive approximate rate of change

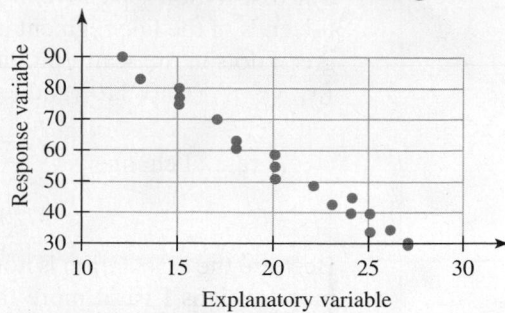

Figure 40 Negative association and negative approximate rate of change

WARNING   If the association between two variables is *nonlinear*, then the rate of change of the response variable with respect to the explanatory variable is *not* constant. For example, consider the company Uber, which enables people to use an app to request a ride from the company's closest driver. Table 22 and the scatterplot in Fig. 41 compare the number of cities where Uber operates with various years.

Table 22 Numbers of Cities where Uber Operates

Year	Number of Cities
2010	1
2011	7
2012	19
2013	67
2014	276

Source: *Uber*

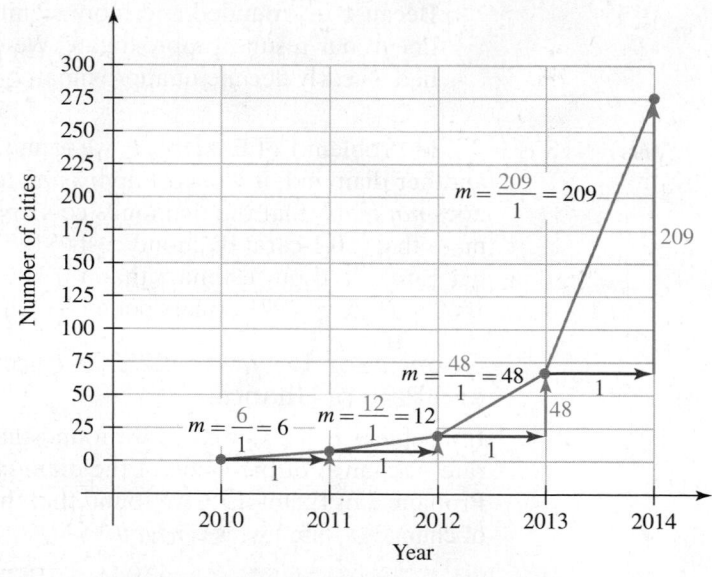

Figure 41 Numbers of cities where Uber operates

From the slopes of the line segments, we see that the rates of change (in number of cities per year) from one year to the next are *not* constant: 6, 12, 48, 209. In fact, they increase by increasing amounts!

Slope of a Line

We can find the rate of change of y with respect to x for a nonvertical line whether or not it is a model. We call the rate of change the *slope* of the line and use the letter m to represent it.

▶ **Definition**   Slope of a nonvertical line

Assume (x_1, y_1) and (x_2, y_2) are two distinct points of a nonvertical line. The **slope** of the line is the rate of change of y with respect to x. In symbols:

$$m = \text{slope} = \frac{y_2 - y_1}{x_2 - x_1} = \frac{\text{rise}}{\text{run}}$$

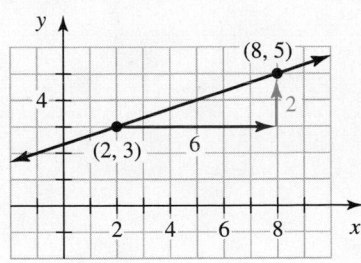

Figure 42 The slope is $\dfrac{2}{6} = \dfrac{1}{3}$

▶ **Example 6** Finding the Slope of a Line

Find the slope of the line that contains the points $(2, 3)$ and $(8, 5)$.

Solution

Using the slope formula with $(x_1, y_1) = (2, 3)$ and $(x_2, y_2) = (8, 5)$, we have

$$m = \frac{y_2 - y_1}{x_2 - x_1} = \frac{5 - 3}{8 - 2} = \frac{2}{6} = \frac{1}{3}$$

By plotting points, we find that if the run is 6, then the rise is 2 (see Fig. 42). So, the slope is $m = \dfrac{\text{rise}}{\text{run}} = \dfrac{2}{6} = \dfrac{1}{3}$, which is the same as our result from using the formula.

In Example 6, we calculated the slope of a line for $(x_1, y_1) = (2, 3)$ and $(x_2, y_2) = (8, 5)$. Here, we switch the roles of the two points to find the slope when $(x_1, y_1) = (8, 5)$ and $(x_2, y_2) = (2, 3)$:

$$m = \frac{y_2 - y_1}{x_2 - x_1} = \frac{3 - 5}{2 - 8} = \frac{-2}{-6} = \frac{1}{3}$$

The result is the same as that in Example 6. In general, when we use the slope formula with two points on a line, it doesn't matter which point we choose to be first, (x_1, y_1), and which point we choose to be second, (x_2, y_2).

WARNING   It is a common error to make incorrect substitutions into the slope formula. Carefully consider why the middle and right-hand formulas are incorrect:

Correct	**Incorrect**	**Incorrect**
$m = \dfrac{y_2 - y_1}{x_2 - x_1}$	$m = \dfrac{y_2 - y_1}{x_1 - x_2}$	$m = \dfrac{x_2 - x_1}{y_2 - y_1}$

▶ **Example 7** Finding the Slope of a Line

Find the slope of the line that contains the points $(3, 4)$ and $(7, 2)$.

Solution

Using the formula for slope with $(x_1, y_1) = (3, 4)$ and $(x_2, y_2) = (7, 2)$, we have

$$m = \frac{y_2 - y_1}{x_2 - x_1} = \frac{2 - 4}{7 - 3} = \frac{-2}{4} = -\frac{1}{2}$$

By plotting points, we find that when the run is 4, the rise is -2 (see Fig. 43). So, the slope is $\dfrac{-2}{4} = -\dfrac{1}{2}$, which is the same as our result from using the formula.

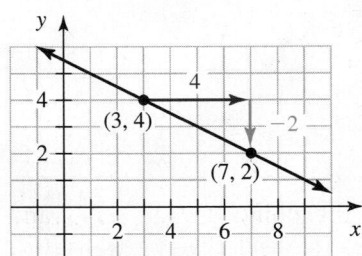

Figure 43 The slope is $\dfrac{-2}{4} = -\dfrac{1}{2}$

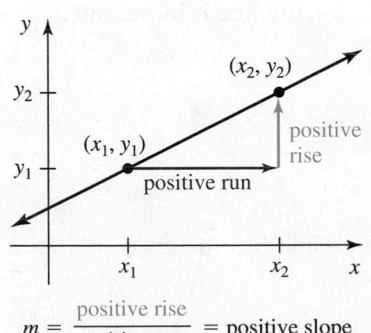

$$m = \frac{\text{positive rise}}{\text{positive run}} = \text{positive slope}$$

Figure 44 Increasing lines have positive slope

Slopes of Increasing and Decreasing Lines

Since the line in Fig. 44 goes upward from left to right, we say the line (and the graph) is **increasing.** A sign analysis of the rise and run in Fig. 44 shows that the slope of the line is positive. This is consistent with the fact that if an association between two variables is positive, then the approximate rate of change of one variable with respect to the other is positive.

Since the line in Fig. 45 goes downward from left to right, we say the line (and the graph) is **decreasing.** A sign analysis of the rise and run in Fig. 45 shows that the slope of the line is negative. This is consistent with the fact that if an association between two variables is negative, then the approximate rate of change of one variable with respect to the other is negative.

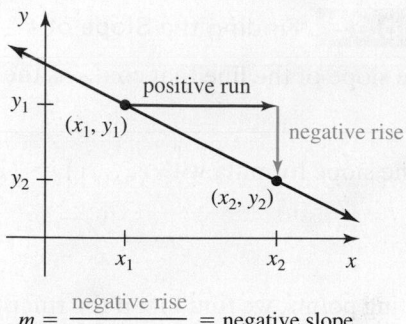

$$m = \frac{\text{negative rise}}{\text{positive run}} = \text{negative slope}$$

Figure 45 Decreasing lines have negative slope

> ### Slopes of Increasing or Decreasing Lines
>
> - An increasing line has positive slope (see Fig. 44).
> - A decreasing line has negative slope (see Fig. 45).

When we compute the slope of two points that have negative coordinates, it can help to write first

$$\frac{(\;) - (\;)}{(\;) - (\;)}$$

and then insert the coordinates of the two points into the appropriate parentheses.

▶ **Example 8** Finding the Slope of a Line

Find the slope of the line that contains the points $(-5, 2)$ and $(3, -4)$.

Solution

$$m = \frac{(-4) - (2)}{(3) - (-5)} = \frac{-4 - 2}{3 + 5} = \frac{-6}{8} = -\frac{6}{8} = -\frac{3}{4}$$

Since the slope is negative, the line is decreasing.

▶ **Example 9** Finding the Slope of a Line

Find the approximate slope of the line that contains the points $(-4.9, -3.5)$ and $(-2.3, 5.8)$. Round the result to the second decimal place.

Solution

$$m = \frac{(5.8) - (-3.5)}{(-2.3) - (-4.9)} = \frac{9.3}{2.6} \approx 3.58$$

So, the slope is approximately 3.58. Since the slope is positive, the line is increasing.

Measuring the Steepness of a Line

In Example 10, we will compare the slopes of two lines.

▶ **Example 10** Comparing the Slopes of Two Lines

Find the slopes of the two lines sketched in Fig. 46. Which line has the greater slope? Which line is steeper?

Solution

In Fig. 47, we see that for line l_1, if the run is 1, the rise is 1. We calculate the slope of line l_1:

$$\text{Slope of line } l_1 = \frac{\text{rise}}{\text{run}} = \frac{1}{1} = 1$$

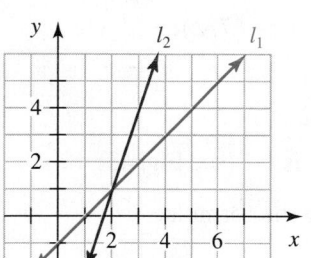

Figure 46 Find the slopes of the two lines

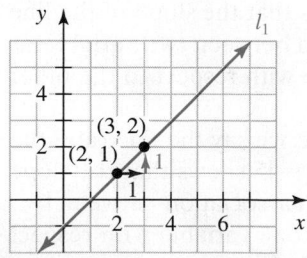

Figure 47 Line with lesser slope

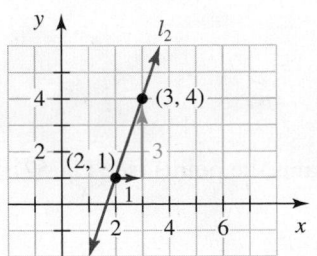

Figure 48 Line with greater slope

In Fig. 48, we see that for line l_2, if the run is 1, the rise is 3. We calculate the slope of line l_2:

$$\text{Slope of line } l_2 = \frac{\text{rise}}{\text{run}} = \frac{3}{1} = 3$$

The slope of line l_2 is greater than the slope of line l_1, and line l_2 is steeper than line l_1. So, the steeper line has the greater slope.

In Fig. 49, we show three decreasing lines and three increasing lines and their slopes. In Fig. 50, we show the same lines and the absolute values of their slopes.

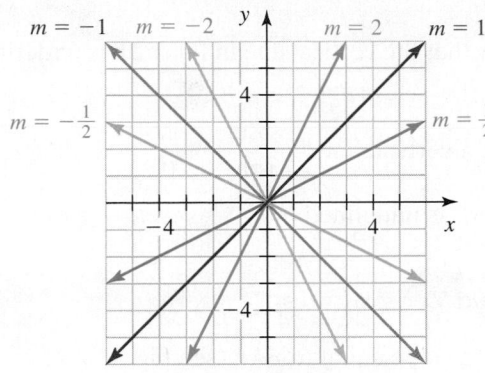

Figure 49 Slopes of some lines

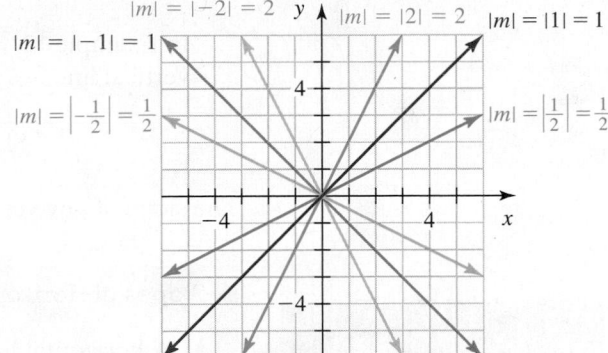

Figure 50 Absolute values of slopes of some lines

Of the six lines, the steepest (blue) lines have the largest absolute value of slope (2). The next steepest (red) lines have the next largest absolute value of slope (1). The least steepest (green) lines have the least absolute value of slope $\left(\frac{1}{2}\right)$.

> **Measuring the Steepness of a Line**
>
> The absolute value of the slope of a line measures the steepness of the line. The steeper the line, the larger the absolute value of its slope will be.

Slopes of Horizontal and Vertical Lines

So far, we have discussed slopes of lines that are increasing or decreasing. What about the slope of a horizontal line or the slope of a vertical line?

> ▶ **Example 11** Finding the Slope of a Horizontal Line

Find the slope of the line that contains the points (3, 2) and (7, 2).

Solution

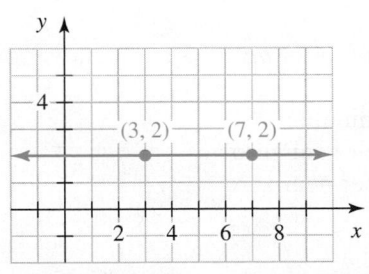

Figure 51 The horizontal line has zero slope

We plot the points (3, 2) and (7, 2) and sketch the line that contains the points (see Fig. 51). The formula for slope gives

$$m = \frac{2 - 2}{7 - 3} = \frac{0}{4} = 0$$

So, the slope of this horizontal line is zero.

In Example 11, we saw that the horizontal line in Fig. 51 has a slope equal to zero. Note that *any* horizontal line has no rise, so

$$\text{Slope of a horizontal line} = \frac{\text{rise}}{\text{run}} = \frac{0}{\text{run}} = 0$$

The slope of any horizontal line is zero.

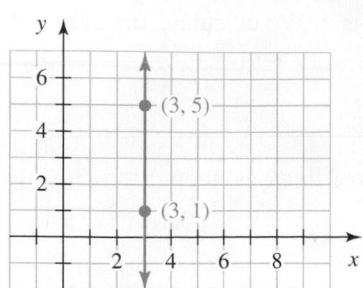

Figure 52 The vertical line has undefined slope

▶ **Example 12** Finding the Slope of a Vertical Line

Find the slope of the line that contains the points $(3, 1)$ and $(3, 5)$.

Solution

We plot the points $(3, 1)$ and $(3, 5)$ and sketch the line that contains the points (see Fig. 52). The formula for slope gives

$$m = \frac{5 - 1}{3 - 3} = \frac{4}{0}$$

Since division by zero is undefined, the slope of the vertical line is *undefined*.

In Example 12, we saw that the vertical line in Fig. 52 has undefined slope. Note that *any* vertical line has zero run, so

$$\text{Slope of a vertical line} = \frac{\text{rise}}{\text{run}} = \left.\frac{\text{rise}}{0}\right\} \quad \text{undefined}$$

The slope of any vertical line is undefined.

▶ **Slopes of Horizontal and Vertical Lines**

- A horizontal line has a slope equal to zero (see Fig. 53).
- A vertical line has undefined slope (see Fig. 54).

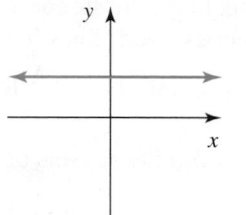

Figure 53 Horizontal lines have a slope equal to zero

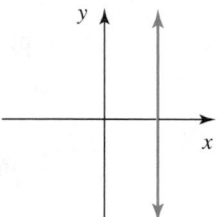

Figure 54 Vertical lines have undefined slope

▲▼ Group Exploration

Section Opener: Graphical significance of *m* for *y = mx*

1. Use technology to graph these equations of the form $y = mx$ in order, and describe what you observe:

$$y = x, \qquad y = 2x, \qquad y = 3x, \qquad y = 4x$$

2. Give an example of an equation of the form $y = mx$ whose graph is a line steeper than the lines you sketched in Problem 1.

3. Use technology to graph these equations in order, and describe what you observe:

$$y = -x, \qquad y = -2x, \qquad y = -3x, \qquad y = -4x$$

4. Describe the graph of $y = mx$ in the following situations:
   a. m is a large positive number
   b. m is a positive number near zero
   c. m is a negative number near zero
   d. m is less than -10
   e. $m = 0$

5. Describe what you have learned in this exploration.

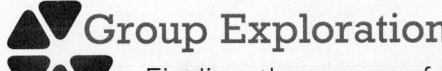

Group Exploration

Finding the mean of rates of change

The numbers (in thousands) of service members in the armed forces diagnosed as overweight are shown in Table 23 for various years.

Table 23 Numbers of Service Members Diagnosed as Overweight

Years	Number of Overweight Service Members (thousands)
2006	56
2007	62
2008	72
2009	79
2010	86

Source: *Defense Department*

1. Estimate the rate of change of the number of overweight service members from 2006 to 2010. [**Hint:** Use the 2006 and 2010 data in Table 23.]

2. Find the rate of change of the number of overweight service members from each year to the next, beginning in 2006. [**Hint:** You should find four results.]

3. Find the mean of the four rates of change you found in Problem 2.

4. Compare the result you found in Problem 3 with the result you found in Problem 1.

5. The following expression is an example of a *telescoping sum*:

$$(62 - 56) + (72 - 62) + (79 - 72) + (86 - 79)$$

Explain why the above sum is equal to $86 - 56$.

6. Use the result you found in Problem 5 to help explain why the following statement is true:

$$\frac{\dfrac{62 - 56}{2007 - 2006} + \dfrac{72 - 62}{2008 - 2007} + \dfrac{79 - 72}{2009 - 2008} + \dfrac{86 - 79}{2010 - 2009}}{4}$$

$$= \frac{86 - 56}{2010 - 2006}$$

Also, explain how this statement is related to the comparison you made in Problem 4.

7. Describe what you have learned from doing this exploration.

Group Exploration

For a line, rise over run is constant

1. A line is sketched in Fig. 55. Plot the points $(-2, -5)$, $(1, 1)$, and $(3, 5)$. (Plotted correctly, these points will lie on the line.)

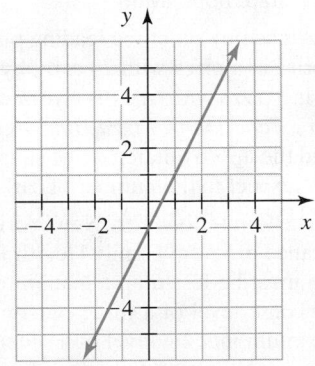

Figure 55 Use different pairs of points to calculate the slope

2. Using the points $(-2, -5)$ and $(1, 1)$, find the slope of the line.

3. Using the points $(1, 1)$ and $(3, 5)$, find the slope of the line.

4. Using the points $(-2, -5)$ and $(3, 5)$, find the slope of the line.

5. Using two other points of your choice, find the slope of the line.

6. What do you notice about the slopes you have calculated? Does it matter which two points on a line are used to find the slope of the line?

Tips for Success   Reread a Problem

After you think you have solved a problem, reread it to make sure you have answered its question(s). Also, reread the problem with the solution in mind. If what you read makes sense, then you've provided another check of your result(s).

Homework 7.2

For extra help ▶ MyMathLab®    Watch the videos in MyMathLab    Download the MyDashboard App

1. If there is an exact linear association between two variables, then the rate of change of one variable with respect to the other is ____.

2. The ____ of a line is the rate of change of y with respect to x.

3. A decreasing line has ____ slope.

4. The absolute value of the slope of a line measures the ____ of the line.

5. The volume of water in a swimming pool increases steadily by 2400 gallons in an 8-hour period. Find the rate of change of volume of water.

6. The height of a ti plant increases steadily by 48 inches in 12 years. Find the rate of change of the ti plant's height.

7. An airplane's altitude declines steadily by 24,750 feet over a 15-minute period. Find the rate of change of the airplane's altitude.

8. The temperature decreases steadily by 15°F over a 3-hour period. Find the rate of change of temperature.

9. The number of female-owned firms increased approximately steadily from 5.4 million firms in 1997 to 8.6 million firms in 2013 (Source: *U.S. Census Bureau*). Find the approximate rate of change of the number of female-owned firms.

10. The number of worldwide subscribers of Netflix increased approximately steadily from 22.9 million users in 2011 to 62.3 million users in 2015 (Source: *Netflix*). Find the approximate rate of change of the number of worldwide Netflix users.

11. The Steller sea lion population decreased approximately steadily from about 300 thousand in 1980 to about 40 thousand in 2012 (Source: *National Marine Mammal Laboratory*). Find the approximate rate of change of the Steller sea lion population.

12. The interest rate for subsidized student loans decreased approximately steadily from 6.8% in 2008 to 3.4% in 2012 (Source: *U.S. Department of Education*). Estimate the approximate rate of change of the interest rate for subsidized student loans.

13. The number of U.S. cable TV subscribers decreased approximately steadily from 61.8 million subscribers in 2009 to 54.3 million subscribers in 2014 (Source: *IHS*).

Find the approximate rate of change of the number of U.S. cable TV subscribers.

14. The percentage of Americans who say they or their spouses have saved money for retirement decreased approximately steadily from 75% in 2009 to 66% in 2012 (Source: *Employee Benefit Research Institute*). Find the approximate rate of change of this percentage.

15. For the fall semester 2014, in-district students at Triton College paid $972 for 9 hours (units) of classes and $1296 for 12 hours of classes (Source: *Triton College*). Find the rate of change of the total cost of classes with respect to the number of hours of classes.

16. In Manhattan, the mean price of a two-bedroom house is $1,694,547, and that of a four-bedroom house is $6,315,496 (Source: *Trulia*). Find the approximate rate of change of mean price with respect to the number of bedrooms.

17. Nebraska has 3 seats in the House of Representatives and a population of 1.83 million. Illinois has 18 seats and a population of 12.83 million (Source: *U.S. Census Bureau*). Find the approximate rate of change of a state's number of seats with respect to the state population.

18. For 29 pizzas made by six of the leading pizza companies, there are 246 calories in a pizza with 28 carbohydrates and there are 549 calories in a pizza with 64 carbohydrates (Source: *Domino's, Little Caesar's, Papa John's, Pizza Hut, DiGiorno Frozen, Kashi Frozen*). Find the approximate rate of change of the number of calories with respect to the number of carbohydrates.

19. In order for a family living in New York to qualify for the health insurance program Family Health Plus, the family's annual income must be less than a maximum level. In 2012, the maximum income level of a three-person family was $28,635 and the maximum income level of a seven-person family was $52,395 (Source: *New York State Department of Health*). Find the approximate rate of change of maximum income level with respect to family size.

20. A person stacks some cups of uniform shape and size (one placed inside the next). The height of 3 stacked cups is 17.5 centimeters and of 5 stacked cups is 23.0 centimeters. Find the rate of change of the height of the stacked cups with respect to the number of cups.

21. Suppose that a stock is worth $5 on October 1 and that its value increases by $3 per week. Let v be the stock's value (in dollars) at t weeks since October 1.

   a. Identify the explanatory and response variables.
   b. Construct a table to describe the association between t and v for the following values of t: 0, 1, 2, 3, 4, and 5.
   c. Use technology to construct a scatterplot for the ordered pairs listed in the table you constructed in part (b).
   d. Describe the four characteristics of the association. Compute and interpret r as part of your analysis. If there is an association, draw an appropriate model in the scatterplot.
   e. Explain how this exercise suggests that if the rate of change of one variable with respect to another variable is constant, then there is an exact linear association between the variables.

22. When a person begins a car trip, there are 12 gallons of gasoline in the car's tank. During the trip, the car consumes 2 gallons of gasoline per hour. Let G be the volume (in gallons) of gasoline after the person has driven t hours.

   a. Identify the explanatory and response variables.
   b. Construct a table to describe the association between t and G for the following values of t: 0, 1, 2, 3, 4, and 5.
   c. Use technology to construct a scatterplot for the ordered pairs listed in the table you constructed in part (b).
   d. Describe the four characteristics of the association. Compute and interpret r as part of your analysis. If there is an association, draw an appropriate model in the scatterplot.
   e. Explain how this exercise suggests that if the rate of change of one variable with respect to another variable is constant, then there is an exact linear association between the variables.

23. **DATA** Before starting a diet, a person weighs 180 pounds. The person's weights (in pounds) are shown in Table 24 for various numbers of weeks since going on the diet.

Table 24 Time versus Person's Weight

Number of Months on Diet	Weight (pounds)
0	180
1	175
2	170
3	165
4	160
5	155

Let w be the person's weight (in pounds) after being on the diet for t months.

   a. Use technology to construct a scatterplot.
   b. Describe the four characteristics of the association. Compute and interpret r as part of your analysis.
   c. Find the rate of change of the person's weight from $t = 2$ to $t = 3$.
   d. Find the rate of change of the person's weight from $t = 3$ to $t = 5$.
   e. Find the rate of change of the person's weight from $t = 0$ to $t = 5$. Compare the result to the results you found in parts (c) and (d).
   f. Explain how this exercise suggests that if there is an exact linear association between two variables, then the

rate of change of one variable with respect to the other is constant.

24. **DATA** The balances (in thousands of dollars) of a person's savings account are shown in Table 25 for various numbers of years since the person opened the account.

Table 25 Years versus Balances in Savings Account

Number of Years Since Account Was Opened	Balance (thousands of dollars)
0	3
1	7
2	11
3	15
4	19
5	23

Let B be the balance (in thousands of dollars) at t years since the account was opened.

   a. Use technology to construct a scatterplot.
   b. Describe the four characteristics of the association. Compute and interpret r as part of your analysis.
   c. Find the rate of change of the balance from $t = 3$ to $t = 4$.
   d. Find the rate of change of the balance from $t = 1$ to $t = 3$.
   e. Find the rate of change of the balance from $t = 0$ to $t = 5$. Compare the result to the results you found in parts (c) and (d).
   f. Explain how this exercise suggests that if there is an exact linear association between two variables, then the rate of change of one variable with respect to the other is constant.

25. The scatterplot and the model in Fig. 56 describe the association between years and the numbers (in millions) of Americans without health insurance.

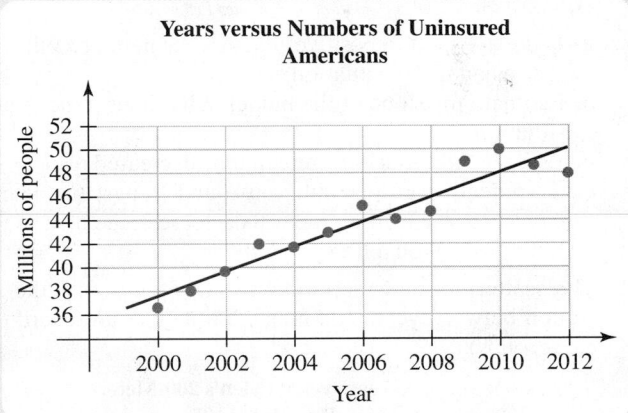

Figure 56 Exercise 25
(**Source:** *U.S. Census Bureau*)

   a. Is the association positive, negative, or neither? What does that mean in this situation?
   b. Estimate the rate of change of the number of Americans without health insurance.

26. The scatterplot and the model in Fig. 57 describe the association between years and the total wealth (in billions of dollars) of the world's 80 richest people.

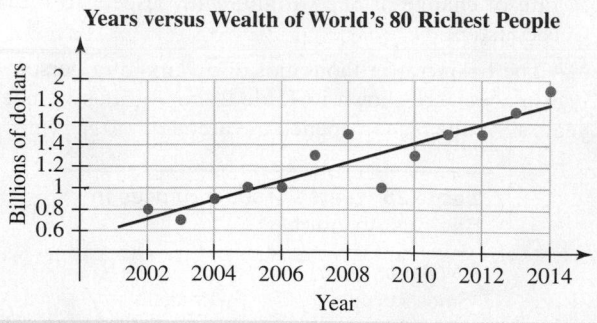

Figure 57 Exercise 26
(**Source:** *Oxfam*)

a. Is the association positive, negative, or neither? What does that mean in this situation?
b. Estimate the rate of change of the total wealth of the world's 80 richest people.

27. The scatterplot and the model in Fig. 58 describe the association between years and women's 200-meter run record times (in seconds).

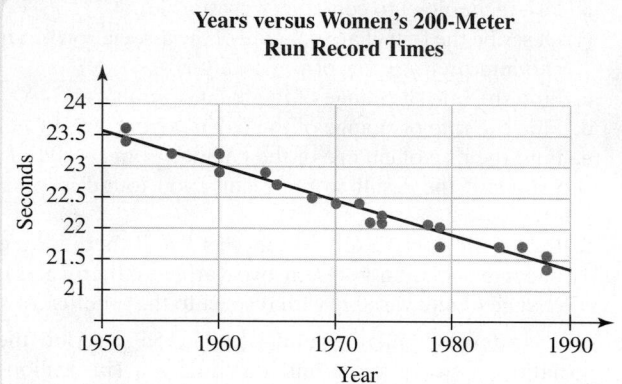

Figure 58 Exercise 27
(**Source:** *International Association of Athletics Federations*)

a. Is the association positive, negative, or neither? What does that mean in this situation?
b. Estimate the slope of the model. What does it mean in this situation?
c. Estimate how much the record time decreased over 15 years.
d. A student says the result you found in part (c) describes how much the record time will decrease in the next 15 years. What would you tell the student?

28. The scatterplot and the model in Fig. 59 describe the association between years and men's 200-meter run record times (in seconds).

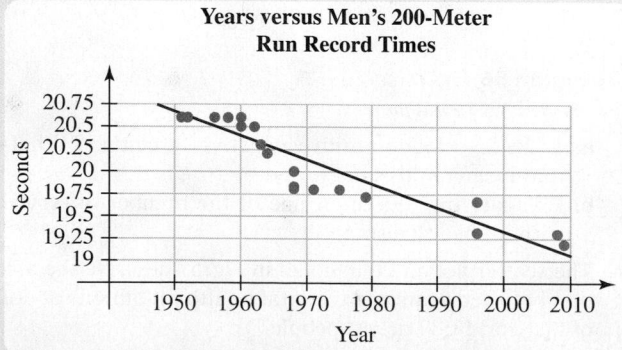

Figure 59 Exercise 28
(**Source:** *International Association of Athletics Federations*)

a. Is the association positive, negative, or neither? What does that mean in this situation?
b. Estimate the slope of the model. What does it mean in this situation?
c. Estimate how much the record time decreased over 25 years.
d. A student says the result you found in part (c) describes how much the record time will decrease in the next 25 years. What would you tell the student?

29. The scatterplot and the model in Fig. 60 describe the association between 2013 U.S. revenues and 2013 worldwide revenues, both in millions of dollars, of the 15 worldwide best-selling video games.

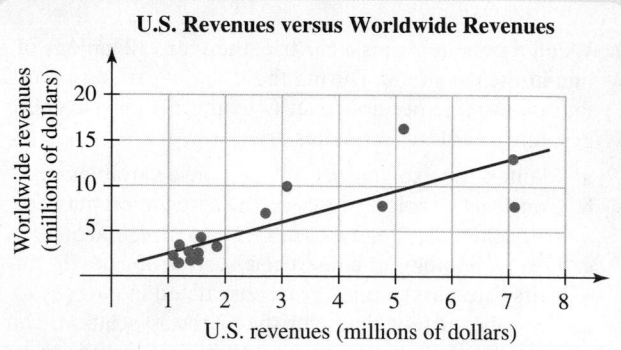

Figure 60 Exercise 29
(**Source:** *VGChartz*)

a. Is the association positive, negative, or neither? What does that mean in this situation?
b. Estimate the rate of change of worldwide revenue with respect to U.S. revenue.

30. Many appliances such as televisions and electronic gadgets such as cell phones contain rare Earth metals. The scatterplot and the model in Fig. 61 compare the annual revenue (in billions of dollars) of U.S. electronics and appliance stores with annual worldwide rare Earth mining (in thousands of metric tons).

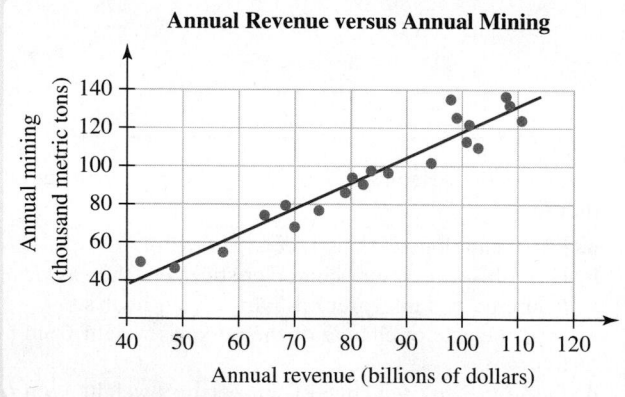

Figure 61 Exercise 30
(**Source:** *U.S Geological Survey*)

a. Is the association positive, negative, or neither? What does that mean in this situation?
b. Estimate the rate of change of annual worldwide rare Earth mining with respect to annual revenue of U.S. electronics and appliance stores.

31. The scatterplot and the model in Fig. 62 compare the maximum weights (in pounds) with the mean life expectancies (in years) for 16 dog breeds.

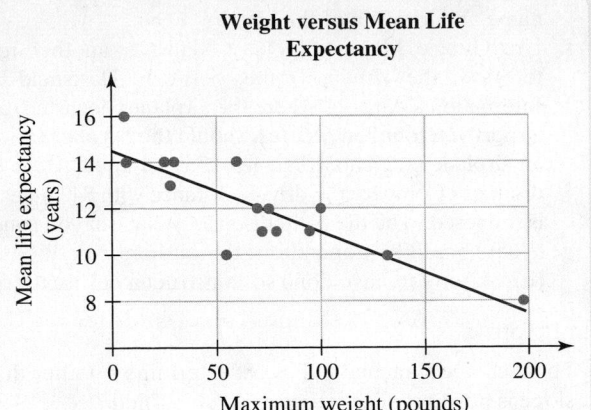

Figure 62 Exercise 31
(**Source:** *FindTheBest*)

a. Is the association positive, negative, or neither? What does it mean in this situation?

b. Estimate the slope of the model. What does it mean in this situation?

c. If the mean weight of a dog breed is 40 pounds greater than the mean weight of another dog breed, estimate how much less the mean life expectancy is for the heavier dog breed than for the lighter dog breed.

d. On the basis of the scatterplot, a student concludes that if a dog gains weight, the dog will not live as long. What would you tell the student?

32. The scatterplot and the model in Fig. 63 compare the annual amounts (in millions of pounds) of herbicides used with the numbers (in millions) of bee colonies in the United States for years between 1939 and 2013, inclusive.

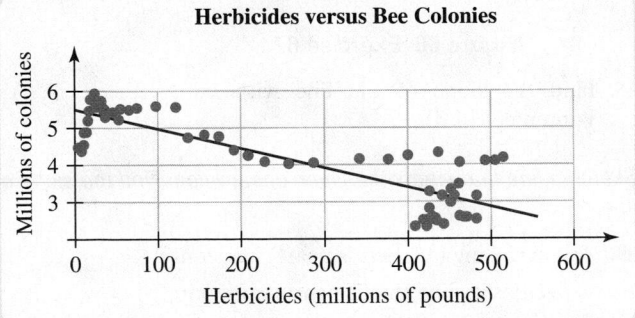

Figure 63 Exercise 32
(**Source:** *U.S. Department of Agriculture*)

a. Is the association positive, negative, or neither? What does it mean in this situation?

b. Estimate the slope of the model. What does it mean in this situation?

c. If the amount of herbicides used in a certain year was 300 million pounds more than in another year, estimate how many fewer colonies of bees there were in the year when more herbicides were used.

d. On the basis of the scatterplot, a student concludes that herbicides kill bees. What would you tell the student?

Use the slope formula to find the slope of the line that passes through the two given points. State whether the line is increasing, decreasing, horizontal, or vertical.

33. $(1, 5)$ and $(3, 9)$ **34.** $(2, 3)$ and $(5, 12)$

35. $(3, 10)$ and $(5, 2)$ **36.** $(5, 8)$ and $(7, 2)$

37. $(2, 5)$ and $(8, 3)$ **38.** $(1, 7)$ and $(9, 5)$

39. $(-2, 4)$ and $(3, -1)$ **40.** $(-3, 4)$ and $(1, -2)$

41. $(5, -2)$ and $(9, -4)$ **42.** $(2, -3)$ and $(8, -6)$

43. $(-7, -1)$ and $(-2, 9)$ **44.** $(-6, -8)$ and $(-4, 2)$

45. $(-6, -9)$ and $(-2, -3)$ **46.** $(-5, -2)$ and $(-1, -3)$

47. $(6, -1)$ and $(-4, 7)$ **48.** $(4, -5)$ and $(-2, 10)$

49. $(0, 0)$ and $(4, -2)$ **50.** $(-6, -9)$ and $(0, 0)$

51. $(3, 5)$ and $(7, 5)$ **52.** $(-4, -6)$ and $(3, -6)$

53. $(-3, -1)$ and $(-3, -2)$ **54.** $(4, 2)$ and $(4, 7)$

For Exercises 55–60, find the approximate slope of the line that contains the two given points. Round your result to the second decimal place. State whether the line is increasing, decreasing, horizontal, or vertical.

55. $(-3.2, 5.1)$ and $(-2.8, 1.4)$ **56.** $(-1.9, 4.8)$ and $(-3.1, 5.5)$

57. $(4.9, -2.7)$ and $(6.3, -1.1)$ **58.** $(9.7, -6.8)$ and $(4.5, -2.7)$

59. $(-4.97, -3.25)$ and $(-9.64, -2.27)$

60. $(-3.22, -8.54)$ and $(-7.29, -6.13)$

61. Find the slope of the line sketched in Fig. 64.

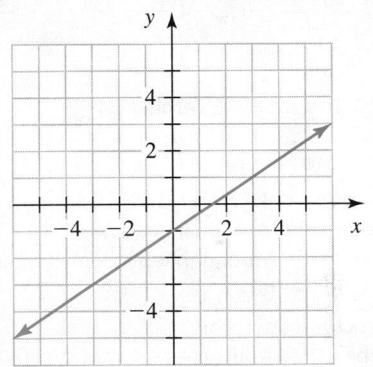

Figure 64 Exercise 61

62. Find the slope of the line sketched in Fig. 65.

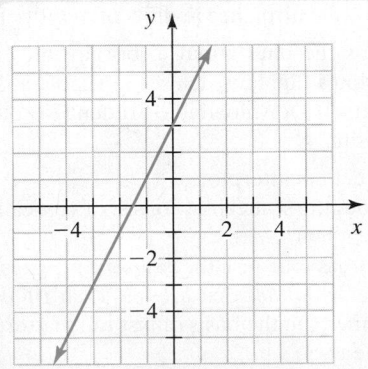

Figure 65 Exercise 62

63. Find the slope of the line sketched in Fig. 66.

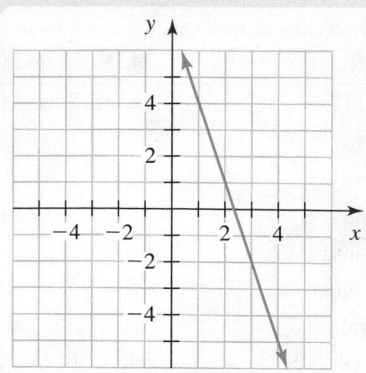

Figure 66 Exercise 63

64. Find the slope of the line sketched in Fig. 67.

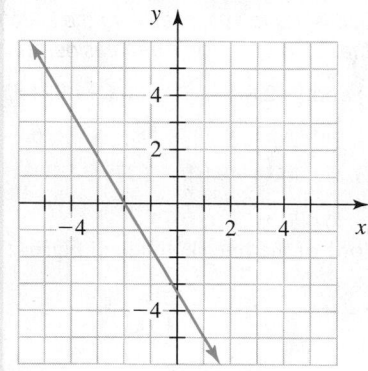

Figure 67 Exercise 64

Large Data Sets

65. **DATA** Access the data about airborne times and distances of Delta Airlines flights, which are available at MyMathLab and at the Pearson Downloadable Student Resources for Math & Stats website.

a. Construct a scatterplot.

b. Describe the four characteristics of the association. Compute and interpret r as part of your analysis.

c. Print your scatterplot and draw a linear model on it.

d. Estimate the slope of the linear model. What does it mean in this situation?

e. Use the reciprocal of the slope to help you estimate the speed of the airplanes in units of miles per *hour*.

66. **DATA** Access the data about airborne times and distances of Delta Airlines flights, which are available at MyMathLab and at the Pearson Downloadable Student Resources for Math & Stats website.

a. Construct a scatterplot.

b. Why does the scatterplot consist of vertically aligned clumps of data points?

c. On the basis of just the scatterplot, guess whether Delta offers more *routes* that are less than 1000 miles or greater 1000 miles. On the basis of just the scatterplot, why is it not possible to be sure?

d. For a given distance other than 849 miles, does an outlier tend to lie above or below the other data points? What does that mean in this situation? From what you know about flights, why does that make sense?

e. For the distance 849 miles, there are a large number of outliers that lie below the other data points. By searching the data set, find the three-letter code for the destination airport for those flights. Research online to determine the name and location of the airport.

f. Use Google Maps, MapQuest Maps, or another resource to view the driving route between Hartsfield-Jackson International Airport (where the airplanes departed) and the airport you found in part (e). Should the distance traveled by an airplane be greater than, less than, or equal to the driving distance? Compare the driving distance with 849 miles, which is supposed to be the flight distance. What can you conclude?

g. What should be done about the outliers described in part (e)? After you have done so, construct a new scatterplot.

Concepts

67. For each line sketched in Fig. 68, determine whether the line's slope is positive, negative, zero, or undefined.

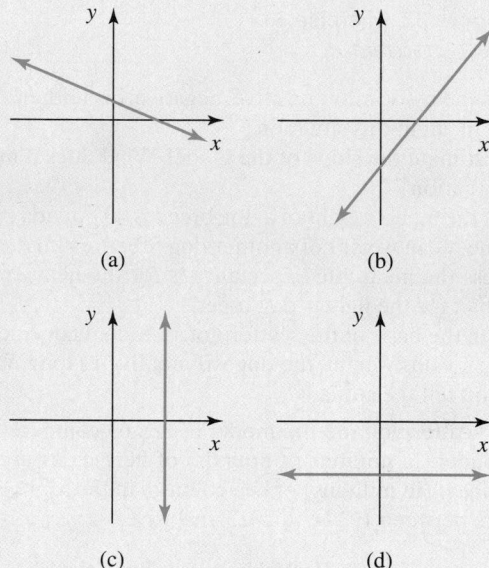

Figure 68 Exercise 67

68. Find the slope of the line with x-intercept $(-3, 0)$ and y-intercept $(0, 4)$.

Sketch a line that meets the given description. Find the slope of the line.

69. An increasing line that is nearly horizontal

70. A decreasing line that is nearly horizontal

71. A decreasing line that is nearly vertical

72. An increasing line that is nearly vertical

For Exercises 73–76, sketch a line that meets the given description.

73. The slope is a large positive number.

74. The slope is a positive number near zero.

75. The slope is a negative number near zero.

76. The slope is less than -5.

77. A student tries to find the slope of the line that contains the points $(1, 3)$ and $(4, 7)$:

$$\frac{4 - 1}{7 - 3} = \frac{3}{4}$$

Describe any errors. Then find the slope correctly.

78. A student tries to find the slope of the line that contains the points $(6, 4)$ and $(3, 9)$:

$$\frac{9 - 4}{6 - 3} = \frac{5}{3}$$

Describe any errors. Then find the slope correctly.

79. A student tries to find the slope of the line that contains the points $(-1, -5)$ and $(3, 8)$:

$$\frac{8 - 5}{3 - 1} = \frac{3}{2}$$

Describe any errors. Then find the slope correctly.

80. A student tries to find the slope of the line that contains the points $(2, 5)$ and $(6, 7)$:

$$\frac{7 - 5}{6 - 2} = \frac{2}{4}$$

Describe any errors. Then find the slope correctly.

81. Sketch a line with a slope of 2 and another line with a slope of 3. Does the steeper line have the greater slope?

82. Sketch a line with a slope of 1 and another line with a slope of $\frac{1}{2}$. Does the steeper line have the greater slope?

83. a. Sketch a line with a slope of -2 and another line with a slope of -3.
   b. Does the steeper line have the greater slope?
   c. Find the absolute value of the slope of each line. Does the steeper line have the greater absolute value of the slope?
   d. Explain why the absolute value of the slope of a line is useful for comparing the steepness of lines.

84. a. Sketch a line with a slope of -2 and another line with a slope of 2.
   b. Is the line with a slope of 2 steeper than the line with a slope of -2?
   c. Find the absolute value of the slope of each line. Compare the results.
   d. Explain why the absolute value of the slope of a line is useful for comparing the steepness of lines.

85. A line contains the points $(2, 1)$ and $(3, 4)$. Find three more points that lie on the line. [**Hint:** Find the slope of the line. Then use the slope and a point on the line to help you find other points on the line.]

86. A line contains the points $(1, 5)$ and $(4, 3)$. Find three more points that lie on the line. [**Hint:** See Exercise 85.]

87. Explain why the slope of a vertical line is undefined.

88. Explain why the slope of a horizontal line is 0.

89. Explain why the slope of an increasing line is positive.

90. Explain why the slope of a decreasing line is negative.

91. a. Carefully graph the given equation by hand. Then find the slope of the line by using the ratio $\dfrac{\text{rise}}{\text{run}}$.
   i. $y = 2x + 1$ **ii.** $y = 3x - 5$ **iii.** $y = -2x + 6$
   b. Compare the slope of each line with the number multiplied by x in the corresponding equation.

92. Here you will explore the relationship among three lines that pass through the origin $(0, 0)$, where the slope of one of the lines is the reciprocal of the slope of one of the other lines.

a. By hand, carefully sketch the lines that pass through the origin $(0, 0)$ and that have slopes $5, 1$, and $\frac{1}{5}$.

b. Sketch the lines that pass through the origin $(0, 0)$ and that have slopes $\frac{2}{5}, 1$, and $\frac{5}{2}$.

c. Sketch the lines that pass through the origin $(0, 0)$ and that have slopes $\frac{3}{4}, 1$, and $\frac{4}{3}$.

d. What pattern do you notice from your graphs in parts (a)–(c)?

e. A line with slope m is sketched in Fig. 69. Sketch a line with slope $\frac{1}{m}$ that passes through the origin $(0, 0)$. Assume both axes are scaled the same.

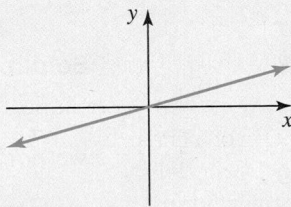

Figure 69 Exercise 93e

93. Suppose that a line with slope $-\frac{2}{3}$ contains a point P. A point Q lies three units to the right and two units down from point P. A point S lies three units to the left and two units up from point P. Does the line contain point Q? point S? Explain.

94. a. A square has vertices at $(3, 1)$ and $(3, 7)$. How many possible positions are there for the other two vertices? Find the coordinates for each possibility.
   b. A parallelogram has vertices at $(-7, -2), (3, 1)$, and $(-4, 2)$. How many possible positions are there for the fourth vertex? Find the coordinates for each possibility. [**Hint:** Try drawing different line segments between the given vertices.]

95. Explain why the absolute value of the slope of a line measures the steepness of the line.

96. Describe the meaning of the slope of a line. Sketch various types of lines and give the slope for each line. For each sketch, explain why the slope assignment makes sense. For example, you could sketch a horizontal line, state that the slope is zero, and explain why it makes sense that the slope of a horizontal line is zero in terms of rise and run.

Hands-On Research

97. By researching blogs, newspapers, magazines, and/or journals, find a statistical diagram that describes a linear association between two numerical variables.

a. Define both variables.
b. What is the point of the article? Does the diagram support that point? Explain.
c. Your association should be linear. Describe the other three characteristics of the association. Compute and interpret r as part of your analysis.
d. Is the diagram a scatterplot? If yes, draw a linear model on the scatterplot. If no, construct a scatterplot that describes the association and then draw a linear model on the scatterplot.
e. Find the slope of the linear model. What does it mean in the situation?

▼7.3 Using Slope to Graph Equations of Lines and Linear Models

Objectives

» Use the slope and the y-intercept of a line to sketch the line.

» Describe the meaning of m for an equation of the form $y = mx + b$.

» Graph an equation of the form $y = mx + b$ by using the line's slope and y-intercept.

» Find an equation of a line from its graph.

» Graph an equation of a linear model by using the model's slope and vertical intercept.

» Use a linear model's slope and vertical intercept to find its equation.

In Section 7.1, we determined that for an linear equation of the form $y = mx + b$, the graph has y-intercept $(0, b)$. In Section 7.2, we found that the slope of a line is the rate of change of y with respect to x. In this section, we will use these concepts to first efficiently graph linear equations in two variables. Then we will find equations of linear models and use the equations to make predictions.

Using the Slope and the y-Intercept to Sketch a Line

We can use the slope and the y-intercept of a line to sketch the line.

▶ **Example 1** Sketching a Line

Sketch the line that has slope $m = -\dfrac{2}{5}$ and y-intercept $(0, 3)$.

Solution

Slope tells us how to go from one point on a line to another point on the line. So, we begin by plotting the y-intercept, $(0, 3)$. The slope is $-\dfrac{2}{5} = \dfrac{-2}{5} = \dfrac{\text{rise}}{\text{run}}$. From $(0, 3)$, we count 5 units to the right and 2 units down, where we plot the point $(5, 1)$. See Fig. 70. We then sketch the line that contains these two points (see Fig. 71).

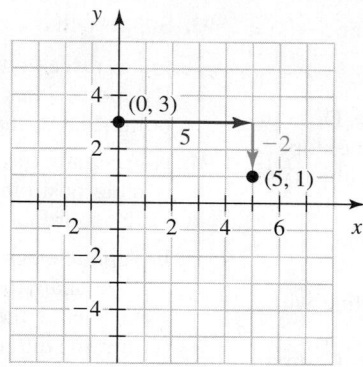

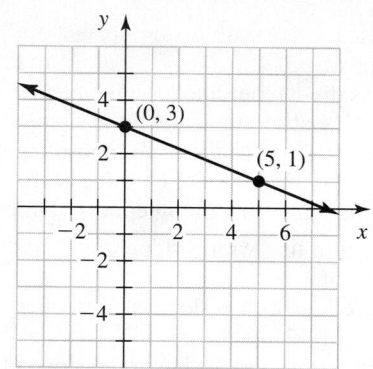

Figure 70 Plot (0, 3). Then count 5 units to the right and 2 units down to plot (5, 1)

Figure 71 Sketch the line containing (0, 3) and (5, 1)

Because the slope is negative, we can verify our work by checking that the line is decreasing.

The Meaning of m for $y = mx + b$

In Example 1, we saw that if we know the slope and the y-intercept of a line, we can sketch that line. Next, we will discuss how to determine the slope and y-intercept of a line from the line's equation.

▶ **Example 2** Finding the Slope of a Line

Find the slope of the line $y = 2x + 1$.

Solution

We list some solutions in Table 26 and sketch the graph of the equation in Fig. 72.

If the run is 1, the rise is 2 (see Fig. 72). So, the slope is

$$m = \frac{\text{rise}}{\text{run}} = \frac{2}{1} = 2$$

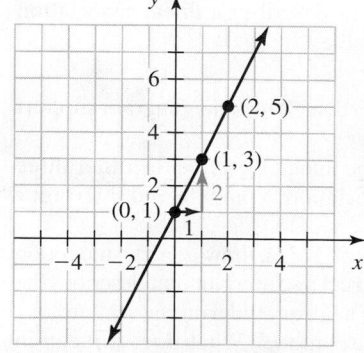

Figure 72 Graph of $y = 2x + 1$

Table 26 Solutions of $y = 2x + 1$

x	y
0	$2(0) + 1 = 1$
1	$2(1) + 1 = 3$
2	$2(2) + 1 = 5$

In Example 2, we found that the line $y = 2x + 1$ has slope 2. Note that 2 is also the number multiplied by x in the equation $y = 2x + 1$. This observation suggests a general property about a linear equation of the form $y = mx + b$.

> **Finding the Slope and *y*-Intercept from a Linear Equation**
>
> For a linear equation of the form $y = mx + b$,
> * the slope of the line is m and
> * the *y*-intercept of the line is $(0, b)$.
>
> We say that this equation is in **slope–intercept form**.

For example, the equation $y = -4x + 5$ is in slope–intercept form with $m = -4$ and $b = 5$. The graph of this equation is a line with slope -4 and *y*-intercept $(0, 5)$. The line $y = 8x - 2$ has slope 8 and *y*-intercept $(0, -2)$.

Graphing Equations of the Form $y = mx + b$

In Example 3, we will graph an equation in slope–intercept form.

▶ **Example 3**   Graphing an Equation

Sketch the graph of $y = 3x - 4$.

Solution

Note that the *y*-intercept is $(0, -4)$ and the slope is $3 = \dfrac{3}{1} = \dfrac{\text{rise}}{\text{run}}$. To graph the line:

1. Plot the *y*-intercept $(0, -4)$.
2. From $(0, -4)$, count 1 unit to the right and 3 units up to plot a second point, which we see by inspection is $(1, -1)$. See Fig. 73.
3. Sketch the line that contains these two points (see Fig. 74).

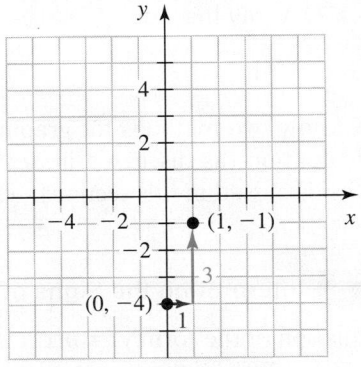   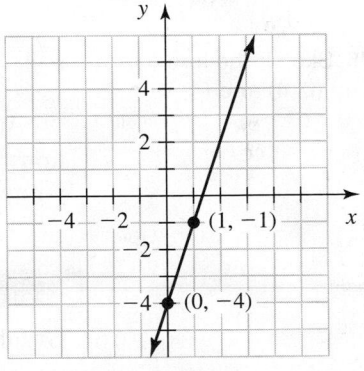

Figure 73 Plot $(0, -4)$. Then count 1 unit to the right and 3 units up to plot $(1, -1)$

Figure 74 Sketch the line containing $(0, -4)$ and $(1, -1)$

Recall from Section 7.1 that every point on the graph of an equation represents a solution of the equation. We can verify our result by checking that both $(0, -4)$ and $(1, -1)$ are solutions of $y = 3x - 4$:

Check that $(0, -4)$ is a solution

$$y = 3x - 4$$
$$-4 \overset{2}{=} 3(0) - 4$$
$$-4 \overset{2}{=} 0 - 4$$
$$-4 \overset{2}{=} -4$$
$$\text{true}$$

Check that $(1, -1)$ is a solution

$$y = 3x - 4$$
$$-1 \overset{2}{=} 3(1) - 4$$
$$-1 \overset{2}{=} 3 - 4$$
$$-1 \overset{2}{=} -1$$
$$\text{true}$$

We check two ordered pairs (rather than just one) because two points determine a line.

> **Graphing an Equation in Slope–Intercept Form**
>
> To graph an equation of the form $y = mx + b$,
> 1. Plot the y-intercept $(0, b)$.
> 2. Use $m = \dfrac{\text{rise}}{\text{run}}$ to plot a second point. For example, if $m = \dfrac{2}{3}$, then count 3 units to the right (from the y-intercept) and 2 units up to plot another point.
> 3. Sketch the line that passes through the two plotted points.

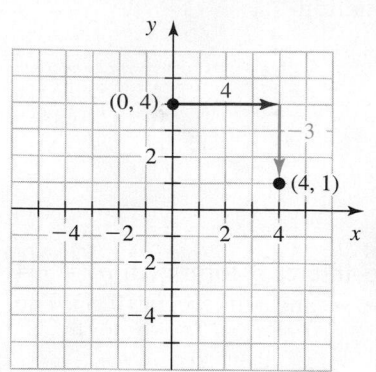

Figure 75 Plot $(0, 4)$. Then count 4 units to the right and 3 units down to plot $(4, 1)$

▶ **Example 4**   Graphing an Equation

Sketch the graph of $y = -\dfrac{3}{4}x + 4$.

Solution

The y-intercept is $(0, 4)$, and the slope is $-\dfrac{3}{4} = \dfrac{-3}{4} = \dfrac{\text{rise}}{\text{run}}$. To graph the line:

1. Plot the y-intercept $(0, 4)$.
2. From $(0, 4)$, count 4 units to the right and 3 units down to plot a second point, which we see by inspection is $(4, 1)$. See Fig. 75.
3. Sketch the line that contains these two points (see Fig. 76).

   We use a TI-84 to verify our work (see Fig. 77). For TI-84 instructions on WINDOW settings, see Appendix A.18.

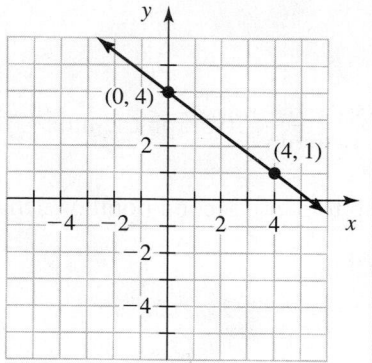

Figure 76 Sketch the line containing $(0, 4)$ and $(4, 1)$

 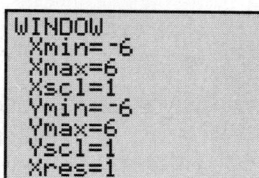 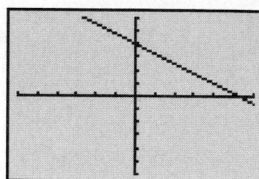

Figure 77 Verify the work

   We now know two methods for graphing a linear equation: We can first find solutions of the equation (as discussed in Section 7.1), or we can first find the slope and y-intercept (as discussed in this section).

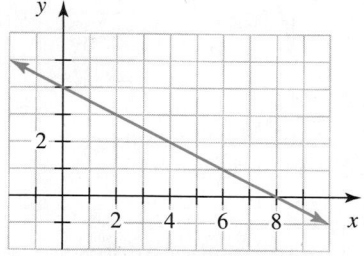

Figure 78 A decreasing line with y-intercept above the origin

▶ **Example 5**   Interpreting the Signs of m and b

Graph an equation of the form $y = mx + b$ where m is negative and b is positive.

Solution

We should sketch a decreasing line, because the slope m is negative. We should draw a line whose y-intercept, $(0, b)$, is above the origin, because b is positive. We sketch such a line in Fig. 78.

   There is nothing special about the line in Fig. 78—any line that is decreasing and has its y-intercept above the origin will do.

Finding an Equation of a Line from Its Graph

Given the slope and y-intercept of a nonvertical line, we can find an equation of the line.

▶ Example 6   Finding an Equation of a Line from Its Slope and *y*-Intercept

Find an equation of the line that has slope $\frac{2}{3}$ and *y*-intercept $(0, -5)$.

Solution

To find an equation, we substitute $\frac{2}{3}$ for *m* and -5 for *b* in the equation $y = mx + b$:

$$y = \frac{2}{3}x + (-5)$$

$$y = \frac{2}{3}x - 5$$

Given an equation in slope–intercept form, $y = mx + b$, we can find the graph of the equation. We can also go backward: Given the graph, we can find an equation for it.

▶ Example 7   Finding an Equation of a Line from Its Graph

Find an equation of the line sketched in Fig. 79.

Solution

From Fig. 80, we see that the *y*-intercept of the line is $(0, -1)$. We also see that if the run is 3, then the rise is -2. So, the slope is $\frac{-2}{3} = -\frac{2}{3}$. By substituting $-\frac{2}{3}$ for *m* and -1 for *b* in the equation $y = mx + b$, we have $y = -\frac{2}{3}x - 1$.

We can verify our equation by checking that both $(0, -1)$ and $(3, -3)$ satisfy the equation. Or we can use a TI-84 to verify our work (see Fig. 81).

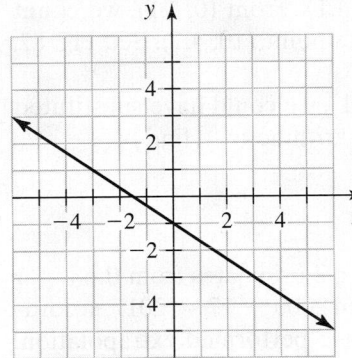

Figure 79 Graph of a line

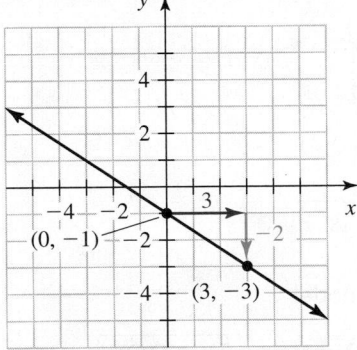

Figure 80 Finding the slope of the line

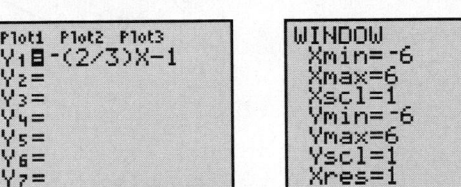
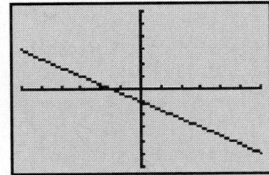

Figure 81 Verify the work

▶ **Finding an Equation of a Line from a Graph**

To find an equation of a line from a graph,
1. Determine the slope *m* and the *y*-intercept $(0, b)$ from the graph.
2. Substitute your values for *m* and *b* into the equation $y = mx + b$.

Graphing an Equation of a Linear Model

We can graph an equation of a linear model by using the model's slope and vertical intercept.

▶ Example 8   Graphing a Model's Equation

The percentages of U.S. food sales that are organic products are shown in Table 27 for various years. Let *p* be the percentage of U.S. food sales that are organic products at *t* years since 2000. A reasonable model is

$$p = 0.3t + 1.1$$

Table 27 Percentages of U.S. Food Sales That Are Organic Products

Year	Percent
2000	1.2
2002	1.6
2004	2.2
2006	2.9
2008	3.6
2010	4.0

Source: *Organic Trade Association*

1. Graph the model.
2. Predict when 6.2% of food sales will be organic products. Did you perform interpolation or extrapolation? Do you have much faith in the prediction? Explain.

Solution

1. The p-intercept is $(0, 1.1)$, and the slope is $0.3 = \dfrac{0.3}{1} = \dfrac{\text{rise}}{\text{run}}$. It will be easier to graph the model if we multiply the slope by $\dfrac{10}{10} = 1$ so the rise and run are larger and both are integers:

$$0.3 = \frac{0.3}{1} \cdot \frac{10}{10} = \frac{3}{10} = \frac{\text{rise}}{\text{run}}$$

To graph the model, we first plot the p-intercept $(0, 1.1)$. From $(0, 1.1)$, we count 10 units to the right and 3 units up, where we plot the point $(10, 4.1)$. See Fig. 82. We then sketch the line that contains the two points.

Instead of using the slope to find the point $(10, 4.1)$, we could have substituted 10 for t in the equation $p = 0.3t + 1.1$ and performed arithmetic to find p:

$$p = 0.3(10) + 1.1 = 4.1$$

So, the point $(10, 4.1)$ is a point on the linear model.

2. The green arrows in Fig. 82 show that the output $p = 6.2$ originates from the input $t = 17$. So, 6.2% of food sales will be organic products in $2000 + 17 = 2017$, according to the model. Because 2017 is in the future, we have performed extrapolation. So, we have little or no faith in our prediction.

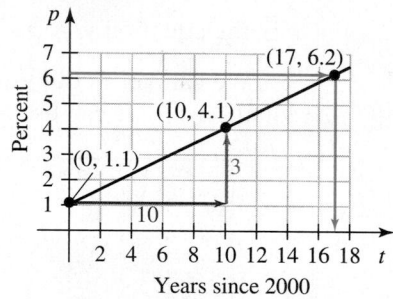

Figure 82 Organic products model

To use a TI-84 to verify our work in Problems 1 and 2, we press WINDOW and set Xmin to be -10 (for 1990), set Xmax to be 25 (for 2025), and use ZoomFit to set the values for Ymin and Ymax automatically (see Fig. 83). Then we use TRACE to check that $(17, 6.2)$ is a point on the linear model. See Appendices A.16, A.17, and A.18 for TI-84 instructions.

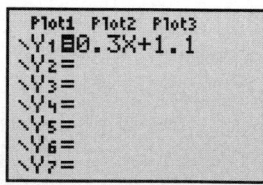

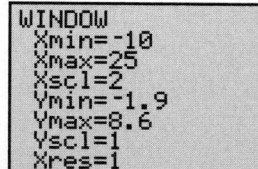
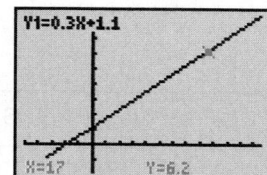

Figure 83 Verify the work

Finding an Equation of a Linear Model

For an equation of the form $y = mx + b$, we know its graph has y-intercept $(0, b)$ and the slope m is the rate of change of y with respect to x. We can use these facts to find an equation of a linear model.

▶ **Example 9**   Finding an Equation of a Model

In 2010, a college's enrollment was 20 thousand students. Each year, the enrollment increases by 2 thousand students for the period 2010–2015. Let E be the enrollment (in thousands of students) at t years since 2010.

1. Is there an exact linear association between t and E? Explain.
2. Find the E-intercept of a linear model. What does it mean in this situation?
3. Find the slope of the linear model. What does it mean in this situation?
4. Find an equation of the linear model.

Table 28 College Enrollments

Years since 2010 t	Enrollment (thousands of students) E
0	20
1	22
2	24
3	26
4	28
5	30

Solution

1. Since the rate of change of enrollment per year is a *constant* 2 thousand students per year, the variables t and E are exactly linearly associated.
2. We list some values of t and E in Table 28. We plot the corresponding points and sketch the line that contains the points in Fig. 84.

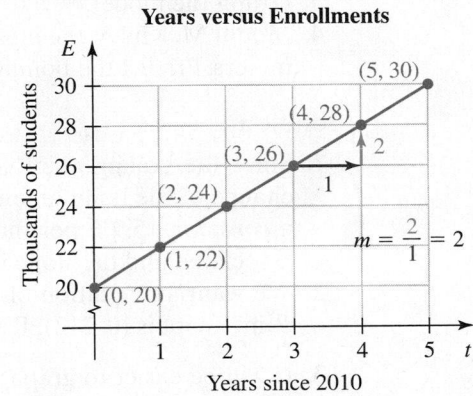

Figure 84 Enrollment scatterplot and model

From the table and the graph, we see that the E-intercept is $(0, 20)$. This means that the enrollment was 20 thousand students in the year 2010.

3. The rate of change of enrollment is 2 thousand students per year. So, the slope of the linear model is 2. This checks with the calculation shown in Fig. 84.
4. An equation of a line can be written in slope–intercept form, $y = mx + b$. Using t and E, we have $E = mt + b$. Since the slope is 2 and the E-intercept is $(0, 20)$, we have $E = 2t + 20$.

   To check our work with a TI-84, we begin by entering our model (see Fig. 85). Then we check that the entries in the TI-84 table in Fig. 86 equal the entries in Table 28. We also check that the graph of our equation contains the points of the scatterplot of the data (see Fig. 87).

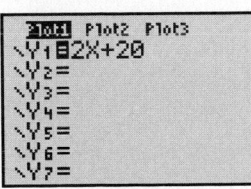

Figure 85 Enter the model

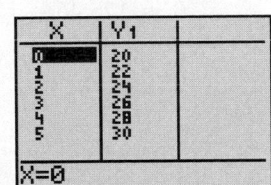

Figure 86 Use a table to verify the model

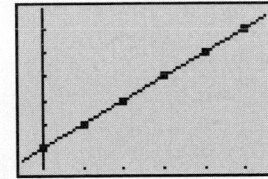

Figure 87 Use a graph to verify the model

For TI-84 instructions on creating tables, see Appendix A.21. For instructions on constructing scatterplots, see Appendices A.10 and A.19.

In Example 9, we found the model $E = 2t + 20$. Here is the connection between parts of the equation and the situation:

$$E = \underbrace{2}_{\substack{\text{rate of} \\ \text{change of} \\ \text{enrollment}}} \cdot\ t + \underbrace{20}_{\substack{\text{enrollment at} \\ t = 0}}$$

▶ **Example 10** Finding and Graphing an Equation of a Model

The temperature at which water boils (the *boiling point*) is explained by elevation: the higher you are, the lower the boiling point will be. At sea level (elevation 0), water boils at 212°F. The boiling point declines by 5.9°F for each thousand-meter increase in elevation (Source: *Thermodynamics, an Engineering Approach* by Cengal & Boles). Let B be the boiling point (in degrees Fahrenheit) at an elevation of E thousand meters.

1. Is there an exact linear association between E and B? If so, find the slope.
2. Find an equation of the model.
3. Graph the model.
4. Mount McKinley, the highest mountain in the United States, reaches 6.194 thousand meters. Predict the boiling point of water at the peak.

Solution

1. Since the boiling point is explained by the elevation, we will consider the rate of change of the boiling point with respect to elevation. Because this rate of change is a *constant* −5.9°F per thousand meters, the variables E and B are exactly linearly associated and the slope is −5.9.

2. We want an equation of the form $B = mE + b$. Since $B = 212$ when $E = 0$, the B-intercept is $(0, 212)$. Recall that $m = -5.9$, so an equation is $B = -5.9E + 212$.

3. It will be easier to graph the model if we multiply the slope −5.9 by $\dfrac{10}{10} = 1$ so the rise and run are larger and both integers:

$$-5.9 = \frac{-5.9}{1} \cdot \frac{10}{10} = \frac{-59}{10} = \frac{\text{rise}}{\text{run}}$$

To graph the model, we first plot the B-intercept $(0, 212)$. From $(0, 212)$, we count 10 units to the right and 59 units down, where we plot the point $(10, 153)$. See Fig. 88. We then sketch the line that contains the two points.

4. To find the boiling point at 6.194 thousand meters, we substitute the input 6.194 for E in the equation $B = -5.9E + 212$:

$$B = -5.9(6.194) + 212 \qquad \textit{Substitute 6.194 for E.}$$
$$\approx 175.46 \qquad\qquad\quad \textit{Compute.}$$

So, the boiling point is 175.46°F at the peak of Mount McKinley. The blue arrows in the coordinate system in Fig. 89 confirm that the input $E = 6.194$ leads to the output $B = 175.46$.

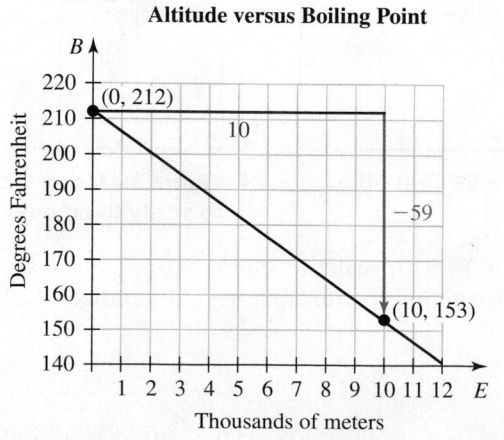

Figure 88 Boiling-point model

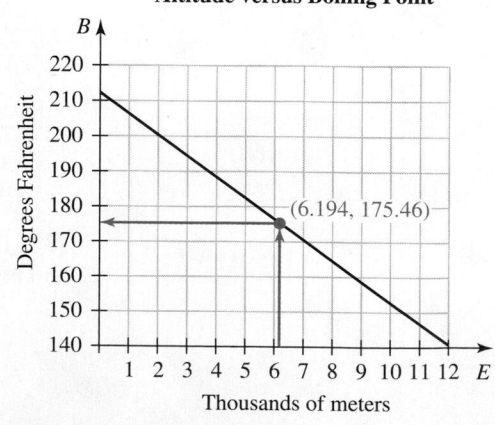

Figure 89 Verify prediction

Table 29 Worldwide Jewelry Sales

Year	Sales (billions of dollars)
2009	202
2010	214
2011	246
2012	278
2013	300

Source: *Euromonitor International*

▶ Example 11   Analyzing a Model and Making an Estimate

Worldwide jewelry sales are shown in Table 29 for various years. Let *s* be annual jewelry sales (in billions of dollars) at *t* years since 2000. A model of the situation is

$$s = 26t - 38$$

1. Use a TI-84 to draw a scatterplot and the model in the same viewing window. Check whether the line comes close to the data points.
2. What is the slope of the model? What does it mean in this situation?
3. Find the rates of change in sales from one year in Table 29 to the next one listed. Compare the rates of change with the result you found in Problem 2.
4. What is the *s*-intercept? What does it mean in this situation?
5. Estimate the sales in 2013.

Solution

1. We draw the scatterplot and the model in the same viewing window (see Fig. 90). For TI-84 instructions, see Appendices A.10 and A.19.

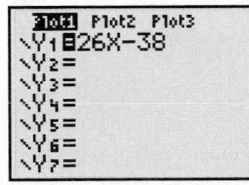 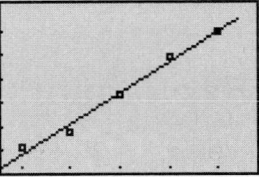

Figure 90 Jewelry scatterplot and model

   The line comes close to the data points, so the model is a reasonable one.

2. The slope is 26, because $s = 26t - 38$ is of the form $y = mx + b$ and $m = 26$. This means that worldwide jewelry sales increased by approximately by \$26 billion per year.
3. The rates of change of sales are shown in Table 30. The rate of change from 2009 to 2010 (\$12 billion per year) is quite different than \$26 billion per year. The other rates of change in Table 30 are fairly close to \$26 billion per year.

Table 30 Rates of Change of Jewelry Sales

Year	Sales (billions of dollars)	Rate of Change of Sales from Previous Year (billions of dollars per year)
2009	202	
2010	214	$(214 - 202) \div (2010 - 2009) = 12$
2011	246	$(246 - 214) \div (2011 - 2010) = 32$
2012	278	$(278 - 246) \div (2012 - 2011) = 32$
2013	300	$(300 - 278) \div (2013 - 2012) = 22$

4. The *s*-intercept is $(0, -38)$, because $s = 26t - 38$ is of the form $y = mx + b$ and $b = -38$. According to the model, jewelry sales were -38 billion dollars in 2000. Model breakdown has occurred because sales cannot be negative.
5. We substitute the input 13 for *t* in the equation $s = 26t - 38$:

$$s = 26(13) - 38 = 300$$

   According to the model, worldwide jewelry sales were \$300 billion in 2013. From Table 29, we see this result equals the actual sales in that year.

WARNING   A common error in describing the meaning of the slope of a model is vagueness. For example, a description such as

The slope means that it is increasing.

neither specifies the quantity that is increasing nor the rate of increase. The following statement includes the missing information:

The slope of 26 means that worldwide jewelry sales increased by approximately \$26 billion per year.

◆ Group Exploration

Section Opener: The meaning of *m* in the equation $Y = mx + b$

1. a. Carefully sketch a graph of the line $y = 2x - 1$.

b. Using the formula $m = \dfrac{\text{rise}}{\text{run}}$, find the slope of the line you sketched.

c. What number is multiplied by x in the equation $y = 2x - 1$? How does it compare with the slope you found in part (b)?

2. a. Carefully sketch a graph of the line $y = -3x + 5$.

b. Using the formula $m = \dfrac{\text{rise}}{\text{run}}$, find the slope of the line you sketched.

c. What number is multiplied by x in the equation $y = -3x + 5$? How does it compare with the slope you found in part (b)?

3. Describe what you have learned in this exploration so far.

4. Without graphing, determine the slope of each line.

a. $y = 4x - 7$ **b.** $y = -2x + 4$

c. $y = \dfrac{2}{5}x - 3$ **d.** $y = x - 2$

e. $y = 3$

◆ Group Exploration

Drawing lines with various slopes

1. Using technology, graph a group of lines (a *family of lines*) to make a starburst like the one in Fig. 91. List the equations of your lines.

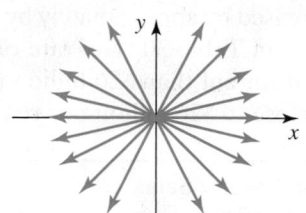

Figure 91 A starburst

2. Using technology, graph a family of lines to make a starburst like the one in Fig. 92. List the equations of your lines.

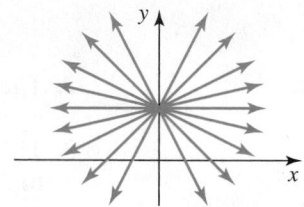

Figure 92 Another starburst

3. Summarize what you have learned about slope from this exploration, this section, and Section 7.2.

◆ Group Exploration

Finding an equation of a line from its graph

In this exploration, you will need to work with one other student.

1. First, both you and your partner should create a linear equation of the form $y = mx + b$, but do not show each other your equations. For m, use a rational number between -3 and 3, inclusive. For b, use an integer between -3 and 3, inclusive.

2. Use ZDecimal to graph the equation.

3. Exchange TI-84s.

4. By inspecting the graph, determine the equation that your partner created. You can use TRACE to find co-ordinates of a point on your line. Or you can use a grid by pressing 2nd [FORMAT]. Press ▽ twice, then press ▷, and then press ENTER. The "GridOn" choice should now be highlighted. Next, press GRAPH. A grid should now be visible.

5. Press Y = to check the equation you found.

Homework 7.3

1. For a linear equation of the form $y = mx + b$, the slope of the line is ____.

2. For a linear equation of the form $y = mx + b$, the y-intercept of the line is ____.

3. A linear equation of the form $y = mx + b$ is in slope-____ form.

4. *True or False:* One way to graph an equation of the form $y = mx + b$ is to first plot the y-intercept and then use the slope to plot another point.

Sketch the line that has the given slope and contains the given point.

5. $m = \frac{2}{3}$, $(0, 1)$

6. $m = \frac{3}{5}$, $(0, 2)$

7. $m = -\frac{5}{2}$, $(0, 4)$

8. $m = -\frac{3}{4}$, $(0, -2)$

9. $m = -\frac{3}{2}$, $(0, 0)$

10. $m = -\frac{1}{2}$, $(0, 0)$

11. $m = 2$, $(0, 1)$

12. $m = 4$, $(0, -3)$

13. $m = -3$, $(0, -2)$

14. $m = -5$, $(0, 4)$

15. $m = 0$, $(4, -5)$

16. $m = 0$, $(6, 3)$

17. m is undefined, $(2, -1)$

18. m is undefined, $(-1, -3)$

For Exercises 19–40, determine the slope and the y-intercept. Then use the slope and the y-intercept to graph the equation by hand.

19. $y = \frac{2}{3}x - 1$

20. $y = \frac{1}{5}x + 2$

21. $y = -\frac{1}{3}x + 4$

22. $y = -\frac{3}{2}x - 1$

23. $y = \frac{4}{3}x + 2$

24. $y = \frac{5}{2}x - 3$

25. $y = -\frac{5}{3}x$

26. $y = -\frac{4}{5}x$

27. $y = 4x - 2$

28. $y = 2x - 4$

29. $y = -2x + 4$

30. $y = -3x + 5$

31. $y = x + 1$

32. $y = -x + 2$

33. $y = -3x$

34. $y = 4x$

35. $y = x$

36. $y = -x$

37. $y = -3$

38. $y = -2$

39. $y = 0$

40. $y = 1$

41. Graphs of four linear equations are shown in Fig. 93. State whether m and b are positive, negative, zero, or undefined for the $y = mx + b$ form of each equation.

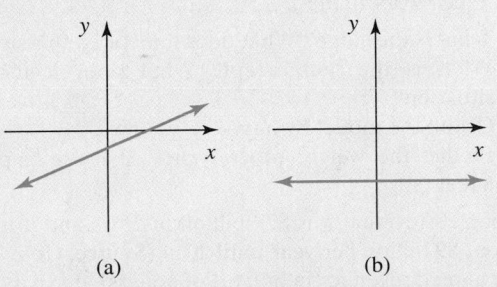

(a)   (b)

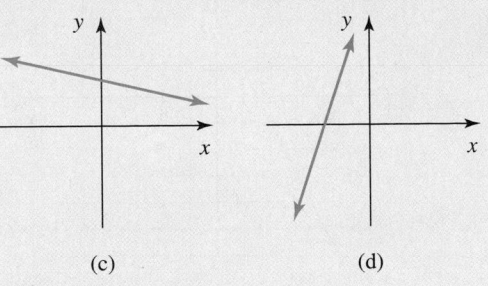

(c)   (d)

Figure 93 Exercise 41

42. Graphs of four linear equations are shown in Fig. 94. State the signs of the constants m and b for the $y = mx + b$ form of each equation.

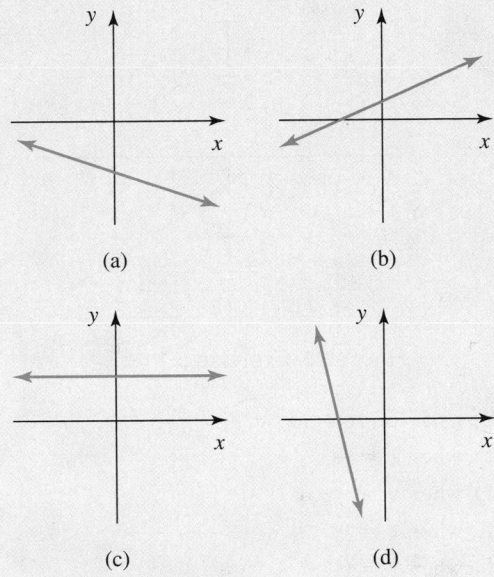

(a)   (b)

(c)   (d)

Figure 94 Exercise 42

Graph by hand an equation of the form $y = mx + b$ that meets the given criteria for m and b. Also, find an equation of each graph.

43. m is positive and b is positive

44. m is positive and b is negative

45. m is negative and b is negative

46. m is negative and $b = 0$

47. $m = 0$ and b is negative

48. $m = 0$ and $b = 0$

For Exercises 49–54, find an equation of a line that has the given slope and contains the given point.

49. $m = 3$, $(0, -4)$

50. $m = -2$, $(0, 5)$

51. $m = -\frac{6}{5}$, $(0, 3)$

52. $m = \frac{3}{4}$, $(0, -2)$

53. $m = -\frac{2}{7}$, $(0, 0)$

54. $m = 0$, $(0, -1)$

55. Find an equation of the line sketched in Fig. 95.

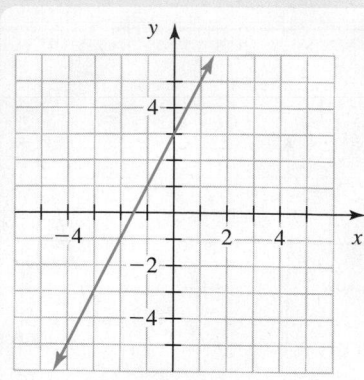

Figure 95 Exercise 55

56. Find an equation of the line sketched in Fig. 96.

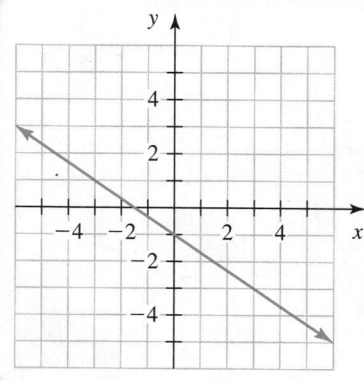

Figure 96 Exercise 56

For Exercises 57–62, refer to Fig. 97.

57. Find y when $x = -3$.

58. Find x when $y = -3$.

59. Find x when $y = 0$.

60. Find y when $x = 0$.

61. Find the slope of the line.

62. Find an equation of the line.

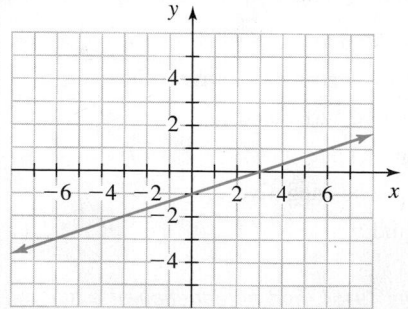

Figure 97 Exercises 57–62

63. Let n be the number (in thousands) of postal employees at t years since 2010. For the period 2010–2014, a reasonable model is $n = -31t + 586$ (Source: *U.S. Postal Service*).

   a. Graph the model by hand.

   b. Estimate the number of postal employees in 2014.

64. Let n be the number of firefighters who died on duty in the year that is t years since 2010. For the period 2010–2014, a reasonable model is $n = -2t + 83$ (Source: *U.S. Fire Administration*).

   a. Graph the model by hand.

   b. Estimate the number of firefighters who died on duty in 2013.

65. A study was based on the records of 6789 patients in 369 hospitals in which the patients' hearts stopped because of conditions that could have been reversed by an electrical shock from a defibrillator. Let p be the percentage of patients who survived at least until being discharged from the hospital, and let d be the delay (in minutes) in the hospital staff giving the shock. For delays from 1 to 6 minutes, inclusive, a reasonable model is $p = -4.4d + 45$ (Source: *New England Journal of Medicine*).

   a. Graph the model by hand.

   b. Predict the delay for patients of which 23% survived at least until being discharged from the hospital. Explain by using arrows on the graph you sketched in part (a).

66. Let H be the price (in dollars) of a hot dog, and let S be the price (in dollars) of a soft drink, both at a Major League Baseball (MLB) stadium. For hot dog prices between $1 and $6.25, inclusive, a reasonable model is $S = 0.8H + 0.4$ (Source: *Team Marketing Report*).

   a. Graph the model by hand.

   b. Predict the price of a hot dog at an MLB stadium where a soft drink costs $4. Explain by using arrows on the graph you sketched in part (a).

67. For monthly recordings since 1909, let B be the mean sea level at The Battery, New York, and let G be the mean sea level at Galveston, Texas, both in feet. For mean sea levels at The Battery between -1.5 and 0.9 feet, inclusive, a reasonable model is $G = 1.43B + 0.46$ (Source: *National Oceanic and Atmospheric Administration*).

   a. What is the slope? What does it mean in this situation?

   b. What is the G-intercept? What does it mean in this situation?

   c. Graph the model by hand.

   d. Estimate the mean sea level at Galveston when the mean sea level at The Battery was 0.6 feet.

   e. On the basis of the positive association between B and G, a student concludes that an increase in the mean sea level at The Battery will cause an increase in the mean sea level at Galveston. What would you tell the student?

68. For the 60 players picked in the 2014 draft for NBA basketball, let h be the height (in inches) of a player and let w be the weight (in pounds) of a player. For heights between 72 and 87 inches, inclusive, a reasonable model is $w = 6.54h - 301.81$ (Source: *nbadraft.net*).

   a. What is the slope? What does it mean in this situation?

   b. What is the w-intercept? What does it mean in this situation?

   c. Graph the model by hand.

   d. Predict the weight of draft-pick Shabazz Napier, who is 6 feet tall.

69. Google's revenue was $29 billion in 2010, and it increased by about $9 billion per year until 2014 (Source: *Google*). Let r be the annual revenue (in billions of dollars) at t years since 2010.

a. Identify the explanatory and response variables.
b. Find the slope and the *r*-intercept of a linear model.
c. Find an equation of the model.
d. Graph the model by hand.
e. Estimate Google's revenue in 2014.

70. As of February 1, a garage band knows 5 songs. Each week, the band members learn 2 more songs. Let *n* be the number of songs that the band knows at *t* weeks since February 1.

a. Identify the explanatory and response variables.
b. Find the slope and the *n*-intercept of a linear model.
c. Find an equation of the model.
d. Graph the model by hand.
e. How many songs will the band know on March 22 (7 weeks after February 1)?

71. A student's savings account has a balance of $4700 on September 1. Each month for 6 months, the balance declines by $650. Let *B* be the balance (in dollars) at *t* months since September 1.

a. Why is there a linear association between *t* and *B*?
b. Find the slope of a linear model. What does it mean in this situation?
c. Find an equation of the model.
d. Graph the model by hand.
e. When was the balance $1450? Explain by using arrows on the graph you sketched in part (d).

72. A person owns a Coleman RoadTrip LXE Propane Grill with a tank that holds 16.4 ounces of propane. The person always sets the burner on high, which uses 0.13 ounce of propane per minute (Source: *Coleman*). Let *p* be the number of ounces of propane that remain in the tank after *t* minutes of cooking since the tank was last filled.

a. Why is there an exact linear association between *t* and *p*?
b. Find the slope of a linear model. What does it mean in this situation?
c. Find an equation of the model.
d. Graph the model by hand.
e. The person fills the tank with propane. After how many hours of use will 6 ounces of propane remain? Explain by using arrows on the graph you sketched in part (d).

73. For fall semester 2014, part-time students at Centenary College paid $575 per credit (unit or hour) for tuition and paid a mandatory part-time student fee of $15 per semester (Source: *Centenary College*). Let *T* be the total cost (in dollars) of tuition and the fee when taking *c* credits of courses.

a. Identify the explanatory and response variables.
b. Find the slope of a linear model. What does it mean in this situation?
c. Find an equation of the model.
d. Graph the model by hand.
e. What was the total one-semester cost of tuition plus part-time student fee for 9 credits of classes?

74. For the spring semester 2014, California residents paid an enrollment fee of $46 per unit (credit or hour) at Santa Barbara City College (Source: *Santa Barbara City College*). Students were also required to pay a $18 health fee each semester.

Let *T* be the total cost (in dollars) of enrollment and health fees when taking *M* units of courses.

a. Identify the explanatory and response variables.

b. Find the slope of a linear model. What does it mean in this situation?
c. Find an equation of the model.
d. Graph the model by hand.
e. What was the total one-semester cost of tuition plus health fee for 15 units of classes?

75. A person drives her Toyota Prius® on a road trip. At the start of the trip, she fills up the 11.9-gallon tank with gasoline. During the trip, the cars uses about 0.02 gallon of gasoline per mile. Let *G* be the number of gallons of gasoline remaining in the tank after driving *d* miles.

a. Why is there an exact linear association?
b. What is the *G*-intercept of a model? What does it mean in this situation?
c. Find an equation of the model.
d. Graph the model by hand.
e. How much gasoline remains in the tank after driving 525 miles?

76. Although the United States and Great Britain use the Fahrenheit (°F) temperature scale, most countries use the Celsius (°C) scale. The temperature reading 0°C is equivalent to the Fahrenheit reading 32°F. An increase of 1°C is equivalent to an increase of 1.8°F. Let *F* be the Fahrenheit reading that is equivalent to a Celsius reading of *C* degrees. Assume that *F* is the response variable.

a. Why is there an exact linear association?
b. What is the *F*-intercept of the model? What does it mean in this situation?
c. Find an equation of the model.
d. Graph the model by hand.
e. If the temperature is 30°C, what is the Fahrenheit reading?

77. **DATA** Digital music sales in the United States are shown in Table 31 for various years.

Table 31 Digital Music Sales in the United States

Year	Sales (billions of dollars)
2005	1.1
2007	2.3
2009	3.0
2011	3.6
2013	4.4

Source: *Recording Industry Association of America*

Let *s* be the annual U.S. digital music sales (in billions of dollars) at *t* years since 2005.

a. Use technology to construct a scatterplot.
b. Describe the four characteristics of the association. Compute and interpret *r* as part of your analysis.
c. Use technology to graph the model $s = 0.4t + 1.3$ on the scatterplot. Does the line come close to the data points?
d. What is the *s*-intercept? What does it mean in this situation?
e. Estimate the sales in 2012.

78. **DATA** The numbers of countries that require picture warnings on cigarette packages are shown in Table 32 for various years.

Table 32 Numbers of Countries Requiring Picture Warning on Cigarette Packages

Year	Number of Countries
2008	18
2010	34
2012	55
2014	70
2015	77

Source: *Canadian Cancer Society*

Let n be the number of countries that require picture warnings on cigarette packages at t years since 2000.

a. Use technology to construct a scatterplot.

b. Describe the four characteristics of the association. Compute and interpret r as part of your analysis.

c. Use technology to graph the model $n = 8.59t - 50.58$ on the scatterplot. Does the line come close to the data points?

d. What is the n-intercept? What does it mean in this situation?

e. Estimate the number of countries that required picture warnings on cigarette packages in 2013.

79. ^{DATA} The percentages of Americans who believe marriages between same-sex couples should be recognized by the law as valid are shown in Table 33 for various age groups.

Table 33 Percentages of Americans Who Support Same-Sex Marriage

Age Group (years)	Age Used to Represent Age Group (years)	Percent
18–29	23.5	71
30–39	34.5	56
40–49	44.5	49
50–59	54.5	46
60–69	64.5	42
70–79	74.5	32
over 79	85.0	21

Source: *The Gallup Organization*

Let p be the percentage of Americans at age a years who believe marriages between same-sex couples should be recognized by the law as valid.

a. Use technology to construct a scatterplot.

b. Describe the four characteristics of the association. Compute and interpret r as part of your analysis.

c. Use technology to graph the model $p = -0.72t + 84.59$ on the scatterplot. Does the line come close to the data points?

d. What is the slope? What does it mean in this situation?

e. Predict the percentage of 27-year-old Americans who believe marriages between same-sex couples should be recognized by the law as valid.

80. ^{DATA} In a study of adults who had received radiation to the brain to treat brain cancer when they were children, researchers found that the cancer survivors tended to develop mild cognitive deficits typically seen in older people. The percentages of childhood brain cancer survivors who have mild cognitive deficits are shown in Table 34 for various ages.

Table 34 Percents of Childhood Brain Cancer Survivors Who Have Mild Cognitive Deficits

Age	Percent
20	22
30	27
40	55
50	77
60	98

Source: *St. Jude's Children's Research Hospital*

Let p be the percentage of childhood brain cancer survivors who have mild cognitive deficits at age a years.

a. Use technology to construct a scatterplot.

b. Describe the four characteristics of the association. Compute and interpret r as part of your analysis.

c. Use technology to graph the model $p = 2.02a - 25$ on the scatterplot. Does the line come close to the data points?

d. What is the slope? What does it mean in this situation?

e. Predict the percentage of 43-year-old childhood brain cancer survivors who have mild cognitive deficits.

81. ^{DATA} In an attempt to raise the low graduation rates of college football and basketball players, Division I institutions require that prospective athletes must meet new grade point average (GPA) standards (based on a maximum of 4.0) to play as freshmen, and enrolled athletes must meet these new standards to keep playing (see Table 35).

Table 35 Core GPAs Needed to Qualify to Play, for Given SAT Scores

SAT Score	Core GPA
620	3.0
700	2.8
780	2.6
860	2.4
940	2.2
1010	2.0

Source: *NCAA*

Let G be the qualifying core GPA for an SAT score of s points.

a. Use technology to construct a scatterplot.

b. Describe the four characteristics of the association. Compute and interpret r as part of your analysis.

c. Use technology to graph the model $G = -0.00254s + 4.58$ on the scatterplot. Does the line come close to the data points?

d. If one athlete's SAT score is 200 points higher than another athlete's SAT score, estimate how much lower the qualifying core GPA is for the student with the higher SAT score.

e. If an athlete's SAT score is 400 points, the lowest possible score, predict the student's qualifying core GPA. The actual qualifying core GPA is 3.55. Compute the error in your prediction. Did you perform extrapolation or interpolation?

82. ^{DATA} The mean selling prices for a Subaru Outback® of various ages are shown in Table 36.

Table 36 Mean Selling Prices of Subaru Outbacks

Age (years)	Mean Price (dollars)
1	27,080
3	21,112
4	16,950
5	13,400
6	9992
7	9159

Source: *AutoTrader®*

Let p be the mean price (in dollars) of a Subaru Outback of age a years.

a. Use technology to construct a scatterplot.

b. Describe the four characteristics of the association. Compute and interpret r as part of your analysis.

c. Use technology to graph the model $p = -3173.8a + 30,035.3$ on the scatterplot. Does the line come close to the data points?

d. What is the slope? What does it mean in this situation?

e. Predict the mean price of a 2-year-old Outback.

83. **DATA** Amazon.com's revenues are shown in Table 37 for various years.

Table 37 Amazon.com's Annual Revenues

Year	Revenue (billions of dollars)
2010	34.2
2011	48.1
2012	61.1
2013	74.5
2014	89.0

Source: *Amazon.com*

Let r be Amazon.com's annual revenues (in billions of dollars) at t years since 2010. A model of the situation is $r = 13.6t + 34.18$.

a. Use technology to construct a scatterplot, and graph the model on the same coordinate system. Does the line come close to the data points?

b. What is the slope? What does it mean in this situation?

c. Use the data in Table 37 to find the approximate rate of change of the revenues for each of the periods 2010–2011, 2012–2014, and 2010–2014. Compare each rate of change with the result you found in part (b).

d. Use the model to predict the revenue in 2019. Have you performed interpolation or extrapolation? Do you have much faith in your prediction?

84. **DATA** Costco's net sales are shown in Table 38 for various years.

Table 38 Costco's Net Sales

Year	Net Sales (billions of dollars)
2009	70
2010	76
2011	87
2012	97
2013	103

Source: *Costco*

Let n be Costco's annual net sales (in billions of dollars) at t years since 2005. A model of the situation is $n = 8.7t + 34.4$.

a. Use technology to construct a scatterplot, and graph the model on the same coordinate system. Does the line come close to the data points?

b. What is the slope? What does it mean in this situation?

c. Use the data in Table 38 to find the approximate rate of change of the net sales for each of the periods 2009–2011, 2011–2013, and 2009–2013. Compare each rate of change with the result you found in part (b).

d. Use the model to predict Costco's net sales in 2020. Did you perform interpolation or extrapolation? Do you have much faith in your prediction? Explain.

Concepts

85. A student says the slope of the line $y = 2x + 1$ is $2x$. Is the student correct? Explain.

86. A student says the y-intercept of the line $y = 3x - 2$ is $(0, 2)$. Is the student correct? Explain.

87. Use the slope and y-intercept of the line $y = 2x - 1$ to graph the equation $y = 2x - 1$ by hand. Then choose two points that lie on the graph, and show that both of the corresponding ordered pairs are solutions of the equation $y = 2x - 1$.

88. Use the slope and y-intercept of the line $y = -2x + 3$ to graph the equation $y = -2x + 3$ by hand. Then choose two points that lie on the graph, and show that both of the corresponding ordered pairs are solutions of the equation $y = -2x + 3$.

89. Recall that we can describe some or all of the solutions of an equation in two variables with an equation, a table, a graph, or words.

a. Describe the solutions of $y = \frac{1}{2}x + 2$ by using a graph.

b. Describe three solutions of $y = \frac{1}{2}x + 2$ by using a table.

c. Describe the solutions of $y = \frac{1}{2}x + 2$ by using words.

90. Recall that we can describe some or all of the solutions of an equation in two variables with an equation, a table, a graph, or words.

a. Describe the solutions of $y = \frac{1}{3}x - 1$ by using a graph.

b. Describe three solutions of $y = \frac{1}{3}x - 1$ by using a table.

c. Describe the solutions of $y = \frac{1}{3}x - 1$ by using words.

91. A student tries to graph the equation $y = -3x + 1$ (see Fig. 98).

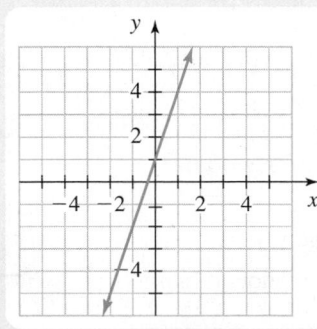

Figure 98 Exercise 91

Choose two points that lie on the line, and show that at least one of these points is not a solution of $y = -3x + 1$. Explain why your work shows that Fig. 98 is incorrect. Then sketch the correct graph.

92. A student tries to graph the equation $y = \dfrac{3}{2}x - 1$ (see Fig. 99).

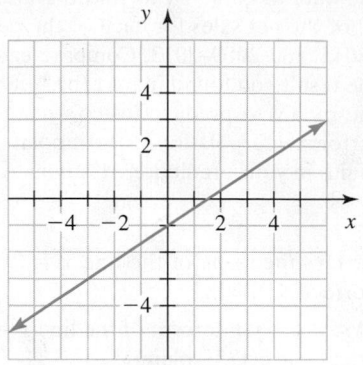

Figure 99 Exercise 92

Choose two points that lie on the line, and show that at least one of these points is not a solution of $y = \dfrac{3}{2}x - 1$. Explain why your work shows that Fig. 99 is incorrect. Then sketch the correct graph.

93. A line passes through the point $(-1, 1)$ and has a slope of 2.

  a. Sketch the line.

  b. Find an equation of the line. [**Hint:** Use part (a) to find b in $y = mx + b$.]

94. A line passes through the point $(1, 2)$ and has a slope of -3.

  a. Sketch the line.

  b. Find an equation of the line. [**Hint:** Use part (a) to find b in $y = mx + b$.]

95. **a.** Find the slope of each line: $y = 3$, $y = -5$, and $y = 0$.

  b. Find the slope of the graph of any linear equation of the form $y = k$, where k is a constant.

96. **a.** Find the slope of each line: $x = 2$, $x = -4$, and $x = 0$.

  b. Find the slope of the graph of any equation of the form $x = k$, where k is a constant.

97. Graphs of the equations $y = mx + b$ and $y = kx + c$ (where m, b, k, and c are constants) are sketched in Fig. 100.

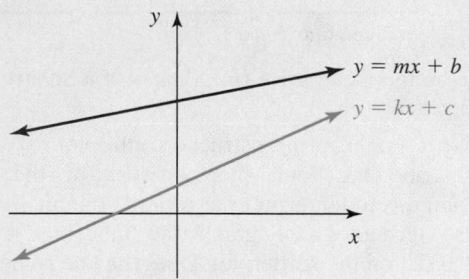

Figure 100 Exercise 97

  a. Which is greater, m or k? Explain.

  b. Which is greater, b or c? Explain.

98. Suppose that m and b are negative constants. Is the x-coordinate of the x-intercept of the line $y = mx + b$ positive or negative? Explain.

99. Create an equation of the form $y = mx + b$. Find two points on the graph of your equation by substituting two values for x. Use the two points to calculate the slope of the line, and compare the result with your chosen value of m.

100. Describe how to use the method discussed in this section to graph an equation of the form $y = mx + b$.

▼ 7.4 Functions

Objectives

» Describe the meanings of *relation, domain, range,* and *function.*

» Identify functions by using the *vertical line test.*

» Explain the meaning of *linear function.*

» Describe the Rule of Four for functions.

» Use the graph of a function to find the function's domain and range.

» Use function notation.

» Use function notation to make predictions.

» Determine the domain and range of a model.

In Chapter 6 and this chapter, we have worked with associations between two variables. In this section, we will describe some of these associations by using an extremely important concept called a *function.*

Relation, Domain, Range, and Function

In Chapter 6 and this chapter, we have used graphs, tables, and equations to describe the association between two variables. To illustrate, Table 39 describes an association between the variables x and y. This association is also described graphically in Fig. 101.

Table 39 A Relationship Described by a Table

x	y
3	2
4	1
5	3
5	4

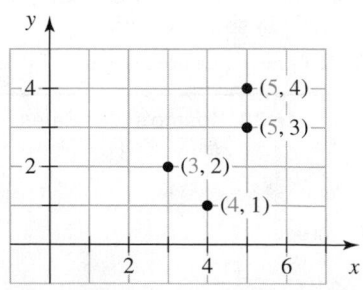

Figure 101 The relationship of Table 39 described by a graph

We call the set of ordered pairs listed in Table 39 a *relation*. This relation consists of the ordered pairs $(3, 2)$, $(4, 1)$, $(5, 3)$, and $(5, 4)$. The *domain* of the relation is the set of all values of *x* (the explanatory variable)—in this case, 3, 4, and 5. The *range* of the relation is the set of all values of *y* (the response variable)—here, 1, 2, 3, and 4.

▶ **Definition   Relation, domain, and range**

A **relation** is a set of ordered pairs. The **domain** of a relation is the set of all values of the explanatory variable, and the **range** of the relation is the set of all values of the response variable.

We can think of a relation as a machine in which values of *x* are "inputs" and values of *y* are "outputs." In general, each member of the domain is an **input,** and each member of the range is an **output**.

For the relation described in Table 39, we can think of the values of *x* as being sent to the values of *y* (see Fig. 102).

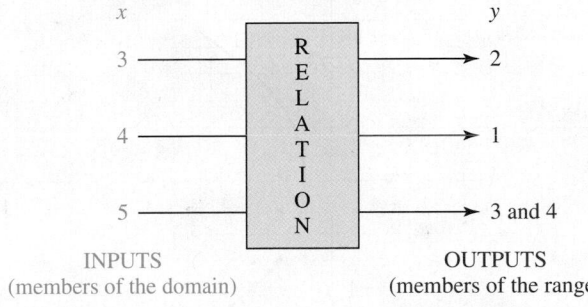

INPUTS
(members of the domain)

OUTPUTS
(members of the range)

Figure 102 Think of a relation as an input–output machine

Note that the input $x = 5$ is sent to *two* outputs: $y = 3$ and $y = 4$. In a special type of relation called a *function*, each input is sent to exactly *one* output. The relation described in Table 39 is not a function.

▶ **Definition   Function**

A **function** is a relation in which each input leads to exactly one output.

The equation $y = x + 2$ describes a relation consisting of an infinite number of ordered pairs. We will determine whether the relation is a function in Example 1.

▶ **Example 1   Deciding whether an Equation Describes a Function**

Is the relation $y = x + 2$ a function? Find the domain and range of the relation.

Solution

Let's consider some input–output pairs (in Fig. 103).

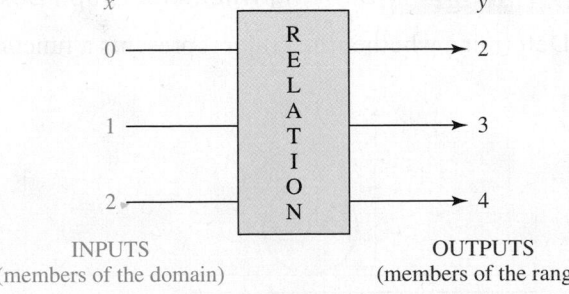

INPUTS
(members of the domain)

OUTPUTS
(members of the range)

Figure 103 The "increasing by 2" relation: $y = x + 2$

Each input leads to just *one* output—namely, the input increased by 2—so the relation $y = x + 2$ is a function.

The domain of the relation $y = x + 2$ is the set of all real numbers because we can add 2 to *any* real number. The range of $y = x + 2$ is also the set of real numbers because any real number is the output of the number that is 2 units less than it.

Table 40 Input–Output Pairs of a Relation

x (input)	y (output)
0	2
1	3
1	5
2	7
3	10

▶ **Example 2**   Deciding whether a Table Describes a Function

Is the relation described by Table 40 a function?

Solution

The input $x = 1$ leads to *two* outputs: $y = 3$ and $y = 5$. So, the relation is not a function.

▶ **Example 3**   Deciding whether a Graph Describes a Function

Is the relation described by the graph in Fig. 104 a function?

Solution

The input $x = 3$ leads to *two* outputs: $y = -4$ and $y = 4$ (see Fig. 105). So, the relation is *not* a function.

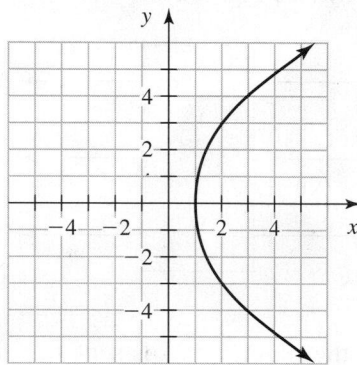

Figure 104 Graph of a relation

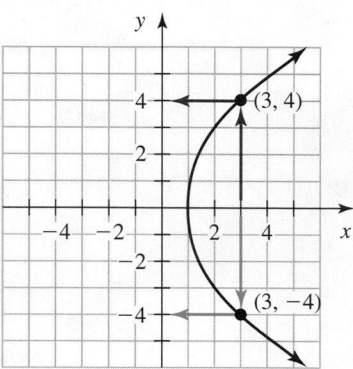

Figure 105 The input $x = 3$ leads to two outputs: $y = -4$ and $y = 4$

Vertical Line Test

The relation described in Example 3 is not a function because some vertical lines would intersect the graph more than once.

▶ **Vertical Line Test**

A relation is a function if and only if each vertical line intersects the graph of the relation at no more than one point. We call this requirement the **vertical line test**.

▶ **Example 4**   Deciding whether a Graph Describes a Function

Determine whether the graph represents a function.

1.

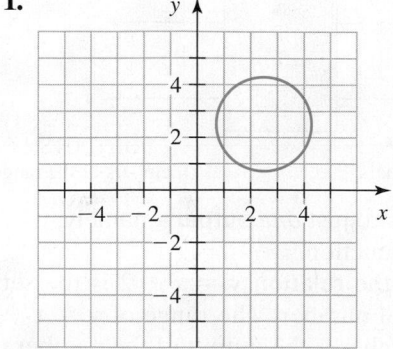

2.

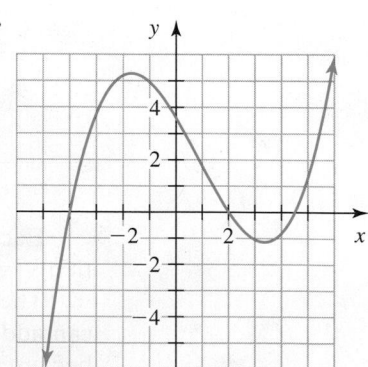

Solution

1. Since the vertical line sketched in Fig. 106 intersects the circle more than once, the relation is not a function.
2. Each vertical line sketched in Fig. 107 intersects the curve at one point. In fact, *any* vertical line would intersect this curve at just one point. So, the relation is a function.

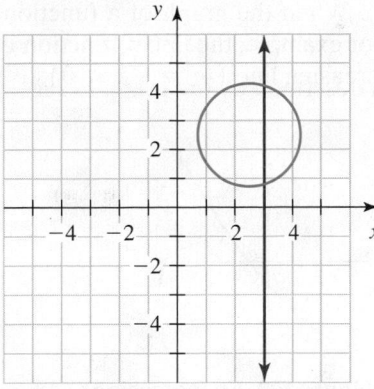

Figure 106 The circle does not describe a function

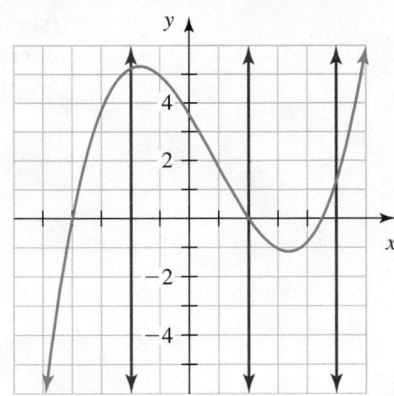

Figure 107 The curve describes a function

▶ **Example 5** Deciding whether an Equation Describes a Function

Is the relation $y = 2x + 1$ a function?

Solution

We begin by sketching the graph of $y = 2x + 1$ in Fig. 108. Each vertical line would intersect the line $y = 2x + 1$ at just one point. So, the relation $y = 2x + 1$ is a function.

Linear Functions

In Example 5, we saw that the line $y = 2x + 1$ is a function. In fact, any nonvertical line is a function because it passes the vertical line test.

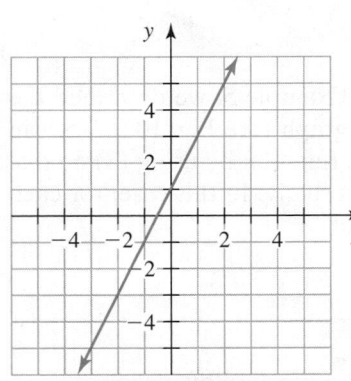

Figure 108 Graph of $y = 2x + 1$

▶ **Definition** Linear function

A **linear function** is a relation whose equation can be put into the form

$$y = mx + b$$

where m and b are constants.

In this chapter, we have made many observations about linear equations in two variables. Since a linear function can be described by a linear equation in two variables, these observations tell us about linear functions. Let's summarize what we know about a linear function $y = mx + b$:

1. The graph of the function is a nonvertical line.
2. The constant m is the rate of change of y with respect to x.
3. The constant m is the slope of the line. The absolute value of the slope is a measure of the line's steepness.
4. If $m > 0$, the graph of the function is an increasing line.
5. If $m < 0$, the graph of the function is a decreasing line.
6. If $m = 0$, the graph of the function is a horizontal line.
7. The y-intercept of the line is $(0, b)$.

Finally, since a linear equation of the form $y = mx + b$ is a *function*, we know each input leads to exactly one output.

If a curve goes upward from left to right, we say that the curve is **increasing** (see Fig. 109). When the graph of a function is increasing, we say that the function is **increasing**. For example, the linear function $f(x) = 3x - 2$ is increasing, because the graph is an increasing line ($m = 3 > 0$).

If a curve goes downward from left to right, we say that the curve is **decreasing** (see Fig. 110). When the graph of a function is decreasing, we say that the function is **decreasing**. For example, the linear function $g(x) = -2x + 5$ is decreasing, because the graph is a decreasing line ($m = -2 < 0$).

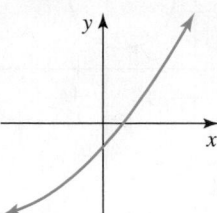

Figure 109 An increasing curve

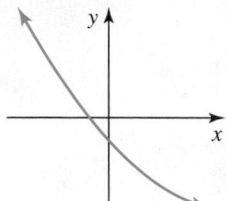

Figure 110 A decreasing curve

For example, the linear function $y = 4x - 7$ is increasing, because the graph is an increasing line ($m = 4 > 0$). The linear function $y = -3x + 6$ is decreasing, because the graph is a decreasing line ($m = -3 < 0$).

Rule of Four for Functions

We can describe functions in four ways. For instance, in Example 5, we described the function $y = 2x + 1$ by using (1) the equation and (2) a graph (see Fig. 108). We can also describe some of the input–output pairs for the same function by using (3) a table (see Table 41). Finally, we can describe the function (4) verbally: In this case, for each input–output pair, the output is 1 more than twice the input.

Table 41 Input–Output Pairs for $y = 2x + 1$

x	y
0	1
1	3
2	5
3	7
4	9

Table 42 Input–Output Pairs of $y = -2x - 1$

x	y
-2	$-2(-2) - 1 = 3$
-1	$-2(-1) - 1 = 1$
0	$-2(0) - 1 = -1$
1	$-2(1) - 1 = -3$
2	$-2(2) - 1 = -5$

> ### Rule of Four for Functions
>
> We can describe some or all of the input–output pairs of a function by means of
>
> **1.** an equation, **2.** a graph,
>
> **3.** a table, or **4.** words.
>
> These four ways to describe input–output pairs of a function are known as the **Rule of Four** for functions.

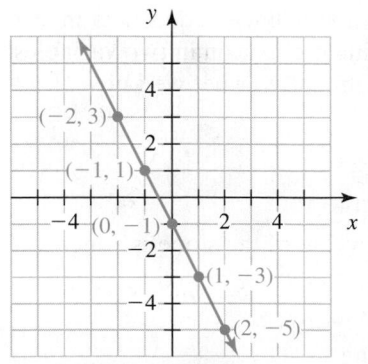

Figure 111 Graph of $y = -2x - 1$

▶ **Example 6** Describing a Function by Using the Rule of Four

1. Is the relation $y = -2x - 1$ a function?
2. List some input–output pairs of $y = -2x - 1$ by using a table.
3. Describe the input–output pairs of $y = -2x - 1$ by using a graph.
4. Describe the input–output pairs of $y = -2x - 1$ by using words.

Solution

1. Because $y = -2x - 1$ is of the form $y = mx + b$, it is a (linear) function.
2. We list five input–output pairs in Table 42.
3. We graph $y = -2x - 1$ in Fig. 111.
4. For each input–output pair, the output is 1 less than -2 times the input.

Using a Graph to Find the Domain and Range of a Function

In Example 7, we will refer to a function's graph to determine its domain and range.

▶ **Example 7**   Finding the Domain and Range

Use the graph of the function to determine the function's domain and range.

1.
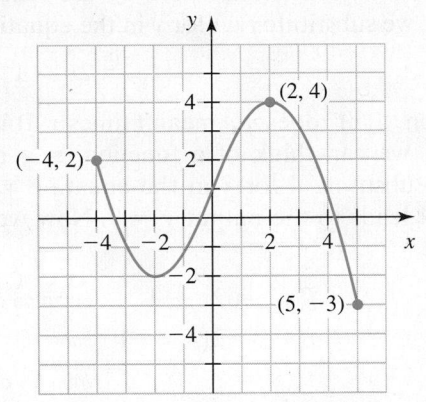

2.
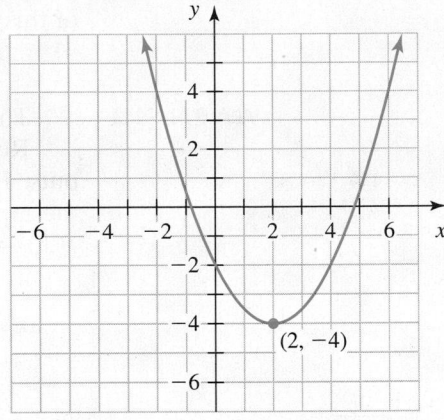

Solution

1. Figure 112 shows arrows pointing from *some* of the inputs (on the x-axis) to *some* of the outputs (on the y-axis). The domain is the set of all x-coordinates of points in the graph. Since there are no breaks in the graph, and since the leftmost point is $(-4, 2)$ and the rightmost point is $(5, -3)$, the domain is $-4 \leq x \leq 5$.

    The range is the set of all y-coordinates of points in the graph. Since the lowest point is $(5, -3)$ and the highest point is $(2, 4)$, the range is $-3 \leq y \leq 4$.

2. Figure 113 shows arrows going from *some* of the inputs (on the x-axis) to *some* of the outputs (on the y-axis). The graph extends to the left and right indefinitely without breaks, so every real number is an x-coordinate of some point in the graph. The domain is the set of all real numbers.

    The output -4 is the smallest number in the range, because $(2, -4)$ is the lowest point in the graph. The graph also extends upward indefinitely without breaks, so every number larger than -4 is also in the range. The range is $y \geq -4$.

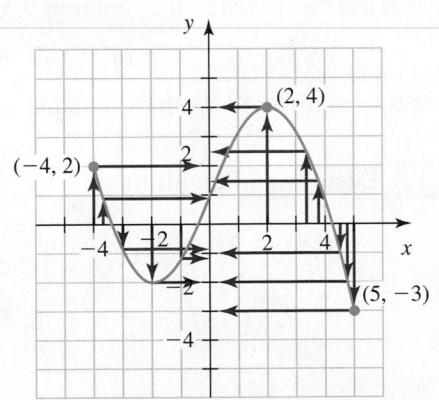

Figure 112 Arrows point from *some* inputs to *some* outputs

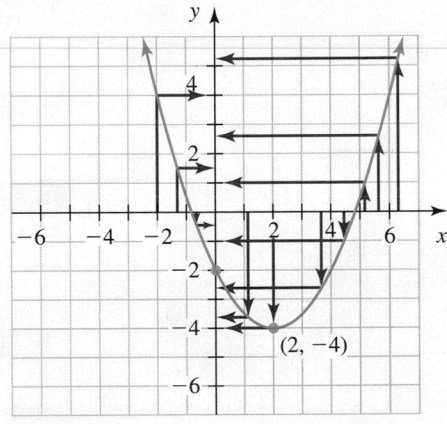

Figure 113 Arrows point from *some* inputs to *some* outputs

Function Notation

Rather than use an equation, table, graph, or words to refer to a function, it would be easier to name the function. For example, to use "f" as the name of the linear function $y = 2x + 1$, we use "$f(x)$" (read "f of x") to represent y:

$$y = f(x)$$

We refer to "$f(x)$" as *function notation*. To use function notation to write the equation of this function, we substitute $f(x)$ for y in the equation $y = 2x + 1$:

$$f(x) = 2x + 1$$

WARNING   The notation "$f(x)$" does *not* mean f times x. It is another variable name for y.

Recall that we can think of a function as a machine that sends inputs to outputs. Here we substitute 4 for x in the equation $y = 2x + 1$: $y = 2(4) + 1 = 9$. So, the input $x = 4$ leads to the output $y = 9$. Now we substitute 4 for x in the equation $f(x) = 2x + 1$:

$$\begin{aligned}
f(x) &= 2x + 1 &&\text{Equation of } f \\
f(4) &= 2(4) + 1 &&\text{Substitute 4 for } x. \\
&= 9 &&\text{Multiply and then add.}
\end{aligned}$$

The equation $f(4) = 9$ means the input $x = 4$ leads to the output $y = 9$. Figure 114 shows the "machine" f sending the input 4 to the output 9.

Notice that $f(4) = 9$ is of the form

$$f(\text{input}) = \text{output}$$

This is true for any function f.

The number $f(4)$ is the value of y when x is 4. To find $f(4)$, we say we **evaluate** the function f at $x = 4$.

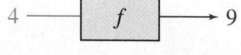

Figure 114 A function "machine"

▶ Example 8   Evaluating a Function

Evaluate $f(x) = -4x + 2$ at 5.

Solution

$$\begin{aligned}
f(x) &= -4x + 2 &&\text{Equation of } f \\
f(5) &= -4(5) + 2 &&\text{Substitute 5 for } x. \\
&= -18 &&\text{Multiply and then add.}
\end{aligned}$$

We can also use "g" to name the function $y = -4x + 2$:

$$g(x) = -4x + 2$$

The most commonly used symbols to name functions are f, g, and h.

▶ Example 9   Evaluating Functions

For $f(x) = 8x - 5$, $g(x) = 2x^2 - 3x$, $h(x) = \dfrac{4x - 2}{5x - 1}$, and $k(x) = 8$, find the following:

1. $f\left(\dfrac{3}{2}\right)$ **2.** $g(-2)$ **3.** $h(3)$ **4.** $k(2)$

Solution

1.
$$\begin{aligned}
f\left(\frac{3}{2}\right) &= 8\left(\frac{3}{2}\right) - 5 &&\text{Evaluate } f \text{ at } \frac{3}{2}. \\
&= 12 - 5 &&\text{Multiply first.} \\
&= 7 &&\text{Subtract.}
\end{aligned}$$

2.

$$g(-2) = 2(-2)^2 - 3(-2) \quad \text{\textit{Evaluate g at} } -2.$$
$$= 2(4) - 3(-2) \quad \text{\textit{Perform exponentiation}}$$
$$\text{\textit{first:} } (-2)^2 = (-2)(-2) = 4$$
$$= 8 + 6 \quad \text{\textit{Multiply.}}$$
$$= 14 \quad \text{\textit{Add.}}$$

3.

$$h(3) = \frac{4 \cdot 3 - 2}{5 \cdot 3 - 1} \quad \text{\textit{Evaluate h at 3.}}$$
$$= \frac{12 - 2}{15 - 1} \quad \text{\textit{Multiply first.}}$$
$$= \frac{10}{14} \quad \text{\textit{Subtract.}}$$
$$= \frac{5}{7} \quad \text{\textit{Simplify.}}$$

4. All outputs of k are 8, so $k(2) = 8$.

So far, we have used equations to evaluate functions. Next, we will use a table to find an output and an input of a function.

Table 43 Input–Output Pairs of g

x	$g(x)$
3	12
4	9
5	8
6	9
7	12

▶ **Example 10** Using a Table to Find an Output and an Input

Some input–output pairs of a function g are shown in Table 43.

1. Find $g(7)$.
2. Find x when $g(x) = 9$.

Solution

1. From Table 43, we see the input $x = 7$ leads to the output $y = 12$. So, $g(7) = 12$.
2. To find x when $g(x) = 9$, we need to find all inputs in the table that lead to the output $y = 9$. From Table 43, we see both inputs $x = 4$ and $x = 6$ lead to the output $y = 9$. So, the values of x are 4 and 6.

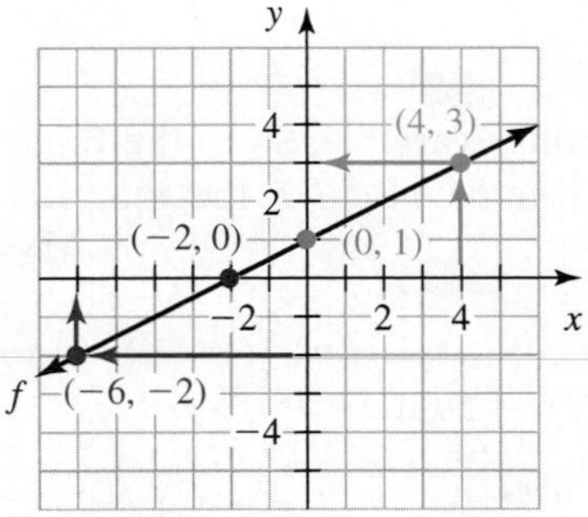

▶ **Example 11** Using a Graph to Find Values of x or $f(x)$

A graph of a function f is sketched in Fig. 115.

1. Find $f(4)$.
2. Find $f(0)$.
3. Find x when $f(x) = -2$.
4. Find x when $f(x) = 0$.

Figure 115 Problems 1–4 of Example 11

Solution

1. Recall that $y = f(x)$. The notation $f(4)$ refers to $f(x)$ when $x = 4$. So, we want the value of y when $x = 4$. The blue arrows in Fig. 115 show that the input $x = 4$ leads to the output $y = 3$. Hence, $f(4) = 3$.
2. To find $f(0)$, we want the value of y when $x = 0$. The line contains the point $(0, 1)$, so $f(0) = 1$.
3. We have $y = f(x) = -2$. Thus, $y = -2$. So, we want the value of x when $y = -2$. The red arrows in Fig. 115 show that the output $y = -2$ originates from the input $x = -6$. Hence, $x = -6$.
4. We have $y = f(x) = 0$. Thus, $y = 0$. The line contains the point $(-2, 0)$, so $x = -2$.

Using Function Notation to Make Predictions

Recall from Section 6.3 that when we are *not* describing an authentic situation, we treat x as the explanatory variable and y as the response variable. Here we label the explanatory variable, response variable, and function name of the equation $y = f(x)$:

response variable ⟶   ⟵ explanatory variable

$$y = f(x)$$

↑
function name

We follow this same format for a function f that *is* a model.

> ▶ **Definition**   Function notation
>
> The response variable of a function f can be represented by the expression formed by writing the explanatory variable name within the parentheses of $f(\ )$:
>
> $$\text{response variable} = f(\text{explanatory variable})$$
>
> We call this representation **function notation**.

In Section 6.3, we defined a linear model as a nonvertical line that describes two quantities in an authentic situation. Because any nonvertical line is a linear function, it follows that every linear model is a linear function.

However, not every linear function is a linear model. Models are used only to describe authentic situations. Functions are used both to describe authentic situations *and* to describe certain *mathematical* relationships between two variables. For example, if the equation $y = 2x$ is not being used to describe a situation, then it is a function, not a model.

▶ **Example 12**   Using Function Notation to Make Predictions

In Exercise 73 of Homework 7.3, you found the model $T = 575C + 15$, where T is the total fall semester 2014 cost (in dollars) of tuition and fees for part-time students who took C credits (units or hours) of courses at Centenary College (Source: *Centenary College*).

1. Rewrite the equation $T = 575C + 15$ with the function name f.
2. Find $f(9)$. What does it mean in this situation?

Solution

1. Here, C is the explanatory variable and T is the response variable. Since the function name is f, we can write $T = f(C)$. Then we substitute $f(C)$ for T in the equation $T = 575C + 15$:

$$f(C) = 575C + 15$$

2. To find $f(9)$, we substitute 9 for C in the equation $f(C) = 575C + 15$:

$$f(9) = 575(9) + 15 \quad \text{Substitute 9 for C.}$$
$$= 5190 \qquad\qquad\quad \text{Multiply and then add.}$$

The model predicts that the total cost of tuition and fees for 9 credits of courses is $5190.

In Example 12, we used f to name the function $f(C) = 575C + 15$, where $f(C)$ represents the total cost (in dollars) of tuition and fees for part-time students who took C credits of courses at Centenary College. When we use more than one function to model situations, naming the functions helps us distinguish among them. For example, we can also use a linear function to model the total cost of tuition and fees at Portland Community College. A good model is $T = 98.2C + 19$, where T is the total cost (in dollars) of tuition and fees for Oregon residents who took C credits of courses in fall semester 2014 (Source: *Portland Community College*). We can distinguish this function from f by using g as its name:

$$g(C) = 98.2C + 19$$

▶ **Example 13** Using Function Notation to Make Predictions

The percentages of American adults who smoke are shown in Table 44 for various years.

Table 44 Percentages of American Adults Who Smoke

Year	Percent
1970	37.4
1980	33.2
1990	25.3
2000	23.1
2010	19.4
2012	18.0

Source: *National Center for Health Statistics*

A model is $p = -0.45t + 36.83$, where p is the percentage of Americans who smoke at t years since 1970.

1. Verify that the graph of $p = -0.45t + 36.83$ comes close to the data points.
2. Rewrite the equation $p = -0.45t + 36.83$ with the function name g.
3. Estimate the percentage of American adults who smoked in 2011.
4. Find the p-intercept of the model. What does it mean in this situation?

Solution

1. We draw the scatterplot and the model in the same viewing window (see Fig. 116). The line comes close to the data points, so the model is a reasonable one.
2. To use the name g, we substitute $g(t)$ for p in the equation $p = -0.45t + 36.83$:

$$g(t) = -0.45t + 36.83$$

3. We represent the year 2011 by $t = 2011 - 1970 = 41$. To find the percentage, we substitute 41 for t in the equation $g(t) = -0.45t + 36.83$:

$$g(41) = -0.45(41) + 36.83 \quad \text{Substitute 41 for t.}$$
$$= 18.38 \quad\quad\quad\quad\quad \text{Compute.}$$

The model estimates that 18.38% of American adults smoked in 2011.

4. Because the model $g(t) = -0.45t + 36.38$ is in slope-intercept form, the p-intercept is $(0, 36.38)$. So, the model estimates that 36.38% of American adults smoked in 1970. This estimate is fairly close to the actual percentage, 37.4% (see Table 44).

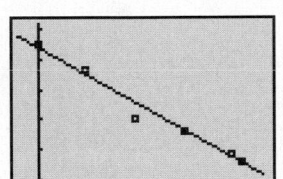

Figure 116 Verifying the smoking model

▶ **Example 14** Finding an Equation of a Model; Using Function Notation

The number of households with cable TV subscriptions was 61.8 million in 2009 and decreased by about 1.5 million per year until 2014 (Source: *IHS*). Let $n = f(t)$ be the number (in millions) of households with cable TV subscriptions at t years since 2009.

1. Find an equation of f.
2. Find $f(4)$. What does it mean in this situation?

Solution

1. Because the number of households with cable TV subscriptions decreased by about a *constant* 1.5 million each year, we will use a linear model f with slope -1.5. The n-intercept is $(0, 61.8)$, because there were 61.8 million households with cable TV in 2009 $(t = 0)$. So, an equation of f is

$$f(t) = -1.5t + 61.8$$

2. To find $f(4)$, we substitute 4 for t in the equation $f(t) = -1.5t + 61.8$:

$$f(4) = -1.5(4) + 61.8 = 55.8$$

So, $n = 55.8$ when $t = 4$. The model estimates that there were 55.8 million households with cable TV in 2013.

Domain and Range of a Model

Recall from earlier in this section that the domain of a function is the set of all inputs and that the range of a function is the set of all outputs. For the **domain** and **range** of a model, we consider input–output pairs only when both the input and the output make sense in the situation. The domain of the model is the set of all such inputs, and the range of the model is the set of all such outputs.

▶ **Example 15**   Finding the Domain and Range of a Model

A store is open from 9 A.M. to 5 P.M., Mondays through Saturdays. Let $I = f(t)$ be an employee's weekly income (in dollars) from working t hours each week at $10 per hour.
1. Find an equation of the model f.
2. Find the domain and range of the model f.

Solution

1. The employee's weekly income (in dollars) is equal to the pay per hour times the number of hours worked per week:

$$f(t) = 10t$$

2. To find the domain and range of the model f, we consider input–output pairs only when both the input and the output make sense in this situation. Time is the input. Since the store is open 8 hours a day, 6 days a week, the employee can work up to 48 hours each week. So, the domain is the set of numbers between 0 and 48, inclusive: $0 \le t \le 48$.

   Income is the output. Since the number of hours worked is between 0 and 48 hours, inclusive, and the pay is $10 per hour, the range is the set of numbers between 0 and $10(48) = 480$, inclusive: $0 \le f(t) \le 480$.

   In Fig. 117, we illustrate the inputs 22, 35, and 48 being sent to the outputs 220, 350, and 480, respectively. We also label the part of the t-axis that represents the domain and the part of the I-axis that represents the range.

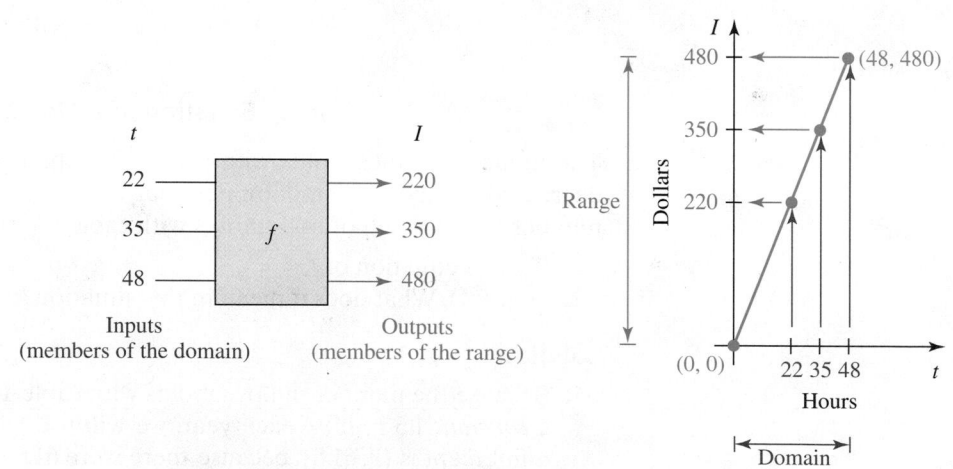

Figure 117 Domain and range of the employee income model

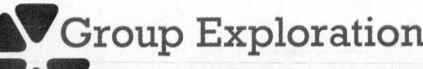

Group Exploration

Use before Vertical line test

1. Consider the relation described by Table 45. Is the relation a function? Explain. Now plot the points on a coordinate system. What do you notice about them?

Table 45 A Relation Described by a Table

x	y
2	1
2	5
2	7

2. Consider the relation described by Table 46. Is the relation a function? Explain. Now plot the points on a coordinate system. What do you notice about them?

Table 46 A Relation Described by a Table

x	y
4	2
4	3
4	6

3. Describe the graph of a relation that is not a function.

4. Determine whether each graph in Fig. 118 is the graph of a function. Explain.

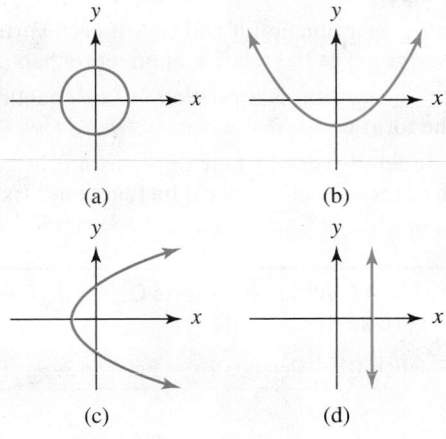

Figure 118 Which graphs describe functions?

Group Exploration

Formula for slope

1. Let $f(x) = 2x + 1$. Find each of the following and compare all three results. [**Hint for part (b):** First, find $f(5)$ and $f(3)$. Then subtract. Finally, divide.]
   a. the slope of the graph of f
   b. $\dfrac{f(5) - f(3)}{5 - 3}$
   c. $\dfrac{f(7) - f(4)}{7 - 4}$

2. Let $g(x) = 3x + 5$. Find each of the following and compare all three results:
   a. the slope of the graph of g
   b. $\dfrac{g(3) - g(1)}{3 - 1}$
   c. $\dfrac{g(4) - g(0)}{4 - 0}$

3. Let f be a function of the form $f(x) = mx + b$. Describe $\dfrac{f(c) - f(d)}{c - d}$, where $c \neq d$. Explain.

▶ **Tips for Success** "Work Out" by Solving Problems

Although there are many things you can do to enhance your learning, there is no substitute for solving problems. Your mathematical ability will respond to solving problems in much the same way that your muscles respond to lifting weights. Muscles increase greatly in strength when you work out intensely, frequently, and consistently.

Just as with building muscles, to learn math, you must be an *active* participant. No amount of watching weight lifters lift, reading about weight-lifting techniques, or conditioning yourself psychologically can replace working out by lifting weights. The same is true of learning math: No amount of watching your instructor do problems, reading your text, or listening to a tutor can replace "working out" by solving problems.

Homework 7.4

For extra help ▶ MyMathLab® 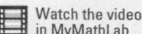 Watch the videos in MyMathLab    Download the MyDashboard App

1. The range of a relation is the set of all values of the ____ variable.

2. A function is a relation in which each input leads to exactly ____ output.

3. A relation is a function if and only if each vertical line intersects the graph of the relation at no more than ____ point.

4. A(n) ____ function is a relation whose equation can be put into the form $y = mx + b$.

5. Some ordered pairs of four relations are listed in Table 47. Which of these relations could be functions? Explain.

Table 47 Which Relations Could Be Functions? (Exercise 5)

Relation 1		Relation 2		Relation 3		Relation 4	
x	y	x	y	x	y	x	y
1	1	3	27	0	4	5	10
2	3	4	24	1	4	6	20
3	5	5	21	2	4	7	30
3	7	6	18	3	4	8	40
4	9	7	15	4	4	8	50

6. Some ordered pairs of four relations are listed in Table 48.

Table 48 Which Relations Could Be Functions? (Exercise 6)

Relation 1		Relation 2		Relation 3		Relation 4	
x	y	x	y	x	y	x	y
1	3	5	27	0	50	3	11
2	4	5	24	1	45	4	13
3	5	5	21	2	40	5	17
3	6	5	18	3	35	6	25
4	7	5	15	4	30	7	40

   a. Which of the relations could be functions? Explain.
   b. Which could be linear functions? Explain.

7. For a certain relation, an input leads to two different outputs. Could the relation be a function? Explain.

8. For a certain relation, two different inputs lead to the same output. Could the relation be a function? Explain.

9. A relation's graph contains the points $(2, 3)$ and $(5, 3)$. Could the relation be a function? Explain.

10. A relation's graph contains the points $(4, 5)$ and $(4, 9)$. Could the relation be a function? Explain.

Determine whether the graph represents a function. Explain.

11.
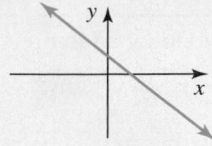

12.
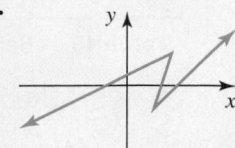

13.
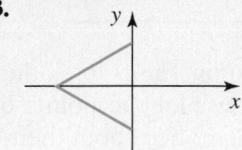

14.
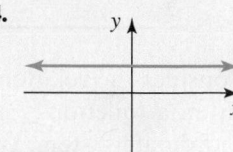

15.
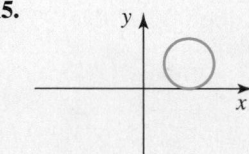

16.
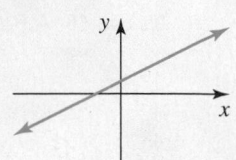

17.
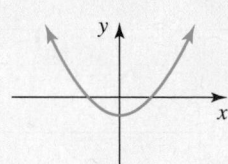

18.

For Exercises 19–24, determine whether the relation is a function. Explain.

19. $y = 5x - 1$
20. $y = -3x + 8$
21. $y = 4$
22. $y = -1$
23. $x = -3$
24. $x = 0$

25. Recall that we can describe some or all of the input–output pairs of a function by means of an equation, a graph, a table, or words.

   a. Describe five input–output pairs of $y = 3x - 2$ by using a table.
   b. Describe the input–output pairs of $y = 3x - 2$ by using a graph.
   c. Describe the input–output pairs of $y = 3x - 2$ by using words.

26. Recall that we can describe some or all of the input–output pairs of a function by means of an equation, a graph, a table, or words.

   a. Describe five input–output pairs of $y = -\dfrac{3}{2}x + 2$ by using a table.
   b. Describe the input–output pairs of $y = -\dfrac{3}{2}x + 2$ by using a graph.
   c. Describe the input–output pairs of $y = -\dfrac{3}{2}x + 2$ by using words.

For Exercises 27–36, use the graph of the function to determine the function's domain and range.

27.

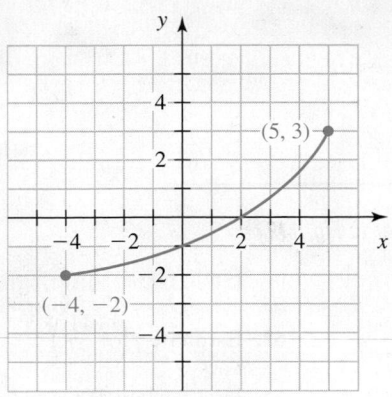

31.

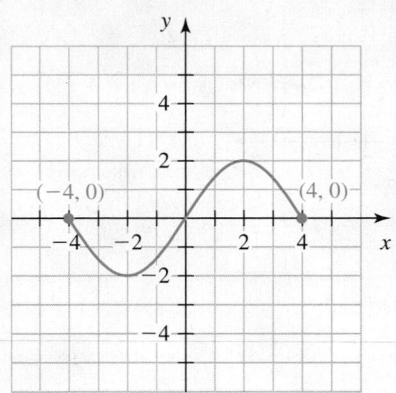

28.

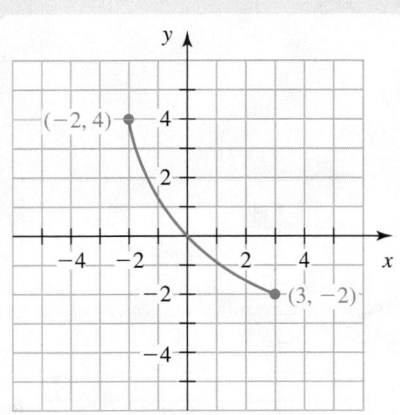

32.

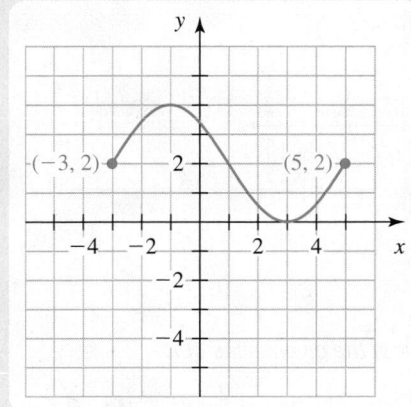

29.

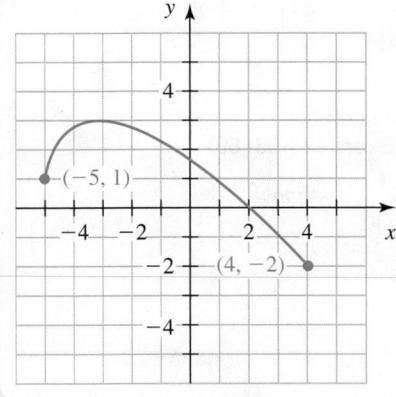

33.

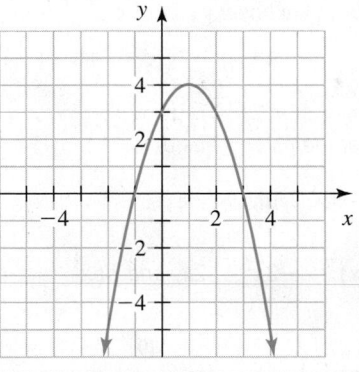

30.

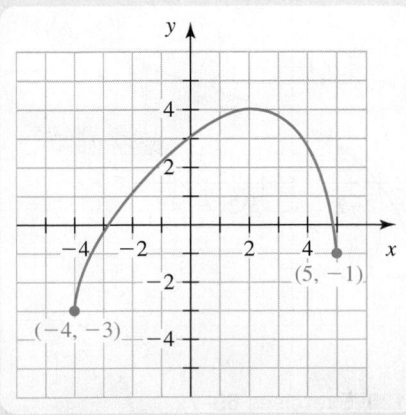

34.

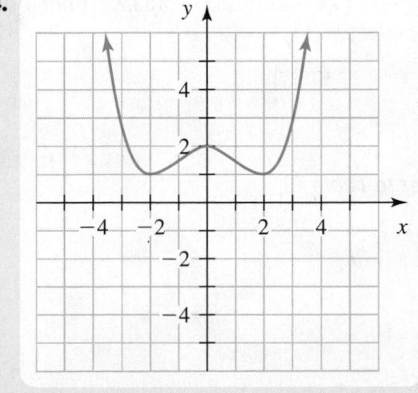

35.

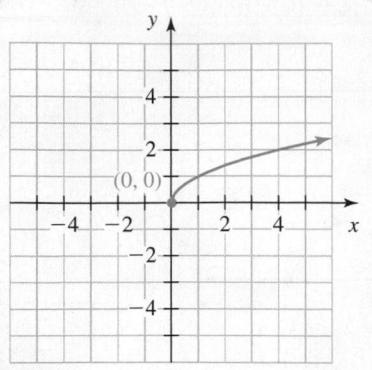

36.

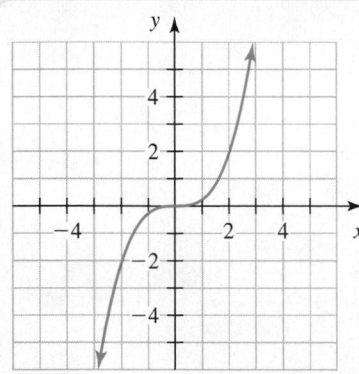

Evaluate $f(x) = 6x - 4$ *at the given value of x.*

37. $f(5)$ **38.** $f(-2)$ **39.** $f\left(\dfrac{2}{3}\right)$ **40.** $f\left(\dfrac{5}{2}\right)$

Evaluate $g(x) = 2x^2 - 5x$ *at the given value of x.*

41. $g(2)$ **42.** $g(3)$ **43.** $g(-3)$ **44.** $g(-2)$

Evaluate $h(x) = \dfrac{3x - 4}{5x + 2}$ *at the given value of x.*

45. $h(2)$ **46.** $h(-4)$

For $f(x) = -2x + 7$, $g(x) = -3x^2 + 2x$, *and* $h(x) = -4$, *find the following.*

47. $g(-2)$ **48.** $g(-1)$ **49.** $f(5)$

50. $f(-4)$ **51.** $h(7)$ **52.** $h(-9)$

For Exercises 53 and 54, let $f(x) = -5.95x + 183.22$. *Round any results to the second decimal place.*

53. Find $f(10.91)$.

54. Find $f(17.28)$.

For Exercises 55–58, refer to Table 49.

55. Find $f(2)$.

56. Find $f(4)$.

57. Find x when $f(x) = 2$.

58. Find x when $f(x) = 4$.

Table 49 Values of f
(Exercises 55–58)

x	f(x)
0	0
1	2
2	4
3	2
4	0

For Exercises 59–68, refer to Fig. 119.

59. Find $f(-6)$.   **60.** Find $f(0)$.

61. Estimate $f(2.5)$.   **62.** Estimate $f\left(-\dfrac{11}{2}\right)$.

63. Find x when $f(x) = 0$.   **64.** Find x when $f(x) = 1$.

65. Estimate x when $f(x) = 3.5$.

66. Estimate x when $f(x) = \dfrac{5}{2}$.

67. Find the domain of f.

68. Find the range of f.

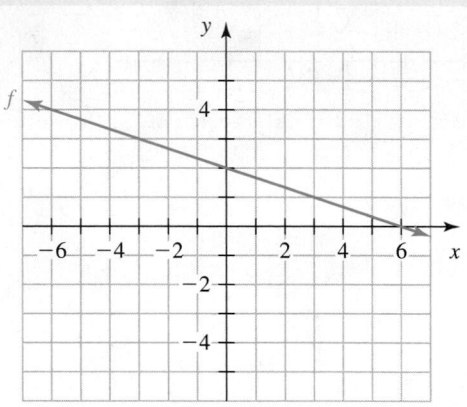

Figure 119 Exercises 59–68

For Exercises 69–72, refer to Fig. 120.

69. Find $g(-2)$.

70. Find x when $g(x) = 3$.

71. Find the domain of g.

72. Find the range of g.

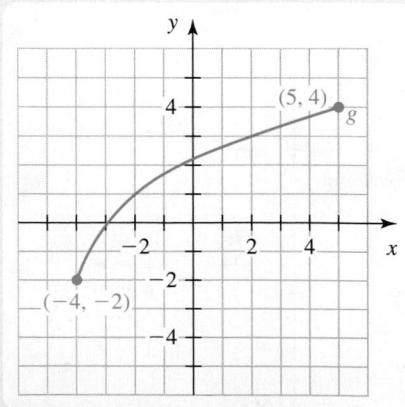

Figure 120 Exercises 69–72

For Exercises 73–76, refer to Fig. 121.

73. Find $h(1)$.

74. Find x when $h(x) = -1$.

75. Find the domain of h.

76. Find the range of h.

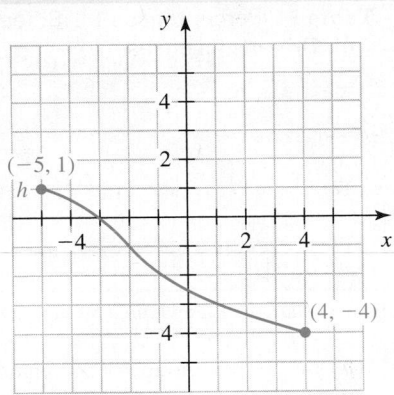

Figure 121 Exercises 73–76

77. Let n be the number of drive-in movie sites in the United States at t years since 2009. The function $n = -4.9t + 381$ models the situation well for the period 2009–2014 (Source: *National Association of Theater Owners*).

   a. Rewrite the equation $n = -4.9t + 381$ using the function name f.

   b. Find $f(3)$. What does it mean in this situation?

   c. Find $f(0)$. What does it mean in this situation?

78. In Exercise 64 of Homework 7.3, you worked with the model $n = -2t + 83$, where n is the number of firefighters who died on duty in the year that is t years since 2010 (Source: *U.S. Fire Administration*). The function models the situation well for the period 2010–2014.

   a. Rewrite the equation $n = -2t + 83$ using the function name f.

   b. Find $f(4)$. What does it mean in this situation?

   c. Find $f(0)$. What does it mean in this situation?

79. In Exercise 75 of Homework 7.3, you found the equation $G = -0.02d + 11.9$, where G is the number of gallons of gasoline remaining in a Toyota Prius tank after driving d miles.

   a. Rewrite the equation $G = -0.02d + 11.9$ using the function name f.

   b. Find $f(450)$. What does it mean in this situation?

   c. Find $f(0)$. What does it mean in this situation?

80. In Example 10 of Section 7.3, we found the model $B = -5.9E + 212$, where B is the boiling point (in degrees Fahrenheit) at an elevation of E thousand meters (Source: *Thermodynamics, an Engineering Approach by Cengal & Boles*).

   a. Rewrite the equation $B = -5.9E + 212$ using the function name f.

   b. Find $f(5)$. What does it mean in this situation?

   c. Find $f(0)$. What does it mean in this situation?

81. **DATA** The percentages of total donations that went to the 10 colleges and universities that received the most in donations are shown in Table 50 for various years.

Table 50 Percentages of Total Donations That Went to the 10 Colleges and Universities That Received the Most in Donations

Year	Percent
2002	15.2
2004	15.3
2006	15.9
2008	16.0
2010	16.3
2012	17.1
2014	17.5

Source: *Council for Aid to Education*

Let p be the percentage of total donations that went to the 10 colleges and universities that received the most in donations at t years since 2000.

   a. Use technology to construct a scatterplot.

   b. Use technology to graph the model $p = 0.19t + 14.63$ on the scatterplot. Does it come close to the data points?

   c. Rewrite the equation $p = 0.19t + 14.63$ using the function name f.

   d. Find $f(11)$. What does it mean in this situation?

   e. In 2013, \$34.7 billion was donated to *all* colleges and universities. Estimate the amount of donations given to the 10 colleges and universities that received the most in donations.

82. **DATA** The annual sales of passenger vehicles in China are shown in Table 51 for various years.

Table 51 Annual Sales of Passenger Vehicles in China

Year	Passenger Vehicle Sales (millions)
2007	5.3
2008	5.7
2009	8.3
2010	11.3
2011	12.2
2012	13.3
2013	16.3

Source: *China Association of Automobile Manufacturers*

Let s be the annual passenger vehicle sales (in millions) in China at t years since 2000.

   a. Use technology to construct a scatterplot.

   b. Use technology to graph the model $s = 1.86t - 8.26$ on the scatterplot. Does it come close to the data points?

   c. Rewrite the equation $s = 1.86t - 8.26$ using the function name f.

   d. Find $f(14)$. What does it mean in this situation? Did you perform interpolation or extrapolation? Do you have much faith in your prediction?

   e. On the basis of your result in part (d), what was the increase in sales from 2013 to 2014? What percentage increase is that? Compare your result with consulting firm IHS's prediction of 10% (Source: *IHS*).

83. **DATA** The *labor force* is made up of people at least 16 years of age who are either employed or actively looking for work.

The *participation rate* is the percentage of people at least 16 years of age who are in the labor force. The *unemployment rate* is the percentage of the labor force who are unemployed. The participation and unemployment rates in March of various years are shown in Table 52.

Table 52 Participation and Unemployment Rates in March

Year	Participation Rate (percent)	Unemployment Rate (percent)	Population at Least 16 Years of Age (millions)
2010	64.9	9.9	237
2011	64.2	9.0	239
2012	63.8	8.2	242.6
2014	63.2	6.6	247.2
2015	62.7	5.5	250.2

(**Source:** *Labor Department*)

Let $P(t)$ and $U(t)$ be the participation and unemployments rates, respectively, both in March at t years since 2010.

a. Use technology to construct a scatterplot that compares the years and the participation rates. Does the graph of $P(t) = -0.41t + 64.74$ come close to the data points?

b. Use technology to construct a scatterplot that compares the years and the unemployment rates. Does the graph of $U(t) = -0.86t + 9.90$ come close to the data points?

c. Find $P(3)$ and $U(3)$. What do your results mean in this situation?

d. Use the results you found in part (c) to estimate the percentage of people at least 16 years of age who were *employed* in March 2013. [**Hint:** Carefully read the definitions in the first paragraph of this exercise.]

e. A student says that even though the unemployment rate has decreased, the number of people working has not increased because the participation rate has decreased. Using the data in Table 52, estimate the *number* of people at least 16 years of age who were employed in 2010. Do the same for 2015. Is the student correct?

84. **DATA** The percentages of elementary and secondary school-teachers who are union members are shown in Table 53 for various years.

Table 53 Percentages of Teachers Who Are Union Members

Year	Percent
1995	58
2000	54
2005	53
2010	52
2013	49

(**Source:** *unionstats.com*)

Let p be the percentage of teachers who are union members at t years since 1995.

a. Use technology to construct a scatterplot.

b. Use technology to graph the model $p = -0.43t + 57.32$ on the scatterplot. Does it come close to the data points?

c. Rewrite the equation $p = -0.43t + 57.32$ using the function name f.

d. Find $f(17)$. What does it mean in this situation?

e. Use f to estimate how much the percentage of teachers who were in unions changed from one year to the next.

85. **DATA** The percentages of U.S. teenagers with driver's licenses are shown in Table 54 for various ages.

Table 54 Percentages of U.S. Teens with Driver's Licenses

Age (years)	Percent
16	56.8
17	61.3
18	72.0
19	79.4
20	81.8

Source: *National Longitudinal Survey of Youth*

Let $p = f(a)$ be the percentage of teenagers at age a years who have driver's licenses.

a. Use technology to construct a scatterplot.

b. Describe the four characteristics of the association. Compute and interpret r as part of your analysis.

c. Graph the model $f(a) = 6.81a - 52.32$ on the scatterplot. Does it come close to the data points?

d. Find $f(20)$. What does it mean in this situation? Find the error in your result.

e. Find $f(15)$. What does it mean in this situation? Did you perform interpolation or extrapolation? Do you have much faith in this prediction? Explain.

86. **DATA** Credit scores measure financial responsibility. Mean credit scores of Americans are shown in Table 55 for various age groups.

Table 55 Mean Credit Scores of Americans

Age Group (years)	Age Used to Represent Age Group (years)	Mean Credit Score (points)
18–29	23.5	637
30–39	34.5	654
40–49	44.5	675
50–59	54.5	697
60–69	64.5	722
70 or more	75	747

Source: *Experian*

Let $c = f(a)$ be the mean credit score (in points) of adults at a years of age.

a. Use technology to construct a scatterplot.

b. Describe the four characteristics of the association. Compute and interpret r as part of your analysis.

c. Graph the model $f(a) = 2.17a + 581.49$ on the scatterplot. Does it come close to the data points?

d. Find $f(23.5)$. What does it mean in this situation? Did you perform interpolation or extrapolation? Find the error in your result.

e. The oldest verified American ever was Sarah Knauss, who died at age 119 years (Source: *Associated Press*). Use f to estimate Knauss's credit score. Did you perform interpolation or extrapolation? Do you have much faith in this estimate? Explain.

87. The numbers of women and people who live alone are shown in Table 56 for various years.

Table 56 Numbers of Women and People Who Live Alone

| Year | Number Living Alone (millions) | |
	Women	People
1990	14.0	23.0
1995	14.6	24.7
2000	15.6	26.8
2005	17.3	30.1
2010	17.4	31.4
2012	17.7	32.0

Source: *U.S. Census Bureau*

Let $W(t)$ be the number of women who live alone and $P(t)$ be the number of people who live alone, both in millions, at t years since 1990. Reasonable models are $W(t) = 0.18t + 13.94$ and $P(t) = 0.43t + 22.84$.

a. Find $W(21)$. What does it mean in this situation?

b. Find $P(21)$. What does it mean in this situation?

c. Use W and P to help estimate the *percentage* of people living alone who were women in 2008.

d. Use the data in Table 56 to find the percentage of people living alone who were women in 1990, 2000, 2010, and 2012.

e. On the basis of your results in part (d), is the association between years and the percentage of people living alone who are women positive, negative, or neither for the period 1990–2012? How is this possible, given that the number of women living alone increased in that period?

88. [DATA] Mean annual U.S. per-person consumption of chicken and red meat (in pounds per person) is described for various years in Table 57.

Table 57 Mean Annual U.S. Per-Person Consumption of Chicken and Red Meat

| Year | Mean Annual Consumption (pounds per person) | |
	Chicken	Red Meat
1970	40.1	145.1
1980	47.4	136.1
1990	60.6	119.4
2000	77.4	120.2
2010	82.8	107.9
2013	83.2	104.4

Source: *U.S. Department of Agriculture*

Let $C(t)$ be the mean annual per-person consumption of chicken and $R(t)$ be the mean annual per-person consumption of red meat, both in pounds per person, at t years since 1970. Equations of C and R are $C(t) = 1.09t + 39.33$ and $R(t) = -0.91t + 143.89$.

a. Use technology to determine whether the graph of C comes close to the points of the appropriate scatterplot.

b. Use technology to determine whether the graph of R comes close to the points of the appropriate scatterplot.

c. Find $C(42)$ and $R(42)$. What do they mean in this situation?

d. Compare the slopes of the two models. What does your comparison mean in this situation?

e. Find the sum of the slopes of the two models. What does your result mean in this situation?

89. To measure the strength of stitching in knitted jerseys, researchers applied pressure perpendicular to the fabric until it burst. The *bursting strength* is the pressure applied just before the fabric bursts. Figure 122 shows a scatterplot that compares the lengths of stitches with the bursting strengths for 26-count yarn and 30-count yarn.

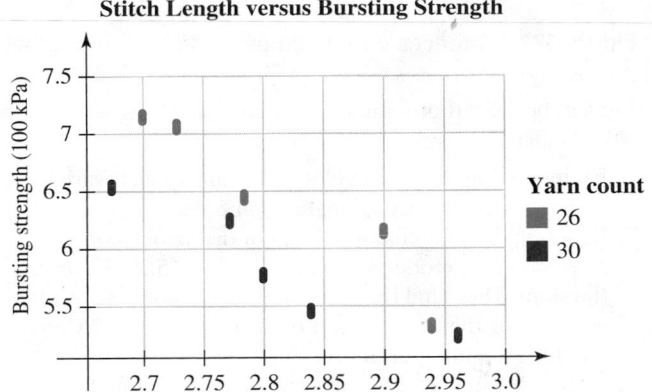

Figure 122 Stitch lengths versus bursting strengths
(**Source:** *Study on the Effect of Stitch Length on Bursting Strength of Knit Fabric, S. A. Shadid and M. F. Rahman*

Let $f(L)$ and $g(L)$ be the bursting strengths (in kPa) of knitted jerseys with yarn counts 26 and 30, respectively, in which the stitch length is L (in mm).

a. Describe the four characteristics of the association between stitch length and bursting strength for 26-count yarn.

b. Describe the four characteristics of the association between stitch length and bursting strength for 30-count yarn.

c. Reasonable models are $f(L) = -7.23L + 26.80$ and $g(L) = -5.13L + 20.27$. Find $f(2.75)$ and $g(2.75)$. What do they mean in this situation?

d. The thinner the yarn, the larger the yarn count. Which type of yarn does this study suggest has greater bursting strength, thinner or thicker yarn? Explain by referring to the scatterplot and the results you found in part (c).

e. For the yarn-count-26 association, r is -0.94. For the yarn-count-30 association, r is also -0.94. But when all the data are combined, r is only -0.83. How is this possible?

90. Figure 123 shows a scatterplot that compares the distances and times for randomly selected Delta Air Lines flights departing from Hartsfield-Jackson International Airport in Atlanta, Georgia.

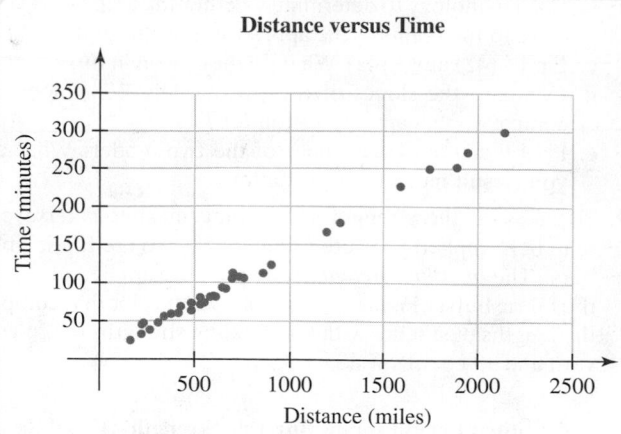

Figure 123 Distances versus times

(**Source:** *Bureau of Transportation Statistics*)

Let $f(d)$ be the airborne time (in minutes) for a flight that traveled d miles.

a. By inspecting the scatterplot, determine which variable is being treated as the explanatory variable.

b. Describe the four characteristics of the association.

c. A reasonable model is $f(d) = 0.14d + 5.39$. Determine the slope. Then find the reciprocal of the slope. Why is it an estimate of the mean speed of all the airplanes? Convert the units to miles per hour.

d. Use f to estimate the airborne time of Flight 833 from Atlanta to Albuquerque, New Mexico, which is a distance of 1267 miles.

e. The actual airborne time of Flight 833 was 179 minutes. Was the airplane's velocity a bit more or a bit less than your result in part (c)? Is the point that represents the flight a bit above or a bit below the graph of f? Explain.

91. The mean number of viewers of Fox prime-time TV shows was 9.1 million viewers in 2010 and decreased by about 0.8 million viewers until 2014. (Source: *Nielsen*). Let $f(t)$ be the mean number (in millions) of Fox prime-time viewers at t years since 2010.

a. Find an equation of f.

b. Find $f(3)$. What does it mean in this situation?

c. Estimate the *percentage* of Americans who were Fox prime-time viewers in 2012. The U.S. population was 312.8 million in that year.

92. The median age of buyers of Harley-Davidson motorcycles was 48.7 years in 2010 and increased by 0.31 year each year until 2014 (Source: *Harley-Davidson Motor Co.*). Let $f(t)$ be the median age (in years) of buyers at t years since 2010.

a. Find an equation of f.

b. Find $f(4)$. What does it mean in this situation?

c. Find $f(0)$. What does it mean in this situation?

93. The number of U.S. brewery openings was 154 openings in 2010 and increased by about 12.4 openings per year until 2014 (Source: *Brewers Association*). Let $f(t)$ be the number of brewery openings in the year that is t years since 2010.

a. Find an equation of f.

b. What is the slope? What does it mean in this situation?

c. Find $f(3) - f(2)$. Compare the result with the slope. Why does your comparison make sense?

d. Estimate the number of openings in 2013.

e. Estimate the mean number of openings *per state* in 2014. Is your result likely an underestimate or an overestimate for Wyoming, which is the state with the smallest population?

94. Annual sales of echinacea were $132 million in 2009 and decreased by about $6.9 million per year until 2014 (Source: *Nutrition Business Journal*). Let $f(t)$ be the annual echinacea sales (in millions of dollars) at t years since 2009.

a. Find an equation of f.

b. What is the slope? What does it mean in this situation?

c. Find $f(3) - f(2)$. Compare the result with the slope. Why does your comparison make sense?

d. Estimate the sales of echinacea in 2013.

e. Estimate the mean sales of echinacea *per state* in 2014. Is your result likely an underestimate or an overestimate for California, which is the state with the largest population?

95. The percentage of the workforce having manufacturing jobs was 10.1% in 2010 and decreased by about 0.43 percentage point per year until 2014 (Source: *U.S. Bureau of Labor Statistics*). Let $p = f(t)$ be the percentage of the workforce who have manufacturing jobs at t years since 2010.

a. Find an equation of f.

b. Find $f(4)$. What does it mean in this situation?

c. Find the slope. What does it mean in this situation?

d. Find the p-intercept. What does it mean in this situation?

96. The number of households that watched the *Miss Universe Pageant* was 4.62 million in 2010 and decreased by about 0.13 million per year until 2014 (Source: *Nielsen Media Research*). Let $n = f(t)$ be the number (in millions) of households that watched the show at t years since 2010.

a. Find an equation of f.

b. Find $f(2)$. What does it mean in this situation?

c. Find the slope. What does it mean in this situation?

d. Find the n-intercept. What does it mean in this situation?

97. A basement is flooded with 640 cubic feet of water. It takes 4 hours to pump out the water at a constant rate. Let $f(t)$ be the number of cubic feet of water that remains in the basement after t hours of pumping.

a. Find a linear equation of f. [**Hint:** You are given information about two points that can be used to find the slope and then an equation.]

b. Graph f by hand.

c. What are the domain and range of the model? Explain.

98. It takes a person 5 minutes to eat all 12 ounces of ice cream in a cup at a constant rate. Let $f(t)$ be the number of ounces of ice cream remaining in the cup t minutes after the person began eating the ice cream.

a. Find a linear equation of f. [**Hint:** You are given information about two points that can be used to find the slope and then an equation.]

b. Graph f by hand.

c. What are the domain and range of the model? Explain.

Concepts

99. Describe the input–output pairs of a function (different from those in this section) by using an equation, a graph, and words. Describe also five input–output pairs of the function by using a table. Explain why your relation is a function.

100. Sketch the graph of a relation for which the input $x = 2$ leads to exactly two outputs and the input $x = 6$ leads to exactly one output. Is the relation a function? Explain.

101. Sketch the graph of a function whose domain is $-3 \le x \le 5$ and whose range is $-2 \le y \le 4$.

102. Sketch the graph of a function whose domain is the set of all real numbers and whose range is $y \le 3$.

103. A student tries to determine whether the relation $y = x^2$ is a function. She finds that both inputs $x = -3$ and $x = 3$ give the same output, $y = 9$. The student concludes the relation is not a function. Is her conclusion correct? Explain.

104. Explain how you can determine whether a relation is a function.

105. Explain why the vertical line test works.

106. Let p be the percentage of adults with income d dollars who own a Mercedes-Benz. A student says that if f is the name of the function that describes the association between d and p, then $f(p) = d$. What would you tell the student?

107. a. For $f(x) = 4x$, find $f(3)$, $f(5)$, and $f(8)$. Is the equation $f(3 + 5) = f(3) + f(5)$ a true statement?

b. For $f(x) = x^2$, find $f(2)$, $f(3)$, and $f(5)$. Is the equation $f(2 + 3) = f(2) + f(3)$ a true statement?

c. For $f(x) = \sqrt{x}$, find $f(9)$, $f(16)$, and $f(25)$. Is $f(9 + 16) = f(9) + f(16)$ a true statement?

d. Is $f(a + b) = f(a) + f(b)$ a true statement for every function f?

108. a. For $f(x) = 3x + 2$, find $f(5) - f(4)$. Compare your result with the slope of the graph of f. [**Hint:** Find $f(5)$ and $f(4)$. Then subtract.]

b. For $f(x) = 2x + 5$, find $f(7) - f(6)$. Compare your result with the slope of the graph of f.

c. For $f(x) = 4x + 1$, find $f(3) - f(2)$. Compare your result with the slope of the graph of f.

d. What do the results that you found in parts (a), (b), and (c) suggest about a linear function of the form $y = mx + b$?

109. A student says $f(4)$ means that $y = 4$, because $y = f(x)$. Is the student correct? Explain.

110. A student tries to find $g(-5)$, where $g(x) = x^2$:

$$g(-5) = -5^2 = -25$$

Describe any errors. Then find $g(-5)$ correctly.

Find the slope of the graph of the linear function f that meets the following conditions:

111. $f(-5) = 2$ and $f(3) = -4$

112. $f(-6) = 1$ and $f(-2) = -5$

Find the intercepts of the graph of the linear function f that meets the following conditions:

113. $f(0) = -3$ and $f(4) = 0$

114. $f(-5) = 0$ and $f(0) = -1$

Hands-On Projects

ᴅᴀᴛᴀ Climate Change Project (continued from Chapter 6)

Scientists estimate that the mean global temperature has increased by about 1°F in the past century (see Table 58). Recall from the Climate Change Lab in Chapter 6 that many scientists believe an increase as small as 3.6°F could be dangerous. Because the planet's mean temperature increased by 1°F in the past century, it might seem it would take about another two-and-a-half centuries for Earth to reach a dangerous temperature level.

However, the Intergovernmental Panel on Climate Change (IPCC), composed of hundreds of scientists around the world, predicted that Earth's mean temperature will rise 2.5°F to 10.4°F in the coming century.* Scientists from the United States and Europe have predicted that there is a 9-out-of-10 chance the global mean temperature will increase 3°F to 9°F, with a range of 4°F to 7°F most likely.†

One result of the last century of global warming is that glaciers are receding and ocean levels are rising. For example, NASA scientist Bill Krabill estimates that Greenland's ice cap, the world's second largest, may be losing ice at a rate of 50 cubic kilometers per year.‡ The ice cap contained 2.85 million cubic kilometers of ice in 2000. Climatologist Jonathan Gregory of the University of Reading, in the

Table 58 Mean Surface Temperatures of Earth

Year	Mean Temperature (degrees Fahrenheit)	Year	Mean Temperature (degrees Fahrenheit)
1900	57.1	1960	57.2
1905	56.8	1965	57.0
1910	56.6	1970	57.3
1915	57.0	1975	57.1
1920	56.9	1980	57.6
1925	56.9	1985	57.3
1930	57.1	1990	57.8
1935	57.0	1995	57.9
1940	57.3	2000	57.8
1945	57.3	2005	58.3
1950	56.9	2010	58.3
1955	57.0	2014	58.4

Source: *NASA–GISS*

*IPCC, 2001: Summary for Policymakers.
†T. M. L. Wigley and S. C. B. Raper, "Interpretation of high projections for global-mean warming," *Science* 293 (5529):451–454, July 20, 2001.
‡W. Krabill et al., "Greenland ice sheet: High-elevation balance and peripheral thinning," *Science* 289:428–430, 2000.

United Kingdom, believes that by 2050 the ice cap may start an irreversible runaway melting. A total meltdown could take 1000 years.[§]

Jonathan Overpeck, director of the Institute for the Study of Planet Earth at the University of Arizona, believes that Greenland's ice cap could melt completely in as little as 150 years and that a partial melting of Greenland's ice cap by 2100 could raise ocean levels by more than 1 meter.[¶] Coastal engineers estimate that a 1-meter rise would translate into a loss of about 100 meters of land.[‖] In some areas of the world, land loss could be much greater.[**] For example, most of Florida is less than 1 meter above sea level.

Most scientists predict that global warming will also cause the extinction of plants and animals, bring severe water shortages, create extreme weather events, and increase the number of heat-related illnesses and deaths.

Analyzing the Situation

1. Let F be the mean global temperature (in degrees Fahrenheit) at t years since 1900. Refer to the scatterplot that you constructed in Problem 3 of the Climate Change Project of Chapter 6. (If you did not construct this scatterplot earlier, construct one now, using the data shown in Table 58.) Does it appear that Earth's mean temperature increased much from 1900 to 1965? Explain.

2. a. Estimate the mean global temperature from 1900 to 1965.

   b. Estimate the mean global temperature from 1990 to 2000.

   c. Use your results in parts (a) and (b) to estimate the change in Earth's mean temperature in the past century. Compare your result with the scientists' estimate of 1°F.

3. In Problem 5 of the Climate Change Project of Chapter 6, you showed that the planet's mean temperature increased approximately linearly from 1965 to 2014. (If you didn't do this earlier, do so now.) Find the approximate rate of change of the mean global temperature from 1965 to 2014.

4. Use the approximate rate of change you found in Problem 3 to predict the change in mean temperature for the coming century. Does your result fall within the 2.5–10.4°F IPCC range? Does it fall within either the 3–9°F or 4–7°F range predicted by the team of scientists from the United States and Europe? If yes, which one(s)?

5. Explain why, by using terminology such as "9 out of 10" and "most likely," scientists have been able to predict narrower ranges of increase in global mean temperatures in the coming century. Explain how this terminology, coupled with narrower ranges, will help policymakers better understand the risks involved in various courses of action or inaction.

6. Let I be the amount of ice (in cubic kilometers) in Greenland's ice cap at t years since 2000. Assuming that Greenland's ice cap continues to melt at the current rate, find an equation that models the situation.

7. a. Use your model to predict the amount of ice in Greenland's ice cap in 1000 years. If a total meltdown occurs in 1000 years, as predicted by Gregory, what must happen to the rate of melting?

   b. If a total meltdown occurs in 150 years, as predicted by Overpeck, what must happen to the rate of melting?

ᴅᴀᴛᴀ Workout Project

In this lab, you will explore your walking or running speed. Check with your instructor about whether you should collect your own data or use the data listed in Table 59.

Table 59 Times for Walking 440-Yard Laps

Lap Number	Distance (yards)	Time (seconds)
0	0	0
1	440	217
2	880	436
3	1320	656
4	1760	878
5	2200	1095
6	2640	1308

Source: *J. Lehmann*

Materials

You will need the following items:

- A timing device
- A pencil or pen
- A small pad of paper

Preparation

Locate a running track on which you can walk or run. For most tracks, one lap is 440 yards; so, four laps are 1760 yards, or 1 mile. You may select some other type of route, provided that you know the distance of one lap and you can easily complete six laps. Or map out a route in your neighborhood and estimate the distance by measuring it with the odometer in a car.

Recording of Data

Start your timing device and begin walking or running. Complete six laps of your course. Each time you complete a lap, record the *total* elapsed time. It will be easier to have a friend record the times for you. You may go slowly or quickly, but try to move at a constant speed throughout this experiment.

[§] J. M. Gregory et al., "Climatology: Threatened loss of the Greenland ice sheet," *Nature* (April 8, 2004):426–616.
[¶] American Geophysical Union meeting, October 2002.
[‖] National Oceanic and Atmospheric Administration.
[**] R. J. Nicholls and F. M. J. Hoozemans, "Vulnerability to sea-level rise with reference to the Mediterranean region," *Medcoast 95*, Vol. II, October 1995.

Analyzing the Data

1. Describe your route and the distance of one lap.

2. Use a table to describe the six total elapsed times for the six laps.

3. Let d be the distance (in yards) after you have walked or run for t seconds. Throughout this lab, treat d as the response variable and t as the explanatory variable. Show the six pairs of values of t and d in a table.

4. Create a scatterplot for the variables t and d.

5. For each of the six laps, calculate your approximate speed. Did you move at a steady rate, slow down, speed up, or engage in a combination of these?

6. Explain how your six approximate speeds are related to the position of the points in your scatterplot.

7. Find the mean of the six approximate speeds you found in Problem 5.

8. Divide the total distance you walked or ran by the total time it took. Compare the result to your result in Problem 7.

9. Convert the result you found in Problem 8 to units of miles per hour.

10. Use your result from Problem 9 to predict how long it would take you to walk or run 2 miles.

11. Use your result from Problem 9 to predict how long it would take you to walk or run a marathon, which is 26.2 miles long. Has model breakdown occurred? Explain.

📊 Balloon Project

In this lab, you will explore how long it takes for air to be released from a balloon when it is inflated with various amounts of air. Check with your instructor about whether you should collect your own data or use the data listed in Table 60.

Table 60 Mean Release Times for a Balloon Inflated with a Single Breath Having a Volume of 20 Ounces

Number of Breaths	Volume (ounces)	Mean Release Time (seconds)
0	0	0
5	100	1.9
10	200	3.5
15	300	6.1
20	400	6.0
25	500	7.7
30	600	10.6

Source: *J. Lehmann*

Materials

You will need one helper and the following items:

- A balloon
- A timing device
- A bucket, sink, or bathtub
- Water to fill the bucket, sink, or bathtub
- A transparent 1-cup or larger measuring cup (a larger cup is more convenient)

Preparation

Inflate and deflate the balloon fully several times to stretch it out. Each time you inflate the balloon, practice blowing into it with uniform-size breaths. There is no need to breathe deeply into the balloon. Medium-size breaths will work well for this lab.

Recording of Data

Perform the following tasks:

1. Count how many medium-size breaths it takes to fill the balloon.

2. For each trial of this experiment, you will time how long it takes for the balloon to release all the air inside it. Run three trials for each of six different volumes. Decide for which volumes you will have trials. For example, if it takes 30 breaths to fill the balloon, you could run three trials for each of the volumes of 5, 10, 15, 20, 25, and 30 breaths. To run a trial, first fill the balloon to the desired volume. Then begin timing as you release the balloon. Stop timing when the balloon is deflated completely. Record the time and the corresponding volume (in number of breaths).

3. To find the volume of a medium-size breath, fill a bucket, sink, or bathtub with water. Have one person fully submerge the measuring cup, so that there is no air in it, and then turn the cup upside down while it is still under water. The cup should not rest on the bottom of the container. Next, have a second person blow once into the balloon and then carefully release the air into the submerged cup. The air from the balloon will displace the water in the cup. The volume of air in the balloon will likely be more than what can fit in the cup, so the second person will have to stop releasing air from the balloon, and then the first person can empty out the air by resubmerging the cup. The second person should continue releasing air from the balloon into the cup until the balloon is empty. Depending on the size of the breath and the cup, the first person may have to resubmerge the cup several times. Then compute the volume of a medium-size breath by measuring the amount of air in the cup and taking into account how many times you filled the cup with air.

Analyzing the Data

1. For each of the volumes of 5, 10, 15, 20, 25, and 30 breaths, compute the mean of the three release times. Then record the numbers of breaths and the mean release times in columns 1 and 3 of a table similar to Table 60. Use the volume of one breath to help you find the entries for the second column of the table.

2. Let T be the time (in seconds) it takes for the balloon to deflate completely when the initial volume is n ounces. Create a scatterplot.

3. Draw a line that comes close to the data.

4. What is the T-intercept of the linear model? What does it mean in this situation? If model breakdown occurs, draw a better model.

5. Find the slope of the linear model. What does it mean in this situation?

6. Find an equation of the model.

7. Use the model to estimate the release time for a volume other than any of the volumes you used for the trials.

8. As you inflated the balloon, did it take about the same amount of effort to breathe in each time, or did it get progressively easier or harder? Thinking about how that effort is related to release times, does it suggest that the data points should lie close to a line or to a curve that bends? Explain. Do the (actual) data points support your theory? Explain.

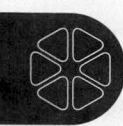 # Chapter Summary

Key Points of Chapter 7

Section 7.1   Graphing Equations of Lines and Linear Models

Solution, satisfy, and solution set of an equation	An ordered pair (a, b) is a **solution** of an equation in terms of x and y if the equation becomes a true statement when a is substituted for x and b is substituted for y. We say (a, b) **satisfies** the equation. The **solution set** of an equation is the set of all solutions of the equation.
Graph	The **graph** of an equation in two variables is the set of points that correspond to all solutions of the equation.
Graph of $y = mx + b$	The graph of an equation of the form $y = mx + b$, where m and b are constants, is a line.
y-intercept of the graph of $y = mx + b$	The graph of an equation of the form $y = mx + b$ has y-intercept $(0, b)$.
Equations of horizontal and vertical lines	If a and b are constants, then • The graph of $y = b$ is a horizontal line. • The graph of $x = a$ is a vertical line.
Equations whose graphs are lines	If an equation can be put into the form $y = mx + b$ or $x = a$, where m, a, and b are constants, then the graph of the equation is a line. We call such an equation a **linear equation in two variables**.
Rule of Four	We can describe some or all of the solutions of an equation in two variables with an equation, a table, a graph, or words. These four ways to describe solutions are known as the **Rule of Four**.

Section 7.2   Rate of Change and Slope of a Line

Formula for rate of change	Suppose that a quantity y changes steadily from y_1, to y_2 as a quantity x steadily changes from x_1 to x_2. Then the **rate of change** of y with respect to x is the ratio of the change in y to the change in x: $$\frac{\text{change in } y}{\text{change in } x} = \frac{y_2 - y_1}{x_2 - x_1}$$
Constant rate of change implies an exact linear association	If the rate of change of one variable with respect to another variable is constant, then there is an exact linear association between the variables.
An exact linear association implies constant rate of change	If there is an exact linear association between two variables, then the rate of change of one variable with respect to the other is constant.
Using run and rise to find rate of change	Assume there is an exact linear association between an explanatory variable x and a response variable y and (x_1, y_1) and (x_2, y_2) are two distinct points of the linear model. Then the rate of change of y with respect to x is the ratio of the rise to the run in going from point (x_1, y_1) to point (x_2, y_2): $$\frac{\text{rate of change of } y}{\text{with respect to } x} = \frac{y_2 - y_1}{x_2 - x_1} = \frac{\text{rise}}{\text{run}}$$

Section 7.2   Rate of Change and Slope of a Line (*Continued*)

Connection between the direction of an association and rate of change	• If an association between two variables is positive, then the approximate rate of change of one variable with respect to the other is positive. • If an association between two variables is negative, then the approximate rate of change of one variable with respect to the other is negative.
Slope of a nonvertical line	Assume (x_1, y_1) and (x_2, y_2) are two distinct points of a nonvertical line. The **slope** of the line is the rate of change of y with respect to x. In symbols: $$m = \text{slope} = \frac{y_2 - y_1}{x_2 - x_1} = \frac{\text{rise}}{\text{run}}$$
Slopes of increasing and decreasing lines	• An increasing line has positive slope. • A decreasing line has negative slope.
Measuring the steepness of a line	The absolute value of the slope of a line measures the steepness of the line. The steeper the line, the larger the absolute value of its slope will be.
Slopes of horizontal and vertical lines	• A horizontal line has a slope equal to zero. • A vertical line has undefined slope.

Section 7.3   Using Slope to Graph Equations of Lines and Linear Models

Slope and y-intercept of a linear equation of the form $y = mx + b$; slope–intercept form	For a linear equation of the form $y = mx + b$, • the slope of the line is m and • the y-intercept of the line is $(0, b)$. We say that this equation is in **slope–intercept form.**
Using slope to graph an equation of the form $y = mx + b$	To graph an equation of the form $y = mx + b$, **1.** Plot the y-intercept $(0, b)$. **2.** Use $m = \dfrac{\text{rise}}{\text{run}}$ to plot a second point. **3.** Sketch the line that passes through the two plotted points.
Find an equation of a line from a graph	To find an equation of a line from a graph, **1.** Determine the slope m and the y-intercept $(0, b)$ from the graph. **2.** Substitute your values for m and b into the equation $y = mx + b$.

Section 7.4   Functions

Relation, domain, and range	A **relation** is a set of ordered pairs. The **domain** of a relation is the set of all values of the explanatory variable, and the **range** of the relation is the set of all values of the response variable.
Input and output	Each member of the domain is an **input**, and each member of the range is an **output**.
Function	A **function** is a relation in which each input leads to exactly one output.
Vertical line test	A relation is a function if and only if each vertical line intersects the graph of the relation at no more than one point.
Linear function	A **linear function** is a relation whose equation can be put into the form $y = mx + b$, where m and b are constants.
Rule of Four for functions	We can describe some or all of the input–output pairs of a function by means of (1) an equation, (2) a graph, (3) a table, or (4) words.
Function notation	The response variable of a function f can be represented by the expression formed by writing the explanatory variable name within the parentheses of $f(\ )$: $$\text{response variable} = f(\text{explanatory variable})$$ We call this representation **function notation**.

Chapter 7 Review Exercises

1. Which of the ordered pairs $(-3, 9)$, $(1, 2)$, and $(4, -5)$ satisfy the equation $y = -2x + 3$?

For Exercises 2–5, refer to the graph sketched in Fig. 124.

2. Find y when $x = -2$.

3. Find y when $x = 0$.

4. Find x when $y = -4$.

5. Find x when $y = 0$.

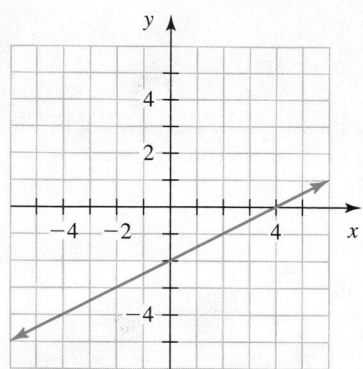

Figure 124 Exercises 2–5

6. *Prepaid cards* can be loaded with cash and used anywhere credit cards or debit cards are accepted. The total amounts of money loaded onto U.S. prepaid cards are shown in Table 61 for various years.

Table 61 Total Amounts of Money Loaded onto Prepaid Cards

Years	Total Amount of Money (billions of dollars)
2008	20
2009	29
2010	41
2011	57
2012	64
2013	84

Source: *Mercator Advisory Group*

Let p be the total amount of money (in billions of dollars) loaded onto prepaid cards at t years since 2005.

a. Identify the explanatory and response variables.

b. Construct a scatterplot by hand.

c. Is the association positive or negative? What does that mean in this situation?

d. Graph the model $p = 12.6t - 20.13$ by hand on the scatterplot. Does the line come close to the data points?

e. Estimate the total amount of money that was loaded onto prepaid cards in 2014. Did you perform interpolation or extrapolation? Mercator Advisory Group, a company that analyzes the payments industry, predicted the amount would be $98.6 billion (Source: *Mercator Advisory Group*). Compare your result with the company's predicted amount.

7. The temperature declines by 6°F over a 4-hour period. Find the average rate of change of temperature.

8. Revenue of Apollo Education decreased approximately steadily from $5.0 billion in 2010 to $3.0 billion in 2014 (Source: *FactSet*). Find the approximate rate of change of revenue.

9. The percentages of American adults who are confident they will retire ahead of their schedule are described by the scatterplot and the model in Fig. 125 for various incomes (in thousands of dollars).

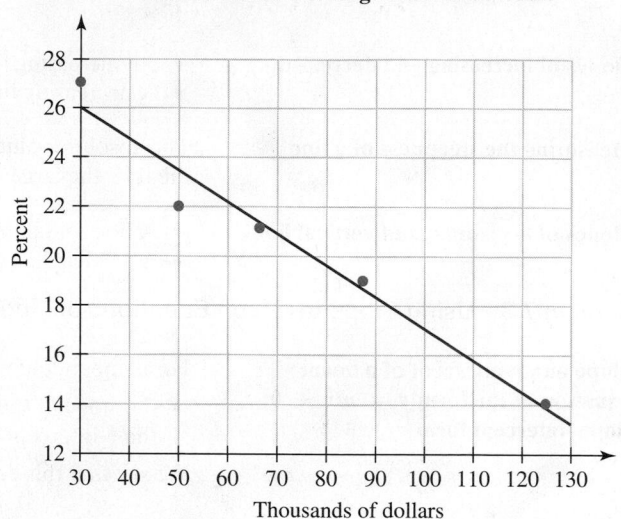

Figure 125 Exercise 9
(**Source:** *Country Financial*)

a. Is the association positive, negative, or neither? What does that mean in this situation?

b. Estimate the slope of the model. What does it mean in this situation?

c. Estimate the difference in the percentages of American adults who are confident they will retire ahead of their schedule for those that earn a certain income and for those that earn $50 thousand more than that income.

For Exercises 10–13, find the slope of the line passing through the two given points. State whether the line is increasing, decreasing, horizontal, or vertical.

10. $(4, -3)$ and $(8, -1)$ **11.** $(-10, -3)$ and $(-4, -5)$

12. $(-4, 7)$ and $(-4, -3)$ **13.** $(-5, 2)$ and $(-1, 2)$

14. Find the approximate slope of the line that contains the points $(-8.74, -2.38)$ and $(-1.16, 4.77)$. Round your result to the second decimal place. State whether the line is increasing, decreasing, horizontal, or vertical.

15. Sketch a line whose slope is a negative number near zero.

Sketch the line that has the given slope and contains the given point.

16. $m = 3$, $(0, -4)$ **17.** $m = \dfrac{4}{3}$, $(0, 1)$

Determine the slope and the y-intercept. Use them to graph the equation by hand.

18. $y = -\dfrac{2}{5}x - 1$ **19.** $y = \dfrac{2}{3}x$

20. $y = -3x + 1$ **21.** $y = x + 2$

For Exercises 22 and 23, graph the equation by hand.

22. $x = -3$ **23.** $y = 2$

24. Recall that we can describe some or all of the solutions of an equation in two variables with an equation, a table, a graph, or words.
   a. Describe three solutions of $y = -2x + 1$ by using a table.
   b. Describe the solutions of $y = -2x + 1$ by using a graph.
   c. Describe the solutions of $y = -2x + 1$ by using words.

25. Find an equation of the line sketched in Fig. 126.

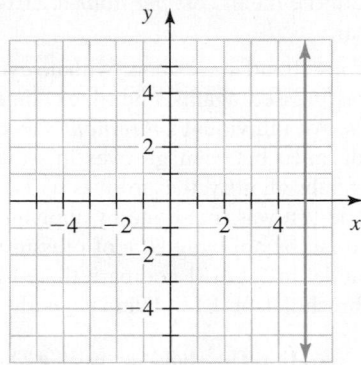

Figure 126 Exercise 25

26. Find an equation of the line sketched in Fig. 127.

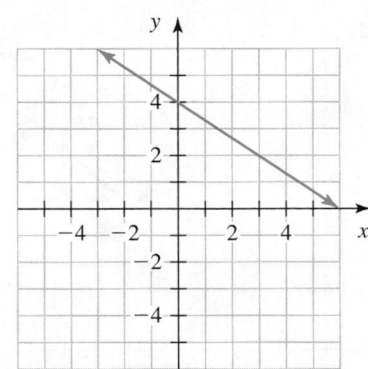

Figure 127 Exercise 26

27. Let c be the charge (in dollars) for a taxi ride for d miles in New York City. A reasonable model is $c = 2d + 2.5$.
   a. Identify the explanatory and response variables.
   b. What is the slope? What does it mean in this situation?
   c. What is the c-intercept? What does it mean in this situation?
   d. Graph the model by hand.
   e. Find the charge for a taxi ride from Battery Park, which is at the south end of Manhattan, to Marble Hill, which is the northernmost neighborhood in Manhattan. The total distance is 16.9 miles.

28. A person weighs 195 pounds when he starts a weight-loss program. For the first 6 months of being in the program, he steadily loses 4 pounds per month. Let w be his weight (in pounds) t months after he has started the program.
   a. Why is there an exact linear association?
   b. Find the slope and the w-intercept of a linear model.
   c. Find an equation of the model.

   d. Graph the model by hand.
   e. Find the person's weight 5 months after starting the program.

29. **DATA** Revenues of the National Football League (NFL) are shown in Table 62 for various years.

Table 62 NFL's Annual Revenues

Years	Revenue (billions of dollars)
2005	5.8
2007	6.6
2009	7.5
2011	8.0
2013	8.9

Source: *Wall Street Journal*

Let r be the NFL's annual revenue (in billions of dollars) at t years since 2000.

   a. Use technology to construct a scatterplot.
   b. Describe the four characteristics of the association. Compute and interpret r as part of your analysis.
   c. Use technology to graph the model $r = 0.38t + 3.94$ on the scatterplot. Does the line come close to the data points?
   d. What is the slope of the model? What does it mean in this situation?
   e. Estimate the revenue in 2014. Did you perform interpolation or extrapolation? The actual revenue was $10.3 billion. Find the error in your result.

30. Some ordered pairs for four relations are listed in Table 63. Which of these relations could be functions? Explain.

Table 63 Which Relations Might Be Functions? (Exercise 30)

Relation 1		Relation 2		Relation 3		Relation 4	
x	y	x	y	x	y	x	y
1	12	3	27	0	7	2	1
2	15	4	24	1	7	2	2
3	18	4	21	2	7	2	3
4	21	5	18	3	7	2	4
5	24	6	15	4	7	2	5

31. Determine whether the graph in Fig. 128 represents a function. Explain.

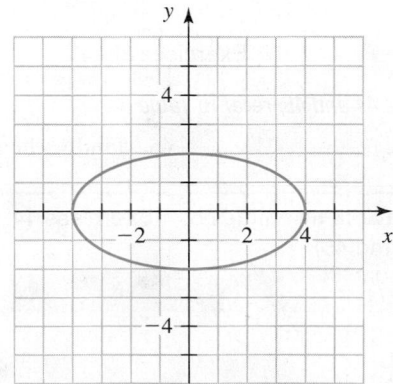

Figure 128 Exercise 31

For Exercises 32 and 33, determine whether each relation is a function. Explain.

32. $y = \dfrac{5}{6}x - 3$ **33.** $x = 9$

34. Use the graph of the function in Fig. 129 to determine the function's domain and range.

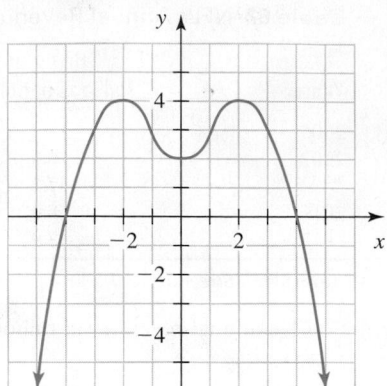

Figure 129 Exercise 34

For $f(x) = -10x - 3$, $g(x) = 3x^2 - 7$, and $h(x) = \dfrac{2x + 5}{3x + 6}$, find the following.

35. $f(-3)$ **36.** $g(-2)$ **37.** $h(4)$

For Exercises 38–43, refer to Fig. 130.

38. Find $f(0)$.   **39.** Estimate $f(-3)$.

40. Find x when $f(x) = 0$.

41. Find x when $f(x) = -1$.

42. Find the domain of f.

43. Find the range of f.

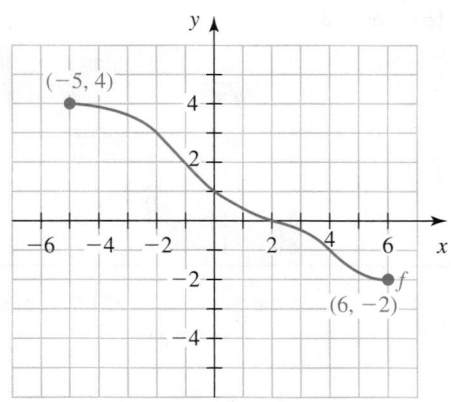

Figure 130 Exercises 38–43

For Exercises 44 and 45, refer to Table 64.

44. Find $f(2)$.   **45.** Find x when $f(x) = 2$.

Table 64 Values of f (Exercises 44 and 45)

x	$f(x)$
0	1
1	2
2	4
3	3
4	0

46. The mean monthly day-care cost was \$972 in 2011, and it increased by approximately \$45.60 per year for the period 2011–2014 (Source: *National Association of Child Care Resource & Referral Agencies*). Costs are based on a three-year-old attending a day-care center eight hours a day, five days a week. Let $f(t)$ be the mean monthly cost (in dollars) at t years since 2011.

  a. Find an equation of f.
  b. What is the slope? What does it mean in this situation?
  c. Find $f(3) - f(2)$. Compare the result with the slope. Why does your comparison make sense?
  d. Estimate the mean monthly cost in 2014.
  e. Estimate the mean cost *per hour* in 2014. There 365 days in 1 year.

47. In a study of 43 blind individuals (with no eyesight), a grooved surface was pressed against the index finger of each of the individuals. An individual's *threshold* was defined to be the smallest distance between grooves in which the individual could correctly identify the grooves as being horizontal or vertical. The test was first done by applying 10 g of pressure and then done by applying 50 g of pressure. The scatterplot and the model in Fig. 131 compare the 10-g thresholds with the 50-g thresholds of the individuals, both in mm.

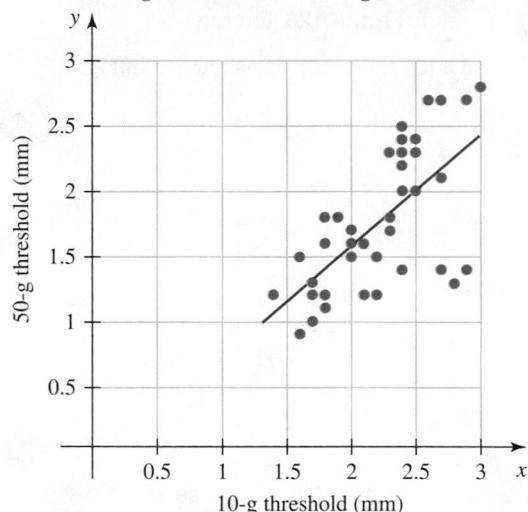

Figure 131 Exercise 47

(**Source:** Tactile Acuity Is Enhanced in Blindness, *Daniel Goldreich and Ingrid M. Kanics*)

Let x be an individual's 10-g threshold, and let y be the individual's 50-g threshold, both in mm. The equation of the model shown in Fig. 131 is $f(x) = 0.86x - 0.16$.

  a. Is the association positive, negative, or neither? What does that mean in this situation?
  b. Find $f(1.5)$ Which is larger, 1.5 or $f(1.5)$? What does your comparison mean in this situation?
  c. Find $f(2)$ Which is larger, 2 or $f(2)$? What does your comparison mean in this situation?
  d. Find $f(3)$ Which is larger, 3 or $f(3)$? What does your comparison mean in this situation?
  e. On the scatterplot, imagine the line $y = x$. Or use a strand of thread, a stretched rubber band, or the edge of a clear ruler to form the line. Do most of the data points lie below, on, or above the line? What does that mean in this situation?

Chapter 7 Test

For Exercises 1–4, refer to the graph sketched in Fig. 132.

1. Find y when $x = -3$.
2. Find x when $y = -1$.
3. Find the y-intercept.
4. Estimate the x-intercept.

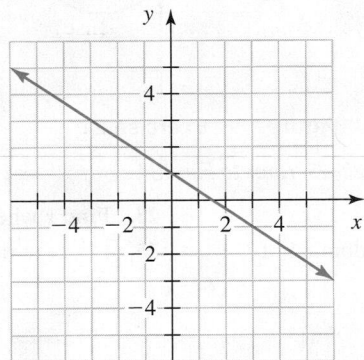

Figure 132 Exercises 1–4

5. The cooking times of a turkey in a 325°F oven are shown in Table 65 for various weights.

Table 65 Cooking Times of a Turkey in a 325°F Oven

Weight Group (pounds)	Weight Used to Represent Weight Group (pounds)	Cooking-Time Group (hours)	Time Used to Represent Cooking-Time Group (hours)
6–8	7	3.0–3.5	3.25
8–12	10	3.5–4.5	4
12–16	14	4.5–5.5	5
16–20	18	5.5–6.5	6
20–24	22	6.5–7.0	6.75

Source: *About, Inc.*

For a 325°F oven, let T be the cooking time (in hours) of a turkey that weighs W pounds.

  a. Identify the explanatory and response variables.
  b. Construct a scatterplot by hand.
  c. Is the association positive or negative? What does that mean in this situation?
  d. Graph the model $T = 0.24W + 1.64$ by hand on the scatterplot. Does the line come close to the data points?
  e. Predict the cooking time for a 20-pound turkey in a 325°F oven.

6. The mean price of Christmas trees increased approximately steadily from \$64.61 in 2010 to \$81.30 in 2013 (Source: *National Christmas Tree Association*). Find the approximate rate of change of the mean price of Christmas trees.

7. The scatterplot and the model in Fig. 133 compare the years and the median square footages of newly built homes.

Years versus Square Footages

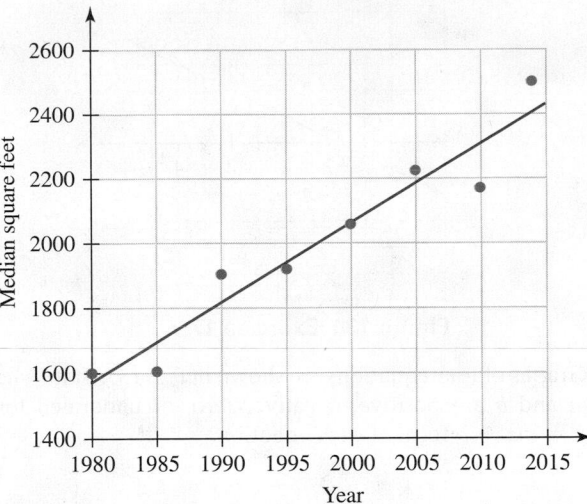

Figure 133 Exercise 7
(**Source:** *National Association of Home Builders*®)

  a. Is the association positive, negative, or neither? What does that mean in this situation?
  b. Estimate the slope of the model. What does it mean in this situation?
  c. Estimate how much the median square footage increased over 15 years.
  d. A student says the result you found in part (c) describes how much the median square footage will increase in the next 15 years. What would you tell the student?
  e. The mean sizes of all types of households were 2.7 people and 2.5 people in 1982 and 2013, respectively. Estimate the median square footage *per person* for newly built homes in 1982. Do the same for 2013 and then compare your two results. Describe any assumptions you have made.

For Exercises 8–11, find the slope of the line passing through the two given points. State whether the line is increasing, decreasing, horizontal, or vertical.

8. $(3, -8)$ and $(5, -2)$
9. $(-4, -1)$ and $(2, -4)$
10. $(-5, 4)$ and $(1, 4)$
11. $(-2, -7)$ and $(-2, 3)$

12. Find the approximate slope of the line that contains the points $(-5.99, -3.27)$ and $(2.83, 8.12)$. Round your result to the second decimal place. State whether the line is increasing, decreasing, horizontal, or vertical.

For Exercises 13–16, determine the slope and the y-intercept. Use them to graph the equation by hand.

13. $y = -\dfrac{3}{2}x + 2$
14. $y = \dfrac{5}{6}x$
15. $y = 2$
16. $y = -2x + 3$

17. Find an equation of the line sketched in Fig. 134.

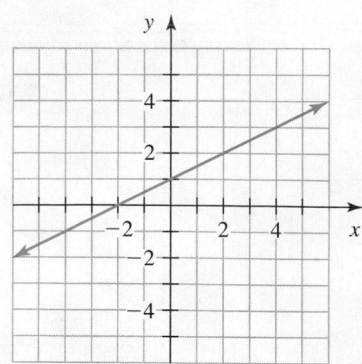

Figure 134 Exercise 17

18. Graphs of four equations are shown in Fig. 135. State whether m and b are positive, negative, zero, or undefined for the $y = mx + b$ form of each equation.

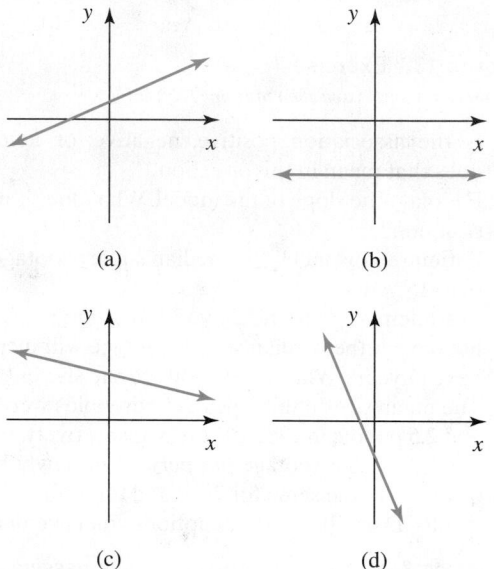

(a)   (b)

(c)   (d)

Figure 135 Exercise 18

19. The median compensation (including salary and bonus) of college presidents was $870 thousand in 2010 and increased by about $37 thousand per year during the period 2010–2014 (Source: *IRS*). Let C be the median annual compensation (in thousands of dollars) of college presidents at t years since 2010.

    a. Is there a linear association between t and C? Explain. If there is a linear association, find the slope and describe what it means in this situation.

    b. What is the C-intercept? What does it mean in this situation?

    c. Find an equation of the model.

    d. Graph the model by hand.

    e. Estimate the median compensation in 2014.

20. Determine whether the relation described by $y = -2x + 5$ is a function. Explain.

21. Use the graph of the relation in Fig. 136 to determine the relation's domain and range. Determine whether the relation is a function.

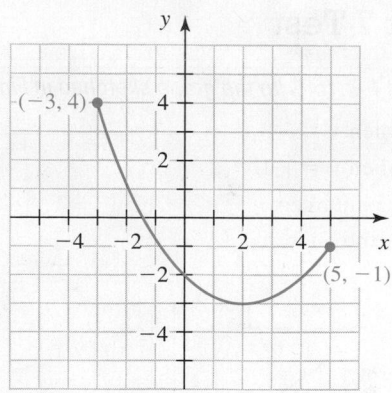

Figure 136 Exercise 21

For Exercises 22–25, refer to Fig. 137.

22. Find $f(-3)$.

23. Find x when $f(x) = 0$.

24. Find the domain of f.

25. Find the range of f.

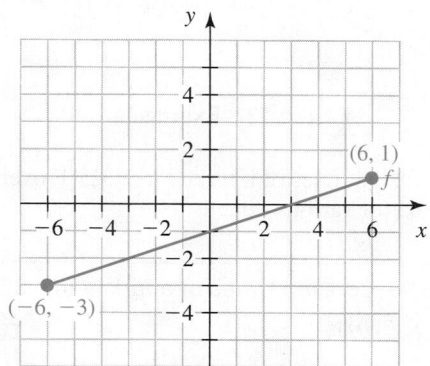

Figure 137 Exercises 22–25

26. Find $f(-3)$, where $f(x) = -4x + 7$.

27. Find $g(-4)$, where $g(x) = -2x^2 - 5x + 1$.

28. DATA The percentages of tax returns in which individuals donated money to presidential elections by checking off boxes (for $1, $2, $3, or $6) on tax returns are shown in Table 66 for various years.

Table 66 Percentages of Tax Returns in which Presidential-Election Donation Boxes Were Checked Off

Years	Percent
1976	27.5
1986	21.7
1996	12.6
2006	10.9
2013	6.0

Source: *Federal Election Commission*

Let p be the percentage of tax returns in which presidential-election donation boxes were checked off at t years since 1970.

    a. Use technology to construct a scatterplot.

    b. Use technology to graph the model $p = -0.57t + 30.31$ on the scatterplot. Does the line come close to the data points?

    c. Rewrite the equation $p = -0.57t + 30.31$ using the function name f.

    d. Find $f(40)$. What does it mean in this situation?

    e. Make an educated guess whether it has become more or less important for candidates to seek huge donations from wealthy Americans. Explain.

Solving Linear Equations and Inequalities to Make Predictions

What was your average grade in high school? The percentage of college freshmen whose average grade in high school was an A has increased greatly since 1970 (see Table 1). Do you think the increase is due to students learning more, teachers lowering their standards, or some other reason? In Exercise 96 of Homework 8.3, you will estimate when 44% of all college freshmen earned an average grade of A in high school.

In this chapter, we will write expressions more simply. Then we will find solutions of equations and inequalities in *one* variable. We will also continue working with formulas. Once we have learned these skills, we will be able to make predictions efficiently about the explanatory variable (or the response variable) in authentic situations, such as the grade data just described.

Table 1 College Freshmen Whose Average Grade in High School Was an A

Year	Percent
1970	19.6
1980	26.6
1985	28.7
1990	29.4
1995	36.1
2000	42.9
2005	46.6
2010	48.4

Source: *Higher Education Research Institute*

▼ 8.1 Simplifying Expressions

Objectives

» Describe the commutative *law*.

» Describe the *associative law*.

» Describe the *distributive law*.

» Describe the meaning of *equivalent expressions*.

» *Combine like terms.*

» *Simplify expressions.*

In this section, we will write expressions more simply.

Commutative Laws

Consider the equations $2 + 6 = 6 + 2$ and $2 \cdot 6 = 6 \cdot 2$. These true statements suggest the *commutative laws*.

> #### Commutative Laws for Addition and Multiplication
>
> **Commutative law for addition:** $a + b = b + a$
> **Commutative law for multiplication:** $ab = ba$
>
> In words: We can add two numbers in either order and get the same result, and we can multiply two numbers in either order and get the same result.

> **Example 1** Using the Commutative Law for Addition

Use the commutative law for addition to write the expression in another form.

1. $3 + x$ **2.** $5 + 3w$

Solution

1. $3 + x = x + 3$ *Commutative law for addition: $a + b = b + a$*
2. $5 + 3w = 3w + 5$ *Commutative law for addition: $a + b = b + a$*

> **Example 2** Using the Commutative Law for Multiplication

Use the commutative law for multiplication to write the expression in another form.

1. cd **2.** $-7x$ **3.** $k \cdot 8 + 2$

Solution

1. $cd = dc$ *Commutative law for multiplication: $ab = ba$*
2. $-7x = x(-7)$ *Commutative law for multiplication: $ab = ba$*
3. $k \cdot 8 + 2 = 8k + 2$ *Commutative law for multiplication: $ab = ba$*

WARNING   There is no commutative law for subtraction. For example, $5 - 2 = 3$ but $2 - 5 = -3$. And there is no commutative law for division. For example, $3 \div 1 = 3$, but $1 \div 3 = \frac{1}{3}$.

A **term** is a constant, a variable, or a product of a constant and one or more variables raised to powers. Here are some terms:

$$4x \qquad -7 \qquad y \qquad -5xy^2 \qquad 97x^6y^3$$

The expression $7xy + 4x - 5y + 3$ has four terms: $7xy$, $4x$, $5y$, and 3. Note that, in the expression, the terms are separated by addition and subtraction symbols.

Variable terms are terms that contain variables. **Constant terms** are terms that do not contain variables. For example, $-5x$ is a variable term, and 8 is a constant term. We usually write a sum of both variable and constant terms with the variable terms to the left of the constant terms. So, for $5 - 8x$, we write

$$5 - 8x = 5 + (-8x) \quad a - b = a + (-b)$$
$$= -8x + 5 \quad \text{Commutative law for addition: } a + b = b + a$$

Associative Laws

As we discussed in Section 1.7, we work from left to right when we perform additions and subtractions. However, if a sum is of the form $a + b + c$, we can get the same result by performing the addition on the left first or the one on the right first. For example,

$$(4 + 2) + 3 = 6 + 3 = 9$$
$$4 + (2 + 3) = 4 + 5 = 9$$

A similar law is true for an expression of the form abc:

$$(4 \cdot 2) \cdot 3 = 8 \cdot 3 = 24$$
$$4 \cdot (2 \cdot 3) = 4 \cdot 6 = 24$$

These examples suggest the *associative laws*.

> **Associative Laws for Addition and Multiplication**
>
> **Associative law for addition:** $a + (b + c) = (a + b) + c$
> **Associative law for multiplication:** $a(bc) = (ab)c$
>
> In words: For an expression $a + b + c$, we get the same result by performing the addition on the right first or the one on the left first. For an expression abc, we get the same result by performing the multiplication on the right first or the one on the left first.

It is important to know the difference between the associative laws and the commutative laws. The associative laws change the order of *operations*, whereas the commutative laws change the order of *terms* or even *expressions*.

For example, consider the expression $(3 + x) + 1$. By the associative law of addition, we can change the order of the additions:

$$(3 + x) + 1 = 3 + (x + 1)$$

By the commutative law of addition, we can change the order of the terms 3 and x:

$$(3 + x) + 1 = (x + 3) + 1$$

We can also use the commutative law to change the order of the expressions $x + 3$ and 1:

$$(x + 3) + 1 = 1 + (x + 3)$$

▶ **Example 3**   Using the Associative Laws

Use an associative law to write the expression in another form.
   1. $(x + 2) + y$ **2.** $w + (9 + k)$ **3.** $3(mp)$ **4.** $(wx)y$

Solution

   1. $(x + 2) + y = x + (2 + y)$ *Associative law for addition:* $(a + b) + c = a + (b + c)$
   2. $w + (9 + k) = (w + 9) + k$ *Associative law for addition:* $a + (b + c) = (a + b) + c$
   3. $3(mp) = (3m)p$ *Associative law for multiplication:* $a(bc) = (ab)c$
   4. $(wx)y = w(xy)$ *Associative law for multiplication:* $(ab)c = a(bc)$

▶

We can use a combination of the commutative law and the associative law, both for addition, to write the terms of $a + b + c$ in different orders:

$$a + c + b \qquad b + a + c \qquad b + c + a \qquad c + a + b \qquad c + b + a$$

We **rearrange the terms** of an expression by writing the terms in a different order.

Likewise, we **rearrange the factors** of an expression by writing the factors in a different order. Here we rearrange the factors of abc to get the following products:

$$acb \qquad bac \qquad bca \qquad cab \qquad cba$$

▶ **Example 4**   Rearranging Terms

Rearrange the terms of $9 - 2x - 5 + 3$ so that the numbers can be added.

Solution

$$
\begin{aligned}
9 - 2x - 5 + 3 &= 9 + (-2x) + (-5) + 3 &&\quad a - b = a + (-b) \\
&= -2x + 9 + (-5) + 3 &&\quad \text{Rearrange terms.} \\
&= -2x + 7 &&\quad \text{Add constant terms.}
\end{aligned}
$$

▶

It is helpful to practice problems such as the one in Example 4 until you can do each problem in one step.

▶ **Example 5**   Using the Commutative and Associative Laws

Use the commutative and associative laws to remove the parentheses. Also, multiply and add numbers when possible.

1. $-3(8x)$ **2.** $(5 + 7p) + 8$

Solution

1. $-3(8x) = (-3 \cdot 8)x$ *Associative law for multiplication:* $a(bc) = (ab)c$
$\quad\quad\quad = -24x$ *Multiply.*

2. $(5 + 7p) + 8 = (7p + 5) + 8$ *Commutative law for addition:* $a + b = b + a$
$\quad\quad\quad\quad\quad = 7p + (5 + 8)$ *Associative law for addition:* $(a + b) + c = a + (b + c)$
$\quad\quad\quad\quad\quad = 7p + 13$ *Add.*

Distributive Law

For the expression $2(3 + 4)$, we perform the addition before the multiplication. However, compare the result with that for computing $2 \cdot 3 + 2 \cdot 4$:

$$2(3 + 4) = 2(7) = 14$$
$$2 \cdot 3 + 2 \cdot 4 = 6 + 8 = 14$$

Since both results are equal to 14, we can write

$$2(3 + 4) = 2 \cdot 3 + 2 \cdot 4$$

The blue curves indicate what numbers we multiply by 2. We say that we *distribute* the 2 to both the 3 and the 4. This example suggests the **distributive law**.

▶ **Distributive Law**

$$a(b + c) = ab + ac$$

In words: To find $a(b + c)$, distribute a to both b and c.

▶ **Example 6**   Using the Distributive Law

Find the product.

1. $3(x + 5)$ **2.** $5(2x + 4y)$

Solution

1. $3(x + 5) = 3x + 3 \cdot 5$ *Distributive law:* $a(b + c) = ab + ac$
$\quad\quad\quad\quad = 3x + 15$ *Multiply.*

2. $5(2x + 4y) = 5 \cdot 2x + 5 \cdot 4y$ *Distributive law:* $a(b + c) = ab + ac$
$\quad\quad\quad\quad\quad = 10x + 20y$ *Multiply.*

WARNING   In Problem 1 of Example 6, we found that $3(x + 5) = 3x + 15$. In applying the distributive law to an expression such as $3(x + 5)$, remember to distribute the 3 to *every* term in the parentheses. For example, the expression $3x + 5$ is an incorrect result.

Since subtracting a number is the same as adding the opposite of the number, it is also true that

$$a(b - c) = ab - ac$$

We can use the distributive law and the commutative law for multiplication to show that we can "distribute from the right" as well as from the left (see Exercises 103 and 104):

$$(a + b)c = ac + bc$$

$$(a - b)c = ac - bc$$

▶ **Example 7**   Using the Distributive Law

Find the product.

1. $4(3x - 2y)$ **2.** $(x - 6)(3)$ **3.** $-2(6 + t)$ **4.** $-5(w - 3)$

Solution

1. $4(3x - 2y) = 4 \cdot 3x - 4 \cdot 2y$ *Distributive law: $a(b - c) = ab - ac$*
$\qquad\qquad = 12x - 8y$ *Multiply.*

2. $(x - 6)(3) = x \cdot 3 - 6 \cdot 3$ *Distributive law: $(a - b)c = ac - bc$*
$\qquad\qquad = 3x - 18$ *Commutative law for multiplication: $ab = ba$; multiply.*

3. $-2(6 + t) = -2(6) + (-2)t$ *Distributive law: $a(b + c) = ab + ac$*
$\qquad\qquad = -12 + (-2t)$ *Multiply.*
$\qquad\qquad = -2t + (-12)$ *Commutative law for addition: $a + b = b + a$*
$\qquad\qquad = -2t - 12$ *$a + (-b) = a - b$*

4. $-5(w - 3) = -5w - (-5)(3)$ *Distributive law: $a(b - c) = ab - ac$*
$\qquad\qquad = -5w + 15$ *Multiply.*

It is extremely helpful to practice problems such as those in Example 7 until you can do them in one step.

We can also use the distributive law when there are more than two terms inside the parentheses.

▶ **Example 8**   Distributive Law

Find the product $3(2t - 5w + 4)$.

Solution

$$3(2t - 5w + 4) = 3 \cdot 2t - 3 \cdot 5w + 3 \cdot 4 \quad \textit{Distributive law}$$
$$= 6t - 15w + 12 \quad \textit{Multiply.}$$

The products $-1 \cdot 2 = -2, -1 \cdot 5 = -5$, and $-1 \cdot 6 = -6$ suggest the following property:

▶ **Multiplying a Number by -1**

$$-1a = -a$$

In words: -1 times a number is equal to the opposite of the number.

To remove parentheses in $-(x + 3)$, we use the fact that $-a = -1a$ to write

$$
\begin{aligned}
-(x + 3) &= -1(x + 3) && \text{$-a = -1a$} \\
&= -1x + (-1)3 && \text{Distributive law: $a(b + c) = ab + ac$} \\
&= -x - 3 && \text{$-1a = -a$}
\end{aligned}
$$

Example 9   Simplifying an Expression

Remove parentheses in $-(x - 4y + 7)$.

Solution

$$
\begin{aligned}
-(x - 4y + 7) &= -1(x - 4y + 7) && \text{$-a = -1a$} \\
&= -1x - (-1)4y + (-1)7 && \text{Distributive law: $a(b - c) = ab - ac$} \\
&= -x + 4y - 7 && \text{$-1a = -a$}
\end{aligned}
$$

It is helpful to compare the distributive law with the associative law for multiplication to avoid confusing the two laws. For the expression $a(b + c)$, we distribute a to both of the terms b and c:

$$
a(b + c) = ab + ac \quad \text{Distributive law}
$$

WARNING   However, the expression $a(bc)$ is *not* equivalent to $(ab)(ac)$. By the associative law, we can write

$$
a(bc) = (ab)c \quad \text{Associative law for multiplication}
$$

Equivalent Expressions

In Problem 1 of Example 6, we found that $3(x + 5) = 3x + 15$. In Example 10, we will explore the meaning of this statement.

Example 10   Evaluating Expressions

Evaluate both of the expressions $3(x + 5)$ and $3x + 15$ for the given values of x.
1. $x = 2$ 2. $x = 4$
3. $x = 0, x = 1, x = 2, x = 3, x = 4, x = 5,$ and $x = 6$

Solution

1. First, we evaluate $3(x + 5)$ for $x = 2$:

$$
3(2 + 5) = 3(7) = 21
$$

Then we evaluate $3x + 15$ for $x = 2$:

$$
3(2) + 15 = 6 + 15 = 21
$$

Both results are equal to 21.

2. First, we evaluate $3(x + 5)$ for $x = 4$:

$$
3(4 + 5) = 3(9) = 27
$$

Then we evaluate $3x + 15$ for $x = 4$:

$$
3(4) + 15 = 12 + 15 = 27
$$

Both results are equal to 27.

3. We use a TI-84 to create a table for the equations $y = 3(x + 5)$ and $y = 3x + 15$ (see Fig. 1). See Appendix A.22 for TI-84 instructions.

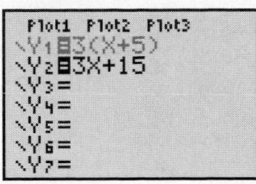  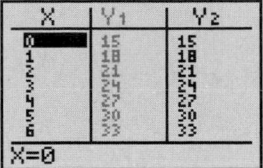

Figure 1 Evaluate the expressions $3(x + 5)$ and $3x + 15$ for several values of x

For each value of x in the table, the values of y for the two equations are equal. So for each of these values of x, the two results of evaluating the two expressions are equal.

In Example 10, we found that for each value that we used to evaluate the expressions $3(x + 5)$ and $3x + 15$, the two results were equal.

▶ **Definition   Equivalent expressions**

Two or more expressions are **equivalent expressions** if, when each variable is evaluated for *any* real number (for which all the expressions are defined), the expressions all give equal results.

Combining Like Terms

The **coefficient** of a term is the constant factor of the term. For example, the coefficient of the term $-3x$ is -3. Since $x = 1 \cdot x$, the coefficient of the term x is 1. Since $-x = -1x$, the coefficient of the term $-x$ is -1. For the term 4, the coefficient is 4.

Like terms are either constant terms or variable terms that contain the same variable(s) raised to exactly the same power(s). For example, 5 and 9 are like terms; so are $3x$ and $8x$. Also, $2y^3$, $5y^3$, and $-6y^3$ are like terms.

If terms are not like terms, we say that they are **unlike terms**. For example, $3x$ and $8y$ are unlike terms, because $3x$ contains an x but $8y$ does not and because $8y$ contains a y but $3x$ does not. The terms $4x^2$ and $7x^3$ are unlike terms, because the exponents of x are different.

For a sum of like terms, such as $2x + 4x$, we can use the distributive law to write the sum as one term:

$$2x + 4x = (2 + 4)x \quad \text{Distributive law: } ac + bc = (a + b)c$$
$$= 6x \quad \text{Add.}$$

When we write a sum or difference of like terms as one term, we say we have **combined like terms**.

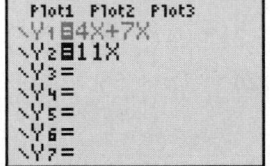

▶ **Example 11**   Combining Two Like Terms

Combine like terms.

1. $4x + 7x$ **2.** $8x - 5x$ **3.** $6a + a$

Solution

1. $4x + 7x = (4 + 7)x \quad \text{Distributive law: } ac + bc = (a + b)c$
$$= 11x \quad \text{Add.}$$

We use a TI-84 table to verify our work (see Fig. 2).

Figure 2 Verify the work

2. $8x - 5x = (8 - 5)x$ *Distributive law: $ac - bc = (a - b)c$*

$\qquad\qquad = 3x$ *Subtract.*

3. $6a + a = 6a + 1a$ *$a = 1a$*

$\qquad\qquad = (6 + 1)a$ *Distributive law: $ac + bc = (a + b)c$*

$\qquad\qquad = 7a$ *Add.*

In Problem 1 of Example 11, we found that $4x + 7x = 11x$. We can find the sum $4x + 7x$ in one step by adding the coefficients of $4x$ and $7x$ (that is, 4 and 7) to get the coefficient of the result, 11:

$$4x + 7x = 11x$$

$$4 + 7 = 11$$

In Problem 2 of Example 11, we found that $8x - 5x = 3x$. Note that if we write $8x - 5x = 8x + (-5x)$, then we can add the coefficients of $8x$ and $-5x$ (that is, 8 and -5) to get the coefficient of the result, 3:

$$8x - 5x = 3x$$

$$8 + (-5) = 3$$

> ### Combining Like Terms
>
> To combine like terms, add the coefficients of the terms and keep the same variable factors.

Simplifying an Expression

We simplify an expression by removing parentheses and combining like terms. The result of simplifying an expression is a **simplified expression**, which is equivalent to the original expression.

Simplifying an expression often makes it easier to evaluate the expression and to graph an equation that contains the expression. We will see other benefits of simplifying expressions throughout the text.

▶ **Example 12** Simplifying Expressions

Simplify.

1. $2x + 5y - 6x + 2 + 3y + 7$ $\qquad\qquad$ **2.** $5(3x - 2) - 4x$

3. $-2(x + 5y) - 3(2x - 4y)$ $\qquad\qquad$ **4.** $4(x - 2) - (x + 3)$

Solution

1. $2x + 5y - 6x + 2 + 3y + 7 = 2x - 6x + 5y + 3y + 2 + 7$ *Rearrange terms.*

$\qquad\qquad\qquad\qquad\qquad\quad = -4x + 8y + 9$ *Combine like terms.*

2. $5(3x - 2) - 4x = 15x - 10 - 4x$ *Distributive law*

$\qquad\qquad\qquad = 15x - 4x - 10$ *Rearrange terms.*

$\qquad\qquad\qquad = 11x - 10$ *Combine like terms.*

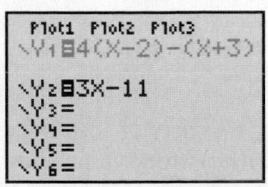

Figure 3 Verify the work

3. $-2(x + 5y) - 3(2x - 4y) = -2x - 10y - 6x + 12y$ *Distributive law*

$$= -2x - 6x - 10y + 12y \quad \text{\textit{Rearrange terms.}}$$

$$= -8x + 2y \qquad\qquad\quad \text{\textit{Combine like terms.}}$$

4. We write $4(x - 2) - (x + 3)$ as $4(x - 2) - 1(x + 3)$ so that we can distribute the -1:

$$4(x - 2) - (x + 3) = 4(x - 2) - 1(x + 3) \quad a - b = a - 1b$$

$$= 4x - 8 - 1x - 3 \qquad \text{\textit{Distributive law}}$$

$$= 4x - 1x - 8 - 3 \qquad \text{\textit{Rearrange terms.}}$$

$$= 3x - 11 \qquad\qquad\;\; \text{\textit{Combine like terms.}}$$

We use a TI-84 table to verify the work (see Fig. 3).

▶ **Example 13**   Translating from English to Mathematics

Let x be a number. Translate the phrase "3 times the number, minus 4, plus twice the number" to an expression. Then simplify the expression.

Solution

The expression is

3 times the number,	minus 4,	plus	twice the number
$3x$	-4	$+$	$2x$

Next, we simplify the expression:

$$3x - 4 + 2x = 3x + 2x - 4 \quad \text{\textit{Rearrange terms.}}$$

$$= 5x - 4 \qquad\;\; \text{\textit{Combine like terms.}}$$

In Example 13, we translated an English phrase into a mathematical expression. In Example 14, we will translate a mathematical expression into an English phrase.

▶ **Example 14**   Translating from Mathematics to English

Let x be a number. Translate the expression $x - 5 \cdot (x - 2)$ into an English phrase. Then simplify the expression.

Solution

Here is the translation:

$x\ -$	$5\ \cdot$	$(x - 2)$
the number, minus	5 times	the difference of the number and 2

One of many possible correct translations is "the number, minus 5 times the difference of the number and 2." Next, we simplify the expression:

$$x - 5 \cdot (x - 2) = x - 5x + 10 \quad \text{\textit{Distributive law}}$$

$$= -4x + 10 \qquad \text{\textit{Combine like terms.}}$$

 **Group Exploration**

Section Opener: Laws of operations

1. a. Evaluate $a(b + c)$ for $a = 2, b = 3$, and $c = 5$.
   b. Evaluate $ab + ac$ for $a = 2, b = 3$, and $c = 5$.
   c. Compare the results you found in Problems 1 and 2.
   d. Evaluate both $a(b + c)$ and $ab + ac$ for $a = 4$, $b = 2$, and $c = 6$, and then compare the results.
   e. Evaluate both $a(b + c)$ and $ab + ac$ for values of your choosing for a, b, and c, and then compare the results.
   f. Make an educated guess as to whether the statement $a(b + c) = ab + ac$ is true for all numbers a, b, and c.

2. Evaluate both $a(bc)$ and $(ab)(ac)$ for values of your choosing for a, b, and c. Is the statement $a(bc) = (ab)(ac)$ true for all numbers a, b, and c?

3. Evaluate both $a(bc)$ and $(ab)c$ for values of your choosing for a, b, and c. Make an educated guess as to whether the statement $a(bc) = (ab)c$ is true for all numbers a, b, and c.

4. Evaluate both $a + b$ and $b + a$ for values of your choosing for a and b. Make an educated guess as to whether the statement $a + b = b + a$ is true for all numbers a and b.

5. What are the main points of this exploration?

 **Group Exploration**

Simplifying expressions

For the work shown, carefully describe any errors.

1. $3(x + 5) + 4x = 3x + 5 + 4x$
$$= 3x + 4x + 5$$
$$= 7x + 5$$

2. $3(xw) = (3x)(3w)$
$$= 3(3)xw$$
$$= 9xw$$

3. $5 + x - 3 = 5 + 3 - x$
$$= 8 - x$$
$$= x - 8$$

▶ **Tips for Success**   **Use 3-by-5 Cards**

Do you have trouble memorizing definitions and laws? If so, try writing a word or phrase on one side of a 3-by-5 card. On the other side, put its definition or state a law and how it can be applied. For example, you could write "distributive law" on one side of a card and "$a(b + c) = ab + ac$" on the other side. You could also describe, in your own words, the meaning of the law and how you can apply it. Once you have completed a card for each definition and law, shuffle the cards and quiz yourself until you are confident you know the definitions and laws and how to apply them. Quiz yourself again later to make sure you have retained the information.

   In addition to memorizing definitions and laws, it is important that you continue to strive to understand their meanings and how to apply them.

Homework 8.1

For extra help ▶   Watch the videos in MyMathLab   Download the MyDashboard App

1. Commutative law for addition: $a + b =$ _____.

2. Associative law for multiplication: $a(bc) =$ _____.

3. The _____ of a term is the constant factor of the term.

4. _____ terms are either constant terms or variable terms that contain the same variable(s) raised to exactly the same power(s).

Use the commutative law for addition to write the expression in another form.

5. $5 + x$ 6. $x + 8$ 7. $2p + 7$ 8. $6 + 8w$

Use the commutative law for multiplication to write the expression in another form.

9. xy 10. pw 11. $15 + 4m$ 12. $7n + 12$

Use an associative law to write the expression in another form.

13. $(x + 4) + y$ 14. $(8 + x) + y$

15. $(4b)c$ 16. $(7p)w$

17. $x + (y + 3)$ 18. $9 + (k + d)$

19. $a(bc)$ 20. $k(pw)$

Use the commutative and associative laws to simplify the expression.

21. $2(5x)$ 22. $3(6x)$

23. $(p + 4) + 3$ 24. $(k + 8) + 1$

25. $2 + (3b + 9)$ 26. $1 + (8m + 4)$

27. $\frac{1}{2}(-8x)$ 28. $\frac{2}{3}(-12x)$

29. $7\left(\frac{x}{4}\right)$ 30. $6\left(\frac{x}{5}\right)$

Use the distributive law to simplify the expression.

31. $3(x + 9)$ 32. $6(x + y)$

33. $(x - 5)(7)$ 34. $(x - 8)(4)$

35. $-2(t + 5)$ 36. $-4(w + 3)$

37. $-5(6 - 2x)$ 38. $-8(5 - 3x)$

39. $5(5x + 3y - 8)$ 40. $6(2x - 4y + 5)$

41. $-(5x + 8y - 1)$ 42. $-(3x - 7y + 4)$

43. $-\frac{3}{5}(45 - 35x)$ 44. $-\frac{5}{6}(30 + 12x)$

Simplify.

45. $2x + 5x$ 46. $3x + 6x$

47. $9x - 4x$ 48. $6x - 5x$

49. $-w - 5w$ 50. $p - 8p$

51. $\frac{2}{3}x + \frac{5}{3}x$

52. $\frac{9}{5}x - \frac{2}{5}x$

53. $2 + 4x - 5 - 7x$

54. $-8x - 1 + 3x - 4$

55. $3y + 5x - 2y - 2x + 1$

56. $5 - 3x + 7y + 6x$

57. $-4.6x + 3.9y + 2.1 - 5.3x - 2.8y$

58. $8.7 - 3.5y + 4.4x - 6.2y + 1.9x$

59. $-3(a - 5) + 2a$

60. $5(k - 8) - 6k$

61. $5.2(8.3x + 4.9) - 2.4$

62. $-3.8(2.7x - 5.5) - 8.4$

63. $2b - 3(5b + 2) + 1$

64. $3a - 2(3a + 4) + 5$

65. $6x - (4x - 3y) - 5y$

66. $8x - (3x + 7y) - 2y$

67. $6(2x - 3y) - 4(9x + 5y)$

68. $2(7x - 5y) - 5(3x + y)$

69. $-(x - 1) - (1 - x)$

70. $-(6x - 7) - (7 - 6x)$

71. $2x - 5y - 3(2x - 4y + 7)$

72. $4x + 3y - 2(5x + 2y + 8)$

73. $\frac{2}{7}(a + 1) - \frac{4}{7}(a - 1)$

74. $\frac{3}{5}(t - 1) + \frac{1}{5}(t + 1)$

75. $5x - \frac{1}{2}(4x + 6)$

76. $7x - \frac{1}{3}(6x + 9)$

Let x be a number. Translate the English phrase into a mathematical expression. Then simplify the expression.

77. The number plus the product of 5 and the number

78. The number minus the product of the number and 3

79. 4 times the difference of the number and 2

80. -6 times the sum of the number and 4

81. The number, plus 3 times the difference of the number and 7

82. The number, minus 5 times the sum of the number and 2

83. Twice the number, minus 4 times the sum of the number and 6

84. Twice the number, plus 9 times the difference of the number and 4

For Exercises 85–92, let x be a number. Translate the expression into an English phrase. Then simplify the expression.

85. $2x + 6x$

86. $3x - 8x$

87. $7(x - 5)$

88. $-2(x + 4)$

89. $x + 5(x + 1)$

90. $x - 8(x - 3)$

91. $2x - 3(x - 9)$

92. $2x + 6(x - 4)$

93. Find the sum of $3x - 7$ and $5x + 2$.

94. Find the sum of $6x - 3$ and $8x - 9$.

95. Find the difference of $4x + 8$ and $7x - 1$.

96. Find the difference of $5x - 9$ and $2x + 6$.

Concepts

97. A student tries to simplify $3(x + 4)$:

$$3(x + 4) = 3x + 4$$

Describe any errors. Then simplify the expression correctly.

98. A student tries to simplify $2(xy)$:

$$2(xy) = (2x)(2y) = 4xy$$

What would you tell the student?

99. A student works 7 hours on Monday, 3 hours on Tuesday, and 6 hours on Wednesday. The student earns $10 per hour. Show two ways to compute the student's total earnings. Use the distributive law to explain why both methods give the same result.

100. When we write $-(x + 5)$ as $-x - 5$, what *number* have we distributed? Show the missing steps.

101. Explain what law was used in each step to rearrange terms of $a + b + c$ as follows:

$$
\begin{aligned}
a + b + c &= (a + b) + c \quad \text{Add from left to right.}\\
&= (b + a) + c\\
&= b + (a + c)\\
&= b + (c + a)\\
&= (b + c) + a\\
&= b + c + a \quad \text{Add from left to right.}
\end{aligned}
$$

102. Explain what law was used in each step to rearrange factors of abc as follows:

$$
\begin{aligned}
abc &= (ab)c \quad \text{Multiply from left to right.}\\
&= a(bc)\\
&= a(cb)\\
&= (ac)b\\
&= (ca)b\\
&= cab \quad \text{Multiply from left to right.}
\end{aligned}
$$

103. Explain what law was used in each step to show that $(a + b)c = ac + bc$:

$$
\begin{aligned}
(a + b)c &= c(a + b)\\
&= ca + cb\\
&= ac + bc
\end{aligned}
$$

104. Use the distributive law $a(b - c) = ab - ac$ and other laws to show that $(a - b)c = ac - bc$. [**Hint:** See Exercise 103.]

105. In Section 1.6, we learned that the product of two numbers with the same sign is positive. Explain what property or law was used in each step to show why $(-2)(-3) = 6$:

$$
\begin{aligned}
(-2)(-3) &= (-1 \cdot 2)(-3)\\
&= -1(2 \cdot (-3))\\
&= -1(-6)\\
&= -(-6)\\
&= 6
\end{aligned}
$$

106. **a.** Pick a number. Next, subtract 3. Then double the result. Then subtract the original number. Finally, add 6. Compare your result with the original number.

   b. Pick another number and follow the instructions in part (a), including comparing your result with the original number.

   c. Let x be a number and follow the instructions in part (a). Your result should be an expression. Simplify the expression to explain the observations you made in parts (a) and (b).

107. Which of the following expressions are equivalent?

$$-2x - 3 \qquad -2(x - 3) \qquad 2(3 - x) \qquad -2x - 6$$
$$-3(x - 2) + x \qquad -2x + 6$$

108. Which of the following expressions are equivalent?

$$-(2x - 3) \qquad 3 - 2x \qquad -(3 - 2x) \qquad -2x + 3$$
$$-(3x - 2) + x + 1 \qquad -(x - 1) - (x - 2)$$

109. Give three examples of expressions equivalent to the expression $2(x - 5) + 3(x + 1)$.

110. Give three examples of expressions equivalent to x.

For Exercises 111–114, simplify the right-hand side of the equation. Then graph the equation by hand.

111. $y = 9x - 4 - 7x$

112. $y = 4x + 3 - 6x$

113. $y = 4(2x - 1) - 5x$

114. $y = -2(3x - 2) + 5x$

115. Give an example of combining like terms. Then use the distributive law to show why we can do this.

116. Describe the meaning of *equivalent expressions*. (See page 12 for guidelines on writing a good response.)

▼ 8.2 Solving Linear Equations in One Variable

Objectives

» Describe the meaning of *linear equation in one variable.*

» For an equation in one variable, describe *satisfy, solution, solution set,* and *solve.*

» Describe *equivalent equations.*

» Describe the *addition property of equality.*

» Describe the *multiplication property of equality.*

» Solve a linear equation in one variable.

» Solve a percentage problem.

» Use a graph to solve a linear equation in one variable.

» Use a table to solve a linear equation in one variable.

Linear Equation in One Variable

In Section 8.1, we simplified expressions. In this section and Section 8.3, we will use that skill to help us work with equations. So far, we have discussed how to graph linear equations in *two* variables, such as

$$y = x + 2 \qquad y = -5x \qquad y = 4x - 7 \qquad y = x - 35.7$$

In this section and Section 8.3, we will work with *linear equations in one variable*, such as:

$$0 = x + 2 \qquad 9 = -5x \qquad 4x - 7 = 3 \qquad 2.3 = \frac{x - 35.7}{6.2}$$

We will see in this section and in Section 8.3 that we can put each of these four equations in the form $mx + b = 0$, where m and b are constants and $m \neq 0$.

▶ **Definition**   **Linear equation in one variable**

A **linear equation in one variable** is an equation that can be put into the form

$$mx + b = 0$$

where m and b are constants and $m \neq 0$.

Working with linear equations in one variable will enable us to make efficient estimates and predictions about authentic situations.

Meaning of Satisfy, Solution, Solution Set, and Solve

Consider the linear equation

$$x + 1 = 6$$

This equation becomes a false statement if we substitute 2 for x:

$$x + 1 = 6 \qquad \text{Original equation}$$
$$(2) + 1 \stackrel{?}{=} 6 \qquad \text{Substitute 2 for } x.$$
$$3 \stackrel{?}{=} 6 \qquad \text{Add.}$$
$$\text{false}$$

However, the equation $x + 1 = 6$ becomes a true statement if we substitute 5 for x:

$$x + 1 = 6 \qquad \text{Original equation}$$
$$(5) + 1 \stackrel{?}{=} 6 \qquad \text{Substitute 5 for } x.$$
$$6 \stackrel{?}{=} 6 \qquad \text{Add.}$$
$$\text{true}$$

We say 5 *satisfies* the equation $x + 1 = 6$ and 5 is a *solution* of the equation. In fact, 5 is the only solution of $x + 1 = 6$ because 5 is the only number that, when increased by 1, is equal to 6. We call the set containing only this number the *solution set* of the equation.

▶ **Definition**   *Solution, satisfy, solution set,* and *solve* for an equation in one variable

A number is a **solution** of an equation in one variable if the equation becomes a true statement when the number is substituted for the variable. We say the number **satisfies** the equation. The set of all solutions of the equation is called the **solution set** of the equation. We **solve** the equation by finding its solution set.

▶ **Example 1**   Identifying Solutions of an Equation

1. Is 3 a solution of the equation $5(x - 1) = 10 + 2x$?
2. Is 5 a solution of the equation $5(x - 1) = 10 + 2x$?

Solution

1. We begin by substituting 3 for x in $5(x - 1) = 10 + 2x$:

$$5(x - 1) = 10 + 2x \qquad \textit{Original equation}$$
$$5(3 - 1) \stackrel{?}{=} 10 + 2(3) \qquad \textit{Substitute 3 for x.}$$
$$5(2) \stackrel{?}{=} 10 + 6 \qquad \textit{Simplify.}$$
$$10 \stackrel{?}{=} 16 \qquad \textit{Simplify.}$$
$$\text{false}$$

So, 3 is not a solution of the equation $5(x - 1) = 10 + 2x$.

2. We begin by substituting 5 for x in $5(x - 1) = 10 + 2x$:

$$5(x - 1) = 10 + 2x \qquad \textit{Original equation}$$
$$5(5 - 1) \stackrel{?}{=} 10 + 2(5) \qquad \textit{Substitute 5 for x.}$$
$$5(4) \stackrel{?}{=} 10 + 10 \qquad \textit{Simplify.}$$
$$20 \stackrel{?}{=} 20 \qquad \textit{Simplify.}$$
$$\text{true}$$

So, 5 is a solution of the equation $5(x - 1) = 10 + 2x$.

Equivalent Equations

Consider the equation $x = 2$. We add 5 to both sides of the equation:

$$x = 2 \qquad \textit{Original equation}$$
$$x + 5 = 2 + 5 \qquad \textit{Add 5 to both sides.}$$
$$x + 5 = 7 \qquad \textit{Add.}$$

Note that 2 satisfies all three equations:

Equation	Does 2 satisfy the equation?	
$x = 2$	$(2) \stackrel{?}{=} 2$	true
$x + 5 = 2 + 5$	$(2) + 5 \stackrel{?}{=} 2 + 5$	true
$x + 5 = 7$	$(2) + 5 \stackrel{?}{=} 7$	true

In fact, 2 is the *only* number that satisfies any of these equations. So, the equations $x = 2$, $x + 5 = 2 + 5$, and $x + 5 = 7$ have the same solution set. We say that the three equations are *equivalent*.

▶ **Definition**   Equivalent equations

Equivalent equations are equations that have the same solution set.

Addition Property of Equality

The fact that the equations $x = 2$ and $x + 5 = 2 + 5$ have the same solution set suggests that adding a number to both sides of an equation does not change an equation's solution set. This property is called the **addition property of equality**.

> **Addition Property of Equality**

If A and B are expressions and c is a number, then the equations $A = B$ and $A + c = B + c$ are equivalent.

To solve an equation in one variable, x, we can sometimes use the addition property of equality to get x alone on one side of the equation. Then we can identify solutions of the equation. For example, for the equation $x = 3$, we can see that the solution is 3.

▶ **Example 2**   Solving an Equation by Adding a Number to Both Sides

Solve $x - 2 = 3$.

Solution

To get x alone on the left side, we undo the subtraction of 2 by adding 2 to both sides:

$$x - 2 = 3 \qquad \text{\textit{Original equation}}$$
$$x - 2 + 2 = 3 + 2 \qquad \text{\textit{Addition property of equality: Add 2 to both sides.}}$$
$$x + 0 = 5 \qquad \text{\textit{Simplify: } } -a + a = 0$$
$$x = 5 \qquad \text{\textit{Simplify: } } a + 0 = a$$

Next, we check that 5 satisfies the original equation, $x - 2 = 3$:

$$x - 2 = 3 \qquad \text{\textit{Original equation}}$$
$$(5) - 2 \stackrel{?}{=} 3 \qquad \text{\textit{Substitute 5 for x.}}$$
$$3 \stackrel{?}{=} 3 \qquad \text{\textit{Subtract.}}$$
$$\text{true}$$

So, the solution is 5.

▶

After solving an equation, check that all of your results satisfy the equation.
In Example 2, we worked with the equations $x - 2 = 3$, $x - 2 + 2 = 3 + 2$, $x + 0 = 5$, and $x = 5$. Although we could find the solution set with any of these equivalent equations, it is easiest to determine the solution set from the equation $x = 5$, which has the variable alone on the left side. **Our strategy in solving linear equations in one variable will be to use properties to get the variable alone on one side of the equation.**

Assume that A and B are expressions and that c is a number. Since subtracting a number is the same as adding the opposite of the number, the addition property of equality implies that the equations $A = B$ and $A - c = B - c$ are equivalent.

▶ **Example 3**   Solving an Equation by Subtracting a Number from Both Sides

Solve $x + 4 = 6$.

Solution

To get x alone on the left side, we undo the addition of 4 by subtracting 4 from *both* sides:

$$x + 4 = 6 \qquad \text{\textit{Original equation}}$$
$$x + 4 - 4 = 6 - 4 \qquad \text{\textit{Subtract 4 from both sides.}}$$
$$x = 2 \qquad \text{\textit{a} } + 0 = a$$

Next, we check that 2 satisfies the original equation, $x + 4 = 6$:

$$x + 4 = 6 \qquad \text{\textit{Original equation}}$$
$$(2) + 4 \stackrel{?}{=} 6 \qquad \text{\textit{Substitute 2 for x.}}$$
$$6 \stackrel{?}{=} 6 \qquad \text{\textit{Add.}}$$
$$\text{true}$$

The solution is 2.

▶

Multiplication Property of Equality

Consider the equation $x = 3$. We multiply both sides of $x = 3$ by 7:

$$x = 3 \qquad \textit{Original equation}$$
$$7 \cdot x = 7 \cdot 3 \qquad \textit{Multiply both sides by 7.}$$
$$7x = 21 \qquad \textit{Multiply.}$$

Note that 3 satisfies all three equations:

Equation	Does 3 satisfy the equation?	
$x = 3$	$(3) \stackrel{?}{=} 3$	true
$7 \cdot x = 7 \cdot 3$	$7 \cdot (3) \stackrel{?}{=} 7 \cdot 3$	true
$7x = 21$	$7(3) \stackrel{?}{=} 21$	true

In fact, 3 is the *only* number that satisfies any of these equations. So, the equations $x = 3$, $7 \cdot x = 7 \cdot 3$, and $7x = 21$ are equivalent. The fact that the equations $x = 3$ and $7 \cdot x = 7 \cdot 3$ are equivalent suggests the **multiplication property of equality**.

> ### Multiplication Property of Equality
>
> If A and B are expressions and c is a nonzero number, then the equations $A = B$ and $Ac = Bc$ are equivalent.*

The multiplication property of equality is often helpful in solving equations that contain fractions. It is useful to know that a fraction times its reciprocal is equal to 1. For example,

$$\frac{2}{7} \cdot \frac{7}{2} = \frac{14}{14} = 1$$

Solving a Linear Equation in One Variable

In Example 4, we will use the multiplication property of equality to solve an equation.

> ▶ **Example 4** Solving an Equation by Multiplying Both Sides by a Number

Solve $\frac{4}{5}x = 8$.

Solution

To get x alone on the left side, we multiply *both* sides by the reciprocal of $\frac{4}{5}$, which is $\frac{5}{4}$:

$$\frac{4}{5}x = 8 \qquad \textit{Original equation}$$
$$\frac{5}{4} \cdot \frac{4}{5}x = \frac{5}{4} \cdot 8 \qquad \textit{Multiplication property of equality: Multiply both sides by } \frac{5}{4}.$$
$$1x = 10 \qquad \textit{Simplify.}$$
$$x = 10 \qquad \textit{1 a = a}$$

---

*If $c = 0$, the equations may not be equivalent. For example, $x = 5$ and $x \cdot 0 = 5 \cdot 0$ (or $0 = 0$) are not equivalent. The only solution of $x = 5$ is 5, but the solution set of $x \cdot 0 = 5 \cdot 0$ is the set of all real numbers because no matter what real number we substitute for x, the equation $x \cdot 0 = 5 \cdot 0$ will become true.

The solution is 10. We use a TI-84 table to check that, when 10 is substituted for x in the expression $\frac{4}{5}x$, the result is 8 (see Fig. 4). For TI-84 instructions on using "Ask" in a table, see Appendix A.23.

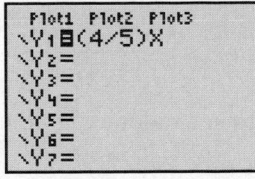  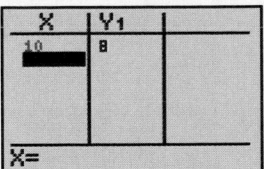

Figure 4 Verify the work

Assume that A and B are expressions and c is a nonzero number. Because dividing by a number is the same as multiplying by its reciprocal, the multiplication property of equality implies that the equations $A = B$ and $\frac{A}{c} = \frac{B}{c}$ are equivalent.

▶ **Example 5**   Solving an Equation by Dividing Both Sides by a Number

Solve $-12 = -3t$.

Solution

The variable term $-3t$ is on the right-hand side this time. To get t alone on this side, we undo the multiplication by -3 by dividing *both* sides by -3:

$$-12 = -3t \quad \text{\textit{Original equation}}$$
$$\frac{-12}{-3} = \frac{-3t}{-3} \quad \text{\textit{Divide both sides by} } -3.$$
$$4 = t \quad \text{\textit{Simplify.}}$$

The solution is 4. We can check that 4 satisfies the original equation (try it).

The last step of Example 5 is $4 = t$. Note that the equation $t = 4$ is equivalent to $4 = t$ and that we can see from either equation that the solution is 4.

▶ **Example 6**   Solving an Equation by Dividing Both Sides by -1

Solve $-w = 5$.

Solution

Recall that $-w = -1w$. To get w alone on the left side, we undo the multiplication by -1 by dividing both sides by -1:

$$-w = 5 \quad \text{\textit{Original equation}}$$
$$-1w = 5 \quad \text{\textit{-w = -1w}}$$
$$\frac{-1w}{-1} = \frac{5}{-1} \quad \text{\textit{Divide both sides by} } -1.$$
$$w = -5 \quad \text{\textit{Divide.}}$$

Next, we check that -5 satisfies the original equation, $-w = 5$:

$$-w = 5 \quad \text{\textit{Original equation}}$$
$$-(-5) \stackrel{2}{=} 5 \quad \text{\textit{Substitute} } -5 \text{ \textit{for w.}}$$
$$5 \stackrel{2}{=} 5 \quad \text{\textit{-(-a) = a}}$$
$$\text{true}$$

The solution is -5.

▶ **Example 7** Solving an Equation by Multiplying Both Sides by a Number

Solve $\dfrac{2x}{3} = \dfrac{5}{6}$.

Solution

Since $\dfrac{2x}{3} = \dfrac{2}{3} \cdot \dfrac{x}{1} = \dfrac{2}{3}x$, we first write $\dfrac{2x}{3}$ as $\dfrac{2}{3}x$. Then we multiply both sides by the reciprocal of $\dfrac{2}{3}$, which is $\dfrac{3}{2}$, to get x alone on the left side:

$$\dfrac{2x}{3} = \dfrac{5}{6} \qquad \text{\textit{Original equation}}$$

$$\dfrac{2}{3}x = \dfrac{5}{6} \qquad \text{\textit{Write } } \dfrac{2x}{3} \text{ \textit{as} } \dfrac{2}{3}x.$$

$$\dfrac{3}{2} \cdot \dfrac{2}{3}x = \dfrac{3}{2} \cdot \dfrac{5}{6} \qquad \text{\textit{Multiply both sides by }} \dfrac{3}{2}.$$

$$1x = \dfrac{3 \cdot 5}{2 \cdot 2 \cdot 3} \qquad \text{\textit{Simplify: }} \dfrac{a}{b} \cdot \dfrac{c}{d} = \dfrac{ac}{bd}$$

$$x = \dfrac{5}{4} \qquad \text{\textit{Simplify: }} \dfrac{3}{3} = 1$$

The solution is $\dfrac{5}{4}$. We use a TI-84 table to check that, when $\dfrac{5}{4} = 1.25$ is substituted for x in the expression $\dfrac{2x}{3}$, the result is $\dfrac{5}{6} \approx 0.83333$ (see Fig. 5).

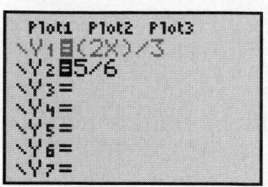

Figure 5 Verify the work

In Examples 2–7, we used either the addition property of equality or the multiplication property of equality to solve a linear equation. In Example 8, we will use a combination of these properties to solve a linear equation.

▶ **Example 8** Using the Addition and Multiplication Properties of Equality

Solve $2x - 3 = 5$.

Solution

We begin by adding 3 to both sides to get $2x$ alone on the left side:

$$2x - 3 = 5 \qquad \text{\textit{Original equation}}$$

$$2x - 3 + 3 = 5 + 3 \qquad \text{\textit{Add 3 to both sides.}}$$

$$2x = 8 \qquad \text{\textit{Combine like terms.}}$$

$$\dfrac{2x}{2} = \dfrac{8}{2} \qquad \text{\textit{Divide both sides by 2.}}$$

$$x = 4 \qquad \text{\textit{Simplify.}}$$

Next, we check that 4 satisfies the equation $2x - 3 = 5$:

$$2x - 3 = 5 \qquad \text{\textit{Original equation}}$$

$$2(4) - 3 \overset{?}{=} 5 \qquad \text{\textit{Substitute 4 for x.}}$$

$$8 - 3 \overset{?}{=} 5 \qquad \text{\textit{Multiply.}}$$

$$5 \overset{?}{=} 5 \qquad \text{\textit{Subtract.}}$$

$$\text{true}$$

So, the solution is 4.

In Example 8, we used the addition property of equality so that a variable term was alone on one side of the equation and a constant term was on the other side. Then we used the multiplication property of equality to get the variable alone on one side of the equation. We will do the same thing in Example 9, but first we will simplify one side of the equation.

▶ **Example 9** Combining Like Terms to Help Solve an Equation

Solve $4x - 7x + 2 = 17$.

Solution

First, we combine like terms on the left side:

$$4x - 7x + 2 = 17 \qquad \textit{Original equation}$$
$$-3x + 2 = 17 \qquad \textit{Combine like terms.}$$
$$-3x + 2 - 2 = 17 - 2 \qquad \textit{Subtract 2 from both sides to get } -3x \textit{ alone on left side.}$$
$$-3x = 15 \qquad \textit{Combine like terms.}$$
$$\frac{-3x}{-3} = \frac{15}{-3} \qquad \textit{Divide both sides by } -3 \textit{ to get } x \textit{ alone on left side.}$$
$$x = -5 \qquad \textit{Simplify.}$$

The solution is -5. We use a TI-84 table to check that if -5 is substituted for x in the expression $4x - 7x + 2$, the result is 17 (see Fig. 6).

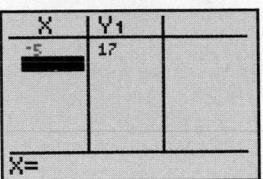

Figure 6 Verify the work

In Example 10, we will solve an equation that contains both variable terms and constant terms on each side of the equation. To solve such an equation, we first use the addition property of equality to write the variable terms on one side of the equation and constant terms on the other side.

▶ **Example 10** Solving an Equation with Variable Terms on Both Sides

Solve $-x + 7 = 2x - 2$.

Solution

First, we subtract $2x$ from both sides to get all of the variable terms on the left side:*

$$-x + 7 = 2x - 2 \qquad \textit{Original equation}$$
$$-x + 7 - 2x = 2x - 2 - 2x \qquad \textit{Subtract } 2x \textit{ from both sides.}$$
$$-3x + 7 = -2 \qquad \textit{Combine like terms.}$$
$$-3x + 7 - 7 = -2 - 7 \qquad \textit{Subtract 7 from both sides.}$$
$$-3x = -9 \qquad \textit{Combine like terms.}$$
$$\frac{-3x}{-3} = \frac{-9}{-3} \qquad \textit{Divide both sides by } -3.$$
$$x = 3 \qquad \textit{Simplify.}$$

The solution is 3. We use a TI-84 table to check that if 3 is substituted for x in the expressions $-x + 7$ and $2x - 2$, the two results are equal (see Fig. 7).

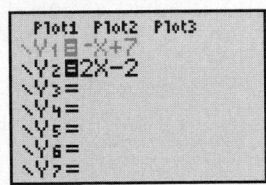
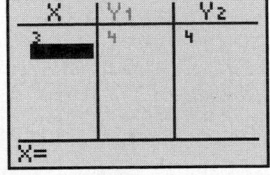

Figure 7 Verify the work

Solving Percentage Problems

Recall that to find the percentage of a quantity, we multiply the decimal form of the percentage and the quantity (Section 1.6).

---

*The addition property of equality implies that we can add a number to both sides of an equation without changing the equation's solution set. We can also add (or subtract) a variable term of the form mx, where m is a constant, to both sides of an equation without changing the equation's solution set.

▶ **Example 11** Solving a Percentage Problem

In a survey, 36% of adults said police departments do a poor job of holding officers accountable (**Source:** *USA Today and Pew Research Center poll*). If the number of adults in the survey who said this is 540, what is the total number of adults surveyed?

Solution

We let n be the total number of adults surveyed. Since 36% of the adults surveyed is equal to 540 adults, we write

36% of adults surveyed	is equal to	540.
$0.36n$	$=$	540

Now we solve the equation:

$$0.36n = 540$$
$$\frac{0.36n}{0.36} = \frac{540}{0.36} \quad \textit{Divide both sides by 0.36.}$$
$$n = 1500 \quad \textit{Simplify: divide.}$$

So, a total of 1500 adults were surveyed.

◀

In many application problems in this text so far, variable names and their definitions have been provided. In Example 11, a key step was to create the variable name n and define it.

Using Graphing to Solve an Equation in One Variable

In Example 2, we showed that the solution of the equation $x - 2 = 3$ is 5. How can we use graphing to solve this equation? Here are three steps:

Step 1. We set y equal to the left side, $x - 2$, to form the equation $y = x - 2$, and we set y equal to the right side, 3, to form the equation $y = 3$. Then we graph the two equations $y = x - 2$ and $y = 3$ in the same coordinate system (see Fig. 8).

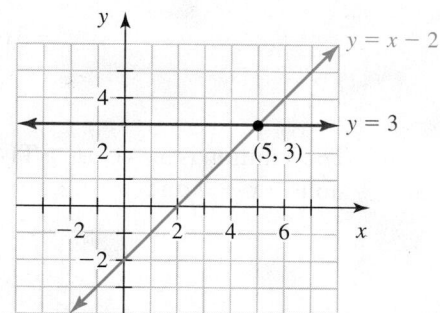

Figure 8 The intersection point is (5, 3)

Step 2. We find the intersection point of the two graphs, which is the point $(5, 3)$.
Step 3. The solution of the original equation $x - 2 = 3$ is the x-coordinate of the intersection point $(5, 3)$. So, the solution is 5.

In general, the solutions are all the x-coordinates of any intersection points.

Using Graphing to Solve an Equation in One Variable

To use graphing to solve an equation $A = B$ in one variable, x, where A and B are expressions,

1. Graph the equations $y = A$ and $y = B$ in the same coordinate system. (For example, if the original equation is $5x - 9 = 3x + 7$, then we would graph the equations $y = 5x - 9$ and $y = 3x + 7$.)
2. Find all intersection points.
3. The x-coordinates of those intersection points are the solutions of the equation $A = B$.

▶ **Example 12** Solving an Equation in One Variable by Graphing

The graphs of $y = \dfrac{3}{2}x + 1$, $y = 4$, and $y = -5$ are shown in Fig. 9. Use these graphs to solve the given equations.

1. $\dfrac{3}{2}x + 1 = 4$ 2. $\dfrac{3}{2}x + 1 = -5$

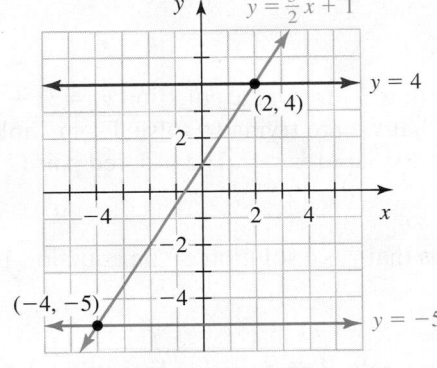

Figure 9 Solving $\dfrac{3}{2}x + 1 = 4$ and $\dfrac{3}{2}x + 1 = -5$

Solution

1. The graphs of $y = \dfrac{3}{2}x + 1$ and $y = 4$ intersect only at the point $(2, 4)$. The intersection point, $(2, 4)$, has x-coordinate 2. So, 2 is the solution of $\dfrac{3}{2}x + 1 = 4$. We can verify our work by checking that 2 satisfies the equation (try it).

2. The graphs of $y = \dfrac{3}{2}x + 1$ and $y = -5$ intersect only at the point $(-4, -5)$. The intersection point, $(-4, -5)$, has x-coordinate -4. So, -4 is the solution of $\dfrac{3}{2}x + 1 = -5$. We can verify our work by checking that -4 satisfies the equation (try it).

▶ **Example 13** Solving an Equation in One Variable by Graphing

Use "intersect" on a TI-84 to solve the equation $-2x + 4 = x - 5$.

Solution

We use a TI-84 to graph the equations $y = -2x + 4$ and $y = x - 5$ in the same coordinate system and then use "intersect" to find the intersection point, which turns out to be $(3, -2)$. See Fig. 10. See Appendix A.24 for TI-84 instructions.

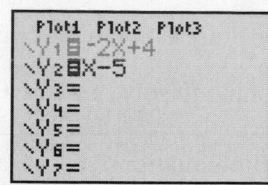 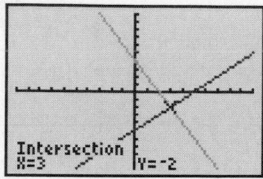

Figure 10 Using "intersect" to solve an equation in one variable

The intersection point, $(3, -2)$, has x-coordinate 3. So, the solution of the original equation is 3. We can verify our work by checking that 3 satisfies the equation $-2x + 4 = x - 5$ (try it).

Using a Table to Solve an Equation

We can use a table of solutions of an equation in two variables to help us solve an equation in one variable.

▶ **Example 14**   Solving an Equation in One Variable by Using a Table

Some solutions of $y = 3x - 5$ are shown in Table 2. Use the table to solve the equation $1 = 3x - 5$.

Solution

If we substitute 1 for y in the equation $y = 3x - 5$, the result is the equation $1 = 3x - 5$, which is what we are trying to solve. From Table 2, we see that the output $y = 1$ originates from the input $x = 2$. The ordered pair $(2, 1)$ satisfies the equation $y = 3x - 5$:

$$1 = 3(2) - 5$$

This means that 2 is a solution of the equation $1 = 3x - 5$.

Table 2 Solutions of
$y = 3x - 5$

x	y
-1	-8
0	-5
1	-2
2	1
3	4

We have solved an equation by getting the variable alone on one side of the equation, by using graphing, and by using a table. All three methods give the same results.

◢◣ Group Exploration
Locating an error in solving an equation

A student tries to solve the equation $x - 3 = 5$:

$$x - 3 = 5$$
$$x - 3 - 3 = 5 - 3$$
$$x - 0 = 2$$
$$x = 2$$

Substitute 2 for x in each of the four equations to determine which equations have 2 as a solution. Explain how your work shows that the student made an error and how your work helps you pinpoint the step in which the error was made.

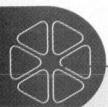

Homework 8.2

For extra help ▶   ▦ Watch the videos in MyMathLab   ◐ Download the MyDashboard App

1. A linear equation in one variable is an equation that can be put into the form _____ = 0, where m and b are constants and $m \neq 0$.

2. A number is a _____ of an equation in one variable if the equation becomes a true statement when the number is substituted for the variable.

3. _____ equations are equations that have the same solution set.

4. *True or False:* If A and B are expressions and c is a number, then the equations $A = B$ and $A + c = B + c$ are equivalent.

Determine whether 2 is a solution of the equation.

5. $3x + 1 = 7$

6. $-2x + 5 = 4$

7. $5(2x - 1) = 0$

8. $-3(x - 2) = 0$

9. $12 - x = 2(4x - 3)$

10. $5 - x = 3(2x - 1)$

For Exercises 11–70, solve.

11. $x - 3 = 2$
12. $x - 4 = 9$
13. $x + 6 = -8$
14. $x + 1 = -5$
15. $-5 = x - 2$
16. $-3 = x - 4$
17. $x - 3 = 0$
18. $x - 5 = 0$
19. $6r = 18$
20. $4w = 24$
21. $-3x = 12$
22. $-5x = 20$
23. $15 = 3x$
24. $24 = 2x$
25. $6x = 8$
26. $4x = 6$
27. $-10x = -12$
28. $-14x = -6$
29. $-2x = 0$
30. $-5x = 0$
31. $\frac{1}{3}t = 5$
32. $\frac{1}{2}w = 4$
33. $-\frac{2}{7}x = -3$
34. $-\frac{4}{9}x = -5$
35. $-9 = \frac{3x}{4}$
36. $-8 = \frac{2x}{5}$
37. $\frac{2}{5}p = -\frac{4}{3}$
38. $\frac{4}{7}b = -\frac{5}{21}$
39. $-\frac{3x}{8} = -\frac{9}{4}$
40. $-\frac{5x}{4} = -\frac{15}{8}$
41. $-x = 3$
42. $-x = 2$
43. $-\frac{1}{2} = -x$
44. $-\frac{3}{4} = -x$
45. $x + 4.3 = -6.8$
46. $x + 7.5 = -2.8$
47. $25.17 = x - 16.59$
48. $5.27 = x - 28.85$
49. $-3.7r = -8.51$
50. $-2.9w = 13.34$
51. $3x - 2 = 13$
52. $5x - 1 = 9$
53. $-4x + 6 = 26$
54. $-2x + 7 = 23$
55. $-5 = 6x + 3$
56. $-7 = 4x - 1$
57. $8 - x = -4$
58. $2 - x = -9$
59. $8x - 5x - 4 = 11$
60. $5x - 2x - 3 = 6$
61. $2x + 6 - 7x = -4$
62. $4x + 3 - 9x = -22$
63. $1 = 3x - 5 - 9x$
64. $10 = 3x - 6 - 7x$
65. $5x + 4 = 3x + 16$
66. $7x + 5 = 4x + 17$

67. $-3r - 1 = 2r + 24$
68. $-6w - 3 = 4w + 17$
69. $9 - x - 5 = 2x - x$
70. $8 - 2x - 2 = 3x + x$

71. In a survey about the drinking age, the number of adults against lowering the drinking age to 18 was 750, which was about 74% of the adults in the survey (Source: *The Gallup Organization*). What was the total number of adults surveyed?

72. In 2013, the sales of Monster® energy drink were $3.1 billion, which was about 35% of all energy drink sales (Source: *Euromonitor*). What were the total sales of all energy drinks?

73. In a study of small-business owners, the number of such owners who said their children always have a place to work in the business was 280, which was 28% of the owners in the study (Source: *Bank of America Small Business Owner Report*). What was the total number of owners surveyed?

74. In a study of graduating college students, the number of students who said they are expecting to stay at their next full-time job for 3–5 years was 5645, which was about 43% of the students surveyed (Source: *Achievers survey*). What was the total number of students surveyed?

75. In 2009, there were 186 thousand students who earned bachelor's degrees in business, which was 20% of the students who earned bachelor's degrees in any field in that year (Source: *U.S. Center for Education Statistics*). How many students earned bachelor's degrees in 2009?

76. In 2011, there were 34.1 million satellite television subscribers, which was about 33% of all television subscribers in that year (Source: *Bernstein Research*). What was the number of television subscribers in 2011?

77. In 2011, Americans consumed an average of 506 servings of carbonated sweetened drinks per person, which was about 69.7% of the average per-person consumption of all sweetened drinks (both carbonated and noncarbonated) in that year (Source: *Centers for Disease Control and Prevention*). What was the average number of servings of sweetened drinks consumed per person in 2011?

78. In 2010, there were 223 thousand new cases of lung cancer, which was about 14.6% of the new cases of all types of cancer in that year (Source: *U.S. National Institutes of Health*). How many new cases of cancer were there in 2010?

For Exercises 79–82, use the graph of $y = -\frac{1}{2}x + 1$, shown in Fig. 11, to solve the given equation.

79. $-\frac{1}{2}x + 1 = 3$

80. $-\frac{1}{2}x + 1 = 2$

81. $-\frac{1}{2}x + 1 = -1$

82. $-\frac{1}{2}x + 1 = -2$

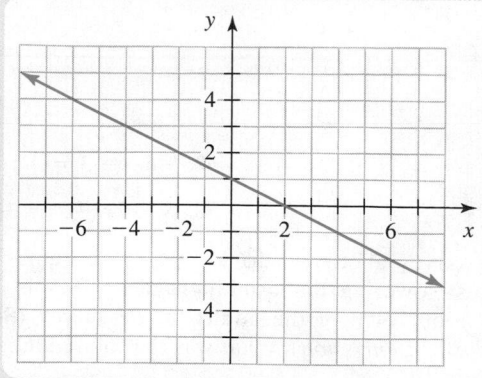

Figure 11 Exercises 79–82

Use "intersect" on a TI-84 to solve the equation.

83. $x + 2 = 7$

84. $x - 3 = 4$

85. $2x - 3 = 5$

86. $-3x + 5 = -7$

87. $-4(x - 1) = -8$

88. $2(x + 5) = 6$

89. $\frac{2}{3}t - \frac{3}{2} = -\frac{7}{2}$

90. $\frac{5}{2}w - \frac{5}{3} = \frac{10}{3}$

For Exercises 91–94, solve the given equation by referring to the solutions of $y = 5x - 3$ shown in Table 3.

91. $5x - 3 = 12$

92. $5x - 3 = 7$

93. $5x - 3 = -13$

94. $5x - 3 = -8$

Table 3 Exercises 91–94

x	y
−3	−18
−2	−13
−1	−8
0	−3
1	2
2	7
3	12

Concepts

95. A student tries to solve the equation $2x = 10$:

$$2x = 10$$
$$2x - 2 = 10 - 2$$
$$x = 8$$

Describe any errors. Then solve the equation correctly.

96. A student tries to solve the equation $x + 6 = 9$:

$$x + 6 = 9$$
$$x + 6 - 6 = 9$$
$$x + 0 = 9$$
$$x = 9$$

Describe any errors. Then solve the equation correctly.

97. A student solves the equation $4x = 12$:

$$4x = 12$$
$$\frac{4x}{4} = \frac{12}{4}$$
$$x = 3$$

Show that 3 satisfies each of the three equations.

98. A student solves the equation $x - 4 = 7$:

$$x - 4 = 7$$
$$x - 4 + 4 = 7 + 4$$
$$x + 0 = 11$$
$$x = 11$$

Show that 11 satisfies each of the four equations.

99. A student tries to solve the equation $x + 2 = 7$:

$$x + 2 = 7$$
$$x + 2 - 7 = 7 - 7$$
$$x - 5 = 0$$
$$x - 5 + 5 = 0 + 5$$
$$x = 5$$

Is the work correct? If yes, show a way to solve the equation in fewer steps. If no, describe the student's error(s) and then solve the equation correctly.

100. A student tries to solve the equation $\frac{3}{7}x = 2$:

$$\frac{3}{7}x = 2$$
$$7 \cdot \frac{3}{7}x = 7 \cdot 2$$
$$\frac{7}{1} \cdot \frac{3}{7}x = 14$$
$$3x = 14$$
$$\frac{3x}{3} = \frac{14}{3}$$
$$x = \frac{14}{3}$$

Is the work correct? If yes, show a way to solve the equation in fewer steps. If no, describe the student's error(s) and then solve the equation correctly.

101. Are the equations $\frac{x}{3} = 2$ and $x - 1 = 4$ equivalent?

102. Are the equations $2x = 10$ and $x + 2 = 7$ equivalent?

103. Give three examples of equations that have 4 as their solution.

104. Give three examples of equations that have −2 as their solution.

105. Give an example of three equations that are equivalent. Explain.

106. Give an example of two equations that are not equivalent. Explain.

107. Are the equations $\frac{x - 5}{2} = \frac{3}{x + 1}$ and $\frac{x - 5}{2} + 6 = \frac{3}{x + 1} + 6$ equivalent? Explain.

108. Are the equations $x(x + 1) = 6$ and $2x(x + 1) = 12$ equivalent? Explain.

109. **a.** Solve $x + 2 = 7$.
     b. Solve $x + 5 = 9$.
     c. Solve $x + b = k$, where b and k are constants. [**Hint:** Do steps similar to the ones you did in parts (a) and (b). The solution will be in terms of b and k.]

110. **a.** Solve $2x = 7$.
     b. Solve $5x = 9$.
     c. Solve $mx = p$, where m and p are constants and m is non-zero. [**Hint:** Do steps similar to the ones you did in parts (a) and (b). The solution will be in terms of m and p.]

111. Describe in your own words what the addition property of equality means.

112. Describe in your own words what the multiplication property of equality means.

113. Explain why it is *not* necessary to state a subtraction property of equality: If A and B are expressions and c is a number, then the equations $A = B$ and $A - c = B - c$ are equivalent.

114. Explain why it is *not* necessary to state a division property of equality: If A and B are expressions and c is a nonzero number, then the equations $A = B$ and $\dfrac{A}{c} = \dfrac{B}{c}$ are equivalent.

▼ 8.3 Solving Linear Equations to Make Predictions

Objectives

» Use combinations of properties of equality to solve a linear equation in one variable.

» Compare the meanings of *simplify an expression* and *solve an equation*.

» Use graphing to solve a linear equation in one variable.

» Use two tables to solve a linear equation in one variable.

» Find the input of a linear function for a given output.

» Make estimates and predictions by solving linear equations.

In this section, we will solve more complicated linear equations in one variable than those in Section 8.2. We will solve such equations to help us make predictions about authentic situations.

Solving Linear Equations in One Variable

When possible, we simplify each side of an equation before using properties such as the addition property of equality or the multiplication property of equality.

▶ **Example 1**   Using the Distributive Law to Help Solve an Equation

Solve $-2(a - 3) + 4 = 4(a - 2)$.

Solution

First, we use the distributive law on each side:

$-2(a - 3) + 4 = 4(a - 2)$	*Original equation*
$-2a + 6 + 4 = 4a - 8$	*Distributive law*
$-2a + 10 = 4a - 8$	*Combine like terms.*
$-2a + 10 - 4a = 4a - 8 - 4a$	*Subtract $4a$ from both sides.*
$-6a + 10 = -8$	*Combine like terms.*
$-6a + 10 - 10 = -8 - 10$	*Subtract 10 from both sides.*
$-6a = -18$	*Combine like terms.*
$\dfrac{-6a}{-6} = \dfrac{-18}{-6}$	*Divide both sides by -6.*
$a = 3$	*Simplify; divide.*

Next, we check that 3 satisfies $-2(a - 3) + 4 = 4(a - 2)$:

$-2(a - 3) + 4 = 4(a - 2)$	*Original equation*
$-2(3 - 3) + 4 \overset{?}{=} 4(3 - 2)$	*Substitute 3 for a.*
$-2(0) + 4 \overset{?}{=} 4(1)$	*Simplify.*
$4 \overset{?}{=} 4$	*Simplify.*
true	

So, 3 is the solution.

▶

A key step in solving an equation that contains fractions is to multiply both sides of the equation by the LCD so there are no fractions on either side of the equation. An equation that is "cleared of fractions" will be easier to solve than the original equation.

▶ **Example 2** Solving an Equation That Contains Fractions

Solve $\dfrac{2}{3}x + \dfrac{1}{6} = \dfrac{3}{4}$.

Solution

To find the LCD of the three fractions in the equation, we list the multiples of 3, the multiples of 6, and the multiples of 4:

$$\text{Multiples of 3:}\quad 3, 6, 9, 12, 15, 18, 21, \ldots$$
$$\text{Multiples of 6:}\quad 6, 12, 18, 24, 30, 36, 42, \ldots$$
$$\text{Multiples of 4:}\quad 4, 8, 12, 16, 20, 24, 28, \ldots$$

The LCD is 12. Next, we multiply both sides of $\dfrac{2}{3}x + \dfrac{1}{6} = \dfrac{3}{4}$ by 12 to clear the equation of fractions:

$$\frac{2}{3}x + \frac{1}{6} = \frac{3}{4} \qquad \textit{Original equation}$$

$$12 \cdot \left(\frac{2}{3}x + \frac{1}{6}\right) = 12 \cdot \frac{3}{4} \qquad \textit{Multiply both sides by the LCD, 12.}$$

$$12 \cdot \frac{2}{3}x + 12 \cdot \frac{1}{6} = 12 \cdot \frac{3}{4} \qquad \textit{Distributive law}$$

$$8x + 2 = 9 \qquad \textit{Simplify.}$$

$$8x + 2 - 2 = 9 - 2 \qquad \textit{Subtract 2 from both sides.}$$

$$8x = 7 \qquad \textit{Combine like terms.}$$

$$\frac{8x}{8} = \frac{7}{8} \qquad \textit{Divide both sides by 8.}$$

$$x = \frac{7}{8} \qquad \textit{Simplify.}$$

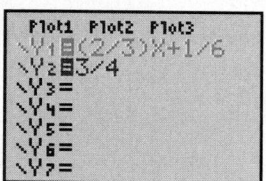

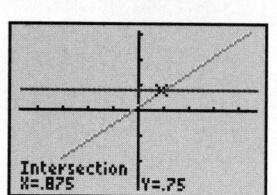

Figure 12 Verify the work

The solution is $\dfrac{7}{8}$. We check our work by using ZDecimal to graph the equations $y = \dfrac{2}{3}x + \dfrac{1}{6}$ and $y = \dfrac{3}{4}$ in the same coordinate system and by using "intersect" to find the intersection point, $(0.875, 0.75)$. See Fig. 12. So, the solution of the original equation is $\dfrac{7}{8} = 0.875$, which checks.

▶

▶ **Example 3** Solving an Equation That Contains Fractions

Solve $\dfrac{3x - 1}{2} = \dfrac{4x + 2}{3}$.

Solution

We multiply both sides of $\dfrac{3x - 1}{2} = \dfrac{4x + 2}{3}$ by the LCD, 6, to clear the equation of fractions:

$$\frac{3x - 1}{2} = \frac{4x + 2}{3} \qquad \textit{Original equation}$$

$$6 \cdot \frac{3x - 1}{2} = 6 \cdot \frac{4x + 2}{3} \qquad \textit{Multiply both sides by the LCD, 6.}$$

$$3(3x - 1) = 2(4x + 2) \qquad \textit{Simplify.}$$

$$9x - 3 = 8x + 4 \qquad \textit{Distributive law}$$

$$9x - 3 - 8x = 8x + 4 - 8x \qquad \textit{Subtract 8x from both sides.}$$

$$x - 3 = 4 \qquad \textit{Combine like terms.}$$

$$x - 3 + 3 = 4 + 3 \qquad \textit{Add 3 to both sides.}$$

$$x = 7 \qquad \textit{Combine like terms.}$$

We can use "intersect" on a TI-84 to check our work.

▶ **Example 4** Solving an Equation That Contains Decimal Numbers

Solve $1.96 = \dfrac{x - 9.7}{2.8}$. Round any solutions to two decimal places.

Solution

$$1.96 = \frac{x - 9.7}{2.8} \qquad \textit{Original equation}$$

$$2.8 \cdot 1.96 = 2.8 \cdot \frac{x - 9.7}{2.8} \qquad \textit{Multiply both sides by 2.8.}$$

$$5.488 = x - 9.7 \qquad \textit{Multiply; simplify.}$$

$$5.488 + 9.7 = x - 9.7 + 9.7 \qquad \textit{Add 9.7 to both sides.}$$

$$15.188 = x \qquad \textit{Combine like terms.}$$

The solution is approximately 15.19. We can check that 15.19 approximately satisfies the original equation (try it).

Comparing Expressions and Equations

It is important to be able to identify expressions and equations. Recall that an expression is a constant, a variable, or a combination of constants, variables, operation symbols, and grouping symbols, such as parentheses. Recall that an equation consists of an equality sign "=" with expressions on both sides. In Example 5, we will compare the processes of simplifying an expression and solving an equation.

▶ **Example 5** Comparing Expressions and Equations

1. Simplify $\dfrac{2}{5}x + \dfrac{1}{3}x$.

2. Solve $\dfrac{2}{5}x + \dfrac{1}{3}x = \dfrac{4}{5}$.

3. Compare the first steps of your work in Problems 1 and 2.

Solution

1. $\dfrac{2}{5}x + \dfrac{1}{3}x = \dfrac{3}{3} \cdot \dfrac{2}{5}x + \dfrac{5}{5} \cdot \dfrac{1}{3}x$ *The LCD of $\dfrac{2}{5}$ and $\dfrac{1}{3}$ is 15.*

   $= \dfrac{6}{15}x + \dfrac{5}{15}x$ *Multiply numerators and multiply denominators: $\dfrac{a}{b} \cdot \dfrac{c}{d} = \dfrac{ac}{bd}$*

   $= \dfrac{11}{15}x$ *Combine like terms.*

2.
   $\dfrac{2}{5}x + \dfrac{1}{3}x = \dfrac{4}{5}$ *Original equation*

   $15 \cdot \left(\dfrac{2}{5}x + \dfrac{1}{3}x\right) = 15 \cdot \dfrac{4}{5}$ *Multiply both sides by the LCD, 15.*

   $15 \cdot \dfrac{2}{5}x + 15 \cdot \dfrac{1}{3}x = 15 \cdot \dfrac{4}{5}$ *Distributive law*

   $6x + 5x = 12$ *Simplify.*

   $11x = 12$ *Combine like terms.*

   $\dfrac{11x}{11} = \dfrac{12}{11}$ *Divide both sides by 11.*

   $x = \dfrac{12}{11}$ *Simplify.*

3. To start simplifying the expression in Problem 1, we multiplied the term $\dfrac{2}{5}x$ by $\dfrac{3}{3} = 1$ and multiplied the term $\dfrac{1}{3}x$ by $\dfrac{5}{5} = 1$. To start solving the equation in Problem 2, we multiplied both sides of the equation by 15, which is *not* equal to 1.

WARNING   In Example 5, we simplified an expression and we solved an equation. In general, it is *incorrect* to say we "solve an expression." It is also incorrect to say we "simplify an equation," although "simplifying one (or both) sides of an equation" can make sense because each side of an equation is an expression.

From our work in Example 5, we can make two important observations: First, the result of simplifying the *expression* $\dfrac{2}{5}x + \dfrac{1}{3}x$ is $\dfrac{11}{15}x$, an *expression*; second, the result of solving the *equation* $\dfrac{2}{5}x + \dfrac{1}{3}x = \dfrac{4}{5}$ is $\dfrac{12}{11}$, a *number*.

Results of Simplifying an Expression and Solving an Equation

The result of simplifying an expression is an expression. The result of solving a linear equation in one variable is a number.

Problem 3 in Example 5 suggests the following comparison of multiplying an expression by a number and multiplying both sides of an equation by a number:

Multiplying Expressions and Both Sides of Equations by Numbers

In simplifying an expression, the only number that we can multiply the expression or part of it by is 1. In solving an equation, we can multiply both sides of the equation by *any* number except 0.

By the property $a \cdot 1 = a$, we know multiplying an expression by 1 gives an equivalent expression. By the multiplication property of equality, we know multiplying both sides of an equation by *any* nonzero number gives an equivalent equation.

Using Graphing to Solve an Equation in One Variable

In Section 8.2, we used graphing to solve some linear equations in one variable. We can use graphing to solve more complicated linear equations in one variable.

▶ **Example 6**   Solving an Equation in One Variable by Graphing

The graphs of $y = -\dfrac{1}{2}x - 1$ and $y = -\dfrac{5}{4}x + 2$ are shown in Fig. 13. Use these graphs to solve the equation $-\dfrac{1}{2}x - 1 = -\dfrac{5}{4}x + 2$.

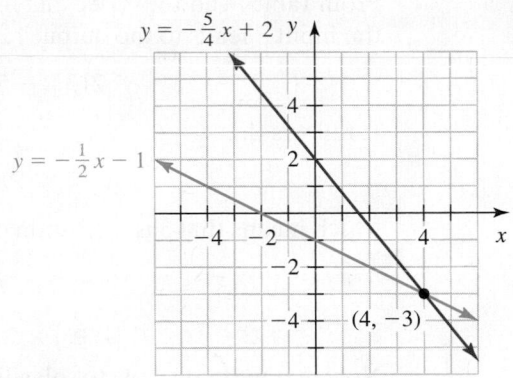

Figure 13 The intersection point is $(4, -3)$

Solution

The two lines intersect only at $(4, -3)$, whose x-coordinate is 4. So, 4 is the solution of the equation $-\dfrac{1}{2}x - 1 = -\dfrac{5}{4}x + 2$.

▶

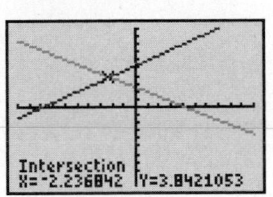

Figure 14 Using "intersect" to solve an equation in one variable

▶ **Example 7**   Solving an Equation in One Variable by Graphing

Use "intersect" on a TI-84 to solve the equation $-\dfrac{3}{5}x + \dfrac{5}{2} = \dfrac{2}{3}x + \dfrac{16}{3}$, with the solution rounded to the second decimal place.

Solution

We use a TI-84 to graph the equations $y = -\dfrac{3}{5}x + \dfrac{5}{2}$ and $y = \dfrac{2}{3}x + \dfrac{16}{3}$ on the same coordinate system and then use "intersect" to find the approximate intersection point, $(-2.24, 3.84)$. See Fig. 14. See Appendix A.24 for TI-84 instructions.

The approximate intersection point $(-2.24, 3.84)$ has x-coordinate -2.24. So, the approximate solution of the original equation is -2.24.

▶

Using Tables to Solve an Equation in One Variable

Recall from Section 8.2 that we can also solve equations by using tables.

▶ **Example 8**   Solving an Equation in One Variable by Using Tables

The solutions of $y = 2x - 5$ and $y = -4x + 13$ are shown in Tables 4 and 5, respectively. Use the tables to solve the equation $2x - 5 = -4x + 13$.

Table 4 Solutions of $y = 2x - 5$	
x	y
0	-5
1	-3
2	-1
3	1
4	3

Table 5 Solutions of $y = -4x + 13$	
x	y
0	13
1	9
2	5
3	1
4	-3

Solution

From Tables 4 and 5, we see that for both of the equations $y = 2x - 5$ and $y = -4x + 13$, the input 3 leads to the output 1:

$$2(3) - 5 = 1 \quad \text{and} \quad -4(3) + 13 = 1$$

It follows that

$$2(3) - 5 = -4(3) + 13$$

which means that 3 is a solution of the equation $2x - 5 = -4x + 13$.

Finding the Input of a Linear Function for a Given Output

Now that we know how to solve linear equations in one variable, we can use an equation of a function to find the input for a given output.

▶ **Example 9**   Using an Equation to Find an Output and an Input

Let $f(x) = \dfrac{3}{2}x - 1$.

1. Find $f(4)$. **2.** Find x when $f(x) = -4$.

Solution

Figure 15 Putting table into "Ask" mode

1.
$$f(4) = \frac{3}{2}(4) - 1 \quad \textit{Substitute 4 for x.}$$
$$= 6 - 1 \qquad \frac{3}{2}(4) = \frac{3}{2} \cdot \frac{4}{1} = 6$$
$$= 5 \qquad \textit{Subtract.}$$

2. We substitute -4 for $f(x)$ in $f(x) = \dfrac{3}{2}x - 1$ and solve for x:

$$-4 = \frac{3}{2}x - 1 \qquad \textit{Substitute −4 for f(x).}$$

$$2 \cdot (-4) = 2 \cdot \left(\frac{3}{2}x - 1 \right) \qquad \textit{Multiply both sides by LCD, 2.}$$

$$2 \cdot (-4) = 2 \cdot \frac{3}{2}x - 2 \cdot 1 \qquad \textit{Distributive law.}$$

$$-8 = 3x - 2 \qquad \textit{Multiply; simplify.}$$

$$-6 = 3x \qquad \textit{Add 2 to both sides.}$$

$$-2 = x \qquad \textit{Divide both sides by 3.}$$

Figure 16 Verify the work

We can verify our work in Problems 1 and 2 by putting a TI-84 table into Ask mode (see Figs. 15 and 16). For TI-84 instructions, see Appendix A.23.

WARNING   Example 9 asks for both a value of *y* (in Problem 1) and a value of *x* (in Problem 2). Be sure you know which value you need to find. When you are asked for $f(x)$, what you are looking for is a value of *y*, not a value of *x*.

Making Estimates and Predictions by Solving Equations

Suppose there is an association between two variables. In statistics, we often predict values for the response variable, but we do not make predictions for the explanatory variable. That is, we do not substitute a value for the response variable and solve for the explanatory variable. However, we often do this in algebra. We will make such a prediction (estimate) in Example 10.

▶ Example 10   Using a Model to Make Estimates

The number of deportations from the United States was 246 thousand in 2005, and it increased by about 24.3 thousand people per year until 2013 (Source: *U.S. Department of Homeland Security*). Let *n* be the number (in thousands) of deportations in the year that is *t* years since 2005.

1. Find a model of the situation.
2. Estimate the number of deportations in 2013.
3. Estimate in which year 400 thousand people were deported.

Solution

1. Because the number of deportations increased by about a *constant* 24.3 thousand people per year, the variables *t* and *n* are linearly associated. We want an equation of the form $n = mt + b$. Because the slope is 24.3 and the *n*-intercept is $(0, 246)$, a reasonable model is

$$n = 24.3t + 246$$

2. We substitute 8 for *t* in the equation $n = 24.3t + 246$ and solve for *n*:

$$n = 24.3(8) + 246 = 440.4$$

About 440.4 thousand people were deported in 2013, according to the model.

3. We substitute 400 for *n* in the equation $n = 24.3t + 246$ and solve for *t*:

$400 = 24.3t + 246$	Substitute 400 for *n*.
$400 - 246 = 24.3t + 246 - 246$	Subtract 246 from both sides.
$154 = 24.3t$	Combine like terms.
$\dfrac{154}{24.3} = \dfrac{24.3t}{24.3}$	Divide both sides by 24.3.
$6.34 \approx t$	

In $2005 + 6.34 \approx 2011$, 400 thousand people were deported, according to the model. We verify our work in Problems 2 and 3 by using a TI-84 table (see Fig. 17).

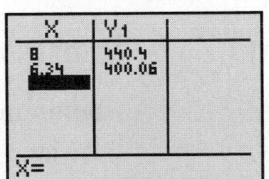

Figure 17 Verify the work

In Problem 2 of Example 10, we made an estimate about the response variable *n*, and in Problem 3 we made an estimate about the explanatory variable *t*. Here we summarize how to use an equation of a model to make such estimates and predictions.

> **Using an Equation of a Model to Make Predictions**

- When making a prediction about the response variable of a linear model, substitute a chosen value for the explanatory variable in the model. Then solve for the response variable.

- When making a prediction about the explanatory variable of a linear model, substitute a chosen value for the response variable in the model. Then solve for the explanatory variable.

Recall that in statistics, we don't usually find a value of the explanatory variable for a given value of the response variable, but we certainly do this in algebra. So, to practice algebra, we will continue to find values of the explanatory variable in the rest of the course.

> **Example 11**   Using a Model to Make an Estimate

The sales of musical instruments were \$6.39 million in 2010, and they decreased by about \$0.28 million per year until 2014 (Source: *National Arts Index*). Estimate in which year sales were \$5.3 million.

Solution

We let s be the annual sales (in millions of dollars) at t years since 2010. Because the sales decreased by about a *constant* \$0.28 million per year, there is a linear association between t and s. We want an equation of the form $s = mt + b$. Because the slope is -0.28 and the s-intercept is $(0, 6.39)$, a reasonable model is

$$s = -0.28t + 6.39$$

To find when the annual sales were \$5.3 million, we substitute 5.3 for s in the equation $s = -0.28t + 6.39$ and solve for t:

$$
\begin{aligned}
5.3 &= -0.28t + 6.39 &&\text{Substitute 5.3 for s.} \\
5.3 - 6.39 &= -0.28t + 6.39 - 6.39 &&\text{Subtract 6.39 from both sides.} \\
-1.09 &= -0.28t &&\text{Combine like terms.} \\
\frac{-1.09}{-0.28} &= \frac{-0.28t}{-0.28} &&\text{Divide both sides by } -0.28. \\
3.89 &\approx t
\end{aligned}
$$

In $2010 + 3.89 \approx 2014$, sales of musical instruments were \$5.3 million, according to the model.

> **Example 12**   Solving a Percentage Problem

In 2014, worldwide sales of digital cameras were 43.4 million cameras, down 31% from 2013 (Source: *Camera & Imaging Products Association*). What were the sales in 2013?

Solution

We let s be the worldwide sales (in millions of digital cameras) in 2013. Because the sales decreased by 31%, we write

Millions of cameras sold in 2013		Decrease in sales (in millions of cameras)		Millions of cameras sold in 2014
s	$-$	$0.31s$	$=$	43.4

Now we solve the equation:

$$s - 0.31s = 43.4$$

$$0.69s = 43.4 \quad \text{Combine like terms.}$$

$$\frac{0.69s}{0.69} = \frac{43.4}{0.69} \quad \text{Divide both sides by 0.69.}$$

$$s \approx 62.90$$

Worldwide sales in 2013 were about 62.90 million digital cameras.

▶ **Example 13** Interpreting Function Notation When Making Predictions

Many appliances such as televisions and electronic gadgets such as cell phones contain rare Earth metals. Annual U.S. electronics and appliance store sales and annual worldwide production of rare Earth metals are shown in Table 6 for the period 1992–2012.

Table 6 Annual Electronics and Appliance Stores Sales and Annual Rare-Earth Productions

Annual Electronics and Appliance Store Sales (billions of dollars)	Annual Worldwide Production of Rare Earth Metals (thousand metric tons)	Annual Electronics and Appliance Store Sales (billions of dollars)	Annual Worldwide Production of Rare Earth Metals (thousand metric tons)
42.631	50.1	86.689	97.1
48.614	46.7	94.416	102
57.266	55.1	101.340	122
64.770	74.3	107.989	137
68.363	79.7	110.673	124
70.061	68.3	108.663	132
74.527	77.1	98.030	135
78.977	86.6	99.128	126
82.206	90.9	100.792	113
80.240	94.5	102.998	110
83.740	98.2		

Source: *U.S. Geological Survey*

Let $p = f(s)$ be the annual worldwide production (in thousands of metric tons) of rare Earth metals, and let s be the annual U.S. electronics and appliance store sales (in billions of dollars) in the same year.

1. Construct a scatterplot.
2. Describe the four characteristics of the association.
3. Graph the model $f(s) = 1.34s - 16.31$ on the scatterplot. Does it come close to the data points?
4. Find $f(100)$. What does it mean in this situation?
5. Find s when $f(s) = 100$. What does it mean in this situation?

Solution

1. We construct a scatterplot in Fig. 18.

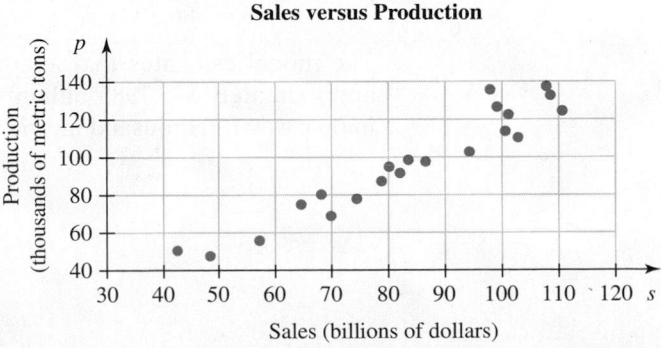

Sales versus Production

Figure 18 Sales versus Production

2. There are no outliers, and the association is linear and positive. The value of r is approximately 0.96 (see Fig. 19). On the basis of the scatterplot and r being so close to 1, we conclude the association is strong.

Simple linear regression results:
Dependent Variable: Production
Independent Variable: Sales
Production $= -16.30848 + 1.3404805$ Sales
Sample size: 21
R (correlation coefficient) $= 0.96074952$
R-sq $= 0.92303964$
Estimate of error standard deviation: 7.9420448

Figure 19 Sales versus Production

3. Figure 20 shows that the model comes close to the data points.

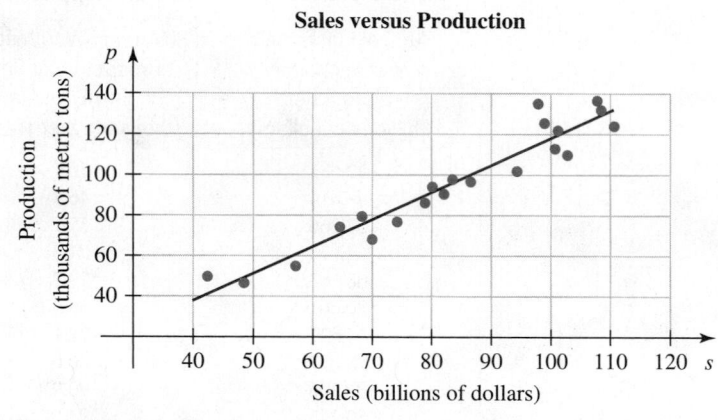

Sales versus Production

Figure 20 Rare Earth model

4. We substitute 100 for s in the equation $f(s) = 1.34s - 16.31$:

$$f(100) = 1.34(100) - 16.31 = 117.69$$

The model estimates that annual worldwide production of rare Earth metals was 117.69 thousand metric tons when annual electronics and appliance store sales were $100 billion.

5. We substitute 100 for $f(s)$ in the equation $f(s) = 1.34s - 16.31$ and solve for s:

$$100 = 1.34s - 16.31 \qquad \text{Substitute 100 for f(s).}$$
$$100 + 16.31 = 1.34s - 16.31 + 16.31 \qquad \text{Add 16.31 to both sides.}$$
$$116.31 = 1.34s \qquad \text{Combine like terms.}$$
$$\frac{116.31}{1.34} = \frac{1.34s}{1.34} \qquad \text{Divide both sides by 1.34.}$$
$$86.7985 \approx s \qquad \text{Simplify.}$$

The model estimates that annual U.S. electronics and appliance store sales were approximately $86.7985 billion when annual worldwide production of rare Earth metals was 100 thousand metric tons.

Group Exploration

Any linear equation in one variable has exactly one solution.

1. Solve $7x + 5 = 0$.

2. Solve $4x + 3 = 0$.

3. Solve $5x + 2 = 0$.

4. Solve $mx + b = 0$, where m and b are constants and $m \neq 0$. [**Hint:** Perform steps similar to those you followed for Problems 1–3.]

5. Describe a linear equation in one variable.

6. Does using the addition property of equality or the multiplication property of equality change an equation's solution set? Explain.

7. Explain why your responses to Problems 4–6 show that any linear equation in one variable has exactly one solution.

8. Use your result from Problem 4 to solve $8x + 3 = 0$ in one step.

▶ **Tips for Success**   **Plan for the Final Exam**

Don't wait until the last minute to begin studying for your final exam. Look at your finals schedule and decide how you will allocate your time to prepare for each final.

It is important that you are well rested so you can fully concentrate during your final exam. Plan to do some fun activities that involve exercise—a great way to neutralize stress.

Homework 8.3

For extra help ▶ **MyMathLab®** 📹 Watch the videos in MyMathLab 🔵 Download the MyDashboard App

1. *True or False:* A key step in solving an equation that contains fractions is to multiply both sides of the equation by the LCD so that there are no fractions on either side of the equation.

2. *True or False:* The result of simplifying a linear expression in one variable is an equation in one variable.

3. In simplifying an expression, the only number that we can multiply the expression or part of it by is ____.

4. *True or False:* When making a prediction about the response variable of a linear model, we substitute a chosen value for the explanatory variable in the model. Then we solve for the response variable.

Solve.

5. $2(x + 3) = 5x - 3$

6. $-3(x - 4) = 2x + 2$

7. $1 - 3(5b - 2) = 4 - (7b + 3)$

8. $3 - 4(3p + 2) = 7 - (9p - 1)$

9. $3(4x - 5) - (2x + 3) = 2(x - 4)$

10. $-2(5x + 3) - (4x - 1) = 5(x + 2)$

11. $\dfrac{x}{2} - \dfrac{3}{4} = \dfrac{1}{2}$

12. $\dfrac{x}{9} - \dfrac{1}{3} = \dfrac{2}{9}$

13. $\dfrac{5x}{6} + \dfrac{2}{3} = 2$

14. $\dfrac{3x}{8} - \dfrac{1}{2} = 1$

15. $\dfrac{5}{6}k = \dfrac{3}{4}k + \dfrac{1}{2}$

16. $\dfrac{3}{8}t = \dfrac{5}{6}t - \dfrac{1}{4}$

17. $\dfrac{7}{12}x - \dfrac{5}{3} = \dfrac{7}{4} + \dfrac{5}{6}x$

18. $\dfrac{5}{4} + \dfrac{9}{2}x = \dfrac{3}{8}x - \dfrac{1}{4}$

19. $\dfrac{4}{3}x - 2 = 3x + \dfrac{5}{2}$

20. $\dfrac{2}{5}x - 4 = 2x - \dfrac{3}{4}$

21. $\dfrac{3(x - 4)}{5} = -2x$

22. $\dfrac{5(x - 2)}{3} = -4x$

23. $\dfrac{4x + 3}{5} = \dfrac{2x - 1}{3}$

24. $\dfrac{3x + 2}{2} = \dfrac{6x - 3}{5}$

25. $\dfrac{4m - 5}{2} - \dfrac{3m + 1}{3} = \dfrac{5}{6}$

26. $\dfrac{2p + 4}{3} - \dfrac{5p - 7}{6} = \dfrac{11}{12}$

For Exercises 27–36, solve. Round the result to the second decimal place.

27. $0.3x + 0.2 = 0.7$

28. $0.6x - 0.1 = 0.4$

29. $5.27x - 6.35 = 2.71x + 9.89$

30. $8.25x - 17.56 + 4.38x = 25.86$

31. $0.4x - 1.6(2.5 - x) = 3.1(x - 5.4) - 11.3$

32. $3.2x + 0.5(7.3 - x) = 4.7 - 6.4(x - 2.1)$

33. $1.960 = \dfrac{x - 25.9}{7.3}$

34. $1.645 = \dfrac{x - 87.1}{12.7}$

35. $-1.282 = \dfrac{x - 9.56}{2.44}$

36. $-2.326 = \dfrac{x - 50.92}{8.39}$

For Exercises 37–44, simplify the expression or solve the equation, as appropriate.

37. $5x - 4 - 3x + 16$

38. $3(x - 2) - (7x + 2) - 4(3x + 1)$

39. $5x - 4 = 3x + 16$

40. $3(x - 2) - (7x + 2) = 4(3x + 1)$

41. $\dfrac{2x}{3} - \dfrac{5x}{2} - 1 = 0$

42. $\dfrac{7}{2}x - \dfrac{5}{6} = \dfrac{1}{3} + \dfrac{3}{4}x$

43. $\dfrac{2x}{3} - \dfrac{5x}{2} - 1$

44. $\dfrac{7}{2}x - \dfrac{5}{6} - \dfrac{1}{3} + \dfrac{3}{4}x$

For Exercises 45–50, solve the given equation by referring to the graphs of $y = -\dfrac{3}{2}x + 2$ and $y = \dfrac{1}{2}x - 2$ shown in Fig. 21.

45. $-\dfrac{3}{2}x + 2 = \dfrac{1}{2}x - 2$ **46.** $-\dfrac{3}{2}x + 2 = -4$

47. $-\dfrac{3}{2}x + 2 = 5$ **48.** $\dfrac{1}{2}x - 2 = -3$

49. $\dfrac{1}{2}x - 2 = 0$ **50.** $-\dfrac{3}{2}x + 2 = 2$

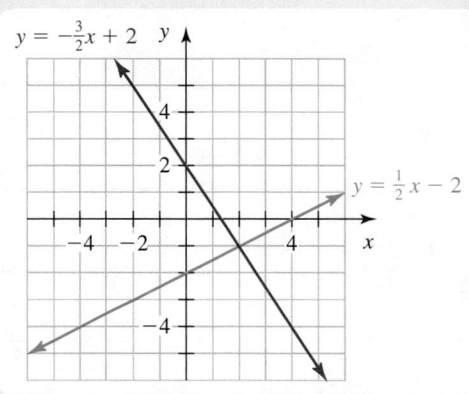

Figure 21 Exercises 45–50

For Exercises 51–56, solve the given equation by referring to the graphs of $y = -\dfrac{1}{3}x + \dfrac{5}{3}$ and $y = \dfrac{3}{2}x + \dfrac{7}{2}$ shown in Fig. 22.

51. $-\dfrac{1}{3}x + \dfrac{5}{3} = \dfrac{3}{2}x + \dfrac{7}{2}$

52. $-\dfrac{1}{3}x + \dfrac{5}{3} = 3$

53. $-\dfrac{1}{3}x + \dfrac{5}{3} = 1$

54. $\dfrac{3}{2}x + \dfrac{7}{2} = -4$

55. $\dfrac{3}{2}x + \dfrac{7}{2} = -1$

56. $-\dfrac{1}{3}x + \dfrac{5}{3} = 0$

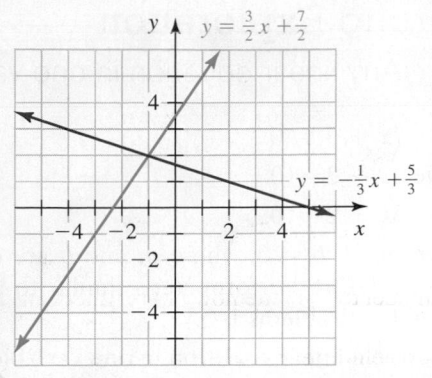

Figure 22 Exercises 51–56

Use "intersect" on a TI-84 to solve the equation. Round the solution to the second decimal place.

57. $-4x + 8 = 2x - 9$

58. $-2x - 7 = x + 1$

59. $2.5x - 6.4 = -1.7x + 8.1$

60. $-1.5x - 9.3 = 3.1x + 2.1$

61. $\dfrac{23}{75}x - \dfrac{99}{38} = -\dfrac{52}{89}x - \dfrac{67}{9}$

62. $-\dfrac{54}{35}x + \dfrac{26}{21} = \dfrac{67}{95}x - \dfrac{72}{31}$

For Exercises 63–66, solve the given equation by referring to the solutions of $y = -3x + 7$ and $y = 5x + 15$ shown in Tables 7 and 8, respectively.

63. $-3x + 7 = 5x + 15$ **64.** $-3x + 7 = 4$

65. $5x + 15 = 5$ **66.** $5x + 15 = 25$

Table 7 Solutions of $y = -3x + 7$		**Table 8** Solutions of $y = 5x + 15$	
x	y	x	y
-2	13	-2	5
-1	10	-1	10
0	7	0	15
1	4	1	20
2	1	2	25

For Exercises 67–72, let $f(x) = -3x + 7$.

67. Find $f(6)$.   **68.** Find $f(2)$.

69. Find x when $f(x) = 6$.   **70.** Find x when $f(x) = 2$.

71. Find x when $f(x) = \dfrac{5}{2}$.

72. Find x when $f(x) = -\dfrac{4}{3}$.

For Exercises 73–76, let $f(x) = -5.85x + 183.22$. Round any results to the second decimal place.

73. Find $f(10.91)$.   **74.** Find $f(17.28)$.

75. Find x when $f(x) = 99.34$.   **76.** Find x when $f(x) = 72.06$.

77. The number of Americans without health insurance was 46.7 million in 2010, and it increased by about 1.04 million per year until 2013 (Source: *Centers for Disease Control and Prevention*). Let n be the number (in millions) of Americans without health insurance at t years since 2010.

   a. Find an equation of a model.

   b. Estimate when 49 million Americans did not have health insurance.

   c. A new law went into effect in 2014 that requires people to get health insurance or pay a penalty if they don't. Use the model to estimate how many Americans in 2014 would have been uninsured if the law hadn't been passed. Do you have much faith in your estimate? Explain.

78. The minimum salary for a hockey player in the National Hockey League was $500 thousand in 2010, and it increased by about $14 thousand per year until 2015 (Source: *National Hockey League Players Association*). Let s be the minimum salary (in thousands of dollars) at t years since 2010.

   a. Find an equation of a model.

   b. Estimate the minimum salary in 2015.

   c. Estimate when the minimum salary was $550 thousand.

79. Martha Stewart went to prison for selling nearly 4000 shares of ImClone Systems stock because she had inside information that the company was about to announce bad news that would result in the stock's value plummeting. Surprisingly, the value of a share of Martha Stewart Living Omnimedia stock increased by about $3.36 per month after she was sentenced. The value of a share of her stock was $8.16 just before she was sentenced (Source: *Bloomberg Financial Markets*). Let v be the value (in dollars) of a share of the stock at t months since Stewart was sentenced.

   a. Find an equation of a model to describe the situation.

   b. Estimate the value of a share of the stock 3 months after the sentencing, which is about when Stewart reported to prison.

   c. Estimate when the value of a share of the stock reached $28.32.

80. The number of Radio Shack stores was 5.74 thousand in 2010, and it decreased by about 0.21 thousand stores per year until 2013 (Source: *Radio Shack*). Let n be the number (in thousands) of Radio Shack stores at t years since 2010.

   a. Find an equation of a model.

   b. Estimate when there were 5.1 thousand Radio Shack stores.

   c. Estimate the number of Radio Shack stores in 2012.

   d. In 2014, there were a large number of Radio Shack store closures, leaving just 3.96 thousand stores. Use the model to help you estimate the number of closures that happened in addition to the usual number of closures per year.

81. There are many Native American tribes that are not recognized by the U.S. government, which means they cannot seek the same health and educational opportunities as federal tribes. Because of difficult requirements to achieve recognition, many tribes have never completed the process. The number of tribes that sought recognition was 11 in 2005, and it decreased by about 1.2 tribes per year until 2014 (Source: *Bureau of Indian Affairs*). Let $f(t)$ be the number of tribes who seek recognition in the year that is t years since 2005.

   a. Find an equation of f.

   b. Find $f(8)$. What does it mean in this situation?

   c. Find t when $f(t) = 8$. What does it mean in this situation?

82. General Motors (GM) struggled from 2005 to 2012. In fact, GMs share of new-vehicle sales was 25.59% in 2005, and it decreased by about 1.15 percentage points per year until 2012 (Source: *GM*). Let $f(t)$ be GM's share of new-vehicle sales at t years since 2005.

   a. Find an equation of f.

   b. Find $f(7)$. What does it mean in this situation?

   c. Find t when $f(t) = 20$. What does it mean in this situation?

83. In 2010, the percentage of private-sector workers who were in a union was 6.95%, and it decreased by about 0.25 percentage point per year until 2014 (Source: *Bloomberg BNA*).

   a. Find an equation of a model to describe the situation. Explain what your variables represent.

   b. Estimate when the percentage was 6.20%.

   c. Estimate the percentage of private-sector workers who were *not* in a union in 2014.

84. In 2010, the total number of overnight visits to national parks was 13.91 million visits, and it decreased by about 0.21 million visits per year until 2014 (Source: *National Park Service*).

   a. Find an equation of a model to describe the situation. Explain what your variables represent.

   b. Estimate in which year there were 13.0 million visits.

85. Chicago taxis charge $2.25 plus $1.80 for each mile traveled.

   a. Find an equation of a linear model to describe the situation. Explain what your variables represent.

   b. What would be the cab fare for the 8.8-mile trip from South Side, Chicago, to Lincoln Park Zoo?

   c. If a person paid $25.65 for a cab fare in Chicago, how far was the ride?

86. Houston taxis charge $2.55 plus $2.20 for each mile traveled.

   a. Find an equation of a linear model to describe the situation. Explain what your variables represent.

   b. What would be the cab fare for the 3.5-mile trip from Minute Maid Park to the Museum of Fine Arts, Houston?

   c. If a person paid $37.75 for a cab fare in Houston, how far was the ride?

87. In 2013, the mean annual per-person consumption of butter was 5.5 pounds, up 12.2% from 2010 (Source: *USDA National Agriculture Statistics Service*). What was the mean annual per-person consumption of butter in 2010?

88. In 2013, the mean annual per-person consumption of cheese was 33.7 pounds, up 13.1% from 2000 (Source: *USDA National Agriculture Statistics Service*). What was the mean annual per-person consumption of cheese in 2000?

89. In 2013, total Visa® debit card purchases were $1.187 trillion, up 7.3% from 2012 (Source: *The Nilson Report*). What were the total Visa debit card purchases in 2012?

90. In 2013, total MasterCard® debit card purchases were $491.1 billion, up 9.5% from 2012 (Source: *The Nilson Report*). What were the total MasterCard debit card purchases in 2012?

91. In 2012, the mean office space per worker was 176 square feet, down 21.8% from 2010 (Source: *CoreNet Global*). What was the mean office space per worker in 2010?

92. In 2013, worldwide revenue for Ivory® bar soap was $94 million, down 10.5% from 2010 (Source: *Euromonitor*). What was the worldwide revenue in 2010?

93. In 2014, the number of U.S. bank failures was 18 failures, down 88.5% from 2010 (Source: *Federal Deposit Insurance Corporation*). What was the number of U.S. bank failures in 2010?

94. The mean ratio of students per teacher in U.S. public schools was 15.1 students per teacher in 2013, down 5.6% from 2000 (Source: *National Center for Education Statistics*). What was the mean ratio of students per teacher in 2000?

95. **DATA.** The mean numeric grades given by full-time and part-time faculty at Siena College are shown in Table 9 for various years.

Table 9 Mean Numeric Grades Given by Full-Time and Part-Time Faculty at Siena College

Year	Mean Numeric Grade Given by Faculty (points)	
	Full Time	Part Time
2006	2.96	3.23
2007	3.00	3.22
2008	3.01	3.19
2009	3.04	3.18
2010	3.03	3.18
2011	3.06	3.20
2012	3.10	3.20

Source: *Siena College*

Let F and P be the mean numeric grades (in points) given by full-time faculty and part-time faculty, respectively, both at t years since 2000.

a. Construct a scatterplot that describes the association between t and F. Do the same for the association between t and P. Show both scatterplots in the same coordinate system if you know how to use colors or shapes to identify which points are for which type of faculty.

b. Graph the model $F = 0.02t + 2.85$ on the scatterplot for full-time faculty. Does it comes close to the data points?

c. Find the mean of the values of P shown in Table 9.

d. Use the model for full-time faculty to predict when the mean numeric grade they give will equal the result you found in part (c). What does that mean in this situation? Do you have much faith in your prediction? Explain.

e. Describe at least two possible lurking variables for the full-time-faculty association.

"According to my calculations, if we wait 20 years to have a child, we'll be guaranteed he or she will get straight As in high school!"

96. **DATA.** The percentages of college freshmen whose average grade in high school was an A are shown in Table 10 for various years.

Table 10 College Freshmen Whose Average Grade in High School Was an A

Year	Percent
1970	19.6
1980	26.6
1985	28.7
1990	29.4
1995	36.1
2000	42.9
2005	46.6
2010	48.4

Source: *Higher Education Research Institute*

Let p be the percentage of college freshmen whose average grade in high school was an A at t years since 1970.

a. Construct a scatterplot.

b. Describe the four characteristics of the association. Compute and interpret r as part of your analysis.

c. A model of the situation is $p = 0.76t + 18.06$. Graph the model on the scatterplot. Does it come close to the data points?

d. Estimate when 44% of all college freshmen earned an average grade of A in high school.

e. Describe at least two possible lurking variables.

97. **DATA.** American child death rates (per 100,000 children ages 5–14 years) are shown in Table 11 for various years.

Table 11 U.S. Child Death Rates

Year	Child Death Rate (number of deaths per 100,000 children)
1980	30.6
1990	24.0
1995	22.5
2000	18.0
2005	16.3
2010	12.9
2013	13.0

Source: *Child Trends Databank*

Let $C = f(t)$ be the U.S. child death rate (per 100,000 children) at t years since 1980.

a. Graph the model $f(t) = -0.55t + 30.09$ on a scatterplot. Does it come close to the data points?

b. What is the C-intercept? What does it mean in this situation?

c. Use f to estimate the U.S. child death rate in 2012.

d. Use f to estimate when the child death rate was 15 deaths per 100,000 children.

e. Approximately one-third of all child deaths are due to unintentional injuries (Source: *Child Trends Databank*). Estimate the child death rate due to unintentional injuries in 2011.

98. **DATA.** Banks charge an *overdraft fee* for writing a check for an amount that is more than the balance in the account. The mean overdraft fees are shown in Table 12 for various years.

Table 12 Mean Overdraft Fees

Year	Mean Overdraft Fee (dollars)
2004	25.81
2006	27.40
2008	28.95
2010	30.47
2012	31.26
2014	32.74

Source: *Bankrate.com*

Let $F = g(t)$ be the mean overdraft fee (in dollars) at t years since 2000.

a. Graph the model $g(t) = 0.68t + 23.30$ on a scatterplot. Does it come close to the data points?

b. What is the slope? What does it mean in this situation?

c. What is the F-intercept? What does it mean in this situation? Do you have much faith in your estimate?

d. Use g to estimate when the mean overdraft fee was $30.

e. Because of inflation, $1 in 2005 had the same buying power as $1.19 in 2013. Use g to determine whether the mean overdraft fee has kept pace with, fallen behind, or gone beyond inflation. Is this good news, bad news, or neutral news for bank customers?

99. ^{DATA} The mean selling prices of a home sold in San Bruno, California, are shown in Table 13 for various square footages.

Table 13 Mean Selling Prices of Homes

Number of Square Feet	Number Used to Represent Square Feet	Mean Selling Price (thousands of dollars)
500–1000	750	632
1001–1500	1250	733
1501–2000	1750	814
2001–2500	2250	894
2501–3000	2750	1025

Source: *Green Banker*

Let p be the mean selling price (in thousands of dollars) of a home measuring s square feet.

a. Construct a scatterplot.

b. Describe the four characteristics of the association. Compute and interpret r as part of your analysis.

c. Graph the model $p = 0.19s + 488.15$ on the scatterplot. Does it come close to the data points?

d. What is the slope? What does it mean in this situation?

e. Use the model to estimate the square footage for which the mean selling price is $950 thousand.

100. ^{DATA} The percentages of American adults who have earned an annual salary over $200,000 at least once in their lives are shown in Table 14 for various years. Let p be the percentage of American adults at age a years who have earned an annual salary over $200,000 at least once in their lives.

a. Construct a scatterplot.

b. Is the association positive, negative, or neither? What does that mean in this situation? Why does that make sense?

Table 14 Percentage of American Adults Who Have Earned over $200,000 at Least Once

Age	Percent
25	1
30	4
35	8
40	13
45	18
50	24
55	28
60	32

Source: *Chasing the American Dream, Thomas A. Hirschl*

c. Graph the model $p = 0.93a - 23.46$ on the scatterplot. Does it come close to the data points?

d. Use the model to estimate the percentage of 32-year-old Americans who have earned an annual salary over $200,000 at least once.

e. Use the model to predict at what age one-fourth of American adults have earned an annual salary over $200,000 at least once.

101. ^{DATA} The percentages of elementary school children who receive free or reduced-fee lunches and the percentages of bicycle riders who wear helmets are shown in Table 15 for 13 Californian neighborhoods.

Table 15 Children Who Receive Free or Reduced-Fee School Lunches and Bicycle Riders Who Wear Helmets

Neighborhood	Percent Receiving Free or Reduced-Fee Lunches	Percent Wearing Bicycle Helmets
Fair Oaks	50	22.1
Strandwood	11	35.9
Walnut Acres	2	57.9
Discovery Bay	19	22.2
Belshaw	26	42.4
Kennedy	73	5.8
Cassell	81	3.6
Miner	51	21.4
Sedgewick	11	55.2
Sakamoto	2	33.3
Toyon	19	32.4
Lietz	25	38.4
Los Arboles	84	46.6

Source: *Perales, D., and Gerstman, B. B. (March 27–29, 1995). A Bi-County Comparative Study of Bicycle Helmet Knowledge and Use by California Elementary School Children*

Let L be the percentage of children who receive free or reduced-fee school lunches, and let $H = f(L)$ be the percentage of bicycle riders who wear helmets.

a. Construct a scatterplot, treating the percentage of children who receive reduced-fee or free school lunches as the explanatory variable.

b. Graph the model $f(L) = -0.33L + 43.64$ on the scatterplot. Does it come close to most of the data points? Should it come close to the point for Los Arboles? Explain.

c. Find $f(40)$. What does it mean in this situation?

d. Find L when $f(L) = 40$. What does it mean in this situation?

e. Does an increase in L *cause* a decrease in H? Explain. Describe at least one possible lurking variable.

102. DATA The maximum weights and mean life expectancies of dogs are shown in Table 16 for various dog breeds.

Table 16 Dog Weights versus Mean Life Expectancies

Dog Breed	Maximum Weight (pounds)	Mean Life Expectancy (years)
Labrador Retriever	79	11
Rottweiler	132	10
German Shepherd	95	11
Dachshund	28	13
Golden Retriever	75	12
Doberman Pinscher	100	12
Beagle	25	14
French Bulldog	30	14
Yorkshire Terrier	7	14
Bulldog	55	10
Boxer	72	11
Siberian Husky	60	14
Great Dane	198	8
Chihuahua	6	16
Pomeranian	8	14

Source: *FindTheBest*

Let $L = f(W)$ be the mean life expectancy (in years) of a dog breed with maximum weight W pounds.

a. Construct a scatterplot.
b. Describe the four characteristics of the association. Compute and interpret r as part of your analysis.
c. Graph the model $f(W) = -0.035W + 14.52$ on the scatterplot. Does it come close to most of the data points?
d. Find $f(8)$. What does it mean in this situation?
e. Find W when $f(W) = 8$. What does it mean in this situation?

Concepts

103. A student tries to solve $2(x - 5) = x - 3$:

$$2(x - 5) = x - 3$$
$$2x - 10 = x - 3$$
$$2x - 10 + 10 = x - 3 + 10$$
$$2x = x + 7$$
$$\frac{2x}{2} = \frac{x + 7}{2}$$
$$x = \frac{x + 7}{2}$$

Describe any errors. Then solve the equation correctly.

104. A student tries to solve the equation $\frac{1}{2}x + 3 = \frac{5}{2}$:

$$\frac{1}{2}x + 3 = \frac{5}{2}$$
$$2 \cdot \frac{1}{2}x + 3 = 2 \cdot \frac{5}{2}$$
$$x + 3 = 5$$
$$x + 3 - 3 = 5 - 3$$
$$x = 2$$

Describe any errors. Then solve the equation correctly.

105. Two students try to solve the equation $5x + 3 = 3x + 7$:

Student A	Student B
$5x + 3 = 3x + 7$	$5x + 3 = 3x + 7$
$5x + 3 - 3x = 3x + 7 - 3x$	$5x + 3 - 3 = 3x + 7 - 3$
$2x + 3 = 7$	$5x = 3x + 4$
$2x + 3 - 3 = 7 - 3$	$5x - 3x = 3x + 4 - 3x$
$2x = 4$	$2x = 4$
$\frac{2x}{2} = \frac{4}{2}$	$\frac{2x}{2} = \frac{4}{2}$
$x = 2$	$x = 2$

Compare the methods of solving the equation. Is one method better than the other? Explain.

106. Two students try to solve the equation $2x + 4 = 0$:

Student 1	Student 2
$2x + 4 = 0$	$2x + 4 = 0$
$2x + 4 + 6 = 0 + 6$	$2x + 4 - 4 = 0 - 4$
$2x + 10 = 6$	$2x = -4$
$2x + 10 - 10 = 6 - 10$	$\frac{2x}{2} = \frac{-4}{2}$
$2x = -4$	$x = -2$
$2x \cdot 3 = -4 \cdot 3$	
$6x = -12$	
$\frac{6x}{6} = \frac{-12}{6}$	
$x = -2$	

Did either student, both students, or neither student solve the equation correctly? Explain.

107. Two students try to solve the equation $\frac{3}{4}x - \frac{5}{6} = \frac{1}{3}$:

Student A	Student B
$\frac{3}{4}x - \frac{5}{6} = \frac{1}{3}$	$\frac{3}{4}x - \frac{5}{6} = \frac{1}{3}$
$\frac{3}{4}x = \frac{1}{3} + \frac{5}{6}$	$12 \cdot \left(\frac{3}{4}x - \frac{5}{6}\right) = 12 \cdot \frac{1}{3}$
$\frac{3}{3} \cdot \frac{3}{4}x = \frac{4}{4} \cdot \frac{1}{3} + \frac{2}{2} \cdot \frac{5}{6}$	$12 \cdot \frac{3}{4}x - 12 \cdot \frac{5}{6} = 4$
$\frac{9}{12}x = \frac{4}{12} + \frac{10}{12}$	$9x - 10 = 4$
$\frac{9}{12}x = \frac{14}{12}$	$9x = 14$
$\frac{3}{4}x = \frac{7}{6}$	$x = \frac{14}{9}$
$\frac{4}{3} \cdot \frac{3}{4}x = \frac{4}{3} \cdot \frac{7}{6}$	
$1x = \frac{2 \cdot 2 \cdot 7}{3 \cdot 2 \cdot 3}$	
$x = \frac{14}{9}$	

Compare the methods of solving the equation. Is one method better than the other? Explain.

108. A student tries to simplify the expression $\frac{1}{4}x + \frac{1}{3}x$:

$$\frac{1}{4}x + \frac{1}{3}x = 12 \cdot \left(\frac{1}{4}x + \frac{1}{3}x\right)$$

$$= 12 \cdot \frac{1}{4}x + 12 \cdot \frac{1}{3}x$$

$$= 3x + 4x$$

$$= 7x$$

Describe any errors. Then simplify the expression correctly.

109. Solve $5(x - 2) = 2x - 1$. Show that your result satisfies each of the equations in your work.

110. Solve $3(x + 1) + 2 = x + 7$. Show that your result satisfies each of the equations in your work.

111. Consider the following equations:

$$x = 3$$
$$x + 2 = 3 + 2$$
$$x + 2 = 5$$
$$3(x + 2) = 3 \cdot 5$$
$$3(x + 2) = 15$$
$$3(x + 2) + 1 = 15 + 1$$
$$3(x + 2) + 1 = 16$$

What is the solution of the equation $3(x + 2) + 1 = 16$? What is the solution of the equation $3(x + 2) = 15$? Explain how you can respond to these questions without doing any work besides the work already included in this problem.

112. If an equation contains fractions, why is it helpful to multiply both sides of the equation by the LCD?

▼ 8.4 Solving Formulas

Objectives

» Find a quantity by substituting values for all but one variable in a formula and then solving for the remaining variable.

» Find a quantity by using a formula with summation notation.

» Solve a formula for one of its variables.

» Solve a formula with a square root.

» Graph a linear equation in two variables by solving for y.

Recall from Section 4.1 that a **formula** is an equation that contains two or more variables. We have worked with many formulas in this course, including formulas for the mean, the standard deviation, probabilities, and the slope of a line. In this section, we will continue to work with these formulas as well as new ones. We will solve some of these formulas for a variable that is not alone on one side of the equation.

Substituting Values for Some Variables in a Formula and Then Solving for a Variable

Recall from Section 5.2 that if E and F are disjoint events, then $P(E \text{ OR } F) = P(E) + P(F)$.

▶ **Example 1**   Substituting Values for Some Variables in a Formula and Then Solving

At Apple, 7% of employees are African American and 18% of employees are African American OR Hispanic (Source: *Apple*). Find the probability of randomly selecting an Apple employee who is Hispanic.

Solution

We use A for the event African American and H for the event Hispanic. The events are disjoint because African Americans and Hispanics have no members in common. So, we can use the formula $P(A \text{ OR } H) = P(A) + P(H)$.

We substitute 0.18 for $P(A \text{ OR } H)$ and 0.07 for $P(A)$ in the formula $P(A \text{ OR } H) = P(A) + P(H)$:

$$0.18 = 0.07 + P(H)$$

Then we solve the equation $0.18 = 0.07 + P(H)$ for $P(H)$:

$$0.18 = 0.07 + P(H)$$
$$0.18 - 0.07 = 0.07 + P(H) - 0.07 \quad \text{Subtract 0.07 from both sides.}$$
$$0.11 = P(H)$$

So, the probability of randomly selecting a Hispanic employee is 0.11.

▶

In Example 1, we substituted values for $P(A \text{ OR } H)$ and $P(A)$ in the formula $P(A \text{ OR } H) = P(A) + P(H)$ and then solved for the probability $P(H)$. In general, **to find a single value of a variable in a formula, we often substitute numbers for all of the other variables and then solve for the remaining variable.**

Recall from Section 5.5 that an observation x with z-score z is given by $x = \bar{x} + zs$.

▶ **Example 2** Substituting Values into a Formula and Then Solving

Recall that the scores on the Wechsler IQ test are normally distributed with mean 100 and standard deviation 15 points. Actress Sharon Stone is reported to have an IQ of 154 points (Source: Chicago Tribune). Use the formula $x = \bar{x} + zs$ to find the z-score for an IQ of 154 points.

Solution

To begin, we substitute 154 for x, 100 for $\bar{x}$, and 15 for s in the formula $x = \bar{x} + zs$:

$$154 = 100 + z(15)$$

Then we solve the equation $154 = 100 + 15z$ for z:

$$154 = 100 + 15z$$
$$154 - 100 = 100 + 15z - 100 \qquad \text{Subtract 100 from both sides.}$$
$$54 = 15z \qquad \text{Combine like terms.}$$
$$\frac{54}{15} = \frac{15z}{15} \qquad \text{Divide both sides by 15.}$$
$$3.6 = z$$

So, the z-score is 3.6. Recall that this means Sharon Stone's IQ is 3.6 standard deviations greater than the mean. Because 99.7% of Americans' IQs lie within three standard deviations of the mean, we conclude that Stone is extremely intelligent.

In Example 2, we found a z-score by using the formula $x = \bar{x} + zs$. But in Section 5.4, we used the formula $z = \dfrac{x - \bar{x}}{s}$ to find z-scores. Later in this section, we will show that for any given values of x, $\bar{x}$, and s, the two formulas will give the same value for the z-score.

Statistics Formulas with Summation Notation

Recall from Section 4.1 that when working with 5 observations, the summation notation Σx_i means $x_1 + x_2 + x_3 + x_4 + x_5$. In Example 3, we will work with the formula $\mu = \Sigma x_i P(x_i)$, which gives the mean for a certain type of distribution that you will study in a statistics course.

▶ **Example 3** Evaluating a Formula

Substitute the following values in the formula $\mu = \Sigma x_i P(x_i)$ and solve for the remaining variable:

$$x_1 = 0, x_2 = 1, x_3 = 2, x_4 = 3, x_5 = 4,$$
$$P(x_1) = 0.0625, P(x_2) = 0.25, P(x_3) = 0.375, \ P(x_4) = 0.25, P(x_5) = 0.0625$$

Solution

First, we write the summation for 5 terms:

$$\mu = \Sigma x_i P(x_i) \qquad \text{Original formula}$$
$$\mu = x_1 P(x_1) + x_2 P(x_2) + x_3 P(x_3) + x_4 P(x_4) + x_5 P(x_5) \qquad \text{Definition of } \Sigma$$

Next, we substitute $x_1 = 0, x_2 = 1, x_3 = 2, x_4 = 3, x_5 = 4, P(x_1) = 0.0625, P(x_2) = 0.25,$ $P(x_3) = 0.375, P(x_4) = 0.25,$ and $P(x_5) = 0.0625,$ and solve for μ:

$\mu = 0(0.0625) + 1(0.25) + 2(0.375) + 3(0.25) + 4(0.0625)$ *Substitute the given values for the variables.*

$\mu = 2$ *Compute.*

In Example 4, we will work with a more complicated formula that describes a certain type of variance that you will also study in a statistics course.

▶ Example 4   Evaluating a Formula

Substitute the following values in the formula $\text{MSE} = \dfrac{\Sigma[(n_i - 1)s_i^2]}{n - k}$ and solve for the remaining variable:

$$n_1 = 22, n_2 = 28, n_3 = 21, s_1 = 2, s_2 = 5, s_3 = 4, n = 71, \text{ and } k = 3$$

Round the result to the first decimal place.

Solution

First, we write the summation for 3 terms:

$$\text{MSE} = \frac{\Sigma[(n_i - 1)s_i^2]}{n - k} \qquad \textit{Original formula}$$

$$\text{MSE} = \frac{(n_1 - 1)s_1^2 + (n_2 - 1)s_2^2 + (n_3 - 1)s_3^2}{n - k} \qquad \textit{Definition of } \Sigma$$

Next, we substitute $n_1 = 22, n_2 = 28, n_3 = 21, s_1 = 2, s_2 = 5, s_3 = 4, n = 71,$ and $k = 3,$ and solve for MSE:

$$\text{MSE} = \frac{(22 - 1)2^2 + (28 - 1)5^2 + (21 - 1)4^2}{71 - 3} \qquad \textit{Substitute the given values for the variables.}$$

$$\text{MSE} = \frac{1079}{68} \qquad \textit{Compute.}$$

$$\text{MSE} \approx 15.9 \qquad \textit{Compute.}$$

Solving a Formula for a Variable

In each of Examples 1 and 2, we substituted values for all but one variable in a formula and then solved the equation for the remaining variable. Sometimes it will be better to reverse the two steps. That is, we can first solve a formula for a variable and then substitute values for the other variables.

▶ Example 5   Solving a Formula for a Variable

1. Solve the formula $P(A \text{ OR } H) = P(A) + P(H)$ for $P(H)$.
2. Substitute 0.18 for $P(A \text{ OR } H)$ and 0.07 for $P(A)$ in the formula found in Problem 1.

Solution

1. We can solve the formula $P(A \text{ OR } H) = P(A) + P(H)$ for $P(H)$ in much the same way that we solved the equation $0.18 = 0.07 + P(H)$ in Example 1:

$0.18 = 0.07 + P(H)$	$P(A \text{ OR } H) = P(A) + P(H)$	*Addition for disjoint events*
$0.18 - 0.07 = 0.07 + P(H) - 0.07$	$P(A \text{ OR } H) - P(A) = P(A) + P(H) - P(A)$	*Subtract P(A) from both sides.*
$0.11 = P(H)$	$P(A \text{ OR } H) - P(A) = P(H)$	*Combine like terms.*

The result is $P(A \text{ OR } H) - P(A) = P(H)$. By switching the sides of the equation, we have $P(H) = P(A \text{ OR } H) - P(A)$.

2. We substitute 0.18 for $P(A \text{ OR } H)$ and 0.07 for $P(A)$ in the formula $P(H) = P(A \text{ OR } H) - P(A)$:

$$P(H) = 0.18 - 0.07 = 0.11$$

This is the same result we found in Example 1.

▶

In Problem 1 of Example 5, the result of solving the formula $P(A \text{ OR } H) = P(A) + P(H)$ for $P(H)$ is the formula $P(H) = P(A \text{ OR } H) - P(A)$. This means that the formulas give the same result when we substitute values for all but one probability and solve for the remaining probability. For example, we found the same value (0.11) for $P(H)$ in Example 1 and Problem 2 of Example 5. In general, **solving for a variable in a formula will not change the association between the variables in the formula.**

▶ Example 6   Solving a Formula for a Variable

1. Solve the formula $x = \bar{x} + zs$ for z.
2. For students who graduated from high school in 2013, their SAT math scores were approximately normally distributed with a mean of 514 points and a standard deviation of 118 points (Source: *College Board*). Use the formula $x = \bar{x} + zs$ or the formula you found in Problem 1 to find the z-scores for the following students' SAT scores (all in points): 621, 457, 555, 748, 242, and 562.
3. Which of the students' scores given in Problem 2 is most unusual? Explain.

Solution

1. We can solve the formula $x = \bar{x} + zs$ for z in much the same way that we solved the equation $154 = 100 + 15z$ in Example 2:

$154 = 100 + 15z$	$x = \bar{x} + zs$ *z-score formula*
$154 - 100 = 100 + 15z - 100$	$x - \bar{x} = \bar{x} + zs - \bar{x}$ *Subtract $\bar{x}$ from both sides.*
$54 = 15z$	$x - \bar{x} = zs$ *Combine like terms.*
$\dfrac{54}{15} = \dfrac{15z}{15}$	$\dfrac{x - \bar{x}}{s} = \dfrac{zs}{s}$ *Divide both sides by s.*
$3.6 = z$	$\dfrac{x - \bar{x}}{s} = z$ *Simplify.*

By switching the sides of the equation $\dfrac{x - \bar{x}}{s} = z$, we have $z = \dfrac{x - \bar{x}}{s}$, which is the formula we used in Section 5.4 to find z-scores.

2. To find the z-scores, we will use the formula $z = \dfrac{x - \bar{x}}{s}$ because z is alone on one side of the equation. First, we substitute 514 for $\bar{x}$ and 118 for s:

$$z = \frac{x - 514}{118}$$

Then we use a TI-84 to substitute $621, 457, 555, 748, 242,$ and 562 for x in the formula $z = \dfrac{x - 514}{118}$ (see Fig. 23). By viewing the second column of the TI-84 table, we see that the values of z are about 0.91, −0.48, 0.35, 1.98, −2.31, and 0.41. So, the SAT scores (in points) 621, 457, 555, 748, 242, and 562 have approximate z-scores 0.91, −0.48, 0.35, 1.98, −2.31, and 0.41, respectively.

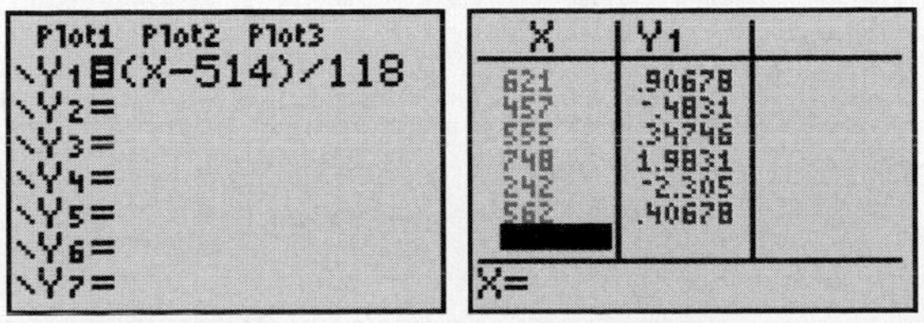

Figure 23 Find the *z*-scores of students' SAT scores

3. The SAT score 242 points is the most unusual score because its approximate *z*-score has the largest absolute value (2.31) of the six students' SAT scores. The SAT score 242 points is about 2.31 standard deviations below the mean, so the score is very low.

In Example 2, we found a *z*-score by substituting Sharon Stone's IQ for *x*, 100 for $\bar{x}$, and 15 for *s* in the formula $x = \bar{x} + zs$ and then solving for *z*. This was efficient because we wanted to find only one *z*-score. But to find the *z*-scores of the six SAT scores in Example 6, substituting before solving would require solving *six* linear equations! Solving the formula $x = \bar{x} + zs$ for *z* and then constructing a table is much more efficient (see Example 6). In general, **to find several values of a variable in a formula, we usually solve the formula for that variable before we make any substitutions**.

In Example 7, we will work with yet another formula that you will study in statistics.

► Example 7 Solving a Formula for One of Its Variables

Solve $u_r = \dfrac{2n_1 n_2}{n} + 1$ for n_1.

Solution

$$u_r = \frac{2n_1 n_2}{n} + 1 \qquad \text{Original formula}$$

$$n \cdot u_r = n \cdot \left(\frac{2n_1 n_2}{n} + 1 \right) \qquad \text{Multiply both sides by n.}$$

$$nu_r = n \cdot \frac{2n_1 n_2}{n} + n \cdot 1 \qquad \text{Distributive law}$$

$$nu_r = 2n_1 n_2 + n \qquad \text{Simplify.}$$

$$nu_r - n = 2n_1 n_2 + n - n \qquad \text{Subtract n from both sides.}$$

$$nu_r - n = 2n_1 n_2 \qquad \text{Combine like terms.}$$

$$\frac{nu_r - n}{2n_2} = \frac{2n_1 n_2}{2n_2} \qquad \text{Divide both sides by } 2n_2.$$

$$\frac{nu_r - n}{2n_2} = n_1 \qquad \text{Simplify.}$$

The result is $n_1 = \dfrac{nu_r - n}{2n_2}$.

► Example 8 Solving a Formula for One of Its Variables

Let *F* be the Fahrenheit reading corresponding to a Celsius reading of *C* degrees. A formula that describes the exact linear association between *F* and *C* is $C = \dfrac{5}{9}(F - 32)$.

1. Solve the Fahrenheit-Celsius formula for *F*.
2. Convert 10°C to the equivalent Fahrenheit temperature.
3. Use a TI-84 to convert 10°C, 15°C, 20°C, 25°C, and 30°C to the equivalent Fahrenheit temperatures.

Solution

1.

$$C = \frac{5}{9}(F - 32)$$ *Fahrenheit-Celsius formula*

$$\frac{9}{5} \cdot C = \frac{9}{5} \cdot \frac{5}{9}(F - 32)$$ *Multiply both sides by $\frac{9}{5}$.*

$$\frac{9}{5}C = F - 32$$ *Simplify.*

$$\frac{9}{5}C + 32 = F - 32 + 32$$ *Add 32 to both sides.*

$$\frac{9}{5}C + 32 = F$$ *Combine like terms.*

The result is $F = \frac{9}{5}C + 32$.

2. We substitute 10 for C in the formula $F = \frac{9}{5}C + 32$ and solve for F:

$$F = \frac{9}{5}(10) + 32 = 50$$

So, 10°C is equivalent to 50°F.

3. We use a TI-84 table to substitute 10, 15, 20, 25, and 30 for C in the formula $F = \frac{9}{5}C + 32$ (see Fig. 24). By viewing the second column of the TI-84 table, we see that the values of F are 50, 59, 68, 77, and 86. So, 10°C, 15°C, 20°C, 25°C, and 30°C are equivalent to 50°F, 59°F, 68°F, 77°F, and 86°F, respectively.

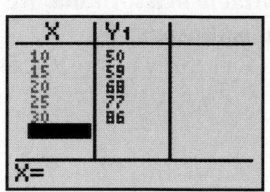

Figure 24 Finding Fahrenheit temperatures

▶ **Example 9**   Solving a Linear Model for One of Its Variables

The mean fee banks charged noncustomers to use their ATMs was $1.40 in 2003 and increased by approximately $0.12 per year until 2014 (Source: *Bankrate.com*). Let F be the mean ATM fee (in dollars) at t years since 2003.

1. Find an equation of a linear model to describe the situation.
2. Solve the equation found in Problem 1 for t.
3. Estimate when the mean ATM fee was $1.60.
4. Use a TI-84 table to estimate in which years the mean ATM fee was $1.80, $2.00, $2.20, $2.40, and $2.60.

Solution

1. Since the mean ATM fee increased by about $0.12 each year, we can model the situation by using a linear model with slope 0.12. Since the mean ATM fee was $1.40 in 2003, the F-intercept is (0, 1.4). So, a reasonable model is

$$F = 0.12t + 1.4$$

2.

$$F = 0.12t + 1.4$$ *Original equation*

$$F - 1.4 = 0.12t + 1.4 - 1.4$$ *Subtract 1.4 from both sides.*

$$F - 1.4 = 0.12t$$ *Combine like terms.*

$$\frac{F - 1.4}{0.12} = \frac{0.12t}{0.12}$$ *Divide both sides by 0.12.*

$$\frac{F - 1.4}{0.12} = t$$ *Simplify.*

The result is the formula $t = \frac{F - 1.4}{0.12}$.

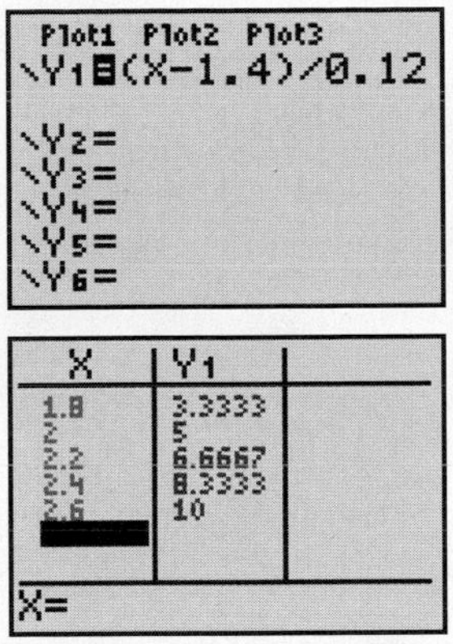

3. To find the year, we use the formula $t = \dfrac{F - 1.4}{0.12}$ because t is alone on one side of the equation. So, we substitute 1.60 for F in $t = \dfrac{F - 1.4}{0.12}$:

$$t = \frac{1.60 - 1.4}{0.12} \approx 1.67$$

The mean ATM fee was $1.60 in 2003 + 1.67 ≈ 2005, according to the model.

4. We use a TI-84 table to substitute $1.80, 2.00, 2.20, 2.40,$ and 2.60 for F in the equation $t = \dfrac{F - 1.4}{0.12}$ (see Fig. 25). By viewing the second column of the TI-84, we see that the approximate values of t (rounded to the ones place) are $3, 5, 7, 8,$ and 10. The model estimates that the mean ATM fee was $1.80, $2.00, $2.20, $2.40, and $2.60 in the years $2006, 2008, 2010, 2011,$ and 2013, respectively.

Figure 25 Finding the years

Solving Formulas with Square Roots

Recall that the standard-deviation formula contains a square root, as many statistics formulas do. Before we solve such a formula for a variable, it will help to explore what happens when we square a principal square root:

$(\sqrt{9})^2 = 3^2 = 9$ The square of the principal square root of 9 is 9.
$(\sqrt{25})^2 = 5^2 = 25$ The square of the principal square root of 25 is 25.
$(\sqrt{36})^2 = 6^2 = 36$ The square of the principal square root of 36 is 36.

These computations suggest that the square of the principal square root of a number is equal to the number, which is true, provided the number is nonnegative. This makes sense because squaring undoes the process of finding a square root.

> **Squaring a Principal Square Root**
>
> If x is nonnegative, then
> $$(\sqrt{x})^2 = x$$
> In words: The square of the principal square root of a nonnegative number is the number.

We can use this property to solve the equation $\sqrt{x} = 3$ by squaring both sides:

$$\sqrt{x} = 3 \quad \textit{Original equation}$$
$$(\sqrt{x})^2 = 3^2 \quad \textit{Square both sides.}$$
$$x = 9 \quad (\sqrt{x})^2 = x$$

In Example 10, we will solve a more complicated equation with a square root by eventually squaring both sides of an equation.

> **Example 10** Solving a Formula with a Square Root

In statistics, you will learn about a *margin of error formula* $E = z \cdot \dfrac{s}{\sqrt{n}}$, where E has to do with the error in making a certain type of estimate. Solve the formula for n, which is the number of individuals who are randomly selected.

Solution

$$E = z \cdot \frac{s}{\sqrt{n}} \qquad \text{\textit{Margin-of-error formula}}$$

$$E \cdot \sqrt{n} = z \cdot \frac{s}{\sqrt{n}} \cdot \sqrt{n} \qquad \text{\textit{Multiply both sides by} } \sqrt{n}.$$

$$E\sqrt{n} = zs \qquad \text{\textit{Simplify.}}$$

$$\frac{E\sqrt{n}}{E} = \frac{zs}{E} \qquad \text{\textit{Divide both sides by E.}}$$

$$\sqrt{n} = \frac{zs}{E} \qquad \text{\textit{Simplify.}}$$

$$(\sqrt{n})^2 = \left(\frac{zs}{E}\right)^2 \qquad \text{\textit{Square both sides.}}$$

$$n = \left(\frac{zs}{E}\right)^2$$

Graphing Linear Equations by Solving for y

In Chapter 7, we graphed linear equations in the slope-intercept form $y = mx + b$. In Example 11, we will graph an equation that is not in this form, but that can be put into it.

▶ **Example 11**   Graphing a Linear Equation by Solving for y

Sketch the graph of $2x + 3y = 9$.

Solution

We will put the equation in $y = mx + b$ form so that we can use the y-intercept and the slope to help us graph the equation. To begin, we get y alone on one side of the equation:

$$2x + 3y = 9 \qquad \text{\textit{Original equation}}$$

$$2x + 3y - 2x = 9 - 2x \qquad \text{\textit{Subtract 2x from both sides to get 3y alone on left-hand side.}}$$

$$3y = -2x + 9 \qquad \text{\textit{Combine like terms; rearrange right-hand side.}}$$

$$\frac{3y}{3} = \frac{-2x}{3} + \frac{9}{3} \qquad \text{\textit{Divide both sides by 3 to get y alone on left-hand side.}}$$

$$y = -\frac{2}{3}x + 3 \qquad \text{\textit{Simplify.}}$$

Since $2x + 3y = 9$ can be put into the form $y = mx + b$ (as $y = -\frac{2}{3}x + 3$), we know that the graph of $2x + 3y = 9$ is a line. The y-intercept is $(0, 3)$ and the slope is $-\frac{2}{3}$. The graph is shown in Fig. 26.

We can verify our result by checking that both $(0, 3)$ and $(3, 1)$ are solutions of $2x + 3y = 9$.

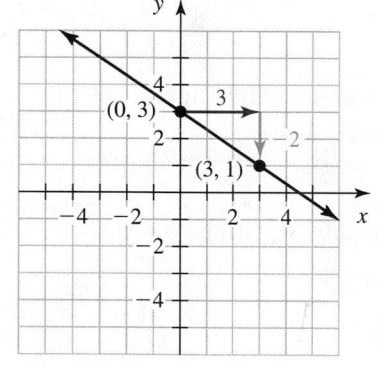

Figure 26 Graph of $2x + 3y = 9$

WARNING

It is a common error to think that the slope of an equation such as $2x + 3y = 9$ is 2, because 2 is the coefficient of x. However, by solving for y (and getting $y = -\frac{2}{3}x + 3$), we see that the slope is $-\frac{2}{3}$ (see Example 11).

> **Graphing a Linear Equation in Two Variables**
>
> To graph a linear equation in two variables,
> 1. If the equation is in slope-intercept form $y = mx + b$, skip to Step 2. If the equation is not in slope-intercept form, put it in that form.
> 2. Plot the y-intercept $(0, b)$.
> 3. Use $m = \dfrac{\text{rise}}{\text{run}}$ to plot a second point. For example, if $m = \dfrac{3}{4}$, count 4 units to the right (from the y-intercept) and 3 units up to plot another point.
> 4. Sketch the line that passes through the two plotted points.

▶ **Example 12** Graphing a Linear Equation by Solving for y

Sketch the graph of $2(3x - 2y) = 9x - 5y - 1$.

Solution

First, we use the distributive law on the left side of the equation:

$$
\begin{array}{lll}
2(3x - 2y) = 9x - 5y - 1 & & \text{Original equation} \\
6x - 4y = 9x - 5y - 1 & & \text{Distributive law} \\
6x - 4y - 6x = 9x - 5y - 1 - 6x & & \text{Subtract } 6x \text{ from both sides.} \\
-4y = 3x - 5y - 1 & & \text{Combine like terms.} \\
-4y + 5y = 3x - 5y - 1 + 5y & & \text{Add } 5y \text{ to both sides.} \\
y = 3x - 1 & & \text{Combine like terms.}
\end{array}
$$

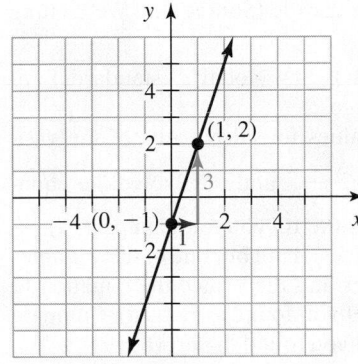

Figure 27 Graph of $2(3x - 2y) = 9x - 5y - 1$

Since $2(3x - 2y) = 9x - 5y - 1$ can be put into the slope-intercept form $y = mx + b$ (as $y = 3x - 1$), we know that the graph of $2(3x - 2y) = 9x - 5y - 1$ is a line. The y-intercept is $(0, -1)$ and the slope is 3. The graph is shown in Fig. 27.

▶

▲▼ Group Exploration

DATA Determining when it's better to solve a formula for a variable

In a survey of 1058 adults, the percentages of people who are confident that they will retire ahead of their schedule are shown in Table 17 for various levels of incomes.

Table 17 Levels of Income versus Percentages of Adults Who Are Confident That They Will Retire Ahead of Schedule

Income Group (thousands of dollars)	Income Used to Represent Income Group (thousands of dollars)	Percent
20–39	29.5	27
40–59	49.5	22
60–75	67.5	21
75–100	87.5	19
over 100	125.0	14

Source: *Country Financial Survey*

Let P be the percentage of adults at income level I thousand dollars who are confident that they will retire ahead of schedule.

1. Construct a scatterplot of the data.
2. Describe the four characteristics of the association.

3. Why is it ironic (surprising) that there is a negative association?
4. Graph the model $P = -0.13I + 29.67$ in the scatterplot. Does it come close to the data points?
5. At what level of income is the given percentage of adults confident that they will retire ahead of schedule?
   a. 15%   **b.** 20%   **c.** 25%
6. Solve the equation $P = -0.13I + 29.67$ for I.
7. Use your result from Problem 6 to predict at what level of income the given percentage of adults is confident that they will retire ahead of schedule.
   a. 15%   **b.** 20%   **c.** 25%
8. In Problem 5, you used the equation $P = -0.13I + 29.67$ to make some predictions. In Problems 6 and 7, you first solved the equation $P = -0.13I + 29.67$ for I before making the same predictions. In your opinion, what number of predictions for the adults' ages would make it worth taking the time and effort to first solve the equation $P = -0.13I + 29.67$ for I before using the result to make the predictions?

Homework 8.4

1. A(n) _____ is an equation that contains two or more variables.

2. *True or False:* Solving for a variable in a formula will change the association between the variables in the formula.

3. *True or False:* To find several values of a variable in a formula, we usually solve the formula for that variable before we make any substitutions.

4. If x is nonnegative, then $(\sqrt{x})^2 = $ _____.

For Exercises 5–20, substitute the given values for the variables and then solve the equation for the remaining variable. Round approximate results to the second decimal place.

5. $\mu = np$; $\mu = 6$, $p = 0.2$ (population mean of binomial distribution)

6. $P(E \text{ AND } F) = P(E)P(F)$; $P(E \text{ AND } F) = 0.14$, $P(F) = 0.2$ (multiplication rule for independent events)

7. $P(E|F) = \dfrac{P(E \text{ AND } F)}{P(F)}$; $P(E|F) = 0.6$, $P(F) = 0.4$ (conditional probability)

8. $\hat{p} = \dfrac{x}{n}$; $\hat{p} = 0.65, n = 1040$ (proportion)

9. $P(E \text{ OR } F) = P(E) + P(F)$; $P(E \text{ OR } F) = 0.95$, $P(E) = 0.42$ (addition rule for disjoint events)

10. $P(E \text{ OR } F) = P(E) + P(F) - P(E \text{ AND } F)$; $P(E \text{ OR } F) = 0.63, P(E) = 0.39, P(E \text{ AND } F) = 0.21$ (general addition rule)

11. $\sigma_{\bar{x}} = \dfrac{\sigma}{\sqrt{n}}$; $\sigma_{\bar{x}} = 7.98$, $n = 36$ (standard deviation of a sampling distribution)

12. $E = t \cdot \dfrac{s}{\sqrt{n}}$; $E = 2.95, s = 4.83, n = 9$ (margin of error)

13. $x = \bar{x} + zs$; $x = 15.56$, $\bar{x} = 8.7$, $s = 3.5$ (value of an observation)

14. $x = \bar{x} + zs$; $x = 58.98$, $\bar{x} = 54.7$, $s = 2.6$ (value of an observation)

15. $y - y_1 = m(x - x_1)$; $y = 8$, $y_1 = 2$, $m = 3$, $x_1 = 4$ (equation of a line)

16. $y - y_1 = m(x - x_1)$; $y = 4, y_1 = 7, x = 1, x_1 = 5$ (equation of a line)

17. $\dfrac{x}{a} + \dfrac{y}{b} = 1$; $a = 2, y = 6, b = 5$ (equation of a line)

18. $\dfrac{x}{a} + \dfrac{y}{b} = 1$; $x = 3, a = 4, b = 7$ (equation of a line)

19. $z = \dfrac{x - \bar{x}}{s}$; $z = 1.96, \bar{x} = 13.1, s = 3.9$ (z-score)

20. $z_0 = \dfrac{r - \mu_r}{\sigma_r}$; $z_0 = -0.39, \mu_r = 19.162, \sigma_r = 2.943$ (test statistic in the runs test for randomness)

21. Let $\bar{x}$ be the mean (in points) of five test scores, x_1, x_2, x_3, x_4, and x_5, all in points.
    a. Find a formula for the mean of the test scores.

b. If a student has test scores of 74, 81, 79, and 84, find the score he needs on the fifth test so that his five-test mean score is 80 points, which is the cutoff for a B.

22. Let $\bar{x}$ be the mean (in points) of four test scores, x_1, x_2, x_3, and x_4, all in points.
    a. Find a formula for the mean of the test scores.
    b. If a student has test scores of 87, 92, and 86, find the score she needs on the fourth test so that her four-test mean score is 90 points, which is the cutoff for an A.

23. Today, Americans' scores on the Wechsler IQ test are normally distributed with mean 100 points and standard deviation 15 points. However, the typical IQ of Americans has been increasing about 3 points per decade (Source: *Are We Getting Smarter?, James R. Flynn*).
    a. Estimate the typical IQ (by today's standards) of Americans two decades ago.
    b. Substitute appropriate values for all but one of the variables in the formula $z = \dfrac{x - \bar{x}}{s}$ and then solve the equation to find the z-score for the IQ you found in part (a).
    c. Substitute appropriate values for all but one of the variables in the formula $x = \bar{x} + zs$ and then solve the equation to find the z-score for the IQ you found in part (a). Compare the result with the z-score you found in part (b).

24. The scores on the Wechsler IQ test are normally distributed with mean 100 and standard deviation 15 points. Actor James Woods is reported to have an IQ of 180 points (Source: Chicago Tribune).
    a. Substitute appropriate values for all but one of the variables in the formula $z = \dfrac{x - \bar{x}}{s}$ and then solve the equation to find the z-score for James Woods's IQ.
    b. Substitute appropriate values for all but one of the variables in the formula $x = \bar{x} + zs$ and then solve the equation to find the z-score for James Woods's IQ. Compare the result with the z-score you found in part (a).

25. In a survey about work, employees were asked for what personal expenses they *most* wanted their company to reimburse them. Let L be the event lunch expenses and G be the event gym membership.
    a. Write a formula for $P(L \text{ OR } G)$.
    b. Of 1000 randomly selected employees, 11% said they most wanted to be reimbursed for lunch expenses and 26% said they most wanted to be reimbursed for lunch expenses OR gym membership (Source: *Accounting Principals*). Use the formula you found in part (a) to find the probability that an individual randomly selected from the study said he or she most wanted to be reimbursed for gym membership.

26. In a survey about work, employees were asked for what personal expenses they *most* wanted their company to reimburse them. Let C be the event cell phone charges and T be the event transportation expenses.

a. Write a formula for $P(C \text{ OR } T)$.

b. Of 1000 randomly selected employees, 10% said they most wanted to be reimbursed for cell phone charges and 52% said they most wanted to be reimbursed for cell phone charges OR transportation expenses (Source: *Accounting Principals*). Use the formula you found in part (a) to find the probability that an individual randomly selected from the study said he or she most wanted to be reimbursed for transportation expenses.

For Exercises 27–32, substitute the given values for the variables and then solve the equation for the remaining variable. Round approximate results to the second decimal place.

27. $\mu = \Sigma x_i P(x_i); x_1 = 0, x_2 = 1, x_3 = 2, x_4 = 3, P(x_1) = 0.027,$ $P(x_2) = 0.189, P(x_3) = 0.441, P(x_4) = 0.343$ (mean of binomial distribution)

28. $\mu = \Sigma x_i P(x_i); \; x_1 = 0, \; x_2 = 1, \; x_3 = 2, \; x_4 = 3, \; x_5 = 4,$ $P(x_1) = 0.240, P(x_2) = 0.412, P(x_3) = 0.265, P(x_4) = 0.076,$ $P(x_5) = 0.008$ (mean of binomial distribution)

29. $\chi^2 = \Sigma \dfrac{(O_i - E_i)^2}{E_i}; O_1 = 15, O_2 = 28, O_3 = 19, E_1 = 11.3,$ $E_2 = 32.5, E_3 = 18.2$ (test statistic for goodness-of-fit test)

30. $\text{MST} = \dfrac{\Sigma n_i(\bar{x}_i - \bar{x})^2}{k - 1}; \; n_1 = 19, \; n_2 = 24, \; n_3 = 15, \; \bar{x}_1 = 35,$ $\bar{x}_2 = 39, \; \bar{x}_3 = 30, \; \bar{x} = 33, \; k = 3$ (mean square due to treatment)

31. $\sigma = \sqrt{\Sigma[x_i^2 P(x_i)] - \mu^2}; \; x_1 = 0, \; x_2 = 1, \; x_3 = 2, \; x_4 = 3,$ $P(x_1) = 0.4, P(x_2) = 0.1, P(x_3) = 0.2, P(x_4) = 0.3, \mu = 1.4$ (standard deviation of a discrete random variable)

32. $s_e = \sqrt{\dfrac{\Sigma(y_i - \hat{y}_i)^2}{n - 2}}; \; y_1 = 2, \; y_2 = 7, \; y_3 = 9, \; \hat{y}_1 = 3, \; \hat{y}_2 = 4,$ $\hat{y}_3 = 11, n = 14$ (standard error of the estimate)

For Exercises 33–54, solve the formula for the specified variable.

33. $\mu = np$, for p

34. $\mu = \lambda t$, for t

35. $\hat{p} = \dfrac{x}{n}$, for x

36. $\sigma_{\bar{x}} = \dfrac{\sigma}{\sqrt{n}}$, for σ

37. $P(E \text{ AND } F) = P(E)P(F)$, for $P(F)$

38. $P(E|F) = \dfrac{P(E \text{ AND } F)}{P(F)}$, for $P(E \text{ AND } F)$

39. $P(\text{NOT } E) = 1 - P(E)$, for $P(E)$

40. $P(E \text{ OR } F) = P(E) + P(F)$, for $P(E)$

41. $P(E \text{ OR } F) = P(E) + P(F) - P(E \text{ AND } F)$, for $P(F)$

42. $P(E \text{ OR } F) = P(E) + P(F) - P(E \text{ AND } F)$, for $P(E \text{ AND } F)$

43. $x = \mu + z\sigma$, for z

44. $x = \mu + z\sigma$, for σ

45. $y - y_1 = m(x - x_1)$, for x

46. $y - y_1 = m(x - x_1)$, for x_1

47. $z_0 = \dfrac{r - \mu_r}{\sigma_r}$, for μ_r

48. $z_0 = \dfrac{r - \mu_r}{\sigma_r}$, for r

49. $u_r = \dfrac{2n_1 n_2}{n} + 1$, for n_1

50. $u_r = \dfrac{2n_1 n_2}{n} + 1$, for n_2

51. $\dfrac{x}{a} + \dfrac{y}{b} = 1$, for y

52. $\dfrac{x}{a} + \dfrac{y}{a} = 1$, for x

53. $\sigma_{\bar{x}} = \dfrac{\sigma}{\sqrt{n}}$, for n

54. $r_s = \dfrac{z_0}{\sqrt{n - 1}}$, for n

55. a. Solve the formula $P(E \text{ OR } F) = P(E) + P(F) - P(E \text{ AND } F)$ for $P(E)$.

b. Substitute 0.8 for $P(F)$, 0.9 for $P(E \text{ OR } F)$, and 0.2 for $P(E \text{ AND } F)$ in the formula that you found in part (a) and solve for $P(E)$.

56. a. Solve the formula $P(E \text{ OR } F) = P(E) + P(F) - P(E \text{ AND } F)$ for $P(E \text{ AND } F)$.

b. Substitute 0.3 for $P(E)$, 0.4 for $P(F)$, and 0.6 for $P(E \text{ OR } F)$ in the formula that you found in part (a) and solve for $P(E \text{ AND } F)$.

57. Let $\bar{x}$ be the mean (in points) of four test scores, $x_1, x_2, x_3,$ and x_4, all in points.

a. Write a formula for the mean of the test scores.

b. Solve the formula you found in part (a) for x_4.

c. If a student has test scores of 85, 93, and 89, all in points, use the formula you found in part (b) to find the score she needs on the fourth test so that her four-test mean is 90 points, which is the cutoff for an A.

58. Let $\bar{x}$ be the mean (in points) of five test scores, $x_1, x_2, x_3, x_4,$ and x_5, all in points.

a. Write a formula for the mean of the test scores.

b. Solve the formula you found in part (a) for x_5.

c. If a student has test scores of 71, 75, 88, and 81, all in points, use the formula you found in part (b) to find the score he needs on the fifth test so that his five-test mean is 80 points, which is the cutoff for a B.

59. The price of an adult one-day ticket to Walt Disney World was $46 in 2000, and it increased by about $3.75 per year until 2012 (Source: *The Walt Disney Company*). Let p be the price (in dollars) of a ticket at t years since 2000.

a. Find an equation of a linear model to describe the situation.

b. Solve the equation you found in part (a) for t.

c. Use the equation you found in part (b) to estimate in which years the prices of tickets were $70, $75, $80, $85, and $90.

60. The number of cases of unruly passenger behavior on board aircraft was 339 in 2007, and it increased by about 1346 cases per year until 2013 (Source: *International Air Transport Association*). Let n be the number of cases of unruly passenger behavior in the year that is t years since 2007.

a. Find an equation of a linear model to describe the situation.

b. Solve the equation you found in part (a) for t.

c. Use the equation you found in part (b) to estimate in which years there were 3000, 4500, 6000, 7500, and 9000 cases of unruly passenger behavior.

61. Due to the airline industry charging more fees, the percentage of revenue from fares was only 71% in 2010, and it decreased by 0.8 percentage point per year until 2014 (Source: *Bureau of Transportation Statistics*). Let p be the percentage of revenue from fares at t years since 2010.

a. Find an equation of a linear model to describe the situation.
b. Use the equation you found in part (a) to estimate in which year 68.5% of revenue came from fares.
c. Solve the equation you found in part (a) for t.
d. Use the equation you found in part (c) to estimate in which year 68.5% of revenue came from fares.
e. Compare the results you found in parts (b) and (d). Which equation was easier to use? Explain.

62. Some 213.9 million pounds of fireworks were used in the United States in 2009, and that amount decreased by about 20.1 million pounds per year until 2013 (Source: *American Pyrotechnics Association*). Let F be the amount (in millions of pounds) of fireworks used in the year that is t years since 2009.

a. Find an equation of a linear model to describe the situation.
b. Use the equation you found in part (a) to estimate in which year 150 million pounds of fireworks were used.
c. Solve the equation you found in part (a) for t.
d. Use the equation you found in part (c) to estimate in which year 150 million pounds of fireworks were used.
e. Compare the results you found in parts (b) and (d). Which equation was easier to use? Explain.

63. The scatterplot in Fig. 28 compares the sizes of people's heads with the weights of their brains.

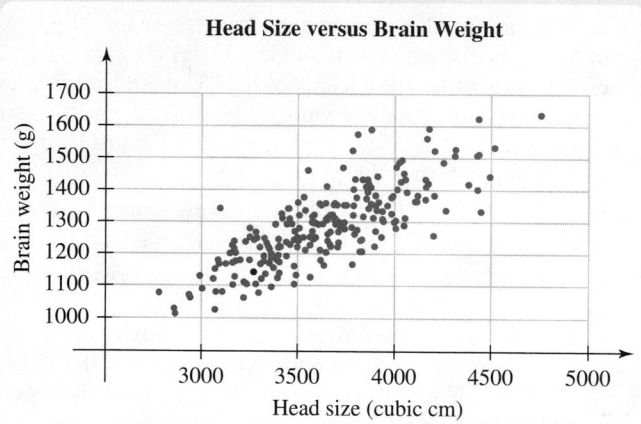

Head Size versus Brain Weight

Figure 28 Exercise 63
(**Source:** A Study of the Relations of the Brain to the Size of the Head, *R. J. Gladstone*)

Let h be the size (in cubic centimeters) of a person's head, and let w be the weight (in grams) of the person's brain.

a. Describe the four characteristics of the association.
b. A reasonable model is $w = 0.26h + 335.43$. Solve the equation for h.
c. Use the equation you found in part (b) to predict the head size of a person whose brain weighs 1142 grams. Find the error in this prediction for the person represented by the red dot in the scatterplot.
d. Predict the head sizes of people whose brains weigh 1100, 1200, 1300, 1400, and 1500, all in grams.
e. The researchers measured the heads and brains of people who had died at the Middlesex Hospital in London. A large proportion of the patients had died of cancer and the rest had died from accidents. Although the researchers found that subjects who had died from cancer had smaller brain-weight to head-size ratios than subjects who had died from accidents, the difference was small. Explain why it is important that the researchers compared the

ratios if the model $w = 0.26h + 335.43$ was intended to be used to predict the brain weights of healthy people.

64. In an experiment of 40 college soccer players (20 women and 20 men), the total distance each participant could cover in three single-leg hops was recorded. The players also jumped vertically, and their jump heights were recorded. The scatterplot in Fig. 29 compares the triple-hop distances of the players with their vertical-jump heights.

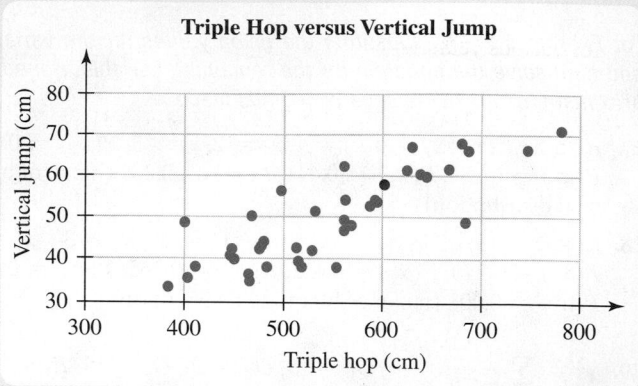

Triple Hop versus Vertical Jump

Figure 29 Exercise 64
(**Source:** Triple-Hop Distance as a Valid Predictor of Lower Limb Strength and Power, *R. T. Hamilton et al.*)

Let T be a college soccer player's triple-hop distance and V be the player's vertical-jump height, both in centimeters.

a. Describe the four characteristics of the association.
b. A reasonable model is $V = 0.093T - 1.57$. Solve the equation for T.
c. Use the equation you found in part (b) to predict the triple-hop distance for a player whose vertical-jump height is 57.6 centimeters. Find the error in this prediction for the player represented by the red dot in the scatterplot.
d. Predict the triple-hop distances for players whose vertical-jump heights are 45, 50, 55, 60, and 65, all in centimeters.
e. Participants were allowed 1 to 3 practice trials of both the three-hop and vertical-jump events. Guess how the strength of the association would have been affected if practice trials had not been allowed. Explain.

65. DATA The numbers of cremations in the United States are shown in Table 18 for various years.

Table 18 Percentages of Bodies That Are Cremated

Year	Percent
1995	19.2
1998	24.1
2000	26.2
2003	29.5
2005	32.3
2007	34.3
2010	40.6
2013	45.3

Source: *Cremation Association of North America*

Let p be the percentage of bodies that are cremated in the year that is t years since 1990.

a. Construct a scatterplot.

b. Describe the four characteristics of the association. Compute and interpret r as part of your analysis.

c. Graph the model $p = 1.40t + 11.96$ on the scatterplot. Does it come close to the data points?

d. What is the slope? What does it mean in this situation?

e. Solve the equation $p = 1.40t + 11.96$ for t.

f. What is the slope of the model you found in part (e)? What does it mean in this situation?

66. **DATA** The percentages of new-vehicle buyers who use the Internet during the shopping process are shown in Table 19 for various years.

Table 19 Percentages of New-Vehicle Buyers Who Use the Internet during the Shopping Process

Year	Percent
2000	54
2002	60
2004	64
2006	68
2008	75
2010	79
2012	79

Source: *J. D. Power and Associates*

Let p be the percentage of new-vehicle buyers who use the Internet during the shopping process at t years since 2000.

a. Construct a scatterplot.

b. Describe the four characteristics of the association. Compute and interpret r as part of your analysis.

c. Graph the model $p = 2.21t + 55.14$ on the scatterplot. Does it come close to the data points?

d. What is the slope? What does it mean in this situation?

e. Solve the equation $p = 2.21t + 55.14$ for t.

f. What is the slope of the model you found in part (e)? What does it mean in this situation?

67. **DATA** A person travels d miles at a constant speed s (in miles per hour) for t hours.

a. Complete Table 20 to help find a formula that describes the association among s, t, and d. Show the arithmetic to help you see a pattern.

Table 20 Speed, Time, and Distance

Speed (miles per hour) s	Time (hours) t	Distance (miles) d
50	4	
70	3	
65	2	
55	5	
s	t	

b. Solve the formula you found in part (a) for t.

c. Use the formula you found in part (b) to find the value of t when $d = 315$ and $s = 70$. What does your result mean in this situation?

d. A student plans to drive from Albuquerque Technical Vocational Institute in Albuquerque, New Mexico, to Denver, Colorado. The speed limit is 75 mph in New Mexico and 65 mph in Colorado. The trip involves 229.9 miles of travel in New Mexico followed by 219.2 miles in Colorado. If the student drives at the posted speed limits, estimate the driving time.

68. **DATA** An employee is paid a total of E dollars for working T hours at P dollars per hour.

a. Complete Table 21 to help find a formula that describes the association among T, P, and E. Show the arithmetic to help you see a pattern.

Table 21 Time, Hourly Pay, and Total Pay

Time (hours) T	Hourly Pay (dollars) P	Total Pay (dollars) E
9	10	
3	15	
6	11	
5	13	
T	P	

b. Solve the formula you found in part (a) for T.

c. Use the formula you found in part (b) to find the value of T when $P = 14$ and $E = 420$. What does your result mean in this situation?

d. An employee is paid $12 per hour for working Monday through Friday, earning a total of $480. She is paid $18 per hour for working on the weekend, earning another $270. What is the total number of hours she worked for the entire week?

For Exercises 69–90, determine the slope and the y-intercept. Use the slope and the y-intercept to graph the equation by hand.

69. $y + 2x = 4$

70. $y - 3x = 2$

71. $3y = 2x$

72. $2y = -5x$

73. $5y = 4x - 15$

74. $4y = 7x - 20$

75. $3x - 4y = 8$

76. $4x + 3y = 9$

77. $6x - 15y = 30$

78. $6x - 8y = 16$

79. $4x + y + 2 = 0$

80. $2x - y - 3 = 0$

81. $6x - 4y + 8 = 0$

82. $15x + 12y - 36 = 0$

83. $y - 3 = 0$

84. $y + 5 = 0$

85. $3(x - 2y) = 9$

86. $2(y - 3x) = 8$

87. $4x - 5y + 3 = 2x - 2y - 3$

88. $3y - 6x + 2 = 7y - x - 6$

89. $1 - 3(y - 2x) = 7 + 3(x - 3y)$

90. $8 - 2(y - 3x) = 2 + 4(x - 2y)$

Concepts

91. Let $\bar{x}$ be the mean (in points) and s be the standard deviation (in points) of three test scores, x_1, x_2, and x_3, all in points.

a. Write a formula for the mean of the three test scores.

b. Use the formula you found in part (a) to show that if all three test scores are equal, then the mean score is x_1.

c. Write a formula for the standard deviation of the three test scores.

d. Use the formula you found in part (c) to show that if all three test scores are equal, then the standard deviation of the three test scores is 0 points.

92. Let M be the median (in points) and R be the range (in points) of four test scores (all in points), x_1, x_2, x_3, and x_4, where the scores are listed from smallest to largest.

  a. Write a formula for the median of the test scores.
  b. Use the formula you found in part (a) to help you show that if the two middle scores are equal, then the median score is x_2.
  c. Write a formula for the range of the test scores.
  d. Use the formula you found in part (c) to help you show that if all four test scores are equal, then the range is 0 points.

93. a. Solve the equation $mx + b = 0$, where $m \neq 0$, for x.

  b. Explain why a linear equation in one variable has exactly one solution.

94. a. Find the slope of the line $3x + 5y = 7$.
  b. Find the slope of the line $2x + 7y = 3$.
  c. Find the slope of the line $ax + by = c$, where a, b, and c are constants and b is nonzero. [**Hint:** Do steps similar to the ones you did in parts (a) and (b).]

95. A student says the graph of $3y + 2x = 6$ has slope 2 because the coefficient of the $2x$ term is 2. Is the student correct? If yes, explain why. If no, find the slope.

96. A student says the y-intercept of the line $2y = 3x + 4$ is $(0, 4)$ because the constant term is 4. Is the student correct? If yes, explain why. If no, find the y-intercept.

97. Recall that we can describe some or all of the solutions of an equation in two variables with an equation, a table, a graph, or words.

  a. Describe the solutions of the equation $3x - 5y = 10$ by using a graph.

  b. Describe three solutions of the equation $3x - 5y = 10$ by using a table.
  c. Describe the solutions of the equation $3x - 5y = 10$ by using words.

98. Recall that we can describe some or all of the solutions of an equation in two variables with an equation, a table, a graph, or words.

  a. Describe the solutions of the equation $5x + 2y = 4$ by using a graph.
  b. Describe three solutions of the equation $5x + 2y = 4$ by using a table.
  c. Describe the solutions of the equation $5x + 2y = 4$ by using words.

99. a. Solve the equation $2y - 6 = 4x$ for y.

  b. Find two solutions of the equation $y = 2x + 3$.
  c. Show that the two ordered pairs you found in part (b) satisfy both of the equations $2y - 6 = 4x$ and $y = 2x + 3$.
  d. Explain why it makes sense that the graphs of $2y - 6 = 4x$ and $y = 2x + 3$ are the same.

100. If the graph of a linear equation has a defined slope, describe how to find the slope and the y-intercept and how to use this information to graph the equation.

101. In what cases would you first solve a formula for a variable and then make substitutions for any other variables, and in what cases would you first make substitutions for all but one variable and then solve for the remaining variable?

102. Give an example of an equation in one variable and an example of a formula. How are these equations different?

8.5 Solving Linear Inequalities to Make Predictions

Objectives

» Describe the *addition property of inequalities*.

» Describe the *multiplication property of inequalities*.

» Describe the meaning of *satisfy*, *solution*, and *solution set* for a *linear inequality in one variable*.

» Solve a linear inequality in one variable, and graph the solution set.

» Substitute values for variables in a compound inequality.

» Solve a *compound inequality in one variable*, and graph the solution set.

» Use linear inequalities to make predictions about authentic situations.

In Section 1.1, we discussed the meaning of inequalities and intervals, and we graphed them. In Section 8.3, we predicted when a quantity would equal a certain amount. In this section, we will put these two concepts together. That is, we will use a model to predict when a quantity will be more than (or less than) a certain amount. For instance, in Example 12, we will estimate the years when the annual U.S. digital music revenue was more than $3.7 billion.

Addition Property of Inequalities

What happens if we add 3 to both sides of the inequality $4 < 6$?

$$4 < 6 \qquad \textit{Original inequality}$$
$$4 + 3 \overset{?}{<} 6 + 3 \qquad \textit{Add 3 to both sides.}$$
$$7 \overset{?}{<} 9 \qquad \textit{Simplify.}$$
$$\text{true}$$

What happens if we add -3 to both sides of the inequality $4 < 6$?

$$4 < 6 \qquad \textit{Original inequality}$$
$$4 + (-3) \overset{?}{<} 6 + (-3) \qquad \textit{Add } -3 \textit{ to both sides.}$$
$$1 \overset{?}{<} 3 \qquad \textit{Simplify.}$$
$$\text{true}$$

These examples suggest the following property:

> **Addition Property of Inequalities**
>
> $$\text{If } a < b, \text{ then } a + c < b + c.$$
>
> Similar properties hold for $\leq$, $>$, and $\geq$.

Similar rules hold for subtraction because subtracting a number is the same as adding the opposite of the number.

The addition property of inequalities is similar to the addition property of equality.

We can use a number line to illustrate that if $a < b$, then $a + c < b + c$. From Figs. 30 and 31, we see that if a lies to the left of b ($a < b$), then $a + c$ lies to the left of $b + c$ ($a + c < b + c$). In Fig. 30, c is negative; in Fig. 31, c is positive.

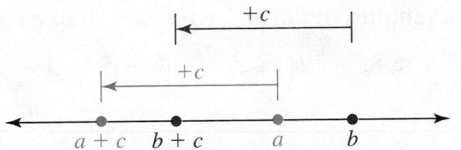

Figure 30 Adding c where c is negative

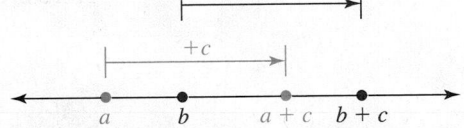

Figure 31 Adding c where c is positive

Multiplication Property of Inequalities

What if we multiply both sides of the inequality $4 < 6$ by 3?

$$4 < 6 \qquad \textit{Original inequality}$$
$$4(3) \overset{?}{<} 6(3) \qquad \textit{Multiply both sides by 3.}$$
$$12 \overset{?}{<} 18 \qquad \textit{Simplify.}$$
$$\text{true}$$

Finally, what happens if we multiply both sides of $4 < 6$ by -3?

$$4 < 6 \qquad \textit{Original inequality}$$
$$4(-3) \overset{?}{<} 6(-3) \qquad \textit{Multiply both sides by } -3.$$
$$-12 \overset{?}{<} -18 \qquad \textit{Simplify.}$$
$$\text{false}$$

The result is the false statement $-12 < -18$. We can get a *true* statement if we *reverse the inequality symbol* when we multiply both sides of $4 < 6$ by -3:

$$4 < 6 \qquad \textit{Original inequality}$$
$$4(-3) \overset{?}{>} 6(-3) \qquad \textit{Reverse inequality symbol.}$$
$$-12 \overset{?}{>} -18 \qquad \textit{Simplify.}$$
$$\text{true}$$

So, when we multiply both sides of an inequality by a *negative* number, we *reverse* the inequality symbol.

> **Multiplication Property of Inequalities**
>
> - For a *positive* number c, if $a < b$, then $ac < bc$.
> - For a *negative* number c, if $a < b$, then $ac > bc$.
>
> Similar properties hold for $\leq$, $>$, and $\geq$. In words: When we multiply both sides of an inequality by a positive number, we keep the inequality symbol. When we multiply both sides by a negative number, we reverse the inequality symbol.

Similar rules apply for division because dividing by a nonzero number is the same as multiplying by its reciprocal. Therefore, **when we multiply or divide both sides of an inequality by a negative number, we reverse the inequality symbol**.

Consider multiplying both sides of $a < b$ by -1:

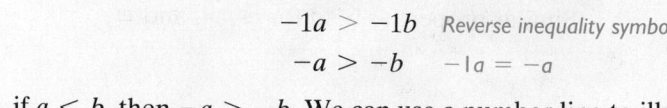

$$a < b \qquad \textit{Original inequality}$$
$$-1a > -1b \qquad \textit{Reverse inequality symbol.}$$
$$-a > -b \qquad \textit{-1a = -a}$$

So, if $a < b$, then $-a > -b$. We can use a number line to illustrate this fact. To plot the point for $-a$, we move the point for a to the other side of the origin so that the points for $-a$ and a are the same distance from the origin (see Fig. 32).

From Fig. 32 we see that if the point for a lies to the *left* of the point for b ($a < b$), then the point for $-a$ lies to the *right* of the point for $-b$ ($-a > -b$).

Figure 32 The points for a, b, $-a$, and $-b$

Satisfying Inequalities, Solutions, and Solution Sets

Here are some examples of *linear inequalities in one variable*:

$$3x + 5 < 8, \qquad 2x \le 5, \qquad w - 5 > 4 - 2w, \qquad 5(x - 3) \le 1$$

> ▶ **Definition**   Linear inequality in one variable
>
> A **linear inequality in one variable** is an inequality that can be put into one of the forms
>
> $$mx + b < 0, \qquad mx + b \le 0, \qquad mx + b > 0, \qquad mx + b \ge 0$$
>
> where m and b are constants and $m \ne 0$.

We say a number **satisfies** an inequality in one variable if the inequality becomes a true statement after we have substituted the number for the variable.

▶ **Example 1**   Identifying Solutions of an Inequality

1. Does the number 4 satisfy the inequality $2x - 3 < 7$?
2. Does the number 6 satisfy the inequality $2x - 3 < 7$?

Solution

1. We substitute 4 for x in the inequality $2x - 3 < 7$:

$$2(4) - 3 \overset{?}{<} 7 \qquad \textit{Substitute 4 for x.}$$
$$8 - 3 \overset{?}{<} 7 \qquad \textit{Multiply.}$$
$$5 \overset{?}{<} 7 \qquad \textit{Subtract.}$$
$$\text{true}$$

So, 4 satisfies the inequality $2x - 3 < 7$.

2. We substitute 6 for x in the inequality $2x - 3 < 7$:

$$2(6) - 3 \overset{?}{<} 7 \qquad \textit{Substitute 6 for x.}$$
$$12 - 3 \overset{?}{<} 7 \qquad \textit{Multiply.}$$
$$9 \overset{?}{<} 7 \qquad \textit{Subtract.}$$
$$\text{false}$$

So, 6 does not satisfy the inequality $2x - 3 < 7$.

> **Definition** *Solution, solution set,* and *solve* for an inequality in one variable

We say a number is a **solution** of an inequality in one variable if it satisfies the inequality. The **solution set** of an inequality is the set of all solutions of the inequality. We **solve** an inequality by finding its solution set.

Solving a Linear Inequality

To solve a linear inequality in one variable, we apply properties of inequalities to get the variable alone on one side of the inequality. The steps are similar to solving a linear equation in one variable, but we must remember to reverse the inequality when we multiply or divide both sides of an inequality by a negative number.

> **Example 2**   Solving a Linear Inequality

Solve $2x - 3 < 7$. Describe the solution set as an inequality, in interval notation, and on a graph.

Solution

We get x alone on one side of the inequality:

$$2x - 3 < 7 \qquad \text{\textit{Original inequality}}$$
$$2x - 3 + 3 < 7 + 3 \qquad \text{\textit{Add 3 to both sides to get 2x alone on left side.}}$$
$$2x < 10 \qquad \text{\textit{Combine like terms.}}$$
$$\frac{2x}{2} < \frac{10}{2} \qquad \text{\textit{Divide both sides by 2 to get x alone on left side.}}$$
$$x < 5 \qquad \text{\textit{Simplify.}}$$

The solution set is the set of all numbers less than 5, which we describe in interval notation as $(-\infty, 5)$. We graph the solution set on a number line in Fig. 33.

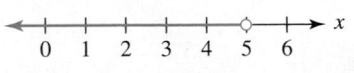

Figure 33 Graph of $x < 5$

In Example 1, we found that 4 is a solution of the inequality $2x - 3 < 7$ but 6 is not. This checks with our work in Example 2, because 4 is on the graph in Fig. 33 and 6 is not.

> **Example 3**   Solving a Linear Inequality

Solve the inequality $-3x \geq -12$. Describe the solution set as an inequality, in interval notation, and on a graph.

Solution

We divide both sides of the inequality by -3, a negative number:

$$-3x \geq -12 \qquad \text{\textit{Original inequality}}$$
$$\frac{-3x}{-3} \leq \frac{-12}{-3} \qquad \text{\textit{Divide both sides by} -3; \textit{reverse inequality symbol.}}$$
$$x \leq 4 \qquad \text{\textit{Simplify.}}$$

Since we divided both sides of the inequality by a negative number, we reversed the inequality symbol. The solution set is $(-\infty, 4]$. We graph the solution set in Fig. 34.

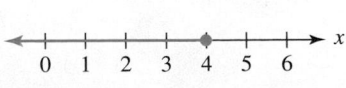

Figure 34 Graph of $x \leq 4$

WARNING   It is a common error to forget to reverse an inequality symbol when multiplying or dividing both sides of an inequality by a negative number. For instance, in Example 3, it is important that we reversed the inequality symbol $\geq$ when we divided both sides of the inequality $-3x \geq -12$ by -3.

▶ **Example 4** Solving a Linear Inequality

Solve the inequality $3x \geq -12$. Describe the solution set as an inequality, in interval notation, and on a graph.

Solution

We divide both sides of the inequality by 3, a positive number:

$$3x \geq -12 \qquad \textit{Original inequality}$$

$$\frac{3x}{3} \geq \frac{-12}{3} \qquad \textit{Divide both sides by 3.}$$

$$x \geq -4 \qquad \textit{Simplify.}$$

Since we divided both sides of the inequality by a positive number, we did *not* reverse the inequality symbol. We graph the solution set, $[-4, \infty)$, in Fig. 35.

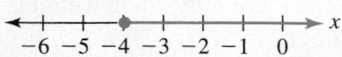

Figure 35 Graph of $x \geq -4$

▶ **Example 5** Solving a Linear Inequality

Solve $2x - 5 > 6x + 3$. Describe the solution set as an inequality, in interval notation, and on a graph.

Solution

$$2x - 5 > 6x + 3 \qquad \textit{Original inequality}$$

$$2x - 5 - 6x > 6x + 3 - 6x \qquad \textit{Subtract 6x from both sides.}$$

$$-4x - 5 > 3 \qquad \textit{Combine like terms.}$$

$$-4x - 5 + 5 > 3 + 5 \qquad \textit{Add 5 to both sides.}$$

$$-4x > 8 \qquad \textit{Combine like terms.}$$

$$\frac{-4x}{-4} < \frac{8}{-4} \qquad \textit{Divide both sides by } -4; \textit{ reverse inequality symbol.}$$

$$x < -2 \qquad \textit{Simplify.}$$

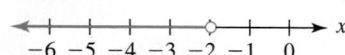

Figure 36 Graph of $x < -2$

We graph the solution set, $(-\infty, -2)$, in Fig. 36.

    To verify our result, we check that for some values of x less than -2, the value of $2x - 5$ is greater than the value of $6x + 3$ (see Fig. 37). We do this by setting up a TI-84 table so x begins at -2 and increases by 1. Then we scroll up 3 rows so we can view values of x that are less than -2 and values of x that are greater than -2.

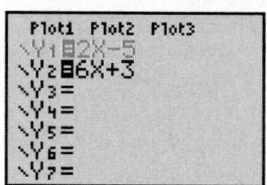 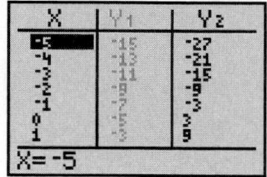

Figure 37 Verify the work

▶ **Example 6** Solving a Linear Inequality

Solve $3(x - 2) < 5x - 3$. Describe the solution set as an inequality, in interval notation, and on a graph.

Solution

$$3(x - 2) < 5x - 3 \qquad \textit{Original inequality}$$

$$3x - 6 < 5x - 3 \qquad \textit{Distributive law}$$

$$3x - 6 - 5x < 5x - 3 - 5x \qquad \textit{Subtract } 5x \textit{ from both sides.}$$

$$-2x - 6 < -3 \qquad \textit{Combine like terms.}$$

$$-2x - 6 + 6 < -3 + 6 \qquad \textit{Add 6 to both sides.}$$

$$-2x < 3 \qquad \textit{Combine like terms.}$$

$$\frac{-2x}{-2} > \frac{3}{-2} \qquad \textit{Divide both sides by } -2; \textit{ reverse inequality symbol.}$$

$$x > -\frac{3}{2} \qquad \textit{Simplify.}$$

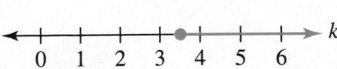

Figure 38 Graph of $x > -\dfrac{3}{2}$

We graph the solution set, $\left(-\dfrac{3}{2}, \infty\right)$, in Fig. 38.

To verify our result, we check that for some values of x greater than $-\dfrac{3}{2}$, the value of $3(x - 2)$ is less than the value of $5x - 3$ (see Fig. 39). We do this by setting up a TI-84 table so x begins at -1.5 and increases by 1. Then we scroll up 3 rows so we can view values of x that are less than -1.5 and values of x that are greater than -1.5.

 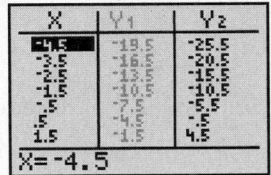

Figure 39 Verify the work

▶

▶ Example 7 Solving a Linear Inequality That Contains Decimals

Solve $-2.9k + 4.1 \le -6.05$. Describe the solution set as an inequality, in interval notation, and on a graph.

Solution

$$-2.9k + 4.1 \le -6.05 \qquad \textit{Original inequality}$$

$$-2.9k + 4.1 - 4.1 \le -6.05 - 4.1 \qquad \textit{Subtract 4.1 from both sides.}$$

$$-2.9k \le -10.15 \qquad \textit{Combine like terms.}$$

$$\frac{-2.9k}{-2.9} \ge \frac{-10.15}{-2.9} \qquad \textit{Divide both sides by } -2.9; \textit{ reverse inequality symbol.}$$

$$k \ge 3.5 \qquad \textit{Simplify.}$$

Figure 40 Graph of $k \ge 3.5$

We graph the solution set, $[3.5, \infty)$, in Fig. 40.

We can use TI-84 tables to check our work.

▶

▶ Example 8 Solving a Linear Inequality

Solve $\dfrac{3x + 5}{4} - \dfrac{2x - 7}{6} < \dfrac{11}{3}$. Describe the solution set as an inequality, in interval notation, and on a graph.

Solution

First, we multiply both sides of the inequality by the LCD, 12:

$$12 \cdot \left(\frac{3x + 5}{4} - \frac{2x - 7}{6} \right) < 12 \cdot \frac{11}{3} \qquad \text{Multiply both sides by LCD, 12.}$$

$$12 \cdot \frac{3x + 5}{4} - 12 \cdot \frac{2x - 7}{6} < 12 \cdot \frac{11}{3} \qquad \text{Distributive law}$$

$$3(3x + 5) - 2(2x - 7) < 44 \qquad \text{Simplify.}$$

$$9x + 15 - 4x + 14 < 44 \qquad \text{Distributive law}$$

$$5x + 29 < 44 \qquad \text{Combine like terms.}$$

$$5x + 29 - 29 < 44 - 29 \qquad \text{Subtract 29 from both sides.}$$

$$5x < 15 \qquad \text{Combine like terms.}$$

$$\frac{5x}{5} < \frac{15}{5} \qquad \text{Divide both sides by 5.}$$

$$x < 3 \qquad \text{Simplify.}$$

We graph the solution set, $(-\infty, 3)$, in Fig. 41.

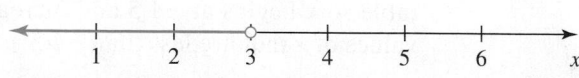

Figure 41 Graph of $x < 3$

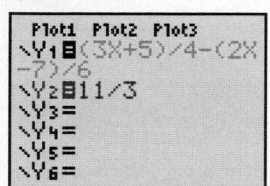

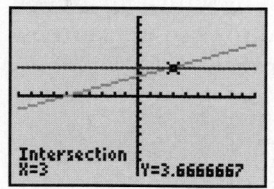

Figure 42 Verify the work

To verify our result, we use a TI-84 to check that, for values of x less than 3, the graph of the equation $y = \frac{3x + 5}{4} - \frac{2x - 7}{6}$ is below the horizontal line $y = \frac{11}{3}$. See Fig. 42.

Substituting Values for Variables in a Compound Inequality

In Section 1.1, we discussed the meaning of compound inequalities and how to graph them. In Example 9, we will substitute values for all but one of the variables in a compound inequality to find a compound inequality in one variable.

▶ **Example 9**   Substituting Values for Variables in a Compound Inequality

Substitute 20.3 for $\bar{x}$, 1.708 for t, 7.6 for s, and 28 for n in the compound inequality

$$\bar{x} - t \cdot \frac{s}{\sqrt{n}} < \mu < \bar{x} + t \cdot \frac{s}{\sqrt{n}}$$

to find a compound inequality that describes the variable μ. Round to the first decimal place.

Solution

We substitute 20.3 for $\bar{x}$, 1.708 for t, 7.6 for s, and 28 for n in the inequality:

$$20.3 - 1.708 \cdot \frac{7.6}{\sqrt{28}} < \mu < 20.3 + 1.708 \cdot \frac{7.6}{\sqrt{28}} \qquad \text{Substitute.}$$

$$20.3 - 2.5 < \mu < 20.3 + 2.5 \qquad \text{Compute.}$$

$$17.8 < \mu < 22.8 \qquad \text{Add/subtract.}$$

So, the values of μ are between 17.8 and 22.8. We can graph the solution set on a number line (see Fig. 43), or we can describe the solution set in interval notation as (17.8, 22.8).

$$\xleftarrow{\qquad} \underset{17 \quad 18 \quad 19 \quad 20 \quad 21 \quad 22 \quad 23}{\overset{\circ \qquad \qquad \qquad \qquad \qquad \circ}{\vert \quad \vert \quad \vert \quad \vert \quad \vert \quad \vert \quad \vert}} \xrightarrow{\quad} \mu$$

Figure 43 Graph of $17.8 < \mu < 22.8$

In statistics, you will find many compound intervals similar to the one you found in Example 9.

Solving Compound Inequalities

In Example 10, we will solve a compound inequality by taking steps much like the ones we took in previous examples.

▶ Example 10 Solving a Compound Inequality

Solve $-5 < 2x - 1 < 7$.

Solution

We can get x alone in the "middle part" of the inequality by applying the same operations to all three parts of the inequality:

$$-5 < 2x - 1 < 7 \qquad \text{\textit{Original inequality}}$$

$$-5 + 1 < 2x - 1 + 1 < 7 + 1 \qquad \text{\textit{Add 1 to all three parts.}}$$

$$-4 < 2x < 8 \qquad \text{\textit{Combine like terms.}}$$

$$\frac{-4}{2} < \frac{2x}{2} < \frac{8}{2} \qquad \text{\textit{Divide all three parts by 2.}}$$

$$-2 < x < 4 \qquad \text{\textit{Simplify.}}$$

So, the solution set is the set of numbers between -2 and 4. We can graph the solution set on a number line (see Fig. 44), or we can describe the solution set in interval notation as $(-2, 4)$.

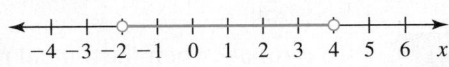

Figure 44 Graph of $-2 < x < 4$

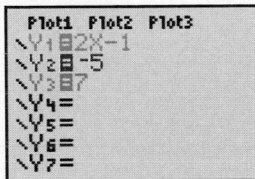

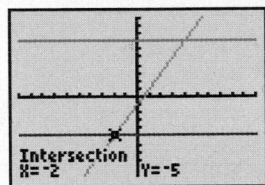

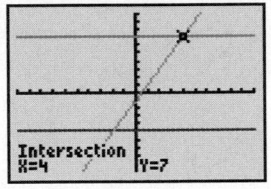

Figure 45 Verify the work

To verify our result, we use a TI-84 to check that, for values of x between -2 and 4, the graph of $y = 2x - 1$ is between the horizontal lines $y = -5$ and $y = 7$ (see Fig. 45).

▶ Example 11 Solving a Compound Inequality

Solve $\dfrac{1}{2} \le 5 - \dfrac{3}{2}w \le 4$.

Solution

$$\frac{1}{2} \le 5 - \frac{3}{2}w \le 4 \qquad \text{\textit{Original inequality}}$$

$$2 \cdot \frac{1}{2} \le 2 \cdot 5 - 2 \cdot \frac{3}{2}w \le 2 \cdot 4 \qquad \text{\textit{Multiply all three parts by LCD, 2.}}$$

$$1 \le 10 - 3w \le 8 \qquad \text{\textit{Simplify.}}$$

$$1 - 10 \le 10 - 3w - 10 \le 8 - 10 \qquad \text{\textit{Subtract 10 from all three parts.}}$$

$$-9 \le -3w \le -2 \qquad \text{\textit{Combine like terms.}}$$

$$\frac{-9}{-3} \ge \frac{-3w}{-3} \ge \frac{-2}{-3} \qquad \text{\textit{Divide all three parts by -3; reverse inequality symbols.}}$$

$$3 \ge w \ge \frac{2}{3} \qquad \text{\textit{Simplify.}}$$

So, the solution set is the set of numbers between $\dfrac{2}{3}$ and 3, inclusive. The compound inequality $\dfrac{2}{3} \le w \le 3$ describes the solution set. We can graph the solution set on a number line (see Fig. 46), or we can describe the solution set in interval notation as $\left[\dfrac{2}{3}, 3\right]$.

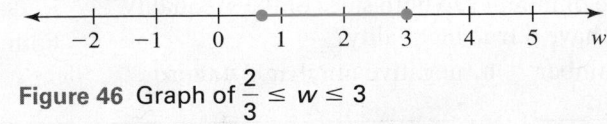

Figure 46 Graph of $\dfrac{2}{3} \le w \le 3$

Using Linear Inequalities to Make Predictions

When working with a linear model, we can make certain types of predictions by solving a linear inequality that is related to the linear model.

▶ **Example 12** Making a Prediction

Digital music revenues in the United States are shown in Table 22 for various years. Let r be the annual revenue (in billions of dollars) at t years since 2000.

Table 22 U.S. Digital Music Revenues

Year	Annual Revenue (billions of dollars)
2006	1.86
2008	2.91
2010	2.89
2012	4.05
2013	4.36

Source: *Recording Industry Association of America*

Stage-fright control. It plays the entire solo of "Free Bird" in a pinch.

What's the button for?

A reasonable model is $r = 0.34t - 0.11$. Estimate when the annual revenue was more than \$3.7 billion.

Solution

To estimate when the annual revenue was more than \$3.7 billion, we find the values of t such that the expression $0.34t - 0.11$ is more than 3.7:

$$0.34t - 0.11 > 3.7 \qquad \text{\textit{Original inequality}}$$
$$0.34t - 0.11 + 0.11 > 3.7 + 0.11 \qquad \text{\textit{Add 0.11 to both sides.}}$$
$$0.34t > 3.81 \qquad \text{\textit{Combine like terms.}}$$
$$\frac{0.34t}{0.34} > \frac{3.81}{0.34} \qquad \text{\textit{Divide both sides by 0.34.}}$$
$$t > 11.21 \qquad \text{\textit{Simplify; divide.}}$$

According to the model, the annual revenue was more than \$3.7 billion after 2011 ($t > 11$). But we have little faith in our result for the years after 2013, because we have performed extrapolation for those years.

▶

◣◥ Group Exploration

Section Opener: Properties of inequalities

1. Decide whether each inequality is true.
   a. $4 < 9$ **b.** $3 > 8$ **c.** $-1 < -7$ **d.** $-2 > -6$

2. Decide whether performing the given operation on both sides of the true inequality $4 < 6$ will give a true statement.
   a. Add 2 to both sides.
   b. Add -2 to both sides.
   c. Multiply both sides by 2.
   d. Multiply both sides by -2.

3. For any true inequality (such as $3 < 7$), can you add the given type of number to both sides of the inequality and then still have a true inequality?
   a. positive number   **b.** negative number   **c.** zero

4. For any true inequality, can you multiply both sides of the inequality by the given type of number and then still have a true inequality?
   a. positive number   **b.** negative number   **c.** zero

5. Recall from Section 1.5 that subtracting a number is the same as adding the opposite of that number. What does this tell you about subtracting a number from both sides of an inequality?

6. Recall from Section 1.3 that dividing by a nonzero number is the same as multiplying by the reciprocal of that number. What does this tell you about dividing both sides of an inequality by a nonzero number?

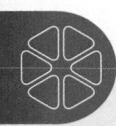Group Exploration

Meaning of the solution set of an inequality

We solve the inequality $-3x + 7 < 1$:

$$-3x + 7 < 1$$
$$-3x + 7 - 7 < 1 - 7$$
$$-3x < -6$$
$$\frac{-3x}{-3} > \frac{-6}{-3}$$
$$x > 2$$

1. Choose a number greater than 2. Check that your number satisfies the inequality $-3x + 7 < 1$.

2. Choose two more numbers greater than 2. Check that both of these numbers satisfy the inequality $-3x + 7 < 1$.

3. Choose three numbers that are *not* greater than 2. Show that each of these numbers does *not* satisfy the inequality $-3x + 7 < 1$.

4. Explain what it means when we write $x > 2$ as the last step in solving the inequality $-3x + 7 < 1$.

Homework 8.5

For extra help ▶ MyMathLab®   ▦ Watch the videos in MyMathLab   Download the MyDashboard App

1. *True or False:* When we add a negative number to both sides of an inequality, we reverse the inequality symbol.

2. *True or False:* When we multiply both sides of an inequality by a negative number, we reverse the inequality symbol.

3. We say a number _____ an inequality in one variable if the inequality becomes a true statement after we have substituted the number for the variable.

4. We say a number is a(n) _____ of an inequality in one variable if it satisfies the inequality.

Which of the given numbers satisfy the given inequality?

5. $3x + 5 \geq 14$;   2, 3, 6
6. $-4x - 7 > 1$; $-3, -2, 0$
7. $2x < x + 2$; $-4, 2, 3$
8. $x - 9 \leq 4x$; $-3, 2, 5$

Solve the inequality. Describe the solution set as an inequality, in interval notation, and on a graph.

9. $x + 2 > 3$
10. $x + 5 \leq 9$
11. $x - 1 < -4$
12. $x - 3 \geq -1$
13. $2x \leq 6$
14. $3x > 9$
15. $4x \geq -8$
16. $2x < -10$
17. $-3t \geq 6$
18. $-2w \leq 2$
19. $-2x > 1$
20. $-4x < -2$
21. $5x \leq 0$
22. $-3x > 0$
23. $-x < 2$
24. $-x \leq -1$
25. $-\frac{2}{3}x \geq 2$
26. $-\frac{5}{2}x \leq 10$
27. $3x - 1 \geq 2$
28. $4x + 7 < 15$
29. $5 - 3x < -7$
30. $8 - 2x \leq 6$
31. $3c - 6 \leq 5c$
32. $7w + 4 > 3w$
33. $5x \geq x - 12$
34. $4x < 6 - 2x$
35. $-3.8x + 1.9 > -7.6$
36. $-2.4x + 5.8 \leq 8.92$
37. $3b + 2 > 7b - 6$
38. $5k - 1 \leq 2k + 8$
39. $4 - 3x < 9 - 2x$
40. $8 - x \geq 2 - 3x$

41. $2(x + 3) \leq 8$
42. $5(x - 2) \geq 15$
43. $-(a - 3) > 4$
44. $-(t + 5) < -2$
45. $3(2x - 1) \leq 2(2x + 1)$
46. $4(3x - 5) > 5(2x - 3)$
47. $4(2x - 3) + 1 \geq 3(4x - 5) - x$
48. $-2(5x + 3) - 2x < -3(2x - 4) + 2$
49. $4.3(1.5 - x) \geq 13.76$
50. $3.1(2.7 - x) > -1.55$
51. $\frac{1}{2}y + \frac{2}{3} \geq \frac{3}{2}$
52. $\frac{3}{4}t - \frac{1}{2} \leq \frac{1}{4}$
53. $\frac{5}{3} - \frac{1}{6}x < \frac{1}{2}$
54. $\frac{1}{4} - \frac{2}{3}x > \frac{7}{12}$
55. $-\frac{1}{2}x - \frac{5}{6} \geq \frac{1}{3} + \frac{3}{2}x$
56. $-\frac{3}{4}x + \frac{5}{2} < \frac{7}{8} - \frac{5}{2}x$
57. $\frac{4c - 5}{6} \leq \frac{3c + 7}{4}$
58. $\frac{6p + 2}{8} \geq \frac{4p - 1}{6}$
59. $\frac{3x + 1}{6} - \frac{5x - 2}{9} > \frac{2}{3}$
60. $\frac{4x - 7}{15} + \frac{2x + 3}{10} < \frac{2}{5}$

Substitute the given values for the variables in the compound inequality

$$\bar{x} - t \cdot \frac{s}{\sqrt{n}} < \mu < \bar{x} + t \cdot \frac{s}{\sqrt{n}}$$

to find a compound inequality that describes the variable μ. Describe the solution set as an inequality, in interval notation, and on a graph. Round to the first decimal place.

61. $\bar{x} = 18.5, t = 2.042, s = 3.1, n = 30$
62. $\bar{x} = 26.9, t = 2.528, s = 4.9, n = 20$
63. $\bar{x} = 87.2, t = 1.676, s = 9.8, n = 50$
64. $\bar{x} = 192.3, t = 2.131, s = 24.1, n = 15$

Substitute the given values for the variables in the compound inequality

$$\hat{p} - z \cdot \sqrt{\frac{\hat{p}(1 - \hat{p})}{n}} < p < \hat{p} + z \cdot \sqrt{\frac{\hat{p}(1 - \hat{p})}{n}}$$

to find a compound inequality that describes the variable p. Describe the solution set as an inequality, in interval notation, and on a graph. Round to the second decimal place.

65. $\hat{p} = 0.27, z = 1.96, n = 850$

66. $\hat{p} = 0.45, z = 1.645, n = 930$

67. $\hat{p} = 0.83, z = 2.33, n = 1005$

68. $\hat{p} = 0.62, z = 1.96, n = 1375$

For Exercises 69–76, solve the inequality. Describe the solution set as an inequality, in interval notation, and on a graph.

69. $4 < x + 3 < 8$

70. $-2 < x - 4 < 3$

71. $-15 \leq 2x - 5 \leq 7$

72. $-5 \leq 3x + 1 \leq 13$

73. $-17 < 3 - 4x \leq 15$

74. $7 < 5 - 2x \leq 13$

75. $\frac{1}{3} \leq 4 - \frac{2}{3}x < 2$

76. $\frac{3}{4} \leq 1 - \frac{1}{4}x < 3$

77. ᴰᴬᵀᴬ The percentages of osteopathic doctors who are women are shown in Table 23 for various years.

Table 23 Percentages of Osteopathic Doctors Who Are Women

Year	Percent
1985	9.7
1990	13.9
1995	17.9
2000	22.8
2005	26.7
2010	32.0
2013	34.6

Source: *American Medical Association*

Let p be the percentage of osteopathic doctors who are women at t years since 1980.

a. Construct a scatterplot.

b. Graph the model $p = 0.89t + 4.92$ on the scatterplot. Does it come close to most of the data points?

c. What is the slope? What does it mean in this situation?

d. Determine the years when more than 30% of osteopathic doctors were women.

78. ᴰᴬᵀᴬ The numbers of married-couple households (in millions) and the percentages of households that are married-couple households are shown in Table 24 for various years. Let p be the percentage of households that are married-couple households at t years since 2000.

a. Construct a scatterplot.

b. Graph the model $p = -0.24t + 52.19$ on the scatterplot. Does it come close to most of the data points?

c. What is the slope? What does it mean in this situation?

d. In which years were less than 52% of households married-couple households?

Table 24 Married-Couple Households

Year	Number of Married-Couple Households (millions)	Percentage of Households That Are Married-Couple Households
2003	57.3	51.5
2005	58.0	51.2
2008	58.3	50.0
2010	58.4	49.7
2014	61.9	49.0

Source: *U.S. Census Bureau*

e. Although the number of married-couple households *increased* from 2003 to 2014, the percentage of households that were married-couple households *decreased*. Explain how this is possible.

79. ᴰᴬᵀᴬ The percentages of Americans who think they will live comfortably when they retire are shown in Table 25 for various years.

Table 25 Percentages of Americans Who Think They Will Live Comfortably when They Retire

Year	Percent
2002	59
2004	59
2006	50
2008	46
2010	46
2012	38

Source: *Gallup Organization*

Let p be the percentage of Americans at t years since 2000 who think they will live comfortably when they retire.

a. Construct a scatterplot.

b. Describe the four characteristics of the association. Compute and interpret r as part of your analysis.

c. Graph the model $p = -2.11t + 64.47$ on the scatterplot. Does it come close to the data points?

d. What is the p-intercept? What does it mean in this situation? Do you have much faith in your result? Explain.

e. In which years did more than 41% of Americans think they will live comfortably when they retire?

80. ᴰᴬᵀᴬ The teenage birthrate in the United States has declined since 1990 (see Table 26).

Table 26 American Teenage Birthrate

Year	Birthrate (number of births per 1000 women ages 15–19)
1990	59.9
1995	56.0
2000	47.7
2005	40.5
2010	34.3
2013	26.5

Source: *U.S. National Center for Health Statistics*

Let B be the American teenage birthrate (number of births per 1000 women ages 15–19) at t years since 1990.

a. Construct a scatterplot.

b. Describe the four characteristics of the association. Compute and interpret r as part of your analysis.

c. Graph the model $B = -1.44t + 61.67$ on the scatterplot. Does it come close to the data points?

d. Estimate the teenage birthrate in 2012. Then estimate the *number* of births to women ages 15–19 in 2012. Use the U.S. Census Bureau's estimate that there were 10,526,000 women ages 15–19 in that year.

e. In which years were there fewer than 50 births per 1000 women ages 15–19?

81. DATA A dollar store is a no-frills retailer that sells many goods for one dollar. The percentages of households in various income groups that shop at dollar stores are shown in Table 27.

Table 27 Households That Shop at Dollar Stores

Household Income Group (thousands of dollars)	Income Used to Represent Income Group (thousands of dollars)	Percentage of Households That Shop at Dollar Stores
0–19.999	10	74
20–29.999	25	71
30–39.999	35	67
40–49.999	45	64
50–69.999	60	58
70+	100	45

Source: *ACNielsen Homescan Panel*

Let p be the percentage of households with an income of d thousand dollars that shop at dollar stores.

a. Construct a scatterplot.

b. Describe the four characteristics of the association. Compute and interpret r as part of your analysis.

c. Graph the model $p = -0.33t + 78.39$ on the scatterplot. Does it come close to the data points?

d. What is the p-intercept? What does it mean in this situation? Do you have much faith in your result? Explain.

e. At what incomes do more than half of households shop at dollar stores?

82. DATA The clarity of Lake Tahoe in California is measured by the depth at which a white disk with diameter 10 inches remains visible when lowered beneath the water's surface. The clarities of Lake Tahoe in the winter and summer are shown in Table 28 for various years.

Table 28 Clarity of Lake Tahoe

Clarity (feet)	
Winter	Summer
99.4	93.5
90.8	74.9
84.6	75.5
70.6	64.1
73.0	51.9

Source: *UC Davis Tahoe Environmental Research Center*

Let w be the clarity of Lake Tahoe in the winter and s be the clarity of Lake Tahoe in the summer, both in feet.

a. Construct a scatterplot, treating winter clarity as the explanatory variable.

b. Describe the four characteristics of the association. Compute and interpret r as part of your analysis.

c. Graph the model $s = 1.16w - 25.47$ on the scatterplot. Does it come close to the data points?

d. What is the slope? What does it mean in this situation?

e. Predict the summer clarity in years when the winter clarity is less than 80 feet.

83. DATA The percentages of Americans who went to the movies at least once in the past year are shown in Table 29 for various age groups.

Table 29 Percentages of Americans Who Go to the Movies

Age Group (years)	Age Used to Represent Age Group (years)	Percent
18–24	21.0	88
25–34	29.5	79
35–44	39.5	73
45–54	49.5	65
55–64	59.5	46
65–74	69.5	38
over 74	80	28

Source: *U.S. National Endowment for the Arts*

Let p be the percentage of Americans at age a years who go to the movies.

a. Construct a scatterplot.

b. Describe the four characteristics of the association. Compute and interpret r as part of your analysis.

c. Graph the model $p = -1.04a + 111.56$ on the scatterplot. Does it come close to the data points?

d. What is the slope? What does it mean in this situation?

e. Estimate at what ages more than half of Americans go to the movies.

84. DATA In a study of adults who had received radiation to the brain to treat brain cancer when they were children, researchers found that the cancer survivors tended to develop mild cognitive deficits typically seen in older people. The percentages of childhood brain cancer survivors who have mild cognitive deficits are shown in Table 30 for various ages.

Table 30 Percents of Childhood Brain Cancer Survivors Who Have Mild Cognitive Deficits

Age	Percent
20	22
30	27
40	55
50	77
60	98

Source: *St. Jude's Children's Research Hospital*

Let p be the percentage of childhood brain cancer survivors at age a years who have mild cognitive deficits.

a. Construct a scatterplot.
b. Describe the four characteristics of the association. Compute and interpret r as part of your analysis.
c. Graph the model $p = 2.02t - 25.00$ on the scatterplot. Does it come close to the data points?
d. What is the slope? What does it mean in this situation?
e. Estimate at what ages less than half of childhood brain cancer survivors have mild cognitive deficits.

Concepts

85. A student tries to solve $-3x < 15$:

$$-3x < 15$$
$$\frac{-3x}{-3} < \frac{15}{-3}$$
$$x < -5$$

Describe any errors. Then solve the inequality correctly.

86. A student tries to solve $4x < -24$:

$$4x < -24$$
$$\frac{4x}{4} > \frac{-24}{4}$$
$$x > -6$$

Describe any errors. Then solve the inequality correctly.

87. a. List three numbers that satisfy $3x - 7 < 5$.
b. List three numbers that do not satisfy $3x - 7 < 5$.

88. a. List three numbers that satisfy $2(x + 3) > 17$.
b. List three numbers that do not satisfy $2(x + 3) > 17$.

89. a. Solve $x + 1 = -2x + 10$.
b. Solve $x + 1 < -2x + 10$.
c. Solve $x + 1 > -2x + 10$.
d. Graph the solutions in parts (a), (b), and (c) on the same number line. Use three colors to identify the different solutions. Make observations about the solutions.

90. a. Is the following statement true?

$$\text{If } a < b, \text{ then } a - c < b - c.$$

Explain.
b. Is the following statement true?

$$\text{If } a < b \text{ and } c \neq 0, \text{ then } \frac{a}{c} < \frac{b}{c}.$$

Explain.

91. Use the number line to show that if $a < b$, then $2a < 2b$.

92. Use the number line to show that if $a < b$, then $-2a > -2b$.

93. Give an example of a linear inequality of the form $mx + b \leq c$, where m, b, and c are constants. Then solve the inequality. Describe the solution set as an inequality, in interval notation, and in a graph.

94. Describe how to solve a linear inequality in one variable. Describe when you need to reverse an inequality symbol. Explain why it is necessary to reverse the symbol in this case. Finally, explain what you have accomplished by solving an inequality.

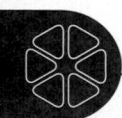

Chapter Summary

Key Points of Chapter 8

Section 8.1   Simplifying Expressions

Commutative law for addition	$a + b = b + a$
Commutative law for multiplication	$ab = ba$
Term	A **term** is a constant, a variable, or a product of a constant and one or more variables raised to powers.
Associative law for addition	$a + (b + c) = (a + b) + c$
Associative law for multiplication	$a(bc) = (ab)c$
Distributive law	$a(b + c) = ab + ac$
Multiplying a number by -1	$-1a = -a$
Equivalent expressions	Two or more expressions are **equivalent expressions** if, when each variable is evaluated for *any* real number (for which all the expressions are defined), the expressions all give equal results.
Like terms	**Like terms** are either constant terms or variable terms that contain the same variable(s) raised to exactly the same power(s).

Section 8.1   Simplifying Expressions (*Continued*)

Combining like terms	To combine like terms, add the coefficients of the terms and keep the same variable factors.
Simplifying an expression	We simplify an expression by removing parentheses and combining like terms.

Section 8.2   Solving Linear Equations in One Variable

Linear equation in one variable	A **linear equation in one variable** is an equation that can be put into the form $mx + b = 0$, where m and b are constants and $m \neq 0$.
Solution, satisfy, solution set, and solve for an equation in one variable	A number is a **solution** of an equation in one variable if the equation becomes a true statement when the number is substituted for the variable. We say that the number **satisfies** the equation. The set of all solutions of the equation is called the **solution set** of the equation. We **solve** the equation by finding its solution set.
Equivalent equations	**Equivalent equations** are equations that have the same solution set.
Addition property of equality	If A and B are expressions and c is a number, then the equations $A = B$ and $A + c = B + c$ are equivalent.
Checking results after solving an equation	After solving an equation, check that all of your results satisfy the equation.
Multiplication property of equality	If A and B are expressions and c is a nonzero number, then the equations $A = B$ and $Ac = Bc$ are equivalent.
Using graphing to solve an equation in one variable	To use graphing to solve an equation $A = B$ in one variable, x, where A and B are expressions, **1.** Graph the equations $y = A$ and $y = B$ in the same coordinate system. **2.** Find all intersection points. **3.** The x-coordinates of those intersection points are the solutions of the equation $A = B$.

Section 8.3   Solving Linear Equations to Make Predictions

Solving an equation that contains fractions	A key step in solving an equation that contains fractions is to multiply both sides of the equation by the LCD so that there are no fractions on either side of the equation.
Results of simplifying an expression and solving an equation	The result of simplifying an expression is an expression. The result of solving a linear equation in one variable is a number.
Multiplying expressions and both sides of equations by numbers	In simplifying an expression, the only number that we can multiply the expression or part of it by is 1. In solving an equation, we can multiply both sides of the equation by *any* number except 0.
Using an equation of a model to make predictions	• When making a prediction about the response variable of a linear model, substitute a chosen value for the explanatory variable in the model. Then solve for the response variable. • When making a prediction about the explanatory variable of a linear model, substitute a chosen value for the response variable in the model. Then solve for the explanatory variable.

Section 8.4   Solving Formulas

Formula	A **formula** is an equation that contains two or more variables.
Finding a single value of a variable in a formula	To find a single value of a variable in a formula, we often substitute numbers for all of the other variables and then solve for the remaining variable.
Finding several values of a variable in a formula	To find several values of a variable in a formula, we usually solve the formula for that variable before we make any substitutions.
Squaring a principal square root	If x is nonnegative, then $(\sqrt{x})^2 = x$.

Section 8.4   Solving Formulas (*Continued*)

Graphing a linear equation in two variables	To graph a linear equation in two variables,
	1. If the equation is in slope-intercept form $y = mx + b$, skip to Step 2. If the equation is not in slope-intercept form, put it in that form.
	2. Plot the y-intercept $(0, b)$.
	3. Use $m = \dfrac{\text{rise}}{\text{run}}$ to plot a second point. For example, if $m = \dfrac{3}{4}$, count 4 units to the right (from the y-intercept) and 3 units up to plot another point.
	4. Sketch the line that passes through the two plotted points.

Section 8.5   Solving Linear Inequalities to Make Predictions

Addition property of inequalities	If $a < b$, then $a + c < b + c$. Similar properties hold for $\leq$, $>$, and $\geq$.
Multiplication property of inequalities	• For a *positive* number c, if $a < b$, then $ac < bc$. • For a *negative* number c, if $a < b$, then $ac > bc$. Similar properties hold for $\leq$, $>$, and $\geq$.
Linear inequality in one variable	A **linear inequality in one variable** is an inequality that can be put into one of the forms $$mx + b < 0, \quad mx + b \leq 0, \quad mx + b > 0, \quad mx + b \geq 0$$ where m and b are constants and $m \neq 0$.
Satisfy an inequality in one variable	We say a number **satisfies** an inequality in one variable if the inequality becomes a true statement after we have substituted the number for the variable.
Solution, solution set, and solve for an inequality in one variable	We say a number is a **solution** of an inequality in one variable if it satisfies the inequality. The **solution set** of an inequality is the set of all solutions of the inequality. We **solve** an inequality by finding its solution set.

Chapter 8 Review Exercises

1. Use the commutative law for addition to write the expression $5w + 9$ in another form.
2. Use the commutative law for multiplication to write the expression $8 + wp$ in another form.

Use an associative law to write the expression in another form.

3. $2 + (k + y)$ 4. $(bx)w$

Simplify.

5. $-3(8x + 4)$

6. $\dfrac{4}{5}(15y - 35)$

7. $-(3x - 6y - 8)$

8. $5a + 2 - 13b - a + 4b - 9$

9. $-5y - 3(4x + y) - 6x$

10. $-2.6(3.1x + 4.5) - 8.5$

11. $-(2m - 4) - (3m + 8)$

12. $4(3a - 7b) - 3(5a + 4b)$

For Exercises 13 and 14, let x be a number. Translate the English phrase to a mathematical expression. Then simplify the expression.

13. -4 times the difference of the number and 7

14. -7, plus 2 times the sum of the number and 8

15. Which of the following expressions are equivalent?

$$-5x - 20 \qquad -5(x - 4) \qquad 5(4 - x) \qquad -5x - 4$$
$$-2(x - 10) - 3x \qquad -5x + 20$$

16. Determine whether 3 is a solution of the linear equation $2 - 5x = -3(4x - 7)$.

For Exercises 17–25, solve.

17. $a + 5 = 12$

18. $-4x = 20$

19. $-p = -3$

20. $-\dfrac{7}{3}a = 14$

21. $8m - 3 - m = 2 - 4m$

22. $8x = -7(2x - 3) + x$

23. $6(4x - 1) - 3(2x + 5) = 2(5x - 3)$

24. $\dfrac{w}{8} - \dfrac{3}{4} = \dfrac{5}{6}$

25. $\dfrac{3p - 4}{2} = \dfrac{5p + 2}{4} + \dfrac{7}{6}$

26. A student tries to solve the equation $x - 5 = 2$:

$$x - 5 = 2$$
$$x - 5 + 5 = 2$$
$$x + 0 = 2$$
$$x = 2$$

Describe any errors. Then solve the equation correctly.

27. Solve $-2.5(3.8x - 1.9) = 83.7$. Round the result to two decimal places.

For Exercises 28 and 29, simplify the expression or solve the equation, as appropriate.

28. $\frac{5}{6}r - \frac{3}{4} = \frac{1}{6} + \frac{7}{2}r$

29. $\frac{5}{6}r - \frac{3}{4} - \frac{1}{6} + \frac{7}{2}r$

30. A student tries to simplify the expression $\frac{2}{3}x + \frac{7}{5}$:

$$\frac{2}{3}x + \frac{7}{5} = 15 \cdot \left(\frac{2}{3}x + \frac{7}{5}\right)$$
$$= 15 \cdot \frac{2}{3}x + 15 \cdot \frac{7}{5}$$
$$= 10x + 21$$

Describe any errors. Then simplify the expression correctly.

31. Use "intersect" on a TI-84 to solve the equation $\frac{1}{2}x + \frac{5}{3} = -\frac{2}{3}x - \frac{1}{4}$. Round the solution to the second decimal place.

For Exercises 32 and 33, solve the given equation by referring to the solutions of $y = -2x + 17$ and $y = 5x - 4$ shown in Tables 31 and 32.

32. $-2x + 17 = 5x - 4$ **33.** $-2x + 17 = 15$

Table 31 Solutions of $y = -2x + 17$

x	y
0	17
1	15
2	13
3	11
4	9

Table 32 Solutions of $y = 5x - 4$

x	y
0	-4
1	1
2	6
3	11
4	16

For $f(x) = 2x + 3$, find the value of x that corresponds to the given value of f(x).

34. $f(x) = -5$ **35.** $f(x) = \frac{2}{3}$

36. In 2011, the mean salary of public school teachers was $56,643, and it increased by about $1199 per year until 2014 (Source: *National Education Association*). Let s be the mean salary (in dollars) at t years since 2011.
a. Find an equation of a model.
b. Estimate the mean salary in 2014.
c. Estimate when the mean salary was $59,041.

37. The total annual ad spending for the NCAA basketball tournament (known as "March Madness") was $239.1 million in 1998, and it increased by about $31.2 million per year until 2013 (Source: *TNS Media Intelligence*). Let $f(t)$ be the total annual ad spending (in millions of dollars) in the year that is t years since 1998.
a. Find an equation of f.
b. Find $f(14)$. What does it mean in this situation?
c. Find t when $f(t) = 600$. What does it mean in this situation?

38. In 2014, the number of Little League baseball participants was 2.4 million participants, up 14.3% from 2011 (source: *Little League International*). What was the number of participants in 2011?

39. **DATA** The percentages of men with high blood pressure are shown in Table 33 for various ages.

Table 33 Percentages of Men with High Blood Pressure

Age Group (years)	Age Used to Represent Age Group (years)	Percent
20–34	27	9.1
35–44	39.5	24.4
45–54	49.5	37.7
55–64	59.5	52.0
65–74	69.5	63.9
over 75.0	80.0	72.1

Source: *National Center for Health Statistics*

Let p be the percentage of men at age a years who have high blood pressure.
a. Construct a scatterplot.
b. Describe the four characteristics of the association. Compute and interpret r as part of your analysis.
c. Graph the model $p = 1.23a - 23.27$ on the scatterplot. Does it come close to the data points?
d. Predict the percentage of 30-year-old men who have high blood pressure.
e. Predict at what age half of men have high blood pressure.

40. Substitute 47.3 for x, 52.9 for $\bar{x}$, and 2.5 for s in the formula $x = \bar{x} + zs$, and then solve for z.

41. Let $\bar{x}$ be the mean (in points) of three test scores, x_1, x_2, and x_3, all in points.
a. Find a formula for the mean of the test scores.
b. If a student has test scores of 72 and 85, find the score she needs on the third test so that her three-test mean score is 80 points, which is the cutoff for a B.

42. Substitute the following values for the indicated variables in the formula $\chi^2 = \sum \frac{(O_i - E_i)^2}{E_i}$, and then solve for χ^2:
$O_1 = 8$, $O_2 = 15$, $O_3 = 12$, $E_1 = 6.5$, $E_2 = 17.8$, and $E_3 = 10.7$. Round the result to the second decimal place.

43. a. Solve $E = t \cdot \frac{s}{\sqrt{n}}$ for s.
b. Substitute 1.746 for t, 16 for n, and 3.623 for E in the formula that you found in part (a), and then solve for s. Round the result to the first decimal place.

44. The number of visits to U.S. libraries was 1.59 billion in 2009, and it increased by about 0.05 billion visits per year until 2014 (Source: *Institute of Museum and Library Services*). Let v be the number of visits (in billions) to U.S. libraries in the year that is t years since 2009.
a. Find an equation of a linear model to describe the situation.
b. Solve the equation you found in part (a) for t.
c. Use the equation you found in part (a) to estimate in which year there were 1.8 billion visits.
d. Use the equation you found in part (b) to estimate in which year there were 1.8 billion visits.
e. Compare your results from parts (c) and (d). Which equation was easier to use? Explain.

f. Which equation is easier to use to estimate the number of U.S. library visits in 2011? Explain. Then find that number of visits.

For Exercises 45 and 46, determine the slope and the y-intercept. Use the slope and the y-intercept to graph the equation by hand.

45. $3x - 2y = -6$ **46.** $-3(y + 2) = 2x + 9$

For Exercises 47–52, solve the inequality. Describe the solution set as an inequality, in interval notation, and on a graph.

47. $x - 3 \geq -4$

48. $-4x < 8$

49. $5w - 3 > 3w - 9$

50. $-3(2a + 5) + 5a \geq 2(a - 3)$

51. $\dfrac{2b - 4}{3} \leq \dfrac{3b - 4}{4}$

52. $1 \leq 2x + 5 < 11$

53. Substitute 75.9 for $\bar{x}$, 1.686 for t, 12.1 for s, and 38 for n in the compound inequality

$$\bar{x} - t \cdot \frac{s}{\sqrt{n}} < \mu < \bar{x} + t \cdot \frac{s}{\sqrt{n}}$$

to find a compound inequality that describes the variable μ. Describe the solution set as an inequality, in interval notation, and on a graph. Round to the first decimal place.

54. **DATA** Violent-crime rates in the United States are shown in Table 34 for various years.

Table 34 Violent-Crime Rates

Year	Number of Violent Crimes per 100,000 People
2004	463
2006	474
2008	458
2010	405
2012	388
2013	368

Source: *FBI*

Let v be the violent-crime rate (number of violent crimes per 100,000 people) at t years since 2000.

a. Construct a scatterplot.

b. Graph the model $v = -11.97t + 531.71$ on the scatterplot. Does it come close to the data points?

c. What is the slope? What does it mean in this situation?

d. In which years were the violent crime rates greater than 400 violent crimes per 100,000 people?

Chapter 8 Test

1. Use the commutative law for addition to write the expression $4 + 3p$ in another form.

2. Use an associative law to write the expression $3(xy)$ in another form.

For Exercises 3–5, simplify.

3. $-\dfrac{2}{3}(6x - 9)$

4. $-5(2w - 7) - 3(4w - 6)$

5. $-(3a + 7b) - (8a - 4b + 2)$

For Exercises 6–11, solve.

6. $6x - 3 = 19$

7. $\dfrac{3}{5}x = 6$

8. $9a - 5 = 8a + 2$

9. $8 - 2(3t - 1) = 7t$

10. $3(2x - 5) - 2(7x + 9) = 49$

11. $\dfrac{7}{8}x + \dfrac{3}{10} = \dfrac{1}{4}x - \dfrac{1}{2}$

12. Solve $8.21x = 3.9(4.4x - 2.7)$. Round your result to two decimal places.

For Exercises 13 and 14, simplify the expression or solve the equation, as appropriate.

13. $9(3x + 2) - (4x - 6)$

14. $9(3x + 2) - (4x - 6) = x$

15. A student believes that $x - 3$ is the solution of an equation. Is the student correct? Explain.

16. Give three examples of an expression equivalent to the expression 4.

For Exercises 17 and 18, solve the given equation by referring to the graphs of $y = \dfrac{3}{2}x - 4$ and $y = \dfrac{1}{2}x - 2$ shown in Fig. 47.

17. $\dfrac{3}{2}x - 4 = \dfrac{1}{2}x - 2$ **18.** $\dfrac{1}{2}x - 2 = -3$

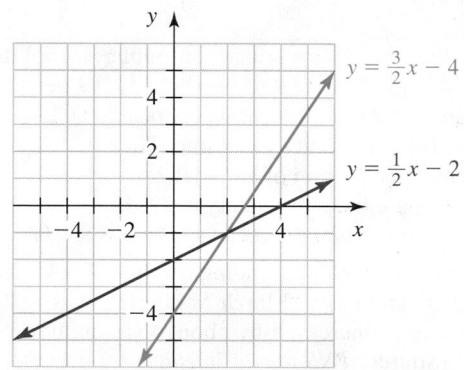

Figure 47 Exercises 17 and 18

19. For $f(x) = 4x - 7$, find the value of x when $f(x) = -1$.

20. The number of U.S. patent applications was 520 thousand in 2010, and it increased by about 21.7 thousand per year until 2014 (Source: *U.S. Patent and Trademark Office*). Let n be the number of patent applications (in thousands) during the year that is t years since 2010.

a. Find an equation of a model.

b. Estimate the number of patent applications in 2014.

c. Estimate in which year there were 580 thousand patent applications.

21. In 2013, the number of complaints of online crime was 263 thousand complaints, down 9.3% from 2012 (Source: *Internet Crime Complaint Center*). What was the number of complaints in 2012?

22. Substitute 0.9 for $P(E \text{ OR } F)$, 0.4 for $P(E)$, and 0.7 for $P(F)$ in the formula $P(E \text{ OR } F) = P(E) + P(F) - P(E \text{ AND } F)$, and then solve for $P(E \text{ AND } F)$.

23. Solve the formula $z = \dfrac{G - \mu_G}{\sigma_G}$ for G.

24. **DATA** The First Amendment says: "Congress shall make no law respecting an establishment of religion or prohibiting the free exercise thereof, or abridging the freedom of speech or of the press, or the right of the people peaceably to assemble, and to petition the government for a redress of grievances." The percentages of Americans who feel the amendment goes too far in the rights it guarantees are shown in Table 35 for various years.

Table 35 Percentages of Americans Who Feel the First Amendment Goes Too Far in the Rights It Guarantees

Year	Percent
2001	39
2003	34
2005	23
2007	25
2009	19
2011	18

Source: *State of the First Amendment 2011*

Let p be the percentage of Americans who think the First Amendment goes too far in the rights it guarantees at t years since 2000.

a. Construct a scatterplot.

b. Graph the model $p = -2.11t + 39.02$ on the scatterplot. Does it come close to the data points?

c. What is the slope? What does it mean in this situation?

d. Solve the equation $p = -2.11t + 39.02$ for t.

e. Use the result you found in part (d) to estimate the years when the percentages of Americans who felt the First Amendment goes too far in the rights it guarantees were equal to 18%, 22%, 26%, 30%, and 34%.

25. Determine the slope and the y-intercept of the line $2x - 3y = 12$. Use the slope and the y-intercept to graph the equation by hand.

Solve the inequality. Describe the solution set as an inequality, in interval notation, and on a graph.

26. $3(2x + 1) \le 4(x + 2) - 1$ **27.** $\dfrac{1}{2} < 3 - \dfrac{5}{2}x \le 13$

28. **DATA** The Internal Revenue Service (IRS) standard mileage rate is a way of computing an automobile expense deduction on a tax return. Mileage rates for businesses are provided in Table 36 for various years.

Table 36 IRS Standard Mileage Rates for Businesses

Year	Standard Mileage Rate (cents per mile)
2000	32.5
2002	36.5
2004	37.5
2006	44.5
2008	54.5
2010	50.0
2012	55.5
2014	56.0

Source: *IRS*

Let $M = f(t)$ be the standard mileage rate (in cents per mile) at t years since 2000.

a. Construct a scatterplot.

b. Graph the model $f(t) = 1.83t + 33.08$ on the scatterplot. Does it come close to the data points?

c. Find $f(13)$. What does it mean in this situation?

d. Find t when $f(t) = 53$. What does it mean in this situation?

e. In which years were the standard mileage rates less than 46 cents per mile?

Finding Equations of Linear Models

If you could stop time and live forever in good health at a particular age, what age would you choose? The mean ideal ages chosen by various age groups are shown in Table 1. What patterns do you notice? In Exercise 33 of Homework 9.3, you will find an equation of a linear model that can be used to predict the mean ideal age people of any current age between 21 and 75 years would choose.

Table 1 Mean Ideal Ages

Age Group (years)	Age Used to Represent Age Group (years)	Mean Ideal Age (years)
18–24	21	27
25–29	27	31
30–39	34.5	37
40–49	44.5	40
50–64	57	44
over 64	75	59

Source: *Harris Poll*

In Chapters 7 and 8, we used the *y*-intercept and the slope of a line to find an equation of the line. In this chapter, we will use the more flexible approach of using two points to find an equation of a line. Then we will use the same approach to find an equation of a linear model. Similar to the slope method, this approach gives a reasonable model but not necessarily the best one. Finally, we will use a statistical approach that involves *all* the data points in a scatterplot to find an equation of the linear model that fits the data points better than any other linear model.

9.1 Using Two Points to Find an Equation of a Line

Objectives

» Find an equation of a line by using the slope-intercept form of a linear equation.

» Find an equation of a line by using the point-slope form of a linear equation.

In this section, we will use two methods to find an equation of a line. We will first use the slope-intercept form. Then we will use another form of an equation of a line called the *point-slope form*.

Method 1: Using the Slope-Intercept Form

In Example 1, we will use the concept that a point that lies on a line satisfies an equation of the line.

▶ **Example 1** Using the Slope and a Point to Find a Linear Equation

Find an equation of the line that has slope $m = 2$ and contains the point $(4, 3)$.

Solution

An equation of a nonvertical line can be put into the form $y = mx + b$. Since $m = 2$, we have

$$y = 2x + b$$

To find b, recall from Section 7.1 that any point on the graph of an equation represents a solution of that equation. In particular, the point $(4, 3)$ should satisfy the equation $y = 2x + b$:

$$y = 2x + b \qquad \text{Slope is } m = 2.$$
$$3 = 2(4) + b \qquad \text{Substitute 4 for } x \text{ and 3 for } y.$$
$$3 = 8 + b \qquad \text{Multiply.}$$
$$3 - 8 = 8 + b - 8 \qquad \text{Subtract 8 from both sides.}$$
$$-5 = b \qquad \text{Combine like terms.}$$

Now we substitute -5 for b in $y = 2x + b$:

$$y = 2x - 5$$

To verify our work, we check that the coefficient of the variable term $2x$ is 2 (the given slope) and we see whether $(4, 3)$ satisfies the equation $y = 2x - 5$:

$$y = 2x - 5 \qquad \text{The equation we found}$$
$$3 \overset{?}{=} 2(4) - 5 \qquad \text{Substitute 4 for } x \text{ and 3 for } y.$$
$$3 \overset{?}{=} 8 - 5 \qquad \text{Multiply.}$$
$$3 \overset{?}{=} 3 \qquad \text{Subtract.}$$
$$\text{true}$$

Finding an Equation of a Line by Using the Slope, a Point, and the Slope-Intercept Form

To find an equation of a line by using the slope and a point,

1. Substitute the given value of the slope m into the equation $y = mx + b$.

2. Substitute the coordinates of the given point into the equation you found in step 1 and solve for b.

3. Substitute the value of b you found in step 2 into the equation you found in step 1.

4. Check that the graph of your equation contains the given point.

▶ **Example 2**   Using the Slope and a Point to Find a Linear Equation

Find an equation of the line that has slope $m = -\dfrac{2}{5}$ and contains the point $(-4, 1)$.

Solution

Since the slope is $-\dfrac{2}{5}$, the equation has the form $y = -\dfrac{2}{5}x + b$. To find b, we substitute the coordinates of the point $(-4, 1)$ into the equation $y = -\dfrac{2}{5}x + b$ and solve for b:

$$y = -\frac{2}{5}x + b \qquad \text{Slope is } -\frac{2}{5}.$$

$$1 = -\frac{2}{5}(-4) + b \qquad \text{Substitute } -4 \text{ for } x \text{ and 1 for } y.$$

$$1 = \frac{8}{5} + b \qquad \text{Simplify.}$$

$$5 = 8 + 5b \qquad \text{Multiply both sides by LCD, 5, to clear equation of fractions.}$$

$$5 - 8 = 8 + 5b - 8 \qquad \text{Subtract 8 from both sides.}$$

$$-3 = 5b \qquad \text{Combine like terms.}$$

$$-\frac{3}{5} = b \qquad \text{Divide both sides by 5.}$$

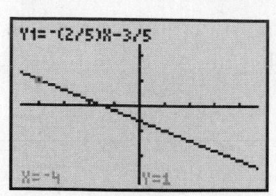

Figure 1 Check that the line contains $(-4, 1)$

So, the equation is $y = -\dfrac{2}{5}x - \dfrac{3}{5}$. In Fig. 1, we use ZDecimal followed by TRACE on a TI-84 to check that the line $y = -\dfrac{2}{5}x - \dfrac{3}{5}$ contains the point $(-4, 1)$.

In Examples 1 and 2, we found an equation of a line by using the slope of the line and a point. We can also find an equation of a line by using two points.

▶ **Example 3**   Using Two Points to Find a Linear Equation

Find an equation of the line that passes through $(-1, 4)$ and $(2, -5)$.

Solution

First, we find the slope of the line:

$$m = \frac{y_2 - y_1}{x_2 - x_1} = \frac{-5 - 4}{2 - (-1)} = \frac{-5 - 4}{2 + 1} = \frac{-9}{3} = -3$$

So, we have $y = -3x + b$. Next, we will find the value of b. Since the line contains the point $(-1, 4)$, we substitute -1 for x and 4 for y:

$$
\begin{aligned}
y &= -3x + b & &\text{Slope is } m = -3.\\
4 &= -3(-1) + b & &\text{Substitute } -1 \text{ for } x \text{ and 4 for } y.\\
4 &= 3 + b & &\text{Multiply.}\\
4 - 3 &= 3 + b - 3 & &\text{Subtract 3 from both sides.}\\
1 &= b & &\text{Combine like terms.}
\end{aligned}
$$

The equation is $y = -3x + 1$. To verify our equation, we check that both $(-1, 4)$ and $(2, -5)$ satisfy the equation $y = -3x + 1$:

Check that $(-1, 4)$ is a solution	**Check that $(2, -5)$ is a solution**
$y = -3x + 1$	$y = -3x + 1$
$4 \overset{?}{=} -3(-1) + 1$	$-5 \overset{?}{=} -3(2) + 1$
$4 \overset{?}{=} 3 + 1$	$-5 \overset{?}{=} -6 + 1$
$4 \overset{?}{=} 4$	$-5 \overset{?}{=} -5$
true	true

In Example 3, we substituted the coordinates of the given point $(-1, 4)$ into the equation $y = -3x + b$ to help us find the constant b. If we had used the other given point, $(2, -5)$, we would have found the same value of b. (Try it.)

> **Finding an Equation of a Line by Using Two Points and the Slope–Intercept Form**
>
> To find an equation of the line that passes through two given points whose x-coordinates are different,
>
> 1. Use the formula $m = \dfrac{y_2 - y_1}{x_2 - x_1}$ to find the slope of the line containing the two points.
> 2. Substitute the m value you found in step 1 into the equation $y = mx + b$.
> 3. Substitute the coordinates of one of the given points into the equation you found in step 2 and solve for b.
> 4. Substitute the b value you found in step 3 into the equation you found in step 2.
> 5. Check that the graph of your equation contains the two given points.

▶ **Example 4** Using Two Points to Find a Linear Equation

Find an equation of the line that passes through the points $(-9, -2)$ and $(-3, 7)$.

Solution

We begin by finding the slope of the line:

$$m = \frac{y_2 - y_1}{x_2 - x_1} = \frac{7 - (-2)}{-3 - (-9)} = \frac{9}{6} = \frac{3}{2}$$

So, we have $y = \frac{3}{2}x + b$. To find b, we substitute the coordinates of $(-3, 7)$ into the equation $y = \frac{3}{2}x + b$ and solve for b:

$$y = \frac{3}{2}x + b \qquad \text{Slope is } \frac{3}{2}.$$

$$7 = \frac{3}{2}(-3) + b \qquad \text{Substitute } -3 \text{ for } x \text{ and } 7 \text{ for } y.$$

$$7 = -\frac{9}{2} + b \qquad \text{Simplify.}$$

$$14 = -9 + 2b \qquad \text{Multiply both sides by LCD, 2, to clear equation of fractions.}$$

$$14 + 9 = -9 + 2b + 9 \qquad \text{Add 9 to both sides.}$$

$$23 = 2b \qquad \text{Combine like terms.}$$

$$\frac{23}{2} = b \qquad \text{Divide both sides by 2.}$$

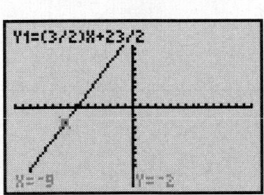

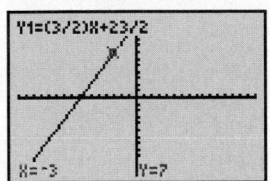

Figure 2 Check that the line contains both $(-9, -2)$ and $(-3, 7)$

The equation is $y = \frac{3}{2}x + \frac{23}{2}$. We use ZStandard followed by ZSquare to check that the line $y = \frac{3}{2}x + \frac{23}{2}$ contains the points $(-9, -2)$ and $(-3, 7)$. See Fig. 2. For TI-84 instructions, see Appendices A.14–A.17.

▶

In Example 5, we will work with decimal numbers to prepare us to find equations of models in Section 9.2.

▶ **Example 5** Finding an Approximate Equation of a Line

Find an approximate equation of the line that contains the points $(-3.1, 5.7)$ and $(1.6, -4.8)$.

Solution

First, we find the slope of the line:

$$m = \frac{y_2 - y_1}{x_2 - x_1} = \frac{-4.8 - 5.7}{1.6 - (-3.1)} = \frac{-10.5}{4.7} \approx -2.23$$

So, the equation has the form $y = -2.23x + b$. To find b, we substitute the coordinates of the point $(-3.1, 5.7)$ into the equation $y = -2.23x + b$ and solve for b:

$$y = -2.23x + b \qquad \text{Slope is approximately } -2.23.$$

$$5.7 = -2.23(-3.1) + b \qquad \text{Substitute } -3.1 \text{ for } x \text{ and } 5.7 \text{ for } y.$$

$$5.7 = 6.913 + b \qquad \text{Multiply.}$$

$$5.7 - 6.913 = 6.913 + b - 6.913 \qquad \text{Subtract 6.913 from both sides.}$$

$$-1.21 \approx b \qquad \text{Combine like terms.}$$

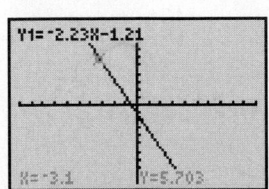

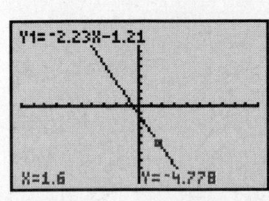

Figure 3 Check that the line comes very close to $(-3.1, 5.7)$ and $(1.6, -4.8)$

The approximate equation is $y = -2.23x - 1.21$.

We use a TI-84 to check that the line $y = -2.23x - 1.21$ comes very close to the points $(-3.1, 5.7)$ and $(1.6, -4.8)$. See Fig. 3.

▶

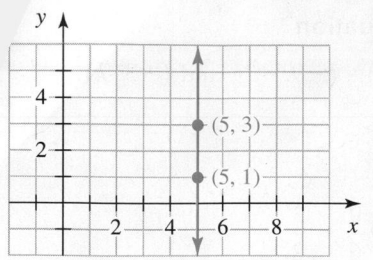

Figure 4 The line that contains (5, 1) and (5, 3)

▶ Example 6   Finding an Equation of a Vertical Line

Find an equation of the line that contains the points $(5, 1)$ and $(5, 3)$.

Solution

Since the x-coordinates of the given points are equal (both 5), the line that contains the points is vertical (see Fig. 4). An equation of the line is $x = 5$. ◀

Method 2: Using the Point-Slope Form

We can find an equation of a line by another method. Suppose a nonvertical line has slope m and contains the point (x_1, y_1). Then if (x, y) represents a different point on the line, the slope of the line is

$$\frac{y - y_1}{x - x_1} = m$$

Multiplying both sides of the equation by $x - x_1$ gives

$$\frac{y - y_1}{x - x_1} \cdot (x - x_1) = m(x - x_1)$$

$$y - y_1 = m(x - x_1)$$

We say this linear equation is in **point-slope form**.*

▶ **Point-Slope Form**

If a nonvertical line has slope m and contains the point (x_1, y_1), then an equation of the line is

$$y - y_1 = m(x - x_1)$$

▶ Example 7   Using the Point-Slope Form to Find an Equation of a Line

Use the point-slope form to find an equation of the line that has slope $m = -3$ and contains the point $(-4, 2)$. Then write the equation in slope-intercept form.

Solution

We begin by substituting $x_1 = -4$, $y_1 = 2$, and $m = -3$ into the point-slope form $y - y_1 = m(x - x_1)$:

$y - y_1 = m(x - x_1)$	*Point-slope form*
$y - 2 = -3(x - (-4))$	*Substitute $x_1 = -4$, $y_1 = 2$, and $m = -3$.*
$y - 2 = -3(x + 4)$	*Simplify.*
$y - 2 = -3x - 12$	*Distributive law*
$y - 2 + 2 = -3x - 12 + 2$	*Add 2 to both sides.*
$y = -3x - 10$	*Combine like terms.*

So, the equation is $y = -3x - 10$. ◀

---

*Although we assumed that (x, y) is different from (x_1, y_1), note that (x_1, y_1) is a solution of the equation $y - y_1 = m(x - x_1)$: $y_1 - y_1 = m(x_1 - x_1)$, or $0 = 0$, a true statement.

▶ **Example 8** Using the Point-Slope Form to Find an Equation of a Line

Use the point-slope form to find an equation of the line that contains the points $(2, -6)$ and $(5, -4)$. Then write the equation in slope-intercept form.

Solution

We begin by finding the slope of the line:

$$m = \frac{-4 - (-6)}{5 - 2} = \frac{2}{3}$$

Then we substitute $x_1 = 2, y_1 = -6,$ and $m = \frac{2}{3}$ into the equation $y - y_1 = m(x - x_1)$:

$$y - y_1 = m(x - x_1) \quad \text{Point-slope form}$$

$$y - (-6) = \frac{2}{3}(x - 2) \quad \text{Substitute, } x_1 = 2, y_1 = -6, \text{ and } m = \frac{2}{3}.$$

$$y + 6 = \frac{2}{3}x - \frac{4}{3} \quad \text{Simplify; distributive law}$$

$$y + 6 - 6 = \frac{2}{3}x - \frac{4}{3} - 6 \quad \text{Subtract 6 from both sides.}$$

$$y = \frac{2}{3}x - \frac{22}{3} \quad \text{Combine like terms; } -\frac{4}{3} - 6 = -\frac{4}{3} - \frac{18}{3} = -\frac{22}{3}$$

So, the equation is $y = \frac{2}{3}x - \frac{22}{3}$.

▶

Group Exploration

Deciding which points to use to find an equation of a line

1. **a.** Use the method shown in Example 3 to find an equation of the line that contains the points $(1, 2)$ and $(3, 8)$.
   b. In part (a), you used either point $(1, 2)$ or point $(3, 8)$ to find the constant b for the equation $y = 3x + b$. Now use the other point to find b.
   c. Does it matter which point is used to find the constant b? Explain.

2. Imagine any line that is not parallel to either axis. Choose four points on the line. Name the points A, B, C, and D.
   a. Use points A and B to find an equation of the line. Write your equation in slope-intercept form.
   b. Use points C and D to find an equation of the line. Write your equation in slope-intercept form.
   c. Are the equations you found in parts (a) and (b) the same? Explain.

Group Exploration

DATA Finding equations of lines

The objective of this game is to earn 19 credits. You earn credits by finding equations of lines that pass through one or more of the following points:

$$(-3, 2), (-3, 0), (-2, -7), (-2, -1), (-1, 4), (0, 2),$$
$$(1, -1), (2, -2), (3, 1), (3, 3)$$

If a line passes through exactly one point, then you earn one credit. If a line passes through exactly two points, then you earn three credits. If a line passes through exactly three points, then you earn five credits. You may use five equations. You may use points more than once. [**Hints:** First, construct a scatterplot. After finding your equations, use technology to check that they are correct.]

▶ **Tips for Success**   **Show What You Know**

Even if you don't know how to do one step of a problem, you can still show the instructor that you understand the other steps of the problem. Depending on how your instructor grades tests, you may earn partial credit even though you pick an incorrect number to be the result for a particular step, as long as you then show what you would do with that number in the remaining steps of the solution. Check with your instructor first.

For example, suppose you want to find an equation of the line that passes through the points $(1, 5)$ and $(2, 8)$, but you have forgotten how to find the slope. You could still write,

I've drawn a blank on finding slope. However, assuming that the slope is 2, then

$$y = 2x + b$$
$$5 = 2(1) + b$$
$$5 = 2 + b$$
$$5 - 2 = 2 + b - 2$$
$$3 = b$$

Therefore, $y = 2x + 3$.

You could point out that you know your result is incorrect, because the graph of $y = 2x + 3$ does *not* pass through the point $(2, 8)$. Also, seeing your result (with the graph) may jog your memory about finding the slope and allow you to go back and do the problem correctly.

Homework 9.1

For extra help ▶ MyMathLab®    Watch the videos in MyMathLab    Download the MyDashboard App

1. *True or False:* If a line has slope 3 and contains the point $(2, 5)$, then an equation of the line is $y = 3x + 5$.

2. *True or False:* If a line has slope 4 and contains the point $(0, 7)$, then an equation of the line is $y = 4x + 7$.

3. An equation written in the form $y - y_1 = m(x - x_1)$ is in ____–slope form.

4. *True or False:* The graph of the equation $y - 2 = 6(x - 4)$ is a line that has slope 6 and contains the point $(-4, -2)$.

Find an equation of the line that has the given slope and contains the given point. If possible, write your equation in slope–intercept form.

5. $m = 2$, $(3, 5)$	6. $m = 3$, $(2, 4)$
7. $m = -3$, $(1, -2)$	8. $m = -5$, $(3, -8)$
9. $m = -6$, $(-2, -3)$	10. $m = -1$, $(-7, -4)$
11. $m = \dfrac{2}{5}$, $(3, 1)$	12. $m = \dfrac{1}{2}$, $(5, 3)$
13. $m = -\dfrac{4}{5}$, $(2, -3)$	14. $m = -\dfrac{7}{3}$, $(-1, -4)$
15. $m = -\dfrac{3}{4}$, $(-2, -5)$	16. $m = -\dfrac{5}{3}$, $(-4, -2)$
17. $m = 0$, $(5, 3)$	18. $m = 0$, $(-1, -3)$
19. m is undefined, $(-2, 4)$	20. m is undefined, $(3, -2)$

Find an approximate equation of the line that has the given slope and contains the given point. Write your equation in slope–intercept form. Round the constant term to two decimal places.

21. $m = 2.1$, $(3.7, 5.9)$	22. $m = 1.3$, $(6.6, 3.8)$
23. $m = -6.59$, $(2.48, -1.61)$	24. $m = -2.07$, $(-4.73, 9.60)$
25. $m = -13.9$, $(-85.6, -254.8)$	
26. $m = -25.3$, $(-42.8, -93.3)$	

Find an equation of the line that passes through the two given points. If possible, write your equation in slope–intercept form.

27. $(3, 2)$ and $(5, 6)$	28. $(1, 4)$ and $(2, 1)$
29. $(-1, -7)$ and $(2, 8)$	30. $(-2, -10)$ and $(3, 5)$
31. $(-5, -4)$ and $(-2, -10)$	32. $(-3, -2)$ and $(-1, -8)$
33. $(0, 9)$ and $(2, 1)$	34. $(-3, 1)$ and $(0, -5)$
35. $(3, 2)$ and $(5, 2)$	36. $(-5, -3)$ and $(-1, -3)$
37. $(-4, -1)$ and $(-4, 3)$	38. $(7, 1)$ and $(7, 6)$
39. $(4, 3)$ and $(8, 5)$	40. $(2, 3)$ and $(6, 1)$
41. $(-3, 2)$ and $(3, 1)$	42. $(2, -2)$ and $(6, -5)$
43. $(-2, 1)$ and $(5, -1)$	44. $(-6, 5)$ and $(-2, 2)$
45. $(-4, -2)$ and $(6, 4)$	46. $(-1, -2)$ and $(5, 6)$
47. $(-4, -8)$ and $(-2, -5)$	48. $(-6, -9)$ and $(-2, -4)$

Find an approximate equation of the line that passes through the two given points. Write your equation in slope–intercept form. Round the slope and the constant term to two decimal places.

49. (4.5, 2.2) and (1.2, 7.5)

50. (8.1, 5.3) and (3.3, 2.7)

51. (−4.57, 8.29) and (7.17, −2.69)

52. (8.99, −4.82) and (−5.85, 3.92)

53. (−398.1, −125.9) and (−258.2, −499.5)

54. (−287.6, −883.7) and (−483.8, −650.2)

55. Find an equation of the line sketched in Fig. 5. [**Hint:** Choose two points whose coordinates appear to be integers.]

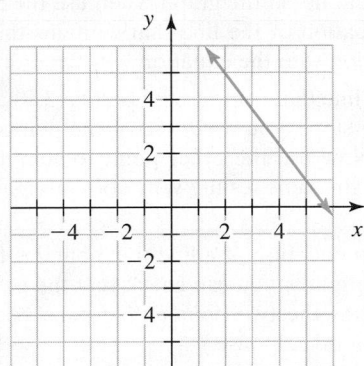

Figure 5 Exercise 55

56. Find an equation of the line sketched in Fig. 6. [**Hint:** Choose two points whose coordinates appear to be integers.]

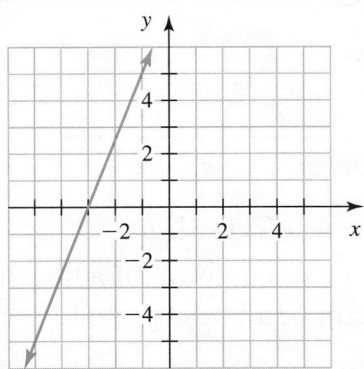

Figure 6 Exercise 56

Concepts

57. A student thinks that if a line has slope 2 and contains the point (3, 5), then the equation of the line is $y = 2x + 5$, because the slope is 2 (the coefficient of x) and the y-coordinate of (3, 5) is 5 (the constant term). What would you tell the student?

58. A student tries to find an equation of the line that contains the points (1, 5) and (3, 9). The student believes an equation of the line is $y = 4x + 1$. The student then checks whether (1, 5) satisfies $y = 4x + 1$:

$$y = 4x + 1$$
$$5 \stackrel{?}{=} 4(1) + 1$$
$$5 \stackrel{?}{=} 5$$
$$\text{true}$$

The student concludes that $y = 4x + 1$ is an equation of the line. Find any errors. Then find an equation correctly.

59. Consider the line that contains the points (−2, −6) and (4, 3).
  a. Find an equation of the line.
  b. Sketch the line on a coordinate system.
  c. Use a table to list five ordered pairs that correspond to points on the line.

60. Consider the line that contains the points (−3, 4) and (6, −2).
  a. Find an equation of the line.
  b. Sketch the line on a coordinate system.
  c. Use a table to list five ordered pairs that correspond to points on the line.

61. Decide whether it is possible for a line to have the indicated number of x-intercepts. If it is possible, find an equation of such a line. If it is not possible, explain why not.
  a. no x-intercepts
  b. exactly one x-intercept
  c. exactly two x-intercepts
  d. an infinite number of x-intercepts

62. Decide whether it is possible for a line to have the indicated number of y-intercepts. If it is possible, find an equation of such a line. If it is not possible, explain why not.
  a. no y-intercepts
  b. exactly one y-intercept
  c. exactly two y-intercepts
  d. an infinite number of y-intercepts

63. <u>DATA</u> Let E be the enrollment (in thousands of students) at a college t years after the college opens. Some pairs of values of t and E are listed in Table 2.

Table 2 Enrollments

Age of College (years) t	Enrollment (thousands of students) E
2	9
4	13
7	19
9	23
12	29

  a. Construct a scatterplot.
  b. Describe the four characteristics of the association.
  c. Find an equation that describes the association between t and E.
  d. Graph the equation you found in part (c) on the scatterplot.

64. <u>DATA</u> Let s be a person's savings (in thousands of dollars) at t years since 2000. Some pairs of values of t and s are listed in Table 3.

Table 3 A Person's Savings

Years t	Savings (thousands of dollars) s
1	5
4	14
5	17
7	23
10	32

  a. Construct a scatterplot.
  b. Describe the four characteristics of the association.

c. Find an equation that describes the association between t and s.

d. Graph the equation you found in part (c) on the scatterplot.

65. a. Use each of the following forms to find an equation of the line that contains the points (2, 1) and (4, 7). Write each result in slope–intercept form.

   i. slope–intercept form
   ii. point–slope form

   b. Compare the results you found in parts (ai) and (aii).

66. a. Use each of the following forms to find an equation of the line that contains the points $(-4, 3)$ and $(2, -5)$. Write each result in slope–intercept form.

   i. slope–intercept form
   ii. point–slope form

   b. Compare the results you found in parts (ai) and (aii).

67. a. Find an equation of a line with slope -2. [**Hint:** There are *many* correct answers.]

   b. Find an equation of a line with y-intercept (0, 4).

   c. Find an equation of a line that contains the point (3, 5).

   d. Determine whether there is a line that has slope -2 and y-intercept (0, 4) and contains the point (3, 5). Explain.

68. Suppose you are trying to find the equation of a line that contains two given points, and you find that the slope is undefined. What type of line is it? Explain.

69. Create a table of seven pairs of values of x and y for which

   a. each point lies on the line $y = 3x - 6$.

   b. each point lies close to, but not on, the line $y = 3x - 6$.

c. the points do not lie close to the line $y = 3x - 6$, but all of them lie close to another line. In addition to creating the table, provide an equation of the other line.

70. Suppose a set of points all lie 0.5 unit above the line $y = -4x + 3$. Find an equation of the line that passes through the points of the set.

71. Sketch a vertical line on a coordinate system. Find an equation of the line. What is the slope of the line?

72. Sketch a decreasing line that is nearly horizontal on a coordinate system. Find an equation of the line.

73. Find an equation of a line that has no solutions in Quadrant I.

74. a. Graph $y = 2x - 3$ by hand.

   b. Choose two points that lie on the graph. Then use the two points to find an equation of the line that contains them. Compare your equation with the equation $y = 2x - 3$.

75. Find an equation of the line that contains the points (3, 7) and (4, 9). After finding the slope, you used one of the points to complete the process. Now use the other point to complete the process. Did you get the same result? Why does this makes sense?

76. Describe how to find an equation of a line that contains two given points. Also, explain how you can check that the graph of your equation contains the two points. (See page 12 for guidelines on writing a good response.)

9.2 Using Two Points to Find an Equation of a Linear Model

Objectives

» Use two points to find an equation of a linear model.

Use Two Points to Find an Equation of a Linear Model

In Section 9.1, we used two given points to find an equation of a line. In this section, we use this skill to find an equation of a linear model.

Table 4 Numbers of Apple Stores

Year	Number of Apple Stores
2003	51
2005	95
2007	169
2009	249
2011	321
2012	358

Source: *Apple, Inc.*

▶ Example 1 Finding an Equation of a Linear Model

The numbers of Apple stores are shown in Table 4 for various years. Let n be the number of Apple stores at t years since 2000. Find an equation of a model that comes close to the points in the scatterplot of the data.

Solution

We begin by viewing the positions of the data points in the scatterplot (see Fig. 7). It appears that there is a linear association, so we will find an equation of a linear model. It is not necessary to use two *data* points to find such an equation, although it is often convenient and satisfactory to do so.

The red line that contains points (3, 51) and (5, 95) does *not* come close to the other data points (see Fig. 7). However, the green line that passes through points (9, 249) and (12, 358) appears to come close to the rest of the points. We will find an equation of this line.

An equation of a line can be written in the form $y = mx + b$, where x is the explanatory variable. Because the variables t and n are linearly associated and t is the

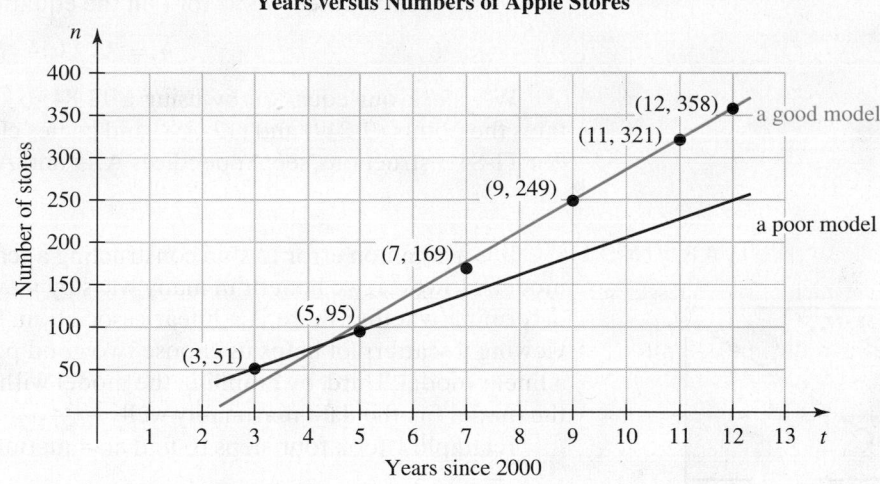

Figure 7 Apple stores scatterplot, a good model, and a poor model

explanatory variable, we will find an equation of the form $n = mt + b$. To find the equation, we first use the points $(9, 249)$ and $(12, 358)$ to find m (see Fig. 8):

$$m = \frac{358 - 249}{12 - 9} = \frac{109}{3} \approx 36.33$$

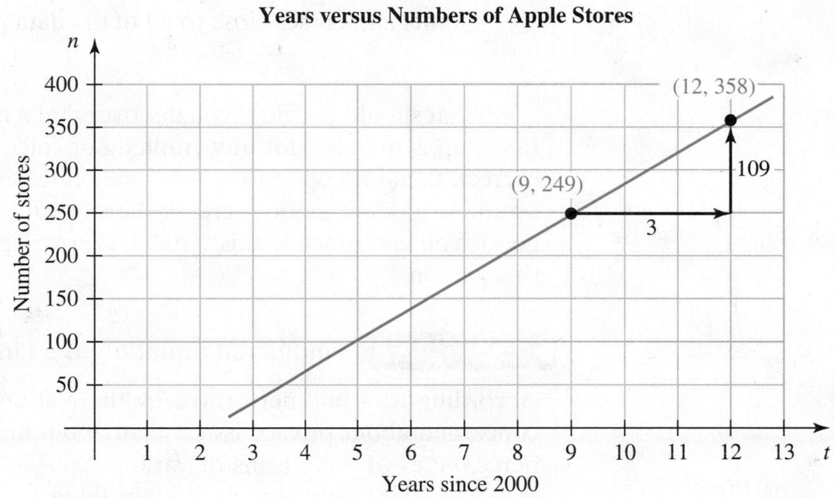

Years versus Numbers of Apple Stores

Figure 8 Two points of the Apple stores scatterplot

Then we substitute 36.33 for m in the equation $n = mt + b$:

$$n = 36.33t + b$$

To find the constant b, we substitute the coordinates of the point $(9, 249)$ into the equation $n = 36.33t + b$ and then solve for b:

$$
\begin{array}{ll}
249 = 36.33(9) + b & \text{\textit{Substitute 9 for t and 249 for n.}} \\
249 = 326.97 + b & \text{\textit{Multiply.}} \\
249 - 326.97 = 326.97 + b - 326.97 & \text{\textit{Subtract 326.97 from both sides.}} \\
-77.97 = b & \text{\textit{Combine like terms.}}
\end{array}
$$

Now we substitute -77.97 for b in the equation $n = 36.33t + b$:

$$n = 36.33t - 77.97$$

We check our equation by using a TI-84 to verify that our line approximately contains the points $(9, 249)$ and $(12, 358)$ and comes close to the other data points (see Fig. 9). For TI-84 instructions, see Appendices A.19 and A.20.

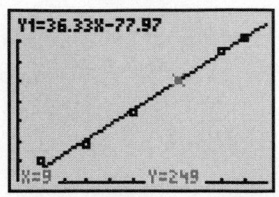

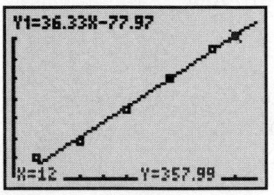

Figure 9 Verify the model

WARNING

It is a common error to skip constructing a scatterplot when we find an equation of a model. However, we benefit in many ways by viewing a scatterplot of data. First, we can determine whether there is a linear association. Second, if there is a linear association, viewing a scatterplot helps us choose two good points with which to find an equation of a linear model. Third, by graphing the model with the scatterplot, we can assess whether the model fits the data reasonably well.

Example 1 took four steps to find an equation of a linear model.

> **Finding an Equation of a Linear Model**
>
> To find an equation of a linear model, given some data,
> 1. Construct a scatterplot of the data.
> 2. Determine whether there is a line that comes close to the data points. If so, choose two points (not necessarily data points) that you can use to find an equation of a linear model.
> 3. Find an equation of the line.
> 4. Use technology to verify that the graph of your equation contains the two chosen points and comes close to all of the data points of the scatterplot.

What should you do if you discover that a model does not fit a data set well? A good first step is to check for any graphing or calculation errors. If your work appears to be correct, then one option is to try using different points to find your equation. Another option is to increase or decrease the slope m and/or the constant term b until the fit is good. You can practice this "trial-and-error" process by completing the exploration in this section.

> **Example 2**   Finding an Equation of a Linear Model

According to a poll performed by the National Consumers League, people are more concerned about privacy issues than about health care, education, crime, or taxes. The percentages of Americans of various age groups who consider their home address to be personal information are listed in Table 5.

Table 5 Americans Who Consider Their Home Address to Be Personal Information

Age Group	Age Used to Represent Age Group	Percent
18–24	21	84
25–34	29.5	80
35–44	39.5	74
45–54	49.5	74
55–64	59.5	67
over 64	75	60

Source: *American Demographics*

Let p be the percentage of Americans at age a years who consider their home address to be personal information.

1. Construct a scatterplot.
2. Describe the four characteristics of the association.
3. Find an equation of a linear model to describe the data.
4. Rewrite the equation with the function name f.
5. Find $f(25)$. What does it mean in this situation?

Solution

1. A scatterplot of the data is shown in Fig. 10.

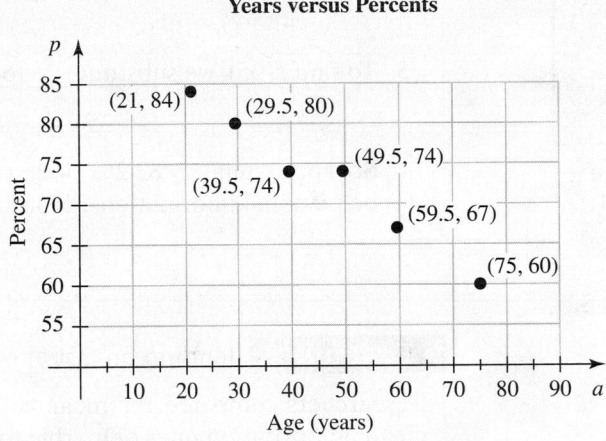

Figure 10 Personal information scatterplot

2. From the scatterplot in Fig. 10, we see there are no outliers. The association is linear, very strong, and negative.
3. We see from the scatterplot in Fig. 11 that a line containing the data points $(21, 84)$ and $(59.5, 67)$ comes close to the rest of the data points.

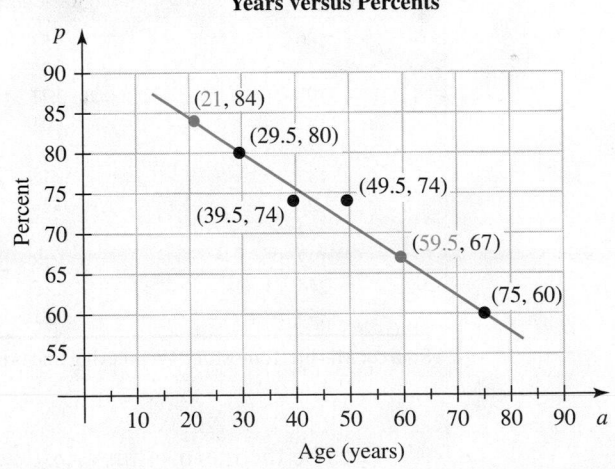

Figure 11 Personal information scatterplot and linear model

For an equation of the form $p = ma + b$, we first use the points $(21, 84)$ and $(59.5, 67)$ to find m:

$$m = \frac{67 - 84}{59.5 - 21} \approx -0.44$$

So, the equation has the form

$$p = -0.44a + b$$

To find b, we use the point $(21, 84)$ and substitute 21 for a and 84 for p:

$$84 = -0.44(21) + b \qquad \textit{Substitute 21 for a and 84 for p.}$$
$$84 = -9.24 + b \qquad \textit{Multiply.}$$
$$84 + 9.24 = -9.24 + b + 9.24 \qquad \textit{Add 9.24 to both sides.}$$
$$93.24 = b \qquad \textit{Combine like terms.}$$

So, the equation is $p = -0.44a + 93.24$.

We use a TI-84 to check that the linear model approximately contains the points $(21, 84)$ and $(59.5, 67)$ and comes close to the other data points (see Fig. 12).

4. To use the function name f, we substitute $f(a)$ for p into $p = -0.44a + 93.24$:

$$f(a) = -0.44a + 93.24$$

5. To find $f(25)$, we substitute 25 for a into the equation $f(a) = -0.44a + 93.24$:

$$f(25) = -0.44(25) + 93.24 = 82.24$$

So, approximately 82.2% of 25-year-old Americans consider their home address to be personal information, according to the model.

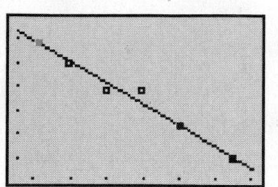

Figure 12 Verify the information model

▶ **Example 3**   Finding an Equation of a Linear Model

Researchers compared the mean sulfur dioxide concentrations (air pollution) and the mean deterioration rates of marble tombstones in 21 U.S. cities for the period 1893–1993 (see Table 6). A microgram (μg) is equal to 0.000001 gram.

Table 6 Mean Sulfur Dioxide Concentrations and Mean Tombstone Deterioration Rates

Mean Sulfur Dioxide Concentration (μg/m^3)	Mean Tombstone Deterioration Rate (mm per century)	Mean Sulfur Dioxide Concentration (μg/m^3)	Mean Tombstone Deterioration Rate (mm per century)
180	1.53	239	2.51
12	0.27	48	0.84
197	2.71	94	1.21
142	1.01	102	1.09
234	1.61	142	1.90
117	1.72	91	1.78
20	0.14	178	1.98
323	3.16	20	0.33
122	1.18	224	2.41
244	2.15	92	1.08
46	0.81		

Source: Marble Tombstone Weathering and Air Pollution in North America, *T. C. Meierding*

Let s be the mean sulfur dioxide concentration (in μg/m^3) and d be the mean deterioration rate (in mm per century) of marble tombstones, both for the same city.

1. Describe the four characteristics of the association. Compute and interpret r as part of your analysis.
2. Find an equation of a model to describe the situation.
3. Estimate the mean deterioration rate of marble tombstones during the period 1983–1993 in a city where the mean sulfur dioxide concentration was 150 μg/m^3 for that period.

4. A city's mean deterioration of marble tombstones for the period 1893–1993 is 2 mm. Estimate the city's mean sulfur dioxide concentration for that period.
5. Because the association is strong, a student concludes that air pollution causes tombstone deterioration. What would you tell the student?

Solution

1. First, we construct a scatterplot (see Fig. 13).

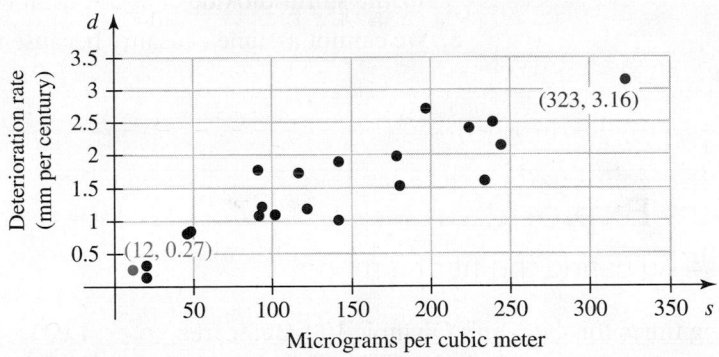

Mean Sulfur Dioxide Concentration versus Mean Tombstone Deterioration Rate

Figure 13 Tombstone deterioration scatterplot

There are no outliers. The association appears to be linear, strong, and positive. The value of r is 0.90 (try it), which confirms that the association is strong.
2. Because the variables s and d are linearly associated, and s is the explanatory variable, we will find an equation of the form $d = ms + b$. Because the line that contains the two data points $(12, 0.27)$ and $(323, 3.16)$ appears to come close to the other data points (see Fig. 14), we will use those two points to find the values of m and b.

First, we use $(12, 0.27)$ and $(323, 3.16)$ to find the slope:

$$m = \frac{3.16 - 0.27}{323 - 12} \approx 0.0093$$

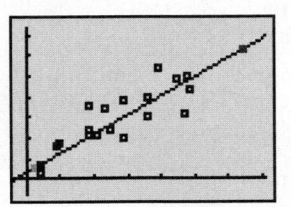

Figure 14 Verify the model

Then we substitute 0.0093 for m in the equation $d = ms + b$:

$$d = 0.0093s + b$$

To find the constant b, we substitute the coordinates of the point $(323, 3.16)$ into the equation $d = 0.0093s + b$ and then solve for b:

$$3.16 = 0.0093(323) + b \qquad \text{Substitute 323 for s and 3.16 for d.}$$
$$3.16 = 3.0039 + b \qquad \text{Multiply.}$$
$$3.16 - 3.0039 = 3.0039 + b - 3.0039 \qquad \text{Subtract 3.0039 from both sides.}$$
$$0.16 \approx b \qquad \text{Combine like terms.}$$

Next, we substitute 0.16 for b in the equation $d = 0.0093s + b$:

$$d = 0.0093s + 0.16$$

We verify our equation by using a TI-84 to check that our line approximately contains the points $(12, 0.27)$ and $(323, 3.16)$ and comes close to the other data points (see Fig. 14).
3. We substitute 150 for s in the equation $d = 0.0093s + 0.16$ and solve for d:

$$d = 0.0093(150) + 0.16 = 1.555$$

So, the mean tombstone deterioration was 1.555 mm, according to the model.

4. We substitute 2 for d in the equation $d = 0.0093s + 0.16$ and solve for s:

$$2 = 0.0093s + 0.16 \qquad \textit{Substitute 2 for d.}$$
$$2 - 0.16 = 0.0093s + 0.16 - 0.16 \qquad \textit{Subtract 0.16 from both sides.}$$
$$1.84 = 0.0093s \qquad \textit{Combine like terms.}$$
$$\frac{1.84}{0.0093} = \frac{0.0093s}{0.0093} \qquad \textit{Divide both sides by 0.0093.}$$
$$197.8 \approx s \qquad \textit{Simplify.}$$

So, the sulfur dioxide concentration was $197.8\,\mu g/m^3$, according to the model.

5. We cannot assume causality because the study is observational.

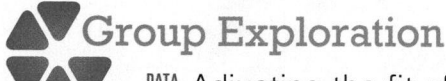

Group Exploration

DATA Adjusting the fit of a model

The winning times for the men's Olympic 100-meter freestyle swimming event are shown in Table 7 for various years.

Table 7 Winning Times for the Men's Olympic 100-Meter Freestyle

Year	Swimmer	Country	Winning Time (seconds)
1980	Jörg Woithe	E. Germany	50.40
1984	Rowdy Gaines	USA	49.80
1988	Matt Biondi	USA	48.63
1992	Aleksandr Popov	Unified Team	49.02
1996	Aleksandr Popov	Russia	48.74
2000	Pieter van den Hoogenband	Netherlands	48.30
2004	Pieter van den Hoogenband	Netherlands	48.17
2008	Alain Bernard	France	47.21
2012	Nathan Adrian	USA	47.52

Source: *The New York Times Almanac*

Let w be the winning time (in seconds) at t years since 1980.

1. Construct a scatterplot of the data.

2. Describe the four characteristics of the association.

3. The linear model $w = -0.0325t + 48.95$ can be found by using the data points (20, 48.30) and (24, 48.17). Graph the model on the scatterplot. Check that the line contains these two points.

4. The model $w = -0.0325t + 48.95$ does not fit the seven data points very well. Adjust the equation by increasing or decreasing the slope -0.0325 and/or the constant term 48.95 so that your new model will fit the data better. Keep adjusting the model until it fits the data points reasonably well.

5. Use your improved model to estimate the winning time in the 2004 Olympics. Calculate the error in your estimate.

> **Tips for Success   Verify Your Work**
>
> Remember to use technology to verify your work. In this section, for example, you can use technology to check your equations. Checking your work increases your chances of catching errors and thus will likely improve your performance on homework assignments, quizzes, and tests.

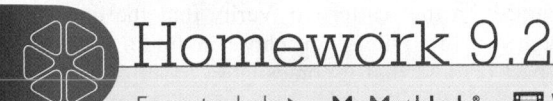

Homework 9.2

For extra help ▶       Watch the videos in MyMathLab   Download the MyDashboard App

1. *True or False:* Because two points determine a line, we can use any two data points to find a reasonable linear model.

2. *True or False:* Viewing a scatterplot of the data can help determine which two points to use to find an equation of a reasonable linear model.

3. *True or False:* It is more desirable to find a linear model that contains several, but not all, data points than it is to find a linear model that does not contain any data points but comes close to all data points.

4. *True or False:* If the variables w and h are linearly associated and w is the explanatory variable, then the situation can be modeled well by an equation of the form $h = mw + b$.

5. **DATA** Find an equation of a line that comes close to the points listed in Table 8. Then use technology to check that your line comes close to the points.

Table 8 Find an Equation

x	y
4	8
5	10
6	13
7	15
8	18

6. **DATA** Find an equation of a line that comes close to the points listed in Table 9. Then use technology to check that your line comes close to the points.

Table 9 Find an Equation

x	y
1	6
5	13
7	14
10	17
12	22

7. **DATA** Find an equation of a line that comes close to the points listed in Table 10. Then use technology to check that your line comes close to the points.

Table 10 Find an Equation

x	y
3	18
7	14
9	9
12	7
16	4

8. Find an equation of a line that comes close to the points listed in Table 11. Then use technology to check that your line comes close to the points.

Table 11 Find an Equation

x	y
2	14
5	10
6	8
8	4
11	1

9. **DATA** The prices of ski rental packages from Gold Medal Sports® are shown in Table 12 for various numbers of days.

Table 12 Prices of Ski Rental Packages

Number of Days	Price of Package (dollars)
1	15.00
2	30.00
3	45.00
4	56.00
5	70.00
6	78.00

Source: *Gold Medal Sports*

Let p be the price (in dollars) of a ski rental package for n days.

a. Construct a scatterplot.
b. Describe the four characteristics of the association. Compute and interpret r as part of your analysis.
c. Find an equation of a linear model.
d. Graph the model on the scatterplot. Verify that the line passes through the two points you chose in finding the equation in part (c) and that it comes close to all of the data points.

10. **DATA** The enrollments in an elementary algebra course at the College of San Mateo are shown in Table 13 for various Tuesdays leading up to the first day of class on Tuesday, January 17.

Table 13 Enrollments in an Elementary Algebra Course

Date	Number of Weeks since November 22	Enrollment
November 22	0	8
November 29	1	9
December 6	2	12
December 13	3	13
December 20	4	16
December 27	5	18
January 3	6	19

Source: *J. Lehmann*

Let E be the enrollment in the elementary algebra course at t weeks since November 22.

a. Construct a scatterplot.
b. Describe the four characteristics of the association. Compute and interpret r as part of your analysis.
c. Find an equation of a linear model.
d. Graph the model on the scatterplot. Verify that the line passes through the two points that you chose in finding the equation in part (c) and that it comes close to all of the data points.

11. **DATA** The percentages of births outside marriage in the United States are shown in Table 14 for various years.

Table 14 Births Outside Marriage

Year	Percentage of Births outside Marriage
1990	28.0
1995	32.2
2000	33.2
2005	36.9
2010	40.8
2013	40.6

Source: *National Center for Health Statistics*

Let p be the percentage of births outside marriage in the United States at t years since 1900.

a. Construct a scatterplot.
b. Describe the four characteristics of the association. Compute and interpret r as part of your analysis.
c. Find an equation of a linear model.
d. Graph the model on the scatterplot. Verify that the line passes through the two points that you chose in finding the equation in part (c) and that it comes close to all of the data points.

12. **DATA** Repeat Exercise 11, but let p be the percentage of births outside marriage in the United States at t years *since 1970*. Compare the slope of your model with the slope of the model you found in Exercise 11. Compare the p-intercepts. Explain why your comparisons make sense.

13. **DATA** Table 15 lists world record times for the women's 400-meter run.

Table 15 Women's 400-Meter Run Record Times

Year	Runner	Country	Record Time (seconds)
1957	Marlene Mathews	Australia	57.0
1959	Maria Itkina	USSR	53.4
1962	Shin Geum Dan	North Korea	51.9
1969	Nicole Duclos	France	51.7
1972	Monika Zehrt	E. Germany	51.0
1976	Irena Szewinska	Poland	49.29
1979	Marita Koch	E. Germany	48.60
1983	Jarmila Kratochvílová	Czechoslovakia	47.99
1985	Marita Koch	E. Germany	47.60

Source: *International Association of Athletics Federations*

Let r be the record time (in seconds) at t years since 1900.

a. Construct a scatterplot.
b. Describe the four characteristics of the association. Compute and interpret r as part of your analysis.
c. Find an equation of a linear model.

d. Graph the model on the scatterplot. Verify that the line passes through the two points you chose in finding the equation in part (c) and that it comes close to all of the data points.

14. **DATA** Table 16 lists world record times for the men's 400-meter run.

Table 16 Men's 400-Meter Run Record Times

Year	Runner	Country	Record Time (seconds)
1900	Maxie Long	USA	47.8
1916	Ted Meredith	USA	47.4
1928	Emerson Spencer	USA	47.0
1932	Bill Carr	USA	46.2
1941	Graver Klemmer	USA	46.0
1950	George Rhoden	Jamaica	45.8
1960	Carl Kaufmann	Germany	44.9
1968	Lee Evans	USA	43.86
1988	Harry Reynolds	USA	43.29
1999	Michael Johnson	USA	43.18

Source: *International Association of Athletics Federations*

Let r be the record time (in seconds) at t years since 1900.

a. Construct a scatterplot.
b. Describe the four characteristics of the association. Compute and interpret r as part of your analysis.
c. Find an equation of a linear model.
d. Graph the model on the scatterplot. Verify that the line passes through the two points you chose in finding the equation in part (c) and that it comes close to all of the data points.

15. **DATA** In Exercises 13 and 14, you found equations for the women's and men's 400-meter run record times. Equations that model the data well are

$$r = -0.27t + 70.45 \quad \text{(women's model)}$$
$$r = -0.053t + 48.08 \quad \text{(men's model)}$$

where r represents the record time (in seconds) at t years since 1900.

a. Graph both models for the period 1900–2050 on the same coordinate system.
b. Do the models predict that the women's record time will ever equal the men's record time? If so, what is that record time, and when will the record be set? Do you have much faith in this prediction? Explain.
c. Do the models predict that the women's record time will ever be less than the men's record time? If so, in what years? Do you have much faith in this prediction? Explain.

16. DATA The mean annual U.S. per-person consumptions of milk and soft drinks are shown in Table 17 for various years.

Table 17 Mean Annual U.S Per-Person Consumptions of Milk and Soft Drinks

Year	Mean Annual Consumption of Milk (gallons per person)	Year	Mean Annual Consumption of Soft Drinks (gallons per person)
1970	31.3	1950	10.8
1980	27.5	1960	13.4
1985	26.7	1970	24.3
1990	25.7	1980	35.1
1995	23.9	1990	46.2
2000	22.5	2000	49.3
2005	21.0	2005	51.5

Source: *USDNEconomic Research Service*

a. Let $M(t)$ be the mean annual per-person consumption of milk and $S(t)$ be the mean annual per-person consumption of soft drinks, both in gallons per person, in the year that is t years since 1900. Find linear equations of M and S.

b. Graph both models for the period 1970–2005 on the same coordinate system.

c. Do the models estimate that the mean annual per-person consumption of milk was ever equal to the mean annual per-person consumption of soft drinks? If so, what is that mean annual per-person consumption, and when did it occur?

d. Do the models estimate that the mean annual per-person consumption of milk was ever greater than the mean annual per-person consumption of soft drinks? If so, in what years?

17. DATA The percentages of American adults whose favorite sport is racing are shown in Table 18 for various years.

Table 18 Percentages of American Adults Whose Favorite Sport Is Racing

Year	Percent
2005	6.1
2006	5.7
2007	5.4
2008	5.0
2009	4.7
2010	4.4
2011	3.5
2012	3.5
2013	3.7

Source: ESPN

Let p be the percentage of American adults whose favorite sport is racing at t years since 2000.

a. Construct a scatterplot.

b. Describe the four characteristics of the association. Compute and interpret r as part of your analysis.

c. Find an equation of a linear model.

d. Find the p-intercept. What does it mean in this situation? Do you have much faith in your estimate? Explain.

e. What is the slope? What does it mean in this situation?

18. DATA Worldwide sales of Alfa Romeo cars are shown in Table 19 for various years.

Table 19 Alfa Romeo Worldwide Sales

Year	Sales (thousands of cars)
2003	175
2005	138
2007	153
2009	117
2011	137
2012	100

Source: *IHS Automotive*

Let s be the annual worldwide sales (in thousands) of Alfa Romeo cars at t years since 2000.

a. Construct a scatterplot.

b. Describe the four characteristics of the association. Compute and interpret r as part of your analysis.

c. Find an equation of a linear model.

d. Find the s-intercept. What does it mean in this situation? How much faith do you have in this estimate? Explain.

e. Alfa Romeo cars have not been sold in the United States since the mid-1990s. Why is it surprising that Fiat Chrysler Automobiles will reintroduce Alfa Romeo cars to the United States?

19. DATA The numbers of conflicts in New York City involving police firearm discharges are shown in Table 20 for various years.

Table 20 Numbers of Conflicts in New York City Involving Police Firearm Discharges

Year	Numbers of Conflicts
2004	61
2006	59
2008	46
2010	33
2011	36

Source: *NYPD*

Let $f(t)$ be the number of conflicts in New York City involving police firearm discharges in the year that is t years since 2000.

a. Construct a scatterplot.

b. Describe the four characteristics of the association. Compute and interpret r as part of your analysis.

c. Find an equation of f.

d. Find $f(9)$. What does it mean in this situation?

e. Find t when $f(t) = 50$. What does it mean in this situation?

20. DATA The percentages of households that consist of married couples with children are shown in Table 21 for various years.

Table 21 Percentages of Households That Consist of Married Couples with Children

Year	Percent
1980	30.9
1990	26.3
1995	25.5
2000	24.1
2005	22.9
2010	20.9
2012	19.6

Source: *U.S. Census Bureau*

Let $f(t)$ be the percentage of households that consist of married couples with children at t years since 1980.

  a. Construct a scatterplot.

  b. Describe the four characteristics of the association. Compute and interpret r as part of your analysis.

  c. Find an equation of f.

  d. Find $f(28)$. What does it mean in this situation?

  e. Find t when $f(t) = 28$. What does it mean in this situation?

21. DATA A runner's *stride rate* is the number of steps per second. The average stride rates of the top female runners are shown in Table 22 for various speeds.

Table 22 Top Female Runners' Speeds and Mean Stride Rates

Speed (feet per second)	Mean Stride Rate (number of steps per second)
15.86	3.05
16.88	3.12
17.50	3.17
18.62	3.25
19.97	3.36
21.06	3.46
22.11	3.55

Source: *Biomechanical Comparison of Male and Female Runners, R. C. Nelson et al.*

Let $f(v)$ be the mean stride rate of a woman running at v feet per second.

  a. Construct a scatterplot.

  b. Describe the four characteristics of the association. Compute and interpret r as part of your analysis.

  c. Find an equation of f.

  d. Find $f(17)$. What does it mean in this situation?

  e. Find v when $f(v) = 3.4$. What does it mean in this situation?

22. DATA The percentages of Americans who have been diagnosed with diabetes are shown in Table 23 for various age groups.

Table 23 Percentages of Americans Diagnosed with Diabetes, by Age Group

Age Group (years)	Age Used to Represent Age Group (years)	Percent
35–39	37	2
40–44	42	4
45–49	47	5
50–54	52	8
55–59	57	10
60–64	62	13
65–69	67	14

Source: *National Health Interview Survey*

Let $f(a)$ be the percentage of Americans at age a years who have been diagnosed with diabetes at some point in their lives.

  a. Construct a scatterplot.

  b. Find an equation of f.

  c. Find $f(50)$. What does it mean in this situation?

  d. Find a when $f(a) = 12$. What does it mean in this situation?

  e. The chance of any one person being diagnosed with diabetes at some point in their past increases as the person grows older. However, 13% of all Americans over the age of 70 have been diagnosed with diabetes at some point in their lives—less than the percentage for ages 65–69 years. How is this possible?

23. DATA Five human subjects were injected with various concentrations of LSD, and with each concentration they solved as many simple arithmetic problems that they could in three minutes. Their work was recorded as the percentage of the number of problems they could do before being injected with LSD. The LSD concentrations and the test percentages are shown in Table 24.

Table 24 LSD Concentrations and Test Percents

LSD Concentration (nanograms per milliliter of plasma)	Mean Test Percent
1.17	78.93
2.97	58.20
3.26	67.47
4.69	37.47
5.83	45.65
6.00	32.92
6.41	29.97

Source: Correlation of Performance Test Scores with "Tissue Concentration" of Lysergic Acid Diethylamide in Human Subjects, *Wagner et al.*

Let p be the mean test percentage when the LSD concentration was c nanograms per milliliter of plasma.

  a. Construct a scatterplot.

  b. Find an equation of a linear model.

  c. What is the slope? What does it mean in this situation?

  d. What is the p-intercept? What does it mean in this situation? Has model breakdown occurred? Explain.

  e. Describe the control group in this experiment.

24. DATA Fibromyalgia (FMS) and chronic fatigue syndrome (CFS) are two illnesses that often coexist. Patients who have both illnesses tend to suffer from fatigue, poor sleep, lack of mental clarity, and body aches (Source: *See Table 25*). Researchers randomly assigned 69 patients with both FMS and CFS and 3 patients with only FMS to a treatment group and a control group. The treatment group took low dosages of a large number of medications and the control group took placebos. At each of four monthly visits, the individuals rated their energy, sleep, mental clarity, degree of feeling pain-free, and overall sense of well-being each on a scale of 0 to 100, for a total possible "well-being" score of 500 points. The mean well-being scores for both groups are shown in Table 25 for the four visits.

Table 25 Numbers of Visits versus Mean Well-Being Scores

Visit	Mean Well-Being Score (points)	
	Treatment Group	Control Group
1	176.1	177.1
2	249.7	187.5
3	264.1	189.2
4	295.7	221.3

Source: Effective Treatment of Chronic Fatigue Syndrome and Fibromyalgia: A Randomized, Double-Blind, Placebo-Controlled, Intent to Treat Study, *Jacob E. Teitelbaum, MD et al.*

a. Construct a scatterplot that describes the association between the visit number and the mean well-being score (in points) for the treatment group. Do the same for the control group. Show both scatterplots in the same coordinate system if you know how to use colors or shapes to identify which points are for which group.

b. During the study, did the treatment group's mean well-being score increase more, less, or about the same as the control group?

c. Let s be the mean well-being score for the treatment group at the nth visit. Find an equation of a linear model that describes the association between n and s.

d. What is the slope of the treatment-group model? What does it mean in this situation?

e. Use the model to predict what the treatment group's well-being score would have been if it had visited 10 times. Has model breakdown occurred? Explain.

25. **DATA** To enroll in intermediate algebra, a student at the College of San Mateo (CSM) must score at least 21 points (out of 50) on a placement test. Using four semesters of data, the CSM Mathematics Department computed the percentages of students who succeeded in intermediate algebra (grade of A, B, or C) for various groups of scores on the placement test (see Table 26).

Table 26 Percentages of Intermediate Algebra Students Who Succeeded

Placement Score Group	Score Used to Represent Score Group	Percentage Who Succeeded in Intermediate Algebra
21–25	23	34
26–30	28	47
31–35	33	55
36–40	38	71
41–45	43	84
46–50	48	*

Source: *College of San Mateo Mathematics Department*

*There were not enough students in this group to give useful data.

Let p be the percentage of intermediate algebra students succeeding in the course who scored x points on the placement test.

a. Construct a scatterplot.

b. Find an equation of a linear model.

c. Students who score below 21 points (out of 50) on the placement test cannot enroll in intermediate algebra. Use the model to predict how high the cutoff score would have to be to ensure that all students succeed in the course. Do you have much faith in your prediction? Explain.

d. Use the model to predict for which scores no students would succeed in the course. Do you have much faith in your prediction? Explain.

e. If, in one semester, 145 students scored in the 16–20-point range on the placement test, predict how many of these students would have succeeded in the course if they had been allowed to enroll in it. Assuming your prediction is accurate, would you advise CSM to lower the placement score cutoff to 16? Explain.

26. **DATA** The "crime index" refers to the number of incidents of crime. The numbers of burglaries, aggravated assaults, and all types of crime per 100,000 Americans are shown in Table 27 for various years.

Table 27 Crime Indexes

	Crime Index (number of incidents per 100,000 people)		
Year	Burglary	Aggravated Assault	All Types of Crime
1993	1100	441	10,974
1995	987	418	10,550
1997	919	382	9855
1999	770	334	8533
2001	742	319	8325
2003	741	295	8134
2005	727	291	7801
2007	726	287	7496
2009	718	265	6947
2011	701	242	6585
2012	670	242	6492

Source: *FBI*

a. Let A be the crime index of aggravated assaults for the year that is t years since 1990. Can the data be modeled well by a linear model? If yes, find such a model. If no, explain why not.

b. Let B be the crime index of burglaries for the year that is t years since 1990. Can the data be modeled well by a linear model? If yes, find such a model. If no, explain why not.

c. Let C be the crime index of all types of crime for the year that is t years since 1990. Can the data be modeled well by a linear model? If yes, find such a model. If no, explain why not.

d. Economist Steven D. Levitt believes crime and the economy are not related. The U.S. economy performed well from 1993 to 2000 and from 2003 to 2008 and did poorly from 2000 to 2003 and from 2008 to 2012. Explain why the aggravated-assault data support Levitt's theory.

e. Explain why the burglary data are less supportive (than the aggravated-assault data) of Levitt's theory.

f. Explain why the data for all types of crime support Levitt's theory.

27. The percentages of the workforce in unions and the percentages of worker hours lost due to strikes are compared by the scatterplot in Fig. 15 for the 50 states, except Alaska and Nebraska. Let U be the percentage of a state's workforce in unions and L be the percentage of the state's worker hours lost due to strikes.

Percent in Unions versus Percent Hours Lost

Figure 15 Exercise 27

(**Source:** *A Statistical Analysis of the Right to Work Conflict, D. Gilbert*)

a. Describe the four characteristics of the association.

b. Imagine the line that contains the two red dots, which represent the data for West Virginia (left dot) and Louisiana (right dot). Or use a strand of thread, a stretched rubber band, or the edge of a clear ruler to form the line. Does the line appear to be a reasonable model? Explain.

c. Find an equation of the line that contains the two red dots.

d. In 2013, 23.2% of the Hawaiian workforce was in unions (Source: *Bureau of Labor Statistics*). For Hawaiians who worked 40 hours per week for 50 weeks per year, estimate the mean per-person *number* of hours lost due to strikes in 2013.

e. The data described by the scatterplot are for the period 1957–1962. Do you have much faith in the result you found in part (d)? Explain.

28. Figure 16 displays a scatterplot that compares the mean lengths and mean heights of 51 types of dinosaurs. Let L be the mean length and H be the mean height, both in meters, of a dinosaur.

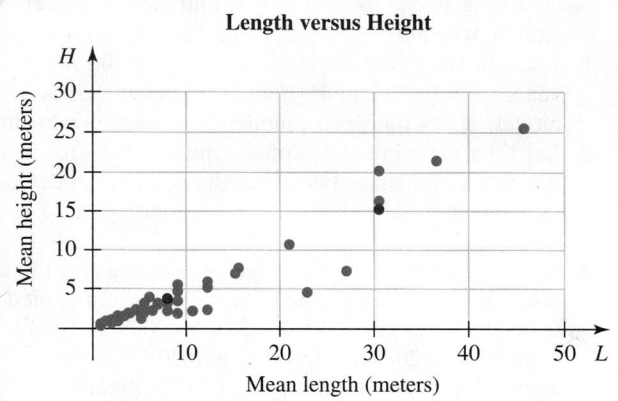

Figure 16 Exercise 28
(**Source:** *DinoDatabase.com*)

a. Describe the four characteristics of the association.

b. Imagine the line that contains the two red dots, which represent the Megalosaurus (left dot) and the Brachiosaurus (right dot). Or use a strand of thread, a stretched rubber band, or the edge of a clear ruler to form the line. Does the line appear to be a reasonable model? Explain.

c. Find an equation of the line that contains the two red dots.

d. Use the model to estimate the mean height of the Triceratops, which had a mean length of 45.7 meters. How many times greater is your result than 69.4 inches, which is the mean height of American men? There are 1.09 yards in 1 meter.

29. The lengths of cruise ships and the sizes of the crews are compared by the scatterplot shown in Fig. 17. Let S be the crew size (in hundreds) for a cruise ship with length L hundred feet.

a. Describe the four characteristics of the association.

b. For the scatterplot, imagine a reasonable linear model. Or use a strand of thread, a stretched rubber band, or the edge of a clear ruler to form the line. Choose two points (not

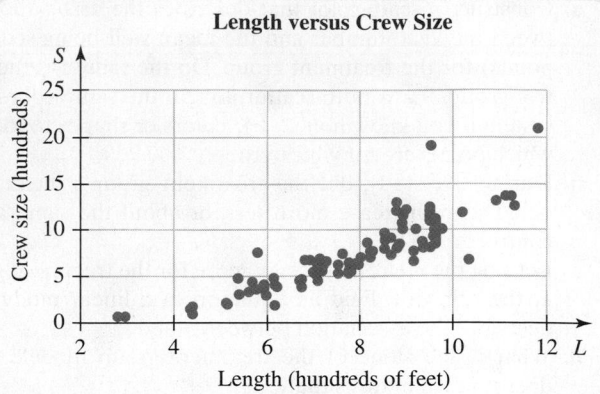

Figure 17 Exercise 29
(**Source:** *True Cruise*)

necessarily data points) that lie on the line. Use the two points to find an equation of a linear model.

c. Use your model to determine whether two 500-foot-long cruise ships would require a crew size less than, equal to, or greater than that for one 1000-foot-long cruise ship.

d. A student says that two 500-foot ships should require the same number of crew members as a 1000-foot ship, because 500 + 500 = 1000. What would you tell the student?

30. When a golfer hits the ball at the starting point, the shot is called a *drive* and the *driving distance* is the distance that the ball travels. The *fairway* is the closely mown area that runs from the starting point almost all the way to the hole. The *fairway accuracy* is the percentage of drives in which the ball comes to rest on the fairway (not on longer grass, sand traps, and ponds). The mean driving distances and the fairway accuracies for Professional Golf Association (PGA) male players are compared by the scatterplot in Fig. 18. Let d be the mean driving distance (in yards) by a PGA male golf player whose fairway accuracy is p percent.

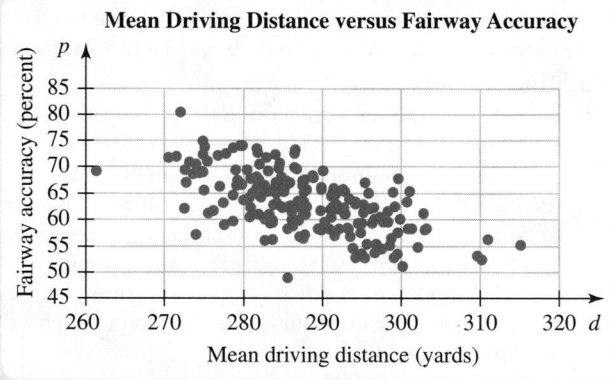

Figure 18 Exercise 30
(**Source:** *PGA*)

a. Is the association positive, negative, or neither? Why does that make sense?

b. In part (a), you described the direction of the association. Now describe the other three characteristics of the association.

c. For the scatterplot, imagine a reasonable linear model. Or use a strand of thread, a stretched rubber band, or the edge of a clear ruler to form the line. Choose two points (not necessarily data points) that lie on the line. Use the two points to find an equation of a linear model.

d. From January 1 to September 14, 2014, David Toms led the PGA in fairway accuracy, with 653 out of 865 drives coming to rest on fairways. Estimate his mean driving distance.

e. It turns out that $r = -0.61$. For women in the Ladies Professional Golf Association (LPGA), $r = -0.424$. Is the association stronger for men or for women? Give at least one possible reason.

Concepts

31. **DATA** Three students are to find a linear model for the data in Table 28. Student A uses points $(1, 5.9)$ and $(2, 6.4)$, student B uses points $(3, 9.0)$ and $(4, 11.0)$, and student C uses points $(5, 12.1)$ and $(6, 15.5)$. Which student seems to have made the best choice of points?

Table 28 Three Students Model Data

x	y
1	5.9
2	6.4
3	9.0
4	11.0
5	12.1
6	15.5
7	16.5

32. **DATA** Three students are to find a linear model for the data in Table 29. Student A uses points $(3, 13.8)$ and $(4, 10.1)$, student B uses points $(6, 7.8)$ and $(7, 4.3)$, and student C uses points $(5, 9.1)$ and $(8, 3.1)$. Which student seems to have made the best choice of points? Explain.

Table 29 Three Students Model Data

x	y
3	13.8
4	10.1
5	9.1
6	7.8
7	4.3
8	3.1
9	1.1

For Exercises 33 and 34, consider the scatterplot of data and the graph of the model $y = mx + b$ in the indicated figure. Sketch the graph of a linear model that describes the data better. Then explain how you would adjust the values of m and b of the original model so it would describe the data better.

33. See Fig. 19.   34. See Fig. 20.

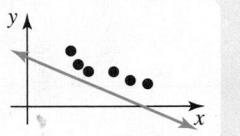

Figure 19 Exercise 33

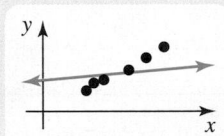

Figure 20 Exercise 34

35. Explain how to find an equation of a linear model for a given situation. Also, explain how to verify that the model describes the situation reasonably well.

36. A student comes up with a shortcut for modeling a situation described by a table that contains several rows of data. Instead of creating a scatterplot of the data, the student chooses two data points at random and uses them to find an equation of a line. Give at least two examples to illustrate what can go wrong with this shortcut.

▼ 9.3 Linear Regression Model

In Section 9.2, we used two points to find an equation of a linear model. In this section, we will find a linear model by a method that takes into account *all* of the data points.

Residuals

In Example 1, we will find the difference between the actual price of a Honda Accord and a model's prediction of the price. This is a key step toward finding a linear model by this section's method.

▶ **Example 1 Computing the Difference Between an Actual Price and a Predicted Price**

The ages and asking prices for 8 Honda Accords at dealerships in the Boston area are shown in Table 30 on page 598. Let x be the age (in years) and y be the price (in thousands of dollars), both for a Honda Accord.

1. Describe the four characteristics of the association.
2. By viewing a scatterplot, we can determine that the line that contains the data points $(9, 9.9)$ and $(5, 13.0)$ comes fairly close to the data points. By using the technique

Objectives

» Compute and interpret residuals.

» Compute and interpret the sum of squared residuals.

» Find an equation of a linear regression model and use it to make predictions.

» Interpret residual plots.

» Use a residual plot to help determine whether a regression line is an appropriate model.

» Identify influential points.

» Compute and interpret the coefficient of determination.

Table 30 Ages and Prices of Some Honda Accords

Age (in years)	Price (thousands of dollars)
8	13.0
7	9.2
12	7.0
9	9.9
5	13.0
11	10.0
3	17.0
5	14.2

Source: *Edmunds.com*

discussed in Section 9.2, we can find that an equation of the line is $y = -0.78x + 16.88$ (try it). Verify that the line comes fairly close to the data points.

3. Use the model to predict the price of a 8-year-old Honda Accord.
4. Find the difference of the actual price and the predicted price for an 8-year-old Honda Accord.
5. Find the difference of the actual price and the predicted price for a 7-year-old Honda Accord.

Solution

1. From the scatterplot shown in Fig. 21, we see there are no outliers. The association is linear, fairly strong, and negative. The value of r turns out to be -0.87, which supports that the association is fairly strong.

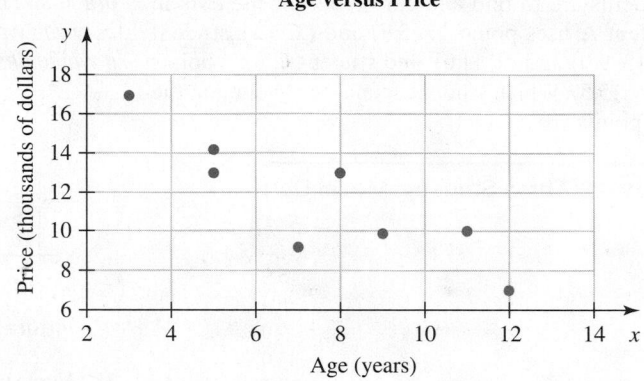

Age versus Price

Figure 21 Ages and Prices of Some Honda Accords

2. The model fits the data fairly well (see Fig. 22).

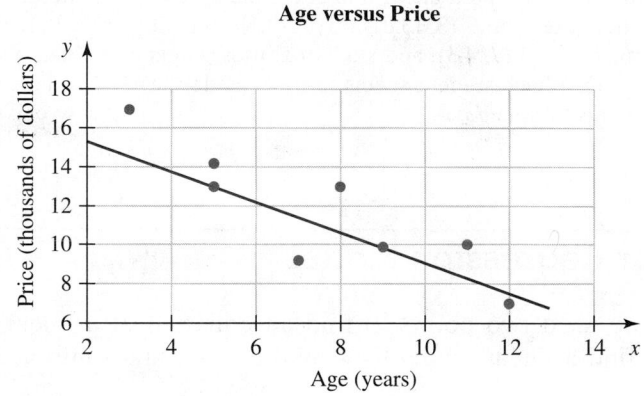

Age versus Price

Figure 22 Honda Accord model

3. We substitute 8 for x in the equation $y = -0.78x + 16.88$:

$$y = -0.78(8) + 16.88 = 10.64$$

So, the price of an 8-year-old Accord is $10.64 thousand, according to the model.

4. There is exactly one 8-year-old Accord included in Table 30. Its price is $13.0 thousand. We find the difference of the actual price and the predicted price we found in Problem 3:

actual price − predicted price = 13.0 − 10.64 = 2.36 thousand dollars

So, the difference is $2.36 thousand.

5. First, we substitute 7 for a in the equation $y = -0.78x + 16.88$:

$$y = -0.78(7) + 16.88 = 11.42$$

So, the price of a 7-year-old Accord is $11.42 thousand, according to the model.

Then we find the difference of the actual price, $9.2 thousand, and the predicted price, $11.42 thousand:

actual price − predicted price = 9.2 − 11.42 = −2.22 thousand dollars

So, the difference is −2.22 thousand dollars.

In Problem 3 of Example 1, we predicted that the price of the 8-year-old Accord is $10.64 thousand. We say that 10.64 is the *predicted value of y* and write $\hat{y} = 10.64$. But the actual price of the 8-year-old Accord is $13.0 thousand (see Table 30). We say the *observed value of y* is 13.0.

In general, for a data point (x, y), the **observed value of y** is y and the **predicted value of y** [written $\hat{y}$ or $\widehat{f(x)}$] is the value obtained by using a model to predict y.

If we had used the letter "*p*" to stand for price (in thousands of dollars) in Example 1, the predicted value of p would be written $\hat{p}$, but the notation $\hat{p}$ is used in statistics to stand for the proportion of a sample. To avoid this double meaning and others like it, we agree to use x for the explanatory variable and y for the response variable for all associations from now on.

Why was the predicted price of the 8-year-old Accord less than the observed price? There are many possible reasons. Maybe the car has low mileage, has lots of features, or is in excellent condition. The seller of the Accord might have unrealistic expectations of how much the car can sell for. Or perhaps we could have found a better model.

In Problem 4 of Example 1, we found that for the 8-year-old Accord, the difference of the observed price and the predicted price is $2.36 thousand:

observed price − predicted price = 13.0 − 10.64 = 2.36

We call the difference $2.36 thousand a *residual*. The residual is indicated in Fig. 23.

Age versus Price

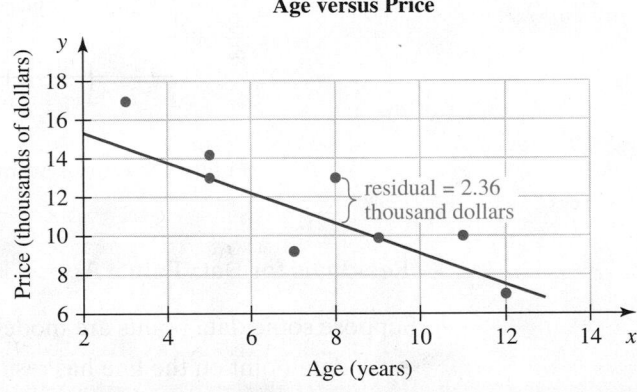

Figure 23 The residual $2.36 thousand

▶ Definition   Residual

For a given data point (x, y), the **residual** is the difference of the observed value of y and the predicted value of y:

Residual = Observed value of y − Predicted value of $y = y - \hat{y}$

If the predicted value of y for a data point is equal to the observed value of y, then the residual is 0. A nonzero residual can be due to lurking variables and/or randomness.

Note that the residual for the 8-year-old Accord is *positive* and the data point lies *above* the model (see Fig. 24). This makes sense, because if a residual is positive, then the observed value of y must be greater than the predicted value of y, which means the data point is above the model.

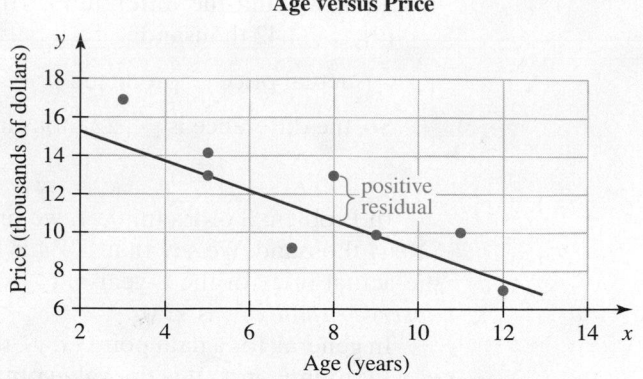

Figure 24 A data point above the line has positive residual

On the basis of our work in Problem 5 of Example 1, we conclude that for the 7-year-old Accord, the residual is −2.22 thousand dollars. Note that the residual is *negative* and the data point lies *below* the model (see Fig. 25). This makes sense, because if a residual is negative, then the observed value of y must be less than the predicted value of y, which means the data point is below the model.

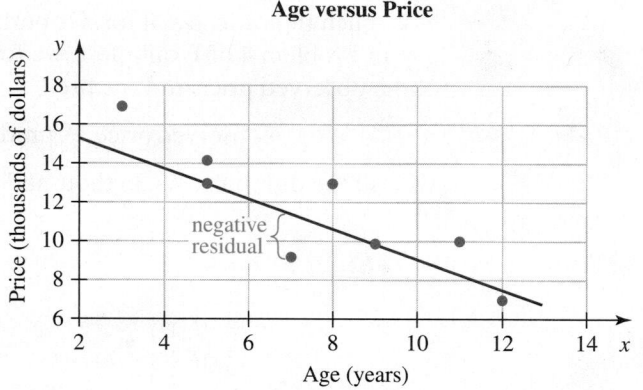

Figure 25 A data point below the line has negative residual

▶ **Residuals for Data Points Above, Below, or on a Line**

Suppose some data points are modeled by a line.
- A data point on the line has residual equal to 0.
- A data point above the line has positive residual (see Fig. 24).
- A data point below the line has negative residual (see Fig. 25).

Sum of Squared Residuals

We can measure how close a linear model comes to some data points by calculating the residuals for the data points. Loosely speaking, the smaller the residuals, the better the line will fit the data. To obtain a single-number measure, it is tempting to total the residuals, but this would not be informative because a part of the negative residuals might undo a part of the positive residuals. For example, even though the line in Fig. 26 fits the data points terribly, the sum of residuals is equal to 0 because the negative residuals (in red) and the positive residuals (in blue) sum to 0. The line is no match for the line in Fig. 27, where the sum of the residuals equals 0 because the line contains all the data points.

Exact Linear Association

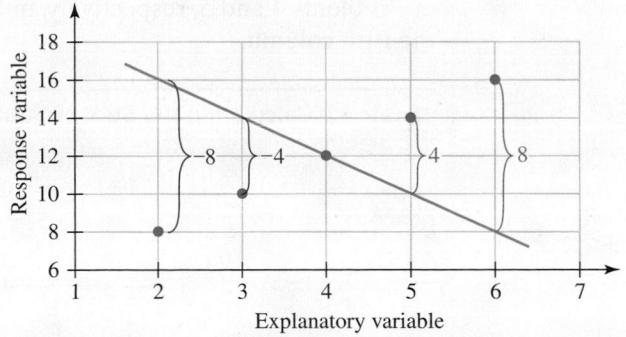

Figure 26 The sum of the residuals is 0 for the line with terrible fit

Exact Linear Association

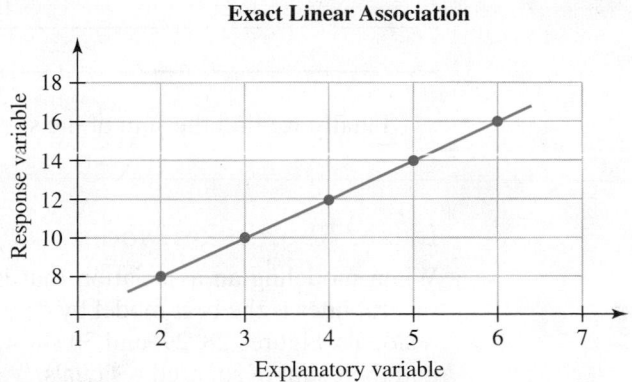

Figure 27 The sum of the residuals is 0 for the line with perfect fit

Similar to standard deviation and variance, we address this issue by totaling the *squared* residuals.

> **Sum of Squared Residuals**
>
> We measure how well a line fits some data points by calculating the sum of the squared residuals:
>
> $$\sum (y_i - \hat{y}_i)^2$$
>
> The smaller the sum of squared residuals, the better the line will fit the data. If the sum of squared residuals is 0, then there is an exact linear association.

Table 31 Ages and Prices of Some Honda Accords

Age (in years)	Price (thousands of dollars)
8	13.0
7	9.2
12	7.0
9	9.9
5	13.0
11	10.0
3	17.0
5	14.2

Source: *Edmunds.com*

▶ **Example 2**   Finding a Sum of Squared Residuals

Here we continue to work with the Honda Accord data (see Table 31). Let x be the age (in years) and y be the price (in thousands of dollars), both for a Honda Accord. Find the sum of squared residuals for the linear model we worked with in Example 1, which is $\hat{y} = -0.78x + 16.88$.

Solution

We list the values of x and y in the first and second column of Table 32. Then we substitute the values of x in the equation $\hat{y} = -0.78x + 16.88$ to find the predicted prices and list them in the third column. In the fourth column, we find the values of the residuals. Note that the first two residuals, 2.36 and −2.22, are the ones we found in Problems 4 and 5, respectively, in Example 1. Next, we list the squared residuals in the fifth column.

Table 32 Calculating the Sum of Squared Residuals for the Model $\hat{y} = -0.78x + 16.88$

Age x_i	Price y_i	Predicted Price Using $\hat{y} = -0.78x + 16.88$ $\hat{y}_i$	Residual $y_i - \hat{y}_i$	Residual^2 $(y_i - \hat{y}_i)^2$
8	13.0	10.64	2.36	5.57
7	9.2	11.42	−2.22	4.93
12	7.0	7.52	−0.52	0.27
9	9.9	9.86	0.04	0.00
5	13.0	12.98	0.02	0.00
11	10.0	8.30	1.70	2.89
3	17.0	14.54	2.46	6.05
5	14.2	12.98	1.22	1.49

$$\sum (y_i - \hat{y}_i)^2 = 21.20$$

Finally, we find the sum of the squared residuals, which is 21.20.

Linear Regression Model

When modeling an association that is approximately linear, we can identify which of several lines is the best model by determining which one has the lowest sum of squared residuals. Figures 28, 29, and 30 show three linear models of the age-price association and their sum of squared residuals.

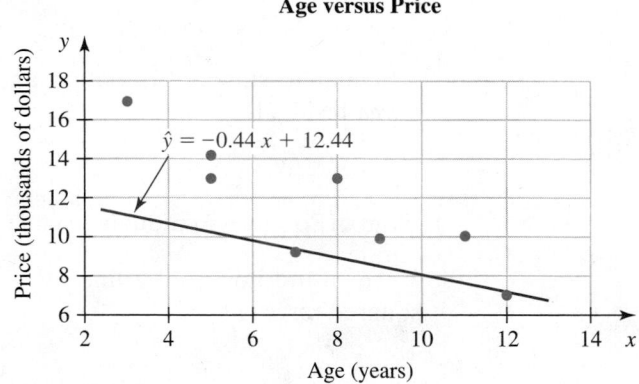

Figure 28 Sum of squared residuals is 82.36

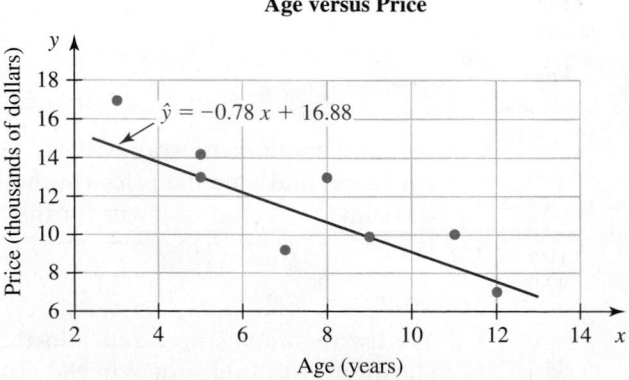

Figure 29 Sum of squared residuals is 21.20

Age versus Price

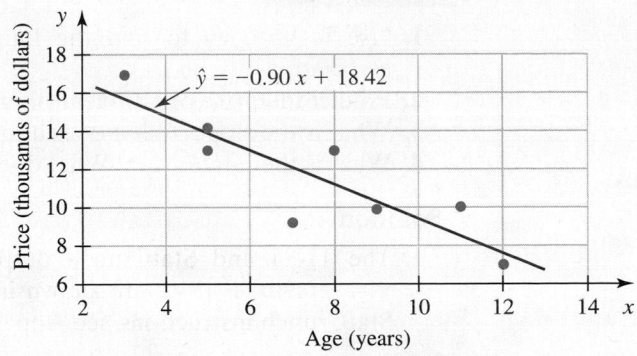

Figure 30 Sum of squared residuals is 17.01

Of the three models, the model $\hat{y} = -0.90x + 18.42$ is the best, because it has the least sum of squared residuals (17.01). In fact, it turns out that the model has the least sum of squared residuals of *all* linear models. This can be shown using calculus. So, the model $\hat{y} = -0.90x + 18.42$ fits the data better than any other linear model. We call it the *linear regression model* and call the equation $\hat{y} = -0.90x + 18.42$ the *linear regression equation*.

> **Definition   Linear regression function, line, equation, and model**
>
> For a group of points, the **linear regression function** is the linear function with the least sum of squared residuals. Its graph is called the **regression line** and its equation is called the **linear regression equation**, written
>
> $$\hat{y} = b_1 x + b_0$$
>
> where b_1 is the slope and $(0, b_0)$ is the y-intercept. The **linear regression model** is the linear regression function for a group of *data* points.

It might surprise you that we use b_1 instead of m for the slope and b_0 instead of b for the constant term. This sets up the pattern to write more complicated regression equations such as $\hat{y} = b_2 x_2 + b_1 x_1 + b_0$, which are useful in statistics courses. Such an equation is usually written in the form $y = b_0 + b_1 x_1 + b_2 x_2$.

Instead of using calculus to find the linear regression equation for some data points, we will use technology. In Fig. 31, we use StatCrunch to find the linear regression equation for the Honda Accord situation. The slope (-0.90) is shown to the right of "Slope" and below "Estimate," and the y-intercept (18.42) is shown to the right of "Intercept" and below "Estimate." The sum of squared residuals (17.01) is shown to the right of "Error" (residual) and below "SS" (sum of squares). For StatCrunch instructions, see Appendix B.16.

Parameter estimates:

Parameter	Estimate	Std. Err.	Alternative	DF	T-Stat	P-value
Intercept	18.418015	1.6429545	$\neq 0$	6	11.210301	< 0.0001
Slope	−0.90073529	0.20417645	$\neq 0$	6	−4.4115533	0.0045

Analysis of variance table for regression model:

Source	DF	SS	MS	F-stat	P-value
Model	1	55.170037	55.170037	19.461803	0.0045
Error	6	17.008713	2.8347855		
Total	7	72.17875			

Figure 31 The slope, the y-intercept, and the sum of squared residuals

▶ **Example 3**   Using a Linear Regression Equation to Make Predictions

1. Use technology to find the linear regression equation for the Honda Accord association.
2. Predict the price of a 10-year-old Accord.
3. What is the slope? What does it mean in this situation?
4. What is the y-intercept? What does it mean in this situation?

Solution

1. The TI-84 and StatCrunch outputs for finding the linear regression equation $\hat{y} = -0.90x + 18.42$ are shown in Fig. 32 and Fig. 33, respectively. For TI-84 and StatCrunch instructions, see Appendices A.26 and B.17.

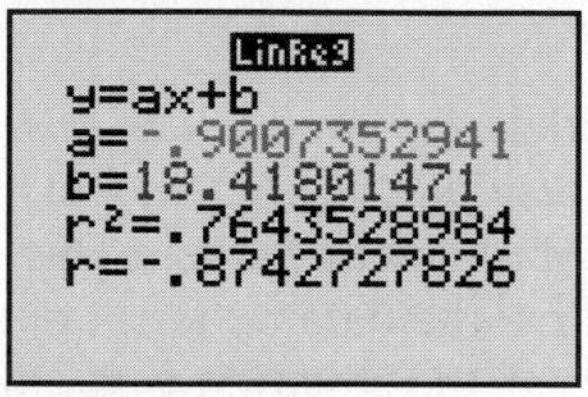

Figure 32 Linear regression equation found by TI-84

Simple linear regression results:
Dependent Variable: Price
Independent Variable: Age
Price = 18.418015 - 0.90073529 Age
Sample size: 8
R (correlation coefficient) = -0.87427278
R-sq = 0.7643529
Estimate of error standard deviation: 1.6836821

Figure 33 Linear regression equation found by StatCrunch

2. We substitute 10 for x in the equation $\hat{y} = -0.90x + 18.42$:

$$\hat{y} = -0.90(10) + 18.42 = 9.42$$

So, the price is $9.42 thousand, according to the model.

3. The slope of the line $\hat{y} = -0.90x + 18.42$ is -0.90. According to the model, the price decreases by $0.90 thousand ($900) per year in age of an Accord.

4. The constant term of $\hat{y} = -0.90x + 18.42$ is 18.42. So, the y-intercept is $(0, 18.42)$. The model predicts that the price of a new Accord is $18.42 thousand. However, we have performed extrapolation, so we have little faith in this prediction. In fact, some research would show that the mean price of new Accords even with no optional equipment is much higher: $21.0 thousand (Source: *Edmunds.com*).

Residual Plots

In Chapters 6–8, we used scatterplots and r to determine whether an association was approximately linear. In this section, we have discussed that for the linear regression model, the sum of squared residuals is lowest. We can picture *individual* residuals by constructing a **residual plot**, which is a graph that compares data values of the explanatory variable with the data points' residuals.

▶ **Example 4**   Constructing and Analyzing a Residual Plot

1. Construct a residual plot for the Honda Accord linear regression model.
2. How many dots are above the zero residual line (the red, dashed, horizontal line) of the residual plot? What do they mean in this situation?
3. How many dots are below the zero residual line of the residual plot? What do they mean in this situation?
4. Which dot is farthest from the zero residual line? What does that mean in this situation?
5. Does the residual plot support that the regression line is a reasonable model? Explain.

Solution

1. First, we construct a table that describes the residuals (see Table 33).

Table 33 Residuals for the Linear Regression Model $\hat{y} = -0.90x + 18.42$

Age x_i	Price y_i	Predicted Price Using $\hat{y} = -0.90x + 18.42$ $\hat{y}_i$	Residual $y_i - \hat{y}_i$
8	13.0	11.22	1.78
7	9.2	12.12	−2.92
12	7.0	7.62	−0.62
9	9.9	10.32	−0.42
5	13.0	13.92	−0.92
11	10.0	8.52	1.48
3	17.0	15.72	1.28
5	14.2	13.92	0.28

Then we use StatCrunch to construct a residual plot that compares the ages of the cars with the cars' residuals (see Fig. 34). For StatCrunch instructions, see Appendix B.18. We add a red, dashed, horizontal line called the *zero residual line*, which represents residuals equal to 0.

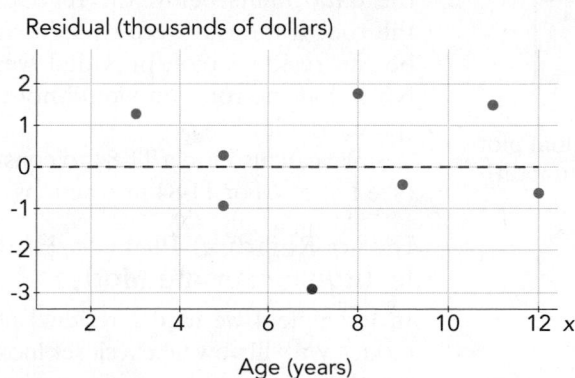

Age versus Residual

Figure 34 StatCrunch Residual Plot

2. There are four dots above the zero residual line. Because their residuals are positive, we conclude that the corresponding four data points are above the linear regression model.
3. There are four dots below the zero residual line. Because their residuals are negative, we conclude that the corresponding four data points are below the linear regression model.
4. The red dot $(7, -2.92)$ is farthest from the zero residual line. This means that the corresponding data point $(7, 9.2)$ is farthest from the regression line. But on the basis of the scatterplot in Fig. 30 on page 603, the data point does not appear to be quite far enough from the linear model to be an outlier.
5. The residual plot supports that the regression line is a reasonable model because not only are there the same number of dots above the zero residual line as below it, the dots generally alternate between being above and below the zero residual line. Also, none of the points are quite far enough from the zero residual line to be outliers.

In Fig. 35, we use green vertical line segments to indicate the residuals in the scatterplot and the residual plot for the Honda Accord situation. Note that for each car, the green vertical lines have the same lengths in both figures. So the same information about residuals is contained in both plots, but it is easier to compare the residuals in a residual plot, because the green line segments all start on the same horizontal line, rather than on a "tilted" regression line.

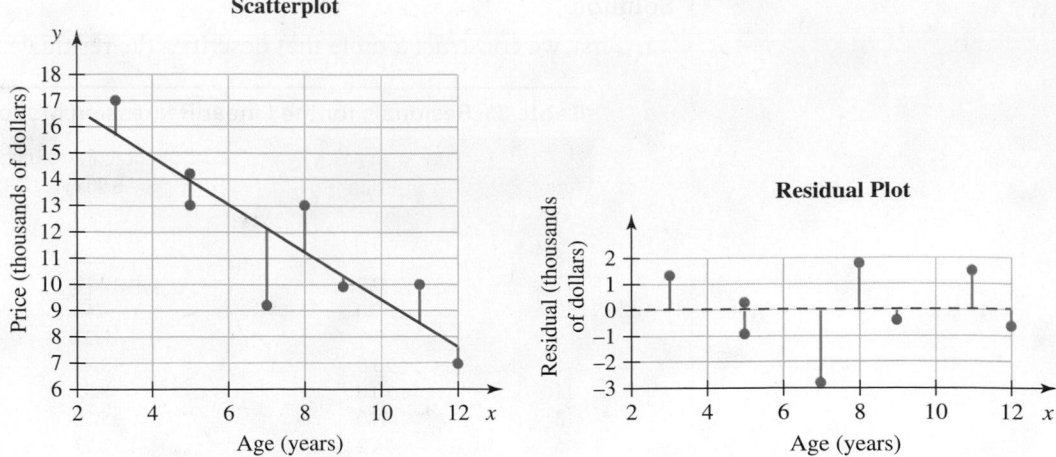

Figure 35 The green vertical line segments, which indicate the residuals, have the same lengths in both plots

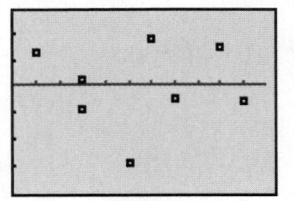

Figure 36 TI-84 residual plot for Honda Accord situation

For the scatterplot, imagine that the data points above the regression line are tiny helium balloons tied by green strings to a tilted curtain rod. Also imagine that the data points below the regression line are marbles hung by green strings from the rod. If the rod were rotated counterclockwise to be horizontal, the result would be the residual plot, provided we used appropriate scaling on the horizontal axis. Note that the rotation would not change the lengths of the strings (which represent the residuals).

We can also use a TI-84 to construct a residual plot for the Honda Accord situation (see Fig. 36). For TI-84 instructions, see Appendix A.27.

Using Residual Plots to Help Determine whether a Regression Line Is an Appropriate Model

In Example 4 we used a residual plot to confirm that a regression line is a reasonable model. We will now take a closer look at how a residual plot can be used to help determine whether a regression line is an appropriate model.

The scatterplot and residual plot in Fig. 37 suggest that if there is a nonlinear association, then there will be a nonlinear pattern in the residual plot.

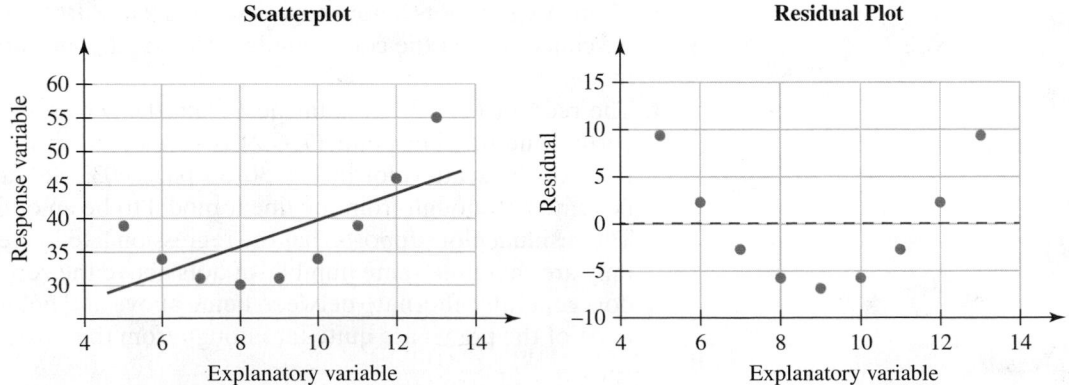

Figure 37 Scatterplot and residual plot for a nonlinear association

Because outliers are much farther away from the regression line than other data points, the outliers' dots in residual plots are much farther from the zero residual line than the other dots (see the red dots in the scatterplot and the residual plot in Fig. 38).

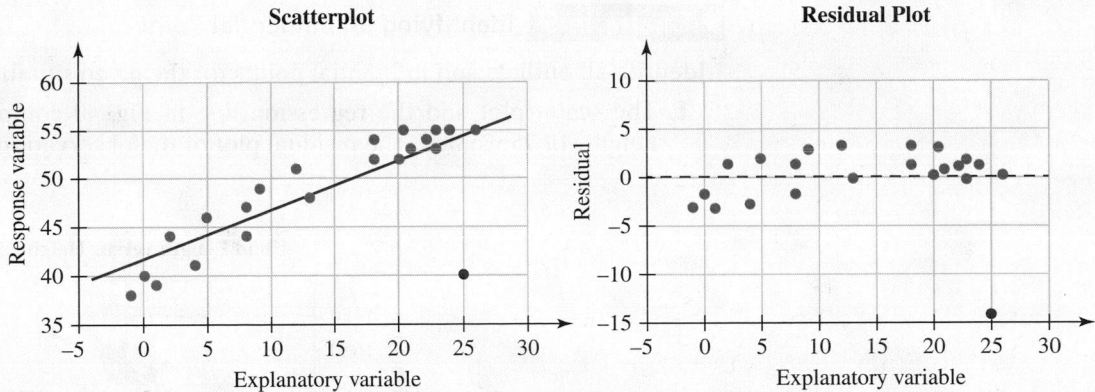

Figure 38 An outlier is described by the red dots in the scatterplot and residual plot

To use certain theories in statistics, we need to add one more requirement for a regression line to be used as a model: for each value of the explanatory variable, the variation of the response variable should be approximately equal. This means that the *vertical* spread of a residual plot should be about the same for each value of the explanatory variable. For example, because the residual plot in Fig. 39 becomes more spread out *vertically* as we view the plot from left to right, we cannot use the regression line shown in the scatterplot to model the situation.

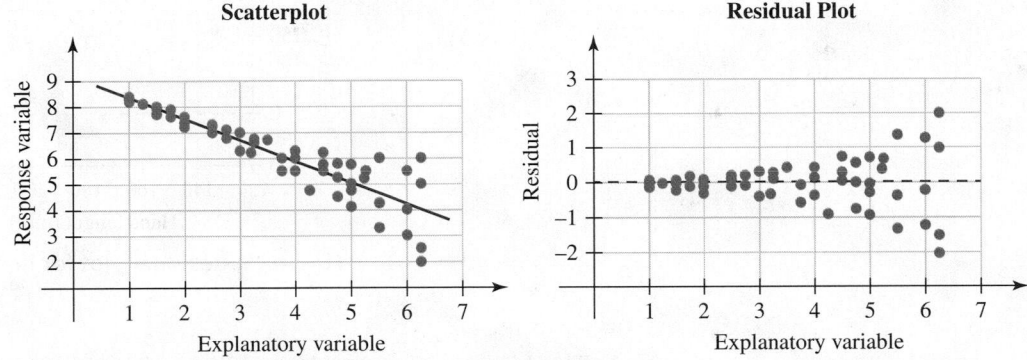

Figure 39 Each plot becomes more spread out vertically as we view the plot from left to right.

> **Using a Residual Plot to Help Determine whether a Regression Line Is an Appropriate Model**

The following statements apply to a residual plot for a regression line.

- If the residual plot has a pattern where the dots do not lie close to the zero residual line, then there is either a nonlinear association between the explanatory and response variables or there is no association.

- If a dot lies much farther away from the zero residual line than most or all of the other dots, then the dot corresponds to an outlier. If the outlier is neither adjusted nor removed, the regression line may *not* be an appropriate model.

- The *vertical* spread of the residual plot should be about the same for each value of the explanatory variable.

Influential Points

The absence of an influential student from a club meeting can greatly affect the meeting's dynamics. If the slope of a regression line is greatly affected by the removal of a data point, we say the data point is an **influential point**.

▶ Example 5   Identifying an Influential Point

Identify all outliers and influential points for the given situation.

1. The scatterplot and the regression line in Fig. 40 compare the hand lengths and heights of 75 women. The residual plot of the observations is shown in Fig. 41.

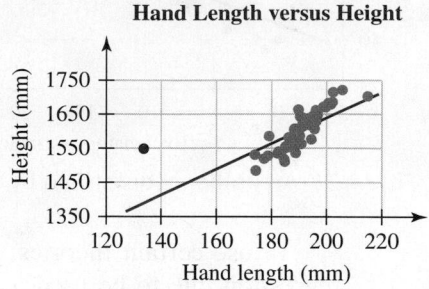

Figure 40 Scatterplot for hand-height data
(**Source:** Stature Estimation Based on Hand Length and Foot Length, S. G. Sani)

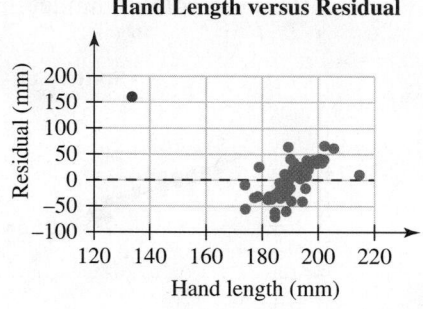

Figure 41 Residual plot for hand-height data

2. The lengths of cruise ships and the sizes of the crews are compared by the scatterplot and regression line shown in Fig. 42. The residual plot of the observations is shown in Fig. 43.

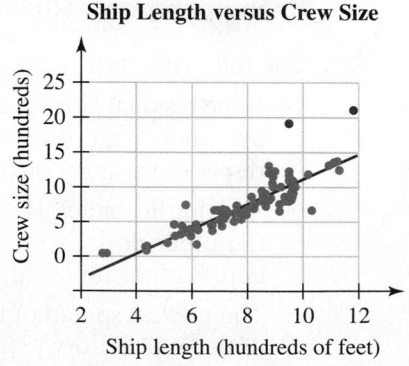

Figure 42 Scatterplot for cruise data
(**Source:** True Cruise)

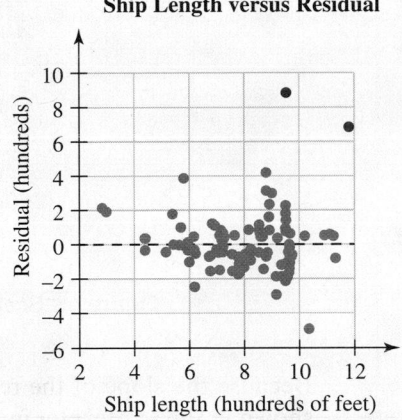

Figure 43 Residual plot for cruise data

Solution

1. Because the red data point in the scatterplot is much farther from the regression line than the other data points and because the red dot in the residual plot is much farther from the zero residual line than the other dots, the red data point in the scatterplot is an outlier. We remove it and construct a scatterplot with the new regression line and construct a residual plot in Figs. 44 and 45, respectively.

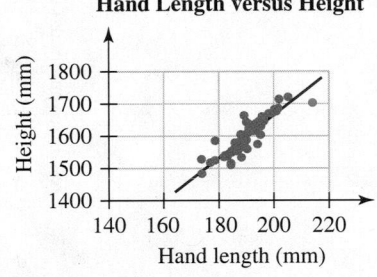

Figure 44 Scatterplot with outlier removed

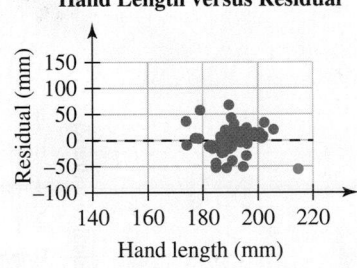

Figure 45 Residual plot with outlier removed

Because the regression line's slope changed significantly, the red data point in Fig. 40 is an influential point. Suppose we perform more research and discover that the influential point is the result of a recording error, so we remove it. Because this changes the regression line's slope, other data points might be outliers. In fact, the green data point shown in the scatterplot in Fig. 44 *might* be an outlier. However, the residual plot in Fig. 45 shows that several other data point dots are as low as the green dot, so the green data point in the scatterplot is *not* an outlier. To determine whether the data point is influential, we remove it and construct a scatterplot and sketch the new regression line in Fig. 46.

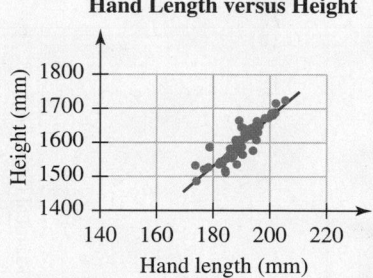

Figure 46 Scatterplot with green data point removed

Because the slope of the regression line did not change much, the green data point shown in the scatterplot in Fig. 44 is *not* an influential point.

2. Because the two red data points in the scatterplot in Fig. 42 are quite a bit farther from the regression line than the other data points and because the red dots in the residual plot in Fig. 43 are quite a bit farther from the zero residual line than the other dots, the red data points in the scatterplot are outliers. We remove them and construct a scatterplot with the new regression line and construct a residual plot in Figs. 47 and 48, respectively.

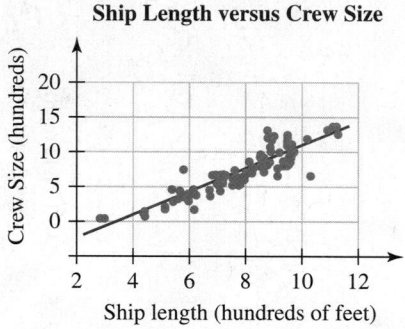

Figure 47 Scatterplot with outliers removed

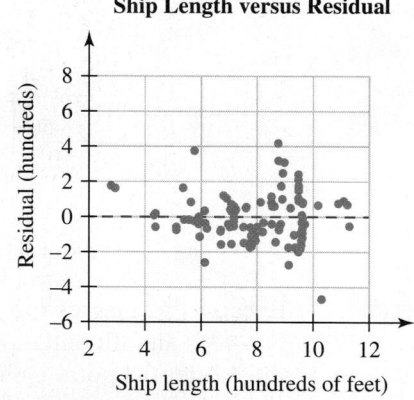

Figure 48 Residual plot with outliers removed

Because the slope of the regression line hardly changed, the two outliers are not influential points.

Why is the outlier for the hand-height data in Problem 1 of Example 5 an influential point but the two outliers for the cruise data in Problem 2 are not? Note that the outlier for the hand-height data is horizontally far from the other data points but the outliers for the cruise data are not. In general, **outliers tend to be influential points when they are horizontally far from the other data points**.

If an influential point is due to measurement or recording error, its coordinates should be corrected, if possible. If an influential point is removed, the researcher should state that it was removed in the report. If an influential point is neither adjusted nor removed, the regression line is probably *not* an appropriate model.

Coefficient of Determination

The heights and weights of the 60 players picked in the 2014 draft for NBA basketball teams are described by the scatterplot in Fig. 49. There appears to be a fairly strong linear association. The correlation coefficient of $r = 0.75$ confirms this.

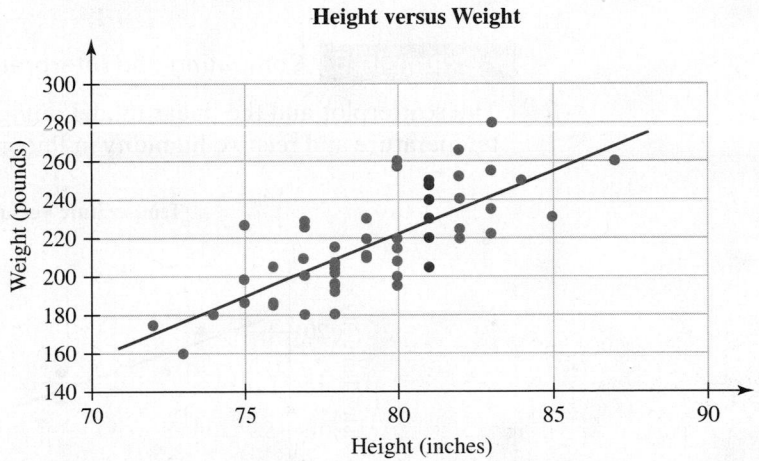

Figure 49 Scatterplot of players' heights and weights
(**Source:** *nbadraft.net*)

The red dots represent all the draft picks who are 81 inches tall. There are 8 such players, but there are only 6 dots because two pairs of players have the same weights. The dotplot in Fig. 50 shows only the weights of the 60 players. The 8 red dots represent the same players represented by the red dots in the scatterplot. Because those 8 players have the same height, the variation in their weights is due to randomness and factors other than height, such as diet and exercise. But the variation of the 8 players' weights is much smaller than the variation of all 60 players's weights. In fact, the variation of the 8 players' weights is 242 pounds^2, which is much smaller than the variation of the 60 players' weights, which is 659 pounds^2. That is because the variable weight tends to increase as the variable height increases. In other words, it is because of the linear association between height and weight.

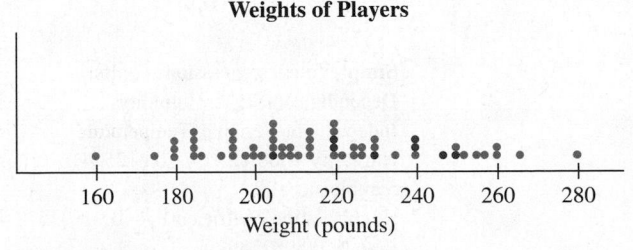

Figure 50 Dotplot of players' heights

But what proportion of the variation of the weights is due to the linear association? The proportion turns out to be equal to the square of the correlation coefficient: r^2. To find r^2, we substitute 0.75 for r in r^2:

$$r^2 = 0.75^2 \approx 0.56$$

So, about 56% of the variation in the players' weights is explained by the regression line. In general, we call r^2 the **coefficient of determination**.

> **Coefficient of Determination**
>
> The coefficient of determination, r^2, is the proportion of the variation in the response variable that is explained by the regression line.

▶ **Example 6**   Computing and Interpreting the Coefficient of Determination

The scatterplot and the linear model in Fig. 51 describe the linear association between temperature and relative humidity in Phoenix, Arizona, on June 1, 2014.

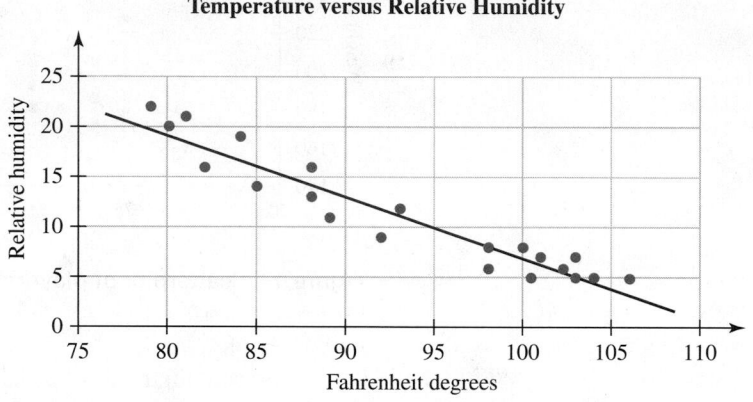

Figure 51 Scatterplot and model for temperature and relative humidity in Phoenix
(**Source:** *Weatherbase*)

Compute the coefficient of determination and describe what it means in this situation.

Solution

We use StatCrunch to compute that $r^2 \approx 0.90$ (see Fig. 52). TI-84 output is shown in Fig. 53. The result $r^2 \approx 0.90$ means that about 90% of the variation in relative humidity is explained by the regression line.

Simple linear regression results:
Dependent Variable: Humidity
Independent Variable: Temperature
Humidity = 68.123034 − 0.61252416 Temperature
Sample size: 22
R (correlation coefficient) = −0.94777578
R-sq = 0.89827893
Estimate of error standard deviation: 1.8621932

Figure 52 StatCrunch output for r^2

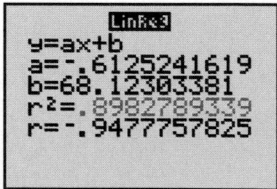

Figure 53 TI-84 output for r^2

Group Exploration
Meaning of residual plots

1. Scatterplots and regression lines are shown for data sets A and B in Figs. 54 and 55, respectively. The residual plots for the two data sets are the same (see Fig. 56). Explain why they are the same.

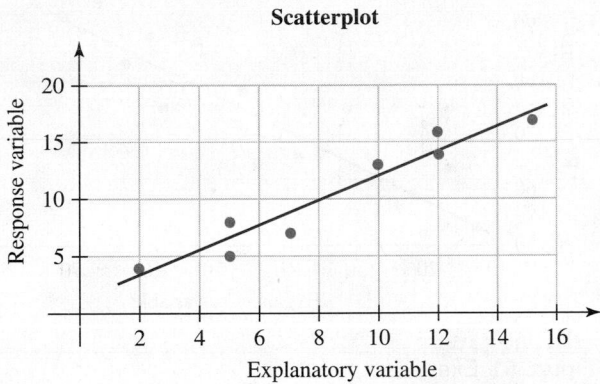

Figure 54 Scatterplot and model for data set A

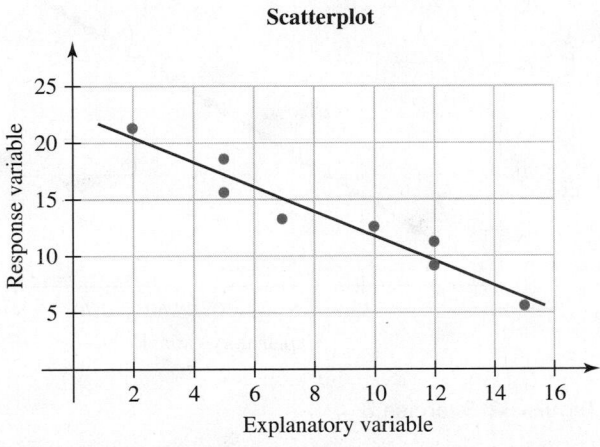

Figure 55 Scatterplot and model for data set B

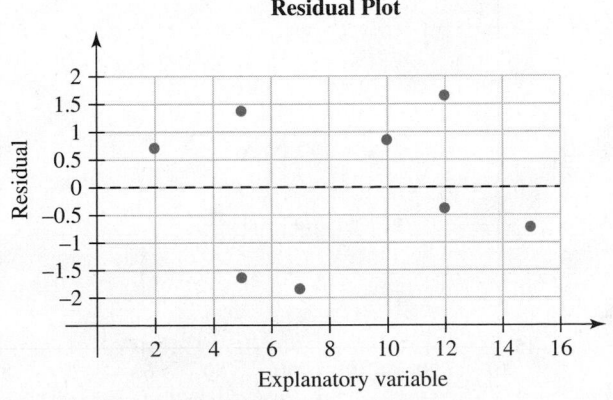

Figure 56 Residual plot

2. Construct a scatterplot and an *increasing* line so that the residual plot for the line is the one shown in Fig. 57.

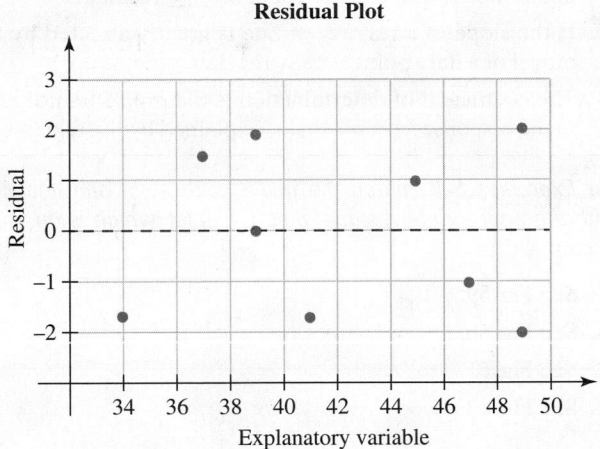

Figure 57 Residual plot

3. Construct a scatterplot and a *decreasing* line so that the residual plot for the line is the one shown in Fig. 57.

4. The linear regression equation for some data is $\hat{y} = 0.53x + 17.07$. The residual plot is shown in Fig. 58. Estimate the values of the data pairs.

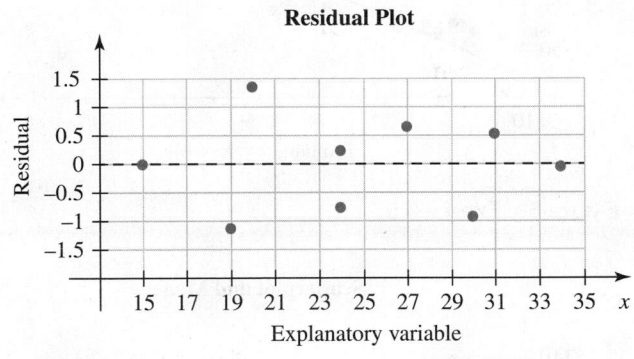

Figure 58 Residual plot

Homework 9.3

For extra help ▶ MyMathLab®   Watch the videos in MyMathLab   Download the MyDashboard App

1. *True or False:* For a given data point (x, y), the residual is equal to the predicted value of y minus the observed value of y.

2. For a group of data points, the linear regression model is the linear model with the least sum of _____ residuals.

3. If the slope of a regression line is greatly affected by the removal of a data point, we say the data point is a(n) _____ point.

4. The coefficient of determination is the proportion of the _____ in the response variable that is explained by the regression line.

For Exercises 5–8, match the given scatterplot and linear model with the appropriate residual plot in Fig. 63, which is on this page and page 615.

5. See Fig. 59.

6. See Fig. 60.

7. See Fig. 61.

8. See Fig. 62.

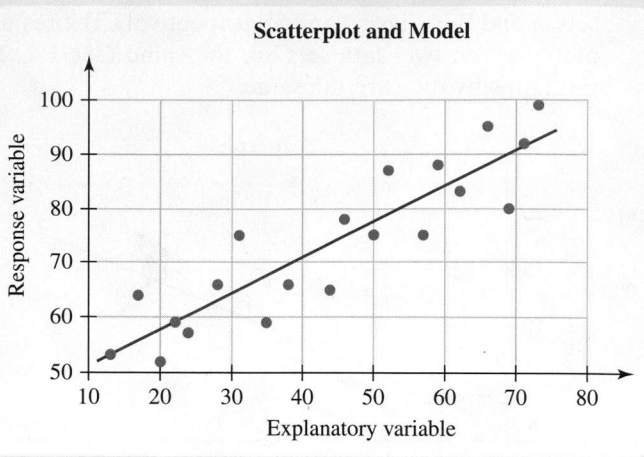

Figure 61 Exercise 7

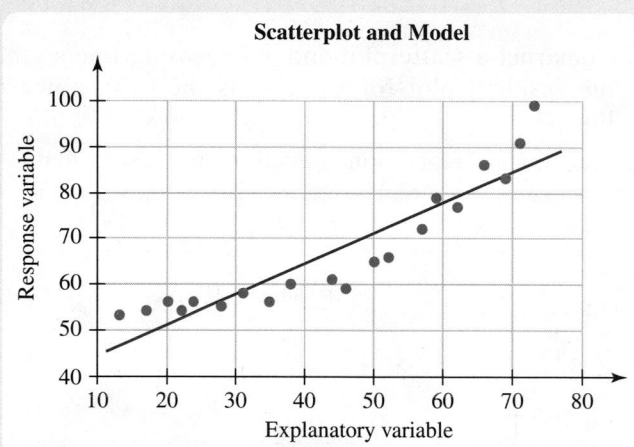

Figure 59 Exercise 5

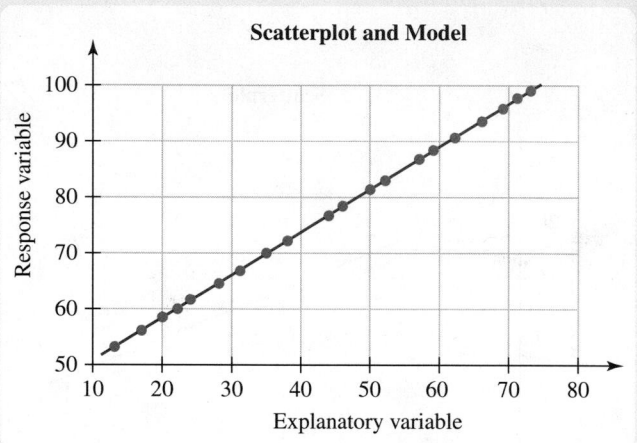

Figure 62 Exercise 8

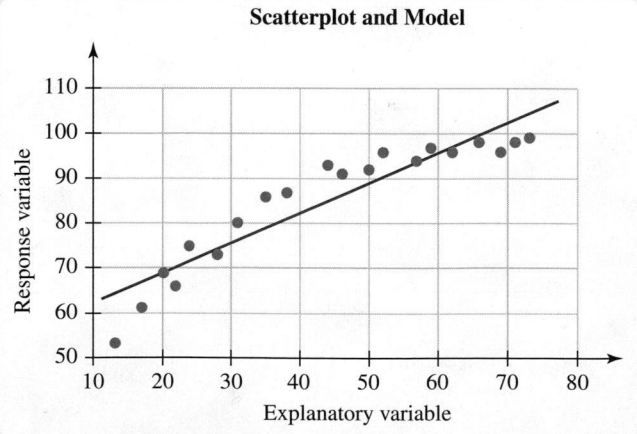

Figure 60 Exercise 6

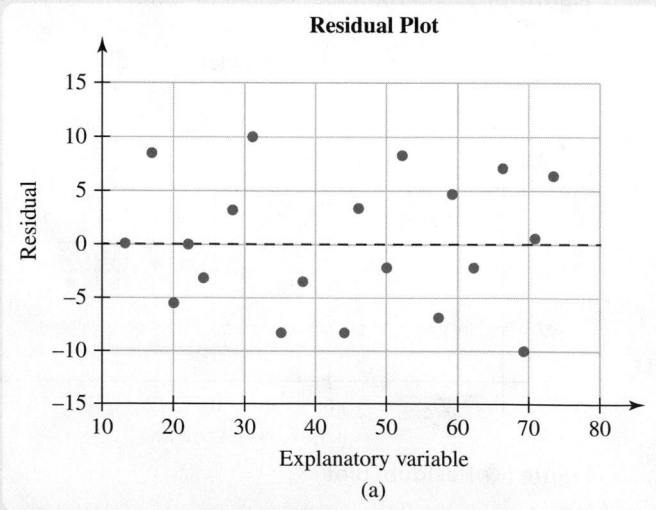

Figure 63 Exercises 5–8 (*continued on page 615*)

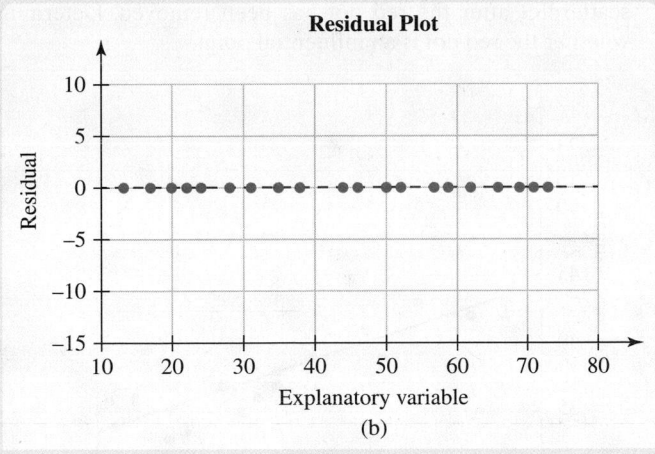

(b)

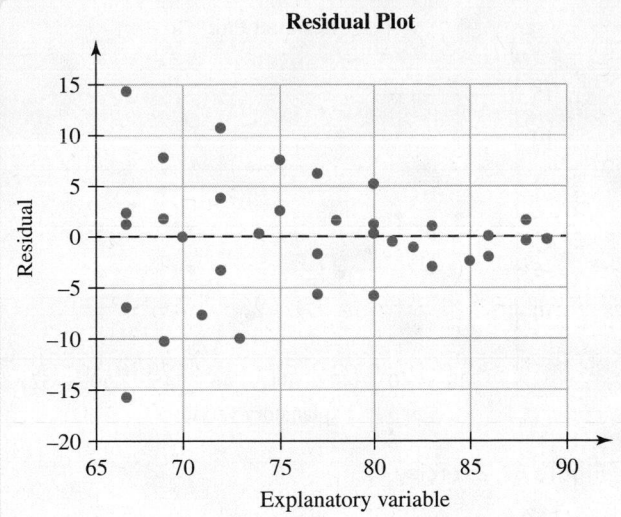

Figure 64 Exercise 9

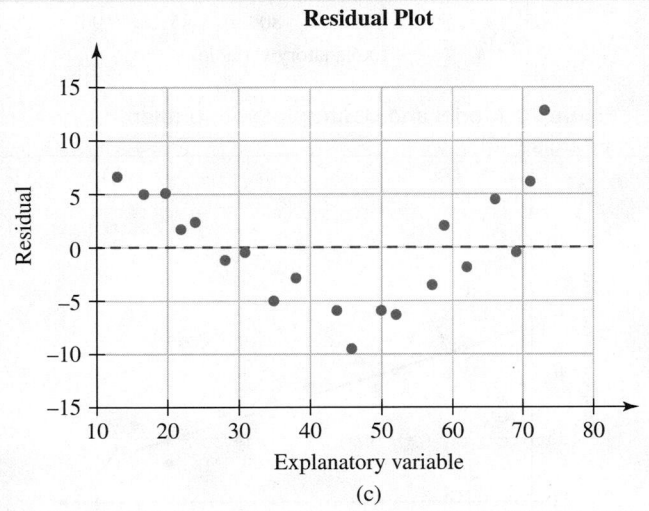

(c)

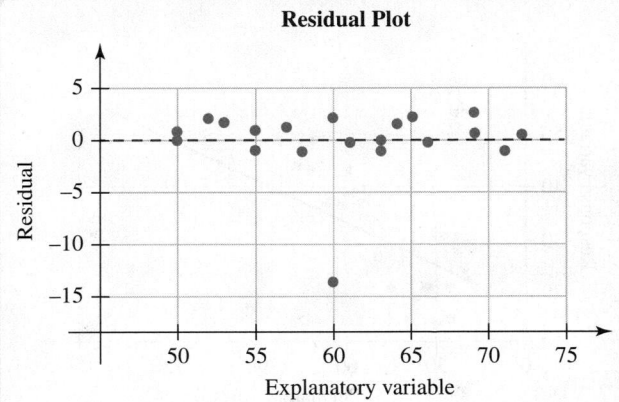

Figure 65 Exercise 10

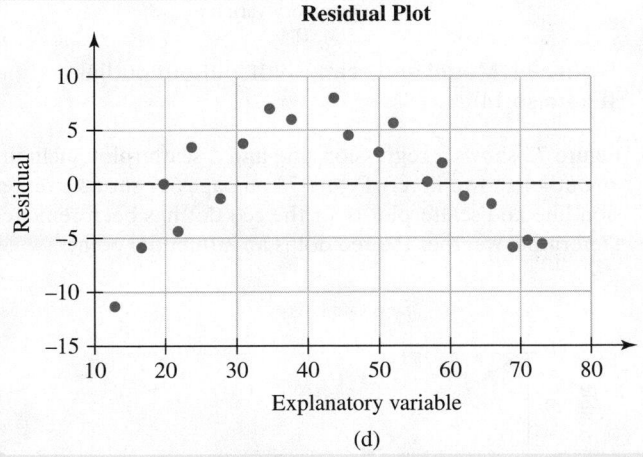

(d)

Figure 63 (*Continued*)

For Exercises 9–12, refer to the given residual plot and identify which conditions, if any, for a regression line are not met.

9. See Fig. 64.
10. See Fig. 65.
11. See Fig. 66.
12. See Fig. 67 on page 616.

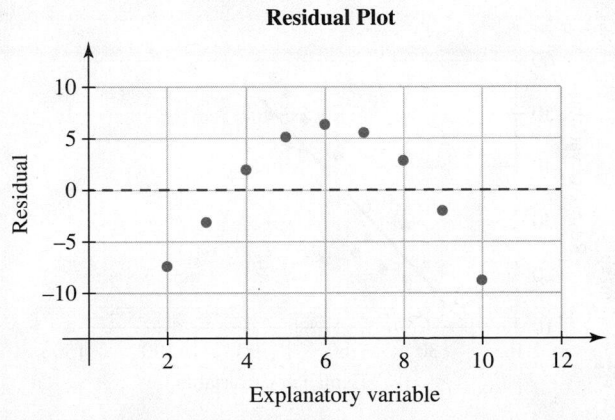

Figure 66 Exercise 11

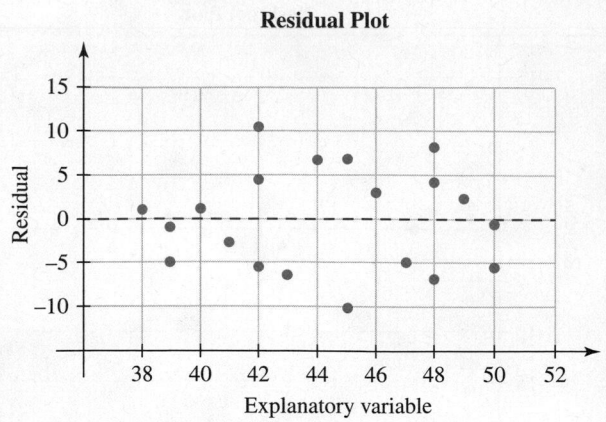

Figure 67 Exercise 12

13. Figure 68 shows a regression line and a scatterplot, including an outlier shown in red. Figure 69 shows a regression line and scatterplot after the red dot has been removed. Determine whether the red dot is an influential point.

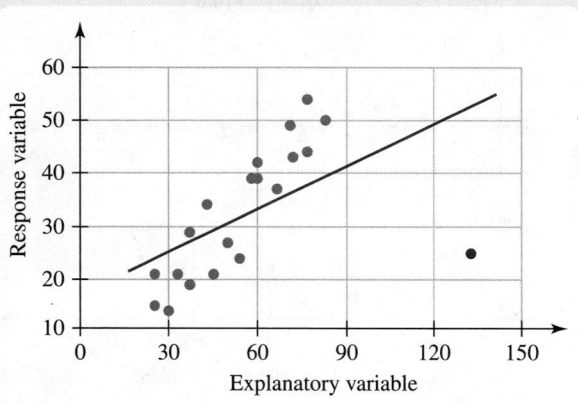

Figure 68 Model and scatterplot with outlier (Exercise 13)

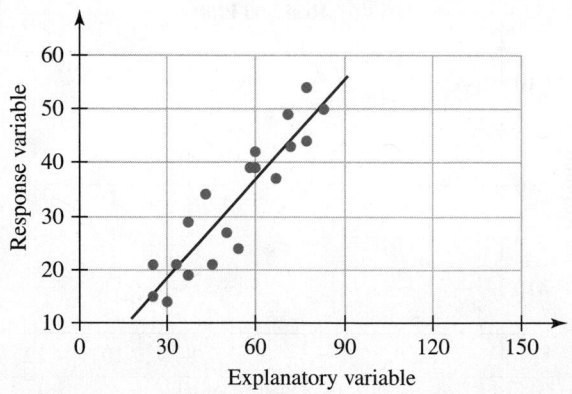

Figure 69 Model and scatterplot without outlier (Exercise 13)

14. Figure 70 shows a regression line and a scatterplot, including an outlier shown in red. Figure 71 shows a regression line and scatterplot after the red dot has been removed. Determine whether the red dot is an influential point.

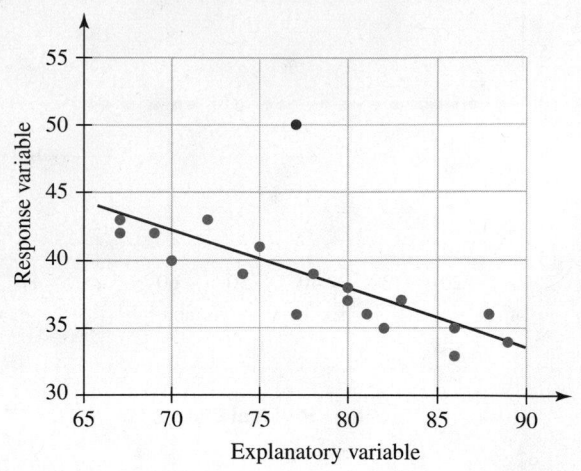

Figure 70 Model and scatterplot with outlier (Exercise 14)

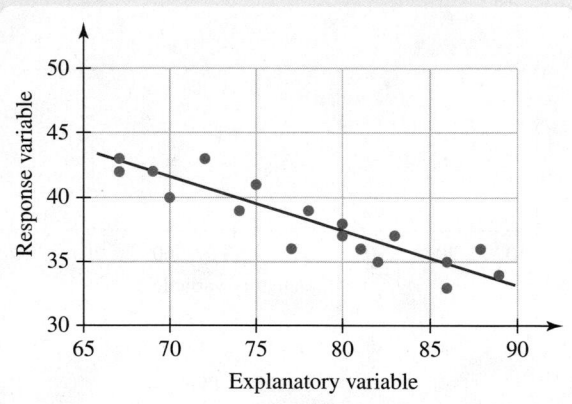

Figure 71 Model and scatterplot without outlier (Exercise 14)

15. Figure 72 shows a regression line and a scatterplot, including an outlier shown in red. Figure 73 on page 617 shows a regression line and scatterplot after the red dot has been removed. Determine whether the red dot is an influential point.

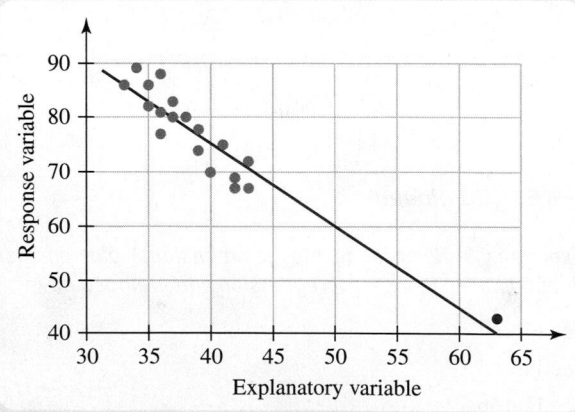

Figure 72 Model and scatterplot with outlier (Exercise 15)

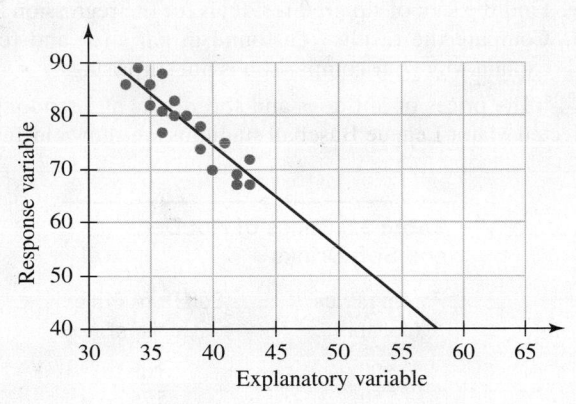

Figure 73 Model and scatterplot without outlier (Exercise 15)

16. Figure 74 shows a regression line and a scatterplot, including an outlier shown in red. Figure 75 shows a regression line and scatterplot after the red dot has been removed. Determine whether the red dot is an influential point.

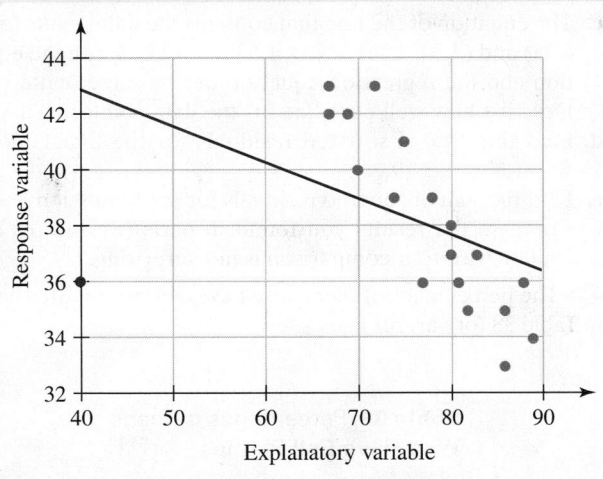

Figure 74 Model and scatterplot with outlier (Exercise 16)

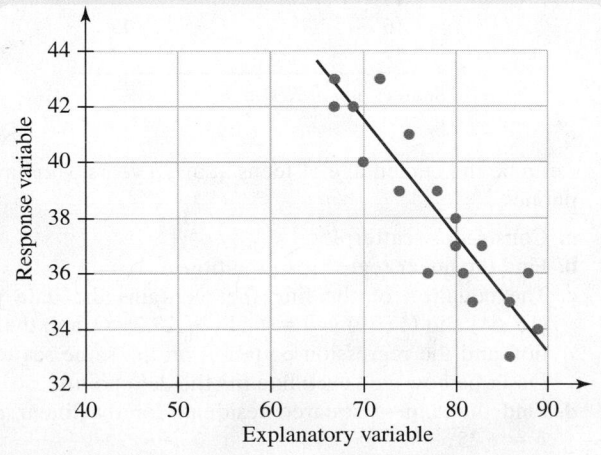

Figure 75 Model and scatterplot without outlier (Exercise 16)

For Exercises 17–20, match the given information to the appropriate scatterplot in Fig. 76.

17. $r^2 = 0.81$ **19.** $r^2 = 1$

18. $r^2 = 0.36$ **20.** $r^2 = 0$

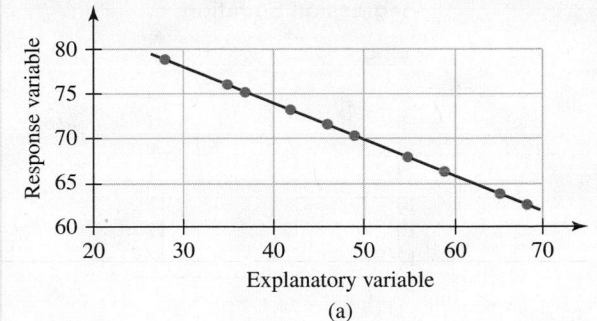

(a)

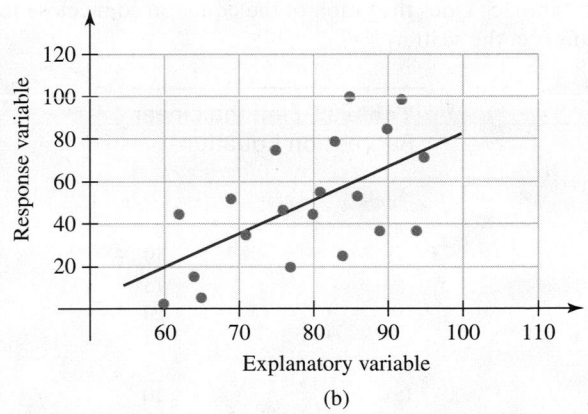

(b)

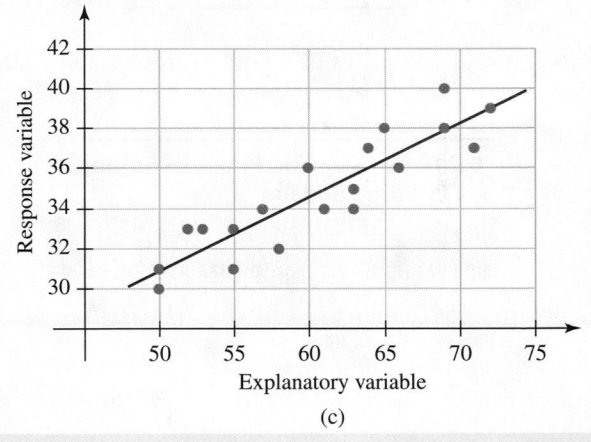

(c)

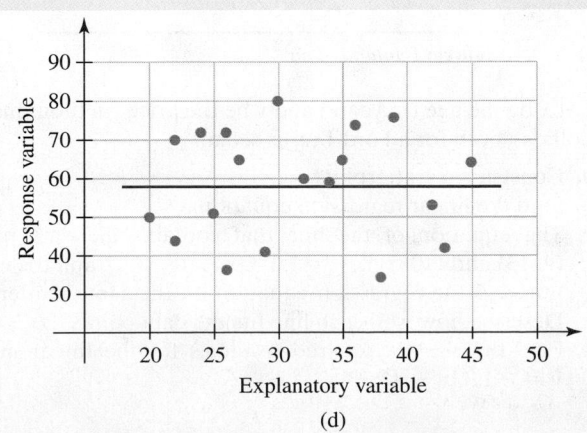

(d)

Figure 76 Exercises 17–20

21. DATA Find the linear regression equation for the points listed in Table 34. Does the graph of the equation come close to the points in the scatterplot?

Table 34 Find the Linear Regression Equation

x	y
2	4
5	7
5	9
7	11
10	15
10	12
12	17

22. DATA Find the linear regression equation for the points listed in Table 35. Does the graph of the equation come close to the points in the scatterplot?

Table 35 Find the Linear Regression Equation

x	y
4	24
6	19
9	15
9	20
12	14
15	6
15	10
19	2

23. DATA The ages and prices of 6 Ford Taurus sedans at dealerships in the Nashville area are shown in Table 36.

Table 36 Ages and Prices of Some Ford Taurus Sedans

Age (in years)	Price (thousands of dollars)
4	13
10	6
5	11
7	9
4	12
10	5

Source: *Edmunds.com*

Let x be the age (in years) and y be the price (in thousands of dollars), both for a Ford Taurus sedan.

a. Construct a scatterplot.
b. Find the linear regression equation.
c. The equation of the line that contains the data points (4, 13) and (10, 5) is $\hat{y} = -1.33x + 18.33$. Graph the equation and the regression equation on the same scatterplot. Describe how well each line fits the data points.
d. Find the sum of squared residuals for the linear model $\hat{y} = -1.33x + 18.33$.

e. Find the sum of squared residuals for the regression line.
f. Compare the results you found in parts (d) and (e) and explain why your comparison is not surprising.

24. DATA The prices of hot dogs and soft drinks at 5 randomly selected Major League Baseball stadiums are shown in Table 37.

Table 37 Prices of Hot Dogs and Soft Drinks

Hot Dog Price (dollars)	Soft Drink Price (dollars)
3.00	3.00
5.50	5.75
4.50	4.25
1.50	1.50
3.50	2.50

Source: *Team Marketing Report*

Let x be the price of a hotdog and y be the price of a soft drink, both in dollars, at a Major League Baseball stadium.

a. Construct a scatterplot, treating x as the explanatory variable.
b. Find the linear regression equation.
c. The equation of the line that contains the data points (5.50, 5.75) and (3.50, 2.50) is $\hat{y} = 1.63x - 3.19$. Graph the equation and the regression equation on the same scatterplot. Describe how well each line fits the data points.
d. Find the sum of squared residuals for the linear model $\hat{y} = 1.63x - 3.19$.
e. Find the sum of squared residuals for the regression line.
f. Compare the results you found in parts (d) and (e) and explain why your comparison is not surprising.

25. DATA The percentages of teens who have cell phones are shown in Table 38 for various ages.

Table 38 Percentages of Teens Who Have Cell Phones

Age (years)	Percent
13	54
14	68
15	68
16	73
17	79

Source: *Jupiter Research*

Let p be the percentage of teens at age a years who have cell phones.

a. Construct a scatterplot.
b. Find the linear regression equation.
c. The equation of the line that contains the data points (13, 54) and (17, 79) is $\hat{y} = 6.25x - 27.25$. Graph the equation and the regression equation on the same scatterplot. Describe how well each line fits the data points.
d. Find the sum of squared residuals for the linear model $\hat{y} = 6.25x - 27.25$.

e. Find the sum of squared residuals for the regression line.

f. Compare the results you found in parts (d) and (e) and explain why your comparison is not surprising.

26. DATA. The percentages of Americans who have been diagnosed with diabetes are shown in Table 39 for various age groups.

Table 39 Percentages of Americans Diagnosed with Diabetes by Age Group

Age Group (years)	Age Used to Represent Age Group (years)	Percent
35–39	37	2
40–44	42	4
45–49	47	5
50–54	52	8
55–59	57	10

Source: *National Health Interview Survey*

Let y be the percentage of Americans at age x years who have been diagnosed with diabetes at some point in their lives.

a. Construct a scatterplot.

b. Find the linear regression equation.

c. The equation of the line that contains the data points $(37, 2)$ and $(47, 5)$ is $\hat{y} = 0.3x - 9.1$. Graph the equation and the regression equation on the same scatterplot. Describe how well each line fits the data points.

d. Find the sum of squared residuals for the line $\hat{y} = 0.3x - 9.1$.

e. Find the sum of squared residuals for the regression line.

f. Compare the results you found in parts (d) and (e) and explain why your comparison is not surprising.

27. DATA. A racquetball is dropped from various heights, and the bounce height is recorded each time (see Table 40).

Table 40 Drop and Bounce Heights of a Racquetball

Drop Height (inches)	Bounce Height (inches)
6	5.0
12	9.3
18	15.0
24	19.6
30	24.0
36	27.6
42	32.8
48	38.0

Source: *J. Lehmann*

Let $f(x)$ be the bounce height (in inches) of the racquetball after it is dropped from an initial height of x inches.

a. Construct a scatterplot.

b. Find the linear regression equation for f. Does the graph of f come close to the data points?

c. Find $f(18)$. What does it mean in this situation?

d. Find the residual for the prediction you made in part (c). What does it mean in this situation?

e. Find x when $f(x) = 30$. What does it mean in this situation?

28. DATA. The heights of objects and the lengths of their shadows are shown in Table 41.

Table 41 Heights of Objects and Lengths of Their Shadows

Object	Height (inches)	Length of Shadow (inches)
Wine bottle	15.5	10.3
Toy putter	20.8	13.0
Box	36.5	23.6
Mop	47.8	31.0
Person's shoulder	55.0	36.5
Person	63.0	41.5

Source: *J. Lehmann*

Let x be the height (in inches) of an object and $f(x)$ be the length (in inches) of the object's shadow.

a. Construct a scatterplot.

b. Find the linear regression equation for f. Does the graph of f come close to the data points?

c. Find $f(55)$. What does it mean in this situation?

d. Find the residual for the prediction you made in part (c). What does it mean in this situation?

e. Find x when $f(x) = 20$. What does it mean in this situation?

29. DATA. Mean gasoline prices in Michigan are shown in Table 42 for various years.

Table 42 Mean Gasoline Prices in Michigan

Year	Mean Gasoline Price (dollars)
2000	1.568
2002	1.385
2004	1.861
2006	2.555
2008	3.275
2010	2.801
2012	3.689
2013	3.589

Source: *AAA*

Let $f(x)$ be the mean gasoline price (in dollars) at x years since 2000.

a. Construct a scatterplot.

b. Find the linear regression equation for f. Does the graph of f come close to the data points?

c. Find $f(13)$. What does it mean in this situation?

d. Find the residual for the estimate you made in part (c). What does it mean in this situation?

e. Find x when $f(x) = 3$. What does it mean in this situation?

30. DATA. Due to inflation, an item that cost $10,000 in 1980 cost $28,486 in 2015. Costs comparable to $10,000 in 1980 are shown in Table 43 for various years.

Table 43 Costs Comparable to $10,000 in 1980

Year	Comparable Cost (dollars)
1980	10,000
1985	13,058
1990	15,862
1995	18,495
2000	20,898
2005	23,701
2010	26,463
2015	28,486

Source: *Bureau of Labor Statistics*

Let $f(x)$ be the cost (in dollars) at x years since 1980 that is comparable to $10,000 in 1980.

a. Construct a scatterplot.

b. Find the linear regression equation for f. Does the graph of f come close to the data points?

c. Find $f(30)$. What does it mean in this situation?

d. Find the residual for the estimate you made in part (c). What does it mean in this situation?

e. Find x when $f(x) = 25{,}000$. What does it mean in this situation?

31. **DATA** The number of cases and deaths from Ebola in some African countries as of October 14, 2014, are shown in Table 44.

Table 44 Numbers of Cases and Deaths from Ebola

Country	Number of Cases	Number of Deaths
Guinea	1472	843
Liberia	4249	2458
Nigeria	20	8
Senegal	1	0
Sierra Leone	3252	1183

Source: *World Health Organization (WHO)*

Let y be the number of Ebola deaths in a country that has had x Ebola cases.

a. What is the explanatory variable? Explain.

b. Construct a scatterplot.

c. Find the linear regression equation. Does the graph of the equation come close to the data points?

d. From September 17 to October 17, 2014, there were about 1000 new cases per week (Source: *WHO*). Substitute 1000 for x in the linear regression equation to estimate the number of Ebola deaths from those new cases per week. Why is that substitution for x not in the spirit of how x is defined? In your opinion, does that matter?

e. WHO predicted that if drastic measures were not taken from October 17 to December 17, 2014, there would have been 10,000 new cases per week after that period. Estimate the number of Ebola deaths per week from that many new cases per week. Do you have much faith in this estimate? Explain.

32. **DATA** The rate at which a cricket chirps depends on the temperature of the surrounding air. You can estimate the air temperature by counting chirps! Some data randomly selected from the study's data are provided in Table 45.

Table 45 Rates of Cricket Chirping

Temperature (°F)	Chirp Rate (number of chirps in 13 seconds)
80.5	44
68	33
65	28
74	37
59	24
62	22
50.75	13
78	44
55	16

Source: The Sound of Crickets: Using Evidence-Based Reasoning to Measure Temperature Using Cricket Chirps, *Peggy LeMone*

Let y be the number of chirps a cricket makes in 13 seconds when the temperature is x degrees Fahrenheit.

a. Construct a scatterplot.

b. Find the linear regression equation. Does the graph of the equation come close to the data points?

c. Predict the number of times a cricket will chirp in 13 seconds if it is 70°F.

d. Find the linear regression equation again, but this time treat y as the explanatory variable. [**Hint:** Do *not* find the equation by solving the equation you found in part (b) for x. This would give a slightly different result.]

e. The researcher says that as a rough estimate, you can predict the current temperature by counting the number of chirps in 13 seconds and then adding 40 to the result. Write an equation for this rule of thumb. Compare the result to the equation you found in part (d).

33. **DATA** If you could stop time and live forever in good health at a particular age, what age would you choose? The mean ideal ages chosen by various age groups are shown in Table 46.

Table 46 Mean Ideal Ages

Age Group (years)	Age Used to Represent Age Group (years)	Mean Ideal Age (years)
18–24	21	27
25–29	27	31
30–39	34.5	37
40–49	44.5	40
50–64	57	44
over 64	75	59

Source: *Harris Poll*

Let y be the mean ideal age (in years) chosen by people whose actual age is x years.

a. Construct a scatterplot.

b. Find the linear regression equation. Graph it on the scatterplot.

c. Construct a residual plot.

d. Find the age and the residual represented by the lowest dot in the residual plot. What does this tell you about the corresponding data point in the scatterplot and the regression line?

e. Find the age and the residual represented by the highest dot in the residual plot. What does this tell you about the corresponding data point in the scatterplot and the regression line?

34. **DATA.** The percentages of Americans who gamble online are shown in Table 47 for various age groups.

Table 47 Percentages of Americans Who Gamble Online

Age Group (years)	Age Used to Represent Age Group (years)	Online
21–29	25	9
30–39	34.5	14
40–49	44.5	18
50–59	54.5	20
over 59	70	37

Source: *American Gaming Association*

Let y be the percentage of Americans who gamble online at age x years.

a. Construct a scatterplot.
b. Find the linear regression equation. Graph it on the scatterplot.
c. Construct a residual plot.
d. Find the age and the residual represented by the lowest dot in the residual plot. What does this tell you about the corresponding data point in the scatterplot and the regression line?
e. Find the age and the residual represented by the highest dot in the residual plot. What does this tell you about the corresponding data point in the scatterplot and the regression line?

35. **DATA.** In Exercise 33 on page 620 you found the linear regression equation $\hat{y} = 0.55x + 15.90$, where y is the mean ideal age (in years) chosen by people whose actual age is x years.

a. Use the model to predict the mean ideal age chosen by 25-year-olds.
b. What is the slope? What does it mean in this situation?
c. What is the y-intercept? What does it mean in this situation?
d. Refer to Table 46 on page 620 to find the coefficient of determination. What does it mean in this situation?
e. What is the age of people who chose a mean ideal age equal to their actual age?

36. **DATA.** In Exercise 34 you found the linear regression equation $\hat{y} = 0.57x - 6.68$, where y is the percentage of Americans who gamble online at age x years.

a. Use the model to predict the percentage of 30-year-old Americans who gamble online.
b. What is the slope? What does it mean in this situation?
c. Use the model to predict the age at which 16% of Americans gamble.
d. Refer to Table 47 to find the coefficient of determination. What does it mean in this situation?

37. **DATA.** Mean scores on tests that evaluate general knowledge and vocabulary, and mean scores on tests that evaluate memory and information-processing speed are shown in Table 48 for various ages. (The higher the mean score, the better the mental function will be. The mean score for everyone in the study is 0.)

Table 48 Age versus Mental Functioning

Age Group (years)	Age Used to Represent Age Group (years)	Mean General Knowledge and Vocabulary Score (points)	Mean Memory and Information-Processing Speed Score (points)
20–30	25	−0.4	1.0
30–40	35	−0.3	0.7
40–50	45	0	0.3
60–70	65	0.2	−0.2
70–80	75	0.3	−0.4

Source: *Denise Park, University of Illinois, Champaign-Urbana*

Let y be the mean score (in points) on general knowledge and vocabulary tests for people at age x years.

a. Construct a scatterplot that compares values of x and y.
b. Describe the four characteristics of the association. Compute and interpret r as part of your analysis.
c. Find the linear regression equation.
d. Construct a residual plot. Does it support any of the conclusions you made in part (b)? If so, which one(s)?
e. Find the coefficient of determination. What does it mean in this situation?

38. **DATA.** Mean scores on tests that evaluate general knowledge and vocabulary, and mean scores on tests that evaluate memory and information-processing speed are shown in Table 48 for various ages. (The higher the mean score, the better the mental function will be. The mean score for everyone in the study is 0.) Let y be the mean score (in points) on memory and information-processing speed tests for people at age x years.

a. Construct a scatterplot that compares values of x and y.
b. Describe the four characteristics of the association. Compute and interpret r as part of your analysis.
c. Find the linear regression equation.
d. Construct a residual plot. Does it support any of the conclusions you made in part (b)? If yes, which one(s)?
e. Find the coefficient of determination. What does it mean in this situation?

39. **DATA.** The levels of rainfall in Tuscola County, Michigan, and the numbers of murders by pushing from high places in the United States are shown in Table 49 for various years.

Table 49 Annual Levels of Rainfall in Tuscola County, Michigan, and Numbers of Murders per Year by Pushing from High Places

Year	Level of Rainfall (mm)	Number of Murders
1999	2.07	17
2000	2.47	18
2001	2.19	17
2002	2.03	16
2003	2.10	15
2004	2.32	17
2005	2.22	18
2006	2.86	19
2007	2.04	15
2008	2.63	20
2009	2.32	18

Source: *U.S. Centers for Disease Control and Prevention*

Let x be the level of rainfall (in mm) in Tuscola County, Michigan, and y be the number of murders by pushing from high places in the United States, both in the same year.

a. Construct a scatterplot, treating the annual level of rainfall as the explanatory variable.
b. Describe the four characteristics of the association. Compute and interpret r as part of your analysis.
c. Find the linear regression equation.
d. Construct a residual plot. Does it support any of the conclusions you made in part (b)? If yes, which one(s)?
e. Find the coefficient of determination. What does it mean in this situation?
f. On the basis of your responses to parts (a) through (e), can you conclude that murders by pushing from high places in the United States were caused by rainfall in Tuscola County, Michigan? Explain.

40. DATA The total revenues of skiing facilities and the numbers of people who died from becoming tangled in their bedsheets are shown in Table 50 for various years.

Table 50 Annual Total Revenues of Skiing Facilities and Numbers of People Who Died from Becoming Tangled in Their Bedsheets

Year	Total Revenue of Skiing Facilities (billions of dollars)	Fatalities from Tangled Bedsheets
2000	1.55	327
2001	1.64	456
2002	1.80	509
2003	1.83	497
2004	1.96	596
2005	1.99	573
2006	2.18	661
2007	2.26	741
2008	2.48	809
2009	2.44	717

Source: *U.S. Census Bureau, Centers for Disease Control and Prevention*

Let x be the total revenue (in billions of dollars) of skiing facilities, and let y be the number of fatalities from people becoming tangled in their bedsheets, both in the same year.

a. Construct a scatterplot, treating the total annual revenue of skiing facilities as the explanatory variable.
b. Describe the four characteristics of the association. Compute and interpret r as part of your analysis.
c. Find the linear regression equation.
d. Construct a residual plot. Does it support any of the conclusions you made in part (b)? If yes, which one(s)?
e. Find the coefficient of determination. What does it mean in this situation?
f. On the basis of your responses to parts (a) through (e), can you conclude that an increase in the revenue of skiing facilities causes an increase in the number of people who die from becoming tangled in their bedsheets? Explain.

41. DATA If you are driving and spot an object in the road, the distance it will take you to stop is equal to the sum of the following:

- The **reaction distance** is the distance you will continue to travel before you hit the brakes.
- The **braking distance** is the distance you will travel as you are braking.

The reaction and braking distances are shown in Table 51 for various driving speeds.

Table 51 Reaction and Braking Distances

Driving Speed (miles per hour)	Reaction Distance (feet)	Braking Distance (feet)
20	44	25
30	66	57
40	88	101
50	110	158
60	132	227
70	154	310
80	176	404

Source: *National Highway Traffic Safety Administration*

Let B be the braking distance (in feet) when driving at s miles per hour.

a. Construct a scatterplot that compares the values of s and B.
b. Describe the four characteristics of the association between s and B.
c. Construct a residual plot for the regression line that models the association between s and B. Does it support any of the conclusions you made in part (b)? If yes, which one(s)?
d. Estimate the driving speed and the residual represented by the lowest point in the residual plot. What does this tell you about the corresponding data point in the scatterplot and the regression line?
e. A student computes that $r = 0.99$ and concludes that there is a strong, linear association. What would you tell the student?

42. DATA Refer to Exercise 41 for a description of reaction and braking distances. Let R be the reaction distance and let B be the braking distance, both in feet, when driving at s miles per hour.

a. Find the stopping distances (in feet) for the driving speeds 20, 30, 40, 50, 60, 70, and 80, all in miles per hour.
b. Let D be the stopping distance (in feet) when driving at s miles per hour. Construct a scatterplot for the association between s and D.
c. Find the linear regression equation for the association between s and D.
d. The linear regression equation for the association between s and R is $\hat{R} = 2.2s$. The linear regression equation for the association between s and B is $\hat{B} = 6.32s - 132.75$. Add the expressions $2.2s$ and $6.32s - 132.75$ and compare the result to the right-hand side of the regression equation you found in part (c). Why did this happen?

43. DATA Refer to Exercise 41 for a description of reaction and braking distances. Let R be the reaction distance, B be the braking distance, and D be the stopping distance, all in feet, when driving at s miles per hour.

a. The residual plot for using the regression line to model the association between s and R is shown in Fig. 77. What can you conclude about the association?

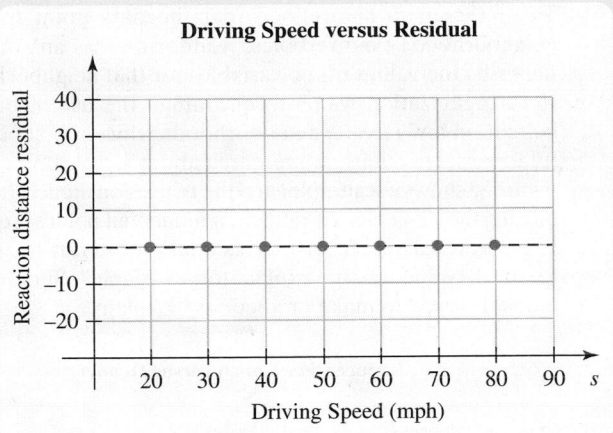

Figure 77 Exercise 43a

b. Construct a scatterplot that compares the values of s and R. Does it confirm your response to part (a)?

c. The linear regression equations for reaction, braking, and stopping distances are $\hat{R} = 2.2s$, $\hat{B} = 6.32s - 132.75$, and $\hat{D} = 8.52s - 132.75$, respectively. Show that the sum of the right-hand sides of the first two equations is equal to the right-hand side of the third equation.

d. The residual plots for using regression lines to model the speed-braking and speed-stopping associations are shown in Fig. 78. What do you notice about the residual plots? Why did this happen?

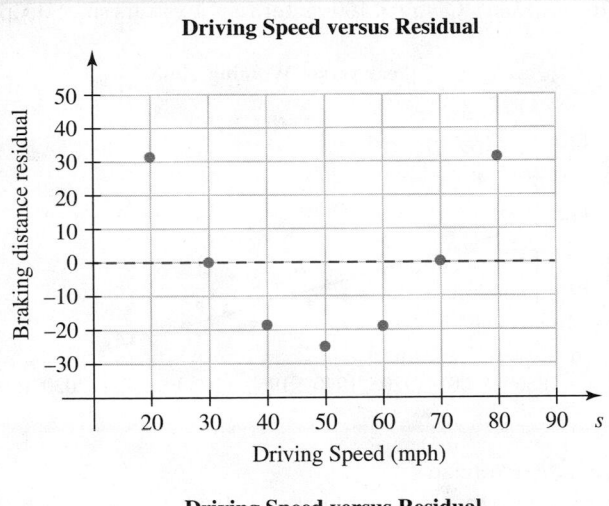

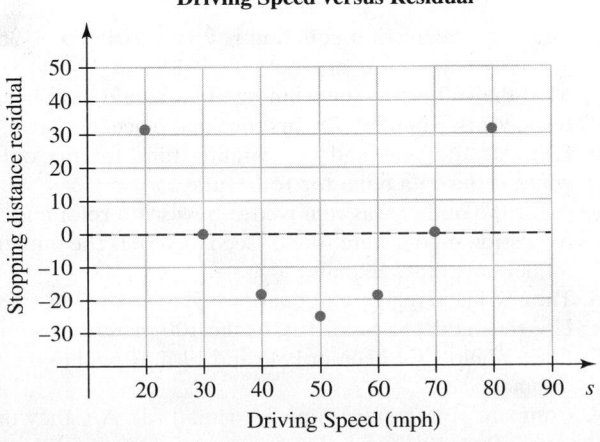

Figure 78 Exercise 43d

44. **DATA** Table 52 compares the SAT scores and acceptance rates of applicants to 10 of the most selective U.S. colleges and universities. The data do not include students who participated in the early-decision application process.

Table 52 SAT Scores and Acceptance Rates

SAT Score Group (points)	Score Used to Represent SAT Score Group (points)	Percent
1100–1190	1145	10
1200–1290	1245	17
1300–1390	1345	31
1400–1490	1445	48
1500–1600	1550	72

Source: *Professor Christopher Avery, Kennedy School of Government, Harvard University*

For students who score x points, let y be the percentage of applicants who are accepted.

a. Construct a scatterplot.

b. Describe the four characteristics of the association. Compute and interpret r as part of your analysis.

c. Construct a residual plot for the regression line that models the association. Does it support any of the conclusions you made in part (b)? If so, which one(s)?

d. Estimate the SAT score and the residual represented by the lowest dot in the residual plot. What does this tell you about the corresponding data point in the scatterplot and the regression line?

e. Because r is so close to 1, a student concludes that there is a strong, linear association. What would you tell the student?

45. The residual plot in Fig. 79 describes the residuals for using the regression line to describe the association between the percentage of people who voted for President Obama in the 2008 presidential election and the percentage of people who voted for him in the 2012 presidential election, where the dots represent the 50 states.

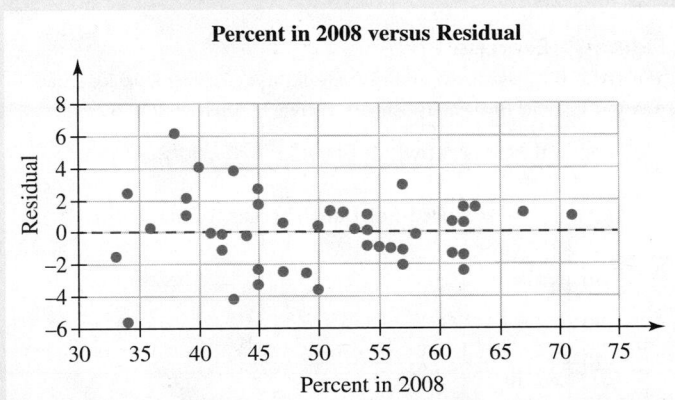

Figure 79 Exercise 45

a. In 2008, 40% of Louisiana residents voted for President Obama. Estimate the residual for that state. What does it tell you about the data point and the regression line?

b. In 2008, 55% of Nevada residents voted for President Obama. Estimate the residual for that state. What does it tell you about the data point and the regression line?

c. Is the vertical spread of the residual plot about the same for each value of the explanatory variable? Can we use the regression line to make predictions? Explain.

d. Is it easier to draw the conclusions you made in part (c) by referring to the residual plot in Fig. 79 or the scatterplot in Fig. 80. Explain.

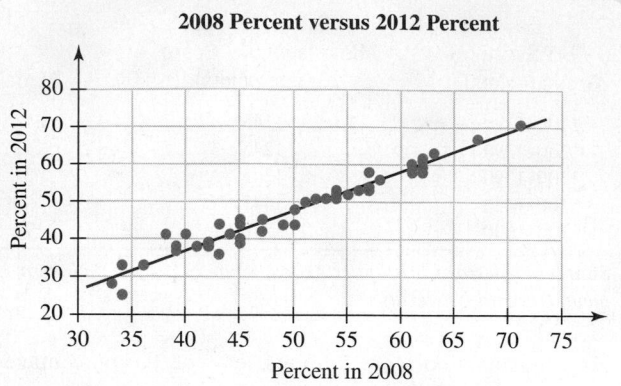

Figure 80 Exercise 45d

(**Source:** *Federal Election Commission report*)

46. The scatterplot and the model in Fig. 81 describe the association between the percentage of children who receive free or reduced-fee lunches at school and the percentage of people who wear bicycle helmets, for each of 12 neighborhoods.

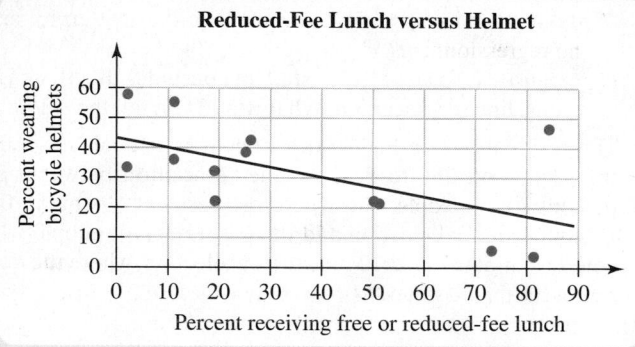

Figure 81 Exercise 46

(**Source:** *A Bi-County Comparative Study of Bicycle Helmet Knowledge and Use by California Elementary School Children, D. Perales et al.*)

A residual plot is shown in Fig. 82.

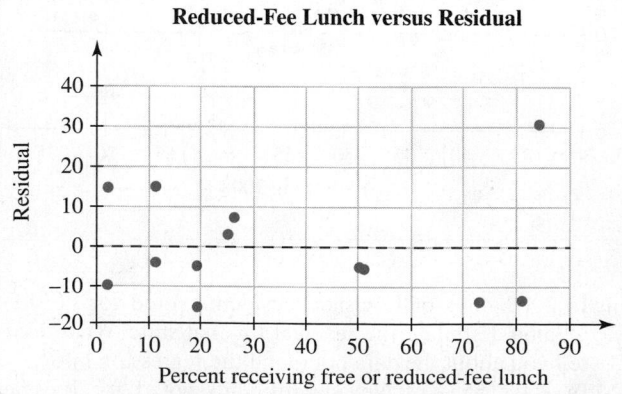

Figure 82 Residual plot for lunch-helmet data (Exercise 46)

a. The researchers determined that the data point for the neighborhood Los Arboles, California, is an outlier. Describe the values of the variables for that neighborhood.

b. If an organization wants to encourage the use of bicycle helmets in low-income neighborhoods, why might it be useful to perform additional research about Los Arboles?

c. Figure 83 shows a scatterplot and the regression model after removing the outlier. Is the outlier an influential point? Explain.

d. Is the vertical spread of the residual plot about the same for each value of the explanatory variable? Should the model be used to make predictions? Explain.

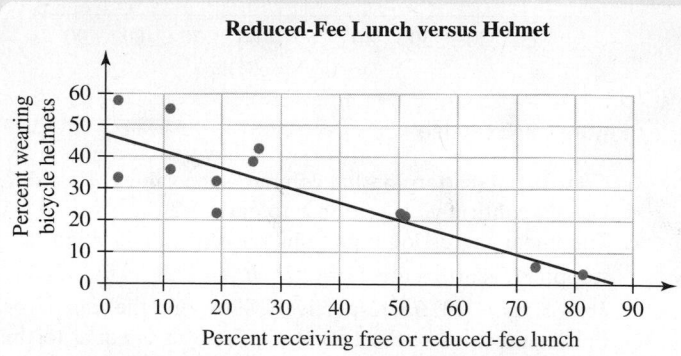

Figure 83 Exercise 46c

47. The scatterplot and the model in Fig. 84 describe the association between the years and the winning times for the men's Olympic 100-meter run. Let y be the winning time (in seconds) for the men's Olympic 100-meter run at x years since 0 A.D.

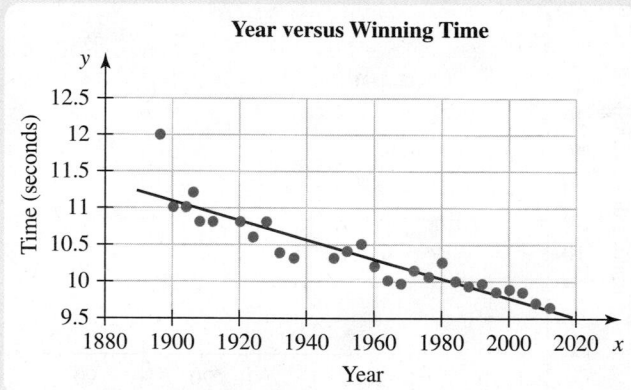

Figure 84 Exercise 47

(**Source:** *The Complete Book of the Olympics, David Wallechinsky and Jaime Loucky*)

a. The linear regression equation is $\hat{y} = -0.01332x + 36.41$. The Olympics were not held in 1940 and 1944. Predict what the 100-meter-run winning times would have been in those years. Round to the first decimal place.

b. Estimate the year and the winning time for the outlier, which is the data point for Tom Burke.

c. After the outlier was removed, a new scatterplot and new regression model were found (see Fig. 85). Is the outlier an influential point? Explain.

d. The new linear regression equation is $\hat{y} = -0.01175x + 33.30$. Use this model to predict what the 100-meter-run winning times would have been in 1940 and 1944. Round to the first decimal place.

e. Compare your results in parts (a) and (d). Are they fairly close? Does this support your response to part (c)?

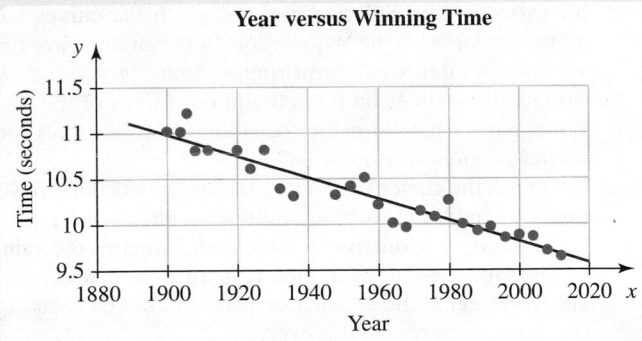

Year versus Winning Time

Figure 85 Exercise 47c

48. DATA The numbers of commercial airline boardings on domestic flights are shown in Table 53 for various years.

Table 53 Numbers of Commercial Airline Boardings on Domestic Flights

Year	Number of Boardings (millions)
1991	452
1993	487
1995	547
1997	599
1999	635
2001	622

Source: Bureau of Transportation Statistics

Let n be the number (in millions) of commercial airline boardings on domestic flights for the year that is t years since 1990.

a. Construct a scatterplot.

b. Identify the outlier. Explain how the terrorist attacks in 2001 are related to the outlier.

c. Determine whether the outlier is an influential point.

d. Find the linear regression equation after the data point for 2001 is removed. Use the equation to estimate the number of boardings in 2001. Estimate how much the number of boardings were affected due to the terrorist attacks.

e. By making the following assumptions, estimate the amount of money airlines lost in 2001.

- All trips were round trips.
- The average number of boardings for a round trip was four (two flights out, two back).
- The average round-trip fare was $340.

Large Data Sets

49. DATA Access the data about roller coasters, which are available at MyMathLab and at the Pearson Downloadable Student Resources for Math & Stats website.

a. Construct a pie chart for the countries and a pie chart for the states. Are all the roller coasters in the world included in the data set? If no, were the roller coasters most likely randomly selected? Explain.

b. Construct a scatterplot that describes the association between the heights and lengths of the wooden roller coasters. Do the same for the steel roller coasters. Show both scatterplots in the same coordinate system if you know how to use colors or shapes to identify which points are for which scatterplot.

c. Describe the four characteristics of the wooden-roller-coaster association. Compute and interpret r as part of your analysis. Do the same for the steel-roller-coaster association.

d. Construct residual plots for the wooden and the steel associations. Describe important features of each residual plot. How do your observations relate to the observations you made in part(c)?

e. For each of the two associations, find the coefficient of determination. What do your results mean in this situation?

f. On this basis of the results you found in parts (b)–(e), does it make sense to use a regression line to model the wooden association? If yes, find an equation of the line. Does it make sense to use a regression line to model the steel association? If yes, find an equation of the line.

g. For any regression lines you found in part (f), predict the length if the height is 30 feet.

50. DATA When a golfer hits a golf ball at the starting point, the shot is called a *drive* and the *driving distance* is the distance that the ball travels. The *fairway* is the closely mown area that runs from the starting point almost all the way to the hole. The *fairway accuracy* is the percentage of drives in which the ball comes to rest on the fairway (not on longer grass, sand traps, and ponds). Access the data about professional golfers, which are available at MyMathLab and at the Pearson Downloadable Student Resources for Math & Stats website.

a. Construct a scatterplot that describes the association between the driving distances and fairway accuracies for female professional golfers. Do the same for the male professional golfers. Show both scatterplots in the same coordinate system if you know how to use colors or shapes to identify which points are for which gender.

b. Which gender tends to have larger driving distances? Explain how you can tell this from the scatterplots.

c. Describe the four characteristics of the female association. Compute and interpret r as part of your analysis. Do the same for the male association.

d. For each of the two associations, find the coefficient of determination. What do your results mean in this situation?

e. Even though the female association and the male association are weak, find the linear regression equation for each association. Also find the linear regression equation for all golfers.

f. On the basis of the scatterplots, explain why it makes sense that the regression line for all golfers is less steep than the regression lines for female golfers and for male golfers.

g. From January 1 to September 14, 2014, Tim Clark had the second-best fairway accuracy in the PGA, with 775 out of 1047 drives coming to rest on fairways. Estimate his mean driving distance.

Concepts

51. Explain why a data point above the regression line has positive residual.

52. Explain why a data point below the regression line has negative residual.

53. A student shopping for a used Fender Jazz Bass finds a regression line to describe the association between the ages and asking prices of basses offered on eBay. The asking price of a 40-year-old bass is $4700 and the predicted price is $4750. The student calculates that the residual is $4750 − 4700 = 50$ dollars. Describe the error the student made and find the correct result. Also explain how the student's result and the correct result are related and why this makes sense.

54. Explain why we use the sum of squared residuals to measure how well a line fits some data. In particular, explain why the smaller the sum of squared residuals is, the better the line fits the data points.

55. What does it mean if some data can be modeled by a line in which the sum of squared residuals is 0? Explain why this makes sense.

56. Explain how to determine whether a data point is an influential point.

57. Is the slope of the regression line for some data sensitive to or resistant to an influential point? Explain.

58. If the regression line is used to model some data pairs, explain why it is impossible for all the dots in the residual plot to be above the zero residual line.

59. Describe three characteristics of residual plots that indicate that a regression line should not be used to make predictions.

60. Assume that for a regression line used to model some data, the coefficient of determination is 0.80. A student concludes that 80% of the data points lie near the line. What would you tell the student?

Hands-On Research

61. Go to the website Spurious Correlations by either searching for "Spurious Correlations" or entering the address http://www.tylervigen.com.

    a. The website displays several time-series plots. Each plot displays two curves that describe the values of two variables

for various years. Select a plot for which the curves look quite similar and the association between the non-time variables is interesting, surprising, or amusing.

    b. Inspect the scalings on the left and right sides of the time-series plot. What does the fact that the two curves look similar mean in this situation?

    c. If you put the cursor near a pair of data points that are vertically aligned, the website will display the coordinates of the data points. Construct a table that compares the values of the two *non-time* variables. It is up to you which variable you treat as the explanatory variable, but be consistent throughout the rest of this exercise.

    d. Construct a scatterplot of the two non-time variables.

    e. Describe the four characteristics of the association between the two non-time variables. Compute and interpret r as part of your analysis. Does the value of r you found equal the value of r provided at the website?

    f. If you treated the response variable as the explanatory variable (and the explanatory variable as the response variable), how would the value of r change?

    g. The website gives the value of r and also expresses it as a percentage. Explain why the percentage is *not* the coefficient of determination but why it could be mistaken for it.

    h. Compute the coefficient of determination. What does it mean in this situation?

    i. A student says a change in one of the non-time variables would *cause* a change in the other because the linear association is so strong. What would you tell the student?

Hands-On Projects

Climate Change Project (continued from Chapter 7)

Recall from the Climate Change Lab in Chapter 7 that many scientists believe that an increase in mean global temperature as small as 3.6°F could be a dangerous climate change (see Table 54).

Most scientists believe global warming is largely the result of carbon dioxide emissions from the burning of fossil fuels such as oil, coal, and natural gas. Carbon dioxide emissions in the United States and in the world have increased greatly since 1950 (see Table 55).

Table 54 Mean Surface Temperatures of Earth

Year	Mean Temperature (degrees Fahrenheit)	Year	Mean Temperature (degrees Fahrenheit)
1900	57.1	1960	57.2
1905	56.8	1965	57.0
1910	56.6	1970	57.3
1915	57.0	1975	57.1
1920	56.9	1980	57.6
1925	56.9	1985	57.3
1930	57.1	1990	57.8
1935	57.0	1995	57.9
1940	57.3	2000	57.8
1945	57.3	2005	58.3
1950	56.9	2010	58.3
1955	57.0	2014	58.4

Source: *NASA-GISS*

Table 55 Carbon Dioxide Emissions from Burning of Fossil Fuels

Year	Carbon Dioxide Emissions (billions of metric tons)	
	United States	World
1950	2.4	5.8
1955	2.7	7.2
1960	2.9	9.4
1965	3.5	11.2
1970	4.3	14.7
1975	4.4	16.5
1980	4.8	19.1
1985	4.6	19.4
1990	5.0	22.7
1995	5.3	23.6
2000	5.9	25.4
2005	5.9	29.4
2010	5.5	33.0
2013	5.3	35.3

Source: *U.S. Department of Energy*

In 1997, the United Nations negotiated a treaty called the Kyoto Protocol. The treaty's goal was to reduce annual greenhouse gas emissions to about 5% to 7% below 1990 levels by 2012. In November 2004, Russia cast the deciding vote to ratify the protocol, which took effect on February 16, 2005.

To create some flexibility in the treaty's requirements, a country that had exceeded its emissions limit could buy emissions credits from a country that was below its emissions limit. Also, a country could receive emissions credits by financing a project to help lower emissions in another country.

The treaty was legally binding for the 128 countries that ratified it. The United States declined to ratify the Kyoto Protocol, saying that reducing emissions to the point called for by the treaty would cripple the U.S. economy. If it had ratified the treaty, the United States would have had to reduce its 2012 greenhouse gas emissions to 7% below its emissions level in 1990.

The treaty's requirements were a crucial first goal, but the Intergovernmental Panel on Climate Change (IPCC) is calling for carbon dioxide emissions in 2050 to be 60% less than carbon dioxide emissions in 1990.

Many countries are condemning the United States because the Bush and Obama administrations refused to ratify the Kyoto Protocol. Although the U.S. population in 2013 was only 4% of the world population, 15% of world carbon dioxide emissions that year were produced by the United States (see Table 56).

Table 56 United States and World Populations

| Year | Population (billions) | |
	United States	World
1960	0.18	3.04
1970	0.21	3.71
1980	0.23	4.45
1990	0.25	5.29
2000	0.28	6.09
2010	0.31	6.85
2013	0.32	7.14

Source: *U.S. Census Bureau*

Critics of the Kyoto Protocol say a fairer pact would be for all countries to commit to the same level of carbon dioxide emissions *per person*. Using the IPCC's recommendation for 2050 carbon dioxide emissions, coupled with the United Nations' prediction of 9.3 billion people in 2050, carbon dioxide emissions should be about 0.9 metric ton per person that year.[*]

In 2010, annual carbon dioxide emissions were about 4.8 metric tons of carbon dioxide per person. Mean annual carbon dioxide emissions for developing countries was 2.7 metric tons per person, which was three times IPCC's recommendation. Even worse, mean annual carbon dioxide

[*] United Nations Population Division.

emissions for developed countries was 10.2 metric tons per person, more than ten times IPCC's recommendation (see Table 57).

Table 57 GDP Ranks, per-Person GDPs, and per-Person Carbon Dioxide Emissions

Country	2010 GDP Rank	2010 per-Person GDP (thousands of international dollars)	2010 per-Person Carbon Emissions (metric tons)
Sweden	22	39.0	5.3
Switzerland	19	46.4	5.4
France	5	34.1	5.5
Italy	8	32.0	6.7
United Kingdom	6	35.7	7.9
Austria	25	40.0	8.1
Denmark	31	40.2	8.4
Japan	3	33.7	8.9
Germany	4	37.4	9.3
Belgium	21	37.6	9.9
Norway	23	57.2	10.5
Australia	15	38.2	16.0
United States	1	47.2	18.1
Netherlands	16	42.2	31.9

Source: *World Bank; Carbon Dioxide Information Analysis Center*

GDP, the gross domestic product, is a measure of a country's economic strength.[†]

With 2013 annual carbon dioxide emissions of 16.6 metric tons per person, the United States would have to reduce emissions by 95% to meet the standard of 0.9 metric ton per person. This means Americans could emit only 5% of the carbon dioxide they currently emit. Imagine driving your car, heating and cooling your home, using your home appliances (including your refrigerator), using your computer, using your lights, and watching your television only 5% (one-twentieth) of the time you currently do.

Now that the Kyoto Protocol has expired, it is time for a new climate deal. The United Nations has agreed to the Durban Platform for Enhanced Action. The details of the plan are still being worked out, and the agreement won't come into effect until 2020. Unlike the Kyoto Protocol, the Durban Platform will be legally binding to all countries.[‡]

Progress on working out the details is slow, mostly because many countries fear that reducing greenhouse emissions will hurt their economies.

Some experts, however, such as the engineer Alan Pears, codirector of the environmental consultancy Sustainable Solutions, believe that it is possible for emissions to be significantly reduced without harming a country's economy. Norway, for instance, has a larger per-person GDP than the United States has, as well as significantly lower per-person carbon dioxide emissions. In fact, with the exception of the Netherlands and Australia, all of the countries listed in

[†] International Energy Agency, "CO_2 Emissions from Fuel Combustion Highlights," 2011.
[‡] United Nations, "Establishment of an Ad Hoc Working Group on the Durban Platform for Enhanced Action," December, 11, 2011.

Table 57 have strong economies and significantly lower per-person carbon dioxide emissions than the United States has. A scatterplot of the data would show that countries with higher per-person GDP do not necessarily have higher carbon dioxide emissions.

Many states have taken the matter into their own hands by adopting policies to reduce carbon dioxide emissions. And by using alternative sources of energy, many countries have slowed or reversed the growth of carbon dioxide emissions.[§] In fact, U.S. carbon emissions in 2013 were 10% less than in 2005, although some or all of the decline may have been due to the struggling economy in 2013.

In addition to national, state, and even corporate actions, individuals can help lower carbon dioxide emissions by purchasing electric and hybrid automobiles, major appliances with the Energy Star logo, and solar thermal systems to help provide hot water. Individuals can also car pool or use public transportation.

Analyzing the Situation

1. **a.** Let F be the mean global temperature (in degrees Fahrenheit) at t years since 1900. Construct a scatterplot of the data in Table 54 and then find an equation of a model *for the years 1965 to 2013*. Then verify that your model fits the data well for those years.
   b. In Problem 2a of the Climate Change Project in Chapter 7, you found the mean global temperature from 1900 to 1965. If you didn't do this, do so now. Use your model to predict when the planet's mean temperature will have increased by 3.6°F—a potentially dangerous climate change.

2. Use Tables 55 and 56 to verify the claims that although the U.S. population in 2013 was only 4% of world population, 15% of annual world carbon dioxide emissions were produced by the United States in that year.

3. Use the United Nations' prediction that the world population will be 9.3 billion in 2050 to verify the claim that per-person carbon dioxide emissions that year should be about 0.9 metric ton per person for the IPCC recommendation of a 60% reduction by then.

4. Let P be the U.S. population (in billions) at t years since 1950. Construct a scatterplot of the data. Describe the four characteristics of the association. Compute and interpret r as part of your analysis. Find an equation of a model. Verify that your model comes close to the data points.

5. Let C be U.S. annual carbon dioxide emissions (in billions of metric tons) at t years since 1950. Construct a scatterplot of the data. Describe the four characteristics of the association. Compute and interpret the coefficient of determination as part of your analysis. Construct a residual plot. Find an equation of a model. Verify that your model comes close to the data points.

6. **a.** Use your population model of Problem 4 to estimate U.S. population in 2011.
   b. Use your emission model of Problem 5 to estimate U.S. carbon dioxide emissions in 2011.
   c. Use your results from parts (a) and (b) to estimate U.S. *per-person* carbon dioxide emissions in 2011.
   d. The actual U.S. per-person carbon dioxide emissions in 2011 were 18.0 metric tons. Is the result you found in part (c) an underestimate or an overestimate? Explain why this can be explained at least in part by the poor economy during the period 2008–2011.

7. Let G be the per-person GDP (in international dollars) of a country with per-person carbon emissions c (in metric tons). Construct a scatterplot of the data. Do countries with higher per-person GDPs always have higher per-person carbon emissions? Explain.

DATA. Golf Ball Project

In this lab, you will explore the association between the height of a golf ball before dropping it and its height after one bounce.*

Materials

You will need at least three people and the following items:

1. a tape measure
2. a golf ball

Recording the Data

The same person should drop the golf ball each time. A second person should measure the height of the golf ball (from the bottom of the ball) before the first person drops it. The ball should be dropped from an initial height of 12 inches. A spotter should estimate the bounce height of the golf ball. Repeat this process three times. Then do the same for initial heights of 24 inches, 36 inches, 48 inches, 60 inches, and 72 inches. If your instructor prefers, use the data listed in Table 58.

Table 58 Drop Heights versus Bounces Heights of a Golf Ball

Drop Height (inches)	Bounce Height (inches)	Drop Height (inches)	Bounce Height (inches)
12	11.3	48	44.5
12	9.2	48	43.4
12	10.0	48	44.5
24	20.3	60	50.8
24	21.7	60	52.0
24	19.0	60	53.0
36	32.5	72	64.0
36	31.0	72	64.7
36	29.7	72	63.4

Source: *J. Lehmann*

[§] Pamela Person, "Reducing Greenhouse Gas Emissions," Maine Center for Economic Policy, *Choices*, VII(9), Oct. 11, 2001.

---

* Golf Ball Project from Lab written by Jim Ryan. Copyright © by James Ryan.

Analyzing the Data

1. Display your golf ball data in a table.

2. Let y be the bounce height (in inches) after the ball was dropped from an initial height of x inches. Construct a scatterplot.

3. Describe the four characteristics of the association. Compute and interpret r as part of your analysis.

4. Compute the coefficient of determination. What does it mean in this situation?

5. Construct a residual plot. Does it support any of the conclusions you made in Problem 3? If so, which one(s)?

6. Find an equation of a linear model to describe the situation.

7. Find the y-intercept of your model. What does it mean in this situation? If you can find a linear model with a better y-intercept, do so.

8. Graph your model on the scatterplot. Does the model come close to the data points?

9. Use your model to estimate the bounce height for a drop height of 54 inches.

10. On a golf course, a golf ball is hit to a maximum height of 50 feet. What does your model estimate the bounce height to be after one bounce? Do you think this estimate is accurate? If not, will it be an underestimate or an overestimate? Explain.

11. Find the slope of your model. What does the slope mean in this situation? Explain.

12. Estimate the bounce height after three bounces for a drop height of 66 inches.

Taking It One Step Further

13. Redo the experiment with a rubber ball and then with a tennis ball. Then repeat Parts 1–8. Finally, compare the slopes of your three linear models and explain why the comparison makes sense.

DATA Rope Project

In this lab, you will explore the relationship between the number of knots tied in a rope and the rope's length.

Check with your instructor whether you should collect your own data or use the data listed in Table 59.

Table 59 Lengths of a Rope with Diameter about 7 Millimeters

Number of Knots	Length of Rope (centimeters)
0	60.0
1	53.2
2	45.8
3	38.3
4	30.6

Source: J. Lehmann

Materials

1. A 60-centimeter-long piece of rope with diameter about 7 millimeters

2. A meterstick or other measuring device with units of millimeters

Recording of Data

Pull the rope taut and measure its length (in centimeters). Then tie a knot close to one end of the rope and measure the length of the rope again. Next, tie another knot next to the first one and measure the length of the rope. Continue tying knots, working your way along the rope and measuring the rope's length after you have tied each knot. Tie a total of four knots.

Analyzing the Situation

1. Display your data in a table or use the data in Table 59.

2. Let y be the length (in centimeters) of the rope with x knots. Construct a scatterplot.

3. Is the association positive, negative, or neither? Why does that make sense?

4. In Problem 3, you determined the direction of the association. Now describe the three other characteristics of the association. Compute and interpret r as part of your analysis.

5. Compute the coefficient of determination. What does it mean in this situation?

6. Construct a residual plot. Does it support any of the conclusions you made in Problems 3 and 4? If so, which one(s)?

7. Find an equation of a model.

8. Graph the model on the scatterplot. Does the model come close to the data points?

9. Find the y-intercept of your model. What does it mean in this situation?

10. Find the slope of the model. What does it mean in this situation?

11. Use the model to estimate the length of the rope with five knots. Did you perform interpolation or extrapolation? Do you have much faith in your estimate?

12. Check whether the result you found in Problem 11 is an under-estimate or an overestimate by tying a fifth knot in the rope and then measuring the rope's length.

13. Continue tying knots in the rope. When does model breakdown first occur? Explain.

DATA Shadow Project

In this lab, you will compare the relationship between an object's height and the length of its shadow.

Check with your instructor whether you should collect your own data or use the data listed in Table 60.

Table 60 Heights of Objects and the Lengths of Their Shadows

Object	Height (inches)	Length of Shadow (inches)
Nothing	0	0
Wine bottle	15.5	10.3
Toy putter	20.8	13.0
Box	36.5	23.6
Mop	47.8	31.0
Person's shoulder	55.0	36.5
Person	63.0	41.5

Source: *J. Lehmann*

Materials

1. Six objects of various heights up to 7 feet
2. A building, pole, tree, or other tall object with height greater than 15 feet
3. A tape measure or other measuring device

Recording of Data

Run the experiment when the objects (including the tall object) have noticeable and measurable shadows. For each object, measure its height and the length of its shadow. Record the beginning and ending time of the experiment. Also, record the length of the shadow of the tall object. It is important that you record all the data quickly.

Analyzing the Situation

1. Display your data in a table similar to Table 60. Those data were collected from 2:10 P.M. to 2:20 P.M., and the tall object is a tree whose shadow has a length of 49.5 feet.
2. Let x be the height (in inches) of an object and y be the length (in inches) of the object's shadow. Construct a scatterplot.
3. Is the association positive, negative, or neither? Why does that make sense?
4. In Problem 3, you determined the direction of the association. Now describe the three other characteristics of the association. Compute and interpret r as part of your analysis.
5. Compute the coefficient of determination. What does it mean in this situation?
6. Construct a residual plot. Does it support any of the conclusions you made in Problems 3 and 4? If so, which one(s)?
7. Find an equation of a model.
8. Graph the model on the scatterplot. Does the model come close to the data points?
9. Find the y-intercept of your model. What does it mean in this situation?
10. Find the slope of the model. What does it mean in this situation?

11. Use the length of the shadow of the tall object to predict the object's height. Do you have much faith in your prediction? Explain.
12. Explain why it was important that you ran the experiment quickly.
13. Suppose you had run the experiment half an hour later. How would that have affected the slope of your model? Explain. (If the Sun would have set by then, describe the impact on the slope if the experiment had been performed half an hour earlier.) Would this change in time have resulted in a different estimate of the height of the tall object? Explain.

Linear Project: Topic of Your Choice

Your objective in this lab is to use a linear model to describe some authentic situation. Find some data on two quantities that describe a situation that has not been discussed in this text. Blogs, newspapers, magazines, and scientific journals are good resources. Or you can conduct a survey or physical experiment. Choose something that interests you!

Analyzing the Situation

1. What two quantities did you explore? Define variables for the quantities. Include units in your definitions.
2. Which variable is the explanatory variable? Which variable is the response variable? Explain.
3. Describe how you found your data. If you performed a survey or a physical experiment, provide a careful description with specific details of how you conducted it. If you didn't conduct a survey or a physical experiment, state the source of your data.
4. Include a table of your data.
5. Construct a scatterplot of your data. (If your data are not linearly associated, find some data that are.)
6. In Problem 5, you determined that the association is linear. Now describe the three other characteristics of the association. Compute and interpret r as part of your analysis.
7. Compute the coefficient of determination. What does it mean in this situation?
8. Construct a residual plot. Does it support any of the conclusions you made in Problems 5 and 6? If so, which one(s)?
9. Find an equation of a model.
10. Graph the model on the scatterplot. Does the model come close to the data points?
11. What is the slope of your linear model? What does it mean in this situation?
12. Choose a value for your explanatory variable. On the basis of that chosen value, use your model to find a value for your response variable. Describe what your result means in the situation you are modeling.
13. Choose a value for your response variable. On the basis of that chosen value, use your model to find a value for

your explanatory variable. Describe what your result means in the situation you are modeling.

14. Find the vertical intercept of your linear model. What does it mean in the situation you are modeling? Has model breakdown occurred?

15. Comment on your project experience.

    a. For example, you might address whether the project was enjoyable, insightful, and so on.

b. Were you surprised by any of your findings? If so, which ones?

c. How would you improve your process for this project if you were to do it again?

d. How would you improve your process if you had more time and money?

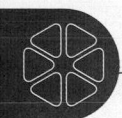

Chapter Summary

Key Points of Chapter 9

Section 9.1   Using Two Points to Find an Equation of a Line

Finding an equation of a line by using the slope, a point, and the slope-intercept form	To find an equation of a line by using the slope and a point, 1. Substitute the given value of the slope m into the equation $y = mx + b$. 2. Substitute the coordinates of the given point into the equation you found in step 1 and solve for b. 3. Substitute the value of b you found in step 2 into the equation you found in step 1. 4. Check that the graph of your equation contains the given point.
Finding an equation of a line by using two points and the slope-intercept form	To find an equation of the line that passes through two given points whose x-coordinates are different, 1. Use the formula $m = \dfrac{y_2 - y_1}{x_2 - x_1}$ to find the slope of the line containing the two points. 2. Substitute the m value you found in step 1 into the equation $y = mx + b$. 3. Substitute the coordinates of one of the given points into the equation you found in step 2 and solve for b. 4. Substitute the b value you found in step 3 into the equation you found in step 2. 5. Check that the graph of your equation contains the two given points.
Point-slope form	If a nonvertical line has slope m and contains the point (x_1, y_1), then an equation of the line is $y - y_1 = m(x - x_1)$.

Section 9.2   Using Two Points to Find an Equation of a Linear Model

Finding an equation of a linear model	To find an equation of a linear model, given some data, 1. Construct a scatterplot of the data. 2. Determine whether there is a line that comes close to the data points. If so, choose two points (not necessarily data points) that you can use to find an equation of a linear model. 3. Find an equation of the line. 4. Use technology to verify that the graph of your equation contains the two chosen points and comes close to all of the data points of the scatterplot.

Section 9.3   Linear Regression Model

Observed and predicted values	For a data point (x, y), the **observed value of y** is y and the **predicted value of y** (written $\hat{y}$) is the value obtained by using a model to predict y.
Residual	For a given data point (x, y), the **residual** is the difference of the observed value of y and the predicted value of y: $$\text{Residual} = \text{Observed value of } y - \text{Predicted value of } y = y - \hat{y}$$
Residuals for data points above, below, or on a line	Suppose some data points are modeled by a line. • A data point on the line has residual equal to 0. • A data point above the line has positive residual. • A data point below the line has negative residual.
Sum of squared residuals	We measure how well a line fits some data points by calculating the sum of the squared residuals: $$\sum (y_i - \hat{y}_i)^2$$ The smaller the sum of squared residuals, the better the line will fit the data. If the sum of squared residuals is 0, then there is an exact linear association.
Linear regression function, line, equation, and model	For a group of points, the **linear regression function** is the linear function with the least sum of squared residuals. Its graph is called the **regression line** and its equation is called the **linear regression equation**, written $$\hat{y} = b_1 x + b_0$$ where b_1 is the slope and $(0, b_0)$ is the y-intercept. The **linear regression model** is the linear regression function for a group of *data* points.
Residual plot	A **residual plot** is a graph that compares data values of the explanatory variable with the data points' residuals.
Using a residual plot to help determine whether a regression line is an appropriate model	The following statements apply to a residual plot for a regression line. • If the residual plot has a pattern where the dots do not lie close to the zero residual line, then there is either a nonlinear association between the explanatory and response variables or there is no association. • If a dot lies much farther away from the zero residual line than most or all of the other dots, then the dot corresponds to an outlier. If the outlier is neither adjusted nor removed, the regression line may *not* be an appropriate model. • The *vertical* spread of the residual plot should be about the same for each value of the explanatory variable.
Influential point	If the slope of a regression line is greatly affected by the removal of a data point, we say the data point is an **influential point**.
Identifying influential points	Outliers tend to be influential points when they are horizontally far from the other data points.
Coefficient of determination	The coefficient of determination, r^2, is the proportion of the variation in the response variable that is explained by the regression line.

Chapter 9 Review Exercises

Find an equation of the line that has the given slope and contains the given point. If possible, write your equation in slope–intercept form.

1. $m = -4, (2, -1)$

2. $m = -\dfrac{2}{3}, (-6, -4)$

3. m is undefined, $(2, 5)$

4. $m = 0, (-1, -4)$

Find an approximate equation of the line that has the given slope and contains the given point. Write your equation in slope–intercept form. Round the constant term to the second decimal place.

5. $m = -5.29, (-4.93, 8.82)$

6. $m = 1.45, (-2.79, -7.13)$

Find an equation of the line that passes through the two given points. If possible, write your equation in slope–intercept form.

7. $(-2, -7)$ and $(1, 2)$

8. $(2, -5)$ and $(4, 5)$

9. $(-3, 9)$ and $(6, -6)$

10. $(-4, -10)$ and $(-2, -7)$

11. $(5, -3)$ and $(5, 2)$

12. $(-4, -3)$ and $(-1, -3)$

For Exercises 13 and 14, find an approximate equation of the line that passes through the two given points. Write your equation in slope–intercept form. Round the slope and the constant term to two decimal places.

13. $(3.5, 9.2)$ and $(8.7, 4.8)$

14. $(-5.22, 2.49)$ and $(1.83, -3.99)$

15. Find an equation of the line sketched in Fig. 86.

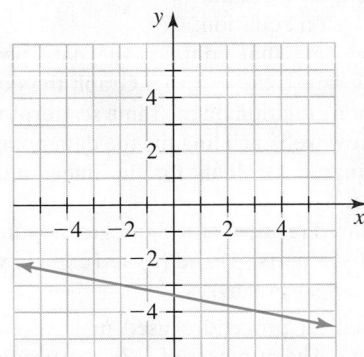

Figure 86 Exercise 15

16. **DATA** Use two points to find an equation of a line that comes close to the points listed in Table 61. Then use technology to check that your line comes close to those points.

Table 61 Find an Equation

x	y
1	28
4	23
6	16
9	13
10	8

17. **DATA** The ratios of the median wealth of upper-income households to the median wealth of middle-income households are shown in Table 62 for various years.

Table 62 Ratios of the Median Wealth of Upper-Income Households to the Median Wealth of Middle-Income Households

Year	Ratio
1995	3.6
1998	4.1
2001	4.4
2004	5.0
2007	4.5
2010	6.2
2013	6.6

Source: *Pew Research Center*

Let y be the ratio of the median wealth of upper-income households to the median wealth of middle-income households at x years since 1990.

a. Construct a scatterplot.

b. Describe the four characteristics of the association. Compute and interpret r as part of your analysis.

c. Use two points to find an equation of a model.

d. Use the model to estimate when the median wealth of upper-income households was 6 times the median wealth of middle-income households.

e. In 2012, the median wealth of middle-income households was $96,500. Use the model to help you estimate the median wealth of upper-income households in 2012.

18. **DATA** The percentages of American adults in favor of banning smoking in public places are shown in Table 63 for various years.

Table 63 Percentages of American Adults in Favor of Banning Smoking in Public Places

Year	Percent
2003	31
2005	39
2007	40
2011	59
2013	55
2014	56

Source: *The Gallup Organization*

Let y be the percentage of American adults who favor banning smoking in public places at x years since 2000.

a. Construct a scatterplot.

b. Use two points to find an equation of a model.

c. What is the slope? What does it mean in this situation?

d. What is the y-intercept. What does it mean in this situation. Do you have much faith in your result?

e. Estimate the percentage of American adults who favored banning smoking in public places in 2009.

19. Refer to the residual plot shown in Fig. 87 and identify which conditions, if any, for a regression line are not met.

Residual Plot

Figure 87 Exercise 19

20. Figure 88 shows a regression line and a scatterplot, including an outlier shown in red. Figure 89 shows a regression line and scatterplot after the red dot has been removed. Determine whether the red dot is an influential point.

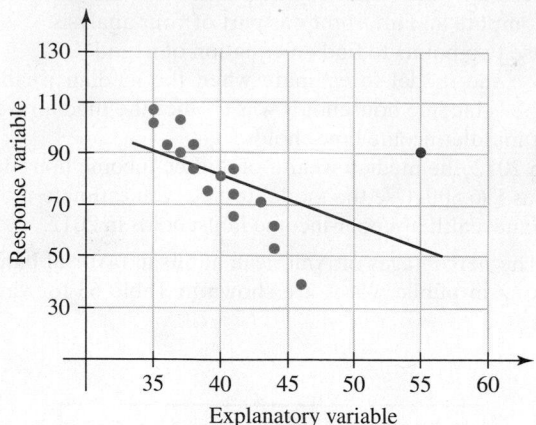

Figure 88 Model and scatterplot with outlier (Exercise 20)

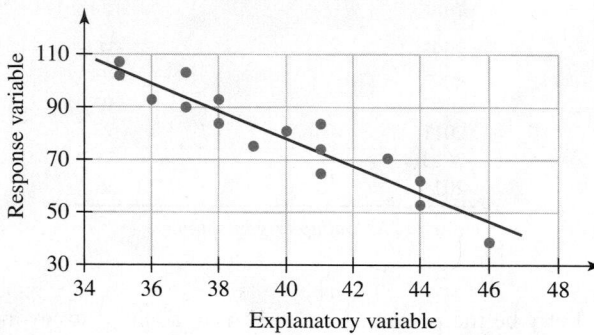

Figure 89 Model and scatterplot without outlier (Exercise 20)

21. Which of the following values is a reasonable estimate of the coefficient of determination for the scatterplot shown in Fig. 90: $-1, -0.9, -0.3, 0, 0.3, 0.9, 1$.

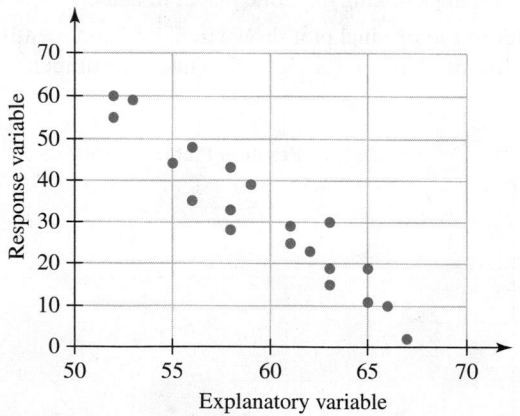

Figure 90 Exercise 21

22. **DATA** The ages and prices of 9 Chevrolet Impala sedans at dealerships in Fort Myers, Florida, are shown in Table 64.

Table 64 Ages and Prices of Some Chevrolet Impala Sedans

Age (in years)	Price (thousands of dollars)
7	13
2	15
5	11
11	6
9	9
3	14
7	8
11	7
3	17

Source: *Edmunds.com*

Let x be the age (in years) and let y be the price (in thousands of dollars), both for a Chevrolet Impala sedan.

a. Construct a scatterplot of the data.
b. Find the linear regression equation.
c. The equation of the line that contains the data points $(3, 17)$ and $(7, 8)$ is $\hat{y} = -2.25x + 23.75$. Graph the equation and the regression equation on the same scatterplot of the data. Describe how well each line fits the data points.
d. Find the sum of squared residuals for the linear model $\hat{y} = -2.25x + 23.75$.
e. Find the sum of squared residuals for the regression line.
f. Compare your results in parts (d) and (e) and explain why your comparison is not surprising.

23. **DATA** An ERCP is a medical procedure used most often to treat tumors or gallstones. The numbers of ERCP procedures in the United States are shown in Table 65 for various years.

Table 65 Numbers of ERCP Procedures in the United States

Year	Number of ERCP Procedures (thousands)
2010	584
2011	600
2012	622
2013	642
2014	669

Source: *U.S. Food and Drug Administration*

Let y be the number (in thousands) of ERCP procedures in the year that is x years since 2010.

a. Construct a scatterplot of the data.
b. Find the linear regression equation.
c. Construct a residual plot.
d. Find the year and the residual represented by the highest point in the residual plot. What does this tell you about the corresponding data point in the scatterplot and the regression line?
e. An ERCP involves a duodenoscope, which is a medical scope that is difficult to clean. In fact, the scope has been tied to deadly superbug outbreaks in hospitals in more than five cities. Research shows that 10% to 30% of duodenoscopes remain contaminated after attempted cleaning (Source: *USA Today*). Assuming the exact percentage

of contaminated duodenoscopes is 20%, predict the *number* of ERCP procedures that will involve contaminated duodenoscopes in 2015. Do you have much faith in your prediction? Explain.

24. **DATA** Researchers studied the impact on children's future incomes due to moving to wealthier neighborhoods. For each child in the study, researchers computed the percentile of the child's future income (at age 24 years) for incomes between the poorer neighborhood and the wealthier neighborhood. So, an income at the 0th percentile is equal to the poorer neighborhood's median income and an income at the 100th percentile is equal to the wealthier neighborhood's median income. The future-income percentiles are shown in Table 66 for various ages when children moved. Let y the future-income percentile for a child who at age x years moved to a wealthier neighborhood.

a. Construct a scatterplot of the data.

b. Is the association positive, negative, or neither? Why does that make sense in this situation?

c. Find the linear regression equation.

d. Find the coefficient of determination. What does it mean in this situation?

e. Use the model to predict the future-income percentile for a child who at age 15 years moved to a wealthier neighborhood. Compute the residual. What does it mean in this situation?

Table 66 Moving Ages versus Future-Income Percentiles

Age when Moved (years)	Future-Income Percentile
9	54
10	50
11	55
12	51
13	43
14	38
15	42
16	29
17	29
18	25
19	19
20	14
21	13
22	8
23	3

Source: The Impacts of Neighborhoods on Intergenerational Mobility: Childhood Exposure Effects and County-Level Estimates, *Raj Chetty and Nathaniel Hendren*

Chapter 9 Test

For Exercises 1 and 2, find an equation of the line that has the given slope and contains the given point. If possible, write your equation in slope-intercept form.

1. $m = 7$, $(-2, -4)$

2. $m = -\dfrac{2}{3}$, $(6, -1)$

3. Find an equation of the line that passes through the points $(-4, 6)$ and $(2, 3)$.

4. Find an approximate equation of the line that passes through the points $(-3.4, 2.9)$ and $(1.8, -7.1)$. Write your equation in slope-intercept form. Round the slope and the constant term to two decimal places.

5. Find an equation of the line sketched in Fig. 91.

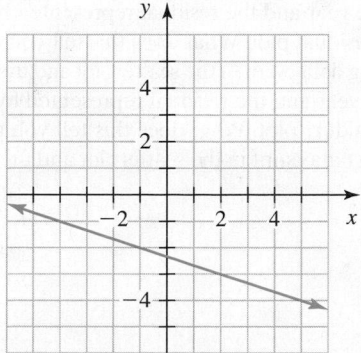

Figure 91 Exercise 5

6. Consider the scatterplot of data and the graph of the model $\hat{y} = mx + b$ in Fig. 92. Sketch a linear model that describes

the data better, and then explain how you would adjust the values of m and b of the original model so that it would describe the data better.

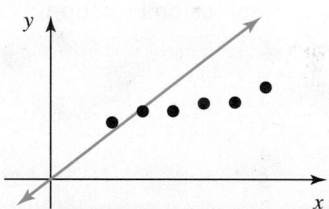

Figure 92 Exercise 6

7. **DATA** The percentages of Fortune 500 companies that still offer traditional pensions to new hires are shown in Table 67 for various years.

Table 67 Percentages of Companies That Offer Traditional Pensions

Year	Percent
1998	50
2001	41
2004	31
2007	21
2010	11
2013	7

Source: *Towers Watson*

Let $f(x)$ be the percentage of Fortune 500 companies that offer traditional pensions to new hires at x years since 1990.

a. Construct a scatterplot.
b. Use two points to find an equation of a model.
c. What is the slope? What does it mean in this situation?
d. Find $f(22)$. What does it mean in this situation?
e. Find x when $f(x) = 22$. What does it mean in this situation?

8. Refer to the residual plot shown in Fig. 93 and identify which conditions, if any, for a regression line are not met.

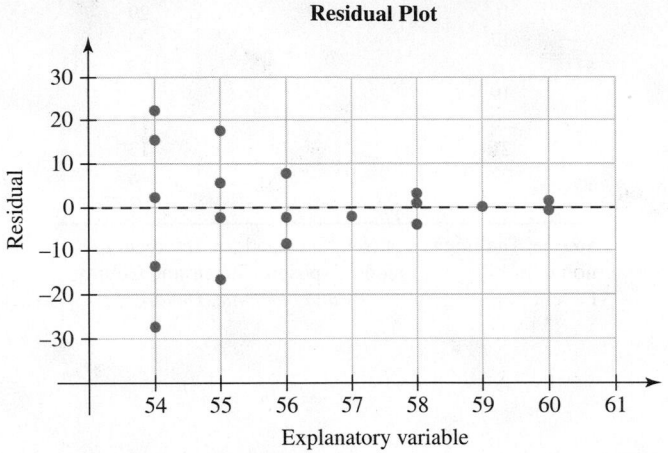

Residual Plot

Figure 93 Exercise 8

9. **DATA** Find the linear regression equation for the points listed in Table 68. Does the graph of the equation come close to the points in the scatterplot?

Table 68 Find the Linear Regression Equation

x	y
3	39
4	28
4	35
5	25
7	27
7	15
7	19
9	14
10	11
10	16

10. **DATA** The mean prices of high school lunches are shown in Table 69 for various years.

Table 69 Mean Prices of High School Lunches

Year	Mean Price (dollars)
2003	1.66
2005	1.77
2007	1.90
2009	2.13
2011	2.20
2013	2.42

Source: *School Nutrition Association*

Let y be the mean price (in dollars) of high school lunches at x years since 2000.

a. Construct a scatterplot.
b. Describe the four characteristics of the association. Compute and interpret r as part of your analysis.
c. Find the linear regression equation. Graph it on the scatter plot.
d. Construct a residual plot. Does it support any of the conclusions you made in part (b)? If so, which one(s)?
e. Find the coefficient of determination. What does it mean in this situation?

11. **DATA** The mean numbers of viewers of *American Idol* are shown in Table 70 for various years.

Table 70 Mean Numbers of Viewers of *American Idol*

Year	Mean Number of Viewers (million)
2006	30.7
2008	28.3
2010	22.5
2012	19.1
2014	11.7
2015	10.9

Source: *Nielsen*

Let y be the mean number (in millions) of viewers in the year that is x years since 2000.

a. Construct a scatterplot.
b. Find the linear regression equation. Graph it on the scatterplot.
c. Construct a residual plot.
d. Find the year and the residual represented by the lowest dot in the residual plot. What does this tell you about the corresponding data point in the scatterplot and the regression line?
e. Find the year and the residual represented by the highest dot in the residual plot. What does this tell you about the corresponding data point in the scatterplot and the regression line?

Using Exponential Models to Make Predictions

Pointing a laser at aircraft, which can temporarily blind pilots, is a serious offense with a maximum punishment of 20 years in prison and a $250,000 fine. The numbers of laser incidents involving aircraft are shown in Table 1 for various years. In Example 4 in Section 10.5, we will estimate the percentage increase of laser incidents per year.

In Section 1.7, we described large and small numbers with scientific notation, which involves integer exponents. In this chapter, we will simplify expressions first involving integer exponents and then involving *rational exponents*. In Chapters 6–9, we worked with linear functions. We will now work with *exponential functions*, including graphing them and finding equations of them. Finally, we will use *exponential models* to describe authentic situations such as the mean tuition at four-year colleges (see Exercise 50 of Homework 10.5).

Table 1 Numbers of Laser Incidents Involving Aircraft

Year	Number
2005	283
2006	446
2007	675
2008	988
2009	1527
2010	2836

Source: *Federal Aviation Administration*

▼ 10.1 Integer Exponents

Objectives

» Describe the *product property* for exponents.

» Simplify expressions involving nonnegative-integer exponents.

» Describe the following properties for exponents: quotient, raising a product to a power, raising a quotient to a power, and raising a power to a power.

» Use combinations of properties of exponents to simplify expressions involving nonnegative-integer exponents.

» Simplify expressions involving negative-integer exponents.

In Section 1.7, we performed exponentiation, such as $2^3 = 2 \cdot 2 \cdot 2 = 8$. In this section, we will determine several properties of exponents and use them to simplify expressions involving exponents.

Product Property for Exponents

Consider the product $b^2 \cdot b^3$. We can write this product as a single power:

$$b^2 \cdot b^3 = (b \cdot b)(b \cdot b \cdot b) \quad \text{Write factors without exponents.}$$
$$= b \cdot b \cdot b \cdot b \cdot b \quad \text{Remove parentheses.}$$
$$= b^5 \quad \text{Simplify.}$$

Note that we can find the same product by adding the exponents:

$$b^2 \cdot b^3 = b^{2+3} = b^5$$

> ### Product Property for Exponents
>
> If n and m are counting numbers, then
> $$b^m b^n = b^{m+n}$$
> In words: To multiply two powers of b, keep the base and add the exponents.

637

» Describe the meaning of *exponential function.*

» Work with models whose equations involve integer exponents.

For example, $b^4 b^8 = b^{4+8} = b^{12}$. Also, $b^5 b = b^5 b^1 = b^{5+1} = b^6$.

▶ **Example 1**   Product Property

Find the product.

1. $3b^4(5b^2)$ **2.** $3b^2 c(2b^3 c^2)$.

Solution

1. We rearrange factors so that the coefficients are adjacent and the powers of b are adjacent:

$$3b^4(5b^2) = (3 \cdot 5)(b^4 b^2) \quad \text{Rearrange factors.}$$
$$= 15b^6 \quad \text{Add exponents: } b^m b^n = b^{m+n}$$

2. We rearrange factors so the coefficients are adjacent, the powers of b are adjacent, and the powers of c are adjacent:

$$3b^2 c(2b^3 c^2) = (3 \cdot 2)(b^2 b^3)(c^1 c^2) \quad \text{Rearrange factors; } b = b^1.$$
$$= 6b^5 c^3 \quad \text{Add exponents: } b^m b^n = b^{m+n}$$

Simplifying Expressions Involving Nonnegative-Integer Exponents

We can use the product property as well as other properties of exponents that we will discuss in this section, to *simplify expressions involving exponents.*

> ▶ **Simplifying an Expression Involving Nonnegative-Integer Exponents**
>
> An expression involving nonnegative-integer exponents is simplified if
> **1.** It includes no parentheses.
> **2.** Each variable or constant appears as a base as few times as possible. For example, for nonzero b, we write $b^3 \, b^5$ as b^8.
> **3.** Each numerical expression (such as 5^2) has been calculated, and each numerical fraction has been simplified.

Quotient Property for Exponents

Consider the quotient $\dfrac{b^5}{b^2}$. We can simplify this quotient by first writing the expression without exponents:

$$\frac{b^5}{b^2} = \frac{b \cdot b \cdot b \cdot b \cdot b}{b \cdot b} \quad \text{Write quotient without exponents.}$$
$$= \frac{b \cdot b}{b \cdot b} \cdot \frac{b \cdot b \cdot b}{1} \quad \frac{AC}{BD} = \frac{A}{B} \cdot \frac{C}{D}$$
$$= 1 \cdot b \cdot b \cdot b \quad \text{Simplify: } \frac{b \cdot b}{b \cdot b} = 1$$
$$= b^3 \quad \text{Simplify.}$$

Note that we can get the same result by subtracting the exponents 5 and 2:

$$\frac{b^5}{b^2} = b^{5-2} = b^3$$

> ▶ **Quotient Property for Exponents**
>
> If m and n are counting numbers and b is nonzero, then
> $$\frac{b^m}{b^n} = b^{m-n}$$
> In words: To divide two powers of b, keep the base and subtract the exponents.

For example, $\dfrac{b^9}{b^4} = b^{9-4} = b^5$.

▶ **Example 2**   Quotient Property

Simplify.

1. $\dfrac{6b^9}{4b^5}$ **2.** $\dfrac{12b^6c^8}{4bc^7}$

Solution

1.
$$\dfrac{6b^9}{4b^5} = \dfrac{6}{4} \cdot \dfrac{b^9}{b^5} \qquad \dfrac{AC}{BD} = \dfrac{A}{B} \cdot \dfrac{C}{D}$$

$$= \dfrac{3}{2} \cdot b^{9-5} \qquad \text{Simplify; subtract exponents: } \dfrac{b^m}{b^n} = b^{m-n}$$

$$= \dfrac{3}{2} \cdot \dfrac{b^4}{1} \qquad \text{Simplify.}$$

$$= \dfrac{3b^4}{2} \qquad \text{Multiply numerators; multiply denominators: } \dfrac{A}{B} \cdot \dfrac{C}{D} = \dfrac{AC}{BD}$$

2.
$$\dfrac{12b^6c^8}{4bc^7} = \dfrac{12}{4} \cdot \dfrac{b^6}{b^1} \cdot \dfrac{c^8}{c^7} \qquad \dfrac{AC}{BD} = \dfrac{A}{B} \cdot \dfrac{C}{D}$$

$$= 3 \cdot b^{6-1} \cdot c^{8-7} \qquad \text{Simplify; subtract exponents: } \dfrac{b^m}{b^n} = b^{m-n}$$

$$= 3b^5c^1 \qquad \text{Simplify.}$$

$$= 3b^5c \qquad c^1 = c$$

Raising a Product to a Power

Consider the power $(bc)^3$. We can write this power without exponents:

$$(bc)^3 = (bc)(bc)(bc) \qquad \text{Write power without exponents.}$$

$$= (b \cdot b \cdot b)(c \cdot c \cdot c) \qquad \text{Rearrange factors.}$$

$$= b^3c^3 \qquad \text{Simplify.}$$

Note that we can get this same result by raising each factor of the base bc to the 3rd power:

$$(bc)^3 = b^3c^3$$

▶ **Raising a Product to a Power**

If n is a counting number, then

$$(bc)^n = b^nc^n$$

In words: To raise a product to a power, raise each factor to the power.

For example, $(bc)^6 = b^6c^6$.

▶ **Example 3**   Raising a Product to a Power

Perform the indicated operation.

1. $(bc)^7$ **2.** $(7b)^2$ **3.** $(-2bc)^4$

Solution

1. $(bc)^7 = b^7c^7$

2. $(7b)^2 = 7^2b^2 = 49b^2$

3. $(-2bc)^4 = (-2)^4b^4c^4 = 16b^4c^4$

WARNING   The expressions $5b^2$ and $(5b)^2$ are *not* equivalent expressions:

$$5b^2 = 5b \cdot b$$
$$(5b)^2 = (5b)(5b) = 25b \cdot b$$

For $5b^2$, the base is the variable b. For $(5b)^2$, the base is the product $5b$.

Here we show a typical error and the correct work performed to find the power $(5b)^2$:

$$(5b)^2 = 5b^2 \qquad \text{\textit{Incorrect}}$$
$$(5b)^2 = 5^2b^2 = 25b^2 \quad \text{\textit{Correct}}$$

Since the base $5b$ is a product, we need to raise *both* factors 5 and b to the 2nd power.

In general, when finding a power of the form $(bc)^n$, don't forget to raise *both* factors b and c to the nth power.

Raising a Quotient to a Power

Consider the expression $\left(\dfrac{b}{c}\right)^4$. We begin to write this expression in another form by writing it without exponents:

$$\left(\frac{b}{c}\right)^4 = \frac{b}{c} \cdot \frac{b}{c} \cdot \frac{b}{c} \cdot \frac{b}{c} \qquad \text{\textit{Write expression without exponents.}}$$

$$= \frac{b \cdot b \cdot b \cdot b}{c \cdot c \cdot c \cdot c} \qquad \text{\textit{Multiply numerators; multiply denominators:}} \ \frac{A}{B} \cdot \frac{C}{D} = \frac{AC}{BD}$$

$$= \frac{b^4}{c^4} \qquad \text{\textit{Simplify.}}$$

Note that we can get the same result by raising both the numerator and the denominator of the base $\dfrac{b}{c}$ to the 4th power:

$$\left(\frac{b}{c}\right)^4 = \frac{b^4}{c^4}$$

This method is called *raising a quotient to a power*.

> **Raising a Quotient to a Power**
>
> If n is a counting number and c is nonzero, then
>
> $$\left(\frac{b}{c}\right)^n = \frac{b^n}{c^n}$$
>
> In words: To raise a quotient to a power, raise both the numerator and the denominator to the power.

▶ **Example 4   Raising a Quotient to a Power**

Simplify.

1. $\left(\dfrac{b}{c}\right)^8$

   2. $\left(\dfrac{b}{2}\right)^3$

Solution

1. $\left(\dfrac{b}{c}\right)^8 = \dfrac{b^8}{c^8}$ \qquad *Raise numerator and denominator to eighth power:* $\left(\dfrac{b}{c}\right)^n = \dfrac{b^n}{c^n}$

2. $\left(\dfrac{b}{2}\right)^3 = \dfrac{b^3}{2^3}$ \qquad *Raise numerator and denominator to third power:* $\left(\dfrac{b}{c}\right)^n = \dfrac{b^n}{c^n}$

   $= \dfrac{b^3}{8}$ \qquad *Simplify.*

Raising a Power to a Power

Consider the expression $(b^2)^3$. We can write this expression as the power b^6:

$$(b^2)^3 = b^2 \cdot b^2 \cdot b^2 \quad \text{\textit{Write expression without exponent 3.}}$$
$$= b^{2+2+2} \quad \text{\textit{Add exponents: } } b^m b^n = b^{m+n}$$
$$= b^6 \quad \text{\textit{Simplify.}}$$

Note that we can get the same result by multiplying the exponents 2 and 3:

$$(b^2)^3 = b^{2 \cdot 3} = b^6$$

This method is called *raising a power to a power.*

> ### Raising a Power to a Power
>
> If m and n are counting numbers, then
>
> $$(b^m)^n = b^{mn}$$
>
> In words: To raise a power to a power, keep the base and multiply the exponents.

▶ **Example 5**   Raising a Power to a Power

Simplify.

1. $(b^2)^7$ **2.** $(b^5)^8$

Solution

1. $(b^2)^7 = b^{2 \cdot 7}$ *Multiply exponents: $(b^m)^n = b^{mn}$*
$$ = b^{14} \quad \text{\textit{Simplify.}}$$

2. $(b^5)^8 = b^{5 \cdot 8}$ *Multiply exponents: $(b^m)^n = b^{mn}$*
$$ = b^{40} \quad \text{\textit{Simplify.}}$$

Using Combinations of Properties of Exponents

Recall from Section 1.7 that if b is nonzero, then $b^0 = 1$. For example, $5^0 = 1$.
Next, we use more than one property of exponents to simplify an expression.

▶ **Example 6**   Simplifying Expressions Involving Exponents

Simplify.

1. $(2b^8c^5)^3$ **2.** $2b^7(4b^4)^2$ **3.** $\dfrac{8b^3b^4}{2b^7}$

Solution

1. $(2b^8c^5)^3 = 2^3(b^8)^3(c^5)^3$ *Raise each factor to third power: $(bc)^n = b^nc^n$*
$$ = 8b^{24}c^{15} \quad \text{\textit{Multiply exponents: } } (b^m)^n = b^{mn}$$

2. $2b^7(4b^4)^2 = 2b^7\left[4^2(b^4)^2\right]$ *Raise each factor to second power: $(bc)^n = b^nc^n$*
$$ = 2b^7\left[16b^8\right] \quad \text{\textit{Multiply exponents: } } (b^m)^n = b^{mn}$$
$$ = (2 \cdot 16)(b^7b^8) \quad \text{\textit{Rearrange factors.}}$$
$$ = 32b^{15} \quad \text{\textit{Add exponents: } } b^m b^n = b^{m+n}$$

3. $\dfrac{8b^3 b^4}{2b^7} = \dfrac{4b^7}{b^7}$ *Simplify; add exponents:* $b^m b^n = b^{m+n}$

$\qquad\qquad = 4b^{7-7}$ *Subtract exponents:* $\dfrac{b^m}{b^n} = b^{m-n}$

$\qquad\qquad = 4b^0$ *Simplify.*

$\qquad\qquad = 4$ $b^0 = 1.$

▶ **Example 7**   Simplifying Expressions Involving Exponents

Simplify.

1. $\left(\dfrac{2b^4}{3c^7}\right)^3$

2. $\dfrac{(2b^3 c)^5}{b^8 c}$

Solution

1. $\left(\dfrac{2b^4}{3c^7}\right)^3 = \dfrac{(2b^4)^3}{(3c^7)^3}$ *Raise numerator and denominator to third power:* $\left(\dfrac{b}{c}\right)^n = \dfrac{b^n}{c^n}$

$\qquad\qquad = \dfrac{2^3 (b^4)^3}{3^3 (c^7)^3}$ *Raise factors to third power:* $(bc)^n = b^n c^n$

$\qquad\qquad = \dfrac{8b^{12}}{27c^{21}}$ *Multiply exponents:* $(b^m)^n = b^{mn}$

2. $\dfrac{(2b^3 c)^5}{b^8 c} = \dfrac{2^5 (b^3)^5 c^5}{b^8 c}$ *Raise factors to fifth power:* $(bc)^n = b^n c^n$

$\qquad\qquad = \dfrac{32b^{15} c^5}{b^8 c^1}$ *Multiply exponents:* $(b^m)^n = b^{mn}; c = c^1$

$\qquad\qquad = 32b^{15-8} c^{5-1}$ *Subtract exponents:* $\dfrac{b^m}{b^n} = b^{m-n}$

$\qquad\qquad = 32b^7 c^4$ *Simplify.*

Simplifying Expressions Involving Negative-Integer Exponents

In Section 1.7, we defined a negative-integer exponent.

▶ **Definition**   Negative-integer exponent

If n is a counting number and $b \neq 0$, then

$$b^{-n} = \dfrac{1}{b^n}$$

In words, to find b^{-n}, take its reciprocal and change the sign of the exponent.

For example, $5^{-2} = \dfrac{1}{5^2} = \dfrac{1}{25}$.

Simplifying an expression involving negative-integer exponents includes writing it so that each exponent is positive.

▶ **Example 8**   Simplifying Expressions Involving Integer Exponents

Simplify.

1. $9b^{-7}$

2. $3^{-1} + 4^{-1}$

Solution

1. $9b^{-7} = 9 \cdot \dfrac{1}{b^7} = \dfrac{9}{b^7}$

2. $3^{-1} + 4^{-1} = \dfrac{1}{3} + \dfrac{1}{4} = \dfrac{4}{12} + \dfrac{3}{12} = \dfrac{7}{12}$

Next, we write $\dfrac{1}{b^{-n}}$ in another form, where b is nonzero and n is a counting number:

$$\dfrac{1}{b^{-n}} = 1 \div b^{-n} \qquad \dfrac{a}{b} = a \div b$$

$$= 1 \div \dfrac{1}{b^n} \qquad \text{\textit{Write power so that exponent is positive: }} b^{-n} = \dfrac{1}{b^n}$$

$$= 1 \cdot \dfrac{b^n}{1} \qquad \text{\textit{Multiply by reciprocal of }} \dfrac{1}{b^n}, \text{\textit{ which is }} \dfrac{b^n}{1}.$$

$$= b^n \qquad \text{\textit{Simplify.}}$$

So, $\dfrac{1}{b^{-n}} = b^n$.

▶ **Negative-Integer Exponent in a Denominator**

If n is a counting number and $b \neq 0$, then

$$\dfrac{1}{b^{-n}} = b^n$$

In words, to find $\dfrac{1}{b^{-n}}$, find its reciprocal and change the sign of the exponent.

▶ **Example 9**   Simplifying Expressions Involving Integer Exponents

Simplify.

1. $\dfrac{1}{2^{-3}}$

2. $\dfrac{5}{b^{-3}}$

Solution

1. $\dfrac{1}{2^{-3}} = 2^3 = 8$

2. $\dfrac{5}{b^{-3}} = 5 \cdot \dfrac{1}{b^{-3}} = 5b^3$

In Example 10, we will simplify some quotients of two powers.

▶ **Example 10**   Simplifying Expressions Involving Integer Exponents

Simplify.

1. $\dfrac{b^{-4}}{c^2}$

2. $\dfrac{b^3}{c^{-7}}$

Solution

1. $\dfrac{b^{-4}}{c^2} = b^{-4} \cdot \dfrac{1}{c^2} \qquad \text{\textit{Write quotient as a product: }} \dfrac{A}{B} = A \cdot \dfrac{1}{B}$

$$= \dfrac{1}{b^4} \cdot \dfrac{1}{c^2} \qquad \text{\textit{Write powers so that exponents are positive: }} b^{-n} = \dfrac{1}{b^n}$$

$$= \dfrac{1}{b^4 c^2} \qquad \text{\textit{Multiply numerators; multiply denominators: }} \dfrac{A}{B} \cdot \dfrac{C}{D} = \dfrac{AC}{BD}$$

2. $\dfrac{b^3}{c^{-7}} = b^3 \cdot \dfrac{1}{c^{-7}}$ *Write quotient as a product:* $\dfrac{A}{B} = A \cdot \dfrac{1}{B}$

$\qquad = b^3 c^7$ *Write powers so that exponents are positive:* $\dfrac{1}{b^{-n}} = b^n$

In Problem 1 of Example 10, we simplified $\dfrac{b^{-4}}{c^2}$. When simplifying such expressions, we usually skip the first two steps shown in the example and write

$$\frac{b^{-4}}{c^2} = \frac{1}{b^4 c^2}$$

Likewise, we can skip the first step of our work in Problem 2 of Example 10 by simplifying $\dfrac{b^3}{c^{-7}}$ as follows:

$$\frac{b^3}{c^{-7}} = b^3 c^7$$

▶ **Example 11**   Simplifying Expressions Involving Integer Exponents

Simplify $\dfrac{4a^{-2}b^4}{7c^{-5}}$.

Solution

$$\frac{4a^{-2}b^4}{7c^{-5}} = \frac{4c^5 b^4}{7a^2}$$ *Write powers so that exponents are positive:* $b^{-n} = \dfrac{1}{b^n}; \dfrac{1}{b^{-n}} = b^n$

So far, we have discussed various properties of counting-number exponents. It turns out that these properties are also true for all negative-integer exponents and the zero exponent.

▶ **Properties of Integer Exponents**

If m and n are integers, $b \neq 0$, and $c \neq 0$, then

- $b^m b^n = b^{m+n}$ *Product property for exponents*

- $\dfrac{b^m}{b^n} = b^{m-n}$ *Quotient property for exponents*

- $(bc)^n = b^n c^n$ *Raising a product to a power*

- $\left(\dfrac{b}{c}\right)^n = \dfrac{b^n}{c^n}$ *Raising a quotient to a power*

- $(b^m)^n = b^{mn}$ *Raising a power to a power*

We can use these properties to expand our rules for simplifying expressions involving exponents to include those which involve negative-integer exponents.

▶ **Simplifying Expressions Involving Integer Exponents**

An expression involving integer exponents is simplified if
1. It includes no parentheses.
2. Each variable or constant appears as a base as few times as possible.
3. Each numerical expression (such as 7^2) has been calculated, and each numerical fraction has been simplified.
4. Each exponent is positive.

▶ **Example 12** Simplifying Expressions Involving Integer Exponents

Simplify.

    1. $2^{-1003}2^{1000}$ **2.** $(b^{-4})^3$ **3.** $(8b^6)(-3b^{-10})$

Solution

    1. $2^{-1003}2^{1000} = 2^{-1003+1000}$ Add exponents: $b^m b^n = b^{m+n}$

                      $= 2^{-3}$ Simplify.

                      $= \dfrac{1}{2^3}$ Write powers so that exponents are positive: $b^{-n} = \dfrac{1}{b^n}$

                      $= \dfrac{1}{8}$ Simplify.

    2. $(b^{-4})^3 = b^{-12}$ Multiply exponents: $(b^m)^n = b^{mn}$

                 $= \dfrac{1}{b^{12}}$ Write power so that exponent is positive: $b^{-n} = \dfrac{1}{b^n}$

    3. $(8b^6)(-3b^{-10}) = 8(-3)b^6 b^{-10}$ Rearrange factors.

                            $= -24b^{-4}$ Add exponents: $b^m b^n = b^{m+n}$

                            $= -\dfrac{24}{b^4}$ Write power so that exponent is positive: $b^{-n} = \dfrac{1}{b^n}$

▶ **Example 13** Simplifying Expressions Involving Integer Exponents

Simplify.

    1. $\dfrac{b^4}{b^9}$ **2.** $\dfrac{4b^7}{3b^{-2}}$ **3.** $\dfrac{6b^{-3}c^{-2}}{9b^4c^{-5}}$

Solution

    1. $\dfrac{b^4}{b^9} = b^{4-9}$ Subtract exponents: $\dfrac{b^m}{b^n} = b^{m-n}$

                  $= b^{-5}$ Subtract.

                  $= \dfrac{1}{b^5}$ Write power so that exponent is positive: $b^{-n} = \dfrac{1}{b^n}$

    2. $\dfrac{4b^7}{3b^{-2}} = \dfrac{4b^{7-(-2)}}{3}$ Subtract exponents: $\dfrac{b^m}{b^n} = b^{m-n}$

                   $= \dfrac{4b^9}{3}$ Simplify.

    3. $\dfrac{6b^{-3}c^{-2}}{9b^4c^{-5}} = \dfrac{2b^{-3-4}c^{-2-(-5)}}{3}$ Simplify; subtract exponents: $\dfrac{b^m}{b^n} = b^{m-n}$

                      $= \dfrac{2b^{-7}c^3}{3}$ Simplify.

                      $= \dfrac{2c^3}{3b^7}$ Write powers so that exponents are positive: $b^{-n} = \dfrac{1}{b^n}$

In the first step of Problem 2 of Example 13, we found that

$$\frac{4b^7}{3b^{-2}} = \frac{4b^{7-(-2)}}{3}$$

WARNING   Note that we need a subtraction symbol *and* a negative symbol in the exponent on the right-hand side. It is a common error to omit writing one of these two symbols in problems of this type.

▶ **Example 14** Simplifying Expressions Involving Integer Exponents

Simplify.

1. $(2b^{-5})^{-3}$ **2.** $\left(\dfrac{2b^{-4}}{c^6}\right)^{-5}$

Solution

1.

$$(2b^{-5})^{-3} = 2^{-3}(b^{-5})^{-3} \qquad \text{Raise factors to power } -3: (bc)^n = b^n c^n$$

$$= 2^{-3}b^{15} \qquad \text{Multiply exponents: } (b^m)^n = b^{mn}$$

$$= \frac{b^{15}}{2^3} \qquad \text{Write powers so that exponents are positive: } b^{-n} = \frac{1}{b^n}$$

$$= \frac{b^{15}}{8} \qquad \text{Simplify.}$$

2.

$$\left(\frac{2b^{-4}}{c^6}\right)^{-5} = \frac{(2b^{-4})^{-5}}{(c^6)^{-5}} \qquad \text{Raise numerator and denominator to power } -5: \left(\frac{b}{c}\right)^n = \frac{b^n}{c^n}$$

$$= \frac{2^{-5}(b^{-4})^{-5}}{(c^6)^{-5}} \qquad \text{Raise each factor to power } -5: (bc)^n = b^n c^n$$

$$= \frac{2^{-5}b^{20}}{c^{-30}} \qquad \text{Multiply exponents: } (b^m)^n = b^{mn}$$

$$= \frac{b^{20}c^{30}}{2^5} \qquad \text{Write powers so that exponents are positive: } b^{-n} = \frac{1}{b^n}; \; \frac{1}{b^{-n}} = b^n$$

$$= \frac{b^{20}c^{30}}{32} \qquad \text{Simplify.}$$

▶ **Example 15** Simplifying Expressions Involving Integer Exponents

Simplify.

1. $\dfrac{(3bc^5)^2}{(2b^{-2}c^2)^3}$ **2.** $\left(\dfrac{18b^{-4}c^7}{6b^{-3}c^2}\right)^{-4}$

Solution

1.

$$\frac{(3bc^5)^2}{(2b^{-2}c^2)^3} = \frac{3^2 b^2 (c^5)^2}{2^3(b^{-2})^3(c^2)^3} \qquad \text{Raise factors to a power: } (bc)^n = b^n c^n$$

$$= \frac{9b^2 c^{10}}{8b^{-6}c^6} \qquad \text{Multiply exponents: } (b^m)^n = b^{mn}$$

$$= \frac{9b^{2-(-6)}c^{10-6}}{8} \qquad \text{Subtract exponents: } \frac{b^m}{b^n} = b^{m-n}$$

$$= \frac{9b^8 c^4}{8} \qquad \text{Simplify.}$$

2.

$$\left(\frac{18b^{-4}c^7}{6b^{-3}c^2}\right)^{-4} = (3b^{-4-(-3)}c^{7-2})^{-4} \qquad \text{Subtract exponents: } \frac{b^m}{b^n} = b^{m-n}$$

$$= (3b^{-1}c^5)^{-4} \qquad \text{Simplify.}$$

$$= 3^{-4}(b^{-1})^{-4}(c^5)^{-4} \qquad \text{Raise factors to nth power: } (bc)^n = b^n c^n$$

$$= 3^{-4}b^4 c^{-20} \qquad \text{Multiply exponents: } (b^m)^n = b^{mn}$$

$$= \frac{b^4}{3^4 c^{20}} \qquad \text{Write powers so that exponents are positive: } b^{-n} = \frac{1}{b^n}$$

$$= \frac{b^4}{81c^{20}} \qquad \text{Simplify.}$$

Definition of an Exponential Function

In Chapters 6–9, we modeled authentic situations using linear functions. In this chapter, we will do the same with *exponential functions*. Here are some examples of such functions:

$$f(x) = 2(3)^x, \qquad g(x) = -7\left(\frac{1}{2}\right)^x, \qquad h(x) = 5^x$$

Notice that, in exponential functions, the variable appears as an exponent.

▶ **Definition   Exponential function**

An **exponential function** is a function whose equation can be put into the form

$$f(x) = ab^x$$

where $a \neq 0, b > 0$, and $b \neq 1$. The constant b is called the **base.**

▶ **Example 16**   Evaluating Exponential Functions

For $f(x) = 3^x$ and $g(x) = 3(2)^x$, find the following.

  1. $f(4)$ **2.** $f(-2)$ **3.** $g(3)$ **4.** $g(-4)$ **5.** $g(0)$

Solution

  1. $f(4) = 3^4 = 81$

  2. $f(-2) = 3^{-2} = \dfrac{1}{3^2} = \dfrac{1}{9}$

  3. $g(3) = 3(2)^3 = 3 \cdot 8 = 24$

  4. $g(-4) = 3(2)^{-4} = \dfrac{3}{2^4} = \dfrac{3}{16}$

  5. $g(0) = 3(2)^0 = 3(1) = 3$

WARNING   It is a common error to confuse exponential functions such as $E(x) = 2^x$ with linear functions such as $L(x) = 2x$. For the *exponential* function $E(x) = 2^x$, the variable x is an *exponent*. For the *linear* function $L(x) = 2x^1$, the variable x is a *base*.

Models Whose Equations Contain Integer Exponents

Many authentic situations can be modeled by equations that contain exponents. Some examples of such situations are the sound level of a guitar, the intensity of illumination by a light bulb, and the gravitational force of the Sun acting on Earth.

▶ **Example 17**   Using a Model Whose Equation Involves an Integer Exponent

The intensity $f(d)$ (in watts per square meter) of a television signal at a distance d kilometers from the transmitter is described by the equation

$$f(d) = 250d^{-2}$$

  1. Simplify the right-hand side of the equation.
  2. Find $f(5)$. What does the result mean in this situation?

Solution

1. We use the definition of a negative-integer exponent to write the expression on the right-hand side without any negative exponents:

$$f(d) = 250d^{-2} \quad \text{\textit{Original equation}}$$

$$f(d) = \frac{250}{d^2} \quad \text{\textit{Write power so exponent is positive: }} b^{-n} = \frac{1}{b^n}$$

2. We substitute 5 for d in the equation $f(d) = \frac{250}{d^2}$:

$$f(5) = \frac{250}{5^2} = \frac{250}{25} = 10$$

So, the intensity is 10 watts per square meter at a distance of 5 kilometers from the transmitter.

 # Group Exploration

Properties of positive-integer exponents

1. For the statement $b^2 b^3 = b^5$, a student wants to know why there are two b variables on the left-hand side of the equation and only one b on the right-hand side. What would you tell the student?

2. A student tries to simplify the expression $(3b^3)^2$:

$$(3b^3)^2 = (3 \cdot 2)b^{3 \cdot 2} = 6b^6$$

Describe any errors. Then simplify the expression correctly.

3. In simplifying expressions involving exponents, a student is confused about when to add exponents and when to multiply exponents. What would you tell the student?

4. A student tries to simplify $\dfrac{(2b^5)^4}{b^2}$:

$$\frac{(2b^5)^4}{b^2} = (2b^{5-2})^4 = (2b^3)^4 = 2^4(b^3)^4 = 16b^{12}$$

Describe any errors. Then simplify the expression correctly.

5. Simplify $b^3 + b^3$. Then simplify $b^3 b^3$. Explain why your two results have different exponents.

Group Exploration

Properties of negative-integer exponents

1. Since $b^{-5} = \dfrac{1}{b^5}$ for nonzero b, does it follow that $-5 = \dfrac{1}{5}$? Explain.

2. A student tries to simplify $7b^{-2}$:

$$7b^{-2} = \frac{1}{7b^2}$$

Describe any errors. Then simplify the expression correctly.

3. A student tries to simplify $\dfrac{b^8}{b^{-5}}$:

$$\frac{b^8}{b^{-5}} = b^{8-5} = b^3$$

Describe any errors. Then simplify the expression correctly.

4. A student tries to simplify $(5b^3)^{-2}$:

$$(5b^3)^{-2} = 5(b^3)^{-2} = 5b^{-6} = \frac{5}{b^6}$$

Describe any errors. Then simplify the expression correctly.

▶ **Tips for Success**   **Form a Study Group to Prepare for the Final Exam**

To prepare for the final exam, it may be helpful to form a study group. The group could list important concepts of the course and discuss the meanings of these concepts and how to apply them. You could also list important techniques learned in the course and practice those techniques. Set aside some solo study time after the group study session to make sure you can do the mathematics without the help of other members of the study group.

Homework 10.1

1. If m and n are integers and $b \neq 0$, then $b^m b^n =$ _____.

2. If m and n are integers and $b \neq 0$, then $(b^m)^n =$ _____.

3. *True or False:* If an expression involving exponents is simplified, then each exponent is positive.

4. A(n) _____ function is a function whose equation can be put into the form $f(x) = ab^x$, where $a \neq 0, b > 0$, and $b \neq 1$.

Simplify.

5. $b^3 b^6$

6. $b^2 b^7$

7. $(6b^2 c^5)(9b^4 c^3)$

8. $(10b^7 c^2)(3b^5 c)$

9. $(2bc)^5$

10. $(3bc)^3$

11. $\dfrac{b^5}{b^2}$

12. $\dfrac{b^8}{b^6}$

13. $\dfrac{15b^6 c^8}{12b^3 c}$

14. $\dfrac{14b^5 c^9}{21b^3 c^6}$

15. $\left(\dfrac{b}{c}\right)^7$

16. $\left(\dfrac{b}{c}\right)^4$

17. $(b^2)^4$

18. $(b^3)^8$

19. $(3b^2 b^7)^3$

20. $(2b^3 b^4)^5$

21. $(9b^8 c^4)^0$

22. $(5b^3 c^9)^0$

23. $(b^2 c^3)^4 b^5 c^8$

24. $(bc^6)^3 b^2 c^4$

25. $5b^4(3b^6)^2$

26. $4b^2(2b^5)^3$

27. $\dfrac{10b^5 b^7}{8b^4}$

28. $\dfrac{15b^9 b^2}{20b^8}$

29. $\left(\dfrac{2b^4 c}{3d^2}\right)^3$

30. $\left(\dfrac{4bc^6}{7d^3}\right)^2$

31. $\left(\dfrac{3b^4}{5c^7}\right)^0$

32. $\left(\dfrac{5c^2}{9b^4}\right)^0$

33. $\dfrac{(4b^5 c^8)^2}{(2b^2 c^3)^3}$

34. $\dfrac{(2b^4 c^3)^3}{(6b^5 c^4)^2}$

Simplify.

35. b^{-4}

36. b^{-2}

37. $\dfrac{1}{b^{-2}}$

38. $\dfrac{1}{b^{-5}}$

39. $\dfrac{b^{-3}}{c^5}$

40. $\dfrac{b^2}{c^{-7}}$

41. $\dfrac{4b^{-9}}{-6c^4 d^{-1}}$

42. $\dfrac{-9b^{-1}}{6c^{-5} d^7}$

43. $(b^{-2})^7$

44. $(b^5)^{-8}$

45. $(-4b^3 c^{-7})(-b^{-5} c^4)$

46. $(-3b^{-2} c^{-6})(-b^9 c^{-1})$

47. $7^{-902} 7^{900}$

48. $4^{2003} 4^{-2000}$

49. $13^{500} 13^{-500}$

50. $(130^{-1})^{-1}$

51. $\dfrac{b^{-3}}{b^5}$

52. $\dfrac{b^{-4}}{b^7}$

53. $\dfrac{b^3}{b^{-2}}$

54. $\dfrac{b^6}{b^{-1}}$

55. $\dfrac{7b^{-3}}{4b^{-9}}$

56. $\dfrac{4b^{-2}}{9b^{-5}}$

57. $\dfrac{5^{-6}}{5^{-4}}$

58. $\dfrac{7^{-5}}{7^{-6}}$

59. $\dfrac{3^4 b^{-8}}{3^2 b^{-3}}$

60. $\dfrac{2^5 b^{-7}}{2^2 b^{-2}}$

61. $(2b^{-1})^{-5}$

62. $(2b^{-4})^{-4}$

63. $3(b^5 c)^{-2}$

64. $-6(bc^4)^{-3}$

65. $(2b^4 c^{-2})^5 (3b^{-3} c^{-4})^{-2}$

66. $(7b^{-4} c^{-1})^{-2}(2b^3 c^{-2})^5$

67. $\dfrac{15b^{-7} c^{-3} d^8}{-45c^2 b^{-6} d^8}$

68. $\dfrac{18b^5 c^3 d^{-7}}{24b^{-6} c^3 d^{-2}}$

69. $\dfrac{(24b^3 c^{-6})(49b^{-1} c^{-2})}{(28b^2 c^4)(14b^{-5} c)}$

70. $\dfrac{(16b^{-2} c)(25b^4 c^{-5})}{(15b^5 c^{-1})(8b^{-7} c^{-2})}$

71. $\dfrac{(2b^{-4} c)^{-3}}{(2b^2 c^{-5})^2}$

72. $\dfrac{(3bc^{-2})^{-2}}{(3b^{-3} c)^{-1}}$

73. $\left(\dfrac{6b^5 c^{-2}}{7b^2 c^4}\right)^2$

74. $\left(\dfrac{2bc^{-7}}{5b^{-1} c^{-2}}\right)^3$

75. $\left(\dfrac{5b^4 c^{-3}}{15b^{-2} c^{-1}}\right)^{-4}$

76. $\left(\dfrac{8b^{-2} c^2}{12b^{-5} c^{-3}}\right)^{-3}$

77. $b^{-1} c^{-1}$

78. $\dfrac{1}{b^{-1}} \cdot \dfrac{1}{c^{-1}}$

79. $\dfrac{1}{b^{-1}} + \dfrac{1}{c^{-1}}$

80. $b^{-1} + c^{-1}$

For Exercises 81–88, let $f(x) = 4^x$ and $g(x) = 2(3)^x$. Find the following.

81. $f(3)$

82. $f(2)$

83. $f(-2)$

84. $f(-3)$

85. $g(2)$

86. $g(0)$

87. $g(-2)$

88. $g(-3)$

89. **a.** Complete Table 2 with output values of the function $f(x) = 2^x$.

b. Plot the ordered pairs you found in part (a). Then guess the graph of f and sketch it by hand.

Table 2 Input-Output Pairs of $f(x) = 2^x$ (Exercise 89)

x	f(x)	x	f(x)
-3		1	
-2		2	
-1		3	
0		4	

c. Use your graph to estimate $2^{\frac{1}{2}}$. Round to the first decimal place.

90. a. Complete Table 3 with output values of the function $f(x) = \left(\frac{1}{2}\right)^x$.

Table 3 Input-Output Pairs of $f(x) = \left(\frac{1}{2}\right)^x$ (Exercise 90)

x	f(x)	x	f(x)
-4		0	
-3		1	
-2		2	
-1		3	

b. Plot the ordered pairs you found in part (a). Then guess the graph of f and sketch it by hand.

c. Use your graph to estimate $\left(\frac{1}{2}\right)^{\frac{1}{2}}$. Round to the first decimal place.

For Exercises 91–94, simplify the right-hand side of the statistics formula.

91. $\hat{p} = xn^{-1}$

92. $t = \mu_X \lambda^{-1}$

93. $F_0 = s_1^2 s_2^{-2}$

94. $\chi_0^2 = (n-1)s^2 \sigma_0^{-2}$

95. A person invests \$5000 in an account at 3% interest compounded annually. The value V (in dollars) of the account after t years is described by $V = 5000(1.03)^t$. Find the value of the account after 10 years.

96. A person invests \$8000 in an account at 7% interest compounded annually. The value V (in dollars) of the account after t years is described by $V = 8000(1.07)^t$. Find the value of the account after 6 years.

97. The amount of money P (in dollars) that must be invested in an account at 2% interest compounded annually so that the balance is A dollars after t years is described by $P = A(1.02)^{-t}$. How much money would a person need to invest in an account at 2% interest compounded annually in order for the balance to be \$10,000 after 9 years?

98. The amount of money P (in dollars) that must be invested in an account at 4% interest compounded annually so that the balance is A dollars after t years is described by $P = A(1.04)^{-t}$. How much money would a person need to invest in an account at 4% interest compounded annually in order for the balance to be \$7000 after 6 years?

99. If an object is moving at a constant speed s (in miles per hour), then $s = dt^{-1}$, where d is the distance traveled (in miles) and t is the time (in hours).

   a. Simplify the right-hand side of the equation.

   b. Substitute 186 for d and 3 for t in the equation you found in part (a), and solve for s. What does your result mean in this situation?

100. The force $f(L)$ (in pounds) you must exert on a wrench handle of length L inches to loosen a bolt is described by the equation $f(L) = 720L^{-1}$.

   a. Simplify the right-hand side of the equation.

   b. Find $f(12)$. What does your result mean in this situation?

101. If a person plays an electric guitar outside, the sound level $f(d)$ (in decibels) at a distance d yards from the amplifier is described by the equation $f(d) = 5760d^{-2}$.

   a. Simplify the right-hand side of the equation.

   b. Find $f(8)$. What does your result mean in this situation?

102. The amount of light $f(d)$ (in milliwatts per square centimeter) of a 50-watt light bulb at a distance d centimeters from the bulb is described by the equation $f(d) = 8910d^{-2}$.

   a. Simplify the right-hand side of the equation.

   b. Find $f(80)$. What does your result mean in this situation?

103. **DATA** The mean costs (in current dollars) of tuition, fees, room, and board at four-year public colleges are shown in Table 4 for various years.

Table 4 Mean Costs (in Current Dollars) of Tuition, Fees, Room, and Board

Year	Mean Cost (dollars)
1975	1646
1980	2328
1985	3682
1990	4715
1995	6620
2000	8080
2005	11,376
2010	15,235
2015	18,943

Source: *The College Board*

Let y be the mean cost (in current dollars) of tuition, fees, room, and board at four-year public colleges at x years since 1970.

   a. Construct a scatterplot.

   b. Is there a linear association, a nonlinear association, or no association?

   c. Use technology to graph the equation $\hat{y} = 1338(1.063)^x$ on the scatterplot. Does the model come close to the data points?

   d. Use the model to estimate the cost in 2015. Compute the residual for your estimate.

   e. Some public four-year colleges in Indiana, Iowa, Minnesota, and Missouri froze their tuitions in 2015 (Source: USA Today). Why is it not surprising that the residual you found in part (d) is negative?

104. **DATA** The revenues of Amazon are shown in Table 5 for various years.

Table 5 Annual Revenues of Amazon

Year	Revenue (billions of dollars)
2001	3.12
2003	5.26
2005	8.49
2007	14.84
2009	24.51
2011	48.08
2013	74.45

Source: *Amazon*

Let y be the annual revenue (in billions of dollars) of Amazon at x years since 2000.

a. Construct a scatterplot.

b. Is there a linear association, a nonlinear association, or no association?

c. Use technology to graph the equation $\hat{y} = 2.32(1.31)^x$ on the scatterplot. Does the model come close to the data points?

d. Use the model to estimate the revenue in 2011. Compute the residual for your estimate.

e. Use the model to estimate the revenue in 2013. Compute the residual for your estimate.

Concepts

105. A student tries to simplify b^3b^5:

$$b^3b^5 = b^{3\cdot5} = b^{15}$$

Describe any errors. Then simplify the expression correctly.

106. A student tries to simplify $\left(b^4\right)^6$:

$$\left(b^4\right)^6 = b^{4+6} = b^{10}$$

Describe any errors. Then simplify the expression correctly.

107. A student tries to simplify $\left(5b^3\right)^2$:

$$\left(5b^3\right)^2 = 5\left(b^3\right)^2 = 5b^6$$

Describe any errors. Then simplify the expression correctly.

108. A student tries to simplify $\left(2b^2\right)^4$:

$$\left(2b^2\right)^4 = (2\cdot4)b^{2\cdot4} = 8b^8$$

Describe any errors. Then simplify the expression correctly.

109. Two students try to simplify $\left(5b^2\right)^{-1}$:

Student A	Student B
$\left(5b^2\right)^{-1} = -5b^{-2}$	$\left(5b^2\right)^{-1} = 5^{-1}\left(b^2\right)^{-1}$
$= \dfrac{-5}{b^2}$	$= 5^{-1}b^{-2}$
	$= \dfrac{1}{5b^2}$

Did either student simplify the expression correctly? Describe any errors.

110. Two students try to simplify an expression:

Student 1	Student 2
$\dfrac{7b^8}{b^{-3}} = 7b^{8-(-3)}$	$\dfrac{7b^8}{b^{-3}} = 7b^{8-3}$
$= 7b^{11}$	$= 7b^5$

Did either student simplify the expression correctly? Describe any errors.

111. A student tries to simplify $\dfrac{3b^{-2}c^4}{d^7}$:

$$\frac{3b^{-2}c^4}{d^7} = \frac{c^4}{3b^2d^7}$$

Describe any errors. Then simplify the expression correctly.

112. a. Simplify $\left(\dfrac{b}{c}\right)^{-2}$.

b. Simplify $\left(\dfrac{b}{c}\right)^{-n}$.

c. Use your result from part (b) to simplify $\left(\dfrac{b}{c}\right)^{-5}$ in one step.

113. It is common to confuse expressions such as $2^2, 2^{-1}, 2(-1)$, $\left(\dfrac{1}{2}\right)^2, \left(\dfrac{1}{2}\right)^{-1}, -2^2, (-2)^2$, and $\dfrac{1}{2}$. List these numbers from least to greatest. Are there any "ties"?

114. Simplify each expression.

a. b^{-1}

b. $\left(b^{-1}\right)^{-1}$

c. $\left(\left(b^{-1}\right)^{-1}\right)^{-1}$

d. $\left(\left(\left(b^{-1}\right)^{-1}\right)^{-1}\right)^{-1}$

e. $\underbrace{\left(\left(\left(\left(b^{-1}\right)^{-1}\right)^{-1}\right)^{-1}\cdots\right)^{-1}}_{n \text{ exponents}}$

115. It is a common error to confuse the properties $b^mb^n = b^{m+n}$ and $\left(b^m\right)^n = b^{mn}$. Explain why each property makes sense, and compare the properties. Give examples to illustrate your comparison. (See page 12 for guidelines on writing a good response.)

116. Describe what it means to use exponential properties to simplify an expression. Include several examples in your description. (See page 12 for guidelines on writing a good response.)

▼10.2 Rational Exponents

Objectives

» Describe the meaning of *rational exponents*.

» Simplify expressions involving rational exponents.

» Describe properties of rational exponents.

In Section 10.1, we worked with integer exponents. In this section, we will work with exponents that are rational numbers.

Definitions of Rational Exponents

How should we define $b^{1/n}$, where n is a counting number? If the exponential property $(b^m)^n = b^{mn}$ is to be true for $m = \dfrac{1}{2}$ and $n = 2$, then

$$(9^{\frac{1}{2}})^2 = 9^{\frac{1}{2} \cdot 2} = 9^1 = 9$$

Since $(-3)^2 = 9$ and $3^2 = 9$, the statement suggests that a good meaning of $9^{1/2}$ is -3 or 3. We define $9^{1/2} = 3$. We call the nonnegative number 3 the *principal second root,* or **principal square root,** of 9, written $\sqrt{9}$.

Similarly, if the property $(b^m)^n = b^{mn}$ is to be true for $m = \dfrac{1}{3}$ and $n = 3$, then

$$(8^{\frac{1}{3}})^3 = 8^{\frac{1}{3} \cdot 3} = 8^1 = 8$$

Since $2^3 = 8$, the statement suggests that a good meaning of $8^{1/3}$ is 2. The number 2 is called the *third root,* or **cube root,** of 8, written $\sqrt[3]{8}$.

For $(-8)^{1/3}$, a good meaning is -2, since $(-2)^3 = -8$. We do not assign a real-number value to $(-9)^{1/2}$, since no real number squared is equal to -9.

> ▶ **Definition** $b^{1/n}$
>
> For the counting number n, where $n \neq 1$,
>
> - If n is odd, then $b^{1/n}$ is the number whose nth power is b, and we call $b^{1/n}$ the **nth root of b.**
> - If n is even and $b \geq 0$, then $b^{1/n}$ is the nonnegative number whose nth power is b, and we call $b^{1/n}$ the **principal nth root of b.**
> - If n is even and $b < 0$, then $b^{1/n}$ is not a real number.
>
> $b^{1/n}$ may be represented by $\sqrt[n]{b}$.

Simplifying Expressions Involving Rational Exponents

▶ **Example 1** Simplifying Expressions Involving Rational Exponents

Simplify.

1. $25^{1/2}$
2. $64^{1/3}$
3. $(-64)^{1/3}$
4. $16^{1/4}$
5. $-16^{1/4}$
6. $(-16)^{1/4}$

Solution

1. $25^{1/2} = 5$, since $5^2 = 25$.
2. $64^{1/3} = 4$, since $4^3 = 64$.
3. $(-64)^{1/3} = -4$, since $(-4)^3 = -64$.
4. $16^{1/4} = 2$, since $2^4 = 16$.
5. $-16^{1/4} = -(16^{1/4}) = -2$.
6. $(-16)^{1/4}$ is not a real number, since the fourth power of any real number is nonnegative.

TI-84 checks for Problems 2 and 3 are shown in Fig. 1. For example, to find $25^{1/2}$, press **25** ⌈∧⌉ ⌈(⌉ **1** ⌈÷⌉ **2** ⌈)⌉ ⌈ENTER⌉.

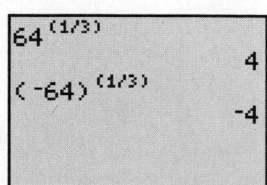

Figure 1 Checks for Problems 2 and 3

What would be a reasonable definition of $b^{m/n}$? If the properties of exponents we discussed in Section 10.1 are to hold true for rational exponents, we have

$$8^{\frac{2}{3}} = 8^{\frac{1}{3}\cdot 2} = (8^{\frac{1}{3}})^2 = 2^2 = 4 \quad \text{or} \quad 8^{\frac{2}{3}} = 8^{2\cdot\frac{1}{3}} = (8^2)^{\frac{1}{3}} = 64^{\frac{1}{3}} = 4$$

Likewise,

$$32^{\frac{3}{5}} = 32^{\frac{1}{5}\cdot 3} = (32^{\frac{1}{5}})^3 = 2^3 = 8 \quad \text{or} \quad 32^{\frac{3}{5}} = 32^{3\cdot\frac{1}{5}} = (32^3)^{\frac{1}{5}} = 32{,}768^{\frac{1}{5}} = 8$$

Also,

$$32^{-\frac{3}{5}} = \frac{1}{32^{\frac{3}{5}}} = \frac{1}{8}$$

> ### Definition   Rational exponent
>
> For the counting numbers m and n, where $n \neq 1$ and b is any real number for which $b^{1/n}$ is a real number,
>
> - $b^{m/n} = (b^{1/n})^m = (b^m)^{1/n}$
>
> - $b^{-m/n} = \dfrac{1}{b^{m/n}}, \quad b \neq 0$
>
> A power of the form $b^{m/n}$ or $b^{-m/n}$ is said to have a **rational exponent.**

> ### Example 2   Simplifying Expressions Involving Rational Exponents

Simplify.

1. $25^{3/2}$ **2.** $(-27)^{2/3}$ **3.** $32^{-2/5}$ **4.** $(-8)^{-5/3}$

Solution

1. $25^{3/2} = (25^{1/2})^3 = 5^3 = 125$

2. $(-27)^{2/3} = ((-27)^{1/3})^2 = (-3)^2 = 9$

3. $32^{-2/5} = \dfrac{1}{32^{2/5}} = \dfrac{1}{(32^{1/5})^2} = \dfrac{1}{2^2} = \dfrac{1}{4}$

4. $(-8)^{-5/3} = \dfrac{1}{(-8)^{5/3}} = \dfrac{1}{((-8)^{1/3})^5} = \dfrac{1}{(-2)^5} = \dfrac{1}{-32} = -\dfrac{1}{32}$

TI-84 checks for Problems 2 and 3 are shown in Fig. 2.

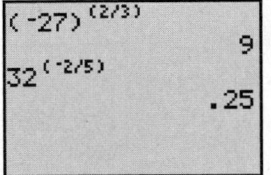

Figure 2 Checks for Problems 2 and 3

> ### Example 3   Evaluating an Exponential Function

For $f(x) = 64^x$, $g(x) = 3(16)^x$, and $h(x) = -5(9)^x$, find the following.

1. $f\left(\dfrac{2}{3}\right)$ **2.** $g\left(\dfrac{3}{4}\right)$ **3.** $h\left(-\dfrac{1}{2}\right)$

Solution

1. $f\left(\dfrac{2}{3}\right) = 64^{2/3} = (64^{1/3})^2 = 4^2 = 16$

2. $g\left(\dfrac{3}{4}\right) = 3(16)^{3/4} = 3(16^{1/4})^3 = 3(2)^3 = 3 \cdot 8 = 24$

3. $h\left(-\dfrac{1}{2}\right) = -5(9)^{-1/2} = \dfrac{-5}{9^{1/2}} = -\dfrac{5}{3}$

Properties of Rational Exponents

The properties of exponents we discussed in Section 10.1 are valid for *rational* exponents.

> **Properties of Rational Exponents**
>
> If m and n are rational numbers and b and c are any real numbers for which b^m, b^n, and c^n are real numbers, then
>
> - $b^m b^n = b^{m+n}$ *Product property for exponents*
> - $\dfrac{b^m}{b^n} = b^{m-n}, b \neq 0$ *Quotient property for exponents*
> - $(bc)^n = b^n c^n$ *Raising a product to a power*
> - $\left(\dfrac{b}{c}\right)^n = \dfrac{b^n}{c^n}, c \neq 0$ *Raising a quotient to a power*
> - $(b^m)^n = b^{mn}$ *Raising a power to a power*

We can use properties of exponents to help us simplify expressions involving rational exponents.

> ▶ **Example 4**   Simplifying Expressions Involving Rational Exponents

Simplify. Assume b is positive.

1. $(4b^6)^{3/2}$ **2.** $\dfrac{b^{2/7}}{b^{-3/7}}$

Solution

1. $(4b^6)^{3/2} = 4^{3/2}(b^6)^{3/2}$ *Raise factors to nth power:* $(bc)^n = b^n c^n$

$= (4^{1/2})^3 b^{\frac{6}{1}\cdot\frac{3}{2}}$ $b^{m/n} = (b^{1/n})^m;$ *multiply exponents:* $(b^m)^n = b^{mn}$

$= 2^3 b^9$ $4^{\frac{1}{2}} = 2;$ *multiply.*

$= 8b^9$ *Simplify.*

2. $\dfrac{b^{2/7}}{b^{-3/7}} = b^{\frac{2}{7} - \left(-\frac{3}{7}\right)}$ *Subtract exponents:* $\dfrac{b^m}{b^n} = b^{m-n}$

$= b^{\frac{2}{7}+\frac{3}{7}}$ *Simplify.*

$= b^{5/7}$ *Add.*

> ▶ **Example 5**   Simplifying Expressions Involving Rational Exponents

Simplify. Assume b is positive.

1. $b^{2/3}b^{1/2}$ **2.** $\left(\dfrac{32b^2}{b^{12}}\right)^{2/5}$

Solution

1. $b^{2/3}b^{1/2} = b^{\frac{2}{3}+\frac{1}{2}}$ *Add exponents:* $b^m b^n = b^{m+n}$

$= b^{\frac{4}{6}+\frac{3}{6}}$ *Find common denominator.*

$= b^{7/6}$ *Add.*

2. $\left(\dfrac{32b^2}{b^{12}}\right)^{2/5} = (32b^{2-12})^{2/5}$ *Subtract exponents:* $\dfrac{b^m}{b^n} = b^{m-n}$

$= (32b^{-10})^{2/5}$ *Subtract.*

$= \left(\dfrac{32}{b^{10}}\right)^{2/5}$ *Write powers so exponents are positive:* $b^{-n} = \dfrac{1}{b^n}$

$$= \frac{32^{2/5}}{(b^{10})^{2/5}}$$

Raise numerator and denominator to nth power: $\left(\dfrac{b}{c}\right)^n = \dfrac{b^n}{c^n}$

$$= \frac{(32^{1/5})^2}{b^{10 \cdot \frac{2}{5}}}$$

$b^{m/n} = (b^{1/n})^m$; *multiply exponents:* $(b^m)^n = b^{mn}$

$$= \frac{2^2}{b^4}$$

$32^{1/5} = 2$; *multiply.*

$$= \frac{4}{b^4}$$

Simplify.

▶ **Example 6**   Simplifying an Expression Involving Rational Exponents

Simplify $\dfrac{(81b^6c^{20})^{1/2}}{(27b^{12}c^9)^{2/3}}$. Assume b and c are positive.

Solution

$$\frac{(81b^6c^{20})^{1/2}}{(27b^{12}c^9)^{2/3}} = \frac{81^{1/2}(b^6)^{1/2}(c^{20})^{1/2}}{27^{2/3}(b^{12})^{2/3}(c^9)^{2/3}}$$

Raise factors to a power: $(bc)^n = b^nc^n$

$$= \frac{9b^{6 \cdot \frac{1}{2}}c^{20 \cdot \frac{1}{2}}}{(27^{1/3})^2 b^{12 \cdot \frac{2}{3}}c^{9 \cdot \frac{2}{3}}}$$

$81^{1/2} = 9$; $b^{m/n} = (b^{1/n})^m$; *multiply exponents:* $(b^m)^n = b^{mn}$

$$= \frac{9b^3c^{10}}{3^2 b^8 c^6}$$

$27^{1/3} = 3$; *multiply.*

$$= \frac{9b^{-5}c^4}{9}$$

Subtract exponents: $\dfrac{b^m}{b^n} = b^{m-n}$

$$= \frac{c^4}{b^5}$$

Write powers so exponents are positive: $b^{-n} = \dfrac{1}{b^n}$

▼▲ Group Exploration

Section Opener: Definition of $b^{1/n}$

Throughout this exploration, assume that $(b^m)^n = b^{mn}$ for rational numbers m and n.

1. First, you will explore the meaning of $b^{1/2}$, where b is nonnegative.
   a. For now, do not use a calculator. You will explore how you should define $9^{1/2}$. You can determine a reasonable value of $9^{1/2}$ by first finding its *square*:

   $$(9^{1/2})^2 = 9^{\frac{1}{2} \cdot 2} = 9^1 = 9$$

   What would be a good meaning of $9^{1/2}$? [**Hint:** Can you think of a positive number whose square equals 9?]
   b. What would be a good meaning of $16^{1/2}$? Of $25^{1/2}$?
   c. Now use a calculator to find $9^{1/2}$, $16^{1/2}$, and $25^{1/2}$. For example, to use a TI-84 to find $9^{1/2}$, press 9 $\boxed{\wedge}$ $\boxed{(}$ 1 $\boxed{\div}$ 2 $\boxed{)}$ $\boxed{\text{ENTER}}$. Is the calculator interpreting $b^{1/2}$ as you would expect?

   d. What would be a good meaning of $b^{1/2}$, where b is nonnegative?

2. Now you will explore the meaning of $b^{1/3}$.
   a. For now, do not use a calculator. You will explore how you should define $8^{1/3}$. You can first find the *cube* of $8^{1/3}$:

   $$(8^{1/3})^3 = 8^{\frac{1}{3} \cdot 3} = 8^1 = 8$$

   What would be a good meaning of $8^{1/3}$? Explain.
   b. What would be a good meaning of $27^{1/3}$? Of $64^{1/3}$?
   c. Use a calculator to find $8^{1/3}$, $27^{1/3}$, and $64^{1/3}$. Is the calculator interpreting $b^{1/3}$ as you would expect?
   d. What would be a good meaning of $b^{1/3}$?

3. What would be a good meaning of $b^{1/n}$, where n is a counting number and b is nonnegative?

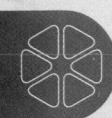

Homework 10.2

1. If n is odd, then _____ is the number whose nth power is b.

2. *True or False:* For the counting numbers m and n, where $n \neq 1$ and b is any real number for which $b^{1/n}$ is a real number, $b^{m/n} = (b^{1/m})^n$.

3. If m and n are rational numbers and b is any nonzero real number for which b^m and b^n are real numbers, then $\dfrac{b^m}{b^n} = $ _____.

4. If m and n are rational numbers and b is any real number for which b^m and b^n are real numbers, then $(b^m)^n = $ _____.

Simplify without using a calculator.

5. $16^{1/2}$ **6.** $27^{1/3}$ **7.** $1000^{1/3}$

8. $32^{1/5}$ **9.** $49^{1/2}$ **10.** $81^{1/4}$

11. $125^{1/3}$ **12.** $64^{1/6}$ **13.** $8^{4/3}$

14. $16^{3/4}$ **15.** $9^{3/2}$ **16.** $64^{2/3}$

17. $32^{2/5}$ **18.** $27^{4/3}$ **19.** $4^{5/2}$

20. $81^{3/4}$ **21.** $27^{-1/3}$ **22.** $16^{-1/4}$

23. $-36^{-1/2}$ **24.** $-32^{-1/5}$ **25.** $4^{-5/2}$

26. $9^{-3/2}$ **27.** $(-27)^{-4/3}$ **28.** $(-32)^{-3/5}$

Simplify without using a calculator.

29. $2^{1/4}2^{3/4}$ **30.** $3^{7/5}3^{3/5}$ **31.** $(3^{1/2}2^{3/2})^2$

32. $(2^{2/3}5^{1/3})^3$

33. $\dfrac{7^{1/3}}{7^{-5/3}}$ **34.** $\dfrac{5^{4/3}}{5^{1/3}}$

For $f(x) = 81^x$, $g(x) = 4(27)^x$, and $h(x) = -2(4)^x$, find the following.

35. $f\left(\dfrac{3}{4}\right)$ **36.** $f\left(\dfrac{1}{4}\right)$

37. $g\left(\dfrac{1}{3}\right)$ **38.** $g\left(\dfrac{2}{3}\right)$

39. $g\left(-\dfrac{1}{3}\right)$ **40.** $g\left(-\dfrac{2}{3}\right)$

41. $h\left(\dfrac{3}{2}\right)$ **42.** $h\left(\dfrac{5}{2}\right)$

43. Without using a calculator, complete Table 6 with values of the function $f(x) = 16^x$.

Table 6 Values of the Function $f(x) = 16^x$

x	$f(x)$	x	$f(x)$
$-\dfrac{3}{4}$		$\dfrac{1}{4}$	
$-\dfrac{1}{2}$		$\dfrac{1}{2}$	
$-\dfrac{1}{4}$		$\dfrac{3}{4}$	
0		1	

44. Without using a calculator, complete Table 7 with values of the function $f(x) = 64^x$.

Table 7 Values of the Function $f(x) = 64^x$

x	$f(x)$	x	$f(x)$
$-\dfrac{5}{6}$		$\dfrac{1}{6}$	
$-\dfrac{2}{3}$		$\dfrac{1}{3}$	
$-\dfrac{1}{2}$		$\dfrac{1}{2}$	
$-\dfrac{1}{3}$		$\dfrac{2}{3}$	
$-\dfrac{1}{6}$		$\dfrac{5}{6}$	
0		1	

Simplify. Assume b and c are positive.

45. $b^{7/6}b^{5/6}$ **46.** $b^{1/5}b^{3/5}$

47. $b^{3/5}b^{-13/5}$ **48.** $b^{2/7}b^{-6/7}$

49. $(16b^8)^{1/4}$ **50.** $(27b^{27})^{1/3}$

51. $4(25b^8c^{14})^{-1/2}$ **52.** $-(8b^{-6}c^{12})^{2/3}$

53. $(b^{3/5}c^{-1/4})(b^{2/5}c^{-7/4})$ **54.** $(b^{-4/3}c^{1/2})(b^{-2/3}c^{-3/2})$

55. $(5bcd)^{1/5}(5bcd)^{4/5}$ **56.** $(6bc^2)^{5/7}(6bc^2)^{2/7}$

57. $\left[(3b^5)^3(3b^9c^8)\right]^{1/4}$ **58.** $\left[(4b^3)^2(b^2c^{12})\right]^{1/4}$

59. $\dfrac{b^{-2/5}c^{11/8}}{b^{18/5}c^{-5/8}}$ **60.** $\dfrac{b^{3/4}c^{1/2}}{b^{-1/4}c^{-1/2}}$

61. $\left(\dfrac{9b^3c^{-2}}{25b^{-5}c^4}\right)^{-1/2}$ **62.** $\left(\dfrac{16b^{12}c^2}{2b^{-3}c^{-4}}\right)^{-1/3}$

63. $32^{1/5}b^{3/7}b^{2/5}$ **64.** $16^{1/4}b^{1/4}b^{1/3}$

65. $\dfrac{b^{5/6}}{b^{1/4}}$ **66.** $\dfrac{b^{-2/3}}{b^{1/7}}$

67. $\dfrac{(9b^5)^{3/2}}{(27b^4)^{2/3}}$ **68.** $\dfrac{(32b^3)^{3/5}}{(16b^3)^{3/2}}$

69. $\left(\dfrac{8b^{2/3}}{2b^{4/5}}\right)^{3/2}$ **70.** $\left(\dfrac{27b^{1/3}c^{3/4}}{8b^{-2/3}c^{1/2}}\right)^{4/3}$

71. $\dfrac{(8bc^3)^{1/3}}{(81b^{-5}c^3)^{3/4}}$ **72.** $\dfrac{(1000b^{-7}c^8)^{2/3}}{(32b^{15}c^4)^{3/5}}$

73. $b^{2/5}(b^{8/5} + b^{3/5})$ **74.** $c^{1/3}(c^{8/3} - c^{5/3})$

For Exercises 75 and 76, simplify the right-hand side of the statistics formula.

75. $\sigma_{\bar{x}} = \sigma n^{-1/2}$

76. $E = tsn^{-1/2}$

77. DATA The amounts (in megawatts) of new solar power installed in the United States are shown in Table 8 for various years.

Table 8 Amounts of New Solar Power Installed in the United States

Year	Amount of New Solar Power Installed (megawatts)
2000	4
2002	23
2004	58
2006	105
2008	298
2010	852
2012	3369
2014	6201

Source: *GTM Research*

Let y be the amount (in megawatts) of new solar power installed in the year that is x years since 2000.

a. Construct a scatterplot.

b. Is there a linear association, a nonlinear association, or no association?

c. Use technology to graph the equation $\hat{y} = 5.79(1.66)^x$ on the scatterplot. Does the model come close to the data points?

d. Predict the amount of new solar power that will be installed in 2015. Do you have much faith in your prediction? Explain.

e. GTM Research predicted that the amount of new solar power installed will increase by 31% from 2014 to 2015. On the basis of this prediction, how much new solar power will be installed in 2015? Compare your result with the prediction you made in part (d).

78. DATA The numbers of AP tests administered are shown in Table 9 for various years.

Table 9 Numbers of AP Tests Administered

Year	Number of AP Tests Administered (millions)
1980	0.2
1985	0.3
1990	0.5
1995	0.9
2000	1.3
2005	2.1
2010	3.3
2014	4.2

Source: *The College Board*

Let y be the number (in millions) of AP tests administered in the year that is x years since 1980.

a. Construct a scatterplot.

b. Is there a linear association, a nonlinear association, or no association?

c. Use technology to graph the equation $\hat{y} = 0.2(1.096)^x$ on the scatterplot. Does the model come close to the data points?

d. Estimate the number of AP tests administered in 2010. Find the residual. What does it mean in this situation?

e. Estimate the number of AP tests administered in 2014. Find the residual. What does it mean in this situation?

Concepts

79. Why can we represent $\sqrt{5}$ by $5^{1/2}$?

80. Explain why $(-1)^{1/4}$ is not a real number.

81. A student tries to simplify $(36x^{36})^{1/2}$:

$$(36x^{36})^{1/2} = 36^{1/2}(x^{36})^{1/2} = 18x^{18}$$

Describe any errors. Then simplify the expression correctly.

82. A student tries to simplify $64^{2/3}$:

$$64^{2/3} = (64^{1/2})^3 = 8^3 = 512$$

Describe any errors. Then simplify the expression correctly.

83. Two students simplify $9^{3/2}$:

Student A	Student B
$9^{3/2} = (9^{1/2})^3 = 3^3 = 27$	$9^{3/2} = (9^3)^{1/2} = 729^{1/2} = 27$

Explain why it makes sense that the students' results are equal. Which method is easier? Explain.

84. A student tries to simplify $25^{-1/2}$:

$$25^{-1/2} = -25^{1/2} = -5$$

Describe any errors. Then simplify the expression correctly.

85. a. Identify which of the following are *not* real numbers:

$$(-9)^{1/2}, (-27)^{1/3}, (-81)^{1/4}, (-32)^{1/5}, (-1)^{1/6}, (-1)^{1/7}$$

b. Describe all of the values of b and counting number n for which $b^{1/n}$ is not a real number.

86. Use a calculator to determine which is larger, $\left(\dfrac{1}{2}\right)^{1/3}$ or $\left(\dfrac{1}{3}\right)^{1/2}$. Explain why this makes sense. [**Hint:** Use the property $\left(\dfrac{b}{c}\right)^n = \dfrac{b^n}{c^n}$.]

87. Describe how to compute a numerical expression of the form $b^{m/n}$, assuming it is a real number.

88. List the exponent definitions and properties that are discussed in this section and Section 10.1. Explain how you can recognize which definition or property will help you simplify a given expression.

▼ 10.3 Graphing Exponential Models

Objectives

» Graph an exponential function.

» Describe the *base multiplier property* and the *increasing or decreasing property*.

» Find the *y*-intercept of the graph of an exponential function.

» Describe the *reflection property for exponential functions*.

» Graph an equation of an exponential model.

» Interpret the coefficient and the base of an exponential model.

Recall from Section 10.1 that an exponential function is a function whose equation can be put into the form $f(x) = ab^x$, where $a \neq 0, b > 0$, and $b \neq 1$. In this section, we will discuss how to use the values of a and b to help us graph an exponential function.

Graphing Exponential Functions

When graphing a certain type of function for the first time, we often begin by finding outputs for integer inputs near zero.

▶ **Example 1** Graphing an Exponential Function with $b > 1$

Graph $f(x) = 2^x$ by hand.

Solution

First, we list input–output pairs of the function f in Table 10. Note that as the value of x increases by 1, the value of y is multiplied by 2 (the base).

Next, we plot the solutions from Table 10 in Fig. 3 and sketch an increasing curve that contains the plotted points. The graph shows that as the value of x increases by 1, the value of y is doubled.

Table 10 Input–Output Pairs of $f(x) = 2^x$

x	$f(x)$
-3	$2^{-3} = \dfrac{1}{2^3} = \dfrac{1}{8}$
-2	$2^{-2} = \dfrac{1}{2^2} = \dfrac{1}{4}$
-1	$2^{-1} = \dfrac{1}{2^1} = \dfrac{1}{2}$
0	$2^0 = 1$
1	$2^1 = 2$
2	$2^2 = 4$
3	$2^3 = 8$

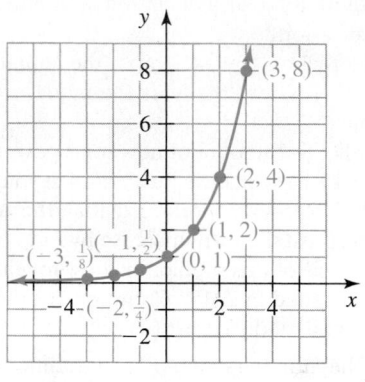

Figure 3 Graph of $f(x) = 2^x$

We can set up a window to verify our graph (see Figs. 4–6).

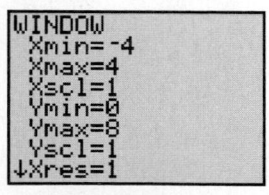

Figure 4 Enter the function

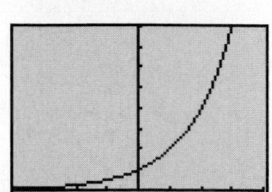

Figure 5 Set up the window

The smooth curve sketched in Fig. 3 implies that 2^x has meaning for *any* real-number exponent x. This is indeed true. In general, for $b > 0$, b^x has meaning for any real-number exponent x. The exponents are defined so the graph of any exponential function is a smooth graph. Also, the exponential properties we have discussed for rational exponents apply to real-number exponents as well. We can use a calculator to find *real-number powers* of numbers.

Recall from Section 7.1 that every point on the graph of an equation represents a solution of the equation. Every point *not* on the graph represents an ordered pair that is *not* a solution. The graph of an exponential function is called an **exponential curve.**

▶ **Example 2** Graphing an Exponential Function with $0 < b < 1$

Graph $g(x) = 4\left(\dfrac{1}{2}\right)^x$ by hand.

Solution

Input–output pairs of g are listed in Table 11. For example,

$$g(-1) = 4\left(\frac{1}{2}\right)^{-1} = 4\left(\frac{1}{2^{-1}}\right) = 4(2^1) = 8$$

Figure 6 Graph the function

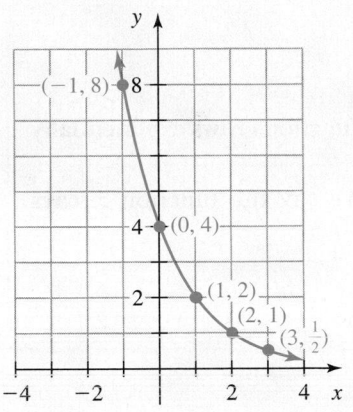

Figure 7 Graph of $y = 4\left(\dfrac{1}{2}\right)^x$

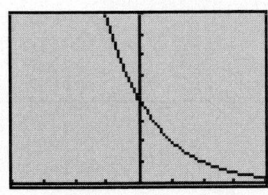

Figure 8 Graph of $y = 4\left(\dfrac{1}{2}\right)^x$

Table 11 Input–Output Pairs of $g(x) = 4\left(\dfrac{1}{2}\right)^x$

x	g(x)
−1	8
0	4
1	2
2	1
3	$\dfrac{1}{2}$

So, $(-1, 8)$ is an input–output pair. Note that as the value of x increases by 1, the value of y is multiplied by $\dfrac{1}{2}$.

We plot the found points in Fig. 7 and sketch a decreasing exponential curve that contains the plotted points. The graph shows that as the value of x increases by 1, the value of y is halved.

We can use a TI-84 to verify our graph (see Fig. 8).

Base Multiplier Property; Increasing or Decreasing Property

Examples 1 and 2 suggest the *base multiplier property* and the *increasing or decreasing property*.

> ▶ **Base Multiplier Property**
>
> For an exponential function of the form $y = ab^x$, if the value of x increases by 1, then the value of y is multiplied by b.

We have seen two examples of this property in Examples 1 and 2. Here are two more examples of the base multiplier property:

1. For the function $f(x) = 2(3)^x$, if the value of x increases by 1, the value of y is multiplied by 3.

2. For the function $f(x) = 5\left(\dfrac{3}{4}\right)^x$, if the value of x increases by 1, the value of y is multiplied by $\dfrac{3}{4}$.

To prove the base multiplier property for the exponential function $f(x) = ab^x$, we compare outputs for the inputs k and $k + 1$, which differ by 1:

$$f(k) = ab^k \qquad f(k + 1) = ab^{k+1}$$
$$= ab^k b^1$$
$$= f(k)b$$

Since $f(k + 1) = f(k)b$, we conclude that if the value of the independent variable increases by 1, the value of the dependent variable is multiplied by b, which is what we set out to show.

For the increasing or decreasing property, we note in Example 1 that the base b is greater than 1 and the graph is increasing. In Example 2, the positive base is less than 1 and the graph is decreasing. For $f(x) = ab^x$ with $a > 0$, in general, we have the property that each multiplication by a base greater than 1 gives a larger value of y, whereas each multiplication by a positive base less than 1 gives a smaller value of y.

▶ Increasing or Decreasing Property

Let $f(x) = ab^x$, where $a > 0$. Then

- If $b > 1$, then the function f is increasing. We say the function **grows exponentially** (see Fig. 9).
- If $0 < b < 1$, then the function f is decreasing. We say the function **decays exponentially** (see Fig. 10).

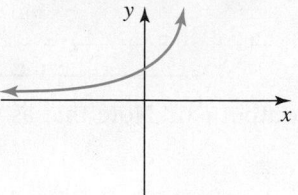

Figure 9 Typical graph of $f(x) = ab^x$, where $a > 0$ and $b > 1$

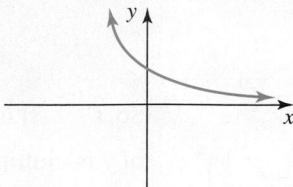

Figure 10 Typical graph of $f(x) = ab^x$, where $a > 0$ and $0 < b < 1$

y-Intercept of the Graph of an Exponential Function

When we sketch the graph of an exponential function, it is helpful to plot the y-intercept first. Substituting 0 for x in the general equation $y = ab^x$ gives

$$y = ab^0 = a(1) = a$$

So, the y-intercept is $(0, a)$.

▶ y-Intercept of the Graph of an Exponential Function

For the graph of an exponential function of the form

$$y = ab^x,$$

the y-intercept is $(0, a)$.

For the graph of the function $y = 5(8)^x$, the y-intercept is $(0, 5)$. For the graph of the function $y = 4\left(\dfrac{1}{7}\right)^x$, the y-intercept is $(0, 4)$.

WARNING   For the graph of an exponential function of the form $y = b^x$ (rather than $y = ab^x$), the y-intercept is *not* $(0, b)$. By writing $y = b^x = 1b^x$, we see the y-intercept is $(0, 1)$. For example, for the graph of $y = 2^x$, the y-intercept is $(0, 1)$. See Example 1.

▶ Example 3   Intercepts and Graph of an Exponential Function

Let $f(x) = 6\left(\dfrac{1}{2}\right)^x$.

1. Find the y-intercept of the graph of f.
2. Find the x-intercept of the graph of f.
3. Graph f by hand.

Solution

1. Since $f(x) = 6\left(\dfrac{1}{2}\right)^x$ is of the form $f(x) = ab^x$, the y-intercept is $(0, a)$, or $(0, 6)$.

Table 12 Input-Output Pairs of $f(x) = 6\left(\dfrac{1}{2}\right)^x$

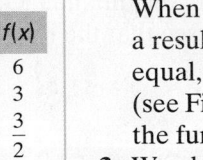

x	$f(x)$
0	6
1	3
2	$\dfrac{3}{2}$
3	$\dfrac{3}{4}$
4	$\dfrac{3}{8}$

2. By the base multiplier property, as the value of x increases by 1, the value of y is multiplied by $\dfrac{1}{2}$ (see Table 12).

When we halve a number, it becomes smaller. But no number of halvings will give a result that is zero. So, as x grows large, y will become extremely close to, but never equal, 0. Likewise, the graph of f gets arbitrarily close to, but never reaches, the x-axis (see Fig. 11). In this case, we call the x-axis a **horizontal asymptote.** We conclude that the function f has no x-intercepts.

3. We plot five solutions from Table 12 and sketch a decreasing exponential curve that contains the five points (see Fig. 11). If we had not already found a table of solutions, we could have plotted the y-intercept and plotted additional solutions by increasing the value of x by 1 and going half as high for the value of y each time.

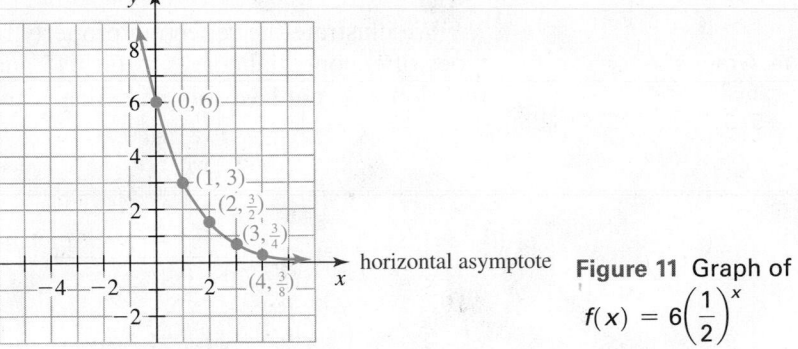

Figure 11 Graph of $f(x) = 6\left(\dfrac{1}{2}\right)^x$

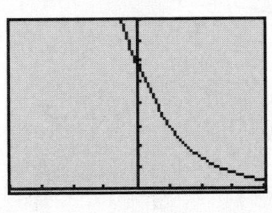

Figure 12 Graph of $y = 6\left(\dfrac{1}{2}\right)^x$

As a check, we note that according to the increasing or decreasing property, the function f is decreasing, since $a > 0$ and the base, $\dfrac{1}{2}$, is between 0 and 1. For a more thorough check, we can use a TI-84 to verify our graph (see Fig. 12).

In Fig. 13, the graphs of both exponential functions get closer and closer to, but never reach, the x-axis. For both graphs, the x-axis is a horizontal asymptote.

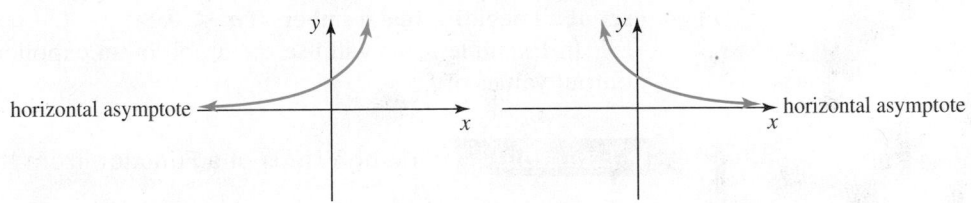

Figure 13 For both exponential functions, the x-axis is a horizontal asymptote

Reflection Property

In Example 4, we will graph two related exponential functions that will help us understand the *reflection property for exponential functions*.

▶ **Example 4** Graphs of Functions of the Form $y = ab^x$ and $y = -ab^x$

1. Sketch and compare the graphs of $f(x) = 5(3)^x$ and $g(x) = -5(3)^x$.
2. Find the domain and range of f.
3. Find the domain and range of g.

Table 13 Input-Output Pairs of $f(x) = 5(3)^x$ and $g(x) = -5(3)^x$

x	$f(x)$	$g(x)$
0	5	−5
1	15	−15
2	45	−45

Solution

1. Input-output pairs of f and g are listed in Table 13 and plotted in Fig. 14. In Table 13, we see that for each value of x, the outputs of g are the opposites of the outputs of f. Because of this, the graph of g is the reflection of the graph of f, with the mirror along the x-axis. We can find the graph of g by reflecting the graph of f across the x-axis.

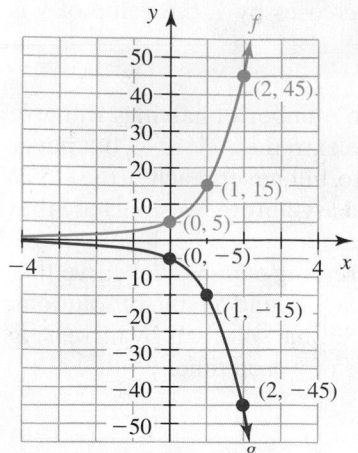

Figure 14 Graphs of $f(x) = 5(3)^x$ and $g(x) = -5(3)^x$

2. The expression $5(3)^x$ is defined for any real number x. So, the domain of f is the set of all real numbers. From Fig. 14, we see that the range of f (the set of all outputs of f) is the set of all positive real numbers.

3. The expression $-5(3)^x$ is defined for any real number x. So, the domain of g is the set of all real numbers. From Fig. 14, we see that the range is the set of all negative real numbers.

▶ **Reflection Property for Exponential Functions**

The graphs of $f(x) = -ab^x$ and $g(x) = ab^x$ are reflections of each other across the x-axis.

We illustrate the reflection property for exponential functions and summarize four types of exponential curves in Figs. 15 and 16. In general, **the graph of an exponential function does not have any x-intercepts and the x-axis is a horizontal asymptote.**

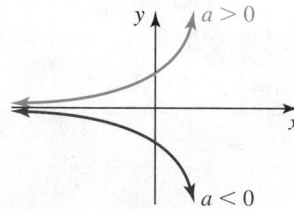

Figure 15 Typical graphs of $f(x) = ab^x$, $b > 1$

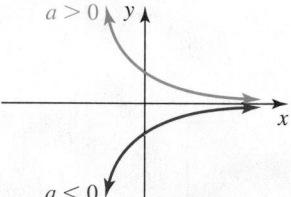

Figure 16 Typical graphs of $f(x) = ab^x$, $0 < b < 1$

Recall that for $b > 0$, b^x has meaning for any real-number exponent x. So, **the domain of any exponential function $f(x) = ab^x$ is the set of all real numbers.**

Further, Figs. 15 and 16 show that an exponential function $f(x) = ab^x$ has positive outputs if $a > 0$ and negative outputs if $a < 0$. Therefore, **the range of an exponential function $f(x) = ab^x$ is the set of all positive real numbers if $a > 0$, and the range is the set of all negative real numbers if $a < 0$.**

In Example 5, we will use the graph of an exponential function f to find input and output values of f.

▶ **Example 5**   Finding Values of a Function from Its Graph

The graph of an exponential function f is shown in Fig. 17.

1. Find $f(2)$.
2. Find x when $f(x) = 2$.
3. Find x when $f(x) = 0$.

Solution

1. The blue arrows in Fig. 17 show that the input $x = 2$ leads to the output $y = 8$. We conclude that $f(2) = 8$.
2. The red arrows in Fig. 17 show that the output $y = 2$ originates from the input $x = -2$. We conclude that $x = -2$ when $f(x) = 2$.
3. Recall that the graph of an exponential function gets close to, but never reaches, the x-axis. So, there is no value of x where $f(x) = 0$.

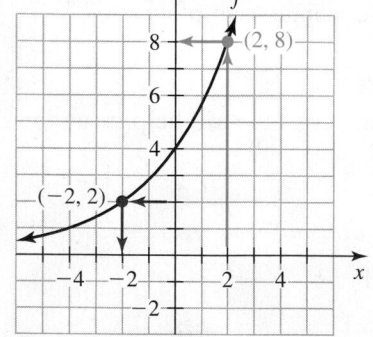

Figure 17 Graph of f

Graphing an Equation of an Exponential Model

In Chapters 6-9, we used linear functions to model authentic situations. In this chapter, we will model authentic situations using exponential functions

> **Definition** **Exponential model**

An **exponential model** is an exponential function, or its graph, that describes an authentic association.

Now that we have discussed how to graph exponential equations in two variables, we can graph equations of exponential models.

> **Example 6** Graphing an Equation of an Exponential Model

The prices of one ounce of gold on the New York Stock Exchange (NYSE) at closing on January 2 are shown in Table 14 for various years. If the NYSE was closed on January 2, then the closing price on the first date it was open was used.

Table 14 Gold Prices

Year	Price (dollars per ounce)
2000	289
2002	278
2004	417
2006	519
2008	856
2010	1102
2012	1587

Source: *Kitco Metals, Inc.*

Let y be the price (in dollars per ounce) of gold on January 2 at x years since 2000.

1. Construct a scatterplot.
2. Describe the shape and the direction of the association.
3. Graph the model $\hat{y} = 289(1.15)^x$ on the scatterplot. Does the model come close to the data points?
4. What is the coefficient a of the model $\hat{y} = 289(1.15)^x$? What does it mean in this situation?
5. What is the base b of the model $\hat{y} = 289(1.15)^x$? What does it mean in this situation?

Solution

1. We construct a scatterplot in Fig. 18.

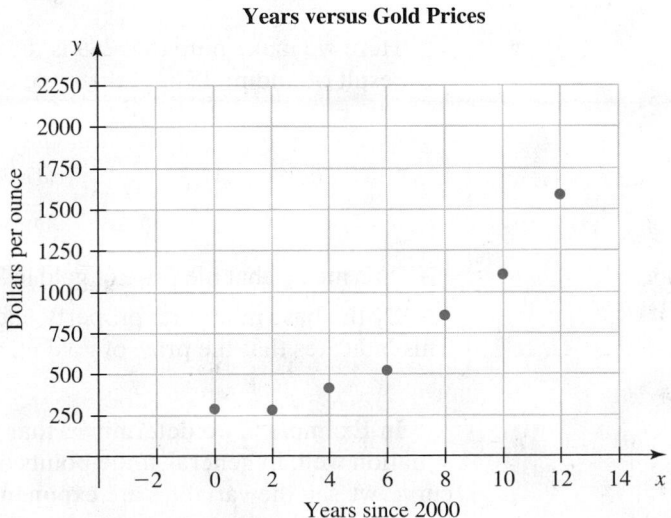

Years versus Gold Prices

Figure 18 Gold scatterplot

2. The association is nonlinear and positive. We suspect an exponential curve will model the situation well, which we will investigate in Problem 3.

3. The x-coordinates of the data points are between 0 and 12, inclusive (see Table 15 and Fig. 18). We will extend the model 2 units (years) on both sides of the data points. So, we will graph the model for values of x between -2 and 14, inclusive. We substitute the values $-2, 2, 6, 10,$ and 14 for x in the equation $\hat{y} = 289(1.15)^x$ and list the solutions in Table 15. Then we plot the (red) points and draw the exponential model from $(-2, 219)$ to $(14, 2045)$ in Fig 19.

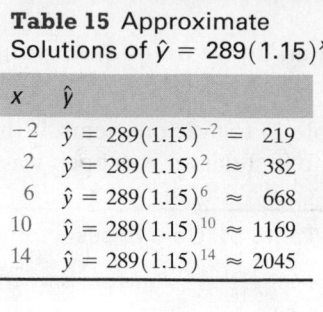

Table 15 Approximate Solutions of $\hat{y} = 289(1.15)^x$

x	$\hat{y}$
-2	$\hat{y} = 289(1.15)^{-2} = 219$
2	$\hat{y} = 289(1.15)^2 \approx 382$
6	$\hat{y} = 289(1.15)^6 \approx 668$
10	$\hat{y} = 289(1.15)^{10} \approx 1169$
14	$\hat{y} = 289(1.15)^{14} \approx 2045$

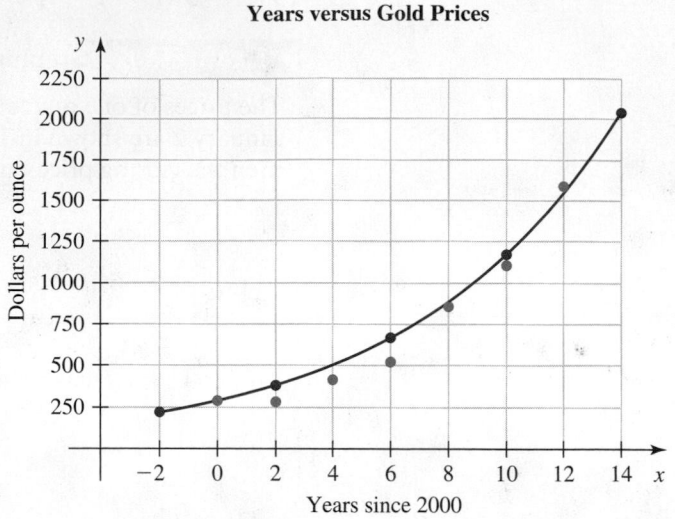

Years versus Gold Prices

Figure 19 Gold scatterplot and exponential model

As we suspected, our model comes close to the data points.

4. The coefficient a of $\hat{y} = 289(1.15)^x$ is 289. So, the y-intercept is $(0, 289)$. This means the price of gold was $289 per ounce on January 2 in 2000. In fact, that was the actual price (see Table 14 on page 663).

5. First, we note that in Problem 4, we found that the price of gold in 2000 was $289 per ounce. Next, we substitute 1 for x in the equation $\hat{y} = 289(1.15)^x$:

$$\hat{y} = 289(1.15)^1 \qquad \textit{Substitue 1 for x.}$$
$$\hat{y} = 289(1.15) \qquad \textit{b}^1 = b$$
$$\hat{y} = 289(1 + 0.15) \qquad 1.15 = 1 + 0.15$$
$$\hat{y} = 289 + 289(0.15) \qquad \textit{Distributive law}$$

Here we make note that 289 is the number of incidents in 2000 and $289(0.15)$ is the result of finding 15% of the price of gold in 2000:

$$\hat{y} = \underbrace{289}_{\substack{\textit{Price of} \\ \textit{gold in 2000}}} + \underbrace{289(0.15)}_{\substack{\textit{15\% of price of} \\ \textit{gold in 2000}}}$$

This means that the price of gold in 2001 was 15% more than the price of gold in 2000.

By the base multiplier property, each time x is increased by 1, y is multiplied by 1.15. This indicates that the price of gold increased by 15% per year, according to the model.

In Example 6, we determined that the exponential curve in Fig. 19 models the gold situation well. In general, if the points of a scatterplot lie close to (or on) an exponential curve, we say the variables are **exponentially associated** and that there is an **exponential association**.

Interpreting the Coefficient and the Base of an Exponential Model

In Problem 4 of Example 6, we found that for the gold model $\hat{y} = 289(1.15)^x$, the coefficient 289 means that the price of gold was $289 per ounce when x is 0.

> **Meaning of the Coefficient of an Exponential Model**

Assume that an association between the variables x and y can be described exactly by the exponential model $\hat{y} = ab^x$. Then the coefficient a equals the quantity y when the quantity x is 0.

In Problem 5 of Example 6, we found that for the gold model $\hat{y} = 289(1.15)^x$, the price of gold grew exponentially by 15% per year. Note that subtracting 1 from the base 1.15 gives the percentage growth rate in decimal form:

$$b - 1 = 1.15 - 1 = 0.15 = \text{percentage growth rate (in decimal form)}$$

> **Meaning of the Base of an Exponential Model**

Assume that an association between the variables x and y can be described exactly by the exponential model $\hat{y} = ab^x$, where $a > 0$. Then the following statements are true.
- If $b > 1$, then the quantity y grows exponentially at a rate $b - 1$ percent (in decimal form) per unit increase of the quantity x.
- If $0 < b < 1$, then the quantity y decays exponentially at a rate of $1 - b$ percent (in decimal form) per unit increase of the quantity x.

It turns out the number (in thousands) of teachers with x years of experience can be modeled well by the exponential model $\hat{y} = 220(0.949)^x$. Because $1 - 0.949 = 0.051$, the base 0.949 indicates that the number of teachers decays exponentially by 5.1% per year of teaching experience.

> **Example 7**   Graphing an Equation of an Exponential Model

Malathion is an insecticide that has been sprayed from helicopters on suburban areas to control mosquitoes and Mediterranean fruit flies. The California Department of Health Services concluded that some people may be sensitive to malathion (see the source shown in Table 16). In an experiment conducted in 1993, researchers sprayed some tomatoes with malathion and recorded how much of the insecticide persisted. The mean malathion concentrations on the surfaces of four groups of tomatoes are shown in Table 16 for various amounts of time.

Table 16 Elapsed Times versus Mean Concentrations of Malathion

Time (hours)	Mean Concentration (micrograms per centimeter squared)
0	0.169
12	0.167
24	0.178
48	0.120
96	0.073
239	0.070
504	0.008

Source: *Assessment of Malathion and Malaoxon Concentration and Persistence in Water, Sand, Soil and Plant Matrices Under Controlled Exposure Conditions,* Neal et al.

Let y be the mean malathion concentration (in micrograms per centimeter squared) on the tomatoes at x hours.

1. Construct a scatterplot.
2. Describe the shape and the direction of the association.

3. Graph the model $\hat{y} = 0.18(0.994)^x$ on the scatterplot Then describe the strength of the association.
4. What is the coefficient of the model $\hat{y} = 0.18(0.994)^x$? What does it mean in this situation?
5. What is the base of the model $\hat{y} = 0.18(0.994)^x$? What does it mean in this situation?
6. Predict the percentage of the malathion that remained on the tomatoes 5 days after spraying.

Solution

1. We construct a scatterplot in Fig. 20.

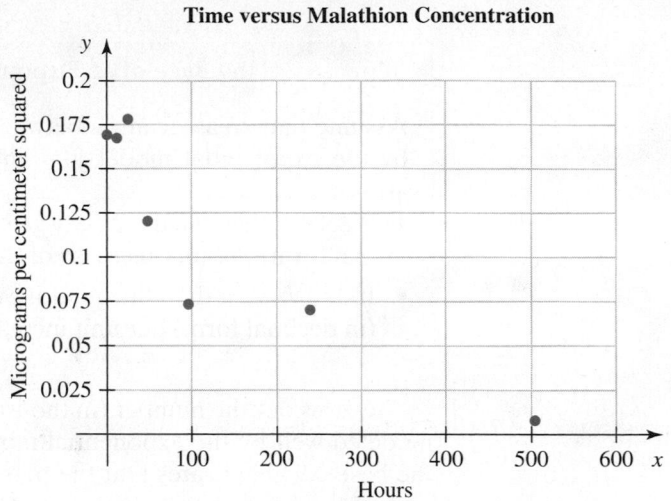

Figure 20 Malathion scatterplot

2. Although the regression line would fit the data points fairly well, it appears that a well-chosen decreasing exponential model might fit them even better (see Fig. 20). We will investigate this further in Problem 3. The association is negative.
3. The x-coordinates of the data points are between 0 and 504, inclusive (see Table 16 and Fig. 20). We often extend a model a bit beyond the data points on both sides, but points to the left of the data point $(0, 0.169)$ have negative x-coordinates, which are not relevant to the experiment. So, we will graph the model for values of x between 0 and 600, inclusive, which is what was done in the study. We substitute the values 0, 100, 200, 300, 400, 500, and 600 for x in Table 17. Then we plot the (red) points and draw the exponential model from $(0, 0.180)$ to $(600, 0.005)$. See Fig. 21.

Table 17 Approximate Solutions of $\hat{y} = 0.18(0.994)^x$

x	$\hat{y}$
0	$\hat{y} = 0.18(0.994)^0 = 0.180$
100	$\hat{y} = 0.18(0.994)^{100} \approx 0.099$
200	$\hat{y} = 0.18(0.994)^{200} \approx 0.054$
300	$\hat{y} = 0.18(0.994)^{300} \approx 0.030$
400	$\hat{y} = 0.18(0.994)^{400} \approx 0.016$
500	$\hat{y} = 0.18(0.994)^{500} \approx 0.009$
600	$\hat{y} = 0.18(0.994)^{600} \approx 0.005$

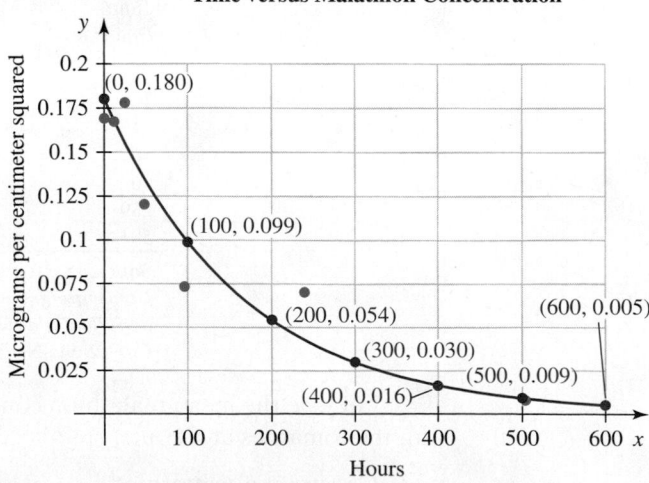

Figure 21 Exponential model

The exponential association is fairly strong.

4. The coefficient of $\hat{y} = 0.18(0.994)^x$ is 0.18. So, the mean malathion concentration was 0.18 microgram per centimeter squared when the tomatoes were sprayed, according to the model. The result is fairly close to the actual mean concentration of 0.169 microgram per centimeter squared (see Table 16 on page 665).

5. The base of $\hat{y} = 0.18(0.994)^x$ is $b = 0.994$. Because $1 - b = 1 - 0.994 = 0.006$, we conclude that the mean malathion concentration decreased by 0.6% per hour.

6. The number of hours in 5 days is $5(24) = 120$ hours. So, we substitute 120 for x in the equation $\hat{y} = 0.18(0.994)^x$:

$$\hat{y} = 0.18(0.994)^{120} \approx 0.0874$$

The mean malathion concentration was about 0.0874 microgram per centimeter squared, according to the model.

This means the proportion of malathion that remained was $\dfrac{0.0874}{0.169} \approx 0.517$. Thus, 51.7% of the malathion remained after 5 days, according to the model.

Group Exploration

Section Opener: Numerical significance of *a* and *b* for $f(x) = ab^x$

In this exploration, you will investigate the nature of exponential functions of the form $f(x) = ab^x$.

1. Use technology to create a table of ordered pairs for $f(x) = 2(3)^x$, $g(x) = 64\left(\dfrac{1}{2}\right)^x$, and a third exponential function of your choice. Use the following values for the *x*-coordinates: $0, 1, 2, 3, 4, 5$, and 6.

2. **a.** What connection do you notice between the *y*-coordinates of each function and the base *b* of the function $y = ab^x$?

   b. Test the connection you described in part (a) by choosing yet another exponential function, and check whether it behaves as you think it should.

c. For $f(x) = ab^x$, we have $f(0) = a$, $f(1) = ab$, $f(2) = abb$, and $f(3) = abbb$. Explain why these results suggest that your response to part (a) is correct.

3. **a.** What connection do you notice between the *y*-coordinates of each function and the coefficient *a* of the function $y = ab^x$?

   b. Test the connection you described in part (a) by choosing yet another exponential function, and check whether it behaves as you think it should.

   c. Use pencil and paper to find $f(0)$, where $f(x) = ab^x$. Explain why your result shows that your response to part (a) is correct.

Group Exploration

Section Opener: Graphical significance of *a* and *b* for $y = ab^x$

1. Use technology to graph these equations of the form $y = b^x$ in order, and describe what you observe:

$$y = 1.2^x, \quad y = 1.5^x, \quad y = 2^x, \quad \text{and} \quad y = 5^x$$

Do the same with the equations

$$y = 0.3^x, \quad y = 0.5^x, \quad y = 0.7^x, \quad \text{and} \quad y = 0.9^x$$

2. Use technology to graph these equations of the form $y = a(1.1)^x$ in order, and describe what you observe:

$$y = 2(1.1)^x, \quad y = 3(1.1)^x, \quad y = 4(1.1)^x,$$
$$\text{and} \quad y = 5(1.1)^x$$

Do the same with the equations

$$y = -2(1.1)^x, \quad y = -3(1.1)^x, \quad y = -4(1.1)^x,$$
$$\text{and} \quad y = -5(1.1)^x$$

3. So far, you have sketched the graphs of equations of only the forms $y = b^x$ (where $a = 1$) and $y = a(1.1)^x$ (where $b = 1.1$). Graph more equations of the form $y = ab^x$, until you are confident you know the graphical significance of the constants *a* and *b*, for any possible combination of values of *a* and *b*. If you have any new insights into the graphical significance of *a* and *b*, describe those insights.

4. Describe the graph of $y = ab^x$ in the following situations.
a. a is positive.   **b.** a is negative.
c. $b > 1$ **d.** $0 < b < 1$
e. $b = 1$ **e.** b is negative.

5. Describe the connection between the y-intercept of the graph of $y = ab^x$ and the values of a and b.

Group Exploration

Drawing families of exponential curves

For each problem, use technology to graph a family of curves.

1. List the equations of a family of exponential curves like the ones shown in Fig. 22.

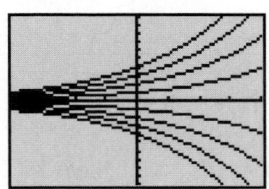

Figure 22 A family of exponential curves

2. List the equations of a family of exponential curves like the ones shown in Fig. 23. All of these curves pass through the point $(0, 2)$.

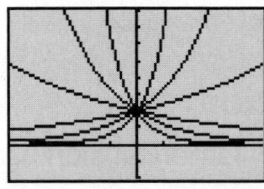

Figure 23 A family of exponential curves passing through $(0, 2)$

3. Summarize what you have learned from this exploration and this section about the coefficient a and the base b in functions of the form $f(x) = ab^x$.

Homework 10.3

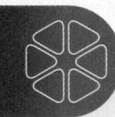

For extra help ▶ MyMathLab®   Watch the videos in MyMathLab   Download the MyDashboard App

1. For an exponential function of the form $y = ab^x$, if the value of x increases by 1, then the value of y is multiplied by _____.

2. Let $f(x) = ab^x$, where $a > 0$. If $0 < b < 1$, then the function f is _____.

3. For the graph of an exponential function of the form $y = ab^x$, the y-intercept is _____.

4. An exponential _____ is an exponential function, or its graph, that describes an authentic association.

Graph the equation by hand.

5. $y = 3^x$ **6.** $y = 4^x$ **7.** $y = 10^x$

8. $y = 5^x$ **9.** $y = 3(2)^x$ **10.** $y = 2(3)^x$

11. $y = 6(3)^x$ **12.** $y = 3(5)^x$

13. $y = 15\left(\dfrac{1}{3}\right)^x$

14. $y = 20\left(\dfrac{1}{4}\right)^x$ **15.** $y = 12\left(\dfrac{1}{2}\right)^x$ **16.** $y = 18\left(\dfrac{1}{3}\right)^x$

Graph both functions by hand on the same coordinate system.

17. $f(x) = 2^x, g(x) = -2^x$

18. $f(x) = 3^x, g(x) = -3^x$

19. $f(x) = 4(3)^x, g(x) = -4(3)^x$

20. $f(x) = 2(10)^x, g(x) = -2(10)^x$

21. $f(x) = 8\left(\dfrac{1}{2}\right)^x, g(x) = -8\left(\dfrac{1}{2}\right)^x$

22. $f(x) = 6\left(\dfrac{1}{3}\right)^x, g(x) = -6\left(\dfrac{1}{3}\right)^x$

For Exercises 23–26, graph the function by hand. Find the domain and range of the function.

23. $f(x) = 5(2)^x$

24. $f(x) = -3(3)^x$

25. $f(x) = -8\left(\dfrac{1}{4}\right)^x$

26. $f(x) = 9\left(\dfrac{1}{3}\right)^x$

27. Recall that we can describe some or all of the input–output pairs of a function by means of an equation, a graph, a table, or words. Let $f(x) = 4(2)^x$.

   a. Describe five input–output pairs of f by using a table.
   b. Describe the input–output pairs of f by using a graph.
   c. Describe the input–output pairs of f by using words.

28. Recall that we can describe some or all of the input–output pairs of a function by means of an equation, a graph, a table, or words. Let $g(x) = 16\left(\frac{1}{2}\right)^x$.

   a. Describe five input–output pairs of g by using a table.
   b. Describe the input–output pairs of g by using a graph.
   c. Describe the input–output pairs of g by using words.

29. Input–output pairs of four exponential functions are listed in Table 18. Complete the table.

Table 18 Complete the Table (Exercise 29)

x	$f(x)$	$g(x)$	$h(x)$	$k(x)$
0	162	3	2	800
1	54	12	10	400
2	18	48		
3	6			
4				

30. Input–output pairs of four exponential functions are listed in Table 19. Complete the table.

Table 19 Complete the Table (Exercise 30)

x	$f(x)$	$g(x)$	$h(x)$	$k(x)$
0	3	64	2	100
1	6	32	6	10
2	12	16		
3	24			
4				

31. Input–output pairs of four exponential functions are listed in Table 20. Complete the table.

Table 20 Complete the Table (Exercise 31)

x	$f(x)$	$g(x)$	$h(x)$	$k(x)$
0	5			
1		80	54	
2	20			
3		20		192
4			2	768

32. Input–output pairs of four exponential functions are listed in Table 21. Complete the table.

Table 21 Complete the Table (Exercise 32)

x	$f(x)$	$g(x)$	$h(x)$	$k(x)$
0			3	400
1		3		
2	25			
3		147		
4	1		30,000	25

For Exercises 33–40, refer to Fig. 24.

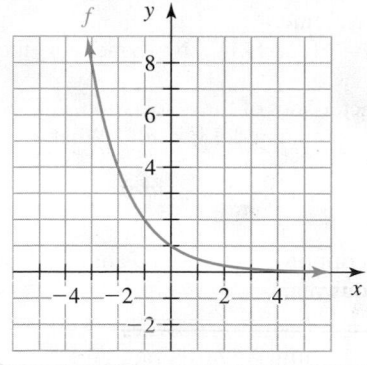

Figure 24 Graph of f—Exercises 33–40

33. Find $f(-3)$.
34. Find $f(-1)$.
35. Find $f(0)$.
36. Estimate $f(1)$.
37. Find x when $f(x) = 4$.
38. Find x when $f(x) = 2$.
39. Find x when $f(x) = 1$.
40. Find x when $f(x) = -2$.

For Exercises 41–48, refer to Table 22.

41. Find $f(3)$. 42. Find $f(6)$.
43. Find $f(5)$. 44. Find $f(0)$.
45. Find x when $f(x) = 3$. 46. Find x when $f(x) = 6$.
47. Find x when $f(x) = 24$. 48. Find x when $f(x) = 96$.

Table 22 Some Values of an Exponential Function f (Exercises 41–48)

x	$f(x)$
0	3
1	6
2	12
3	24
4	48
5	96
6	192

49. The numbers of firearms discovered at TSA checkpoints at U.S. airports are shown in Table 23 for various years.

Table 23 Numbers of Firearms Discovered at TSA Checkpoints

Year	Number of Firearms
2005	660
2007	803
2009	976
2011	1320
2013	1813
2014	2212

Source: *Transportation Security Administration*

Let y be the number of firearms that were discovered at TSA checkpoints in the year that is x years since 2005.

a. Construct a scatterplot by hand.

b. Graph the model $\hat{y} = 618(1.144)^x$ by hand on the scatterplot.

c. Describe the four characteristics of the association.

d. What is the coefficient a of the model $\hat{y} = ab^x$? What does it mean in this situation?

e. What is the base b of the model $\hat{y} = ab^x$? What does it mean in this situation?

50. The advertising and subscription revenues of Pandora® are shown in Table 24 for various years.

Table 24 Pandora's Annual Advertising and Subscription Revenues

| Year | Revenue (millions of dollars) | |
	Advertising	Subscriptions
2006	3.25	0.93
2007	13.31	0.99
2008	18.25	1.09
2009	50.15	5.04
2010	119.33	18.43
2011	239.96	34.38
2012	375.22	51.93

Source: *Pandora*

Let y be the annual advertising revenue (in millions of dollars) at x years since 2005.

a. Construct by hand a scatterplot that describes the association between x and y.

b. Graph the model $\hat{y} = 2(2.19)^x$ by hand on the scatterplot.

c. Describe the four characteristics of the association.

d. What is the y-intercept? What does it mean in this situation? Do you have much faith in your estimate?

e. The annual subscription revenue can be modeled fairly well by the equation $\hat{s} = 0.24(2.19)^x$, where s is the annual subscription revenue (in millions of dollars) at x years since 2005. Compare the bases of the subscription and advertising models. What does your comparison mean in this situation?

51. The numbers of Russian children adopted by American families are shown in Table 25 for various years.

Table 25 Numbers of Russian Children Adopted by American Families

Year	Number of Children
2004	5862
2006	3702
2008	1857
2010	1079
2011	962

Source: *U.S. Department of State*

Let $f(x)$ be the number of Russian children adopted by American families in the year that is x years since 2000.

a. Construct a scatterplot by hand.

b. Graph the model $\widehat{f(x)} = 17{,}555(0.76)^x$ by hand on the scatterplot. Then describe the shape and the strength of the association.

c. What is the base of the model $\widehat{f(x)} = 17{,}555(0.76)^x$? What does it mean in this situation?

d. Find $f(13)$. What does it mean in this situation? Do you have much faith in your result? Explain.

e. In 2013, Russian banned the adoption of Russian children by American families. How does this relate to your responses to part (d)?

52. The numbers of men's colleges, not including seminaries, are shown in Table 26 for various years.

Table 26 Numbers of Men's Colleges

Year	Number of Men's Colleges
1967	145
1975	80
1985	27
1995	11
2012	4

Source: *National Association of Independent Colleges and Universities*

Let $f(x)$ be the number of men's colleges at x years since 1960.

a. Construct a scatterplot by hand.

b. Graph the model $\widehat{f(x)} = 242(0.92)^x$ by hand on the scatterplot. Then describe the shape and the strength of the association.

c. What is the coefficient of the model $\widehat{f(x)} = 242(0.92)^x$? What does it mean in this situation? Do you have much faith in your estimate?

d. What is the base of the model $\widehat{f(x)} = 242(0.92)^x$? What does it mean in this situation?

e. Find $f(47)$. What does it mean in this situation?

53. The acceptance rates by 10 of the most selective colleges and universities are shown in Table 27 for various groups of SAT scores.

Table 27 Acceptance Rates by 10 of the Most Selective Colleges and Universities

SAT Score Group (points)	Score Used to Represent SAT Score Group (points)	Percent
1100–1190	1150	10
1200–1290	1250	17
1300–1390	1350	31
1400–1490	1450	48
1500–1600	1550	72

Source: *Professor Christopher Avery, Kennedy School of Government, Harvard University*

For students who score x points, let y be the percentage of applicants who are accepted.

a. Construct a scatterplot by hand.

b. Graph the equation $\hat{y} = 0.034(1.005)^x$ by hand on the scatterplot.

c. What is the base of the model $\hat{y} = 0.034(1.005)^x$? What does it mean in this situation?

d. What percentage of applicants who score 1425 points get accepted?

54. The percentages of seniors with severe memory impairment (based on memory tests) are shown in Table 28.

Table 28 Percentages of Seniors with Severe Memory Impairment

Age Group (years)	Age Used to Represent Age Group (years)	Percent
65-69	67	1.1
70-74	72	2.5
75-79	77	4.5
80-84	82	6.4
over 84	88	12.9

Source: *Federal Interagency Forum on Aging-Related Statistics*

Let y be the percentage of seniors at age x years with severe memory impairment.

a. Construct a scatterplot by hand.

b. Graph the equation $\hat{y} = 0.00067(1.119)^x$ by hand on the scatterplot.

c. Describe the four characteristics of the association.

d. What is the base b of the model $\hat{y} = ab^x$? What does it mean in this situation?

e. In Exercise 37 of Homework 9.3, you may have found that there is a linear, negative association between adults' ages and mean scores on a test measuring memory and information processing speed. Would that *linear* association necessarily conflict with an *exponential* association between a senior's age and the percentages of seniors with severe memory impairment? Explain.

Find the x- and y-intercepts of the graph of the function.

55. $y = 7^x$

56. $y = 2(5)^x$

57. $y = 3\left(\dfrac{1}{5}\right)^x$

58. $y = -9\left(\dfrac{2}{3}\right)^x$

For Exercises 59–62, let $f(x) = 2^x + 3^x$.

59. Find $f(2)$.

60. Find $f(0)$.

61. Find $f(-2)$.

62. Find $f(-1)$.

For Exercises 63–66, let $f(x) = 3^x$.

63. Find x when $f(x) = 3$.

64. Find x when $f(x) = 9$.

65. Find x when $f(x) = 1$.

66. Find x when $f(x) = \dfrac{1}{3}$.

For Exercises 67–78, use TI-84 tables or other technology to compare each pair of functions f and g. What do you observe? Use exponential properties to show why this is so. [TI-84: For 2^{3x}, press 2 $\boxed{\wedge}\,\boxed{(}\,\boxed{3}\,\boxed{X, T, \Theta, n}\,\boxed{)}\,\boxed{ENTER}$.]

67. $f(x) = 3^x 3^x, g(x) = 3^{2x}$

68. $f(x) = \dfrac{3^{2x}}{3^x}, g(x) = 3^x$

69. $f(x) = 2^{3x}, g(x) = 8^x$

70. $f(x) = 2^{-x}, g(x) = \left(\dfrac{1}{2}\right)^x$

71. $f(x) = 2^{x+3}, g(x) = 8(2)^x$

72. $f(x) = \dfrac{6^x}{3^x}, g(x) = 2^x$

73. $f(x) = 2^0, g(x) = 3^0$

74. $f(x) = 2^x 3^x, g(x) = 6^x$

75. $f(x) = 5^{x/3}, g(x) = (5^{1/3})^x$

76. $f(x) = 2^x, g(x) = 8^{x/3}$

77. $f(x) = x^{1/2}, g(x) = \sqrt{x}$ [**TI-84:** For $\sqrt{x}$, press $\boxed{\text{2nd}}$ $\boxed{x^2}$ $\boxed{X, T, \Theta, n}$ $\boxed{ENTER}$.]

78. $f(x) = 25^{x/2} \cdot 5^x, g(x) = 25^x$

Concepts

79. Graphs of four functions of the form $y = ab^x$ are shown in Fig. 25. Describe the constants a and b of each function.

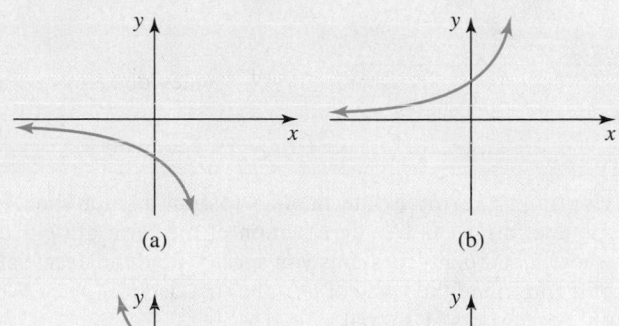

(a)   (b)

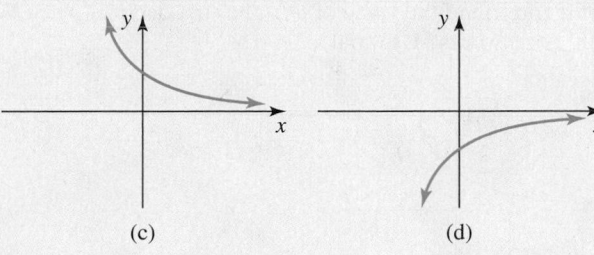

(c)   (d)

Figure 25 Exercise 79

80. The graphs of functions $f(x) = ab^x$ and $g(x) = cd^x$ are shown in Fig. 26.

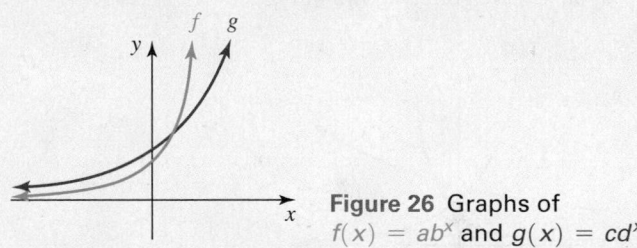

Figure 26 Graphs of $f(x) = ab^x$ and $g(x) = cd^x$

a. Which coefficient is greater, a or c? Explain.

b. Which base is greater, b or d? Explain.

81. Use technology to graph a family of exponential curves similar to the family graphed in Fig. 27. List the equations of that family.

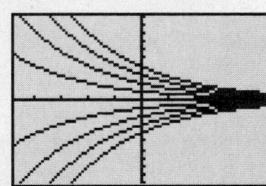

Figure 27 A family of exponential curves

82. Use technology to graph a family of exponential curves similar to the family graphed in Fig. 28. All of these curves pass through the point $(0, -2)$. List the equations of that family.

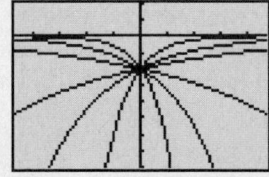

Figure 28 A family of exponential curves passing through $(0, -2)$

83. Use trial and error with technology to draw a graph similar to the one in Fig. 29. Use an equation of the form $f(x) = ab^x$, where a and b are constants you specify. Find the exact value of a and round the value of b to the first decimal place. What equation works?

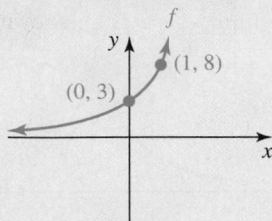

Figure 29 Exercise 83

84. Use trail and error with technology to draw a graph similar to the one in Fig. 30. Use an equation of the form $g(x) = ab^x$, where a and b are constants you specify. Find the exact value of a and round the value of b to the first decimal place. What equation works? Use trial and error.

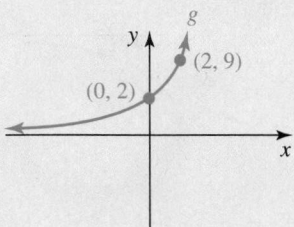

Figure 30 Exercise 84

85. Find equations of exponential functions that could correspond to the graphs shown in Fig. 31.

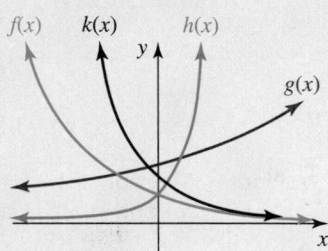

Figure 31 Exercise 85

86. a. For each part that follows, simplify and compare both expressions.

   i. $4(3)^2$ and 12^2 **ii.** $4^2 \cdot 3^2$ and 12^2

b. For each part that follows, build an input–output table that shows the same input values for both functions. Explain in terms of order of operations or exponential properties why the tables are the same or different.

   i. $f(x) = 4(3)^x$ and $g(x) = 12^x$
   ii. $f(x) = 4^x \cdot 3^x$ and $g(x) = 12^x$

87. In this exercise, you will compare the function $f(x) = 100(2)^x$ with the function $g(x) = 5(3)^x$.

a. Find the y-intercept of the graph of each function.

b. What does the base multiplier property tell you about each function?

c. On the basis of your comments in parts (a) and (b), which function's outputs will eventually be much greater than the other's outputs? Explain.

d. Use technology to construct an input–output table to verify your responses to parts (a)–(c).

88. What are the x-intercepts and y-intercepts of the graph of a function of the form $y = ab^x$, where $b > 0$?

89. Is the statement true for $f(x) = 2^x$?

a. $f(3 + 4) = f(3) + f(4)$ **b.** $f(x + y) = f(x) + f(y)$

90. Is the statement true for $g(x) = 3^x$?

a. $g(2 + 5) = g(2) \cdot g(5)$ **b.** $g(4 + 6) = g(4) \cdot g(6)$
c. $g(2 + 4) = g(2) \cdot g(4)$ **d.** $g(x + y) = g(x) \cdot g(y)$

91. Let $f(x) = ab^x$, where $a > 0$. Explain why f is increasing if $b > 1$ and f is decreasing if $0 < b < 1$.

92. In an exponential function $f(x) = b^x$, the base b is a positive number not equal to 1. In this exercise, you will explore what happens if we try to define a function whose base is negative. Consider $f(x) = (-4)^x$.

a. Explain why $f\left(\dfrac{1}{2}\right)$ is undefined.

b. Explain why $f\left(\dfrac{1}{4}\right)$ is undefined.

c. List three more values of x that result in undefined outputs.

93. The graphs of the exponential functions $f(x) = -ab^x$ and $g(x) = ab^x$ are reflections of each other across the x-axis. Explain why this makes sense.

94. Explain how to sketch the graph of a function of the form $f(x) = ab^x$, where $b > 0$. Include the effect of a value of a or b on the graph.

▼ 10.4 Using Two Points to Find an Equation of an Exponential Model

Objectives

» Solve an equation of the form $ab^n = k$ for the base b.

» Use two points to find an equation of an exponential curve.

» Use two points to find an equation of an exponential model.

In Section 10.3, we graphed exponential equations in two variables. In this section, we will go backward. That is, we will find equations of exponential curves. We will then use that skill to find equations of exponential models.

Solving Equations of the Form $ab^n = k$ for b

Before we can find equations of exponential curves, we need to determine how to solve equations of the form $ab^n = k$ for the base b.

▶ **Example 1** One-Variable Equations Involving Exponents

Find all real-number solutions.

1. $b^2 = 25$ **2.** $b^3 = 8$ **3.** $2b^4 = 32$
4. $10b^5 = 90$ **5.** $b^6 = -28$

Solution

1. $b^2 = 25$ *Original equation*

$b = -5$ or $b = 5$ $(-5)^2 = 25$ *and* $5^2 = 25$

So, the solutions are -5 and 5. We can use the notation ± 5 to stand for the numbers -5 and 5.

2. $b^3 = 8$ *Original equation*

$b = 2$ $2^3 = 8$

3. $2b^4 = 32$ *Original equation*

$b^4 = 16$ *Divide both sides by 2.*

$b = \pm 2$ $(-2)^4 = 16$ *and* $2^4 = 16$

We can check that both -2 and 2 satisfy the equation $2b^4 = 32$ (try it).

4. $10b^5 = 90$ *Original equation*

$b^5 = 9$ *Divide both sides by 10.*

$b = 9^{1/5}$ $9^{1/5}$ *is the number whose 5th power is 9.*

$b \approx 1.55$ $1.55^5 \approx 9$

We can check that 1.55 approximately satisfies the equation $10b^5 = 90$ by using a TI-84 to verify that $10(1.55)^5 \approx 90$ (see Fig. 32).

5. The equation $b^6 = -28$ has no real-number solutions, since an even-numbered exponent gives a positive number.

```
10(1.55)⁵
        89.46609688
```

Figure 32 Checking that 1.55 approximately satisfies $10b^5 = 90$

The problems in Example 1 suggest how to solve equations of the form $b^n = k$ for b.

▶ **Solving Equations of the Form $b^n = k$ for b**

To solve an equation of the form $b^n = k$ for b,

1. If n is odd, the real-number solution is $k^{1/n}$.

2. If n is even and $k \geq 0$, the real-number solutions are $\pm k^{1/n}$.

3. If n is even and $k < 0$, there is no real-number solution.

▶ **Example 2** One-Variable Equations Involving Exponents

Find all real-number solutions. Round any results to the second decimal place.

1. $5.42b^6 - 3.19 = 43.74$ **2.** $\dfrac{b^9}{b^4} = \dfrac{70}{3}$

Solution

1. $5.42b^6 - 3.19 = 43.74$ *Original equation*

$5.42b^6 = 43.74 + 3.19$ *Add 3.19 to both sides.*

$5.42b^6 = 46.93$ *Add.*

$b^6 = \dfrac{46.93}{5.42}$ *Divide both sides by 5.42.*

$b = \pm \left(\dfrac{46.93}{5.42}\right)^{1/6}$ *The solutions of $b^6 = k$ are $\pm k^{1/6}$ if $k \geq 0$.*

$b \approx \pm 1.43$ *Compute.*

2. $\dfrac{b^9}{b^4} = \dfrac{70}{3}$ *Original equation*

    $b^5 = \dfrac{70}{3}$ *Subtract exponents:* $\dfrac{b^m}{b^n} = b^{m-n}$

    $b = \left(\dfrac{70}{3}\right)^{1/5}$ *The solution of $b^5 = k$ is $k^{1/5}$.*

    $b \approx 1.88$ *Compute.*

Using Two Points to Find an Equation of an Exponential Curve

Now that we have discussed how to solve equations of the form $ab^n = k$ for b, we can find an equation of an exponential curve.

▶ Example 3   Finding an Equation of an Exponential Curve

Find an approximate equation $y = ab^x$ of the exponential curve that contains the points $(0, 3)$ and $(4, 70)$. Round the value of b to two decimal places.

Solution

Because the y-intercept is $(0, 3)$, the equation has the form $y = 3b^x$. Next, we substitute $(4, 70)$ in the equation $y = 3b^x$ and solve for b:

$70 = 3b^4$ *Substitute 4 for x and 70 for y.*

$3b^4 = 70$ *If $c = d$, then $d = c$.*

$b^4 = \dfrac{70}{3}$ *Divide both sides by 3.*

$b = \pm\left(\dfrac{70}{3}\right)^{1/4}$ *The solutions of $b^4 = k$ are $\pm k^{1/4}$ if $k \geq 0$.*

$b \approx 2.20$ *Compute; base of an exponential function is positive.*

So, our equation is $y = 3(2.20)^x$; its graph contains the given point $(0, 3)$. Because we rounded the value b, the graph of the equation comes close to, but does not pass through, the given point $(4, 70)$.

We use a TI-84 to verify our work (see Fig. 33).

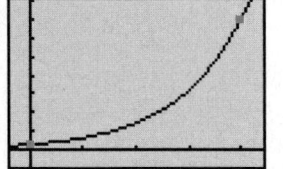

Figure 33 Verify the work

In Example 4, we will find an equation of a curve that approximates the exponential curve containing two given points. Neither point will be the y-intercept. To do this, we will use the following property.

▶ Dividing Left Sides and Right Sides of Two Equations

If $a = b$, $c = d$, $c \neq 0$, and $d \neq 0$, then

$$\frac{a}{c} = \frac{b}{d}$$

In words, the quotient of the left sides of two equations is equal to the quotient of the right sides.

For example, if we divide the left sides and divide the right sides of the equations $2 = 2$ and $3 = 3$, we obtain the true statement $\dfrac{2}{3} = \dfrac{2}{3}$.

▶ **Example 4** Finding an Equation of an Exponential Curve

Find an approximate equation $y = ab^x$ of the exponential curve that contains $(2, 5)$ and $(5, 63)$. Round the values of a and b to two decimal places.

Solution

Since both of the ordered pairs $(2, 5)$ and $(5, 63)$ must satisfy the equation $y = ab^x$, we have the following system of equations:

$$5 = ab^2 \quad \textit{Substitute 2 for x and 5 for y.}$$
$$63 = ab^5 \quad \textit{Substitute 5 for x and 63 for y.}$$

It will be slightly easier to solve this system if we switch the equations to list the equation with the greater exponent of b first:

$$63 = ab^5$$
$$5 = ab^2$$

We divide the left sides and divide the right sides of the two equations to get the following result for nonzero a and b:

$$\frac{63}{5} = \frac{ab^5}{ab^2}$$

By then applying the properties $\dfrac{b^m}{b^n} = b^{m-n}$ and $\dfrac{a}{a} = 1$, where a and b are nonzero, to the right-hand side of the equation, we have an equation in terms of b (and not a):

$$\frac{63}{5} = b^3$$

We can now solve for b by finding the cube root of $\dfrac{63}{5}$:

$$b^3 = \frac{63}{5} \qquad \textit{If c = d, then d = c.}$$

$$b = \left(\frac{63}{5}\right)^{1/3} \qquad \textit{The solution of } b^3 = k \textit{ is } k^{1/3}.$$

$$\approx 2.33 \qquad \textit{Compute.}$$

So, we can substitute 2.33 for the constant b in the equation $y = ab^x$:

$$y = a(2.33)^x$$

To find a, we substitute the coordinates of the given point $(2, 5)$ into $y = a(2.33)^x$:

$$5 = a(2.33)^2 \quad \textit{Substitute 2 for x and 5 for y.}$$

$$\frac{5}{2.33^2} = a \qquad \textit{Divide both sides by } 2.33^2.$$

$$a \approx 0.92 \qquad \textit{Compute.}$$

So, an equation that approximates the exponential curve that passes through $(2, 5)$ and $(5, 63)$ is $y = 0.92(2.33)^x$.

We use a TI-84 to verify our work (see Fig. 34).

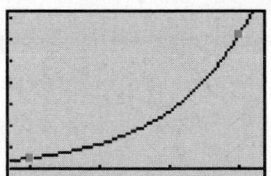

Figure 34 Check that the curve approximately contains $(2, 5)$ and $(5, 63)$

Using Two Points to Find an Equation of an Exponential Model

In Example 5, we will find an equation of an exponential model by using the model's y-intercept and another point.

Table 29 Tuition Rates at Princeton

Year	Tuition Rate (thousands of dollars)
1950	0.6
1960	1.5
1970	2.4
1980	5.6
1990	14.4
2000	24.6
2010	35.3
2015	41.8

Source: *Princeton University*

▶ Example 5   Finding an Equation of an Exponential Model

The tuition rates at Princeton University are shown in Table 29 for the academic years ending in the displayed year. Let y be the tuition rate (in thousands of dollars) at x years since 1950.

1. Construct a scatterplot.
2. Find an equation of a model.
3. What is the base b of the model $\hat{y} = ab^x$? What does it mean in this situation?
4. Estimate the tuition in 2012.

Solution

1. We construct a scatterplot in Fig. 35.

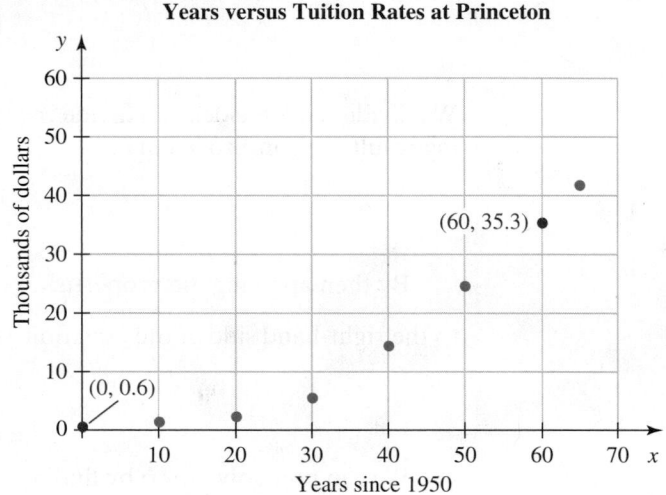

Figure 35 Princeton tuition rate scatterplot

2. If we imagine an exponential curve that contains the data points $(0, 0.6)$ and $(60, 35.3)$, it appears that the curve might come close to the other data points (see Fig. 35). We will find an equation of the curve. Because the y-intercept is $(0, 0.6)$, the equation has the form $\hat{y} = 0.6b^x$. Next, we substitute $(60, 35.3)$ in the equation $\hat{y} = 0.6b^x$ and solve for b:

$$35.3 = 0.6b^{60} \qquad \text{Substitute 60 for x and 35.3 for y.}$$

$$0.6b^{60} = 35.3 \qquad \text{If } c = d, \text{ then } d = c.$$

$$b^{60} = \frac{35.3}{0.6} \qquad \text{Divide both sides by 0.6.}$$

$$b = \pm\left(\frac{35.3}{0.6}\right)^{1/60} \qquad \text{The solution of } b^{60} = k \text{ is } \pm k^{1/60}.$$

$$b \approx 1.07 \qquad \text{Compute; the base of an exponential function is positive.}$$

So the equation is $\hat{y} = 0.6(1.07)^x$. We use a TI-84 to check that our exponential curve contains the point $(0, 0.6)$, approximately contains the point $(60, 35.3)$, and comes close to the other data points (see Fig. 36).

3. The base of $\hat{y} = 0.6(1.07)^x$ is $b = 1.07$. Because $1.07 - 1 = 0.07$, the base indicates that the tuition rate increased by 7% per year.

4. To estimate the tuition rate in 2012, we substitute 62 for x in the equation $\hat{y} = 0.6(1.07)^x$:

$$\hat{y} = 0.6(1.07)^{62} = 39.81$$

In 2012, the tuition rate was $39.81 thousand, according to the model.

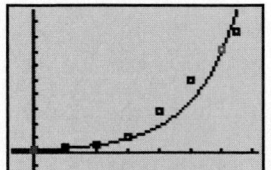

Figure 36 Princeton tuition rate scatterplot and model

In Example 6, we will again use two points to find an equation of an exponential model, but this time neither point will be the *y*-intercept.

▶ Example 6   Finding an Equation of an Exponential Model

The numbers of viewers of the Major League Baseball (MLB) All-Star Game are shown in Table 30 for various years. Let *y* be the number (in millions) of viewers at *x* years since 1980.

1. Construct a scatterplot.
2. Is the association positive, negative, or neither? What does it mean in this situation?
3. Find an equation of a model.
4. What is the coefficient *a* of the model $\hat{y} = ab^x$? What does it mean in this situation?
5. What is the base *b* of the model $\hat{y} = ab^x$? What does it mean in this situation?

Table 30 Numbers of Viewers of MLB All-Star Game

Year	Numbers of Viewers (millions)
1982	34
1985	28
1990	24
1995	20
2000	15
2005	12
2010	12
2014	11

Source: *Nielsen Media Research, Inc.*

Solution

1. We construct a scatterplot in Fig. 37.

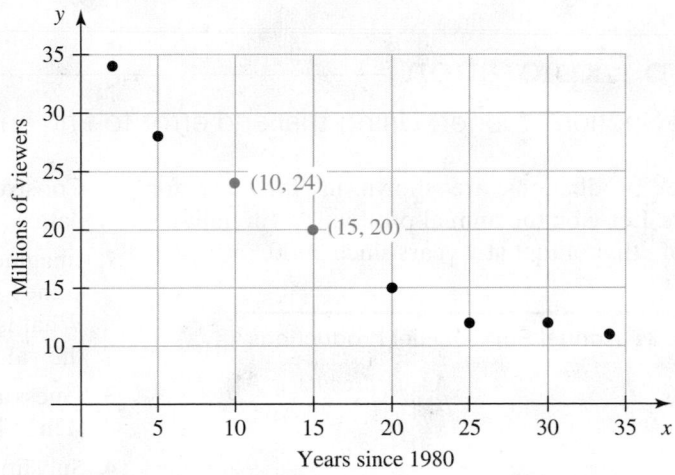

Years versus Numbers of Viewers

Figure 37 MLB All-Star Game scatterplot

2. The association is negative, which means the number of viewers decreased.
3. The association appears to be exponential (see Fig. 37). If we imagine an exponential curve that contains the data points $(10, 24)$ and $(15, 20)$, it appears that the curve might come close to the other data points. To find an equation of this curve, we substitute the coordinates of the points $(10, 24)$ and $(15, 20)$ into the equation $\hat{y} = ab^x$:

$$20 = ab^{15}$$
$$24 = ab^{10}$$

Next, we divide the two left sides and divide the two right sides and solve for *b*:

$$\frac{20}{24} = \frac{ab^{15}}{ab^{10}} \qquad \text{Divide left sides and divide right sides.}$$

$$\frac{20}{24} = b^5 \qquad \text{Simplify; subtract exponents: } \frac{b^m}{b^n} = b^{m-n}$$

$$b = \left(\frac{20}{24}\right)^{1/5} \qquad \text{The solution of } b^5 = k \text{ is } k^{1/5}.$$

$$\approx 0.964 \qquad \text{Compute; base of an exponential function is positive.}$$

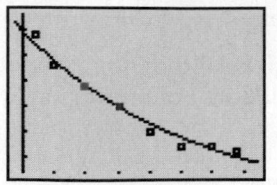

Figure 38 Verify the model

So, an equation is $\hat{y} = a(0.964)^x$. To find a, we substitute the coordinates of $(15, 20)$ into the equation:

$$20 = a(0.964)^{15} \quad \text{Substitute 15 for } x \text{ and 20 for } \hat{y}.$$

$$a = \frac{20}{0.964^{15}} \quad \text{Divide both sides by } 0.964^{15}.$$

$$\approx 34.66 \quad \text{Compute.}$$

The equation is $\hat{y} = 34.66(0.964)^x$. We verify our equation by using a TI-84 to check that our exponential curve approximately contains the points $(10, 24)$ and $(15, 20)$ and comes close to the other data points (see Fig. 38).

4. The coefficient of $\hat{y} = 34.66(0.964)^x$ is $a = 34.66$. So, the y-intercept is $(0, 34.66)$. This means there were about 34.66 million viewers in 1980. Because we have performed extrapolation, we would usually have little or no faith that this estimate is correct. However, a little research would show that the actual number of viewers was 36 million, so the model's estimate is fairly accurate.

5. The base of $\hat{y} = 34.66(0.964)^x$ is $b = 0.964$. Because $1 - 0.964 = 0.036$, we conclude that the number of viewers is decaying exponentially by 3.6% per year.

Group Exploration

DATA Section Opener: Using trial and error to find an equation of a model

Ethanol fuel productions are shown in Table 31 for various years. Let y be the annual production (in millions of gallons) of ethanol fuel at x years since 2000.

Table 31 Annual Ethanol Fuel Productions

Year	Annual Production (millions of gallons)
2000	1.6
2001	1.8
2002	2.1
2003	2.8
2004	3.4
2005	3.9
2006	4.9
2007	6.5
2008	9.0
2009	10.6
2010	13.2
2011	13.9

Source: *Renewable Fuels Association*

1. Construct a scatterplot. Would it be better to model the data with a linear or an exponential function? Explain.

2. Imagine an exponential function $\hat{y} = ab^x$ whose graph comes close to the data points in your scatterplot. What is the y-intercept? What does this tell you about the value of a or b? Explain.

3. Guess a reasonable value of b for your function $\hat{y} = ab^x$. [**Hint:** The base multiplier property may help.]

4. Substitute the values of a and b you found in Problems 2 and 3 into the equation $\hat{y} = ab^x$.

5. Graph your model and the scatterplot in the same viewing window to see how well your model fits the data.

6. Now find better values of a and b through trial and error. When you are satisfied with your values of a and b, write the equation that you have found.

Homework 10.4

For extra help ▶　**MyMathLab®**　 Watch the videos in MyMathLab　 Download the MyDashboard App

1. If n is even and $k \geq 0$, then the real-number solutions of an equation of the form $b^n = k$ are _____.

2. If n is odd, then the real-number solution of an equation of the form $b^n = k$ is _____.

3. *True or False:* If n is even and $k < 0$, then an equation of the form $b^n = k$ has one real-number solution.

4. *True or False:* If $a = b, c = d, c \neq 0$, and $d \neq 0$, then $\dfrac{a}{c} = \dfrac{b}{d}$.

Find all real-number solutions. Round your result(s) to the second decimal place.

5. $b^2 = 16$

6. $b^4 = 81$

7. $b^3 = 27$

8. $b^5 = 100{,}000$

9. $3b^5 = 96$

10. $5b^2 = 45$

11. $35b^4 = 15$

12. $44b^3 = 12$

13. $3.6b^3 = 42.5$

14. $1.7b^4 = 86.4$

15. $32.7b^6 + 8.1 = 392.8$

16. $2.1b^5 - 8.2 = 237.5$

17. $\dfrac{1}{4}b^3 - \dfrac{1}{2} = \dfrac{9}{4}$

18. $\dfrac{1}{6}b^4 + \dfrac{5}{3} = \dfrac{11}{2}$

19. $\dfrac{b^6}{b^2} = 81$

20. $\dfrac{b^{10}}{b^3} = 2187$

21. $\dfrac{b^8}{b^3} = \dfrac{79}{5}$

22. $\dfrac{b^9}{b^6} = \dfrac{2}{9}$

Find all real-number solutions or simplify, whichever is appropriate. Round your solution(s) to the second decimal place.

23. $\dfrac{b^7}{b^2}$

24. $\dfrac{b^8}{b^4}$

25. $\dfrac{b^7}{b^2} = 76$

26. $\dfrac{b^8}{b^4} = \dfrac{65}{3}$

27. $\dfrac{8b^3}{6b^{-1}}$

28. $\dfrac{10b^{-7}}{15b^{-2}}$

29. $\dfrac{8b^3}{6b^{-1}} = \dfrac{3}{7}$

30. $\dfrac{10b^{-7}}{15b^{-2}} = \dfrac{4}{7}$

Find an approximate equation $y = ab^x$ of the exponential curve that contains the given pair of points. Round the value of b to two decimal places.

31. $(0, 4)$ and $(1, 8)$

32. $(0, 5)$ and $(1, 15)$

33. $(0, 3)$ and $(5, 100)$

34. $(0, 8)$ and $(4, 79)$

35. $(0, 87)$ and $(6, 14)$

36. $(0, 256)$ and $(7, 23)$

37. $(0, 5.5)$ and $(2, 73.9)$

38. $(0, 2.1)$ and $(5, 9.7)$

39. $(0, 7.4)$ and $(3, 1.3)$

40. $(0, 97.2)$ and $(4, 17.1)$

41. $(0, 39.18)$ and $(15, 3.66)$

42. $(0, 12.94)$ and $(20, 357.03)$

Find an approximate equation $y = ab^x$ of the exponential curve that contains the given pair of points. Round the values of a and b to two decimal places.

43. $(1, 4)$ and $(2, 12)$

44. $(2, 5)$ and $(3, 10)$

45. $(3, 4)$ and $(5, 9)$

46. $(2, 1)$ and $(5, 7)$

47. $(10, 329)$ and $(30, 26)$

48. $(11, 492)$ and $(17, 8)$

49. $(5, 8.1)$ and $(9, 2.4)$

50. $(1, 3.5)$ and $(5, 1.3)$

51. $(13, 24.71)$ and $(21, 897.35)$

52. $(4, 6.3)$ and $(10, 250.8)$

53. $(2, 73.8)$ and $(7, 13.2)$

54. $(8, 39.43)$ and $(12, 6.52)$

55. DATA World populations are shown in Table 32 for various years.

Table 32 World Population

Year	Population (billions)
1930	2.070
1940	2.295
1950	2.500
1960	3.050
1970	3.700
1980	4.454
1990	5.279
2000	6.080
2010	6.916
2015	7.248

Source: *U.S. Census Bureau*

Let y be the world's population (in billions) at x years since 1930.

a. Construct a scatterplot.

b. Find an equation of a model.

c. Graph the model on the scatterplot. Verify that the model passes through the two points you chose in finding the equation in part (b) and that it comes close to all of the data points.

d. What is the coefficient a of the model $\hat{y} = ab^x$? What does it mean in this situation?

e. Estimate the increase in world population from 2009 to 2013. Compare your result to the 2013 U.S. population of 317 million (Source: *U.S. Census Bureau*).

56. DATA The numbers of firearms discovered at TSA checkpoints in U.S. airports are shown in Table 33 for various years.

Table 33 Numbers of Firearms Discovered at TSA Checkpoints

Year	Number of Firearms
2005	660
2007	803
2009	976
2011	1320
2013	1813
2014	2212

Source: *Transportation Security Administration*

Let y be the number of firearms that were discovered at TSA checkpoints in the year that is x years since 2005.

a. Construct a scatterplot.

b. Find an equation of a model.

c. Graph the model on the scatterplot. Verify that the model passes through the two points you chose in finding the equation in part (b) and that it comes close to all of the data points.

d. What is the coefficient a of the model $\hat{y} = ab^x$? What does it mean in this situation?

e. In 2014, 1835 of the discovered firearms were loaded. Assuming the percentage of discovered firearms that were loaded in 2012 was the same as in 2014, estimate the *number* of loaded firearms discovered in that year.

57. **DATA** The numbers of Twitter employees are shown in Table 34 for various years.

Table 34 Numbers of Twitter Employees

Year	Number of Employees
2008	8
2009	29
2010	130
2011	350
2012	900
2013	2712

Source: *Twitter*

Let *y* be the number of Twitter employees at *x* years since 2005.

a. Construct a scatterplot.
b. Find an equation of a model.
c. Graph the model on the scatterplot. Verify that the model passes through the two points you chose in finding the equation in part (b) and that it comes close to all of the data points.
d. What is the base *b* of the model $\hat{y} = ab^x$? What does it mean in this situation?
e. Use the model to estimate the number of employees in 2012. Compute and interpret the residual.

58. **DATA** A building with a LEED certification has met a wide range of green requirements. The numbers of LEED-certified projects are shown in Table 35 for various years.

Table 35 Numbers of LEED-Certified Projects

Year	Number of Projects
2002	18
2004	118
2006	337
2008	994
2010	3337
2012	4605

Source: *U.S. Green Building Council*

Let *y* be the number of LEED-certified projects at *t* years since 2000.

a. Construct a scatterplot.
b. Find an equation of a model.
c. Graph the model on the scatterplot. Verify that the model passes through the two points you chose in finding the equation in part (b) and that it comes close to all of the data points.
d. What is the base *b* of the model $\hat{y} = ab^x$? What does it mean in this situation?
e. Estimate the number of LEED-certified projects in 2011.

59. **DATA** The numbers of polio cases in the world are shown in Table 36 for various years.

Table 36 Numbers of Polio Cases Worldwide

Year	Number of Polio Cases (thousands)
1988	350
1992	138
1996	33
2000	4
2004	1.3
2008	1.7
2011	0.7
2014	0.4

Source: *World Health Organization*

Let *y* be the number (in thousands) of polio cases at *x* years since 1980.

a. Construct a scatterplot.
b. Find an equation of a model and graph it on the scatterplot.
c. Describe the four characteristics of the association.
d. What is the base *b* of the model $\hat{y} = ab^x$? What does it mean in this situation?
e. The Polio Eradication and Endgame Strategic Plan 2013–2018 is a strategy to have a polio-free world by 2018 (Source: *The Global Polio Eradication Initiative*). Does your model suggest that the goal will be achieved? Do you have much faith in your result? Explain.

60. **DATA** The infant mortality rate is the number of deaths of infants (not including fetuses) under 1 year old per 1000 births. Infant mortality rates in the United States are shown in Table 37 for various years.

Table 37 Infant Mortality Rates in the United States

Year	Rate (number of deaths per 1000 infants)
1915	99.9
1920	85.8
1930	64.6
1940	47.0
1950	29.2
1960	26.0
1970	20.0
1980	12.6
1990	9.2
2000	6.9
2010	6.2
2014	6.2

Source: *National Center for Health Statistics*

Let *y* be the U.S. infant mortality rate (number of deaths per 1000 infants) at *x* years since 1900.

a. Construct a scatterplot.
b. Find an equation of a model and graph it on the scatterplot. Then describe the shape and the strength of the association.
c. What is the base *b* of the model $\hat{y} = ab^x$? What does it mean in this situation?

d. Estimate the U.S. infant mortality rate in 2007.

e. Find the ratio of the 2005 U.S. infant mortality rate to the 2005 Singapore infant mortality rate, which was 2.1 deaths per 1000 infants (Source: *European Perinatal Health Report*). What does your result mean in this situation?

61. **DATA.** If you place your hand on a piano and play a note, you will feel the piano vibrate. The number of vibrations per second (hertz) of a note is called its *frequency*. If you strike the piano keys from left to right, the frequencies of the notes increase. We use combinations of letters of the alphabet, numbers, and sometimes the "sharp" symbol # to refer to these notes (see Fig. 39). The frequencies of 13 notes in a row are listed in Table 38.

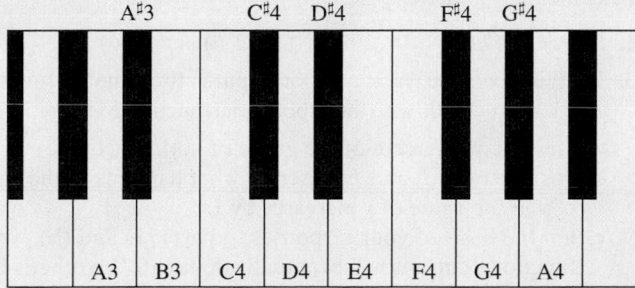

Figure 39 Some notes on a piano

Table 38 Frequencies of Some Notes on a Piano

Note	Number of Notes above A3	Frequency (in hertz)
A3	0	220.0
A#3	1	233.1
B3	2	246.9
C4	3	261.6
C#4	4	277.2
D4	5	293.7
D#4	6	311.1
E4	7	329.6
F4	8	349.2
F#4	9	370.0
G4	10	392.0
G#4	11	415.3
A4	12	440.0

Source: *sengpielaudio.com*

Let y be the frequency (in hertz) of the note that is x notes above the note A3.

a. Construct a scatterplot.

b. Find an equation of a model and graph it on the scatterplot. Then describe the shape and the strength of the association.

c. What is the y-intercept? What does it mean in this situation?

d. Use the model to predict the frequencies of the notes E4 and F4.

e. Find the ratio of your predicted frequency of the note F4 to your predicted frequency of the note E4. Compare the ratio to the base of your model. Why does your comparison make sense?

62. **DATA.** Saks Fifth Avenue® offered a promotional sale in which customers could receive a gift card. The values of the gift cards are shown in Table 39 for various expenditures.

Table 39 Saks Fifth Avenue Gift Card Values

Expenditure Group (dollars)	Expenditure Used to Represent Expenditure Group (dollars)	Gift Card Value (dollars)
250–499	375	25
500–999	750	50
1000–1999	1500	100
2000–2999	2500	200
3000 or more	3500	450

Source: *Saks Fifth Avenue*

Let y be the value (in dollars) of a gift card that a customer who spends x dollars will receive.

a. Construct a scatterplot.

b. Find an equation of a model and graph it on the scatterplot. Then describe the shape and the strength of the association.

c. What is the coefficient a of the model $\hat{y} = ab^x$? What does it mean in this situation?

d. What is the base b of the model $\hat{y} = ab^x$? What does it mean in this situation?

e. Customer A spends $2000, customer B spends $2500, and customer C spends $2999. Use your model to predict the values of the gift cards these customers will receive. Compare your results with the actual gift card values they will receive.

Concepts

63. Find an equation of the exponential curve sketched in Fig. 40. [**Hint:** Choose two points whose coordinates appear to be integers.]

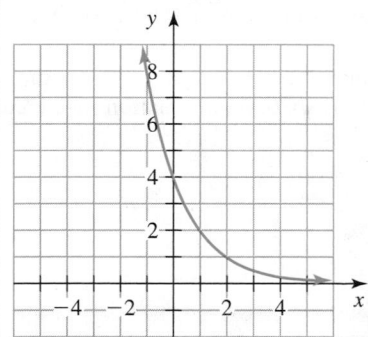

Figure 40
Exercise 63

64. Find an equation of the exponential curve sketched in Fig. 41. [**Hint:** Choose two points whose coordinates appear to be integers.].

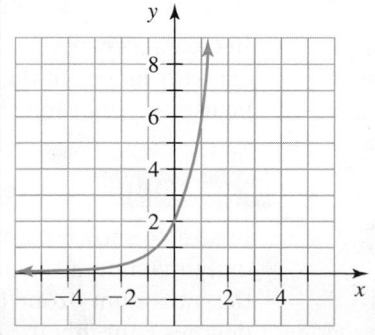

Figure 41
Exercise 64

65. Find an equation of the exponential curve sketched in Fig. 42.

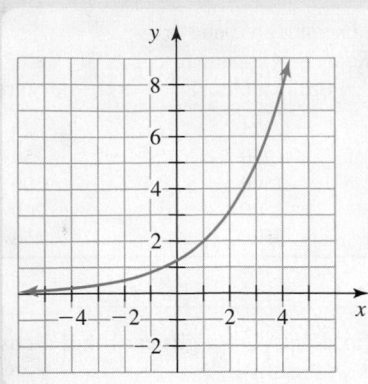

Figure 42
Exercise 65

66. Find an equation of the exponential curve sketched in Fig. 43.

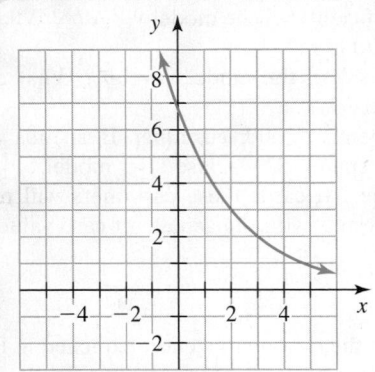

Figure 43
Exercise 66

67. a. Is there an exponential curve that contains the given point? If so, find an equation of the curve. If not, explain.
   i. $(0, 2)$ **ii.** $(2, 0)$
b. Is there an exponential curve that contains the points $(2, -1)$ and $(3, 1)$? If so, find an equation of the curve. If not, explain.

68. a. Is there an exponential curve that passes through the given points? If so, find an equation of the curve. If not, explain.
   i. $(0, 5)$ and $(1, 5)$ **ii.** $(0, 3)$ and $(7, 3)$
b. Is there an exponential curve that passes through two given points that have the same y-coordinate? Explain.

Find equations of a linear function L and an exponential function E such that the graph of each equation contains the given points.

69. $(0, 2)$ and $(1, 6)$ **70.** $(2, 8)$ and $(5, 2)$

For Exercises 71 and 72, the graph of a function contains the given pair of points. Could the function be linear, exponential, either linear or exponential, or neither? Explain.

71. $(5, 3)$ and $(7, 6)$ **72.** $(2, 6)$ and $(4, 6)$

73. In this exercise, you will compare the linear function $L(x) = 2x + 100$ with the exponential function $E(x) = 3(2)^x$.
a. Find the y-intercept of the graph of both functions.
b. For functions L and E, describe what happens to the value of y as the value of x increases by 1.
c. On the basis of your responses to parts (a) and (b), which function's outputs will eventually dominate the other's outputs? Explain.

74. Is it possible for a linear function and an exponential function to have the indicated number of intersection points? If so, give equations of the two functions. If not, explain. [**Hint:** First sketch some graphs.]
a. 3 intersection points
b. 2 intersection points
c. 1 intersection point
d. 0 intersection points

75. Describe how to find an equation of an exponential curve that contains two given points. Include both the case in which one of the points is the y-intercept and the case in which neither of the points is the y-intercept.

▼ 10.5 Exponential Regression Model

Objectives

» Compute and interpret the *exponential correlation coefficient*.

» Compute and interpret *residuals* when working with an exponential model.

» Find the sum of squared residuals for an *exponential model*.

» Find an equation of an *exponential regression model*.

In Section 9.3, we worked with linear regression models. In this section, we will work with *exponential regression models*.

Exponential Correlation Coefficient

In Chapters 6–9, we used a (linear) correlation coefficient to measure the strength of a linear association. In this section, we will work with another correlation coefficient that can be used to determine how well an association can be modeled with an exponential curve.

▶ **Example 1** Determining Four Characteristics of an Exponential Association

Figure 44 shows a scatterplot of daily maximum temperatures in New York City versus daily maximum ozone levels, which are measured in parts per billion (ppb). Let y be the maximum ozone (in ppb) for a day in New York City with maximum temperature x degrees Fahrenheit.

» Use a residual plot to help determine whether an exponential regression curve is an appropriate model.

» Identify influential points for an exponential regression model.

» Compute and interpret the *exponential coefficient of determination*.

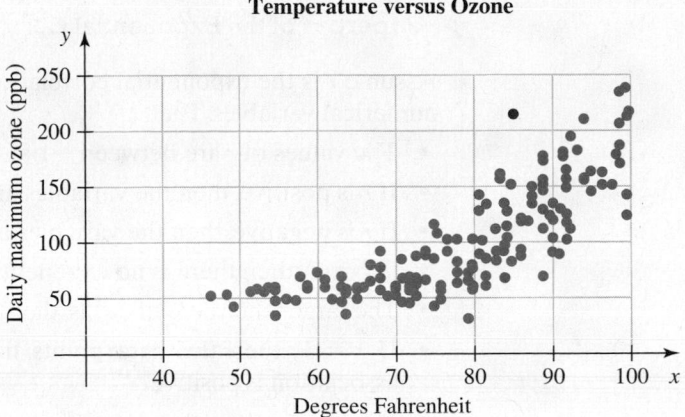

Figure 44 Daily maximum temperatures versus daily maximum ozone levels
(**Source:** *NAST*)

Describe the four characteristics of the association.

Solution

The red dot could be considered an outlier (see Fig. 44). The association is nonlinear and positive. In fact, the increasing exponential curve in Fig. 45 fits the data fairly well, which means the association is exponential and fairly strong.

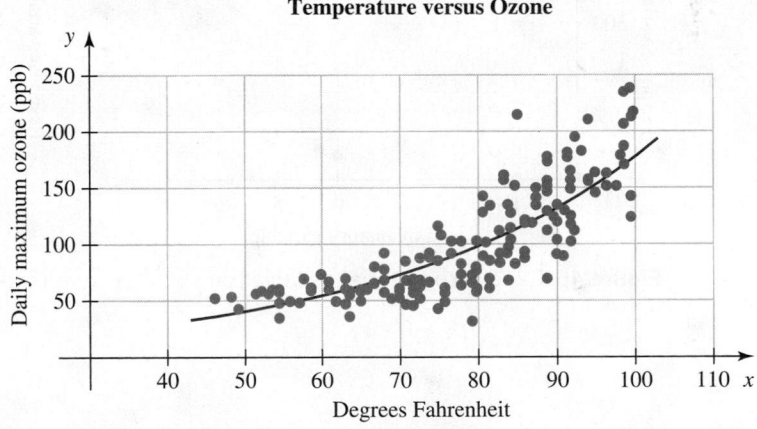

Figure 45 Temperature-ozone scatterplot and exponential curve

WARNING   In Example 1, we determined that the temperature-ozone association is positive, which means that the maximum ozone level tends to be larger on days when the maximum temperature is larger. Because the study is observational, this does *not* necessarily mean that an increase in temperature *causes* an increase in ozone level. We could determine this only by performing an experiment.

In Chapters 6–9, we used a correlation coefficient to measure the strength of a linear association. We can measure the strength of an exponential association by using a different correlation coefficient. To keep the two measures straight, we will refer to them as the *linear correlation coefficient* and the *exponential correlation coefficient*. We will use *r* to represent both correlation coefficients, so it is important that we state which one we are using.

The properties of the exponential correlation coefficient are the same as the ones for the linear correlation coefficient, except that we replace the word *linear* with *exponential*.

> **Properties of the Exponential Correlation Coefficient**
>
> Assume r is the exponential correlation coefficient for the association between two numerical variables. Then
>
> - The values of r are between -1 and 1, inclusive.
> - If r is positive, then the variables are positively associated.
> - If r is negative, then the variables are negatively associated.
> - If $r = 0$, then there is no exponential association.
> - The larger the value of $|r|$ is, the stronger the exponential association will be.
> - If $r = 1$, then the data points lie exactly on an exponential curve and the association is positive.
> - If $r = -1$, then the data points lie exactly on an exponential curve and the association is negative.

Figures 46–53 display the values of the exponential correlation coefficient r for various associations.

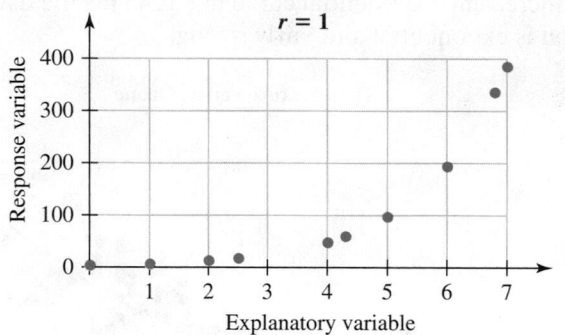

Figure 46 Exact exponential association

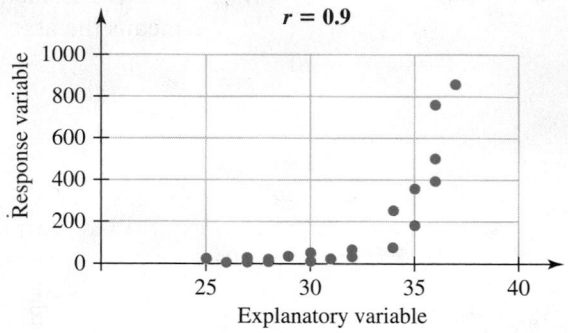

Figure 47 Strong exponential association

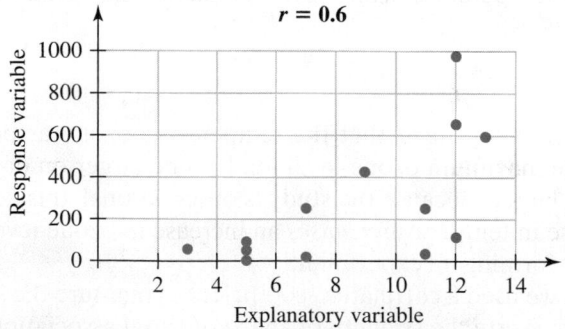

Figure 48 Weak exponential association

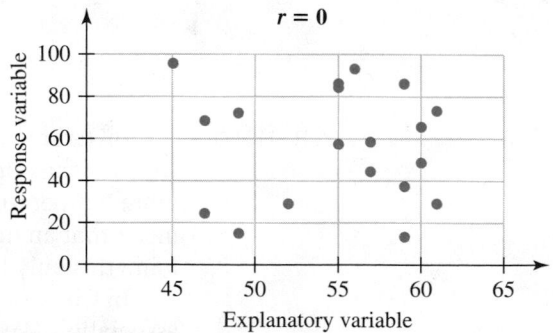

Figure 49 No exponential association

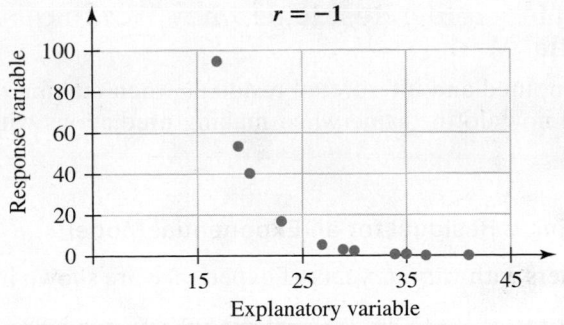

Figure 50 Exact exponential association

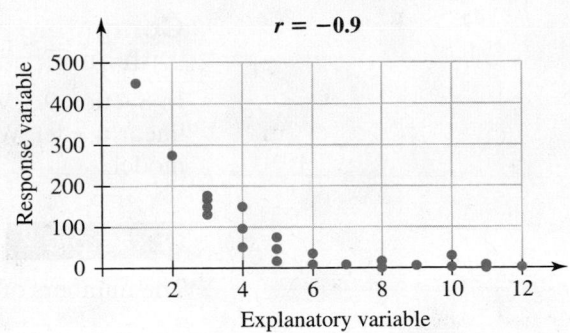

Figure 51 Strong exponential association

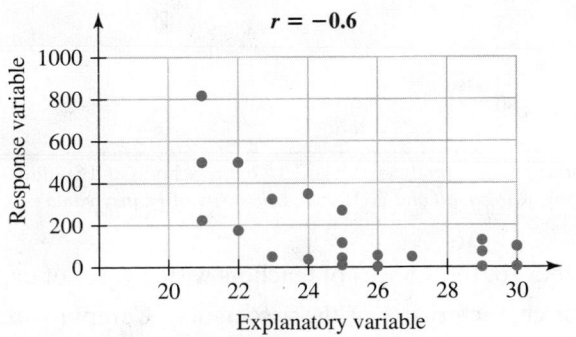

Figure 52 Weak exponential association

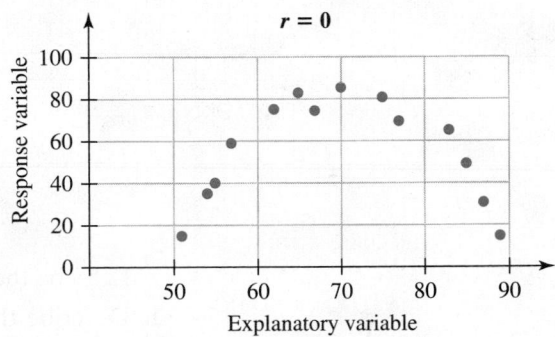

Figure 53 No exponential association

WARNING

In Fig. 53, $r = 0$ tells us that there is no *exponential* association. However, there *is* an association of another type. In general, if the exponential correlation coefficient is 0, then there is no exponential association, but there may be some other type of association.

WARNING

A scatterplot is shown in Fig. 54. It turns out that the *exponential* correlation coefficient is 0.97, which suggests there is a strong exponential association. However, by inspecting the scatterplot, we see there is a strong linear association. In fact, the *linear* correlation coefficient is 0.99, which confirms that the linear association is strong. We conclude that the association is linear (and not exponential). Recall that **to determine the strength and type of an association, we should inspect the scatterplot of the data as well as compute one or more correlation coefficients.**

Scatterplot

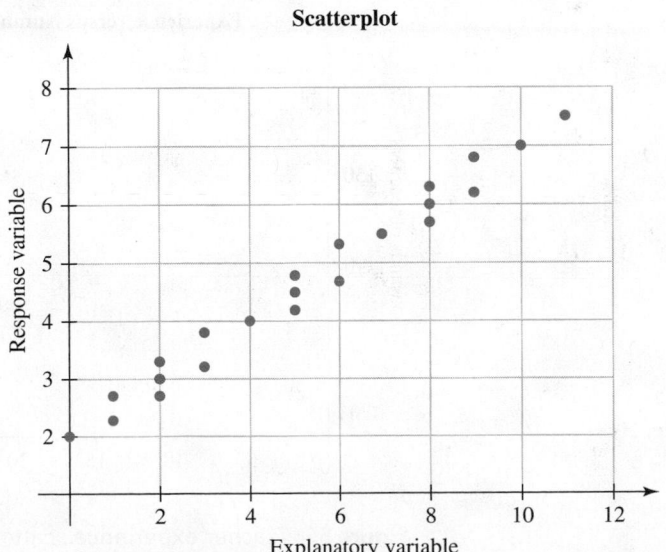

Figure 54 Association appears to be linear

Computing and Interpreting Residuals when Working with an Exponential Model

In Section 9.3, we computed and interpreted residuals when making predictions with a linear model. We will now do the same when making predictions with an exponential model.

▶ **Example 2** Finding a Residual for an Exponential Model

The numbers of teachers with various years of experience are shown in Table 40.

Table 40 Teachers' Years of Experience

Years of Experience	Number of Teachers (thousands)
5	185
10	145
15	90
20	63
25	54
30	50
35	38

Source: *Original analyses for NCTAF of the* Schools and Staffing Survey, *R.Ingersoll and E. Merrill, University of Pennsylvania*

Let y be the number (in thousands) of teachers with x years of experience.

1. Describe the four characteristics of the association. Compute and interpret an appropriate correlation coefficient r as part of your analysis.
2. We can use the data points (10, 145) and (25, 54) to find the exponential model $\hat{y} = 280(0.936)^x$ (try it). How well does the model fit the data?
3. Use the model to predict the number of teachers who have 5 years of experience.
4. Find the residual for the prediction you made in Problem 3. What does it mean in this situation? What does it mean about the positions of the data point and the exponential curve?

Solution

1. We construct a scatterplot in Fig. 55. There are no outliers. The data points appear to lie close to a curve that "bends.'" So, we check whether the variables are exponentially associated by using a TI-84 to compute the exponential correlation coefficient, which is $r = -0.98$ (see Fig. 56). For TI-84 instructions, see Appendix A.12.

Figure 56 Compute the exponential correlation coefficient

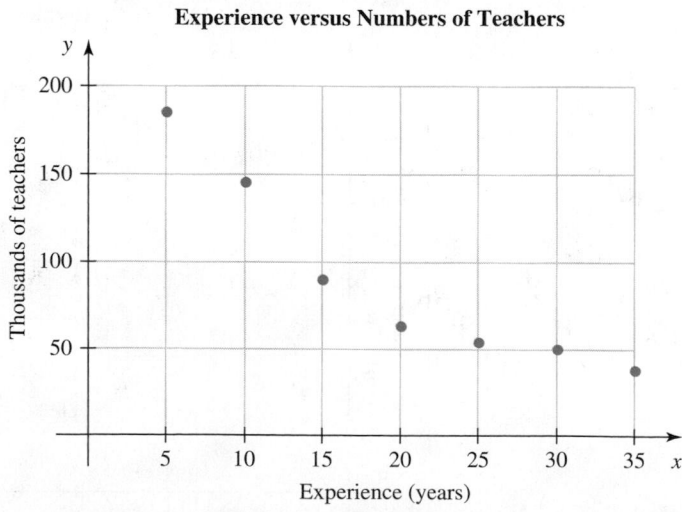

Figure 55 Teacher experience scatterplot

Because the points lie close to a curve that bends and the exponential correlation coefficient is very close to −1, we conclude that there is a strong exponential association. The association is negative, which means that the greater the amount of experience, the fewer teachers there will be.

2. The model comes close to the data points (see Fig. 57).

Experience versus Numbers of Teachers

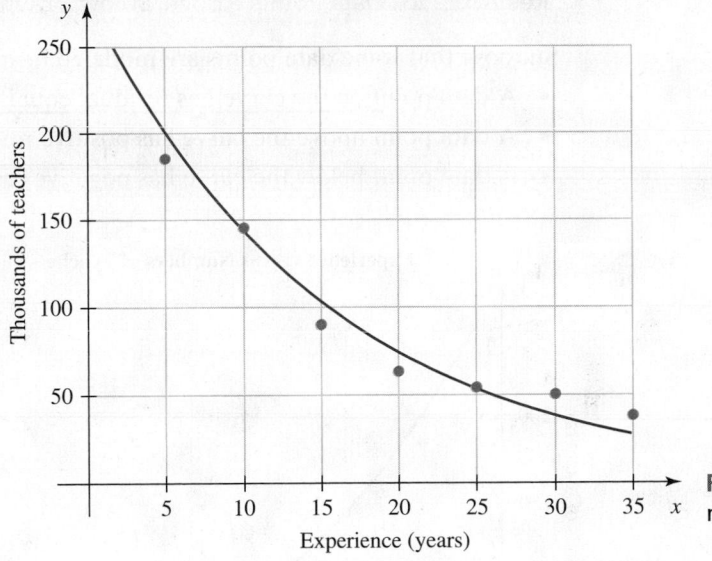

Experience (years)

Figure 57 The teacher model

3. We substitute 5 for x in the equation $\hat{y} = 280(0.936)^x$:

$$\hat{y} = 280(0.936)^5 \approx 201.2$$

So, there are 201.2 thousand teachers with 5 years of experience, according to the model.

4. For teachers with 5 years of experience, the observed value of y is 185 (see Table 40 on page 686). In Problem 3, we found that the predicted value of y is 201.2. We substitute 185 for y and 201.2 for $\hat{y}$ in the formula Residual $= y - \hat{y}$:

$$\text{Residual} = y - \hat{y}$$
$$= 185 - 201.2$$
$$= -16.2$$

So, the residual is −16.2 thousand teachers. This means that the actual number of teachers with 5 years of experience is 16.2 thousand teachers less than the predicted number of teachers. So, the data point $(5, 185)$ is 16.2 thousand teachers below the exponential curve (see Fig. 58).

Experience versus Numbers of Teachers

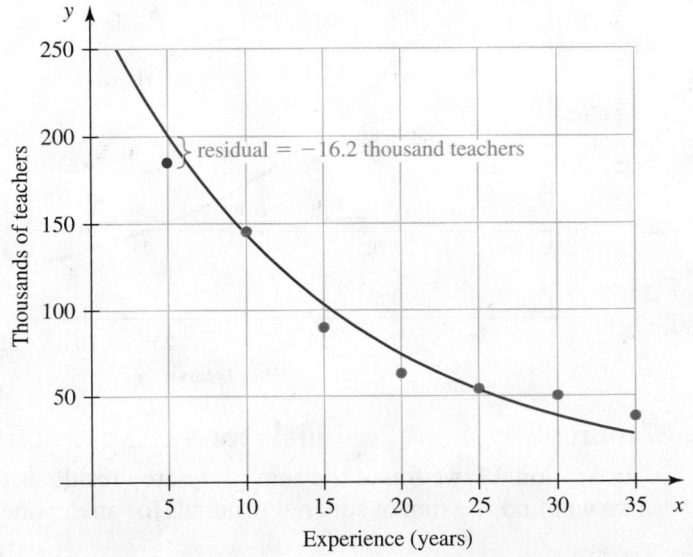

residual $= -16.2$ thousand teachers

Experience (years)

Figure 58 The data point (5, 185) is below the exponential curve

When using a line to model a situation, a data point on the line has residual equal to 0, a data point above the line has positive residual, and a data point below the line has negative residual (Section 9.3). Similarly, in Problem 4 of Example 2, we found that the data point (5, 185) lies below the exponential curve $\hat{y} = 280(0.936)^x$ and has negative residual, -16.2 thousand teachers.

Residuals for Data Points Above, Below, or On a Line

Suppose that some data points are modeled by an exponential curve.

- A data point on the curve has residual equal to 0.
- A data point above the curve has positive residual (see Fig. 59).
- A data point below the curve has negative residual (see Fig. 60).

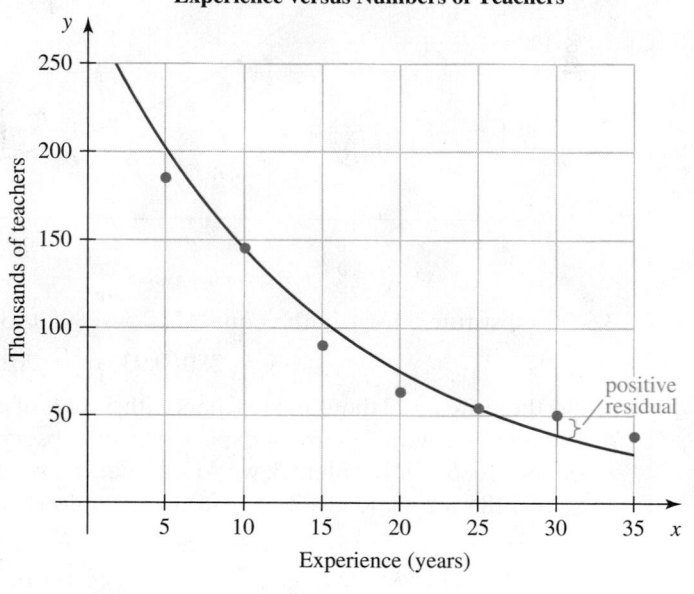

Figure 59 A data point above the curve has positive residual

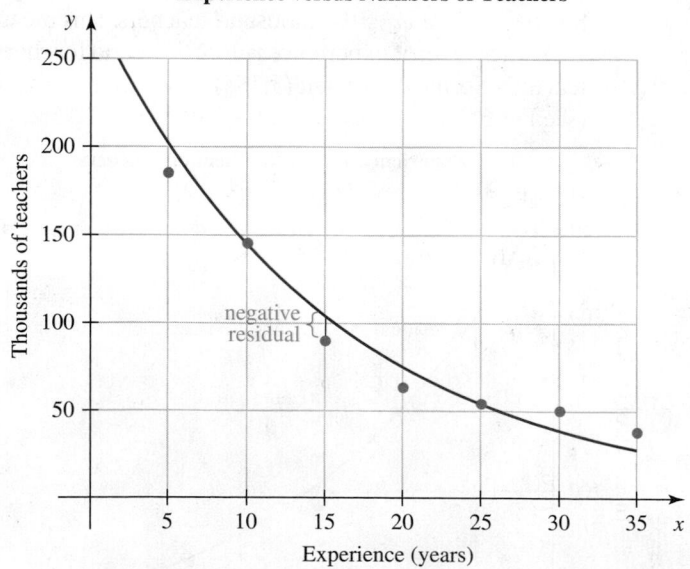

Figure 60 A data point below the curve has negative residual

Sum of Squared Residuals for an Exponential Model

In Section 9.3, we found the sum of squared residuals for a linear model. In Example 3, we will find the sum of squared residuals for an exponential model.

▶ **Example 3**   Sum of Squared Residuals for an Exponential Model

In Example 2, we modeled the teacher association with the function $\hat{y} = 280(0.936)^x$. Find the sum of squared residuals for the model.

Solution

We list the values of x and y in the first and second columns of Table 41. Then we substitute the values of x in the equation $\hat{y} = 280(0.936)^x$ to find the predicted numbers of teachers and list them in the third column. In the fourth column, we find the values of the residuals. Note that the first residual, -16.2 thousand teachers, is the one we found in Problem 4 of Example 2. Next, we list the squared residuals in the fifth column and total them to find the sum of the squared residuals, 826.19 thousand teachers^2.

Table 41 Calculating the Sum of Squared Residuals for the Model $\hat{y} = 280(0.936)^x$

Years x_i	Thousands of Teachers y_i	Predicted Number (in thousands) of Teachers Using $\hat{y} = 280(0.936)^x$ $\hat{y}_i$	Residual (thousands of teachers) $y_i - \hat{y}_i$	Residual^2 (thousands of teachers^2) $(y_i - \hat{y}_i)^2$
5	185	201.2	-16.2	262.44
10	145	144.5	0.5	0.25
15	90	103.8	-13.8	190.44
20	63	74.6	-11.6	134.56
25	54	53.6	0.4	0.16
30	50	38.5	11.5	132.25
35	38	27.7	10.3	106.09

$$\Sigma(n_i - \hat{y}_i)^2 = 826.19$$

Exponential Regression Model

For a given data set, the linear regression model has the least sum of squared residuals of all linear models and therefore fits the data points the best of all linear models (Section 9.3). Similarly, the *exponential regression model* has the least sum of squared residuals of all exponential models and fits the data points the best of all exponential models.

For a given data set, we can use a TI-84 to find an equation whose graph would be very close to the graph of the exponential regression model. In fact, the graphs would be so similar, we will refer to the TI-84's equation as the exponential regression model.

For the teacher data set, the TI-84 gives the exponential regression model $\hat{y} = 220(0.949)^x$ (see Fig. 61). For TI-84 instructions, see Appendix A.26. If we use all the digits for a and b shown in Fig. 61, then the sum of squared residuals for the equation is 814.44 thousand teachers^2, which is less than the sum of squared residuals 826.19 thousand teachers^2 we found in Example 3 for the equation $\hat{y} = 280(0.936)^x$.

The graph of the exponential regression model and the scatterplot are shown in Fig. 62.

Figure 61 Find the exponential regression equation

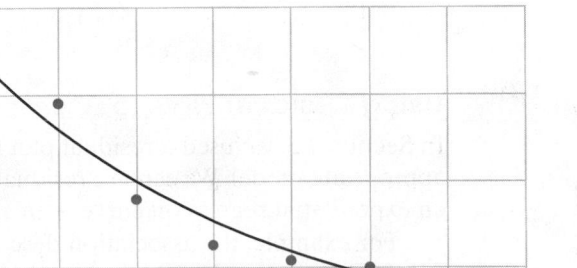

Experience versus Numbers of Teachers

Thousands of teachers

Experience (years)

Figure 62 The exponential regression model

▶ Definition   Exponential regression function, curve, equation, and model

For a group of points, the **exponential regression function** is the exponential function with the least sum of squared residuals of all exponential functions. Its graph is called the **exponential regression curve** and its equation is called the **exponential regression equation**, written

$$\hat{y} = ab^x$$

where $a \neq 0$, $b > 0$, and $b \neq 1$ The **exponential regression model** is the exponential regression function for a group of *data* points.

▶ **Example 4**   Finding an Exponential Regression Equation

Pointing a laser at aircraft, which can temporarily blind pilots, is a serious offense with a maximum punishment of 20 years in prison and a $250,000 fine. The numbers of laser incidents involving aircraft are shown in Table 42 for various years. Let y be the number of laser incidents involving aircraft in the year that is x years since 2000.

Table 42 Numbers of Laser Incidents Involving Aircraft

Year	Number
2005	283
2006	446
2007	675
2008	988
2009	1527
2010	2836

Source: *Federal Aviation Administration*

1. Construct a scatterplot.
2. Find a regression equation to describe the data. How well does the model fit the data points?
3. What is the y-intercept? What does it mean in this situation?
4. Estimate the percentage increase in laser incidents per year.

Solution

1. We use a TI-84 to construct a scatterplot (see Fig. 63).
2. A curve that "bends" will fit the data points better than a line, so we will try to model the situation with the exponential regression equation. We use a TI-84 to find the equation $\hat{y} = 30(1.56)^x$ See Fig. 64. The model fits the data very well (see Fig. 65).

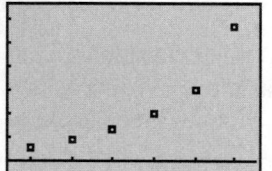

Figure 63 Laser scatterplot

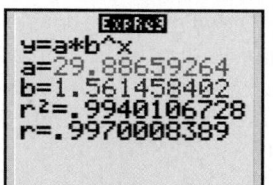

Figure 64 The exponential regression equation

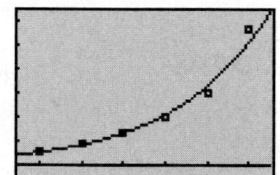

Figure 65 Check the fit of the model

3. The y-intercept of the graph of $\hat{y} = 30(1.56)^x$ is $(0, 30)$. This indicates that there were 30 laser incidents involving aircraft in 2000. We have little faith in this estimate, because 0 is not in the scope of the data values for x.
4. The base of the model $\hat{y} = 30(1.56)^x$ is $b = 1.56$. Because $b - 1 = 1.56 - 1 = 0.56$, we conclude that the number of laser incidents involving aircraft increased by 56% per year.

Using Residual Plots to Assess an Exponential Regression Curve

In Section 9.3, we used a residual plot to help determine whether a regression line is an appropriate model. We can use residual plots in a similar way to help determine whether an exponential regression curve is an appropriate model.

For example, the association described by the scatterplot in Fig. 66 is not exponential. This is indicated by the residual plot in Fig. 66; which has a pattern where the dots do not lie close to the zero residual line.

For another example, because an outlier is much farther from an exponential regression curve than the other data points, the outlier's dot in the residual plot is much farther from the zero residual line than the other dots (see the red dots in Fig. 67)

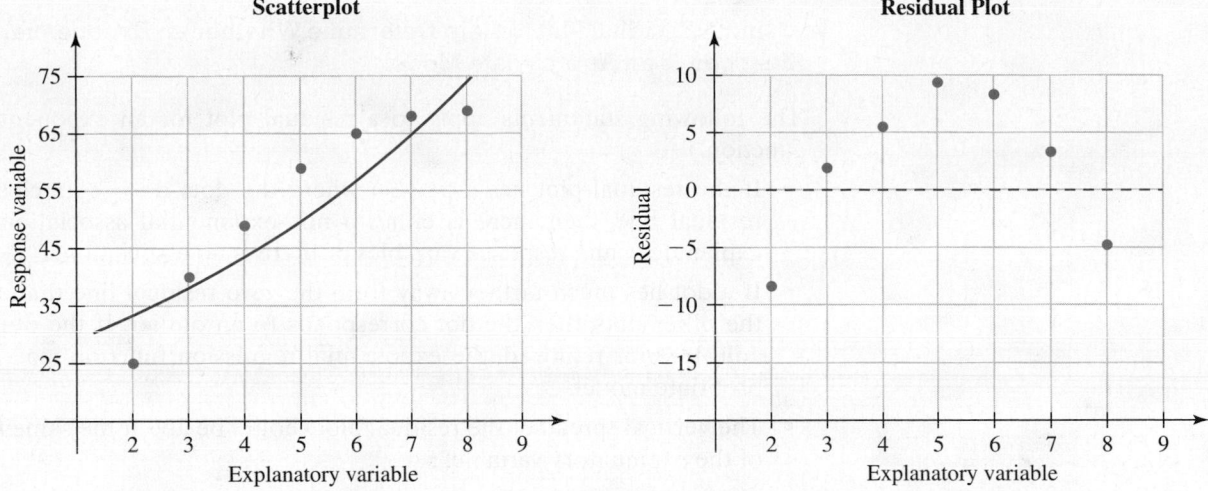

Figure 66 Scatterplot and residual plot for an association that is not exponential

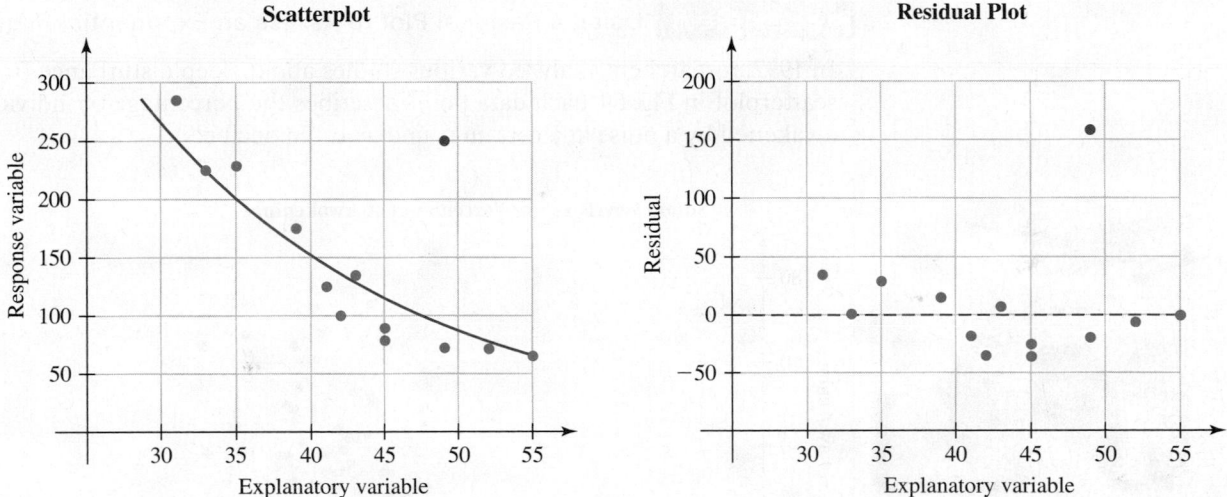

Figure 67 An outlier is described by the red dots in the scatterplot and residual plot

Finally, recall that for each value of the explanatory variable, the variation of the response variable should be about the same. So, because the residual plot in Fig. 68 becomes more spread out *vertically* as we view the plot from left to right, we cannot use the exponential regression curve shown in the scatterplot to model the situation.

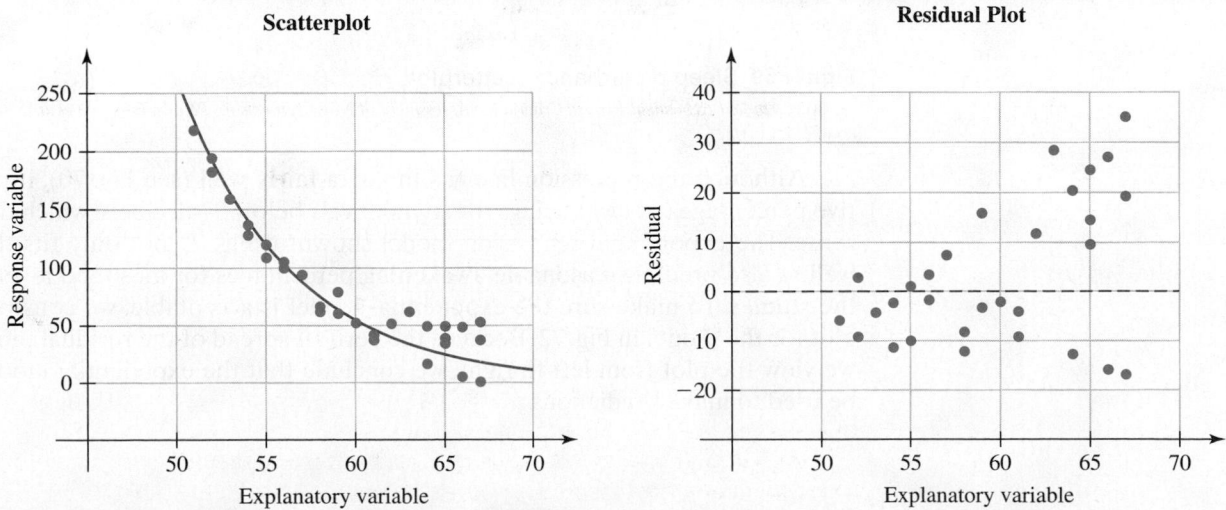

Figure 68 The residual plot becomes more spread out vertically as we view the plot from left to right

> **Using a Residual Plot to Help Determine Whether an Exponential Regression Function Is an Appropriate Model**

The following statements apply to a residual plot for an exponential regression function.

- If the residual plot has a pattern where the dots do not lie close to the zero residual line, then there is either a non-exponential association between the explanatory and response variables or there is no association.

- If a dot lies much farther away from the zero residual line than most or all of the other dots, then the dot corresponds to an outlier. If the outlier is neither adjusted nor removed, the exponential regression function may *not* be an appropriate model.

- The vertical spread of the residual plot should be about the same for each value of the explanatory variable.

▶ **Example 5** Using a Residual Plot to Assess an Exponential Regression Model

In 1992, researchers analyzed various studies about sleep disturbance from noise. In the scatterplot in Fig. 69, each data point describes the percentage of individuals who were awakened by a noise at a certain sound level (in decibels).

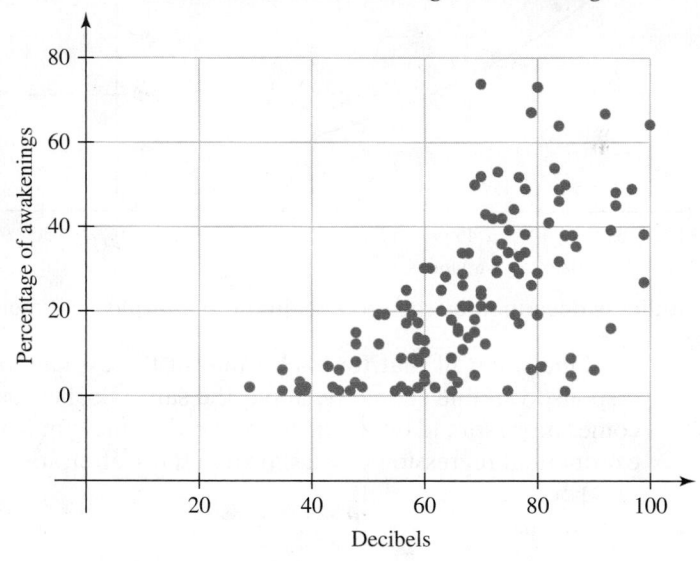

Sound Levels versus Percentages of Awakenings

Figure 69 Sleep disturbance scatterplot

(**Source:** *Applied Acoustical Report: Criteria for Assessment of Noise Impacts on People, Finegold et at.*)

Although the regression line fits the data fairly well (see Fig. 70), it predicts negative percentages of awakenings for sound levels below 37 decibels, which does not make sense. The exponential regression model shown in Fig. 71 not only fits the data fairly well, it also predicts reasonable awakening percentages for all sound levels recorded in the studies. To make sure the exponential model is acceptable, we construct a residual plot for the model in Fig. 72. Because the vertical spread of the residual plot increases as we view the plot from left to right, we conclude that the exponential model should *not* be used to make predictions.

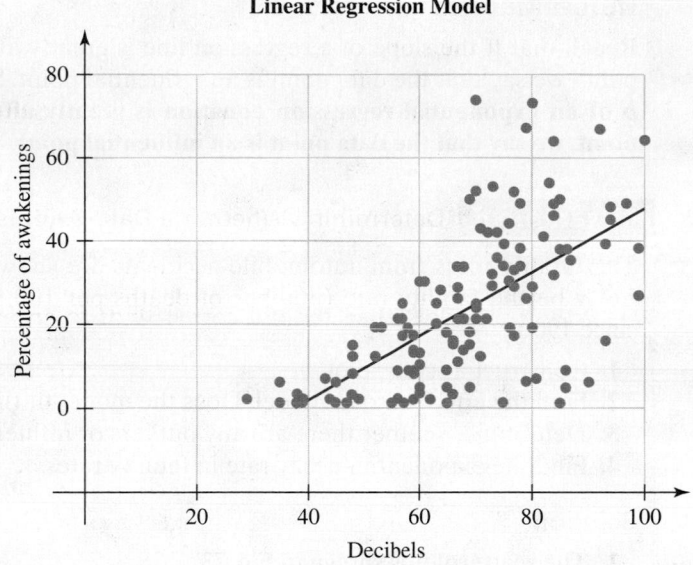

Figure 70 Linear sleep disturbance model

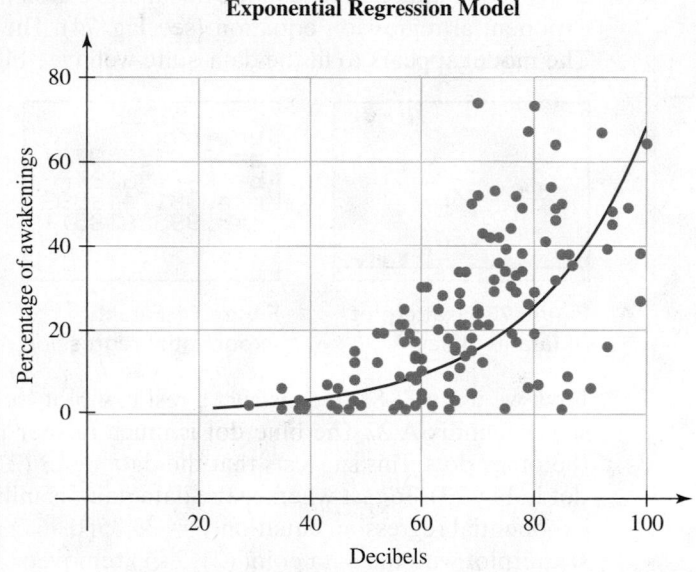

Figure 71 Exponential sleep disturbance model

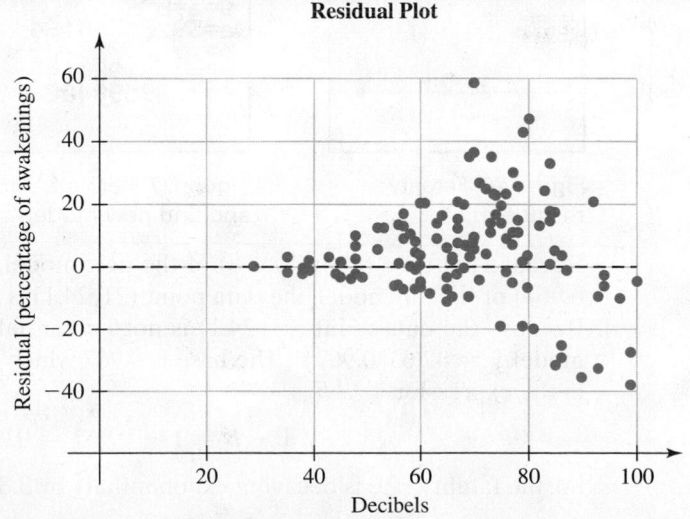

Figure 72 Residual plot for the exponential model

Influential Points

Recall that if the slope of a regression line is greatly affected by the removal of a data point, we say that the data point is an influential point. Similarly, **if the value of the base b of an exponential regression equation is greatly affected by the removal of a data point, we say that the data point is an influential point.**

▶ **Example 6**   Determining Whether a Data Point Is Influential

Table 43 Years versus Fatality Rates from Automobile Accidents

Year	Fatality Rate (deaths per 100 million miles)
1921	24.1
1930	15.1
1940	10.9
1950	7.2
1960	5.1
1970	4.7
1980	3.3
1990	2.1
2000	1.5
2010	1.1

Source: *National Highway Traffic Safety Administration*

The fatality rates from automobile accidents are shown in Table 43 for various years. Let y be the fatality rate (number of deaths per 100 million miles driven) at x years since 1900.

1. Construct a scatterplot.
2. Find the equation of a model. Does the model fit the data points well?
3. Determine whether there are any outliers or influential points.
4. Find the exponential decay rate in fatality rates.

Solution

1. The scatterplot is shown in Fig. 73.
2. Because the data points suggest a curve that "bends," the exponential model will likely fit the data better than the linear regression model. We use a TI-84 to find the exponential regression equation (see Fig. 74). The equation is $\hat{y} = 42.63(0.967)^x$. The model appears to fit the data quite well (see Fig. 75).

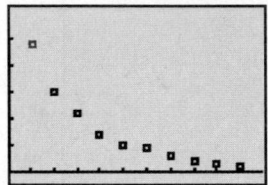

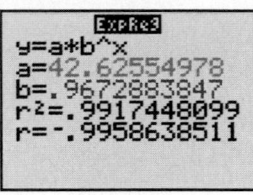

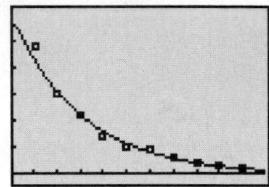

Figure 73 Scatterplot of fatality data

Figure 74 Fatality exponential regression model

Figure 75 Check the fit

3. First, we use a TI-84 to construct a residual plot (see Fig. 76). For TI-84 instructions, see Appendix A.27. The blue dot is much farther from the zero residual line than the other dots. This suggests that the data point $(21, 24.1)$ is an outlier (see the red dot in Fig. 73). To test whether the data point is influential, we remove it, find a new exponential regression equation $\hat{y} = 39.05(0.968)$ (see Fig. 77), and graph it on the scatterplot with the data point $(21, 24.1)$ removed (see Fig. 78).

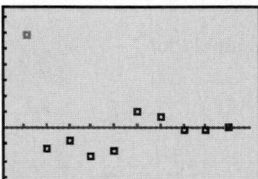

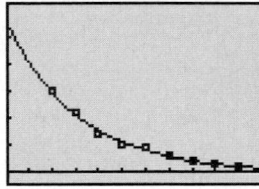

Figure 76 Fatality residual plot

Figure 77 Remove outlier and find new model

Figure 78 Check the fit of the new model

Because the base (about 0.968) of the new model is very close to the base (about 0.967) of the old model, the data point $(21, 24.1)$ is *not* influential.

4. Because the data point $(21, 24.1)$ is not influential, we will work with the original model $\hat{y} = 42.63(0.967)^x$. The base is 0.967, which is less than 1, so we substitute it in the expression $1 - b$:

$$1 - b = 1 - 0.967 = 0.033$$

So, the fatality rate is decaying exponentially by 3.3% per year.

Exponential Coefficient of Determination

In Section 9.3, we used the linear coefficient of determination to measure the percentage of the variation in the response variable that could be explained by the regression line. Recall that the coefficient of determination for a regression line is equal to r^2, where r is the linear correlation coefficient. Similarly, if r is the *exponential* correlation coefficient, we call r^2 the **exponential coefficient of determination.**

> ### Exponential Coefficient of Determination
>
> Let r be the exponential correlation coefficient for a group of data points. The exponential coefficient of determination, r^2, is the proportion of the variation in the response variable that is explained by the exponential regression curve.

WARNING

When we report a coefficient of determination, it is important that we state whether it is for a linear regression line or an exponential regression curve.

> ▶ **Example 7** Coefficient of Determination for an Exponential Regression Curve

In Example 1, we analyzed a scatterplot of daily maximum temperatures in New York City versus daily maximum ozone levels (see Fig. 79). Let y be the maximum ozone (in parts per billion) for a day in New York City with maximum temperature x degrees Fahrenheit.

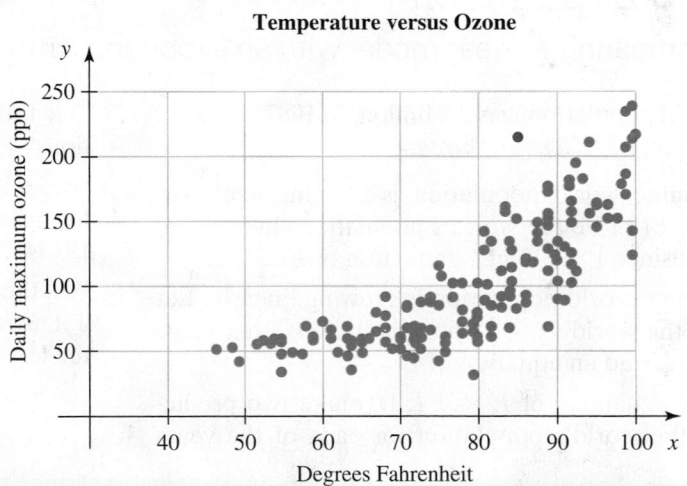

Figure 79 Daily maximum temperatures versus daily maximum ozone levels

(Source: *NAST)*

1. In Example 1, we assumed the association is exponential. Verify this by computing the exponential correlation coefficient.
2. Find an equation of the exponential regression model.
3. Compute and interpret the exponential coefficient of determination.

Solution

1. We use a TI-84 to compute the exponential correlation coefficient, which is approximately 0.83, which confirms that the exponential association is fairly strong (see Fig. 80).
2. From Fig. 80, we see that the exponential regression equation is $\hat{y} = 9.22(1.03)^x$. The model appears to fit the data points fairly well (see Fig. 81).

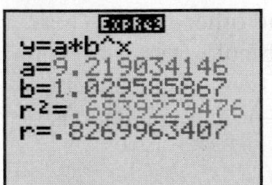

Figure 80 Exponential correlation coefficient, regression equation, and coefficient of determination

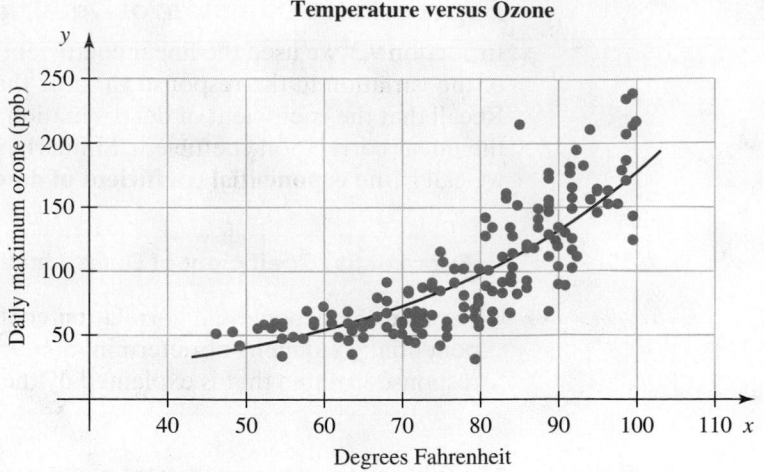

Figure 81 Temperature-ozone scatterplot and exponential regression model

3. The exponential coefficient of determination is approximately 0.68 (see Fig. 80). This means that 68% of the variation in the daily maximum ozone in New York City is explained by the exponential regression curve.

Group Exploration

Comparing a linear model with an exponential model

In 1950, world population was 2.5 billion. In 1987, it was 5.0 billion (Source: *U.S. Census Bureau*).

1. First, assume world population is growing exponentially. Let $E(x)$ be the world's population (in billions) at x years since 1950. Find an equation of E.

2. Now assume world population is growing linearly. Let $L(x)$ be the world's population (in billions) at x years since 1950. Find an equation of L.

3. Use your equations of E and L to make two predictions of the world's population for each of the years

that follow. Do you have much faith in these predictions? Explain.
   a. 2020   b. 2050   c. 2150

4. Compare the graphs of E and L for the period 1950–2100.

5. Will there be much difference in the world's population if it grows exponentially or linearly in the short run? in the long run? Explain.

▶ **Tips for Success**   **Retake Quizzes and Exams to Prepare for the Final Exam**

To study for your final exam, consider retaking your quizzes and other exams. These quizzes and exams can reveal your weak areas. If you have difficulty with a certain concept, you can refer to homework exercises that address that concept. Reflect on *why* you are having such difficulty, rather than just doing more homework exercises that address the concept.

Homework 10.5

For extra help ▶ MyMathLab®   ▣ Watch the videos in MyMathLab   ◉ Download the MyDashboard App

1. Assume r is the exponential correlation coefficient for the association between two numerical variables. If r is negative, then the variables are _____ associated

2. *True or False:* A data point above an exponential curve has negative residual.

3. For a group of data points, the exponential _____ model is the exponential function with the least sum of squared residuals of all exponential models.

4. The exponential coefficient of determination is the proportion of the variation in the _____ variable that is explained by the exponential regression curve.

For Exercises 5–8 match the given information to the appropriate scatterplot in Fig. 82.

5. $r = -0.9$ **6.** $r = 1$ **7.** $r = -0.7$ **8.** $r = 0.9$

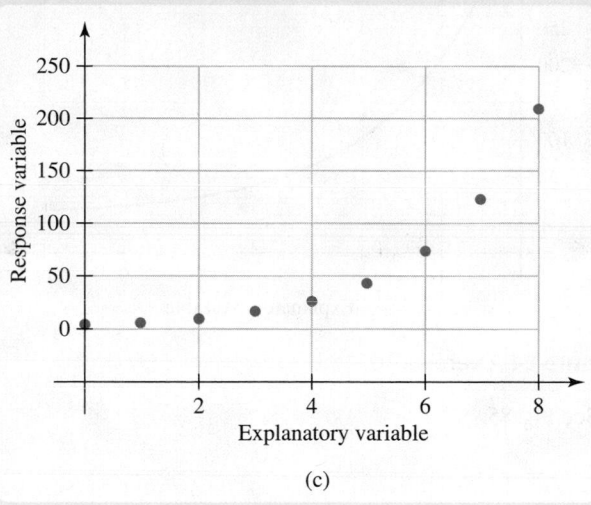

(c)

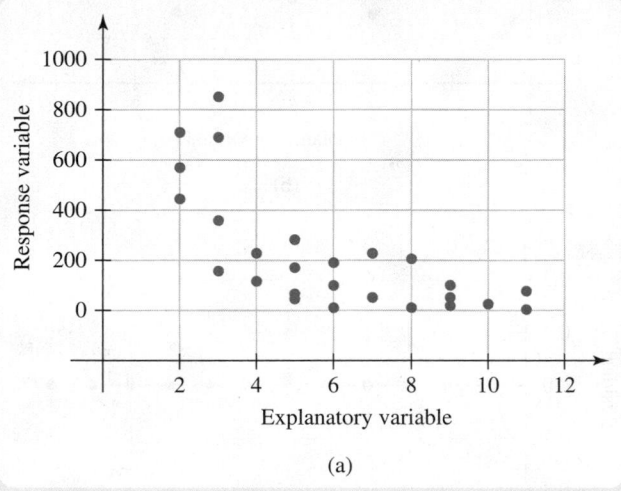

(a)

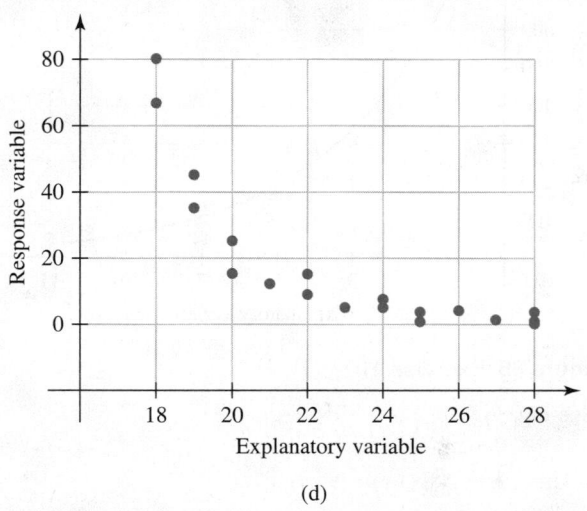

(d)

Figure 82 Exercises 5–8

For Exercises 9–12, match the given scatterplot and exponential model with the appropriate residual plot in Fig. 87.

9. See Fig. 83.

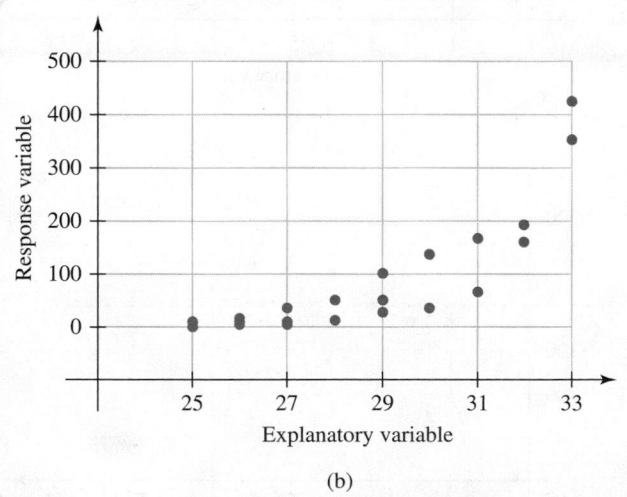

(b)

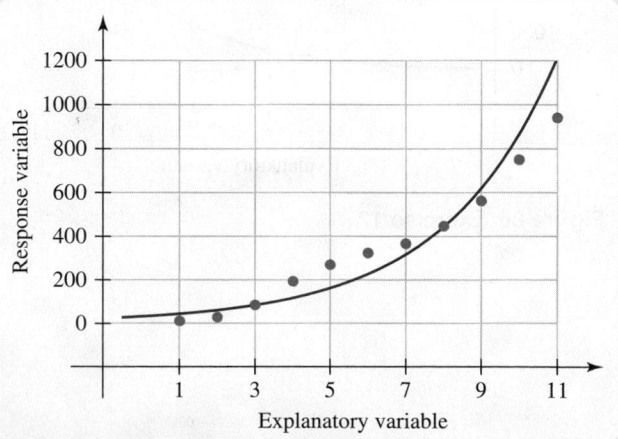

Figure 83 Exercise 9

10. See Fig. 84.

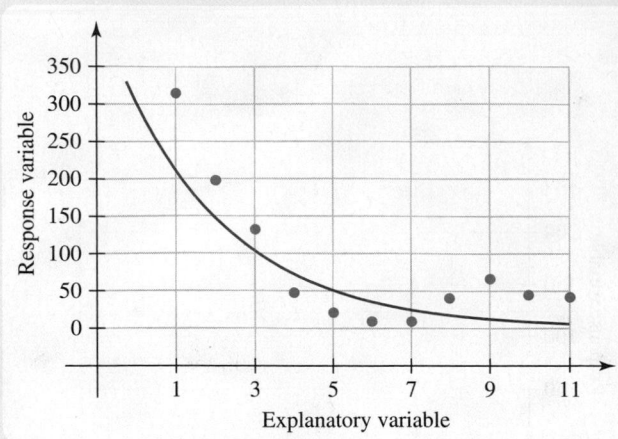

Figure 84 Exercise 10

11. See Fig. 85.

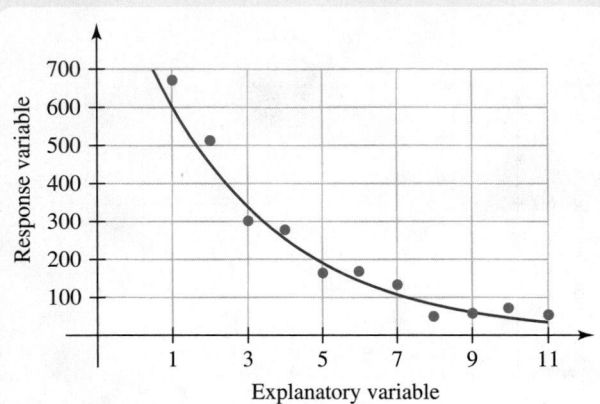

Figure 85 Exercise 11

12. See Fig. 86.

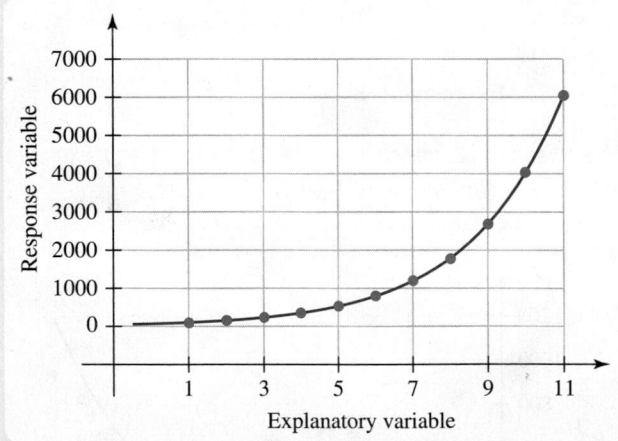

Figure 86 Exercise 12

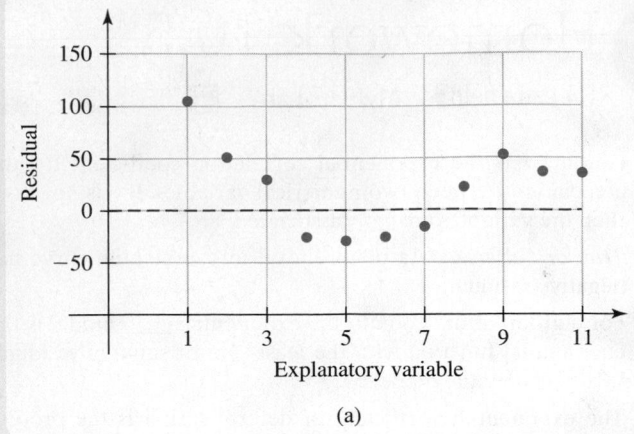

(a)

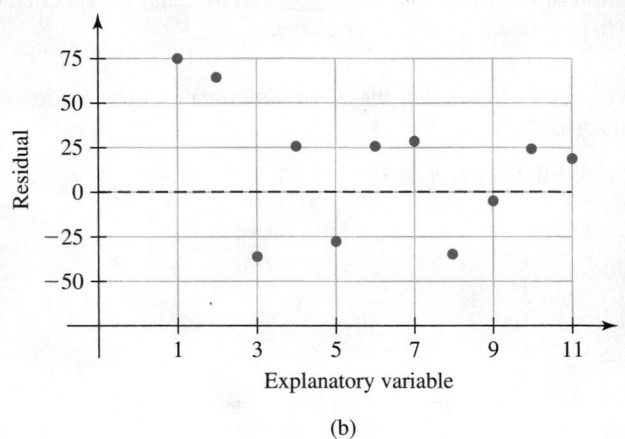

(b)

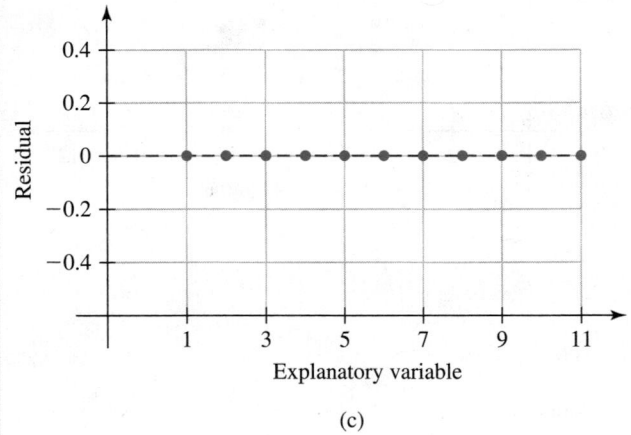

(c)

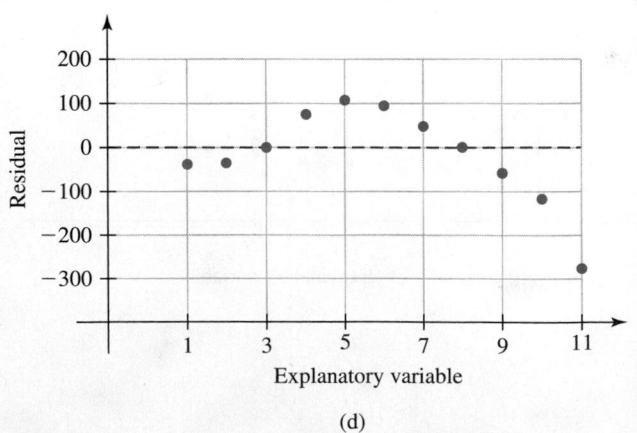

(d)

Figure 87 Exercises 9–12

For Exercises 13–16, refer to the given residual plot and identify which conditions, if any, for a regression exponential curve are not met.

13. See Fig. 88.

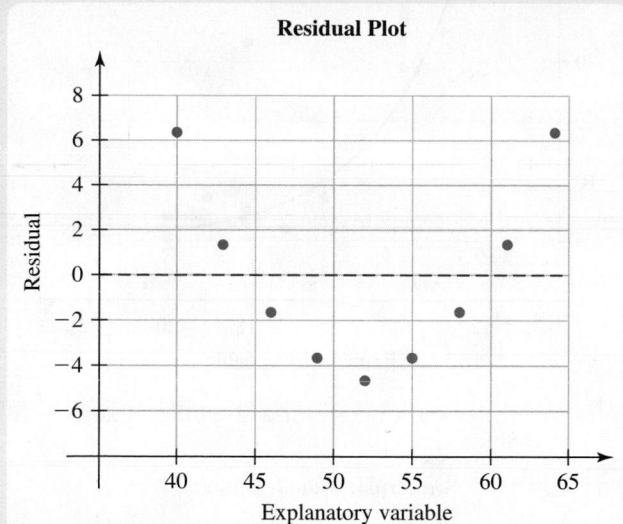

Figure 88 Exercise 13

14. See Fig. 89.

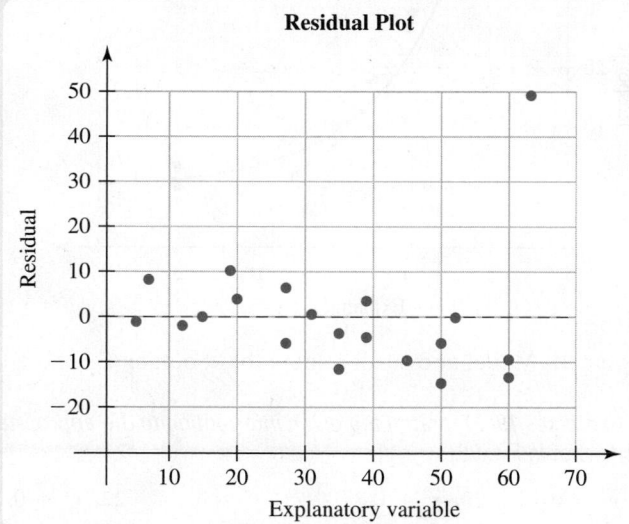

Figure 89 Exercise 14

15. See Fig. 90.

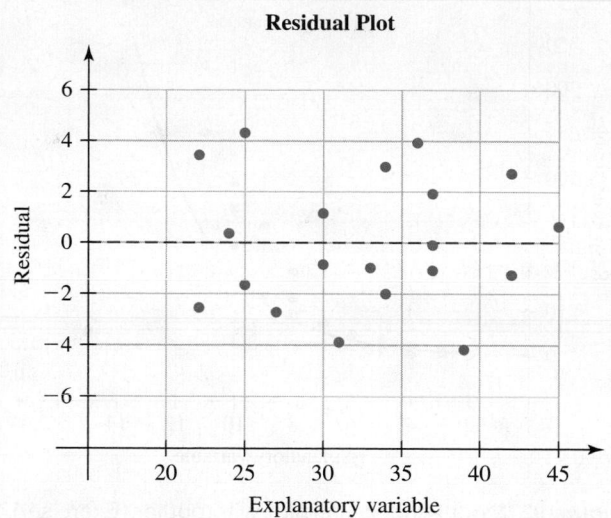

Figure 90 Exercise 15

16. See Fig. 91.

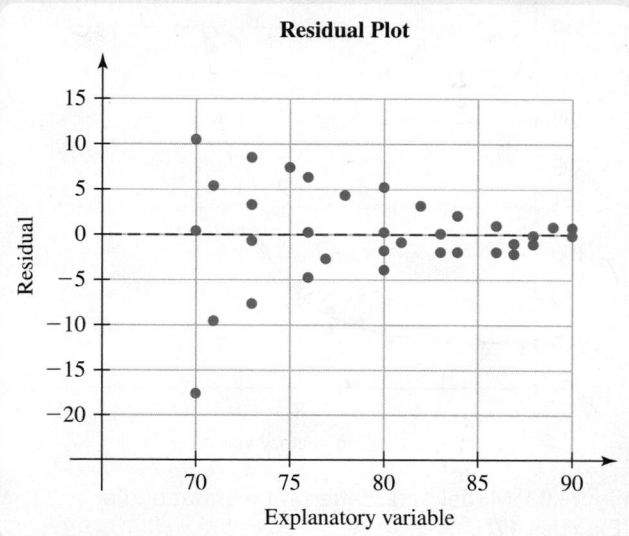

Figure 91 Exercise 16

17. Figure 92 shows an exponential regression curve and a scatterplot, including an outlier shown in red. Figure 93 shows an exponential regression curve and a scatterplot after the outlier has been removed. Determine whether the outlier is an influential point.

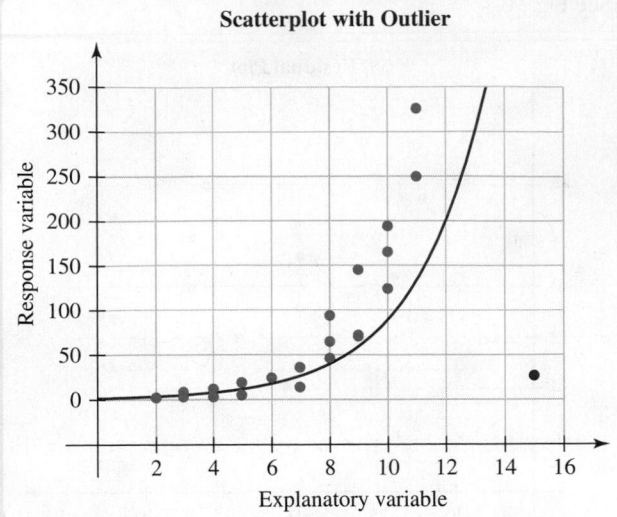

Figure 92 Model and scatterplot with outlier (Exercise 17)

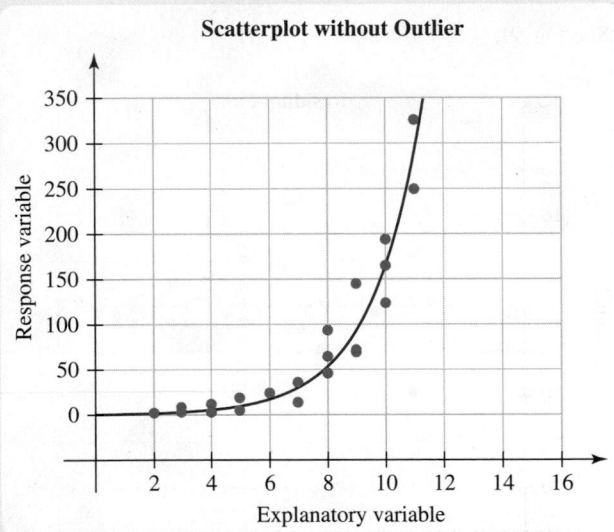

Figure 93 Model and scatterplot without outlier (Exercise 17)

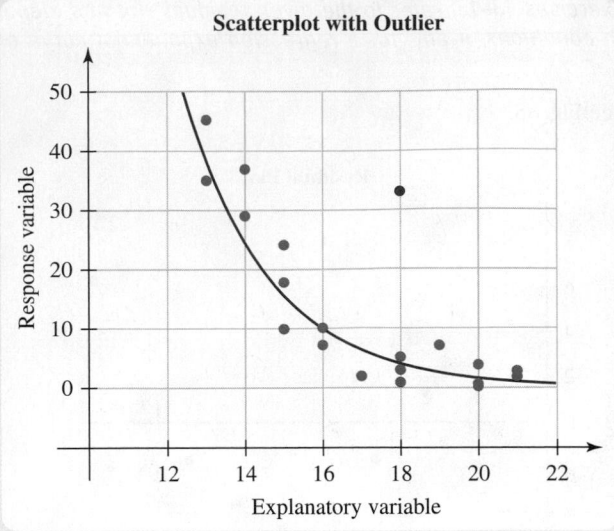

Figure 94 Model and scatterplot with outlier (Exercise 18)

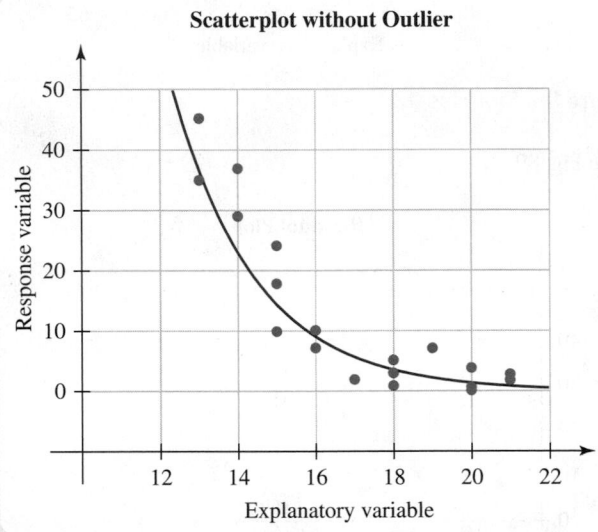

Figure 95 Model and scatterplot without outlier (Exercise 18)

18. Figure 94 shows an exponential regression curve and a scatterplot, including an outlier shown in red. Figure 95 shows an exponential regression curve and scatterplot after the outlier has been removed. Determine whether the outlier is an influential point.

For Exercises 19–22, match the given information to the appropriate scatterplot in Fig. 96.

19. $r^2 = 0.6$ **20.** $r^2 = 0.8$ **21.** $r^2 = 1$ **22.** $r^2 = 0$

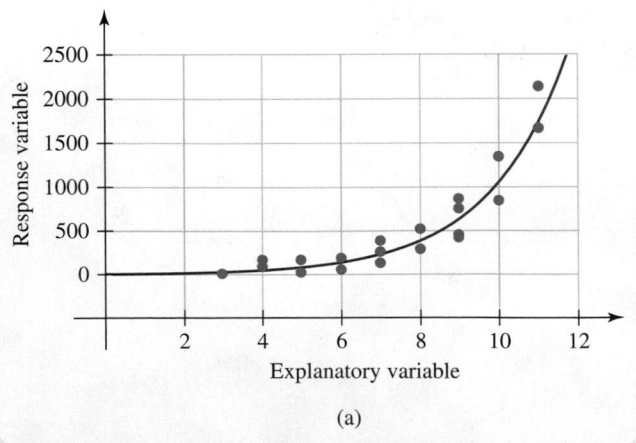

(a)

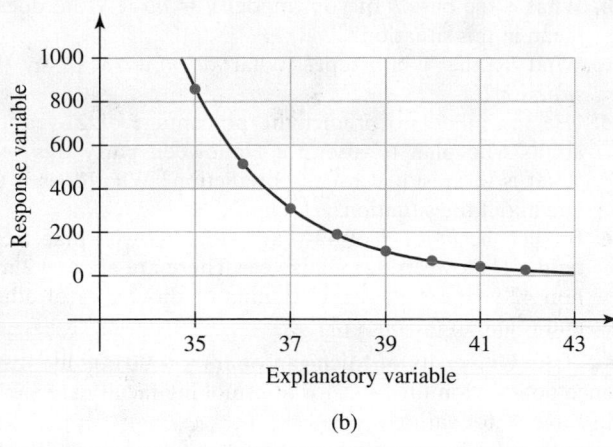

(b)

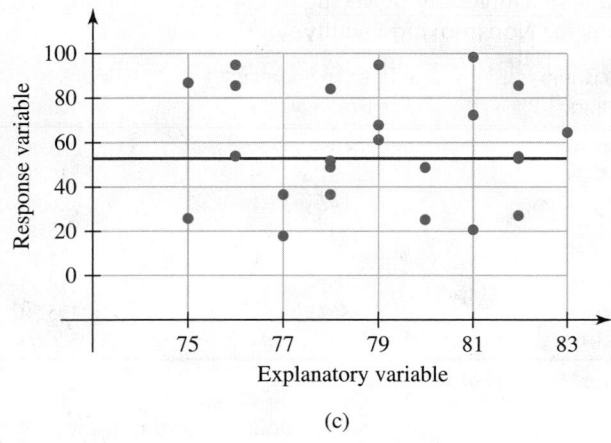

(c)

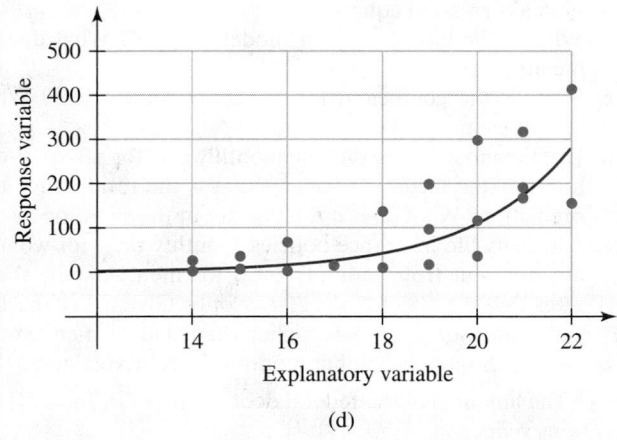

(d)

Figure 96 Exercises 19–22

23. **DATA** Find the exponential regression equation for the data set shown in Table 44. Does the graph of the equation come close to the data points in the scatterplot?

Table 44 Find the Exponential Regression Equation

x	y
3	2
4	3
4	5
7	8
11	9
13	15
17	20
17	25
20	35

24. **DATA** Find the exponential regression equation for the data set shown in Table 45. Does the graph of the equation come close to the data points in the scatterplot?

Table 45 Find the Exponential Regression Equation

x	y
2	172
5	50
8	27
8	18
11	9
14	4
14	2
17	1

25. **DATA** The numbers of subscribers to Showtime are listed in Table 46 for various years.

Table 46 Numbers of Showtime Subscribers

Year	Number of Showtime Subscribers (millions)
2004	13
2006	14
2008	16
2010	19
2012	22

Source: *SNL Kagan*

Let y be the number (in millions) of Showtime subscribers at x years since 2000.

a. Construct a scatterplot.
b. Find a regression equation.
c. Describe the four characteristics of the association. Compute and interpret an appropriate correlation coefficient as part of your analysis.
d. Use the model to estimate the number of subscribers in 2012. What is the residual for the estimate? What does it tell you about the positions of the data point and the model?

e. AT&T charged $10 per month for Showtime in 2011. Estimate the total amount of money subscribers paid for Showtime in the *entire* year 2011, assuming all Showtime subscribers paid $10 per month and they subscribed for the entire year. Use units of billions of dollars and round to the second decimal place.

26. **DATA** The numbers of North American cruise ship passengers are shown in Table 47 for various years.

Table 47 Numbers of North American Cruise Ship Passengers

Year	Number of Passengers (in millions)
1980	1.4
1985	2.2
1990	3.8
1995	4.7
2000	7.2
2005	11.2
2010	14.2

Source: *Cruise Lines International Association*

Let y be the number (in millions) of North American cruise ship passengers in the year that is x years since 1980.

a. Construct a scatterplot.
b. Find a regression equation.
c. Describe the four characteristics of the association. Compute and interpret an appropriate correlation coefficient as part of your analysis.
d. Estimate the number of North American cruise ship passengers in 2010. What is the residual for the estimate? What does it tell you about the positions of the data point and the model?
e. The average amount of money spent per passenger was $1770 per week in 2013 (Source: *American Association of Port Authorities*). Estimate the total amount of money spent by all passengers on North American cruise ships in 2013, assuming the average time passengers were on cruises was one week. Do you have much faith in this estimate? Explain.

27. **DATA** Percentages of adults surveyed who plan to attend a Halloween party this year are shown in Table 48 for various age groups.

Table 48 Percentages of Adults Who Plan to Attend a Halloween Party

Age Group (years)	Age Used to Represent Age Group (years)	Percent
18–24	21.0	44
25–34	29.5	34
35–44	39.5	25
45–54	49.5	14
55–64	59.5	10
65 or over	70.0	6

Source: *International Mass Retail Association*

Let y be the percentage of adults at age x years who plan to attend a Halloween party this year.

a. Find an exponential regression equation.

b. What is the base b of your model $\hat{y} = ab^x$? What does it mean in this situation?
c. What is the y-intercept? What does it mean in this situation?
d. Use the model to predict the percentage of 21-year-old adults who plan to attend a Halloween party this year. What is the residual for the prediction? What does it tell you about the situation?
e. Predict the *number* of 42-year-old adults who plan to attend a Halloween party this year. There are about 4.2 million 42-year-old adults. Use units of thousands of adults and round to the ones place.

28. **DATA** The University of Michigan offers a $500,000 life insurance policy. Monthly rates for nonsmoking faculty are shown in Table 49 for various ages.

Table 49 University of Michigan Life Insurance Monthly Rates for Nonsmoking Faculty

Age Group (years)	Age Used to Represent Age Group (years)	Monthly Rate (dollars)
30–34	32	15.00
35–39	37	18.50
40–44	42	26.00
45–49	47	46.00
50–54	52	75.50
55–59	57	118.00
60–64	62	195.50
65–69	67	326.60

Source: *University of Michigan*

Let y be the monthly rate (in dollars) for a nonsmoking faculty member at x years of age.

a. Find a regression equation.
b. What is the base b of your model $\hat{y} = ab^x$? What does it mean in this situation?
c. What is the coefficient a of your model $\hat{y} = ab^x$? What does it mean in this situation?
d. Use the model to predict the monthly rate for a 37-year-old nonsmoking faculty member. What is the residual for the prediction? What does it tell you about the situation?
e. For many life insurance policies, monthly rates for women are different from monthly rates for men. Assume these rates depend on life expectancy only. Given that the life expectancy of women is higher than that of men, would women or men pay higher monthly rates? Explain.

29. **DATA** The amounts of the federal debt are listed in Table 50 for various years.

Table 50 Federal Debt Amounts

Year	Federal Debt (billions of dollars)
1960	291
1970	381
1980	909
1990	3206
2000	5629
2010	13,529
2015	18,152

Source: *U.S. Office of Management and Budget*

Let y be the federal debt (in billions of dollars) at x years since 1960.

a. Find a regression equation.

b. Estimate the federal debt in 2014. Use units of trillions of dollars and round to the second decimal place.

c. If the federal debt had been paid off in 2014 by each U.S. citizen contributing an equal amount of money, how much would each person have had to pay? Round to the ones place. The population was 319 million in 2014.

d. Predict the federal debt in 2050. Use units of trillions of dollars and round to the second decimal place. Do you have much faith in this prediction? Explain.

e. Assuming the result you found in part (d) is correct, if the federal debt were paid off in 2050 by each U.S. citizen contributing an equal amount of money, how much would each person have to pay? Round to the ones place. Assume the population will be 439 million in 2050. Explain why some people want to reduce or eliminate the debt now rather than later.

30. **DATA** A person's heart attack risk can be estimated by using *Framingham point scores*, which are based on such factors as age, cholesterol level, blood pressure, and smoking habits. Men's risks of having a heart attack in the next 10 years are shown in Table 51 for various scores.

Table 51 Risks of Having a Heart Attack

Framingham Point Scores	Risk (percent)
0	1
5	2
10	6
15	20
17	30

Source: *Journal of the American Medical Association; Framingham Heart Study*

Let y be a man's risk (in percent) of having a heart attack in the next 10 years if his score is x points.

a. Find a regression equation.

b. Find the coefficient a of your model $\hat{y} = ab^x$. What does it mean in this situation?

c. A 47-year-old man with high cholesterol has high blood pressure but does not smoke. His score is 11 points. What is the risk he will have a heart attack in the next 10 years?

d. Another 47-year-old man has the same cholesterol level and blood pressure as the man described in part (c). However, this man's score is 5 points higher, because he smokes. What is the risk that he will have a heart attack in the next 10 years?

31. **DATA** World population is shown in Table 52 for various years.

Table 52 World Population

Year	Population (billions)
1930	2.070
1940	2.295
1950	2.500
1960	3.050
1970	3.700
1980	4.454
1990	5.279
2000	6.080
2010	6.916
2015	7.248

Source: *U.S. Census Bureau*

Let y be the world's population (in billions) at x years since 1930.

a. Construct a scatterplot of the data.

b. Find a regression equation. Write the equation using the function name f.

c. Find $f(85)$. What does it mean in this situation? Find the residual.

d. Construct a scatterplot for the period 1970–2015.

e. Find a regression equation for the period 1970–2015. Write the equation using the function name g.

f. Find $g(85)$. What does it mean in this situation? Find the residual. Compare your result with the residual you found in part (c). By referring to the shape of the association, explain why your comparison makes sense.

32. **DATA** The numbers of men's colleges, not including seminaries, are shown in Table 53 for various years.

Table 53 Numbers of Men's Colleges

Year	Number of Men's Colleges
1967	145
1975	80
1985	27
1995	11
2012	4

Source: *National Association of Independent Colleges and Universities*

Let $f(x)$ be the number of men's colleges at x years since 1960.

a. Construct a scatterplot.

b. Find a regression equation.

c. Describe the four characteristics of the association. Compute and interpret an appropriate correlation coefficient as part of your analysis.

d. Find $f(45)$. What does it mean in this situation? Have you performed interpolation or extrapolation?

e. Find $f(65)$. What does it mean in this situation? Have you performed interpolation or extrapolation? How much faith do you have in your result? Explain.

33. <u>DATA</u> Percentages of adults surveyed who plan to attend a Halloween party this year are shown in Table 54 for various age groups.

Table 54 Percentages of Adults Who Plan to Attend a Halloween Party

Age Group (years)	Age Used to Represent Age Group (years)	Percent
18–24	21.0	44
25–34	29.5	34
35–44	39.5	25
45–54	49.5	14
55–64	59.5	10
65 or over	70.0	6

Source: *International Mass Retail Association*

Let y be the percentage of adults at age x years who plan to attend a Halloween party this year.

a. Find the exponential regression equation.

b. The equation of the exponential curve that contains the data points (39.5, 25) and (49.5, 14) is $\hat{y} = 247(0.944)^x$. Graph the equation and the regression equation on the same scatterplot of the data. Describe how well each curve fits the data points.

c. Find the sum of the squared residuals for the model $\hat{y} = 247(0.944)^x$.

d. Find the sum of the squared residuals for the exponential regression model.

e. Compare your results in parts (c) and (d) and explain why your comparison is not surprising.

34. <u>DATA</u> The University of Michigan offers a $500,000 life insurance policy. Monthly rates for nonsmoking faculty are shown in Table 55 for various ages.

Table 55 University of Michigan Life Insurance Monthly Rates for Nonsmoking Faculty

Age Group (years)	Age Used to Represent Age Group (years)	Monthly Rate (dollars)
30–34	32	15.00
35–39	37	18.50
40–44	42	26.00
45–49	47	46.00
50–54	52	75.50
55–59	57	118.00
60–64	62	195.50
65–69	67	326.60

Source: *University of Michigan*

Let y be the monthly rate (in dollars) for a nonsmoking faculty member at x years of age.

a. Find the exponential regression equation.

b. The equation of the exponential curve that contains the data points (42, 26.00) and (67, 326.60) is $\hat{y} = 0.37(1.1065)^x$. Graph the equation and the regression equation on the same scatterplot of the data. Describe how well each curve fits the data points.

c. Find the sum of the squared residuals for the model $\hat{y} = 0.37(1.1065)^x$.

d. Find the sum of the squared residuals for the exponential regression model.

e. Compare your results in parts (c) and (d) and explain why your comparison is not surprising.

35. <u>DATA</u> In a study about beer froth, a researcher poured beer into a cylindrical beer mug and recorded the heights of the froth at the start and then every minute for the first 6 minutes. For each of three brands of beer, the researcher ran the experiment several times and computed the mean height of beer froth at each of the 7 times (see Table 56).

Table 56 Times and Mean Heights of Beer Froth

Time (minutes)	Mean Height of Froth (centimeters)		
	Erdinger Weissbier	Augustinerbräu München	Budweiser Budvar
0	17.0	14.0	14.0
1	13.2	8.5	9.3
2	10.7	6.0	7.0
3	8.9	4.4	5.5
4	7.5	2.9	3.5
5	6.3	1.3	2.0
6	5.2	0.7	0.9

Source: *Demonstration of the Exponential Decay Law Using Beer Froth*, A. Leike

Let y be the mean height (in centimeters) of froth for an Erdinger Weissbier x minutes after having been poured.

a. Construct a scatterplot that compares values of x and y.

b. Find the regression equation that models the association between x and y.

c. Describe the four characteristics of the association between x and y. Compute and interpret an appropriate correlation coefficient as part of your analysis.

d. Use the model to predict the mean height of froth 1 minute after an Erdinger Weissbier is poured.

e. Find the residual for your prediction in part (d). What does it mean in this situation?

36. <u>DATA</u> Refer to Exercise 35 for a description of a study about beer froth. Let y be the mean height (in centimeters) of froth for an Augustinerbräu München x minutes after having been poured.

a. Construct a scatterplot that compares values of x and y.

b. Find the regression equation that models the association between x and y.

c. Describe the four characteristics of the association between x and y. Compute and interpret an appropriate correlation coefficient as part of your analysis.

d. Use the model to predict the mean height of froth 4 minutes after an Augustinerbräu München is poured.

e. Find the residual for your prediction in part (d). What does it mean in this situation?

37. <u>DATA</u> Refer to Exercise 35 for a description of a study about beer froth. Let y be the mean height (in centimeters) of froth for a Budweiser Budvar x minutes after having been poured.

a. Construct a scatterplot that compares values of x and y.

b. Find the regression equation that models the association between x and y.

c. Describe the four characteristics of the association between x and y. Compute and interpret an appropriate correlation coefficient as part of your analysis.

d. Compute the coefficient of determination for the model. What does it mean in this situation?

e. Imagine a scatterplot that compares the values of x to the *individual* heights of froth, rather than the mean heights of froth. Would there be more or less scatter (spread) than in the scatterplot you constructed in part (a)? Would the coefficient of determination be larger or smaller? Explain.

38. **DATA** Refer to Exercise 35 on page 704 for a description of a study about beer froth. Let $E, A,$ and B be the heights (in centimeters) of froth for an Erdinger Weissbier, an Augustinerbräu München, and a Budweiser Budvar, all at x minutes after having been poured.

a. Find exponential regression equations for x and E, x, and A and x and B.

b. Find the vertical intercept for each of the three models. On the basis of your results, which beer had the most froth just after the beer was poured? Is that what actually happened? Explain.

c. What is the percentage rate of decay per minute for the mean height of froth of each of the three beers? Which beer had the greatest mean-height decay of froth? How you can tell this by inspecting Table 56 on page 704?

d. The researcher used a cylindrical beer mug. Why is it important that a cylindrical mug was used rather than a glass whose width varies from top to bottom?

e. The volume of a cylinder is given by $V = \pi r^2 h$, where r is the radius of the circular base and h is the height. The radius of the circular base of the beer mug was 3.6 centimeters. Predict the mean volume of froth 5.5 minutes after a Budweiser Budvar is poured.

39. **DATA** In 1797, a researcher shot a gun with various amounts of gunpowder and measured the forces of the fired gunpowder (see Table 57).

Table 57 Amounts of Gunpowder and Forces

Gunpowder (milligrams)	Force (newtons)
65	255
194	747
324	1838
454	2659
583	5081
713	7268
842	10,754
972	15,467
1166	35,953

Source: *Experiments to Determine the Force of Fired Gunpowder,* Count

Let y be the force (in newtons) of x milligrams of fired gunpowder.

a. Construct a scatterplot.

b. Find an equation of a regression model.

c. Describe the four characteristics of the association. Compute and interpret an appropriate correlation coefficient as part of your analysis.

d. What is the base b of your model $\hat{y} = ab^x$? What does it mean in this situation?

e. What is the coefficient a of the model $\hat{y} = ab^x$ What does it mean in this situation?

40. **DATA** In 1797, a researcher shot a gun with various amounts of gunpowder and measured the forces of the fired gunpowder (see Table 57). Let y be the force (in newtons) of x milligrams of fired gunpowder.

a. Find an equation of the exponential regression equation.

b. Use the model to predict the force of 454 milligrams of fired gunpowder. What is the residual? What does it tell you about the data point and the graph of the model?

c. Use the model to predict the force of 842 milligrams of fired gunpowder. What is the residual? What does it tell you about the data point and the graph of the model?

d. Compute the coefficient of determination for the model. What does it mean in this situation?

e. Construct a residual plot. Even though the coefficient of determination is quite close to 1, explain why the residual plot suggests that there might be a model that fits the data better than the one you found in part (a).

41. **DATA** The fuel consumptions of a 3000-TEU ship that travels from Singapore to Jakarta, Indonesia, are shown in Table 58 for various speeds.

Table 58 Ship Speeds and Fuel Consumptions

Speed (knots)	Fuel Consumption (tons per day)
15.4	38
15.7	39
16.3	43
16.6	45
17.1	50
17.6	56

Source: *Sailing Speed Optimization for Container Ships in a Liner Shipping Network,* S. Wang, Q. Meng

Let y be fuel consumed (in tons per day) by a ship that travels at a rate of x knots.

a. Construct a scatterplot.

b. Is the association positive, negative, or neither? What does that mean in this situation?

c. Find the exponential regression equation.

d. In part (b), you described the direction of the association. Now describe the other three characteristics of the association.

e. Predict the fuel consumption of a 3000-TEU ship that travels at 18 miles per hour. A knot is equivalent to 1 nautical mile per hour, there are 2025 yards in 1 nautical mile, and there are 5280 feet in 1 mile.

42. **DATA** The fuel consumptions of a 3000-TEU ship that travels from Singapore to Jakarta, Indonesia, are shown in Table 58 for various speeds. Let y be fuel consumed (in tons per day) by a ship that travels at a rate of x knots.

a. Find an exponential regression model.

b. Use the model to predict the fuel consumption of a 3000-TEU ship that travels at 16.6 knots.

c. Compute the residual for your prediction in part (b). What does it mean in this situation?

d. Compute the coefficient of determination for the model. What does it mean in this situation?

e. Construct a residual plot. Explain why an exponential regression equation might not be the best model even though its coefficient of determination is so close to 1.

43. **DATA** The percentages of Earth's land surface that is forested and worldwide carbon dioxide emissions are shown in Table 59 for various years.

Table 59 Percentages of Earth's Land Surface That Is Forested and Worldwide Carbon Dioxide Emissions

Year	Percentage of Earth's Land Surface That Is Forested	Carbon Dioxide Emissions (billion tonnes)
1992	31.9	22.3
1994	31.8	22.7
1996	31.7	23.7
1998	31.6	24.3
2000	31.4	24.8
2002	31.4	25.7
2004	31.3	28.6
2006	31.2	30.7
2008	31.1	32.2
2010	31.0	33.6

Source: *World Bank*

Let x be the percentage of Earth's land surface that is forested and y be worldwide carbon emissions (in billion tonnes).

a. Construct a scatterplot.
b. Find the exponential regression equation.
c. Graph the exponential regression equation on the scatterplot. Explain why the curve is below so many of the data points.
d. Remove the data points (31.4, 24.8) and (31.4, 25.7) and then find the exponential regression equation.
e. Determine whether the data points (31.4, 24.8) and (31.4, 25.7) are influential. Explain.

44. **DATA** The percentages of Earth's land surface that is forested and worldwide carbon dioxide emissions are shown in Table 59 for various years. Let x be the percentage of Earth's land surface that is forested and y be worldwide carbon emissions (in billion tonnes). After removing two influential points, the exponential regression equation is $\hat{y} = 106{,}140{,}092(0.617)^x$.

a. What is the y-intercept of the model? What does it mean in this situation?
b. What is the percentage rate of decay of carbon emissions per 1 percentage point increase in forestation, according to the model?
c. Remove the data points (31.4, 24.8) and (31.4, 25.7) and then find the coefficient of determination for the model. What does it mean in this situation?
d. On the basis of the coefficient of determination being so close to 1, a student concludes that deforestation causes an increase in worldwide carbon emissions. What would you tell the student?
e. Describe at least one possible lurking variable.

45. **DATA** From 1790 to 1860, U.S. population grew rapidly (see Table 60).

Table 60 U.S. Population

Year	Population (millions)	Population Ratio (current to previous)
1790	3.9	—
1800	5.3	1.36
1810	7.2	
1820	9.6	
1830	12.9	
1840	17.1	
1850	23.2	
1860	31.4	

Source: *U.S. Census Bureau*

a. Complete the third column of Table 60. The first entry is 1.36, since $\dfrac{1800\,\text{population}}{1790\,\text{population}} = \dfrac{5.3}{3.9} \approx 1.36$.

b. What do you observe about the ratios in the third column?
c. On the basis of your observation in part (b), is the association linear or exponential?
d. Let y be the U.S. population (in millions) at x years since 1790. Find the equation of the exponential regression curve that models the data from 1790 to 1860.
e. Complete the third column of Table 61.

Table 61 U.S. Population

Year	Population (millions)	Population Ratio (current to previous)
1860	31.4	—
1870	39.8	
1880	50.2	
1890	62.9	
1900	76.0	

Source: *U.S. Census Bureau*

f. Is it likely the model you found in part (d) gives reasonable population estimates after 1860? Explain.
g. Use the model you found in part (d) to estimate the population in 2012. The actual population was 313.9 million (Source: *U.S. Census*). What is the residual of your estimate?

46. **DATA** The total attendances at Major League Baseball (MLB) games are shown in Table 62 for various years.

Table 62 Total Attendances at MLB Games

Year	Attendance (millions)
1900	1.83
1910	5.93
1920	9.12
1930	10.13
1940	9.82
1950	17.15
1960	19.93
1970	28.75
1980	43.01
1990	54.82
2000	72.70
2010	73.17
2014	73.74

Source: *ballparksofbaseball.com*

Let y be the total attendance (in millions) at MLB games in the year that is x years since 1900.

a. Construct a scatterplot.
b. Find the linear regression equation and the exponential regression equation. Which model describes the association better?
c. Use the exponential model to estimate the total attendance in 1940. Find the residual. On the basis of the fact that World War II was happening in 1940, explain why it is not surprising that the residual is negative.
d. Use the exponential model to estimate the total attendance in 2010. Find the residual. On the basis of the fact

that the economy was suffering in 2010, explain why it is not surprising that the residual is negative.

e. Use the exponential model to estimate the total attendance in 2014. Find the residual. The economy was stronger in 2014 than in 2010. Does this suggest that MLB attendance is strongly affected by the economy?

f. On the basis of the data alone, can we be sure that World War II and/or changes in the economy caused changes in total attendance? Explain.

47. ▼**DATA** The mean numbers of lightning deaths per million people per year are shown in Table 63 for various decades, and the percentages of Americans who live in rural areas are shown in the table for various years.

Table 63 Mean Numbers of Lightning Deaths per Million People per Year; Percentages of Americans Who Live in Rural Areas

Decade	Year Used to Represent Decade	Lightning Fatality Rate	Year	Percentage of Americans Who Live in Rural Areas
1940-1949	1945	2.4	1950	36.0
1950-1959	1955	1.1	1960	30.1
1960-1969	1965	0.7	1970	26.3
1970-1979	1975	0.5	1980	26.3
1980-1989	1985	0.4	1990	24.8
1990-1999	1995	0.2	2000	22.0
2000-2010	2005	0.1	2010	21.0

Source: Storm Data, *López and Holle*; *U.S. Census Bureau*

Let y be the lightning fatality rate (mean number of lightning deaths per million people per year) at x years since 1900.

a. Find a regression equation.

b. Estimate the lightning fatality rate in 2000.

c. Use your result in part (b) to estimate the *number* of lightning deaths in 2000. The U.S. population was 281 million in that year (Source: *U.S. Census*).

d. The ratio of injuries to deaths, both from lightning, is about 10 to 1 (Source: *Cherington et al. 1999*). Use your result in part (c) to estimate the number of injuries from lightning in 2000.

e. Let A be the percentage of Americans who live in rural areas at x years since 1900. Find a regression equation.

f. It has been hypothesized that the lightning fatality rate has decreased due to the migration of Americans from rural areas to urban ones (Source: Annual Rates of Lightning Fatalities by Country, *Holle 2008*). Explain why this is probably not the only reason by comparing the meaning of the bases of the models you found in parts (a) and (e).

48. ▼**DATA** Table 64 describes countries' percentages of populations who are involved in producing agriculture and countries' per-person *gross national product* (*GNP*), which is a measure of the amount of goods and services a country produces.

Table 64 Percentages of Populations Involved in Agriculture versus Per-Person GNPs

Country	Percent of Population Involved in Agriculture	GNP per Person (dollars)
United States	1	43,743
Great Britain	1	37,632
France	2	35,854
Canada	2	32,546
Australia	3	32,170
Italy	3	29,999
Japan	5	38,984
New Zealand	9	25,942
Slovenia	9	17,352
Korea (North and South)	12	10,975
Latvia	16	6757
Chile	17	5865
Panama	23	4626
Brazil	25	3455
Columbia	32	2292
Bolivia	36	1009
Bangladesh	50	467
Vietnam	60	623

Source: *United Nations*

Let y be the GNP per person (in dollars) for a country in which x percent of the population was involved in agriculture.

a. Construct a scatterplot.

b. Is the association positive, negative, or neither? What does that mean in this situation? Explain why this makes sense in terms of productivity.

c. Find a regression equation. Describe the shape and strength of the association.

d. What is the base b of your model $\hat{y} = ab^x$? What does it mean in this situation?

e. Which country's data point has the largest residual? What is the residual? What does the position of the point in relation to the other data points and the regression curve suggest about this situation?

Large Data Sets

49. ▼**DATA** Access the data about heart rates and lactate concentrations, which are available at MyMathLab and at the Pearson Downloadable Student Resources for Math & Stats website. During intense exercise, the body produces lactate. In a 2013 study, female soccer players walked and ran on a treadmill at various levels of intensity and their heart rates and lactate concentrations in their bloodstreams were recorded (Source: Mathematical Modeling of Physiological Characteristics in Female Soccer Athletes, *Thomas S. Goeppinger*). Let y be the lactate concentration (in millimole per liter) for a player with heart rate x beats per minute.

a. Construct a scatterplot.

b. Is the association positive, negative, or neither? What does it mean in this situation? Why does it make sense?

c. Find a regression equation.

d. In part (b), you described the direction of the association. Now describe the three other characteristics of the association. Compute and interpret an appropriate correlation coefficient as part of your analysis.

e. What is the base b of your model $\hat{y} = ab^x$? What does it mean in this situation?

f. Find the change in the predicted lactate concentration for a heart rate that increases from 140 to 141 beats per minute. Do the same for a heart rate that increases from 160 to 161 beats per minute. Which result is greater? How does this relate to your response to part (e)?

50. **DATA** Access the data about *mean* tuitions at colleges, which are available at MyMathLab and at the Pearson Downloadable Student Resources for Math & Stats website. Do *not* access the other data about individual tuitions. Let y be the mean tuition at private, nonprofit, four-year colleges and w be the mean tuition at public, four-year colleges, both in dollars at x years since 1970.

a. Construct a scatterplot that describes the association between years and mean tuitions at private, nonprofit, four-year colleges. Also construct a scatterplot that describes the association between years and mean tuitions at public, four-year colleges.

b. Find a regression equation for each of the associations described in part (a).

c. Describe the four characteristics of each of the associations described in part (a). Compute and interpret an appropriate correlation coefficient as part of your analysis.

d. For each of the regression equations that you found in part (b), find and interpret the base.

e. Work with the entries in the appropriate column of the data table to find the mean one-year percentage change in mean tuition for private, nonprofit, four-year colleges for the period 1971–2015. Do the same for public, four-year colleges. Compare the two results with the two results you found in part (d).

f. Which type of college had a higher mean tuition in the academic year 2014–2015? Which type of college had the larger mean one-year percentage change in mean tuition for the period 1971–2015? Why do your two responses suggest that the two types of colleges will have equal mean tuitions at some point in the future? Do you have much faith that this will happen? Explain.

Concepts

Consider the scatterplot and the graph of the model $f(\widehat{x}) = ab^x$ in the figure. Sketch the graph of an exponential model that describes the data better; then explain how you would adjust the values of a and b of the original model to describe the data better.

51. See Fig. 97.

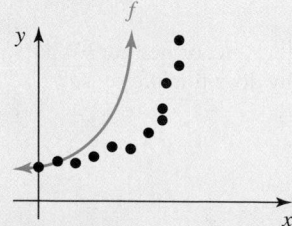

Figure 97 Exercise 51

52. See Fig. 98.

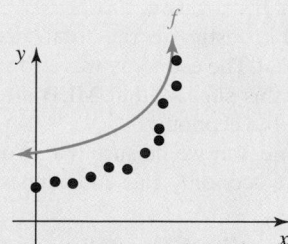

Figure 98 Exercise 52

53. When using an exponential regression curve to model some data, what is the residual of a data point? Why is it an important concept?

54. The better that an exponential model fits some data points, the lower the sum of squared residuals will be. Explain why this makes sense.

55. How can you determine whether to model a data set with a regression line, an exponential regression curve, or neither? In particular, describe how you would use scatterplots, correlation coefficients, graphs of models, residual plots, and coefficients of determinations to help you decide.

56. How are linear regression models and exponential regression models similar? How are they different?

57. For an exponential model $\hat{y} = ab^x$, explain why the coefficient a is equal to the quantity y when $x = 0$.

58. For an exponential model $\hat{y} = ab^x$ with $a > 0$ and $b > 1$, explain why the quantity y grows exponentially at a rate $b - 1$ percent (in decimal form) per unit increase of the quantity x.

59. For an exponential model $\hat{y} = ab^x$ with $a > 0$ and $0 < b < 1$, explain why the quantity y decays exponentially at a rate $1 - b$ percent (in decimal form) per unit increase of the quantity x.

60. How can a residual plot be used to help determine whether an exponential regression curve is an appropriate model?

61. When modeling some data with an exponential regression curve, explain how to determine whether a data point is an influential point. Compare this process with modeling some data with a regression line.

62. A student says that the larger the coefficient of determination is for an exponential model, the more likely a change in the explanatory variable *causes* a change in the response variable. What would you tell the student?

Hands-On Projects

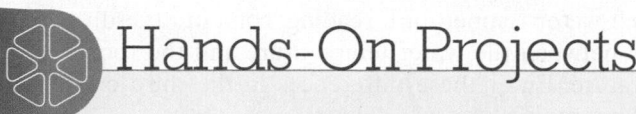

Stringed Instrument Project

Many stringed instruments (such as guitars, banjos, and basses) have thin metal strips called *frets* across the neck and underneath the strings. Frets are precisely placed so that the instruments produce the 12 chromatic notes of our Western musical scale. In this lab, you will discover where the frets of a stringed instrument must be placed to produce the 12 chromatic notes. You can apply what you learn to determine where violinists and cellists must put their fingers to produce these 12 notes.

Materials

You will need the following materials:

1. a stringed instrument with frets
2. a meterstick or tape measure

Recording of Data

Measure the length (in centimeters) of one of the strings of the instrument from the *nut* to the *bridge* (see Fig. 99). This is the length of an *open string*. Then measure the length (in centimeters) of the same string from the 12th fret to the bridge.

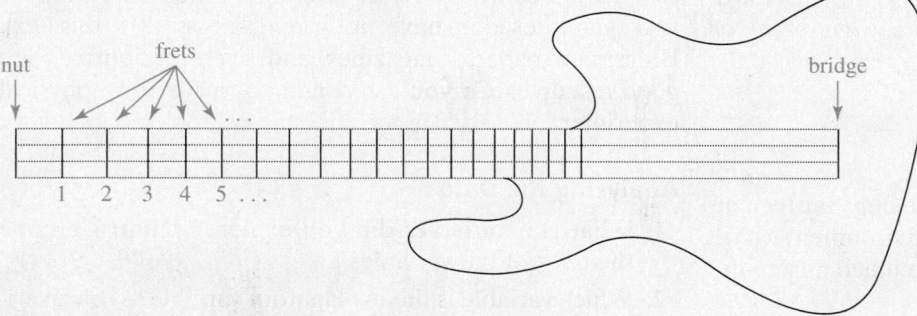

Figure 99 A bass guitar

Analyzing the Data

1. Compare the sound produced by plucking the open string with that produced by plucking the string when it is pressed just before the 12th fret (toward the nut). The higher pitched note is called the *octave* of the lower note. How do distances between the nut and the bridge and between the 12th fret and the bridge compare?

2. You probably found that when you halve the length of an open string, you can produce the octave of the open string. Since the Western chromatic musical scale has 12 notes from each note to its octave, this means that the 12th fret should be placed in the middle of the nut and the bridge. If there were a 24th fret where would it have to be placed to achieve the next octave?

3. Complete Table 65.

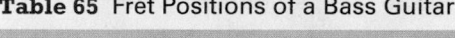

Table 65 Fret Positions of a Bass Guitar

Number of Frets	Length of String from Fret to Bridge (centimeters)
0	
12	
24	
36	
48	
n	

4. Let y be the distance (in centimeters) from the xth fret to the bridge. On the basis of the entries in Table 65, is the association linear, exponential, or neither? Explain.

5. Is the association positive, negative, or neither? What does that mean in this situation?

6. Find an equation of a model.

7. Use your model to find the distance between the fifth fret and the bridge.

8. Use a TI-84 table or other technology to find the appropriate distances for all the frets of the instrument. Compare these values with the actual distances.

9. Construct a residual plot. How does the plot relate to the comparison you made in Problem 8?

10. Compute the coefficient of determination for your model. What does it mean in this situation?

11. On a 1974 Fender® Jazz bass, one of the strings has an open-string length of 86.5 cm. Explain how the Fender music technician knew where to put the frets.

12. In this lab, you observed that an octave is achieved by halving a string. Explain why it follows that the frets of an instrument are closer together when they are closer to the bridge.

^{DATA} Cooling Water Project

In this lab, you will relate the temperature of some heated water to the amount of time that the water has been cooling.

In case the necessary measuring devices are not available for this experiment, some sample data are listed in Table 66.

Table 66 Temperatures of Water

Time (minutes)	Water Temperature (°C)	Difference Between Water Temperature and Room Temperature (°C)
0	83.49	
5	71.22	
10	63.09	
15	57.23	
20	52.65	
25	48.83	
30	45.63	

Source: *J. Lehmann*
*Room temperature = 21.7°C

Materials

To do this lab, you will need the following materials:

1. hot water
2. a coffee cup
3. a temperature probe that measures temperatures in degrees Celsius. Or measure temperatures in degrees Fahrenheit and use the formula $C = \dfrac{F - 32}{1.8}$ to convert from degrees Fahrenheit to degrees Celsius. *If you are using a thermometer as the temperature probe, make sure that it can handle hot water; otherwise it might break or even explode.*
4. a timing device

Preparation

Heat some water and pour it into a coffee cup. Set the cup on a counter or table to cool. Choose an environment where the temperature of the air will not change much during the 30-minute experiment.

Recording of Data

Record the temperature of the air at the start of the experiment and then again at the end. If the room temperatures at the beginning and end of the experiment are significantly different, redo the experiment in an environment where the room temperature will not change much. Also, record the water temperature every five minutes for 30 minutes.

Analyzing the Data

1. Find the average of the room temperature in degrees Celsius at the start and end of the experiment. This average will be referred to as *the* room temperature. If you are using the sample data in Table 66, the room temperature when it was collected was 21.7°C.
2. Display the water temperature data in the first two columns of a table like Table 66.

3. For each water temperature reading, compute the difference between the water temperature and the room temperature. Enter these differences in the third column of your table.
4. Let *y* be the difference between the water and room temperatures (in degrees Celsius) at *x* minutes after the water is allowed to cool. Construct a scatterplot, comparing *x* with *y*.
5. Find an equation of a model.
6. Graph your model on the scatterplot.
7. Describe the four characteristics of the association. Compute and interpret an appropriate correlation coefficient as part of your analysis.
8. If your model is linear, what is the meaning of the slope? If your model is exponential, what is the meaning of the base?
9. Use your model to predict the water temperature at 22 minutes.
10. What will be the temperature of the water when it stops cooling? Does your model predict when this temperature will be reached? According to your model, how much time will it take for this to happen? Is this a reasonable prediction? Explain.

Exponential Project: Topic of Your Choice

Your objective in this lab is to use an exponential model to describe some authentic situation. Find some data on two quantities that have not been discussed in this text. Blogs, newspapers, magazines, and scientific journals are good resources. Or you can conduct a survey or a physical experiment.

Analyzing the Data

1. What two variables did you explore? [**Hint:** Describe the units of the variables.]
2. Which variable is the explanatory variable? Which variable is the response variable? Explain.
3. Does it make sense to you why an exponential model would describe your situation well? Explain.
4. State the source of your data. If you conducted a survey or a physical experiment, provide a careful description with specific details of how you conducted it.
5. Include a table of your data.
6. Construct a scatterplot.
7. Find the exponential regression equation.
8. Graph your exponential model on the scatterplot.
9. Describe the four characteristics of the association. Compute and interpret the exponential correlation coefficient as part of your analysis.
10. Compute the coefficient of determination for your model. What does it mean in this situation?

11. What is the base b of your exponential model $\hat{y} = ab^x$? What does it mean in this situation?

12. What is the coefficient a of your exponential model $\hat{y} = ab^x$? What does it mean in this situation?

13. Find any intercepts of your exponential model. What do they mean in this situation?

14. Choose an input of your model. Find the output that comes from the input. What does your result mean in this situation?

15. Comment on your lab experience.

a. For example, you might address whether this lab was enjoyable, insightful, and so on.

b. Were you surprised by any of your findings? If so, which ones?

c. How would you improve your lab procedure if you did this project again?

d. How would you improve your procedure if you had more time and money?

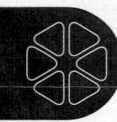

Chapter Summary

Key Points of Chapter 10

Section 10.1   Integer Exponents

Negative-integer exponent	If n is a counting number and $b \neq 0$, then $b^{-n} = \dfrac{1}{b^n}$.
Negative-integer exponent in a denominator	If n is a counting number and $b \neq 0$, then $\dfrac{1}{b^{-n}} = b^n$.
Properties of integer exponents	If m and n are integers, $b \neq 0$, and $c \neq 0$, then

- $b^m b^n = b^{m+n}$ **Product property for exponents**

- $\dfrac{b^m}{b^n} = b^{m-n}$ **Quotient property for exponents**

- $(bc)^n = b^n c^n$ **Raising a product to a power**

- $\left(\dfrac{b}{c}\right)^n = \dfrac{b^n}{c^n}$ **Raising a quotient to a power**

- $(b^m)^n = b^{mn}$ **Raising a power to a power**

Simplifying expressions involving integer exponents	An expression involving integer exponents is simplified if

1. It includes no parentheses.
2. Each variable or constant appears as a base as few times as possible.
3. Each numerical expression has been calculated, and each numerical fraction has been simplified.
4. Each exponent is positive.

Exponential function	An **exponential function** is a function whose equation can be put into the form $f(x) = ab^x$, where $a \neq 0$, $b > 0$, and $b \neq 1$. The constant b is called the **base**.

Section 10.2   Rational Exponents

Definition of $b^{1/n}$	For the counting number n, where $n \neq 1$,

- If n is odd, then $b^{1/n}$ is the number whose nth power is b, and we call $b^{1/n}$ the **nth root of b**.
- If n is even and $b \geq 0$, then $b^{1/n}$ is the nonnegative number whose nth power is b, and we call $b^{1/n}$ the **principal nth root of b**.
- If n is even and $b < 0$, then $b^{1/n}$ is not a real number.

$b^{1/n}$ may be represented by $\sqrt[n]{b}$.

Section 10.2   Rational Exponents (*Continued*)

Rational exponent	For the counting numbers m and n, where $n \neq 1$ and b is any real number for which $b^{1/n}$ is a real number, • $b^{m/n} = (b^{1/n})^m = (b^m)^{1/n}$ • $b^{-m/n} = \dfrac{1}{b^{m/n}}, b \neq 0$
Properties of rational exponents	If m and n are rational numbers and b and c are any real numbers for which b^m, b^n, and c^n are real numbers, then • $b^m b^n = b^{m+n}$ **Product property for exponents** • $\dfrac{b^m}{b^n} = b^{m-n}, b \neq 0$ **Quotient property for exponents** • $(bc)^n = b^n c^n$ **Raising a product to a power** • $\left(\dfrac{b}{c}\right)^n = \dfrac{b^n}{c^n}, c \neq 0$ **Raising a quotient to a power** • $(b^m)^n = b^{mn}$ **Raising a power to a power**

Section 10.3   Graphing Exponential Models

Base multiplier property	For an exponential function of the form $y = ab^x$, if the value of x increases by 1, then the value of y is multiplied by b.
Increasing or decreasing property	Let $f(x) = ab^x$, where $a > 0$. Then • If $b > 1$, then the function f is increasing. We say the function **grows exponentially.** • If $0 < b < 1$, then the function f is decreasing. We say the function **decays exponentially**.
y-intercept of the graph of an exponential function	For the graph of an exponential function of the form $y = ab^x$, the y-intercept is $(0, a)$.
Reflection property	The graphs of $f(x) = -ab^x$ and $g(x) = ab^x$ are **reflections** of each other across the x-axis.
x-intercept and horizontal asymptote of an exponential function	The graph of an exponential function does not have any x-intercepts and the x-axis is a horizontal asymptote.
Domain	The domain of any exponential function $f(x) = ab^x$ is the set of real numbers.
Range	The range of an exponential function $f(x) = ab^x$ is the set of all positive real numbers if $a > 0$, and the range is the set of all negative real numbers if $a < 0$.
Exponential model	An **exponential model** is an exponential function, or its graph, that describes an authentic association.
Meaning of the coefficient of an exponential model	Assume that an association between the variables x and y can be described exactly by the exponential model $\hat{y} = ab^x$. Then the coefficient a equals the quantity y when the quantity x is 0.
Meaning of the base of an exponential model	Assume that an association between the variables x and y can be described exactly by the exponential model $\hat{y} = ab^x$, where $a > 0$. Then the following statements are true. • If $b > 1$, then the quantity y grows exponentially at a rate $b - 1$ percent (in decimal form) per unit increase of the quantity x. • If $0 < b < 1$, then the quantity y decays exponentially at a rate of $1 - b$ percent (in decimal form) per unit increase of the quantity x.

Section 10.4   Using Two Points to Find an Equation of an Exponential Model

Solving $b^n = k$ for b	To solve an equation of the form $b^n = k$ for b, • If n is odd, the real-number solution is $k^{1/n}$. • If n is even and $k \geq 0$, the real-number solutions are $\pm k^{1/n}$. • If n is even and $k < 0$, there is no real-number solution.
Dividing left sides and right sides of two equations	If $a = b$, $c = d$, $c \neq 0$, and $d \neq 0$, then $\dfrac{a}{c} = \dfrac{b}{d}$.

Section 10.5   Exponential Regression Model

Properties of the exponential correlation coefficient	Assume r is the exponential correlation coefficient for the association between two numerical variables. Then • The values of r are between -1 and 1, inclusive. • If r is positive, then the variables are positively associated. • If r is negative, then the variables are negatively associated. • If $r = 0$, then there is no exponential association. • The larger the value of $\lvert r \rvert$ is, the stronger the exponential association will be. • If $r = 1$, then the data points lie exactly on an exponential curve and the association is positive. • If $r = -1$, then the data points lie exactly on an exponential curve and the association is negative.
Residuals for data points above, below, or on a line	Suppose that some data points are modeled by an exponential curve. • A data point on the curve has residual equal to 0. • A data point above the curve has positive residual. • A data point below the curve has negative residual.
Exponential regression function, curve, equation, and model	For a group of points, the **exponential regression function** is the exponential function with the least sum of squared residuals of all exponential functions. Its graph is called the **exponential regression curve** and its equation is called the **exponential regression equation,** written $\hat{y} = ab^x$, where $a \neq 0$, $b > 0$, and $b \neq 1$. The **exponential regression model** is the exponential regression function for a group of *data* points.
Using a residual plot to assess an exponential regression function	The following statements apply to a residual plot for an exponential regression function. • If the residual plot has a pattern where the dots do not lie close to the zero residual line, then there is either a non-exponential association between the explanatory and response variables or there is no association. • If a dot lies much farther away from the zero residual line than most or all of the other dots, then the dot corresponds to an outlier. If the outlier is neither adjusted nor removed, the exponential regression function may *not* be an appropriate model. • The vertical spread of the residual plot should be about the same for each value of the explanatory variable.
Exponential coefficient of determination	Let r be the exponential correlation coefficient for a group of data points. The exponential coefficient of determination, r^2, is the proportion of the variation in the response variable that is explained by the exponential regression curve.

Chapter 10 Review Exercises

Simplify. Assume b and c are positive.

1. $(2b^5 b^4)^3$

2. $(b^2 c)^4 (bc^5)^2$

3. $\dfrac{2^{-400}}{2^{-405}}$

4. $(8b^{-3} c^5)(6b^{-9} c^{-2})$

5. $\dfrac{4b^{-3} c^{12}}{16b^{-4} c^3}$

6. $(2b^{-5} c^{-2})^3 (3b^4 c^{-6})^{-2}$

7. $(37b^{-3} c^4)^{-97} (37b^{-3} c^4)^{97}$

8. $\dfrac{(20b^{-2} c^{-9})(27b^5 c^3)}{(18b^3 c^{-1})(30b^{-1} c^{-4})}$

9. $32^{4/5}$

10. $16^{-3/4}$

11. $b^{-4/5} b^{2/3}$

12. $\dfrac{b^{-1/3}}{b^{4/3}}$

13. $\dfrac{(16b^8 c^{-4})^{1/4}}{(25b^{-6} c^4)^{3/2}}$

14. $\left(\dfrac{32b^2 c^5}{2b^{-6} c^1} \right)^{1/4}$

15. $(8^{2/3} b^{-1/3} c^{3/4})(64^{-1/3} b^{1/2} c^{-5/2})$

For $f(x) = 4^x$ and $g(x) = 3(5)^x$, find the following.

16. $f(3)$

17. $g(-2)$

For $f(x) = 49^x$ and $g(x) = 2(81)^x$, find the following.

18. $f\left(\dfrac{1}{2}\right)$

19. $g\left(-\dfrac{3}{4}\right)$

Graph the equation by hand.

20. $f(x) = 2(3)^x$

21. $k(x) = -18\left(\dfrac{1}{3}\right)^x$

Graph the function by hand. Find its domain and range.

22. $h(x) = -3(2)^x$

23. $g(x) = 12\left(\dfrac{1}{2}\right)^x$

24. Input–output pairs of four exponential functions are shown in Table 67. Complete the table.

Table 67 Values of Four Functions (Exercise 24)

x	f(x)	g(x)	h(x)	k(x)
2	3		2	96
3	15			
4	75	18		
5			128	
6		2		6

25. The mean ticket prices to Major League Baseball (MLB) games are shown in Table 68 for various years.

Table 68 Mean Ticket Prices to MLB Games

Year	Mean Ticket Price (dollars)
1950	1.54
1960	1.96
1970	2.72
1980	4.45
1991	8.84
2000	16.22
2010	26.74
2014	27.93

Sources: *The Sporting News and the Sporting News Baseball Dope Book, 1950–85; Team Marketing Report, 1991–2014*

Let y be the mean ticket price (in dollars) to MLB games for the year that is x years since 1950.

a. Construct a scatterplot by hand.
b. Graph the equation $\hat{y} = 1.22(1.05)^x$ by hand on the scatterplot. Then describe the shape and strength of the association.
c. What is the y-intercept? What does it mean in this situation?
d. Use the model to estimate the mean ticket prices in 2013 and 2014.
e. Find the ratio of your estimated 2014 mean ticket price to your estimated 2013 mean ticket price. Round your result to the second decimal place. Compare the ratio with the base 1.05 of the model $\hat{y} = 1.22(1.05)^x$. Why does your comparison make sense?

For Exercises 26–31, find all real-number solutions of the equation. Round any result(s) to the second decimal place.

26. $b^3 = 8$

27. $2b^5 = 60$

28. $3.9b^7 = 283.5$

29. $5b^4 - 13 = 67$

30. $\dfrac{1}{3}b^2 - \dfrac{1}{5} = \dfrac{2}{3}$

31. $\dfrac{b^7}{b^2} = \dfrac{83}{6}$

For Exercises 32–35, find an approximate equation $y = ab^x$ of the exponential curve that contains the given pair of points. Round the values of a and b to two decimal places.

32. $(0, 2)$ and $(5, 3)$

33. $(0, 3.8)$ and $(4, 113.2)$

34. $(3, 30)$ and $(9, 7)$

35. $(5, 6.9)$ and $(20, 78.6)$

36. ^{DATA} Zimride is a website where people at least 18 years old can offer and get paid for shared car rides. The numbers of users of the site are shown in Table 69 for various years.

Table 69 Numbers of Users of Zimride

Year	Number of Users (thousands)
2007	6
2008	10
2009	35
2010	105
2011	200
2012	400

Source: *Zimride*

Let y be the number (in thousands) of users of Zimride at x years since 2000.

a. Construct a scatterplot.
b. Use two points to find an equation of a model. Graph the model on the scatterplot.
c. Describe the four characteristics of the association.
d. What is the base b of the model $y = ab^x$? What does it mean in this situation?
e. What is the coefficient a of the model $\hat{y} = ab^x$? What does it mean in this situation?

37. Refer to the residual plot in Fig. 100 and identify which conditions, if any, for a regression exponential curve that are not met.

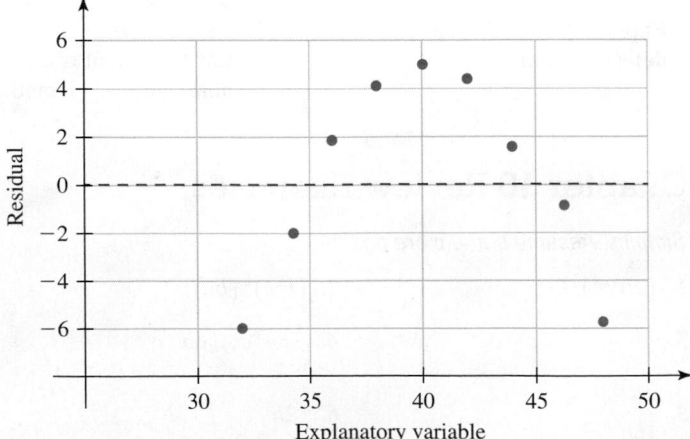

Figure 100 Exercise 37

38. **DATA** Table 70 shows the unemployment rates for people with various full-time-equivalent years of education.

Table 70 Years of Education versus Unemployment Rate

Grade-Level Completion or Degree	Full-Time-Equivalent Years of Education	Unemployment Rate (percent)
9th to 12th Grade, no diploma	10	7.3
High school	12	4.2
Some college	13	3.5
Associate	14	2.9
Bachelor's	16	2.5
Master's	18	2.1
Doctoral	20	1.1

Source: *U.S. Census Bureau*

Let *y* be the unemployment rate (in percent) for people with a full-time equivalent of *x* years of education.

a. Construct a scatterplot.

b. Find an equation of a regression model.

c. Describe the four characteristics of the association. Compute and interpret an appropriate correlation coefficient as part of your analysis.

d. Compute and interpret the coefficient of determination.

e. Predict the unemployment rate for people who have 16 full-time-equivalent years of education. Compute and interpret the residual for your result.

39. **DATA** New York Life offers a $250,000 life insurance policy. Quarterly rates for women and men are shown in Table 71 for various ages.

Table 71 New York Life Quarterly Rates for a $250,000 Policy

Age (years)	Quarterly Rate (dollars)	
	Women	Men
35	25.00	28.75
40	33.75	35.75
45	51.25	57.50
50	70.00	87.50
55	104.50	145.00
60	145.75	230.75
64	220.00	355.00

Source: *New York Life*

Let *W(x)* and *M(x)* be the quarterly rates (in dollars) for women and men, respectively, both at *x* years of age.

a. Find exponential regression equations of *W* and *M*.

b. Compare the bases of your two models. What does your comparison mean in this situation?

c. For a $250,000 policy, how much would a 52-year-old woman pay per quarter? How much would a 52-year-old man pay per quarter?

d. Due to Montana insurance regulations, both sexes must pay the same quarterly rates. So, New York Life uses the male rates for all residents of Montana. Estimate how much more a 62-year-old woman would pay per quarter for a $250,000 policy if she lived in Montana rather than in some other state.

40. **DATA** The Yellowstone cutthroat trout population has decreased significantly in Yellowstone Lake in part due to increasing numbers of lake trout, which are not native to the park. The National Park Service has been removing lake trout in hopes that the population will collapse so the Yellowstone cutthroat trout can make a comeback. The *catch rate* is the number of lake trout that are caught per 100 meters of net per night. The numbers of lake trout removed and the catch rates are shown in Table 72 for various years.

Table 72 Numbers of Lake Trout Removed and Catch Rates

Year	Number of Lake Trout Removed (thousands)	Catch Rate (number of lake trout caught per 100 meters of net per night)
2002	16	0.9
2003	24	1.0
2004	22	1.8
2005	37	1.9
2006	64	2.4
2007	76	2.6
2008	79	4.7
2009	102	4.8
2010	148	4.9
2011	223	8.2
2012	302	

Source: *Yellowstone Center for Resources*

Let *n* be the total number (in thousands) of lake trout removed and *c* be the catch rate (number of lake trout caught per 100 meters of net per night), both in the year that is *x* years since 2000.

a. Find a regression equation that models the association between *x* and *n*.

b. Find a regression equation that models the association between *x* and *c*.

c. Which of the two models is the better indicator of the number of lake trout that remain in the lake? Explain.

d. Compare the bases of the two models. What does your comparison mean in this situation?

e. Use the catch-rate model to predict the catch rate in 2012. The actual catch rate was 6.7 lake trouts per 100 meters of net per night. Find the residual for your prediction.

f. On the basis of the comparison you made in part (d), explain why it is not surprising that the residual you found in part (e) is negative?

41. Consider the scatterplot of data and the graph of the model $\hat{y} = ab^x$ in Fig. 101. Sketch the graph of an exponential model that describes the data better; then explain how you would adjust the values of *a* and *b* of the original model to describe the data better.

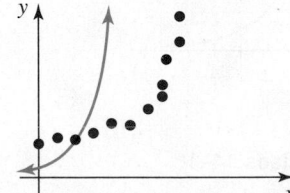

Figure 101 Exercise 41

Chapter 10 Test

Simplify.

1. $32^{2/5}$

2. $-8^{-4/3}$

Simplify. Assume b and c are positive.

3. $(2b^3c^8)^3$

4. $\left(\dfrac{4b^{-3}c}{25b^5c^{-9}}\right)^0$

5. $\dfrac{b^{1/2}}{b^{1/3}}$

6. $\dfrac{25b^{-9}c^{-8}}{35b^{-10}c^{-3}}$

7. $\left(\dfrac{6b(b^3c^{-2})}{3b^2c^5}\right)^2$

8. $\dfrac{(25b^8c^{-6})^{3/2}}{(7b^{-2})(2c^3)^{-1}}$

For f(x) = 4ˣ, find the following.

9. $f(-2)$

10. $f\left(-\dfrac{3}{2}\right)$

For Exercises 11 and 12, graph the function by hand. Find its domain and range.

11. $f(x) = -5(2)^x$

12. $f(x) = 18\left(\dfrac{1}{3}\right)^x$

13. For each graph in Fig. 102, find an equation of an exponential function that could fit the graph.

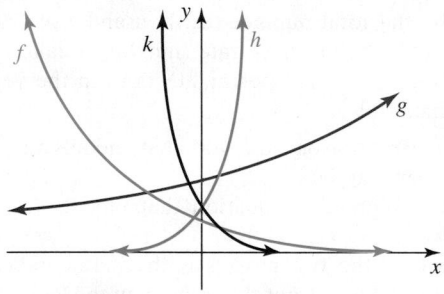

Figure 102 Exercise 13

For Exercises 14–16, refer to Fig. 103.

14. Find $f(0)$.

15. Find x when $f(x) = 3$.

16. Find an equation of f.

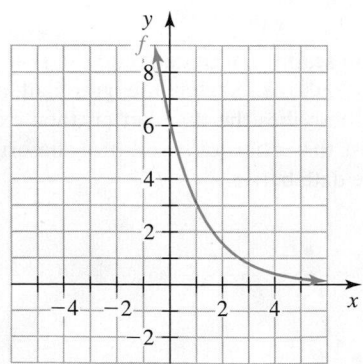

Figure 103 Exercises 14–16

17. The numbers of hours of video uploaded to YouTube per minute are shown in Table 73 for various years.

Table 73 Video Upload Rates to YouTube

Year	Number of Hours of Video Uploaded per Minute
2007	6
2008	10
2009	20
2010	24
2011	48
2012	60
2013	100

Source: *YouTube*

Let y be the video upload rate (number of hours of video uploaded per minute) to YouTube at x years since 2005.

   a. Construct a scatterplot by hand.
   b. Graph the model $\hat{y} = 2.6(1.59)^x$ by hand on the scatterplot. Does the model come close to the data points?
   c. What is the base b of the model? What does it mean in this situation?
   d. Use your model to estimate the video upload rate to YouTube in 2014. The actual rate was 300 hours of video per minute. Compute and interpret the residual.
   e. A student says your model should estimate the 2014 video upload rate well because the model fits the data shown in Table 73 so well. What would you tell the student?

18. Find all real-number solutions of $3b^6 + 5 = 84$. Round any result(s) to the second decimal place.

For Exercises 19 and 20, find an approximate equation $y = ab^x$ of an exponential curve that contains the given pair of points. Round the values of a and b to two decimal places.

19. $(0, 70)$ and $(6, 20)$

20. $(4, 9)$ and $(7, 50)$

21. **DATA** The numbers of U.S. bank failures are shown in Table 74 for various years.

Table 74 Numbers of U.S. Bank Failures

Year	Number of Bank Failures
2010	157
2011	92
2012	51
2013	24
2014	18

Source: *Federal Deposit Insurance Corporation*

Let y be the number of bank failures in the year that is x years since 2010.

   a. Construct a scatterplot.
   b. Use two points to find an equation of an exponential model.
   c. Graph the model on the scatterplot. Verify that the model passes through the two points you chose in finding the equation in part (b) and that it comes close to all of the data points.

d. What is the coefficient a of your model $\hat{y} = ab^x$? What does it mean in this situation?

e. Use the model to estimate the number of banks that failed in 2014. Compute and interpret the residual.

22. Figure 104 shows an exponential regression curve and a scatterplot, including an outlier shown in red. Figure 105 shows an exponential regression curve and scatterplot after the outlier has been removed. Determine whether the outlier is an influential point.

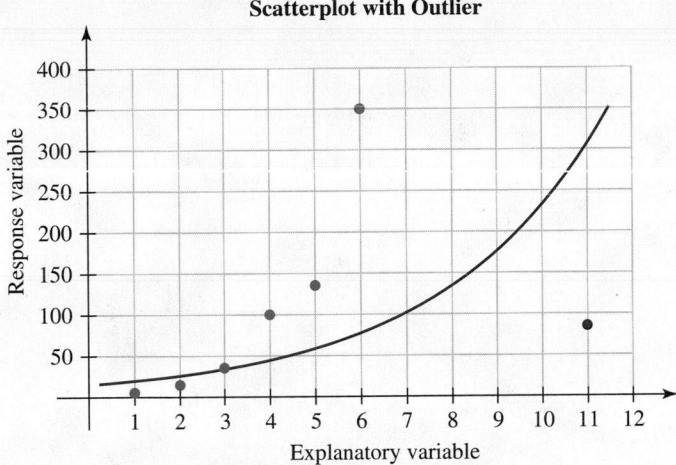

Figure 104 Model and scatterplot with outlier (Exercise 22)

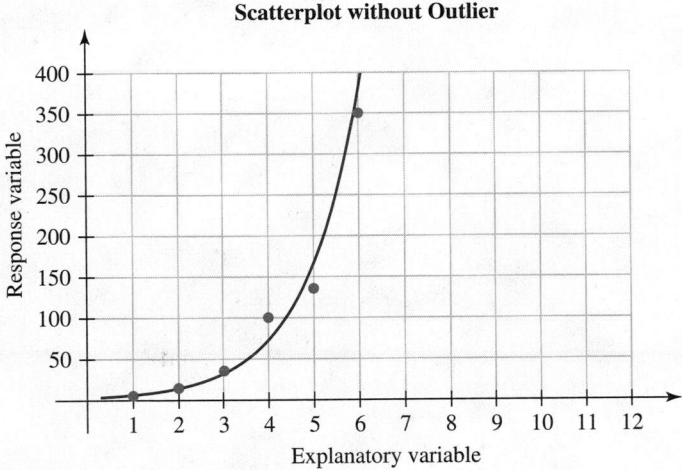

Figure 105 Model and scatterplot without outlier (Exercise 22)

23. The minimum salaries for major league baseball players are shown in Table 75 for various years.

Table 75 Minimum Salaries for Major League Baseball Players

Year	Minimum Salary (thousands of dollars)
1970	12
1975	16
1980	30
1985	60
1990	100
1995	109
2000	200
2005	316
2010	400
2012	480

Source: *baseball-reference.com*

Let y be the minimum salary (in thousands of dollars) at x years since 1970.

a. Construct a scatterplot.

b. Find an equation of a regression model.

c. Compute and interpret the coefficient of determination for the model.

d. Estimate the minimum salary in 2011.

e. Construct a residual plot. Identify two data points that might be considered outliers. Do you have much faith in your result in part (d)? Explain.

Using a TI-84 Graphing Calculator

The more you experiment with a TI-84, the more comfortable and efficient you will become with it.

A TI-84 can detect several types of errors and display an error message. When this occurs, refer to Appendix A.28 for explanations of some common error messages and how to fix these types of mistakes. Errors do not hurt the calculator. In fact, you can't hurt the calculator regardless of the order in which you press its keys. So, the more you experiment with the calculator, the better off you will be.

To access a command written in blue above a key, first press ⃞2ND⃞, then the key. Whenever a key must follow the ⃞2ND⃞ key, this appendix will use brackets for the key. For example, "Press ⃞2ND⃞ [OFF]" means to press ⃞2ND⃞ and then press ⃞ON⃞ (because "OFF" is written in blue above the ⃞ON⃞ key).

▼ A.1 Turning a Graphing Calculator On or Off

To turn a graphing calculator on, press ⃞ON⃞. To turn it off, first press ⃞2ND⃞. Then press [OFF].

▼ A.2 Making the Screen Lighter or Darker

To make the screen darker, first press ⃞2ND⃞ (then release it); then hold the ⃞△⃞ key down for a while. To make the screen lighter, first press ⃞2ND⃞ (then release it); then hold the ⃞▽⃞ key down for a while.

▼ A.3 Selecting Numbers Randomly

To randomly select 6 numbers from the numbers 53 through 98,

1. Press ⃞2ND⃞ [QUIT] to reach the home screen.
2. Pick any counting number for the seed. Because it was 7:42 a.m. when this sentence was written, we will use 742.

When we show two or more buttons in a row, press them one at a time and in order.

3. Press **742** ⃞STO ▷⃞ ⃞MATH⃞ to reach the Math screen (see Fig. 1). Next, press ⃞▷⃞ three times to reach the PRB (probability) screen (see Fig. 2). Then press **1** ⃞ENTER⃞ to store 742 in the "rand" command (see Fig. 3).

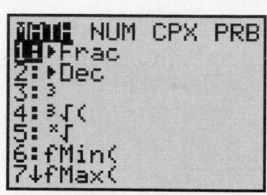

Figure 1 MATH screen

Figure 2 Probability screen

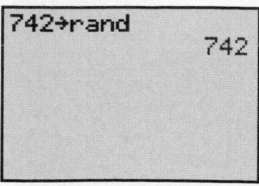

Figure 3 Storing 742 in "rand" command

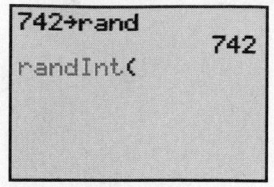

Figure 4 Selecting "randInt" command

4. Press MATH to get back to the Math screen (see Fig. 1). Next, press ▷ three times to get back to the PRB screen (see Fig. 2). Then press **5** to select the "randInt" command (see Fig. 4).

5. To use the "randInt" command, you must enter the smallest number that might be selected (53), the largest number that might be selected (98), and the number of random numbers to be selected. To begin, we select only 4 numbers so that they are all visible. Press **53**, **98**, **4**). See Fig. 5.

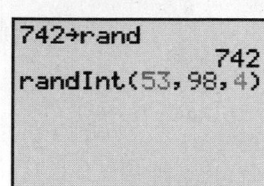

Figure 5 Setting up "randInt" command

6. Press ENTER. The first 4 numbers are 62, 56, 62, 60 (see Fig. 6).

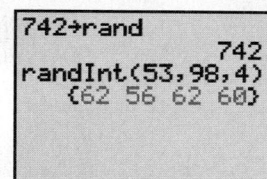

Figure 6 Selecting the first 4 numbers

7. Press ENTER. The next 4 numbers are 77, 53, 71, and 73 (see Fig. 7).

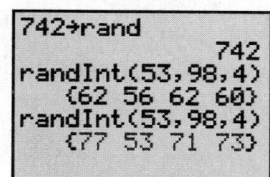

Figure 7 Selecting the next 4 numbers

8. If the numbers are to be selected *with* replacement, the 6 numbers are 62, 56, 62, 60, 77, and 53. If the numbers are to be selected *without* replacement, the 6 numbers are 62, 56, 60, 77, 53, and 71.

▼ **A.4** Entering Data for a Single Variable

To enter the data 23, 29, 31, 34, 39, 40, 45, and 69 in the STAT LIST editor,

1. Press STAT **1**.

To clear a column, make sure you press CLEAR rather than DEL. If you press DEL, the column will vanish. If you ever do this by mistake, press STAT 5 ENTER to get back the missing column.

2. If there are numbers listed in the first column (list L1), clear the column by pressing ◁ as many times as necessary to reach list L1. Next, press △ once to reach the top of list L1. L1 should now be highlighted. Then press CLEAR ENTER.

3. Press **23** ENTER **29** ENTER **31** ENTER **34** ENTER **39** ENTER **40** ENTER **45** ENTER **69** ENTER to enter the data in list L1 (see Fig. 8). The observations 23

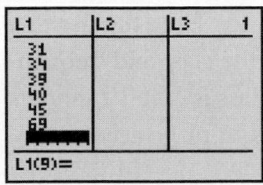

Figure 8 Entering data

and 29 are not visible because the editor scrolled down to fit the other observations. You can view them by pressing △ several times or by holding △ down for a while.

4. If you enter a number incorrectly, use △ or ▽ as many times as necessary to move the cursor over the number, type the correct number, and press ENTER. To delete a number, press DEL. To insert a number, press 2ND [INS], type the number, and press ENTER.

▼ A.5 Constructing a Frequency Histogram

To construct a frequency histogram for the observations 23, 29, 31, 34, 39, 40, 45, and 69,

1. Enter the data in the STAT LIST editor (see Appendix A.4).
2. Press 2ND [STAT PLOT] **1** to reach the settings for Plot 1.
3. Press ENTER to turn Plot 1 on.
4. Press ▽. Next, press ▷ twice. Then press ENTER to choose the histogram mode.
5. Press ▽ so the cursor is at Xlist. Then press 2ND [L1]. See Fig. 9.

Figure 9 Setting up Plot 1

6. Press 2ND [Quit] to reach the home screen.
7. To let the calculator determine the classes, press ZOOM **9** (see Fig. 10).

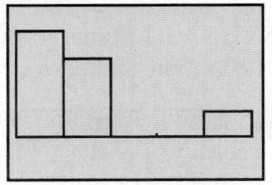

Figure 10 Using Zoom 9 to construct a histogram

8. To adjust the classes, press WINDOW. To begin with lower class limit 10, press **10** to set Xmin to be 10. To change the class width to 20, press ▽ twice and press **20** to set Xscl to be 20. To have tick marks spaced apart 1 unit vertically, press ▽ three times and press **1** to set Yscl to be 1. (see Fig. 11). Then press GRAPH (see Fig. 12).

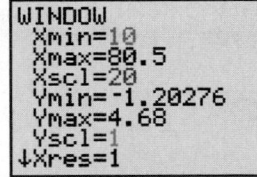

Figure 11 Adjusting the classes

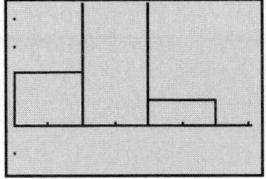

Figure 12 Histogram with adjusted classes

9. In Fig. 12, the middle bar does not fit on the screen. To view it completely, press WINDOW. Next, press ▽ four times. Then press **6** to set Ymax to be 6 (see Fig. 13). Finally, press GRAPH (see Fig. 14).

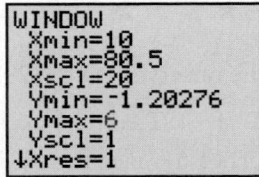

Figure 13 Increase Ymax to 6

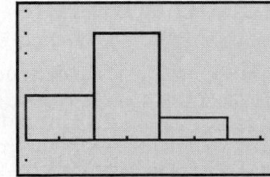

Figure 14 View the middle bar completely

10. To view the frequencies of the classes, press TRACE (see Fig. 15). Note the flashing "×" at the top of the leftmost bar in Fig. 15. This indicates that the class and frequency for that bar are shown at the bottom of the screen. So, the class is $[10, 30)$ and its frequency is 2. To find the frequencies of the other classes, press ◁ or ▷ accordingly.

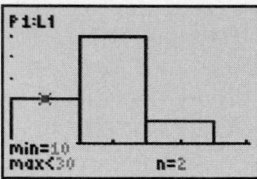

Figure 15 Viewing frequencies of classes

▼ A.6 Computing Median, Mean, Standard Deviation, and other Measures

To find the median, sum, mean, standard deviation, smallest value, Q_1, Q_3, and largest value of the observations 23, 29, 31, 34, 39, 40, 45, and 69,

1. Enter the data in the STAT LIST editor (see Appendix A.4).
2. Go to 1-Var Stats by pressing STAT ▷ 1 (see Fig. 16).

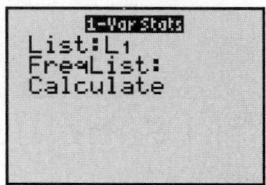

Figure 16 Setting up 1-Var Stats

3. Press ENTER three times. From Fig. 17, we see that the mean ($\bar{x}$) is 38.75, the sum ($\sum x$) is 310, and the standard deviation ("Sx") is 14.04838577.

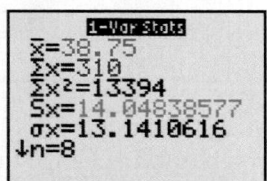

Figure 17 Computing mean, sum, and standard deviation

4. To see more measures, press ▽ five times. From Fig. 18, we see that the smallest observation ("minX") is 23, Q_1 is 30, the median ("Med") is 36.5, Q_3 is 42.5, and the largest observation ("maxX") is 69.

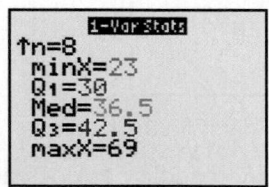

Figure 18 Computing median and other measures

▼ A.7 Constructing a Boxplot

To construct a boxplot for the observations 23, 29, 31, 34, 39, 40, 45, and 69,

1. Enter the data in the STAT LIST editor (see Appendix A.4).
2. Press 2ND [STAT PLOT] 1 to reach the settings for Plot 1.
3. Press ENTER to turn Plot 1 on. Then press ▽.

4. To construct a boxplot that does not display outliers, press ▷ four times and press ENTER (see Fig. 19). To construct a boxplot that displays outliers, press ▷ three times and press ENTER (see Fig. 20).

Figure 19 Selecting boxplot that does not display outliers

Figure 20 Selecting boxplot that displays outliers

5. Press ▽ so the cursor is at Xlist. Then press 2ND [L1]. The settings for a boxplot that does not display outliers is shown in Fig. 19. If you are constructing a boxplot that displays outliers, press ▽ twice. Then choose whether to represent any outliers using squares, plus signs, or dots by pressing ◁ or ▷ the appropriate number of times and then pressing ENTER. In Fig. 20, the square has been selected.

6. Press 2ND [QUIT] to reach the home screen.

7. Press ZOOM 9. A boxplot that does not display any outliers is shown in Fig. 21. A boxplot that displays an outlier is shown in Fig. 22.

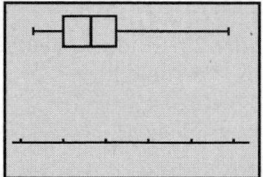

Figure 21 Boxplot that does not display any outliers

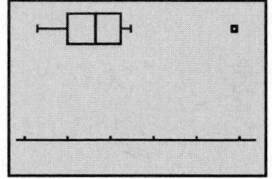

Figure 22 Boxplot that displays an outlier

8. To view the measures of a boxplot, press TRACE. Notice the flashing "×" on the left end of the left whisker of each of Figs. 23 and 24. At the bottom of the screen, we see that the smallest observation ("minX") is 23. To find the other measures, press ◁ or ▷ the appropriate number of times.

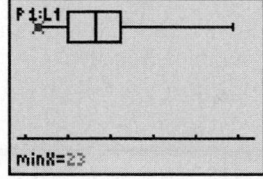

Figure 23 Finding the smallest observation

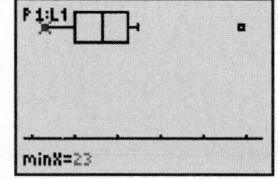

Figure 24 Finding the smallest observation

▼A.8 Computing Probabilities for a Normal Distribution

To find a probability for a normal curve with mean 50 and standard deviation 10,

1. Press 2ND [QUIT] to get to the home screen.

2. Press 2ND [DISTR] 2 to select the command "normalcdf" (see Fig. 25). It does not matter if the numbers on your screen are different than those in Fig. 25 because you will enter new numbers in the following steps. Across from "lower," we enter the left end of the region. If there is no left end, we use −999999. Across from "upper," we enter the right end of the region. If there is no right end, we use 999999. The symbols "μ" and "σ" stand for the (population) mean and the (population) standard deviation, respectively.

Figure 25 Selecting "normalcdf"

3. Recall that we are assuming that the mean is 50 and the standard deviation is 10. To find $P(X < 43)$, enter the values shown in Fig. 26 and then press ENTER several times so that your screen looks like Fig. 27. The probability is about 0.2420.

Figure 26 Entering values into "normalcdf"

Figure 27 Computing $P(X < 43)$

4. To find $P(45 < X < 59)$, press 2ND [DISTR] **2** and enter the values shown in Fig. 28. Then press ENTER several times so that your screen looks like Fig. 29. The probability is about 0.5074.

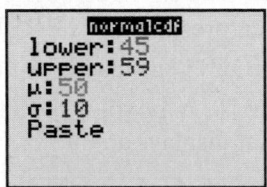

Figure 28 Entering values into "normalcdf"

Figure 29 Computing $P(45 < X < 59)$

5. To find $P(X > 64)$, press 2ND [DISTR] **2** and enter the values shown in Fig. 30. Then press ENTER several times so that your screen looks like Fig. 31. The probability is about 0.0808.

Figure 30 Entering values into "normalcdf"

Figure 31 Computing $P(X > 64)$

▼ A.9 Finding a Value of a Variable for a Normal Distribution

Suppose that a distribution is normally distributed with mean 30 and standard deviation 5. To find a number that is greater than or equal to 70% of the observations and less than 30% of the observations,

1. Press 2ND [QUIT] to get to the home screen.

2. Press 2ND [DISTR] **3** to select the command "invNorm" (see Fig. 32). The numbers on your screen may be different than in Fig. 32. This does not matter because you will enter new numbers in the following steps.

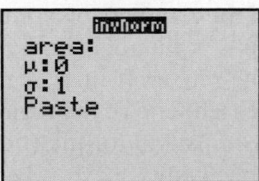

Figure 32 Selecting the command "invNorm

3. The number across from "area" should be the area to the *left* (0.70) of the number we're trying to find. So, enter 0.7 across from "area," 30 across from "μ" (the population mean), and 5 across from "σ" (the population standard deviation). See Fig. 33.

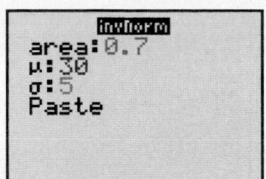

Figure 33 Entering values into "invNorm"

4. Press ENTER several times so that your screen looks like Fig. 34. The number we want to find is about 32.6.

Figure 34 Finding the number

▼ A.10 Constructing a Time-Series Plot or Scatterplot

Table 1 Some Data

x	y
2	5
3	6
4	10
5	11

To construct a time-series plot or a scatterplot of the data shown in Table 1,

1. Press STAT 1 and enter the numbers 2, 3, 4, and 5 in list L1 (see Appendix A.4). Then enter the numbers 5, 6, 10, and 11 in list L2 (see Fig. 35).

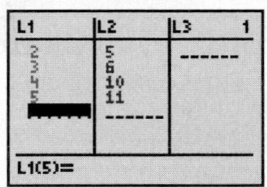

Figure 35 Entering data in lists L1 and L2

2. Press 2ND [STAT PLOT] 1 to reach the settings for Plot 1.

3. Press ENTER to turn Plot 1 on. Press ▽.

4. To construct a scatterplot, press ENTER. To construct a time-series plot, press ▷ and then press ENTER.

5. Press ▽ so the cursor is at Xlist. Next, press 2ND [L1]. Then press ▽ so the cursor is at Ylist. Finally, press 2ND [L2].

6. Press ▽ and choose whether to represent the data points using squares, plus signs, or dots by pressing ◁ or ▷ the appropriate number of times and then pressing ENTER. Figure 36 displays a screen in which squares have been selected for a scatterplot. Figure 37 displays a screen in which squares have been selected for a time-series plot.

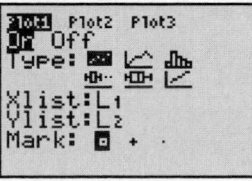

Figure 36 Settings for a scatterplot

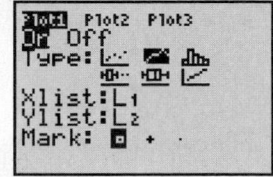

Figure 37 Settings for a time-series plot

7. Press $\boxed{\text{2ND}}$ [Quit] to reach the home screen.

8. Press $\boxed{\text{ZOOM}}$ 9. A scatterplot and a time-series plot are shown in Figs. 38 and 39, respectively.

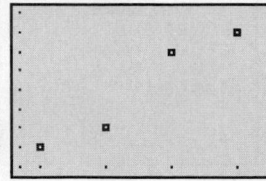
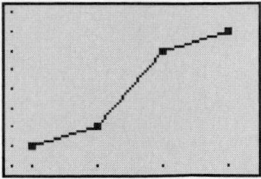

Figure 38 A scatterplot   **Figure 39** A time-series plot

9. To view the coordinates of the data points, press $\boxed{\text{TRACE}}$. Notice the flashing "×" on the data point on the far left of each of Figs. 40 and 41. At the bottom of each screen, we see that the coordinates of the point are $(2, 5)$. To find the coordinates of the other points, press $\boxed{\triangleleft}$ or $\boxed{\triangleright}$ accordingly.

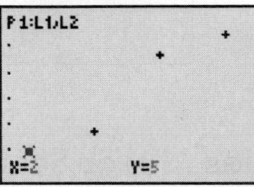
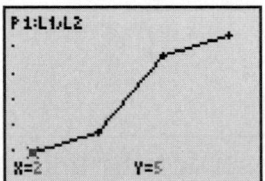

Figure 40 Finding the coordinates of a point in a scatterplot   **Figure 41** Finding the coordinates of a point in a time-series plot

▼ A.11 Constructing Two Scatterplots That Share the Same Axes

To construct scatterplots of the data in Table 2 and Table 3 that share the same axes,

Table 2 A First Data Set

x	y
2	5
3	6
4	10
5	11

Table 3 A Second Data Set

x	y
2	11
2	9
3	6
5	4

1. Construct a scatterplot of the data in Table 2, using squares as marks for the data points (see Appendix A.10).

2. Press $\boxed{\text{STAT}}$ 1. Then for Table 3, enter the values of x in list L3 and the values of y in list L4 (see Fig. 42).

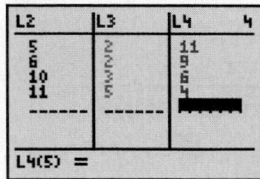

Figure 42 Entering data in lists L3 and L4

3. Press $\boxed{\text{2ND}}$ [STAT PLOT] 2 to reach the settings for Plot 2.

4. Press $\boxed{\text{ENTER}}$ to turn Plot 2 on. Press $\boxed{\triangledown}$.

5. Press $\boxed{\text{ENTER}}$ to select a scatterplot.

6. Press ▽ so the cursor is at Xlist. Next, press 2ND [L3]. Then press ▽ so the cursor is at Ylist. Finally, press 2ND [L4].

7. Press ▽ and choose plus signs to represent data points shown in Table 3 (see Fig. 43).

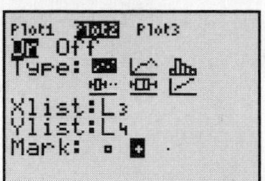

Figure 43 Settings for second scatterplot

8. Press 2ND [QUIT] to reach the home screen.

9. Press ZOOM 9 to construct the scatterplots (see Fig. 44).

Figure 44 Two scatterplots that share the same axes

10. To view the coordinates of the data points plotted by Plot1, press TRACE (see Fig. 45). To view the coordinates of the data points plotted by Plot2, press ▽ (see Fig. 46). You can return to finding coordinates of the data points plotted by Plot1 by pressing △. Notice that the labels "P1:L1,L2" and "P2:L3,L4" lie in the upper left corners of the screens when you view the coordinates of the data points plotted by Plot 1 and Plot 2, respectively.

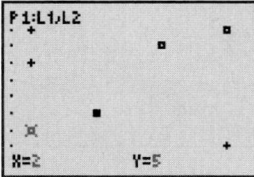

Figure 45 Finding the coordinates of a point plotted by Plot 1

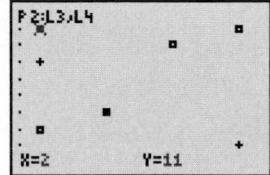

Figure 46 Finding the coordinates of a point plotted by Plot 2

▼A.12 Computing Correlation Coefficients and Coefficients of Determination

To find the linear and exponential r and r^2,

1. Press 2ND [CATALOG] and scroll down to DiagnosticON (see Fig. 47). Then press ENTER twice (see Fig. 48). You will never have to do this step again, unless you ever select DiagnosticOff.

Table 4 Some Data

x	y
2	5
3	6
4	10
5	11

Figure 47 Scroll down to DiagnosticOn

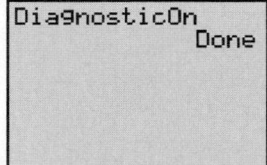

Figure 48 Diagnostic is now on

2. See Appendix A.10 to construct a scatterplot of the data in Table 4.

3. Press $\boxed{\text{STAT}}$ $\boxed{\triangleright}$ to choose the CALC menu (see Fig. 49).

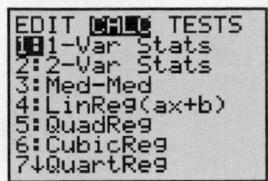

Figure 49 CALC menu

4. To find the linear r and r^2, press **4** to get to the LinReg($ax + b$) menu (see Fig. 50). Or to find the exponential r and r^2, press **0** to get to the ExpReg menu (see Fig. 51).

Figure 50 LinReg($ax + b$) menu   **Figure 51** ExpReg menu

5. Whether you are at the LinReg($ax + b$) or the ExpReg menu, press $\boxed{\text{ENTER}}$ five times. The linear r (about 0.965) and the linear r^2 (about 0.931) are shown in Fig. 52. The exponential r (about 0.966) and the exponential r^2 (about 0.933) are shown in Fig. 53.

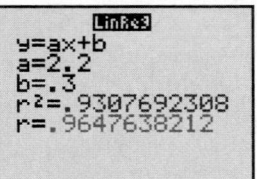   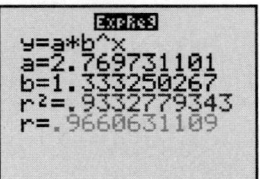

Figure 52 Linear r and r^2 **Figure 53** Exponential r and r^2

▼A.13 Turning a Plotter On or Off

To change the on/off status of the plotter,

1. Press $\boxed{\text{Y} =}$ to reach the equation editor (see Fig. 54).

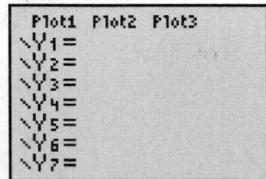

Figure 54 Equation editor

2. Press $\boxed{\triangle}$. A flashing rectangle will be on "Plot 1."

3. Press $\boxed{\triangleright}$ if necessary to move the flashing rectangle to the plotter you wish to turn on or off.

4. Press $\boxed{\text{ENTER}}$ to turn your plotter on or off. The plotter is on if the plotter icon is highlighted.

▼ A.14 Entering an Equation

Figure 55 Entering an equation

To enter the equation $y = 2x + 1$,

1. Press $\boxed{Y =}$.
2. If necessary, press $\boxed{\text{CLEAR}}$ to erase a previously entered equation.
3. Press **2** $\boxed{\text{X, T, } \Theta, n}$ $\boxed{+}$ **1** (see Fig. 55).
4. If you want to enter another equation, press $\boxed{\text{ENTER}}$. Then type in the next equation.
5. Use $\boxed{\triangle}$ or $\boxed{\triangledown}$ to get from one equation to another.

▼ A.15 Graphing an Equation

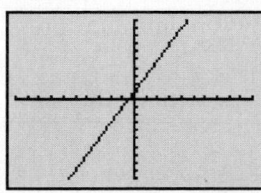

Figure 56 Graphing an equation

To graph the equation $y = 2x + 1$,

1. Turn Plot 1, Plot 2, and Plot 3 off (see Appendix A.13).
2. Enter the equation $y = 2x + 1$ (see Appendix A.14).
3. Press $\boxed{\text{ZOOM}}$ **6** to draw a graph of your equation between the values of -10 and 10 for both x and y (see Fig. 56).
4. See Appendix A.17 if you want to zoom in or zoom out to get another part of the graph to appear on the calculator screen. Or see Appendix A.18 to change the window format manually.

▼ A.16 Tracing a Curve without a Scatterplot

To *trace a curve,* we find coordinates of points on the curve. To trace the line $y = 2x + 1$,

1. Graph $y = 2x + 1$ (see Appendix A.15).
2. Press $\boxed{\text{TRACE}}$.
3. If you see a flashing "×" on the curve, the coordinates of that point will be listed at the bottom of the screen. If you don't see the flashing "×," press $\boxed{\text{ENTER}}$, and your calculator will adjust the viewing window so you can see it.
4. To find coordinates of points on your curve that are off to the right, press $\boxed{\triangleright}$.
5. To find coordinates of points on your curve that are off to the left, press $\boxed{\triangleleft}$.
6. Find the y-coordinate of a point by entering the x-coordinate. For example, to find the y-coordinate of the point that has x-coordinate 3, press **3** $\boxed{\text{ENTER}}$ (see Fig. 57). This feature works for values of x between Xmin and Xmax, inclusive (see Appendix A.18).

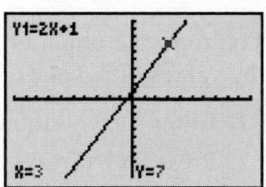

Figure 57 Tracing a curve

7. If more than one equation has been graphed, press $\boxed{\triangledown}$ to trace the second equation. Continue pressing $\boxed{\triangledown}$ to trace the third equation, and so on. Press $\boxed{\triangle}$ to return to the previous equation. Notice that the equation of the curve being traced is listed in the upper left corner of the screen.

▼ A.17 Zooming

The $\boxed{\text{ZOOM}}$ menu has several features that allow you to adjust the viewing window. Some of the features adjust the values of x that are used in tracing.

- **Zoom In** magnifies the graph around the cursor location. The following instructions are for zooming in on the graph of $y = 2x + 1$:

    1. Graph $y = 2x + 1$ (see Appendix A.15).
    2. Press $\boxed{\text{ZOOM}}$ **2**.

    If you lose sight of the line, you can always press $\boxed{\text{TRACE}}$ $\boxed{\text{ENTER}}$.

    3. Use $\boxed{\triangleleft}$, $\boxed{\triangleright}$, $\boxed{\triangle}$, and $\boxed{\triangledown}$ to position the cursor on the portion of the line that you want to zoom in on.
    4. To zoom in, press $\boxed{\text{ENTER}}$.
    5. To zoom in on the graph again, you have two options:
        a. To zoom in at the same point, press $\boxed{\text{ENTER}}$.
        b. To zoom in at a new point, move the cursor to the new point; then press $\boxed{\text{ENTER}}$.

    When zooming out, you will return to the original graph only if you did not move the cursor while zooming in.

    6. To return to your original graph, zoom out (see the next instruction) the same number of times you zoomed in.

- **Zoom Out** does the reverse of Zoom In: It allows you to see *more* of a graph. To zoom out, follow the preceding instructions, but press $\boxed{\text{ZOOM}}$ **3** instead of $\boxed{\text{ZOOM}}$ **2** in step 2.

- **ZStandard** will change your viewing screen so both x and y will go from -10 to 10. To use ZStandard, press $\boxed{\text{ZOOM}}$ **6**.

- **ZDecimal** lets you trace a curve by using the numbers $0, \pm 0.1, \pm 0.2, \pm 0.3, \ldots$ for x. ZDecimal will change your viewing screen so x will go from -4.7 to 4.7 and y will go from -3.1 to 3.1. To use ZDecimal, press $\boxed{\text{ZOOM}}$ **4**.

- **ZInteger** allows you to trace a curve by using the numbers $0, \pm 1, \pm 2, \pm 3, \ldots$ for x. ZInteger can be used for any viewing window, although it will usually change the view. To use ZInteger, press $\boxed{\text{ZOOM}}$ **8** $\boxed{\text{ENTER}}$.

- **ZSquare** will change your viewing window so the spacing of the tick marks on the x-axis is the same as that on the y-axis. To use ZSquare, press $\boxed{\text{ZOOM}}$ **5**.

- **ZoomStat** will change your viewing window so you can see a time-series plot or a scatterplot of points that you have entered in the STAT LIST editor. To use ZoomStat, press $\boxed{\text{ZOOM}}$ **9**. For complete instructions, see Appendix A.10.

- **ZoomFit** will adjust the dimensions of the y-axis to display as much of a curve as possible. The dimensions of the x-axis will remain unchanged. To use ZoomFit, press $\boxed{\text{ZOOM}}$ **0**.

▼ A.18 Setting the Window Format

To graph the equation $y = 2x + 1$ between the values of -2 and 3 for x and between the values of -5 and 7 for y,

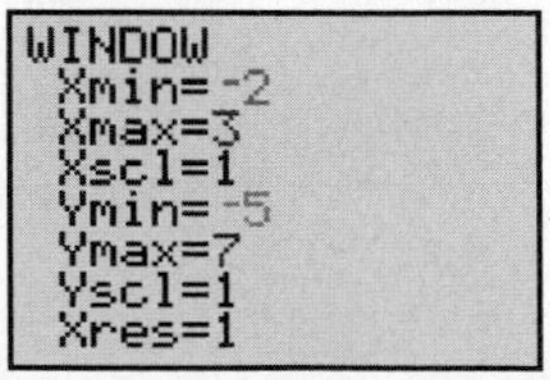

```
WINDOW
 Xmin=-2
 Xmax=3
 Xscl=1
 Ymin=-5
 Ymax=7
 Yscl=1
 Xres=1
```

Figure 58 Window settings

1. Enter the equation $y = 2x + 1$ (see Appendix A.14).
2. Press $\boxed{\text{WINDOW}}$. Then change the window settings so the window looks like the one displayed in Fig. 58 after you have used steps 3–8.
3. Press $\boxed{(-)}$ **2** $\boxed{\text{ENTER}}$ to set the smallest value of x to -2.
4. Press **3** $\boxed{\text{ENTER}}$ to set the largest value of x to 3.
5. Press **1** $\boxed{\text{ENTER}}$ to set the scaling for the x-axis to increments of 1.
6. Press $\boxed{(-)}$ **5** $\boxed{\text{ENTER}}$ to set the smallest value of y to -5.
7. Press **7** $\boxed{\text{ENTER}}$ to set the largest value of y to 7.

If you press ZOOM 6 or ZOOM 9 or zoom in or zoom out, your window settings will change accordingly.

8. Press 1 ENTER to set the scaling for the y-axis to increments of 1.
9. Press GRAPH to view the graph of $y = 2x + 1$ (see Fig. 59).

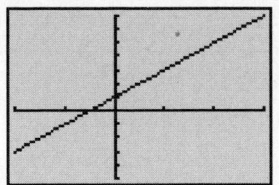

Figure 59 Graph of $y = 2x + 1$

▼A.19 Graphing Equations with a Scatterplot

Table 5 Data for a Scatterplot

x	y
2	5
3	6
4	10
5	11

To graph the equation $y = 2x + 1$ with a scatterplot of the data displayed in Table 5,

1. Enter the equation $y = 2x + 1$ (see Appendix A.14).
2. Follow the instructions in Appendix A.10 to construct the scatterplot. The graph of the equation will also be drawn, because you turned the equation on (see Fig. 60).

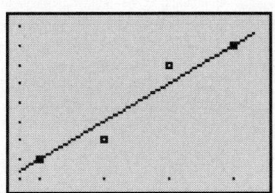

Figure 60 Graphing an equation and a scatterplot

▼A.20 Tracing a Curve with a Scatterplot

To trace a curve with a scatterplot,

1. Graph an equation with a scatterplot (see Appendix A.19).

Recall from Appendix A.16 that if you do not see the flashing "×", press ENTER, and the calculator will adjust the viewing window so you can see it.

2. Press TRACE to trace points that make up the scatterplot. Press ▽ to trace points that lie on the curve. If other equations are graphed, continue pressing ▽ to trace the second equation, and so on. Press △ to begin to return to the scatterplot. Notice that the label "P1:L1, L2" is in the upper left corner of the screen when Plot 1's points are being traced and that the equation entered in the Y= mode is listed in the upper left corner of the screen when the curve is being traced.

▼A.21 Constructing a Table

To create a table of ordered pairs for the equation $y = 2x + 1$, where the values of x are 3, 4, 5, ... (see Fig. 61),

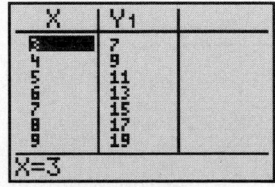

Figure 61 Table of ordered pairs for $y = 2x + 1$

1. Enter the equation $y = 2x + 1$ for Y_1 (see Appendix A.14).
2. Press 2ND [TBLSET].
3. Press 3 ENTER to tell the TI-84 that the first x value in your table is 3.

Figure 62 Table setup

4. Press **1** ENTER to tell the TI-84 that the x values in your table increase by 1.
5. Press ENTER ▽ ENTER to highlight "Auto" for both "Indpnt" and "Depend" (see Fig. 62).
6. Press 2ND [TABLE]. To make the message "Press + for △ Tbl" disappear, press ENTER (see Fig. 61).

▼ A.22 Constructing a Table for Two Equations

Figure 63 Table for two equations

To construct a table of ordered pairs for the equations $y = 2x + 1$ and $y = -2x + 7$, where the values of x are 3, 4, 5, ... (see Fig. 63),

1. Enter the equation $y = 2x + 1$ for Y_1, and enter the equation $y = -2x + 7$ for Y_2 (see Appendix A.14).
2. Follow steps 2–6 of Appendix A.21.

▼ A.23 Using "Ask" in a Table

Table 6 Using "Ask" in a Table with $y = 2x + 1$

x	y
2	
2.9	
5.354	
7	
100	

To use the Ask option in the Table Setup mode to complete Table 6 for $y = 2x + 1$,

1. Enter the equation $y = 2x + 1$ for Y_1 (see Appendix A.14).
2. Press 2ND [TBLSET].
3. Press ENTER twice. Next, press ▷. Then press ENTER. The Ask option for "Indpnt" will now be highlighted. Make sure the Auto option for "Depend" is highlighted (see Fig. 64). The values for Tbl Start and △ Tbl do not matter.
4. Press 2ND [TABLE].
5. To construct the table, press **2** ENTER **2.9** ENTER **5.354** ENTER **7** ENTER **100** ENTER (see Fig. 65).

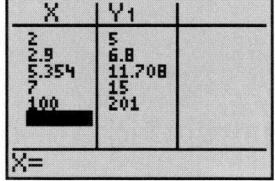

Figure 64 Table Setup   **Figure 65** Using "Ask" for a table with $y = 2x + 1$

▼ A.24 Finding the Intersection Point(s) of Two Curves

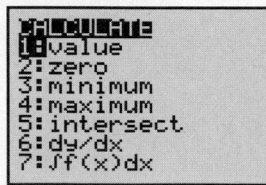

Figure 66 CALCULATE menu

To find the intersection point of the lines $y = 2x + 1$ and $y = -2x + 7$,

1. Enter the equation $y = 2x + 1$ for Y_1, and enter the equation $y = -2x + 7$ for Y_2 (see Appendix A.14).
2. By zooming in or out or by changing the window settings, draw a graph of both curves so you can see an intersection point. For our example, press ZOOM **6**.
3. Press 2ND [CALC] to reach the CALCULATE menu (see Fig. 66).
4. Press **5** to select "intersect."

5. You will now see a flashing cursor on your first curve. If there is more than one intersection point on your display screen, move the cursor by pressing ◁ or ▷ so it is much closer to the intersection point you want to find. The screen will look something like the one displayed in Fig. 67.

6. Press ENTER to put the cursor on the second curve. Press ENTER again to display "Guess?" Press ENTER once more (see Fig. 68). The intersection point is (1.5, 4).

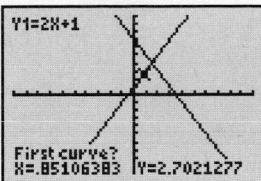

Figure 67 Putting cursor near intersection point

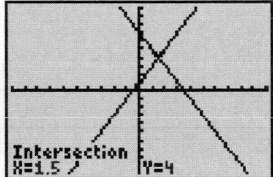

Figure 68 Finding intersection point

▼A.25 Turning an Equation On or Off

You can graph an equation only if its equals sign is highlighted. (The equation is then "on"). Up to 10 equations can be graphed at one time. To change the on–off status of an equation,

1. Press Y =.

2. Move the cursor to the equation whose status you want to change.

3. Use ◁ to place the cursor over the " = " sign of the equation.

4. Press ENTER to change the status.

▼A.26 Finding a Regression Equation

Table 7 A Data Set

x	y
2	5
3	6
4	10
5	11

To find a regression equation for the data displayed in Table 7,

1. Press STAT 1 and enter the numbers 2, 3, 4, and 5 in list L1 (see Appendix A.4). Then enter the numbers 5, 6, 10 and 11 in list L2.

2. Press STAT ▷ to reach the CALC menu (see Fig. 69).

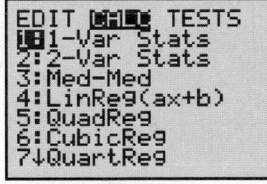

Figure 69 CALC menu

3. To find the linear regression equation, press **4** to get to the LinReg($ax + b$) menu (see Fig. 70). Or to find the exponential regression equation, press **0** to get to the ExpReg menu (see Fig. 71).

Figure 70 LinReg($ax + b$) menu

Figure 71 ExpReg menu

4. Whether you are at the LinReg($ax + b$) or the ExpReg menu, press ENTER five times. The linear regression equation ($\hat{y} = 2.2x + 0.3$) and the exponential regression equation [approximately $\hat{y} = 2.77(1.33)^x$] are shown in Figs. 72 and 73, respectively.

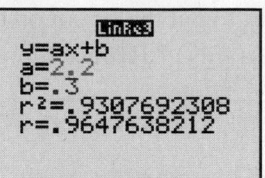

Figure 72 Linear regression equation

Figure 73 Exponential regression equation

▼ A.27 Constructing a Residual Plot

Table 8 Some Data

x	y
2	5
3	6
4	10
5	11

Figure 74 Setting up Plot 1

To construct a residual plot,

1. Find the linear regression equation or the exponential regression equation for the data shown in Table 8 (see Appendix A.26).
2. Press 2ND [STAT PLOT] **1**. Then press ENTER to turn Plot 1 on. Press ▽.
3. Press ENTER to select the scatterplot icon.
4. Press ▽ twice to move the cursor to Ylist. Then press 2ND [STAT] **7**.
5. Press ▽ and select the square, plus sign, or dot icon. The square icon is selected in Fig. 74.
6. Press 2ND [Quit] to reach the home screen.
7. Press ZOOM **9**. If you found the linear regression equation in Step 1, then your screen should display the linear residual plot shown in Fig. 75. If you found the exponential regression equation in Step 1, then your screen should display the exponential residual plot shown in Fig. 76.

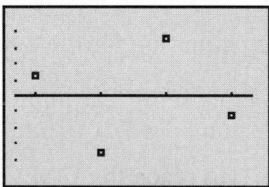

Figure 75 Linear residual plot

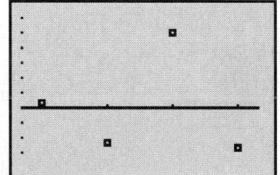

Figure 76 Exponential residual plot

▼ A.28 Responding to Error Messages

Here are several common error messages and how to respond to them:

- The **Syntax** error (see Fig. 77) means you have misplaced one or more parentheses, operations, or commas. The calculator will find this type of error if you choose "Goto" by pressing ▽, then ENTER. Your error will be highlighted by a flashing black rectangle.

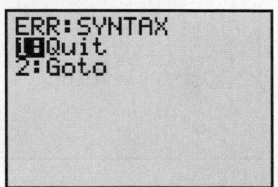

Figure 77 "Syntax" error message

The most common "Syntax" error is pressing $\boxed{(\text{-})}$ when you should have pressed $\boxed{-}$, or vice versa:

1. Press the $\boxed{(\text{-})}$ key when you want to take the opposite of a number or are working with negative numbers. To compute $-5(-2)$, press $\boxed{(\text{-})}\,\mathbf{5}\,\boxed{(}\boxed{(\text{-})}\,\mathbf{2}\,\boxed{)}$.

2. Press the $\boxed{-}$ key when you want to subtract two numbers. To compute $5 - 2$, press $\mathbf{5}\,\boxed{-}\,\mathbf{2}$.

- The **Invalid** error (see Fig. 78) means you have tried to enter an inappropriate number, expression, or command. The most common "Invalid" error is to try to enter a number that is not between Xmin and Xmax, inclusive, when you use a command such as $\boxed{\text{TRACE}}$.

- The **Invalid dimension** error (see Fig. 79) means you have the plotter turned on (see Fig. 80) but have not entered any data points in the STAT LIST editor (see Fig. 81). In this case, first press $\boxed{\text{ENTER}}$ to exit the error message display. Then either turn the plotter off or enter data in the STAT LIST editor.

Figure 78 "Invalid" error message

Figure 79 "Invalid dimension" error message

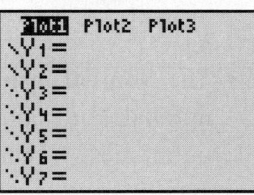

Figure 80 Plotter is on

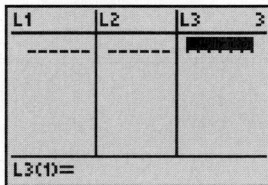

Figure 81 STAT LIST editor's columns are empty

- The **Dimension mismatch** error (see Fig. 82) is fixed in two ways:

1. In the STAT LIST editor, one column you are using to plot has more numbers than the other column has (see Fig. 83). In this case, first press $\boxed{\text{ENTER}}$ to exit the error message display. Then add or delete numbers so the two columns have the same length.

Figure 82 "Dimension mismatch" error message

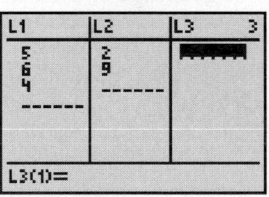

Figure 83 Columns of unequal length in STAT LIST editor

2. In the $\boxed{\text{STAT}}$ LIST editor, one column you are using to plot has more numbers than the other column has, but you didn't notice the difference in length because you deleted one or both of the columns by mistake. You can find the missing column(s) by pressing STAT 5 $\boxed{\text{ENTER}}$.

- The **Window range** error (see Fig. 84) means you made an error in setting up your window. This usually means you entered a larger number for Xmin than for Xmax or you entered a larger number for Ymin than for Ymax. In this case, first press $\boxed{\text{ENTER}}$ to exit the error message display. Then change your window settings accordingly (see Appendix A.18).

Figure 84 "Window range" error message

- The **No sign change** error (see Fig. 85) means one of two things:

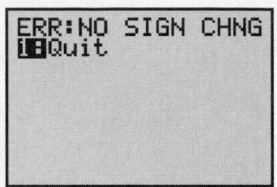

Figure 85 "No sign change" error message

1. You are trying to locate a point that does not appear on the screen. For example, you may be trying to find an intersection point of two curves that intersect offscreen. In this case, press ENTER and change your window settings so the point you are trying to locate is on the screen.

2. You are trying to locate a point that does not exist. For example, you may be trying to find an intersection point of two lines that do not intersect. In this case, press ENTER and stop looking for the point that doesn't exist!

- The **Nonreal answer** error (see Fig. 86) means your computation did not yield a real number. For example, $\sqrt{-4}$ is not a real number. The calculator will locate this computation if you choose "Goto" by pressing ▽, then ENTER.

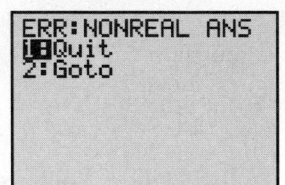

Figure 86 "Nonreal answer" error message

- The **Divide by 0** error (see Fig. 87) means you asked the calculator to perform a calculation that involves a division by zero. For example, $3 \div (5-5)$ will yield the error message shown in Fig. 87.

Figure 87 "Divide by zero" error message

The calculator will locate the division by zero if you choose "Goto" by pressing ▽, then ENTER.

Using StatCrunch

StatCrunch produces excellent statistical diagrams and does not require much time to learn how to use. The more you experiment with StatCrunch, the more comfortable and efficient you will become with it.

▼ B.1 Selecting Numbers Randomly

To randomly select some numbers,

1. Click on the button Data. Next, select "Simulate" in the drop-down menu. Then click on "Discrete Uniform" in the side menu.
2. To indicate the number of random numbers you want to appear in each column of StatCrunch's data table, enter a counting number in the box labeled "Rows," which is under the heading "**Number of rows and columns**."
3. To indicate the number of columns of random numbers you want to appear in Stat-Crunch's data table, enter a counting number in the box labeled "Columns," which is under the heading "**Number of rows and columns**."
4. Of all the numbers you want StatCrunch to choose from, enter the smallest one in the box labeled "Minimum," which is under the heading "**Discrete Uniform parameters**."
5. Of all the numbers you want StatCrunch to choose from, enter the largest one in the box labeled "Maximum," which is under the heading "**Discrete Uniform parameters**."
6. To enter a seed, select "Use fixed seed," which is under the heading "**Seeding**" and enter the seed in the box labeled "Seed."
7. Click on the button Compute!. The random numbers should appear in the left-most column(s) free of data in StatCrunch's data table. A small window should also appear, saying that the random numbers have been added to StatCrunch's data table.

▼ B.2 Entering Data

To manually enter data, click on the cell at the top of the first column, delete "var1," and enter the name of the variable. Next, enter the values of the variable below the name of the variable in the first column. If you have data for other variables, enter those observations in other columns similarly.

To perform a cut and paste of data from other spreadsheet software such as Excel, copy the data, click on the top of the first column of StatCrunch's data table, and paste the data. In some cases, you can add data by simply dragging and dropping a file onto StatCrunch's data table.

There are many other ways to import data, including a feature called "StatCrunchThis," which imports data tables from the Internet. See the instructions in StatCrunch or ask your professor for details.

▼B.3 Constructing Frequency and Relative Frequency Tables

To construct one or more frequency and relative frequency tables,

1. Click on the button Stat. Next, select "Tables" in the drop-down menu. Then click on "Frequency" in the side menu.

2. Under the heading "**Select column(s)**," select one or more variables. For each variable you select, StatCrunch will construct a table.

3. To immediately display the table(s), click on the button Compute! at the bottom of the window. To display the table(s) and then make adjustments, click on the button Compute!, then click on the button Options at the top of the window, and in the drop-down menu click on "Edit." To take advantage of more features, do some or all of the following steps.

4. Suppose that a numerical variable in your data set is Height of Student and a categorical variable is Ethnicity. To construct a table for the heights of each ethnicity, you would select "Height of Student" under the heading "**Select column(s)**" and you would select "Ethnicity" in the drop-down menu under the heading "**Group by**."

5. To adjust how the table entries are ordered, click on the appropriate choice in the drop-menu under the heading "**Order by**."

6. To display the table(s), click on the button Compute! at the bottom of the window.

7. Click on the button Options at the top of the window and select from the drop-down menu to edit, save, copy, print, or download the results.

▼B.4 Constructing Bar Graphs and Multiple Bar Graphs

To construct one or more bar graphs,

1. Click on the button Graph. Next, select "Bar Plot" in the drop-down menu. If you have entered the categories of a variable in one column and the frequencies in another column, select the categorical variable in the drop-down menu under the heading "Categories in," and select the column that contains the frequencies in the drop-down menu under the heading "Counts in." Then skip to Step 3 in these instructions. If you have entered the categories with repetitions in one column such as "female," "female," "male," "female," "male," and "male," click on "With Data" in the side menu.

2. Under the heading "**Select column(s)**," select one or more categorical variables. For each variable you select, StatCrunch will construct a bar graph.

3. To immediately construct the bar graphs, click on the button Compute! at the bottom of the window. If you have constructed more than one bar graph, you can view the other one(s) by clicking one or more times on the buttons < or > at the bottom right corner of the window. To view the bar graph and then make adjustments, click on the button Compute!, then click on the button Options at the top of the window, and in the drop-down menu click on "Edit." To take advantage of more features, do some or all of the following steps.

4. This step only applies if you clicked on "With Data" in Step 1. Suppose that a categorical variable in your data set is Eye Color of Student and another categorical variable is Gender. To construct a *multiple* bar graph of eye colors broken down further into each gender, you would select "Eye Color of Student" under the heading "**Select column(s)**" and you would select "Gender" in the drop-down menu under the heading "**Group by**."

5. In the drop-down menu under the heading "**Type**," select which type of bar graph you want to construct.

6. Under the heading "**Display**," click on the box labeled "Value above bar" to instruct StatCrunch to display the frequencies, relative frequencies, or percents above the bars in the bar graph.

7. StatCrunch will automatically display the variable name on the horizontal axis, but you can enter something more complete by making the appropriate entry in the box labeled "X-axis label," which is under the heading "**Graph properties**."

8. To title the bar graph plot(s), enter the title in the box labeled "Title," which is under the heading "**Graph properties**."

9. To display the bar graph(s), click on the button ⬚Compute!⬚ at the bottom of the window.

10. Click on the button ⬚Options⬚ at the top of the window and select from the drop-down menu to edit, save, copy, print, or download the bar graph(s).

▼ B.5 Constructing Pie Charts

To construct one or more pie charts,

1. Click on the button ⬚Graph⬚. Next, select "Pie Chart" in the drop-down menu. If you have entered the categories of a variable in one column and the frequencies in another column, select the categorical variable in the drop-down menu under the heading "**Categories in,**" and select the column that contains the frequencies in the drop-down menu under the heading "**Counts in.**" Then skip to Step 3 in these instructions. If you have entered the categories with repetitions in one column such as "female," "female," "male," "female," "male," and "male," click on "With Data" in the side menu.

2. Under the heading "**Select column(s)**," select one or more categorical variables. For each variable you select, StatCrunch will construct a pie chart.

3. To immediately construct the pie chart(s), click on the button ⬚Compute!⬚ at the bottom of the window. If you have constructed more than one pie chart, you can view the other one(s) by clicking one or more times on the buttons ⬚<⬚ or ⬚>⬚ at the bottom right corner of the window. To view the pie chart(s) and then make adjustments, click on the button ⬚Compute!⬚, then click on the button ⬚Options⬚ at the top of the window, and in the drop-down menu click on "Edit." To take advantage of more features, do some or all of the following steps.

4. This step only applies if you clicked on "With Data" in Step 1. Suppose that a categorical variable in your data set is Eye Color of Student and another categorical variable is Gender. To construct a pie chart of eye colors for each gender, you would select "Eye Color of Student" under the heading "**Select column(s)**" and you would select "Gender" in the drop-down menu under the heading "**Group by**."

5. To list the categories' frequencies, categories' percents, or both, select the appropriate choices under the heading "**Display**."

6. To title the pie chart(s), enter the title in the box labeled "Title," which is under the heading "**Graph properties**."

7. To display the pie chart(s), click on the button ⬚Compute!⬚ at the bottom of the window.

8. Click on the button ⬚Options⬚ at the top of the window and select from the drop-down menu to edit, save, copy, print, or download the time-series plot.

▼ B.6 Constructing Two-Way Tables

To construct a two-way table,

1. Click on the button Stat . Next, select "Tables" in the drop-down menu. Then select "Contingency" in the side menu. Finally, click on "With Data" in the second side menu.

2. In the drop-down menu under the heading "**Row variable**," select a categorical variable.

3. In the drop-down menu under the heading "**Column variable**," select another categorical variable.

4. To view the two-way table, click on the button Compute! .

5. Click on the button Options at the top of the window and select from the drop-down menu to edit, save, copy, print, or download the two-way table.

▼ B.7 Constructing Dotplots

To construct one or more dotplots,

1. Click on the button Graph . Then click on "Dotplot" in the drop-down menu.

2. Under the heading "**Select column(s)**," select one or more numerical variables. For each variable you select, StatCrunch will construct a dotplot.

3. To immediately construct the dotplot(s), click on the button Compute! at the bottom of the window. To view the dotplot(s) and then make adjustments, click on the button Compute! , then click on the button Options at the top of the window, and in the drop-down menu click on "Edit." To take advantage of more features, do some or all of the following steps.

4. Suppose that a numerical variable in your data set is Height of Student and a categorical variable is Ethnicity. To construct a dotplot of heights for each ethnicity, you would select "Height of Student" under the heading "**Select column(s)**" and you would select "Ethnicity" in the drop-down menu under the heading "**Group by**."

5. To adjust the size of the dots, select a size in the drop-down menu under the heading "**Point size**."

6. To display the mean and/or median on the dotplot(s), click on the appropriate boxes under the heading "**Markers**."

7. StatCrunch will automatically display the variable name on the horizontal axis, but you can enter something more complete such as the variable name and the units by making the appropriate entry in the box labeled "X-axis label," which is under the heading "**Graph properties**."

8. To title the dotplot(s), enter the title in the box labeled "Title," which is under the heading "**Graph properties**."

9. To show light horizontal and/or vertical lines, click on one or both of the boxes labeled "Horizontal lines" and "Vertical lines," which are under the heading "**Graph properties**." StatCrunch will display horizontal lines only if two or more dotplots are displayed in the same diagram.

10. To display the dotplot(s), click on the button Compute! at the bottom of the window.

11. If all the dotplots are not shown in the window, you can view the other one(s) by clicking one or more times on the buttons < or > at the bottom right corner of the window.

12. You can adjust the labels on the axes, the scaling on the horizontal axis, and the title by clicking on the small icon consisting of three horizontal lines near the bottom left corner of the window.

13. Click on the button ⬚Options at the top of the window and select from the drop-down menu to edit, save, copy, print, or download the dotplot(s).

▼ B.8 Constructing Stemplots and Split Stems

After you perform the following steps, on the basis of the complexity of the data, StatCrunch will automatically determine whether to use a stemplot or a split stem. To simplify the following instructions, we will refer to either diagram as a stemplot. To construct one or more stemplots,

1. Click on the button ⬚Graph. Then click on "Stem and Leaf" in the drop-down menu.

2. Under the heading "**Select column(s)**," select one or more numerical variables. For each variable you select, StatCrunch will construct a stemplot.

3. To immediately construct the stemplot(s), click on the button ⬚Compute! at the bottom of the window. If you have constructed more than one stemplot, you can view the other one(s) by clicking one or more times on the buttons ⬚< or ⬚> at the bottom right corner of the window. To view the stemplot(s) and then make adjustments, click on the button ⬚Compute!, then click on the button ⬚Options at the top of the window, and in the drop-down menu click on "Edit." To take advantage of more features, do some or all of the following steps.

4. Suppose that a numerical variable in your data set is Height of Student and a categorical variable is Ethnicity. To construct a stemplot of heights for each ethnicity, you would select "Height of Student" under the heading "**Select column(s)**" and you would select "Ethnicity" in the drop-down menu under the heading "**Group by.**"

5. StatCrunch will automatically select the leaf unit, but you can change it by using the drop-down menu under the heading "**Leaf unit.**"

6. For the choices under the heading "**Outlier trimming**," select "Mild and extreme" if you want all outliers to be listed (across from the label "High") under the stemplot(s) but not included in the stemplot(s). Select "Extreme only" if you want only the outliers that are very different than the rest of the data to be listed under the stemplot(s) but not included in the stemplot(s). If you do not want outliers to be excluded from the stemplot(s), select "None."

7. To display the stemplot(s), click on the button ⬚Compute! at the bottom of the window.

8. Click on the button ⬚Options at the top of the window and select from the drop-down menu to edit, save, copy, print, or download the stemplot(s).

▼ B.9 Constructing Time-Series Plots

To construct a time-series plot,

1. Click on the button ⬚Graph. Then click on "Scatter Plot" in the drop-down menu.

2. In the drop-down menu under the heading "**X column**," select a variable that represents time.

3. In the drop-down menu under the heading "**Y column**," select a numerical variable other than the one you selected in Step 2.

4. Under the heading "**Display**," select "Points" and "Lines."

5. To immediately construct a time-series plot, click on the button $\boxed{\text{Compute!}}$ at the bottom of the window. To view the time-series plot and then make adjustments, click on the button $\boxed{\text{Compute!}}$, then click on the button $\boxed{\text{Options}}$ at the top of the window, and in the drop-down menu click on "Edit." To take advantage of more features, do some or all of the following steps.

6. Suppose that you chose the variables Years since 2000 and Population for Steps 2 and 3, respectively, and that a categorical variable in your data set is Country. To construct a time-series plot comparing years and populations for each country, you would select "Country" in the drop-down menu under the heading "**Group by.**"

7. To adjust the thickness of the line segments, select a size in the drop-down menu under the heading "**Line size.**"

8. To adjust the size of the dots, select a size in the drop-down menu under the heading "**Point size.**"

9. StatCrunch will automatically display the variable names on the axes, but you can enter something more complete such as the variable names and the units by making the appropriate entries in the boxes labeled "X-axis label" and "Y-axis label," which are under the heading "**Graph properties.**"

10. To title the time-series plot, enter the title in the box labeled "Title," which is under the heading "**Graph properties.**"

11. To show light horizontal and/or vertical lines, click on one or both of the boxes labeled "Horizontal lines" and "Vertical lines," which are under the heading "**Graph properties.**"

12. To display the time-series plot, click on the button $\boxed{\text{Compute!}}$ at the bottom of the window.

13. You can adjust the labels on the axes, the scaling on the axes, and the title by clicking on the small icon consisting of three horizontal lines near the bottom left corner of the window.

14. Click on the button $\boxed{\text{Options}}$ at the top of the window and select from the drop-down menu to edit, save, copy, print, or download the time-series plot.

▼ B.10 Constructing Histograms

To construct one or more histograms,

1. Click on the button $\boxed{\text{Graph}}$. Then click on "Histogram" in the drop-down menu.

2. Under the heading "**Select column(s),**" select one or more numerical variables. For each variable you select, StatCrunch will construct a histogram.

3. To immediately construct the histogram(s), click on the button $\boxed{\text{Compute!}}$ at the bottom of the window. If you have constructed more than one histogram, you can view the other one(s) by clicking on the buttons $\boxed{<}$ or $\boxed{>}$ at the bottom right corner of the window. To view the histogram(s) and then make adjustments, click on the button $\boxed{\text{Compute!}}$, then click on the button $\boxed{\text{Options}}$ at the top of the window, and in the drop-down menu click on "Edit." To take advantage of more features, do some or all of the following steps.

4. Suppose that a numerical variable in your data set is Height of Student and a categorical variable is Ethnicity. To construct a histogram of heights for each ethnicity, you would select "Height of Student" under the heading "**Select column(s)**" and you would select "Ethnicity" in the drop-down menu under the heading "**Group by.**"

5. In the drop-down menu under the heading "**Type,**" select which type of histogram you want to construct.

6. StatCrunch will automatically determine the classes, but if you want to determine them, enter the lower limit of the first class in the box labeled "Start at," which is under the heading "**Bins**" and enter the class width in the box labeled "Width."

7. To display the frequencies, relative frequencies, or densities above the bars of the histogram(s), click on the box labeled "Value above bar," which is under the heading "**Display options**."

8. To display the mean and/or median on the histogram(s), click on the appropriate boxes under the heading "**Markers**."

9. StatCrunch will automatically display the variable name on the horizontal axis, but you can enter something more complete such as the variable name and the units by making the appropriate entry in the box labeled "X-axis label," which is under the heading "**Graph properties**."

10. To title the histogram(s), enter the title in the box labeled "Title," which is under the heading "**Graph properties**."

11. To show light horizontal and/or vertical lines, click on one or both of the boxes labeled "Horizontal lines" and "Vertical lines," which are under the heading "**Graph properties**."

12. To display the histogram(s), click on the button Compute! at the bottom of the window.

13. You can adjust the labels on the axes, the scaling on the axes, and the title by clicking on the small icon consisting of three horizontal lines near the bottom left corner of the window.

14. Click on the button Options at the top of the window and select from the drop-down menu to edit, save, copy, print, or download the histogram(s).

▼ B.11 Computing Medians, Means, Standard Deviations, and Other Measures

To compute the median, mean, standard deviation, and other measures of one or more numerical variables,

1. Click on the button Stat. Next, select "Summary Stats" in the drop-down menu. Then click on "Columns" in the side menu.

2. Under the heading "**Select column(s)**," select one or more numerical variables. For each variable you select, StatCrunch will perform computations.

3. StatCrunch will automatically compute several measures including the median, mean, and standard deviation, but you can select exactly which measures you want it to compute by selecting choices under the heading "**Statistics**." To see the entire list, scroll down.

4. To immediately display the results of the computations, click on the button Compute! at the bottom of the window. To display the results and then make adjustments, click on the button Compute!, then click on the button Options at the top of the window, and in the drop-down menu click on "Edit." To take advantage of one more feature, do the following step.

5. Suppose that a numerical variable in your data set is Height of Student and a categorical variable is Ethnicity. To compute measures of heights for each ethnicity, you would select "Height of Student" under the heading "**Select column(s)**" and you would select "Ethnicity" in the drop-down menu under the heading "**Group by**."

6. To display the results of the computations, click on the button Compute! at the bottom of the window.

7. Click on the button Options at the top of the window and select from the drop-down menu to edit, save, copy, print, or download the results.

▼ B.12 Constructing Boxplots

To construct one or more boxplots,

1. Click on the button $\boxed{\text{Graph}}$. Next, click on "Boxplot" in the drop-down menu.

2. Under the heading "**Select column(s)**," select one or more numerical variables. For each variable you select, StatCrunch will construct a boxplot.

3. To immediately construct the boxplot(s), click on the button $\boxed{\text{Compute!}}$ at the bottom of the window. To view the boxplot(s) and then make adjustments, click on the button $\boxed{\text{Compute!}}$, then click on the button $\boxed{\text{Options}}$ at the top of the window, and in the drop-down menu click on "Edit." To take advantage of more features, do some or all of the following steps.

4. Suppose that a numerical variable in your data set is Height of Student and a categorical variable is Ethnicity. To construct a boxplot of heights for each ethnicity, you would select "Height of Student" under the heading "**Select column(s)**" and you would select "Ethnicity" in the drop-down menu under the heading "**Group by**."

5. To instruct StatCrunch to display outliers and/or to construct the boxplot(s) horizontally (rather than vertically), click on the appropriate boxes under the heading "**Other options**."

6. To display the mean and/or median on the boxplot(s), click on the appropriate boxes under the heading "**Markers**."

7. If the boxplot is constructed vertically, StatCrunch will automatically display the variable name on the vertical axis, but you can enter something more complete such as the variable name and the units by making the entry in the box labeled "Y-axis label," which is under the heading "**Graph properties**." If the boxplot is displayed horizontally, then make the entry in the box labeled "X-axis label."

8. To title the boxplot(s), enter the title in the box labeled "Title," which is under the heading "**Graph properties**."

9. To show light horizontal and/or vertical lines, click on one or both of the boxes labeled "Horizontal lines" and "Vertical lines," which are under the heading "**Graph properties**." StatCrunch will display both horizontal and vertical lines only if two or more boxplots are displayed in the same diagram.

10. To display the boxplot(s), click on the button $\boxed{\text{Compute!}}$ at the bottom of the window.

11. If all the boxplots are not shown in the window, you can view the other one(s) by clicking one or more times on the buttons $\boxed{<}$ or $\boxed{>}$ at the bottom right corner of the window.

12. You can adjust the labels on the axes, the scaling on the appropriate axis, and the title by clicking on the small icon consisting of three horizontal lines near the bottom left corner of the window.

13. Click on the button $\boxed{\text{Options}}$ at the top of the window and select from the drop-down menu to edit, save, copy, print, or download the boxplot.

▼ B.13 Computing Probabilities for a Normal Distribution

To compute probabilities for a normal distribution,

1. Click on the button $\boxed{\text{Stat}}$. Next, select "Calculators" in the drop-down menu. Then click on "Normal" in the side menu.

2. To find $P(X \leq 43)$ where the mean is 50 and the standard deviation is 10, enter 50 for the mean, 10 for the standard deviation ("Std. Dev."), and 43 in the box just to the right of the inequality symbol "$\leq$". Leave the box to the right of the equal sign blank. Click on the button $\boxed{\text{Compute}}$. The probability 0.24196365 should appear in the box to the right of the equal sign.

3. To find $P(X \geq 43)$ where the mean is 50 and the standard deviation is 10, select "$\geq$" in the drop-down menu for the inequality. If you competed Step 2, the probability 0.75803635 should appear in the box to the right of the equal sign. If you did not complete Step 2, enter 50 for the mean, 10 for the standard deviation ("Std. Dev."), and 43 in the box just to the right of the inequality symbol "$\geq$." Then click on the button $\boxed{\text{Compute}}$. The probability should appear in the box to the right of the equal sign.

4. To find $P(43 \leq X \leq 59)$ where the mean is 50 and the standard deviation is 10, click on the button "Between" near the top of the window. Next, enter 50 for the mean, 10 for the standard deviation, 43 in the box to the left of "$\leq X \leq$," and 59 in the box just to the right of "$\leq X \leq$." Leave the box to the right of the equal sign blank. Then click on the button $\boxed{\text{Compute}}$. The probability 0.57397622 should appear in the box to the right of the equal sign.

▼B.14 Finding Values of a Variable for a Normal Distribution

To find values of a variable for a normal distribution,

1. Click on the button $\boxed{\text{Stat}}$. Next, select "Calculators" in the drop-down menu. Then click on "Normal" in the side menu.

2. Suppose a distribution is normally distributed with mean 30 and standard deviation 5. To find a number that is greater than 70% of the observations and less than 30% of the observations, enter 30 for the mean, enter 5 for the standard deviation ("Std. Dev."), select "$\leq$" in the drop-down menu to the right of "P(X," and enter 0.70 in the box to the right of the equal sign.

3. Click on the button $\boxed{\text{Compute}}$. The number we want to find, 32.622003, should appear in the box just to the left of the equal sign.

▼B.15 Constructing Scatterplots

To construct a scatterplot,

1. Click on the button $\boxed{\text{Graph}}$. Then click on "Scatter Plot" in the drop-down menu.

2. In the drop-down menu under the heading "**X column**," select a numerical variable to be the explanatory variable.

3. In the drop-down menu under the heading "**Y column**," select a numerical variable to be the response variable.

4. Under the heading "**Display**," select "Points."

5. To immediately construct a scatterplot, click on the button $\boxed{\text{Compute!}}$ at the bottom of the window. To view the scatterplot and then make adjustments, click on the button $\boxed{\text{Compute!}}$, then click on the button $\boxed{\text{Options}}$ at the top of the window, and in the drop-down menu click on "Edit." To take advantage of more features, do some or all of the following steps.

6. Suppose that you chose the variables Height and Weight of students for Steps 2 and 3, respectively, and that a categorical variable in your data set is Ethnicity. To construct a scatterplot comparing heights and weights for each ethnicity, you would select "Ethnicity" in the drop-down menu under the heading "**Group by**."

7. To adjust the size of the dots, select a size in the drop-down menu under the heading "**Point size**."

8. StatCrunch will automatically display the variable names on the axes, but you can enter something more complete such as the variable names and the units by making the appropriate entries in the boxes labeled "X-axis label" and "Y-axis label," which are under the heading "**Graph properties**."

9. To title the scatterplot, enter the title in the box labeled "Title," which is under the heading "**Graph properties**."

10. To show light horizontal and/or vertical lines, click on one or both of the boxes labeled "Horizontal lines" and "Vertical lines," which are under the heading "**Graph properties**."

11. To display the scatterplot, click on the button $\boxed{\text{Compute!}}$ at the bottom of the window.

12. You can adjust the labels on the axes, the scaling on the axes, and the title by clicking on the small icon consisting of three horizontal lines near the bottom left corner of the window.

13. Click on the button $\boxed{\text{Options}}$ at the top of the window and select from the drop-down menu to edit, save, copy, print, or download the scatterplot.

▼B.16 Computing Linear Correlation Coefficients, Coefficients of Determination, and Sum of Squared Residuals

To compute the linear correlation coefficient and the coefficient of determination,

1. Click on the button $\boxed{\text{Stat}}$. Next, select "Regression" in the drop-down menu. Then click on "Simple Linear" in the side menu.

2. In the drop-down menu under the heading "**X variable**," select a numerical variable to be the explanatory variable.

3. In the drop-down menu under the heading "**Y variable**," select a numerical variable to be the response variable.

4. Click on the button $\boxed{\text{Compute!}}$ at the bottom of the window. Under the heading "**Simple linear regression results**," the correlation coefficient is listed across from "R (correlation coefficient)" and the coefficient of determination is listed across from "R-sq." Under the heading "**Analysis of variance table for regression model**," the sum of squared residuals is listed across from "Error" and below "SS."

▼B.17 Finding Linear Regression Equations

To find a linear regression equation,

1. Click on the button $\boxed{\text{Stat}}$. Next, select "Regression" in the drop-down menu. Then click on "Simple Linear" in the side menu.

2. In the drop-down menu under the heading "**X variable**," select a numerical variable to be the explanatory variable.

3. In the drop-down menu under the heading "**Y variable**," select a numerical variable to be the response variable.

4. To immediately find the linear regression equation, click on the button $\boxed{\text{Compute!}}$ at the bottom of the window. The equation is displayed under the heading "**Simple linear regression results**." To view the regression line on the scatterplot, click on the button $\boxed{>}$ at the bottom right corner of the window. To view the equation and the regression line on the scatterplot and then make adjustments, click on the button $\boxed{\text{Compute!}}$, next click on the button $\boxed{>}$, then click on the button $\boxed{\text{Options}}$ at the top of the window, and in the drop-down menu click on "Edit." See Steps 6 and 8–13 of Appendix B.15 for adjustments you can then make to the scatterplot.

▼ B.18 Constructing Residual Plots for Linear Regression Models

To construct a residual plot for a linear regression model,

1. Click on the button Stat . Next, select "Regression" in the drop-down menu. Then click on "Simple Linear" in the side menu.

2. In the drop-down menu under the heading "**X variable**," select a numerical variable to be the explanatory variable.

3. In the drop-down menu under the heading "**Y variable**," select a numerical variable to be the response variable.

4. In the window under the heading "**Graphs**," scroll down and click on "Residuals vs. X-values."

5. To construct the residual plot, click on the button Compute! at the bottom of the window. Then click on the button > at the bottom right corner of the window. To view the residual plot and then make adjustments, click on the button Compute! , next click on the button > , then click on the button Options at the top of the window, and in the drop-down menu click on "Edit." See Steps 6 and 8–13 of Appendix B.15 for adjustments you can then make to the residual plot.

Appendix C

Standard Normal Distribution Table

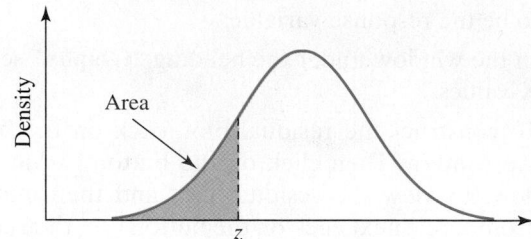

Table 1

	Standard Normal Distribution									
z	.00	.01	.02	.03	.04	.05	.06	.07	.08	.09
−3.4	0.0003	0.0003	0.0003	0.0003	0.0003	0.0003	0.0003	0.0003	0.0003	0.0002
−3.3	0.0005	0.0005	0.0005	0.0004	0.0004	0.0004	0.0004	0.0004	0.0004	0.0003
−3.2	0.0007	0.0007	0.0006	0.0006	0.0006	0.0006	0.0006	0.0005	0.0005	0.0005
−3.1	0.0010	0.0009	0.0009	0.0009	0.0008	0.0008	0.0008	0.0008	0.0007	0.0007
−3.0	0.0013	0.0013	0.0013	0.0012	0.0012	0.0011	0.0011	0.0011	0.0010	0.0010
−2.9	0.0019	0.0018	0.0018	0.0017	0.0016	0.0016	0.0015	0.0015	0.0014	0.0014
−2.8	0.0026	0.0025	0.0024	0.0023	0.0023	0.0022	0.0021	0.0021	0.0020	0.0019
−2.7	0.0035	0.0034	0.0033	0.0032	0.0031	0.0030	0.0029	0.0028	0.0027	0.0026
−2.6	0.0047	0.0045	0.0044	0.0043	0.0041	0.0040	0.0039	0.0038	0.0037	0.0036
−2.5	0.0062	0.0060	0.0059	0.0057	0.0055	0.0054	0.0052	0.0051	0.0049	0.0048
−2.4	0.0082	0.0080	0.0078	0.0075	0.0073	0.0071	0.0069	0.0068	0.0066	0.0064
−2.3	0.0107	0.0104	0.0102	0.0099	0.0096	0.0094	0.0091	0.0089	0.0087	0.0084
−2.2	0.0139	0.0136	0.0132	0.0129	0.0125	0.0122	0.0119	0.0116	0.0113	0.0110
−2.1	0.0179	0.0174	0.0170	0.0166	0.0162	0.0158	0.0154	0.0150	0.0146	0.0143
−2.0	0.0228	0.0222	0.0217	0.0212	0.0207	0.0202	0.0197	0.0192	0.0188	0.0183
−1.9	0.0287	0.0281	0.0274	0.0268	0.0262	0.0256	0.0250	0.0244	0.0239	0.0233
−1.8	0.0359	0.0351	0.0344	0.0336	0.0329	0.0322	0.0314	0.0307	0.0301	0.0294
−1.7	0.0446	0.0436	0.0427	0.0418	0.0409	0.0401	0.0392	0.0384	0.0375	0.0367
−1.6	0.0548	0.0537	0.0526	0.0516	0.0505	0.0495	0.0485	0.0475	0.0465	0.0455
−1.5	0.0668	0.0655	0.0643	0.0630	0.0618	0.0606	0.0594	0.0582	0.0571	0.0559
−1.4	0.0808	0.0793	0.0778	0.0764	0.0749	0.0735	0.0721	0.0708	0.0694	0.0681
−1.3	0.0968	0.0951	0.0934	0.0918	0.0901	0.0885	0.0869	0.0853	0.0838	0.0823
−1.2	0.1151	0.1131	0.1112	0.1093	0.1075	0.1056	0.1038	0.1020	0.1003	0.0985
−1.1	0.1357	0.1335	0.1314	0.1292	0.1271	0.1251	0.1230	0.1210	0.1190	0.1170
−1.0	0.1587	0.1562	0.1539	0.1515	0.1492	0.1469	0.1446	0.1423	0.1401	0.1379
−0.9	0.1841	0.1814	0.1788	0.1762	0.1736	0.1711	0.1685	0.1660	0.1635	0.1611
−0.8	0.2119	0.2090	0.2061	0.2033	0.2005	0.1977	0.1949	0.1922	0.1894	0.1867
−0.7	0.2420	0.2389	0.2358	0.2327	0.2296	0.2266	0.2236	0.2206	0.2177	0.2148
−0.6	0.2743	0.2709	0.2676	0.2643	0.2611	0.2578	0.2546	0.2514	0.2483	0.2451
−0.5	0.3085	0.3050	0.3015	0.2981	0.2946	0.2912	0.2877	0.2843	0.2810	0.2776
−0.4	0.3446	0.3409	0.3372	0.3336	0.3300	0.3264	0.3228	0.3192	0.3156	0.3121
−0.3	0.3821	0.3783	O.3745	0.3707	0.3669	0.3632	0.3594	0.3557	0.3520	0.3483
−0.2	0.4207	0.4168	0.4129	0.4090	0.4052	0.4013	0.3974	0.3936	0.3897	0.3859
−0.1	0.4602	0.4562	0.4522	0.4483	0.4443	0.4404	0.4364	0.4325	0.4286	0.4247
−0.0	0.5000	0.4960	0.4920	0.4880	0.4840	0.4801	0.4761	0.4721	0.4681	0.4641

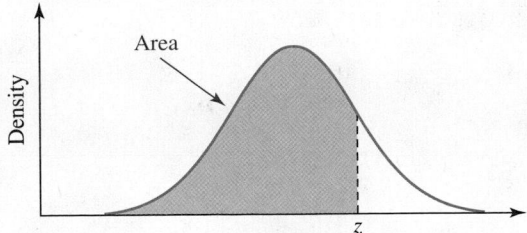

Table 1   (*Continued*)

Standard Normal Distribution

z	.00	.01	.02	.03	.04	.05	.06	.07	.08	.09
0.0	0.5000	0.5040	0.5080	0.5120	0.5160	0.5199	0.5239	0.5279	0.5319	0.5359
0.1	0.5398	0.5438	0.5478	0.5517	0.5557	0.5596	0.5636	0.5675	0.5714	0.5753
0.2	0.5793	0.5832	0.5871	0.5910	0.5948	0.5987	0.6026	0.6064	0.6103	0.6141
0.3	0.6179	0.6217	0.6255	0.6293	0.6331	0.6368	0.6406	0.6443	0.6480	0.6517
0.4	0.6554	0.6591	0.6628	0.6664	0.6700	0.6736	0.6772	0.6808	0.6844	0.6879
0.5	0.6915	0.6950	0.6985	0.7019	0.7054	0.7088	0.7123	0.7157	0.7190	0.7224
0.6	0.7257	0.7291	0.7324	0.7357	0.7389	0.7422	0.7454	0.7486	0.7517	0.7549
0.7	0.7580	0.7611	0.7642	0.7673	0.7704	0.7734	0.7764	0.7794	0.7823	0.7852
0.8	0.7881	0.7910	0.7939	0.7967	0.7995	0.8023	0.8051	0.8078	0.8106	0.8133
0.9	0.8159	0.8186	0.8212	0.8238	0.8264	0.8289	0.8315	0.8340	0.8365	0.8389
1.0	0.8413	0.8438	0.8461	0.8485	0.8508	0.8531	0.8554	0.8577	0.8599	0.8621
1.1	0.8643	0.8665	0.8686	0.8708	0.8729	0.8749	0.8770	0.8790	0.8810	0.8830
1.2	0.8849	0.8869	0.8888	0.8907	0.8925	0.8944	0.8962	0.8980	0.8997	0.9015
1.3	0.9032	0.9049	0.9066	0.9082	0.9099	0.9115	0.9131	0.9147	0.9162	0.9177
1.4	0.9192	0.9207	0.9222	0.9236	0.9251	0.9265	0.9279	0.9292	0.9306	0.9319
1.5	0.9332	0.9345	0.9357	0.9370	0.9382	0.9394	0.9406	0.9418	0.9429	0.9441
1.6	0.9452	0.9463	0.9474	0.9484	0.9495	0.9505	0.9515	0.9525	0.9535	0.9545
1.7	0.9554	0.9564	0.9573	0.9582	0.9591	0.9599	0.9608	0.9616	0.9625	0.9633
1.8	0.9641	0.9649	0.9656	0.9664	0.9671	0.9678	0.9686	0.9693	0.9699	0.9706
1.9	0.9713	0.9719	0.9726	0.9732	0.9738	0.9744	0.9750	0.9756	0.9761	0.9767
2.0	0.9772	0.9778	0.9783	0.9788	0.9793	0.9798	0.9803	0.9808	0.9812	0.9817
2.1	0.9821	0.9826	0.9830	0.9834	0.9838	0.9842	0.9846	0.9850	0.9854	0.9857
2.2	0.9861	0.9864	0.9868	0.9871	0.9875	0.9878	0.9881	0.9884	0.9887	0.9890
2.3	0.9893	0.9896	0.9898	0.9901	0.9904	0.9906	0.9909	0.9911	0.9913	0.9916
2.4	0.9918	0.9920	0.9922	0.9925	0.9927	0.9929	0.9931	0.9932	0.9934	0.9936
2.5	0.9938	0.9940	0.9941	0.9943	0.9945	0.9946	0.9948	0.9949	0.9951	0.9952
2.6	0.9953	0.9955	0.9956	0.9957	0.9959	0.9960	0.9961	0.9962	0.9963	0.9964
2.7	0.9965	0.9966	0.9967	0.9968	0.9969	0.9970	0.9971	0.9972	0.9973	0.9974
2.8	0.9974	0.9975	0.9976	0.9977	0.9977	0.9978	0.9979	0.9979	0.9980	0.9981
2.9	0.9981	0.9982	0.9982	0.9983	0.9984	0.9984	0.9985	0.9985	0.9986	0.9986
3.0	0.9987	0.9987	0.9987	0.9988	0.9988	0.9989	0.9989	0.9989	0.9990	0.9990
3.1	0.9990	0.9991	0.9991	0.9991	0.9992	0.9992	0.9992	0.9992	0.9993	0.9993
3.2	0.9993	0.9993	0.9994	0.9994	0.9994	0.9994	0.9994	0.9995	0.9995	0.9995
3.3	0.9995	0.9995	0.9995	0.9996	0.9996	0.9996	0.9996	0.9996	0.9996	0.9997
3.4	0.9997	0.9997	0.9997	0.9997	0.9997	0.9997	0.9997	0.9997	0.9997	0.9998

Glossary

Absolute value (1.4) The absolute value of a number is the distance the number is from 0 on the number line.

Addition property of equality (8.2) Adding a number to both sides of an equation does not change that equation's solution set.

Addition property of inequality (8.5) If $a < b$, then $a + c < b + c$. Similar properties hold for $\leq$, $>$, and $\geq$.

Addition rule for disjoint events (5.2) If two events have no outcomes in common, then the probability of one event OR the other event occurring is equal to the sum of their probabilities. So, if E and F are disjoint events, then $P(E \text{ OR } F) = P(E) + P(F)$.

Area (1.1) The area (in square inches) of a flat surface is the number of square inches that it takes to cover the surface.

Area-probability equality property (5.4) The area under a normal curve for an interval is equal to the probability of randomly selecting an observation that lies in the interval. The total area under a normal curve is equal to 1.

Association (2.3) When the response variable changes as the explanatory variable changes, there is an association between the two variables.

Associative law for addition (8.1) For an expression $a + b + c$, we get the same result by performing the addition on the right first or the one on the left first: $a + (b + c) = (a + b) + c$.

Associative law for multiplication (8.1) For an expression abc, we get the same result by performing the multiplication on the right first or the one on the left first: $a(bc) = (ab)c$.

Axes (1.1) The perpendicular number lines that divide a specific graphing system into four quadrants, creating a coordinate system.

Base (1.7) In the expression b^n, b is the base.

Base multiplier property (10.3) For an exponential function of the form $y = ab^x$, if the value of x increases by 1, then the value of y is multiplied by b.

Base of an exponential model (10.3) Assume that an association between the variables x and y can be described exactly by the exponential model $\hat{y} = ab^x$, where $a > 0$. If $b > 1$, then quantity y grows exponentially at a rate $b - 1$ percent (in decimal form) per unit increase of the quantity x. If $0 < b < 1$, then quantity y decays exponentially at a rate $1 - b$ percent (in decimal form) per unit increase of the quantity x.

Bias (2.1) A consistent underemphasis or overemphasis of some characteristic(s) of a population by a sampling method.

Bimodal distribution (3.4) A distribution with two mounds.

Boxplot that describes distributions without outliers (4.3) A diagram that consists of four parts: the left whisker, the left part of the box, the right part of the box, and the right whisker. Each part represents 25% of the observations.

Categorical variable (3.1) A variable consisting of names or labels of groups of individuals. Also known as the **qualitative variable**.

Causation (2.3) When the change in the explanatory variable is determined to be the cause of the change in the response variable.

Census (2.1) A census is a study in which data are collected about all members of a given population.

Cluster sampling (2.2) A sampling process that first divides a population into subgroups called clusters, which are then subjected to simple random sampling to select some of the clusters.

Coefficient (8.1) The coefficient of a term is the constant factor of the term. For example, the coefficient of the term $-3x$ is -3.

Coefficient of an exponential model (10.3) Assume that an association between the variables x and y can be described exactly by the exponential model $\hat{y} = ab^x$. Then the coefficient a equals the quantity y when the quantity x is 0.

Combining like terms (8.1) When we write a sum or difference of like terms as one term, we say we have combined like terms; to combine like terms, add the coefficients of the terms and keep the same variable factors.

Commutative law for addition (8.1) Two numbers can be added in any order, and the result will be the same: $a + b = b + a$.

Commutative law for multiplication (8.1) Two numbers can be multiplied in any order, and the result will be the same: $ab = ba$.

Complement (5.2) The complement of an event E is the event that consists of all the outcomes not in E. We write "NOT E" to stand for the complement of E.

Complement rule (5.2) For any event E, the probability of the complement of E is equal to 1 minus the probability of E: $P(\text{NOT } E) = 1 - P(E)$

Conditional probability (5.3) If E and F are events, then the conditional probability $P(E|F)$ is the probability that E occurs, given that F occurs.

Confounding variable (2.3) A variable other than the explanatory variable that causes or helps cause the response variable to change during the study.

Constant (1.1) A constant is a symbol that represents a specific number (a quantity that does not vary).

Constant term (8.1) A term that does not contain a variable.

Continuous variable (3.3) A variable that can take on any value between two possible values.

Control group (2.3) A collection of individuals who do not receive treatment (or do not have the characteristic of a treatment group).

Convenience sampling (2.2) Gathering data that is easy to collect without bothering with random sampling.

Coordinate (1.1) One of the numbers in an ordered pair.

Coordinate system (1.1) A graphing system containing two perpendicular number lines, called axes, that divide the system into four quadrants.

Correlation (6.1) An association of numerical explanatory and response variables is often called a correlation.

Counting numbers (1.1) The counting numbers, also called the **natural numbers**, are the numbers 1, 2, 3, 4, 5, and so on.

Data (1.1) Data are quantities or categories that describe people, animals, or things.

Decreasing line (7.2) A line that goes downward from left to right.

Decreasing property of an exponential function (10.3) Let $f(x) = ab^x$, where $a > 0$. If $0 < b < 1$, then the function f is decreasing. We say the function decays exponentially.

Denominator (1.1) The denominator is the number at the bottom of a given fraction.

Density histogram (3.4) A histogram in which the area of each bar in the graph represents the relative frequency of the bar's class.

Dependent events (5.3) Events E and F are dependent if $P(E|F)$ and $P(E)$ are not equal.

Dependent variable (2.3) The variable that researchers hope to show will change in response to the change of another variable.

Descriptive statistics (2.1) Descriptive statistics is the practice of using tables, graphs, and calculations about a sample to draw conclusions about only the sample.

Discrete variable (3.3) A discrete variable is a variable that has gaps between successive, possible values.

Disjoint events (5.2) Two events are disjoint (or mutually exclusive) if they have no outcomes in common.

Distributive law (8.1) To find a $(b + c)$, distribute a to both b and c: $a(b + c) = ab + ac$.

Domain of a relation (7.4) The domain of a relation is the set of all values of the explanatory variable.

Dotplot (3.3) To construct a dotplot, for each observation, we plot a dot above the number line, stacking dots as necessary.

Double-blind study (2.3) A study in which neither the participants nor the researchers know who is in the control group or the treatment group.

Empirical rule (4.2) For unimodal symmetric distributions, the following are true: approximately 68% of observations lie within one standard deviation of the mean; approximately 95% lie within two standard deviations of the mean; and approximately 99.7% lie within three standard deviations of the mean.

Equally likely probability formula (5.1) If the sample space of a random experiment consists of a finite number of equally likely outcomes, then the probability of an event is equal to the number of outcomes in the event divided by the number of outcomes in the sample space.

Equivalent equations (8.2) Equations that have the same solution set.

Equivalent expressions (8.1) Two or more expressions are equivalent expressions if, when each variable is evaluated for any real number (for which all the expressions are defined), the expressions all give equal results.

Error (6.3) The error in an estimate is the amount by which the estimate differs from the actual value. For an overestimate, the error is positive. For an underestimate, the error is negative.

Evaluating an expression (1.2) We evaluate an expression by substituting a number for each variable in the expression and then calculating the result. If a variable appears more than once in the expression, the same number is substituted for that variable each time.

Event (5.1) An event is some of the outcomes in the sample space, all of them, or none of them.

Exact association (6.2) If a curve passes through all the points of a scatterplot, we say there is an exact association with respect to the curve.

Experiment (2.3) In this type of study, researchers determine which individuals are in the treatment group(s) and the control group, often by using random assignment.

Explanatory variable (2.3) The variable that researchers hope to show affects another variable.

Exponent (1.7) The exponent is a raised integer or variable just to the right of another integer or variable (the base) that indicates the number of times the base should be multiplied by itself. In the expression b^n, b is the base and n is the exponent. We refer to b^n as the power, the nth power of b, or b raised to the nth power.

Exponential association (10.3) If the points of a scatterplot lie close to (or on) an exponential curve, we say the variables are exponentially associated.

Exponential coefficient of determination (10.5) Let r be the exponential correlation coefficient for a group of data points. The exponential coefficient of determination, r^2, is the proportion of the variation in the response variable that is explained by the exponential regression curve.

Exponential curve (10.3) The graph of an exponential function.

Exponential function (10.1) An exponential function is a function whose equation can be put into the form $f(x) = ab^x$, where $a \neq 0$, $b > 0$, and $b \neq 1$. The constant b is called the base.

Exponential model (10.3) An exponential function, or its graph, that describes an authentic association.

Exponential regression curve (10.5) The graph of an exponential regression function.

Exponential regression equation (10.5) The equation of an exponential regression function, written $\hat{y} = ab^x$, where $a \neq 0$, $b > 0$, and $b \neq 1$.

Exponential regression function (10.5) For a group of points, the exponential regression function is the exponential function with the least sum of squared residuals of all exponential functions.

Exponential regression model (10.5) The exponential regression model is the exponential regression function for a group of *data* points.

Exponentiation (1.7) The act of raising a number to a given power.

Expression (1.2) An expression is a constant, a variable, or a combination of constants, variables, operation symbols, and grouping symbols, such as parentheses.

Extrapolation (6.3) For a situation that can be described by a model whose explanatory variable is x, we perform extrapolation when we use a part of the model whose x-coordinates are not between the x-coordinates of any two data points.

Factor (1.2) We call a and b factors of ab.

Fences (4.3) The left fence is given by $Q_1 - 1.5$ IQR; the right fence is given by $Q_3 + 1.5$ IQR.

Formula (8.4) An equation that contains two or more variables.

Fraction bar (1.1) The dash between the numerator and the denominator is called the fraction bar.

Frame (2.1) A numbered list of all the individuals in the population.

Frequency bar graph (3.1) A graph that uses heights of bars to describe frequencies of categories.

Frequency distribution of a categorical variable (3.1) The frequency distribution of a categorical variable is the categories of the variable together with their frequencies.

Frequency distribution of a numerical variable (3.3 and 3.4) The frequency distribution of a numerical variable is the observations together with their frequencies. When using classes, the frequency distribution of a numerical variable is the classes together with their frequencies.

Frequency of a category (3.1) The frequency of a category is the number of observations in that category.

Frequency of a class (3.3) The frequency of a class is the number of observations in the class.

Frequency of an observation (3.3) The frequency of an observation of a numerical variable is the number of times the observation occurs in a group of data.

Frequency table for a categorical variable (3.1) A frequency table for a numerical variable is a table that lists all the categories and their frequencies.

Frequency table for a numerical variable (3.4) A frequency table for a categorical variable is a table that lists all the classes and their frequencies.

Function (7.4) A relation in which each input leads to exactly one output.

Function notation (7.4) The response variable of a function f can be represented by the expression formed by writing the explanatory variable name within the parentheses of $f(\ )$: response variable $= f$(explanatory variable).

General addition rule (5.2) The probability of one event OR another event occurring is equal to the sum of their probabilities minus the probability of one event AND the other event occurring. So, $P(E \text{ OR } F) = P(E) + P(F) - P(E \text{ AND } F)$.

Histogram (3.4) A graph with bars that can be used to describe any numerical variable, constructed on a horizontal axis that represents classes, and a vertical axis that represents frequencies of the classes.

Impossible events (5.1) An impossible event does not contain any outcomes of the sample space.

Increasing line (7.2) A line that goes upward from left to right.

Increasing property of an exponential function (10.3) Let $f(x) = ab^x$, where $a > 0$. If $b > 1$, then the function f is increasing. We say the function grows exponentially.

Independent events (5.3) Two events E and F are independent if $P(E \mid F) = P(E)$.

Independent variable (2.3) The variable that researchers hope to show affects another variable.

Individual (2.1) Individuals are the people or objects a study wants to learn about.

Inequality (1.1) An inequality contains one of the symbols $<, >, \leq,$ or $\geq$ with a constant or variable on one side and a constant or variable on the other side.

Inequality symbol (1.1) Symbols used to compare the values of two quantities are called inequality symbols, and include $<$ (is less than), $>$ (is greater than), $\leq$ (is less than or equal to), and $\geq$ (is greater than or equal to).

Inferences (2.1) The conclusions reached about an entire population by performing inferential statistics.

Inferential statistics (2.1) Inferential statistics is the practice of using information from a sample to draw conclusions about the entire population.

Influential point for a linear association (9.3) If the slope of a regression line is greatly affected by the removal of a data point, we say the data point is an influential point.

Influential point for an exponential association (10.5) If the base of an exponential regression curve is greatly affected by the removal of a data point, we say the data point is an influential point.

Input (6.3) An input is a permitted value of the explanatory variable that leads to at least one output.

Integers (1.1) The integers are the numbers $\ldots -3, -2, -1, 0, 1, 2, 3, \ldots$.

Intercept of a line (6.3) An intercept of a line is any point where the line and an axis (or axes) of a coordinate system intersect.

Interpolation (6.3) For a situation that can be described by a model whose explanatory variable is x, we perform interpolation when we use a part of the model whose x-coordinates are between the x-coordinates of two data points.

Interquartile range (IQR) (4.3) A measure of the spread of the middle 50% of observations, calculated by finding the difference between the third quartile and the first quartile: $\text{IQR} = Q_3 - Q_1$.

Interval notation (1.1) A system of symbols used to describe sets of numbers. For example, all real numbers greater than 3 uses the interval notation $(3, \infty)$.

Irrational numbers (1.1) The irrational numbers are real numbers that cannot be written in the form $\frac{n}{d}$, where n and d are integers and $d \neq 0$.

Least common denominator (LCD) (1.3) The least common denominator (LCD) of a group of fractions is the LCM of the denominators of all of the fractions.

Least common multiple (LCM) (1.3) When a given value is multiplied, the resulting numbers are called its multiples. The least common multiple of a group of numbers is the smallest number that is a multiple of all of the numbers in the group.

Like terms (8.1) Either constant terms or variable terms that contain the same variable(s) raised to exactly the same power(s).

Linear association (6.2) If the points of a scatterplot lie close to (or on) a line, we say there is a linear association.

Linear coefficient of determination (9.3) Let r be the linear correlation coefficient for a group of data points. The linear coefficient of determination, r^2, is the proportion of the variation in the response variable that is explained by the regression line.

Linear equation in one variable (8.2) An equation that can be put into the form $mx + b = 0$, where m and b are constants and $m \neq 0$.

Linear equation in two variables (7.1) If an equation can be put into the form $y = mx + b$ or $x = a$, where $m, a,$ and b are constants, then the graph of the equation is a line and we call the equation a linear equation in two variables.

Linear function (7.4) A relation whose equation can be put into the form $y = mx + b$ where m and b are constants.

Linear inequality in one variable (8.5) A linear inequality in one variable is an inequality that can be put into one of these forms: $mx + b < 0, mx + b \leq 0, mx + b > 0, mx + b \geq 0$, where m and b are constants and $m \neq 0$.

Linear model (6.3) A non-vertical line that describes the association between two quantities in an authentic situation.

Linear regression equation (9.3) The equation of the linear regression model, written $\hat{y} = b_1 x + b_0$, where b_1 is the slope and $(0, b_0)$ is the y-intercept.

Linear regression function (9.3) For a group of data points, the linear regression function is the linear function with the least sum of squared residuals.

Linear regression model (9.3) The linear regression function for a group of *data* points.

Lurking variable (2.3) A variable not officially part of the study that causes both the explanatory and response variables to change during the study.

Mean (4.1) The mean, also called the **arithmetic mean**, is the sum of the observations divided by the number of observations.

Median (4.1) If the number of observations is odd, then the median is the middle observation. If the number of observations is even, then the median is the average of the two middle observations.

Mode (4.1) The mode of some data is an observation with the greatest frequency. There can be more than one mode, but if all the observations have frequency 1, there is no mode.

Model (3.4) A model is a mathematical description of an authentic situation.

Model breakdown (6.3) When a model gives a prediction that does not make sense or an estimate that is not a good approximation.

Multimodal distribution (3.4) A distribution with more than two mounds.

Multiple bar graph (3.1) A graph that has two or more bars for each category of the variable described on the horizontal axis.

Multiplication property of equality (8.2) If A and B are expressions and c is a nonzero number, then the equations $A = B$ and $Ac = Bc$ are equivalent.

Multiplication property of inequalities (8.5) When we multiply both sides of an inequality by a positive number, we keep the inequality symbol. When we multiply both sides by a negative number, we reverse the inequality symbol.

Multiplication rule for independent events (5.3) If two events are independent, then the probability of one event AND the other event occurring is equal to the product of their probabilities: $P(E \text{ AND } F) = P(E) \cdot P(F)$.

Natural numbers (1.1) The natural numbers, also called the **counting numbers**, are the numbers $1, 2, 3, 4, 5$, and so on.

Negative association (6.1) If two numerical variables are the explanatory and response variables of a study, and the response variable tends to decrease as the explanatory variable increases, there is a negative association (or **negative correlation**).

Negative correlation (6.1) If two numerical variables are the explanatory and response variables of a study, and the response variable tends to decrease as the explanatory variable increases, there is a negative correlation (or **negative association**).

Negative integers (1.1) The numbers $-1, -2, -3, \ldots$.

Negative numbers (1.1) The negative numbers are the real numbers less than 0.

Negative-integer exponent (1.7) If n is a counting number and $b \neq 0$, then $b^{-n} = \frac{1}{b^n}$.

Negative-integer exponent in a denominator (10.1) If n is a counting number and $b \neq 0$, then $\frac{1}{b^{-n}} = b^n$.

No association (6.2) If no curve comes close to all the points of a scatterplot, we say there is no association.

Nonlinear association (6.2) If the points of a scatterplot lie close to (or on) a curve that is not a line, we say there is a nonlinear association.

Nonresponse bias (2.1) Nonresponse bias happens if individuals refuse to be part of the study or if the researcher cannot track down individuals identified to be in the sample.

Nonsampling error (2.1) An error caused by using biased sampling, recording data incorrectly, and analyzing data incorrectly.

Normal curve (5.4) Also called a **bell-shaped curve**, a normal curve describes a symmetrical, unimodal distribution whose z-scores are described by the z-table in Appendix C. Distributions that come close to this are called approximately normal or approximately normally distributed.

nth root of b (10.2) For the counting number n, where $n \neq 1$, if n is odd, then $b^{1/n}$ is the number whose nth power is b, and we call b/n the nth root of b.

Numerator (1.1) The numerator is the number at the top of a given fraction.

Numerical variable (3.1) A variable consisting of measurable quantities that describe individuals. Also called the **quantitative variable**.

Observational study (2.3) In this type of study, researchers do not determine which individuals are in the treatment group(s) or the control group.

Observations (2.1) The data observed for a variable are called observations.

Observed value (9.3) In general, for a data point (x, y), the observed value of y is y.

Opposite numbers (1.4) Two numbers are called opposites of each other if they are the same distance from 0 on the number line on opposite sides of 0.

Ordered pair (1.1) Two variables whose values occur together in a particular coordinate system. Each number is called a coordinate.

Origin (1.1) The origin is the intersection point of the axes in a coordinate system.

Outlier (4.3) A value that lies outside the left and right fences of a distribution.

Output (6.3) A permitted value of the response variable to which a value of the explanatory variable leads.

Percentage (1.6) Percent means "for each hundred" and is written as %: $a\% = \frac{a}{100}$.

Percentile (3.3) The kth percentile of some data is a value (not necessarily a data value) that is greater than or equal to approximately $k\%$ of the observations and is less than approximately $(100 - k)\%$ of the observations. In general, the 50th percentile can be used to measure the center of the distribution.

Perfect square (1.7) A number whose principal square root is rational.

Pie chart (3.2) A disk-shaped chart with slices that represent categories. The proportion of the total area for one slice is equal to the relative frequency for the category represented by the slice.

Placebo (2.3) A fake drug or treatment given to a control group.

Placebo effect (2.3) Occurs when the characteristic of interest changes in individuals due to their belief that the characteristic should change.

Point-slope form (9.1) If a non-vertical line has slope m and contains the point (x_1, y_1), then an equation of the line is $y - y_1 = m(x - x_1)$.

Population (2.1) A population is the entire group of individuals about which a study wants to learn.

Positive association (6.1) If two numerical variables are the explanatory and response variables of a study, and the response variable tends to increase as the explanatory variable increases, there is a positive association (or positive correlation).

Positive correlation (6.1) If two numerical variables are the explanatory and response variables of a study, and the response variable tends to increase as the explanatory variable increases, there is a positive association (or positive correlation).

Positive integers (1.1) The numbers $1, 2, 3, \ldots$.

Positive numbers (1.1) The positive numbers are the real numbers greater than 0.

Predicted value (9.3) In general, for a data point (x, y), the predicted value of y, (written $\hat{y}$) is the value obtained by using a model to predict y.

Prime factorization (1.3) When a value is written as a product of its prime factors.

Prime number (1.3) A prime number, or prime, is any counting number larger than 1 whose only positive factors are itself and 1.

Principal nth root of b (10.2) For the counting number n, where $n \neq 1$, if n is even and $b \geq 0$, then $b^{1/n}$ is the nonnegative number whose nth power is b, and we call $b^{1/n}$ the principal nth root of b.

Principal square root (1.7) If a is a nonnegative number, then the square root of a is the nonnegative number we square to get a. For example, the principal square root of 16 is 4.

Probability (5.1) The probability of an outcome from a random experiment is the relative frequency of the outcome if the experiment were run an infinite number of times.

Product (1.2) The product of a and b is ab.

Product property for exponents (10.2) If m and n are rational numbers, and b is any real number for which b^m and b^n are real numbers, then $b^m b^n = b^{m+n}$.

Proportion (1.3) In statistics, a proportion is a fraction of the whole. A proportion can also be written as a decimal but is never negative or larger than 1.

Quadrants (1.1) The four areas that the two perpendicular axes divide a coordinate system into are called quadrants.

Qualitative variable (3.1) A variable consisting of names or labels of groups of individuals. Also known as the **categorical variable**.

Quantitative variable (3.1) A variable consisting of measurable quantities that describe individuals. Also called the **numerical variable**.

Quartile (4.3) The first quartile is the 25th percentile; the second quartile is the 50th percentile; and the third quartile is the 75th percentile.

Quotient (1.2) The quotient of a and b is $\frac{a}{b}$, where b is not zero.

Quotient property for exponents (10.2) If m and n are rational numbers, b is any real number for which b^m and b^n are real numbers, and $b \neq 0$, then $\frac{b^m}{b^n} = b^{m-n}$.

Radical sign (1.7) The symbol used to mean "the square root of." It is written as $\sqrt{}$.

Radicand (1.7) The expression under the radical, or square root, sign.

Raising a power to a power (10.2) If m and n are rational numbers, and b is any real number for which b^m and b^n are real numbers, then $(b^m)^n = b^{mn}$.

Raising a product to a power (10.2) If n is a rational number and b and c are any real numbers for which b^n and c^n are real numbers, then $(bc)^n = b^n c^n$.

Raising a quotient to a power (10.2) If n is a rational number and b and c are any real numbers for which b^n and c^n are real numbers, and $c \neq 0$, then $\left(\frac{b}{c}\right)^n = \frac{b^n}{c^n}$.

Random assignment (2.3) The process of randomly assigning participants to the control group and the treatment group(s).

Random variable (5.1) A numerical measure of an outcome from a random experiment. We often use a capital letter such as X to stand for a random variable.

Range (4.2) A measure of spread equal to the difference between the largest observation and the smallest observation.

Range of a relation (7.4) The range of a relation is the set of all values of the response variable.

Rate of change (7.2) If a quantity y changes steadily from y_1 to y_2 as a quantity x changes steadily from x_1 to x_2, then the rate of change of y with respect to x is the ratio of the change in y to the change in x: $\frac{\text{change in } y}{\text{change in } x} = \frac{y_2 - y_1}{x_2 - x_1}$.

Rational exponent (10.2) For the counting numbers m and n, where $n \neq 1$ and b is any real number for which $b^{1/n}$ is a real number, $b^{m/n} = (b^{1/n})^m = (b^m)^{1/n}$. A power of the form $b^{m/n}$ or $b^{-m/n}$ is said to have a rational exponent.

Rational numbers (1.1) The numbers that can be written in the form $\frac{n}{d}$, where n and d are integers and $d \neq 0$.

Real numbers (1.1) The real numbers are all of the numbers represented on the number line.

Reciprocal (1.3) To find the reciprocal of a given fraction, write a new fraction with the numerator and denominator switched. So, the reciprocal of $\frac{a}{b}$ is $\frac{b}{a}$.

Reflection property for exponential functions (10.3) The graphs of $f(x) = -ab^x$ and $g(x) = ab^x$ are reflections of each other across the x-axis.

Regression line (9.3) The graph of the linear regression model.

Relation (7.4) A set of ordered pairs.

Relative frequency bar graph (3.1) A graph that uses heights of bars to describe relative frequencies of categories.

Relative frequency distribution of a categorical variable (3.1) The categories of the variable together with their relative frequencies.

Relative frequency distribution of a numerical variable (3.4) The classes together with their relative frequencies.

Relative frequency of a category (3.1) The proportion of all the observations that fall in that category. This is found by dividing the frequency of the category by the total number of observations.

Relative frequency of a class (3.4) The proportion of all the observations that fall in the class. This is found by dividing the frequency of the class by the total number of observations.

Residual (9.3) For a given data point (x, y), the residual is the difference of the observed value of y and the predicted value of y.

Residual plot (9.3) A graph that compares data values of the explanatory variable with the data points' residuals.

Response bias (2.1) Response bias occurs if surveyed people's answers do not match with what they really think.

Response variable (2.3) The variable that researchers hope to show will change in response to the change of another variable.

Rise (7.2) The rise is the vertical change in going from one point on the line to another point on the line.

Rule of Four (7.1) A rule that states we can describe some or all of the solutions of an equation in two variables in one of four ways: an equation, a table, a graph or words.

Rule of Four for functions (7.4) A rule that states there are four ways to describe input–output pairs of a function: an equation, a table, a graph or words.

Run (7.2) The horizontal change in going from one point on the line to another point on the line.

Sample (2.1) The part of a population from which data are collected.

Sample space (5.1) The sample space of a random experiment is the group of all possible outcomes.

Sampling bias (2.1) Sampling bias occurs if the sampling technique favors one group of individuals over another.

Sampling error (2.1) The error involved in using a sample to estimate information about a population due to randomness in the sample.

Sampling with replacement (2.1) This type of sampling allows an individual to be selected more than once.

Sampling without replacement (2.1) This type of sampling does not allow an individual to be selected more than once.

Scientific notation (1.7) A number is written in scientific notation if it has the form $N \times 10^k$, where k is an integer and the absolute value of N is between 1 and 10 or is equal to 1.

Simple random sample (2.1) A sample of size n chosen by a process in which every sample of size n has the same chance of being chosen from a given population.

Simple random sampling (2.1) A process of selecting a sample of size n in which every sample of size n has the same chance of being chosen.

Simplify a fraction (1.3) To simplify a fraction, we write it as an equivalent fraction in which the numerator and the denominator do not have any common positive factors other than 1.

Simplifying an expression (8.1) We simplify an expression by removing parentheses and combining like terms. The result is called a **simplified expression**.

Simulation (5.1) The use of technology to imitate a random experiment.

Single-blind study (2.3) A study in which the participants don't know whether they're in the control group or the treatment group(s).

Skewed distribution (3.4) If the left tail of a unimodal distribution is longer than the right tail, then the distribution is skewed left; if it is longer on the right, the distribution is skewed right.

Slope (m) (7.2) If (x_1, y_1) and (x_2, y_2) are two distinct points on a nonvertical line, the slope of the line is the rate of change of y with respect to x: $m = \text{slope} = \frac{y_2 - y_1}{x_2 - x_1} = \frac{\text{rise}}{\text{run}}$.

Slope-intercept form (7.3) The linear equation $y = mx + b$, where m is the slope and the y-intercept is $(0, b)$.

Solution (7.1) An ordered pair (a, b) is a solution of an equation in terms of x and y if the equation becomes a true statement when a is substituted for x and b is substituted for y. We say (a, b) satisfies the equation.

Solution for an equation in one variable (8.2) A number is a solution of an equation in one variable if the equation becomes a true statement when the number is substituted for the variable. We say the number satisfies the equation.

Solution for an inequality in one variable (8.5) A number is a solution of an inequality in one variable if it satisfies the inequality.

Solution set (7.1) The set of all solutions of an equation.

Solution set for an equation in one variable (8.2) The set of all solutions for an equation in one variable.

Solution set for an inequality in one variable (8.5) The set of all solutions for an inequality in one variable.

Solving for an equation in one variable (8.2) We solve an equation by finding its solution set.

Solving for an inequality in one variable (8.5) We solve an inequality by finding its solution set.

Split stem (3.3) A stemplot that lists the leaves from 0 to 4 in one row and lists the leaves from 5 to 9 in the next.

Standard deviation (4.2) A measure of spread, calculated by taking the square root of the following: the sum of the squared distances between the data values and the mean, all divided by 1 less than the number of data values.

Standard normal distribution (5.4) The distribution of z-scores in a normal distribution, with mean 0 and standard deviation 1.

Statistics (2.1) The practice of raising a precise question about one or more variables, creating a plan to answer the question, collecting and analyzing data, and drawing a conclusion about the question.

Stemplot (3.3) A diagram that breaks up each data value into two parts: the leaf, which is the rightmost digit, and the stem, which is the other digits. Also called a **stem-and-leaf plot**.

Stratified sampling (2.2) A sampling process that first divides a population into subgroups called **strata** (singular: **stratum**), each of which is then subjected to simple random sampling.

Strength of an association (6.2) If a curve passes through all the points of a scatterplot, we say there is an exact association with respect to the curve. If a curve comes quite close to all the points, we say there is a strong association with respect to the curve. If a curve comes somewhat close to all the points, we say there is a weak association with respect to the curve.

Strong association (6.2) If a curve comes quite close to all the points, we say there is a strong association with respect to the curve.

Sum of squared residuals (9.3 and 10.5) A calculation that measures how well a curve fits some data points: $\sum(y_i - \hat{y}_i)^2$.

Summation notation (4.1) For the data values $x_1, x_2, x_3, \ldots, x_n$, the summation notation $\sum x_i$ stands for the sum of those data values. So, $\sum x_i = x_1 + x_2 + x_3 + \cdots + x_n$.

Symmetric distribution (3.4) A distribution in which the left half is roughly the mirror image of the right half.

Systematic sampling (2.2) To perform systematic sampling, we randomly select an individual out of the first k individuals and also select every kth individual after the first selected individual.

Tails (3.4) For a unimodal distribution, the left tail is the part of the histogram to the left of the 50th percentile and the right tail is the part of the histogram to the right of the 50th percentile.

Term (8.1) A constant, a variable, or a product of a constant and one or more variables raised to powers.

Time-series plot (3.3) A diagram that plots points in a coordinate system in which the horizontal axis represents time and the vertical axis represents some other quantity.

Treatment group (2.3) A collection of individuals who receive a certain treatment (or have a certain characteristic of interest).

Two-way table (3.2) A table in which frequencies correspond to two categorical variables. The categories of one variable are listed vertically on the left side of the table, and the categories of the other variable are listed along the top.

Typical observation (3.4) An observation at or near the 50th percentile of a unimodal distribution tends to be a typical observation.

Unimodal distribution (3.4) A distribution with one mound.

Unit ratio (1.6) A unit ratio is a ratio written in reference to 1. It is written as either $\frac{a}{b}$ with $b = 1$ or $a : b$ with $b = 1$.

Unlike terms (8.1) Constant or variable terms that are not like terms.

Unusual observation (4.2) An observation more than two standard deviations from the mean of a distribution.

Unusual observation in a normal distribution (5.5) Assume that some data are normally or approximately normally distributed. Under these circumstances, an observation is unusual if it is more than 1.96 standard deviations away from the mean

Value of an observation (5.5) A data value is equal to the mean plus the product of its z-score and the standard deviation. So, an observation x with z-score z is given by $x = \bar{x} + zs$.

Variable (2.1) In statistics, a variable is a characteristic of the individuals to be measured or observed.

Variable term (8.1) A term that contains a variable.

Variance (4.2) The variance is the square of the standard deviation.

Venn diagram (5.2) A figure that consists of one or more shaded circles that are all bordered by a rectangle. The region inside the rectangle represents the sample space, and the region inside each circle represents an event.

Vertical line test (7.4) A relation is a function if and only if each vertical line intersects the graph of the relation at no more than one point.

Voluntary response sampling (2.2) Sampling that occurs voluntarily on the part of individuals from the population.

Weak association (6.2) If a curve comes somewhat close to all the points, we say there is a weak association with respect to the curve.

Whisker (4.3) A line segment of a boxplot.

x-intercept of a line (6.3) A point where the line and the x-axis intersect. The y-coordinate of an x-intercept is 0.

y-intercept of a line (6.3) A point where the line and the y-axis intersect. The x-coordinate of a y-intercept is 0.

Zero exponent (1.7) For any nonzero number b, $b^0 = 1$.

z-score (5.4) The z-score of an observation is the number of standard deviations that the observation is from the mean. If the observation lies to the left of the mean, then its z-score is negative. If the observation lies to the right of the mean, then its z-score is positive.

Answers to Odd-Numbered Exercises

Answers to most discussion exercises and to exercises in which answers may vary have been omitted.

Chapter 1

Homework 1.1 **1.** variable **3.** real **5.** 25 thousand fans attended the concert. **7.** In 2013, 70 million iPads were sold.
9. The company lost $45 thousand that year. **11.** The statement $t = 9$ represents the year 2009. **13.** h; 67, 72; $-5, 0$; answers may vary.
15. p; 50, 60; $-2, -8$; answers may vary. **17.** T; 15, 40; 240, -10; answers may vary. **19.** s; 25, 32; $-15, -9$; answers may vary.
21. a. The rectangles are not drawn to scale. Answers may vary. 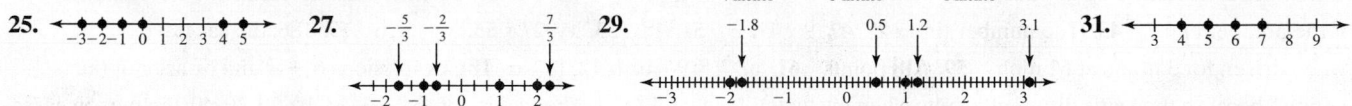 **b.** W, L **c.** A
23. a. The rectangles are not drawn to scale. Answers may vary. 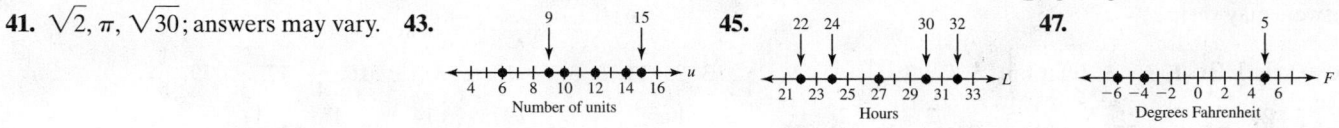 **b.** W, L, A **c.** None
25.   **27.**   **29.**   **31.**
33.   **35.**   **37.** $-8, -9, -27$; answers may vary. **39.** $\dfrac{1}{2}, \dfrac{17}{5}, -\dfrac{7}{3}$; answers may vary.
41. $\sqrt{2}, \pi, \sqrt{30}$; answers may vary. **43.**   **45.**   **47.**
49. a.   **b.** increase **c.** increase **51. a.**   **b.** increase
c. decrease **53–67 odd.**   **69.** 2 **71.** $A(-4, -3), B(-5, 0), C(-2, 4), D(1, 3), E(0, -2), F(5, -4)$
73. true **75.** true **77.**   **79.**
81.   **83.**

85.

In Words	Inequality	Graph	Interval Notation
numbers greater than or equal to 4	$x \geq 4$		$[4, \infty)$
numbers less than or equal to -2	$x \leq -2$		$(-\infty, -2]$
numbers less than 1	$x < 1$		$(-\infty, 1)$
numbers greater than -5	$x > -5$		$(-5, \infty)$

87.   **89.**   **91.**

93.

In Words	Inequality	Graph	Interval Notation
numbers between 1 and 5	$1 < x < 5$		$(1, 5)$
numbers between -5 and 2, as well as -5	$-5 \leq x < 2$		$[-5, 2)$
numbers between -2 and 4, as well as 4	$-2 < x \leq 4$		$(-2, 4]$
numbers between 0 and 4, inclusive	$0 \leq x \leq 4$		$[0, 4]$

95. w; the average daily coffee consumption is greater than 8 ounces.
97. $h \geq 70$; $[70, \infty)$;
99. $0 \leq t \leq 170$; $[0, 170]$;

101. w; a hamburger served at a fast-food restaurant weighs between 1 ounce and 3 ounces, inclusive.
103. $15 \leq d \leq 20$; $[15, 20]$; d **105. a.** -5 **b.** no; a variable alone can represent negative numbers.

107. A variable represents a quantity that can vary, but a constant represents a quantity that does *not* vary.   **109.** infinitely many; 2.1, 3, 3.682   **111.** Actually, the sentence means $x \geq 5$.   **113.** Answers may vary.

Homework 1.2   **1.** expression   **3.** ab **5.** 8   **7.** 3   **9.** 42   **11.** 2   **13.** 12   **15.** 36   **17.** 36; the total cost of 4 albums is $36.   **19.** 11.4 million video subscribers   **21. a.**

Tuition (dollars)	Total Cost (dollars)
400	400 + 20
401	401 + 20
402	402 + 20
403	403 + 20
t	$t + 20$

$t + 20$ **b.** $437   **23. a.**

Number of Hours of Courses	Total Cost (dollars)
1	$101 \cdot 1$
2	$101 \cdot 2$
3	$101 \cdot 3$
4	$101 \cdot 4$
n	$101n$

$101n$

b. 1515; the total cost for 15 credit hours of classes is $1515.   **25.** $x + 4$; 12   **27.** $x \div 2$; 4   **29.** $x - 5$; 3   **31.** $7x$; 56   **33.** $16 \div x$; 2   **35.** The number divided by 2   **37.** 7 minus the number   **39.** The number plus 5   **41.** The product of 9 and the number   **43.** The difference of the number and 7   **45.** The number times 2   **47.** 9   **49.** 3   **51.** 18   **53.** xy; 27   **55.** $x - y$; 6   **57.** 186; the car traveled 186 miles when driven for 3 hours at 62 mph.   **59.** 518 points   **61. a.** 7; 8; 9   **b.** 6, 12, 18   **c.** The expressions $6 + x$ and $6x$ are not the "same thing," because they give different results when evaluated for $x = 1$ (and other values of x).   **63. a.** 10, 20, 30, 40; the person earns $10, $20, $30, and $40 for working 1, 2, 3, and 4 hours, respectively.   **b.** $10 per hour   **c.** Answers may vary.   **65.** Answers may vary.   **67.** Answers may vary.

Homework 1.3   **1.** b **3.** fraction   **5.** 7   **7.** $2 \cdot 2 \cdot 5$ **9.** $2 \cdot 2 \cdot 3 \cdot 3$ **11.** $3 \cdot 3 \cdot 5$ **13.** $2 \cdot 3 \cdot 13$ **15.** $\frac{3}{4}$ **17.** $\frac{3}{5}$ **19.** $\frac{1}{5}$ **21.** $\frac{5}{6}$ **23.** $\frac{2}{15}$ **25.** $\frac{3}{10}$ **27.** $\frac{5}{3}$ **29.** $\frac{5}{6}$ **31.** $\frac{2}{3}$ **33.** $\frac{2}{15}$ **35.** $\frac{3}{4}$ **37.** $\frac{1}{3}$ **39.** $\frac{3}{4}$ **41.** $\frac{19}{12}$ **43.** $\frac{14}{3}$ **45.** $\frac{1}{9}$ **47.** $\frac{17}{63}$ **49.** $\frac{11}{5}$ **51.** 1   **53.** 599   **55.** Undefined   **57.** 0   **59.** 1   **61.** 0   **63.** $\frac{1}{3}$ **65.** $\frac{9}{5}$ **67.** $\frac{1}{3}$ **69.** 0.17   **71.** 1.33   **73.** 0.43   **75.** Answers may vary.   **77.** $\frac{1}{10}$ square mile   **79.** 1   **81.** $\frac{7}{9}$ **83.** $\frac{5}{7}$ **85.** $\frac{13}{21}$ **87.** $\frac{5}{12}$ **89.** $\frac{1}{5}$ **91.** $\frac{1}{4}$ of the course points   **93. a. i.** 0.484   **ii.** 0.516   **iii.** 0.227   **b.** 1, the sum of the proportions for all categories always equals 1.

95.

Number of People	Cost per Person (dollars)
2	$\frac{19}{2}$
3	$\frac{19}{3}$
4	$\frac{19}{4}$
5	$\frac{19}{5}$
n	$\frac{19}{n}$

$\frac{19}{n}$

97. 5.32 feet   **99.** 56.76 liters   **101.** 101.88 milligrams   **103.** 5.30 gallons per day   **105. a. i.** $\frac{5}{9}$ **ii.** $\frac{5}{4}$ **iii.** $\frac{3}{2}$ **iv.** $\frac{1}{6}$ **b.** Answers may vary.   **107.** Answers may vary.   **109.** Answers may vary.

Homework 1.4   **1.** opposites   **3.** true   **5.** 4   **7.** -7 **9.** 3   **11.** 8   **13.** -4 **15.** -7 **17.** -5 **19.** -5 **21.** 2   **23.** -3 **25.** -10 **27.** -3 **29.** 0   **31.** 0   **33.** -13 **35.** -22 **37.** -1145 **39.** 0   **41.** -6.7 **43.** -4.8 **45.** -97.3 **47.** $\frac{2}{7}$ **49.** $-\frac{1}{4}$ **51.** $-\frac{3}{4}$ **53.** $\frac{7}{12}$ **55.** 6221.4   **57.** $-97,571.14$ **59.** -0.11 **61.** $175

63.

Check No.	Date	Description of Transaction	Payment	Deposit	Balance
					-89.00
	7/18	Transfer		300.00	211.00
3021	7/22	State Farm	91.22		119.78
3022	7/22	MCI	44.26		75.52
	7/31	Paycheck		870.00	945.52

65. −2871 dollars   **67.** −1633 dollars   **69.** 4°F   **71.** negative   **73.** The numbers are equal in absolute value and opposite in sign, or the numbers are both 0.   **75. a.** 3   **b.** 4   **c.** 6   **d.** no; answers may vary.

Homework 1.5   **1.** change   **3.** false   **5.** −2   **7.** −6   **9.** 9   **11.** −1   **13.** −3   **15.** 11   **17.** −6   **19.** −79   **21.** 420   **23.** −5.4   **25.** −11.3   **27.** 5.7   **29.** 15.98   **31.** −1   **33.** $\dfrac{1}{2}$ **35.** $\dfrac{3}{4}$ **37.** $-\dfrac{13}{24}$ **39.** 2   **41.** −2   **43.** $-\dfrac{1}{4}$ **45.** −2.7   **47.** −7   **49.** −2   **51.** −3128.17   **53.** 112,927.91   **55.** −0.95   **57.** −12°F   **59.** 11°F   **61. a.** −12°F   **b.** −6°F   **c.** Answers may vary.   **63.** 20,602 feet   **65. a.** 0.7 percentage point, −2.5 percentage points, 4.5 percentage points, −7.7 percentage points, 0.5 percentage point, 9.1 percentage points, −0.2 percentage point, −6.1 percentage points   **b.** 9.1 percentage points   **c.** no   **67. a.** 144 thousand cars   **b.** From 2006 to 2007, from 2009 to 2010, from 2011 to 2012   **c.** From 2007 to 2009, from 2010 to 2011, from 2012 to 2013   **69.** −3   **71.** −7   **73.** 9   **75.** −3 − x:2   **77.** x − 8: −13   **79.** x − (−2): −3   **81.** Answers may vary.   **83. a. i.** 2   **ii.** 8   **iii.** 5   **b.** Answers may vary.   **85. a.** 3   **b.** −3   **c.** They are equal in absolute value and opposite in sign.   **d.** −6, 6; they are equal in absolute value and opposite in sign.   **e.** Answers may vary.   **f.** They are equal in absolute value and opposite in sign.

Homework 1.6   **1.** 100   **3.** false   **5.** 0.63   **7.** 8%   **9.** 0.09   **11.** 5.2%   **13.** In 2013, the proportion of toys and sporting goods that were purchased online was 0.12.   **15.** The approximate proportion of students' college costs that are paid by their parents borrowing money is 0.07.   **17.** In June 2014, 41% of Americans approved of the way President Obama was doing his job.   **19.** In 2012, 8.8% of plastics were recycled.   **21.** 81.0% of 273 thousand people living with spinal cord injuries are men.   **23.** 2450 cars   **25.** $13.80   **27.** 21,829 undergraduates   **29.** −12   **31.** 18   **33.** −5   **35.** 8   **37.** 555   **39.** −39   **41.** 0.08   **43.** −0.3   **45.** −9   **47.** 4   **49.** $-\dfrac{1}{10}$ **51.** $\dfrac{1}{15}$ **53.** $-\dfrac{9}{14}$ **55.** $\dfrac{15}{14}$ **57.** −3   **59.** 13   **61.** 6   **63.** −100   **65.** $-\dfrac{29}{9}$ **67.** $-\dfrac{6}{7}$ **69.** $-\dfrac{4}{5}$ **71.** $\dfrac{3}{4}$ **73.** 10,252.84   **75.** −6.78   **77.** 0.48   **79.** −8.07   **81.** $\dfrac{3}{4}$ **83.** $\dfrac{1.89}{1}$; the number of U.S. billionaires in 2014 is 1.89 times the number of U.S. billionaires in 2001.   **85. a.** $\dfrac{0.8\,\text{red bell pepper}}{1\,\text{black olive}}$; for each black olive used, 0.8 bell pepper is needed.   **b.** $\dfrac{1.25\,\text{black olives}}{1\,\text{red bell pepper}}$; for each red bell pepper used, 1.25 olives are required.   **87. a.** $\dfrac{12.77}{1}$; the FTE enrollment at Texas A&M University is 12.77 times larger than that at St. Olaf College.   **b.** $\dfrac{2.93}{1}$; the number of FTE faculty at University of Massachusetts Amherst is 2.93 times greater than that at Butler University.   **c.** Butler University: $\dfrac{12.44}{1}$; St. Olaf College: $\dfrac{11.87}{1}$; Stonehill College: $\dfrac{13.21}{1}$; University of Massachusetts Amherst: $\dfrac{17.32}{1}$; Texas A&M University: $\dfrac{21.1}{1}$ **d.** Texas A&M University; St. Olaf College   **e.** Answers may vary.   **89. a.** $\dfrac{2.39}{1}$ **b.** For each $1 the person pays to her MasterCard account, she should pay about $2.39 to her Discover account.   **91.** −3162 dollars   **93.** −29.52 dollars   **95. a.** −6   **b.** 8   **c.** A negative number times a negative number is equal to a positive number.   **d.** Answers may vary.   **97.** $\dfrac{a}{b} = \dfrac{-a}{-b}, \dfrac{-a}{b} = \dfrac{a}{-b} = -\dfrac{a}{b} = -\dfrac{-a}{-b}$ **99.** One number is positive and one number is negative.

Homework 1.7   **1.** $\dfrac{1}{b^n}$ **3.** false   **5.** 64   **7.** 32   **9.** −64   **11.** 64   **13.** $\dfrac{36}{49}$ **15.** 1   **17.** $\dfrac{1}{8}$ **19.** $\dfrac{1}{81}$ **21.** $\dfrac{1}{100,000}$ **23.** 2   **25.** −6   **27.** not a real number   **29.** not a real number   **31.** irrational; 5.48   **33.** rational; 8   **35.** 12   **37.** −18   **39.** 4   **41.** $-\dfrac{2}{3}$ **43.** 20   **45.** −17   **47.** −50   **49.** −10   **51.** −14   **53.** 15   **55.** −27   **57.** $\dfrac{1}{2}$ **59.** 27   **61.** −48   **63.** 2   **65.** 3   **67.** 48   **69.** $\dfrac{26}{3}$ **71.** 13   **73.** −7   **75.** 21.04   **77.** 0.08   **79.** 4.17   **81.** 8   **83.** −2   **85.** $-\dfrac{13}{7}$ **87.** 26   **89.** $\dfrac{4}{9}$ **91.** 4.16   **93.** 5 + (−6)x: 29   **95.** $\dfrac{x}{-2} - 3; -1$ **97. a.**

Years since 2000	Congressional Pay (thousands of dollars)
0	3.6 · 0 + 141.3
1	3.6 · 1 + 141.3
2	3.6 · 2 + 141.3
3	3.6 · 3 + 141.3
4	3.6 · 4 + 141.3
t	3.6t + 141.3

3.6t + 141.3   **b.** 166.5; congressional pay was about $166.5 thousand in 2007.   **c.** $175.4 thousand

99. a.

Years since 2000	Population (thousands)
0	$-2.0 \cdot 0 + 102.7$
1	$-2.0 \cdot 1 + 102.7$
2	$-2.0 \cdot 2 + 102.7$
3	$-2.0 \cdot 3 + 102.7$
4	$-2.0 \cdot 4 + 102.7$
t	$-2.0t + 102.7$

$-2.0t + 102.7$ **b.** 74.7; Gary's population was 74.7 thousand in 2014.   **101.** 326.3 million connections   **103.** \$15.0 billion   **105.** 0.52   **107.** 0.38   **109.** 49,000   **111.** 0.00859   **113.** -0.000295 **115.** 4.57×10^7 **117.** 6.59×10^{-5} **119.** -1×10^{-6} **121.** 0.0000063; 0.00013; 3,200,000; 64,000,000   **123.** 3,600,000,000 years   **125.** 0.000000063 mole per liter   **127.** 1.008×10^7 gallons   **129.** 4.7×10^{-7} meter   **131.** Answers may vary; 25.   **133. a.** -2 **b.** 3   **c.** Answers may vary; -2.   **135. a. i.** 1   **ii.** -1 **iii.** 1   **iv.** -1 **v.** -1 **vi.** 1   **b.** even   **c.** odd

Chapter 1 Review Exercises

1. The total box office gross was \$10.02 billion in 2014.   **2.** p; 60, 70 (Answers may vary.); -12, 107 (Answers may vary.)

3. 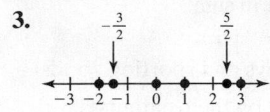   **4.** 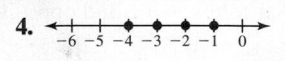   **5.** 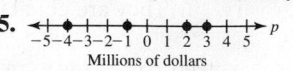   **6.**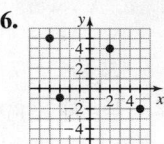

7. 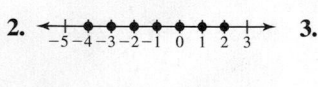 x, numbers less than -3; $(-\infty, -3)$ **8.** 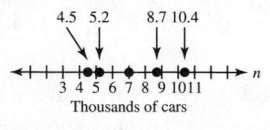 t; the teenager plays the video game between 2 and 5 hours, inclusive.   **9.** $\dfrac{17}{30}$ **10.** $\dfrac{1}{12}$ **11.** 5.30 cups per day   **12.** -3 **13.** 8   **14.** -45 **15.** -4 **16.** -3 **17.** 12   **18.** 5   **19.** $-\dfrac{3}{2}$ **20.** -16 **21.** -20 **22.** -12 **23.** 14   **24.** 4   **25.** 10.9   **26.** $\dfrac{5}{6}$ **27.** $\dfrac{7}{9}$ **28.** -64 **29.** $\dfrac{27}{64}$ **30.** $\dfrac{1}{64}$ **31.** -7 **32.** -54 **33.** 3   **34.** 0   **35.** $-\dfrac{8}{11}$ **36.** -19 **37.** 58   **38.** 7   **39.** -8.68 **40.** 0.03   **41.** $\dfrac{7}{10}$ **42.** -4095.49 dollars   **43. a.** $-12°F$ **b.** $-4°F$ **c.** Answers may vary.   **44. a.** \$5.6 million   **b.** -39 million dollars   **c.** from 2004 to 2008; \$239 million   **d.** from 2000 to 2004; \$167.7 million   **45.** 1.40; the number of messages sent or received per day in 2011 is 1.40 times larger than the number of messages sent or received per day in 2009.   **46.** 59.0% of 1501 adults think that Iran's nuclear program is the biggest threat to the United States.   **47.** 4986 people   **48.** -4394.40 dollars   **49.** -11 **50.** $-\dfrac{1}{2}$ **51.** 22   **52.** $-7 - x$; -4 **53.** $1 + \dfrac{-24}{x}$; 9   **54.** 50; if the total cost is \$650 and there are 13 players on the team, the cost is \$50 per player.

55. a.

Time (hours)	Volume of Water (cubic feet)
0	$-50 \cdot 0 + 400$
1	$-50 \cdot 1 + 400$
2	$-50 \cdot 2 + 400$
3	$-50 \cdot 3 + 400$
4	$-50 \cdot 4 + 400$
t	$-50t + 400$

$-50t + 400$ **b.** 50; there will be 50 cubic feet of water in the basement after 7 hours of pumping.   **56.** 77.7 thousand employees   **57.** 0.34   **58.** 0.000385   **59.** 5.4×10^7

Chapter 1 Test

1. a. The rectangles are not drawn to scale. Answers may vary.   6 feet ☐ 4 feet ☐ 2 feet ▭   6 feet   9 feet   18 feet   **b.** W, L **c.** A

2. 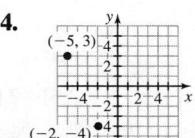   **3.** 4.5 5.2 8.7 10.4 ... 3 4 5 6 7 8 9 10 11 n Thousands of cars   **4.** $(-5, 3)$ ... $(-2, -4)$ **5.** $c \geq 1450$; $[1450, \infty)$; 1400 1450 1500 c Dollars

6. $\dfrac{13}{30}$ **7.** 9.92 ounces   **8.** -6 **9.** -8 **10.** $\dfrac{1}{2}$ **11.** 18   **12.** 3   **13.** -25 **14.** -0.08 **15.** $-\dfrac{45}{4}$ **16.** $\dfrac{13}{40}$ **17.** $\dfrac{1}{32}$ **18.** 6   **19.** -17 **20.** $\dfrac{58}{3}$ **21.** $-\dfrac{21}{4}$ **22. a.** 0.7 audit per 1000 tax returns   **b.** -1.5 audits per 1000 returns   **c.** from 2003 to 2005; 3.2 audits per 1000 returns   **23.** 1.9   **24.** 2.95; the average ticket price in 2012 was 2.95 times larger than the average ticket price in 1991.   **25.** -33 **26.** 11   **27.** -3 **28.** $2x - 3x$; 5   **29.** $\dfrac{-10}{x} - 6$; -4

30. a.

Years since 2008	First-Class Mail Volume (billions of pieces)
0	$-4.9(0) + 91.7$
1	$-4.9(1) + 91.7$
2	$-4.9(2) + 91.7$
3	$-4.9(3) + 91.7$
4	$-4.9(4) + 91.7$
t	$-4.9t + 91.7$

$-4.9t + 91.7$ **b.** 67.2; the first-class mail volume was 67.2 billion pieces in 2013.   **31.** 1.40 million couples   **32.** 6.78×10^{-5}

Chapter 2

Homework 2.1   1. variable   **3.** inferential   **5. a.** Pelee, Kelut, Lamington, El Chichon, Ruiz   **b.** country, year, caused a tsunami, number of deaths   **c.** country: Martinique, Indonesia, Papua New Guinea, Mexico, Columbia; year: 1902, 1919, 1951, 1982, 1985; caused a tsunami: yes, no, no, no, no; number of deaths: 28,000, 5110, 2942, 1879, 23,080   **d.** Pelee   **e.** 26,121 deaths   **7. a.** Student 1, Student 2, Student 3, Student 4, Student 5   **b.** major, daily exercise (in minutes), statistics study time (in hours), read statistics textbook   **c.** major: psychology, undecided, film, business, journalism; daily exercise (in minutes): 0, 2, 30, 60, 60; statistics study time (in hours): 4, 6.5, 8, 10, 5; read statistics textbook: yes, yes, yes, no, no   **d.** 2 minutes; people do not usually exercise for just 2 minutes; the student might exercise for 2 hours and assumed the units were hours, rather than minutes; the student might exercise for 20 minutes, but the author entered 2, rather than 20.   **e.** If a student says they read a statistics textbook that does not make it clear how often or for how long the student reads it; answers may vary.   **9. a.** whether American adults strongly disapprove of the way Congress is doing its job   **b.** the 1000 adults   **c.** all American adults   **11. a.** whether the drinking age should be lowered to age 18   **b.** the 1013 adults   **c.** all American adults   **13. a.** whether the police can protect people from violent crime   **b.** the 218 non-Caucasians   **c.** all non-Caucasians   **15. a.** whether a student plans to stay at their next full-time job for 3–5 years after graduating from college   **b.** the 13,127 students   **c.** all graduating college students   **17. a.** Does the flu shot help hospital employees avoid getting the flu?   **b.** the 179 hospital staff   **c.** all hospital staff   **d.** The flu shot does not help hospital staff avoid getting the flu; inferential statistics   **19. a.** Does music improve sleep quality?   **b.** the 94 students   **c.** all students ages 19 to 29 years   **d.** Listening to relaxing classical music for 45 minutes at bedtime improves sleep quality in students ages 19 to 29 years; inferential statistics   **21. a.** Do low-skilled college students learn more from completing online homework than textbook homework?   **b.** the 75 students   **c.** all low-skilled college algebra students   **d.** Low-skilled college algebra students learn more from completing online homework than textbook homework; inferential statistics   **23. Using a TI-84:** Nubia, Paul, Dawn; **Using StatCrunch:** Jabra, Dawn, Becky   **25. a. Using a TI-84:** Harold, Amy, Frank, Jason, Angelica, Keith; **Using StatCrunch:** Gerardo, Anas, Calvin, Keith, Frank, Harold   **b. Using a TI-84:** $\frac{1}{3}$; **Using StatCrunch:** $\frac{1}{3}$; descriptive statistics   **c. Using a TI-84:** Damon, Keith, Jason, Kat, Anas, Calvin; **Using StatCrunch:** Gerardo, Marissa, Calvin, Damon, Anas, Kat   **d. Using a TI-84:** $\frac{1}{3}$; **Using StatCrunch:** $\frac{1}{2}$ **e. Using TI-84:** yes; no; **Using StatCrunch:** no; because we used simple random sampling, each student had the same chance of being selected. So, different samples happened to be collected in the two samplings.   **27. a. Using a TI-84:** Hee-Sang, Hallie, Claudia, Sean, Jackson, Ben; **Using StatCrunch:** Jackson, Chris, Jasmine, Charles, Hallie, Hee-sang   **b. Using a TI-84:** $\frac{1}{6}$; **Using StatCrunch:** $\frac{1}{6}$ **c.** $\frac{3}{14}$ **d.** no; sampling error   **e.** no; sampling error   **29. a.** 0.350   **b.** 0.38   **c.** no; sampling error   **d.** inferential statistics   **31.** Do you drink smoothies or not drink them?   **33.** Do you pay taxes or not pay them?   **35.** sampling bias   **37.** sampling bias, response bias   **39.** sampling bias, response bias, nonresponse bias   **41.** response bias, nonresponse bias   **43.** sampling bias, response bias   **45.** sampling bias   **47.** sampling bias, response bias   **49. a.** 5.5   **b.** no; the scores 7 and 8 are *both* larger than the middle of the scores.   **c.** There are many better interpretations. For example, customers who chose scores between 0 and 4 points, inclusive, could be interpreted to be unhappy with the product. Customers who chose either of the scores 5 or 6 could be interpreted to feel neutral about the product. Customers who chose scores between 7 and 11 points, inclusive, could be interpreted to be happy with the product.   **51.** Answers may vary.   **53.** Answers may vary.   **55.** Answers may vary.   **57.** Answers may vary.   **59.** Answers may vary.

Homework 2.2   1. kth   **3.** true   **5.** cluster   **7.** systematic   **9.** convenience   **11.** stratified   **13.** simple random   **15.** cluster   **17.** stratified   **19. a.** 4   **b. Using a TI-84:** 3; **Using StatCrunch:** 2   **c. Using a TI-84:** 3, 7, 11, 15, 19; **Using StatCrunch:** 2, 6, 10, 14, 18

21. a. 102   **b. Using a TI-84:** 85; **Using StatCrunch:** 35   **c. Using a TI-84:** 85, 187, 289, 391, 493; **Using StatCrunch:** 35, 137, 239, 341, 443   **23.** 45 employees from the police department, 23 employees from the fire department, 2 employees from the justice department   **25.** 30 students from Notre Dame De La Baie Academy, 5 students from Northeastern Wisconsin Lutheran High School, 4 students from Bay City Baptist School, 1 student from Beth Haven Academy   **27.** 6 students from music and human learning, 99 students from music performance, 4 students from music theory, 11 students from musicology/ethnomusicology   **29.** 46 WUE undergraduate students, 505 resident undergraduate students, 72 resident graduate students, 229 nonresident undergraduates, 48 nonresident graduate students   **31. Using a TI-84:** Kelsey, Southerland, Stevens, Ramsey, Campfield, Massey, Johnson, Gardenhire, Burks, Finney; **Using StatCrunch:** Gardenhire, Johnson, Summerville, Southerland, Haile, Tracy, Ketron, Norris, Finney, Tate   **33. Using a TI-84:** all the active players on the Marlins, Rockies, and Cubs; **Using StatCrunch:** all the active players on the Cubs, Rockies, Dodgers   **35.** stratified   **37.** systematic; systematic is the best choice because they couldn't form a frame of drivers; the police could decide to pull over every 4th car. To determine which car to first pull over, they should randomly select a number between 1 and 4, inclusive.   **39.** stratified; stratified would be best because Republicans, Democrats, and Independents will likely have different opinions. And most individuals within each strata will have the same opinion; the newspaper should determine a total number of people to be surveyed and then use the percentages of each political party to determine how many people of each party to survey.   **41. a.** Some of the classes could be randomly selected and all of the students in the selected classes could be surveyed.   **b.** A sample size could be determined and then the percentages of the ethnic groups could be used to determine the number of students of each ethnic group that should be surveyed.   **c.** cluster; the cluster sampling would require fewer classroom visitations.   **d.** stratified; stratified samplings often require smaller sample sizes than the other sampling methods.   **43. a.** systematic   **b.** 0.592   **c.** The proportion of cars that the police had planned to pull over was 0.333, but the proportion of cars they actually pulled over was 0.592.   **45.** Answers may vary.   **47.** Answers may vary.   **49.** Answers may vary.   **51.** Answers may vary.

Homework 2.3   1. control   **3.** false   **5. a.** treatment group: the 9 cyclists when they took the caffeine pill; control group: the 9 cyclists when they took the Metamucil pill.   **b.** experiment; it is an experiment because the researchers determined which cyclists were in the treatment and control groups of each session.   **c.** Random assignment means that the cyclists were randomly assigned to the treatment and control groups. Random assignment could be accomplished by creating a frame of the 9 cyclists and randomly selecting 5 of the cyclists to take the caffeine pill on the second session. The other 4 cyclists would take the Metamucil pill. Then for the third session, each cyclist would take the other pill.   **d.** sample: the 9 cyclists; population: all highly-trained cyclists   **7. a.** A sugar pill might give cyclists a small boost in energy and then a small dip in energy, but a Metamucil pill would not affect their performance.   **b.** Both the cyclists and the researcher in communication with the cyclists did not know which cyclists were in which group for both sessions; one researcher could label the pill vials with numbers to identify the pills, but the researcher would not tell the code to another researcher who would be in contact with the cyclists.   **c.** explanatory variable: taking a caffeine pill; response variable: a cyclist's time in a hot environment   **d.** A small dose of caffeine improves a cyclist's time in a hot environment; causality; we can assume there is causality because the cyclists were randomly assigned to the control and treatment groups.   **e.** 1 cup; yes   **9. a.** observational study; the researchers did not determine who was subject to a law that bans the use of handheld cell phones while driving and who was not subject to such a law.   **b.** explanatory variable: whether there is a law that bans the use of handheld cell phones while driving; response variable: whether people talk on handheld cell phones while driving   **c.** The proportion of drivers who use handheld cell phones while driving has decreased since the law was passed; an association; causality cannot be concluded from an observational study.   **d.** no; it is possible that before and after the law was passed, fewer people used handheld cell phones in Wellington than in Auckland.   **e.** 6 years; it is possible that fewer drivers used handheld cell phones in 2012 than in 2006 because of other reasons than the law. For example, hands-free devices for cell phones might have been more available and affordable in 2012 than in 2006.   **11. a.** observational study; the researchers did not determine the amount of presidential advertisements voters were exposed to.   **b.** explanatory variable: amount of exposure to presidential advertisements; response variable: voters' opinions about the candidates   **c.** Neighboring states would not have as much "on-the-ground" interaction, which could be a confounding variable.   **d.** A voter who had been exposed to more advertisements for a certain candidate tended to have a more positive opinion about that candidate; an association; causality cannot be concluded from an observational study.   **13. a.** treatment group: the patients who received the drug; control group: the patients who received the placebo   **b.** experiment; the patients were randomly assigned to the treatment and control groups.   **c.** Random assignment means that the patients were randomly assigned to the treatment and control groups. Random assignment could have been accomplished by creating a frame of the 942 patients and randomly selecting half of the patients to take the drug. The other half would take the placebo.   **d.** sample: the 942 patients; population: RRMS patients ages 18 to 50 years   **15. a.** Both the patients and the researcher in communication with the patients did not know which patients were in which group; one researcher could label the pill vials with numbers to identify the pills, but the researcher would not tell the code to another researcher who would be in contact with the patients.   **b.** explanatory variable: whether a patient took the drug; response variable: the annual relapse rate   **c.** The drug reduced the relapse rate; causality; we can assume there is causality because there was random assignment.   **d.** Researchers might have been influenced, consciously or unconsciously, by receiving money from the company that manufactures natalizumab. By reporting that they had received such money, other researchers who have not received money from the company could run a similar experiment and see if they get similar results.   **17. a.** The researchers did not determine which adults were on which diet.   **b.** It would be unethical to require some adults to change their diet to one that might cause them to die sooner than if they had eaten their usual diet.   **c.** sample: the 7316 Danish adults; the entire Danish population   **d.** explanatory variable: whether a person eats a high-fiber diet or a Western diet; response variable: the death rate   **e.** Those on a high-fiber diet had lower death rates than those on the Western diet; association; causality cannot be concluded from an observational study.   **19. Using a TI-84:** Microsoft Lumia 535, Google Nexus 6, ZTE Grand X, Huawei Ascend Y550, Kyocera DuraForce, CAT B15Q; **Using StatCrunch:** HTC Desire 510, Posh Memo S580, LG G3 Vigor, Google Nexus 6 Nokia Lumia 735, CAT B15Q; yes; it is random assignment because the smartphones were randomly assigned to the treatment and control groups.   **21. Using a TI-84:** Mercedes-Benz ML350, Chevrolet Camero, Hyundai Elantra, Lexus LS 460 L, Audi 55; **Using StatCrunch:** Chevrolet Camero, Lexus LS 460 L, Chevrolet Silrerado 15, Hyundai Elantra, KIA Sportage; yes; it is random assignment because the cars were randomly assigned to the treatment and control groups.   **23. a.** Because the study did not use random assignment, it cannot conclude that their program works. Because the 30 members chose to be in the treatment group, they might have been more motivated to lose weight than a typical adult. So, motivation might be a lurking variable. Because the study paid members of the control group but not the members of the treatment group, the treatment group might have had a different financial standing than the control group. So, financial standing might also be a lurking variable.   **b.** The researchers could find 60 volunteers for the study. If necessary, the researchers could offer a monetary incentive to all 60 individuals. Then the researchers should randomly assign 30 of the individuals to the treatment group and the rest of the individuals to the control group. Although it would be impossible for the study to be double-blind, the researcher in contact with the individuals should not know who is in which group (blind study).   **25. a.** Because the student did not use random assignment, he cannot conclude that Tide is more effective than Wisk Deep Clean. It could be the stains came out better when using Tide because Kenmore washing machines are more effective than Samsung washing machines. So, the type of washing machine could be a lurking variable. The towels cleaned with Tide might have had weaker stains than the towels cleaned with Wisk Deep Clean. So, the strength of the stain could be a lurking variable.   **b.** The stained towels should be randomly assigned to be washed with Tide or Wisk Deep Clean. Both groups of towels should be washed using two identical new washing machines. The experiment should be repeated but with the detergents used in the other washing machines.   **27. a.** Because the trainer did not randomly assign the weight lifters to the treatment and control groups, she cannot conclude that the shake works. Weight lifters who drink the shake might work out longer and harder than weight lifters who do not drink the shake. So, commitment might be a lurking variable.   **b.** The trainer could randomly assign weight lifters to the treatment and control groups. The treatment group could be given the protein shake and the control group could be given shakes that do not contain the key ingredients of the protein shake. The study could be double-blind.   **29.** Volunteers with severe ache could be randomly assigned to the treatment and control groups. For six months, individuals in the treatment could be given the experimental drug and the control group could be given sugar pills. The study could be double-blind.   **31.** The runners could be randomly assigned to the treatment and control groups. For one month of daily workouts, the runners in the treatment group could train with the weight belt and the runners in the control group could train without a weight belt. The coach could have an assistant oversee the training, so the coach is blind to which runners are in which group.   **33.** Answers may vary.   **35.** Answers may vary.   **37.** Answers may vary.   **39.** Neither the researchers nor the individuals know which individuals are in the treatment and control groups; in double-blind studies, there is a better chance that the researchers will treat all the individuals the same way (except for the treatment) and the individuals in the two groups will be affected by the placebo effect, if any, by about the same amount.   **41.** The terminologies *explanatory variable* and *response variable* do not necessarily mean there is causation. For example, the terminologies are used in observational studies, which cannot conclude there is causation.

Chapter 2 Review Exercises

1. a. Bahrain, Iraq, Israel, Kuwait, Saudi Arabia   **b.** government, population (in millions), 2012 military expenditure (in billion dollars), oil production (billion barrels per day)   **c.** government: monarchy, republic, republic, monarchy, monarchy; population (millions): 1.3, 31.9, 7.7, 2.7, 26.9; 2012 military expenditure (billion dollars): 0.92, 5.69, 15.54, 5.95, 54.22; oil production (billion barrels per day): 0.05, 2.99, 0.20, 2.69, 9.90   **d.** Bahrain: 18.4, Iraq: 1.903, Israel: 77.7, Kuwait: 2.212, Saudi Arabia: 5.477   **e.** Israel   **2. a.** whether adults say they experience a lot of happiness and enjoyment   **b.** the 500 adults   **c.** all American adults   **3. a. Using a TI-84:** Antoine, Jacob, Ruben, Sandra, Dante, Jose; **Using StatCrunch:** Mario, Sandra, Antoine, Jacob, Alyssa, John   **b. Using a TI-83:** $\frac{1}{2}$; **Using StatCrunch:** $\frac{1}{2}$; descriptive statistics   **c.** $\frac{5}{12}$ **d. Using a TI-83:** no; **Using StatCrunch:** no   **e.** no; the proportions for the samples would not all equal the proportion for all 12 students due to sampling error.   **f.** no; the inferences may not be the same because the samples are not the same.   **4.** Choose the number of questions you usually ask during one hour of your prestatistics class: 0, 1, 2, 3, more than 3.

5. sampling bias; the sampling method favors those who visit the militia group website.   **6.** sampling bias, response bias, nonresponse bias; the sampling method favors those who are often in the financial district, some individuals may exaggerate their salary, and 55 people refused to take part in the study.   **7.** Answers may vary.   **8.** cluster   **9.** simple random   **10.** convenience   **11.** stratified   **12.** systematic   **13. a.** 131   **b. Using a TI-84:** 57; **Using StatCrunch:** 47   **c. Using a TI-84:** 57, 188, 319, 450, 581; **Using StatCrunch:** 47, 178, 309, 440, 571   **14.** commercial airplanes: 33; defense, space, and security: 23; corporate: 12; engineering, operations, and technology: 8; shared services group: 3; other: 1   **15. Using a TI-84:** Gerratana, Ayala, Duff, Frantz, Linares; **Using StatCrunch:** Looney, Stillman, Bartolomeo, Guglielmo, Welch   **16.** cluster; if the clusters are the city blocks, then cluster sampling would require less driving than simple, stratified, and systematic sampling; the city hall should create a frame of all city blocks, randomly select some of the blocks, and survey all of the residents who live on the selected blocks.   **17. a.** treatment groups: the three groups that took different dosages of the drug; control group: the group that took the placebo   **b.** experiment; there was random assignment   **c.** The researchers randomly assigned the individuals to the groups; create a frame of the 560 MDD adults. Then for each treatment group, randomly select 140 of the adults to be in the group. The remaining 140 adults should be in the control group.   **d.** sample: the 560 MDD adults; population: all MDD adults   **18. a.** a sugar pill   **b.** Neither the individuals nor the researcher in contact with the individuals knew who was in which group; one researcher could have labeled the pill vials with numbers to identify the pills but not tell the code to another researcher who was in contact with the individuals.   **c.** explanatory variable: the dosage of the drug; response variable: HRSD score   **d.** The drug successfully lowers MDD adults' HRSD scores; causality; we can assume causality because there was random assignment.   **e.** Although the drug tends to lower MDD adults' HRSD scores, that might not mean that the drug tends to lower depression in MDD adults.   **19. a.** The mothers were not randomly assigned to the group with eating disorders and the group without eating disorders.   **b.** People could not start having an eating disorder or stop having an eating disorder because the researchers told them to.   **c.** sample: the mothers observed by the researchers; population: mothers with first-born infants   **d.** explanatory variable: whether a mother has an eating disorder; response variable: the number of negative emotions mothers expressed to their first-born infant during mealtimes   **e.** Mothers with eating disorders expressed more negative emotions toward their first-born infants during mealtimes than mothers without eating disorders; an association; causality cannot be concluded from an observational study.   **20. Using a TI-84:** Elon University, Campbell University, Wellesley College, Columbia College, University of Mount Union; **Using StatCrunch:** Nichols College, Columbia College, Mills College, Rider University, Villanova University; yes; the colleges were randomly assigned to the two groups.   **21. a.** The coordinator did not use random assignment, so she cannot conclude causality; The variable attending the math center should include a time requirement — a student who attends the math center for only five minutes once in the entire semester should not be considered a student who uses the center; also, students who attend the math center might be more motivated and study harder in a variety of ways than other students; motivation might be a lurking variable.   **b.** The coordinator could randomly assign some students to a treatment group and some students to a control group. Students in the treatment group would attend the math center for one hour per weekday throughout the semester, and students in the control group would not attend the math center. After the semester is over, the coordinator could compare the proportion of the treatment group who passed their math classes that semester with the proportion of the control group who passed their math classes that semester.   **22.** The company could randomly assign some bald people to a treatment group and a control group. The treatment group would take the drug and the control group would take sugar pills. The study could be double-blind. The researchers would measure the extent of the individuals' hair growth after 8 months.

Chapter 2 Test

1. a. Delaware, Hawaii, Mississippi, Texas, Wisconsin   **b.** region, number of workers (in thousands), number of workers in unions (in thousands)   **c.** region: East, West, South, South, Midwest; number of workers (in thousands): 370, 549, 1040, 10,877, 2569; number of workers in unions (in thousands): 38, 121, 38, 518, 317   **d.** Delaware: 10.3%, Hawaii: 22.0%, Mississippi: 3.7%, Texas: 4.8%, Wisconsin: 12.3%   **e.** Hawaii   **2. a.** whether an adult intends to buy wearable technology   **b.** the 2011 adults   **c.** all American adults   **3.** response bias and nonresponse bias; the complex question may cause response bias, and there was a 92% nonresponse rate.   **4. Using a TI-84:** Jamie, Jared, Isabel, Lisa; **Using StatCrunch:** Jamie, Brianna, Dan, Michael   **5.** cluster   **6. a.** 6   **b. Using a TI-84:** 1; **Using StatCrunch:** 2   **c. Using a TI-84:** 1, 7, 13, 19, 25; **Using StatCrunch:** 2, 8, 14, 20, 26   **7.** 113 female undergraduate students, 42 female graduate students, 226 male undergraduate students, 119 male graduate students   **8. a.** treatment groups: the 4 groups taking different drug dosages; control group: the placebo group   **b.** experiment; there was random assignment   **c.** The researchers randomly assigned the patients to the groups; create a frame of the 361 patients. Then for each of the 4 treatment groups, randomly select 72 of the adults to be in the group. The remaining 73 adults should be in the control group.   **d.** sample: the 361 patients; population: all Japanese patients with type 2 diabetes.   **9. a.** a sugar pill   **b.** Neither the patients nor the researcher in contact with the patients knew which patients were in which group; one researcher could have labeled the pill vials with numbers to identify the pills but not tell the code to another researcher who

was in contact with the individuals.   **c.** explanatory variable: the dosage of the drug; response variable: glycated hemoglobin level
d. The drug successfully lowers glycated hemoglobin levels in Japanese patients with type 2 diabetes; causality; we can assume causality because there was random assignment.   **e.** Researchers might be influenced, consciously or unconsciously, by earning a salary from the company that manufactures ipragliflozin. By reporting that they work for the company, other researchers who do not work for the company could run a similar experiment and see if they get similar results.   **10. a.** The researcher did not use random assignment. The players who run every day may also practice basketball longer and harder than players who do not run every day; motivation may be a lurking variable.   **b.** The researcher could randomly assign players to a treatment group and a control group. Players in the treatment group would run for one hour every day, and players in the control group would not run. The researcher could have an assistant oversee the training, so the researcher would be blind to which players are in which group. After the players have run daily for one month, the researcher would compare the scoring of the two groups in the next month, while the players in the treatment group continued to run daily.   **11.** The owner could randomly assign his sales force to a treatment group and a control group. The treatment group would attend a workshop about emotions for a weekend. The control group would not attend the workshop. The owner could be blind to which employees attended the workshop. One month later, the owner would compare the monthly sales by the treatment and control groups.

Chapter 3

Homework 3.1   1. categorical   **3.** frequency   **5.** numerical   **7.** categorical   **9.** categorical   **11.** numerical   **13. a.** how often Instagram users visit the website; categorical   **b.** 140 users   **c.** 189 users   **d.** 107 users   **e.** 17%   **15. a.** 51 employees
b. 109 employees   **c.** 276 employees   **d.** 0.07   **e.** sampling bias, response bias   **17.** 3, 10, 17, 24   **19.** 8, 9, 10, 11, 12, 13, 14
21. 3, 8, 9, 10, 11, 12, 13, 14, 17, 24   **23.** 10   **25. a.** C   **b.** 0.498   **c.** by subtracting from 1: 0.987; by adding relative frequencies: 0.986   **d.** 0.131   **e.** 2072 loans   **27. a.** 0.19   **b.** by subtracting from 1: 0.81; by adding relative frequencies: 0.82   **c.** 0.35
29. a. 291 adults   **b.** The sum of the parts equals the whole; 1.01; the result is not 1 due to rounding.   **c.** Even if the survey was carried out well, the proportion for the population might be a bit different than the proportion for the sample due to sampling error.   **d.** sampling bias; response bias   **31. a.** women; although an equal proportion of women and men have a great deal of confidence in Congress, a greater proportion of women than men have some confidence.   **b.** 0.93   **c.** 362 women   **d.** Even if the survey was carried out well, the proportion for the population might be a bit different than the proportion for the sample due to sampling error.   **e.** The student is incorrect. Actually, 50 women and 41 men had a great deal of confidence in Congress.

33. a.

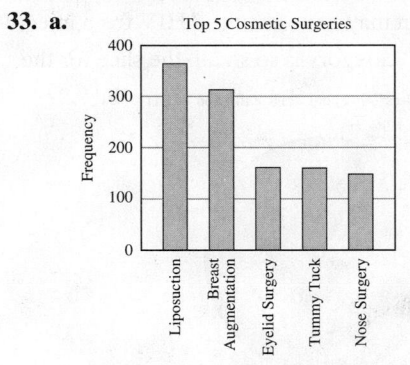

b.

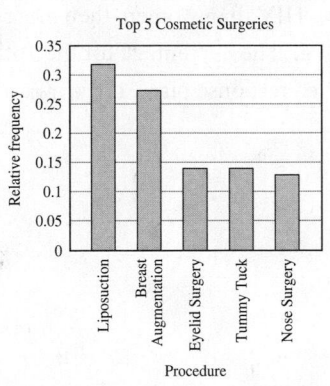

c. descriptive statistics; the bar graphs describe the data that were collected.   **d.** 0.458   **e.** 0.86

35. a.

Category	Frequency	Relative Frequency
Japan	2	$\frac{2}{15} \approx 0.133$
Kenya	5	$\frac{1}{3} \approx 0.333$
U.S.	2	$\frac{2}{15} \approx 0.133$
Uganda	2	$\frac{2}{15} \approx 0.133$
Other	4	$\frac{4}{15} \approx 0.267$
Total	15	$\frac{15}{15} \approx 0.999$

b.

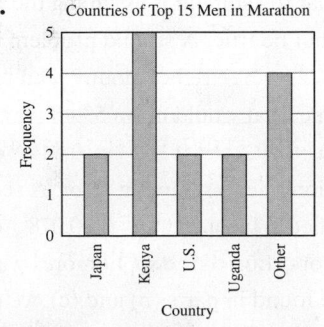

c.

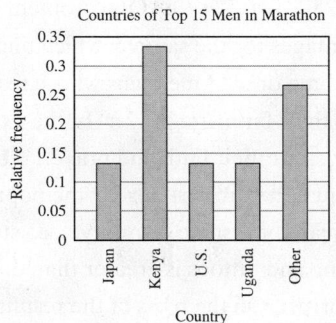

d. 86.7%   **e.** 46.7%

37. a.

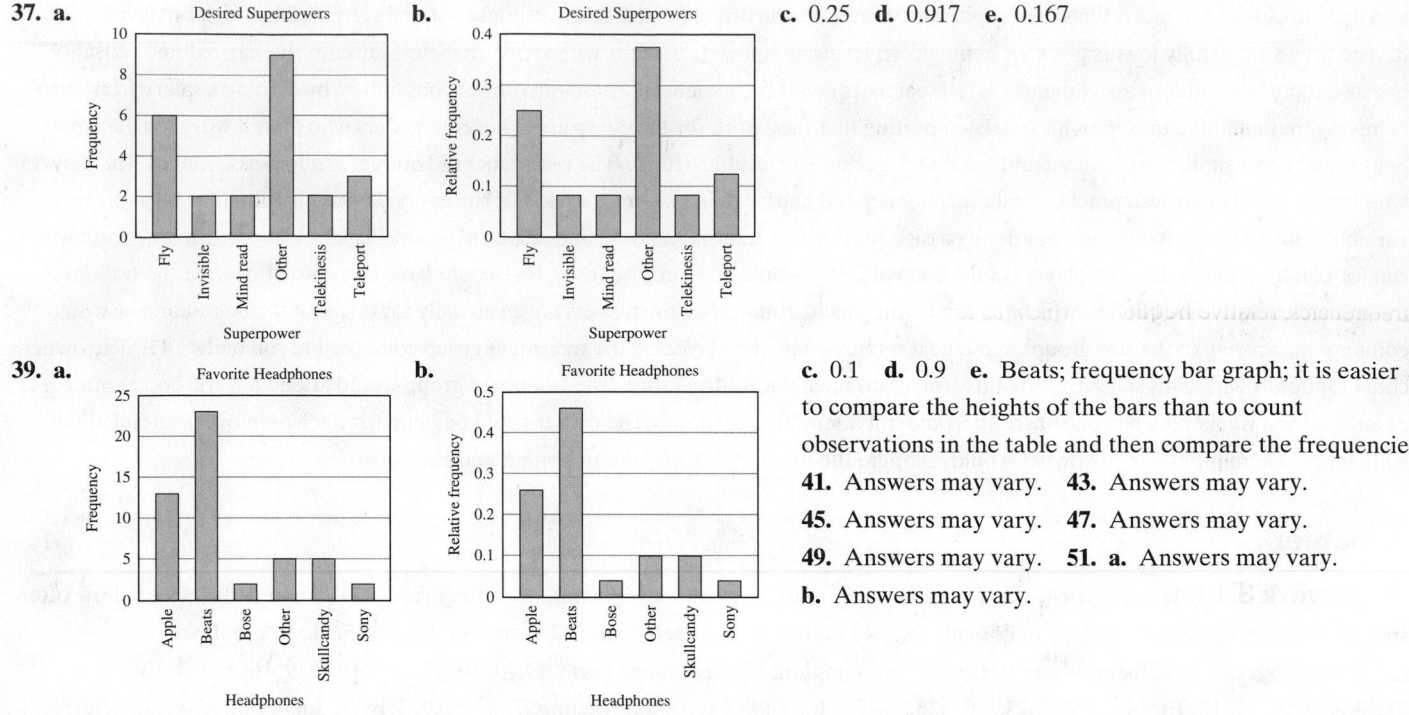

b. **c.** 0.25   **d.** 0.917   **e.** 0.167

39. a. **b.** **c.** 0.1   **d.** 0.9   **e.** Beats; frequency bar graph; it is easier to compare the heights of the bars than to count observations in the table and then compare the frequencies. **41.** Answers may vary.   **43.** Answers may vary. **45.** Answers may vary.   **47.** Answers may vary. **49.** Answers may vary.   **51. a.** Answers may vary. **b.** Answers may vary.

Homework 3.2   **1.** relative   **3.** categorical   **5. a.** activities of students on weekdays; categorical   **b.** 0.8583   **c.** 0.0917   **d.** 17 hours   **e.** The student's estimate of 14.17% might have a lot of error. A typical college student might behave very differently on the weekend than on weekdays.   **7. a.** 80.6%   **b.** The percentages of the categories add to 100%, so two weapons in different categories were never counted for one murder.

c.

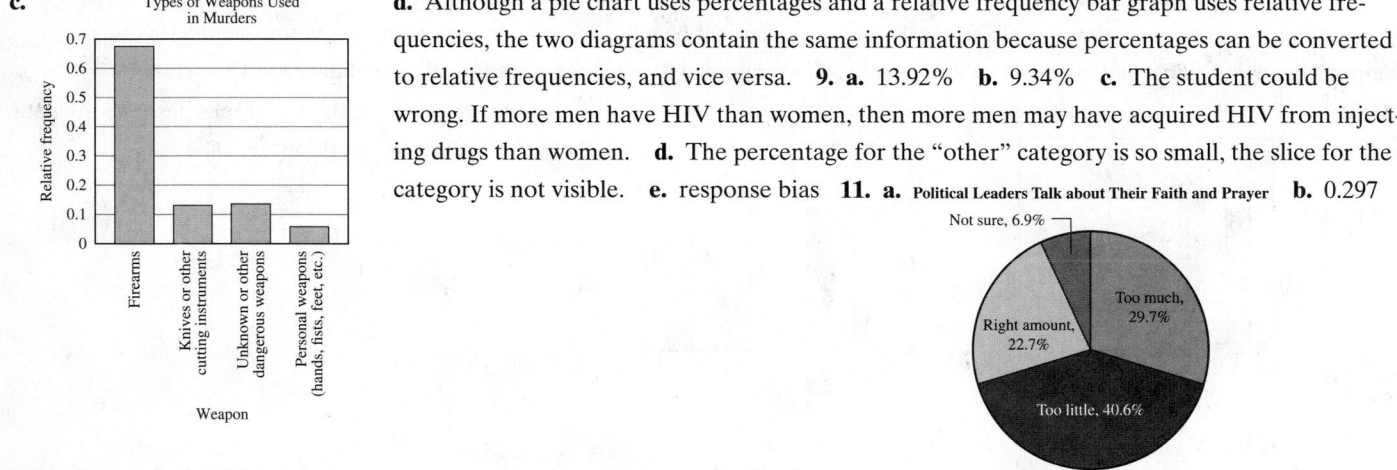

d. Although a pie chart uses percentages and a relative frequency bar graph uses relative frequencies, the two diagrams contain the same information because percentages can be converted to relative frequencies, and vice versa.   **9. a.** 13.92%   **b.** 9.34%   **c.** The student could be wrong. If more men have HIV than women, then more men may have acquired HIV from injecting drugs than women.   **d.** The percentage for the "other" category is so small, the slice for the category is not visible.   **e.** response bias   **11. a.** Political Leaders Talk about Their Faith and Prayer   **b.** 0.297

c. 0.525   **d.** 0.773   **e.** One problem is that the student is assuming the percentages for all Americans would be the same as the percentages for the sample, which might not be true. A second problem is that if politicians talk more about their faith and prayer, then some or all Americans whose responses were "right amount" and "not sure" might then think that politicians talk too much about their faith and prayer. In the extreme, that would mean 59.4% (more than half) of Americans would think politicians talk too much about their faith and prayer.   **13. a.** 0.192   **b.** 0.320   **c.** 0.444   **d.** One problem is that the proportions for the sample may not equal the proportions for the population due to sampling error. A second problem is that the survey is an observational study, so we cannot assume causality.   **e.** stratified   **15. a.** 0.090   **b.** 0.078   **c.** 0.089; the proportion of the juniors who have both body piercing and tattoos is greater than the proportion of the sophomores who have both body piercing and tattoos.   **d.** The student is incorrect. On the basis of the results we found in parts (b) and (c), we conclude that the surveyed juniors are more likely to have both body piercings and tattoos than the surveyed sophomores.   **e.** The result for all college students may not be 11% due to likely sampling bias and sampling error.   **17. a.** 0.667   **b.** 0.376   **c.** 0.598   **d.** 0.521   **e.** 0.131

19.

	Run Lights	Don't Run Lights	Total
Smoke	1	1	2
Don't Smoke	2	6	8
Total	3	7	10

21. a. 0.304   **b.** 0.043   **c.** 0.25   **d.** 0.158; The proportion of the students who run red lights, given that they smoke is greater than the proportion of the students who run red lights, given that they do not smoke. **e.** Nonresponse bias, response bias; the result would be larger; for each of the 6 students, it is not clear whether the student smokes but doesn't run red lights, runs red lights but doesn't smoke, or smokes and runs red lights.   **23.** two-way table, multiple bar graph   **25.** frequency and relative frequency table, frequency bar graph, relative frequency bar graph, pie chart   **27.** two-way table, multiple bar graph   **29.** relative frequencies; relative frequencies allow us to meaningfully compare categories of individuals even if more individuals in some categories were included in the study than individuals in other categories.   **31.** relative frequency bar graph; same ease; relative frequency bar graph; pie chart; it is easier to visually compare the area of a slice to the rest of a pie chart than it is to visually compare the height of a bar to the sum of the heights of the other bars.   **33. a.** Answers may vary.   **b.** Answers may vary.   **c.** Answer may vary.

Homework 3.3   1. discrete   **3.** 45th   **5.** discrete   **7.** continuous   **9.** continuous   **11.** discrete   **13. a.** 0.697   **b.** 0.152   **c.** $435 thousand, $445 thousand   **d.** Homes with large square footages tend to have high asking prices.   **e.** sampling bias; the ratio of the number of dots on the left side to the right side would increase.   **15. a.** 3; the percentage of residents who are Protestant is 58% in 3 states.   **b.** 77%   **c.** 52%   **d.** 11%; Utah   **17. a.** 6500 stores   **b.** 2006   **c.** decreased   **d.** The student's estimate could have a lot of error. We *cannot* assume the number of stores will continue to decrease by about 200 stores per year.   **e.** yes; a decrease of 1100 stores in one year is much greater than past decreases of about 200 stores per year.   **19. a.**

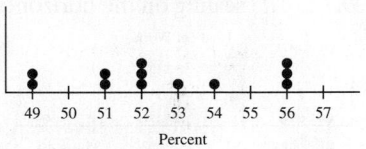

Percent Who Exercise, by Midwestern State

b. 3; There are three Midwestern states where the percentage of adults who exercise often is 52%.   **c.** 49%   **d.** 4 states   **e.** 0.583   **21. a.** yes; the student might have thought the units were minutes, or the student might have given the number of hours per week; response bias

b. 7 hours   **c.**

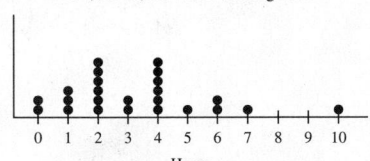

TV, Movie, and Video Viewing Time

d. 0.583   **e.** 0.458   **23. a.** numbers of restaurants in Alameda County, California, that received yellow cards per day; discrete   **b.**

Restaurants Given Yellow Grades

c. 4; there were 4 days when exactly 6 restaurants received yellow grades.   **d.** 1 restaurant receiving a yellow grade per day; 11; there were 11 days when exactly 1 restaurant received a yellow grade.   **e.** 3 restaurants receiving yellow grades per day   **25. a.**

Stem (ones)	Leaf (tenths)
1	3 5
2	6 7 9
3	4 7
4	1 2 3

b. 5 cities   **c.** 0.3   **d.** 0.6   **e.** 4.3 miles per square mile

27. a. continuous   **b.**

Stem (tens)	Leaf (ones)
2	4 4 4 4
2	6 6 8 8 9 9 9
3	0 0 0 1 1 1 4 4 4 4
3	5 6 6 7 7 9
4	0

c. 3; there are 3 Kia car models whose gas mileages are 30 miles per gallon.   **d.** 40 miles per gallon   **e.** 30 miles per gallon

29. a.

Stem (tens)	Leaf (ones)
12	8 9
13	0 0 1 1 1 1 2 2 2 3 3
13	7 8
14	2 2 3 3 3 3 4 4 4
14	5 5 6 6 7 8 8 9
15	0 0 0 0 1 2 2 2 2 2 3 3 4 4
15	5 5 5 5

b. 4; 4 of the times were 131 minutes.   **c.** 152 minutes; 5   **d.** 17   **e.** 18

31. a.

Cities with Open Streets Initiatives

b. generally increased   **c.** 10 cities   **d.** 29 cities; 59 cities; no; we cannot assume the change in the number of large cities with Open Streets Initiatives from 2012 to 2018 will be the same as from 2006 to 2012.

33. a.

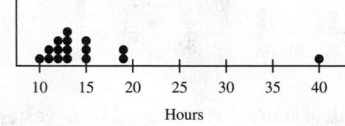

Deaths from Collisions at Highway-Railroad Crossings

b. decreased   **c.** −9 deaths; the number of deaths from collisions at highway-railroad crossings decreased by 9 deaths from 2010 to 2013.   **d.** −108 deaths; 143 deaths; no; we cannot assume the change in the number of deaths from collisions from 2013 to 2021 will be the same as from 2005 to 2013.   **35. a.** 99 years, 160 years, 236 years, 360 years   **b.** The observation 76 years is also an outlier and the frequency of the outlier 99 is 2 (not 1).   **c.** The units of the following are years: 1, 1, 1, 3, 160, 236, and 360; 1-year-old and 3-year-old babies would not get stopped by police, and people don't live to be 160 years, 236 years, or 360 years in age; response bias   **d.** All three actions should be taken; first, review the original reports completed by the police officers and correct as much of the data as possible. If some of the data are still in question, then interview police officers and correct more of the data. If some of the data are still in question, then remove them, but state in the study's article which data were removed.

37. a.

United Airline Departure Delays
Up to 22 values per dot

b. −15 minutes; an airplane departed 15 minutes early.   **c.** 286 minutes; an airplane departed 286 minutes late.   **d.** −3; the most frequent departure delay is −3 minutes (3 minutes early).   **e.** A total of 150 flights had delays of at least 39 minutes; such a delay is much greater than a typical delay.

39. a. The scaling on the horizontal axis should be uniform.

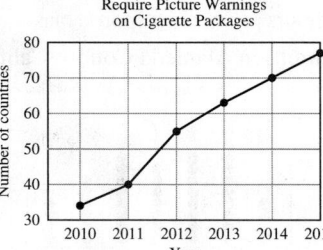

Hours of Work

b. no; the observation 40 hours is an outlier. **c.** yes; using nonuniform scaling distorts the view (shape) of the distribution.   **41.** stemplot; the digits of the observations are displayed in a stemplot, and the precise values of observations can be hard to estimate from the scaling of the horizontal axis of some dotplots.

43. a.

Require Picture Warnings on Cigarette Packages

b. 8.6 countries per year   **c.** 2011–2012   **d.** 2010–2011, 2012–2013, 2013–2014, 2014–2015

Homework 3.4   **1.** 1   **3.** bimodal   **5. a.** commute distance to college (in miles); continuous   **b.** 2 commute distances
c. 7 commute distances   **d.** 0.818   **e.** The following are all in miles: 4, 6, 6, 9, 11, 11, 12, 13, 14, 22, and 34 (answers may vary).
7. a. 1125 murders, 1875 murders; for each estimate the largest possible error is 125 murders.   **b.** skewed right   **c.** 16 states
d. 0–249 murders; yes; over half of the observations are in the class 0–250 murders.   **e.** The ratio of the number of murders in a state to the state's population is a better measure of the risk of getting murdered.   **9. a.** skewed right; the ages of people stopped by police that are older than the 50th percentile are much more spread out than the ages of people stopped by police that are younger than the 50th percentile.   **b.** 0.31   **c.** We cannot assume the proportion of people stopped by police who were under 20 years would be the same amount on other days. In particular, the proportion on weekdays might be quite different than on weekends.   **d.** 20–24 years; yes; 27% of the observations lie in the class 20–24 years, which is quite a bit more than all the other classes except the class 15–19 years.   **11. a.** bimodal; for a large number of days Obama's approval ratings were close to 49%, and for a large number of days Obama's approval ratings were close to 57%.   **b.** no; most of Obama's approval ratings are not close to 54%. Rather, they are close to 49% or close to 57%.   **c.** greater; Obama's approval ratings tended to be much higher for the first half of 2009 then the second half.   **d.** The distribution for the first half of 2009 is more spread out than the distribution for the second half of 2009.
e. The distribution for 2009 is bimodal because Obama's approval rating tended to be much larger for the first half of 2009 than the second half.   **13. a.** The areas of the bars, from left to right, are 0.1, 0.2, 0.4, 0.2, and 0.1.   **b.** 1   **c.** For any density histogram, the area of a bar is equal to the relative frequency of the bar's class. And the sum of the relative frequencies of all the classes is equal to 1.

d.

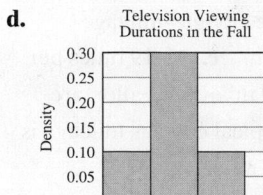

15. a. 7–13 days; yes; 33% of the observations are in the class 7–13 days, which is a lot more than all the other classes except one. **b.** 0.03 **c.** We cannot assume the distribution in 2015 will be the same as it was in 2012 and 2013. **d.** 0.34 **e.** 0.67 **17. a.** the tuition (in thousands of dollars) of a 4-year, private, not-for-profit university or college; discrete **b.** 20–24 thousand dollars; yes; there are more observations in the class 20–24 thousand dollars than any other class. **c.** 0.75 **d.** 0.13 **e.** 0.065 **19. a.** 28th percentile **b.** 58th percentile **c.** $10 thousand **d.** $20 thousand

21. a.

Class	Frequency	Relative Frequency
22–23.9	3	$\frac{3}{21} \approx 0.143$
24–25.9	4	$\frac{4}{21} \approx 0.190$
26–27.9	1	$\frac{1}{21} \approx 0.048$
28–29.9	9	$\frac{9}{21} \approx 0.429$
30–31.9	4	$\frac{4}{21} \approx 0.190$
Total	21	$\frac{21}{21} = 1$

b.

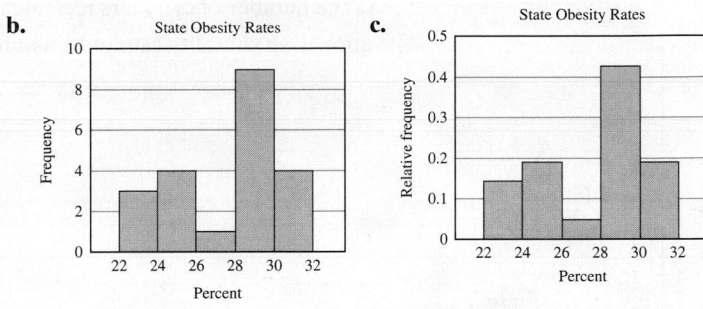

c.

d. skewed left **e.** The distribution is bimodal.

23. a.

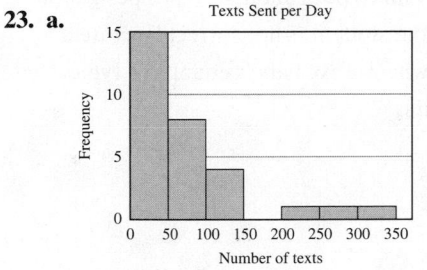

b.

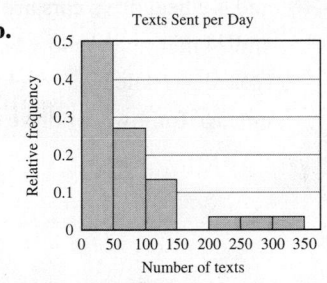

c. 0.233 **d.** skewed right **e.** The bars would be thinner, and all but one would be far off to the left.

25. a.

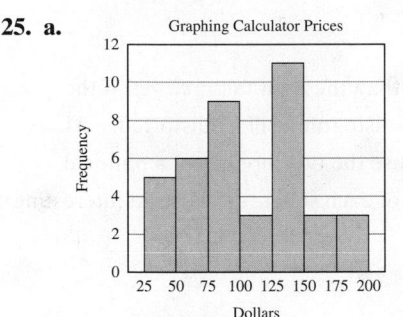

b.

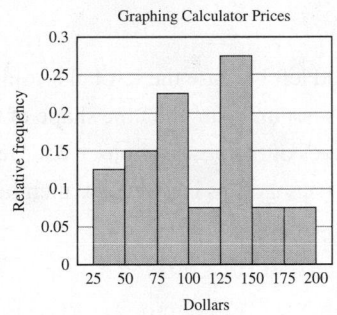

c. 0.225 **d.** bimodal **e.** The observation $115 is in the class 100–124.99 dollars, which has a frequency (3) quite a bit lower than many of the other classes.

27. a.

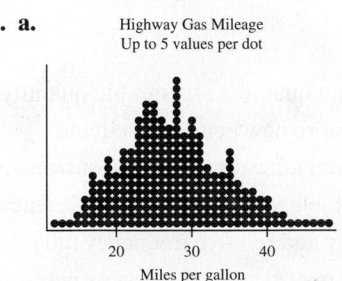

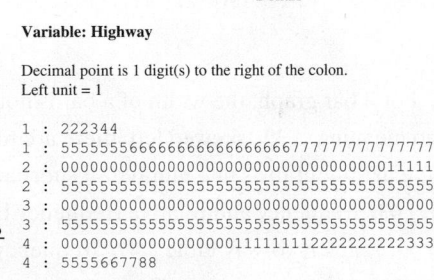

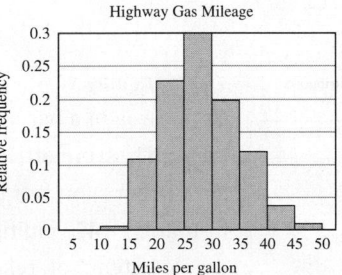

The dotplot provides the most detail, but it does not make it easy to estimate relative frequencies. The stemplot shown is truncated, but the actual one is difficult to use because it extends so far to the right. The relative frequency histogram provides less detail than the dotplot, but it is the easiest to use to estimate relative frequencies.

b.

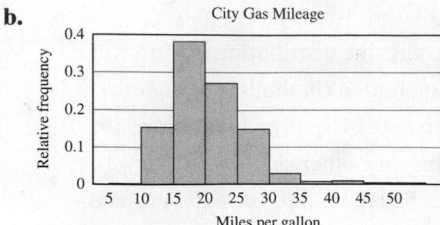

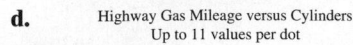

By comparing classes with relative frequencies of at least 0.03, the city-gas-mileage distribution is narrower than the highway-gas-mileage distribution. **c.** 15–19 miles per gallon; 25–29 miles per gallon; the observations in the class 15–19 miles per gallon are less than the observations in the class 25–29 miles per gallon; a typical city gas mileage is approximately 10 miles per gallon less than a typical highway gas mileage.

d.

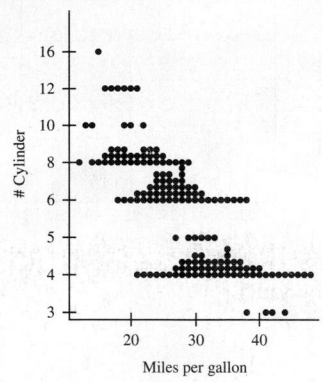

As the number of cylinders increases, the gas mileage tends to decrease; no; the study is observational, so causality cannot be assumed.

e.

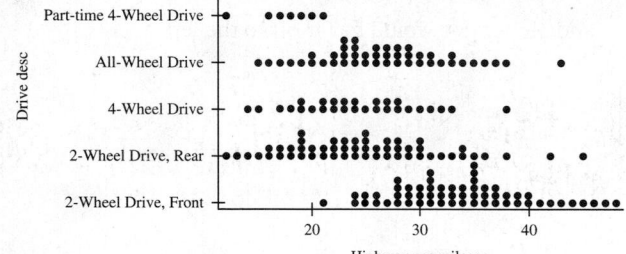

The 50th percentiles of front-wheel drive cars, rear-wheel drive cars, and 4-wheel drive cars are 33 miles per gallon, 24 miles per gallon, and 24 miles per gallon. So, the student is not correct because a typical gas mileage for rear-wheel-drive cars is equal to a typical gas mileage for 4-wheel-drive cars.

29. a. The scaling on the horizontal axis is not uniform.

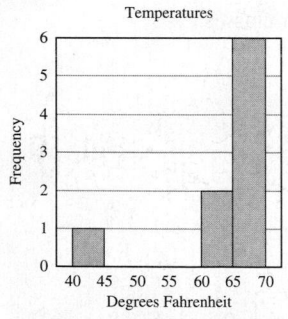

b. no; the distribution is skewed left because the left tail is longer than the right tail. **c.** yes; if the scaling on the horizontal axis is not uniform, then the shape of the histogram will be distorted. **31.** A stemplot displays the exact values of the observations. **33.** Because the two mounds of a bimodal distribution often represent two subgroups in which the members of each subgroup share an interesting characteristic.

35.

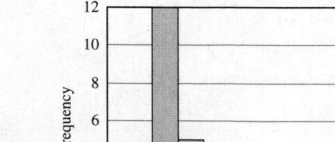

Answers may vary. **37.** For a bar graph, the width of a bar is not equal to a measurable quantity, so the area of a bar has no meaning. **39.** skewed left; there are more new pennies in circulation. **41.** symmetric; there are about the same number of short and tall women, and most women are neither short nor tall. **43.** frequency and relative frequency table, dotplot, relative frequency histogram **45.** multiple bar graph, two-way table **47.** frequency and relative frequency table, dotplot, stemplot, relative frequency histogram **49. a.** Answers may vary. **b.** Answers may vary. **c.** Answers may vary. **d.** Answers may vary.

Homework 3.5   **1.** false   **3.** time-series   **5. a.** the histogram with class widths of 3°F.   **b.** the histogram with class widths of 3°F; 3 days   **c.** the histogram with class widths of 3°F; the class from 91°F to 93.99°F;   **d.** the histogram with class widths of 7°F; the histogram with class widths of 7°F makes it seem possible for there to have been days with high temperatures of 80°, but we can tell from the histogram with class widths 3°F that the high temperature was never between 79°F and 81.99°F; yes.   **7. a.** the bar graph with the vertical axis starting at 40%; by having the vertical axis start at 40%, the bar graph emphasizes that the resale rate of the FJ Cruiser is more than the resale rates of the other cars.   **b.** the bar graph with the vertical axis starting at 0; the bar graph de-emphasizes the difference in the resale rate of the GL 350 and the ES 350.   **c.** the bar graph with the vertical axis starting at 40%; the graph is more "zoomed in" on the vertical axis; 53%   **d.** $12,519.01   **e.** $43,775; $14,583.65; the difference in the base prices is quite a bit more than the difference in the resale values.   **9. a.** the time-series plot with vertical axis starting at $0 thousand; the differences between the tuitions are de-emphasized because the scaling increases by a larger amount than in the other time-series plot.   **b.** the time-series plot with the vertical axis starting at $9.0 thousand; it is easier to estimate the tuition because the scaling on the vertical axis increases by a smaller amount than in the other time-series plot; $11.4 thousand   **c.** from 2009 to 2010; $1.5 thousand   **d.** $2.9 thousand; $14.9 thousand; no; we cannot assume the change in tuition from 2012 to 2016 will be the same as from 2008 to 2012.   **11. a.** the time-series plot with vertical axis starting at 4 million square kilometers; the differences between the sea ice extents are emphasized because the scaling on the vertical axis increases by a smaller amount than in the other time-series plot.   **b.** the time-series plot with vertical axis starting at 0 million square kilometers; the differences between the sea ice extents are de-emphasized because the scaling on the vertical axis increases by a larger amount than in the other time-series plot.   **c.** the time-series plot with vertical axis starting at 4 million square kilometers; it is easier to estimate the sea ice extent because the scaling on the vertical axis increases by a smaller amount than in the other time-series plot; 4.9 million square kilometers   **d.** −1.4 million square kilometers; 3.5 million square kilometers; no; we cannot assume the change in the sea ice extent from 2010 to 2020 will be the same as from 2000 to 2010.   **e.**

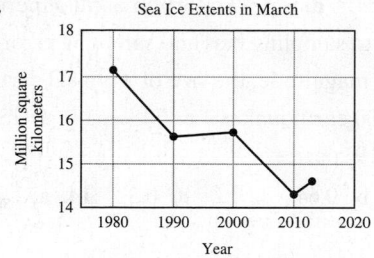

13. a. The scaling on the horizontal axis is not uniform.   **b.**

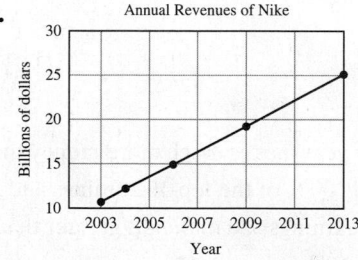

The scaling on both axes is uniform.

c. the bar graph; the bar graph makes it seem like the annual revenue is increasing by greater and greater amounts because the scaling on the horizontal axis is not uniform.   **d.** $25 billion   **e.** $14 billion; $39 billion; no; we cannot assume the change in the revenue from 2013 to 2023 will be the same as from 2003 to 2013.   **15. a.** It is difficult to tell how the tops of the boxes line up with the scaling on the vertical axis.   **b.** relative frequency bar graph

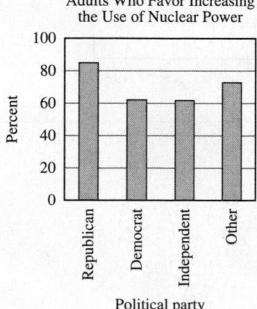

c. The percentages do not describe the percentages of adults who belong to the political parties. Rather, they describe the percentages of adults in the political parties who favor increasing the use of nuclear power.   **d.** 37.8%   **17.** the changes in the variable described by the vertical axis   **19.** The observations represented by the two mounds of a histogram using small class width would be in the same class of a histogram using large enough class width; no; if a distribution is bimodal, then the histogram should display the two mounds.   **21.** It can be difficult to tell how the tops of three-dimensional objects line up with the scaling on the vertical axis; bar graph

Chapter 3 Review Exercises

1. numerical **2.** categorical **3. a.** 0.05 **b.** 0.95 **c.** 0.68 **d.** 0.2 **e.** Some surveyed adults might underestimate how often they are late to work because they are in denial. Or if the survey is not anonymous, some survey adults might try to impress the data collector; smaller

4. a.

Category	Frequency	Relative Frequency
Alternative	1	$\frac{1}{20} = 0.05$
Country	3	$\frac{3}{20} = 0.15$
Hip-hop	3	$\frac{3}{20} = 0.15$
Pop	11	$\frac{11}{20} = 0.55$
Rap	1	$\frac{1}{20} = 0.05$
Soul	1	$\frac{1}{20} = 0.05$
Total	20	$\frac{20}{20} = 1$

b.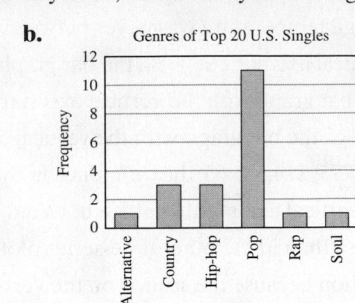

c.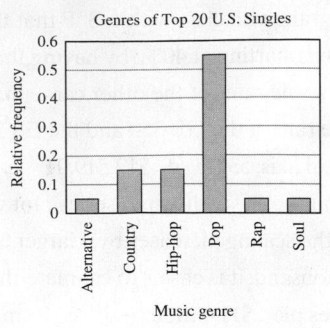

d. 0.85 **e.** 0.2 **5. a.** 0.07 **b.** 0.93 **c.** 0.41 **d.** 20 restaurants
6. a. 0.333 **b.** 0.360 **c.** 0.572 **d.** 0.312; the proportion of Republicans who think the government should reduce the income difference is less than the proportion of Democrats who feel that way. **e.** 0.177 **f.** The student is incorrect because in part (d) we are treating all Republicans as the whole and finding a certain fraction of that whole, but in part (e) we are treating the surveyed adults who think the government should reduce the income difference as the whole and finding a certain fraction of that whole. **7. a.** 0.736 **b.** 0.113 **c.** 0.580 **d.** The proportion of all Americans who think the government should reduce the income difference may not equal 47% due to sampling bias and sampling error. **e.** stratified **8.** continuous **9.** discrete
10. a. symmetric **b.** 4–4.99 surface-wave magnitude; the size of a typical earthquake is between 4 and 4.99 surface-wave magnitudes. **c.** 2000 earthquakes **d.** 1600 earthquakes **e.** 34% **11. a.** 15% **b.** 21% **c.** 0.25 **d.** 0.6 **12. a.** bimodal
b. 40–49 species **c.** 92nd percentile **d.** 19 species

13. a.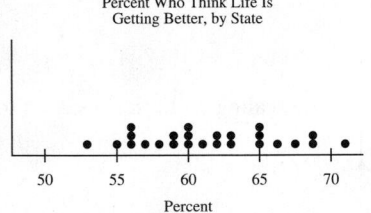

b. 0.64 **c.** 0.2 **d.** 0.32 **14. a.**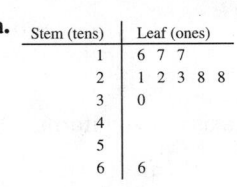

Stem (tens)	Leaf (ones)
1	6 7 7
2	1 2 3 8 8
3	0
4	
5	
6	6

b. 66; the DJ who earns $66 million per year makes much more money than the other 9 DJs. **c.** $17 million; the earnings $17 million is greater than or equal to approximately 30% of the top-10 earnings and less than approximately 70% of the top-10 earnings.
d. $22 million **e.** 90th percentile; the earnings $30 million is greater than or equal to approximately 90% of the top-10 earnings and less than approximately 10% of the top-10 earnings. **15. a.** 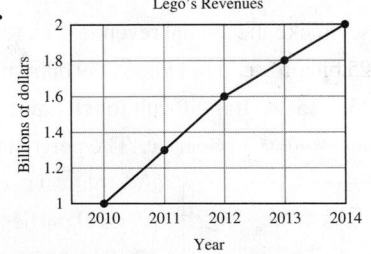 **b.** increased

c. $0.2 billion; the revenue increased by $0.2 billion from 2013 to 2014. **d.** $1.0 billion; $3.0 billion; no; we cannot assume the change in revenue from 2014 to 2018 will be the same as from 2010 to 2014.

16. a.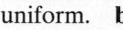

Class	Frequency	Relative Frequency
0–9	8	$\frac{8}{27} \approx 0.296$
10–19	12	$\frac{12}{27} \approx 0.444$
20–29	3	$\frac{3}{27} \approx 0.111$
30–39	3	$\frac{3}{27} \approx 0.111$
110–119	1	$\frac{1}{27} \approx 0.037$
Total	27	$\frac{27}{27} \approx 0.999$

b.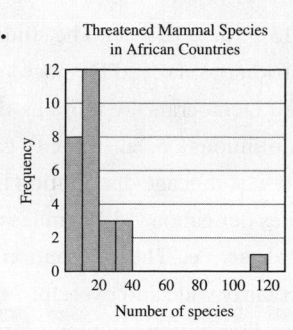

c.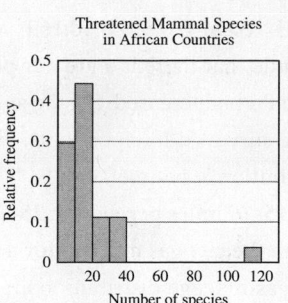

d. skewed right **e.** 10–19 threatened species **f.** 114 threatened species
17. a. the bar graph with the vertical axis starting at 25 thousand cars; the differences between the numbers of stolen cars are emphasized because the scaling on the vertical axis increases by a smaller amount than in the other bar graph. **b.** the bar graph with the vertical axis starting at 0 thousand cars; the differences between the numbers of stolen cars are de-emphasized because the scaling on the vertical axis increases by a larger amount than in the other bar graph. **c.** the bar graph with the vertical axis starting at 25 thousand cars; it is easier to make the estimation because the scaling on the vertical axis increases by a smaller amount than in the other bar graph. **d.** 17 thousand cars **18. a.** The scaling on the horizontal axis is not uniform. **b.**

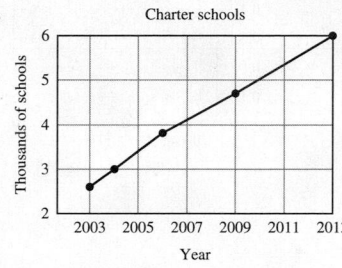

The scalings on both axes are uniform. **c.** the bar graph; the bar graph makes it seem like the annual revenue is increasing by greater and greater amounts because the scaling on the horizontal axis is not uniform. **d.** 6 thousand charter schools **e.** 3.4 thousand charter schools; 9.4 thousand charter schools; no; we cannot assume that the change in the number of charter schools from 2013 to 2023 will be the same as from 2003 to 2013. **19. a.** It is difficult to line up the tops of the boxes with the scaling on the vertical axis.

b. relative frequency bar graph

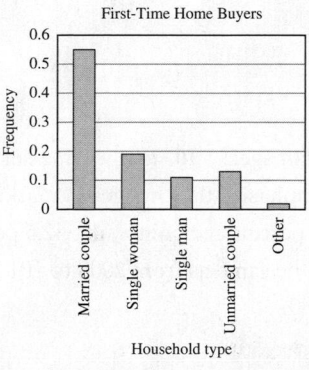

c. 0.45 **d.** 0.3

Chapter 3 Test

1. numerical **2. a.**

Category	Frequency	Relative Frequency
Democrat	6	$\frac{6}{15} = 0.4$
Independent	2	$\frac{2}{15} \approx 0.133$
Libertarian	1	$\frac{1}{15} \approx 0.067$
None	2	$\frac{2}{15} \approx 0.133$
Republican	4	$\frac{4}{15} \approx 0.267$
Total	15	$\frac{15}{15} = 1$

b.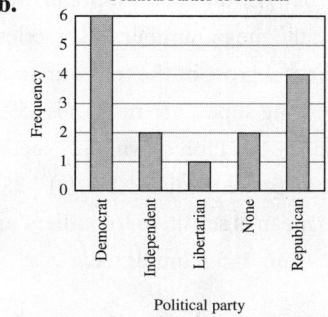

c.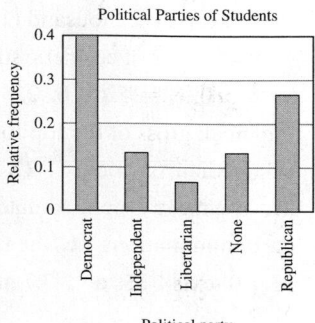

d. 0.733   **e.** 0.533   **3. a.** 0.632   **b.** 0.169   **c.** 0.135   **d.** 0.127   **e.** The student is incorrect. On the basis of the results from parts (c) and (d), we conclude that there is a greater proportion of surveyed Democrats than surveyed Independents that think the number of immigrants should be increased. So, the surveyed Democrats are more likely to think that the number of immigrants should be increased than the surveyed Independents.   **4.** continuous   **5. a.** The city-gas-mileage distribution is skewed right, and the highway-gas-mileage distribution is symmetric.   **b.** The city-gas-mileage distribution is slightly narrower than the highway-gas-mileage distribution.   **c.** 15–19 miles per gallon; 25–29 miles per gallon   **d.** 10 miles per gallon; the highway gas mileage is about 10 miles per gallon greater than the city gas mileage for a typical car.   **e.** The distribution is unimodal because the city-gas-mileage distribution and the highway-gas-mileage distribution are both fairly wide and overlap.   **f.** Even though the distribution of city and highway gas mileages together is unimodal, the city-gas-mileage distribution is quite different than the highway-gas-mileage distribution.
6. a. skewed right   **b.** yes; a fare of $460 is much larger than at least 99% of the other fares.   **c.** 0.38   **d.** 0.95   **e.** 52,429 fares
7. a. 88th percentile   **b.** $10   **c.** $9,699,328   **8. a.**

Stem (tens)	Leaf (ones)
5	4 9
6	0 4 6
7	6 8 8
8	1

   b. symmetric   **c.** 66 years   **d.** 0.333   **e.** 33rd percentile

9. a.

Class	Frequency	Relative Frequency
0–9	19	$\frac{19}{32} \approx 0.594$
10–19	8	$\frac{8}{32} \approx 0.250$
20–29	3	$\frac{3}{32} \approx 0.094$
30–39	0	$\frac{0}{32} = 0$
40–49	1	$\frac{1}{32} \approx 0.031$
100–109	1	$\frac{1}{32} \approx 0.031$
Total	32	$\frac{32}{32} = 1$

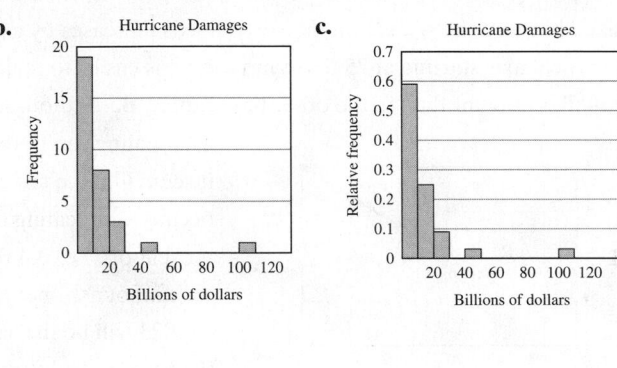

d. $106 billion; the damage $106 billion is much greater than the other damages.   **10. a.** the time-series plot with vertical axis starting at 0%; the large increases in the scaling of the vertical axis de-emphasize the changes in market share.   **b.** the time-series plot with vertical axis starting at 11.5%; 13.2%   **c.** from 2010 to 2011; 1 percentage point   **d.** −2.5 percentage points; 9.3%; no; we cannot assume the change in market share from 2013 to 2018 will be the same as from 2008 to 2013.

Chapter 4

Homework 4.1   1. outliers   **3.** less   **5.** $\bar{x} = 7.41$ **7. a.** $x_1 = 290, x_2 = 221, x_3 = 216, x_4 = 200, x_5 = 133, x_6 = 112$ **b.** 1172; there were 1172 thousand (1, 172, 000) of the six most popular cosmetic surgeries performed in 2013.   **c.** 195.3; for the six most popular types of cosmetic surgery, the mean number of surgeries was 195.3 thousand surgeries per type.   **9. a.** $x_1 = 641, x_2 = 622,$ $x_3 = 580, x_4 = 526$ **b.** 2369; the total gross of the top four grossing superhero movies is $2369 million ($2.369 billion).   **c.** 592.3; the mean gross of the top four grossing superhero movies is $592.3 million.   **11.** $\bar{x} = 8; M = 9$ **13.** $\bar{x} = 29; M = 30.5$ **15.** mean: 5.3; median: 6; mode: 6   **17.** mean: 5.44 million downloads; median: 5.5 million downloads   **19.** The mean is greater than the median.   **21.** The mean is approximately equal to the median.   **23.** (c)   **25.** (d)   **27. a.** 8.3 nominations   **b.** 6.5 nominations   **c.** 6 nominations   **d.** 19 nominations   **e.** mean; the mean is sensitive to outliers, and the median is resistant to outliers.   **29.** 7.4 hours   **31.** 19 years   **33.** 70 texts   **35. a.** 143.7 minutes   **b.** 145 minutes   **c.**

d. The distribution is skewed left.   **e.** For some skewed distributions, the mean and median are not in the class with the largest frequency.   **37. a.** 20–29 years   **b.** 48th percentile   **c.** 49 years   **d.** bimodal; teenagers may break their fingers a lot because they are so active. Adults in their forties might break their fingers a lot because they are both young enough to still be fairly active but old enough to be quite vulnerable to breaking their fingers.   **e.** First group: people ages 0–29 years; second group: people ages 30–79 years; for the first group, the class 10–19 years contains the median; for the second group, the class 40–49 years contains the median.   **39. a.** 70–74.9 degrees Fahrenheit   **b.** The student is incorrect. The median temperature might be quite different the next day.   **41. a.** It is more appropriate to measure the center of the other seven states and report that South Carolina was not included in making the estimate.   **b.** mean: $7.047 per hour; median: $7.25 per hour   **c.** $7.335 per hour; $7.25 per hour   **d.** The mean became larger, but the median did not change.   **43. a.** skewed right   **b.** mean; median; for skewed distributions, the median measures the center better than the mean.   **c.** skewed right   **d.** People were not sure of the exact durations and estimated using multiples of 5; response bias   **e.** yes; by inspecting only Fig. 35, we can tell that the distribution is skewed right, but by inspecting both figures, we can tell there is right-skewness within the right-skewness.   **45. a.** skewed right; mean; median; for skewed distributions, the median is a better measure of the center than the mean.   **b.** 30–34 years   **c.** 40–44 years   **d.** yes; the median age for actors who win the best-actor award is greater than the median age for actresses who win the best-actress award; this comparison is true, assuming the characteristic "highly successful" is measured in terms of winning best-actress and best-actor awards.   **e.** Next year's awards for best actress and best actor may not fit the pattern of past years.   **47. a.** All states with approximately equal populations should have approximately equal numbers of House seats.   **b.** mean; the mean is larger than the median for skewed-right distributions.   **c.** 1.41; 18 representatives   **d.** $6.16(1.41) = 8.6856 \approx 8.7$ **e.** population distribution's ratio: 1.39; seat distribution's ratio: 1.45; the ratios should be approximately equal because the two distributions have about the same shape.

49. a.

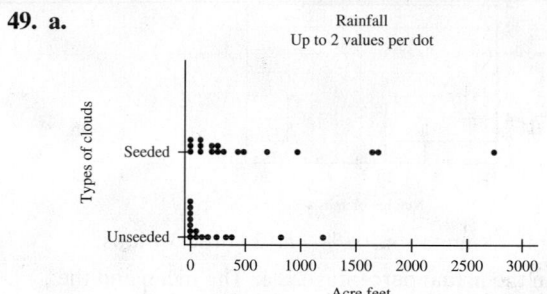

b. medians; the median is a better measure of the center than the mean for skewed distributions.   **c.** median rainfall for seeded clouds: 221.6 acre-feet; median rainfall for unseeded clouds: 44.2 acre-feet   **d.** greater   **e.** sampling error; there is probably no sampling bias because the experiment was probably carried out well, but even with well-designed experiments, there is sampling error.   **51. a.** experiment; it is an experiment because there was random assignment.   **b.** explanatory variable: seeding the clouds; response variable: the amount of rainfall   **c.** Random assignment was used to determine which clouds were in the treatment group and which clouds were in the control group; the researchers could construct a frame of the 52 clouds and randomly select 26 of the clouds to be in the treatment group. The other clouds would be in the control group.   **d.** sample: the 52 clouds; population: all isolated cumulus clouds in south Florida.   **e.** causality; we can determine whether there is causality when conducting experiments.

53. a.

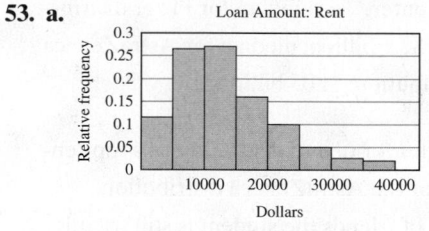 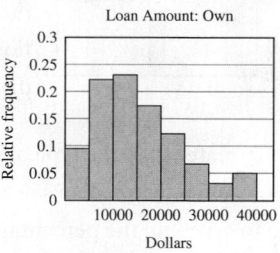 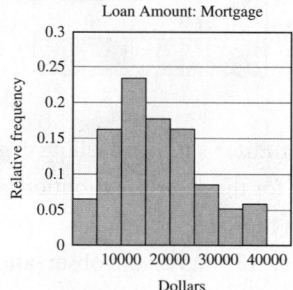

b. All three distributions are unimodal and skewed right.   **c.** For all three distributions, the median is the better measure of the center; the median measures the center better than the mean when a distribution is skewed.

d. rent: the mean is $12,930.8 and the median is $11,337.5; own: the mean is $14,692.8 and the median is $12,850; mortgage: the mean is $16,721.3 and the median is $15,000.   **e.** For all three distributions, the mean is larger; this makes sense because the distributions are skewed right.   **f.** From smallest to largest: rent distribution, own distribution, mortgage distribution; This means that the typical loan for someone who owns a home is larger than the typical loan for someone who rents a home, and the typical loan for someone who has a mortgage is the largest of all three types of borrowers; answers may vary.   **55.** Answers may vary.   **57.** $52 thousand   **59.** stay the same   **61. a.** $6.8 thousand   **b.** $6 thousand   **c.** $166.8 thousand   **d.** $6 thousand   **e.** The outlier increased the mean by a lot ($160 thousand) but did not affect the median.   **63.** the mean weight of 5 randomly selected human adults; the mean measures the center and is unaffected by the number of observations.   **65.** Answers may vary.   **67. a.** Answers may vary.   **b.** Answers may vary.   **c.** Answers may vary.   **d.** Answers may vary.   **e.** Answers may vary.

Homework 4.2   **1.** spread   **3.** two   **5.** $\bar{x} = 2.2$ **7.** $s = 1.5$ **9.** $\bar{x} = 7, M = 7, R = 6, s = 2, s^2 = 4$ **11.** $\bar{x} = 15, M = 15,$ $R = 25, s = 8.6, s^2 = 73.1$ **13.** range: 2.1 million downloads; standard deviation: 0.89 million downloads   **15.** 5.1 hours   **17.** 300 texts

19. 6.6 hours **21. a.** $2735.9 **b.** $1132.2 **c.** 10 public colleges; the mean tuition for the sample of public colleges is less than the mean tuition for the sample of private colleges. **d.** 10 public colleges; the standard deviation of tuitions for the sample of public colleges is less than the standard deviation of tuitions for the sample of private colleges. **e.** The student cannot draw such a conclusion because the spread of all 2-year, public colleges might be different than the randomly selected 2-year, public colleges due to sampling error. Also, the spread of all 2-year, private colleges might be different than the randomly selected 2-year, private colleges due to sampling error. **23.** (b) **25.** (c) **27. a.** 0.68 **b.** 0.96 **c.** 1 **d.** 4 inches **29. a.** 70 points **b.** 70 points **c.** 10 points **d.** 100 points^2 **31. a.** The areas of the bars, from left to right, are 0.05, 0.15, 0.3, 0.3, 0.15, and 0.05. **b.** 1 **c.** The area of a bar is equal to the relative frequency of the bar's class, so the total area of the bars is equal to the total of the relative frequencies, which is always equal to 1. **d.** Answers may vary. **33. a.** Professor A; the student does very well in math and Professor A's students' scores have the same center but more spread than Professor B's students' scores. **b.** yes; the scores on Professor A's tests might have larger spread due to very low scores, not high scores, so the top student in Professor B's class might have scored higher than all the students in Professor A's class. Another possibility is that the students in Professor A's class might have been stronger students due to what they learned in previous math courses, and Professor B might actually be the more effective professor. **35. a.** Event B **b.** Event A **c.** The distribution for Event A is unimodal because the spreads of the women's times and the men's times are relatively large and, hence, their times are intermixed quite a bit. The distribution for Event B is bimodal because the spreads of the women's times and the men's times are relatively small and, hence, do not intermix much. **d.** The standard deviations should also be compared to help determine how much the times intermix. **37. a.**

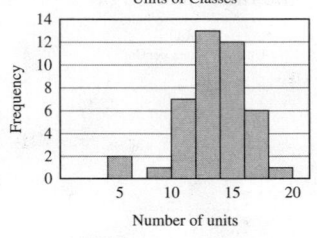

unimodal and approximately symmetric; yes **b.** mean: 13.0 units; standard deviation: 2.9 units **c.** 68% **d.** 29 observations; 69%; the percentage estimated by the Empirical Rule is 1 percentage point less than the actual percentage. **e.** The mean and the standard deviation for all students might be different than the sample's mean and standard deviation, respectively, due to sampling bias and sampling error. **39. a.**

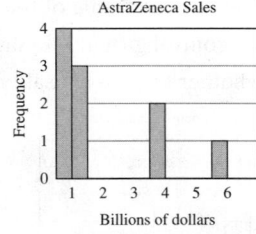

Both distributions are skewed right. **b.** median; for a distribution that is skewed right, the median should be used to measure the center. **c.** median for Pfizer distribution: $2.1 billion; median for AstraZeneca distribution: $1.05 billion

d. Pfizer; the typical sales for one of Pfizer's top-ten-selling drugs is greater than the typical sales for one of AstraZeneca's top-ten-selling drugs. **e.** The median sales for the Pfizer distribution is larger than the median sales for the AstraZeneca distribution.
41. a. If a student had no friends in high school, it makes no sense to compute the percentage of friends the student is still friends with. **b.** 17 FRRs **c.** 24 FRRs **d.** 100%; yes; the observation is more than two standard deviations away from the mean. **e.** The percentage for all college students is probably different than 42.7% due to sampling bias and sampling error. **43. a.** the mean of the stuffed-crust data is larger than the mean of the Thin 'N Crispy data; stuffed-crust pizza has one or more additional ingredients such as cheese baked into the crust. **b.** 55.6 calories; they are equal; answers may vary. **c.** 1.6 calories; the result is less than the standard deviation of the paired differences. **d.** The standard deviation of the paired differences is less than the standard deviation of the stuffed-crust data; the standard deviation of the paired differences is less than the standard deviation of the Thin 'N Crispy data; deciding which toppings to order **45. a.**

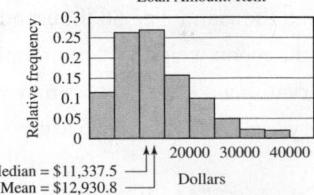
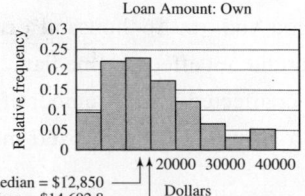
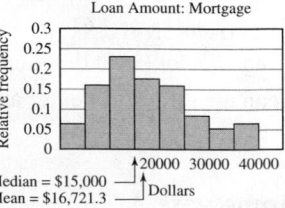

b. For all three distributions, the range is the better measure of the spread; the range measures the spread better than the standard deviation when a distribution is skewed. **c.** For each distribution, the range is $34,000. On the basis of just the range, the spreads of the three distributions are the same.   **d.** For each distribution, the mean and the median appear to be close together (see the histograms found in part (a)).   **e.** Because the mean and the median are fairly close in value, the standard deviation might be a fairly good measure of the spread.   **f.** rent: $7693.7; own: $8704.6; mortgage: $8932.1; the standard deviation of the own distribution is larger than the standard deviation of the rent distribution, and the standard deviation of the mortgage distribution is the largest of all three standard deviations.   **g.** Although the ranges of the three distributions are equal, the standard deviations of the distributions suggest that the own distribution has more spread than the rent distribution, and the mortgage distribution has the most spread of the three distributions. This matches with the histograms because the right tail of the own distribution is higher than the right tail of the rent distribution, and the right tail of the mortgage distribution is the highest of all three distributions' right tails.   **47.** stay the same **49.** increase **51.** decrease **53.** Answers may vary.   **55.** the weights of 5 randomly selected human adults; the weights of human adults are spread out more than the weights of cats.   **57. a.** 99.7% **b.** 6 **c.** If 99.7% of the data are between 30 and 90 points, then the smallest observation might be approximately 30 points and the largest observation might be approximately 90 points. If that is true, then the range is approximately the distance between 30 and 90 points, which is also equal to 6 standard deviations, or 6s. So, the range is approximately 6s.   **59. a.** 75 years **b.** 12.5 years **c.** 0.1 year **61. a.** small-cap stock; because the values of the small-cap stock have greater standard deviation than the values of a blue-chip dividend stock, the absolute value of the decreases in the value of the small-cap stock tend to be greater than the absolute value of the decreases in the value of the blue-chip dividend stock.   **b.** The value of the small-cap stock tends to have larger increases than the value of the blue-chip dividend stock.

Homework 4.3   **1.** 75th **3.** 25th **5.** (b) **7.** (c) **9. a.** $61 thousand **b.** 25% **c.** 75% **d.** 100%; the sum of the parts is equal to the whole.   **e.** 25 states **11. a.** skewed right; the smallest 25% of the observations are less spread out than the largest 25% of the observations.   **b.** 17%; the interest rate 17% is greater than or equal to approximately 75% of the interest rates and less than approximately 25% of the interest rates.   **c.** 25th percentile; the interest rate 10.2% is greater than or equal to approximately 25% of the interest rates and less than approximately 75% of the interest rates.   **d.** interest rate 26% **e.** greater **13. a.** A: 90 points; B: 70 points, C: 60 points **b.** 8 students **c.** Each side of the box represents 25% of the scores, so there were the same number of Cs and Bs.   **d.** the cutoffs described in part (a); students who scored between 70 and 80 points would have earned Cs with the normal cutoffs but earned Bs with the cutoffs described in part (a). And students who scored between 60 and 70 points would have earned Ds with the normal cutoffs but earned Cs with the cutoffs described in part (a).   **15. a.** skewed right; median; the median is better than the mean at measuring the center of a skewed distribution.   **b.** 8 moons; approximately half of the planets have less than 8 moons and approximately half of the planets have more than 8 moons.   **c.** 45 moons; approximately 75% of the planets have less than 45 moons and approximately 25% of the planets have more than 45 moons.   **d.** The student is incorrect. The left whisker together with the left side of the box represent a total of 50% of the observations, but the right side of the box represents only 25% of the observations.   **e.** 4 planets **17.** Answers may vary.   **19. a.** smallest observation: 3 miles; first quartile: 5 miles; median: 7 miles; third quartile: 13 miles; largest observation: 30 miles **b.** 30 miles **c.**

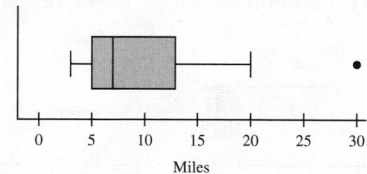

Distances to College

d. skewed right; the right side of the box is longer than the left side, and the right whisker is longer than the left whisker.   **e.** median; the median is a better measure of the center than the mean for a skewed distribution.

21. a.

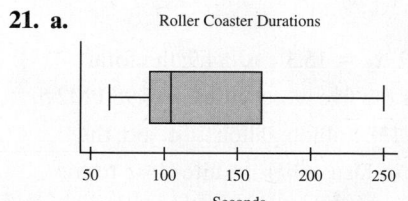

Roller Coaster Durations

b. 166.5 seconds; the value 166.5 seconds is greater than or equal to approximately 75% of the roller coaster durations and less than approximately 25% of the roller coaster durations.   **c.** 76.5 seconds; it is a measure of the spread of the middle 50% of the roller coaster durations.   **d.** no **e.** sampling bias **23. a.**

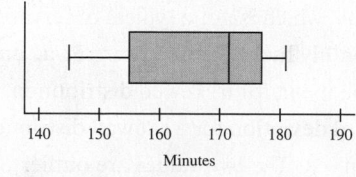

NYC Marathon Top 50 Women's Times

b. 171.5 minutes; the value 171.5 minutes is greater than or equal to approximately 50% of the top-50 women's times and less than approximately 50% of the top-50 women's times.   **c.** 22 minutes; it is a measure of the spread of the middle 50% of the top-50 women's times.   **d.** the running times in the bottom 25% of the sample; the left whisker is longer than the right whisker.   **e.** On the basis of the boxplot, at most 25% of the observations lie in the class 160–169 and at least 25% of the observations lie in the class 170–179, so the student is probably incorrect. In fact, only 9 of the observations lie in the class 160–169 minutes, and 24 observations lie in the class 170–179 minutes, so the student is definitely incorrect.   **25.**

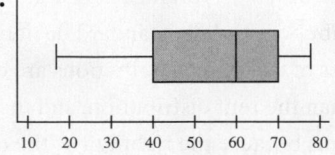

27. a. hardcover prices; a typical hardcover price is greater than a typical paperback price.   **b.** hardcover prices; the hardcover prices are more spread out than the paperback prices.

c. paperback prices; the middle 50% of paperback prices have more spread than the middle 50% of hardcover prices.   **d.** The range takes into account only the smallest and largest observations, but the standard deviation takes into account all of the observations. So, the standard deviation of the paperback prices might be greater than the standard deviation of the hardcover prices due to the middle 50% of the paperback prices having more spread than the middle 50% of the hardcover prices.   **e.** The range of paperback prices of all elementary statistics textbooks might be different than the range of the paperback prices in the sample due to sampling error. And the range of hardcover prices of all elementary statistics textbooks might be different than the range of the hardcover prices in the sample due to sampling error. So, the student might be incorrect.   **29. a.** skewed right; median; the median is a better measure of the center than the mean for a skewed distribution.   **b.** median pre-MBA salary: $49 thousand; median post-MBA salary: $116 thousand; a typical pre-MBA salary is $49 thousand and a typical post-MBA salary is $116 thousand.   **c.** $67 thousand; a typical post-MBA salary is $67 thousand greater than a typical pre-MBA salary.   **d.** $83 thousand   **e.** 5.9 years   **31. a.** The tuition increased by $8 thousand, the median pre-MBA salary increased by $1 thousand, and the median post-MBA salary increased by $2 thousand.   **b.** The tuition increased by $8 thousand but the difference in the median post-MBA salary and the median pre-MBA salary only increased by $1 thousand.   **33. a.**

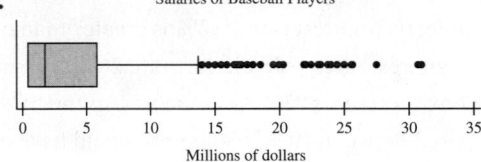

skewed right   **b.** median; the median is a better measure of the center than the mean for a skewed distribution; $1.8 million

c.

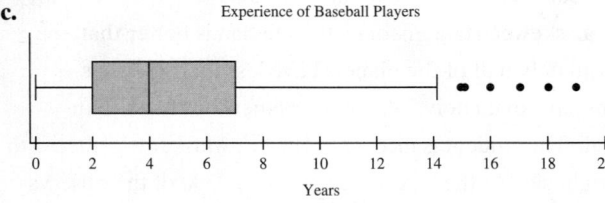

skewed right   **d.** median; the median is a better measure of the center than the mean for a skewed distribution; 4 years   **e.** $7.2 million; overestimate; the player's salary in the first three years would probably have been less than $1.8 million.   **35.** $Q_1 = 50$; $Q_3 = 50$ **37.** IQR $= 20$; $Q_1 = 70$, $Q_3 = 90$ **39. a.** increase; the smallest value and the largest value would become farther apart.   **b.** stay the same; the values of Q_1 and Q_3 would not change, so the IQR would not change.   **41.** The student is incorrect. Each whisker represents 25% of the observations.   **43. a.** The median is between 30 and 50, inclusive.   **b.** The range is at least 20.   **45.** The student did not use uniform scaling on the horizontal axis.

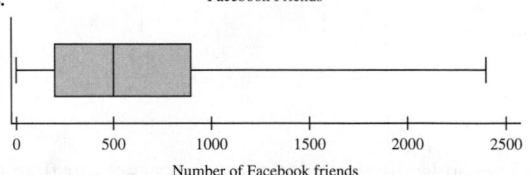

47. frequency and relative frequency table, relative frequency bar graph, pie chart   **49.** frequency and relative frequency table, dotplot, relative frequency histogram, boxplot

Chapter 4 Review Exercises   **1.** $\bar{x} = 8$ **2. a.** $x_1 = 18.2, x_2 = 18.2, x_3 = 17.0, x_4 = 16.2, x_5 = 15.3$ **b.** 84.9; the total level of pollution for the five worst cities is 84.9 PM2.5.   **c.** 16.98; the mean level of pollution for the five worst cities is 16.98 PM2.5.   **3.** mean: 54.8; median: 56.5; mode: 25   **4. a.** 125.3 million gallons   **b.** 125.5 million gallons   **c.** 138 million gallons   **d.** no; the mode is the largest observation, which is not a typical observation.   **e.** no; the fuel consumption in December is quite close to the median, which is a typical monthly fuel consumption.   **5. a.** unimodal and skewed right   **b.** mean; median; the median is a better measure of the center than the mean for a skewed distribution.   **c.** 0–4.9 billion dollars   **d.** range; the range is a better measure of the spread than the standard deviation for a skewed distribution.   **e.** $49 billion   **6. a.** 89th percentile   **b.** $14.9 billion   **c.** $4.9 billion   **d.** $3.9 billion   **e.** The two values are outliers.   **7.** stay the same   **8.** stay the same   **9.** mean: 47.6; median: 45; range: 32; standard deviation: 10.3; variance: 106.3   **10.** 2.0 units   **11.** 7.3 novels^2 **12.** 3 hours per day   **13.** (d)   **14.** (a)   **15.** (b)   **16.** (c)   **17. a.** The distribution is unimodal and symmetric.   **b.** 68%   **c.** 95%   **d.** 98%   **e.** All of the observations between 30 and 75 years lie in the class 27.3–75.3 years.   **18. a.** symmetric   **b.** 398 points   **c.** 106 points; Q_1 and Q_3 are 106 points apart.

d. 598 points **e.** 75th percentile **19. a.**

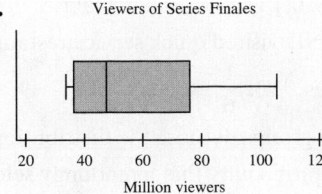

Viewers of Series Finales

Million viewers

b. skewed right; the upper 50% of the observations have more spread than the lower 50%. **c.** median; the median is a better measure of the center than the mean for a skewed distribution. **d.** 48 million viewers; the observation 48 million viewers is greater than or equal to approximately 50% of the observations and less than approximately 50% of the observations. **e.** Each whisker represents 25% of the observations.

20. a.

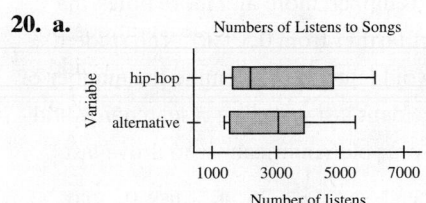

Numbers of Listens to Songs

Number of listens

b. hip-hop **c.** median for alternative distribution: 3067 listens; median for hip-hop distribution: 2365 listens; the median for the alternative distribution is larger than the median for the hip-hop distribution; the typical number of listens of an alternative song is greater than the typical number of listens of a hip-hop song. **d.** IQR for alternative distribution: 2360 listens; IQR for hip-hop distribution: 3101 listens; the IQR for the alternative distribution is less than the IQR for the hip-hop distribution; the middle 50% of the observations of the alternative distribution has less spread than the middle 50% of the observations of the hip-hop distribution. **e.** Q_3 for alternative distribution: 3887 listens; Q_3 for hip-hop distribution: 4792 listens; Q_3 for the alternative distribution is less than Q_3 for the hip-hop distribution; the 75th percentile for the alternative distribution is less than the 75th percentile for the hip-hop distribution.

Chapter 4 Test **1.** less than; the distribution is skewed left. **2. a.** skewed right **b.** mean; median; the median is a better measure of the center than the mean for a skewed distribution. **c.** 45–49 years **d.** The typical age of a female professor is less than the typical age of a male professor. **e.** 96th percentile **3.** $40,000 **4.** mean: 55.7; median: 55; mode: 80; range: 76; standard deviation: 25.5; variance: 651.7 **5. a.** 42.26 thousand albums **b.** 17.80 thousand albums **c.** 316.84 thousand albums^2 **6. a.** 99.7% **b.** 68% **c.** 38 students **d.** The center of the first-test scores is greater than the center of the second-test scores. The first-test scores are less spread out than the second-test scores. **7.** $s = 14$ **8. a.** skewed right; the upper 50% of the observations have more spread than the lower 50% of the observations. **b.** 16 deaths per 1000 births **c.** 117 deaths per 1000 births; yes **d.** 25th percentile **e.** 23,593 infant deaths **9. a.** numerical **b.**

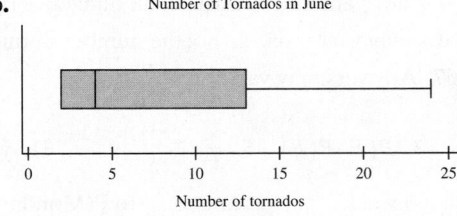

Number of Tornados in June

Number of tornados

c. skewed right; the right side of the box is longer than the left side, and the right whisker is longer than the left whisker. **d.** 4 tornados **e.** The lowest 25% of the observations are each equal to 2 tornados.

10. a. Additional research should be performed to determine why the observations are negative. If the six pitches all hit the ground before reaching the batter, then each negative observation should be replaced with 0 feet. If the negative observations were due to camera error, the observations should be removed. If the negative observations were due to data entry, the values should be corrected, or if this is not possible, they should be removed. If the observations are removed, then this should be stated in the report. **b.** 3 feet; the observation 3 feet is greater than or equal to 75% of the observations and less than 25% of the observations. **c.** Some of the outliers have the same value or are so close in value that they appear to be one dot. **d.** right-hand-hitter distribution; a typical pitch height to a right-hand hitter was less than a typical pitch height to a left-hand hitter. **e.** right-hand-hitter distribution; the middle 50% of pitch heights to right-hand hitters were less spread out than the middle 50% of pitch heights to left-hand hitters.

Chapter 5

Homework 5.1 **1.** 1 **3.** random **5. a.** THTHT **b.** 0.4; no; the estimate 0.4 is not that close to the probability 0.5. **c.** 0.4965; yes; the estimate 0.4965 is very close to the probability 0.5. **7.** $\frac{5}{12}$ **9.** 1 **11.** $\frac{1}{2}$ **13.** $\frac{1}{38}$ **15.** $\frac{9}{19}$ **17.** 0 **19.** lose; the probability of the ball landing on red is less than the probability of the ball landing on black OR green. **21.** $\frac{1}{2}$ **23.** $\frac{1}{8}$ **25.** $\frac{5}{8}$ **27.** $\frac{1}{2}$ **29.** $\frac{5}{8}$

31. The student has made a mistake. The sum of the probabilities of all genres should equal 1. **33.** The values $\frac{9}{2}$ and -0.65 cannot be probabilities because they are not between 0 and 1, inclusive. **35.** $\frac{2}{15}$ **37.** $\frac{4}{15}$ **39.** $\frac{7}{30}$ **41. a.** 0.065 **b.** 0.298 **c.** 0.702

d. For the 30 days prior to the study, there are three possibilities: (1) They visited casual restaurants 2 times AND visited quick service restaurants 2 times, (2) they visited casual restaurants 2 times AND visited quick service restaurants 3 times, and (3) they visited casual restaurants 3 times AND visited quick service restaurants 3 times. **43.** $\frac{1}{6}$ **45.** $\frac{2}{3}$ **47.** $\frac{1}{2}$ **49.** $\frac{2}{3}$ **51.** $\frac{2}{3}$ **53. a.** 0.26 **b.** 0.55 **c.** 0.45 **d.** bimodal **e.** The data should be separated into two groups because the distribution is bimodal. **55. a.** The proportion of mean response times between 0 and 34 days is 0.90. **b.** The probability that a randomly selected response time is between 0 and 34 days is 0.90. **c.** 0.67 **d.** 0.53 **e.** 0.21 **57. a.** 0.95 **b.** 0.997 **c.** 0.68 **59.** The probability that a randomly selected American lives in Ohio is 0.04. **61.** The proportion of Instagram users who are Hispanic is 0.17. **63.** The student is incorrect because a street light is not green, yellow, and red equal amounts of time. **65.** The student is incorrect. As a coin is flipped more and more times, the proportion of heads will generally get closer and closer to 0.5, but at times the proportion may get farther from 0.5. **67.** The student must have made a mistake because a probability cannot be greater than 1. **69.** The researcher would need to determine the number of 20-year-old Americans who drove last year and the number of those drivers who were in a car accident last year. The researcher would then need to divide the number of those drivers who were in a car accident by the number of 20-year-old Americans who drove last year. **71.** Answers may vary. **73. a.** Answers may vary. **b.** Answers may vary. **c.** It would be close to $\frac{1}{2}$. **75. a.** Answers may vary. **b.** Answers may vary. **c.** Answers may vary. **d.** Answers may vary.

Homework 5.2 **1.** E **3.** $P(E) + P(F)$ **5.** 0.3 **7.** 0.7 **9.** 0.9 **11.** $\frac{1}{4}$ **13.** $\frac{3}{4}$ **15.** $\frac{3}{4}$ **17.** $\frac{18}{19}$ **19.** $\frac{10}{19}$ **21.** $\frac{5}{38}$ **23.** $\frac{11}{19}$ **25.** $\frac{1}{2}$ **27.** $\frac{1}{3}$ **29.** $\frac{5}{6}$ **31.** 0 **33.** $\frac{1}{8}$ **35.** $\frac{7}{8}$ **37.** $\frac{5}{8}$ **39.** $\frac{3}{4}$ **41. a.** 0.675 **b.** 0.325 **c.** 0.733 **d.** The student is incorrect. The percentage of weapons used in murders that are personal weapons might be quite different than 5.8% in other years. **43. a.** 0.05 **b.** 0.95 **c.** 0.08 **d.** a great deal: 8%, don't know: 3%, none at all: 56%, some: 33% **e.** response bias, nonresponse bias **45. a.** 0.99 **b.** 0.01 **c.** 0.34 **d.** The percentage of Americans between 10 and 19 years, inclusive, of age who run could be quite different than the percentage of Americans between 10 and 19 years, inclusive, of age who ran in the race. **e.** response bias **47. a.** A flight departed 15 minutes early. **b.** 0.38 **c.** 0.13 **d.** 0.62 **e.** The percentage of United flights departing early from O'Hare Airport could vary from day to day. **49.** $\frac{5}{6}$ **51.** $\frac{2}{5}$ **53.** $\frac{1}{30}$ **55.** $\frac{4}{15}$ **57. a.** 0.576 **b.** 1 **c.** 0 **d.** 0.514 **e.** 0.040 **59. a.** 0.701 **b.** 0 **c.** 0.873 **d.** 0.132 **e.** 0.544 **61.** E; the event that consists of all outcomes that are not in the event that consists of all outcomes not in E is the event E. **63.** Answers may vary. **65. a.** no; the number of outcomes in E AND F is at most the number of outcomes in E. **b.** yes; answers may vary **67.** Answers may vary.

Homework 5.3 **1.** conditional **3.** $P(E)P(F)$ **5.** $\frac{3}{4}$ **7.** $\frac{1}{5}$ **9.** $\frac{1}{2}$ **11.** yes **13.** $\frac{1}{2}$ **15.** $\frac{5}{9}$ **17.** $\frac{1}{6}$ **19.** no **21.** $\frac{2}{3}$ **23.** $\frac{1}{5}$ **25.** $\frac{1}{5}$ **27.** no; $P(\text{Monday AND second week}) = \frac{1}{30}$ is not equal to $P(\text{Monday}) \cdot P(\text{second week}) = \frac{2}{15} \cdot \frac{7}{30} = \frac{7}{225}$. **29.** $P(L\,|\,W) = 0.45$ **31. a.** 0.324 **b.** 0.261 **c.** no; $P(\text{agree}\,|\,\text{between 30 and 49 years}) = 0.410$, $P(\text{agree}\,|\,\text{over 64 years}) = 0.480$; surveyed adults ages 30–49 years are less likely to agree with the statement than surveyed adults over 64 years of age. **33. a.** 0.456 **b.** 0.729 **c.** 0.399 **d.** the result from part (b) is larger; a student who began at a 4-year private institution in fall 2007 was more likely to complete a degree or certificate than a student who began at a 2-year public institution in fall 2007. **e.** The student is incorrect. The student should compare relative frequencies (probabilities), not frequencies. On the basis of the results from parts (b) and (c), we conclude that 2-year public institutions were less effective than 4-year private institutions in fall 2007. **35. a.** 0.488 **b.** 0.464 **c.** 0.364 **d.** 0.306 **e.** 0.257 **f.** The probabilities from parts (a) through (e) decrease; as the classes increase in age, the probability that an individual will smile decreases; no; the study is observational. **g.** response bias, sampling bias; response bias likely occurred because the researchers probably estimated some of the individuals' ages incorrectly; sampling bias likely occurred because some people are more likely to be in public places than others. **37.** dependent **39.** independent **41.** 0.06 **43.** $\frac{1}{32}$ **45.** 0.00583 **47.** $\frac{1}{625}$ **49. a.** 0.01874 **b.** 0.84247 **c.** sampling bias **51. a.** 0.00019 **b.** 0.63152 **c.** nonresponse bias, response bias **53. a.** 0.432 **b.** 0.330 **c.** no; the probabilities found in parts (a) and (b) are not equal. **55. a.** 0.242 **b.** 0.439 **c.** no; the results found in parts (a) and (b) are not equal. **d.** nonresponse bias **57. a.** $\frac{1}{1296}$ **b.** $\frac{1}{1296}$ **c.** They are equal. **d.** The student is incorrect because in both parts (a) and (b) exactly one of six outcomes is specified for each roll. **59. a.** $\frac{1}{128}$ **b.** $\frac{1}{2}$ **c.** The result found in part (a) is much smaller than the result found in part (b). **d.** The student is incorrect. In part (a), the probability describes getting tails on *all* seven flips, but in part (b), the probability describes what is flipped only on the seventh flip (tails). **61.** 1 **63.** The student might be incorrect.

For $P(E\,|\,F)$, the space is narrowed to F, but for $P(F\,|\,E)$, the space is narrowed to E.   **65. a.** yes; answers may vary.   **b.** yes; answers may vary.   **67.** no; the study is observational.   **69.** Answers may vary.

Homework 5.4   **1.** 1   **3.** mean   **5.** approximately normal   **7.** not approximately normal   **9.** (d)   **11.** (c)   **13.** 50 points   **15.** between 40 and 45 points   **17.**

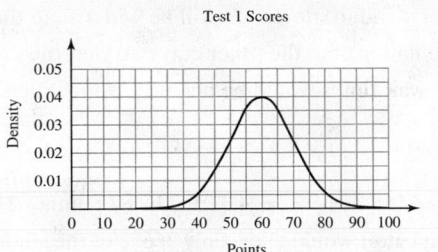

19.

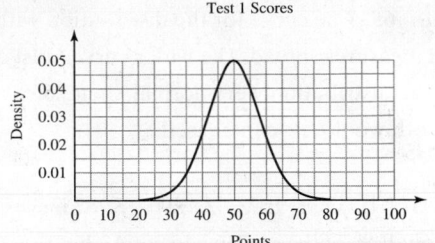

21.

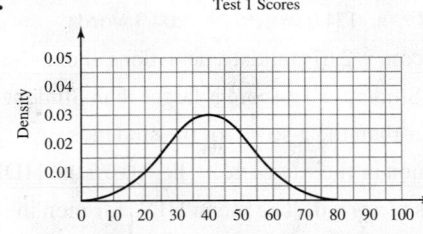

23.

Test 1 Scores

25. a. The areas of the bars, from left to right, are 0.03, 0.09, 0.33, 0.36, 0.18, and 0.03.   **b.** 1.02; due to rounding

c.

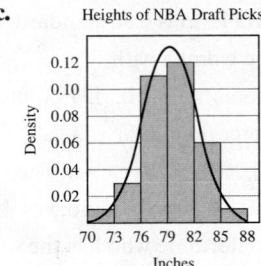

Heights of NBA Draft Picks

d. The total area of the regions where the curve undercounted are equal to the total area of the regions where the curve overcounted.   **e.** 1; 1   **27.** 1.05; the song length is 1.05 standard deviations greater than the mean.   **29.** -2.49; the lowest temperature is 2.49 standard deviations less than the mean.
31. 0.0268   **33.** 0.9995   **35.** 0.7549   **37.** 0.0158   **39.** 0.9130   **41.** 0.4582   **43. a.** 0.7437   **b.** 0.2525
c. 0.2525   **d.** 0.0038; people could die due to a smoke alarm that does not last as long as it has been
certified.   **45. a.** 0.0685   **b.** 0.0228   **c.** 0.9772   **47. a.** 0.1756   **b.** 42nd percentile; it is highly likely that
the percentiles for individual psychology majors vary.   **c.** 63rd percentile; 71st percentile; it is surprising
that the percentile for engineering majors' mean SAT math score is not that much higher than the percentile
for social-science majors.   **49. a.** 17th percentile, 77th percentile, 99th percentile

b.

Math Score	Percentile (Assuming Normality)	Actual Percentile
300	3	3
400	17	16
500	45	45
600	77	75
700	94	93
800	99	99

c. The percentiles assuming normality are close to the actual percentiles; approximately normally distributed   **51. a.** 0.3677   **b.** 0.3677   **c.** 0.4768
d. 835 women   **53. a.** yes   **b.** no   **c.** 38%   **d.** bimodal; no; if Frank measured his blood glucose levels only just before and just after meals, he would tend to miss measuring his blood glucose levels at times when the readings would be between 79 and 121 mg/dL.   **55. a.** The probability that a randomly selected woman from the study has a pregnancy that lasts less than 270 days is 0.05.   **b.** The proportion of pregnancies that last less than 270 days is
0.05.   **c.** A pregnancy that lasts 270 days is at the 5th percentile.   **d.** decreased; increased   **57. a.** 0.3452   **b.** 0.2014   **c.** The mean
and the standard deviation of salt intake are larger for the African American girls than for the Caucasian girls.   **59. a. i.** 0.6316
ii. 0.9280   **iii.** 0.0472   **b.** unimodal and symmetric   **c. i.** 0.62   **ii.** 0.93   **iii.** 0.05   **d.** The results are approximately equal;
approximately normal   **61. a.**

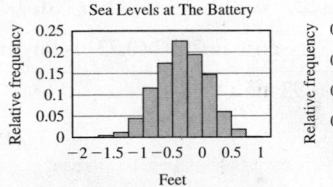

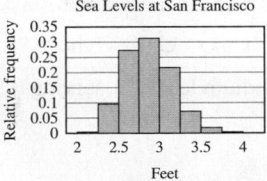

b. Both distributions are unimodal and symmetric. They might be approximately normal.   **c.** The Battery: -0.3392 feet; San Francisco: 2.8590 feet; The mean of The Battery distribution is less than the mean of the San Francisco distribution; the center of The Battery distribution is less than the center of the San Francisco distribution.   **d.** The Battery: 0.4257 foot; San Francisco: 0.2932 foot; The standard deviation of The Battery distribution is greater than the standard deviation of the San Francisco distribution; The Battery distribution is more spread out than the San Francisco distribution.   **e.** yes; the greater increase in the mean annual sea level at The Battery would contribute more to the spread than the smaller increase in the mean annual sea level at San Francisco.   **f.** The annual mean sea level at The Battery was quite a bit lower than the annual mean sea level at San Francisco in 1900.   **63.** If a distribution is normally distributed, then in using a normal curve to approximate a density histogram for the distribution, the total area of regions undercounted by the normal curve is equal to the total area of regions overcounted by the normal curve. So, the area under the normal curve is equal to the total area of the bars of the density curve, which is 1. Therefore, the area under the normal curve is equal to 1.   **65.** The curve for the distribution with larger standard deviation will be wider than the other curve because standard deviation measures spread. The wider curve must also be flatter than the other curve so that the area under the curve equals 1.   **67.** 1   **69. a.** If the distribution were normal, then 7.66% of the wait times would be negative, which does not make sense.   **b.** skewed right   **c.** They both would have increased.

Homework 5.5   **1.** true   **3.** 1.96   **5.** 87 inches   **7.** 110 seconds   **9. a.** 4.32%   **b.** 87 points   **11.** 90th percentile; the 90th percentile would allow 10% of the students to get As, but the 90-point cutoff would allow only 7.66% of the students to get As.   **13. a.** 77 inches   **b.** 0.08%   **c.** 98.79%   **d.** part (a): larger; part (b): impossible to say; part (c): smaller   **15. a.** 31 years   **b.** 55 years   **c.** 0.50%   **d.** 1 inmate   **17. a.** Only half of the boxes of cereal would meet the weight requirement.   **b.** 11.7 ounces   **19. a.** 174.0 words   **b.** 204.3 words   **c.** 30.3 words   **d.** part (a): smaller; part (b): smaller; part (c): larger   **21. a.** 2.67; Student A's score is 2.67 standard deviations above the mean.   **b.** 2.25; Student B's score is 2.25 standard deviations above the mean.   **c.** Student A; Student A's z-score is larger than Student B's z-score.   **23.** the student's ACT score; the z-score for the student's ACT score (1.53) is larger than the z-score for the student's SAT score (1.36).   **25. a.** -0.40; the HDL reading is 0.40 standard deviation below the mean for men in their twenties.   **b.** -0.29; the HDL reading is 0.29 standard deviation below the mean for men in their fifties.   **c.** the man in his fifties   **d.** The mean HDL for men in their twenties is larger than the mean HDL for men in their fifties. Also, the standard deviation of the HDL readings for men in their twenties is less than the standard deviation of the HDL readings for men in their fifties.   **27. a.** -0.30; the BMD level is 0.30 standard deviation less than the mean.   **b.** -2.30; the BMD level is 2.30 standard deviations less than the mean.   **c.** The woman with BMD 666 mg/cm^2 has four times the risk of a hip fracture than the woman with BMD 986 mg/cm^2.   **29. a.** 125 points; no   **b.** 131 points; yes   **c.** yes; no   **31. a.** 3.86   **b.** 3.74   **c.** Cambage and Griner   **d.** yes; yes; all three people have z-scores greater than 1.96.   **33. a.** no; the z-score (1.61) is not more than 1.96.   **b.** yes; the z-score (2.25) is more than 1.96.   **c.** 745 points   **35. a.** 0.0030   **b.** yes   **c.** The mean sodium level per serving is 141 mg.   **d.** no; yes   **37. a.** 0.0062   **b.** yes   **c.** The mean delivery time is 40 minutes.   **d.** no; yes   **39. a.** 0.0038   **b.** yes   **c.** The mean fiber level is 4.6 g.   **d.** no; yes   **41.** There is not enough information to determine who has the greater relative height.   **43.** Answers may vary.   **45.** Answers may vary.   **47.** yes; finding z-scores of observations preserves the order of the observations because a z-score measures the number of standard deviations an observation is from the mean.

Chapter 5 Review Exercises   **1.** The student is incorrect because the outcomes winning and losing are not equally likely events.   **2.** $\frac{1}{38}$ **3.** $\frac{9}{19}$ **4.** $\frac{10}{19}$ **5.** 0   **6.** $\frac{10}{19}$ **7.** $\frac{2}{19}$ **8.** $\frac{12}{19}$ **9.** yes   **10.** no   **11. a.** 0.411   **b.** 0.550   **c.** 0.602   **d.** 0.752   **e.** The probabilities from parts (a) through (d) increase; as the classes increase in age, the probability that a randomly selected individual drinks coffee increases; no; the study is observational.   **12. a.** 0.302   **b.** 0.411   **c.** no; the results of parts (a) and (b) are not equal.   **d.** 0.217   **e.** 0.433   **13. a.** 0.145   **b.** 0.997   **14.** -0.82; the inmate's age is 0.82 standard deviation less than the mean age.   **15.** 0.9492   **16. a.** 0.0203   **b.** 0.2510   **c.** 0.0705   **17.** 34°F   **18.** 90th percentile   **19. a.** yes; the z-score (2.13) is more than 1.96.   **b.** yes; the z-score (2.28) is more than 1.96.   **c.** the student from Florida; the z-score for the student from Florida is greater than the z-score for the student from Texas.   **20. a.** 0.0062   **b.** yes   **c.** The mean fat level per serving is 3.8 g.   **d.** no; yes

Chapter 5 Test   **1.** $\frac{1}{6}$ **2.** $\frac{1}{3}$ **3.** $\frac{1}{6}$ **4.** $\frac{5}{6}$ **5.** $\frac{2}{3}$ **6.** yes   **7. a.** 0.055   **b.** 0.117   **c.** 0.527   **d.** 0.056   **e.** no   **8. a.** 0.3142   **b.** 0.0823   **c.** 0.7062   **9.** 67 years   **10. a.** no   **b.** no   **c.** 2007; the U.S. z-score in 2007 (0.73) is larger than the U.S. z-score in 2011 (0.70).   **11. a.** 0.0038   **b.** yes   **c.** The mean sodium level per serving is 93 mg.   **d.** no; yes

Chapter 6

Homework 6.1   **1.** explanatory   **3.** association   **5.** explanatory: n; response: s; n; s **7.** explanatory: a; response: h; a; h **9.** explanatory: I; response: G; I; G **11.** explanatory: d; response: c; d; c **13.** explanatory: n; response: T; n; T **15.** explanatory: t; response: h; t; h **17.** $(98, 29.4)$; C; T **19.** $(3, 4)$; B; n **21.** $(4, 18.3)$; t; n **23.** After drinking 16 ounces of the protein shake for six months, a weight lifter can bench-press 200 pounds 8 times.   **25.** 38% of Americans at age 21 years say they volunteer.   **27.** In 2014, 45% of Americans streamed television shows at least once a month.   **29.** In 2009, 33% of Americans believed travel websites did a good job of presenting travel choices.   **31.** negative   **33.** neither   **35. a.** explanatory: number of nominees; response: number of winners

b. positive; as the number of nominees increases, the number of winners tends to increase.   **c.** 74 winners; 34 winners; 14 winners

d. 17.9%; 16.5%; 14.3%; for drama, romance, and comedy genres, the larger the number of nominees, the larger the percentage of nominees that will be winners.   **37. a.** At least one of the dots represents more than one data pair.   **b.** explanatory: temperature (Fahrenheit degrees); response: relative humidity (percent)   **c.** negative; as the temperature increases, the humidity tends to decrease.   **d.** The humidity tends to decrease as the *temperature* (not time) increases.   **e.** increase; as temperature decreases, the humidity tends to increase because the association is negative.   **39.**

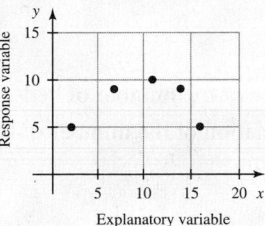

41. a. explanatory: b; response: p **b.** b; p

c.

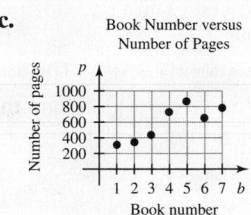

Book Number versus Number of Pages

d. The fifth book   **e.** The third book to the fourth book; the vertical distance between the data points for the third book and the fourth book is greater than the vertical distance between the data points for any other two successive books.   **43. a.** explanatory: t; response: E **b.**

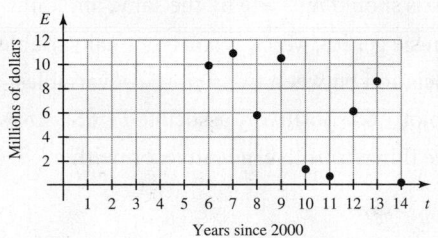

Years versus Tournament Earnings

c. 2014; $0.1 million ($100 thousand)   **d.** 2007; $10.9 million   **e.** no; answers may vary.

45. a.

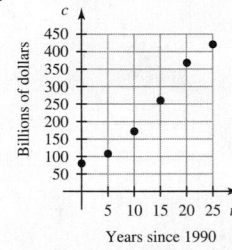

Years versus Total Cost

b. positive; the total cost to prepare taxes has generally increased.   **c.** greater

47. a.

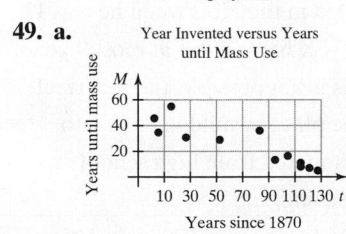

Age versus Accident Rate

b. 60–69-year-old drivers   **c.** 16-year-old drivers   **d.** 16 years and 17 years; we cannot know for sure because accident rates are not given for individual ages greater than 19 years.   **e.** The accident rate for teenagers is much higher than for people who are older.

49. a.

Year Invented versus Years until Mass Use

b. negative; it has taken less time for recent inventions to reach mass use; answers may vary.   **c.** no; it took longer for the microwave to reach mass use than it did for several other earlier inventions.   **d.** People's fears about microwaves slowed their acceptance of the new technology.   **e.** It took longer for the automobile to reach mass use than it did for the earlier inventions of electricity and the telephone; answers may vary.

51. a.

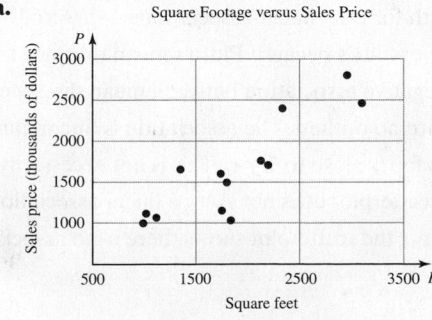

Square Footage versus Sales Price

b. positive; the larger the square footage, the larger the sales price tends to be.   **c.** Driscoll Pl; we can limit our search to points that are low and off to the right.   **d.** La Jennifer Wy; we can limit our search to points that are high up and off to the left.

53. a.

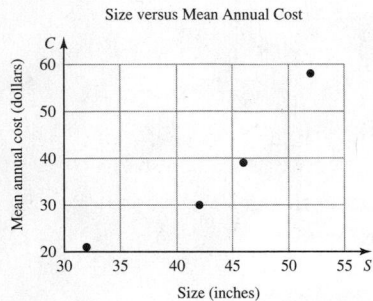

Beds versus Baths

b. positive; the greater the number of beds, the greater the number of baths tends to be.   **c.** Half a bathroom usually means there is a sink and toilet but no shower or bathtub.   **d.** 1 bath; 2.1 baths; 3 baths; 3.5 baths

e.

Number of Beds versus Mean Number of Baths

The association of the number of beds and the mean number of baths is much stronger than the association of the number of beds and the number of baths.

55. The explanatory variable, S, should be described by the horizontal axis, and the response variable, C, should be described by the vertical axis. Also, the numbers used for the scaling on an axis should increase by the same amount.

Size versus Mean Annual Cost

57. The bar graph describes favorite music genres, which is a categorical variable. *Negative association* is used to describe the association between two *numerical* variables.   **59.** multiple bar graph, two-way table   **61.** scatterplot   **63.** positively associated   **65.** Answers may vary.   **67.** An association must be one of *three* things: positive, negative, or neither.   **69.** Answers may vary.

Homework 6.2   1. linear   **3.** -1 **5.** linear   **7.** nonlinear   **9.** d   **11.** b   **13.** f   **15. a.**

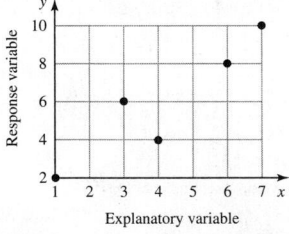

b. linear   **c.** 0.93   **d.** strong association   **17. a.** There are no outliers. The association is linear, very strong, and positive.   **b.** The greater the population, the greater the number of seats tends to be.   **c.** 53 seats; 27 seats; 14 seats   **d.** 1.4 seats per million people; 1.4 seats per million people; 1.4 seats per million people; yes   **e.** large; small   **19. a.** The red dot does not fit the pattern of most of the other points; (24, 12); a 24-year-old player has 12 years of experience; the player would have had to have started in the NFL when he was 12 years of age, which would mean he would have had to have graduated from high school or started college when he was at most 9 years old, which is highly unlikely; the researcher should try to identify the player and correct the error. If this is not possible, the incorrect age should be removed from the data set.   **b.** (25, 8); a 25-year-old player has 8 years of experience; the player would have had to have started in the NFL when he was 17 years of age, which would mean he would have had to have graduated from high school or started college when he was at most 14 years old, which is unlikely.   **c.** The association is linear, fairly strong, and positive.

21. a.

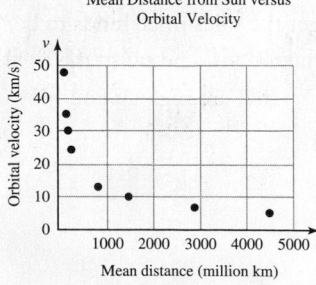

Mean Distance from Sun versus Orbital Velocity

b. There are no outliers. The association is nonlinear, strong, and negative.   **c.** The correlation coefficient r is a meaningful measure of the strength for only linear associations.   **d.** 66,621 mph   **e.** less; Pluto's orbital velocity is less than Neptune's because Pluto's mean distance from the Sun is greater than Neptune's and there is a negative association between mean distance from the Sun and orbital velocity.   **23. a.** There are no outliers. The association is linear, fairly strong, and positive.   **b.** yes; the value $r = 0.83$ is fairly close to 1.   **c.** This is not necessarily true because we cannot assume causation.   **d.** no; the scatterplot does not show a linear association, and the value $r = 0.27$ is not at all close to 1.   **25.** no; the scatterplot shows there is no association and the value $r = 0.20$ is not at all close to 1.

27. a.

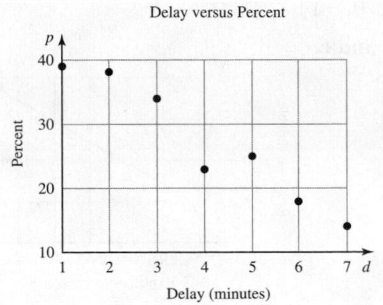

Delay versus Percent

b. linear   **c.** −0.97; strong association   **d.** the survival percentage drops by only 1 percentage point from a 1-minute delay to a 2-minute delay, but the survival percentage drops by 4 percentage points from a 2-minute delay to a 3-minute delay.   **e.** 1568 patients

29. a.

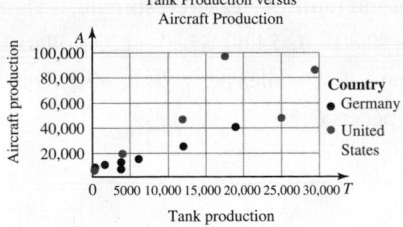

Tank Production versus Aircraft Production

b. linear; linear   **c.** 0.97; strong association; 0.79; weak association   **d.** United States

31. a.

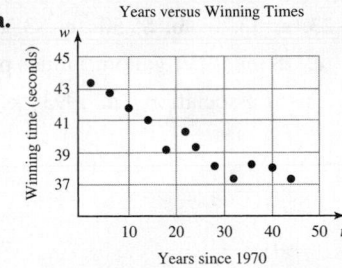

Years versus Winning Times

b. linear association   **c.** −0.94; strong association   **d.** negative association; the winning times have been generally decreasing.   **e.** The winning times for 2006 and 2010 are larger than the winning time in 2002, and the winning time in 2014 is approximately equal to the winning time in 2002.

33. a.

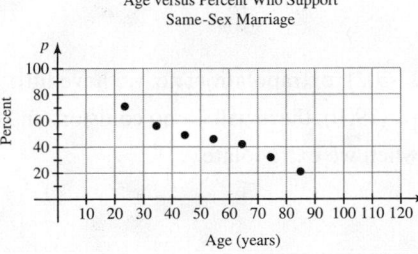

Age versus Percent Who Support Same-Sex Marriage

b. linear association   **c.** −0.98; strong association   **d.** negative association; the older an American, the smaller the percentage of Americans who believe marriages between same-sex couples should be recognized by the law will be.   **e.** We cannot assume people have changed their opinions about same-sex marriage.   **35. a.** negative association; the mean sea ice extent has generally decreased.   **b.** negative association; the mean sea ice extent has generally decreased.   **c.** 9 million square kilometers   **d.** yes; the center of the heights of the points appears to be approximately 9 million square kilometers.   **e.** positive association; the larger the mean sea ice extent in March, the larger the mean sea ice extent tends to be in September; because the mean sea ice extent in September tends to be approximately 9 million square kilometers less than in March, it makes sense that the larger the mean sea ice extent in March, the larger the mean sea ice extent tends to be in September.

37. a.

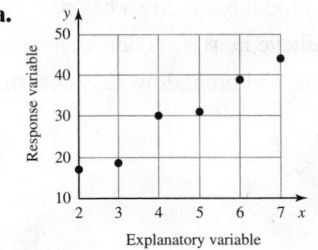

0.98 **b.**

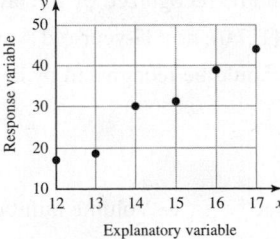

0.98; the points in the scatterplot are 10 units to the right of the points in the scatterplot constructed in part (a); the values of r are equal; the points in the scatterplot have moved 10 units to the right because we added 10 to the x-coordinates of the points; the values of r are equal because if a line that comes close to the points in the scatterplot constructed in part (a) is moved to the right 10 units, it will fit the points in the other scatterplot just as well.

c.

0.98; the points in the scatterplot are 3 units above the points in the scatterplot constructed in part (a); the values of r are equal; the points in the scatterplot are 3 units higher because we added 3 to the y-coordinates of the points; the values of r are equal because if a line that comes close to the points in the scatterplot constructed in part (a) is moved up 3 units, it will fit the points in the other scatterplot just as well.

39.

Explanatory Variable	Response Variable
x	y
10	60
11	47
12	49
13	31
14	34
15	21

Answers may vary.   **41.** The correlation coefficient does not measure causation.   **43.** If $r = 0$, we cannot conclude there is no relationship because there might be a nonlinear association.   **45.** First find any outliers, next determine the shape, then determine the strength, and finally, determine the direction; answers may vary.   **47. a.** Answers may vary.   **b.** Answers may vary.   **c.** Answers may vary.   **d.** Answers may vary.   **e.** Answers may vary.

Homework 6.3 **1.** linear **3.** extrapolation **5.** 2 **7.** 6 **9.** $(8, 0)$ **11.** -2 **13.** 6 **15.** $(0, -1)$

17. a. and **c.**   **b.** linear association **d.** $(8, 3.2)$ **e.** $(5.9, 6)$ **19. a.** and **c.**

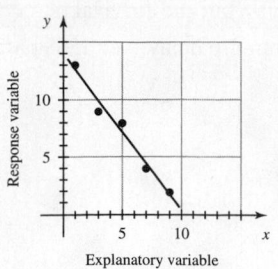

b. linear association **d.** $(-3.8, 0)$ **e.** $(0, 4.8)$ **21. a.** 2.0 thousand injuries **b.** 1.6 thousand injuries **c.** overestimate; the line is above the data point; 0.4 thousand injuries **23. a.** $3.70 **b.** $5.50 **c.** $3.30; $2.50; $0.80 **d.** $4.00; $5.30 -1.30 dollars **25. a.** 14 miles per gallon **b.** 28 miles per gallon **c.** 18 miles per gallon; 3 miles per gallon **d.** 29 miles per gallon; -7 miles per gallon **27. a.** and **c.**   **b.** linear association **d.** 1993 **e.** 1500 species

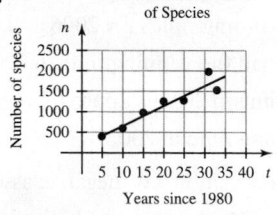

29. a–b.

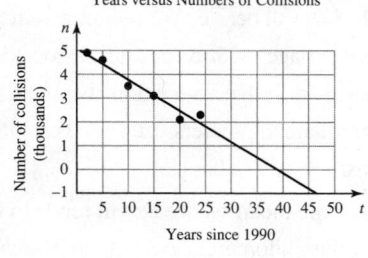

c. 2.2 thousand collisions; interpolation **d.** 2021; extrapolation; no; we have little or no faith in our results when we extrapolate. **e.** $(39, 0)$; there will be no collisions in 2029; no; we have little or no faith in our results when we extrapolate.

31. a–b.

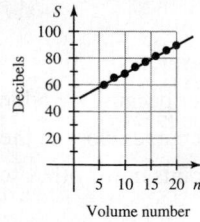

c. 27 years **d.** $(0, 85)$; 85% of newborns believe marriages between same-sex couples should be recognized by the law as valid; model breakdown has occurred. **e.** $(117, 0)$; no 117-year-old Americans believe marriages between same-sex couples should be recognized by law as valid; model breakdown has occurred.

33. a. and **c.**   **b.** linear association **d.** 88 decibels **e.** Volume number 11

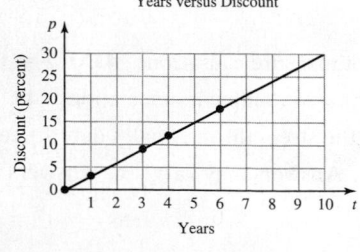

35. a.   **b.** 30%; extrapolation **c.** and **e.**   **d.** 12 percentage points; answers may vary.

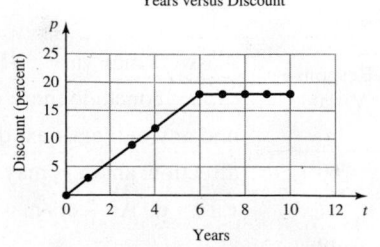

37. a.

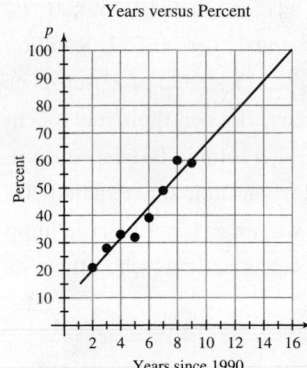

Years versus Percent

b. 100%; extrapolation   **c. and e.**

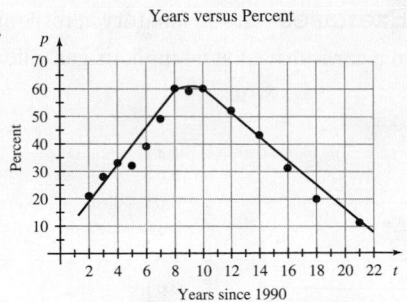

Years versus Percent

d. 69 percentage points; answers may vary.

39. a. $(2, 0)$; the profit was 0 dollars in 2004.   **b.** $(0, -68)$; the profit was -68 million dollars in 2002.

41. a. and c.

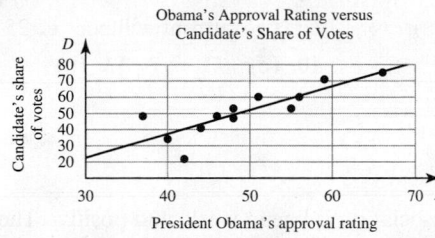

Obama's Approval Rating versus Candidate's Share of Votes

b. There are no outliers. The association is linear, fairly strong, and positive. The value of r is 0.84, which confirms that the association is fairly strong.   **d.** 27%; if Senator Johnson's share of the votes was 27%, he would not win the election.   **e.** extrapolation; no; we have little or no faith in our results when we extrapolate.

43. a. and c.

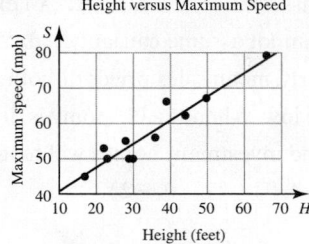

Height versus Maximum Speed

b. There are no outliers. The association is linear, strong, and positive. The value of r is 0.95, which confirms that the association is strong.   **d.** 61 mph

45. a. and c.

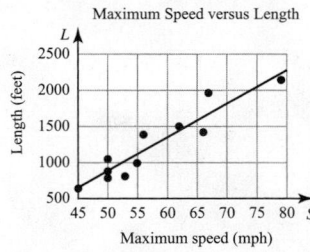

Maximum Speed versus Length

b. There are no outliers. The association is linear, strong, and positive. The value of r is 0.94, which confirms that the association is strong.   **d.** 1350 feet

47. a. and d.

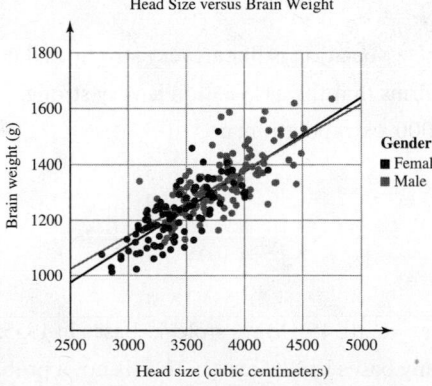

Head Size versus Brain Weight

b. There are no outliers for the women's association. The association is linear, fairly strong, and positive. The value of r is 0.76, which confirms that there is a fairly strong association; there are no outliers for the men's association. The association is linear, fairly strong, and positive. The value of r is 0.71, which confirms that the association is fairly strong.   **c.** Both associations are linear, fairly strong, and positive. We can perform interpolation with the women's model for head sizes between 2720 and 4200 cubic centimeters, but we can perform interpolation for the men's model for head sizes between 3100 and 4750 cubic centimeters.   **d.** The linear models are quite similar (see the scatterplots with the models).   **e.** 3.2 pounds

f. no; the study was performed over 100 years ago, so typical head sizes and brain weights may have changed.   **49.** Answers may vary.
51. interpolation; answers may vary.   **53.** not necessarily; answers may vary.   **55. a. i.** Answers may vary; one.   **ii.** Answers may vary; one.
iii. Answers may vary; one.   **b.** one; answers may vary.   **c.** one; answers may vary.   **57.** no; the x-coordinate of a y-intercept must be 0.
59. yes; answers may vary; $(0, 0)$ **61.** Answers may vary.

Chapter 6 Review Exercises

1. explanatory: t; response: s; t; s **2.** explanatory: c; response: w; c; w **3.** $(110, 139)$; M; T **4.** $(4, 21.4)$; t; s **5.** When a car is driven at 64 mph, its gas mileage is 42 miles per gallon.   **6.** In 2014, there were 1645 U.S. billionaires.

7.

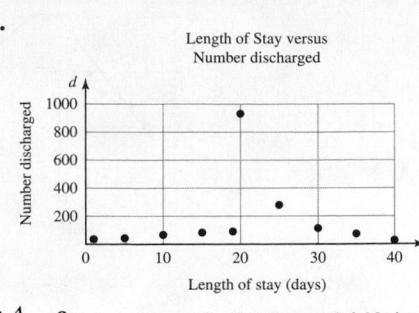

8. a.

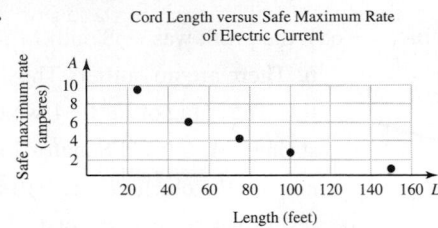

b. neither   **c.** $(40, 27)$; a total of 27 sepsis patients were discharged when their stay reached 40 days.   **d.** $(20, 931)$; a total of 931 sepsis patients were discharged when their stay reached 20 days.   **e.** There was an extremely large jump in discharges of sepsis patients when their stay reached 20 days.

9. a. explanatory: L; response: A **b.** L; A **c.**

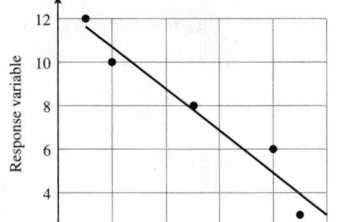

d. negative; the longer the cord, the smaller the safe maximum rate of electric current will be.   **e.** 25 feet
10. (c)   **11.** (a)   **12.** (b)

13. a.

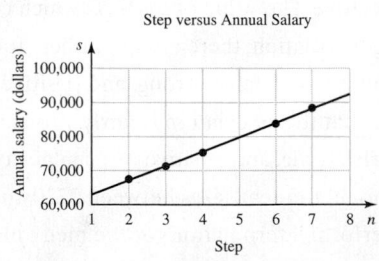

b. There are no outliers. The association is linear, weak, and positive. The value of r is 0.75, which confirms that the association is weak.   **c.** An experiment has not been conducted, so we cannot assume causality.   **d.** the economy; when the economy is doing poorly, mean sales prices of both vacation and investment homes will tend to be low. When the economy is doing well, mean sales prices of both vacation and investment homes will tend to be high.   **14.** $y = -1$ **15.** $x = -6$ **16.** $(-4, 0)$ **17.** $(0, -2)$

18. a. and **c.**

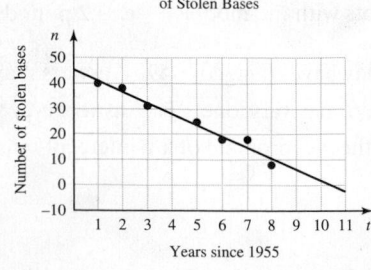

b. linear association   **d.** 8.8   **e.** 5.8

19. a. and **c.**

Step versus Annual Salary

b. There are no outliers. The association is linear, very strong, and positive. The value of r is 0.999. This confirms that the association is very strong.   **d.** $80,000; interpolation; yes   **e.** $92,000; extrapolation; no

20. a–b.

Years versus Numbers of Stolen Bases

c. 28 bases; interpolation; 1 base   **d.** $(0, 45)$; Mays stole 45 bases in 1955; yes   **e.** $(10.4, 0)$; Mays did not steal any bases in 1965; yes   **21.** r is not a probability. It is a measure of the direction and strength of a linear association between two numerical variables.

Chapter 6 Test   **1.** explanatory: a; response: p; a; p **2.** $(6, 255)$; n; c **3.** In 2014, LeBron James's annual salary was $72.3 million.

4. a. explanatory: a; response: s **b.**

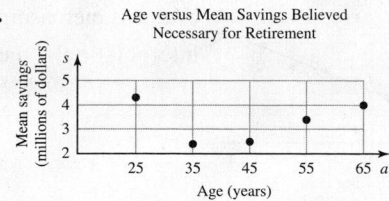

Age versus Mean Savings Believed Necessary for Retirement

c. $(25, 4.3)$; 25-year-old adults think that $4.3 million is enough to have at retirement.   **d.** $(35, 2.4)$; 35-year-old adults think that $2.4 million is enough to have at retirement.   **e.** the 20–29.99 age group; for the other age groups, the older the age group, the more savings the group believes will be necessary to have at retirement.   **5.** no association

6. nonlinear association   **7.** linear association

8. a.

Age versus Percent Who Have High Blood Pressure

b. There are no outliers. The association is linear, very strong, and positive. The value of r is 0.99, which confirms that the association is very strong.   **c.** There are no outliers. The association is linear, very strong, and positive. The value of r is 0.997. This confirms that the association is very strong.   **d.** A very strong association does not guarantee causation.   **e.** 59.5 years; men; women

9. a. and c.

b. linear association   **d.** $(11.2, 0)$ **e.** $(0, 24.3)$

10. a–b.

Years versus Numbers of Space Debris

c. 8.4 thousand debris; interpolation   **d.** 11.9 thousand debris; extrapolation; no; we have little or no faith in our results when we perform extrapolation.

e. 4.1 thousand; we are assuming that if the two collisions had not occurred, then the linear model would have predicted the exact number of debris in 2010. We have little or no faith this is true.

Chapter 7

Homework 7.1   **1.** satisfies   **3.** line   **5.** $(-3, -10), (2, 0)$ **7.** $(0, 7), (4, -5)$

9. $(0, 2)$ **11.** $(0, -4)$ **13.** $(0, 0)$ **15.** $(0, 0)$ **17.** $(0, 0)$ **19.** $(0, 0)$ **21.** $(0, 0)$

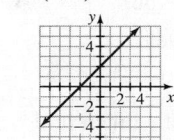

23. $(0, 1)$ **25.** $(0, -3)$ **27.** $(0, 5)$ **29.** $(0, -3)$ **31.** $(0, -3)$ **33.** $(0, 1)$ **35.**

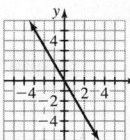
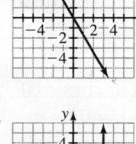

37.   **39.**   **41.**   **43.**

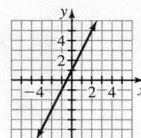

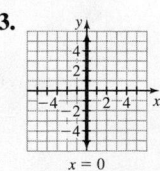

$x = 0$

45. a. explanatory: t; response: n **b.** and **d.**

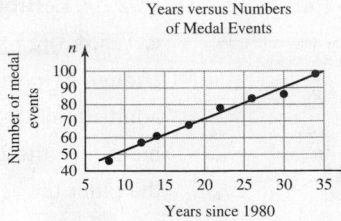
Years versus Numbers of Medal Events

c. positive; the number of winter Olympic medal events is increasing.   **d.** yes   **e.** 90 medal events; interpolation; 4 medal events

47. a. explanatory: t; response: s **b–c.**

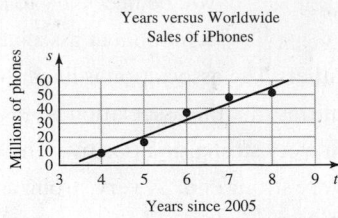
Years versus Worldwide Sales of iPhones

c. yes   **d.** 67 million iPhones; the linear association between t and s may have led analysts to make a prediction close to 67 million iPhones, which is less than 74.5 million iPhones.   **e.** 33,741 iPhones

49. a. explanatory: C; response: F **b.** and **d.**

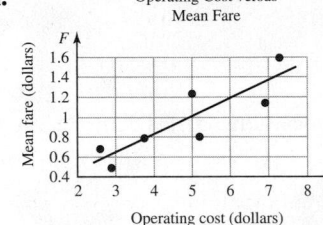
Operating Cost versus Mean Fare

c. positive; the larger the operating cost, the larger the mean fare will tend to be; the larger the operating the cost, the larger the mean fare will tend to be to pay for the cost.   **d.** yes   **e.** $0.875

51. a.

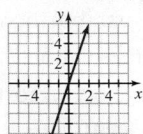

x	y
0	-3
1	-1
2	1

Answers may vary.   **b.**

c. For each solution, the y-coordinate is 3 less than twice the x-coordinate.
53. a. 7; one   **b.** one; answers may vary.   **c.** Answers may vary; one
d. one; answers may vary.   **e.** one; answers may vary.

55. a. i. x-intercept: $(0,0)$; y-intercept: $(0,0)$ **ii.** x-intercept: $(0,0)$; y-intercept: $(0,0)$ **iii.** x-intercept: $(0,0)$; y-intercept: $(0,0)$

b. x-intercept: $(0,0)$; y-intercept: $(0,0)$ **57.** Answers may vary.   **59.** 3   **61.** 0   **63.** 4   **65.** -2 **67.** C, D, E   **69.** Answers may vary; infinitely many   **71.** $x = -3$ **73.** $y = x + 3$ **75.** $y = x$ **77. a.** Answers may vary.   **b.** $y = 3x$ **79.**

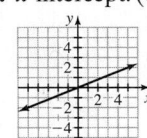

81. Answers may vary.   **83.** vertical line; answers may vary.   **85.** no; answers may vary.   **87.** Answers may vary.

Homework 7.2

1. constant   **3.** negative   **5.** 300 gallons per hour   **7.** -1650 feet per minute   **9.** 0.2 million firms per year (200 thousand firms per year)   **11.** -8.125 thousand Steller sea lions per year   **13.** -1.5 million subscribers per year   **15.** $108 per hour of classes   **17.** 1.36 seats per million people   **19.** $5940 per person   **21. a.** explanatory: t; response: v

b.

Time (number of weeks) t	Value (dollars) v
0	5
1	8
2	11
3	14
4	17
5	20

c–d.

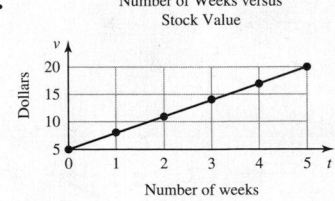
Number of Weeks versus Stock Value

d. There are no outliers; the association is exactly linear and positive; $r = 1$ **e.** Answers may vary.

23. a.

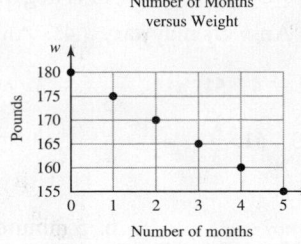

Number of Months versus Weight

b. There are no outliers; the association is exactly linear and negative; $r = -1$ **c.** 5 pounds per month   **d.** 5 pounds per month   **e.** 5 pounds per month; the three results are equal.   **f.** Answers may vary.   **25. a.** positive; the number of Americans without health insurance generally increased.   **b.** 1.0 million Americans per year   **27. a.** negative; the record times generally decreased.   **b.** -0.06 second per year; the record times decreased by about 0.06 second per year.   **c.** 0.9 second   **d.** We have little or no faith the result is correct because we have little or no faith that the rate of change will be the same in the future.   **29. a.** positive; the larger U.S. sales, the larger worldwide sales will tend to be.   **b.** $1.75 worldwide sales per $1 U.S. sales

31. a. negative; the larger the maximum weight of a dog breed, the shorter the mean life expectancy tends to be.   **b.** -0.03; a dog breed that has a maximum weight 1 pound greater than another dog breed tends to have a mean life expectancy 0.03 year less than the other dog breed.   **c.** 1.2 years   **d.** The association describes the association between maximum weights of dog breeds and their mean life expectancies, not the weight gain of a single dog.   **33.** 2; increasing   **35.** -4; decreasing   **37.** $-\dfrac{1}{3}$; decreasing

39. -1; decreasing   **41.** $-\dfrac{1}{2}$; decreasing   **43.** 2; increasing   **45.** $\dfrac{3}{2}$; increasing   **47.** $-\dfrac{4}{5}$; decreasing   **49.** $-\dfrac{1}{2}$; decreasing

51. 0; horizontal   **53.** undefined slope; vertical   **55.** -9.25; decreasing   **57.** 1.14; increasing   **59.** -0.21; decreasing   **61.** $\dfrac{2}{3}$ **63.** -3

65. a. and c.

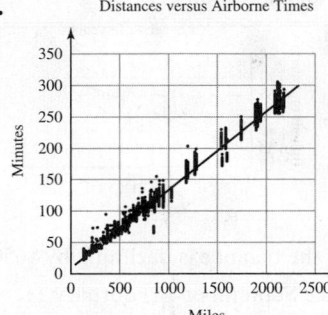

Distances versus Airborne Times

b. Many data points could be interpreted to be outliers, but the most extreme candidates are $(356, 95)$, $(1197, 203)$, and the clump of data points with distance 849 miles and airborne times between 65 and 76 minutes, inclusive; the association is linear, strong, and positive; $r = 0.99$ **d.** 0.12; a route that is 1 mile longer than another route tends to have an airborne time that is 0.12 minute greater.   **e.** 500 miles per hour

67. a. negative   **b.** positive   **c.** undefined   **d.** zero   **69.** Answers may vary.

71. Answers may vary.   **73.** Answers may vary.   **75.** Answers may vary.

77. The numerator and denominator of $\dfrac{4-1}{7-3}$ should be switched; $\dfrac{4}{3}$

79. The student should have kept the signs of -1 and -5: $\dfrac{8-(-5)}{3-(-1)}; \dfrac{13}{4}$ **81.** Answers may vary; yes   **83. a.** Answers may vary.

b. no   **c.** yes   **d.** Answers may vary.   **85.** Answers may vary.   **87.** Answers may vary.   **89.** Answers may vary.

91. a. i.

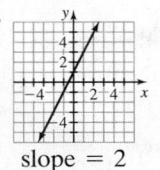

slope = 2

ii.

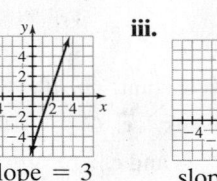

slope = 3

iii.

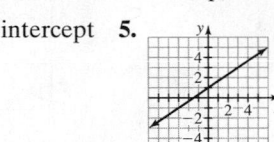

slope = -2

b. They are equal.   **93.** yes; yes; answers may vary.

95. Answers may vary.   **97. a.** Answers may vary.

b. Answers may vary.   **c.** Answers may vary.

d. Answers may vary.   **e.** Answers may vary.

Homework 7.3   1. m **3.** intercept   **5.**

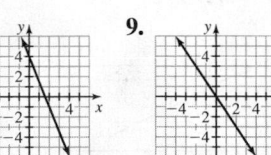

7.

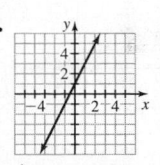

9. **11.** **13.**

15.

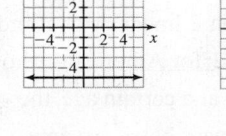

17.

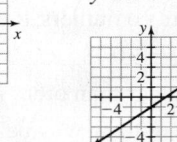

19. slope: $\dfrac{2}{3}$; y-intercept: $(0, -1)$

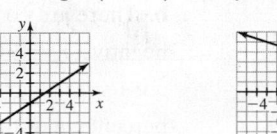

21. slope: $-\dfrac{1}{3}$; y-intercept: $(0, 4)$

23. slope: $\dfrac{4}{3}$; y-intercept: $(0, 2)$

25. slope: $-\dfrac{5}{3}$; y-intercept: $(0, 0)$

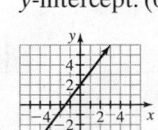

27. slope: 4; y-intercept: $(0, -2)$

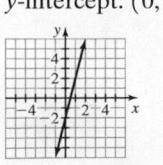

29. slope: -2; y-intercept: $(0, 4)$

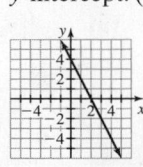

31. slope: 1; y-intercept: $(0, 1)$

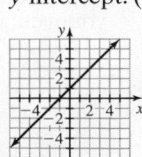

33. slope: -3; y-intercept: $(0, 0)$

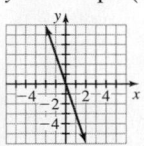

35. slope: 1; y-intercept: $(0, 0)$

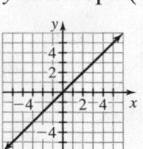

37. slope: 0; y-intercept: $(0, -3)$

39. slope: 0; y-intercept: $(0, 0)$

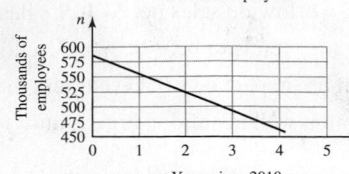

$y = 0$

41. a. m is positive; b is negative   **b.** m is zero; b is negative   **c.** m is negative; b is positive   **d.** m is positive; b is positive   **43.** Answers may vary.   **45.** Answers may vary.   **47.** Answers may vary.   **49.** $y = 3x - 4$ **51.** $y = -\dfrac{6}{5}x + 3$ **53.** $y = -\dfrac{2}{7}x$ **55.** $y = 2x + 3$ **57.** -2 **59.** 3   **61.** $\dfrac{1}{3}$

63. a.

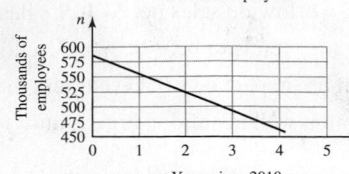

Years versus Numbers of Postal Employees

b. 462 thousand employees   **65. a.**

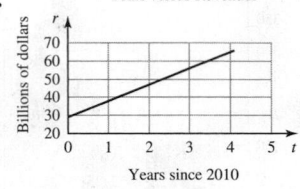

Defibrillator Delay versus Survival Rate

b. 5 minutes

67. a. 1.43; in a year when the mean sea level at The Battery is 1 foot higher than some other year, the mean sea level at Galveston is 1.43 feet higher than in that other year.   **b.** $(0, 0.46)$; in a year when the mean sea level at The Battery was 0 feet, the mean sea level at Galveston was 0.46 foot.   **c.**

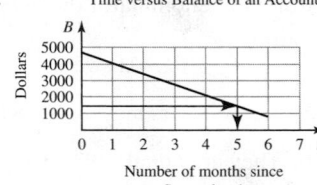

Sea Levels at The Battery and Galveston

d. 1.3 feet   **e.** The study is observational so we cannot assume causation.   **69. a.** explanatory: t; response: r **b.** slope: 9; r-intercept: $(0, 29)$ **c.** $r = 9t + 29$ **d.**

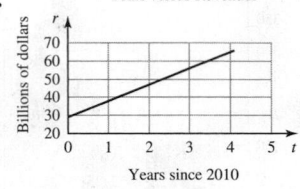

Years versus Revenues

e. $65 billion   **71. a.** The balance is declining by the same amount each month.   **b.** -650; the balance is declining by $650 per month.   **c.** $B = -650t + 4700$ **d.** and **e.**

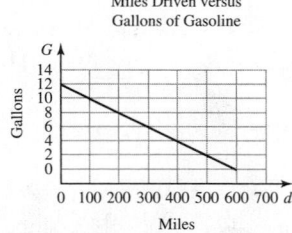

Time versus Balance of an Account

e. 5 months since September 1 (February 1)

73. a. explanatory: c; response: T **b.** 575; the total cost of tuition and the fee increases by $575 per credit.   **c.** $T = 575c + 15$

d.

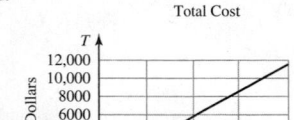

Number of Credits versus Total Cost

e. $5190   **75. a.** The rate of change of gasoline is constant.   **b.** $(0, 11.9)$; at the start of the trip, there are 11.9 gallons of gasoline in the tank.   **c.** $G = -0.02d + 11.9$

d.

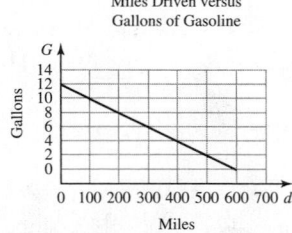

Miles Driven versus Gallons of Gasoline

e. 1.4 gallons   **77. a.** and **c.**

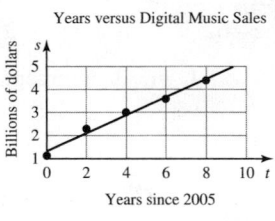

Years versus Digital Music Sales

b. There are no outliers; the association is linear, strong, and positive; $r = 0.99$ **c.** yes   **d.** $(0, 1.3)$; the sales were $1.3 billion in 2005.   **e.** $4.1 billion   **79. a.** and **c.**

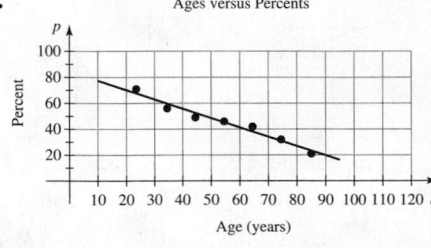

Ages versus Percents

b. There are no outliers; the association is linear, strong, and negative; $r = -0.98$ **c.** yes   **d.** -0.72; for Americans who are 1 year older than other Americans at a certain age, the percentage of them who believe marriages between same-sex couples should be recognized by the law as valid is 0.72 percentage point less than for Americans who are 1 year younger.   **e.** 65.2%

81. a. and c.

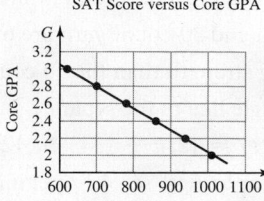

SAT Score versus Core GPA

b. There are no outliers; the association is linear, very strong, and negative; $r = 0.9998$ **c.** yes **d.** 0.51 point **e.** 3.56; 0.01; extrapolation

83. a.

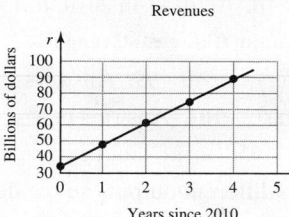

Years versus Amazon.com's Revenues

b. 13.6; the revenue increased by \$13.6 billion per year. **c.** \$13.9 billion per year; \$13.95 billion per year; \$13.7 billion per year; all three rates of change are greater than the slope, but they are close in value. **d.** \$156.58 billion; extrapolation; no **85.** no; the slope is the number multiplied by x, which is 2. **87.** Answers may vary. **89. a.**

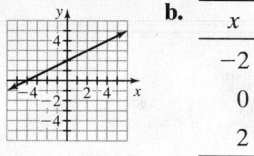

b.

x	y
-2	1
0	2
2	3

Answers may vary. **c.** For each solution, the y-coordinate is two more than half the x-coordinate.

91. Answers may vary;

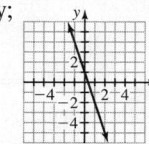

93. a.

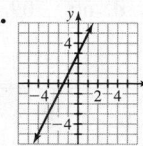

b. $y = 2x + 3$ **95. a.** The slope of each line is zero. **b.** 0 **97. a.** k; the blue line is the steeper of the two increasing lines. **b.** b; the red line's y-intercept is above the blue line's y-intercept. **99.** Answers may vary.

Homework 7.4

1. response **3.** one **5.** relations 2 and 3 **7.** no **9.** yes **11.** yes **13.** no **15.** no **17.** yes **19.** yes; the equation $y = 5x - 1$ is of the form $y = mx + b$, so the relation is a (linear) function. **21.** yes; the equation $y = 4$ is of the form $y = mx + b$, so the relation is a (linear) function. **23.** no; a vertical line intersects the graph of $x = -3$ at more than one point.

25. a. Answers may vary. **b.**

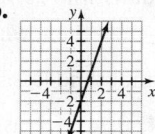

c. For each input–output pair, the output is 2 less than 3 times the input. **27.** domain: $-4 \le x \le 5$; range: $-2 \le y \le 3$ **29.** domain: $-5 \le x \le 4$; range: $-2 \le y \le 3$ **31.** domain: $-4 \le x \le 4$; range: $-2 \le y \le 2$ **33.** domain: all real numbers; range: $y \le 4$

35. domain: $x \ge 0$; range: $y \ge 0$ **37.** 26 **39.** 0 **41.** -2 **43.** 33 **45.** $\frac{1}{6}$ **47.** -16 **49.** -3 **51.** -4 **53.** 118.31 **55.** 4 **57.** 1, 3 **59.** 4 **61.** 1.2 **63.** 6 **65.** -4.5 **67.** all real numbers **69.** 1 **71.** $-4 \le x \le 5$ **73.** -3 **75.** $-5 \le x \le 4$ **77. a.** $f(t) = -4.9t + 381$ **b.** 366.3; there were 366 drive-in movie sites in 2012. **c.** 381; there were 381 drive-in movies in 2009. **79. a.** $f(d) = -0.02d + 11.9$ **b.** 2.9; there were 2.9 gallons of gasoline in the tank after driving 450 miles. **c.** 11.9; there were 11.9 gallons of gasoline in the tank at the start of the trip. **81. a–b.**

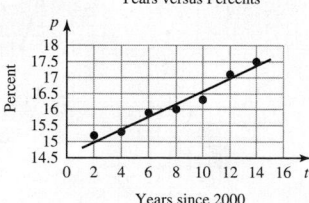

Years versus Percents

b. yes **c.** $f(t) = 0.19t + 14.63$ **d.** 16.72; in 2011, 16.72% of total donations went to the 10 colleges and universities that received the most in donations. **e.** \$5.93 billion

83. a.

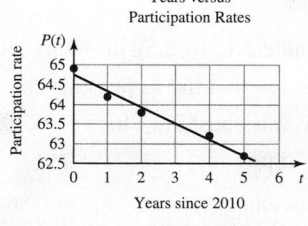

Years versus Participation Rates

b.

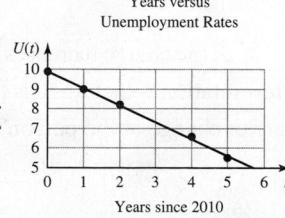

Years versus Unemployment Rates

c. 63.51; 7.32; in 2013, the participation rate was 63.51% and the unemployment rate was 7.32%. **d.** 58.86% **e.** 138.6 million people; 148.2 million people; no

85. a. and c.

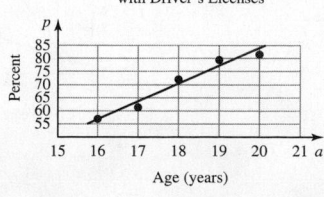

Age versus Percentage of Teens with Driver's Licenses

b. There are no outliers; the association is linear, strong, and positive; $r = 0.98$ **c.** yes **d.** 83.88; 83.88% of 20-year-old teenagers have a driver's license; 2.08 percentage points **e.** 49.83; 49.83% of 15-year-old teenagers have a driver's license; extrapolation; no, model breakdown has occurred. **87. a.** 17.72; in 2011, 17.72 million women lived alone. **b.** 31.87; in 2011, 31.87 million people lived alone. **c.** 56.18% **d.** 60.87%; 58.21%; 55.41%; 55.31% **e.** negative; the number of men living alone increased at a greater rate than the number of women.

89. a. There are no outliers; the association is linear, strong, and negative.   **b.** There are no outliers; the association is linear, strong, and negative.   **c.** 6.92; 6.16; for a stitch length of 2.75 mm, the bursting strengths of 26-count yarn and 30-count yarn are 6.92 hundred kPA and 6.16 hundred kPA, respectively.   **d.** thicker yarn; the 26-count yarn has greater bursting strength than the 30-count yarn, so thicker yarn has greater bursting strength.   **e.** The combined data points do not lie as close to a single line as the separate data points do.
91. a. $f(t) = -0.8t + 9.1$ **b.** 6.7; in 2013, there were 6.7 million viewers.   **c.** 2.40%   **93. a.** $f(t) = 12.4t + 154$ **b.** 12.4; there were 12.4 openings per year.   **c.** 12.4; the result is equal to the slope; answers may vary.   **d.** 191 openings   **e.** 4.1 openings per state; overestimate   **95. a.** $f(t) = -0.43t + 10.1$ **b.** 8.38; in 2014, 8.38% of the workforce had manufacturing jobs.   **c.** -0.43; the percentage of the workforce having manufacturing jobs decreased by 0.43 percentage point per year.   **d.** $(0, 10.1)$; in 2010, 10.1% of the workforce had manufacturing jobs.   **97. a.** $f(t) = -160t + 640$ **b.**

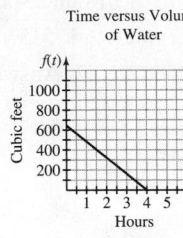
Time versus Volume of Water

c. domain: $0 \le t \le 4$; range; $0 \le f(t) \le 640$ **99.** Answers may vary.   **101.** Answers may vary.   **103.** no; no input corresponds to two different outputs, so the definition of a function is not violated.

105. Answers may vary.   **107. a.** 12; 20; 32; yes   **b.** 4; 9; 25; no   **c.** 3; 4; 5; no   **d.** no   **109.** no; $f(4)$ stands for the value of y when $x = 4$.
111. $-\dfrac{3}{4}$ **113.** x-intercept: $(4, 0)$; y-intercept: $(0, -3)$

Chapter 7 Review Exercises   **1.** $(-3, 9), (4, -5)$ **2.** -3 **3.** -2 **4.** -4 **5.** 4   **6. a.** explanatory: t; response: p

b. and d.

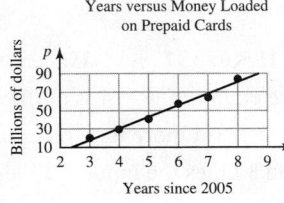
Years versus Money Loaded on Prepaid Cards

c. positive; the total amount of money loaded onto prepaid cards increased.   **d.** yes   **e.** $93.3 billion; extrapolation; the result $93.3 billion is $5.3 billion less than the company's predicted amount.   **7.** $-1.5°F$ per hour   **8.** -0.5 billion dollars per year   **9. a.** negative; the greater the income that an American adult earns, the less confident the adult will be that he or she will retire ahead of schedule.   **b.** -0.13; for American adults who earn $1 thousand more than American adults with a certain income, the percentage of them who are confident they will retire ahead of their schedule is 0.13 percentage point less than for Americans who earn $1 thousand less.   **c.** 6.5 percentage points

10. $\dfrac{1}{2}$; increasing   **11.** $-\dfrac{1}{3}$; decreasing   **12.** undefined slope; vertical   **13.** 0; horizontal   **14.** 0.94; increasing   **15.** Answers may vary.

16.
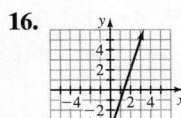

17.
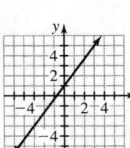

18. slope: $-\dfrac{2}{5}$; y-intercept: $(0, -1)$

19. slope: $\dfrac{2}{3}$; y-intercept: $(0, 0)$

20. slope: -3; y-intercept: $(0, 1)$

21. slope: 1; y-intercept: $(0, 2)$

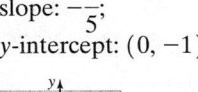

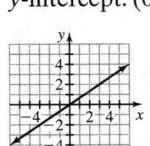

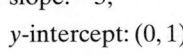

22.
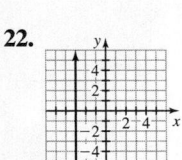

23.
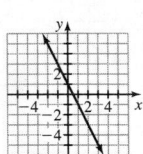

24. a.

x	y
-1	3
0	1
1	-1

Answers may vary.   **b.**

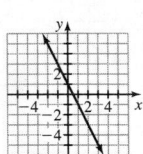

c. For each solution, the y-coordinate is 1 more than -2 times the x-coordinate.

25. $x = 5$ **26.** $y = -\dfrac{2}{3}x + 4$ **27. a.** explanatory: d; response: c **b.** 2; the charge increases $2 per mile.   **c.** $(0, 2.5)$; the charge for a taxi ride for 0 miles is $2.50, which is model breakdown; another interpretation is that $2.50 is the base fare that the $2 per mile charge is added to.   **d.**

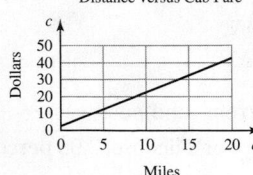
Distance versus Cab Fare

e. $36.30   **28. a.** The rate of change of the person's weight was constant.   **b.** -4; $(0, 195)$
c. $w = -4t + 195$ **d.**

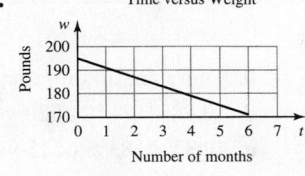
Time versus Weight

e. 175 pounds

29. a. and c.

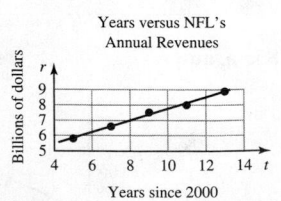
Years versus NFL's Annual Revenues

b. There are no outliers; the association is linear, very strong, and positive; $r = 0.997$ **c.** yes   **d.** 0.38; the revenue increased by $0.38 billion ($380 million) per year.   **e.** $9.26 billion; extrapolation; -1.04 billion dollars   **30.** relations 1 and 3   **31.** no; a vertical line intersects the graph of the relation at more than one point.   **32.** yes; the equation $y = \dfrac{5}{6}x - 3$ is of the form $y = mx + b$, so the relation is a (linear) function.   **33.** no; a vertical line intersects the graph of $x = 9$ at more than one point.   **34.** domain: all real numbers; range: $y \le 4$ **35.** 27   **36.** 5   **37.** $\dfrac{13}{18}$ **38.** 1   **39.** 3.6   **40.** 2   **41.** 4   **42.** $-5 \le x \le 6$ **43.** $-2 \le y \le 4$ **44.** 4   **45.** 1   **46. a.** $f(t) = 45.60t + 972$ **b.** 45.60; the mean monthly cost increased by $45.60 per year.   **c.** 45.60; the result is equal to the slope; answers may vary.   **d.** $1108.80   **e.** $6.40   **47. a.** positive; the larger a person's 10-g threshold, the larger the person's 50-g threshold will be.   **b.** 1.13; 1.5; a person's 10-g threshold of 1.5 mm is larger than the person's 50-g threshold of 1.13 mm.   **c.** 1.56; 2; a person's 10-g threshold of 2 mm is larger than the person's 50-g threshold of 1.56 mm.   **d.** 2.42; 3; a person's 10-g threshold of 3 mm is larger than the person's 50-g threshold of 2.42 mm.   **e.** below; most of the individuals have larger 10-g thresholds than 50-g thresholds; most individuals' ability to feel gaps between grooves is better when more pressure is applied to their index finger.

Chapter 7 Test   **1.** 3   **2.** 3   **3.** $(0, 1)$ **4.** $(1.5, 0)$ **5. a.** explanatory: W; response: T **b.**

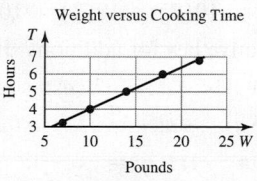

Weight versus Cooking Time

c. positive; the heavier the turkey, the greater the cooking time will be.   **d.** yes   **e.** 6.44 hours   **6.** $5.56 per year   **7. a.** positive; the median square footage was increasing.   **b.** 24; the median square footage increased by 24 square feet per year   **c.** 360 square feet   **d.** We have little or no faith the square footage will increase at the same rate as in the past.   **e.** 600 square feet per person; 950 square feet per person; the square footage per person was lower in 1982 than in 2013; the results were found by assuming the mean household sizes for newly built homes and existing homes were equal for each of the years 1982 and 2013.

8. 3; increasing   **9.** $-\dfrac{1}{2}$; decreasing   **10.** 0; horizontal   **11.** undefined slope; vertical   **12.** 1.29; increasing

13. slope: $-\dfrac{3}{2}$;   **14.** slope: $\dfrac{5}{6}$;   **15.** slope: 0;   **16.** slope: -2;   **17.** $y = \dfrac{1}{2}x + 1$ **18. a.** m is positive; b is
y-intercept: $(0, 2)$ y-intercept: $(0, 0)$ y-intercept $(0, 2)$ y-intercept $(0, 3)$ positive   **b.** m is zero; b is negative
c. m is negative; b is positive   **d.** m is
negative; b is negative

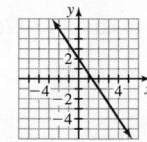 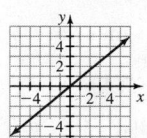 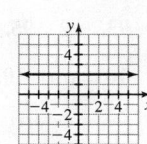 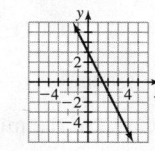

19. a. yes; the median compensation increased by an approximate constant rate; 37; the median compensation increased by about $37 thousand per year.   **b.** $(0, 870)$; the median compensation was $870 thousand in 2010.   **c.** $C = 37t + 870$

d.

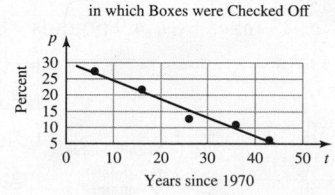

Years versus Median Compensation

e. $1018 thousand ($1.018 million)   **20.** yes; an equation of the form $y = mx + b$ is a (linear) function   **21.** domain: $-3 \le x \le 5$; range: $-3 \le y \le 4$; yes

22. -2 **23.** 3   **24.** $-6 \le x \le 6$ **25.** $-3 \le y \le 1$ **26.** 19   **27.** -11 **28. a–b.**

Years versus Percentages of Tax Returns in which Boxes were Checked Off

b. yes   **c.** $f(t) = -0.57t + 30.31$ **d.** 7.51; in 2010, 7.51% of presidential-election donation boxes were checked off.   **e.** more important; the percentage of presidential-election donation boxes that are checked off has greatly declined since 1976, and campaign spending has greatly increased, so candidates must seek donations from other sources such as wealthy Americans.

Chapter 8

Homework 8.1 **1.** $b + a$ **3.** coefficient **5.** $x + 5$ **7.** $7 + 2p$ **9.** yx **11.** $15 + m \cdot 4$ **13.** $x + (4 + y)$ **15.** $4(bc)$

17. $(x + y) + 3$ **19.** $(ab)c$ **21.** $10x$ **23.** $p + 7$ **25.** $3b + 11$ **27.** $-4x$ **29.** $\dfrac{7x}{4}$ **31.** $3x + 27$ **33.** $7x - 35$ **35.** $-2t - 10$

37. $10x - 30$ **39.** $25x + 15y - 40$ **41.** $-5x - 8y + 1$ **43.** $21x - 27$ **45.** $7x$ **47.** $5x$ **49.** $-6w$ **51.** $\dfrac{7}{3}x$ **53.** $-3x - 3$

55. $3x + y + 1$ **57.** $-9.9x + 1.1y + 2.1$ **59.** $-a + 15$ **61.** $43.16x + 23.08$ **63.** $-13b - 5$ **65.** $2x - 2y$ **67.** $-24x - 38y$

69. 0 **71.** $-4x + 7y - 21$ **73.** $-\dfrac{2}{7}a + \dfrac{6}{7}$ **75.** $3x - 3$ **77.** $x + 5x; 6x$ **79.** $4(x - 2); 4x - 8$ **81.** $x + 3(x - 7); 4x - 21$

83. $2x - 4(x + 6); -2x - 24$ **85.** twice the number, plus six times the number; $8x$ **87.** 7 times the difference of the number and 5; $7x - 35$ **89.** the number, plus 5 times the sum of the number and 1; $6x + 5$ **91.** twice the number, minus 3 times the difference of the number and 9; $-x + 27$ **93.** $8x - 5$ **95.** $-3x + 9$ **97.** The student should have distributed the factor 3 to the constant term 4; $3x + 12$ **99.** $10(7 + 3 + 6) = 160$ dollars; $10(7) + 10(3) + 10(6) = 160$ dollars; distributive law: $10(7 + 3 + 6) = 10(7) + 10(3) + 10(6)$ **101.** commutative law for addition; associative law for addition; commutative law for addition; associative law for addition **103.** commutative law for multiplication; distributive law; commutative law for multiplication **105.** $-a = -1 \cdot a$; associative law for multiplication; the product of two real numbers with different signs is negative; $-1 \cdot a = -a$; $-(-a) = a$ **107.** $-2(x - 3), 2(3 - x), -3(x - 2) + x, -2x + 6$ **109.** Answers may vary.

111. $y = 2x - 4$ **113.** $y = 3x - 4$ **115.** Answers may vary.

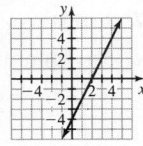 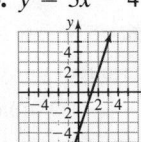

Homework 8.2 **1.** $mx + b$ **3.** Equivalent **5.** yes **7.** no **9.** yes **11.** 5 **13.** -14 **15.** -3 **17.** 3 **19.** 3 **21.** -4 **23.** 5

25. $\dfrac{4}{3}$ **27.** $\dfrac{6}{5}$ **29.** 0 **31.** 15 **33.** $\dfrac{21}{2}$ **35.** -12 **37.** $-\dfrac{10}{3}$ **39.** 6 **41.** -3 **43.** $\dfrac{1}{2}$ **45.** -11.1 **47.** 41.76 **49.** 2.3 **51.** 5 **53.** -5

55. $-\dfrac{4}{3}$ **57.** 12 **59.** 5 **61.** 2 **63.** -1 **65.** 6 **67.** -5 **69.** 2 **71.** 1014 adults **73.** 1000 owners **75.** 930 thousand students

77. 726 servings **79.** -4 **81.** 4 **83.** 5 **85.** 4 **87.** 3 **89.** -3 **91.** 3 **93.** -2 **95.** Answers may vary; 5 **97.** $4(3)$ is equal to 12; $\dfrac{4(3)}{4} = 3$ is equal to $\dfrac{12}{4} = 3, 3$ is equal to 3. **99.** yes; answers may vary. **101.** no **103.** Answers may vary. **105.** Answers may vary.

107. yes; we know the equations are equivalent because of the multiplication property of equality. **109. a.** 5 **b.** 4 **c.** $k - b$

111. Answers may vary. **113.** Answers may vary.

Homework 8.3 **1.** true **3.** 1 **5.** 3 **7.** $\dfrac{3}{4}$ **9.** $\dfrac{5}{4}$ **11.** $\dfrac{5}{2}$ **13.** $\dfrac{8}{5}$ **15.** 6 **17.** $-\dfrac{41}{3}$ **19.** $-\dfrac{27}{10}$ **21.** $\dfrac{12}{13}$ **23.** -7 **25.** $\dfrac{11}{3}$ **27.** 1.67

29. 6.34 **31.** 21.85 **33.** 40.21 **35.** 6.43 **37.** $2x + 12$ **39.** 10 **41.** $-\dfrac{6}{11}$ **43.** $-\dfrac{11}{6}x - 1$ **45.** 2 **47.** -2 **49.** 4 **51.** -1 **53.** 2

55. -3 **57.** 2.83 **59.** 3.45 **61.** -5.43 **63.** -1 **65.** -2 **67.** -11 **69.** $\dfrac{1}{3}$ **71.** $\dfrac{3}{2}$ **73.** 119.40 **75.** 14.34 **77. a.** $n = 1.04t + 46.7$

b. 2012 **c.** 50.86 million Americans; no; we have little or no faith in our results when we perform extrapolation. **79. a.** $v = 3.36t + 8.16$

b. \$18.24 **c.** 6 months after Stewart was sentenced **81. a.** $f(t) = -1.2t + 11$ **b.** 1.4; about 1 tribe sought recognition in 2013.

c. 2.5; 8 tribes sought recognition in 2008. **83. a.** $p = -0.25t + 6.95$, where p is the percentage of private-sector workers who were in a union at t years since 2010. **b.** 2013 **c.** 94.05% **85. a.** $F = 1.80d + 2.25$, where F is the cab fare (in dollars) for traveling d miles. **b.** \$18.09 **c.** 13 miles **87.** 4.9 pounds **89.** \$1.106 trillion **91.** 225 square feet **93.** 157 U.S. bank failures

95. a–b.

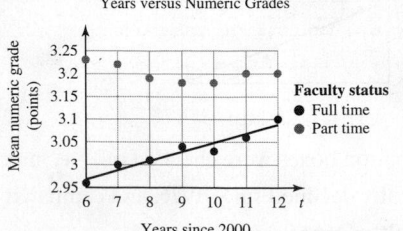

Years versus Numeric Grades

b. yes **c.** 3.2 points **d.** 2018; in 2018, the mean numeric grade given by full-time faculty will equal the mean numeric grade given by part-time faculty; no; we have little or no faith in our results when we extrapolate. **e.** Students' learning is improving; full-time faculty's standards are lowering.

97. a. yes

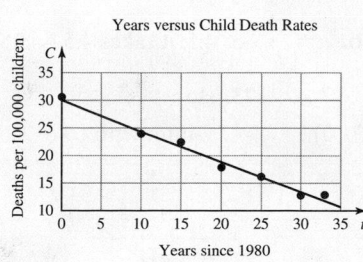

Years versus Child Death Rates

b. $(0, 30.09)$; in 1980, the child death rate was 30.09 deaths per 100,000 children.
c. 12.49 deaths per 100,000 children   **d.** 2007   **e.** 4.35 deaths per 100,000 children

99. a. and c. yes

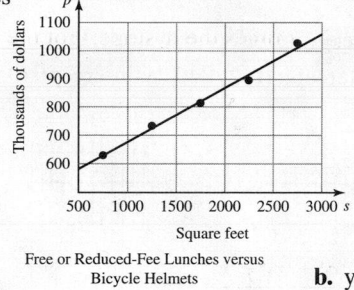

Square Footage versus Mean Selling Price

b. There are no outliers; the association is linear, very strong, and positive; $r = 0.996$ **c.** yes   **d.** 0.19; the mean selling price increases by \$0.19 thousand per square foot.   **e.** 2430.8 square feet

101. a–b.

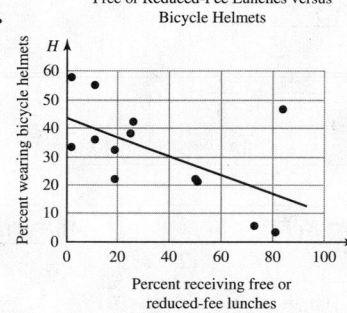

Free or Reduced-Fee Lunches versus Bicycle Helmets

b. yes; no   **c.** 30.44; for a neighborhood where 40% of the children receive free or reduced-fee school lunches, 30.44% of the bicycle riders wear helmets.   **d.** 11.03; for a neighborhood where about 11.03% of the children receive free or reduced-fee school lunches, 40% of the bicycle riders wear helmets.   **e.** no; the study is observational, so we cannot assume causation; answers may vary.   **103.** The student did not get x alone on one side of the equation; 7   **105.** Student A first subtracted $3x$ from both sides of the equation, whereas Student B first subtracted 3 from both sides of the equation; both methods required the same number of steps, so both methods are acceptable.

107. Student A began by multiplying each of the three fractions by 1 so that they would have the same denominator, whereas Student B began by multiplying both sides of the equation by the LCD of all three fractions; Student B's method is better because it requires fewer steps and leads to simpler coefficients and constant terms sooner.   **109.** 3; answers may vary.   **111.** 3; 3; all of the equations are equivalent, and it is clear that the solution of the first equation $x = 3$ is 3.

Homework 8.4   **1.** formula   **3.** true   **5.** $n = 30$ **7.** $P(E \text{ AND } F) = 0.24$ **9.** $P(F) = 0.53$ **11.** $\sigma = 47.88$ **13.** $z = 1.96$

15. $x = 6$ **17.** $x = -0.4$ **19.** $x = 20.74$ **21. a.** $\bar{x} = \dfrac{x_1 + x_2 + x_3 + x_4 + x_5}{5}$ **b.** 82 points   **23. a.** 94 points   **b.** -0.4 **c.** -0.4; the two results are equal.   **25. a.** $P(L \text{ OR } G) = P(L) + P(G)$ **b.** 0.15   **27.** 2.1   **29.** 1.87   **31.** 1.28   **33.** $p = \dfrac{\mu}{n}$ **35.** $x = \hat{p}n$

37. $P(F) = \dfrac{P(E \text{ AND } F)}{P(E)}$ **39.** $P(E) = 1 - P(\text{NOT } E)$ **41.** $P(F) = P(E \text{ OR } F) - P(E) + P(E \text{ AND } F)$

43. $z = \dfrac{x - \mu}{\sigma}$ **45.** $x = \dfrac{mx_1 + y - y_1}{m}$ **47.** $\mu_r = r - z_0\sigma_r$ **49.** $n_1 = \dfrac{nu_r - n}{2n_2}$ **51.** $y = \dfrac{ab - xb}{a}$ **53.** $n = \left(\dfrac{\sigma}{\sigma_{\bar{x}}}\right)^2$

55. a. $P(E) = P(E \text{ OR } F) - P(F) + P(E \text{ AND } F)$ **b.** 0.3   **57. a.** $\bar{x} = \dfrac{x_1 + x_2 + x_3 + x_4}{4}$ **b.** $x_4 = 4\bar{x} - x_1 - x_2 - x_3$
c. 93 points   **59. a.** $p = 3.75t + 46$ **b.** $t = \dfrac{p - 46}{3.75}$ **c.** 2006, 2008, 2009, 2010, 2012   **61. a.** $p = -0.8t + 71$ **b.** 2013
c. $t = \dfrac{p - 71}{-0.8}$ **d.** 2013   **e.** The results are equal; $t = \dfrac{p - 71}{-0.8}$ **63. a.** There are no outliers; the association is linear, fairly strong,
and positive.   **b.** $h = \dfrac{w - 335.43}{0.26}$ **c.** 3102 cubic cm; about -170 cubic cm   **d.** The head sizes, all in cubic cm, are 2941, 3325, 3710, 4095, and 4479.   **e.** If the ratios had not been compared, the researchers would be unable to make any conclusions about healthy people.

65. a. and c.

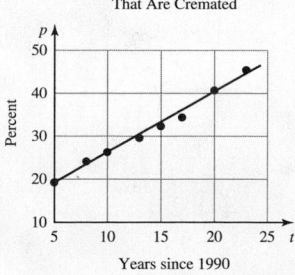

Years versus Percentages of Bodies That Are Cremated

b. There are no outliers; the association is linear, very strong, and positive; $r = 0.995$
c. yes   **d.** 1.40; the percentage of bodies that are cremated is increasing by 1.40 percentage points per year.   **e.** $t = \dfrac{p - 11.96}{1.40}$ **f.** 0.71; every 0.71 year, the percentage of bodies that are cremated increases by 1 percentage point.

67. a. The entries in the third column are $50 \cdot 4, 70 \cdot 3, 65 \cdot 2, 55 \cdot 5$, st, all in miles; $d = st$ **b.** $t = \dfrac{d}{s}$ **c.** 4.5; it takes 4.5 hours to travel 315 miles at 70 miles per hour.   **d.** 6.44 hours   **69.** slope: -2; y-intercept: $(0, 4)$ **71.** slope: $\dfrac{2}{3}$; y-intercept: $(0, 0)$ **73.** slope: $\dfrac{4}{5}$; y-intercept: $(0, -3)$

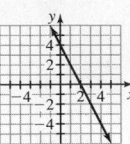
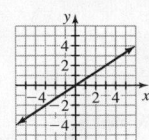
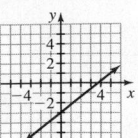

75. slope: $\dfrac{3}{4}$; y-intercept: $(0, -2)$ **77.** slope: $\dfrac{2}{5}$; y-intercept: $(0, -2)$ **79.** slope: -4; y-intercept: $(0, -2)$ **81.** slope: $\dfrac{3}{2}$; y-intercept: $(0, 2)$ **83.** slope: 0; y-intercept: $(0, 3)$

85. slope: $\dfrac{1}{2}$; y-intercept: $\left(0, -\dfrac{3}{2}\right)$ **87.** slope: $\dfrac{2}{3}$; y-intercept: $(0, 2)$ **89.** slope: $-\dfrac{1}{2}$; y-intercept: $(0, 1)$

91. a. $\bar{x} = \dfrac{x_1 + x_2 + x_3}{3}$

b. $\bar{x} = \dfrac{x_1 + x_1 + x_1}{3} = \dfrac{3x_1}{3} = x_1$

c. $s = \sqrt{\dfrac{(x_1 - \bar{x})^2 + (x_2 - \bar{x})^2 + (x_3 - \bar{x})^2}{2}}$

d. $s = \sqrt{\dfrac{(x_1 - x_1)^2 + (x_1 - x_1)^2 + (x_1 - x_1)^2}{2}}$

$= \sqrt{\dfrac{0^2 + 0^2 + 0^2}{2}} = \sqrt{\dfrac{0}{2}} = \sqrt{0} = 0$

93. a. $x = -\dfrac{b}{m}$ **b.** A linear equation can be put in the form $mx + b = 0$, and part (a) shows that such an equation has exactly one solution.   **95.** no; $-\dfrac{2}{3}$ **97. a.** 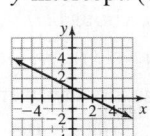   **b.** Answers may vary.   **c.** For each solution, the difference of three times the x-coordinate and five times the y-coordinate is equal to 10.   **99. a.** $y = 2x + 3$ **b.** Answers may vary.   **c.** Answers may vary.   **d.** We used properties of equality to solve $2y - 6 = 4x$ for y, which gave $y = 2x + 3$, and properties of equality do not change the solution set of an equation.   **101.** To find several values of a variable in a formula, we usually solve the formula for that variable before we make any substitutions. To find a single value of a variable in a formula, we often substitute numbers for all of the other variables and then solve for the remaining variable.

Homework 8.5   **1.** false   **3.** satisfies   **5.** 3, 6   **7.** -4 **9.** $x > 1; (1, \infty)$;   **11.** $x < -3; (-\infty, -3)$;

13. $x \le 3; (-\infty, 3]$;   **15.** $x \ge -2; [-2, \infty)$;

17. $t \le -2; (-\infty, -2]$;   **19.** $x < -\dfrac{1}{2}; \left(-\infty, -\dfrac{1}{2}\right)$;   **21.** $x \le 0; (-\infty, 0]$;

23. $x > -2; (-2, \infty)$;   **25.** $x \le -3; (-\infty, -3]$;

27. $x \ge 1; [1, \infty)$;   **29.** $x > 4; (4, \infty)$;   **31.** $c \ge -3; [-3, \infty)$;

33. $x \ge -3; [-3, \infty)$;   **35.** $x < 2.5; (-\infty, 2.5)$;   **37.** $b < 2; (-\infty, 2)$;

39. $x > -5; (-5, \infty)$;   **41.** $x \le 1; (-\infty, 1]$;   **43.** $a < -1; (-\infty, -1)$;

45. $x \le \dfrac{5}{2}; \left(-\infty, \dfrac{5}{2}\right]$;   **47.** $x \le \dfrac{4}{3}; \left(\infty, \dfrac{4}{3}\right]$;   **49.** $x \le -1.7; (-\infty, -1.7]$;

51. $y \ge \dfrac{5}{3}; \left[\dfrac{5}{3}, \infty\right)$;   **53.** $x > 7; (7, \infty)$;

55. $x \le -\dfrac{7}{12}; \left(-\infty, -\dfrac{7}{12}\right]$;   **57.** $c \ge -31; [-31, \infty)$;

59. $x < -5$; $(-\infty, -5)$; 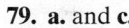 x **61.** $17.3 < \mu < 19.7$; $(17.3, 19.7)$; μ

63. $84.9 < \mu < 89.5$; $(84.9, 89.5)$; μ **65.** $0.24 < p < 0.30$; $(0.24, 0.30)$; p

67. $0.80 < p < 0.86$; $(0.80, 0.86)$; p **69.** $1 < x < 5$; $(1, 5)$; x **71.** $-5 \le x \le 6$; $[-5, 6]$;

x **73.** $-3 \le x < 5$; $[-3, 5)$; x **75.** $3 < x \le \dfrac{11}{2}$; $\left(3, \dfrac{11}{2}\right]$; x

77. a–b.

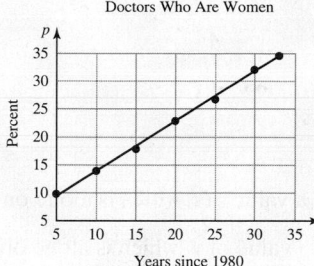

Years versus Percentages of Osteopathic Doctors Who Are Women

b. yes **c.** 0.89; the percentage of osteopathic doctors who are women increased by 0.89 percentage point per year. **d.** after 2008, although we have little or no faith the result is correct for years after 2013

79. a. and **c.**

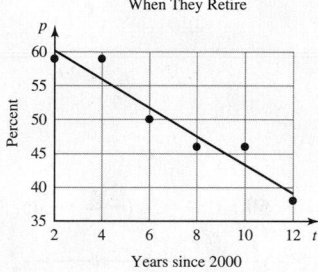

Years versus Percentages of Americans Who Think They Will Live Comfortably When They Retire

b. There are no outliers; the association is linear, strong, and negative; $r = -0.96$ **c.** yes **d.** $(0, 64.47)$; in 2000, 64.47% of Americans thought they would live comfortably when they retire; no; we have little or no faith in our results when we extrapolate. **e.** before 2011, although we have little or no faith the result is correct for years before 2002

81. a. and **c.**

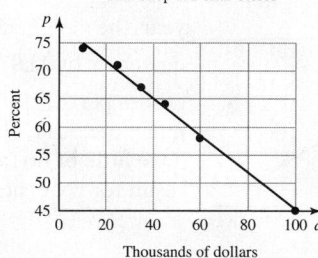

Income versus Percentage of Households That Shop at Dollar Stores

b. There are no outliers; the association is linear, very strong, and negative; $r = -0.998$ **c.** yes **d.** $(0, 78.39)$; about 78.4% of households with no income shop at dollar stores; no; we have little or no faith in our results when we extrapolate. **e.** less than \$86.0 thousand, although we have little or no faith in the result for incomes less than \$10 thousand

83. a. and **c.**

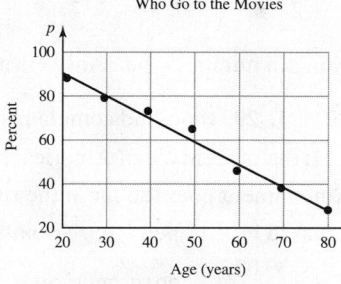

Age versus Percentage of Americans Who Go to the Movies

b. There are no outliers; the association is linear, very strong, and negative; $r = -0.99$ **c.** yes **d.** -1.04; the percentage of Americans at a certain age who go to the movies is 1.04 percentage points less than for Americans who are one year younger. **e.** Americans who are less than 59 years of age, although we have little or no faith in the result for Americans who are less than 21 years of age **85.** The student should have reversed the inequality when the student divided both sides of the inequality by -3; $x > -5$

87. a. Answers may vary. **b.** Answers may vary. **89. a.** $x = 3$ **b.** $x < 3$ **c.** $x > 3$ **d.** Answers may vary.

91. Answers may vary. **93.** Answers may vary.

Chapter 8 Review Exercises **1.** $9 + 5w$ **2.** $8 + pw$ **3.** $(2 + k) + y$ **4.** $b(xw)$ **5.** $-24x - 12$ **6.** $12y - 28$ **7.** $-3x + 6y + 8$ **8.** $4a - 9b - 7$ **9.** $-18x - 8y$ **10.** $-8.06x - 20.2$ **11.** $-5m - 4$ **12.** $-3a - 40b$ **13.** $-4(x - 7)$; $-4x + 28$ **14.** $-7 + 2(x + 8)$; $2x + 9$ **15.** $-5(x - 4)$, $5(4 - x)$, $-2(x - 10) - 3x$, $-5x + 20$ **16.** no **17.** 7 **18.** -5 **19.** 3 **20.** -6 **21.** $\dfrac{5}{11}$ **22.** 1 **23.** $\dfrac{15}{8}$ **24.** $\dfrac{38}{3}$ **25.** $\dfrac{44}{3}$ **26.** The student should have added 5 to *both* sides of the equation; 7 **27.** -8.31 **28.** $-\dfrac{11}{32}$ **29.** $\dfrac{13}{3}r - \dfrac{11}{12}$ **30.** When simplifying an expression, you cannot multiply it by any number other than 1; $\dfrac{2}{3}x + \dfrac{7}{5}$ **31.** -1.64 **32.** 3

33. 1   **34.** −4   **35.** −$\frac{7}{6}$ **36. a.** $s = 1199t + 56{,}643$ **b.** \$60,240   **c.** 2013   **37. a.** $f(t) = 31.2t + 239.1$ **b.** 675.9; in 2012, the total ad spending was \$675.9 million.   **c.** 11.57; in 2010, the total ad spending was \$600 million.   **38.** 2.1 million participants

39. a. and **c.**

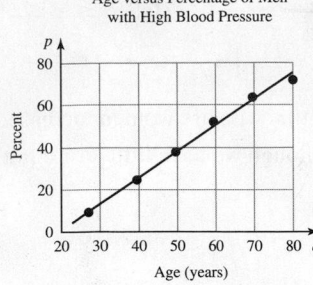

Age versus Percentage of Men with High Blood Pressure

b. There are no outliers; the association is linear, very strong, and positive; $r = 0.997$

c. yes   **d.** 13.63%   **e.** 59.57 years   **40.** −2.24   **41. a.** $\bar{x} = \dfrac{x_1 + x_2 + x_3}{3}$

b. 83 points   **42.** 0.94   **43. a.** $s = \dfrac{E\sqrt{n}}{t}$ **b.** $s = 8.3$ **44. a.** $v = 0.05t + 1.59$

b. $t = \dfrac{v - 1.59}{0.05}$ **c.** 2013   **d.** 2013

e. The results are equal; $t = \dfrac{v - 1.59}{0.05}$; the equation was easier to use because we wanted to find a value of t, which is alone on one side of the equation.   **f.** $v = 0.05t + 1.59$; the equation is easier to use because we want to find a value of v, which is alone on one side of the equation; 1.69 billion visits   **45.** slope: $\dfrac{3}{2}$; y-intercept: $(0, 3)$ **46.** slope: $-\dfrac{2}{3}$; y-intercept: $(0, -5)$ **47.** $x \geq -1$; $[-1, \infty)$;

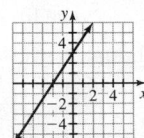   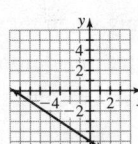

48. $x > -2$; $(-2, \infty)$;   **49.** $w > -3$; $(-3, \infty)$ **50.** $a \leq -3$; $(-\infty, -3]$;

51. $b \geq -4$; $[-4, \infty)$;   **52.** $-2 \leq x < 3$; $[-2, 3)$;

53. $72.6 < \mu < 79.2$; $(72.6, 79.2)$;   **54. a–b.**

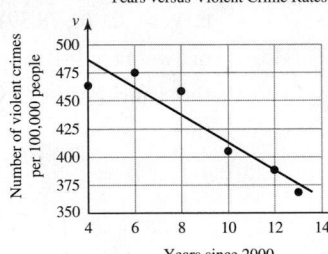

Years versus Violent Crime Rates

b. yes   **c.** −11.97; each year, the violent crime rate decreases by 11.97 violent crimes per 100,000 people.   **d.** before 2011, although we have little or no faith in the result for years before 2004

Chapter 8 Test   1. $3p + 4$ **2.** $(3x)y$ **3.** $-4x + 6$ **4.** $-22w + 53$ **5.** $-11a - 3b - 2$ **6.** $\dfrac{11}{3}$ **7.** 10   **8.** 7   **9.** $\dfrac{10}{13}$

10. $-\dfrac{41}{4}$ **11.** $-\dfrac{32}{25}$ **12.** 1.18   **13.** $23x + 24$ **14.** $-\dfrac{12}{11}$ **15.** no; the solution of an equation is a number.   **16.** Answers may vary.

17. 2   **18.** −2   **19.** $\dfrac{3}{2}$ **20. a.** $n = 21.7t + 520$ **b.** 606.8 thousand applications   **c.** 2013   **21.** 290 thousand complaints

22. 0.2   **23.** $G = \mu_G + z\sigma_G$ **24. a–b.**

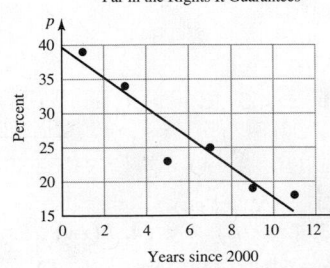

Years versus Percentages of Americans Who Feel the First Amendment Goes Too Far in the Rights It Guarantees

b. yes   **c.** −2.11; the percentage of Americans who feel the First Amendment goes too far in the rights it guarantees decreased by 2.11 percentage points per year.   **d.** $t = \dfrac{p - 39.02}{-2.11}$ **e.** 2010, 2008, 2006, 2004, 2002

25. Slope: $\dfrac{2}{3}$; y-intercept: $(0, -4)$;

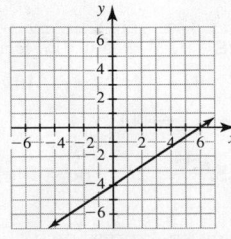

26. $x \leq 2$; $(-\infty, 2]$; 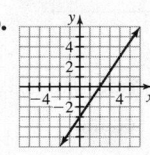   **27.** $-4 \leq x < 1$; $[-4, 1)$;

28. a–b.

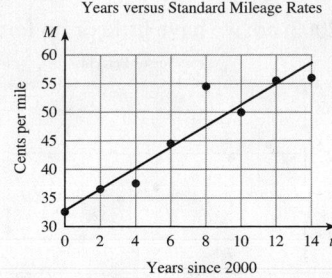

Years versus Standard Mileage Rates

b. yes   **c.** 56.87; in 2013, the standard mileage rate was 56.87 cents per mile.
d. 10.89; in 2011, the standard mileage rate was 53 cents per mile.   **e.** before 2007, although we have little or no faith in the result for years before 2000

Chapter 9

Homework 9.1   **1.** false   **3.** point   **5.** $y = 2x - 1$ **7.** $y = -3x + 1$ **9.** $y = -6x - 15$ **11.** $y = \frac{2}{5}x - \frac{1}{5}$

13. $y = -\frac{4}{5}x - \frac{7}{5}$ **15.** $y = -\frac{3}{4}x - \frac{13}{2}$ **17.** $y = 3$ **19.** $x = -2$ **21.** $y = 2.1x - 1.87$ **23.** $y = -6.59x + 14.73$

25. $y = -13.9x - 1444.64$ **27.** $y = 2x - 4$ **29.** $y = 5x - 2$ **31.** $y = -2x - 14$ **33.** $y = -4x + 9$ **35.** $y = 2$

37. $x = -4$ **39.** $y = \frac{1}{2}x + 1$ **41.** $y = -\frac{1}{6}x + \frac{3}{2}$ **43.** $y = -\frac{2}{7}x + \frac{3}{7}$ **45.** $y = \frac{3}{5}x + \frac{2}{5}$ **47.** $y = \frac{3}{2}x - 2$

49. $y = -1.61x + 9.43$ **51.** $y = -0.94x + 4.02$ **53.** $y = -2.67x - 1189.02$ **55.** $y = -\frac{4}{3}x + \frac{23}{3}$ **57.** The constant term b of
an equation of the form $y = mx + b$ is the y-coordinate of the y-intercept, and the point (3,5) is not the y-intercept.

59. a. $y = \frac{3}{2}x - 3$ **b.** 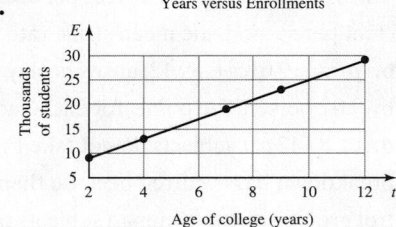   **c.** Answers may vary.   **61. a.** possible; answers may vary.   **b.** possible;
answers may vary.   **c.** not possible; answers may vary.   **d.** possible; $y = 0$

63. a. and **d.**

Years versus Enrollments

b. There are no outliers; the
association is linear, exact, and
positive.   **c.** $E = 2t + 5$

65. a. i. $y = 3x - 5$ **ii.** $y = 3x - 5$ **b.** The results are the same.   **67. a.** Answers may vary.   **b.** Answers may vary.
c. Answers may vary.   **d.** no such line; answers may vary.   **69. a–c.** Answers may vary.   **71.** Answers may vary; answers may vary,
undefined slope   **73.** Answers may vary.   **75.** $y = 2x + 1$; $y = 2x + 1$; yes; answers may vary.

Homework 9.2   **1.** false   **3.** false   **5.** $y = 2.5x - 2.2$; answers may vary   **7.** $y = -1.11x + 20.83$; answers may vary

9. a.

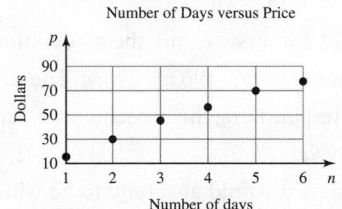

Number of Days versus Price

b. There are no outliers; the association is linear, strong, and positive; $r = 0.996$
c. $p = 12.74n + 4.40$; answers may vary.   **d.**

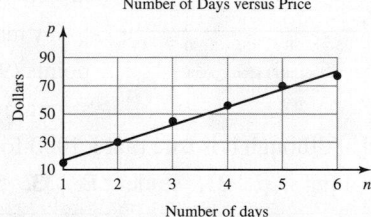

Number of Days versus Price

11. a.

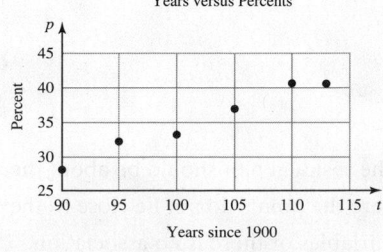

Years versus Percents

b. There are no outliers; the association is linear, strong, and positive; $r = 0.99$
c. $p = 0.56t - 22.39$; answers may vary.

d.

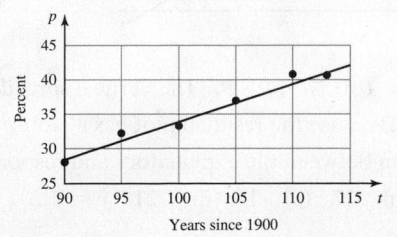

Years versus Percents

13. a.

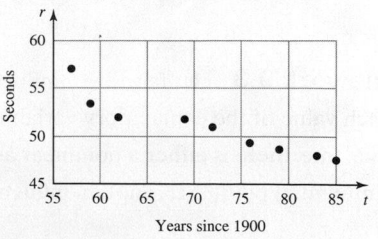

Years versus Record Times

b. There are no outliers; the association is linear, strong, and negative; $r = -0.94$ **c.** $r = -0.27t + 70.45$; answers may vary.

d.
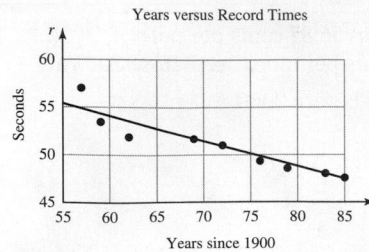
Years versus Record Times

15. a.
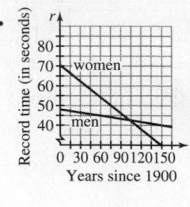

b. yes; 42.62 seconds; 2003; no; we have little or no faith when we extrapolate.   **c.** yes; after 2003; no; we have little or no faith when we extrapolate.   **17. a.**
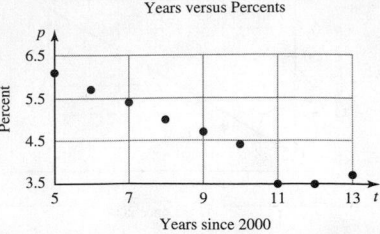
Years versus Percents

b. There are no outliers; the association is linear, fairly strong, and negative; $r = -0.97$ **c.** $p = -0.34t + 7.76$; answers may vary.

d. $(0, 7.76)$; in 2000, 7.76% of American adults preferred racing over other sports; no; we have little or no faith when we extrapolate.

e. -0.34; the percentage of American adults who preferred racing over other sports decreased by 0.34 percentage point per year.

19. a.
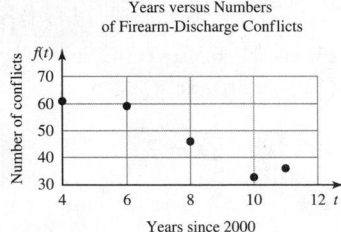
Years versus Numbers of Firearm-Discharge Conflicts

b. There are no outliers; the association is linear, strong, and negative; $r = -0.96$

c. $f(t) = -4.30t + 80.53$; answers may vary.   **d.** 41.83; in 2009, there were 42 conflicts involving police firearm discharges.   **e.** 7.1; in 2007, there were 50 conflicts involving police firearm discharges.   **21. a.**
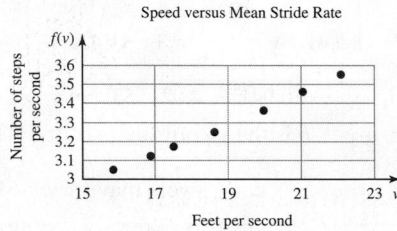
Speed versus Mean Stride Rate

b. There are no outliers; the association is linear, very strong, and positive; $r = 0.999$

c. $f(v) = 0.08v + 1.77$; answers may vary.   **d.** 3.13; for a speed of 17 feet per second, the mean stride rate is 3.13 steps per second.   **e.** 20.375; for a speed of 20.375 feet per second, the mean stride rate is 3.4 steps per second.

23. a.
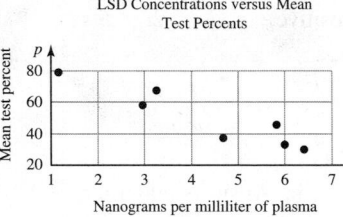
LSD Concentrations versus Mean Test Percents

b. $p = -9.01c + 89.12$; answers may vary.   **c.** -9.01; the mean test percent decreases by 9.01 percentage points for each nanogram of LSD per milliliter of plasma.

d. $(0, 89.12)$; if subjects do not take any LCD, their mean test percent is 89.12%; model breakdown has occurred because their mean test percent would be 100%.   **e.** The control group is the five human subjects taking the test before being injected with LCD.

25. a.
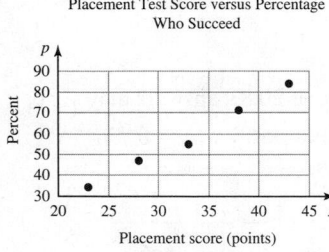
Placement Test Score versus Percentage Who Succeed

b. $p = 2.48x - 23.64$; answers may vary.   **c.** 50 points; no; we have little or no faith in our results when we extrapolate.   **d.** less than 10 points; no; we have little or no faith in our results when we extrapolate.   **e.** 30 students; no; a success rate of 21% is too low.

27. a. There are no outliers; the association is linear, weak, and positive.

b. yes   **c.** $L = 0.041U + 0.17$; answers may vary.   **d.** 22.4 hours   **e.** no; the association today may be different than the association during the period 1957–1962.   **29. a.** The data points $(9.5, 19)$, $(10.3, 7)$, and $(11.8, 21)$ might be considered outliers; the association is linear, strong, and positive.   **b.** $S = 1.75L - 6.43$; answers may vary.

c. less than   **d.** Although it is true that a 1000-foot-long ship is twice as long as a 500-foot-long ship, it would also tend to be wider than a 500-foot-long ship.   **31.** student B   **33.**   increase b **35.** Answers may vary.

Homework 9.3   **1.** false   **3.** influential   **5.** (c)   **7.** (a)   **9.** The vertical spread of the residual plot should be about the same for each value of the explanatory variable.   **11.** Because the residual plot has a pattern where the points do not lie close to the zero residual line, there is either a nonlinear association between the explanatory and response variables or there is no association.

13. influential point   **15.** not an influential point   **17.** (c)   **19.** (a)   **21.** $\hat{y} = 1.23x + 1.77$; yes

23. a. and c.

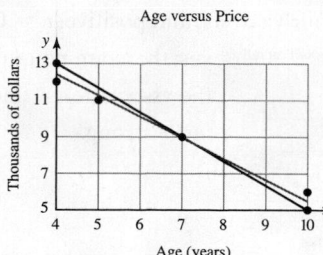

Age versus Price

b. $\hat{y} = -1.15x + 17.02$ **c.** The line $\hat{y} = -1.33x + 18.33$ fits the data points fairly well, but the regression line fits the data points better.   **d.** 2.42   **e.** 1.09
f. The sum of squared residuals is smaller for the regression line than for the line $\hat{y} = -1.33x + 18.33$, which is not surprising because the regression line has the least sum of squared residuals of all lines.

25. a. and c.

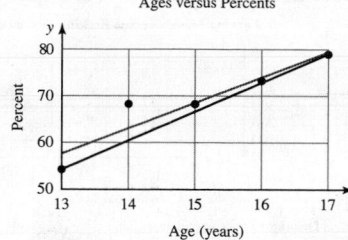

Ages versus Percents

b. $\hat{y} = 5.5x - 14.1$ **c.** The line $\hat{y} = 6.25x - 27.25$ fits the data points fairly well, but the regression line fits the data points better.   **d.** 62.38   **e.** 38.7   **f.** The sum of squared residuals is smaller for the regression line than for the line $\hat{y} = 6.25x - 27.25$, which is not surprising because the regression line has the least sum of squared residuals of all lines.

27. a.

Drop Heights versus Bounce Heights

b. $\widehat{f(x)} = 0.78x + 0.48$; yes   **c.** 14.52; if the racquetball is dropped from a height of 18 inches, the bounce height will be 14.52 inches.   **d.** 0.48 inch; for a drop height of 18 inches, the actual bounce height is 0.48 inch greater than the predicted bounce height.   **e.** 37.8; if the racquetball is dropped from a height of 37.8 inches, the bounce height will be 30 inches.

29. a.

Years versus Mean Gasoline Price

b. $\widehat{f(x)} = 0.18x + 1.34$; yes   **c.** 3.68; in 2013, the mean price of gasoline in Michigan was $3.68.   **d.** −0.091 dollars; the actual mean price of gasoline in Michigan in 2013 was $0.091 less than the predicted price.   **e.** 9.2; in 2009, the mean price of gasoline in Michigan was $3.

31. a. the number of Ebola cases   **b.**

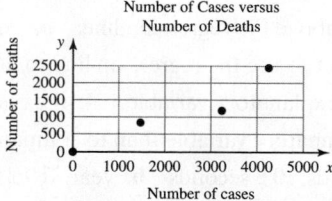

Number of Cases versus Number of Deaths

c. $\hat{y} = 0.51x - 18.58$; yes   **d.** 491 deaths; x is meant to be the number of Ebola cases for a single country; answers may vary.   **e.** 5081 deaths; no; we have little or no faith when we perform extrapolation.

33. a–b.

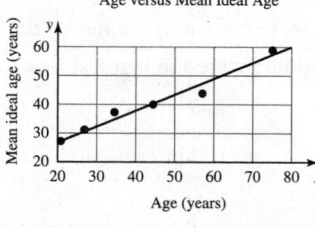

Age versus Mean Ideal Age

b. $\hat{y} = 0.55x + 15.90$ **c.**

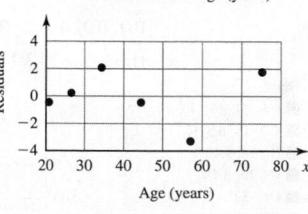

Residuals versus Age (years)

d. age: 57 years; residual: −3.3 years; the corresponding data point in the scatterplot is 3.3 years below the regression line.

e. age: 34.5 years; residual: 2.1 years; the corresponding data point in the scatterplot is 2.1 years above the regression line.
35. a. 29.7 years   **b.** 0.55; the mean ideal age chosen by people at a certain age is 0.55 year more than the mean ideal age chosen by people who are 1 year younger.   **c.** $(0, 15.90)$; the mean ideal age chosen by newborns is 15.9 years; model breakdown has occurred.   **d.** 0.97; 97% of the variation of the chosen mean ideal ages is explained by the regression line.   **e.** 35.3 years
37. a.

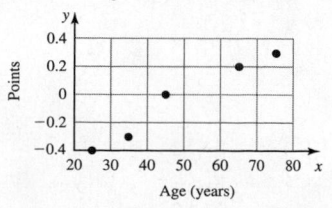

Ages versus Mean Test Scores

b. There are no outliers; the association is linear, strong, and positive; $r = 0.98$.
c. $\hat{y} = 0.014x - 0.75$ **d.**

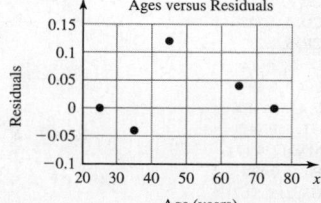

Ages versus Residuals

yes; there are no outliers, and the association is linear and strong.   **e.** 0.96; 96% of the variation in the mean test scores is explained by the regression line.

39. a.

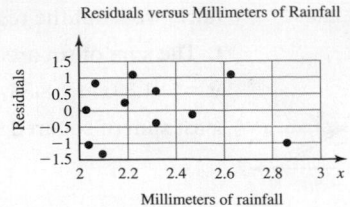

Level of Rainfall versus Number of Murders

b. There are no outliers; the association is linear, fairly strong, and positive; $r = 0.83$.

c. $\hat{y} = 4.86x + 6.11$ **d.**

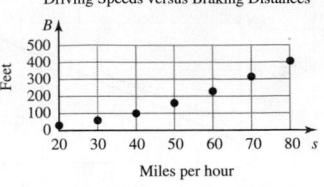

Residuals versus Millimeters of Rainfall

yes; there are no outliers, and the association is linear and fairly strong.

e. 0.69; 69% of the variation in the number of murders per year by pushing from high places in the United States is explained by the regression line.   **f.** no; we cannot conclude causality from an observational study.   **41.   a.**

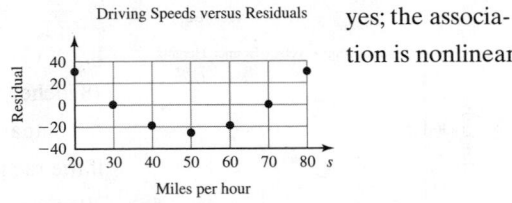

Driving Speeds versus Braking Distances

b. There are no outliers; the association is nonlinear, very strong, and positive.   **c.**

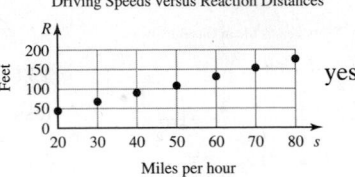

Driving Speeds versus Residuals

yes; the association is nonlinear.

d. driving speed: 50 mph; residual: −25.3 feet; the data point lies 25.3 feet below the regression line.   **e.** The association is nonlinear because the data points "bend upward."   **43. a.** The association is exactly linear.   **b.**

Driving Speeds versus Reaction Distances

yes

c. $2.2s + (6.32s - 132.75) = 8.52s - 132.75$ **d.** The residual plots are the same; the stopping distance is the sum of the reaction distance and the braking distance, and the association between driving speed and reaction distance is exactly linear.

45. a. 4 percentage points; the data point is 4 percentage points above the regression line.   **b.** −0.9 percentage point; the data point is 0.9 percentage point below the regression line.   **c.** no; no; in order to use the regression line to model some data, the vertical spread of the residual plot should be about the same for each value of the explanatory variable.   **d.** the residual plot; it is easier to compare the vertical spreads of the residual plot for various values of the explanatory variable than to compare the vertical spreads of the scatterplot for various values of the explanatory variable.   **47. a.** 10.6 seconds; 10.5 seconds   **b.** year: 1896; winning time: 12 seconds   **c.** no; when the outlier is removed, the slope of the regression line does not change much.   **d.** 10.5 seconds; 10.5 seconds   **e.** yes; yes

49. a.

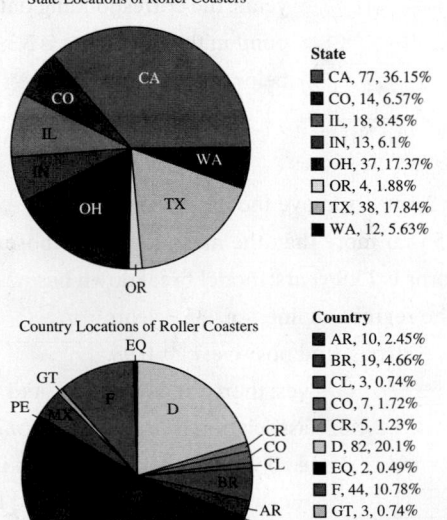

State Locations of Roller Coasters

State
- CA, 77, 36.15%
- CO, 14, 6.57%
- IL, 18, 8.45%
- IN, 13, 6.1%
- OH, 37, 17.37%
- OR, 4, 1.88%
- TX, 38, 17.84%
- WA, 12, 5.63%

Country Locations of Roller Coasters

Country
- AR, 10, 2.45%
- BR, 19, 4.66%
- CL, 3, 0.74%
- CO, 7, 1.72%
- CR, 5, 1.23%
- D, 82, 20.1%
- EQ, 2, 0.49%
- F, 44, 10.78%
- GT, 3, 0.74%
- MX, 17, 4.17%
- PE, 2, 0.49%
- US, 213, 52.21%
- VE, 1, 0.25%

no; no; if the roller coasters were randomly selected, it would be highly unlikely that the 213 U.S. roller coasters selected would be located in just 8 states.

b.

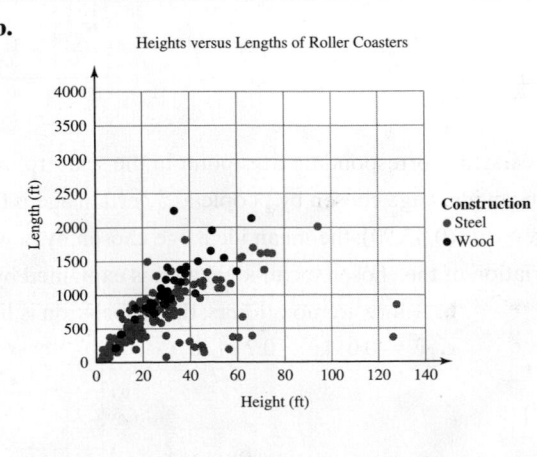

Heights versus Lengths of Roller Coasters

Construction
- Steel
- Wood

c. The Beast at Kings Island in Mason, Ohio, with height 33.528 feet and length 2243.02 feet, might be considered an outlier for the wooden association; the wooden association is linear, fairly strong, and positive with $r = 0.83$; Superman the Escape at Six Flags Magic Mountain in Valencia, California, with height 126.492 feet and length 376.428 feet, might be considered an outlier for the steel association; so might Top Thrill Dragster at Cedar Point in Sandusky, Ohio, with height 128.016 feet and length 853.44 feet; many other roller coasters might be considered outliers; the steel association is linear, weak, and positive with $r = 0.61$.

d.

Heights versus Residuals for Wooden Association

Heights versus Residuals for Steel Association

For the wooden residual plot, the dots lie fairly close to the zero residual line, except for the dot that represents the residual for the Beast; this suggests that the Beast is an outlier and that there is a fairly strong linear association, which agrees with the results we found in part (c); for the steel residual plot, most of the dots are not near the zero residual line and a typical residual is much larger than a typical residual for the wooden association; this suggests that there are outliers and that there is a weak linear association, which agrees with the results we found in part (c).   **e.** wooden association: $r^2 = 0.68$; steel association: 0.37; for the wooden association, 68% of the variation in the lengths is explained by the wooden regression line; for the steel association, 37% of the variation in the lengths is explained by the steel regression line.
f. yes; $\hat{y} = 28.40x + 209.83$, where y is the length (in feet) of a wooden roller coaster with height x feet; no   **g.** 1061.83 feet   **51.** If a data point is above the regression line, then the actual value y is larger than the predicted value $\hat{y}$: $y > \hat{y}$; Subtracting $\hat{y}$ on both sides gives $y - \hat{y} > 0$. So, the residual, $y - \hat{y}$, is positive.   **53.** The student subtracted the two values in the wrong order; −50 dollars; the student's result is the opposite of the correct result; the value of $a - b$ is equal to the opposite of the value of $b - a$: $-(b - a) = -b + a = a - b$ **55.** The association is exactly linear; if the sum of the squared residuals is 0, then all the residuals are 0, which means the data points all lie on the line.   **57.** sensitive; when an influential point is removed, the slope of the regression line is greatly affected.   **59.** Answers may vary.   **61. a–i.** Answers may vary.

Chapter 9 Review Exercises
1. $y = -4x + 7$ **2.** $y = -\dfrac{2}{3}x - 8$ **3.** $x = 2$ **4.** $y = -4$ **5.** $y = -5.29x - 17.26$

6. $y = 1.45x - 3.08$ **7.** $y = 3x - 1$ **8.** $y = 5x - 15$ **9.** $y = -\dfrac{5}{3}x + 4$ **10.** $y = \dfrac{3}{2}x - 4$ **11.** $x = 5$ **12.** $y = -3$

13. $y = -0.85x + 12.16$ **14.** $y = -0.92x - 2.31$ **15.** $y = -\dfrac{1}{5}x - \dfrac{17}{5}$ **16.** $y = -2.13x + 30.38$

17. a.

Years versus Ratios

b. There are no outliers. The association is linear, strong, and positive; $r = 0.93$ **c.** $\hat{y} = 0.16x + 2.70$ **d.** 2011   **e.** \$600,230

18. a.

Years versus Percents

b. $\hat{y} = 2.41x + 25.41$ **c.** 2.41; the percentage of American adults who were in favor of banning smoking in public places increased by 2.41 percentage points per year.

d. (0, 25.41); in 2000, 25.41% of American adults were in favor of banning smoking in public places; no   **e.** 47.1%   **19.** There is an outlier.   **20.** influential point   **21.** 0.9   **22. a.** and **c.**

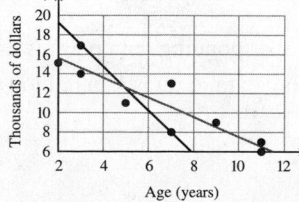

Ages versus Prices

b. $\hat{y} = -1.01x + 17.64$ **c.** The line $\hat{y} = -2.25x + 23.75$ fits the data points poorly; the regression line fits the data points very well.   **d.** 197.56   **e.** 22.21

f. The sum of squared residuals for the regression line is less than the sum of square residuals for the line $\hat{y} = -2.25x + 23.75$; this makes sense because the regression line has the least sum of squared residuals of all lines.   **23. a.**

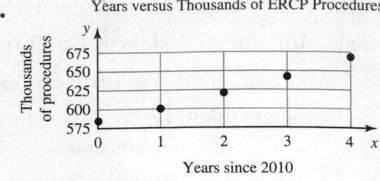
Years versus Thousands of ERCP Procedures

b. $\hat{y} = 21.2x + 581$ **c.**

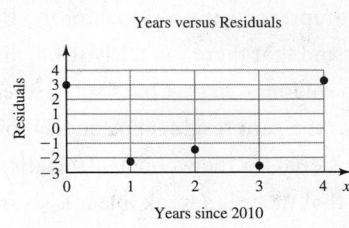
Years versus Residuals

d. year: 2014; residual: 3.2 thousand procedures; for the year 2014, the corresponding data point is 3.2 thousand procedures above the regression line.   **e.** 137.4 procedures; no; we have little or no faith when we perform extrapolation.

24. a.

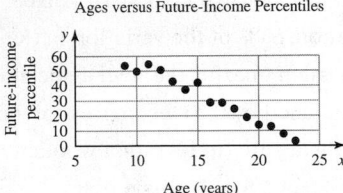
Ages versus Future-Income Percentiles

b. negative; the older a child is when he or she moves, the less time the wealthy neighborhood or lurking variables (if any) will have to impact the child.

c. $\hat{y} = -3.85x + 93.13$ **d.** 0.97; 97% of the variation in the future-income percentiles can be explained by the linear regression line.   **e.** 35th percentile; 7 percentiles; for a child who at age 15 moves to a wealthier neighborhood, the actual future-income percentile is 7 percentiles more than the percentile predicted by the regression line.

Chapter 9 Test   **1.** $y = 7x + 10$ **2.** $y = -\dfrac{2}{3}x + 3$ **3.** $y = -\dfrac{1}{2}x + 4$ **4.** $y = -1.92x - 3.64$ **5.** $y = -\dfrac{1}{3}x - \dfrac{7}{3}$

6.

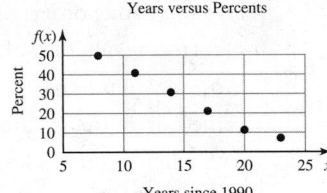

decrease m and increase b **7. a.**

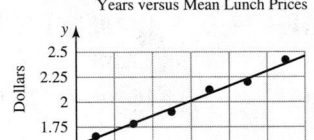
Years versus Percents

b. $\widehat{f(x)} = -3x + 73.33$

c. -3; the percentage of Fortune 500 companies that offered traditional pensions to new hires decreased by 3 percentage points per year.   **d.** 7.3; in 2012, 7.3% of Fortune 500 companies offered traditional pensions to new hires.   **e.** 17.1; in 2007, 22% of Fortune 500 companies offered traditional pensions to new hires.   **8.** For the residual plot, the vertical spread decreases from left to right.

9. $\hat{y} = -3.33x + 44.87$; yes   **10. a. and c.**

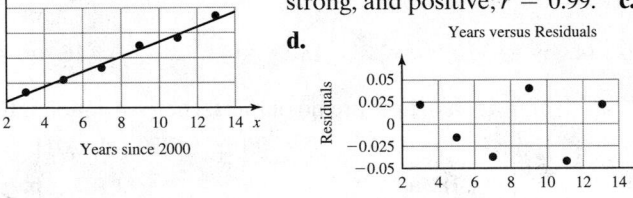
Years versus Mean Lunch Prices

b. There are no outliers; the association is linear, strong, and positive; $r = 0.99$.   **c.** $\hat{y} = 0.076x + 1.41$

d.
Years versus Residuals

yes; there are no outliers, and the association is linear and strong.

e. 0.98; 98% of the variation in the mean lunch prices is explained by the linear regression line.

11. a–b.

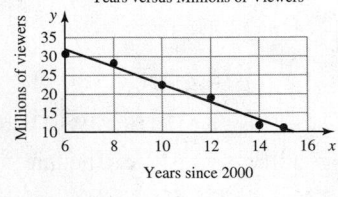
Years versus Millions of Viewers

b. $\hat{y} = -2.34x + 45.92$ **c.**

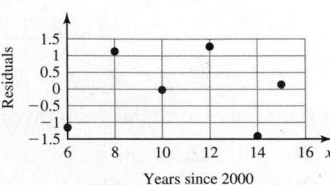
Years versus Residuals

d. year: 2014; residual: -1.46 million viewers; the corresponding data point is 1.46 million viewers below the regression line.   **e.** year: 2012; residual: 1.26 million viewers; the corresponding data point is 1.26 million viewers above the regression line.

Chapter 10

Homework 10.1 **1.** b^{m+n} **3.** true **5.** b^9 **7.** $54b^6c^8$ **9.** $32b^5c^5$ **11.** b^3 **13.** $\dfrac{5b^3c^7}{4}$ **15.** $\dfrac{b^7}{c^7}$ **17.** b^8 **19.** $27b^{27}$ **21.** 1

23. $b^{13}c^{20}$ **25.** $45b^{16}$ **27.** $\dfrac{5b^8}{4}$ **29.** $\dfrac{8b^{12}c^3}{27d^6}$ **31.** 1 **33.** $2b^4c^7$ **35.** $\dfrac{1}{b^4}$ **37.** b^2 **39.** $\dfrac{1}{b^3c^5}$ **41.** $-\dfrac{2d}{3b^9c^4}$ **43.** $\dfrac{1}{b^{14}}$ **45.** $\dfrac{4}{b^2c^3}$ **47.** $\dfrac{1}{49}$

49. 1 **51.** $\dfrac{1}{b^8}$ **53.** b^5 **55.** $\dfrac{7b^6}{4}$ **57.** $\dfrac{1}{25}$ **59.** $\dfrac{9}{b^5}$ **61.** $\dfrac{b^5}{32}$ **63.** $\dfrac{3}{b^{10}c^2}$ **65.** $\dfrac{32b^{26}}{9c^2}$ **67.** $-\dfrac{1}{3bc^5}$ **69.** $\dfrac{3b^5}{c^{13}}$ **71.** $\dfrac{b^8c^7}{32}$ **73.** $\dfrac{36b^6}{49c^{12}}$

75. $\dfrac{81c^8}{b^{24}}$ **77.** $\dfrac{1}{bc}$ **79.** $b+c$ **81.** 64 **83.** $\dfrac{1}{16}$ **85.** 18 **87.** $\dfrac{2}{9}$ **89. a.**

x	$f(x)$	x	$f(x)$
-3	$\dfrac{1}{8}$	1	2
-2	$\dfrac{1}{4}$	2	4
-1	$\dfrac{1}{2}$	3	8
0	1	4	16

b. 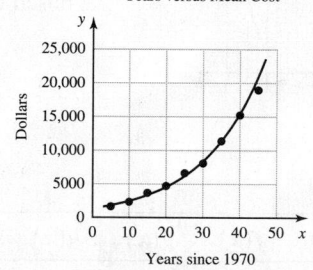 **c.** 1.4

91. $\hat{p}=\dfrac{x}{n}$ **93.** $F_0=\dfrac{s_1^2}{s_2^2}$ **95.** \$6719.58 **97.** \$8367.55 **99. a.** $s=\dfrac{d}{t}$ **b.** 62; an object that travels 186 miles in 3 hours at a

constant speed is traveling at a speed of 62 miles per hour. **101. a.** $f(d)=\dfrac{5760}{d^2}$ **b.** 90; the sound level is 90 decibels at a distance

of 8 yards from the amplifier. **103. a. and c.**

Years versus Mean Cost

b. nonlinear association **c.** yes

d. \$20,914.8; -1971.8 dollars **e.** The 2015 mean cost might be lower than what the model estimates because some public four-year colleges froze their tuitions. **105.** The student should have added the exponents; b^8 **107.** The student should have squared the coefficient $5; 25b^6$ **109.** Student B was correct; Student A should have written 5 in the denominator, not -5 in the numerator.

111. The 3 should stay in the numerator; $\dfrac{3c^4}{b^2d^7}$ **113.** -2^2, which is $-4; 2(-1)$, which is $-2; \left(\dfrac{1}{2}\right)^2$, which is $\dfrac{1}{4}; 2^{-1}=\dfrac{1}{2}$ (tie); $\left(\dfrac{1}{2}\right)^{-1}$,

which is $2; (-2)^2=(2)^2$, which are 4 (tie) **115.** Answers may vary.

Homework 10.2 **1.** $b^{\frac{1}{n}}$ **3.** b^{m-n} **5.** 4 **7.** 10 **9.** 7 **11.** 5 **13.** 16 **15.** 27 **17.** 4 **19.** 32 **21.** $\dfrac{1}{3}$ **23.** $-\dfrac{1}{6}$ **25.** $\dfrac{1}{32}$ **27.** $\dfrac{1}{81}$

29. 2 **31.** 24 **33.** 49 **35.** 27 **37.** 12 **39.** $\dfrac{4}{3}$ **41.** -16 **43.**

x	$f(x)$	x	$f(x)$
$-\dfrac{3}{4}$	$\dfrac{1}{8}$	$\dfrac{1}{4}$	2
$-\dfrac{1}{2}$	$\dfrac{1}{4}$	$\dfrac{1}{2}$	4
$-\dfrac{1}{4}$	$\dfrac{1}{2}$	$\dfrac{3}{4}$	8
0	1	1	16

45. b^2 **47.** $\dfrac{1}{b^2}$ **49.** $2b^2$ **51.** $\dfrac{4}{5b^4c^7}$ **53.** $\dfrac{b}{c^2}$

55. $5bcd$ **57.** $3b^6c^2$ **59.** $\dfrac{c^2}{b^4}$ **61.** $\dfrac{5c^3}{3b^4}$ **63.** $2b^{29/35}$ **65.** $b^{7/12}$ **67.** $3b^{29/6}$ **69.** $\dfrac{8}{b^{1/5}}$ **71.** $\dfrac{2b^{49/12}}{27c^{5/4}}$ **73.** b^2+b **75.** $\sigma_{\bar{x}}=\dfrac{\sigma}{n^{1/2}}$

77. a. and c.

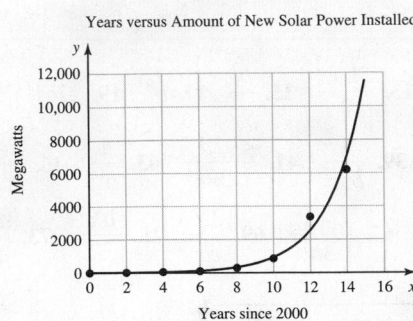

Years versus Amount of New Solar Power Installed

b. nonlinear association   **c.** yes   **d.** 11,595.8 megawatts; no; we have little or no faith in our results when we extrapolate.   **e.** 8123.3 megawatts; GTM Research's prediction is much less than the result found in part (d).
79. Answers may vary.   **81.** The student did not compute $36^{1/2}$ correctly; $6x^{18}$
83. Answers may vary.   **85. a.** $(-9)^{1/2}, (-81)^{1/4}, (-1)^{1/6}$ **b.** b is negative and n is even.   **87.** Answers may vary.

Homework 10.3   1. b **3.** $(0, a)$ **5.**

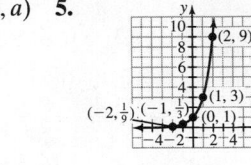   **7.** 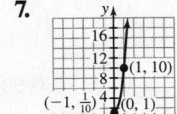   **9.** 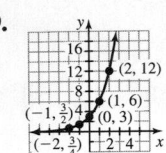   **11.**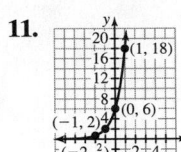

13. 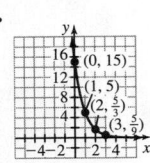   **15.** 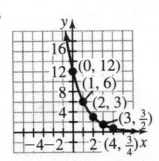   **17.** 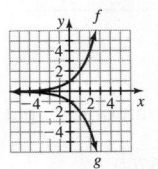   **19.** 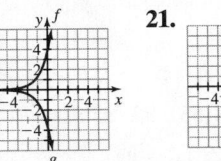   **21.** 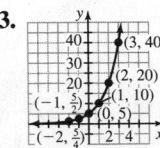   **23.**

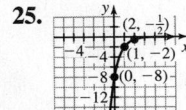   domain: all real numbers; range: $y > 0$

25. 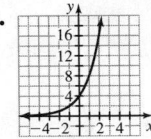   **27. a.** Answers may vary.   **b.** 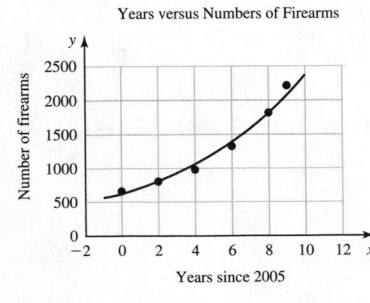   **c.** For each input–output pair, the output is 4 times 2 raised to the power equal to the input.

domain: all real numbers; range: $y < 0$

29.

x	$f(x)$	$g(x)$	$h(x)$	$k(x)$
0	162	3	2	800
1	54	12	10	400
2	18	48	50	200
3	6	192	250	100
4	2	768	1250	50

31.

x	$f(x)$	$g(x)$	$h(x)$	$k(x)$
0	5	160	162	3
1	10	80	54	12
2	20	40	18	48
3	40	20	6	192
4	80	10	2	768

33. 8   **35.** 1   **37.** -2 **39.** 0
41. 24   **43.** 96   **45.** 0   **47.** 3

49. a.–b.

Years versus Numbers of Firearms

c. There are no outliers. The association is exponential, very strong, and positive.
d. 618; in 2005, 618 firearms were discovered at TSA checkpoints, according to the model; this is an underestimate of the actual value, 660 firearms.
e. 1.144; the number of firearms discovered at TSA checkpoints increased by 14.4% per year, according to the model.

51. a.–b.

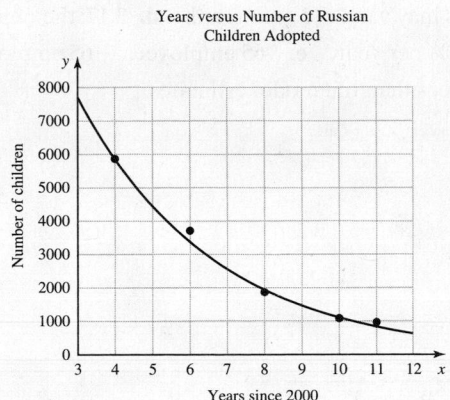

Years versus Number of Russian Children Adopted

b. There is a strong, exponential association.   **c.** 0.76; the number of Russian children adopted by American families decayed by 24% per year.   **d.** 495.4; in 2013, 495 Russian children were adopted by American families; no; we have little or no faith in our results when we extrapolate.   **e.** Model breakdown occurred for the year 2013; this illustrates the danger of extrapolating.

53. a.–b.

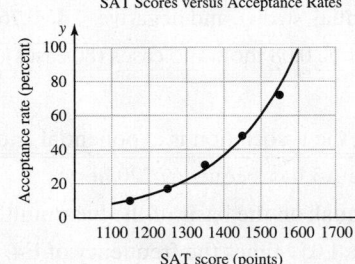

SAT Scores versus Acceptance Rates

c. 1.005; for students who score 1 point more than students with a certain score, the acceptance rate will be 0.5% higher than for students who score 1 point less.
d. 41.5%   **55.** no x-intercept; y-intercept: $(0, 1)$ **57.** no x-intercept;
y-intercept: $(0, 3)$ **59.** 13   **61.** $\dfrac{13}{36}$ **63.** 1   **65.** 0   **67.** $f(x) = g(x)$
69. $f(x) = g(x)$ **71.** $f(x) = g(x)$ **73.** $f(x) = g(x)$ **75.** $f(x) = g(x)$
77. $f(x) = g(x)$ **79. a.** $a < 0, b > 1$ **b.** $a > 0, b > 1$ **c.** $a > 0, 0 < b < 1$
d. $a < 0, 0 < b < 1$ **81.** Answers may vary.   **83.** $f(x) = 3(2.7)^x$

85. Answers may vary.   **87. a.** for f: $(0, 100)$; for g: $(0, 5)$ **b.** For f, as the value of x increases by 1, the value of $f(x)$ is multiplied by 2. For g, as the value of x increases by 1, the value of $g(x)$ is multiplied by 3.   **c.** The outputs of g will eventually be much greater.

d.

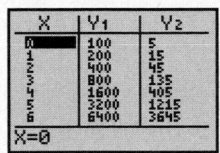 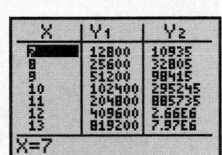

89. a. no   **b.** no   **91.** Answers may vary.   **93.** Answers may vary.

Homework 10.4   **1.** $\pm k^{\frac{1}{n}}$ **3.** false   **5.** ± 4 **7.** 3   **9.** 2   **11.** ± 0.81 **13.** 2.28   **15.** ± 1.51 **17.** 2.22   **19.** ± 3 **21.** 1.74
23. b^5 **25.** 2.38   **27.** $\dfrac{4b^4}{3}$ **29.** ± 0.75 **31.** $y = 4(2)^x$ **33.** $y = 3(2.02)^x$ **35.** $y = 87(0.74)^x$ **37.** $y = 5.5(3.67)^x$
39. $y = 7.4(0.56)^x$ **41.** $y = 39.18(0.85)^x$ **43.** $y = 1.33(3)^x$ **45.** $y = 1.19(1.50)^x$ **47.** $y = 1170.33(0.88)^x$ **49.** $y = 37.05(0.74)^x$
51. $y = 0.07(1.57)^x$ **53.** $y = 146.91(0.71)^x$ **55. a. and c.**

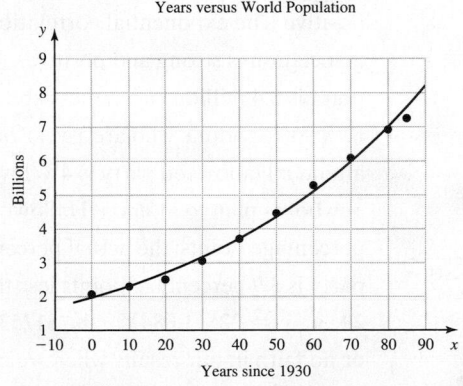

Years versus World Population

b. $\hat{y} = 1.97(1.016)^x$; answers may vary.   **c.** see graph   **d.** 1.97; in 1930, the world population was 1.97 billion.   **e.** 0.4525 billion (452.5 million); the result is quite a bit larger than the U.S. population.

57. a. and c.

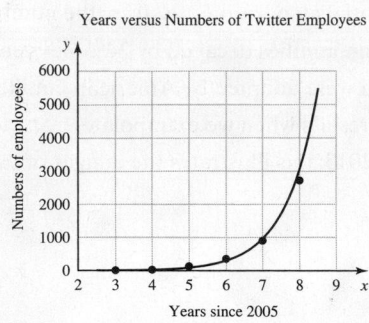

Years versus Numbers of Twitter Employees

b. $\hat{y} = 0.30(3.17)^x$; answers may vary.   **c.** see graph   **d.** 3.17; the number of employees increased by 217% per year.   **e.** 965 employees; −65 employees; there were 65 fewer employees than the model estimates.

59. a.–b.

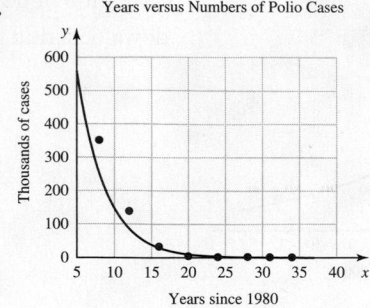

Years versus Numbers of Polio Cases

b. $\hat{y} = 2123(0.766)^x$; answers may vary.   **c.** There are no outliers. The association is exponential, strong, and negative.   **d.** 0.766; the number of polio cases decreased by 23.4% per year.   **e.** no, the model predicts that there will be 0.08 thousand cases (80 cases); no; we have little or no faith in our results when we extrapolate.

61. a.–b.

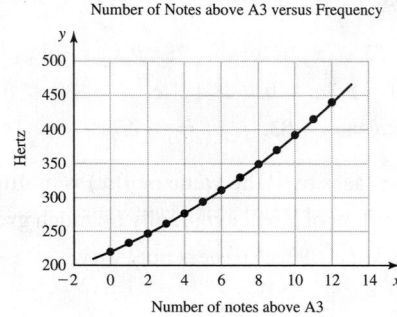

Number of Notes above A3 versus Frequency

b. $\hat{y} = 220(1.059)^x$; answers may vary; the association is exponential and extremely strong.   **c.** $(0, 220)$; the note A3 has frequency 220 hertz.
d. 328.62; 348.01   **e.** 1.059; they are equal; on the basis of the base multiplier property, the frequency of F4 should be 1.059 times the frequency of E4.

63. $y = 4\left(\dfrac{1}{2}\right)^x$ **65.** $y = 1.26(1.58)^x$ **67. a. i.** yes; answers may vary.   **ii.** no; answers may vary.   **b.** no; answers may vary.

69. $L(x) = 4x + 2$; $E(x) = 2(3)^x$ **71.** could be linear or exponential   **73. a.** $L: (0, 100)$; $E: (0, 3)$ **b.** L: y increases by 2; E: y is multiplied by 2.   **c.** The outputs of E will eventually dominate over the outputs of L.   **75.** Answers may vary.

Homework 10.5   **1.** negatively   **3.** regression   **5.** (d)   **7.** (a)   **9.** (d)   **11.** (b)   **13.** The plot has a pattern where the points do not lie close to the zero residual line.   **15.** The conditions are met.   **17.** influential point   **19.** (d)   **21.** (b)   **23.** $\hat{y} = 1.989(1.157)^x$; yes

25. a.

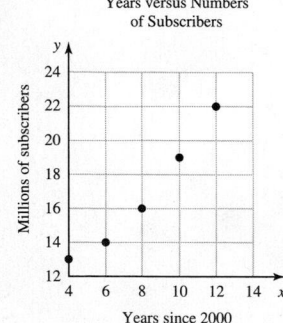

Years versus Numbers of Subscribers

b. $\hat{y} = 9.58(1.07)^x$ **c.** There are no outliers. The association is exponential, strong, and positive. The exponential correlation coefficient is 0.99, which confirms that the exponential association is strong and positive.   **d.** 21.6 million subscribers; 0.4 million subscribers; the data point is 0.4 million subscribers above the model.   **e.** $2.42 billion   **27. a.** $\hat{y} = 113(0.96)^x$
b. 0.96; for adults who are 1 year older than adults at a certain age, the percentage who plan to attend a Halloween party is 4% lower than for the adults 1 year younger.   **c.** $(0, 113)$; 113% of newborns plan to attend a Halloween party; model breakdown has occurred.   **d.** 47.9%; −3.9 percentage points; the actual percentage of 21-year-old adults who plan to attend a Halloween party is 3.9 percentage points less than what the model predicts.   **e.** 853 thousand adults
29. a. $\hat{y} = 225(1.084)^x$ **b.** $17.53 trillion   **c.** $54,953   **d.** $319.76 trillion; no; we have little or no faith in our results when we extrapolate.   **e.** $728,383; the per-person share of the federal debt in 2050 will be over 10 times what it was in 2014.

31. a.

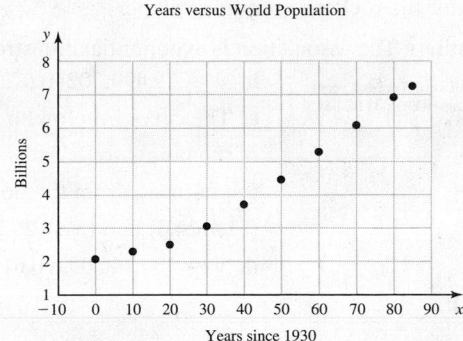

Years versus World Population

b. $\widehat{f(x)} = 1.97(1.016)^x$

c. 7.5933; in 2015, the population is 7.5933 billion; -0.3453 billion

d.

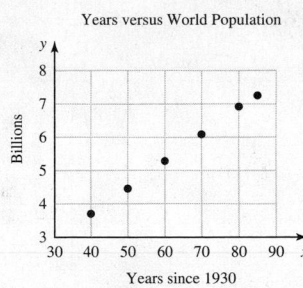

Years versus World Population

e. $\widehat{g(x)} = 0.08x + 0.49$

f. 7.29; in 2015, the population is 7.29 billion; -0.042 billion; the absolute value of the result is less than the absolute value of the result found in part (c); for the period 1970–2015, the association is linear.

33. a. $\hat{y} = 113(0.96)^x$ **b.**

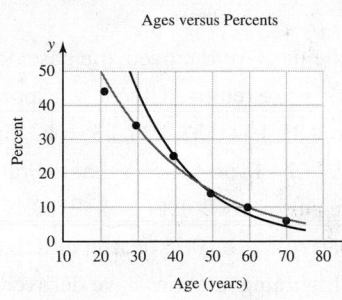

Ages versus Percents

The model $\hat{y} = 247(0.944)^x$ does not fit the data very well; the regression model fits the data very well.   **c.** 1008.9   **d.** 22.9

e. The sum of squared residuals for the exponential regression model is less than for the other model; for a given set of data, the sum of squared residuals for the exponential regression model is always smaller than for any other exponential model.

35. a.

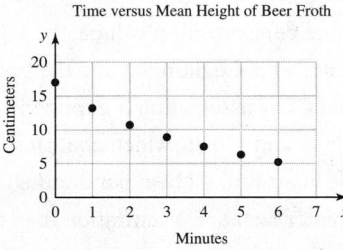

Time versus Mean Height of Beer Froth

b. $\hat{y} = 16.2(0.825)^x$ **c.** There are no outliers. The association is exponential, very strong, and negative. The value of the exponential correlation coefficient is -0.998, which confirms that the exponential association is very strong and negative.

d. 13.37 centimeters   **e.** -0.17 centimeter; one minute after the Erdinger Weissbier was poured, the actual mean height of beer froth was 0.17 centimeter less than the model's prediction.

37. a.

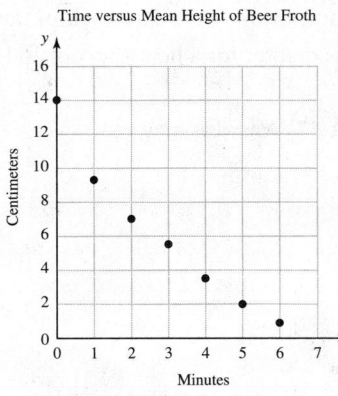

Time versus Mean Height of Beer Froth

b. $\hat{y} = 15.9(0.651)^x$ **c.** There are no outliers. The association is exponential, strong, and negative. The value of the exponential correlation coefficient is -0.98, which confirms that the association is strong and negative.   **d.** 0.96; 96% of the variation in the mean beer froth heights is explained by the regression exponential curve.

e. more scatter; smaller; when there is more scatter, the coefficient of determination is smaller.

39. a.

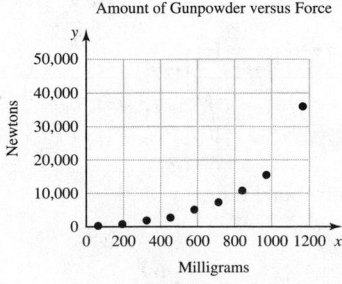

Amount of Gunpowder versus Force

b. $\hat{y} = 336(1.00416)^x$ **c.** There are no outliers. The association is exponential, strong, and positive. The value of the exponential correlation coefficient is 0.98, which confirms that the association is strong and positive.   **d.** 1.00416; for each additional milligram of gunpowder, the force grows by 0.416%.   **e.** 336; if there is no gunpowder, the force is 336 newtons; model breakdown has occurred.

41. a.

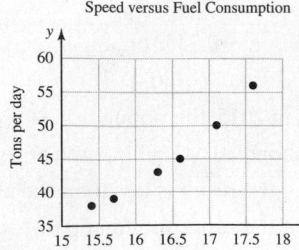

Speed versus Fuel Consumption

b. positive; the greater the speed, the greater the fuel consumption will be.

c. $\hat{y} = 2.43(1.194)^x$ **d.** There are no outliers. The association is exponential and strong.

e. 38.6 tons per day   **43. a.**

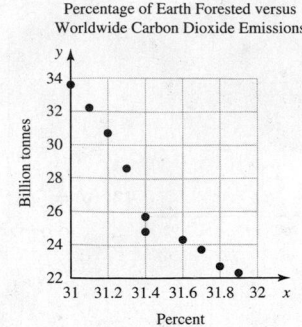

Percentage of Earth Forested versus Worldwide Carbon Dioxide Emissions

b. $\hat{y} = 79,499,302(0.6223)^x$

c. The curve is below so many of the points so that it is not too far away from the points $(31.4, 24.8)$ and $(31.4, 25.7)$.

d. $\hat{y} = 106,140,092(0.6170)^x$

e. One could argue that the two points are not influential because when the they are removed, the base does not change that much; however, we will decide that the two points are influential because when they are removed, the new exponential regression curve is not below so many of the data points.   **45. a.** right-hand column: 1.36, 1.36, 1.33, 1.34, 1.33, 1.36, 1.35   **b.** They are approximately equal. **c.** exponential   **d.** $\hat{y} = 3.94(1.03)^x$ **e.** right-hand column: 1.27, 1.26, 1.25, 1.21   **f.** no; the fact that the ratios found in part (e) are not approximately equal implies that the association is not exponential for the period 1860–1900.   **g.** 2788.43 million (2.78843 billion) people; −2474.53 million (−2.47453 billion) people   **47. a.** $\hat{y} = 18.47(0.9529)^x$ **b.** 0.15 death per million people per year   **c.** 42 lightning deaths   **d.** 420 lightning injuries   **e.** $\hat{A} = 50.68(0.9918)^x$ **f.** Lightning fatalities have decayed exponentially at a greater rate than the number of Americans who live in rural areas.

49. a.

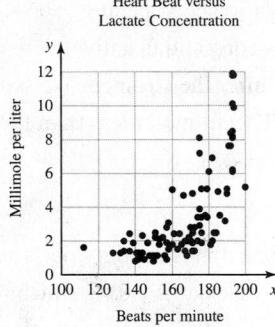

Heart Beat versus Lactate Concentration

b. positive; the higher the heart rate, the greater the lactate concentration will be; the higher the heart rate, the greater the intensity the body will experience and the more lactate the body will produce.   **c.** $\hat{y} = 0.026(1.028)^x$ **d.** There are no outliers. The association is exponential and fairly strong. The value of the exponential correlation coefficient is 0.76, which confirms that the association is fairly strong.   **e.** 1.028; soccer players whose heart rate is 1 beat per minute greater than soccer players with a certain heart rate have 2.8% greater lactate concentration than the soccer players with heart rate 1 beat per minute less.   **f.** 0.04 millimole per liter; 0.06 millimole per liter; 0.06 millimole per liter; the changes in lactate concentration are the result of the same *percentage* growth (2.8%) and the lactate concentration is greater for a heart beat of 160 beats per minute than for a heart beat of 140 beats per minute.

51.

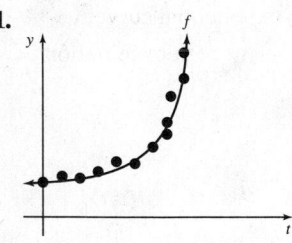

Decrease b.   **53.** Answers may vary.   **55.** Answers may vary.   **57.** Answers may vary.

59. Answers may vary.   **61.** Answers may vary.

Chapter 10 Review Exercises   **1.** $8b^{27}$ **2.** $b^{10}c^{14}$ **3.** 32   **4.** $\dfrac{48c^3}{b^{12}}$ **5.** $\dfrac{bc^9}{4}$ **6.** $\dfrac{8c^6}{9b^{23}}$ **7.** 1   **8.** $\dfrac{b}{c}$ **9.** 16   **10.** $\dfrac{1}{8}$ **11.** $\dfrac{1}{b^{2/15}}$

12. $\dfrac{1}{b^{5/3}}$ **13.** $\dfrac{2b^{11}}{125c^7}$ **14.** $2b^2c$ **15.** $\dfrac{b^{1/6}}{c^{7/4}}$ **16.** 64   **17.** $\dfrac{3}{25}$ **18.** 7   **19.** $\dfrac{2}{27}$ **20.**

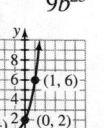

21.

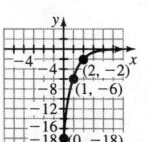

22.

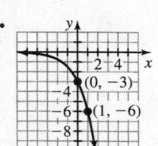

domain: all real numbers; range: $y < 0$ **23.**

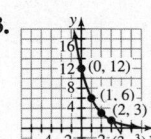

domain: all real numbers; range: $y > 0$

24.

x	$f(x)$	$g(x)$	$h(x)$	$k(x)$
2	3	162	2	96
3	15	54	8	48
4	75	18	32	24
5	375	6	128	12
6	1875	2	512	6

25. a.–b.

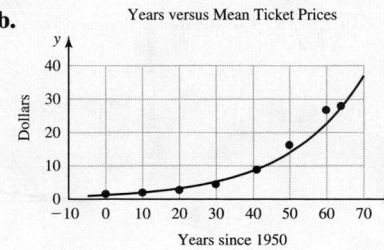

Years versus Mean Ticket Prices

b. The association is exponential and strong.
c. $(0, 1.22)$; in 1950, the mean ticket price was \$1.22.   **d.** \$26.38; \$27.70   **e.** 1.05; they are equal; on the basis of the base multiplier property, if the year is increased by 1, then the mean ticket price is multiplied by the base 1.05.   **26.** 2   **27.** 1.97   **28.** 1.84

29. ± 2 **30.** ± 1.61 **31.** 1.69   **32.** $y = 2(1.08)^x$ **33.** $y = 3.8(2.34)^x$ **34.** $y = 62.11(0.78)^x$ **35.** $y = 3.07(1.18)^x$

36. a.–b.

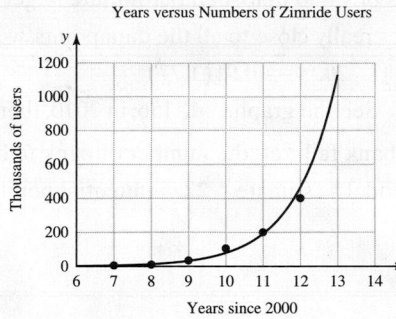

Years versus Numbers of Zimride Users

b. $y = 0.011(2.43)^x$; answers may vary.   **c.** There are no outliers. The association is exponential, strong, and positive.   **d.** 2.43; the number of users increased by 143% per year.   **e.** 0.011; in 2000, there were 0.011 thousand (11) users; model breakdown has occurred because a little research would show that Zimride began in 2007.

37. The plot has a pattern where the points do not lie close to the zero residual line.

38. a.

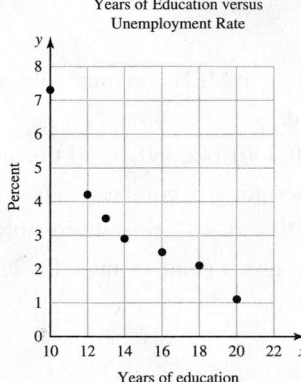

Years of Education versus Unemployment Rate

b. $\hat{y} = 32.66(0.849)^x$ **c.** There are no outliers. The association is exponential, strong, and negative. The value of the exponential correlation coefficient is -0.97, which confirms that the association is strong and negative.   **d.** 0.95; 95% of the variation of the unemployment rate is explained by the exponential regression curve.   **e.** 2.38%; 0.12 percentage point; the unemployment rate for people who have 16 full-time-equivalent years of education is 0.12 percentage point greater than the model's prediction.   **39. a.** $\widehat{W}(x) = 1.79(1.077)^x$; $\widehat{M}(x) = 1.11(1.093)^x$ **b.** The base of W, 1.077, is less than the base of M, 1.093. As ages increase, the quarterly rates for women grow exponentially at a rate less than that for men.   **c.** \$84.74; \$113.13   **d.** \$97.36   **40. a.** $\hat{n} = 9(1.332)^x$ **b.** $\hat{c} = 0.58(1.264)^x$ **c.** the catch-rate model; the larger the catch rate, the larger the lake trout population will tend to be.   **d.** The trout-removal model's base (1.332) is greater than the catch-rate model's base (1.264); the percentage growth rate of the number of trout removed is greater than the percentage growth rate of the catch rate.   **e.** 9.65 lake trout caught per 100 meters of net per night; -2.95 lake trout caught per 100 meters of net per night   **f.** The removal of trout may have become large enough to slow the exponential growth of lake trout or even reduce the lake trout population.   **41.**

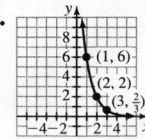

Increase a and decrease b.

Chapter 10 Test   **1.** 4   **2.** $-\dfrac{1}{16}$ **3.** $8b^9c^{24}$ **4.** 1   **5.** $b^{1/6}$ **6.** $\dfrac{5b}{7c^5}$ **7.** $\dfrac{4b^4}{c^{14}}$ **8.** $\dfrac{250b^{14}}{7c^6}$ **9.** $\dfrac{1}{16}$ **10.** $\dfrac{1}{8}$

11.

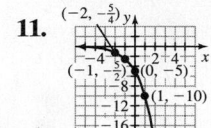

domain: all real numbers; range: $y < 0$ **12.**

domain: all real numbers; range $y > 0$

13. Answers may vary.   **14.** 6   **15.** 1   **16.** $f(x) = 6\left(\dfrac{1}{2}\right)^x$ **17. a.–b.**   **b.** yes

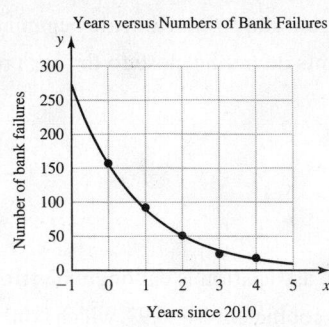

Years versus Video Upload Rate

c. 1.59; the number of hours of video uploaded per minute increased by 59% per year.   **d.** 168.9 hours of video uploaded per minute; 131.1 hours of video uploaded per minute; in 2014, the video upload rate was 131.1 hours of video uploaded per minute larger than what the model estimates. So, the model's estimate is terrible.   **e.** Even if a model comes really close to all the data points, we still have little or no faith in our results when we extrapolate.   **18.** ± 1.72 **19.** $y = 70(0.81)^x$ **20.** $y = 0.91(1.77)^x$

21. a. and c.

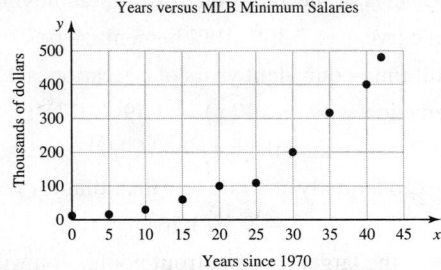

Years versus Numbers of Bank Failures

b. $\hat{y} = 156(0.57)^x$; answers may vary.   **c.** See the graph.   **d.** 156; in 2010, there were 156 bank failures.   **e.** 16 bank failures; 2 bank failures; the number of bank failures in 2014 was 2 bank failures more than the model's estimate.   **22.** influential point

23. a.

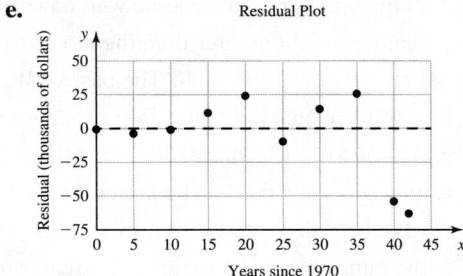

Years versus MLB Minimum Salaries

b. $\hat{y} = 12.7(1.094)^x$ **c.** 0.99; 99% of the variation in MLB minimum salaries is explained by the exponential regression curve.   **d.** $505.3 thousand

e.   Residual Plot

$(40, 400), (42, 480)$; no; if the model does not give good estimates for 2010 and 2012, then it probably does not give a good estimate for 2011, either.

Index